Building Regulations Explained

2000 Revision

Building Regulations Explained

2000 Revision

Sixth edition

John Stephenson

London and New York

First published by E&FN Spon 1986
Second edition 1988
Third edition 1991
Fourth edition 1993
Reprinted 1994
Fifth edition 1995
Reprinted 1997, 1998

This edition published 2001
by Spon Press
11 New Fetter Lane, London EC4P 4EE

Simultaneously published in the USA and Canada
by Spon Press
29 West 35th Street, New York, NY 10001

Spon Press is an imprint of the Taylor & Francis Group

Typeset in Goudy by Wearset, Boldon, Tyne and Wear
Printed and bound in Great Britain by TJ International Ltd, Padstow, Cornwall

British Library Cataloguing in Publication Data
A catalogue record for this book is available from the British Library

Library of Congress Cataloging in Publication Data
Stephenson, John, 1920–
    Building regulations explained : 2000 revision/John Stephenson,– 6th ed,
        p. cm.
    Includes bibliographical references and index.
    ISBN 0–419–23430–6 (hbk.:alk.paper)
        1. Building laws–Great Britain. I. Title.

KD1140 .S853 2001
343.41'07869–dc21

00–041188

ISBN 0–419–23430–6

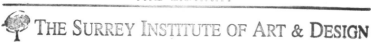

Contents

CONTENTS

Chapter 19   282

Protection from falling, collision and impact
(Approved document K)

Chapter 20   293

Conservation of fuel and power
(Approved document L)

Forewords

FOREWORD TO SIXTH EDITION

Once again most of the chapters have attracted additional comment or adjustment. One facet of building control which has been slow to mature is the matter of approved inspectors. For a number of years only the National House Building Council enjoyed this status but now the position has changed. There is an increasing number of approved inspectors now available, and the body given the responsibility for vetting candidates is the Building Industry Council. The chapter relating to approved inspectors has been reassembled and extended.

Part B relating to fire safety has also received attention. Requirements relating to the spread of fire on linings, structure and externally have been qualified – measures taken are to be 'adequate'; and, in connection with access facilities for the fire service, measures taken are to be 'reasonable'.

There are many alterations and additions to the approved document to Part B, including requirements for habitable rooms up to and above 4.5 m from ground level in dwellings, flats and maisonettes. Guidance is offered on the use of insulating core panels.

The title of Part K has been changed to 'Protection from falling, collision and impact'. It incorporates additional requirements regarding guarding and barriers, protection from collision with open windows, skylights or ventilators, and also the prevention of people being trapped by doors.

A change of title also applies to Part N, which is now called 'Glazing: safety in relation to impact, opening and cleaning'. There is now provision to ensure the safe opening of windows, skylights and ventilators, and also the safe cleaning of windows and other glazed surfaces.

An important addition to the regulations is the extension of access and facilities for disabled people to dwellings. The main points are the need for a safe access to the dwelling, a 'level' threshold, easy circulation within the entrance storey of the dwelling and accessible WCs. The placing of light switches and sockets are controlled. Lifts and common stairs in flats are also subject to requirements.

Another privately prepared, approved document has been published. This relates to the basements of dwellings and is added to Chapter 23. Published by the British Cement Association it brings together all the facets of building regulations affecting basements.

There are additions to the chapter on materials and workmanship, and also more about Radon which has been found to extend to areas not previously reported.

Lists of relevant reading have been revised, as have the sources of information and research. There is a new selection of appeal decisions and determinations made by the Secretary of State.

As for the future, several approved documents are 'under active review' by DETR, including E – resistance to the passage of sound; H – drainage and waste disposal; J – heat producing appliances, and L – conservation of fuel and power, and consultation documents are or have been available.

Various tables and material is Crown Copyright and is reproduced with the permission of the Controller of Her Majesty's Stationery Office.

April 2000
John Stephenson

Performance standards
'Building Control Performance Standards' was published in July 1999; it sets out standards to be applied to a building control service, together with guidance towards achieving these standards.

It is recommended that building control bodies should:

- prepare and make public a policy on the provision of the service;
- have adequate staff resources supplemented when necessary;
- carry out consultations with other agencies;
- give feedback to client upon completion of plans assessment;
- keep records of the plan-checking process;
- have appropriate site inspection regime;
- act promptly in the event of contraventions and keep records;
- give certificates to direct client when work is satisfactorily completed;
- keep records for appropriate time – could be for 15 years, or more;
- arrange for CPD and in-service training;
- arrange for self review of performance;
- adopt quality management principles;
- keep to professional standards and business ethics.

The document is available from DETR Free Literature, PO Box 236, Wetherby L23 7NB.

FOREWORD TO FIRST EDITION

November 1985 was a red letter month for building control. New regulations regarding construction came into operation, and methods of plan checking and inspection were expanded to allow for involvement of people and organisations other than local authorities.

These new regulations arose because of demands for improvements and simplification. They are certainly different – and brief. Implementation arises from the contents of a number of documents approved by the Secretary of State for the Environment.

As with the deemed-to-satisfy requirements in the 1976 Regulations, compliance with the approved documents is not mandatory if other equally suitable materials and methods are employed. However, if something goes wrong the fact that an approved document has been observed is a defence for the developer.

There are additionally some mandatory requirements. They are based on British Standards for means of escape in case of fire, and it is to be expected that British Standards and Codes will play an increasing role in building control in the future. Agrément certificates also have a part to play and some may in due course be regarded as approved documents.

It must be emphasized that, as with the now obsolete regulations, this new version sets down minimum standards which are acceptable. There is no attempt to state standards which ought to be attained to ensure a standard of construction, comfort and protection of health and safety that the 'man in the street' regards as reasonable to the norm as we approach the 21st century.

Many familiar portions of the old regulations have disappeared. Examples are the absence of a minimum ceiling height, the complicated shafts of open space outside windows and the long-standing requirement that there should not be direct access from a living room into a toilet.

Introduction of the scheme for approved inspectors has stimulated local authorities to search for methods they can employ to provide a comparable service. At the time of writing the only approved inspector is the National House Building Council; insurance requirements having delayed the approval of individuals.

In consequence local authorities are cooperating in the formation of consortia to operate house type approval schemes. Basically the idea is that when a builder has a house design that he intends to use repetitively he may send his plans to one local authority. They will arrange to consult building control officers in two other council areas and examine the plans. When approved the plans will be acceptable to all councils embraced in the consortium.

The service of vetting plans being provided is not the subject of a fee; payment of the appropriate building control fees will be made when a proposal to build a particular scheme is initiated.

These arrangements are being supported by a scheme for insurance enabling a developer to provide a ten year warranty for each job upon payment of a premium to the insurance company participating in the scheme.

Early chapters outline events leading to the current situation and the amount of legislation normally associated with building control. The aim to have overall national requirements in England and Wales has been largely achieved, but not completely so. Much of the Inner London legislation remains and in addition to the Building Act 1984 many of the county councils have promoted and secured Acts with requirements limited to proposals within the county boundary. They can deal with such matters as:

- access for fire appliances
- fire precautions for parking places
- minimum areas for habitable rooms
- rooms liable to flooding
- separate foul and surface water sewers
- subsidence.

As with all regulatory situations, success of the 1985 Regulations and its variety of control processes, will hinge on the eagerness, willingness and capacity of all engaged in the construction industry to make the system work. The watchword is cooperation; not conflict.

John Stephenson
Houghton on the Hill
March 1986

FOREWORD TO SECOND EDITION

The largest single change has been the addition of Part M – facilities for disabled people – for Schedule 2 and so Chapter 21 has been completely re-written. Amendments No. 1 to the Manual and Approved Documents was issued by the Department of the Environment during May and these are included in the text as appropriate.

The system of numbering pages has been varied to make each chapter complete and there is now an extensive index.

Additionally more useful information has been provided including a number of determinations made by the Secretary of State and there are several draft checklists, all to be found in Annexes.

June 1988   John Stephenson

FOREWORD TO THIRD EDITION

The work of preparing this new edition was under way during the storms of January and February, and an item in Building Today on 1 February 1990 is most pertinent.

Shoddy workmanship, it was claimed, was to blame for much of the damage inflicted on buildings during the gales. In the opinion of Dr Nick Cook, the Building Research Station's Wind Engineering Consultant 'there is something radically wrong – buildings should have withstood these winds'. The main reason for the damage (which in some incidents meant that entire roofs were ripped off houses) is that roof and walls are too often inadequately tied. The BRE spokesman insisted that if properly followed, the 1985 Regs are adequate.

The 1985 Regs do not aim for a Utopian standard, so it is up to all to ensure that the standards of design, construction and workmanship do not fall short of the standards set out.

This edition was prompted and is much altered following issue of the Building Regulations (Amendment) Regulations 1989, which came into force on 1 April 1990.

In the wake of these amended regulations there have been five new Approved Documents, each labelled '1990 Edition'. They are

Part F — Ventilation and condensation.
Part G — Bathrooms, hot water storage, sanitary and washing accommodation.
Part H — Drainage and solid waste storage.
Part J — Heat producing appliances.
Part L — Conservation of fuel and power.

There are some major changes. In Part F there is now a requirement for mechanical ventilation in kitchens and bathrooms together with 'background ventilation' in other rooms. Not quite a return to the old byelaw airbrick in every room, but recognizing why the old airbrick was there.

John Stephenson

FOREWORD TO FOURTH EDITION

First published in 1972 as 'Building Regulations in Detail' and running to four editions and five supplements, the book assumed its present title with the advent of the Building Regulations 1985, and its place on the bookshelves of those concerned at all levels of the construction industry continues.

Revised building regulations came into use on 1 June 1992 and they reflected the most extensive change since the 1985 Regulations, and the attendant approved documents were published. The amendments are also against a background of new Statutes, which have an influence on building control, with a variety of titles from the National Health Service and Community Care Act 1990 to the Water Industry Act 1991.

The Construction Products Regulation 1991 are a further step towards harmonization throughout Europe so far as materials are concerned, and this type of regulation will play an increasing part in the construction industry.

Among the changes to be found within the new requirements and the attendant approved documents are, particularly, fire safety provisions and the need for satisfactory access to buildings for the fire services.

There is also new guidance on stairs and the risks associated with young children. The thinking links to the new Part N aimed at reducing the risks associated with glazing in buildings, conservatories and porches.

More alterations have been made to the requirements relating to access facilities for disabled people which covers all floors in non-domestic buildings.

On the administrative side there is the introduction of completion certificates which are to be furnished upon application and will indicate that building work has been completed satisfactorily.

There is therefore much that is new in this edition.

The permission of HMSO to reproduce a number of tables is acknowledged and appreciated.

John Stephenson
Houghton on the Hill
April 1992

Shortly after the 1991 Building Regulations and the 1992 Approved Documents became operative, a 33 page publication became available – 'Amendments 1992 to the Approved Documents'. The publishing process has been delayed a little to accommodate these amendments.

25 June 1992

FOREWORD TO FIFTH EDITION

Most of the chapters have additional comment and also incorporated are the alterations brought about by the Building Regulations (Amendment) Regulations 1994. The main purpose of these regulations is to strengthen the requirements of Part L relating to the conservation of fuel and power.

The new requirements are aimed at reducing CO_2 emissions from buildings by involving cost-effective measures without introducing 'unacceptable' technical risks, while at the same time maintaining some flexibility for designers. It has been estimated by BRE that the energy performance of space and water heating in houses will be improved by between 25 and 35%. Not everyone agrees, and the inclusion of fuel costs in calculations has caused quite a rumpus.

Alterations to Part F are designed to improve and simplify provisions for ventilation in dwellings with a wide range of options on how this may be achieved. Non-domestic buildings are for the first time affected by additions to Part F. In Part A the requirements relating to disproportionate collapse have been simplified.

There is also a new regulation, 13A, which has been introduced to allow local authorities to certify building work which should have been subject to control but was previously unauthorized.

There are notes in a new Chapter 23 referring to an approved document prepared by the Timber Development Association; a collection of appeal and determination decisions based on the 1991 Regulations and many additions to the various lists of 'relevant reading' throughout the book.

The first edition of 'Building Regulations 1972 in detail' concluded with the following paragraph. 'If this book assists members of the profession in practice and public employment, builders, tradesmen and students in appreciating the range and aim of the various regulations, then there is a considerable measure of job satisfaction for the author'. Those sentiments still apply today!

The permission of HMSO to reproduce a number of tables and Appendix G to Part L is acknowledged and appreciated.

John Stephenson
Houghton on the Hill
14 January 1995

List of abbreviations

The following abbreviations have been used:

Public Health Act 1875	1875 Act
Public Health Act 1936	1936 Act
Public Health Act 1961	1961 Act
Health and Safety at Work, etc., Act 1974	1974 Act
Local Government (Miscellaneous Provisions) Act 1982	1982 Act
Housing and Building Control Act 1984	1984 HBC Act
Building Act 1984	1984 Act
Town and Country Planning Act 1971	1971 Act
Water Act 1973	1973 Act
Fire Precautions Act 1971	1971 Fire Act
Building Regulations 1976	1976 Regs
Building Regulations 1985	1985 Regs
Building (Approved Inspectors) Regulations 1985	1985 AI Regs
Building (Inner London) Regulations 1985	1985 IL Regs
Water Act 1989	1989 Act
Building Regulations (Amendment) Regulations 1989	1989 Regs
National Health Service and Community Care Act 1990	1990 NH Act
Town and Country Planning Act 1990	1990 Act
Planning (Listed Buildings and Conservation Areas) Act 1990	1990 LBC Act
The Construction Products Regulations 1991	1991 CP Regs
Water Industry Act 1991	1991 Act
Building Regulations 1991	1991 Regs
Building Regulations (Amendment) Regulations 1992	1992 Regs
Building Regulations (Amendment) Regulations 1994	1994 Regs
Party Wall, etc. Act 1996	PW Act
Building Regulations (Amendment) Regulations 1997	1997 Regs
Building (Local Authority Charges) Regulations 1998	1998 LAC Regs
Building Regulations (Amendment) Regulations 1998	1998 Regs
Building Regulations (Approved Inspectors, etc.) (Amendment) Regulations 1998	1998 AI Regs
Building Regulations (Amendment) Regulations 1999	1999 Regs
Building Regulations (Amendment) (No. 2) Regulations 1999	1999 No. 2 Regs

List of statutes mentioned in text

CHAPTER 1

The development of building control

One is frequently impelled to the conclusion that building regulations are a modern impediment to progress, but in effect building control in some form or another, is of great antiquity. The most extensive changes in the extent and methods of control have, however, happened since 1961.

In ancient Greece there was a crude form of constraint largely exercised by the amount of building which could be contained within a city wall. A certain Roman emperor is thought to have burned down a city so that it could be rebuilt in accordance with his new building regulations.

Building control in this country began in London as long ago as 1189 when regulations relating to party walls, rights of light, drainage and related matters were made. Fire-resistant construction and rudimentary means of escape in case of fire were also subject to regulation about the same time.

After the fire of London the first comprehensive building Act came onto the statute book. This included provisions for the appointment of surveyors who were charged with the duty of ensuring that the regulations were observed.

There was a consolidating act in 1774 which gave rise to the appointment of district surveyors in London, and a scale of fees was laid down for their services. This piece of legislation operated for a further 70 years until 1845 when a London Building Act came into force.

PUBLIC HEALTH ACT 1875

Towns in the provinces were at this time also becoming interested in building control and the Liverpool Building Act came about in 1842. But it was not until the Public Health Act 1875 that general powers to control construction by means of byelaws came into effect. Local authorities were empowered to make byelaws aimed at securing the interests of health. This included the sufficiency of air space about buildings, ventilation, drainage, water closets, earth closets, cesspools, etc. This power was extended by an amending Act in 1890 to include the provision for the flushing of water closets, the height of rooms intended for human habitation, the paving of yards and open spaces around dwellings and the provision of back yards intended to facilitate the removal of refuse from a dwelling. Model byelaws were issued by the then Ministry of Health, and byelaws made by councils were normally based upon this model. They were not operative until confirmed by the Ministry of Health. The model byelaws were offered as a guide only, and they were considerably varied in many towns, leading to a lack of uniformity.

PUBLIC HEALTH ACT 1936

The most major piece of legislation was the Public Health Act 1936 which repealed the greater part of earlier legislation, and provided for byelaws with respect to buildings and sanitation. But local authorities were not obliged to make byelaws. The position was that 'every local authority may, and if required by the Minister, shall make byelaws for regulating all or any of the following matters':

- the construction of buildings and the materials to be used;
- space about buildings, the lighting and ventilation of buildings, and the dimensions of, rooms intended for human habitation;
- the height of buildings, the height of chimneys above roofs;
- sanitary conveniences, drainage, including the conveyance of water from roofs and yards; cesspools and other means for the reception or disposal of foul matter;
- ashpits;
- wells, tanks and cisterns for the supply of water for human consumption;
- stoves and other fittings in so far as required for the purposes of health and the prevention of fire;
- private sewers; junctions between drains and sewers and between sewers.

Byelaws made under the 1936 Act could also make provisions requiring:

- the deposit of plans and giving notices;
- the inspection of work; testing of drains and sewers and the taking of samples of materials.

The byelaws also related to structural alterations or extensions of buildings and instances where any material change of use had taken place.

In many areas there have also been Acts promoted by counties or large cities dealing with a specific local requirement. The many local Acts, some of which dated from the beginning of the century, or even earlier, gave rise to yet further variations or inflexibility.

PUBLIC HEALTH ACT 1961

In an effort to achieve uniformity, and to enable the requirements to be more readily amended to take account of advances in knowledge together with changes in techniques and materials, powers were taken in the 1961 Act to make building regulations.

This was a drastic change in the law relating to building control because it removed from local authorities the powers they had possessed for so many years, to make building byelaws. The new procedure was for the Ministry to make building regulations by way of the statutory instrument, having a universal application throughout England and Wales. The regulations were to be made to cover the list of objectives set out in the 1936 Act and mentioned above. Local authorities remained responsible for the enforcement of building regulations (section 4).

There was a new provision in the 1961 Act enabling a local authority to dispense with or relax any requirement of the building regulations, where the authority (or the Minister) considered that the operation of the regulation would be unreasonable (section 6). When a local authority refused an application to relax or dispense with a requirement the applicant could, within a month, appeal to the Minister.

The first building regulations were made in 1965 coming into force on 1 February 1966; there were many amendments before the regulations were metricated in 1972, and there have been numerous alterations and additions.

HEALTH AND SAFETY AT WORK, ETC., ACT 1974

In August 1972 the Government published and circulated to many organizations a consultative document called *Proposals for a Building Bill*. Building organizations, manufacturing and professional interests all had the opportunity of sending their views to the Department of the Environment. Many people naturally expected that after all the representations had been sifted, analysed and collated there would ultimately appear on the statute book a Building Act devoted entirely to those matters referred to in the consultative booklet. But this was not to be; at least not in 1974.

Instead Parliament passed the Health and Safety at Work, etc., Act 1974. It was a long Act of which more than one-third related to building regulations. The legislation concerned with building regulations commenced at section 61. This was a remarkable coincidence as it was section 61 of the 1936 Act which gave the power enabling building byelaws to be made.

The scope, purpose and coverage of building regulations could be increased to an unexpected degree. There was also a number of alterations to procedures claimed to give greater flexibility to building control.

The powers made possible some rationalization of the then system of controls affecting new buildings and the preparation of a more comprehensive code of building regulations covering England and Wales. Scottish regulations were already more extensive than those applied in England and Wales, and there still remains an opportunity for one code of regulations embracing the Scots, Welsh and English.

There was provision for extending the scope of building regulations enabling a much wider range of subjects to be covered. These included electrical and other building services, water fittings, together with access for firemen and disabled people. All types of building services and equipment could be included.

It was possible by a strict interpretation of the 1974 Act for all types of structures and erections – for example, radio and floodlighting masts – to be subjected to building regulation requirements.

Provision was included to enable local authorities for the first time to approve plans in stages. Also for the first time power was given enabling building regulations to impose continuing requirements on owners and occupiers – for example, to ensure that lifts were maintained in good operating condition.

Additional powers were given to local authorities regarding the testing of materials and the Act provided for the introduction of a system of fees in connection with building control.

Not all of the new requirements in the Act ever came into operation, the provisions consisted of 'enabling powers' which means that they did not come into force on the day it received Royal Assent. The coming into force of the new provisions depended upon the Secretary of State making 'Commencement Orders' bringing various parts of the 1974 Act into operation at different times.

BUILDING (PRESCRIBED FEES) REGULATIONS 1980

One of the sections of the 1974 Act which was the subject of 'a commencement order' related to the charging of prescribed fees for and in connection with a council's duties relating to building regulations.

The first regulations became operative on 1 April 1980 and set down scales of fees to be charged by local authorities for the performance of their functions, comprising:

- the passing or rejection of plans;
- the first inspection of any building.

LOCAL GOVERNMENT (MISCELLANEOUS PROVISIONS) ACT 1982

This odd piece of legislation deals with such exciting premises as sex and take-away food shops; acupuncture and ear piercing. However, it is of interest, for there is one section which related to building regulations.

It introduced a measure of flexibility in the arrangements a local authority could make for the passing or rejection of plans.

Previously if deposited plans complied with the regulations they had to be passed within a prescribed period. If plans were defective or showed contraventions in any particular way the local authority was obliged to reject, thereby requiring the developer to submit plans a second time.

This Act enabled a local authority upon receipt of defective plans to reject them as before or pass them subject to either or both of the following conditions:

- the plans must be modified and
- more plans are to be deposited relating to the parts of the building in doubt.

Both parties had to be in agreement that this procedure would operate.

HOUSING AND BUILDING CONTROL ACT 1984

In February 1981 the Secretary of State issued a command paper entitled *The Future of Building Control in England and Wales*. This document contained a whole series of proposals covering a revision of building regulations and mode of control. Wider exemptions and new arrangements for appeals were defined, as was the suggestion that plans could be certified as complying with the building regulations by suitably qualified persons in private practice. The criteria for securing approval as a private certifier would be a professional qualification, practical experience and adequate indemnity insurance.

Upon this basis the Department of Environment set about completely recasting the building regulations and to produce a new Act to amend the law relating to the supervision of building work, the building regulations, sanitation of buildings and building control.

Unfortunately these requirements were linked with some very controversial housing legislation which had a very rough ride in both houses of Parliament and a general election also intervened. Such was the delay that although the Bill was first before Parliament in 1982, it was not until June 1984 that the Housing and Building Control Act 1984 became law.

The 1984 Act contained legislation enabling:

- the supervision of plans and work otherwise than by local authorities;
- certain exemptions and relaxations from procedural requirements for public bodies;
- provision for approved documents giving guidance to building regulations;
- private certification of compliance with building regulations.

However, the parts of the 1984 Act relating to building control only survived for a few months. On 1 December another Act was approved which assumed the mantle of the radical changes for private certification.

BUILDING ACT 1984

This Act consolidated various building control statutes enacted over the last 90 years. It is no longer necessary to search through a confusing array of enactments concerning buildings and related matters;

most are now restated in the 1984 Act and an immense amount of past legislation has in consequence been repealed.

There are altogether 135 sections and 7 schedules to this Act and the various sections are grouped under the following heads:

- Approved documents (sections 6 and 7)
- Appeals in certain cases (sections 39 to 43)
- Application to Crown, etc. (sections 44 and 45)
- Appeal to Crown Court (section 86)
- Appeal against notice requiring works (section 102)
- Breach of building regulations (sections 35 to 38)
- Buildings (sections 69 to 75)
- Consultation (sections 14 and 15)
- Classification of buildings (section 34)
- Compensation and recovery of sums (sections 106 to 111)
- Determination of questions (section 30)
- Drainage (sections 59 to 63)
- Defective premises – demolition (sections 76 to 83)
- Duties of local authorities (section 91)
- Documents (sections 92 to 94)
- Default powers (sections 116 to 118)
- Exemption from building regulations (sections 3 to 5)
- Entry on premises (sections 95 and 96)
- Exemption of works (sections 97 to 101)
- General provisions about appeals and applications (sections 103 to 105)
- Inner London – new regulations apply to Inner London from 1 January 1986 (sections 46 and 88)
- Interpretation (sections 121 to 127)
- Lapse of deposit of plans (section 32)
- Local inquiries (section 119)
- Miscellaneous (sections 89 and 90)
- Obstruction (section 112)
- Orders (section 120)
- Power to make building regulations (sections 1 and 2)
- Passing of plans (sections 16 to 29)
- Proposed departure from plans (section 31)
- Provision for sanitary conveniences (sections 64 to 68)
- Prosecutions (sections 113 and 114)
- Protection of members of authorities (section 115)
- Relaxation of building regulations (sections 9 to 11)
- Supervision of plans and work by approved inspectors (sections 47 to 53)
- Supervision of their own work by public bodies (section 54)
- Supplementary (sections 55 to 58; 132 to 135)
- Savings (sections 128 to 131)
- Type approval of building matter (sections 12 and 13)
- Tests for conformity with building regulations (section 33)
- Yards and passages (sections 84 and 85).

BUILDING REGULATION MATTERS

The matters for which building regulations could be made was considerably extended by the Health and Safety at Work, etc., Act 1974, and this widening of scope has been transferred to the 1984 Act.

A perusal of the following list setting down all the matters for which building regulations may be made reveals that in many cases existing regulations are very brief. In other examples regulations have not been made. It is possible that existing regulations may in the future be modified in some respects. Other topics may never be subject to regulatory control if the current pressure towards a reduction of controls persists.

Broadly the Secretary of State may make regulations controlling the design and construction of buildings and services, fittings and equipment associated with them.

Regulations may be made to:

- secure the health, safety and welfare and convenience of persons in or about buildings and others connected with buildings;
- to conserve fuel and power;
- to prevent waste, misuse or contamination of water (section 1).

The Fire Precautions Act 1971 also enables regulations to be made in connection with fire escapes.

So far as the 1984 Act is concerned, regulations may be made in respect of the following:

- preparation of sites;
- suitability, durability and use of materials and components (including surface finishes);
- structural strength and stability, including

  (a) precautions against overloading, impact and explosion
  (b) measures to safeguard adjacent buildings and services
  (c) underpinning;

- fire precautions, including

  (a) structural measures to resist the outbreak and spread of fire and to mitigate its effects
  (b) services, fittings and equipment designed to mitigate the effects of fire or to facilitate firefighting
  (c) means of escape in case of fire and means for securing that such means of escape can be safely and effectively used at all material times;

- resistance to moisture and decay;
- measures affecting the transmission of heat;
- measures affecting the transmission of sound;
- measures to prevent infestation;
- measures affecting the emission of smoke, gases, fumes, grit or dust or other noxious or offensive substances;
- drainage (including waste-disposal units);
- cesspools and other means for the reception, treatment or disposal of foul matter;
- storage, treatment and removal of waste;
- installations utilizing solid fuel, oil, gas, electricity or any other fuel or power (including appliances, storage tanks, heat exchangers, ducts, fans and other equipment);
- water services (including wells and bore-holes for the supply of water) and fittings and fixed equipment associated therewith;
- telecommunications services (including telephones and radio and television wiring installation);
- lifts, escalators, hoists, conveyors and moving footways;
- plant providing air under pressure;
- standards of heating, artificial lighting, mechanical ventilation and air-conditioning and provision of power outlets;
- open space about buildings and the natural lighting and ventilation of buildings;
- accommodation for specific purposes in or in connection with buildings, and the dimensions of rooms and other spaces within buildings;
- means of access to and egress from buildings and parts of buildings;
- prevention of danger and obstruction to persons in and about buildings (including passers-by);
- matters connected with or ancillary to any of the foregoing matters.

In their broadest sense building regulations may be made to provide for the following:

- guidance on acceptable methods of construction, referred to as approved documents;
- giving documents published by bodies other than the Secretary of State the status of approved documents;
- procedures for the giving of notices;
- the deposit of plans for work proposed; and for work already undertaken;
- the council to keep one copy of deposited plans;
- the inspection and testing of materials and the taking of samples;
- approved inspectors and others to make consultations prior to approving plans;
- the acceptance of a certificate from an approved person;
- the issue of certificates by local authorities that work has been completed;

- the imposition of fees;
- prescribed persons to undertake some functions in parallel with local authorities;
- matters to be provided, or undertaken in connection with buildings;
- the way in which building work may be carried out;
- the manner in which they may be applied to alterations, extensions, services and fittings;
- installation of new services;
- application to certain changes of use;
- application to particular localities, where appropriate;
- the repeal or modification of legislation (section 1 and Schedule 1).

This Act then is the basis of the new building control system. It is supported by the Building Regulations 1985, which is divided into five parts. Compared with the 1976 Regulations it is a very modest document indeed:

Part I   General (Regulations 1 and 2)
Part II   Control of building work (Regulations 3 to 9)
Part III   Relaxation of requirements (Regulation 10)
Part IV   Notices and plans (Regulations 11 to 14)
Part V   Miscellaneous (Regulations 15 to 20)

There are four schedules:

Requirements: Schedule 1
Facilities for disabled people: Schedule 2 (now replaced by part M)
Exempt buildings or work: Schedule 3
Revocations: Schedule 4.

There is one mandatory document with requirements for means of escape in case of fire; a manual giving explanatory notes and 13 approved documents, each containing guidance on the implementation of the regulations:

- materials and workmanship
- structure
- fire
- site preparation
- toxic substances
- sound insulation
- ventilation
- hygiene
- drainage and waste disposal
- heat-producing appliances
- stairways, ramps and guards
- conservation of fuel and power
- facilities for disabled people.

Also, on 11 November 1985 the Building (Approved Inspectors, etc.) Regulations 1985 came into force, and they are concerned with the supervision of building work by approved inspectors instead of the local authority.

These regulations are longer than the 1985 Building Regulations containing as they do 29 regulations and 8 schedules. The regulations are divided into 10 parts:

Part I   General (Regulations 1 and 2)
Part II   Grant and withdrawal of approval (Regulations 3 to 7)
Part III   Supervision of work by approved inspectors (Regulations 8 to 11)
Part IV   Plans certificates (Regulations 12 to 14)
Part V   Final certificates (Regulation 15)
Part VI   Cessation of effect of initial notice (Regulations 16 to 18)
Part VII   Public bodies (Regulations 19 to 26)
Part VIII Certificates relating to deposited plans (Regulation 27)
Part IX   Registers (Regulation 28)
Part X   Effect of contravening building regulations (Regulation 29)

There are eight schedules:
Schedule 1:   Enabling powers
Schedule 2:   Forms
Schedule 3:   Grounds for rejecting an initial notice
Schedule 4:   Grounds for rejecting a plans certificate
Schedule 5:   Grounds for rejecting a final certificate
Schedule 6:   Grounds for rejecting a public body's notice
Schedule 7:   Grounds for rejecting a public body's plans certificate
Schedule 8:   Grounds for rejecting a public body's final certificate

CLASSIFICATION OF BUILDINGS

For the purposes of the building regulations and related matters, buildings may be classified by reference to size, description, design, purpose, location or any other characteristic whatsoever (1984 Act, section 34).

ADDITIONS AND ALTERATIONS

Since the making of the building regulations in 1985 there have been a number of legal developments:

- **The Building (Inner London) Regulations 1985** applied the national system to London from 1 July 1987.
- **The Building (Disabled People) Regulations 1987** replaced Schedule 2 of the Building Regulations 1985 by placing a new Part M, supported by an approved document which has been superseded by a 1992 edition.
- **The Fire Safety and Safety at Places of Sport Act 1987** made amendments to the Fire Precautions Act 1971.
- **The Water Act 1989** set out a framework of new bodies to be responsible for water and sewerage services following privatization. There were revised provisions for requisitioning water mains and sewers and a power under section 79 enabling the undertakers to ask for payment for each property reflecting the cost of infrastructure. There were also numerous amendments to the Building Act 1984.
- **The Water Act 1991** consolidated numerous enactments concerning water supply and sewerage and replaced some of the provisions of the Public Health Act 1936.
- **Water Consolidation (Consequential Provisions) Act 1991** also made some amendments to the Building Act 1984.
- **The Construction Products Regulations 1991** implemented in the UK the provisions of Directive 89/106/EEC; coming into force on 27 December 1991.
- **The Building Regulations 1991** came into force on 1 June 1991. The mandatory document regarding means of escape in case of fire has been abandoned, as has the manual. There are now 14 approved documents, each containing advice on the implementation of regulations:

  - materials and workmanship
  - structure
  - fire safety – much enlarged and containing means of escape
  - site preparation and resistance to moisture
  - toxic substances
  - resistance to passage of sound
  - ventilation
  - hygiene
  - drainage and waste disposal
  - heat-producing appliances
  - stairs, ramps and guards
  - conservation of fuel and power
  - access and facilities for disabled people
  - glazing – materials and protection.

- **The Building Regulations (Amendment) Regulations 1992** contained sundry minor amendments to the Building Regulations 1991.

- **Workplace (Health, Safety and Welfare) Regulations 1992.** These regulations implement provisions in Directive 89/654/EEC regarding minimum safety and health requirements for the workplace.

  Requirements imposed include:

  - maintenance of buildings and equipment; ventilation; temperature; lighting; cleanliness; room dimensions; work stations; condition of floors; falling objects; windows; doors; gates; escalators; sanitary conveniences; and washing facilities

- **The Building Regulations (Amendment) Regulations 1994** amended the Building Regulations 1991, and

  - inserted a new regulation, 13A, to allow regularization certificates to be issued in relation to unauthorized work
  - inserted a new regulation, 14A, requiring energy rating for new dwellings.
  - revoked Requirement A4 relating to disproportionate collapse
  - limited the application of requirements regarding ventilation with a new approved document for Part F
  - introduced further requirements relating to the conservation of fuel and power with a new approved document for Part L.

- **The Building (Prescribed Fees) Regulations 1994.** Revised the basis for charging fees for work in connection with the consideration of plans and inspection of building work.

- **Construction (Design and Management) Regulations 1994.** These regulations cover the minimum health and safety requirements of Directive 92/57/EEC at temporary or mobile construction sites.

  They apply to sites where work lasts for more than 30 days or will involve more than 500 person days of construction work.

  Requirements are:

  - the client must appoint a competent planning supervisor;
  - the planning supervisor prepares a health and safety plan;
  - the designer must co-operate with the planning supervisor, and incorporate health and safety considerations in the design;
  - the principal contractor must ensure that all at works on the project comply with the health and safety plan. Enforcement is the responsibility of the Health and Safety Executive.

    In circumstances where building control authorities serve notices requiring specified work to be carried out on buildings, or issue regulation approvals containing conditions, the authority may be directly concerned with these regulations.

- **Building Regulations (Amendment) Regulations 1995** contained amendments relating to a material change of use and unauthorized building work.

- **Building (Approved Inspectors) (Amendment) Regulations 1995** contains sundry amendments to the Building (Approved Inspectors) Regulations 1985 including reducing the period in connection with initial notices and the rejection of a plans certificate from 10 to 5 days.

- **Construction (Health, Safety and Welfare) Regulations 1996.** These regulations implement the minimum health and safety requirements of Directive 92/57/EEC at temporary construction sites.

  The regulations apply to a site where building or civil engineering construction work is undertaken. A workplace on a site used for other purposes is excluded.

  Principal requirements are:

  - place of work is to be safe with safe access and egress;
  - steps are to be taken to prevent falls;
  - measures are to be taken to prevent any person falling through a fragile material;
  - all practicable steps are to be taken to prevent collapse of the structure during building;
  - demolition or dismantling are to be safely carried out;
  - prevent excavations collapsing;

  - pedestrians and vehicles must be able to move safely;
  - doors and gates are not to present a danger to persons;
  - suitable firefighting equipment is to be on site;
  - where necessary, emergency exits from the site are to be provided;
  - a wide range of welfare services are necessary.

  Enforcement is the responsibility of the Health and Safety Executive, although the fire services may be involved.

- **The Building (Approved Inspectors) (Amendment) Regulations 1996** provided for the insertion of a new regulation 8A relating to amendment notices into the Building (Approved Inspectors, etc.) Regulations 1985.

- **Party Wall, etc. Act 1996.** This Act sets out arrangements for preventing and resolving disputes which may arise between adjoining building owners. It is concerned with the construction and repair of walls on the line of junction between adjoining properties, and also relating to excavations near neighbouring buildings.

  Anyone proposing to carry out work in these circumstances is required to give notice to the adjoining owner who can agree or disagree with what is proposed. There is provision for the resolution of disputes.

  The provisions extend to the whole of England and Wales.

- **The Building Regulations (Amendment) Regulations 1997** came into force on 1 January 1998 and amended the Building Regulations 1991 in the following respects.

  - The requirements of K1 and K2 were extended to apply to stairs, etc., which give access to levels used only for maintenance purposes.
  - There is a requirement to protect people from colliding with vehicles in loading bays (K3).
  - People are also to be protected from collision with open windows, skylights or ventilators (K4).
  - People in buildings are also to be protected from being struck or trapped by sliding or powered doors or gates (K5).
  - There is provision requiring the safe opening and closing, or adjusting of windows, skylights or ventilators (N3).
  - Safe access must be provided for cleaning windows, skylights and other transparent surfaces (N4).

- **The Building Regulations (Amendment) Regulations 1998.** This was a further amendment to the 1991 Regs, making provision for Part M – Access for disabled people – to be applied to new dwellings. In particular, care has to be taken in the provision of access to and into a dwelling; circulation within the entrance storey of a dwelling, switches and socket outlets in a dwelling; the provision of a WC in the entrance storey of a dwelling and passenger lifts or common stairs serving blocks of flats.

- **The Building Regulations (Approved Inspectors, etc.) (Amendment) Regulations 1998.** These regulations substitute a new regulation 3 in place of regulation 3 in the 1985 AI Regs, requiring an individual or corporate body desiring to become an approved inspector to make an application to a designated body.

  This body is the Construction Industry Council.

  There are also several other small amendments to the 1985 AI Regs.

- **The Building (Local Authority Charges) Regulations 1998.** These regulations require local authorities in England and Wales to set their own individual charges for building control work to be based on the cost of the service.

  The new system enables each local authority to set out its charges and administer the system. The DETR has published a circular advising local authorities how to produce its fees scheme. It is DETR circular 09/98 – Welsh Office circular 39/98, and is entitled 'Building and Buildings – The Building (Local Authority Charges) Regulations 1998'.

- **The Building Regulations (Amendment) Regulations 1999.** These regulations amend the 1991 Regs by substituting a revised regulation 7. It requires that building work must be carried out using materials which are proper for the circumstances in which they are being used. Materials are also to be used in a workmanlike manner, adequately mixed or prepared, and they must be used so as to adequately fulfil the function for which they are designed.

  The definitions of 'Construction Products Directive', 'European technical approval' and 'harmonized standard' were removed from the 1991 Regs. 2.

- **The Building Regulations (Amendment) (No. 2) Regulations 1999.** These regulations, made on 17 December 1999, came into force on 1 July 2000, when the 1991 Regulations Part B were amended.

Internal fire spread (linings)
B2(1)(a), now requires linings to *adequately* resist the spread of flame.

Internal fire spread (structure)
B3(2), A wall separating two buildings must *adequately* resist the spread of fire.

External fire spread
B4(1), External walls shall *adequately* resist the spread of fire.
B4(2), Roof of building shall *adequately* resist the spread of fire.
The first four amendments therefore introduce the word 'adequate' to make clear that is the standard of fire resistance required.

Access and facilities for fire service
B5(1), There are to be *reasonable* facilities to assist firefighters, and.
B5(2), *reasonable* provision is to be made for firefighting vehicles.
These amendments are intended to make it clear that facilities for access and facilities for the fire service must be reasonable.

In the wake of these regulations, a new Approved document for Part B (2000) was published, which includes for the first time guidance on the provision of fire alarm and detection systems in all relevant buildings; the provision in dwellings of an emergency exit window to habitable rooms not more than 4.5 m above ground level; schools and insulating core panel construction and sprinkler protection in new single storey retail buildings having compartments of 2000 m² or larger.

CHAPTER 2

Control of building work

Control of building work is empowered by the 1984 Act supported by regulations made by the Secretary of State for the Environment.

The regulations relate to the design and construction of buildings and the provision of services, fittings and equipment. Regulations may only contain provisions directed at:

- securing the health, safety, welfare and convenience of persons in or about buildings and of others who may be affected by buildings or matters connected with buildings;
- furthering the conservation of fuel and power;
- preventing waste, undue consumption, misuse or contamination of water.

Under the new arrangements the following have responsibility for enforcement of building control requirements:

- the Secretary of State
- local authorities
- public bodies
- private individuals or organizations who have been approved for the purpose by a designated body or the Secretary of State.

The 1985 Regulations are much shorter and much simpler than the regulations they replaced. They have been subject to amendment and all have been consolidated and added to in the Building Regulations 1991 which have been operative from 1 June 1992. They are supported by a series of approved documents setting out practical guidance to meeting requirements of the regulations. There is, however, no obligation to adopt particular solutions set out in the documents if a developer wishes to satisfy a requirement in another way.

The importance attached to approved documents arises should a contravention of a regulation be alleged. If a developer has followed the guidance set out in the appropriate approved document, that constitutes evidence tending to show that the regulations have been complied with. Each approved document contains the following advice: 'If you have not followed the guidance then that will be evidence tending to show that you have not complied. It will then be up to you to demonstrate by other means that you have satisfied the requirement.'

Approval of documents for purposes of building regulations is provided for in the 1984 Act, section 6. The Secretary of State may approve or issue any document which, in his opinion, gives guidance with respect to building regulations. An approved document need not of necessity be drafted and issued by the Secretary of State; it may have been prepared by a body designated by him for that purpose or from any other source. Examples throughout the approved documents exist where alternative approaches list a number of British Standards. A number of products may only be used when they are covered by an Agrément certificate.

Compliance or non-compliance of approved documents is covered by the 1984 Act, section 7, and it is there stated that proof of compliance with an approved document 'may be relied on as tending to negate liability'.

The 1991 Regs are supported by Schedule 1 giving the requirements. The various parts are arranged as follows:

Part
A Structure
B Fire safety
C Site preparation and resistance to moisture
D Toxic substances
E Resistance to passage of sound
F Ventilation
G Hygiene
H Drainage and waste disposal
J Heat-producing appliances
K Stairs, ramps and guards
L Conservation of fuel and power
M Access and facilities for disabled people
N Glazing – materials and protection.

These are all supported by approved documents, but not all originate on the same date, although most were issued in 1992. Parts F and L are dated 1995; K–1998; N–1998; M–1999 and B 2000.

INTERPRETATION

Regulations cannot be enforced unless one is clear as to the meaning intended by certain words and phrases. A number of statutory definitions are set out in **Regulation 2**; others will be found in the various approved documents. There is a list of definitions having a bearing on building control in Annexe A.

Sometimes a definition will contain the word 'means', and when this occurs the implication is that interpretation is limited to that particular meaning. On occasions when the word 'includes' is used, it extends the ordinary meaning of the word which is being defined.

Definitions given in **Regulation 2** are as follows:

- **Building** means any permanent or temporary building, but not any other kind of structure or erection. Reference to a building includes part of a building. This definition is much less wide than the definition in the 1984 Act, section 121. This narrower definition has been included to restrict the scope of the regulations to what are generally referred to as buildings. Such things as radio masts and hovercraft are regarded as structures or erections in the 1984 Act definition, but regulations are not to be applied to them.
- **Building notice** means a notice in accordance with Regulations 11 and 12.
- **Building work** is defined in Regulation 3(1).
- **Controlled service and fitting** means fittings to which Parts G, H or J relate.
- **Dwelling** includes a dwellinghouse and a flat.
- **Dwellinghouse** does not include a flat or a building containing a flat.
- **Flat** means separate and self-contained premises constructed or adapted for residential use and forming part of a building divided horizontally from some other part.

This definition enables different requirements to be made for houses and flats. The definition of flat covers maisonettes as well as flats.

- **Floor area** means the aggregate area of every floor in a building or extension. It is calculated to the finished internal surfaces of walls enclosing, or where there is no wall, to the outermost edge of the floor (Fig. 2.1).

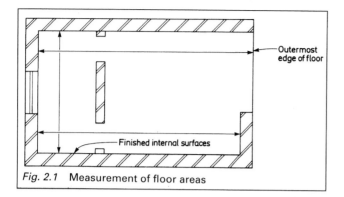

Fig. 2.1   Measurement of floor areas

- **Full plans** means plans deposited in accordance with Regulations 11 and 13.
- **Height** means the height of the building measured from the mean level of the adjoining ground to the level of half the vertical height of the roof, or to the top of the walls or a parapet, if any, whichever is the higher (Fig. 2.2).

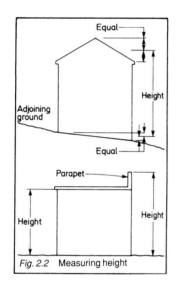

Fig. 2.2   Measuring height

- **Institution** means a hospital, home, school or similar establishment used as living accommodation for, or the treatment, care and maintenance of, persons suffering from disabilities due to illness, old age, physical disability, mental disability or children under the age of five years where such persons sleep on the premises.
- **Shop** includes premises where members of the public may purchase food or drink for consumption on or off the premises. It includes premises used for retail sales; or used by the public as a barber or hairdresser or for hiring any item and where members of the public may deliver goods for repair or other treatment.
- **Public building** means a building consisting of or containing: a theatre hall, public library or other place of public resort, a school or other educational establishment not exempted from the regulations (1984 Act, section 4), a place of worship. But a building is not to be treated as a public building because it consists of or contains a shop, restaurant, store or warehouse, or is a dwelling to which members of the public are occasionally admitted.

The definitions of 'Construction Products Directive', 'European technical approval', and 'harmonized standard' were removed from 1991 Reg. 2 by the 1999 Reg. 3.

ERECTION OF A BUILDING

The 1984 Act, section 123, says that for building control purposes references to 'construct' or 'erect' includes:

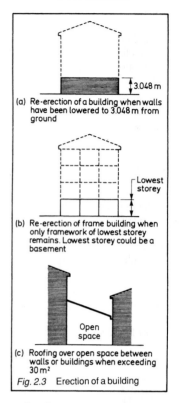

Fig. 2.3   Erection of a building

- the carrying out of such operations as may be set out in building regulations, whether it be the reconstruction of a building or the roofing over of an open space between buildings or walls or otherwise;
- the conversion of a movable object into a building. (1984 Act, section 121, includes vehicle, vessel, hovercraft or other movable object of any kind which may be prescribed by the Secretary of State.)

For other purposes in the 1984 Act contained in Part III, but not for building regulations, the following operations are deemed to be the erection of a building:

- re-erection when outer walls have been pulled or burnt down to within 10 ft of the ground;
- re-erection of a frame building when that building has been pulled or burnt down leaving only the framework of the lowest storey;
- roofing over an open space between buildings.

MEANING OF BUILDING WORK

Regulation 3. (1) 'Building work' means:

- **erection or extension of a building;**
- **material alteration of a building or a controlled service or fitting;**
- **provision, extension of a controlled service or fitting in or in connection with a building;**
- **work when a material change of use is to take place;**
- **insertion of insulating material into a cavity wall;**
- **underpinning of a building.**

(2) A material alteration occurs if work, or any part of it, would adversely affect the existing building if

(a) the building or a controlled service or fitting did not comply with a relevant requirement where previously it did so;

(b) the building or controlled service or fitting did not comply with a relevant requirement prior to the work, but as a result of the work is more unsatisfactory.

(3) 'Relevant requirement' means any of the following:

Part A   Structure
para. B1   Means of escape
       B3   Internal fire spread – structure
       B4   External fire spread
       B5   Access and facilities for fire service.
Part M   Access and facilities for disabled people.

When contemplating alterations to buildings the regulations only apply where the proposal is a 'material alteration', and they apply to the work irrespective of when the building was erected.

Controlled services or fitting is defined and the provision, extension or material alteration of these services constitutes building work.

Later in Regulation 6 are provisions relating to a change of use, and when any work is required by that regulation it is regarded as building work.

It is up to the developer to decide whether his proposals comprise a material alteration, and when in doubt it will be wise to have early discussions with the approved inspector if he is to be engaged, or the building control officer.

Examples of work which might adversely affect an existing building are:

- removing part of a loadbearing wall;
- demolishing part of a non-loadbearing wall, which has a fire protection function;
- altering a three-storey house and in consequence a fire escape provision is necessary.

REQUIREMENTS RELATING TO BUILDING WORK

Regulation 4. (1) Building work is to be carried out:

- **in compliance with those requirements of Schedule 1 which apply.** Not all the technical requirements apply in all situations or buildings. The limitations are set out in the third column of Schedule 1.
- **when fulfilling one requirement other requirements are to be observed.**

(2) When building work has been carried out it must not result in any of the following being adversely affected:

- an extended building or one that has been materially altered;
- a building where a controlled service or fitting has been provided, extended or materially altered;
- a controlled service or fitting must meet requirements of Schedule 1. When there was previously not full compliance the result must be no more unsatisfactory than previously.

MATERIAL CHANGE OF USE

Regulation 5. A material change of use occurs when:

- **a building is used as a dwelling when previously it was not used for that purpose.** The term 'dwelling' includes a dwellinghouse and a flat.
- **a building contains a flat when previously the building was not used for that purpose.** Flat means a separate self-contained premises including a maisonette forming part of a building separated horizontally from the remainder of the building.
- **a building used as a hotel or boarding house when previously the building was not used for this purpose.**
- **a building used as an institution when previously it was not used for that purpose.** 'Institution' has been defined earlier but

there is no definition of 'hotel'. It would have been helpful for a definition of hotel to have been included. This might in some way have related it to the Fire Precautions Act 1971 under which hotels and boarding houses have to obtain fire certificates. One would expect that a motel is embraced by the term 'hotel'.

- **a building used as a public building where previously it was not used for that purpose;**
- **a building is no longer exempted when a change of use has removed it from one of the Classes I to VI in Schedule 2.**

The exempted classes in Schedule 2 are:

Class I   Building controlled under other legislation
Class II   Building not frequented by people
Class III   Greenhouses and agricultural buildings
Class IV   Temporary buildings and mobile homes
Class V   Ancillary buildings
Class VI   Small detached buildings.

Exempted buildings are examined in Chapter 5.

- **a building, which already contains at least one dwelling, contains a greater or lesser number of dwellings.** This clause was inserted by the Building Regulations (Amendments) Regulations 1995.

Town planning requirements relating to a change of use should always be explored as their status does not always coincide with the building control position.

Requirements relating to material change of use

Regulation 6. (1) The requirements of Schedule 1 as listed below are to be satisfied in all cases where a material change of use of the whole building occurs:

(a) B1   Means of escape
    B2   Internal fire spread – linings
    B3   Internal fire spread – structure
    B4(2)   External fire spread – roofs
    B5   Access and facilities for the fire service
    F1   Ventilation
    F2   Condensation in roofs
    G1   Sanitary conveniences and washing facilities
    G2   Bathrooms
    H4   Solid waste storage
    J1   Heat-producing appliances – air
    J2   Discharge of products of combustion
    J3   Protection of building
    L1   Conservation of fuel and power.
(b) Where there is a change of use to a hotel, boarding house, institution or public building or the building is removed from the exempted classes the following requirements are also to be fulfilled:
    A1   Loading
    A2   Ground movement
    A3   Disproportionate collapse.
(c) When the change of use is of a building exceeding 15 m high the following also applies:
    B4(1)   External fire spread – walls.
(d) When the change of use is to a dwelling the following also applies:
    C4   Resistance to weather and ground moisture.
(e) When the change of use is to a dwelling or flat or the building which contained at least one dwelling when altered contains a greater or a lesser number of dwellings than it did previously.
    E1   Airborne sound – walls
    E2   Airborne sound – floors and walls
    E3   Impact sound – floors and stairs.

(2) If the material change of use relates to only part of a building, the work in that part shall be carried out to comply with the following requirements.

(a) In all cases with
   B1   Means of escape
   B2   External fire spread – linings
   B3   External fire spread – structure
   B4(2)   External fire spread – roofs
   B5   Access and facilities for fire service
   F1   Ventilation
   F2   Condensation in roofs
   G1   Sanitary conveniences and washing facilities
   G2   Bathrooms
   H4   Solid waste storage
   J1   Heat-producing appliances
   J2   Discharge of products of combustion
   J3   Protection of building
   L1   Conservation of fuel and power.
(b) When the change of use is to a boarding house, institution, a public building or it was previously an exempted building it must comply with the following requirements:
   A1   Loading
   A2   Ground movement
   A3   Disproportionate collapse.
(c) When a material change of use takes place in a building more than 15 m high the whole building must comply with:
   B4(1)   External fire spread – walls.

Any new bulding work undertaken in connection with a change of use is 'building work' which must comply with all relevant requirements.

APPRAISAL

The approach to the structural appraisal of an existing building differs from that taken in designing the structure of a proposed building. BRS Digest 366 explains the differences and describes a practical sequence for carrying out such an appraisal.

The digest examines the structural appraisal of buildings constructed using rules of thumb and experience for the layout and sizing of members; and also buildings where the structure has been designed, calculated and specified according to engineering principles.

The aim of a structural appraisal is to assess the real condition of a structure and to relate this to some requirement for a present or future use. Digest 366 describes the planning and sequence of an appraisal and offers guidance on procedure under the following headings:

- the brief
- initial inspection and appraisal
- immediate action
- detailed documentary search and review
- detailed investigation of the building
- assessment of the structure
- structural work needed
- report on appraisal
- load testing
- involvement of building control authority.

MATERIALS AND WORKMANSHIP

Regulation 7. All building work is required to be carried out with proper materials and in a workmanlike manner.
   Suitable materials must:

- bear appropriate EC mark; or
- be in accordance with a harmonized standard; or
- have European technical approval; or
- conform to appropriate British Standard; or
- have an Agrément certificate; or

- conform to national technical specification of any member state which gives in use the required protection and performance.

The effect of this regulation is discussed in Chapter 9.

LIMITATIONS ON REQUIREMENTS

Regulation 8. No obligation imposed by the regulations to comply with any requirement in Parts A to K, L and N in, Schedule 1, require anything to be done beyond that necessary to secure a reasonable standard of health and safety.

The Secretary of State is enabled to make regulations for the purpose of health and safety and also for the welfare, convenience of persons in or about buildings, and for furthering the conservation of fuel and power and preventing waste, undue consumption, misuse or contamination of water (1984 Act, section 1). This regulation says that the requirements of Parts A to K, L and N have the aim of securing health and safety. Welfare and convenience, etc., are not mentioned here.

These requirements relate to people in or about the building and others who may be affected by any failure to comply with a requirement.

It is clear the principle always apparent in byelaws and the more recent regulations that the standards prescribed were minimum standards continues. Of course higher standards may be employed but so far as the regulations are concerned one cannot ask for more than will ensure 'reasonable' standards of health and safety.

There is no explanation of the word 'reasonable' so it may be reasonable to assume that materials and design, workmanship and construction given in the various approved documents achieve this target.

TESTS FOR CONFORMITY TO BUILDING REGULATIONS

The 1984 Act, section 33, empowers a local authority to require or carry out tests to establish conformity with the regulations. Expenses are to be met by the person responsible, and not by the council. However, the council may if they think fit, bear the whole or part of the cost themselves.

If a dispute arises it may be considered by a magistrates' court.

The above section of the 1984 Act is to be brought into operation on a day to be appointed by the Secretary of State.

Regulation 16 does, however, enable a local authority to take samples as may be necessary to enable them to ascertain whether such materials comply with the requirements of the regulations.

This regulation does not apply when an approved inspector is being employed or when an approved inspector's final certificate has been accepted.

Testing of drains and private sewers

Regulation 16. A local authority may carry out tests on any drain or private sewer to ascertain whether it complies with any of the requirements of Part H, Drainage and Waste Disposal, of Schedule 1.

This gives a local authority the power to test drains or private sewers to establish their soundness. A drain is used for the drainage of one building or of buildings and yards within the same curtilage. It includes manholes, ventilating shafts, pumps and other accessories belonging to the drain (1984 Act, section 21). The definition of 'private sewer' is one which is not public. A public sewer is a sewer vested in the sewerage undertaker (1984 Act, section 126). See Chapter 17.

This facility does not apply when work is under the supervision of an Approved Inspector. He may, however, arrange to have authority to test within the terms of his contract.

BREACH OF REGULATIONS

The 1984 Act, section 35, says that anyone who contravenes any building regulation to which this section applies may upon summary conviction be fined and maybe subjected to a further fine on a daily basis if the default continues after conviction.

REMOVAL OF OFFENDING WORK

The 1984 Act, section 36, enables a local authority without prejudice of their right to take proceedings to require an owner to pull down or alter work which is in contravention of the regulations. If the developer ignores a local authority notice they may pull down or alter the work. The council's costs must be borne by the person receiving the notice.

Following receipt of a notice to put the work right a developer is allowed 28 days to comply before the local authority may enter to do the work.

These notices are called 'Section 36 notices' and must be given before the end of 12 months from the date of completion of the work in question.

INDEPENDENT REPORTS

The 1984 Act, section 37, says that when a person has received a Section 36 notice he may obtain a report from a suitably qualified person regarding the work to which the notice relates.

Obtaining a report has the immediate effect of extending from 28 days to 70 days the period after which they may enter and remedy the alleged defect.

Should the local authority withdraw their Section 36 notice after considering the independent report they are allowed to meet the expenses of obtaining the report.

CIVIL LIABILITY

The 1984 Act, section 38, enables a breach of a duty imposed by any building regulation to be actionable in civil proceedings unless the regulations provide otherwise. This section will come into force on a day appointed by the Secretary of State.

APPEALS

Appeal against a Section 36 notice

The 1984 Act, section 40, enables an appeal against a Section 36 notice to be made to a magistrates' court. The court may:

- say that the local authority was entitled to give the notice, and confirm it; and
- in every other case give the local authority a direction to withdraw the notice.

The court may require the local authority, if making an order as to costs, to meet the expenses of obtaining the report.

Appeal to higher court

Anyone aggrieved by an appeal decision in a magistrates' court may appeal to the Crown Court. However, this does not confer a right of appeal if the dispute could have been referred to arbitration in the first place (1984 Act, section 14).

HEALTH, SAFETY AND WELFARE

There are several regulations which have been issued arising from directives by the Council of the European Communities.

WORKPLACE (HEALTH, SAFETY AND WELFARE) REGULATIONS 1992

These regulations implement the provisions of directive 89/654/EEC and became operative on 1 January 1993 for premises first coming into use after 31 December 1992. Workplaces which existed at that date were subject to the regulations with effect from 1 January 1996.

Enforcement of these regulations is the responsibility of the Health and Safety Executive, and it should be noted that some of the requirements overlap with building control. However, a health and safety inspector may not serve an improvement notice requiring remedial work which is more onerous than the current building regulations, except in those instances where the relevant statutory provision requires a more onerous provision.

Requirements

There is a duty on every employer; a person in control of a workplace and an occupier of a factory to ensure that the relevant requirements are complied with (Reg. 4).

Maintenance

The workplace and equipment are to be maintained in good working order (Reg. 6).

Environment

- Every enclosed workplace is to be ventilated by a sufficient quantity of fresh or purified air (Reg. 6).
- Temperature in all workplaces is to be 'reasonable' (Reg. 7).
- There is to be suitable and sufficient lighting; whenever possible by natural light. There is to be provision for emergency lighting (Reg. 8).
- The workplace, furniture, furnishings and fittings are to be kept clean (Reg. 9).
- Rooms are to have sufficient floor area, height and unoccupied space to ensure health, safety and welfare (Reg. 10).
- All workstations are to be so arranged as to be suitable for the work to be carried on (Reg. 11).
- Floors and the surface of traffic routes must be suitable for the purpose; have no unguarded hole or slope, or be slippery or uneven. Where necessary they are to be drained (Reg. 12).

Falls

Measures are to be taken to prevent any person falling, or being struck by a falling object. Tanks, pits or structures are, if there is a danger of anyone falling into them, to be securely covered or fenced (Reg. 13).

Windows, skylights, etc.

(see Part N of 1991 Regs, as amended)

- Windows, transparent or translucent doors, gates and walls must be of safety material or be protected against breakage (Reg. 14).
- Windows, skylights and ventilators when being opened, closed or adjusted, must not expose the operator to a risk (Reg. 15).
- All windows and skylights are to be designed and constructed to ensure that they may be cleaned safely (Reg. 16).

Traffic

Ensure that pedestrians and vehicles can move safely about within the workplace (Reg. 17).

Doors and gates

Doors and gates are to have suitable safety devices (Reg. 18).

Escalators

Escalators and moving walkways are to function safely, and be equipped with safety devices and prominent emergency stop controls (Reg. 19).

Welfare

Sanitary conveniences, washing facilities, drinking water, storage for clothing, facilities for changing clothing and facilities for rest and to eat meals are to be provided (Regs 20 to 25).

CONSTRUCTION (DESIGN AND MANAGEMENT) REGULATIONS 1994

These regulations which came into force on 31 March 1995 implement parts of directive 92/57/EEC not already covered by existing UK legislation. They are administered by the Health and Safety Executive and have no direct reference to building control. However, caution follows in the wake of these regulations, as certain aspects of building control may be affected.

The regulations apply to construction work where it lasts for more than 30 days, or will entail more than 500 person days. The regulations also apply to work not falling to be notified to the Health and Safety Executive which involves:

- 5 or more persons on a site at any one time;
- design work irrespective of time involved.

The regulations also apply to work involving demolition with no restriction on time involved or the numbers of persons engaged in the work.

A number of persons have responsibilities under the regulations:

- The Client is any person for whom a project is carried out, whether by another person or in house;
- The Designer is a person who carries on a trade, business or other undertaking which prepares designs or arranges for designs to be prepared;
- The Contractor is a person who carries on a trade or business undertaking to or does carry out or manage work;
- The Planning Supervisor is a person appointed to make sure that designers comply with their responsibilities regarding the avoidance or reduction of risk.

Building control may become involved if, when dealing with full plans, the Local Authority intends to pass them subject to conditions which involve modifications to the plans. Another instance may be when dealing with applications for relaxation or dispensation of a particular regulation. A Local Authority taking enforcement action under the 1984 Act, section 36 involving the removal or alteration of work may also be affected.

Under these circumstances the Local Authority may find themselves embraced by the definitions of client, designer or contractor, and be bound by the requirements of the regulations.

There are also other circumstances in which a Local Authority may issue a notice requiring work to be undertaken; for example the following sections of the 1984 Act where a local authority may also carry out work themselves.

- 76 – repair of defective premises;
- 77 – repair of dangerous buildings;
- 78 – emergency measures in connection with dangerous buildings;
- 79 – work of restoration on a ruinous or dilapidated building; or
- 81 – demolition of a building.

Contents

A brief summary of the contents of the Construction (Design and Management) Regulations 1994 follows.

Reg. 2. Interpretation.
3. Application of the regulations.
4. Clients and Agents of clients.
5. Requirements on developer.
6. Appointment of planning supervisor and principal contractor.
   The client is to appoint a planning supervisor and a principal contractor.
7. Notification of project.
   Notice is to be given to the Executive by the planning supervisor and it is to give the following information:

- date;
- location of construction site;
- type of project;
- name and address of planning supervisor with a declaration confirming the appointment;
- name and address of principal contractor with a declaration confirming the appointment;
- date construction is to start together with duration of the work;
- estimated number of persons at work on the site;
- number of contractors to be engaged on the site;
- names and addresses of contractors already chosen.

8. Competence of planning supervisor.
9. Provision for health and safety.
10. Start of construction phase.
11. Client to ensure information is available.
    The client must make sure that the planning supervisor is provided with all information regarding the site and conditions of premises at or on which the construction work is to be carried out.
12. Client to ensure health and safety file is available for inspection.
13. Requirements on designer.
    Every designer must incorporate design considerations to avoid foreseeable risks to any person carrying out construction work. The designer must combat at source risks to health and safety and give priority to measures which will protect all persons at work.
14. Requirements on planning supervisor.
15. Requirements on the health and safety plan.
    The principal contractor is to take reasonable steps to ensure that all on the site comply with health and safety regulations.
16. Requirements on powers of principal contractors.
    The principal contractor must take steps to ensure co-operation between all contractors in complying with requirements of the regulations.
17. Information and training.
18. Advice from, and views of persons at work.
    There is to be a facility for employees and self-employed persons on the site to discuss matters of health and safety.
19. Requirements and prohibitions on contractor.

CONSTRUCTION (HEALTH, SAFETY AND WELFARE) REGULATIONS 1996

These regulations which implement directive 92/57/EEC set out minimum safety and health requirements at temporary or mobile sites where construction is in progress. The site includes building, civil engineering or engineering construction works, but does not include a workplace on such a site if it is set aside for other purposes (Reg. 2).

The requirements are briefly summarized as follows:

Safe workplaces

When reasonably practicable

- suitable and sufficient access and exits are to be provided;
- the site is to be made and kept safe and there is not to be a risk to the health of any person working there;
- all places of work, having in mind the nature of work carried on, are to have sufficient working space conveniently arranged for the person working there (Reg. 5).

Falls

Measures are to be taken to prevent any person falling (Reg. 6).

Weak material

Measures are to be taken to prevent a person falling through fragile material (Reg. 7).

Stability

Ensure that any new or existing structure does not become unstable, or collapse, during construction work. Temporary supports or structures are to be erected or taken down under the supervision of a competent person (Reg. 9).

Demolition

Demolition or dismantling work is to be carried out under the supervision of a competent person, to prevent as far as practicable the risk of danger to any person (Reg. 10).

Excavations

Ensure that new or existing excavations in a temporary state of instability do not collapse accidentally (Reg. 12).

Traffic

Ensure that pedestrians and vehicles can move about safely within the construction site (Reg. 15).

Doors and gates

Doors, gates and hatches are to have suitable safety devices. This does not include mobile plant and equipment (Reg. 16).

Fire risk

A construction site is to be provided with suitable and sufficient firefighting equipment, fire detectors and alarms. Care is to be taken to ensure that no person suffers injury from fire or explosion (Regs 18 and 21).

Emergencies

Procedures to accommodate foreseeable emergencies are to be prepared and adequate means of escape to a place of safety provided (Regs 19 and 20).

Welfare

Sanitary conveniences, washing facilities, drinking water, storage for clothing and rest facilities are to be provided (Reg. 22).

Enforcement

Enforcement is carried out by staff of the Health and Safety Executive, but the Fire authority may enforce Regs 19, 20 and 21 when persons other than those carrying out the construction work are within the site (Reg. 33).

FIRE PRECAUTIONS (WORKPLACE) REGULATIONS 1997

These regulations, made under the provisions of the Fire Precautions Act 1971, give effect to two Council directives, 89/391/EEC and 89/654/EEC regarding the health and safety of people at work.

The regulations are in five parts.

- Part II gives requirements for fire precautions in workplaces which are not covered by other legislation.
- Part III extends certain requirements of the Management of Health and Safety at Work Regulations 1992 to Part II.

Enforcement is the responsibility of the Fire authority. However, prior to the passing of plans under the 1991 Regs for a new building or a change of use, a local authority must consult with the appropriate Fire authority under section 16 of the 1971 Fire Act.

- Part IV contains provisions for enforcing the requirements.
- Part V contains miscellaneous provisions.

Application

A 'workplace' is premises not being domestic premises used for the purpose of an employer's undertaking and which are made available to the employee as a place of work.

All employers and any person in control of a workplace must see that the requirements are complied with. There are exempted workplaces and these comprise premises where more specific fire safety requirements exist, for example those having a fire certificate under the 1971 Fire Act. Also exempted are construction sites within the meaning of the Construction (Health, Safety and Welfare) Regulations 1996.

Fire protection

In order to ensure the safety of employees in the event of a fire

- a workplace must be equipped with appropriate firefighting equipment, together with fire detectors and alarms, and
- non-automatic firefighting equipment must be easily accessible, easy to use and clearly signed;
- there must be appropriate measures for firefighting in the workplace, having regard to the size of the workplace and taking into account people other than employees who may be in the workplace;
- suitable employees are to be nominated to ensure adequate training and equipment, and to implement the firefighting measures;
- ensure contacts with the appropriate external emergency services (Reg. 4).

Exits

Routes to emergency exits are to be kept clear at all times (Reg. 5).

Enforcement

Fire authorities are responsible for enforcement.

- Intentional or reckless serious breach of fire precaution requirements is a criminal offence.
- Fire authority may issue an enforcement notice relating to serious breaches.
- Breach of enforcement notice is a criminal offence (Regs 9 to 16).

Consequential provisions

- Local authority must consult Fire authority before passing plans (section 16, 1971 Fire Act).
- Section 9A of the 1971 Fire Act is amended requiring means of escape in case of fire and for fighting fire, in premises which are exempt from the requirement for a fire certificate (Reg. 22).

POWER TO ENTER PREMISES

The 1984 Act, section 95, gives an authorized officer of a local authority a right to enter premises at all reasonable hours.

The authorized person must have a duly authenticated document showing his authority.

Admission to premises, other than a factory or workplace, cannot be demanded as of right unless 24 hours' notice of the intended entry has been given to the occupier. If access is refused the authorized person may obtain a warrant.

When the authorized officer enters premises he may be accompanied by such persons as may be necessary (section 96).

BUILDING CONTROL

Until the 1985 Regs came into force there was only one way in which to secure building control in accordance with the regulations. Approved public bodies are now enabled to undertake their own checking and inspection and must issue notices and certificates to that effect. But it is for the private developer that the greatest change has come about. He now has three choices, which will be described in detail in the following chapters, but which are summarized below. He may proceed:

(1) by sending a 'building notice' to the local authority, or
(2) by depositing full plans with a local authority, or
(3) by engaging an approved inspector, who may be a private individual, a company or the NHBC.

The following are the principal stages in each method:

1. Building notice

- A building notice is sent to the local authority.
- This procedure may not be used if certain offices, shops, factories, railway premises, hotels and boarding houses are to be erected.
- A fee must be paid to the local authority.
- After giving the building notice to the local authority, work may start, but only after 2 days' notice of commencement has been given to the council. The usual stage notices must also be given to the council.
- The local authority may ask for more plans they may consider necessary to enable them to discharge their functions.
- If the local authority are of the opinion that building regulations have been contravened they may serve a notice requiring the defective work to be removed.
- An appeal may be made against the local authority's notice.
- The local authority are obliged to give a certificate of completion if one is asked for.

For the purposes of sections 219 to 225 of the 1980 Act, covering the advance payments code:

- the giving of a building notice with appropriate plans shall be treated as the deposit of plans;
- the plans accompanying a building notice shall be treated as the deposit of plans; and
- the receipt of a building notice shall be treated as the passing of those plans.

This is to ensure that the advance payment code is brought into operation when a building notice is given.

2. Full plans

- Full plans and information are sent in duplicate to the local authority.
- A fee must be paid to the local authority.
- Having plans passed gives a developer some protection but one is not required to await for the plans to be passed before commencing work. Work can be started after giving 2 days' notice to the local authority. The usual stage notices must be given to the council.
- The local authority are required to pass or reject the plans within 5 weeks, or if the developer agrees this period may be extended to 8 weeks.
- If the local authority are of the opinion that building regulations

have been contravened they may serve a notice requiring the defective work to be removed. If work is in accordance with the plans they have passed they may not serve a notice requiring defective work to be removed.

- The local authority are required to give a certificate of completion if one is asked for.

3. By employing an approved inspector

- The person having the work done and the approved inspector must jointly send to the authority an initial notice.
- This document must be accompanied by some plans and evidence that insurance cover has been taken out relating to the work.
- The local authority are allowed 10 working days in which to accept or reject the initial notice.
- Once accepted a council's powers to enforce the building regulations are suspended.
- The approved inspector may be required to supply a plans certificate before work is commenced.
- A fee will be payable to the approved inspector. Unlike local authority fees this will not be set out in regulations. The inspector is able to charge a fee by negotiation to cover his services for checking plans, certificates, inspections and travel.
- Work must not be commenced until the council has
  (a) accepted the initial notice, or
  (b) it has been with the council for ten days and has not been rejected.
- If the approved inspector finds the work is in contravention of the regulations he must notify the person having the work done.
- This person must remedy the contravention within 3 months.
- If the contraventions are not corrected during this period the approved inspector must cancel the initial notice.
- The person having the work done then has two options available:
  (a) another approved inspector may be appointed, or
  (b) the local authority will take over responsibility.
- Upon satisfactory completion of the work the approved inspector must provide a final certificate to that effect.

TIMING

The 1984 Act sets down periods which are allowed for certain actions to take place. It is imperative that in the appropriate circumstances these timings are not overlooked as failure to comply will lead to difficulties.

The following periods apply to various actions associated with contraventions of the regulations.

- Period allowed when council requires work to be exposed. A 'reasonable time' must be given (1991 Regs, 14).
- Period to be allowed for the removal of 'offending work'. 28 days (1984 Act, section 36).
- Period allowed for builder to tell council after altering offending work. 'Reasonable time' (1991 Regs, 14).
- Period within which council may serve a Section 36 notice. 12 months from completion of work. (1984 Act, section 36).
- Period in which owner may tell council an independent report is being obtained. 28 days (1984 Act, section 37).
- Period allowed for removal of offending work when an independent report is being obtained. 70 days (1984 Act, section 37).
- Period allowed for making an appeal to magistrates' court against a Section 36 notice. 28 days (1984 Act, section 40).
- Period allowed for making an appeal to a magistrates' court against a Section 36 notice when an independent report is being obtained. 70 days (1984 Act, section 40).

- Period allowed for removal or alteration of offending work when an appeal is lodged.
28 days of determination or withdrawal of appeal (1984 Act, section 40).
- Period within which council may bring prosecution for a contravention of building regulations.
6 months from contravention (Magistrates' Court Act 1980, section 127).

FEES

Since 1985 there has been a series of regulations up to the Building (Prescribed Fees) Regulations 1994 authorizing local authorities to charge fees for discharging their building control functions and the fees have been prescribed in these ongoing regulations.

Local authorities may charge fees for:

- building notices
- passing plans
- inspecting the work
- taking over responsibility from an approved inspector
- regularization in the event of unauthorized building work.

Fees are also payable for determination of questions by the Secretary of State.

No fees are payable in connection with the erection of any building exempted from the building regulations by Schedule 2 of the 1991 Regs.

Where the work only involves the installation of new gas heating appliances carried out by or under the supervision of someone approved under the Gas Safety Regulations no fees are payable.

When the work carried out is solely for the purpose of providing means of access to enable disabled people to get into a building, and to any part of it, or providing facilities intended to secure their health and safety or welfare and convenience, plans must be deposited but no fees are payable.

The level of fees has been set by regulations published from time to time, but this has changed. The Building Act 1984 (Commencement No. 2) Order 1998, brought into operation paragraph 9 of Schedule 9 of the Building Act 1984. This Order was operative from 7 August 1998 and enables building regulations to be made authorizing local authorities to fix and recover charges in connection with their building regulation functions. Thus the local authority will be setting a scale of fees operative only within its own boundaries.

These regulations have been published in the Building (Local Authority Charges) Regulations 1998, and came into force on 1 April 1999.

These Regulations require each local authority to prepare a scheme fixing charges for the performance of their building control functions, and aimed at recovering the cost of the service.

Regulation 2 contains definitions including:

- 'small domestic building' which now includes connected drainage building work, and the total floor area is not to exceed 300 m².

Regulation 3 states that a local authority is authorized to prepare a scheme to fix and recover charges in connection with building control functions.

Regulation 4 says the scheme is to make provision for a plan charge; an inspection charge; a building notice charge; a reversion charge and a regularization charge.

Regulation 5 requires that within three years the fees received are not to be less than the costs directly or indirectly incurred.

Regulation 6 sets out the principles of the scheme for establishing the basis on which charges are made, and provides that when the estimated cost of building work is less than £5000 an inspection charge may be waived.

Regulation 7 states that charges for small domestic buildings and domestic garages, carports and extensions are to be based on floor area. If more than one extension is involved, floor areas are to be added together and charged according to total floor area up to a limit of 60 m².

Regulation 8 gives discretion for a local authority to reduce charges when

- building work contained in an application or building notice concerning two or more buildings or building works which are substantially the same;
- an application or building notice concerning building work which is substantially the same as has been approved or inspected, and the same owner is involved.

Regulation 9 exempts from charges specified work required to provide facilities for disabled persons.

Regulation 10 sets out when a local authority may require payment.

Regulation 11 Requires plans to be accompanied by an estimate in writing, of the cost of building work.

Regulation 12 requires a local authority to publish its scheme within 7 days of being made.

Regulation 13 states that contravention of the regulations is not an offence under the 1984 Act.

Regulation 14 contains transitional provisions, and the fees set under the Building (Prescribed Fees) Regulations 1994 continue to apply to building work where plans were first deposited, or a building or initial notice was given prior to 1 April 1999.

Regulation 15 sets out fees payable to the Secretary of State in respect of determinations, which are to be based upon 50% of the local authority charges, subject to a minimum of £50 and a maximum of £500.

APPROVED INSPECTORS

Regulations do not prescribe fees to be charged by approved inspectors. They are to be negotiated, as for all other professional work.

PENALTIES

Legislation relating to building control provides for fines as penalties for those who transgress. In the past the actual figures were stated in the appropriate sections of the Public Health Act 1936. The passage of time revealed how inadequate some of these fines had become. Some fines had a maximum as little as £5.

This arrangement has been changed by the 1984 Act, which related fines to the Criminal Justice Act 1982. This has been amended by the Criminal Justice Act 1991 and current levels are as follows:

- Level 1   £200
- Level 2   £500
- Level 3   £1000
- Level 4   £2500
- Level 5   £5000.

The following are examples of possible fines relating to building control.

Sections of 1984 Act

Level 1:   fine up to £200

| 19(5) | For not removing a building of short-lived materials on the due date |
| 19(6) | Failure to comply with conditions – short-lived materials |

25(6)   Occupying house without proper water supply
62(5)   For not disconnecting and sealing drains
63(1)   Improper construction or repair of WC or drain
68(2)   Building a public convenience without permission
73(3)   For not raising an adjoining chimney
74(3)   Forming a cellar without consent
74(5)   Contravention of conditions relating to a cellar
77(2)   Not observing an order in connection with a dangerous building
84(4)   For not maintaining entrance to a courtyard

Level 2:   fine up to £500
82(3)   For not notifying council regarding drainage affected by demolition

Level 3:   fine up to £1000
61(3)   Altering drainage without notifying council
96(3)   Disclosing manufacturing or trade secrets. *Penalty:* up to 3 months' imprisonment

Level 4:   fine up to £2500
23(4)   Obstruction of access for removal of refuse
59(1)   Not acting on council notice requiring drainage
60(5)   Not ceasing to use unsuitable pipes as soil pipes
65(3)   For not providing adequate sanitary conveniences at a workplace
66(5)   Failure to replace an earth closet
68(2)   Not removing a public convenience when required to do so
70(2)   For not providing food storage accommodation
71(2)   For not providing entrances and exits
71(2)   For failing to keep entrances free from obstruction
72(3)   Failing to provide means of escape in case of fire
74(4)   For not acting on a notice requiring alteration of a cellar
79(3)   Not complying with requirements regarding a ruinous or dilapidated building
80(4)   Not giving notice of intended demolition

Level 5:   fine up to £5000
3(3)   Contravention of condition regarding exemption of a particular class of building
11(6)   Contravention of a condition concerning a type relaxation

20(7)   Contravention of condition regarding materials not suitable for permanent building
35   *Breach of any requirement of Building Regulations*
52(4)   Failure to cancel initial notice

1991 CP Regulations
5   False declaration of conformity to EC mark requirements. *Penalty:* up to 3 months' imprisonment and/or fine of up to £2000.

PERFORMANCE STANDARDS

'Building Control Performance Standards' was published in July 1999. It sets out standards to be applied to a building control service, together with guidance towards achieving these standards.

It is recommended that building control bodies should:

- prepare and make public a policy on the provision of the service;
- have adequate staff resources, supplemented when necessary;
- carry out consultations with other agencies;
- give feedback to clients upon completion of plans assessment;
- keep records of the plan checking process;
- have appropriate site inspection regimes;
- act promptly in the event of contraventions and keep records;
- give certificates to direct client when work is satisfactorily completed;
- keep records for appropriate time – could be 15 years, or more;
- arrange for CPD and in-service training;
- arrange for self review of performance;
- adopt quality management principles;
- keep to professional standards and business ethics.

RELEVANT READING

BRE books and reports
Building Regulations and Health (D. C. Mant and J. A. Muir Gray)
BRE Digest 366. Structural appraisal of existing buildings for change of use

CHAPTER 3

Application of building regulations to Inner London

The decision to apply the 1985 Regulations to Inner London was expected to coincide with the demise of the Greater London Council, but when the Building (Inner London) Regulations were published they came into force on 6 January 1986.

This gave three months' familiarization experience and practice of the new system before the London borough councils took over from the GLC in April 1986.

The purpose of the 1985 IL Regs is to arrange for building control procedures to become essentially the same in Inner London as elsewhere in England and Wales. The regulations extend to 15 pages but the effect is complex as they make many repeals and amendments to building legislation affecting Inner London.

Requirements that are special to Inner London remain, particularly with regard to fire precautions in large buildings and in assembly buildings. Provisions relating to the rights of building and adjoining owners continue in force.

From 1 April 1986 responsibility for building control in Inner London rested with the Inner London boroughs, the Temples and the City Corporation.

The Inner London boroughs are:

Camden
City of London
City of Westminster
Greenwich
Hackney
Hammersmith and Fulham
Islington
Lambeth
Lewisham
Royal Borough of Kensington and Chelsea
Southwark
Tower Hamlets
Wandsworth

APPLICATION

In addition to the provisions in the 1984 Act and the 1985 Regulations the Building (Inner London) Regulations 1985; the Building Act 1984 (Commencement No. 1) Order 1985 and the Building Act Appointed Day and Repeal Order 1985; the Local Government (Miscellaneous Provisions) (No. 4) Order 1986 and the Building (Inner London) Regulations 1981 have all had an effect on the extent to which the legislation and regulations now relate to the Inner London boroughs and also the Inner and Middle Temples.

Building Act 1984

This applied to Inner London from 1 January 1986, but not all the sections were in force. A number of existing sections have been repealed and taking all the alterations arising from the previous paragraph into account the following are the sections which apply to Inner London.

1. Power to make building regulations.
2. Continuing requirements.
3. Exemptions of particular classes of building.
4. Exemption of educational buildings and buildings of statutory undertakers.
5. Exemption of public bodies from procedural requirements of building regulations.
6. Approval of documents for the purpose of building regulations.
7. Compliance or non-compliance with approved documents.
8. Relaxation of building regulations.
9. Application for relaxation.
10. Advertisement of proposal for relaxation of building regulations.
11. Type relaxation of building regulations.
14. Consultation with building regulations advisory committee.
15. Consultation with fire authority.
16. Passing and rejection of plans (except 16(13)).
17. Approval of person to give certificates.
18. Building over a sewer, etc.
19. Use of short-lived materials.
21. Provision for drainage.
22. Drainage of buildings in combination.
23. Provision of facilities for refuse.
24(3). Provision of exits.
25. Provision of water supply.
32. Lapse of deposit of plans.
34. Classification of buildings.
35. Penalty for contravening building regulations.
36. Removal or alteration of offending work.
37. Obtaining a report when Section 36 notice is given.
39. Appeal against refusal to relax building regulations.
40. Appeal against Section 36 notice.
41. Appeal to Crown Court.
42. Appeal and statement of case to High Court.
46. Inner London.
47. Giving and acceptance of initial notice.
48. Effect of initial notice.
49. Approved inspectors.
50. Plans certificates.
51. Final certificates.
52. Cancellation of initial notice.
53. Effect of initial notice ceasing to be in force.
54. Public body's notice.
55. Appeals.
56. Recording and furnishing information.
57. Offences.
58. Meaning of expressions.
59. Drainage of building. (Does not apply to Middle and Inner Temples)
60. Use and ventilation of soil pipes. (Does not apply to Middle and Inner Temples)
61. Repair, etc, of drains. (Does not apply to Middle and Inner Temples)
62. Disconnection of drain.

63. Improper construction or repair of water closet or drain.
64. Provision of closets in buildings.
65. Provision of sanitary conveniences in workplace.
66. Replacement of earth closets, etc.
67. Loan of temporary sanitary conveniences.
68. Erection of public conveniences.
69. Provision of water supply in occupied house.
70. Provision of food storage accommodation in a house.
76. Defective premises.
84. Paving and drainage of yards and passages.
86. Appeals to Crown Court.
87. Application of provisions to Crown Property.
88. Inner London.
89. References in Acts to building byelaws.
91. Duties of local authorities.
92. Form of documents.
93. Authentication of documents.
94. Service of documents.
95. Power to enter premises.
96. Supplementary provisions as to entry.
97. Power to execute work.
98. Power to require occupiers to permit work.
99. Content and enforcement of notice requiring works.
100. Sale of materials.
101. Breaking open streets.
102. Appeal against notice requiring works.
103. Procedure on appeal.
104. Local authority to give effect to appeal.
105. Judge not disqualified by liability to rates.
106. Compensation for damage.
107. Recovery of expenses.
108. Payments by instalments.
109. Inclusion of several sums in one complaint.
110. Liability of agent or trustee.
111. Arbitration.
112. Obstruction.
113. Prosecution of offences.
114. Continuing offences.
115. Protection of members of authorities.
116. Default powers of Secretary of State.
117. Expenses of Secretary of State.
118. Variation or revocation of Order transferring powers.
119. Local inquiries.
120. Orders.
121. Meaning of 'building'.
122. Meaning of 'building regulation'.
123. Meaning of 'construct' and 'erect'.
124. Meaning of 'deposit of plans'.
125. Construction and availability of sewers.
126. General interpretation.
127. Construction of certain references concerning Temples.
128. Protection for dock and railway undertakings.
129. Saving for Local Land Charges Act 1975.
130. Savings for other laws.
131. Restriction of application.
132. Transitional provisions.
133. Consequential amendments and repeals.
134. Commencement.
135. Short title and extent.

Schedule 1 Building regulations
Schedule 2 Relaxation of building regulations
Schedule 3 Inner London
Schedule 4 Provisions consequential upon public body's notice
Schedule 5 Transitional provisions
Schedule 6 Consequential amendments.

Fees

From 1 January 1986 the national fees regulations came into force in Inner London, replacing the London fees byelaws for work initiated after that date.

Plans certificates

The provisions for private certification of building work by approved persons now also apply to Inner London.

Regulations

The 1985 Inner London Regulations (1985 IL Regs) brought into force in Inner London the following:

The Building Regulations 1985 (1985 Regs)
The Building (Approved Inspectors) Regulations 1985 (1985 AI Regs)
The Building (Prescribed Fees) Regulations 1985 (1985 PF Regs).

Therefore the Greater London Council Building Byelaws were replaced by the National Building Regulations, but much of the London Building Acts were retained.

There are a number of amendments to the 1985 Regs for application to Inner London.

The 1985 IL Regs also state that the GLC Building (Constructional) Byelaws 1972 remain applicable to work where before 6 January 1986:

• a fee had become due to the GLC, or
• a building notice was served by the District Surveyor, or
• consent was given under section 30 of London Building Act (Amendment) Act 1939 to the setting up, erection or retention of the building.

Schedule 2 makes the following amendments to the 1985 Regs:

• Regulation 1(2) stated that the 1985 Regs did not apply to Inner London, so it is removed.
• Regulation 1(3) stated that the 1985 AI Regs did not apply to Inner London, so it also is omitted.
• Regulation 20 is about transitional provisions and a new paragraph is to be inserted saying that the regulation does not apply to Inner London.
• In the 1985 PF Regs, Regulation 1(2) saying that the regulation does not apply to Inner London, is removed.
• Also, after Regulation 17 relating to transitional provisions, is to be inserted a new paragraph saying 'This regulation does not apply to Inner London'.

CONSEQUENTIAL MODIFICATIONS OF ENACTMENTS

Regulation 3(1), Schedule 3, contains numerous modifications to the London Building Act (Amendment) Act 1939 and a small number to the London County Council (General Powers) Act 1955.

Schedule 3 also lists a number of partial repeals to the London Building Act 1930; the London Building (Amendment) Act 1939; The London County Council (General Powers) Acts 1954 and 1955 and 1956; Thermal Insulation (Industrial Buildings) Act 1957; the Greater London Council (General Powers) Act 1965 and the Post Office Act 1969.

All the following are repealed. London County Council Byelaws for the regulation of lamps, signs and other structures overhanging the public way 1914; the Greater London Council London (Constructional) Byelaws 1972; the Greater London Council London Building (Constructional) Amending Byelaws 1974 and 1979 and the Greater London Council London (District Surveyor's Fees) Byelaws 1980.

The following legislation does remain in force:

London Building Act 1930

Part XI refers to dangerous and noxious businesses, and requires isolation from other properties.

A 'dangerous business' means the manufacture of matches or other substances liable to sudden explosion, inflammation or ignition or of turpentine, naphtha, varnish, tar, resin or Brunswick black. Other manufactures may be dangerous on account of the liability of the substances employed to cause sudden fire or explosion.

No one is to erect a building within 50 ft of a dangerous business, and such occupations must be at least 40 ft from streets and 50 ft from houses.

Part XII refers to those parts of London which have suffered flooding and requires that a licence be obtained to build houses on a site less than 11.4 ft above Newlyn datum when they cannot be drained into a sewer. In giving consent the council will be mindful of the effect of the now operative Thames Barrier, but conditions regarding floor levels and drainage outlets may be imposed.

London Building Act (Amendment) Act 1939

This substantial piece of legislation, as amended, remains in force.

Sections 1 to 4 deal with introduction and definitions.

Sections 5 and 6 and 8 to 15 are concerned with the naming of streets and numbering of buildings. The functions are the responsibility of borough councils. Section 7 has been repealed.

Sections 16, 17, 18, and 19 have been repealed leaving sections 20 and 21 concerned with the construction of buildings.

Paragraph 2 of Schedule 3 to the 1985 IL Regs provides a complete re-write of section 20.

It applies where a building is to have a storey at a greater height than 30 m or 25 m if the area of the building exceeds 930 m². Height is measured at street level in the centre of a building to the ceiling of the top storey. Where a building is of the warehouse class or is used for trade or manufacture has a volume of more than 7100 m³ it must be divided by division walls so that no part of the building exceeds 7100 m³ (Fig. 3.1).

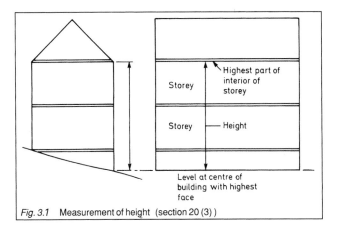

Fig. 3.1 Measurement of height (section 20 (3))

DIVISION WALLS

Division walls (Fig. 3.2) must fulfil the following requirements:

- form a complete vertical separation;
- have a fire resistance of four hours;
- openings are to be protected by four-hour fire-resisting doors or shutters;
- have fire separation at roof level:

1. roof height difference not less than 375 mm, or
2. when roof height difference is less, then there must be a parapet wall 375 mm high, or
3. roofs on both sides within 1.5 m are to be of non-combustible material and have an AA rating.

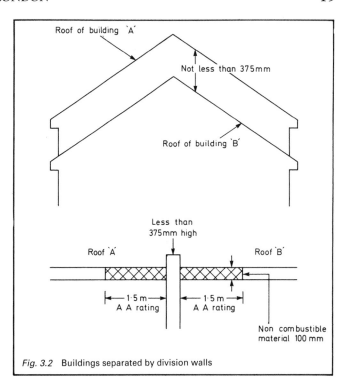

Fig. 3.2 Buildings separated by division walls

Councils may also ask for fire alarms, extinguishing systems, smoke extractors, smoke detectors and access for fire appliances.

In special risk areas councils may restrict the use of any part of the building and also require the provision and maintenance of proper arrangements for reducing the danger of fire in the building.

'Special risk fire areas' may include boiler rooms, electric and refrigeration plant rooms, engine rooms. Also included are places where any flammable or combustible substances are stored or used.

Garages or loading bays in basements or in other storeys and the spaces are not adequately ventilated are also high-risk areas.

Adequate ventilation requires openings or ducts in the external enclosures, and they must be on opposing faces to give cross ventilation. Unrestricted ventilation openings are to equal not less than 5% of the floor area. A local authority has 5 weeks, or such other extended period as may be agreed in writing to give a decision in response to plans deposited for the purposes of section 20. The London Fire Authority must be consulted by the council.

Section 21 has been altered by the 1985 IL Regs. Subsection 4 requires openings in division and party walls to have 4-hour fire-resisting doors or shutters.

A new sub-section (4A) was inserted by the 1987 IL Regs saying that sub-section 4 does not apply when:

- the width of any opening or all openings taken together do not exceed one half the length of the wall, and
- each opening is closed by two steel plate doors, metal covered doors, or steel rolling shutters one on each side of the wall, and
- the doors meet specification 1, 2, 3 or 4 and are installed in accordance with the requirements of the rules for the construction and installation of firecheck doors and shutters of the Fire Offices Committee dated February 1985.

Sections 29 to 31 are about special and temporary structures, and do not apply to buildings subject to control by the 1985 Regs. Buildings exempted by the 1984 Act, section 4, are also excluded. These are certain educational buildings and buildings of statutory undertakers.

Sections 33 to 43 cover means of escape in case of fire. In this respect buildings to which the 1985 Regs, B1, apply, are exempt from section 34. However, these sections apply to a wide range of buildings and have more stringent requirements than the 1985 Regs.

'New buildings' also include one which:

- is rebuilt after being destroyed or pulled down by an amount exceeding half the areas of the roof, floors and walls;
- is extended by an amount greater than the original cubic content;
- is raised so that there is a storey at a greater height than 42 ft.

Under section 34, a council's formal consent is required prior to the erection of:

- public buildings; churches; meeting rooms; halls, schools; concert rooms; ballrooms; places of assembly; and also all other new buildings (Fig. 3.3):
  - (a) when single storey exceed 600 ft² in area;
  - (b) when more than one storey is more than 1000 ft² in area, excluding basements used for storage;
  - (c) with a storey height exceeding 20 ft;
  - (d) when more than 10 persons are employed above the ground storey.

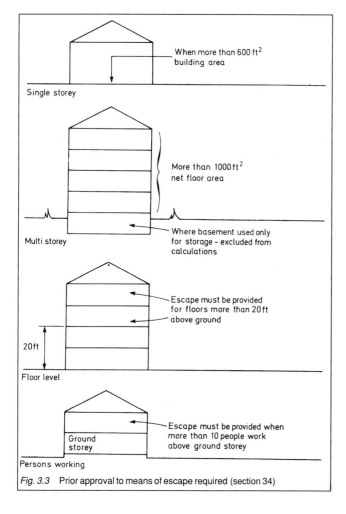

Fig. 3.3 Prior approval to means of escape required (section 34)

Section 35 allows a council, when it is satisfied that sufficient means of escape is not available from an 'old building', to require adequate provision. This section is usually applied when alterations are proposed to an existing building.

'Old buildings' for the purpose of this section (Figs 3.4 and 3.5) include those:

- having a storey at a height of more than 42 ft;
- having sleeping accommodation for more than 20 people;
- having more than 20 occupants or employees;
- where more than 10 persons are employed above the first storey or on any storey more than 20 ft above the ground;

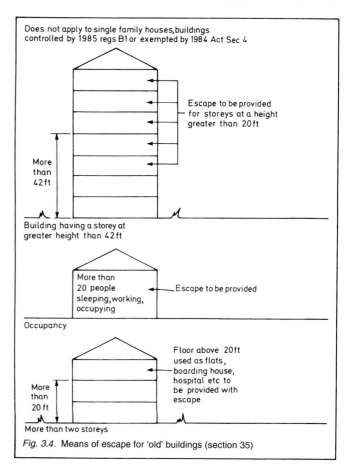

Fig. 3.4. Means of escape for 'old' buildings (section 35)

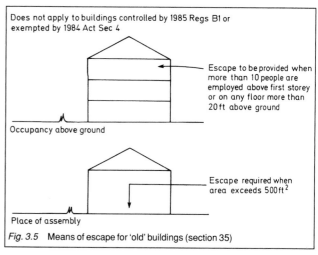

Fig. 3.5 Means of escape for 'old' buildings (section 35)

- which exceed two storeys and have a storey more than 20 ft above ground, used as a flat, inn, hotel, boarding house, hospital, nursing home, boarding school, children's home or other institution.

Also included are:

- restaurants, shops, stores, warehouses having sleeping accommodation above the ground storey;
- places of assembly having an area more than 500 ft².

Section 35 does not apply to houses occupied by one family or to buildings subject to control by the 1985 Regs B1. This latter requirement relates only to:

- a building which
  - (i) is or contains a dwelling house of three or more storeys;
  - (ii) contains a flat and is of three or more storeys;

(iii) is or contains an office;

(iv) is or contains a shop;

- a dwelling house extended or altered to have three or more storeys;
- a building having three or more storeys converted for use as a dwelling house.

Buildings exempted by the 1984 Act, section 4, are also excluded from control by section 35. They are schools and other educational establishments and also certain buildings of statutory undertakers and other public bodies. The exemption in respect of statutory undertakers and other authorities does not apply to houses or buildings used as offices or showrooms unless they are part of a railway station or on an aerodrome owned by the authority.

Part VI of the Act comprised sections 44 to 59 which were concerned with the rights of adjoining owners in regard to party walls. However, this part has been repealed with effect from 1 July 1997 by the Party Wall, etc. Act 1996 (Repeal of Local Enactments) Order 1997. The provisions of the Party Wall, etc. Act 1996 apply to Inner London and are very similar indeed to the provisions previously in place.

The repealed Part VI of the Act continues to apply in respect of work commenced on or before 1 September 1997. All procedures initiated after 1 July 1997 must be in accordance with the 1996 PW Act. (see Annexe C)

Sections 60 to 70 have more speedy and more effective procedures for dealing with dangerous buildings than those in the 1984 Act.

Section 61 Dangerous structures

Section 62 Certification

Section 63 Disputes

Section 64 Enforcing compliance

Section 65 Shoring and hoarding

Section 68 Removing dilapidated and neglected structures

Section 70 Payment of expenses incurred by council.

Sections 73 to 82 related to the district surveyors examination board.

Section 91 is about fees in connection with dangerous structures.

Sections 94 to 96 concern fees for other services.

Section 97 gives power to make byelaws regarding exhibitions.

Sections 101 to 108 are procedures for legal proceedings.

Sections 109 to 120 are about appeal tribunals.

Sections 121 to 126 set out the way notices must be given.

Section 132 is about rebuilding structures of architectural interest.

Sections 133 to 139 and 141 to 146 embrace a number of legal matters regarding means of escape, making alterations, powers of entry and the provision of plans.

Section 148 sets out penalties for contraventions.

Section 149 lists certain exemptions including embankments and the offices and buildings of the Bank of England.

In a number of other Acts, sections have been modified to remove mention of 'byelaws'.

DANGEROUS STRUCTURE FEES

The London Building (Dangerous Structure Fees and Expenses) Byelaws 1976 contains a schedule of fees:

(a) for general purposes;

(b) for surveys, inspections and other services and also fees payable to the council for work done in consequence of default by the owner of a dangerous structure.

MISCELLANEOUS PROVISIONS

The Local Government Reorganization (Miscellaneous Provisions) (No. 4) Order 1986 transferred building control in Inner London to the borough councils. It also made the London Fire and Civil Defence Authority the 'Fire Authority'.

EXTENDING POWERS

The Building (Inner London) Regulations 1987 completed application of the 1985 Regs to Inner London and made appropriate amendments to those regulations. A number of Acts were also modified and are included in the commentary. The City of London Drainage Byelaws were repealed.

Parts G4 and H of the 1985 Regs were applied to Inner London.

G4 is the regulation requiring sufficient sanitary conveniences and Part H is concerned with drainage systems.

Wherever the 1985 Regs were applied to Inner London, they are replaced by the 1991 Regs and subsequent amendments.

CHAPTER 4

Relaxation of building regulations

The Building Regulations 1965 to 1976 made provision for the exercise of the power of dispensation or relaxation of the regulations. The power was initially with the Secretary of State, but as time went by more and more of the various parts of the regulations could be relaxed by local authorities without prior reference to the Secretary of State.

This trend has continued under the current arrangements although the scope for relaxation has diminished.

It is section 8 of the 1984 Act which gives the Secretary of State this power to relax or dispense with requirements in the regulations. However, this power is now exercisable by the local authority (Regulation 10).

Relaxation of a requirement means that a local authority agrees there are special circumstances which allow it to say that all the terms of a particular regulation need not be adhered to.

Dispensation means a local authority is agreeable that a particular regulation may be ignored in the circumstances of the case.

Therefore when it appears to a developer that in the special circumstances of a particular case the requirement of a building regulation is unreasonable then an application may be made to the local authority for relaxation, or dispensation.

Even though the powers of a local authority seem wide in this direction they are in fact quite limited. This arises because all the technical requirements are written in functional form. They ask for something to be undertaken to provide an adequate standard, and to relax such a requirement implies that the result will be inadequate. There may be circumstances when a dispensation will be reasonable.

APPLICATION FOR RELAXATION

The 1984 Act, section 9, says that an application for relaxation or dispensation should contain such particulars as may be prescribed. The 1991 Regs do not prescribe any form of or indicate the information to be given to a local authority when making an application.

These were prescribed in the 1976 Regs and it would seem wise to base any application upon the information required by them. It is clear that any local authority considering an application would require what is really basic information regarding the building, the problem and the solution as seen by the developer.

'I/We hereby apply under section 9 of the Building Act 1984 and Regulation 10 of the Building Regulations 1991 for the requirements of building regulations specified below to be dispensed with or relaxed as specified in connection with the proposed building or works shown on the enclosed forms.'

- State briefly the nature of the proposed building works.
- Give address of the premises or location of site.
- Say if the work has already been carried out.
- State the requirements of building regulations it is considered should be dispensed with or relaxed.
- Set out the grounds upon which the application is based. These should indicate clearly why the operation of specific requirements are unreasonable in the particular circumstances of the case.

In some circumstances it is possible to apply for relaxation of building regulations in respect of work which has already been carried out (1984 Act, section 9, and Schedule 2).

An application for relaxation may not be made

- if the local authority have already served a notice (1984 Act, section 36(3)) requiring that the work be pulled down or altered;
- if there is in force an injunction or other order of the court requiring the work to be pulled down, removed or altered.

An application made to a council for the relaxation or dispensation in connection with work that has already been carried out has the effect of preventing the issue of a notice (1984 Act, section 36) requiring the work to be pulled down, removed or altered, on the ground that it contravenes the regulation referred to in the application.

Withdrawal of the application or a refusal to relax or dispense with the regulation will free the council to issue a notice.

If an application for relaxation or dispensation is made less than 12 months after the work has been done, the council may require it to be pulled down, removed or altered, only if the notice is served within 3 months of the application being withdrawn or finally disposed of.

APPLICATION TO BE ADVERTISED

At least 21 days before making a decision in connection with an application for relaxation or dispensation of a regulation the local authority must place an advertisement in a local newspaper circulating in the area (1984 Act, section 10).

- This must state the nature of the work, its situation and the regulation sought to be relaxed or dispensed with, indicating that any representations concerning the effect the proposal would have on public health and safety be sent to the council within 21 days.

The local authority will require the applicant to pay the cost of the advertisement.

It is not necessary to publish an advertisement if

- the effect on public health and safety is limited to premises adjoining the site of the work. In such a case the owners of adjoining premises must be told of the application;
- the work only affects an internal part of a building.

Any representations received must be taken into account by a local authority in making its decision. Before exercising their powers to relax building regulations relating to structural fire precautions, or concerning means of escape in case of fire, the fire authority must be consulted (1984 Act, section 15).

Should the local authority refuse the application and the developer appeals, copies of any representations must be sent to the Secretary of State.

APPEAL AGAINST REFUSAL

When issuing a notice of refusal the local authority must advise the applicant of the statutory right of appeal. Within one month of a local authority refusing an application to relax or dispense with regulations the applicant may appeal to the Secretary of State (1984 Act, section 39).

Should the local authority not give a decision within two months of the application being made or such extended period as the applicant may agree, then an appeal against the failure to give a decision may be sent to the Secretary of State.

No form of appeal is stipulated but the applicant should send to the Secretary of State a letter giving

- name of local authority concerned;
- plans of proposed work including any calculations;
- statement setting out the matter in dispute;
- reasons for making the appeal;
- grounds for appeal.
- copy of rejection of request for relaxation or dispensation of the statutory requirement.

A copy of the letter of appeal is to be sent to the local authority. They must immediately send to the Secretary of State copies of all the documents furnished by the applicant when asking for relaxation or dispensation.

At the same time the local authority must tell the Secretary of State their views on the application and send a copy to the applicant.

Appeals should be addressed to the

Building Regulations Division,
Department of the Environment, Transport and the Regions
Zone 3/C1
Eland House
Bressenden Place
London SW1E 5DU,

and in Wales to:

Housing Division
Welsh Office
Crown Buildings
Cathays Park
Cardiff CF1 3NQ.

An appeal must have been preceded by a request to a local authority for a relaxation or determination of a statutory requirement and the receipt of a rejection notice. An appeal submitted simply on the grounds of a rejection of the initial plans will not be entertained.

Copies of some appeal decisions are included in Annexe D.

HEALTH AND SAFETY

When giving consideration to an application for relaxation or dispensation of a particular building regulation the local authority should have regard to the Construction (Design and Management) Regulations 1994, as their actions may bring them within the definition of 'client', 'designer' or 'contractor' (see Chapter 2).

APPROVED INSPECTORS

Approved inspectors have no power to give relaxations or dispensations. Where such a course becomes necessary when work is being overseen by an approved inspector the application must be made to the local authority.

TYPE RELAXATION

Should the Secretary of State consider that a particular building regulation would be unreasonable in relation to a particular type of building work, he may give a direction dispensing with or relaxing that requirement either generally or in relation to that type of building matter (1984 Act, section 11).

Thus, after consultation with relevant representative bodies, the Secretary of State may give 'general' or 'class' relaxations.

He may do so on his own initiative or in consequence of a request made to him for relaxation. Fees may be charged for considering an application for type relaxation.

The first type relaxation to be made was under pre-1984 legislation and referred to filling cavity walls with thermal insulation.

TIMING

The 1984 Act sets down periods which are allowed for certain action to take place. These timings are important – failure to meet the deadlines will lead to difficulties.

- Period during which council must consider an application for relaxation.
  2 months, extendable by agreement (1984 Act, section 39).
- Time permitted in which to make representations following an advertisement about relaxation.
  21 days (1984 Act, section 10).
- Time allowed for a neighbour to respond to consultation.
  21 days (1984 Act, section 10).
- Time which must pass between council advertising proposal for relaxation and making decision.
  21 days (1984 Act, section 10).
- Time allowed to make an appeal to Secretary of State against refusal to relax.
  1 month (1984 Act, section 39).
- Time allowed for council to send appeal documents to Secretary of State.
  Must send documents 'at once' (1984 Act, section 39).
- Time allowed for Secretary of State to make a decision on relaxation appeal.
  There is no period stipulated (1984 Act, section 39).

CHAPTER 5

Exempt buildings and works

The majority of buildings are required to comply with the 1991 Regs but some types do not have to conform. In this respect the list of exempted buildings and the conditions attached to their exemption are much simpler than previously in the 1976 Regs.

It is only in recent years that there has been a definition of 'building' in operation. The 1984 Act, section 121, gives a very wide meaning indeed, including moveable objects of any kind. Such moveable objects could be vehicles or even hovercraft. Fortunately the 1991 Regs have ignored this very wide almost all embracing definition and in Regulation 2 narrows the meaning considerably. Gone are the vehicles, masts and flagpoles. 'Building' is now a permanent or temporary building, but not any other type of structure or erection. The word 'Building' must be taken to include 'part of a building'.

Several bodies have exemption now from some or all of the regulations. Reasons for this may vary. Regulations may not be appropriate to the kind of building the body erects or administers, and it may mean that the body is sufficiently responsible to construct buildings to a standard not lower than the 1991 Regs.

EXEMPTIONS

The 1984 Act, section 3, says that building regulations may exempt some buildings from the regulations. The exemption may relate to a particular building or to buildings of a particular class. The exemptions may be unconditional or subject to compliance with any condition. There is provision for a fine and a daily continuing fine for people who contravene one of these conditions when erecting an exempted building. The 1991 Regs contain in Schedule 2 a list of exempted buildings.

Exemption of educational buildings and buildings of statutory undertakings

The 1984 Act, section 4, says that nothing in the 1991 Regs applies in relation to

- buildings for a school or other educational establishment built in accordance with plans which have been approved by the Secretary of State for Education and Science or the Secretary of State for Wales. For many small buildings associated with school or other educational establishments plans do not have to be approved by the Secretary of State, and this exemption applies only to those that are so approved. (The expression 'school' includes a Sunday school or sabbath school (1984 Act, section 126).)
- a building belonging to
 (a) statutory undertakers.
  The 1984 Act, section 126, gives a definition of statutory undertakers. They are bodies who carry out work on buildings for
   railways
   canals
   inland navigation
   docks and harbours
   tramways
   gas
   electricity
   water
   other public undertakings.

(b) The United Kingdom Atomic Energy Authority
 The Civil Aviation Authority.

This exemption does not apply to houses.

The exemption does not apply to offices or showrooms which do not form part of a railway station.

The exemption also does not apply to offices, or showrooms in the case of the Civil Aviation Authority if they are not on an aerodrome owned by the authority in question.

Exemption of public bodies from procedural requirements of the Regulations

The 1984 Act, section 5, enables building regulations to exempt local authorities, county councils, and any other body which is non-profit making and acts for public purposes to be exempt from procedural requirements but not from the technical requirements (see Chapter 8).

By the 1984 Act, section 126, 'local authority' means
 district councils
 London borough councils
 Common council of the City of London
 Sub-Treasurer of the Inner Temple
 Under-Treasurer of the Middle Temple
 Council of the Isles of Scilly.

THE CROWN

The 1984 Act, section 44, provides for the application of the technical requirements of the 1991 Regs to Crown buildings. This reverses the position under previous legislation which was that building regulations did not apply to Crown property. The procedural arrangements still do not apply.

The Crown is given powers of dispensation and relaxation similar to those enjoyed by the local authorities.

This particular section did not become operative when the rest of the 1984 Act came into force but will come into operation when the Secretary of State makes a commencement order.

The situation has further been changed by the introduction of the National Health Service and Community Care Act 1990, section 60. With effect from 1 April 1991, Crown immunity was removed from NHS bodies. This means that from that date health service bodies were subject to the full substantive procedural and enforcement provisions of the 1984 Act and the 1991 Regs.

Health service bodies are defined as:

- Health Authorities as defined in the 1977 NHS Act;
- Family Health Service Authorities (previously Family Practitioner Committees);
- The Dental Practice Board;
- The Public Health Service Board.

This covers regional, district and special health authorities and also other organizations such as the special hospitals. The Department of

Health and the NHS Management Executive continue to be covered by Crown immunity.

Private hospitals and clinics have not enjoyed immunity in the past, and this position has not changed.

EXEMPT BUILDINGS AND WORK

Regulation 9 is short and explicit, saying that, **building regulations do not apply to the erection of or work in connection with any of the buildings described in Schedule 2.**

However, if any work carried out on one of the buildings within the six classes is such that it would no longer fall into one of the descriptions within Schedule 2 then building regulations must be applied. For example, if work or addition to an agricultural store resulted in it being less than $1\frac{1}{2}$ times its height from a dwelling then the regulations would be applied. If a building in one of the classes were to be altered to provide sleeping accommodation, then the regulations would apply.

There is a Class VII in Schedule 2 which states that the regulations do not apply to certain extensions. If building work is carried out on any of the following buildings or extensions and the building still remains listed when the work is complete then that work is also exempt.

The various classes of exempt buildings and work exempted from the regulations by Schedule 2 are as follows:

Class I Buildings which are controlled by other legislation:

• Buildings which are subject to the Explosives Acts 1895 and 1923 (Fig. 5.1).

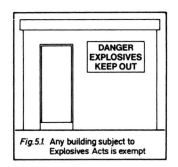

Fig.5.1. Any building subject to Explosives Acts is exempt

• Any building erected on a site subject to a licence under the Nuclear Installations Act 1965. The exemption does not apply to a canteen, a dwelling or an office (Fig. 5.2).

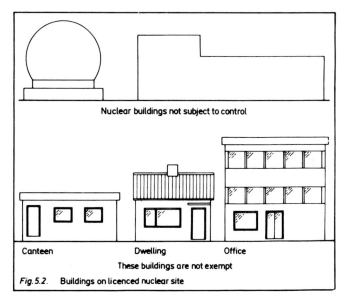

Nuclear buildings not subject to control

Canteen   Dwelling   Office

These buildings are not exempt

Fig. 5.2. Buildings on licenced nuclear site

• A building which is contained in the Schedule of Monuments kept under section 1 of the Ancient Monuments and Archaeological Areas Act 1979 (Fig. 5.3). These buildings are not to be confused with the much larger number of properties 'listed' as being of architectural or historic interest, to which the regulations continue to apply.

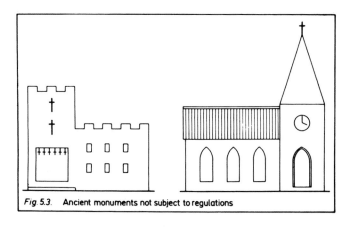

Fig. 5.3. Ancient monuments not subject to regulations

Class II This class embraces those buildings that are not frequented by people.

• A detached building into which people cannot or do not normally go (Fig. 5.4). Such a building is not exempt if it less than $1\frac{1}{2}$ times its height from boundary of curtilage of a building where people can and do normally go.

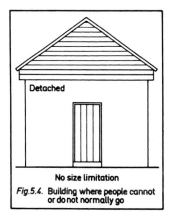

Detached

No size limitation

Fig.5.4. Building where people cannot or do not normally go

• A detached building containing fixed plant or machinery and to which people only go to inspect the plant or machinery (Fig. 5.5). Such a building is not exempt if it is less than $1\frac{1}{2}$ times its height from boundary of curtilage or a building where people can and do normally go.

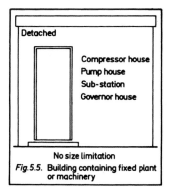

Detached

Compressor house
Pump house
Sub-station
Governor house

No size limitation

Fig.5.5. Building containing fixed plant or machinery

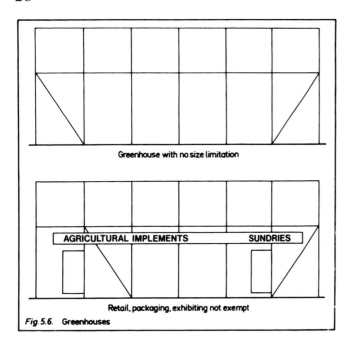

Fig.5.6. Greenhouses

Class III Greenhouses and agricultural buildings.

• A building used as a greenhouse.

The exemption does not apply if the main purpose for which a greenhouse is used for retailing packing or exhibiting (Fig. 5.6).

• A building used for agriculture, or principally for the keeping of animals, which is
  (a) not less than 1½ times its own height from any building containing sleeping accommodation;
  (b) has an exit which can be used in case of fire not more than 30 m from any point in the building;
  (c) has no part that is used as a dwelling.

This exemption does not apply if the main purpose for which the building is used for retailing packing or exhibiting (Fig. 5.7).

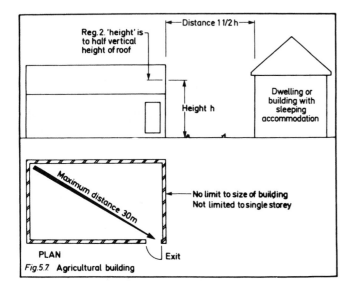

Fig.5.7 Agricultural building

There will be occasions when the 'main purpose' of a building will be difficult to determine or agree with building control staff. A difficult example may well be when a greater area of the building is used as an agricultural store, but a smaller area used for retail purposes generates much activity from members of the public calling to pur-

chase and producing a greater financial benefit to the farmer than the portion used for storage.

For the purpose of this building the definition 'agriculture' includes horticulture, fruit growing, the growing of plants for seed and fish farming.

Class IV Temporary buildings.

• A building which will not remain where it is erected for more than 28 days (Fig. 5.8).

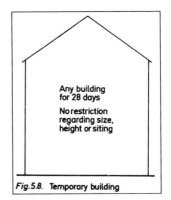

Fig.5.8. Temporary building

Class V Ancillary buildings.

• A building on a site to be used only for the disposal of buildings or building plots on that site (Fig. 5.9).

Fig.5.9. Ancillary buildings

• A building used only by people working in the construction, alteration, extension or repair of a building during the course of that work (Fig. 5.10). Must not contain sleeping accommodation.
• A building erected in connection with a mine or quarry (Fig. 5.11) but only if it does not contain
  (a) a dwelling, or
  (b) is used as an office, or
  (c) is used as a showroom.

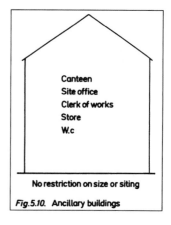

Fig.5.10. Ancillary buildings

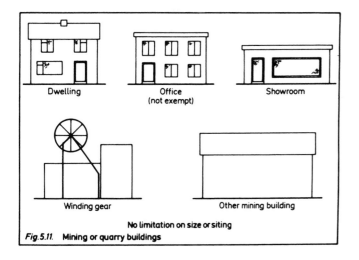

Fig.5.11. Mining or quarry buildings

Dwelling | Office (not exempt) | Showroom

Winding gear | Other mining building

No limitation on size or siting

Class VI Small detached buildings.

- The erection of a detached single-storey building having a floor area of not more than 30 m² and which does not contain sleeping accommodation and is either
  (a) at least 1 m from the boundary of its curtilage (Fig. 5.12) or
  (b) constructed substantially of non-combustible material (Fig. 5.13).
- A detached building to shelter people from the effects of nuclear, chemical, or conventional weapons. It is not to be used for any other purpose (Fig. 5.14):
  (a) the floor area must not be more than 30 m² and

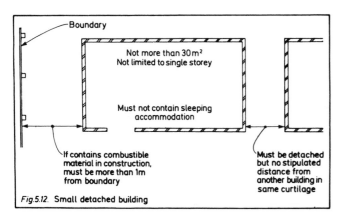

Boundary

Not more than 30 m²
Not limited to single storey

Must not contain sleeping accommodation

If contains combustible material in construction, must be more than 1m from boundary

Must be detached but no stipulated distance from another building in same curtilage

Fig.5.12. Small detached building

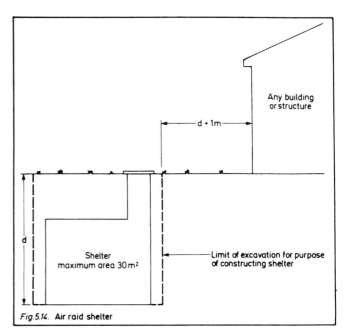

Any building or structure

d + 1m

Shelter maximum area 30 m²

Limit of excavation for purpose of constructing shelter

Fig.5.14. Air raid shelter

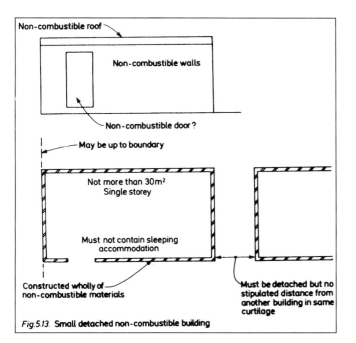

Non-combustible roof

Non-combustible walls

Non-combustible door?

May be up to boundary

Not more than 30 m²
Single storey

Must not contain sleeping accommodation

Constructed wholly of non-combustible materials

Must be detached but no stipulated distance from another building in same curtilage

Fig.5.13. Small detached non-combustible building

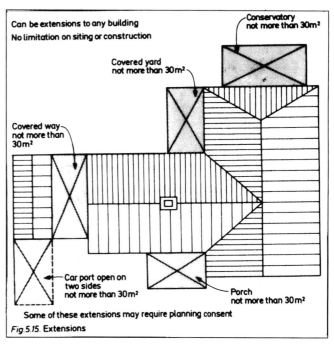

Can be extensions to any building
No limitation on siting or construction

Conservatory not more than 30 m²

Covered yard not more than 30 m²

Covered way not more than 30 m²

Car port open on two sides not more than 30 m²

Porch not more than 30 m²

Some of these extensions may require planning consent

Fig.5.15. Extensions

(b) the excavation for the building must not be nearer to any exposed part of another building or structure than a distance equal to the depth of the excavation plus 1 m.

- A detached building with floor area not more than 15 m² which does not contain sleeping accommodation. There is no requirement regarding materials of construction or distance of detachment.

Class VII Extensions.

- The extension of a building by adding at ground level a
     conservatory
     porch
     covered yard
     covered way
     carport open on at least two sides.

When the floor area of that extension is not more than 30 m² (Fig. 5.15).

To be a 'conservatory' a building must have a transparent or translucent roof.

A wholly or partially glazed conservatory or porch must have glazing that fulfils the requirements of Part N (see Chapter 22).

When an extension is constructed so as to cover a balanced flue the Gas Safety Regulations 1998 (SI 1984, No. 1358) may be relevant.

Addition
Building Regulations (Amendment) Regulations 2000, in force on 3 July 2000, make the Metropolitan Police Authority exempt from the procedural requirements and enforcement procedures of B.Regs. However, the Authority must comply with the substantive requirements.

CHAPTER 6

Notices and plans

Procedural arrangements which must be followed when someone proposes to undertake building work are set out in the 1991 Regs, and several sections of the 1984 Act.

When plans are deposited with a local authority they must be passed unless (1984 Act, section 16)

- they are faulty; or
- show that the intended work would contravene building regulations.

If the plans are faulty or do show that regulations are contravened then there are two options available to the council:

- the plans may be rejected; or
- the plans may be passed if the developer will alter the plans to the local authority's requirements or agree that further plans must be deposited.

In order that the council may impose these conditions the developer must have asked them to do so, or have agreed to the conditions being imposed. The request for, or consent to the imposition of conditions must be in writing.

It is the duty of a local authority to pass plans unless they are defective. A decision that plans have been approved or rejected, or approved subject to conditions must be made within 5 weeks or, if the developer is agreeable, a period of 2 months from the date that plans are deposited. The 1984 Act, section 16, says that a local authority 'SHALL' give a decision within the time allowed. If they do not then they have defaulted. The developer can proceed, but must comply with building regulations and must give 2 days' notice to the council that he intends to commence. He must give all other notices normally required.

Also, it may not generally be understood that a developer can commence work at any time after his plans have been deposited, provided that he gives 2 days' notice of his intention. He is not required to await the council's decision before commencing work, but if he fails to comply with building regulations the council can require the matter to be put right.

Often work subject to building regulations also requires planning approval. Such work under planning legislation should not be commenced prior to the receipt of planning permission.

When a local authority rejects plans they must state the defects or failure to comply with a regulation or section of the 1984 Act upon which they have based their decision.

If plans are passed the local authority must

- set down any condition to which the approval is subject;
- state that the approval is only for the purposes of the 1984 Act and the building regulations.

There are several sections of the 1984 Act that require or authorize a council in certain cases to reject plans. They are:

- where it is proposed to build over a sewer;
- the use of short-lived materials;
- the use of materials unsuitable for a permanent building;
- that provision for drainage is unsatisfactory;
- that satisfactory means of entry and exit are not provided for a building to which the public have access;
- satisfactory provision has not been made for a water supply.

A provision appearing in the 1984 Act allows plans to be accompanied by a certificate from an 'approved person' that parts of the work shown on the deposited plans will comply with the appropriate regulations. The approved person must supply evidence of authorized insurance cover.

The matters in which an approved person may be allowed to give certificates are:

- structure stability (Part A);
- energy conservation (Part L1).

A local authority cannot reject plans on the grounds that they do not conform with a regulation to which an approved person's certificate relates.

An 'approved person' is not the same as an 'approved inspector' (see Chapter 7).

A local authority may not institute proceedings against work carried out in accordance with a certificate from an approved person.

DISPUTES

If there is a dispute between a developer and a local authority regarding plans which are allegedly defective or that work would contravene the building regulations, or concerning the certificate of an approved person, the developer may refer the matter to the Secretary of State for a determination.

The Secretary of State only has the power to determine whether regulations are being complied with. He does not have the power to determine whether the regulations as a whole apply to particular work. This question would ultimately be for the courts to decide.

A determination can only be given on proposed work; work which would be affected by the outcome of a determination should not, therefore, be started before a decision is given. If started, any decision subsequently issued is invalid.

In order to make an application for a determination the following documents should be sent to:

Building Regulations Division,
Department of the Environment, Transport and the Regions
Zone 3/C1
Eland House
Bressenden Place
London SW1E 5DU,

and in Wales to:
Housing Division
Welsh Office
Crown Building
Cathays Park
Cardiff CF1 3NQ.

- **Plans of the proposed work.** These must be the plans (including any calculations, diagrams, etc.) that were rejected by the local authority or a copy of those plans. A determination cannot be given on an amended proposal.

- **A statement setting out the question in dispute.** This must include the regulation(s) or requirement(s) in Schedule 1 which you wish to be determined.
- **A statement setting out your contentions.** In this context it should be noted that a determination will be given on the evidence which has been provided in support of your case. If there is insufficient evidence to show that the proposal would conform with the regulations, the Secretary of State will be obliged to determine against you.
- **A fee.** The fee due is half the plan fee (excluding VAT) and details are set out in the current prescribed fees regulations. The current minimum fee is £50 and the upper limit is £500. Cheques should be made payable to the 'Department of the Environment' and they will be acknowledged.
- **Rejection notice.** A copy of the rejection notice issued by the local authority in respect of the plans submitted for determination. Please note that the plans must be rejected; a determination cannot be given after a conditional approval.

Upon receipt of a valid application the Department will invite the local authority to comment and allow them two weeks to do so. If, after this time, no comments have been received, the Department will proceed with the case, and make a determination under the provisions of the 1984 Act, section 16(10).

It is not necessary to obtain the council's agreement to ask for a determination.

Once the Secretary of State has issued a decision he has no further jurisdiction in the matter. All matters following a decision, including the possibility of devising alternative solutions, should be discussed with the local authority.

Examples of determinations already made are contained in Annexe D.

APPROVAL OF PERSONS TO GIVE CERTIFICATES

We have seen from the above provisions (1984 Act, section 16) approved persons may certify that proposed works if carried out in accordance with the certified plans will comply with those parts of the building regulations covered by the document.

Building regulations may make provisions for the approval of persons by the Secretary of State or a body which he will designate (1984 Act, section 17). The nature of work an approved person undertakes may be limited when he is approved.

The following bodies have been designated to approve private individuals wishing to be approved persons who will certify plans to be deposited with a local authority as complying with energy conservation requirements:

> Architecture and Surveying Institute
> Association of Building Engineers
> Chartered Institute of Building
> Chartered Institution of Building Services
> Institution of Building Control Officers
> Institution of Civil Engineers
> Institution of Structural Engineers
> Royal Institute of British Architects
> Royal Institution of Chartered Surveyors

The Institutions of Civil and Structural Engineers have been designated to approve persons to certify plans as complying with the structural requirements.

A fee is payable to the designated body when application is made to become an approved person. On every occasion where an approved person proposes to act in that capacity he must be able to satisfy the local authority that scheme of insurance which has been approved by the Secretary of State applies in respect to the proposed

work. This is to provide insurance cover for persons suffering damage attributed to negligence on the part of the approved person (1985 AI Regs, 3, 4 and 5).

TERMINATION OF APPROVAL OF APPROVED PERSONS

The designated body's approval of a person will cease to have effect

- after 5 years from the date it was given;
- if a notice in writing is given by the person who approved him.

The Secretary of State is empowered to strike a designated body off the list, but

- this would not affect the operation of any subsisting approval given by the designated body; however,
- a subsisting approval may be withdrawn by the Secretary of State.

Should an approved person be convicted for issuing a false or misleading certificate (1984 Act, section 57) his approval will be withdrawn and may not be renewed for a period of 5 years from the date of conviction (1985 AI Regs, 6).

LISTS OF APPROVALS AND DESIGNATIONS

The Secretary of State is obliged to keep a list of the bodies he has designated for approving persons, and also a list of the persons he may have approved.

He must also supply to every local authority administering the building regulations a list of the designated bodies and of any person he may have approved. Every council is to be told of any bodies or persons removed from or added to the lists.

The Secretary of State has indicated that persons will only be approved by the designated bodies. They must maintain a list of all the persons approved, and notify every local authority when any person has approval withdrawn.

Lists are to set out any limitation placed on the approval or designation, and must indicate in each case when approval will expire.

The designated bodies are not obliged to notify local authorities of the names of persons they approve. The onus is on each approved person to produce evidence of his approval when he issues a certificate to the local authority. This may best be accomplished by submitting a photo copy of the approval document in his possession (1985 AI Regs, 7).

BUILDING OVER SEWERS

A building or an extension is not to be erected over a sewer or drain unless approved by the local authority.

For many years, culminating in the 1984 Act, section 18, the position has been that plans showing a building will be erected over a sewer will be rejected by the local authority unless they are satisfied that no harm will arise to their installations by allowing the work to proceed.

Now, in the Water Act 1989, Schedule 8(6), this attitude has been recognized and the 1984 Act, section 18, has been amended, extending the powers not only to building over but to interference with the use of a sewer or obstructing access to the sewer or drain. Control of this legislation is now placed with sewer authorities who now replace water authorities.

In many instances the council may wish, if levels are adequate, to have the sewer or drain diverted, or some special form of construction above the sewer or drain to be undertaken to offer protection. In these circumstances a legal agreement must be entered into requiring the developer to agree about future access to the installations for maintenance or repair. Such an agreement will usually require the developer, and those owning the property after him to bear costs arising from reaching the sewers or drains through his property.

Disputes between a developer and a local authority may be determined by a magistrates' court.

If anyone builds over a sewer or drain to which the Act refers he will be liable to penalties upon conviction (1984 Act, section 35) and may be required to demolish the building (1984 Act, section 36).

If a public sewer is damaged by the erection of a building or an extension the local authority is in a position to seek damages.

Some caution must be attached to interpretation of the phrase 'over a sewer' in the 1984 Act, section 18. It would seem that some water authorities are concerned that buildings erected adjacent to or close to a sewer, or even within 3 m of a sewer, may in some conditions give rise to damage to the sewer, or not afford them access if necessary.

There does not appear to be any official interpretation of the word 'over', but it is simple and not subject to ambiguity and it must surely be synonymous with 'above'. It is a thought that if a sewer is modern it may be surrounded by granular fill, or an older sewer may be bedded, haunched or surrounded by concrete.

In these circumstances it may be argued that the pipe and the fill, or the pipe and the concrete, comprise the sewer, particularly as for the purpose of section 18 a drain includes a 'pipe including associated works' (Fig. 6.1).

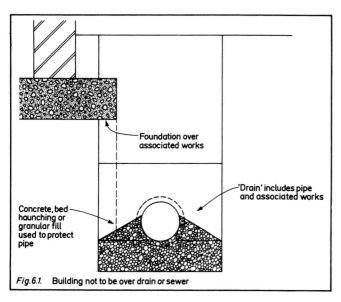

Fig.6.1. Building not to be over drain or sewer

Uncertainty will now surround the different interpretations which will apply locally regarding the extended powers of 'interference with the use of a sewer, or obstructing access to a sewer or drain'. This situation is therefore one in which, should it be found that a building is to be near a sewer, early discussions about the implications contained in the 1989 Act amendment should take place with the local authority and the sewer authority.

The 1984 Act relates the requirements about building over sewers only to circumstances where 'plans, etc., are deposited with a local authority'. It embraces those projects where a building notice, full plans procedures are used, or an approved inspector is appointed, or a public body controls its own development.

See Chapter 17 for details of precautions necessary when drains are in close proximity to buildings.

USE OF SHORT-LIVED MATERIALS

If plans show that a building is to be constructed of materials which are short lived, the council may reject the plans. They may do so even though the plans show conformation with the building regulations. Alternatively the authority may pass the plans and impose conditions which

- give a date when the building must be removed, and
- are appropriate to the nature of materials used.

The simplest condition would be to require regular protective painting of the structure. No condition may be imposed which would conflict with any condition attached to a planning approval (1984 Act, section 19).

Should a local authority discover that a building to which this section applies has been erected without having received their consent, then two courses of action are available to them:

- fix a date when the building must be removed, and impose conditions appropriate to the materials used; and
- take proceedings in the court in respect of any contraventions of the regulations.

It is possible to ask for and receive an extension of the time the building is allowed to remain.

A person aggrieved by the action of a local authority may appeal to a magistrates' court.

At the expiration of the period allowed the building must be removed unless the time period has been extended. If it is not removed the local authority may do so and recover the costs incurred. They may also prosecute in the magistrates' court and, if convicted, the owner may be fined and subjected to a continuing daily penalty.

Building regulations may specify materials which are, in the absence of special care, liable to rapid deterioration or which are in other ways unsuitable for use on permanent buildings.

USE OF MATERIALS UNSUITABLE FOR PERMANENT BUILDING

The section of the 1984 Act immediately following is unusual in that it is designed to replace the 1984 Building Act, section 19. It provides wider powers of control over the use of materials, components and fittings which are short lived or for some other reason are unsuitable for use in permanent buildings.

Upon receipt of plans indicating that it is proposed to construct a permanent building of unsuitable materials the local authority may reject them, notwithstanding that they may conform to the regulations.

Alternatively, the plans may be passed and the council fix a date when the building must be removed. They may also impose conditions considered to be appropriate for the materials to be used. However, no condition may conflict with any permission given under the 1990 Act.

Should work be carried out and no plans have been deposited with the local authority it may take proceedings in the magistrates' court against the owner.

Also the council may fix a period after which the building must be removed and impose conditions with regard to its maintenance (1984 Act, section 20).

The Secretary of State may make regulations listing unsuitable materials for the purposes of this section. He may also specify the type of service fitting or equipment that are likely to be unsuitable to be used in connection with a permanent building in the absence of conditions relating to their use.

This section also provides for the case when building regulations do not require plans to be deposited, but the council sees the building and adjudges a service, fitting or items of equipment as unsuitable for use in connection with a permanent building. They may within 12 months of the work being completed

- fix a period when the work must be removed; and
- if necessary impose conditions and give appropriate notices to the owner.

The time limit may be extended if the condition of the building justifies this action and the council may also vary the conditions when granting a further life to the building.

A person aggrieved by a local authority action

- in rejecting plans;
- in fixing or refusing to extend the time limit, or
- in making or refusing to alter any conditions

is enabled to appeal to the Secretary of State.

The procedure for making an appeal is in the 1984 Act, section 43, and will come into operation when the Secretary of State makes an order to that effect. The Secretary of State may

- arrange for the appellant and the local authority to appear before a person appointed by the Secretary of State;
- when determining this appeal make whatever conditions he considers necessary.

As required in section 19, if the owner does not remove a building when the time is up, the council may remove the offending work, or the building, and recover its expenses from the owner.

A person who offends against the requirements of this section may upon summary conviction be fined with a daily continuing fine until work is removed.

PROVISION OF DRAINAGE

A local authority must reject deposited plans if they do not show satisfactory provision for drainage of the building or of an extension, unless they are satisfied that this requirement may be dispensed with.

For the purposes of this section (1984 Act, section 21), 'drainage' is defined to include the conveyance, by means of a sink and any other necessary appliance, of refuse water and the conveyance of rainwater from the roofs. Thus the drainage system commences with the sink, via a waste pipe through the underground pipeline to its point of connection with a sewer or cesspool. Drainage of rainwater commences with the eaves gutter which is part of the drainage.

Should there be a difference of opinion between the owner and a local authority regarding the need for a drainage system or the adequacy of that shown on plans, the matter may be determined by a magistrates' court.

A local authority may not insist upon a connection being made to a sewer

- if the nearest sewer is more than 100 ft from the building;
- if the levels are such that connection to the sewer is not reasonably practicable;
- if the owner does not have a right to lay his drain through the land between his boundary and the sewers.

However, if the council is prepared to pay the cost of constructing, repairing and maintaining the drain attributable to the distance above 100 ft then they may insist on a connection being made to a sewer.

Where no public sewer is within easy reach of one or more premises the Water Industry Act 1991, section 94, places duty on a sewerage undertaker to provide a public sewer to receive domestic drainage. While the Act was designed to avoid stifling development for lack of infrastructure, the cost is likely to fall on the developer.

One of the words now in common use by planners is 'infrastructure' and this is a collective term for the fixed installations and facilities necessary to support development.

One essential ingredient of an adequate infrastructure is the ability to get sewage away from a site in a satisfactory and efficient manner.

When a planning authority issues a refusal to allow development on the ground that this would overload existing sewers or disposal facilities it must be prepared on appeal to make available the most authentic evidence supported by calculations to justify the decision.

There are two bodies involved. First the council, which is the local planning authority. They will be required to consult the water authority regarding the adequacy of the sewers and the disposal facilities before giving a planning decision. At the same time the council will be acting as agents for the water authority in the matter of sewers.

So the builder can find himself faced with a complex situation covered by legislation relating to town planning and public health. But how can he rid his site of sewage which will arise from the development he wishes to undertake?

The owner or occupier of any premises or the owner of a private sewer is entitled to join his drain or sewer to a public sewer (1991 Act, section 106). To exercise this right he must own any intervening land or have acquired an easement for the purpose of laying his pipeline.

This right is said to be absolute and as long ago as 1897 the courts held that once a connection to a public sewer had been lawfully made under the 1875 Act any type of effluent which was not injurious to health could be discharged into the public sewer.

Anyone wishing to make a connection to a public sewer must follow a prescribed procedure, and also meet the cost of the work.

Once having followed the water authority's requirements and a satisfactory junction with the public sewer has been accomplished a person exercising his right to make a connection is not responsible if the public sewer is not correctly maintained.

There are a number of restrictions regarding the nature of liquids that may be discharged into a public sewer and these concern principally discharges arising from trade premises. A special agreement may be required and there is a system of payment.

Naturally, if separate public sewers are available for storm water and foul water, any connections must be made to the appropriate sewer. Also a person may not cause his drains or sewers to be connected directly to a storm water overflow sewer.

Anyone wishing to make a connection to a public sewer must write to the water authority setting out his intentions. Within 21 days of receiving the notice the water authority may refuse to permit the connection to be made.

The circumstances in which the water authority may refuse are set out in the 1991 Act, section 106, and it must be satisfied 'that the mode of construction or condition of the drain or sewer is such that the making of the connection would be prejudicial to its sewerage system'.

At its simplest interpretation this clause could be used to prevent a sewer being connected to a public sewer of smaller dimension.

Arising from this right of connection it is interesting that an authority may not be liable for damages for nuisance when its public sewers become overloaded or surcharged because premises have been lawfully connected. This was a legal decision arrived at in 1954 and in consequence it is unlikely that an authority has the power under the 1991 Act, section 106, to refuse to allow a connection because it would cause surcharging of the public sewers.

This is one of the situations which cause a fusion between planning law and public health legislation. The planners may be told by a water authority that surcharging of its sewers will take place if certain development is given planning permission. In consequence the planning authority may issue a refusal notice containing a reason to the effect that the existing public sewers are of insufficient capacity to accept further discharge.

The power to connect one's drains to a public sewer is a right: not a duty. Therefore an authority cannot require an owner or occupier of premises to connect his drains to a public sewer unless satisfactory provision has not been made for disposal of sewage from the building. Thus, if there is a private method of disposal available and this is in good order and adequate, the authority cannot require this to be abolished in favour of a connection to a public sewer unless it pays the cost.

Even if the private arrangements may not be wholly satisfactory the site must be within 100 ft of a public sewer before an authority can compel connection. The same distance applies equally to a new property. The 100 ft must be measured not from the boundary but from the building, and of course the owner must have a right to put his pipe through the intervening ground.

There are occasions when no public sewer is within easy reach of one or more premises, and in these circumstances the 1991 Act, section 98, places a duty on a water authority to provide a public sewer.

This provides for the owners or occupiers of existing or new buildings to make a request for the provision of a public sewer to serve their premises. If utilized, this provision could enable a developer to ensure that the public sewerage system will be extended to meet his needs.

Naturally there are conditions attached to any request made under this section; the principal one being the matter of finance, because a water authority will not necessarily make the money available.

The actual charges in respect of the provision of a public sewer in these circumstances are arrived at following knowledge of constructional and operating costs, and will inevitably be the subject of some negotiation. The water authority must be satisfied that the money will be recoverable before they will agree to provision of a public sewer.

If the water authority chooses to lay a new public sewer of larger diameter than the minimum, or than is necessary to accommodate flow from the site, then the owner would expect only to contribute towards the cost an amount equal to expenditure that would be incurred in providing a smaller sewer.

One aspect an owner will contemplate before invoking this power is the position of any building and open land which could make use of the new public sewer passing the property. He would wish for contributions from the owners, but there is no way in which they could be compelled to assist financially in the project.

Switching back to planning legislation the council often, with the agreement of the water authority, thinks that in preference to a refusal on the grounds that the public sewerage system is inadequate the problem can be solved by using the 1990 Act, section 106.

This provision enables the planning authority and the developer to enter into a legal agreement. Such agreements may undertake the provision of finance for some necessary infrastructure including new public sewers or extensions to sewage disposal works. Indeed the developer might even undertake to do the work himself.

Also, when the problem is one where new building would aggravate an existing sewerage overload, permission may be granted subject to an agreement stipulating the occupancy rate of building being dependent upon the completion of the works specified.

DRAINAGE OF BUILDINGS IN COMBINATION

There may be occasions when a local authority decides that the drainage of buildings may be undertaken more economically or advantageously if it is done in combination (1984 Act, section 22).

In this event they may require when the drains of a building are first laid, that they are drained in combination to an existing sewer. This is done by means of a private sewer which may be laid by the owners or by the local authority on behalf of the owners. The requirement must be made in respect of each building before the plans have been passed.

After plans have been passed the arrangement can only be carried out by agreement with the council.

The council must fix

- the proportion of expenses to be borne by people involved;
- where the distance from any of the buildings to the public sewer is more than 100 ft, the cost to be borne by the local authority. If an owner is aggrieved by any of these decisions he may appeal to a magistrates' court.

Should the local authority construct the sewer it does not automatically become a public sewer.

PROVISION OF EXITS

Plans which do not show that a building or an extension will have satisfactory means of ingress and egress must be rejected by a local authority (1984 Act, section 24).

The following buildings are affected by this requirement:

- a theatre, hall or other building which the public visit
- a restaurant, shop, store, or warehouse where the public are admitted and where more than 20 people are employed
- a club controlled by the Licensing Act 1964
- a school which is not exempted from building regulations
- a church, chapel, or other place of public worship.

A local authority must consult with the fire authority before making a decision.

The requirement does not extend to

- a private house when the public are admitted occasionally or exceptionally;
- a church or chapel in use since 1890;
- a building in use before 1 October 1937.

Disagreement over the adequacy of provisions required or to be provided may be referred to a magistrates' court for a determination.

This provision has, however, ceased to apply to work under the 1991 Regs as the limits previously imposed by Part B of Schedule 1 have been removed. The section therefore relates only to full plans deposited before 1 June 1992 in accordance with the 1985 Regs.

FACILITIES FOR REMOVAL OF REFUSE

It is an offence, unless a local authority agrees, to close or obstruct the means of an access used for the removal of refuse from a building (1984 Act, section 23).

If they give permission the council may make conditions involving the improvement of an alternative access or the substitution of another access.

Upon summary conviction for an offence against this section, one may be fined.

PROVISION OF WATER SUPPLY

A local authority must reject plans of a dwellinghouse, whether a private dwellinghouse or not, if they do not show that there are to be satisfactory arrangements to provide the residents with a wholesome supply of water, which is sufficient for their domestic requirements.

This may be provided in any one of three ways:

- by way of a connection to a public water supply;
- if a public supply is not available, then by taking other water into the house by means of a pipe; or
- if the above alternatives are not available then by providing a supply of water within a reasonable distance of the house.

The 1984 Act, section 25, contains this requirement.

There have been many arguments circling around the meaning of 'domestic purposes' and as long ago as 1914 Lord Atkinson said 'water supplied for domestic purposes would mean water supplied to help satisfy the needs, or perform, or help in performing services which, according to ordinary habits of civilized life, are commonly satisfied and performed in people's homes, as distinguished from those deeds and services which are satisfied or performed outside their homes, and are not connected with, nor incident to, the occupation of them'.

Earlier in 1858 water used for washing a carriage was held to be used for domestic purposes, as also in 1907 water used for washing a motorcar.

There was a recent case (1984) where water supplied to a boiler used for heating an office in which the owner did not live was held to be domestic use.

However 'domestic purposes' has now been defined in a statute (1991 Act, section 218) to include drinking, washing, cooking, central heating and sanitary purposes in a house. It also includes the washing of vehicles and watering gardens connected with the house, but not when using a hosepipe.

Any difference of opinion between a council and a developer may be referred to a magistrates' court for a determination.

If the approved scheme is not carried out, or if put into effect has not resulted in a wholesome water supply of sufficient quantity, they may give the owner a notice prohibiting occupation of the house until such time as they issue a certificate saying that the supply is satisfactory.

Contraventions of this section may lead to prosecution, and upon conviction, a fine, and a daily continuing fine.

PROPOSED DEPARTURE FROM PLANS

When plans have been passed under 1984 Act, section 16, further plans may be deposited with a view to gaining approval of departures or deviations from the approved plans (1984 Act, section 31).

They will be dealt with in the same way as the originally deposited plans.

Regulations bringing this section of the Act into operation have not yet been made.

LAPSE OF DEPOSIT OF PLANS

When deposited plans have been approved and the work covered by the plans has not been commenced within three years from the date of deposit, the council may declare that the deposit of those plans is of no effect (1984 Act, section 32).

The person who deposited the plans or the present owner must be told of this decision.

Builders have been known to excavate for and place concrete in part of the approved foundations in order to be able to claim that the work had been 'commenced' and thereby keep the plans alive.

Problems may arise where an approved plan relates to a number of separate buildings if work has commenced on some buildings but not on others. The D of E has offered advice that notices may be served in respect of those buildings where work has not been commenced.

So far in this chapter several sections of the 1984 Act relating to the passing of plans have been outlined.

PROCEDURES

Changes in building control arrangements enable a developer to employ an approved inspector to check his plans, inspect the work in progress and finally give a certificate that all is in order.

Running alongside this arrangement is the facility of sending plans to the local authority. But this is now varied from the way things have been done for many years. A person intending to carry out work, or to make a material change of use, now has the option of giving a local authority a 'building notice' or deposit full plans.

There are five regulations (1991 Regs) in Part IV of the regulations covering notices and plans, the requirements not being nearly so extensive as those contained in earlier regulations.

Regulation 11. (1) A person who intends to undertake building work or make a material change of use must

- send the local authority a 'building notice' (1991 Regs, 12) or;
- deposit 'full plans' with the local authority (1991 Regs, 13).

A number of buildings having a specialized use and sundry small buildings such as buildings not frequented by people, and greenhouses, are completely exempt from the regulations (Schedule 2).

(2) Anyone who is going to carry out building work on a building to be put to a designated use requiring a fire certificate (1971 Fire Act, section 1) must deposit full plans.

Under the Fire Precautions (Hotels and Boarding Houses) Order 1972 and the Fire Precautions (Factories, Offices, Shops and Railway Premises) Order 1989, certain hotel and boarding houses, factories, offices and shops and railway premises are designated under the 1971 Fire Act, section 1.

The Fire Precautions (Workplace) Regulations 1997 adds a 'workplace' as a designated use.

A local authority which is not also the fire authority must consult with the fire authority before passing full plans.

A fire certificate is not required when not more than 20 people are working at any one time and not more than 10 people are at work at any one time, elsewhere than on the ground floor. The exception is not appropriate when explosives or highly inflammable materials are stored or used. The exception does not apply when two or more units would result in the total number of persons being exceeded.

The option to send a building notice is not available.

(3) When building work only consists of the installation of a heat-producing gas appliance to be installed by a person or an employee of a person approved under the Gas Safety (Installation and Use) Regulations, neither a building notice nor the deposit of full plans is required.

The Gas Safety (Installation and Use) Regulations 1994, as amended in 1996 is the version now relevant.

The Council for Registered Gas Installers (CORGI) have a scheme for assessing the competence of gas fitters to carry out safe gas work.

(4) The 1985 AI Regs, 18, sets down local authority powers in relation to partly completed work. When an approved Inspector has been employed, the owner must abide by those requirements instead of this regulation. The requirements are discussed in Chapter 7.

(5) For the purposes of the advance payments code (Highway Act 1980, sections 219 to 225 (see Annexe C))

- **the submission of a building notice and plans shall be treated as a deposit of plans;**
- **plans accompanying a building notice are to be treated as deposited plans; and**
- **the receipt of a building notice shall be treated as the passing of those plans.**

GIVING A BUILDING NOTICE

An outline of the requirements surrounding a building notice is set out below.

However, as we have seen from **Regulation 11**, this procedure cannot be used.

- when the work is on a building intended to be put to a designated use under 1971 Fire Act, section 1 (compulsory fire certificates).

Regulation 12. (1) Anyone deciding to use this procedure is required to send a building notice to the council containing the following information:

- name and address of person intending to carry out work;
- a statement that the notice is given in accordance with Regulation 11(1)(a).
- a description of the building work proposed to be carried out or the nature of a proposed material change of use;
- details of location of the building and the proposed use of the building.

The notice is to be signed by the applicant, or on his behalf.

(2) Included with a building notice relating to the erection or extension of a building must be

- a block plan not less than 1:1250 scale showing

  (i) size and position of the building and its relationship to adjoining boundaries;
  (ii) curtilage of the building: the size location and use of all other buildings in the curtilage;
  (iii) the position and width of any street on, or within the curtilage of the building (Fig. 6.2);

- the number of storeys in the building; including basement storeys;
- particulars relating to the following:

  (i) drainage arrangements;
  (ii) if the building, or an extension, is to be erected over a sewer or drain to which 1984 Act, section 18, applies the manner in which the sewer or drain is to be protected;
  (iii) in the event of there being a local enactment affecting the work, the method to be adopted to comply with its requirements.

Notices or plans are not required to be submitted in duplicate.

An example of a requirement of some local Acts, is that there shall be access to the rear of a building, including houses, to facilitate firefighting.

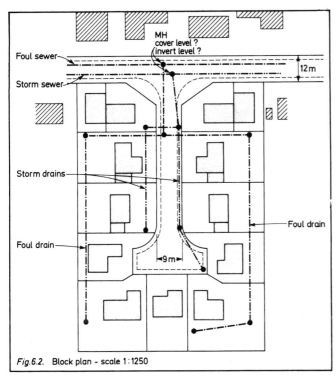

Fig.6.2. Block plan - scale 1 : 1250

(3) When a building notice relates to the insertion of insulating material into cavity walls it is to be accompanied by information regarding

- the type of insulating material and its name;
- approval of the material by the British Board of Agrément or whether it conforms to a British Standard Specification;
- the status of the installer: is he a person holding a certificate of registration from the British Standards Institution or has he the approval of the British Board of Agrément?

C4 of Schedule 1 relates to the installation of insulating material and the approved document gives advice regarding installation.

(4) Where it is proposed to install a hot water storage system to which paragraph G3 of Schedule 1 imposes requirements, a building notice must also include the following information:

- the name, make, model and the type of system it is proposed to install;
- the name of any body which has approved or certified the system;

- the name of any body giving a current registered operative identity card to the installer of the system.

(5) The developer is required to supply plans that may be necessary to enable a local authority to discharge their functions.

(6) The building notice and any accompanying plans are not to be treated as having been deposited.

Upon receipt of a building notice the local authority is able to locate and identify the site and the building. It is also able to examine the possibility that work is the subject of linked powers or any local Act having an effect on the proposal.

The 'linked' matters are

1984 Act, section 18	Building over a sewer
1984 Act, section 21	Provision of drainage
1984 Act, section 25	Water supply

It is always necessary to be cautious if it is found that the building will encroach above a sewer. The local authority may at some time in the future require to dig down to the sewer for repair, maintenance or replacement and unless the legal position in regard to such matters can be resolved in a written agreement the local authority may be able to require removal of the building at some time in the future or require payment for extra expense involved in repairing the sewer. If adequate falls are available it may be possible to divert a sewer around a building (Fig. 6.3) or in the case of small buildings to provide for access through the floor. Whatever is undertaken the prime object is to avoid damage to the sewer with consequent danger to health.

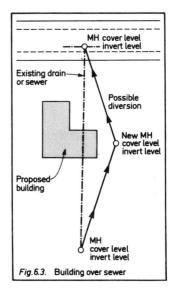

Fig.6.3. Building over sewer

Another situation where caution is essential is when it is proposed to erect or alter a building which will be put to a designated use, the B1 of Schedule 1 may not apply. In this event the persons intending to use the building notice procedure should consult the fire authority before sending in the notice as in their experience it may be necessary to require something different from the proposal before they would issue a fire certificate. Also before constructing or altering buildings which need a licence before they can be used, it will be wise to consult the fire authority.

Safety in fire is becoming an increasingly sensitive aspect of building.

The local authority, upon receipt of a building notice, may ask for plans as may reasonably be required by them in the discharge of their functions. They must be provided within a reasonable time.

A building notice and any accompanying plans are not to be treated as having been deposited. In this respect 1984 Act, section 16, says that a local authority on receipt of deposited plans must approve them or reject them or approve them in part, within five weeks. The effect of the building notice arrangement is to put the local authority in a position where it is not given the power to approve or reject the plans because they will not have been deposited.

Examples embrace many buildings and extensions, including;

- the erection of a dwellinghouse of not more than two storeys;
- a building containing flats but not of more than two storeys;
- a material change in the use of a building of not more than two storeys for use as a dwellinghouse.

The building must of course be erected to comply with the regulations. The person carrying out the building work is required to give the usual notices of commencement and at various stages of work and at completion.

The notices to be given are similar to those now in use. However, whereas at the present time all notices must be in writing, it will be possible in future for notices to be given by such other means as the builder and local authority may agree.

This idea of an optional deposit of plans, or the giving of a building notice allows developers to choose whether they wish to deposit full plans with the council as they have done in the past or whether they wish to avoid the routine of getting approval or rejection to their proposal, by providing only outline information.

If a developer chooses to deposit full plans then the authority must deal with them in the same way as in the past. In this respect passed plans give the builder protection against enforcement notices which have been enjoyed under 1936 Act, section 65, now replaced by 1984 Act, section 36.

The developer may think that his project is so straightforward that he does not need the protection of passed plans. He may also be satisfied that informal discussions with the building control officers had solved any possibility of disagreements arising during the progress of his scheme.

In these circumstances the local authority, on receipt of a building notice, would not be able to insist on the supply of full plans. They will nevertheless be able to ask for any extra information reasonably needed to enable them to discharge their functions in enforcing the regulations, and the listed powers in public and local legislation.

The local authority will still be able to take action against contraventions found in the course of the work. It appears that the arrangements are proposed to reduce the need for unnecessary bureaucratic procedures.

(7) A building notice will cease to have effect three years after the date it was given, if the work to which it relates has not been commenced, or the material change of use has been made.

FULL PLANS

Regulation 13. (1) Full plans must be accompanied by a statement that they are deposited in accordance with 1991 Regs, 11(1)(b).

(2) Full plans must be deposited in duplicate and one copy will be retained by the local authority. Two further copies of plans demonstrating compliance with Part B – fire safety – must be deposited when that Part applies to the proposal. Both copies will be retained by the council.

The additional two copies of plans are not required where the work refers to a dwellinghouse or flat (1992 Regs).

(3) The following make up 'full plans':

- name and address of person intending to carry out the work;
- a description of the building work proposed to be carried out, or the nature of a proposed material change of use;
- details of location of the building and the proposed use of the building.

The notice is to be signed by the applicant or on his behalf. There must also be

- a block plan not less than 1:1250 scale showing:
  (i) size and position of building and its relationship to adjoining boundaries;
  (ii) curtilage of the building; the size location and use of all other buildings in the curtilage;
  (iii) the position and width of any street on or within the curtilage of the building (see Fig. 6.1).
- the number of storeys in the building (each basement level is regarded as one storey);
- particulars relating to the following:
  (i) drainage arrangements;
  (ii) if the building or an extension is to be erected over a drain, sewer or disposal main the manner in which the sewer or drain is to be protected;
  (iii) in the event of there being a local enactment affecting the work, the method to be adopted to comply with its requirements.

If insulating material is to be inserted in cavity walls the plans are to be accompanied by information regarding

- the type of insulating material used and its name;
- approval of the material by the British Board of Agrément or whether it conforms to a British Standard Specification;
- the status of the installer: is he a person holding a certificate of registration from the British Standards Institution or has he the approval of the British Board of Agrément?

C4 of Schedule 1 relates to the installation of insulating material and the approved document gives advice regarding installation.

Where it is proposed to install a hot water storage system to which paragraph G3 of Schedule 1 imposes requirements, full plans must also include the following information:

- the name, make, model and type of system it is proposed to install;
- if approved or certified, the name of the body concerned;
- the name of any body giving a current registered identity card to the installer of the system.

In addition all plans necessary to show that work to be carried out complies with the 1991 Regs.

(4) A statement must accompany full plans saying whether the building is or will be put to a designated use under the Fire Precautions Act 1971.

The premises which may be the subject of designation are when a building is used for

- any purpose involving provision of sleeping accommodation;
- an institution providing treatment and care;
- entertainment, recreation or instruction or for purposes of a club, society or association;
- teaching, training or research;
- purposes involving access by members of the public, whether on payment or otherwise;
- a place of work.

(5) When full plans are deposited they may be accompanied by a request that on completion of the work a completion certificate is requested from the local authority.

The long Schedule 3 of the 1976 Regs has not been reproduced, or even summarized in the 1991 Regs but the purpose of submitting full plans to a local authority is to get them approved and secure the protection that an approval offers.

It is therefore advisable to ensure that deposited plans are complete and enable a local authority to check that compliance with building regulations is intended. It will be recalled that there are two reasons for rejection of full plans:

- that they are defective, meaning incomplete and not giving adequate information;

- that they show the proposed work would contravene any building regulation.

In the case of new buildings full plans will show all the work proposed in connection with the building. Where alterations and extensions are intended it will be necessary to show all new work covered by building regulations and its relationship to the existing building.

CERTIFICATES RELATING TO DEPOSITED PLANS

Full plans may be accompanied by a certificate given by an approved person that certain requirements of the building regulations are satisfied (1984 Act, section 16). A certificate may relate to:

- Part A, structural stability, or
- Part L, energy conservation, and when a valid certificate accompanies full plans the local authority may not reject them in respect of any matter covered by the certificate.

On each occasion that a certificate is issued it must be accompanied by evidence of insurance cover. This must be in the form of a declaration signed by the insurer that the certifier has taken out insurance embraced by a scheme approved by the Secretary of State.

It is also necessary when a certificate relates to work covered by paragraph A3 of Schedule 1 (disproportionate collapse) to enclose a declaration that the certifier does not, and will not until work is complete have a professional or financial interest in the work (1985 AI Regs, 27).

An approved person giving a certificate will have a professional or financial interest in the work covered by the certificate if

- he is or has been responsible for the design or construction of any of the work in any capacity;
- he or his nominee is employed in any capacity with an organization having a professional or financial interest in the work;
- he is a partner or employed by someone having a professional or financial interest in the work (1985 AI Regs, 9).

Professional or financial interest includes people who may be a trustee for the benefit of some other person, or if married and the interest of one spouse is known to the other then the second spouse is regarded as having an interest.

An approved person is not regarded as having a professional or financial interest when he is solely performing the function and responsibilities of an approved person.

PASSING PLANS

When plans are deposited with a local authority it has a duty to pass them unless

- they are defective;
- the proposal would contravene any building regulations (1984 Act, section 16).

If the plans are defective, or show that work would contravene any building regulation, the council is given two options:

- the plans may be rejected; or
- they may be passed subject to conditions:
  a. specifying modifications to be made to the plans; and
  b. more plans are to be deposited.

In these circumstances the person depositing plans must have previously requested the council to pass plans subject to conditions or have agreed with the council that they may do so.

When passing plans the council is to state that it operates only for approval under the building regulations and any section of the 1984 Act that expressly authorizes the council in certain cases to reject plans.

If there is a disagreement between a developer and the council as to whether the proposed work is in conformity with the building regulations the matter may be referred to the Secretary of State for a determination.

The council must issue a decision on the passing or rejection of plans within five weeks of deposit, or such extended period agreed with the developer not being later than 2 months from the date plans were deposited.

When giving consideration to applying conditions to an approval, a local authority should have regard to the Construction (Design and Management) Regulations 1994 as their actions may bring them within the definition of 'client', 'designer' or 'contractor' (see Chapter 2).

Also, prior to the passing of plans for a new building or a change of use, a Fire authority must be consulted if the fire precautions in the premises are not covered by other legislation. (See Fire Precautions (Workplace) Regs. 1997; Chapter 2.)

NOTICES CONCERNING STAGES OF WORK

It is not a breach of the 1984 Act or the 1991 Regs to begin work on a plan which has not yet been approved. Once the full plans have been deposited two days' notice of commencement may be given and then a start made. The developer does, however, put himself in jeopardy if he does not do the work in accordance with all building regulation requirements. Caution is also necessary if a planning permission is awaited.

In very many instances and for a variety of reasons it may not be possible to build exactly as shown on the plans that have been deposited and approved. Any substituted material or work must comply with the building regulations and if this is done then the obligation has been fulfilled. Deviation from plans should, however, be notified to the council, and also be mindful of planning requirements and conditions.

It is always very useful to have council's written approval of plans showing work as built because they are thereby not able to serve notice requiring it to be taken down or altered.

UNAUTHORIZED BUILDING WORK

Regulation 13A

(1) **This regulation applies**

- **when unauthorized work has been commenced on or after 11 November 1985; and**
- **at the time the unauthorized work was carried out there should have been a deposit of plans, or a building notice given; or an initial notice in the case of work supervised by an approved inspector; and**
- **plans were not deposited and no notices were given.**

(2) **In these circumstances the applicant (owner) may in writing ask the local authority for a 'regularization certificate'. Such an application must be accompanied by**

- **a statement that the application is being made in accordance with Regulation 13A;**
- **a description of the unauthorized work;**
- **a plan of the unauthorized work;**
- **a plan showing any additional work necessary to ensure that the work complies with building regulations applicable at the time the unauthorized work was carried out.**

(3) **A local authority may require an applicant to take reasonable steps, including opening unauthorized work for inspection, making tests, or taking samples required to ensure that the relevant requirements are met.**

(4) There is provision for consideration being given to the relaxation or dispensation of a requirement in building regulations (see Chapter 4) applying to the unauthorized work, and the local authority shall notify the applicant

- of the work required to comply with relevant requirements, or requirements dispensed with or relaxed, or
- that they cannot determine what work is required to be done, or
- that no work is required to be done.

(5) When a local authority is satisfied

- **that the relevant requirements have been satisfied,** or
- **that no work is required to ensure that contravening work satisfies the relevant requirements they may issue a regularization certificate.**

(6) A regularization certificate is evidence that the relevant requirements specified in the certificate have been complied with. However, the certificate is not 'conclusive' evidence.

(7) If this regulation applies, Regulations 11 (giving a building notice or deposit of plans) and 13 (full plans) do not apply, and the operation of 13A is not to be treated for the purposes of the 1984 Act, section 16 (passing or rejection of plans).

(8) This regulation is without prejudice to the powers of a local authority under the 1984 Act, section 36 (removal or alteration of offending work).

This regulation was introduced by the 1994 Regs to allow local authorities to certify building work which should have been subject to control when it was undertaken, but had not been authorized at that time. It does not apply to work carried out prior to 11 November 1985 which is the date the 1985 Regs became operative.

Local authorities may charge fees for the consideration and issue of regularization certificates.

Regulation 14

Having deposited full plans or given a building notice it is a requirement that a number of notices must be given to the local authority. These notices must be in writing or by such other means as the developer and council may agree. In view of the many misunderstandings that can arise in consequence of telephone messages it is wise for the developer to ascertain the name of the person receiving his telephone notice and the time of the conversation. It is essential for a local authority to maintain a similar record of telephone notices if they are prepared to agree to such an arrangement.

In previous regulations the amount of notice required has mostly been given in hours. The 1991 Regs changed these to periods of 'days' and define a day as any period of 24 hours commencing at midnight.

Saturdays, Sundays, Bank holidays and public holidays are excluded from the calculations.

Thus the periods relating to notices which must be given to a local authority are as follows:

- at least 2 days after the end of the day on which notice is given that work is to be commenced;
- at least 1 day after the end of the day on which notice is given that
  - (i) excavation for foundations are to be covered;
  - (ii) foundations are to be covered;
  - (iii) damp-proof course is to be covered;
  - (iv) any concrete or other material laid over the site is to be covered;
  - (v) any drain or sewer is to be covered in anyway.
- not more than 5 days following completion of laying haunching or covering any drain or sewer;
- not more than 5 days following completion of building work;
- if to be occupied prior to completion, not less than 5 days before the building or any part of it is to be occupied.

Here is an example of the timing under this arrangement:

1. Notice of commencement given to local authority at any time during Monday.
2. First day of notice commences at midnight on Monday.
3. Second day of notice commences at midnight on Tuesday.
4. Second day of notice ends at midnight on Wednesday.
5. Work may commence at 00.01 on Thursday.

There have been a number of claims for negligence arising from foundations having been placed on poor ground. The requirement that a notice be given to the council before covering the excavation for a foundation gives an opportunity for the subsoil to be examined before concrete is placed. It assists, too, in establishing that care was exercised in preparation for the placing of foundations.

Notice that drainage work has been backfilled must also be given to the council. Many defects are known to arise in shallow drainage systems between the time they are backfilled and the completion of all the building operations taking place. It is therefore to be expected that local authorities will not test the drain immediately upon receipt of this notice, but much later when the building operations are almost complete. Great care should therefore be taken to protect drainage systems as the location of buried defects can be expensive in both time and money.

If a local authority is not given the appropriate notice the developer may be required, by a written notice, to cut into, lay open or pull down, sufficient of the work to enable an inspection to be made to ascertain whether any of the regulations have been contravened. It is necessary to emphasize the effect on a job if the local authority requires any work to be pulled down, and therefore the greatest care must be taken to give correct notices.

A local authority may give the developer a written notice setting out in the way in which work contravenes the regulations. When work has been undertaken to secure compliance with those regulations the developer is required to give a written notice to the local authority that the work has been done. He must give this notice within a reasonable time following completion of the remedy.

In a case **Blaenau Gwent Borough Council v Sabz Ali Khan** (1993) it was held that the person carrying out the work could be the owner if he had authorized and commissioned the work, and did not necessarily apply solely to the person who physically carried out the work.

The erection of a new building may therefore involve up to nine notices.

1. Commencement
2. Excavation for foundations
3. Foundations
4. Damp-proof courses
5. Oversite concrete
6. Drains laid
7. Drains backfilled
8. Occupation and/or
9. Completion.

It must not be assumed that only at these stages is the local authority interested in a building. These are the stages which must be notified in order to ensure that the council has an opportunity to inspect important parts of the work. The authorized officer, who is usually a building control officer, may examine a building any day at any reasonable hour to ensure compliance with all the regulations applicable to a particular project.

ENERGY RATING

Regulation 14A

(1) This regulation applies

- **where a new dwelling is created by building work, or**
- **where a new dwelling is created by a material change of use involving work.**

(2) The energy rating of the dwelling is to be calculated following a procedure approved by the Secretary of State, and the local authority is to be told of this rating.

(3) Notice of the rating is to be given to the local authority

- not later than the period allowed by paragraph (4) of Regulation 14, which is 5 days after the work has been completed, and
- if the building is to be occupied before completion, the energy rating notice is to be given at least 5 days before any part is occupied.

Regulation 14A introduced by the 1994 Regs is part of the Government's commitment to reducing CO_2 emissions and in response to recommendations made by the Select Committee on the Environment and the Select Committee on indoor pollution, and is a further stage in the regulation updating process.

It involves for the first time the calculation of an energy rating for a new building, on a scale of 1 to 100, indicating the cost of running new homes using the Standard Assessment Procedure (SAP). It is aimed at encouraging developers to use more efficient fuels, heating appliances and dwelling types.

The requirements are set out in Part L, and for the first time apply to certain building conversions and material alterations.

Completion certificates

A feature of building control is that a local authority may be required to give a certificate of completion in connection with building work, although the certificate is not intended to be all embracing.

Regulation 15. (1) A local authority is to give a certificate of completion where they receive a notice that building work has been completed or that the building has been partly occupied before completion.

This is dependent upon two procedures having been followed:

- they were requested to issue a completion certificate when full plans were deposited (Regulation 13(5)), or
- there was a statement included when full plans were deposited that the building has a designated use or will have a designated use as set out in the Fire Precautions Act 1971.

(2) If a local authority have taken all reasonable steps to ascertain that 'relevant requirements' have been satisfied they must issue a certificate.

(3) 'Relevant requirements' means

(a) if the deposit of full plans included a statement that the building was having or would have a designated use the requirements of Part B of Schedule 1 – fire safety – and

(b) if a request for a completion certificate was made when full plans were deposited, any applicable requirements of Schedule 1.

(4) A completion certificate is to be regarded as evidence that relevant requirements set out in the certificate have been complied with. However, Regulation 15(4) says that this is not to be regarded as conclusive evidence.

A contravention of Regulation 15 is not an offence (Regulation 21).

It will be noted that completion certificates are not issued when the work has been carried out under a building notice.

Forms

Forms for a building notice, or one to accompany full plans, have not been prescribed, although as is seen in the preceding pages, the content of any notice is set out. It is convenient to use a form for the purpose of notifying a council of one's intentions, and the following are suggested.

Building Regulations 1991
Building Notice

I give notice, in accordance with Regulation 11(1)(a) of the above regulations that the undermentioned work is to be carried out.

Date _____ Signature _____
Address _____
Telephone number _____

Details of proposal
1. Name and address of _____
   person intending to _____
   carry out work if _____
   different from above _____
2. Location _____
3. Description of work to be undertaken _____
   _____
4. Proposed use of building on completion _____
   _____
5. Details
   - number of storeys _____
   - provision for drainage and position of outfall _____
   _____
   - building WILL / WILL NOT be over a sewer
   - any requirement of Local Act _____
6. Filling cavity in an existing building
   - name and type of material._____
   - approved by British Board of Agrément   Yes/No
   - conforms to British Standard Specification   Yes/No
   - installer holds BSS certificate of registration   Yes/No
   - installer approved by British Board of Agrément   Yes/No
7. Unvented hot water system
   - system to be installed _____
   - name of body approving system _____
   - name of body issuing registered operative identity card _____
8. A plan is attached showing the following
   - size and position of building_____Yes/No
   - size and position of extended building_____Yes/No
   - relationship to adjoining boundaries _____Yes/No
   - boundaries of curtilage of building_____Yes/No
   - boundaries of curtilage of extended building_____Yes/No
   - size, position and use of all other buildings within the curtilage _____Yes/No
   - width and position of any street on or within boundaries of curtilage _____Yes/No
9. The appropriate fee of £____ is enclosed.

Building Regulations 1991
Full Plans Notice

I give notice, in accordance with Regulation 11(1)(b) of the above regulations that the undermentioned work is to be carried out. Relevant plans, particulars and statements are deposited herewith.

Date _____ Signature _____
Address _____
Telephone number _____

Details of proposal
1. Name and address of _____
   person intending to _____
   carry out work if _____
   different from above _____
2. Location _____
3. Description of work to be undertaken _____
   _____
4. Proposed use of building on completion _____
   _____

5. Details
   * number of storeys _____
   * provision for drainage and position of outfall _____

   _____
   * building WILL / WILL NOT be over a sewer
   * any requirement of Local Act _____
6. A plan is attached showing the following
   * size and position of building   Yes/No
   * size and position of extended building   Yes/No
   * relationship to adjoining buildings   Yes/No
   * boundaries of curtilage of building   Yes/No
   * boundaries of curtilage of extended building   Yes/No
   * size, position and use of all other buildings within
     the curtilage   Yes/No
   * width and position of any street on or within
     boundaries of curtilage   Yes/No
7. The attached plans show that work is to be carried out in accordance with the Buildings Regulations 1991.
8. Filling cavity wall with insulating material
   * name and type of material _____
   * approved by British Board of Agrément   Yes/No
   * conforms to British Standard Specification   Yes/No
   * installer holds BSS certificate of registration   Yes/No
   * installer approved by British Board of Agrément   Yes/No
9. Unvented hot water system
   * system to be installed _____
   * name of body approving system _____
   * name of body issuing registered
     operative identity card _____
10. The building has a designated use (Fire Precautions Act 1971)   Yes/No
11. I agree to conditional or stage approval within the terms of section 16(2) of the Building Act 1984, should this be considered necessary.
12. I request the issue of a completion certificate in accordance with Regulation 15.   Yes/No
13. The appropriate fee of £____ is enclosed.

NATIONAL TYPE APPROVALS

Local authorities have cooperated in the formation of 12 regional consortiums into a national confederation of building control consortia called LANTAC (Local Authority National Type Approval Confederation). LANTAC operates the national type approval scheme in England and Wales which applies to housing and non-residential buildings which are in a standard format and are likely to be repeated in different areas.

Foundation designs can also be type-approved (subject to local ground conditions) as long as the foundation design is specific to a LANTAC type approved building. A further LANTAC scheme, System Approval, has been introduced for manufacturers of building systems which use standard design details and components in the construction of different building designs. LANTAC has also launched a scheme to assist the retail and leisure sectors when 'fitting out' building shells.

A LANTAC certificate is issued for a design which has been checked in accordance with the building regulations under the rules of the confederation. Once a design has a LANTAC certificate, either type or system approval, when it is used in conjunction with a building regulation submission it will automatically be accepted under the procedures of LANTAC by all other local authorities participating in the scheme. The LANTAC type and system approval schemes speed up the building regulation approval process and guarantee consistent interpretation for LANTAC approved designs.

The Local Authority National Type Approval Confederation is at 35 Great Smith Street, Westminster, London SW1 3BJ. (020 7664 3000).

NEGLIGENCE

The requirement that a local authority should enforce building regulations in its area is clearly set out in the 1984 Act, section 91. From 1972 until a recent judgement given in the House of Lords on 26 July 1990 it had been held that a local authority's responsibilities in building control carried a duty at common law to take reasonable care to see that building regulations were complied with. The 1990 judgement was a dramatic change.

There have been a number of well reported cases where, since Dutton v Bognor Regis UDC in 1972, local authorities through their Building Control Officers have been found negligent in their application of building regulations. If approved inspectors are taking upon themselves the responsibilities of ensuring building regulation compliance, they must be cautious and mindful of the problems which have arisen in the past.

In the Dutton v Bognor Regis case the building inspector saw excavations for the foundations of a house in 1959 when building byelaws were in force. Apparently the work did not comply with the byelaws and as it transpired that the ground was made up the work should not have been accepted. Inspectors, the court declared, must be diligent, visit the work as occasion requires, and carry out inspection with reasonable care.

Then there was Anns v Merton London Borough Council in 1977. It was alleged that the council was negligent as it had not required foundations to be placed at the depth shown on approved plans. The House of Lords, in this case reiterated that when making an inspection there was a duty to take reasonable care, and that was to ensure compliance with the bye laws.

Acrecrest v Hattrell and Partners followed in 1982. In this case defective foundations gave rise to cracks in the building. It was found that as the foundations were in accordance with a building inspector's requirements the local authority was partly responsible for the defects.

In 1982 Dennis v Charnwood Borough Council, it was claimed that reasonable care had not been taken in approving plans. Approved plans showed a house on a concrete raft. After construction defects appeared and it was found that the raft was inadequate. It had been constructed partly on filled and partly on sound subsoil. Settlement was uneven with consequent inevitable damage to the structure.

In another case a builder who had not previously built a bungalow undertook the work and the foundations were not satisfactory. This was Warlock and Saws in 1981. The building inspector did not notice the defect and the building suffered serious damage. As there was no architect involved in supervising the job it was held that the local authority role was important and took 40% of the blame for the defects.

The last case concerned a semi-detached house built in 1969. It was thought that the site presented a risk of settlement and therefore a raft foundation was designed. The local authority engaged an independent firm of consultant engineers to check the design.

The house was purchased in 1970 and cracks appeared in 1977. They were repaired, but further cracks appeared in 1981. Investigations revealed that there had been differential subsidence of the foundation raft. Cracking and distortion occurred and gas and drain pipes fractured.

Thus, Murphy v Brentwood District Council came about and proceeded through the courts to the House of Lords. Here the Lords reconsidered the Anns v London Borough of Merton case and came to the conclusion that it had been wrongly decided, and that all decisions subsequent to the Anns case which purported it should be over-ruled, and the appeal was allowed.

It would appear that a local authority in exercising its building control function is not now under a duty to protect property owners from economic loss. However, it is not clear whether local authorities have no liability at all, or merely no liability for financial loss.

It follows that when making inspections the date, time and the exact extent of the work should be recorded together with any opinion given that work was not in accordance with the building regulations and should be altered or removed. Needless to say, all discrepancies noticed should be reported to the builder in writing.

ROLE OF THE OMBUDSMAN

The Local Government Act 1974 provided for the appointment of two commissioners for local administration – one in England and one in Wales. A local commissioner is generally recognized by the name 'ombudsman'.

An ombudsman's duty is to investigate and report on complaints against the way that local authorities have acted; that is, the way that decisions have been arrived at, attributable to administrative faults; or the influence of improper considerations or conduct.

The majority of complaints investigated relate to planning matters and housing, but any subject may be scrutinized, including building control. The basic concept of his duty is related to procedures and the effect these procedures have on the justice of decision making.

There are two steps to take before actually making an approach to the ombudsman. First the council should be asked to look into the complaint in a letter to the chief executive, and if the response is not satisfactory, a councillor may be approached and requested to enquire into the complaint. If the response is still unsatisfactory a form of complaint may be addressed to the ombudsman, either directly or through a councillor.

The kind of information the ombudsman requires is as follows:

- Your name.
- Your address.
- Your telephone number, at home and at work.
- The name of the council complained about.
- What you think the council did wrong.
- What injustice you think you have suffered in consequence of what the council did.
- Have you already complained to the council? If so when and how?
- List any information you have to support the complaint. Letters from the council and any other documents should be sent to the ombudsman. They will be returned if required.
- Give the date on which you first heard of the action you are complaining about.
- When did you first tell a councillor about it? If it was more than 12 months before you told a councillor give your reasons for the delay.

Any matter in which the complainant has a right of appeal to a tribunal, a Minister, or the courts may not be investigated by the ombudsman. However, there is some discretion allowed if it is not reasonable to expect a complainant to exercise or to have made use of that right.

The 1974 Act is silent when it comes to the meaning of the word 'maladministration' – the only thing that the ombudsman may investigate. In effect it is faulty administration where a decision has been arrived at based on or influenced by improper considerations or conduct. The Commission for Local Administration in England has headquarters at 21 Queen Anne's Gate, London SW1H 9BU. There is a separate one for Wales at Darwen House, Court Road, Bridgend, Mid Glamorgan CF31 1BN.

Transitional provision

The 1991 Regs and the approved documents 1992 came into operation on 1 June 1992, but they do not apply to development where before that date a local authority had received any of the following:

- a building notice
- full plans
- an initial notice, or
- a public bodies notice.

The 1991 Regs also do not relate to building work carried out after that date, following the above procedures, with any notice or plans with or without any departure.

The 1994 Regs added to the transitional provision as follows:

- The new Regulation 13A regarding unauthorized building work came into force on 1 October 1994.
- The new Regulation 14A regarding energy rating came into force on 1 July 1995, and similar transitional arrangements apply.

Deregulation and Contracting Out Act 1994

This Act created new procedures to be applied to enforcement proceedings. They do not prevent or delay urgent enforcement action when this is necessary; nor are existing rights of appeal affected. The new procedures apply to existing legislation and are available to business and individuals (Section 5).

An Order was expected relating Section 5 to building control procedures, but it seems possible that a Code of Practice will be forthcoming.

RELEVANT READING

British Standards
BS 1192: Part 1: 1984. Drawings for the construction industry
BS 1192: Part 2: 1987. Recommendations for architectural and engineering drawings
BS 1192: Part 3: 1987. Recommendations for symbols and other graphic conventions
BS 1192: Part 4: 1984. Recommendation for landscape drawings
BS 1192: Part 5: 1990. Guide for structuring of computer graphic information
BS 1635: 1990. Graphic symbols and abbreviations for fire protection drawings
BS 6100: Part 1: section 1.1: 1987. Glossary of building terms – types of building
BS 6100: sub-section 1.5.2: 1987. Glossary of building terms – jointing
BS 6100: Part 6: section 6.5: 1987. Glossary of building terms – formwork

BRE information paper
3/88   Production drawings – arrangement and content

CHAPTER 7

Approved inspectors

Controversial provisions in the 1984 Act related to the surveillance of building work otherwise than by local authorities; although they are still involved. It is now possible for the supervision of plans and work to be carried out by 'approved inspectors'.

There are fourteen sections in the 1984 Act which relate to the supervision of plans and work by approved inspectors.

Section 47.   Giving and acceptance of initial notice.
Section 48.   Effect of initial notice.
Section 49.   Approved inspectors.
Section 50.   Plans certificate.
Section 51.   Final certificate.
Section 51A.   Variation of work to which initial notice relates.
Section 51B.   Effect of amendment notice.
Section 51C.   Change of person intending to carry out work.
Section 52.   Cancellation of initial notice.
Section 53.   Effect of initial notice ceasing to be in force.
Section 55.   Appeals.
Section 56.   Recording and furnishing of information.
Section 57.   Offences.
Section 58.   Definitions.

Regulatory control of approved inspectors is contained in the Building (Approved Inspectors, etc.) Regulations 1985. (These are abbreviated to 1985 AI Regs.)

The 1985 AI Regs have been amended by the following later regulations.

The Building (Approved Inspectors, etc.) (Amendment) Regulations 1992; and

The Building (Approved Inspectors, etc.) (Amendment) Regulations 1995; and

The Building (Approved Inspectors, etc.) (Amendment) Regulations 1996; and

The Building (Approved Inspectors, etc.) (Amendment) Regulations 1998.

Approved inspectors may be architects, building control officers, chartered builders, engineers and surveyors in private practice. They may, at the option of the person intending to carry out the work, supervise construction instead of a local authority's building control staff.

Several new terms are introduced into building control vocabulary; such as 'initial notice', 'plans certificate' and 'final certificate'.

The procedure for approving inspectors is set out in Regulations 3 to 6. They are to have no professional or financial interest in the work they supervise unless it is minor work (Regulation 9). Their duties are set out in Regulations 10 and 11.

An approved inspector may achieve his/her status in one of two ways. A person may be approved by the Secretary of State, or a body or organization which, in accordance with regulations, is designated by the Secretary of State for that purpose (1984 Act, section 49).

Until D of E circular 10/96 dated 1 July 1996, the bodies designated by the Secretary of State were as follows:

Architecture and Surveying Institute
Association of Building Engineers
Chartered Institute of Building
Institution of Building Control Officers
Institution of Civil Engineers
Institution of Structural Engineers
Royal Institute of British Architects
Royal Institution of Chartered Surveyors

Circular 19/96 added the designation of the Construction Industry Council, and this opens the way for persons who wish to submit applications for approval as an approved inspector as provided for in 1985 Regs 3(2)

An amendment made by the 1998 AI Regs provides that where a body has been designated for approving inspectors, corporate bodies and individuals also wishing to become approved inspectors, are to apply to the designated body. This is currently the Construction Industry Council.

This means that the various professional bodies are replaced by the Construction Industry Council, although they still have a part to play within the Council.

An approved inspector since 1985, the NHBC Building Control Services Ltd has been joined by more companies as corporate approved inspectors for building control work. Conditions of the approval have been revised and there are in fact no limitations placed on the approval of the NHBC, or the following companies:

13 January 1997	BRCS Building Control Ltd
13 January 1997	Butler and Young Associates, Building Control and Fire Safety Consultants Ltd
13 January 1997	TPS Special Services Ltd (now Carillon Special Services Ltd)
28 April 1997	BAA Building Control Ltd
28 April 1997	Maunsell Associates
1 September 1998	STMC (Building Control) Ltd.

It is anticipated that from time to time more names will be added to the list.

Enquiries about individual or corporate approved inspectors should be addressed to the Construction Industry Council, 26 Store Street, London WC1E 7BT (020 7637 8692).

Insurance cover

Regulations impose requirements with respect to the provision of insurance cover against the occurrence of defects, in particular:

• the form and content of policies;
• provision for approved schemes of insurance.

An essential ingredient of the approved inspector scheme is the possession of adequate insurance to cover each job, and the Secretary of State has approved several insurance schemes for this purpose.

INTERPRETATION

The 1985 AI Regs 2 contains definitions of a number of expressions. Where the word 'means' is included in a definition, the interpretation relates only to that particular meaning. Sometimes the word 'includes' is used, and this is intended to expand the ordinary meaning of a word.

- 'the Act' means the Building Act 1984 (the 1984 Act);
- 'building' means any permanent or temporary building but not any other kind of structure or erection. Reference to a building includes part of a building. The definition is identical to that in the 1991 Regs 2(1);
- 'designated use' means designated under the 1971 Fire Act;
- 'flat' means separate and self contained premises in a building divided horizontally and is almost identical to that in the 1991 Regs 2(1);
- 'fire authority' means county councils and metropolitan county fire authorities and the London Fire and Civil Defence authorities;
- 'material alteration and material change of use' have the meaning set out in the 1991 Regs;
- reference to the carrying out of works includes the making of a material change of use.

APPROVED INSPECTORS

Regulations 3 and 4 outline the methods of securing approval. As amended, these regulations require corporate bodies and individuals also wishing to become approved inspectors to apply to the Construction Industry Council.

The 1984 Act, section 49 says that applicants for the status of approved inspector may have to pay a fee, and the CIC will charge a fee calculated to recover their costs.

AUTHORITY TO BE IN WRITING

Regulation 5. Approval of an inspector or designation of a body is to be given in writing

REMOVAL FROM LIST OF APPROVED INSPECTORS

Regulation 6. Approval as an inspector lasts for five years, but may be withdrawn by the designated body if there is a conviction for giving false or misleading notices and certificates.

- An inspector may apply to extend the 5-year period but it must follow on without interruption.
- there is no limit to the number of extensions that may be sought.

Approvals issued by the Secretary of State up to 1 March 1999 will remain indefinite in duration, although it is expected that such approvals will be withdrawn at the end of February 2000. Application may then be made to the CIC. for renewal which will be limited to five years.

An approved inspector removed from the list because of a conviction (1984 Act, section 57) may not be reinstated until 5 years from the date of conviction.

Conviction for issuing notices or certificates recklessly or knowingly when they contain false or misleading information may upon:

- summary conviction mean a fine or up to 6 months' imprisonment;
- conviction on indictment, mean a fine, or up to 2 years' imprisonment, or both.

In a court in 1932 it was held that a written statement may be false because of what it withheld, omitted or implied (R v Kylsant).

'Recklessly' was considered in another court case (R v Lawrence 1981) and it was said there that a person acted recklessly if, before doing the act, he failed to give thought to the possibility of any harmful consequences or having seen that there was a risk, he nevertheless proceeds to go on and do it.

LIST OF APPROVALS AND DESIGNATIONS

Regulation 7. Lists of approvals and designations are to be kept by the Secretary of State and copies of the lists are to be supplied to local authorities with notification of subsequent changes.

The Construction Industry Council is also to maintain a list of inspectors which it has approved, and it must notify every local authority of any withdrawal of approval. Unlike the Secretary of State, the Construction Industry Council is not required to tell local authorities of additions to its list.

An inspector is required to give evidence of his approval when dealing with a local authority.

A local authority is able to check if an inspector is approved by contacting the Construction Industry Council.

FUNCTIONS OF APPROVED INSPECTORS

Regulation 10 says that an approved inspector must exercise professional skill and care.

- to ensure that work will achieve reasonable standards for protection of the health and safety of persons in or about the building, and that the work will conform to the requirements.
- to ensure that the requirements of the 1991 Regs, Schedule 1 are met so far as they relate to any material change of use.
- to ensure that there is satisfactory provision for the removal of foul water, and rainwater from roofs to a sewer, cesspool or other suitable outfall.
- to ensure that the requirements relating to energy ratings are complied with.

Where the building regulations involve the insertion of insulating material in a cavity wall after the wall is constructed the approved inspector is not required to supervise the work, but when issuing a final certificate is to say whether or not the insulation has been placed.

If an approved inspector discovers that work he is supervising is over a public sewer he must notify the local authority as soon as practicable.

ENERGY RATINGS

A regulation introduced by the Building (Approved Inspectors, etc.) (Amendment) Regulations 1995 is concerned with energy ratings.

Regulation 10A places a duty on a person carrying out building work as defined in the 1991 Regs 3(1) and 5 for a new dwelling or a material change of use to calculate the energy rating and give this information to the approved inspector.

The method of calculating energy ratings is set out in Chapter 20.

Contravention of this regulation is an offence listed in the 1984 Act, section 35.

CONSULTATIONS

Regulation 11 states that the responsibilities of an approved inspector include the duty of consulting any stipulated person or body, as for example, the fire authority, before taking any specified step in connection with work or other matters to which building regulations are applicable.

This will involve giving the fire authority sufficient information and plans to enable it to decide whether the work would comply with fire requirements.

Consultation must take place before giving a plans certificate and also before giving a final certificate.

A plans certificate or final certificate are not to be issued until fifteen working days have elapsed from the date of consultation.

INITIAL NOTICES

Approved inspectors may, at the option of a person intending to carry out work, inspect construction instead of a local authority.

The person intending to carry out work and the appointed approved inspector are each required to tell the local authority of the arrangement. This is achieved by completing a form called an 'initial notice' (1984 Act, section 47).

The phrase contained in the regulations 'person intending to carry out the work' is cumbersome, and is clearly intended to embrace anyone, even a DIY person. It is most likely to be an owner or a contractor, but throughout will be called 'the developer'.

Regulation 8 states that when these notices are sent to a council they must be accompanied by some plans and evidence of insurance cover relating to the work.

A standard for the initial notice is contained in AI Regs, **Schedule 2**, and the following information is to be given:

- address and description of work;
- use to which the building is to be put;
- name and address of the approved inspector;
- name and address of the developer;
- a copy of the notice of approval held by the approved inspector;
- a statement signed by the insurer that a named scheme of insurance applies to the work;
- plan, scale not less than 1:1250 indicating location and boundaries of the site;
- position of any connection to be made to sewers;
- if sewers not available, details of disposal method to be employed and the siting of any cesspool or septic tank;
- where no provision is made for drainage, reasons are to be given;
- where it is intended to build over a public sewer, the precautions that will be taken to protect the sewer;
- if any local enactments apply, what means are to be taken to ensure compliance;
- a statement that the work is, or is not, 'minor work'.

Minor work means that the material alterations of a dwelling house which before the work is carried out has no more than two storeys and after the work no more than three storeys; the provision, extension or material alteration of a controlled service or fitting; or work involving the underpinning of a building.

For this particular purpose:

(a) 'dwellinghouse' does not include a flat, or a building containing a flat;
(b) a basement is not a storey;
(c) 'controlled service or fitting' means a service or fitting relating to
   G1 Sanitary conveniences
   G2 Bathrooms
   G3 Hot water storage
   H   Drainage and waste disposal
   J   Heat-producing appliances
   L1 Heating system controls
   L1 Insulation of heating services.

The approved inspector must also give:

- a declaration that he does not have a financial or professional interest in the work. (Regulation 9)

A person will have a financial or professional interest if he

1. designs or constructs any of the work in any capacity, or
2. he or a nominee is a member, officer or employee of a company or body with a professional or financial interest in the work, or
3. he is a partner or an employee of a person with a professional or financial interest;

- a declaration indicating whether or not the Fire Authority is to be consulted;
- declaration that he is aware of his obligations.

The initial notice is to be signed by the developer AND the approved inspector.

If an initial notice is accepted by the council, the approved inspector may then undertake the prescribed functions with respect to the inspection of plans, supervision of the work and the giving of certificates and notices.

TRANSFER TO ANOTHER APPROVED INSPECTOR

Generally if an initial notice is already in force the council is precluded from accepting a second initial notice relating to the same work. The exception is when two approved inspectors agree to a takeover of responsibility. Or the circumstances may not be completely cordial. One approved inspector may in the opinion of the developer not be in a position to carry on with the work, and he appoints another approved inspector.

In these circumstances, if the second initial notice is accompanied by a letter from the first approved inspector promising to cancel his notice immediately the second notice is accepted, the local authority may accept the second notice.

When an arranged takeover is not possible the council will become responsible between the date of cancellation of the first notice and acceptance of the second. If a second notice is imminent it is unlikely that a local authority will wish to become involved in the building control function.

INITIAL NOTICE REJECTION

A local authority may not reject an initial notice unless it fails to comply with a list of requirements set out in AI Regs, Schedule 3. An initial notice must be rejected if it fails to meet any of the prescribed grounds. The council has no discretion in this matter.

The various grounds for the rejection of an initial notice are as follows:

- The notice is not in the proper form; the notice has been served on the wrong authority or the person signing the document is not an approved inspector.
- There is no information about the address of the work, the nature of the work, or the use to which the building is to be put.
- There is inadequate information about the drainage to be provided, or reasons why drainage is not required.
- Satisfactory provision has not been made when it is necessary to build over a sewer.
- Steps to comply with any local enactment are not indicated.
- The inspector has not enclosed a copy of the notice of approval from the designated body.
- There is no declaration from an insurer that proper insurance cover has been obtained.
- There is no undertaking that the Fire authority will be consulted when the nature of the proposal requires that a consultation be made.
- There is no declaration that the approved inspector does not have any professional or financial interest in the work. If a declaration is not necessary the proposal must have been declared to be 'minor work'.
- An earlier initial notice has already been given. This does not apply if:
   (a) the earlier notice is no longer in force and the local authority have not commenced to supervise the work; or
   (b) the approved inspector who gave the earlier notice writes that the earlier notice will be cancelled just as soon as the new initial notice is accepted.

The Public Health Act 1936 enabled a local authority to attach conditions when passing plans, but not actually relating to the implementation of the regulations themselves. They are now listed in the 1984 Act.

Section 21   Provision of drainage
Section 22   Drainage of buildings in combination
Section 23   Obstruction of means of access to remove refuse
Section 25   Provision of water supply

If, when considering an initial notice, it is apparent to a local authority that it is appropriate to impose a condition they may do so.

Most of the reasons a local authority may use for rejecting an initial notice do not require evaluation or judgement; information on the initial notice is as required or it is not adequate.

However, in a number of matters the council may be able to exercise discretion and it may be in the approved inspector's interest to have preliminary discussions with the Building Control Department.

Adequacy of drainage and the availability of public sewerage systems are topics for clarification. It is essential that the point of connection to any public sewer is agreeable to the local authority and also that connections are proposed to the correct sewers, be they combined or separate for both foul and surface water.

When constructing some private form of sewage disposal arrangements the local authority may have information regarding the nature of outfall arrangements found to be satisfactory in the area and point the direction towards any additional consents which may be required from a water authority.

When it appears necessary to build over a sewer, it may be requisite to enter into an agreement with the local authority or undertake such work as they require to protect the sewer from damage.

Local enactments can pose a surprise for an approved inspector operating from an area where these do not exist. They can relate to access arrangements and enquiries should be made. Detailed plans are not required. The initial notice requirements are met if it contains an account of the steps to be taken, although on occasion the giving of a plan will be the ideal explanation of the work proposed.

Local authorities are given a minimal time in which to consider an initial notice. This is not stated in the 1984 Act but it is prescribed in AI Regs, 8(4). The period is 5 working days from the day on which the notice is given. Within this period the local authority must give notice of rejection, specifying the reasons to each of the persons by whom the initial notice was given. If they fail to do so within the time allowed the authority shall be 'conclusively presumed to have accepted' the initial notice and have done so without imposing any requirements.

The Interpretation Act 1978 applies to the posting of notices when dates of receipt are crucial. For example if a notice is sent through the post it is presumed to have been received within the normal time required for the class of postage paid. But even with first-class post one cannot rely upon delivery on the following working day. If a notice must therefore be in the hands of the approved inspector within 5 working days it appears that the local authority have only an effective 3 days to complete their investigations prior to posting first class. It is therefore important that an approved inspector date stamps receipt of all incoming documents in case an out of date rejection should arrive after the date when the local authority are deemed to have approved his notice.

An initial notice comes into effect when it is accepted by a local authority and continues in force until it is cancelled, or six weeks after completion of the work.

Highways Act 1980

When a local authority have accepted an initial notice it is to be regarded as a deposit of plans for the purpose of sections 219 to 225 of the above Act. This relates to the advance payments code and therefore the local authority must tell the street works authority within 1 week. The street works authority, usually a county council, then has 6 weeks in which to require the developer to enter into arrangements for the future street works changes.

AMENDMENT NOTICE

The Deregulation (Buildings) (Initial Notice and Final Certificate) Order 1996 created a new section 51A in the 1984 Act which provided for the amendment of initial notices. This had not previously been possible. A new regulation followed into the 1985 AI Regs making provision for notices to be amended.

Regulation 8A states that an amendment notice has to be prepared on a prescribed form and be accompanied by:

- plans and documents;
- a signed declaration by the insurer that the cover still applies. The notice details the amendments proposed and must be accompanied by a number of documents;
- a copy of the original notice;
- if appropriate, a copy of the inspector's approval by a designated body;
- a statement that all the original plans remain unchanged, or including such plans as are amended;
- a statement by an insurer that a named scheme applies to the work;
- plan to scale of 1:1250 showing boundaries of the site together with information regarding connection to a sewer or cesspool. If no provision is made for drainage, the reason must be given;
- should a building be erected over a sewer, details of precautions to be taken must be given;
- details of any local enactment affecting the work;
- a certificate stating whether or not the work is minor work, and that the approved inspectors will not have a financial or professional interest in the work.

The grounds which a local authority may use to reject an amendment notice are the same as refer to an initial notice (see page 44).

A local authority is allowed only five working days in which to give a notice of rejection.

The acceptance, or rejection, of an amendment notice is regarded as the passing, or rejection, of plans and the new plans supersede those previously deposited. The acceptance of an amendment notice is regarded as the deposit of plans for the purpose of sections 9D and 13 of the 1971 Fire Act. (1984 Act, section 51B)

CHANGE OF PERSON INTENDING TO CARRY OUT WORK

The Deregulation (Building) (Initial Notices and Final Certificates) Order 1996 authorized section 51C to the 1984 Act. It makes provision for giving a written notice to the local authority when a change in the person intending to carry out the work takes place.

There is no prescribed form to be used in this instance, but the notice must be given jointly:

- by the approved inspector who gave the original initial notice and the person who now proposes to carry out the work.

The local authority cannot reject the notice and there is no provision in this section for the name of the approved inspector to be changed.

EFFECT OF INITIAL NOTICE

Acceptance of an initial notice by a local authority means that the developer and the approved inspector may proceed with the preparation and vetting of the proposal.

The actual arrangements for examination and checking of plans is a matter to be settled between the designer and the approved inspector. Clearly it is an opportunity for the approved inspector to be involved in design discussions in order that the finalized plans for his inspection do not present any building regulation problems, and he can proceed to give a plan certificate with the minimum of delay.

So long as an initial notice remains in force the function of enforcing building regulations conferred on a local authority is removed from work which is specified in the notice.

In consequence, a local authority may not (1984 Act, section 48):

- give notice requiring the removal or alteration of any work which contravenes building regulations;
- institute proceedings for any contravention of the building regulations which arises out of carrying out work covered by the initial notice.

The giving of an initial notice covers a developer's responsibilities under several enactments, and for this purpose:

- the giving of an initial notice, together with plans, is to be treated as the deposit of plans;
- plans accompanying any initial notice are to be regarded as deposited plans;
- the acceptance or rejection of an initial notice is to be regarded as the passing or rejection of plans;
- should an initial notice cease to be in force it is to be treated as a declaration that the plans are of no effect.

The enactments referred to are as follows:

(a) the power contained in 1984 Act, section 36(2), which enables a local authority to require work which has been undertaken without deposit of plans, or notwithstanding the rejection of plans, to be pulled down or removed;
(b) the provision contained in 1984 Act, section 36(5), whereby a local authority may not require removal of work in accordance with plans they have approved;
(c) the power of a local authority under the 1984 Act, section 36(6), to obtain an injunction for certain contraventions;
(d) the requirement in 1984 Act, section 18(2), that the water authority be advised if building will be near a public sewer;
(e) the advance payment code: 1980 Act, sections 219 to 225.

For the purpose of the Fire Precautions Act 1971, sections 9D and 13, the acceptance of an initial notice by a local authority is to be treated as a deposit of plans.

INITIAL NOTICE CEASES TO BE IN FORCE

Regulation 16 states that there are a number of circumstances which may lead to an initial notice ceasing to be in force, in addition to those already outlined. There are time limits attaching to a number of circumstances:

- if a final certificate is rejected, the initial notice ceases to be in force 4 weeks after the date of the rejection notice;
- when work involves the erection, extension or material alteration and the premises are occupied and no final certificate has been issued, then the initial notice ceases to be in force

  (a) if the building is put to a designated use, 4 weeks from the date of occupation. The time allowed previously was 1 day, but this was extended to 4 weeks to provide time for a final certificate to be given if a building is to have a designated use (1971 Fire Act) and the construction or materially altered part of a building will be occupied prior to the final certifacte being issued (1998 AI Regs)
  (b) in all other cases 8 weeks from the date of occupation.

However, an initial notice remains valid if part of a building is occupied and a final certificate has been issued in respect of that part:

- when there has been a material change of use and no final certificate issued then the initial notice ceases to be in force 8 weeks from the date of this change of use;

- in all other instances when no final certificate has been issued the initial notice ceases to be in force 8 weeks from the day work was substantially completed.

A local authority may extend any of these periods, either before or after expiry of the periods if they have reason to believe that a final certificate will be given within the extended period.

Responsibility for the work remains with the approved inspector during any extended period.

CANCELLATION OF INITIAL NOTICE

Regulation 17. When an initial notice is in force and the approved inspector

- cannot carry out his functions; or
- is unable adequately to carry out his functions; or
- is of the opinion that there is a contravention of a building regulation

the initial notice must be cancelled by informing the local authority and also the developer (1984 Act, section 52).

There are a variety of reasons why an approved inspector may not be in a position to carry out his duties. The obvious reason is that his approval has been withdrawn for a misdemeanour. Other reasons may arise because of business difficulties, or illness.

The first action of an approved inspector upon discovering a contravention, and certainly before cancelling an initial notice, is to give notice of the fault to the developer. If after three months that person has neither pulled down nor removed the work, nor undertaken any alterations to secure compliance with the building regulations, then the approved inspector has no option but to cancel the initial notice.

If at any time it appears to the developer that the approved inspector is no longer willing or able to carry out his function he must cancel the initial notice. The cancellation must be sent to the local authority and, when practicable, to the approved inspector. Failure to give this notice to a local authority may, upon summary conviction, result in a fine.

There is a prescribed form which is to be used by an approved inspector when cancelling an initial notice. It is to be addressed to the local authority and contain the following information:

- the address of the work, a description of the work and the purpose for which the building is to be used;
- a statement to the effect that an initial notice was given on a stated date referring to the above work;
- a declaration that the person signing the cancellation is the approved inspector;
- a declaration saying: 'I hereby cancel the initial notice';
- if cancellation is because of failure to secure a remedy to a contravention a statement to the effect that notice had been given to the person carrying out the work, and that it has not been put right within the prescribed period, which is three months;
- it is then necessary to state what the contravention is;
- the notice must be signed and dated.

An initial notice may be cancelled by the developer if it appears that the approved inspector is not willing or able to carry out his functions (1984 Act, section 52(3)).

When such a decision is reached the developer must send a written notice to the local authority and the approved inspector. The form to be used is prescribed in 1985 AI Regs, Schedule 2, and requires the following information to be given:

- It must be addressed to the approved inspector and the local authority concerned. The names and addresses are to be on the form.
- It should state the address and description of the work and what the use of the building would be.
- It must contain the date of the initial notice relating to the work.

- It must say whether the signatory is the person carrying out the work, or is the person intending to carry out the work if the job has not been started.
- It must contain the statement: 'I hereby cancel the initial notice'.
- The notice is to be signed and dated.

A local authority may cancel an initial notice if the work has not been started within 3 years of the date it was accepted (1984 Act, section 52(5)).

Should a local authority decide to cancel an initial notice it must send to the approved inspector and the person intending to carry out the work a written notice in the form prescribed in 1985 AI Regs, Schedule 2.

Such a notice is to contain the following information:

- the address and description of the work and what the use of the building would be;
- a statement that the person signing the notice has been authorized to do so by the local authority;
- the date on which the initial notice relating to the work was accepted;
- a statement that the work has not been commenced within 3 years of acceptance and that the local authority thereby cancels the initial notice.

The notice is to be signed and dated.

INITIAL NOTICE NO LONGER IN FORCE

There are provisions which apply when an initial notice ceases to be in force (1984 Act, section 53). They are quite involved and apply when an initial notice is no longer operative because:

- it has been cancelled by the approved inspector saying that he is unable to carry out his functions, or that he is unable to secure a remedy for defective work;
- it has been cancelled by the developer when it appears to him that the approved inspector is no longer willing or able to carry out his functions;
- the period covered by the initial notice has expired.

There are limitations on a local authority's powers regarding work in respect of which a plans certificate or final certificate have been given while the initial notice was still valid.

In a situation when the day before an initial notice ceased to be in force a final certificate was issued relating to part of the work, and was accepted, the final certificate remains valid in respect of the work it covers.

Building regulations are enabled to require that plans be supplied covering work which has been covered by a certificate. This information will facilitate any action a local authority may initiate regarding any work not covered by a certificate.

Normally a local authority may require contravening work to be pulled down or altered within 12 months of completion of the work. But when an approved inspector has been engaged, and the work has been the subject of an initial notice, the period is 12 months from the day the initial notice ceased to be in force.

Likewise, a local authority must initiate any necessary court proceedings within 6 months of the date they assumed responsibility for work which was previously the subject of a cancelled initial notice.

Regulation 18 enumerates a local authority's powers in relation to partly completed work.

The circumstances which enable the local authority to proceed are:

1. Some of the work covered by an initial notice has been done.
2. The initial notice has ceased to be in force or has been cancelled.
3. No initial notice has been accepted.

In this situation the 'owner' (the only time owner is named in the AI Regs) upon being given reasonable notice must provide the local authority with:

(a) plans of the work which has been done and is not covered by a final certificate, and
(b) if a plans certificate had been given and not rejected, a copy of plans relating to the work covered by the certificate; and the owner may be required, given a reasonable time to comply, to cut into, lay open or pull down any work that prevents the council discovering whether any work for which there is no final certificate contravenes the 1991 Regs.

Where it happens that work has been commenced but not completed a person who intends to do further work in connection with the partly completed work must give the local authority some plans. These must indicate that the proposed work will not contravene any building regulation. On occasions it may be necessary to provide plans of work already carried out in order to indicate that the further work can be done without contravening any regulations.

Any plans given to a local authority are not regarded as 'deposited plans'.

PLANS CERTIFICATES

When an approved inspector has examined plans and is satisfied that they are neither defective nor show that the work would contravene building regulations and he has complied with any requirements as to consultation or otherwise, he must, if requested to do so by the developer, give a 'plans certificate' to the local authority and also to that person (1984 Act, section 50).

Any disagreement arising between an approved inspector and a developer regarding compliance with building regulations may be referred by the developer to the Secretary of State for a determination. A fee will be payable.

A plans certificate may be combined with the initial notice, and a form is specified for the purpose. When this is done, and the initial notice is later cancelled the plans certificate continues to be valid.

A plans certificate may relate to all or only part of the work referred to in the initial notice, and it does not come into force until it is accepted by the local authority. It can be 'accepted' by failure of the local authority to reject within 5 working days beginning with the day the certificate is given.

A local authority may only reject a plans certificate on grounds that are set out in the 1985 AI Regs. When rejected both the approved inspector and the developer must be given written notice of the decision.

A plans certificate may be revoked if the work has not been commenced within 3 years of its acceptance by the authority.

FORMS OF PLANS CERTIFICATE

Regulation 12. The form of a plans certificate is set out in Schedule 2 and the approved inspector is required to give the following information:

- a description of the work and its location;
- a declaration that he is the approved inspector appointed for the work referred to in the initial notice and the date of the initial notice;
- a declaration signed by the insurer that an approved named scheme of insurance applies to work covered by the certificate;
- a statement that he has examined plans of the work and he is satisfied that they are not defective; he must also say that work carried out in accordance with the plans would not contravene the building regulations;
- a statement that the work is, or is not, minor work;
- a statement that he has no direct or indirect professional or financial interest;

- an indication of whether the fire authority has been consulted;
- a list of plans he has examined giving the date and reference number of each plan;
- the certificate must be signed and dated.

Regulation 13 refers to Schedule 4 where the grounds and period for rejecting a plans certificate are set out. A rejection must be given within 5 working days otherwise the certificate 'shall be conclusively presumed' to be accepted by the local authority.

The grounds for rejecting a plans certificate are:

- It is not in the proper form.
- The certificate does not specify the work or the plans to which it relates.
- There is no initial notice in force.
- It is not signed by the approved inspector who gave the initial notice or that person is no longer an approved inspector.
- There is no declaration by an insurer that a named scheme of insurance relates to the work.
- The certificate does not, where it is necessary contain a declaration that the fire authority has been consulted.
- There is no declaration that the approved inspector does not have any direct or indirect financial interest.

COMBINED PLANS CERTIFICATE AND INITIAL NOTICE

In submitting a combined plan certificate and initial notice it is necessary to supply information required by the separate forms, and any rejection must be based upon one or more of the grounds listed for rejecting a separate plans certificate or initial notice.

EFFECT OF PLANS CERTIFICATE

Regulation 14. If an initial notice ceases to be in force because it has been cancelled by the approved inspector or the developer and the conditions in 1984 Act, section 53(2), relating to plans certificates given, accepted and not rescinded are fulfilled, the local authority may not

- require removal or alteration of work which contravenes building regulations; or
- institute proceedings for a contravention of building regulations in connection with work described in the certificate and which has been carried out as indicated on the plans.

FINAL CERTIFICATE

A local authority's enforcement powers cannot be exercised in respect of work which has been supervised by an approved inspector.

Consequently, when an approved inspector is satisfied that the work set out in his initial notice has been completed he must give to the local authority a 'final certificate'.

This final certificate will indicate completion of the work and the discharge of his/her functions by the approved inspector. A local authority is allowed 10 working days from receipt in which to reject a final certificate. Only certain grounds for rejection may be used. (1984 Act, section 51 and AI Regs, 15)

Regulation 15 says that the form of final certificate is set out in Schedule 2, and each final certificate must contain the following information:

- a description of the work and its location;
- a declaration that the person completing the form is an approved inspector and that the work was the subject of an initial notice;
- the work has been completed and the functions of an approved inspector carried out;

- when insulation is to be placed in a cavity wall, whether or not this work has been carried out;
- a statement that, as far as he is aware, no building or extension has been erected over a public sewer except:
  (i) that notified in the initial notice, or
  (ii) that discovered during the course of the work and previously notified to the local authority;
- if all the work is complete a statement that final certificates have been issued covering all of the work in the initial notice;
- a declaration by the insurer that an approved named scheme of insurance applies to the work;
- whether the work is or is not minor work;
- a statement that the approved inspector has no direct or indirect financial interest in the work;
- where necessary the fire authority has been consulted;
- the date, and signature of the approved inspector.

The grounds on which a final certificate may be rejected by a local authority are contained in 1985 AI Regs, Schedule 5. They are:

- It is not in the proper form.
- The certificate does not specify the work.
- No initial notice applies to the work.
- The certificate is not signed, or the person is no longer an approved inspector.
- The certificate does not contain a declaration that proper insurance cover exists.
- There is no declaration that the fire office has been consulted where necessary.
- Except when the work is minor, there is no declaration regarding direct or indirect financial or professional interest.

LOCAL AUTHORITY RECORDS

Regulation 28. A local authority is obliged by the 1984 Act, section 56, to keep records and these are to be available to the public for inspection at all reasonable hours. The records are to contain details of receipt and action upon all notices and certificates recorded:

- initial notices, plans certificates, final certificates;
- amendment notices and change of person carrying out work;
- description of work;
- address of work;
- name and address of person signing notice or certificate;
- name and address of insurer;
- date of acceptance, or presumed date of acceptance;
- an index based on address of development.

All information must be in the register within 14 days.

APPEALS

There is a right of appeal to the magistrates' court against a council's decision to reject an initial notice, an amendment notice, a plans or final certificate (1984 Act, section 55).

The court has two options. If it thinks that the notice or certificate was correctly rejected it may confirm the rejection. If it does not think that the notice or certificate was properly rejected it may direct the local authority to accept the document.

An appeal to the magistrates' court must be made within 21 days from the date the authority's rejection was received by the approved inspector.

A person aggrieved by a magistrates' court decision may make an appeal to the Crown Court.

REGULATIONS NOT APPLICABLE

Regulation 18 1991 Regs relates to the supervision of work otherwise than by local authorities and says that **the following building**

regulations do not apply in respect of any work specified in an initial notice given under the 1984 Act, section 47:

11. Giving of a building notice or deposit of plans
14. Notice of commencement or completion of certain stages of work
15. Completion certificates
16. Testing of drains and private sewers
17. Sampling of materials.

The following building regulations do not apply in respect of any work in relation to which a final certificate has been given under the 1984 Act, section 51, and has been accepted by the local authority:

16. Testing of drains and private sewers
17. Sampling of materials.

FEES

Approved inspectors may make charges for carrying out their responsibilities. There is to be no fixed scale of fees as is used by local authorities. Fees are to be agreed between an approved inspector and the person who intends to carry out the work (1984 Act, section 49).

DELEGATION

An approved inspector may arrange for another person to inspect plans or work on his behalf. However, he may not delegate the giving of 'plans certificates' or 'final certificates' to any other person. Delegation will also not affect his liability whether civil or criminal if it arises out of functions conferred upon him by the 1984 Act or by any building regulation.

An approved inspector is also liable for negligence on the part of anyone carrying out any inspection on his behalf as if it were negligence by a servant acting in the course of his employment (1984 Act, section 49(8)).

The approved inspector must always be conscious that when the work is complete he must certify he is satisfied that within the limits of professional skill and care the work complies with building regulations.

CHAPTER 8

Work undertaken by public bodies

A considerable volume of constructional work is undertaken by public bodies. They may be a local authority, county council or any other body that operates for the good of the public and not for its own profit and is named by the Secretary of State for this purpose.

There is a definition of 'local authority' in the 1984 Act, section 126, and it means:

- district councils
- London borough councils
- Common council of City of London
- Sub-Treasurer of the Inner Temple
- Under-Treasurer of the Middle Temple
- Council of the Isles of Scilly.

Such bodies are exempt from compliance with the procedural regulations, but nevertheless must comply with the technical requirements (1984 Act, section 5). There is, however, a part of the 1985 AI Regs which sets out the procedures to be followed by a public body contemplating and carrying out building work.

If a public body considers that the operation of a building regulation would be unreasonable in a particular instance it may give a direction dispensing with or replacing that requirement (1984 Act, section 8). Before exercising this power to relax a building regulation relating to structural fire protection or concerning means of escape in case of fire, the fire authority must be consulted (1984 Act, section 15).

In effect public bodies may supervise their own work without inspection by a local authority or an approved inspector.

This power is contained in the 1984 Act, section 54, which says that

- when a public body is to undertake work to a building belonging to it and the substantive requirements of the building regulations apply and
- the work may be properly supervised by its own employees or agents,

it may give to the local authority a 'public body's notice.'

The procedure in relation to a public body's notice is very similar to that which is followed when an approved inspector serves an initial notice.

To be effective a public body's notice must be accepted by the local authority. To be accepted the notice must contain all the information required by the 1985 AI Regs. If it does not provide all the information required, the local authority does not have any discretion, the notice must be rejected. In accepting a public body's notice the local authority may impose conditions under the various enactments enabling them to do so when passing plans which have been deposited.

The local authority is allowed ten working days in which to reject the notice, and if they do not do so within that period then the council is 'conclusively presumed to have accepted' the public body's notice.

Part VII of the 1985 AI Regs is concerned with public bodies. Regulation 19 says that when the Secretary of State decides certain public bodies should be enabled to supervise their own work, he may approve them either by description of a group or individually. Local authorities are to be informed of his decision.

Approval may be withdrawn by the Secretary of State at any time.

PUBLIC BODY'S NOTICE

Regulation 20 refers to the prescribed form of public body's notice in Schedule 2, and to the grounds under which a local authority may reject a notice contained in Schedule 6. The form is prescribed and it is important that all information requested is given:

- name and address of local authority in whose area the work is to be undertaken;
- the address and description of the work;
- use of any buildings to which the notice relates;
- name and address of the public body; unlike an approved inspector the public body is not required to send proof that it has been approved for this purpose;
- a statement that the work can be adequately supervised by its own employees or agents;
- a plan 1:1250 giving location of site and its boundaries;
- the position of any proposed connection to a sewer;
- when no connection is to be made to a sewer, details of how drainage is to be disposed of, and the siting of any cesspool;
- the reasons when drainage provision is not being made;
- precautions to be taken when any building is to be over a sewer or drain;
- steps to be taken to meet the requirements of any local enactment which may apply to the work;
- whether or not the fire authority will be consulted.

The form must be signed and dated. There is no stipulation regarding the standing of a person signing the form on behalf of the public body and this is not a matter to be investigated by the receiving local authority. Inevitably the public body will limit the number of people authorized to sign and will authorize individuals or post-holders to act either generally or specially in the matter of their responsibilities under the 1985 Regs.

Grounds upon which a local authority may reject a public body's notice are set out in Schedule 6:

- The notice is not in the proper form.
- The notice has been sent to the wrong local authority.
- The body is not one which has been approved by the Secretary of State.
- The notice and plans do not contain the address and description of the work.
- Information is not supplied about drain outfalls, building over sewers and any local enactments that apply.
- There is not an undertaking to consult the fire authority if necessary.
- The authority is not satisfied that a drain or private sewer will be satisfactory.
- The local authority is not satisfied that arrangements for building over a sewer or drain are satisfactory. They may impose conditions.

- The proposed work will not comply with local enactments authorizing the rejection of plans.

EFFECT OF PUBLIC BODY'S NOTICE

The 1984 Act, section 54, applies to the acceptance of a public body's notice by a local authority enabling the body to proceed with the preparation and vetting of the proposal.

The actual arrangements for examination and checking of plans is a matter to be settled by the public body. Clearly it is an opportunity for designs to be controlled by a team including the building control expert in order that the finalized plans do not present any building regulation problems, and a public body's plan certificate can be issued with the minimum of delay.

So long as a public body's notice remains in force the function of enforcing building regulations conferred on a local authority is removed from work which is specified in the notice.

In consequence, a local authority may not (1984 Act, section 48):

- give notice requiring the removal or alteration of any work which contravenes building regulations;
- institute proceedings for any contravention of the building regulations which arises out of carrying out work covered by an initial notice.

The giving of a public body's notice covers a body's responsibilities under several enactments, and for this purpose:

- the giving of an initial notice, together with plans is to be treated as a deposit of plans;
- plans accompanying any public body's notice are to be regarded as deposited plans;
- the acceptance or rejection of a public body's notice is to be regarded as the passing or rejection of plans;
- should a public body's notice cease to be in force it is to be treated as a declaration that the plans are of no effect.

The enactments referred to are as follows:

(a) the power contained in 1984 Act, section 36(2), which enables a local authority to require work which has been undertaken without deposit of plans, or notwithstanding the rejection of plans, to be pulled down or removed;
(b) the provision contained in 1984 Act, section 36(5), whereby a local authority may not require removal of work in accordance with plans they have approved;
(c) the power of the local authority under the 1984 Act, section 36(6), to obtain an injunction for certain contraventions;
(d) the requirement in 1984 Act, section 18(2), that the water authority be advised if building will be near a public sewer;
(e) the advance payment code: 1980 Act, sections 219 to 225.

For the purpose of the Fire Protection Act 1971, section 13, the acceptance of a public body's notice by a local authority is to be treated as a deposit of plans.

PUBLIC BODY'S PLANS CERTIFICATE

When a public body is satisfied that its drawings have been inspected by someone who is competent and the plans are found not to be defective and indicate that work carried out in accordance with them will not contravene the building regulations, they may serve on a local authority a public body's plans certificate.

The public body must, before serving the notice have carried out any consultations which are required (Regulation 21).

The public body's plans may be examined by an employee who is competent to assess the plans. This person may also be an agent, or consultant acting in that capacity and he is not required to have achieved the qualifications or experience necessary for a professional

to become an approved inspector. He is also not debarred from having a professional or financial interest in the proposal.

A public body's plans certificate may refer to all or only part of the work described in the public body's notice, and has no effect until it is accepted by the local authority receiving it. The certificate may be rejected only on grounds listed in 1985 AI Regs, Schedule 7, and unless this is done within 10 days, the local authority is 'conclusively presumed to have accepted' the certificate.

A public body's plans certificate must be in a prescribed form, be signed and dated, and contain the following information:

- address and description of work;
- use to which the building will be put;
- name and address of public body;
- a certificate that the plans have been inspected by a competent employee or agent and that they were found to be satisfactory;
- the date (where necessary) that the fire authority were consulted;
- the date and reference number of all the plans inspected and found to be satisfactory.

As with the public body's notice the status of the person signing the plans certificate is not stipulated.

A public body may, if it wishes, combine the notice and plans certificate in one prescribed form. This should contain all the information required when the forms are completed separately, and be signed and dated.

The only grounds for rejecting a public body's certificate are set out in 1985 AI Regs, Schedule 7:

- The proper form has not been used.
- The work is not correctly described.
- The number and dates of plans are not set out.
- A public body's notice was not in force when the certificate was received (this does not apply when the two forms are combined).
- The certificate was not signed on behalf of the body giving the earlier notice.
- The body has ceased to be entitled to supervise its own work.
- Where the body should have consulted the fire authority it has not done so.

When a public body's notice and plans certificate have been given on the same forms, a local authority may reject on any of the combined grounds.

If work is not commenced within three years of the date of acceptance of a public body's plans certificate, it may be rescinded by the local authority.

While a public body's plans certificate is in force the local authority may not require the removal or alteration of work not meeting regulation requirements or start proceedings to secure a remedy for contravention of building regulations if the work has been carried out in accordance with plans referred to in the certificate.

PUBLIC BODY'S FINAL CERTIFICATE

When a public body is satisfied that any of the work described in its public body's notice has been completed, it may send to the local authority a public body's final certificate. It is not necessary to await completion of the whole building before giving a final certificate. For example, such a certificate could be issued relating solely to foundations or work up to ground level before proceeding further. The effect would be for the public body's notice to cease to have effect in relation to the portion of work described in the public body's final certificate, but to remain in force for the balance of work not yet the subject of a final certificate (1984 Act, Schedule 4).

There is a prescribed form of public body's final certificate in Schedule 1, 1985 AI Regs, and it must contain the following information:

- address and description of the work;
- a statement that the work has been supervised by a servant or an agent to ensure that the substantive requirements of the 1985 Regs have been observed;

- if the work has involved inserting insulating material into a cavity wall, a statement indicating whether or not the process has been carried out;
- a statement that a public body's final notice has been given in respect of all the work;
- a statement that, where necessary, the fire authority has been consulted.

The form must be signed and dated.

A certificate may be rejected by the receiving local authority but only on grounds listed in Schedule 8, 1985 AI Regs. They are as follows:

- The certificate is not in the proper form.
- It does not set out the work to which it relates.
- A public body's notice was not in force.
- The certificate was not signed.
- There is no declaration that the fire authority has been consulted where necessary.

PUBLIC BODY'S NOTICE NO LONGER IN FORCE

There are provisions which apply when a public body's notice ceases to be in force (1984 Act, Schedule 4). They are similar to those involving approved inspectors and apply when a public body's notice is no longer operative because it has not been accepted by a local authority or after a period of three years it has been rescinded.

There are limitations on a local authority's power regarding work in respect of which a public body's plans certificate or a public body's final certificate have been given whilst the public body's notice is still valid.

In a situation when the day before a public body's notice ceased to be in force a public body's certificate was issued relating to part of the work, and was accepted, the public body's final certificate remains valid in respect of the work it covers.

Building regulations are enabled to require that plans be supplied covering work which has been covered by a certificate. This information will facilitate any action a local authority may initiate regarding any work not covered by a certificate.

Normally a local authority may require contravening work to be pulled down or altered within 12 months of completion of the work. But when work is being undertaken by a public body, and the work has been the subject of a public body's notice, the period is 12 months from the day the public body's notice ceased to be in force.

Likewise, a local authority must initiate any necessary court proceedings within 6 months of the date they assumed responsibility for work which was previously the subject of a public body's notice which has ceased to be in force.

There are a number of circumstances which may lead to a public body's notice ceasing to be in force and they are based on 1985 AI

Regs, 16. There are time limits attaching to a number of circumstances:

- if a public body's final certificate is rejected the public body's notice ceases to be in force 4 weeks after date of the rejection notice;
- when work involves the erection, extension or material alteration and
  - (a) the premises are occupied and
  - (b) no public body's final certificate has been issued, then the initial notice ceases to be in force
    - (i) if the building is put to a designated use, one day from the date of occupation, and
    - (ii) in all other cases 8 weeks from the date of occupation.

However, a public body's notice remains valid if part of the building is occupied and a public body's final certificate has been issued in respect of that part:

- when there has been a material change of use and no public body's final certificate issued, then the public body's notice ceases to be in force 8 weeks from the date of this change of use;
- in all other instances when no public body's final certificate has been issued the public body's notice ceases to be in force 8 weeks from the day work was substantially completed.

A local authority may extend any of these periods either before or after expiry of the periods if they have reason to believe that a public body's final certificate will be given within the extended period.

Responsibility for the work remains with the public body during any extended period.

REGULATIONS NOT APPLICABLE

Regulation 18 1991 Regs, relates to the supervision of work otherwise than by local authorities and says that **the following building regulations do not apply in respect of any work specified in a public body's notice** given under the 1984 Act, section 54, and is in force:

11. Giving of a building notice or deposit of plans
14. Notice of commencement or completion of certain stages of work
15. Completion certificates
16. Testing of drains and private sewers
17. Sampling of materials.

The following building regulations do not apply in respect of any work in relation to which a public body's final certificate has been given under the 1984 Act, para. 3 of Schedule 4, and has been accepted by the local authority:

16. Testing of drains or private sewers
17. Sampling of materials.

CHAPTER 9
Materials and workmanship
(Approved document to support Regulation 7)

The requirements for materials and workmanship have been similar for many years, basically requiring that proper materials are to be used in a workmanlike manner. The 1991 Regs, 7 took account of the move towards European harmonization, and the amendment in the 1999 Regs, 7 no longer refers to product approval. Practice guidance on materials and workmanship is given in the 1999 Approved document relating to Reg. 7 and embracing European approvals.

Regulation 7 states that building work is to be carried out using adequate and proper materials:

- **which are of a suitable nature in relation to the purpose and condition of use;**
- **which are adequately mixed or prepared;**
- **to be applied, used or fixed so as to adequately perform the function for which they are intended;**
- **building work is to be carried out in a workmanlike manner.**

The purpose of this standard is to reduce risks to the health and safety of people in or about buildings. It is also aimed at the conservation of energy and the provisions to be made for disabled people.

DEFINITION

To avoid misunderstanding, 'materials' has been defined to include:

- products
- components
- stone, timber, thatch and other 'naturally' occurring materials
- fittings
- items of equipment
- backfilling for excavations.

ENVIRONMENTAL IMPACT

Careful choice of materials may reduce the environmental impact of building work, and where appropriate the use of recycled or recyclable materials is to be considered. Care must be taken to ensure that the use of these materials will not have any adverse consequences to the health and safety requirements of the building work.

RANGE OF APPLICATION

In achieving the level of performance the standards of materials and workmanship are not required to be more than necessary to enable the finished work to be safe, and not a danger to the health or safety of persons in or about the building. This applies to the following parts of Schedule 1:

- A Structure
- B Fire safety
- C Site preparation and resistance to moisture

- D Toxic substances
- E Resistance to passage of sound
- F Ventilation
- G Hygiene
- H Drainage and waste disposal
- J Heat-producing appliances
- K Protection from falling, collision and impact
- N Glazing – safety in relation to impact, opening and cleaning

Standards of materials and workmanship have not to be more than is necessary to limit the calculated heat losses to those in Schedule 1.

- L Conservation of fuel and power.

Standards of materials and workmanship are not required to be in excess of those necessary to serve their purpose in the following instance.

- M Access and facilities for disabled people.

CONTINUING CONTROL

The 1984 Act, section 2, provides for the imposition of continuing requirements on owners and occupiers of buildings, including buildings which were not, when erected, subject to building regulations.

This provision has not been taken up in the 1991 Regs which do not contain any provisions for continuing control over the use of materials after completion of the work.

There is, however, in the 1984 Act, section 19 a provision which enables local authorities to impose conditions when a building is to be constructed of short-lived materials.

This is also of little effect as no materials have been prescribed for its purpose.

SECTION 1. MATERIALS

The approved document mentions materials and products which may be suitable for the purposes of the regulations. The references are not exclusive and other materials or products may be acceptable in particular circumstances if they fulfil the requirements of the regulations.

The earlier version of Regulation 7 included a definition of 'proper materials' but this detail is not now included, reliance being placed upon the various ways of establishing the fitness of materials.

Fitness of materials

The approved document lists several ways in which the fitness of the material may be established.

- **British Standards.** If a material conforms to a British Standard (BS) and it is appropriate for the use and conditions in which it will be placed, then it may be used.

BSI is well known as the recognized body in the United Kingdom for the preparation and promulgation of national standards. The scope of standards work includes glossaries of terms, definitions, quantities, units and symbols, methods of test, specifications for quality, safety, performance or dimensions, preferred sizes and types and codes of practice. The principles observed in the preparation of a BS are that it should be in accordance with the needs of the economy, meet a recognized demand and take into account the interests of producers and users.

Numerous British Standards are listed throughout the approved documents.

If a product has been tested and it is certified as complying with a BS by an approved body in a member state, it is to be accepted by a building control authority and an approved inspector.

If it is not so accepted, then the Trading Standards Officer must be notified and the UK Government may notify the Commission.

Many BS relating to construction products are undergoing revision to tie up with new European Standards known as 'ENs'. The BSI numbering policy is now to adopt the CEN number following BS as for example 'BS EN 459: 1994 – Building lime'.

Inevitably there will be a period when the old BS will apply alongside the new and this may continue for some years. The old BS may still relate to work in progress or even in the situation where an approved document has not been updated. In these circumstances it may be necessary to check the applicability of an old or new standard to the product to be used.

There are two sections in a European-based standard:

(i) A harmonized part containing requirements relevant to building regulations, and
(ii) A voluntary part embracing additional matters of concern to the construction industry.

The approved document references relate only to the harmonized part.

- **CE marking.** Appropriate use of a product which carries the 'CE' symbol as explained in the Construction Products Directive (89/106/EC). A construction product is defined in the directive to mean any product which is produced for incorporation in a permanent manner in construction works, including both building and civil engineering works. A conformity mark may only be issued following involvement of a certification body, an inspection body, and a testing laboratory.

  There has to be compliance with a harmonized European Standard, or with a European Technical approval following the appropriate attestation procedure.

  The symbol is printed below, and is to be accompanied by information concerning the name of the product, the last two digits of the year of manufacture, characteristics of the product, and the number of the certificate.

A material bearing the 'CE' symbol must be accepted as fit for its purpose by a building control authority or approved inspector unless there are reasonable grounds for believing the product does not satisfy the relevant requirements.

The relevant requirements embrace energy economy; heat retention; hygiene, health and the environment; mechanical resistance and stability; protection against noise; safety in case of fire; and safety in use. Some, or all of these requirements may be relevant in any particular situation.

If a material is rejected because its performance is not in accordance with its technical specification the Trading Standards Officer must be notified. A material must not be rejected on grounds

relating to the manufacture or the testing, or the certification organization.

Not all materials will carry the CE mark and not all products will be marked until all the relevant specifications are published. Gas boilers are one product where CE marking must be shown.

- **Independent certification.** There are a number of UK product certification schemes and the most widely used is operated by BSI, utilizing its kitemark symbol. All the schemes are required to certify that when used correctly the material will show compliance with requirements. Many of the bodies operating a certification scheme are accredited to the United Kingdom Accredited Service (UKAS) of 21–47 High Street, Feltham, Middlesex, TW3 4UN (020 8917 8400).

  A product may be tested and certified by an approved body in accordance with Article 16 of the Construction Products Directive. In these circumstances the product must be accepted by the building control authority or an approved inspector. If it is not so accepted, then the Trading Standards Officer must be notified and the UK government may notify the commission. If a product is rejected in these circumstances the onus of proof rests with the building control authority or the approved inspector.

- **National and international specifications.** A material which conforms to other national or international specifications prepared by contracting partners to the European Economic Area may be used. Such specifications must provide an equal level of performance. When proposed to be used, the satisfactory nature of the product must be demonstrated and translations of literature and documents provided in support.

- **Past experience.** This is not a method that should be lightly dismissed. On many occasions it can be shown that a material in a building in use is satisfactory, and is capable of performing the function for which it is intended. There may be geographical limitations in applying this test. A material or mode or workmanship satisfactory on a south slope in Kent may not be adequate to meet climatic conditions on a south slope in Cumbria.

- **Sampling.** 1991 Regs, 17 gives power to a local authority to take whatever samples of materials may be necessary in order to establish whether or not it complies with any provisions of the regulations.

  Approved inspectors do not possess this power to take samples, so these inspectors may think it prudent to make arrangements with the client giving authority to take samples when considered necessary. 1991 Regs, 17 does not apply to any work specified in an initial notice; also it does not apply when an approved inspector has given a final certificate which has been accepted by the local authority.

- **Technical approvals.** Certificates issued by European Technical Approvals issuing bodies may indicate the suitability of a material, which must be used and applied in accordance with conditions set out in the certificate.

  A European Technical Approvals issuing body operates under Article 10 of the Construction Products Directive and currently the only bodies listed in the UK are the British Board of Agrément and WIMLAS.

  Since inception Agrément certificates have carried a paragraph setting out an opinion regarding the particular product in the context of building regulations.

The product must, of course, be manufactured and fitted or used in accordance with the terms of the particular certificate. Each certificate is divided into three parts.

1. Part I is concerned with certification of the product, including marketing, use, assessment, building regulations and conditions of certification.
2. Part II gives a technical specification including description, manufacture, delivery, storage and installation. There is also a section on design data, giving practicability of installation, performance, resistance to loading, behaviour in fire, durability and maintenance. This is usually supported by tables and diagrams.

3. Part III summarizes the technical investigations which have been carried out.

*   **Tests and calculations.** A material may be used if it can be demonstrated by tests, calculations or other methods that it is capable of performing the function for which it is to be used.

    In this connection the Accreditation Scheme for Testing Laboratories run by UKAS provides a route whereby tests are carried out within the prescribed criteria. There are similar schemes run by certification bodies and equivalent testing bodies operating in member states, and this ensures that tests are carried out following recognized criteria.

MATERIALS SUBJECT TO ADVERSE CHANGES

A number of materials when placed in a building may undergo adverse changes from contact with the surroundings. This can occur when high alumina cement is used in concrete. Other materials where adverse changes occur are intumescent paints used for fire resistance purposes, structural silicon sealants and some stainless steels.

These materials may be used in building works and satisfy the requirements of the 1991 Regs. This will occur only when a reliable evaluation indicates that the final residual properties can be achieved prior to installation, and that the building will remain capable of performing the function for which it was erected.

BS 915: Part 2: 1972 gives the composition, manufacture sampling procedures, and tests for fineness, chemical composition and strength, setting time and soundness of high alumina cement.

SHORT-LIVED MATERIALS

The 1984 Act, section 19, says that if plans show a building is to be constructed of materials that are short lived, it is open to the authority to reject them. Alternatively they may fix a period at the end of which the building must be removed and the council may impose conditions which are relevant to the use of those particular materials.

This is to be replaced by section 20 of the same Act on a date when the Secretary of State decides that section 20 shall come into force. Section 20 widens the control over the use of materials that are short lived or unsuitable for use in permanent buildings.

The 1976 Regs gave explicit guidance as to the nature of short-lived materials. The approved document supporting Regulation 7 does not do so. It regards a short-lived material as one which if it is not given special care, may be unsuitable because of rapid deterioration when compared to the life of the building. It does not offer a specific criteria enabling a material to be evaluated against regulation requirements. Materials may be selected on the basis of economic judgements, but such judgements may not be proper as the regulations are concerned with the health and safety of persons in and around the building.

The approved document goes on to say that a short-lived material which is easily accessible for inspection, maintenance or even replacement may be acceptable.

This may be the case when the consequences of failure will not be likely to create circumstances which will affect the health and safety of persons in and around the building.

When it is proposed to use a short-lived material in a position where it will not be accessible for inspection, maintenance or replacement, particularly if failure will lead to conditions likely to be injurious to health or safety, then the material will not be suitable. This is qualified by the advice that if the results of failure of the material will have only a minor effect on health and safety its use may be permitted.

This is an area when the various certification schemes may do much to reduce the possibility of disagreement.

The 1984 Act, section 19 is currently of no effect, as no materials are listed for its purpose.

RESISTANCE TO MOISTURE

Material which may be harmed by condensation or by moisture from the ground or by rain or snow, will meet regulation requirements in the following circumstances:

*   the form of construction used will combat access of moisture to the material;
*   the material is treated, or subjected to a form of protection from moisture.

RESISTANCE TO SUBSOIL

The effect of sulphates upon certain building materials in contact with polluted subsoil and other subsoil problems may be apparent, as described in Chapter 12.

All materials used in contact with the ground must be suitable, and be capable of resisting attack by any deleterious material (Fig. 9.1).

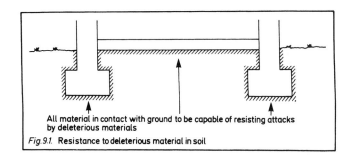

All material in contact with ground to be capable of resisting attacks by deleterious materials

Fig.9.1. Resistance to deleterious material in soil

SPECIAL TREATMENT OF SOFTWOOD TIMBER

There is no mention of the house longhorn beetle in the 1999 edition of the approved document, as it is not now considered appropriate to deal with this topic when discussing materials. However, the guidance contained in the 1992 Approved document to support Regulation 7 regarding the house longhorn beetle continues to apply.

The house longhorn beetle (*Hylotrupes bajulis* L.) has been giving trouble in the south of England for some years. The 1991 Regs included provisions to ensure preservative treatment in the following areas:

*   The district of Bracknell Forest
*   The borough of Elmbridge
*   The borough of Guildford excluding the area of the former borough of Guildford
*   The district of Hart excluding the area of the former urban district of Fleet
*   The district of Runnymede
*   The borough of Spelthorne
*   The borough of Surrey Heath
*   That part of the borough of Rushmoor comprising the former urban district of Farnborough
*   The district of Waverly excluding the parishes of Goldalming and Hazlemere
*   The parishes of Old Windsor, Sunningdale and Sunninghill in the Royal borough of Windsor and Maidenhead
*   The borough of Woking.

The approved document to support Regulation 7 advises that all soft-wood timber

- used for roof construction
- fixed in the roof space
- including ceiling joists in the void spaces of a roof

is to be adequately treated with a suitable preservative to prevent infestation (Fig. 9.2).

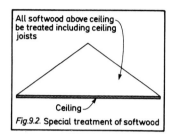

Fig.9.2. Special treatment of softwood

There is no detailed advice on the measures to be taken to protect the timber.

The house longhorn beetle (Fig. 9.3) is primarily a pest of soft-wood, and hardwoods are not attacked by this insect.

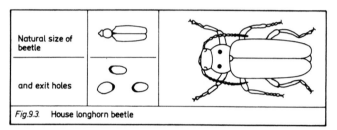

Fig.9.3. House longhorn beetle

The beetle is rather large, being from 8 mm to 25 mm in length, and has been known for many years as a major pest in certain parts of the continent. The requirement there is to protect new properties. Curiously the house longhorn beetle is always found in roof timbers, attacking softwood rafters and purlins. The grub eats up the wood and is capable of destroying large dimension timbers. Even when the beetle is present it is not easy to detect as it leaves a veneer of sound wood on the surface. The exit hole of the adult is not round and clear cut, as in the case of most boring beetles, but an oblique slit, usually filled with powder.

The beetle is generally brown or black. Head and first body segment are covered by grey hairs. There are two shiny black spots behind the head.

There are some 25 British Standards available relating to materials to resist special forms of attack such as rot, fungus, condensation or insects.

One that should be mentioned is BS 1282: 1975, which is a guide to the choice, use and application of wood preservatives. The standard describes conditions under which wood-boring insects, etc., are likely to damage timber and gives guidance on the selection of suitable treatments. Wood preservatives are classified into three main groups: tar oil (TO); organic solvent (OS) and waterborne (WB).

BS 5219: 1975 is about the determination of eradicant action against *Hylotrupes bajulis* L. The method is applicable to organic water-insoluble formulations of water-soluble materials such as salts.

BS 5434: 1977 is about the preventative action against recently hatched larvae of *Hylotrupes bajulis* L. and involves exposure of specimens of surface-treated wood to larvae. The method is applicable to water-soluble chemicals, or organic water-insoluble formulation, or water-soluble materials.

BS 5435: 1977 describes the determination of the toxic values against *Hylotrupes bajulis* L. The method is applicable to water-insoluble chemicals or organic water-insoluble formulations or water-soluble materials.

BS 4072: 1987 specifies requirements for treatment with water-borne preservatives of copper sulphate, sodium or potassium dichromate and arsenic pentoxide. It covers preservative composition, application and resulting retentions.

BS 5268: Part 5: 1977 specifies preservative treatments for constructional timbers, and gives alternative treatments for structural components in different situations related to risk of decay or insect attack.

SECTION 2. WORKMANSHIP

Five ways of establishing the adequacy of workmanship are listed and they are in tune with the manner in which materials may be evaluated. The product may be covered by an equivalent technical approval of a member state. The person carrying out the work must ensure that the method of workmanship is satisfactory.

- **Management systems.** Follow a workmanship scheme complying with recommendations in BS EN ISO 9000: 1994. Quality management and quality assurance standards. One appropriate scheme is the registration of firms of assured capability by BSI. There may also be a suitable BS or other technical standard covering products and processes relating to an approved scheme.
- **Past experience.** It may be shown by examination of a building already in use that a method of workmanship may satisfy requirements for that particular function.
- **Standards.** Following a BS Code of practice appropriate to the purpose and the material will ensure compliance with the regulations. Once again care must be taken to ensure that the conditions under which a material is to be tested is in accordance with the provisions of the code of practice.

   BS 8000: 'Workmanship on building sites', gathers together guidance from a number of BSs and codes (see 'Relevant reading').

   Alternatively workmanship which conforms to an equivalent technical specification including a national technical specification of one of the member states is acceptable.
- **Technical approvals.** Workmanship is adequate if it is covered by

   (i) a certificate issued by a European Technical approvals issuing body; or
   (ii) an equivalent technical approval providing a similar level of performance; or
   (iii) technical approval by member of EOTA.

The person carrying out work is required to show that the workmanship will give an equivalent level of protection and performance.

- **Tests.** Local authorities may undertake tests (1991 Regs, 16) in connection with:

   (a) sanitary pipework and drainage;
   (b) cesspools, septic tanks and settlement tanks; and
   (c) rainwater drainage.

The test results must indicate compliance with the requirements of Schedule 1 (see Chapter 17).

Approved inspectors do not possess these powers so these inspectors may think it prudent to make arrangements with the client giving authority to take samples when considered necessary. 1991 Regs, 16 does not apply to any work specified in an initial notice, also it does not apply when an approved inspector has given a final certificate which has been accepted by the local authority.

STANDARDS AND SPECIFICATIONS

- Building regulations relate to certain specified purposes. They are aimed at the protection of health and safety of people living in or resorting to buildings; energy conservation and the convenience of people having disabilities.
- European Technical Specifications, British Standards and Agrément certificates contain guidance that may be used in achieving

the aims of building regulations. There are, however, matters in these documents which relate to aspects not covered by the regulations.

- There has been agreement between the British Board of Agrément and the Secretary of State regarding the aspects of performance to be assessed before an Agrément certificate is issued demonstrating compliance with the regulations.
- Products holding an Agrément certificate or a European technical approval, will, if they are properly used as described in the certificate, or the approval, meet the relevant requirements of the regulations.
- Approved documents refer to named BS and the relevant version is the one listed at the end of the approved document concerned. Those British Standards are contained in the body of each relevant chapter. There will be occasions when the named BS has been superseded by a later version, and when this is known it will be listed in 'Relevant reading'.

ASBESTOS

The Asbestos (Prohibition) Regulations 1985 stopped with effect from 1 January 1986 the importation of raw crocidolite (blue) asbestos fibre and raw amosite (brown) asbestos fibre into Britain. It also prohibited the supply of crocidolite and amosite, and products containing these minerals, as articles or substances. The use of crocidolite and amosite, and products containing them, are prohibited in the manufacture and repair of any other product. Asbestos spraying and the installation of new asbestos insulation are not allowed.

The prohibition of crocidolite and asbestos spraying is required by a European Directive. The United Kingdom has legislated beyond these directives by additionally stopping amosite and the installation of new asbestos insulation.

Although chrysolite (white) asbestos is not prohibited whenever it is used, it is subject to stringent controls.

COMMUNITY HARMONIZATION

The Construction Products Directive (89/106/EEC) must be one of the European Community's most important directives and it came into effect on 27 June 1991.

The Directive covers all products for use in building and civil engineering works. These products when properly installed must enable works in which they have been placed to satisfy 'relevant requirements' for an economically reasonable working life.

These requirements affect all in the building team – architects, surveyors, engineers, site supervisors and building control.

Although expected to take several years to implement fully, as European procedures and specifications become available, the equivalent national documents will be withdrawn.

The Directive says that member states are responsible for ensuring that building and civil engineering works are designed and executed in a way that does not endanger the safety of persons, domestic animals and property. The position of domestic animals by regulatory means is a new approach – perhaps cat flaps will become obligatory. This is in addition to the health aspects of building control.

Article 3 of the Directive says that, in order to take account of the differences in geographical or climatic conditions, there will be interpretative documents giving guidance on local application.

Essential requirements are set out in Annexe 1 to the Directive. These requirements must, subject to normal maintenance, be satisfied for an economically reasonable working life; and relate to actions which are foreseeable.

1. Mechanical resistance to stability

It is a requirement that loadings during construction and use will not lead to any of the following:

- collapse of whole or part of the work
- major deformations
- damage to other parts of the works
- damage to an extent disproportionate to the original cause.

Looking to the building regulations 1991 these essential requirements may be compared with Part A, Structure; loading; ground movement; and disproportionate collapse.

2. Safety in case of fire

Construction works are to be designed and built in such a way that in the event of fire:

- loadbearing capacity of construction must last a specific period of time;
- generation and spread of fire and smoke must be limited;
- spread of fire to neighbouring construction is to be limited;
- occupants can leave the works, or be rescued by other means;
- the safety of rescue teams is taken into consideration.

Comparison may be made with the 1991 Regs – Fire: Means of escape: Part B, Internal fire spread; External fire spread and some of Part J, Heat-producing appliances.

3. Hygiene, health and environment

Construction works are to be designed and built in a way that will not be a threat to hygiene or health of the occupants, or neighbours, as a result of

- the giving off of toxic gas;
- dangerous particles or gases in the air;
- emission of dangerous radiation;
- pollution or poisoning of water or soil;
- faulty elimination of waste water, smoke, solid or liquid wastes;
- dampness in parts of the works or on surfaces within the works.

1991 Regs do relate in Part C to site preparation and resistance to moisture; Part D, Toxic substances; Part F, Ventilation; Part G, Hygiene; Part H, Drainage and waste disposal and in Part J, Heat-producing appliances.

4. Safety in use

- Design and construction are not to present unacceptable risks of accidents in service or in operation from

  slipping
  falling
  collision
  burns
  electrocution
  injury from explosion.

Protection from accidents in the 1991 Regs seems to be limited to Part K, Stairways, ramps and guards and Part N, Glazing – materials and protection.

5. Protection against noise

Design and construction is to be such that noise perceived by the occupants or people nearby is kept down to a level which will not threaten their health and allow them to sleep, rest and work in satisfactory conditions.

The 1991 Regs contain Part E, Resistance to sound.

6. Energy economy and heat retention

Heating, cooling and ventilation installations are to be designed and built so that the amount of energy required will be low, having regard to climatic conditions, and the occupants.

The 1991 Regs contain Part L, Conservation of fuel and power.

The Directive introduces a system of certification in the form of European Technical Approvals. The British Board of Agrément has a new role as the UK's issuer of these technical approvals. Products to be used anywhere in the community must enable buildings to meet the essential requirements. Relevant products, in order to comply with the Directive, will carry an EC mark. For innovative products where no harmonized standard has been agreed, these will in the UK be issued by the Agrément Board, and in other countries of the community by national members of the European Organization for Technical Approval, most of whom will be former members of the European Union of Agrément.

THE CONSTRUCTION PRODUCTS REGULATIONS 1991

These UK regulations implement the EC Construction Directive and were operative from 27 December 1991. They establish the procedure which manufacturers must follow to legally fix the EC mark on their products. Products bearing this mark may be sold anywhere within the community without having any further technical or procedural requirements.

These regulations are not enforced by building control staff or approved inspectors. Enforcement in England and Wales, the area covered by the Building Regulations 1991 and amendments, is placed in the hands of Trading Standards Officers. Nevertheless, reading through these regulations indicates that designers, building control staff, approved inspectors, purchasers and users of construction products should be aware of the requirements and be wary of specifying, purchasing or building with a product which does not satisfy the requirements of the 1991 CP Regs.

Regulation 1 gives the title of the regulations and states that they came into operation on 27 December 1991. They do not apply to any product which was supplied before that date.

Regulation 2 contains numerous definitions for use in interpreting the regulations and these definitions have been included in Annexe A.

Regulation 3 requires construction products except 'minor part products' to have such characteristics that the works in which they are incorporated can (if properly designed and built) satisfy the regulations.

Regulation 4 says that when a construction product bears the EC mark there is a presumption that the product satisfies the relevant requirements.

Regulation 5. The EC mark may be affixed to a construction product which is not a 'minor part product' by the manufacturer or his agent. The criteria involved require attestation that the product

1. meets relevant national standards, or
2. complies with the applicable European standards, or
3. where no harmonized specifications are available it is to comply with the national technical specifications.

The EC mark must be accompanied by information enabling the manufacturer to be identified, and, where appropriate, giving

(a) the characteristics of the product;
(b) the last two digits of the year of manufacture;
(c) the name of the approved body involved;
(d) the number of EC certificate of conformity.

A person making a declaration of conformity which is not in accordance with the various requirements is guilty of an offence, and liable upon summary conviction to three months' imprisonment, or a fine up to £2000; or both.

Regulation 6 requires people who have affixed an EC mark to have available and produce the certificate or declaration of conformity to an enforcement authority or its officers if required to do so.

Regulation 7 says that a person supplying a construction product not bearing the EC mark is to provide information about the product to the enforcement authority or its officers, if required to do so.

Regulation 8 states that a person shall be guilty of an offence if he supplies a construction product which does not satisfy Regulation 3. It is a defence for a person to show

1. that he reasonably believed that the product would not be used in the community; or
2. that the following conditions are satisfied:

   (a) the product was supplied in the course of carrying on a general retail business, and that he neither knew nor had reasonable grounds for believing that the product did not meet the relevant requirements, or
   (b) that the terms of supply indicated that the product was not supplied as a new product and provided for, or contemplated, the acquisition of an interest in the product by the persons supplied.

A person guilty of an offence against the regulation is subject to the same penalty as mentioned above.

Regulations 9 to 14 are similar to sections 13 to 18 of the Consumer Protection Act 1987.

Regulation 9 enables the Secretary of State to serve a prohibition notice, prohibiting the supply of construction products considered not to satisfy the requirements of Regulation 3. The Secretary of State may also serve on any person a 'notice to warn' requiring that person to provide a notice warning about products considered not to satisfy relevant requirements. Anyone found guilty of contravening these notices is subject to the same penalty as mentioned above.

Regulation 10. An enforcement authority having reasonable grounds for believing that any provision of Regulations 5, 6, 7 or 8 have been contravened may serve a 'suspension notice' preventing the supply of specified goods for a period of up to six months. Enforcement authorities may apply to the court for an order that offending products be forfeited.

A person found guilty of contravening a suspension notice is subject to the same penalty as mentioned above.

If, after service of a suspension notice, there is found to be no contravention of the regulations, or the exercise of the power is not attributable to any neglect or default of that person, the enforcement authority may be liable to pay compensation.

Regulation 11. A person with an interest in any construction product that is the subject of a suspension notice may appeal, in England or Wales, to the magistrates' court. The court has power to set aside a suspension notice only if it is satisfied there has been no contravention of Regulations 5, 6, 7 or 8.

Regulation 12. An enforcement authority may apply to a magistrates' court for an order for the forfeiture of construction products in contravention of Regulations 5, 6, 7 or 8. Any person aggrieved by the decision of the magistrates' court may appeal to the Crown Court.

Products forfeited under this regulation may be destroyed in accordance with directions the court may give.

Regulation 13 relates to forfeiture in Scotland.

Regulation 14 gives the Secretary of State the power to require information to enable him to decide whether to issue, vary or revoke a prohibition notice or serve or revoke a notice to warn. A person is required to provide information requested by the Secretary of State and, upon conviction for failure to do so, penalties may be imposed as mentioned above.

Regulation 15. These regulations are not enforced by building control staff or approved inspectors. This regulation makes it the duty of every weights and measures authority in England and Wales to enforce the requirements of these regulations.

Trading standards officers in Scotland are also responsible for enforcing these regulations. Environmental health officers are responsible in Northern Ireland. DoE Circular 13/91; Scottish Office Circular 19/91; Welsh Office Circular 35/91 and Northern Ireland CP1(91) outline enforcement responsibilities and are available from SO. The reference is ISBN 0-11-7525243.

Regulation 16 enables an enforcement authority to make, or authorize an officer to make, any purchase of construction products in order to ascertain whether requirements of the regulations have been contravened.

When a product has been submitted to test the person likely to be the subject of proceedings should the tests fail, shall be allowed to have the products tested.

Regulation 17. An authorized officer of an enforcement authority may at any reasonable hour enter premises, examine any procedure and carry out tests. If there are reasonable grounds for suspecting that there has been a contravention, records may be examined and copies made. The officer may also seize and detain records and construction products where there are reasonable grounds they may be required as evidence in any proceedings.

Regulation 18. When construction products or records are seized the person from whom they are seized – or, in the case of imported goods, the importer – must be informed.

A justice of the peace, when satisfied that it is necessary, may issue a warrant authorizing the entry of premises, if need be by force. Any premises entered when the occupier is not there must be left effectively secured against unauthorized entry.

Regulation 19 enables a customs officer to detain goods for not more than two working days.

Regulation 20 creates offences for obstructing an officer with enforcement powers, failing to comply with any requirement made under the regulations by such an officer, failing to provide information and making false or reckless statements to an enforcing officer.

Regulation 21 gives a right of appeal against the detention of any product.

Regulation 22 provides for compensation to be payable in certain circumstances in respect of loss or damage arising from the seizure and detention of products.

Regulation 23 says that where a court convicts a person or makes an order for forfeiture of any product, it may order the person concerned to reimburse the expenses of the enforcement authority in connection with the seizure, detention or forfeiting of products.

Regulation 24 allows customs and excise to disclose information relating to imported products to enforcement authorities and their officers.

Regulation 25 makes restrictions on the disclosure of certain information obtained, including secret manufacturing processes or trade secrets.

Regulation 26 applies a defence of due diligence where the alleged offence was due to the act or default of another, or to reliance on information given by another.

Regulation 27 provides for the liability of persons whose act or default leads to the commission of an offence by others.

Regulation 28 gives details regarding the service of any document required or authorized by the regulations.

Regulation 29. Records subject to legal professional privilege may not be required to be produced.

Regulation 30. An enforcement authority is to give immediate notice to the Secretary of State of any suspension notice or any order for forfeiture served by them.

Regulation 31. In England and Wales a magistrates' court may try an information or complaint. Information must be laid within 12 months from the time when the offence is committed.

Regulation 32 says that it is unlawful for public and certain other bodies to make, impose or enforce, or purport to do so or enforce, any rules or conditions which impede the use of construction products fulfilling the requirements of Regulation 3.

Regulation 33 relates to certain cases where there may be an overlap between these regulations and the Health and Safety at Work legislation.

The Regulations are supported by four schedules; principally giving details of procedures.

Schedule 1. This illustrates a specimen of the EC mark to be placed on or to accompany products.

Schedule 2. This is a reproduction of the EC Council Directive of 21 December 1988 on the approximation of laws, regulations and administrative provisions of the member states relating to construction products. Essential requirements are:

1. Mechanical resistance and stability

Works must be designed and built in such a way that during construction and use there will not be collapse of a whole or part; major deformations; damage to other parts of the works; damage by an event disproportionate to the original cause.

2. Safety in fire

Works must be designed and built in such a way that in the event of fire loadbearing capacity can be assumed for a specific time; generation and spread of fire and smoke are limited; spread of fire to neighbouring buildings is limited; occupants can leave or be rescued; safety of rescue teams is considered.

3. Hygiene, health and the environment

Works must be designed and built in such a way that it will not be a threat to health or hygiene, arising from the giving off of toxic gas; presence of dangerous particles or gases in the air; emission of dangerous radiation; pollution or poisoning of water or soil; faulty elimination of waste water smoke, solid or liquid wastes; presence of damp in or on surface of works.

4. Safety in use

Works must be designed and built in such a way to prevent unacceptable risks of accidents such as slipping, falling, collision, burns, electrocution, or injury from explosion.

5. Protection against noise

Works must be designed and built so that noise heard by the occupants or people nearby is not loud enough to threaten their health and allow sleep, rest and work in satisfactory conditions.

6. Energy economy and heat radiation

Works and its heating, cooling and ventilation, installation are to be designed and built so that the amount of energy in use is low, bearing in mind climatic conditions, and the occupants.

Schedule 3. Attestation of conformity to relevant technical specifications.

An EC certificate of conformity may be issued for a system of production control and surveillance or for a product itself. This schedule sets out the requirements to be followed by a manufacturer, and also for the approved body with regard to production control and testing.

Schedule 4. Prohibition notices and notices to warn.

Part I sets out procedures to be followed if the Secretary of State considers that a product does not satisfy the relevant requirements in Regulation 3.

Part II applies if the Secretary of State proposes to serve a notice to warn any person in respect of any construction products and gives details of procedures to be followed.

Part Ill relates to the timing of oral representations.

The General Product Safety Regulations 1994

These regulations apply to all EU member states and came into operation in October 1994.

- Where a product conforms to specific rules there is a presumption that the product is safe.
- If no such rules exist conformity is to be assessed using UK national standards; harmonized European standards or EOTA technical approvals.

When making an assessment it is necessary to take into account the measure of safety that consumers may reasonably expect.

ABBREVIATIONS AND GLOSSARY

The approved document to Regulation 7 includes a short list of abbreviations and a brief glossary.

BBA	British Board of Agrément
BS	British Standard
BSI	British Standards Institute
Building Control Body	Includes local authority building control and also approved inspectors
CE marking	The conformity mark
CEN	Comité Européen de Normalization. This is the European body charged with the preparation of harmonized standards.
CPD	Construction Products Directive reference 89/106/EEC dated 21 December 1988. Is amended by marking directive 93/68/EEC.
EEA	European Economic Area; comprises
	Austria
	Belgium
	Denmark
	Finland
	France
	Germany
	Greece
	Iceland
	Ireland
	Italy
	Luxembourg
	Leichtenstein
	Netherlands
	Norway
	Portugal
	Spain
	Sweden
	United Kingdom
EOTA	European Organization for Technical Approvals
European Commission	Executive Organization of the EU
EN	These letters and a number identify the standard – identical BSs may exist
European Technical Approval	Technical assessment of fitness of product for its intended use
European Technical Approval Issuing Body	Currently in the UK there is the British British Board of Agrément and WIMLAS Ltd.
EU	The European Union comprising the following countries:
	Austria
	Belgium
	Denmark
	Finland
	France
	Germany
	Greece
	Ireland
	Italy
	Luxembourg
	Netherlands
	Portugal
	Spain
	Sweden
	United Kingdom
	Proposals are afoot to admit:
	Cyprus
	Czech Republic
	Estonia
	Hungary
	Poland
	Slovenia.
ISO	International Organization for Standardization – the worldwide organization.
Technical specification	Used to make an assessment of a European Technical Approval.

RELEVANT READING

British Standards and Codes of Practice

BS 12: 1996. Specification for ordinary and rapid-hardening Portland cement
BS 144: Part 1: 1990. Specification using coal tar creosote
BS 144: Part 2: 1990. Methods of timber treatment
BS 493: 1995. Specification for airbricks and gratings
BS 743: 1970. Specification for materials for damp-proof courses
BS 812: Part 101: 1984. Guide to sampling and testing aggregates
BS 812: Part 102: 1984. Methods for sampling aggregates
BS 812: Part 105: 1985. Flakiness index of aggregates
BS 812: Part 106: 1989. Determining shell content of aggregates
BS 812: Part 119: 1985. Method for determination of acid soluble material in fine aggregate
BS 890: 1994. Specification for building limes
BS 915: Part 2: 1972 (1983). High alumina cement
BS 1199: 1976 and BS 1200: 1976. Specification for building sands
BS 1243: 1978. Specification for metal ties
BS 1881: Part 105: 1989. Testing concrete of high workability
BS 1881: Part 114: 1983. Determining density of hardened concrete
BS 1881: Part 115: 1986. Compression testing machines for concrete
BS 1881: Part 116: 1983. Determination of comprehensive strength of concrete cubes
BS 1881: Part 117: 1983. Determination of tensile splitting strength of concrete
BS 1881: Part 118: 1983. Determination of flexural strength of concrete
BS 1881: Part 119: 1983. Determination of compressive strength using portions of beams
BS 1881: Part 120: 1983. Determination of compressive strength of concrete cores
BS 1881: Part 121: 1983. Determination of static modulus of elasticity
BS 1881: Part 122: 1983. Determination of water absorption of concrete
BS 4072: Part 1: 1987. Specification for copper / chromium / arsenic compositions
BS 4072: Part 2: 1987. Methods of timber treatment
BS 4721: 1981 (1986). Specification for ready mixed building materials
BS 4887: Part 1: 1986. Air entraining (plastering) admixtures
BS 5219: 1975 (1993). Wood preservatives
BS 5224: 1995. Specification for masonry cement
BS 5268: Part 5: 1989. Treatments for constructional timbers
BS 5268: Part 7:
  Section 7.1: 1989. Domestic floor joists
  Section 7.2: 1989. Joists for flat roofs
  Section 7.3: 1989. Ceiling joists
  Section 7.4: 1989. Ceiling binders
  Section 7.5: 1990. Domestic rafters
  Section 7.6: 1990. Purlins supporting rafters
  Section 7.7: 1990. Purlins supporting sheeting or decking
BS 5589: 1989. Code of practice for preservation of timber
BS 5606: 1991. Guide to accuracy in building
BS 5666: Part 1: 1987. Guide to sampling and preservation of wood preservatives and treated timber for analysis
BS 5628: Part 3: 1985. Components, design and workmanship in masonry
BS 5642: 1978–83. Sills and copings

BS 5964: Part 1: 1990. Setting out and measurement

BS 6477: 1992. Specification for water repellants for masonry surfaces

BS 6577: 1985. Mastic asphalt for building

BS 6925: 1988. Mastic asphalt for building and civil engineering

BS 5750: 1993. Quality systems

BS 6100: Part 1: Section 1.7, sub-section 1.7.1: 1986 defines terms for performance

BS 6100. Glossary of building and civil engineering terms
Section 1.3.1: 1987. Walls and claddings
Section 1.3.3: 1987. Floors and ceilings

BS 6319: Part 6: 1984. Method for determination of modules of elasticity of compression – resin compositions

BS 6398: 1983. Bitumen damp-proof courses for masonry

BS 6651: 1990. Protection of structure against lightning

BSCP 144: Part 3: 1970. Galvanized corrugated steel

BS 7307: Part 1: 1990. Building tolerances, methods and measurement

BS 7307: Part 2: 1990. Position of measuring points

BS 7308: 1990. Dimensional accuracy data in building construction

BS 7334: Part 1: 1990. Determining accuracy of measuring instruments

BS 7334: Part 2: 1990. Accuracy of measuring tapes

BS 7334: Part 3: 1990. Accuracy of optical measuring instruments

BS 7543: 1992. Guide to durability of buildings and building elements, products and components

BS 8000. Workmanship on building sites

BS 8000: Part 1: 1989. Code of practice for excavation and filling

BS 8000: Part 2: Code of practice for concrete work.
Section 2.1: 1990. Mixing and transporting concrete
Section 2.2: 1990. Sitework with in situ and precast concrete

BS 8000: Part 3: 1989. Code of practice for masonry
Amendment 1: AMD 6195

BS 8000: Part 4: 1989. Code of practice for waterproofing

BS 8000: Part 5: 1990. Code of practice for carpentry, joinery and general fixings

BS 8000: Part 6: 1990. Code of practice for slating and tiling of roofs and claddings

BS 8000: Part 7: 1990. Code of practice for glazing

BS 8000: Part 8: 1994. Code of practice for plasterboard partitions and dry linings
Amendment 1: AMD 6475

BS 8000: Part 9: 1989. Code of practice for cement/sand floor screeds and concrete floor toppings

BS 8000: Part 10: 1995. Code of practice for plastering and rendering
Amendment 1: AMD 6476

BS 8000: Part 11: Code of practice for wall and floor tiling
Section 11.1: 1989. Ceramic tiles, terrazzo tiles and mosaics
Section 11.2: 1990. Natural stone tiles

BS 8000: Part 12: 1989. Code of practice for decorative wallcoverings and painting

BS 8000: Part 13: 1989. Code of practice for above-ground drainage and sanitary appliances

BS 8000: Part 14: 1989. Code of practice for below-ground drainage

BS 8000: Part 15: 1990. Code of practice for hot and cold water services (domestic scale)

BS 8000: Part 16: 1997. Code of practice for sealing joints in buildings

BS 8102: 1990. Protection of building against water from the ground

BS EN ISO 9000: Quality management and quality assurance standards.

BS EN ISO 9001: 1994. Quality systems. Model for quality assurance in design, development, production, installation and servicing

BS EN ISO 1994. Quality systems. Model for quality assurance in production, installation and servicing

BSI also publish a Manual of British Standards in building construction.

BRE digests

69 Durability and application of plasters

89 Sulphate attack on brickwork

160 Mortars for bricklaying

237 Materials for concrete

250 Sulphate-bearing soil

262 Selection of windows by performance

296 Timbers – their natural durability and resistance to preservative treatment

299 Dry rot – its recognition and control

300 Toxic effects of fires

301 Corrosion of metals by wood

304 Preventing decay in external joinery

307 Identifying damage by wood-boring insects

323 Selecting wood-based panel products

325 Concrete Part 1 – materials

326 Concrete Part 2 – quality control of mixes

327 Insecticidal treatments against wood-boring insects

330 Alkali silica reaction

331 GRC

345 Wet rot recognition and control

349 Stainless steel as a building material

362 Building mortar

393 Specifying preservative treatments: new European approach

394 Plywood

397 Standardization in support of European legislation

400 Orientated strand board

408 A guide to attestation of conformity under the Construction Products Directive

415 Reducing the risk of pest infestation in buildings

429 Timbers: durability and resistance to preservative treatment

430 Plastics external glazing

BRE information papers

1/96 Management of construction and demolition wastes

2/96 An assessment of exterior medium density fibreboard

3/85 Wood chipboard – recommendations for use

4/85 Specifying plywood

4/89 Expert systems

4/94 Waterborne coatings for exterior wood

5/90 Preservation of hem-fir timber

6/85 Selection of building management systems

7/95 Bituminous roofing membrane – performance in use

8/89 BREMORTEST – a rapid method of testing fresh mortars for cement content

9/87 Joint primers and sealants: performance between porous claddings

9/96 Preservative treated timber – checking compliance with European standards

10/89 Building management systems: user experience

10/90 Fungicidal paint to control mould growth

11/84 BRE screen tester

12/82 House longhorn beetle survey

12/91 Fibre building board types and uses

13/83 Conformance of some common building products with BS

13/85 Using experience and publications in building design

14/85 Meeting building designers' needs for trade information

14/91 In situ treatment of exterior timber using boron-based implant

17/87 Factory applied priming paints for exterior joinery

18/94 The role of windows in domestic burglary

19/84 Proof testing building components against wind loads

19/86 Controlling death watch beetle

20/87 External joinery: end grain sealers and moisture control

22/81 Assessment of chemical attack on high alumina cement concrete

24/80 Quality control assurance in building components

24/86 Preservative treatment for veneer plywood against decay

28/81 Quality control on building sites

6/99 Preservative treated timber – conformity with European Standards

BRE defects action sheets

  87 Wood entrance doors: discouraging illegal entry
  88 Windows: discouraging illegal entry
103 Wood floors: reducing risk of recurrent rot (Design)
109 Protection against frost
113 Brickwork, prevention of sulphate attack (Design)

BRE leaflet

XL ii. Preservation of building timbers by boron diffusion.

BRE reports or books

Better Briefing Means Better Building (J. J. N. O'Reilly)
Recognizing Wood Rot and Insect Damage in Buildings (Bravery, Berry, Carey and Cooper)
British Softwoods – properties and uses (T. Harding), Forestry Commission, BRE, Ref. 5043

EDC Report

Achieving Quality on Building Sites (from NEDO books, Millfield Tower, Millbank, London SW1P 4QX)

Technical note

39 House Longhorn beetle

Monograph

High alumina cement concrete

Also published

A Survey of Quality and Value in Building (1978)

CHAPTER 10

Structure

(Approved document A)

The 1991 Regs and 1992 approved document A changed the structural requirements in a number of ways. There is provision for landslip and subsidence, and disproportionate collapse. Regulations require that A1 to A3 are to apply where there is a material change of use into a hotel, institution or public building.

There is provision in the sizing of certain roof members to allow for higher snow loading in certain areas of England. There is a simplified procedure when dealing with windspeed and the design of walls and the re-roofing of buildings gets attention requiring that there should be inspection and any necessary strengthening carried out before re-roofing with a heavier material.

The fixing of external wall cladding is also controlled.

Requirement A1 Loading
Buildings are to be constructed to sustain the dead, imposed and wind loads and to transmit them to the ground

- **safely, and**
- **without causing deflection or deformation of any part of the building or movement**
  of the ground to the extent that it will impair the stability of another building.

Assessment of imposed and wind loads are to be those normally to be expected in the ordinary use of the building.

Requirement A2 Ground movement
Buildings are to be constructed in a way that their stability will not be impaired by movements of subsoil arising from

- **swelling**
- **shrinking**
- **freezing**
- **reasonably foreseen subsidence or landslip.**

There are no limits to the application of A1 or A2.

CONTENTS OF APPROVED DOCUMENT

Technical guidance is in five sections:

- houses up to three storeys
- external wall cladding
- re-roofing roofs
- codes standards and other references
- disproportionate collapse

Section 1. Sizes for structural elements for houses and other small buildings
Section 1/A. Basic requirements for stability.
Section 1/B. Sizes of timber floor, ceiling and roof members.
Section 1/C. Thickness of walls in certain buildings.
Section 1/D. Proportions for masonry chimneys.
Section 1/E. Strip foundations.
Section 2. External wall cladding.
Section 3. Re-covering of roofs.
Section 4. Codes and standards.
Section 5. Disproportionate collapse.

SECTION 1. INTERPRETATION

Section 1 commences by giving a number of definitions to aid interpretation of what is to follow:

- Buttressing wall. A wall offering lateral support to another wall for the whole of its height (see Fig. A1).
- Cavity width is restated to be the horizontal distance between two leaves of a cavity wall.
- Compartment wall is a wall separating two compartments and complying with Regulation B3(2).
- Dead loads (see Fig. A1).
- Imposed load is the calculated load arising from the use to which the building is put and are the distributed, concentrated, impact,

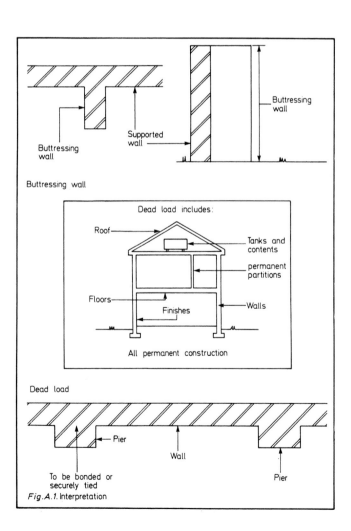

Fig. A.1. Interpretation

inertia and snow loads. Such items as the weight of moveable partitions are included, but wind loads are not.

- Pier. An integral part of a wall and giving lateral support (see Fig. A1).
- Separating wall is a wall separating two adjacent buildings and complying with Regulation B3(2).
- Spacing (see Fig. A2).
- Supported wall is a wall receiving lateral support from buttressing walls, piers, chimneys, together with floors or roof (see Fig. A1).
- Wind load arises from wind pressure or suction.

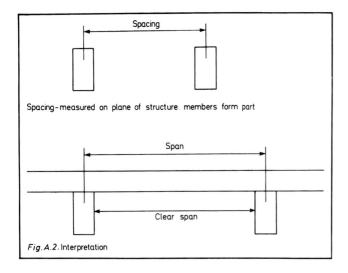

Fig.A.2. Interpretation

SECTION 1/A. REQUIREMENTS FOR STABILITY

Trussed rafters

Trussed rafters are to be braced.

Traditionally constructed roofs of rafters, purlins and ceiling joists that do not have sufficient built-in stability are also to be braced.

Recommendations are contained in BS 5268: Part 3: 1985 which offers advice on the design, fabrication and use of trussed rafters for pitched roofs. It recommends maximum range spans for a range of sizes and roof pitches. Design guidance is also given for several roof configurations together with consideration of bracing requirements and overall roof design.

The function of permanent bracing is to hold the trussed rafters in the position required. Permanent bracing should also provide the necessary lateral restraint required by members subject to compression and should ensure that the roof structure acts as a rigid component capable of transmitting horizontal forces to the buttressing elements. The standard bracing method in the code is limited to spans not exceeding 12 m.

Appendix A of the code which is illustrated is devoted to a standard bracing for domestic roofs.

BS 5268: Part 3: 1985 is being revised following recommendations in BS 5268: Part 2: 1998; the principal changes relate to timber grades used in standard trusses.

Lateral stability – walls

No special provision is necessary to take account of wind loads, when

- the roof structure is correctly braced
- walls are designed and restrained as required in Section 1/C which follows.

SECTION 1/B. SINGLE FAMILY HOUSES

Sizes for timber floors, ceilings and roof construction for houses not exceeding three storeys

There follows a number of tables and in the guidance given in the Approved Documents the following assumptions have been made:

- Dead and imposed loads do not exceed those given in the tables and figures.
- Grade and species of timber are as set out in Table 1, or

New European product requirements may mean that SC3 and SC4 strength classes are not readily available. Acceptable replacements for SC3 and SC4 graded timbers are timbers graded as C16 and C24.

Span tables are to be found in BS 8103: Part 3: 1996 which is a code of practice for timber floors and roofs for housing. There is also additional advice regarding span tables for joists supporting partitions and for trimmer joists.

(a) are in accordance with BS 5268: Part 2: 1991 which is a code of practice for permissible stress design, materials and workmanship.

It gives evidence on the structural use of timber, plywood, glued laminated timber and tempered hardboard in load-bearing members.

Included are recommendations on quality, grade stresses and modification factors. There are also recommendations for the design of nailed, screwed, bolted, connectored and glued joints. There is general advice on workmanship, inspection, and treatments.

There are numerous tables, including comprehensive recommendations regarding strength classes of various species and grades of timber.

A section is devoted to workmanship and requires that the work is carried out in all respects to accepted good practice. Adequate supervision is to be provided and timber which is damaged, crushed or split beyond the limits permitted is to be rejected or repaired. Workmanship embraces machining and preparation, joints, transportation, storage, handling, assembly, treatments and inspection. Regular inspection and maintenance are advocated.

Where a bearing is on the side grain of the timber the permissible stress in compression perpendicular to the grain is dependent on the length and position of the bearing. Bearings should be 150 mm but when they are less than 150 mm long, located 75 mm or more from the end of a member the grade stress is to be multiplied by a modified factor K_4.

The recommended modification factor K_4 for bearing stress are as follows. Interpolation is allowed.

Length of bearing	Value of K_4
10 mm	1.74
15 mm	1.67
25 mm	1.53
40 mm	1.33
50 mm	1.20
75 mm	1.14
100 mm	1.10
150 mm or more	1.00

BS 5268: Part 2: 1991 was updated in 1996 to reflect Eurocode DD ENV 1995–1–1.

(b) floorboards comply with BS 1297: 1987 which is concerned with the grading and sizing of softwood flooring.

The standard is descriptive of the use of tongued and grooved boarding and strip flooring laid on joints, fillets or sub-flooring. All must be free of any sign of decay or insect attack.

Requirements for moisture control are given and the maximum amount of sapwood, fissures, cup, knots and wane are stated. Finished dimensions specified are:

Finished thickness	16 mm	19 mm	21 mm	28 mm
Finished widths	65 mm	90 mm	113 mm	137 mm

Table 1 Common species/grade combinations which satisfy the requirements for the strength classes to which Tables **A1** to **A24** relate

Species	Origin	Grading rules	Grades to satisfy strength class				
			SC3	SC3	SC3	SC4	SC4
Redwood or whitewood	Imported	BS 4978	GS	MGS	M50	SS	MSS
Douglas fir	UK	BS 4978	M50	SS	MSS	–	–
Larch	UK	BS 4978	GS	MGS	M50	SS	MSS
Scotch pine	UK	BS 4978	GS	MGS	M50	SS	MSS
Corsican pine	UK	BS 4978		M50		SS	MSS
European spruce	UK	BS 4978		M75			
Sitka spruce	UK	BS 4978		M75			
Douglas fir–larch	Canada	BS 4978					
Hem–fir			GS	MGS	M50	SS	MSS
Spruce–pine–fir							
Douglas fir–larch	Canada	NLGA	Joist & Plank No. 1 & No. 2			Joist & Plank	Select
Hem–fir			Struct. L.F. No. 1 & No. 2			Struct. L.F.	Select
Spruce–pine–fir							
Douglas fir–larch	Canada	MSR		Machine			Machine
Hem–fir				Street-Rated			Stress-Rated
Spruce–fine–fir				1450f-1.3E			1650f-1.5E
Douglas fir–larch	USA	BS 4978	GS	MGS		SS	MSS
Hem–fir	USA	BS 4978	GS	MGS	M50	SS	MSS
Western whitewoods	USA	BS 4978	SS	MSS		–	–
Southern pine	USA	BS 4978	GS	MGS		SS	MSS
Douglas fir–larch	USA	NGRDL	Joist & Plank No. 1 & No. 2			Joist & Plank	Select
			Struct. L.F. No. 1 & No. 2			Struct. L.F.	Select
Hem–fir	USA	NGRDL	Joist & Plank No. 1 & No. 2			Joist & Plank	Select
			Struct. L.F. No. 1 & No. 2			Struct. L.F.	Select
Western whitewoods	USA	NGRDL	Joist & Plank	Select		–	
			Struct. L.F.	Select			
Southern pine	USA	NGRDL	Joist & Plank	No. 3		Joist & Plank	Select
				Stud grade			
Douglas fir–larch	USA	MSR		Machine			Machine
Hem–fir				Stress-Rated			Stress-Rated
Southern pine				1450f-1.3E			1650f-1.5E

Notes

1. The common species/grade combinations given in this table are for particular use with the other tables in Appendix A and for cross section sizes given in those tables.
2. Definitive and more comprehensive tables for assigning species/grade combinations to strength classes are given in BS 5268: Part 2: 1991.
3. The grading rules for American and Canadian Lumber are those approved by the American Lumber Standards (ALS)

Board of Review and the Canadian Lumber Standards (CLS) Accreditation Board respectively (see BS 5268: Part 2: 1991).
4. NLGA denotes the National Lumber Grading Association.
5. NGRDL denotes the National Grading Rules for Dimension Lumber.
6. MSR denotes the North American Export Standard for Machine Stress-Rated Lumber.

Species and grades

The various species, grades and strength classes are defined in BS 5268: Part 2: 1991 mentioned above.

Sawn sizes

The tables refer to basic sawn sizes defined in BS 4471: Part 1: 1987 and do not relate when the basic sawn dimensions have been reduced by planing. North American Timber is indicated to surfaced sizes, unless the timber has been resawn to BS 4471 requirements.

BS 4471: 1987 species dimensions of sawn softwood sizes and gives tables of reductions upon sawn sizes for various products.

Fig.A3A   Imposed snow loading

Workmanship

Bearing areas and workmanship are to meet the requirements of BS 5268: Part 2: 1991.

Notches and holes

Roof rafters are not to have notches cut in them except for a birdsmouth bearing which must not be in excess of 0.33 the rafter depth (see Fig. A3).

Strutting to joists

Suitable solid or herringbone strutting is to be provided at spans of more than 2.5 m and two rows of strutting where spans are in excess of 4.5 m at one third span positions:

* solid strutting must be not less than 38 mm extending at least 0.75 depth of the joist;

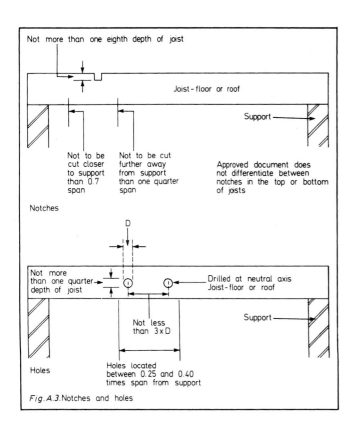

Fig. A.3. Notches and holes

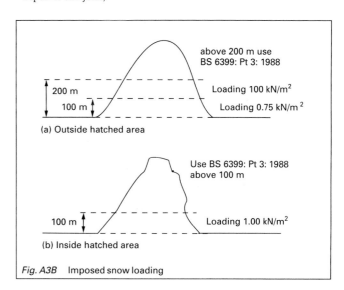

Fig. A3B   Imposed snow loading

- herringbone strutting is to be formed of not less than 38 mm × 38 mm timber.

It is not to be used when distance between joists is greater than the depth of joist × 3.

Snow loading

Snow loading criteria from BS 6399: Part 3: 1988 has now been taken into account in deriving the following tables and this gives rise to additional span tables using an increased imposed load off 1.00 kN/m², which is dependent upon the location of the site using Fig. A3A.

There is a corresponding effect on the maximum clear span (see Fig. A3B).

TABLES

Table 2 and the following tables relating to timber members are reproduced by permission of HMSO. Table 2 forms an index to the following tables. Throughout the tables spans are the clear dimen-

sions between supports (see Fig. A4, etc.) and spacings are between longitudinal centres of members:

- if bath installed, joists supporting bath are to be duplicated;
- notches and drilling not to be greater than in Fig. A3
- section sizes are regularized from BS 4471 basic sawn sizes with allowed tolerances or CLS/ALS sizes with BS 4471.
- where spans are unequal use longer span to choose section sizes
- bracing to be installed where required;
- ceiling joist sizes regularized from BS 4411 basic sawn sizes, with tolerances or CLS/ALS sizes with BS 4411 tolerances
- sizes of binders are either BS 4471 basic sawn sizes or CLS/ALS sizes with BS 4971 tolerances

The tables relate to single family dwellings not exceeding three storeys in height, but with regard to roof construction it seems rea-

Table 2 Summary of Tables **A1** to **A24** relating to timber members

Construction	Timber members	Imposed loading (kN/m²)	Table numbers Strength class SC3	SC4
Floors	Joists		A1	A2
Ceilings	Joists		A3	A3
	Binders		A4	A4
Pitched roofs greater than 15° but less than or equal to 22½°	Rafters	0.75	A5	A5
	Rafters	1.00	A7	A7
	Purlins	0.75	A6	A6
	Purlins	1.00	A8	A8
Pitched roofs greater than 22½° but less than or equal to 30°	Rafters	0.75	A9	A9
	Rafters	1.00	A11	A11
	Purlins	0.75	A10	A10
	Purlins	1.00	A12	A12
Pitched roofs greater than 30° but less than or equal to 45°	Rafters	0.75	A13	A13
	Rafters	1.00	A15	A15
	Purlins	0.75	A14	A14
	Purlins	1.00	A16	A16
Flat roofs access for maintenance only	Joists	0.75	A17	A18
	Joists	1.00	A19	A20
Flat roofs full access allowed	Joists		A21	A22
Sheeted or decked roofs greater than 10° but less than or equal to 35°	Purlins	0.75	A23	A23
	Purlins	1.00	A24	A24

Notes
1. The strength class given in this table assumes that the species and grades of timber to be used are those described in Table 1.
2. The diagrams are only illustrative and do not show all details of construction. Adequate connections between members should be provided as appropriate.
3. These tables do not apply to trussed rafter roofs.

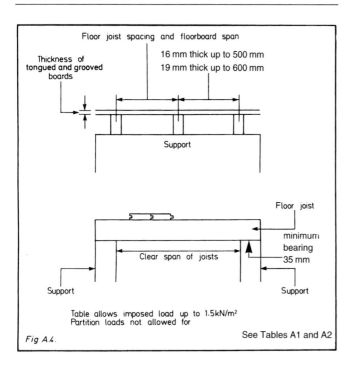

Fig A.4.

Table A1 **Floor joists**

Maximum clear span of joist (m). Timber of strength class **SC3** (see Table 1)

Size of joist (mm × mm)	Dead Load (kN/m²) excluding the self weight of the joist								
	Not more than 0.25			More than 0.25 but not more than 0.50			More than 0.50 but not more than 1.25		
	Spacing of joists (mm)								
	400	450	600	400	450	600	400	450	600
38 × 97	1.83	1.69	1.30	1.72	1.56	1.21	1.42	1.30	1.04
38 × 122	2.48	2.39	1.93	2.37	2.22	1.76	1.95	1.79	1.45
38 × 147	2.98	2.87	2.51	2.85	2.71	2.33	2.45	2.29	1.87
38 × 170	3.44	3.31	2.87	3.28	3.10	2.69	2.81	2.65	2.27
38 × 195	3.94	3.75	3.26	3.72	3.52	3.06	3.19	3.01	2.61
38 × 220	4.43	4.19	3.65	4.16	3.93	3.42	3.57	3.37	2.92
47 × 97	2.02	1.91	1.58	1.92	1.82	1.46	1.67	1.53	1.23
47 × 122	2.66	2.56	2.30	2.55	2.45	2.09	2.26	2.08	1.70
47 × 147	3.20	3.08	2.79	3.06	2.95	2.61	2.72	2.57	2.17
47 × 170	3.69	3.55	3.19	3.53	3.40	2.99	3.12	2.94	2.55
47 × 195	4.22	4.06	3.62	4.04	3.89	3.39	3.54	3.34	2.90
47 × 220	4.72	4.57	4.04	4.55	4.35	3.79	3.95	3.74	3.24
50 × 97	2.08	1.97	1.67	1.98	1.87	1.54	1.74	1.60	1.29
50 × 122	2.72	2.62	2.37	2.60	2.50	2.19	2.33	2.17	1.77
50 × 147	3.29	3.14	2.86	3.13	3.01	2.69	2.81	2.65	2.27
50 × 170	3.77	3.62	3.29	3.61	3.47	3.08	3.21	3.03	2.63
50 × 195	4.31	4.15	3.73	4.13	3.97	3.50	3.65	3.44	2.99
50 × 220	4.79	4.66	4.17	4.64	4.47	3.91	4.07	3.85	3.35
63 × 97	2.32	2.20	1.92	2.19	2.08	1.82	1.93	1.84	1.53
63 × 122	2.93	2.82	2.57	2.81	2.70	2.45	2.53	2.43	2.09
63 × 147	3.52	3.39	3.08	3.37	3.24	2.95	3.04	2.92	2.58
63 × 170	4.06	3.91	3.56	3.89	3.74	3.40	3.50	3.37	2.95
63 × 195	4.63	4.47	4.07	4.44	4.28	3.90	4.01	3.85	3.35
63 × 220	5.06	4.92	4.58	4.91	4.77	4.37	4.51	4.30	3.75
75 × 122	3.10	2.99	2.72	2.97	2.86	2.60	2.68	2.58	2.33
75 × 147	3.72	3.58	3.27	3.56	3.43	3.13	3.22	3.09	2.81
75 × 170	4.28	4.13	3.77	4.11	3.96	3.61	3.71	3.57	3.21
75 × 195	4.83	4.70	4.31	4.68	4.52	4.13	4.24	4.08	3.65
75 × 220	5.27	5.13	4.79	5.11	4.97	4.64	4.74	4.60	4.07
38 × 140	2.84	2.73	2.40	2.72	2.59	2.17	2.33	2.15	1.75
38 × 184	3.72	3.56	3.09	3.53	3.33	2.90	3.02	2.85	2.47
38 × 235	4.71	4.46	3.89	4.43	4.18	3.64	3.80	3.59	3.11

- See Fig. A4

Table A2 **Floor joists**

Maximum clear span of joist(m). Timber of strength class **SC4** (see Table 1)

Size of joist (mm × mm)	Dead Load (kN/m²) excluding the self weight of the joist								
	Not more than 0.25			More than 0.25 but not more than 0.50			More than 0.50 but not more than 1.25		
	Spacing of joists (mm)								
	400	450	600	400	450	600	400	450	600
38 × 97	1.94	1.83	1.59	1.84	1.74	1.51	1.64	1.55	1.36
38 × 122	2.58	2.48	2.20	2.47	2.37	2.08	2.18	2.07	1.83
38 × 147	3.10	2.98	2.71	2.97	2.85	2.59	2.67	2.56	2.31
38 × 170	3.58	3.44	3.13	3.43	3.29	2.99	3.08	2.96	2.68
38 × 195	4.10	3.94	3.58	3.92	3.77	3.42	3.53	3.39	3.07
38 × 220	4.61	4.44	4.03	4.41	4.25	3.86	3.97	3.82	3.46
47 × 97	2.14	2.03	1.76	2.03	1.92	1.68	1.80	1.71	1.50
47 × 122	2.77	2.66	2.42	2.65	2.55	2.29	2.38	2.27	2.01
47 × 147	3.33	3.20	2.91	3.19	3.06	2.78	2.87	2.75	2.50
47 × 170	3.84	3.69	3.36	3.67	3.54	3.21	3.31	3.18	2.88
47 × 195	4.39	4.22	3.85	4.20	4.05	3.68	3.79	3.64	3.30
47 × 220	4.86	4.73	4.33	4.71	4.55	4.14	4.26	4.10	3.72
50 × 97	2.20	2.09	1.82	2.08	1.98	1.73	1.84	1.75	1.54
50 × 122	2.83	2.72	2.47	2.71	2.60	2.36	2.43	2.33	2.06
50 × 147	3.39	3.27	2.97	3.25	3.13	2.84	2.93	2.81	2.55
50 × 170	3.91	3.77	3.43	3.75	3.61	3.28	3.38	3.25	2.94
50 × 195	4.47	4.31	3.92	4.29	4.13	3.75	3.86	3.72	3.37
50 × 220	4.93	4.80	4.42	4.78	4.64	4.23	4.35	4.18	3.80
63 × 97	2.43	2.32	2.03	2.31	2.19	1.93	2.03	1.93	1.71
63 × 122	3.05	2.93	2.67	2.92	2.81	2.55	2.63	2.53	2.27
63 × 147	3.67	3.52	3.21	3.50	3.37	3.07	3.16	3.04	2.76
63 × 170	4.21	4.06	3.70	4.04	3.89	3.54	3.64	3.51	3.19
63 × 195	4.77	4.64	4.23	4.61	4.45	4.05	4.17	4.01	3.65
63 × 220	5.20	5.06	4.73	5.05	4.91	4.56	4.68	4.51	4.11
75 × 122	3.22	3.10	2.83	3.09	2.97	2.71	2.78	2.68	2.43
75 × 147	3.86	3.72	3.39	3.70	3.57	3.25	3.34	3.22	2.93
75 × 170	4.45	4.29	3.91	4.27	4.11	3.75	3.86	3.71	3.38
75 × 195	4.97	4.83	4.47	4.82	4.69	4.29	4.41	4.25	3.86
75 × 220	5.42	5.27	4.93	5.25	5.11	4.78	4.88	4.74	4.35
38 × 140	2.96	2.84	2.58	2.83	2.72	2.47	2.54	2.44	2.17
38 × 184	3.87	3.72	3.38	3.70	3.56	3.23	3.33	3.20	2.90
38 × 235	4.85	4.71	4.31	4.70	4.54	4.12	4.24	4.08	3.70

- See Fig. A4

sonable that these tables may also be used in other types of similar size small buildings.

- clear span of purlin is dimension between supporting struts or walls;
- purlins placed perpendicular to roof slope;
- where spans are unequal use longer span to choose section sizes;
- sizes are either BS 4471 basic sawn sizes with tolerances or CLS/ALS sizes with BS 4471 tolerances;

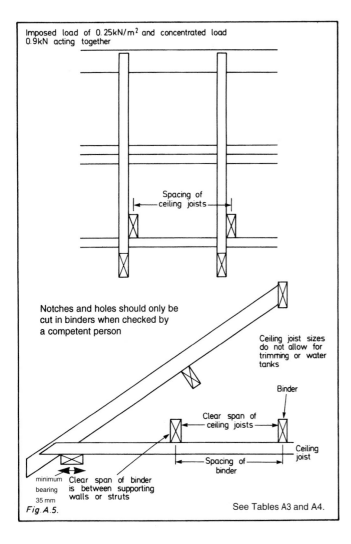

Imposed load of 0.25kN/m² and concentrated load 0.9kN acting together

Spacing of ceiling joists

Notches and holes should only be cut in binders when checked by a competent person

Ceiling joist sizes do not allow for trimming or water tanks

Binder

Clear span of ceiling joists

Ceiling joist

Spacing of binder

minimum bearing 35 mm

Clear span of binder is between supporting walls or struts

See Tables A3 and A4.

Fig. A.5.

Table **A3**   Ceiling joists

Maximum clear span of joist (m). Timber of strength class **SC3** and **SC4** (see Table 1)

Size of joist (mm × mm)	Dead Load (kN/m²) excluding the self weight of the joist											
	Not more than 0.25			More than 0.25 but not more than 0.50			Not more than 0.25			More than 0.25 but not more than 0.50		
	Spacing of joists (mm)											
	400	450	600	400	450	600	400	450	600	400	450	600
38×72	1.15	1.14	1.11	1.11	1.10	1.06	1.21	1.20	1.17	1.17	1.16	1.12
38×97	1.74	1.72	1.67	1.67	1.64	1.58	1.84	1.82	1.76	1.76	1.73	1.66
38×122	2.37	2.34	2.25	2.25	2.21	2.11	2.50	2.46	2.37	2.37	2.33	2.22
38×147	3.02	2.97	2.85	2.85	2.80	2.66	3.18	3.13	3.00	3.00	2.94	2.79
38×170	3.63	3.57	3.41	3.41	3.34	3.16	3.81	3.75	3.58	3.58	3.51	3.32
38×195	4.30	4.23	4.02	4.02	3.94	3.72	4.51	4.43	4.22	4.22	4.13	3.89
38×220	4.98	4.88	4.64	4.64	4.54	4.27	5.21	5.11	4.86	4.86	4.75	4.47
47×72	1.27	1.26	1.23	1.23	1.21	1.17	1.35	1.33	1.30	1.30	1.28	1.24
47×97	1.92	1.90	1.84	1.84	1.81	1.73	2.03	2.00	1.93	1.93	1.90	1.83
47×122	2.60	2.57	2.47	2.47	2.42	2.31	2.74	2.70	2.60	2.60	2.55	2.43
47×147	3.30	3.25	3.11	3.11	3.05	2.90	3.47	3.42	3.27	3.27	3.21	3.04
47×170	3.96	3.89	3.72	3.72	3.64	3.44	4.15	4.08	3.89	3.89	3.81	3.61
47×195	4.68	4.59	4.37	4.37	4.28	4.04	4.90	4.81	4.57	4.57	4.47	4.22
47×220	5.39	5.29	5.03	5.03	4.91	4.63	5.64	5.53	5.25	5.25	5.14	4.84
50×72	1.31	1.30	1.27	1.27	1.25	1.21	1.39	1.37	1.34	1.34	1.32	1.28
50×97	1.97	1.95	1.89	1.89	1.86	1.78	2.08	2.06	1.99	1.99	1.96	1.88
50×122	2.67	2.63	2.53	2.53	2.49	2.37	2.81	2.77	2.66	2.66	2.62	2.49
50×147	3.39	3.34	3.19	3.19	3.13	2.97	3.56	3.50	3.35	3.35	3.29	3.12
50×170	4.06	3.99	3.81	3.81	3.73	3.53	4.25	4.18	3.99	3.99	3.91	3.69
50×195	4.79	4.70	4.48	4.48	4.38	4.13	5.01	4.92	4.68	4.68	4.58	4.32
50×220	5.52	5.41	5.14	5.14	5.03	4.73	5.77	5.66	5.37	5.37	5.25	4.95
38×89	1.54	1.53	1.48	1.48	1.46	1.41	1.63	1.62	1.57	1.57	1.55	1.49
38×140	2.84	2.79	2.68	2.68	2.63	2.50	2.99	2.94	2.82	2.82	2.77	2.63
38×184	4.01	3.94	3.75	3.75	3.68	3.47	4.20	4.13	3.94	3.94	3.85	3.64

- See Fig. A5

Table **A4**   Binders supporting ceiling joists

Maximum clear span of binder(m) Timber of strength class **SC3** and **SC4** (see Table 1)

Size of binder (mm × mm)		Dead Load (kN/m²) excluding the self weight of the binder											
		Not more than 0.25						More than 0.25 but not more than 0.50					
		Spacing of binders (mm)											
		1200	1500	1800	2100	2400	2700	1200	1500	1800	2100	2400	2700
47×150	SC3	2.17	2.05	1.96	1.88	1.81		1.99	1.87				
47×175		2.59	2.45	2.33	2.24	2.15	2.08	2.37	2.23	2.11	2.02	1.94	1.87
50×150		2.22	2.11	2.01	1.93	1.86		2.04	1.92	1.83			
50×175		2.65	2.51	2.39	2.29	2.21	2.13	2.42	2.28	2.16	2.07	1.99	1.91
50×200		3.08	2.91	2.77	2.65	2.55	2.47	2.81	2.64	2.50	2.39	2.29	2.21
63×125		1.97	1.87					1.82					
63×150		2.44	2.31	2.20	2.12	2.04	1.97	2.23	2.11	2.00	1.91	1.84	
63×175		2.90	2.74	2.61	2.51	2.41	2.33	2.65	2.49	2.37	2.26	2.17	2.10
63×200		3.37	3.18	3.03	2.90	2.79	2.69	3.07	2.88	2.74	2.61	2.51	2.42
63×225		3.83	3.61	3.44	3.29	3.16	3.05	3.49	3.27	3.10	2.96	2.84	2.74
75×125		2.12	2.01	1.92	1.85			1.95	1.84				
75×150		2.61	2.47	2.36	2.26	2.18	2.11	2.39	2.25	2.14	2.05	1.97	1.90
75×175		3.10	2.93	2.79	2.68	2.58	2.49	2.83	2.66	2.53	2.42	2.32	2.24
75×200		3.59	3.39	3.23	3.09	2.98	2.88	3.27	3.08	2.92	2.79	2.68	2.58
75×225		4.08	3.85	3.66	3.51	3.37	3.26	3.71	3.50	3.31	3.16	3.03	2.92
47×150	SC4	2.28	2.16	2.06	1.98	1.90	1.84	2.09	1.97	1.87			
47×175		2.72	2.57	2.45	2.34	2.26	2.18	2.48	2.34	2.22	2.12	2.03	1.96
50×150		2.33	2.21	2.11	2.02	1.95	1.89	2.14	2.02	1.92	1.83		
50×175		2.78	2.63	2.51	2.40	2.31	2.23	2.54	2.39	2.27	2.17	2.08	2.01
50×200		3.23	3.05	2.90	2.78	2.67	2.58	2.95	2.77	2.62	2.51	2.40	2.32
63×125		2.07	1.97	1.88	1.81			1.91	1.80				
63×150		2.56	2.42	2.31	2.22	2.14	2.07	2.34	2.21	2.10	2.01	1.93	1.86
63×175		3.04	2.87	2.74	2.62	2.53	2.44	2.78	2.61	2.48	2.37	2.28	2.20
63×200		3.52	3.32	3.16	3.03	2.92	2.82	3.21	3.02	2.86	2.73	2.63	2.53
63×225		4.00	3.77	3.59	3.44	3.31	3.19	3.65	3.42	3.24	3.10	2.97	2.86
75×125		2.22	2.11	2.01	1.94	1.87	1.81	2.04	1.93	1.84			
75×150		2.73	2.59	2.47	2.37	2.28	2.21	2.50	2.36	2.24	2.15	2.06	1.99
75×175		3.24	3.07	2.92	2.80	2.70	2.61	2.96	2.79	2.65	2.53	2.43	2.35
75×200		3.75	3.54	3.37	3.23	3.11	3.00	3.42	3.22	3.05	2.92	2.80	2.70
75×225		4.26	4.02	3.82	3.66	3.52	3.40	3.88	3.65	3.46	3.30	3.17	3.06

- See Fig. A5

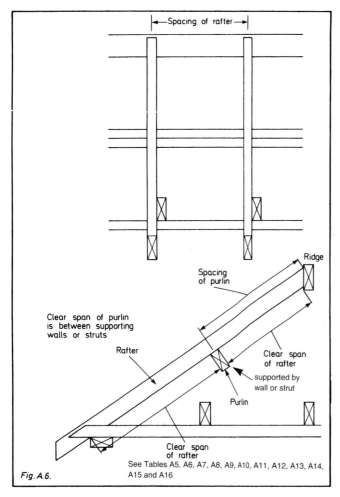

Fig. A.6.

Labels: Spacing of rafter — Ridge — Spacing of purlin — Clear span of purlin is between supporting walls or struts — Rafter — Clear span of rafter — supported by wall or strut — Purlin — Clear span of rafter

See Tables A5, A6, A7, A8, A9, A10, A11, A12, A13, A14, A15 and A16

- minimum bearing: rafters   35 mm
                          purlins   50 mm
- notches or holes should only be cut in purlins when checked by a competent person.

Table A5   **Common or jack rafters** for roofs having a pitch more than 15° but not more than 22.5° with access only for purposes of maintenance or repair. Imposed loading 0.75kN/m²

Maximum clear span of rafter (m). Timber of strength class **SC3** and **SC4** (see Table 1)

Size of rafter (mm x mm)		Dead Load [kN/m²] excluding the self weight of the rafter								
		Not more than 0.50			More than 0.50 but not more than 0.75			More than 0.75 but not more than 1.00		
		___Spacing of rafters (mm)___								
		400	450	600	400	450	600	400	450	600
38 x 100	SC3	2.10	2.05	1.93	1.93	1.88	1.75	1.80	1.75	1.61
38 x 125		2.89	2.79	2.53	2.63	2.55	2.34	2.44	2.35	2.15
38 x 150		3.47	3.34	3.03	3.26	3.14	2.78	3.08	2.96	2.57
47 x 100		2.46	2.40	2.18	2.25	2.19	2.03	2.10	2.03	1.87
47 x 125		3.10	2.99	2.72	2.92	2.81	2.56	2.78	2.67	2.41
47 x 150		3.71	3.57	3.25	3.50	3.36	3.06	3.32	3.20	2.86
50 x 100		2.54	2.45	2.23	2.35	2.29	2.09	2.19	2.12	1.95
50 x 125		3.17	3.05	2.78	2.98	2.87	2.61	2.83	2.73	2.48
50 x 150		3.78	3.64	3.32	3.57	3.43	3.12	3.39	3.26	2.94
38 x 89		1.76	1.72	1.63	1.63	1.59	1.49	1.53	1.49	1.38
38 x 140		3.24	3.12	2.83	3.05	2.93	2.61	2.82	2.72	2.41
38 x 100	SC4	2.42	2.33	2.11	2.28	2.19	1.99	2.16	2.08	1.88
38 x 125		3.01	2.90	2.64	2.83	2.73	2.48	2.69	2.59	2.35
38 x 150		3.60	3.47	3.16	3.39	3.26	2.97	3.22	3.10	2.82
47 x 100		2.59	2.49	2.27	2.44	2.35	2.14	2.32	2.23	2.02
47 x 125		3.22	3.11	2.83	3.04	2.92	2.66	2.89	2.78	2.53
47 x 150		3.85	3.71	3.38	3.63	3.50	3.18	3.45	3.32	3.02
50 x 100		2.64	2.54	2.32	2.49	2.40	2.18	2.37	2.28	2.07
50 x 125		3.29	3.17	2.89	3.10	2.98	2.72	2.95	2.83	2.58
50 x 150		3.93	3.78	3.45	3.70	3.57	3.25	3.52	3.39	3.09
38 x 89		2.16	2.07	1.88	2.03	1.95	1.77	1.92	1.85	1.68
38 x 140		3.37	3.24	2.95	3.17	3.05	2.77	3.01	2.90	2.63

- Provides support for dead load and imposed load of 0.75 kN/m² measured on plan or concentrated load of 0.9 kN
- See Fig. A6.

Table A6   **Purlins** supporting rafters to which Table A5 refers (Imposed loading 0.75 kN/m²).

Maximum clear span of purlin(m) Timber of strength class **SC3** and **SC4** (see Table 1)

Size of purlin (mm x mm)		Dead Load [kN/m²] excluding the self weight of the purlin																	
		Not more than 0.5						More than 0.5 but not more than 0.75						More than 0.75 but not more than 1.0					
		___Spacing of purlins (mm)___																	
		1500	1800	2100	2400	2700	3000	1500	1800	2100	2400	2700	3000	1500	1800	2100	2400	2700	3000
50 x 150	SC3	1.90																	
50 x 175		2.22	2.08	1.96	1.87			2.08	1.95	1.84				1.97	1.84				
50 x 200		2.53	2.37	2.24	2.13	2.02	1.92	2.38	2.22	2.10	1.97	1.85		2.25	2.10	1.95	1.82		
50 x 225		2.84	2.66	2.52	2.40	2.26	2.14	2.67	2.50	2.35	2.20	2.07	1.96	2.53	2.36	2.18	2.03	1.91	1.81
63 x 150		2.06	1.94	1.83				1.94	1.82					1.84					
63 x 175		2.41	2.26	2.13	2.03	1.95	1.87	2.26	2.12	2.00	1.91	1.82		2.14	2.01	1.90	1.80		
63 x 200		2.75	2.58	2.44	2.32	2.22	2.14	2.58	2.42	2.29	2.18	2.08	1.97	2.45	2.29	2.16	2.05	1.93	1.83
63 x 225		3.09	2.89	2.74	2.61	2.50	2.40	2.90	2.72	2.57	2.45	2.33	2.20	2.75	2.58	2.43	2.29	2.16	2.04
75 x 125		1.83																	
75 x 150		2.19	2.06	1.95	1.86			2.06	1.94	1.83				1.96	1.83				
75 x 175		2.56	2.40	2.27	2.17	2.08	2.00	2.41	2.26	2.13	2.03	1.95	1.87	2.28	2.14	2.02	1.92	1.84	
75 x 200		2.92	2.74	2.59	2.47	2.37	2.28	2.75	2.58	2.44	2.32	2.22	2.14	2.61	2.44	2.31	2.20	2.10	2.00
75 x 225		3.28	3.08	2.91	2.78	2.66	2.56	3.09	2.89	2.74	2.61	2.50	2.40	2.93	2.74	2.60	2.47	2.36	2.23
50X 150	SC4	1.99	1.86					1.87											
50 x 175		2.32	2.17	2.05	1.95	1.87		2.18	2.04	1.92	1.83			2.06	1.93	1.82			
50 x 200		2.64	2.48	2.34	2.23	2.14	2.05	2.49	2.33	2.20	2.09	2.00	1.92	2.36	2.20	2.08	1.98	1.89	
50 x 225		2.97	2.78	2.63	2.51	2.40	2.31	2.79	2.62	2.47	2.35	2.25	2.16	2.65	2.48	2.34	2.22	2.12	1.94
63 x 150		2.16	2.02	1.91	1.82			2.03	1.90					1.92	1.80				
63 x 175		2.51	2.36	2.23	2.13	2.04	1.96	2.36	2.22	2.10	2.00	1.91	1.84	2.24	2.10	1.99	1.89	1.81	
63 x 200		2.87	2.69	2.55	2.43	2.33	2.24	2.70	2.53	2.39	2.28	2.18	2.10	2.56	2.40	2.27	2.16	2.06	1.98
63 x 225		3.22	3.02	2.86	2.73	2.61	2.52	3.03	2.84	2.69	2.56	2.45	2.36	2.88	2.70	2.55	2.43	2.32	2.23
75 x 125		1.91																	
75 x 150		2.29	2.15	2.04	1.94	1.86		2.16	2.02	1.91	1.82			2.05	1.92	1.82			
75 x 175		2.67	2.51	2.37	2.26	2.17	2.09	2.51	2.36	2.23	2.13	2.04	1.96	2.39	2.24	2.12	2.02	1.93	1.85
75 x 200		3.05	2.86	2.71	2.58	2.48	2.39	2.87	2.69	2.55	2.43	2.33	2.24	2.72	2.55	2.42	2.30	2.20	2.12
75 x 225		3.42	3.21	3.04	2.90	2.78	2.68	3.22	3.02	2.86	2.73	2.62	2.52	3.06	2.87	2.72	2.59	2.48	2.38

- Provides support for dead load and imposed load of 0.75 kN/m² measured on plan or concentrated load of 0.9 kN.
- See Fig. A6.

Table A7   **Common or jack rafters** for roofs having a pitch more than 15° but not more than 22.5° with access only for purposes of maintenance or repair. Imposed loading of 1.00 kN/m².

Maximum clear span of rafter(m) Timber of strength class **SC3** and **SC4** (see Table 1)

Size of rafter (mm X mm)		Dead Load [kN/m²] excluding the self weight of the rafter								
		Not more than 0.50			More than 0.50 but not more than 0.75			More than 0.75 but not more than 1.00		
		___Spacing of rafters (mm)___								
		400	450	600	400	450	600	400	450	600
38 X 100	SC3	2.10	2.05	1.90	1.93	1.88	1.75	1.80	1.75	1.61
38 x 125		2.73	2.63	2.35	2.59	2.49	2.17	2.44	2.34	2.03
38 x 150		3.27	3.14	2.79	3.10	2.97	2.58	2.94	2.78	2.41
47 x 100		2.35	2.26	2.05	2.23	2.15	1.95	2.10	2.03	1.63
47 x 125		2.93	2.82	2.56	2.78	2.68	2.41	2.66	2.56	2.26
47 x 150		3.50	3.37	3.07	3.33	3.20	2.86	3.18	3.06	2.68
50 x 100		2.40	2.31	2.10	2.28	2.19	1.99	2.18	2.09	1.88
50 x 125		2.99	2.88	2.62	2.84	2.73	2.48	2.71	2.61	2.33
50 x 150		3.57	3.44	3.13	3.40	3.27	2.95	3.25	3.12	2.76
38 x 89		1.76	1.72	1.63	1.63	1.59	1.49	1.53	1.49	1.38
38 x 140		3.05	2.94	2.61	2.90	2.78	2.42	2.76	2.61	2.26
38 X 100	SC4	2.28	2.19	1.99	2.16	2.08	1.89	2.07	1.99	1.80
38 x 125		2.84	2.73	2.48	2.70	2.59	2.35	2.58	2.48	2.25
38 x 150		3.40	3.27	2.97	3.23	3.10	2.82	3.09	2.97	2.69
47 x 100		2.44	2.35	2.14	2.32	2.23	2.03	2.22	2.13	1.94
47 x 125		3.04	2.93	2.67	2.89	2.78	2.53	2.77	2.66	2.42
47 x 150		3.64	3.50	3.19	3.46	3.33	3.03	3.31	3.18	2.89
50 X 100		2.49	2.40	2.18	2.37	2.28	2.07	2.27	2.18	1.98
50 x 125		3.10	2.99	2.72	2.95	2.84	2.58	2.82	2.72	2.47
50 X 150		3.71	3.57	3.26	3.46	3.40	3.09	3.38	3.25	2.95
38 x 89		2.03	1.95	1.77	1.93	1.85	1.68	1.84	1.77	1.60
38 x 140		3.18	3.06	2.78	3.02	2.90	2.63	2.88	2.77	2.52

- Provides support for dead load and imposed load of 1.00 kN/m² measured on plan or concentrated load of 0.9 kN.
- See Fig. A6.

Table A8 **Purlins** supporting rafters to which Table A7 refers (Imposed loading 1.0 kN/m²).

Maximum clear span of purlin(m) Timber of strength class **SC3** and **SC4** (see Table 1)

Size of purlin (mm x mm)		Not more than 0.50						More than 0.50 but not more than 0.75						More than 0.75 but not more than 1.00					
		1500	1800	2100	2400	2700	3000	1500	1800	2100	2400	2700	3000	1500	1800	2100	2400	2700	3000
50 x 175	SC3	2.09	1.95	1.84				1.97	1.85					1.88					
50 x 200		2.38	2.23	2.10	1.97	1.85		2.26	2.11	1.96	1.82			2.15	1.98	1.83			
50 x 225		2.68	2.50	2.36	2.20	2.07	1.96	2.54	2.36	2.18	2.04	1.92	1.81	2.42	2.21	2.04	1.90		
63 x 150		1.94	1.82																
63 x 175		2.27	2.12	2.01	1.91	1.83		2.15	2.01	1.90	1.81			2.05	1.92	1.81			
63 x 200		2.59	2.42	2.29	2.18	2.09	1.98	2.45	2.30	2.17	2.06	1.94	1.83	2.30	2.19	2.06	1.92	1.81	
63 x 225		2.91	2.72	2.58	2.45	2.33	2.21	2.76	2.58	2.44	2.30	2.16	2.05	2.63	2.46	2.30	2.15	2.02	1.91
75 x 150		2.07	1.94	1.83										1.87					
75 x 175		2.41	2.26	2.14	2.04	1.95	1.88	2.29	2.14	2.03	1.93	1.85		2.18	2.04	1.93	1.84		
75 x 200		2.75	2.58	2.44	2.33	2.23	2.14	2.61	2.45	2.31	2.20	2.11	2.01	2.49	2.33	2.20	2.10	1.98	1.88
75 x 225		3.09	2.90	2.74	2.61	2.50	2.41	2.93	2.75	2.60	2.48	2.36	2.24	2.80	2.62	2.48	2.35	2.21	2.09
50 x 150	SC4	1.87																	
50 x 175		2.18	2.04	1.93	1.83			2.07	1.93	1.82				1.97	1.84				
50 x 200		2.49	2.33	2.20	2.10	2.00	1.92	2.36	2.21	2.08	1.98	1.89		2.25	2.10	1.98	1.88		
50 x 225		2.80	2.62	2.48	2.36	2.25	2.16	2.65	2.48	2.34	2.23	2.13	1.95	2.53	2.36	2.23	2.12	1.91	
65 x 150		2.03	1.90	1.80				1.93	1.81					1.84					
63 x 175		2.37	2.22	2.10	2.00	1.91	1.84	2.25	2.10	1.99	1.89	1.81		2.14	2.01	1.90	1.80		
63 x 200		2.70	2.53	2.40	2.28	2.19	2.10	2.57	2.40	2.27	2.16	2.07	1.99	2.45	2.29	2.16	2.06	1.97	1.89
63 x 225		3.04	2.85	2.70	2.57	2.46	2.36	2.88	2.70	2.55	2.43	2.32	2.23	2.75	2.58	2.43	2.31	2.21	2.12
75 x 125		1.80																	
75 x 150		2.16	2.03	1.92	1.83			2.05	1.92	1.82				1.96	1.83				
75 x 175		2.52	2.36	2.24	2.13	2.04	1.96	2.39	2.24	2.12	2.02	1.93	1.86	2.28	2.14	2.02	1.92	1.84	
75 x 200		2.87	2.70	2.55	2.43	2.33	2.24	2.73	2.56	2.42	2.31	2.21	2.12	2.61	2.44	2.31	2.20	2.10	2.02
75 x 225		3.23	3.03	2.87	2.74	2.62	2.52	3.07	2.88	2.72	2.59	2.48	2.39	2.93	2.75	2.60	2.47	2.36	2.27

- Provides support for dead load and imposed load of 1.00 kN/m² measured on plan at concentrated load of 0.9 kN.
- See Fig. A6.

Table A10 **Purlins** supporting rafters to which Table A9 refers (Imposed load 0.75 kN/m²).

Maximum clear span of purlin (m). Timber of strength class **SC3** and **SC4** (see Table 1)

Size of purlin (mm x mm)		Not more than 0.50						More than 0.50 but not more than 0.75						More than 0.75 but not more than 1.00					
		1500	1800	2100	2400	2700	3000	1500	1800	2100	2400	2700	3000	1500	1800	2100	2400	2700	3000
50 x 150	SC3	1.95	1.83					1.83											
50 x 175		2.27	2.12	2.01	1.92	1.83		2.13	1.99	1.88				2.02	1.89				
50 x 200		2.59	2.43	2.30	2.19	2.09	1.99	2.43	2.28	2.15	2.03	1.91	1.81	2.30	2.15	2.01	1.88		
50 x 225		2.92	2.73	2.58	2.46	2.34	2.22	2.74	2.56	2.42	2.27	2.14	2.02	2.59	2.42	2.25	2.10	1.98	1.87
63 x 150		2.12	1.98	1.88				1.99	1.86					1.88					
63 x 175		2.47	2.31	2.19	2.09	2.00	1.92	2.32	2.17	2.05	1.95	1.87		2.19	2.05	1.94	1.85		
63 x 200		2.81	2.64	2.50	2.38	2.28	2.19	2.64	2.48	2.34	2.23	2.13	2.04	2.50	2.35	2.22	2.11	1.99	1.89
63 x 225		3.16	2.97	2.81	2.68	2.56	2.47	2.97	2.78	2.63	2.51	2.40	2.28	2.82	2.64	2.49	2.37	2.23	2.11
75 x 125		1.88																	
75 x 150		2.25	2.11	2.00	1.91	1.83		2.11	1.98	1.87				2.00	1.88				
75 x 175		2.62	2.46	2.33	2.22	2.13	2.05	2.46	2.31	2.19	2.08	1.99	1.92	2.33	2.19	2.07	1.97	1.89	1.81
75 x 200		2.99	2.81	2.66	2.54	2.43	2.34	2.81	2.64	2.50	2.38	2.28	2.19	2.67	2.50	2.36	2.25	2.15	2.07
75 x 225		3.36	3.15	2.99	2.85	2.73	2.63	3.16	2.96	2.80	2.67	2.56	2.46	3.00	2.81	2.66	2.53	2.42	2.31
50 x 150	SC4	2.04	1.91	1.81				1.91						1.31					
50 x 175		2.37	2.22	2.10	2.00	1.92	1.84	2.23	2.09	1.97	1.88			2.11	1.97	1.86			
50 x 200		2.71	2.54	2.40	2.29	2.19	2.11	2.54	2.38	2.25	2.14	2.05	1.97	2.41	2.26	2.13	2.02	1.94	1.84
50 x 225		3.05	2.86	2.70	2.57	2.46	2.37	2.86	2.68	2.53	2.41	2.30	2.21	2.71	2.54	2.39	2.28	2.18	2.07
63 x 150		1.84												1.97	1.84				
63 x 175		2.21	2.07	1.96	1.87			2.08	1.95	1.84				1.97	1.84				
63 x 200		2.57	2.42	2.29	2.18	2.09	2.01	2.42	2.27	2.15	2.04	1.96	1.88	2.29	2.15	2.03	1.93	1.85	
63 x 225		2.94	2.76	2.61	2.49	2.39	2.30	2.76	2.59	2.45	2.33	2.24	2.15	2.62	2.45	2.32	2.21	2.11	2.03
75 x 125		1.96	1.84					1.84											
75 x 150		2.35	2.20	2.09	1.99	1.91	1.84	2.21	2.07	1.96	1.87			2.09	1.96	1.86			
75 x 175		2.73	2.57	2.43	2.32	2.22	2.14	2.57	2.41	2.28	2.18	2.09	2.01	2.44	2.29	2.16	2.06	1.97	1.90
75 x 200		3.12	2.93	2.78	2.65	2.54	2.45	2.93	2.75	2.61	2.49	2.38	2.29	2.79	2.61	2.47	2.35	2.26	2.17
75 x 225		3.50	3.29	3.12	2.98	2.86	2.75	3.30	3.10	2.93	2.80	2.68	2.58	3.13	2.94	2.78	2.65	2.54	2.44

- Provides support for dead load and imposed load of 0.75 kN/m² measured on plan, or concentrated load of 0.9 kN.
- See Fig. A6.

Table A9 **Common or jack rafters** for roofs having a pitch more than 22.5° but not more than 30° with access only for purposes of maintenance or repair. Imposed loading 0.75 kN/m²

Maximum clear span of rafter (m). Timber of strength class **SC3** and **SC4** (see Table 1)

Size of rafter (mm x mm)		Not more than 0.50			More than 0.50 but not more than 0.75			More than 0.75 but not more than 1.00		
		400	450	600	400	450	600	400	450	600
38 x 100	SC3	2.18	2.13	2.01	2.01	1.96	1.82	1.88	1.82	1.68
38 x 125		2.97	2.86	2.60	2.74	2.66	2.44	2.54	2.46	2.25
38 x 150		3.55	3.42	3.11	3.34	3.21	2.92	3.17	3.04	2.72
47 x 100		2.55	2.46	2.23	2.35	2.28	2.10	2.18	2.12	1.95
47 x 125		3.18	3.06	2.79	2.99	2.88	2.62	2.84	2.73	2.48
47 x 150		3.80	3.66	3.33	3.57	3.44	3.13	3.39	3.27	2.97
50 x 100		2.60	2.51	2.28	2.45	2.36	2.14	2.28	2.21	2.03
50 x 125		3.24	3.12	2.84	3.05	2.93	2.67	2.89	2.79	2.53
50 x 150		3.87	3.73	3.40	3.65	3.51	3.20	3.46	3.33	3.03
38 x 89		1.82	1.79	1.69	1.69	1.65	1.55	1.59	1.55	1.44
38 x 140		3.32	3.19	2.90	3.12	3.00	2.72	2.94	2.84	2.55
38 x 100	SC4	2.48	2.38	2.17	2.33	2.24	2.03	2.21	2.12	1.93
38 x 125		3.08	2.97	2.70	2.90	2.79	2.53	2.75	2.65	2.40
38 x 150		3.69	3.55	3.23	3.47	3.34	3.04	3.29	3.17	2.88
47 x 100		2.65	2.55	2.32	2.49	2.40	2.18	2.37	2.28	2.07
47 x 125		3.30	3.18	2.90	3.11	2.99	2.72	2.95	2.84	2.58
47 x 150		3.94	3.80	3.46	3.71	3.58	3.26	3.53	3.40	3.09
50 x 100		2.71	2.61	2.37	2.55	2.45	2.23	2.42	2.32	2.11
50 x 125		3.37	3.24	2.96	3.17	3.05	2.78	3.01	2.90	2.63
50 x 150		4.02	3.87	3.53	3.79	3.65	3.32	3.60	3.46	3.15
38 x 89		2.21	2.12	1.93	2.07	1.99	1.81	1.97	1.89	1.72
38 x 140		3.45	3.32	3.02	3.24	3.12	2.84	3.08	2.96	2.69

- Provides support for dead load and imposed load of 0.75 kN/m² measured on plan or concentrated load of 0.9 kN.
- See Fig. A6.

Table A11 **Common or jack rafters** for roofs having a pitch more than 22.5° but not more than 30° with access only for purposes of maintenance or repair. Imposed loading 1.00 kN/m²

Maximum clear span of rafter (m). Timber of strength class **SC3** and **SC4** (see Table 1)

Size of rafter (mm x mm)		Not more than 0.50			More than 0.50 but not more than 0.75			More than 0.75 but not more than 1.00		
		400	450	600	400	450	600	400	450	600
38 X 100	SC3	2.18	2.13	1.96	2.01	1.96	1.82	1.88	1.82	1.68
38 X 125		2.80	2.69	2.45	2.65	2.55	2.30	2.53	2.44	2.15
38 X 150		3.35	3.22	2.93	3.18	3.06	2.73	3.03	2.92	2.55
47 X 100		2.41	2.32	2.11	2.28	2.20	2.00	2.18	2.10	1.90
47 X 125		3.00	2.89	2.63	2.85	2.74	2.49	2.72	2.62	2.37
47 X 150		3.59	3.46	3.14	3.41	3.28	2.98	3.25	3.13	2.83
50 X 100		2.46	2.37	2.15	2.33	2.24	2.04	2.23	2.14	1.94
50 X 125		3.06	2.95	2.68	2.91	2.80	2.54	2.78	2.67	2.43
50 X 150		3.66	3.52	3.21	3.48	3.34	3.04	3.32	3.20	2.90
38 X 89		1.82	1.79	1.69	1.69	1.65	1.55	1.59	1.55	1.44
38 X 140		3.13	3.01	2.74	2.97	2.85	2.56	2.83	2.72	2.29
38 X 100	SC4	2.34	2.25	2.04	2.21	2.13	1.93	2.11	2.03	1.84
38 X 125		2.91	2.80	2.55	2.76	2.66	2.41	2.64	2.53	2.30
38 X 150		3.48	3.35	3.05	3.30	3.18	2.89	3.16	3.04	2.76
47 X 100		2.51	2.41	2.19	2.38	2.29	2.08	2.27	2.18	1.98
47 X 125		3.12	3.00	2.73	2.96	2.85	2.59	2.83	2.72	2.47
47 X 150		3.73	3.59	3.27	3.54	3.41	3.10	3.38	3.26	2.96
50 X 100		2.56	2.46	2.24	2.42	2.33	2.12	2.32	2.23	2.02
50 X 125		3.18	3.06	2.79	3.02	2.91	2.64	2.89	2.78	2.52
50 X 150		3.80	3.66	3.34	3.61	3.48	3.16	3.45	3.32	3.02
38 X 89		2.08	2.00	1.82	1.97	1.90	1.72	1.88	1.81	1.64
38 X 140		3.25	3.13	2.85	3.09	2.97	2.70	2.95	2.84	2.57

- Provides support for dead load and imposed load of 1.0 kN/m² measured on plan or concentrated load of 0.9 kN.
- See Fig. A6.

Table **A12** **Purlins** supporting rafters to which Table A11 refers (Imposed load 1.00 kN/m²)

Maximum clear span of purlin (m). Timber of strength class **SC3** and **SC4** (see Table 1)

Dead Load (kN/m²) excluding the self weight of the purlin

Size of purlin (mm x mm)		Not more than 0.50						More than 0.50 but not more than 0.75						More than 0.75 but not more than 1.00					
		Spacing of purlins (mm)																	
		1500	1800	2100	2400	2700	3000	1500	1800	2100	2400	2700	3000	1500	1800	2100	2400	2700	3000
50 x 150	SC3	1.84																	
50 x 175		2.14	2.00	1.89				2.03	1.89					1.93	1.80				
50 x 200		2.45	2.29	2.16	2.05	1.93	1.82	2.31	2.16	2.03	1.89			2.20	2.05	1.89			
50 x 225		2.75	2.57	2.43	2.29	2.15	2.04	2.60	2.43	2.26	2.11	1.99	1.88	2.48	2.29	2.11	1.97	1.85	
63 x 150		2.00	1.87					1.89						1.88					
63 x 175		2.33	2.18	2.06	1.96	1.88	1.80	2.20	2.06	1.95	1.85			2.10	1.96	1.85			
63 x 200		2.66	2.49	2.35	2.24	2.14	2.05	2.51	2.35	2.22	2.12	2.00	1.90	2.40	2.24	2.12	1.99	1.87	
63 x 225		2.98	2.80	2.65	2.52	2.41	2.29	2.83	2.65	2.50	2.38	2.24	2.12	2.69	2.52	2.38	2.22	2.09	1.98
75 x 150		2.12	1.99	1.88				2.01	1.88					1.92					
75 x 175		2.47	2.32	2.20	2.09	2.00	1.93	2.34	2.20	2.08	1.98	1.89	1.82	2.24	2.09	1.98	1.88	1.80	
75 x 200		2.82	2.65	2.51	2.39	2.29	2.20	2.68	2.51	2.37	2.26	2.16	2.08	2.55	2.39	2.26	2.15	2.05	1.94
75 x 225		3.17	2.98	2.82	2.68	2.57	2.47	3.01	2.82	2.67	2.54	2.43	2.32	2.87	2.69	2.54	2.42	2.29	2.17
50 x 150	SC4	1.92						1.82											
50 x 175		2.24	2.10	1.98	1.89	1.80		2.12	1.98	1.87				2.02	1.89				
50 x 200		2.56	2.39	2.26	2.15	2.06	1.98	2.42	2.26	2.14	2.03	1.94	1.86	2.31	2.16	2.03	1.93	1.81	
50 x 225		2.87	2.69	2.54	2.42	2.32	2.22	2.72	2.55	2.40	2.29	2.18	2.09	2.59	2.42	2.29	2.17	2.04	1.83
63 x 150		2.09	1.95	1.85				1.98	1.85					1.88					
63 x 175		2.43	2.28	2.16	2.05	1.97	1.89	2.30	2.16	2.04	1.94	1.86		2.20	2.06	1.94	1.85		
63 x 200		2.77	2.60	2.46	2.35	2.25	2.16	2.63	2.46	2.33	2.22	2.12	2.04	2.51	2.35	2.22	2.11	2.02	1.94
63 x 225		3.12	2.92	2.77	2.64	2.52	2.43	2.95	2.77	2.62	2.49	2.39	2.29	2.82	2.64	2.49	2.37	2.27	2.18
75 x 125		1.85																	
75 x 150		2.22	2.08	1.97	1.88			2.10	1.97	1.86				2.01	1.88				
75 x 175		2.58	2.42	2.29	2.19	2.10	2.02	2.45	2.30	2.17	2.07	1.98	1.91	2.34	2.19	2.07	1.97	1.89	1.81
75 x 200		2.95	2.77	2.62	2.50	2.39	2.30	2.80	2.62	2.48	2.36	2.26	2.18	2.67	2.50	2.37	2.25	2.16	2.07
75 x 225		3.31	3.11	2.94	2.81	2.70	2.59	3.14	2.95	2.79	2.66	2.55	2.45	3.00	2.81	2.66	2.53	2.42	2.33

- Provides support for dead load and imposed load of 1.0 kN/m² measured on plan or concentrated load of 0.9 kN.
- See Fig. A6.

Table **A13** **Common or jack rafters** for roofs having a pitch more than 30° but not more than 45° with access only for purposes of maintenance or repair. Imposed loading 0.75 kN/m²

Maximum clear span of rafter (m). Timber of strength class **SC3** and **SC4** (see Table 1)

Dead load (kN/m²) excluding the self weight of the rafter

Size of rafter (mm x mm)		Not more than 0.50			More than 0.50 but not more than 0.75			More than 0.75 but not more than 1.25		
		Spacing of rafters (mm)								
		400	450	600	400	450	600	400	450	600
38 X 100	SC3	2.28	2.23	2.10	2.10	2.05	1.91	1.96	1.91	1.76
38 X 125		3.07	2.95	2.69	2.87	2.77	2.52	2.65	2.56	2.35
38 X 150		3.67	3.53	3.22	3.44	3.31	3.01	3.26	3.14	2.85
47 X 100		2.64	2.54	2.31	2.45	2.38	2.17	2.28	2.21	2.04
47 X 125		3.29	3.17	2.88	3.09	2.97	2.70	2.92	2.81	2.56
47 X 150		3.93	3.78	3.45	3.69	3.55	3.23	3.50	3.37	3.06
50 x 100		2.69	2.59	2.36	2.53	2.43	2.21	2.38	2.30	2.09
50 x 125		3.35	3.23	2.94	3.15	3.03	2.76	2.98	2.87	2.61
50 x 150		4.00	3.86	3.52	3.76	3.62	3.30	3.57	3.44	3.13
38 x 89		1.91	1.87	1.77	1.77	1.73	1.62	1.67	1.62	1.50
38 x 140		3.43	3.30	3.01	3.22	3.10	2.82	3.05	2.93	2.66
38 x 100	SC4	2.56	2.47	2.24	2.40	2.31	2.10	2.28	2.19	1.99
38 x 125		3.19	3.07	2.80	2.99	2.88	2.62	2.84	2.73	2.48
38 x 150		3.81	3.67	3.35	3.58	3.45	3.14	3.39	3.27	2.97
47 x 100		2.74	2.64	2.41	2.58	2.48	2.25	2.44	2.35	2.13
47 x 125		3.41	3.29	3.00	3.21	3.09	2.81	3.04	2.93	2.66
47 x 150		4.08	3.93	3.59	3.83	3.69	3.36	3.64	3.50	3.19
50 x 100		2.80	2.70	2.45	2.63	2.53	2.30	2.49	2.40	2.18
50 x 125		3.48	3.35	3.06	3.27	3.15	2.87	3.10	2.99	2.72
50 x 150		4.16	4.01	3.66	3.91	3.77	3.43	3.71	3.57	3.25
38 x 89		2.28	2.20	2.00	2.14	2.06	1.87	2.03	1.95	1.77
38 x 140		3.56	3.43	3.13	3.35	3.22	2.93	3.17	3.05	2.77

- Provides support for dead load and imposed load of 0.75 kN/m² measured on plan or concentrated load of 0.9 kN.
- See Fig. A6.

Table **A14** **Purlins** supporting rafters to which Table A13 refers (Imposed load 0.75 kN/m²)

Maximum clear span of purlin (m). Timber of strength class **SC3** and **SC4** (see Table 1)

Dead Load (kN/m²) excluding the self weight of the purlin

Size of purlin (mm x mm)		Not more than 0.50						More than 0.50 but not more than 0.75						More than 0.75 but not more than 1.00					
		Spacing of purlins (mm)																	
		1500	1800	2100	2400	2700	3000	1500	1800	2100	2400	2700	3000	1500	1800	2100	2400	2700	3000
50 x 150	SC3	2.02	1.89					1.89											
50 x 175		2.36	2.21	2.09	1.99	1.90	1.83	2.21	2.06	1.95	1.86			2.08	1.95	1.84			
50 x 200		2.69	2.52	2.38	2.27	2.17	2.09	2.52	2.36	2.23	2.12	2.01	1.90	2.38	2.23	2.10	1.97	1.85	
50 x 225		3.02	2.83	2.68	2.55	2.44	2.34	2.83	2.65	2.50	2.38	2.24	2.12	2.68	2.50	2.36	2.20	2.07	1.96
63 x 125		1.83																	
63 x 150		2.19	2.06	1.95	1.85			2.05	1.93	1.82				1.94	1.82				
63 x 175		2.55	2.40	2.27	2.16	2.07	1.99	2.39	2.24	2.12	2.02	1.94	1.86	2.27	2.12	2.01	1.91	1.83	
63 x 200		2.91	2.74	2.59	2.47	2.37	2.28	2.73	2.56	2.42	2.31	2.21	2.13	2.59	2.42	2.29	2.18	2.09	1.98
63 x 225		3.28	3.07	2.91	2.78	2.66	2.56	3.07	2.88	2.73	2.60	2.49	2.39	2.91	2.72	2.58	2.45	2.33	2.21
75 x 125		1.94	1.82					1.82											
75 x 150		2.33	2.19	2.07	1.97	1.89	1.82	2.18	2.05	1.94	1.85			2.07	1.94	1.83			
75 x 175		2.71	2.55	2.41	2.30	2.21	2.12	2.55	2.39	2.26	2.15	2.06	1.99	2.41	2.26	2.14	2.04	1.95	1.87
75 x 200		3.10	2.91	2.75	2.63	2.52	2.43	2.91	2.73	2.58	2.46	2.36	2.27	2.75	2.58	2.44	2.33	2.23	2.14
75 x 225		3.48	3.27	3.10	2.95	2.83	2.73	3.26	3.06	2.90	2.77	2.65	2.55	3.09	2.90	2.74	2.61	2.50	2.41
50 x 150	SC4	2.11	1.98	1.87				1.98	1.85					1.87					
50 x 175		2.46	2.31	2.18	2.08	1.99	1.91	2.31	2.16	2.04	1.94	1.86		2.18	2.04	1.93	1.83		
50 x 200		2.81	2.63	2.49	2.37	2.27	2.19	2.63	2.47	2.33	2.22	2.12	2.04	2.49	2.33	2.20	2.09	2.00	1.92
50 x 225		3.16	2.96	2.80	2.67	2.56	2.46	2.96	2.77	2.62	2.50	2.39	2.30	2.80	2.62	2.48	2.36	2.25	2.16
63 x 125		1.91																	
63 x 150		2.29	2.15	2.03	1.94	1.86		2.15	2.01	1.90	1.81			2.03	1.90	1.80			
63 x 175		2.67	2.50	2.37	2.26	2.17	2.08	2.50	2.35	2.22	2.12	2.03	1.95	2.37	2.22	2.10	2.00	1.91	1.84
63 x 200		3.04	2.86	2.71	2.58	2.47	2.38	2.86	2.68	2.54	2.42	2.31	2.23	2.70	2.53	2.40	2.28	2.19	2.10
63 x 225		3.42	3.21	3.04	2.90	2.78	2.68	3.21	3.01	2.85	2.72	2.60	2.50	3.04	2.85	2.70	2.57	2.55	2.36
75 x 125		2.03	1.90	1.80				1.90						1.80					
75 x 150		2.43	2.28	2.16	2.06	1.98	1.91	2.28	2.14	2.03	1.93	1.85		2.16	2.03	1.92	1.83		
75 x 175		2.83	2.66	2.52	2.40	2.31	2.22	2.66	2.49	2.36	2.25	2.16	2.08	2.52	2.36	2.24	2.13	2.04	1.96
75 x 200		3.23	3.03	2.88	2.74	2.63	2.54	3.03	2.85	2.70	2.57	2.47	2.37	2.88	2.70	2.55	2.43	2.33	2.24
75 x 225		3.63	3.41	3.23	3.08	2.96	2.85	3.41	3.20	3.03	2.89	2.77	2.67	3.23	3.03	2.87	2.74	2.62	2.52

- Provides support for dead load and imposed load of 0.75 kN/m² measured on plan or concentrated load of 0.9 kN.
- See Fig. A6.

Table **A15** **Common or jack rafters** for roofs having a pitch more than 30° but not more than 45° with access only for purposes of maintenance or repair. Imposed loading 1.00 kN/m²

Maximum clear span of rafter (m). Timber of strength class **SC3** and **SC4** (see Table 1)

Dead load (kN/m²) excluding the self weight of the rafter

Size of rafter (mm x mm)		Not more than 0.50			More than 0.50 but not more than 0.75			More than 0.75 but not more than 1.00		
		Spacing of rafters (mm)								
		400	450	600	400	450	600	400	450	600
38 X 100	SC3	2.28	2.23	2.03	2.10	2.05	1.91	1.96	1.91	1.76
38 X 125		2.90	2.79	2.54	2.75	2.64	2.40	2.62	2.52	2.26
38 X 150		3.47	3.34	3.04	3.29	3.16	2.87	3.13	3.01	2.69
47 X 100		2.50	2.40	2.18	2.36	2.27	2.06	2.25	2.17	1.97
47 X 125		3.11	2.99	2.72	2.94	2.83	2.58	2.81	2.70	2.45
47 X 150		3.72	3.58	3.26	3.52	3.39	3.08	3.36	3.23	2.94
50 x 100		2.55	2.45	2.23	2.41	2.32	2.11	2.30	2.21	2.01
50 x 125		3.17	3.05	2.78	3.00	2.89	2.63	2.87	2.76	2.51
50 x 150		3.79	3.65	3.33	3.59	3.46	3.15	3.43	3.30	3.00
38 x 89		1.91	1.87	1.77	1.77	1.73	1.62	1.67	1.62	1.50
38 x 140		3.24	3.12	2.84	3.07	2.95	2.68	2.93	2.82	2.52
38 x 100	SC4	2.42	2.33	2.12	2.29	2.20	2.00	2.18	2.10	1.90
38 x 125		3.02	2.90	2.64	2.86	2.75	2.50	2.72	2.62	2.38
38 x 150		3.61	3.47	3.16	3.42	3.29	2.99	3.26	3.14	2.85
47 x 100		2.60	2.50	2.27	2.46	2.36	2.15	2.34	2.25	2.05
47 x 125		3.23	3.11	2.83	3.06	2.95	2.68	2.92	2.81	2.55
47 x 150		3.86	3.72	3.39	3.66	3.52	3.21	3.49	3.36	3.06
50 x 100		2.65	2.55	2.32	2.51	2.41	2.19	2.39	2.30	2.09
50 x 125		3.30	3.17	2.89	3.12	3.01	2.73	2.98	2.87	2.61
50 x 150		3.94	3.79	3.46	3.73	3.60	3.27	3.57	3.43	3.12
38 x 89		2.16	2.08	1.89	2.04	1.96	1.78	1.95	1.87	1.70
38 x 140		3.37	3.25	2.95	3.19	3.07	2.79	3.05	2.93	2.66

- Provides support for dead load and imposed load of 1.00 kN/m² measured on plan or concentrated load of 0.9 kN.
- See Fig. A6.

Table A16 **Purlins** supporting rafters to which Table A15 refers (Imposed load 1.00 kN/m²)

Maximum clear span of purlin (m). Timber of strength class **SC3** and **SC4** (see Table 1)

		Dead Load [kN/m²] excluding the self weight of the purlin																	
		Not more than 0.50						More than 0.50 but not more than 0.75						More than 0.75 but not more than 1.00					
Size of purlin (mm x mm)		Spacing of purlins (mm)																	
		1500	1800	2100	2400	2700	3000	1500	1800	2100	2400	2700	3000	1500	1800	2100	2400	2700	3000
50 × 150	SC3	1.91						1.80											
50 × 175		2.22	2.08	1.97	1.87			2.10	1.96	1.85				2.00	1.87				
50 × 200		2.54	2.38	2.25	2.14	2.03	1.92	2.40	2.24	2.12	1.99	1.87		2.28	2.13	1.99	1.85		
50 × 225		2.85	2.67	2.53	2.40	2.27	2.15	2.70	2.52	2.38	2.22	2.09	1.98	2.56	2.40	2.22	2.07	1.95	1.84
63 × 150		2.07	1.94	1.84				1.96	1.83					1.86					
63 × 175		2.41	2.26	2.14	2.04	1.95	1.88	2.28	2.14	2.02	1.92	1.84		2.17	2.03	1.92	1.83		
63 × 200		2.76	2.58	2.44	2.33	2.23	2.14	2.61	2.44	2.31	2.20	2.10	2.00	2.48	2.32	2.19	2.09	1.97	1.86
63 × 225		3.10	2.90	2.75	2.62	2.51	2.41	2.93	2.74	2.59	2.47	2.36	2.23	2.79	2.61	2.47	2.33	2.20	2.08
75 × 125		1.84																	
75 × 150		2.20	2.07	1.96	1.86			2.08	1.95	1.85				1.98	1.86				
75 × 175		2.57	2.41	2.28	2.17	2.08	2.00	2.43	2.28	2.15	2.05	1.96	1.89	2.31	2.17	2.05	1.95	1.87	
75 × 200		2.93	2.75	2.60	2.48	2.38	2.29	2.77	2.60	2.46	2.34	2.24	2.16	2.64	2.47	2.34	2.23	2.13	2.04
75 × 225		3.29	3.09	2.92	2.79	2.67	2.57	3.12	2.92	2.76	2.63	2.52	2.43	2.97	2.78	2.63	2.50	2.40	2.28
50 × 150	SC4	1.99	1.87					1.88											
50 × 175		2.32	2.18	2.06	1.96	1.88	1.80	2.20	2.06	1.94	1.85			2.09	1.96	1.85			
50 × 200		2.65	2.49	2.35	2.24	2.14	2.06	2.51	2.35	2.22	2.11	2.02	1.94	2.39	2.23	2.11	2.00	1.91	
50 × 225		2.98	2.79	2.64	2.52	2.41	2.31	2.82	2.64	2.49	2.37	2.27	2.18	2.68	2.51	2.37	2.25	2.15	2.01
63 × 125		1.81																	
63 × 150		2.16	2.03	1.92	1.83			2.05	1.92	1.81				1.95	1.83				
63 × 175		2.52	2.36	2.24	2.13	2.04	1.97	2.39	2.24	2.11	2.01	1.93	1.85	2.27	2.13	2.01	1.91	1.83	
63 × 200		2.88	2.70	2.56	2.44	2.33	2.24	2.72	2.55	2.41	2.30	2.20	2.12	2.59	2.43	2.30	2.19	2.09	2.01
63 × 225		3.23	3.03	2.87	2.74	2.62	2.52	3.06	2.87	2.71	2.59	2.48	2.38	2.92	2.73	2.58	2.46	2.35	2.26
75 × 125		1.92						1.82											
75 × 150		2.30	2.16	2.04	1.95	1.87		2.18	2.04	1.93	1.84			2.07	1.94	1.84			
75 × 175		2.68	2.51	2.38	2.27	2.18	2.10	2.54	2.38	2.25	2.15	2.06	1.98	2.42	2.27	2.14	2.04	1.96	1.88
75 × 200		3.06	2.87	2.72	2.59	2.49	2.39	2.89	2.72	2.57	2.45	2.35	2.26	2.76	2.59	2.45	2.33	2.23	2.15

• Provides support for dead load and imposed load of 1.00 kN/m² measured on plan or concentrated load of 0.9 kN.
• See Fig. A6.

Table A17 **Joists for flat roofs** with access only for purposes of maintenance or repair. Imposed loading 0.75 kN/m²

Maximum clear span of joist (m). Timber of strength class **SC3** (see Table 1)

	Dead load [kN/m²] excluding the self weight of the joist								
	Not more than 0.50			More than 0.50 but not more than 0.75			More than 0.75 but not more than 1.00		
Size of joist (mm × mm)	Spacing of joists (mm)								
	400	450	600	400	450	600	400	450	600
38 × 97	1.74	1.72	1.67	1.67	1.64	1.58	1.61	1.58	1.51
38 × 122	2.37	2.34	2.25	2.25	2.21	2.11	2.16	2.11	2.01
38 × 147	3.02	2.97	2.85	2.85	2.80	2.66	2.72	2.66	2.51
38 × 170	3.63	3.57	3.37	3.41	3.34	3.17	3.24	3.17	2.98
38 × 195	4.30	4.23	3.86	4.03	3.94	3.63	3.81	3.72	3.45
38 × 220	4.94	4.76	4.34	4.64	4.49	4.09	4.38	4.27	3.88
47 × 97	1.92	1.90	1.84	1.84	1.81	1.74	1.77	1.74	1.65
47 × 122	2.60	2.57	2.47	2.47	2.43	2.31	2.36	2.31	2.19
47 × 147	3.30	3.25	3.12	3.12	3.06	2.90	2.96	2.90	2.74
47 × 170	3.96	3.89	3.61	3.72	3.64	3.40	3.53	3.44	3.23
47 × 195	4.68	4.53	4.13	4.37	4.28	3.89	4.14	4.04	3.70
47 × 220	5.28	5.09	4.65	4.99	4.81	4.38	4.75	4.58	4.17
50 × 97	1.97	1.95	1.89	1.89	1.86	1.78	1.81	1.78	1.70
50 × 122	2.67	2.64	2.53	2.53	2.49	2.37	2.42	2.37	2.25
50 × 147	3.39	3.34	3.19	3.19	3.13	2.97	3.04	2.97	2.80
50 × 170	4.06	3.99	3.69	3.81	3.73	3.47	3.61	3.53	3.30
50 × 195	4.79	4.62	4.22	4.48	4.36	3.97	4.23	4.13	3.78
50 × 220	5.38	5.19	4.74	5.09	4.90	4.47	4.85	4.67	4.25
63 × 97	2.19	2.16	2.09	2.09	2.06	1.97	2.01	1.97	1.87
63 × 122	2.95	2.91	2.79	2.79	2.74	2.61	2.66	2.61	2.47
63 × 147	3.72	3.66	3.44	3.50	3.43	3.25	3.33	3.26	3.07
63 × 170	4.44	4.35	3.97	4.16	4.07	3.74	3.95	3.85	3.56
63 × 195	5.14	4.96	4.54	4.86	4.69	4.28	4.61	4.47	4.07
63 × 220	5.77	5.57	5.10	5.46	5.27	4.82	5.21	5.02	4.59
75 × 122	3.17	3.12	3.00	3.00	2.94	2.80	2.86	2.80	2.65
75 × 147	3.98	3.92	3.64	3.75	3.67	3.44	3.56	3.48	3.27
75 × 170	4.74	4.58	4.19	4.44	4.33	3.96	4.21	4.11	3.77
75 × 195	5.42	5.23	4.79	5.13	4.95	4.53	4.89	4.72	4.31
75 × 220	6.07	5.87	5.38	5.76	5.56	5.09	5.50	5.30	4.85
38 × 140	2.84	2.79	2.68	2.68	2.63	2.51	2.56	2.51	2.37
38 × 184	4.01	3.94	3.64	3.76	3.68	3.43	3.56	3.48	3.25

• Provides support for dead and imposed loads of 0.75 kN/m² or concentrated load of 0.9 kN.
• See Fig. A7.

Table A18 **Joists for flat roofs** with access only for purposes of maintenance or repair. Imposed loading 0.75 kN/m²

Maximum clear span of joist (m). Timber of strength class **SC4** (see Table 1)

	Dead Load [kN/m²] excluding the self weight of the joist								
	Not more than 0.50			More than 0.50 but not more than 0.75			More than 0.75 but not more than 1.00		
Size of joist (mm × mm)	Spacing of joists (mm)								
	400	450	600	400	450	600	400	450	600
38 × 97	1.84	1.82	1.76	1.76	1.73	1.66	1.69	1.66	1.59
38 × 122	2.50	2.46	2.37	2.37	2.33	2.22	2.27	2.22	2.11
38 × 147	3.18	3.13	3.00	3.00	2.94	2.79	2.85	2.79	2.64
38 × 170	3.81	3.75	3.50	3.58	3.51	3.30	3.40	3.32	3.12
38 × 195	4.51	4.40	4.01	4.22	4.13	3.78	3.99	3.90	3.59
38 × 220	5.13	4.95	4.51	4.85	4.67	4.25	4.59	4.44	4.04
47 × 97	2.03	2.00	1.94	1.94	1.91	1.83	1.86	1.83	1.74
47 × 122	2.74	2.70	2.60	2.60	2.55	2.43	2.48	2.43	2.30
47 × 147	3.47	3.42	3.26	3.27	3.21	3.04	3.11	3.04	2.87
47 × 170	4.15	4.08	3.76	3.89	3.81	3.54	3.69	3.61	3.35
47 × 195	4.88	4.70	4.29	4.58	4.44	4.05	4.33	4.22	3.85
47 × 220	5.48	5.29	4.83	5.18	5.00	4.56	4.94	4.76	4.33
50 × 97	2.08	2.06	1.99	1.99	1.96	1.88	1.91	1.88	1.79
50 × 122	2.81	2.77	2.66	2.66	2.62	2.49	2.54	2.49	2.36
50 × 147	3.56	3.50	3.32	3.35	3.29	3.12	3.19	3.12	2.94
50 × 170	4.26	4.18	3.83	3.99	3.91	3.61	3.78	3.69	3.43
50 × 195	4.97	4.80	4.38	4.68	4.53	4.13	4.43	4.31	3.93
50 × 220	5.59	5.39	4.93	5.28	5.09	4.65	5.04	4.85	4.42
63 × 97	2.31	2.28	2.20	2.20	2.16	2.07	2.11	2.07	1.97
63 × 122	3.10	3.05	2.93	2.93	2.88	2.74	2.80	2.74	2.59
63 × 147	3.90	3.84	3.58	3.67	3.60	3.38	3.49	3.41	3.21
63 × 170	4.65	4.51	4.12	4.35	4.26	3.89	4.13	4.03	3.70
63 × 195	5.33	5.15	4.71	5.05	4.87	4.45	4.82	4.64	4.24
63 × 220	5.98	5.78	5.30	5.67	5.47	5.00	5.41	5.22	4.76
75 × 122	3.33	3.27	3.14	3.14	3.08	2.93	2.99	2.93	2.77
75 × 147	4.17	4.10	3.78	3.92	3.84	3.57	3.73	3.64	3.40
75 × 170	4.92	4.75	4.35	4.64	4.50	4.11	4.40	4.29	3.92
75 × 195	5.61	5.42	4.97	5.32	5.14	4.70	5.08	4.90	4.48
75 × 220	6.29	6.08	5.59	5.97	5.77	5.28	5.70	5.50	5.04
38 × 140	2.99	2.94	2.82	2.82	2.75	2.63	2.69	2.63	2.49
38 × 184	4.21	4.13	3.79	3.94	3.85	3.57	3.73	3.64	3.39

• Provides support for dead and imposed loads of 0.75 kN/m² or concentrated load of 0.9 kN.
• See Fig. A7.

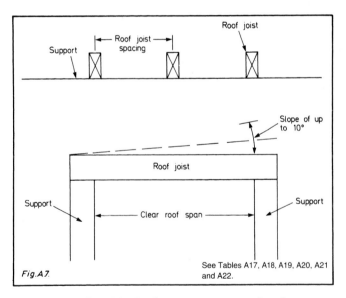

Fig. A7.

See Tables A17, A18, A19, A20, A21 and A22.

• Sizes are either BS4471 basic sawn sizes with tolerances or CLS/ALS sizes with BS 4471 tolerances;
• minimum bearing 35 mm;
• notches or holes should accord with Fig. A3.

Table A19   Joists for flat roofs with access only for purposes of maintenance or repair. Imposed loading 1.0 kN/m²

Maximum clear span of joist (m). Timber of strength class **SC3** (see Table 1)

Size of joist (mm × mm)	Dead Load [kN/m²] excluding the self weight of the joist								
	Not more than 0.50			More than 0.50 but not more than 0.75			More than 0.75 but not more than 1.00		
	Spacing of joists (mm)								
	400	450	600	400	450	600	400	450	600
38 × 97	1.74	1.72	1.67	1.67	1.64	1.58	1.61	1.58	1.51
38 × 122	2.37	2.34	2.25	2.25	2.21	2.11	2.16	2.11	2.01
38 × 147	3.02	2.97	2.75	2.85	2.80	2.61	2.72	2.66	2.49
38 × 170	3.62	3.49	3.17	3.41	3.31	3.01	3.24	3.17	2.88
38 × 195	4.15	3.99	3.63	3.94	3.79	3.45	3.77	3.63	3.29
38 × 220	4.67	4.49	4.09	4.44	4.27	3.88	4.25	4.09	3.71
47 × 97	1.92	1.90	1.84	1.84	1.81	1.74	1.77	1.74	1.65
47 × 122	2.60	2.57	2.45	2.47	2.43	2.31	2.36	2.31	2.19
47 × 147	3.30	3.24	2.95	3.12	3.06	2.80	2.96	2.90	2.68
47 × 170	3.88	3.74	3.40	3.69	3.56	3.23	3.53	3.40	3.09
47 × 195	4.44	4.27	3.89	4.23	4.07	3.70	4.05	3.89	3.54
47 × 220	4.99	4.81	4.38	4.75	4.58	4.17	4.55	4.38	3.99
50 × 97	1.97	1.95	1.89	1.89	1.86	1.78	1.81	1.78	1.70
50 × 122	2.67	2.64	2.50	2.53	2.49	2.37	2.42	2.37	2.25
50 × 147	3.39	3.31	3.01	3.19	3.13	2.86	3.04	2.97	2.73
50 × 170	3.96	3.81	3.47	3.77	3.63	3.30	3.61	3.47	3.16
50 × 195	4.53	4.36	3.97	4.31	4.15	3.78	4.13	3.97	3.61
50 × 220	5.09	4.90	4.47	4.85	4.67	4.25	4.65	4.47	4.07
63 × 97	2.19	2.16	2.09	2.09	2.06	1.97	2.01	1.97	1.87
63 × 122	2.95	2.91	2.70	2.79	2.74	2.57	2.66	2.61	2.46
63 × 147	3.70	3.56	3.25	3.50	3.39	3.09	3.33	3.25	2.95
63 × 170	4.26	4.10	3.74	4.06	3.91	3.56	3.89	3.74	3.41
63 × 195	4.86	4.69	4.28	4.64	4.47	4.07	4.45	4.28	3.90
63 × 220	5.46	5.27	4.82	5.21	5.02	4.59	5.00	4.82	4.39
75 × 122	3.17	3.12	2.86	3.00	2.94	2.72	2.86	2.80	2.60
75 × 147	3.90	3.76	3.44	3.72	3.59	3.27	3.56	3.44	3.13
75 × 170	4.49	4.33	3.96	4.29	4.13	3.77	4.11	3.96	3.61
75 × 195	5.13	4.95	4.53	4.89	4.72	4.31	4.70	4.53	4.13
75 × 220	5.76	5.56	5.09	5.50	5.30	4.85	5.28	5.09	4.65
38 × 140	2.84	2.79	2.62	2.68	2.63	2.48	2.56	2.51	2.37
38 × 184	3.92	3.77	3.43	3.73	3.58	3.25	3.56	3.43	3.11

- Provides support for dead and imposed loads of 1.0 kN/m² or concentrated load of 0.9 kN.
- See Fig. A7.

Table A20   Joists for flat roofs with access only for purposes of maintenance or repair. Imposed loading 1.0 kN/m²

Maximum clear span of joist (m). Timber of strength class **SC4** (see Table 1)

Size of joist (mm × mm)	Dead Load [kN/m²] excluding the self weight of the joist								
	Not more than 0.50			More than 0.50 but not more than 0.75			More than 0.75 but not more than 1.00		
	Spacing of joists (mm)								
	400	450	600	400	450	600	400	450	600
38 × 97	1.84	1.82	1.76	1.76	1.73	1.66	1.69	1.66	1.59
38 × 122	2.50	2.46	2.37	2.37	2.33	2.22	2.27	2.22	2.11
38 × 147	3.18	3.13	2.86	3.00	2.94	2.71	2.85	2.79	2.59
38 × 170	3.77	3.63	3.30	3.58	3.45	3.13	3.40	3.30	2.99
38 × 195	4.31	4.15	3.78	4.10	3.95	3.59	3.93	3.78	3.43
38 × 220	4.85	4.67	4.25	4.61	4.44	4.04	4.42	4.25	3.86
47 × 97	2.03	2.00	1.94	1.94	1.91	1.83	1.86	1.83	1.74
47 × 122	2.74	2.70	2.55	2.60	2.55	2.42	2.48	2.43	2.30
47 × 147	3.47	3.37	3.07	3.27	3.21	2.91	3.11	3.04	2.79
47 × 170	4.03	3.88	3.54	3.84	3.70	3.36	3.68	3.54	3.22
47 × 195	4.61	4.44	4.05	4.39	4.23	3.85	4.21	4.05	3.68
47 × 220	5.18	5.00	4.56	4.94	4.76	4.33	4.73	4.56	4.15
50 × 97	2.08	2.06	1.99	1.99	1.96	1.88	1.91	1.88	1.79
50 × 122	2.81	2.77	2.60	2.66	2.62	2.47	2.54	2.49	2.36
50 × 147	3.56	3.44	3.13	3.35	3.27	2.97	3.19	3.12	2.85
50 × 170	4.11	3.96	3.61	3.92	3.77	3.43	3.75	3.61	3.28
50 × 195	4.70	4.53	4.13	4.48	4.31	3.93	4.29	4.13	3.76
50 × 220	5.28	5.09	4.65	5.04	4.85	4.42	4.83	4.65	4.23
63 × 97	2.31	2.28	2.20	2.20	2.16	2.07	2.11	2.07	1.97
63 × 122	3.10	3.05	2.81	2.93	2.88	2.67	2.80	2.74	2.56
63 × 147	3.84	3.70	3.38	3.66	3.52	3.21	3.49	3.38	3.07
63 × 170	4.42	4.26	3.89	4.21	4.06	3.70	4.04	3.89	3.54
63 × 195	5.05	4.87	4.45	4.81	4.64	4.24	4.62	4.45	4.06
63 × 220	5.67	5.47	5.00	5.41	5.22	4.76	5.19	5.00	4.56
75 × 122	3.33	3.26	2.97	3.14	3.08	2.83	2.99	2.93	2.71
75 × 147	4.05	3.91	3.57	3.86	3.72	3.40	3.71	3.57	3.25
75 × 170	4.66	4.50	4.11	4.45	4.29	3.92	4.27	4.11	3.75
75 × 195	5.32	5.14	4.70	5.08	4.90	4.48	4.88	4.70	4.29
75 × 220	5.97	5.77	5.28	5.70	5.50	5.04	5.48	5.28	4.83
38 × 140	2.99	2.94	2.72	2.82	2.77	2.59	2.69	2.63	2.47
38 × 184	4.07	3.92	3.57	3.87	3.73	3.39	3.71	3.57	3.24

- Provides for dead and imposed loads of 1.0 kN/m² or concentrated load of 0.9 kN.
- See Fig. A7.

Table A21   Joists for flat roofs with access not limited to the purposes of maintenance or repair. Imposed loading 1.50 kN/m²

Maximum clear span of joist (m). Timber of strength class **SC3** (see Table 1)

Size of joist (mm × mm)	Dead load [kN/m²] excluding the self weight of the joist								
	Not more than 0.50			More than 0.50 but not more than 0.75			More than 0.75 but not more than 1.00		
	Spacing of joists (mm)								
	400	450	600	400	450	600	400	450	600
38 × 122	1.80	1.79	1.74	1.74	1.71	1.65	1.68	1.65	1.57
38 × 147	2.35	2.33	2.27	2.27	2.25	2.18	2.21	2.18	2.09
38 × 170	2.88	2.85	2.77	2.77	2.74	2.64	2.68	2.64	2.53
38 × 195	3.47	3.43	3.29	3.33	3.28	3.16	3.21	3.16	3.02
38 × 220	4.08	4.03	3.71	3.90	3.84	3.56	3.75	3.68	3.43
47 × 122	2.00	1.99	1.94	1.94	1.93	1.87	1.89	1.87	1.81
47 × 147	2.60	2.58	2.51	2.51	2.48	2.40	2.44	2.40	2.31
47 × 170	3.18	3.14	3.06	3.06	3.02	2.91	2.95	2.91	2.68
47 × 195	3.82	3.78	3.54	3.66	3.61	3.40	3.52	3.46	3.28
47 × 220	4.48	4.38	3.99	4.27	4.20	3.83	4.10	4.03	3.70
50 × 122	2.06	2.05	2.00	2.00	1.98	1.93	1.95	1.93	1.86
50 × 147	2.68	2.65	2.59	2.59	2.56	2.47	2.51	2.47	2.38
50 × 170	3.27	3.23	3.14	3.14	3.10	2.99	3.04	2.99	2.86
50 × 195	3.93	3.88	3.61	3.76	3.70	3.47	3.62	3.56	3.35
50 × 220	4.60	4.48	4.07	4.38	4.30	3.91	4.21	4.13	3.78
63 × 97	1.67	1.66	1.63	1.63	1.61	1.57	1.59	1.57	1.53
63 × 122	2.31	2.29	2.24	2.24	2.21	2.15	2.17	2.15	2.07
63 × 147	2.98	2.95	2.87	2.87	2.84	2.74	2.78	2.74	2.63
63 × 170	3.62	3.59	3.41	3.48	3.43	3.28	3.36	3.30	3.16
63 × 195	4.34	4.29	3.90	4.15	4.08	3.75	3.99	3.92	3.62
63 × 220	5.00	4.82	4.39	4.82	4.64	4.22	4.62	4.48	4.08
75 × 122	2.50	2.48	2.42	2.42	2.40	2.32	2.35	2.32	2.24
75 × 147	3.23	3.19	3.11	3.11	3.07	2.96	3.00	2.96	2.84
75 × 170	3.91	3.87	3.61	3.75	3.69	3.47	3.61	3.55	3.35
75 × 195	4.66	4.53	4.13	4.45	4.36	3.97	4.28	4.20	3.84
75 × 220	5.28	5.09	4.65	5.09	4.90	4.47	4.92	4.74	4.32
38 × 140	2.19	2.17	2.12	2.12	2.10	2.04	2.07	2.04	1.94
38 × 184	3.21	3.17	3.08	3.08	3.04	2.93	2.98	2.93	2.80

- Provides for dead and imposed loads of 1.5 kN/m² or concentrated load of 0.9 kN.
- See Fig. A7.

Table A22   Joists for flat roofs with access not limited to purposes of maintenance or repair. Imposed loading 1.50 kN/m²

Maximum clear span of joist (m). Timber of strength class **SC4** (see Table 1)

Size of joist (mm × mm)	Dead load [kN/m²] excluding the self weight of the joist								
	Not more than 0.50			More than 0.50 but not more than 0.75			More than 0.75 but not more than 1.00		
	Spacing of joists (mm)								
	400	450	600	400	450	600	400	450	600
38 × 122	1.91	1.90	1.86	1.86	1.84	1.79	1.81	1.79	1.73
38 × 147	2.49	2.46	2.40	2.40	2.38	2.30	2.33	2.30	2.21
38 × 170	3.04	3.01	2.93	2.93	2.89	2.79	2.83	2.79	2.67
38 × 195	3.66	3.62	3.43	3.51	3.46	3.29	3.38	3.33	3.18
38 × 220	4.30	4.25	3.86	4.10	4.04	3.71	3.94	3.87	3.58
47 × 122	2.12	2.10	2.06	2.06	2.04	1.98	2.00	1.98	1.91
47 × 147	2.75	2.73	2.66	2.66	2.62	2.54	2.57	2.54	2.44
47 × 170	3.35	3.32	3.22	3.22	3.18	3.06	3.11	3.06	2.93
47 × 195	4.03	3.98	3.68	3.85	3.80	3.54	3.71	3.64	3.42
47 × 220	4.71	4.56	4.15	4.49	4.39	3.99	4.31	4.23	3.85
50 × 122	2.19	2.17	2.12	2.12	2.10	2.04	2.06	2.04	1.97
50 × 147	2.83	2.81	2.73	2.73	2.70	2.61	2.65	2.61	2.51
50 × 170	3.45	3.41	3.28	3.31	3.27	3.15	3.20	3.15	3.01
50 × 195	4.14	4.09	3.76	3.96	3.90	3.61	3.81	3.74	3.49
50 × 220	4.83	4.65	4.23	4.61	4.47	4.07	4.42	4.32	3.93
63 × 97	1.77	1.75	1.72	1.72	1.71	1.66	1.68	1.66	1.61
63 × 122	2.44	2.42	2.36	2.36	2.34	2.27	2.30	2.27	2.18
63 × 147	3.15	3.12	3.03	3.03	2.99	2.89	2.93	2.89	2.77
63 × 170	3.82	3.78	3.54	3.66	3.61	3.41	3.53	3.47	3.29
63 × 195	4.56	4.45	4.06	4.36	4.29	3.90	4.19	4.11	3.77
63 × 220	5.19	5.00	4.56	5.00	4.82	4.39	4.84	4.66	4.24
75 × 122	2.64	2.62	2.56	2.56	2.53	2.45	2.48	2.45	2.35
75 × 147	3.40	3.36	3.25	3.27	3.23	3.11	3.16	3.11	2.98
75 × 170	4.11	4.07	3.75	3.94	3.88	3.61	3.79	3.73	3.49
75 × 195	4.79	4.70	4.29	4.67	4.53	4.13	4.49	4.38	3.99
75 × 220	5.48	5.28	4.83	5.28	5.09	4.65	5.11	4.93	4.49
38 × 140	2.32	2.30	2.25	2.25	2.22	2.16	2.19	2.15	2.08
38 × 184	3.39	3.35	3.24	3.25	3.21	3.09	3.14	3.09	2.95

- Provides for dead and imposed loads of 1.5 kN/m² or concentrated load of 0.9 kN.
- See Fig. A7.

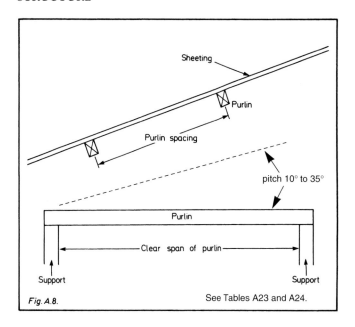

Fig. A.8.   See Tables A23 and A24.

(Figure labels: Sheeting; Purlin; Purlin spacing; pitch 10° to 35°; Purlin; Clear span of purlin; Support; Support)

- sizes are either BS 4471 basic sawn sizes with tolerances or CLS/ALS sizes with BS 4471 tolerances;
- minimum bearing 50 mm;
- notches or holes should accord with Fig. A3.

Table A24   **Purlins** supporting sheeting or decking for roofs having a pitch more than 10° but not more than 35°. Imposed loading 1.0 kN/m²

Maximum clear span of purlin (m). Timber of strength class **SC3** and **SC4** (see Table 1)

Size of purlin (mm × mm)		Dead Load [kN/m²] excluding the self weight of the purlin																
		Not more than 0.25						More than 0.25 but not more than 0.50						More than 0.50 but not more than 0.75					
		Spacing of purlins (mm)																
		900	1200	1500	1800	2100	2400	900	1200	1500	1800	2100	2400	900	1200	1500	1800	2100	2400
50 × 100	SC3	1.67	1.51	1.40	1.31	1.24	1.18	1.55	1.42	1.31	1.22	1.16	1.10	1.45	1.34	1.24	1.16	1.09	1.04
50 × 125		2.08	1.88	1.74	1.64	1.55	1.47	1.95	1.77	1.63	1.53	1.45	1.38	1.85	1.67	1.54	1.44	1.36	1.30
50 × 150		2.49	2.26	2.09	1.96	1.85	1.77	2.34	2.12	1.96	1.83	1.73	1.65	2.22	2.00	1.85	1.73	1.64	1.56
50 × 175		2.90	2.63	2.43	2.28	2.16	2.06	2.73	2.47	2.28	2.14	2.02	1.92	2.58	2.34	2.16	2.02	1.91	1.81
50 × 200		3.31	3.00	2.78	2.61	2.47	2.35	3.11	2.82	2.60	2.44	2.31	2.20	2.95	2.67	2.46	2.31	2.18	2.07
50 × 225		3.72	3.37	3.12	2.93	2.77	2.64	3.49	3.16	2.93	2.74	2.59	2.47	3.31	3.00	2.77	2.59	2.45	2.31
63 × 100		1.80	1.64	1.51	1.42	1.35	1.28	1.69	1.54	1.42	1.33	1.26	1.20	1.60	1.45	1.34	1.26	1.19	1.13
63 × 125		2.25	2.04	1.89	1.77	1.68	1.60	2.11	1.92	1.77	1.66	1.57	1.50	2.00	1.81	1.68	1.57	1.49	1.41
63 × 150		2.69	2.44	2.26	2.13	2.01	1.92	2.53	2.29	2.12	1.99	1.88	1.80	2.40	2.17	2.01	1.88	1.78	1.70
63 × 175		3.13	2.85	2.64	2.48	2.35	2.24	2.95	2.67	2.47	2.32	2.20	2.09	2.80	2.53	2.34	2.20	2.08	1.98
63 × 200		3.57	3.25	3.01	2.83	2.68	2.55	3.36	3.05	2.82	2.65	2.51	2.39	3.19	2.89	2.67	2.51	2.37	2.26
63 × 225		4.01	3.65	3.38	3.18	3.01	2.87	3.77	3.43	3.17	2.98	2.82	2.69	3.58	3.25	3.01	2.82	2.67	2.54
50 × 100	SC4	1.74	1.58	1.46	1.37	1.30	1.24	1.64	1.48	1.37	1.28	1.21	1.16	1.53	1.40	1.30	1.21	1.15	1.09
50 × 125		2.17	1.97	1.82	1.71	1.62	1.54	2.04	1.85	1.71	1.60	1.52	1.44	1.94	1.75	1.62	1.51	1.43	1.36
50 × 150		2.60	2.36	2.19	2.05	1.94	1.85	2.45	2.22	2.05	1.92	1.82	1.73	2.32	2.10	1.94	1.82	1.72	1.63
50 × 175		3.03	2.75	2.55	2.39	2.26	2.16	2.85	2.58	2.39	2.24	2.12	2.02	2.70	2.45	2.26	2.12	2.00	1.90
50 × 200		3.46	3.14	2.91	2.73	2.58	2.46	3.25	2.95	2.73	2.56	2.42	2.30	3.08	2.79	2.58	2.42	2.28	2.17
50 × 225		3.88	3.52	3.27	3.07	2.90	2.77	3.65	3.31	3.06	2.87	2.72	2.59	3.46	3.14	2.90	2.72	2.57	2.44
63 × 100		1.89	1.71	1.58	1.49	1.41	1.34	1.77	1.61	1.49	1.39	1.32	1.26	1.68	1.52	1.41	1.32	1.25	1.19
63 × 125		2.35	2.13	1.98	1.86	1.76	1.68	2.21	2.00	1.85	1.74	1.65	1.57	2.10	1.90	1.76	1.65	1.56	1.48
63 × 150		2.81	2.55	2.37	2.22	2.11	2.01	2.65	2.40	2.22	2.08	1.97	1.88	2.51	2.27	2.10	1.97	1.87	1.78
63 × 175		3.27	2.97	2.76	2.59	2.46	2.34	3.08	2.79	2.59	2.43	2.30	2.19	2.92	2.65	2.45	2.30	2.18	2.07
63 × 200		3.73	3.39	3.15	2.96	2.80	2.67	3.51	3.19	2.95	2.77	2.63	2.50	3.33	3.02	2.80	2.63	2.48	2.37
63 × 225		4.18	3.81	3.53	3.32	3.15	3.01	3.94	3.58	3.32	3.12	2.95	2.81	3.74	3.30	3.15	2.95	2.79	2.66

- Provides for dead and imposed loads measured on plan of 0.75 kN/m² or concentrated load of 0.9 kN.
- See Fig. A8.

Table A23   **Purlins** supporting sheeting or decking for roofs having a pitch more than 10° but not more than 35°. Imposed loading 0.75 kN/m².

Maximum clear span of purlin (m). Timber of strength class **SC3** and **SC4** (see Table 1)

Size of purlin (mm × mm)		Dead Load [kN/m²] excluding the self weight of the purlin																
		Not more than 0.25						More than 0.25 but not more than 0.50						More than 0.50 but not more than 0.75					
		Spacing of purlins (mm)																
		900	1200	1500	1800	2100	2400	900	1200	1500	1800	2100	2400	900	1200	1500	1800	2100	2400
50 × 100	SC3	1.68	1.63	1.51	1.42	1.34	1.28	1.55	1.48	1.40	1.31	1.24	1.18	1.45	1.37	1.31	1.22	1.16	1.10
50 × 125		2.24	2.03	1.88	1.77	1.67	1.60	2.06	1.88	1.74	1.63	1.54	1.47	1.91	1.77	1.63	1.53	1.44	1.37
50 × 150		2.68	2.44	2.26	2.12	2.01	1.91	2.49	2.26	2.09	1.96	1.85	1.76	2.34	2.12	1.96	1.83	1.73	1.65
50 × 175		3.12	2.84	2.63	2.47	2.34	2.23	2.90	2.63	2.43	2.28	2.16	2.06	2.72	2.47	2.28	2.13	2.02	1.92
50 × 200		3.56	3.24	3.00	2.82	2.67	2.55	3.31	3.00	2.78	2.60	2.46	2.35	3.11	2.81	2.60	2.44	2.30	2.19
50 × 225		4.00	3.63	3.37	3.17	3.00	2.86	3.71	3.37	3.12	2.93	2.77	2.64	3.49	3.16	2.92	2.74	2.59	2.47
63 × 100		1.87	1.77	1.64	1.54	1.46	1.39	1.72	1.64	1.51	1.42	1.34	1.28	1.60	1.52	1.42	1.33	1.26	1.20
63 × 125		2.42	2.20	2.04	1.92	1.82	1.73	2.25	2.04	1.89	1.77	1.68	1.60	2.10	1.91	1.77	1.66	1.57	1.50
63 × 150		2.90	2.63	2.44	2.30	2.18	2.08	2.69	2.44	2.26	2.12	2.01	1.92	2.53	2.29	2.12	2.00	1.88	1.79
63 × 175		3.37	3.07	2.85	2.67	2.54	2.42	3.13	2.84	2.63	2.47	2.34	2.23	2.94	2.67	2.47	2.32	2.19	2.09
63 × 200		3.84	3.50	3.25	3.05	2.89	2.76	3.57	3.24	3.01	2.82	2.67	2.55	3.36	3.05	2.82	2.65	2.51	2.39
63 × 225		4.31	3.92	3.64	3.43	3.25	3.10	4.01	3.64	3.38	3.17	3.01	2.87	3.77	3.42	3.17	2.97	2.82	2.68
50 × 100	SC4	1.79	1.71	1.58	1.48	1.40	1.34	1.64	1.57	1.46	1.37	1.30	1.23	1.53	1.45	1.37	1.28	1.21	1.15
50 × 125		2.34	2.13	1.97	1.85	1.75	1.67	2.17	1.97	1.82	1.71	1.62	1.54	2.02	1.85	1.71	1.60	1.51	1.44
50 × 150		2.80	2.55	2.36	2.22	2.10	2.00	2.60	2.36	2.18	2.05	1.94	1.85	2.44	2.21	2.05	1.92	1.81	1.73
50 × 175		3.26	2.97	2.75	2.58	2.45	2.34	3.03	2.75	2.54	2.39	2.26	2.15	2.85	2.58	2.39	2.24	2.12	2.01
50 × 200		3.72	3.38	3.14	2.95	2.79	2.67	3.45	3.13	2.90	2.73	2.58	2.46	3.25	2.94	2.72	2.55	2.42	2.30
50 × 225		4.17	3.80	3.52	3.31	3.14	3.00	3.88	3.52	3.26	3.06	2.90	2.77	3.65	3.31	3.06	2.87	2.72	2.59
63 × 100		1.99	1.84	1.71	1.61	1.52	1.45	1.81	1.71	1.58	1.49	1.41	1.34	1.69	1.60	1.48	1.39	1.32	1.26
63 × 125		2.53	2.30	2.13	2.00	1.90	1.81	2.35	2.13	1.97	1.85	1.76	1.68	2.21	2.00	1.85	1.74	1.65	1.57
63 × 150		3.02	2.75	2.55	2.40	2.28	2.17	2.81	2.55	2.37	2.22	2.10	2.01	2.64	2.40	2.22	2.08	1.97	1.88
63 × 175		3.52	3.20	2.97	2.80	2.65	2.53	3.27	2.97	2.76	2.59	2.46	2.34	3.08	2.79	2.59	2.43	2.30	2.19
63 × 200		4.01	3.65	3.39	3.19	3.03	2.89	3.73	3.39	3.14	2.95	2.80	2.67	3.51	3.19	2.95	2.77	2.62	2.50
63 × 225		4.49	4.10	3.81	3.58	3.40	3.25	4.18	3.80	3.53	3.32	3.15	3.00	3.94	3.58	3.32	3.11	2.95	2.81

- Provides for dead and imposed loads measured on plan of 0.75 kN/m² or concentrated load of 0.9 kN.
- See Fig. A8.

Ground instability

Requirement A2 of Schedule 1 imposes consideration of the possibility of reasonably foreseen landslip or subsidence.

Initially local knowledge is valuable and also walking over the surrounding area may give clues of previous landslips and subsidence. Local information may be available in council offices, libraries, museums and societies concerned with environmental matters.

Ground instability may arise from

- disused mines
- geological faults
- landslides
- unstable strata.

There is in existence a series of reviews of geological condition which have been carried out by the Minerals and Land Reclamation Division of the Directorate of Planning Services of the Department of the Environment. They identify various forms of land instability, the scale of the problems and possible remedies.

The information is contained in a number of regional reports and county maps to a scale of 1:250 000. Reports also cover the nature and causes of instability with implications for development. Methods of investigation and preventive measures are also described.

Currently the following reviews have been prepared and information regarding their availability may be obtained from DPS/2 Department of the Environment, Transport and the Regions, Eland House, Bressenden Place, London SW1E 5DU.

- Review of research into landsliding in Great Britain
- Review of mining instability in Great Britain
- Review of natural underground cavities in Great Britain
- Review of foundation conditions in Great Britain.

Existing buildings

Compliance with Part A in the event of change of use may involve a structural assessment of the existing building prior to schemes being prepared and the following documents will be helpful:

- BRE Digest 366, 'Structural appraisal of existing buildings for change of use'
- Report 'Appraisal of existing structures 1980', by the Institution of Structural Engineers. When BS codes or standards are quoted in this report the latest versions mentioned in the approved document should be used.

SECTION 1/C. THICKNESS OF WALLS

Where used

The requirement regarding thicknesses of certain walls applies only to the following:

- residential buildings up to and including three storeys
- small single-storey non-residential buildings
- small buildings which are annexes to residential buildings, such as outbuildings and garages.

Walls to comply

The walls stated in Table 3 are considered in this section.

Table 3

Residential buildings up to three storeys	Small single-storey non-residential buildings and annexes.
● compartment walls ● external walls ● internal loadbearing walls ● separating walls	● external walls ● internal loadbearing walls

The following have to be taken into account when using this part of the approved document.

- All pertinent design conditions must be satisfied.
- Relevant requirements of BS 5628 : Part 3 : 1985 must be adhered to. This BS deals with materials and components, design and workmanship.
- Guidance in the approved document is said to be based upon the worst combination of circumstances. Minor departures may be permitted when based upon:
  1. calculations, or
  2. judgement and experience.
- The following unit compressive strengths have been used:
  1. bricks 5 N/mm², 7 N/mm² and 15 N/mm²
  2. blocks 2.8 N/mm² and 7 N/mm²
  3. bricks/blocks 7 N/mm²
  BS 5628 : Part 1 : 1978 relates to unreinforced masonry. It makes use of the limit state philisophy in arriving at design methods. It emphasizes that the thickness of walls arrived at from strength considerations may not always satisfy other requirements as, for example, fire regulations and thermal insulation.

This section enables wall thicknesses to be determined easily when the following conditions are applicable.

Buildings

- when they are within the size limits, and
- when they do not exceed maximum areas given, and
- when the imposed load and wind loads are not exceeded.

Walls

- when they do not exceed the maximum height and length given, and
- when workmanship and material requirements are satisfied, and
- when the loading is not exceeded, and
- when proper end restraint is provided, and
- when the requirements for openings, recesses, overhangs and chases are not exceeded, and
- when lateral support is provided for floors and roofs, and
- when in the case of small single-storey non-residential buildings or annexes the advice in Section 1/C is observed.

Where the conditions relating to a wall or building do not fall into the scope of recommendations in this section they are outside the scope of the part and will be subject to individual design.

Thickness of walls

- Walls forming part of a bay window are excluded from the requirements.

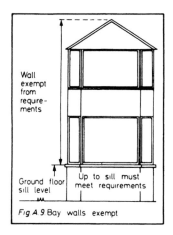

Fig A.9 Bay walls exempt

- Solid walls in coursed brickwork or blockwork.
  (a) compartment walls
  (b) external walls
  (c) separating walls.

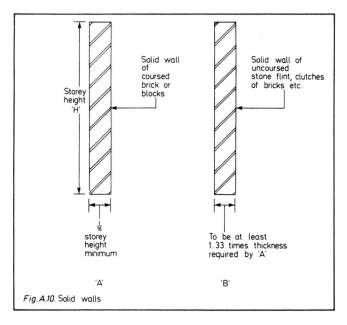

Fig. A.10. Solid walls

- Cavity walls in coursed brickwork or blockwork.
  (a) leaves not less than 90 mm;
  (b) cavities not less than 50 mm.

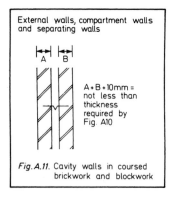

Fig. A.11. Cavity walls in coursed brickwork and blockwork

- Spacing cavity ties.
- Solid walls of uncoursed stone and flints (see Fig. A10).
- Minimum thickness of external walls, compartment walls and separating walls. (Fig. A13).

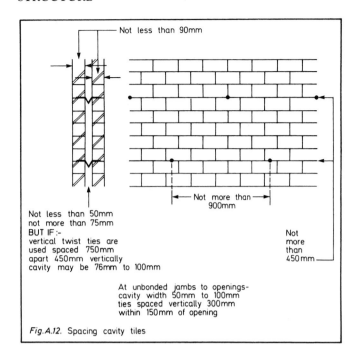

Fig.A.12. Spacing cavity tiles

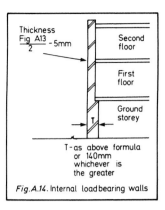

Fig.A.14. Internal loadbearing walls

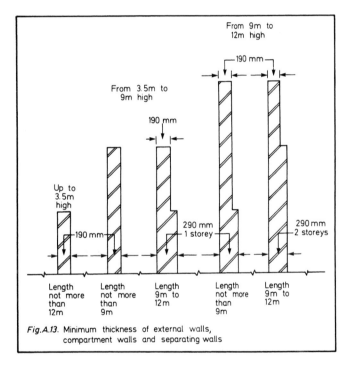

Fig.A.13. Minimum thickness of external walls, compartment walls and separating walls

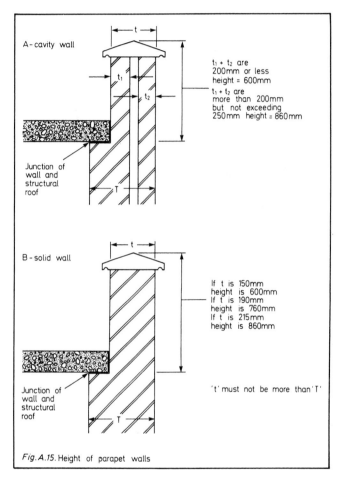

Fig.A.15. Height of parapet walls

- Whatever materials are used a wall giving vertical support to another wall should not be thinner than the wall it supports.
- Internal loadbearing walls in brickwork, but not compartment or separating walls, should have a thickness not less than shown in Fig. A14.

Example:

Thickness from Fig. A13 = 190 mm

$$\frac{190}{2} - 5\,\text{mm} = 90\,\text{mm}$$

if this were to be a wall in the lowest storey of a three-storey building carrying load from both upper storeys, required thickness would be 140 mm.

- Parapet walls. Minimum thicknesses of parapet walls are specified.

- Single leaves of external walls of small single-storey non-residential buildings and annexes may be only 90 mm thick.
- Modular bricks and blocks may be used having dimensions from BS 6750: 1986. This BS gives recommendations for the derivation of the basic sizes for the coordinating dimensions of building components and assemblies for all types of building and all forms of construction. That being so walls referred to in this part using modular bricks or blocks may be reduced by an amount not exceeding the deviation from permitted work size relating to equivalent sized bricks or blocks manufactured from the same material.

When designing residential buildings of not more than three storeys height, they are measured as indicated in Fig. A16. This method of measurement relates only to this part. Elsewhere in the regulations height is measured differently.

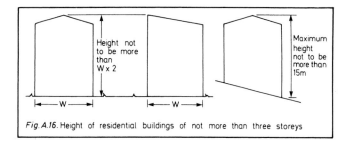

Fig. A.16. Height of residential buildings of not more than three storeys

Height of the wing to a residential building is linked with the amount of projection and width as shown in Fig. A17.

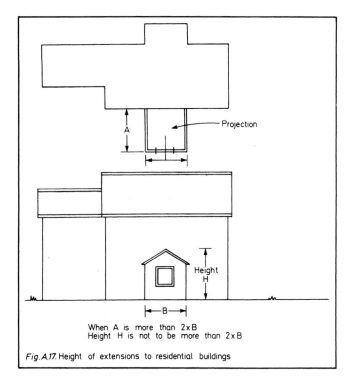

When A is more than 2 x B
Height H is not to be more than 2 x B

Fig. A.17. Height of extensions to residential buildings

There are also restrictions on the height of small single-storey non-residential buildings (Fig. A18)

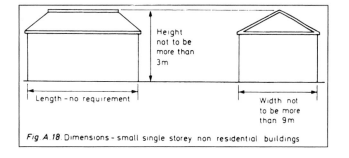

Fig. A.18. Dimensions - small single storey non residential buildings

Annexes are not to be more than 3 m high (Fig. A19).

Walls

This part does not relate to walls longer than 12 m or more than 12 m in height (see Figs A13 and A20).

Rules for the measurement of heights of walls and storeys are shown in Figs A23 and A24. The compressed strengths of bricks and blocks in internal walls of buildings up to three storeys are shown in Figs A25 and A26.

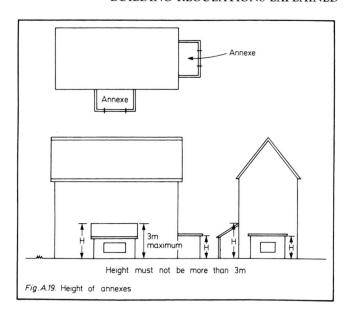

Fig. A.19. Height of annexes

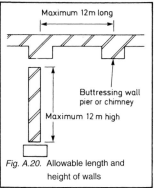

Fig. A.20. Allowable length and height of walls

Floor area

Requirements in the approved document relate to a floor area of up to $70\,m^2$ being enclosed by a structural wall on all sides. The limit in area is $30\,m^2$ when one side is not enclosed by a structural wall (Fig. A21). Above this limit one must resort to calculations.

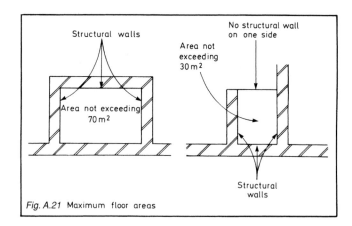

Fig. A.21 Maximum floor areas

Imposed loads

Recommendations in the approved document will prove adequate for the imposed loads as follows.

Roof:

- span not more than $12\,m = 1.00\,kN/m^2$ distributed
- span not more than $6\,m = 1.5\,kN/m^2$ distributed

Floors above ground storey:

- 2.00 kN/m² distributed.

Ceilings:

- 0.25 kN/m² distributed, or
- 0.9 kN concentrated.

Wind load

Wind loads are to be calculated in accordance with BSCP 3: Chapter V: Part 2: 1972 and the design wind speed V_s, is not to be more than 44 m per second. The factor S_3 is not to be less than 1. Most failures attributable to inadequate protection against wind load arise from installation detail, for example:

- roofs stripped off; eaves overhang lifted
- gable ends sucked out
- chimney stacks blown down
- wall cladding panels sucked off.

The formula used is as follows:

$$V_s = V \times S_1 \times S_2 \times S_3$$

when

V_s = design wind speed
V = basic wind speed for location in CP3
S_1 = topography factor
S_2 = factor for ground roughness. Also by reference to CP3 assume a building size B and make allowances for a building close to a cliff or escarpment. (BRE Digest 119 is helpful here)
S_3 = probability factor to be taken as not less than 1.

The code of practice states that wind load is a significant factor affecting the stability of buildings. Factors to take into account are the part of the country in which the building is to be placed, its degree of exposure and nearness to other buildings, all play a part in affecting the actual wind force on a building.

Basic wind speeds are recorded on a map of the UK. In some cases wind suction may be more critical than wind pressure and the code covers this contingency. Tables of modification for ground roughness give many factors in the following classifications:

1. Open country with no obstructions
2. Open country with scattered windbreaks
3. Country with many windbreaks, small towns, outskirts of large cities
4. Surface with large and frequent obstructions such as in city centres.

Recommendations for the maximum height of buildings in these various locations (see Tables 4 and 5) are given in relation to the wind speeds varying between 36 metres per second and 48 metres per second.

An indication of gust speeds which may be exceeded only once in 50 years is shown in Fig. A22. These speeds are in open country at 10 m above ground level.

Table 4   Maximum height of buildings on normal or slightly sloping sites

Basic wind speed m/s	Maximum building height in metres			
	Location			
	Unprotected sites, open countryside with no obstructions	open countryside with scattered windbreaks	Country with many windbreaks,small towns outskirts of large cities	Protected sites city centre
36	15	15	15	15
38	15	15	15	15
40	15	15	15	15
42	15	15	15	15
44	15	15	15	15
46	11	15	15	15
48	9	13	15	15

Table 5   Maximum height of buildings on steeply sloping sites, including hill, cliff and escarpment sites

Basic wind speed m/s	Maximum building height in metres			
	Location			
	Unprotected sites, open countryside with no obstructions	Open countryside with scattered windbreaks	Country with many windbreaks,small towns,outskirts of large cities	Protected sites, city centres
36	8	11	15	15
38	6	9	15	15
40	4	7.5	14	15
42	3	6	12	15
44	calculate	5	10	15
46	calculate	4	8	15
48	calculate	3	6.5	14

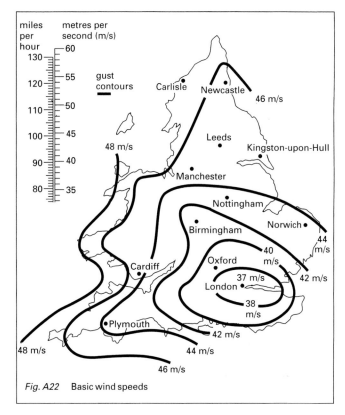

Fig. A22   Basic wind speeds

This section does not relate to walls higher, or longer, than 12 m (see Fig. A13).

BS 6399: Part 2: 1997 – Code of practice for wind loads, will be embraced by the B. Regs soon. It describes a method of calculating wind loads intended to replace the code of practice CP3: Chapter V: 1972. CP3 is not to be withdrawn immediately and there will be an overlap period during which calculations based upon either code of practice will be valid.

BS 6399 describes two ways of calculating wind loads:

- the standard method, and
- the directional method.

The standard method is similar to the CP3 method and has been designed to mirror CP3 for use by manual calculations.

The directional method provides for a more accurate result according to the availability of definite site information. It is more labour intensive, and additionally requires a computer and programmes which have been developed with the BRE.

Wind speeds are given as an hourly average, permitting a more accurate treatment of topography enabling calculations determining fatigue or dynamic exposure of a situation. Altitude is an important requirement for the calculations, and there is information regarding the effect of heights of buildings in city centres. The effect of surrounding buildings on the outskirts of towns is also examined.

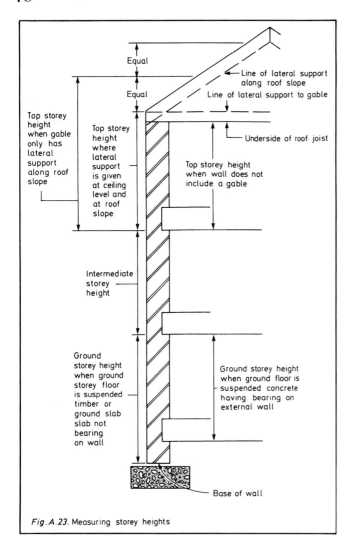

Fig.A.23. Measuring storey heights

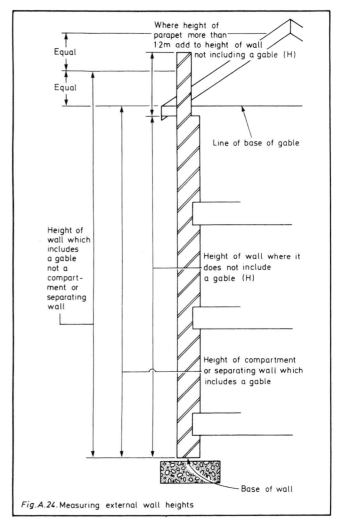

Fig.A.24. Measuring external wall heights

The directional method is more complicated than CP3 but it does provide for more flexibility without compromising safety.

CONSTRUCTION MATERIALS AND WORKMANSHIP

Design advice is also contained in BS 8103: Part 2: 1996 which is a code of practice for masonry walls in housing. Useful information in this BS relates to external walls of small buildings and garages.

• Wall ties must comply with BS 1243: 1978 which describes metal ties for cavity constructions. The standard covers wall ties manufactured from wire or strips of the following materials: low carbon steel strip, low carbon steel wire, copper, copper alloys, stainless steel wire and stainless steel strip. They are classified as butterfly, triangle and vertical-twist strip wall ties. Protective coating is specified and correct coating is important to the life of a tie.

The approved document requires that in severely exposed conditions austenitic stainless steel or other suitable non-ferrous wall ties are to be used. 'Severe exposure' is defined in BS 5628: Part 3: 1985.

The BS 5628 contains a table of classification of exposure to local wind driven rain.

The 'driving rain' index 1976 postulated that the quantity of rain falling on a wall was proportional to the quantity falling on a horizontal surface and to the local wind speed.

Recent computer analysis has been able to produce more realistic values, and the data produced has enabled a local spell index method to be prepared, described in DD 93.

For severe exposure the local spell index lies between 68 and 123. The old exposure category – Laceys Annual mean driving rain index for severe exposure was $>7\,\mathrm{m^2/s}$.

DD 93 is entitled 'Methods of assessing exposure to wind driven rain'.

• Walls of brick or block construction are to be correctly bonded and solidly constructed using mortar and the following materials:

Table 6 (from BS 3921)

Designation (mm)		Work size		
		Length (mm)	Width (mm)	Height (mm)
Bricks	225 × 112.5 × 75	215	102.5	65
Blocks	300 × 62.5 × 225	290	62.5	215
	300 × 75 × 225	290	75	215
	300 × 100 × 225	290	100	215
	300 × 150 × 225	290	150	215

BS 6649 : 1985 relates to clay bricks with modular dimensions and contain the following modular formats.

Table 7 (from BS 6649)

Designation	Work size		
(mm)	Length (mm)	Width (mm)	Height (mm)
200 × 100 × 75	190	90	65
300 × 100 × 75	288	90	65
200 × 100 × 100	190	90	90
300 × 100 × 100	288	90	90

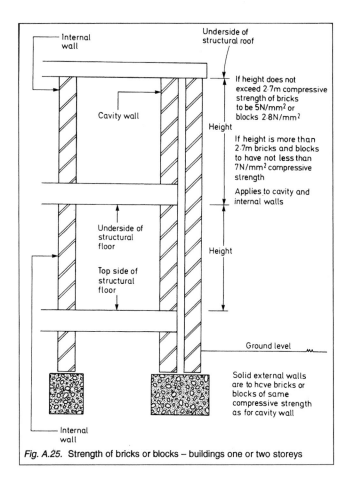

Fig. A.25. Strength of bricks or blocks – buildings one or two storeys

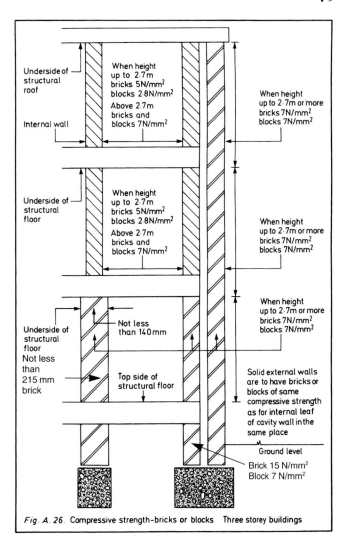

Fig. A.26. Compressive strength-bricks or blocks   Three storey buildings

(a) clay bricks or blocks in accordance with BS 3921: 1974 or BS 6649: 1985, BS 3921: 1974 is in three other sections – general; bricks and blocks for walling; and hollow blocks for structural floors and roofs.

(b) calcium silicate bricks conforming to BS 187: 1978 or BS 6649: 1985 may be used. BS 187: 1978 embraces sand lime and flint lime bricks but not lime based bricks made with calcined shale, slag or other materials made by a similar process.

BS 6649: 1985 relates to calcium silicate bricks with modular dimensions.

(c) concrete bricks or blocks in accordance with BS 6073: Part 1: 1981.

This BS specifies materials, tolerances and minimum performance levels, covering solid, cellular and hollow units not in excess of 650 mm in any work size dimension.

(d) square dressed natural stone complying with BS 5390: 1976 (1984).

This BS is concerned with good design and construction and the selection of materials for stone masonry.

• Mortar must be suitable for the position in which it is to be used and two mixes are specified in the approved document.

1. It must be mortar designation (iii) given in BS 5628: Part 1: 1978 which is relevant to three types of mortar:

   Type of mortar – proportion by volume
   cement 1: lime 1:sand 5 to 6
   masonry cement 1:sand 4 to 5
   cement 1: sand with plasticizers 5 to 6

2. A 1:1:6 mix by volume of dry materials comprising Portland Cement, lime and fine aggregate. Other materials of equal or greater strength may be used when suitable for the type of masonry and its location.

Loading on walls

Floors supported by walls are not to exceed a span of 6 m. See Fig. A27.

Vertical loading on walls is to be distributed (Fig. A28). With concrete floor slabs, precast concrete floors and timber floors following the recommendations in Part 1/B, this is assured.

There are also requirements where there are differences in the level of ground between one side of a wall and the other. Combined dead and imposed loads at the base of the wall are not to be in excess of 70 kN/m.

Walls based on these design assumptions are not to be subjected to

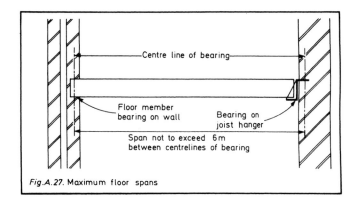

Fig.A.27. Maximum floor spans

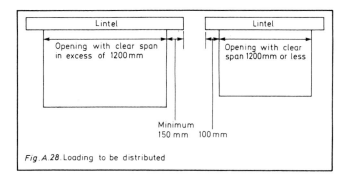

Fig.A.28. Loading to be distributed

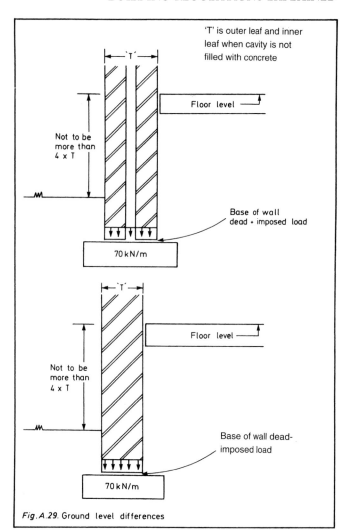

Fig.A.29. Ground level differences

lateral load other than that from wind and that provided for in Fig. A29. Walls outside these stipulations must be designed.

Openings in buttressing walls are not to impair lateral support being offered to the supported wall and each are to be bonded or securely tied together.

(Refer to Figs A23 and A24 for rules for measuring height of supported wall)

• When a buttressing wall is not itself supported as shown in Fig. A30,

   (a) the dimension 'T' must be at least half the thickness required for an external or separating wall of the same height and length minus 5 mm, or

   (b) dimension 'T' can be 75 mm if the wall is in a dwellinghouse and is not more than 10 m long × 6 m total height and

   (c) in every other instance it should not be less than 90 mm;

   (d) ends of loadbearing walls are to be bonded or securely tied in other ways for their full height. This may be to a chimney, buttressing wall or pier (Fig. A31).

The requirement does not relate to single leaf walls less than 2.5 m high and 2.5 m long when they are in a small single-storey non-residential building or annexe.

Openings and recesses

It is essential that the number, size and position of any openings or recesses do not impair the stability of a wall in which they are placed, or adjoining walls. The approved document has produced a formula to control the sizes of openings and recesses depending upon whether timber or concrete floors are supported. Adequate support is required over all openings and recesses but there are no requirements in this respect in the approved document.

The formulae are supported by Fig. A32 and Table 8.

The formula can be utilized to determine the various dimensions which must be preserved in respect of a smaller number of openings or recesses in a wall than those shown in Figs A32 and

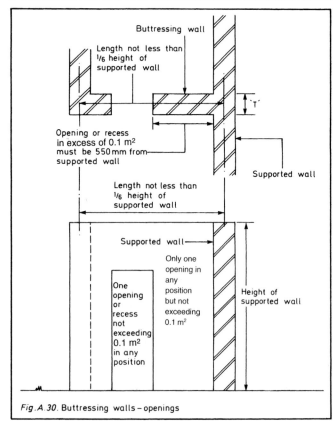

Fig.A.30. Buttressing walls – openings

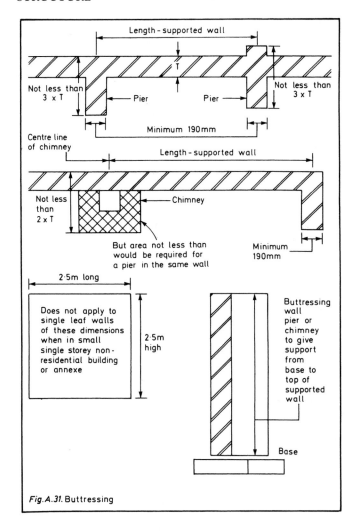

Fig. A.31. Buttressing

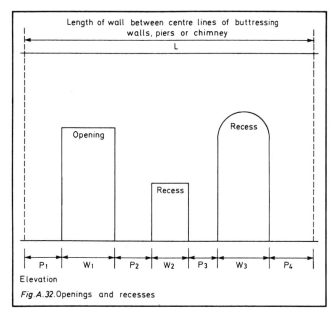

Fig. A.32. Openings and recesses

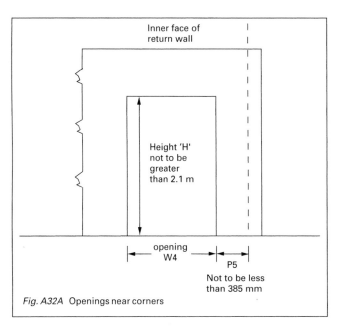

Fig. A32A Openings near corners

Table 8. Factor 'X' to determine sizes of recesses and openings

	Nature of roof span	
	Roof spans parallel to wall X factor	Timber roof spans into wall (maximum roof span 9 m) X factor
Minimum thickness inner leaf of wall 100 mm	0	0
Span of floor parallel to wall	6	6
Span of timber floor into wall:		
4.5 m maximum	6	6
6.0 maximum	6	5
Span of concrete floor into wall:		
4.5 m maximum	6	4
6.0 m maximum	6	3
Minimum thickness inner leaf of wall 90 mm	6	6
Span of timber floor into wall:		
4.5 m maximum	6	4
6.0 m maximum	6	4
Span of concrete floor into wall:		
4.5 m maximum	6	3
6.0 m maximum	5	3

A32A, and the formula does not differentiate between an opening or a recess.

1. To determine the maximum widths of openings which may be placed in a length of wall

$$W_1 + W_2 + W_3 \text{ are not to exceed } 2L/3.$$

Example: Length of wall 12 m

$$= \frac{2 \times 12}{3} = 8 \text{ m total widths of openings or recesses.}$$

2. None of the openings W_1, W_2 or W_3 is to be more than 3 m. Thus for (1) above there could be two openings or recesses each 3 m wide and one 2 m wide.

Widths of uninterrupted wall between openings and recesses are subjected to control by formula using a value 'X' taken from Table 8. The value is 6 in most instances, and it can be given the value 6 if the compressive strength of the bricks or blocks is not less than 7 N/mm^2. When the wall is a cavity wall the value 6 may be used if the loaded leaf is of bricks or blocks not less than 7 N/mm^2.

3. P_1 should not be less than W_1/X. Therefore, if W_1 is 3 m and factor 6 is used,

$$P_1 = \frac{3}{6} = 500 \text{ mm minimum.}$$

4. P_2 should not be less than $(W_1 + W_2)/X$. Therefore, if W_1 is 3 m: W_2 is 2 m and factor 6 is used,

$$P_2 = \frac{3 + 2}{6} = 833 \text{ mm minimum.}$$

5. P_3 should not be less than $(W_2 + W_3)/X$. Therefore, if W_2 is 2 m and W_3 is 2.75 m and factor 6 is used,

$$P_3 = (2 + 2.75)/6 = 4.75/6 = 791 \text{ mm minimum.}$$

6. P_4 is not to be less than W_3/X. Suppose for this example a 4.5 m span concrete floor built into the wall is 90 mm thick. W_3 is taken as 2.75 m and the factor would be 3.

$$P_4 = \frac{2.75}{3} = 916 \text{ mm minimum.}$$

7. P_5 is to be greater than or equal to W_4/X but is not to be less than 385 mm.

Chases

The number and location of chases in walls must not be such as will impair stability and there are limitations on the depth of both horizontal and vertical chases (Fig. A33). The limitation of depth means that for a 75 mm inner leaf the depth of a horizontal chase is limited to 12.8 mm. Care must be exercised when it is necessary to cut a chase in a hollow block wall.

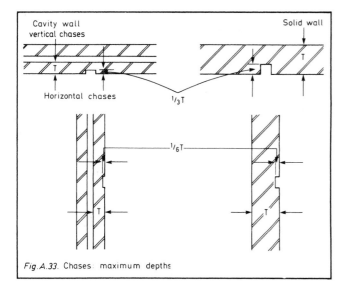

Fig.A.33. Chases: maximum depths

Overhangs

An overhang in a wall must not impair stability (Fig. A34), and the amount of overhang is normally not expected to project more than one-third the thickness of the wall or leaf in the case of a cavity wall.

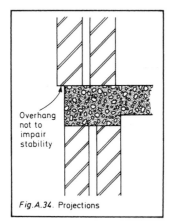

Fig.A.34. Projections

Lateral support by roofs and floors

*   The wall in each storey must reach to the full height of that storey. It is to have horizontal lateral support to limit movement at right angles to its plane.
*   Floors and roofs must achieve transfer of lateral forces from walls to buttressing walls, piers and chimneys.
*   Floors and roofs are to be secured to walls.

A solid external, compartment or separating wall of any length is to be given support by every roof having a junction with it. When length of the wall exceeds 3 m all floors having a junction with it are also required to give lateral support (see Fig. A35).

Roof and floor lateral support is also required at the top of each storey for an internal loadbearing wall of any length.

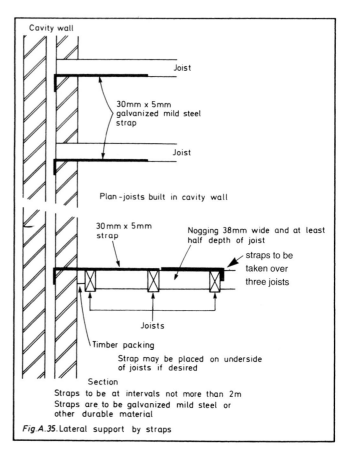

Fig.A.35. Lateral support by straps

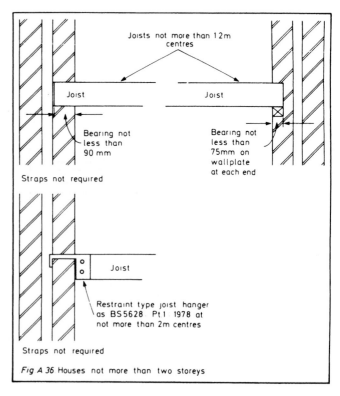

Fig A 36 Houses not more than two storeys

Straps to provide lateral support are not required with some forms of construction in houses of up to two storeys (Fig. A36), with some concrete floor construction and when floors on each side offer support to internal walls (see Figs A37 to A39).

Strapping to roofs

Vertical strapping is to be provided at eaves level (Fig. A40) and to be not less than 1 m long. There are exceptions:

- where the pitch is not less than 15°;
- the roof is slated or tiled;
- if the roof is of a construction known locally to resist gusts of wind;
- main timber members on the supporting wall are spaced at centres not greater than 1.2 m centres.

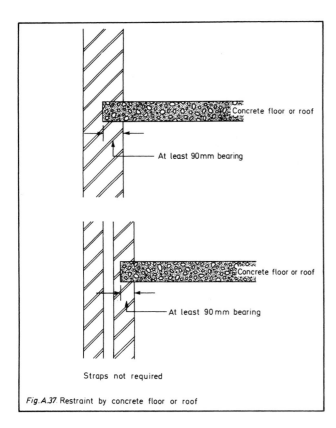

Fig.A.37. Restraint by concrete floor or roof

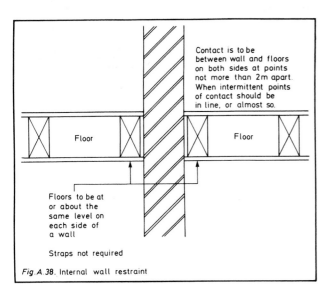

Fig.A.38. Internal wall restraint

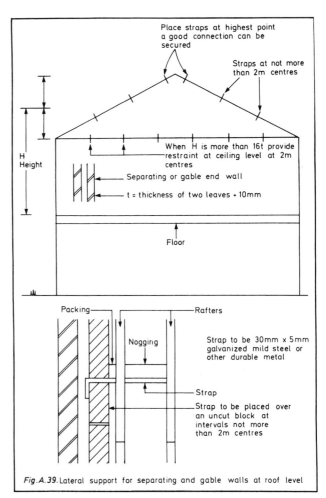

Fig.A.39. Lateral support for separating and gable walls at roof level

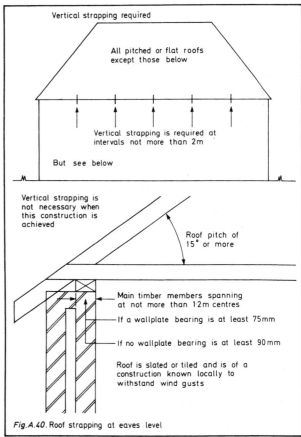

Fig.A.40. Roof strapping at eaves level

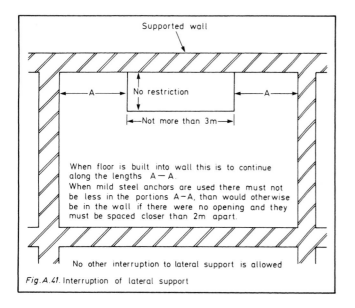

Fig. A.41. Interruption of lateral support

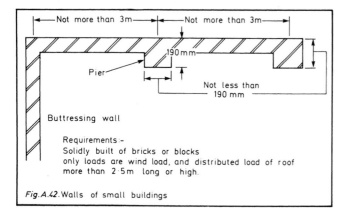

Fig. A.42. Walls of small buildings

Interruption of lateral support from floor

An opening may be provided in a floor, or roof for a stairway or other purpose, and as this will interrupt the lateral support required by an adjoining wall it must meet the limitations shown in Fig. A41.

External walls of small buildings

There is provision for single leaf external walls of small single-storey non-residential buildings and annexes; when properly bonded to buttressing walls and piers they may be built as in Fig. A42.

Floor area enclosed by the single leaf walls is not to be more than 36 m².

SECTION 1/D. PROPORTIONS FOR MASONRY CHIMNEYS

For many years the ratio of height to minimum width of a chimney was 6:1. However, the 1976 Regs, in order to reduce the likelihood of wind damage, decreased the height to width ratio to 4½:1 and this requirement is retained in the approved document.

When a chimney is not supported by ties or securely restrained by other means, the height is measured from the highest point of intersection with the roof, and it includes the chimney pot or other flue terminal (Fig. A43).

Density of masonry in chimney is to be more than 1500 kg/m³.

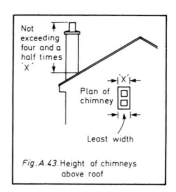

Fig. A.43. Height of chimneys above roof

SECTION 1/E. STRIP FOUNDATIONS OF PLAIN CONCRETE

In deciding upon the nature of foundations required for any particular site some investigation of subsoil must take place.

Sound, solid rock usually only requires to be levelled and cavities or fissures filled with concrete. It may be necessary to determine the need to excavate below any possible slip planes or provide retaining walls where rock outcrops on a slope.

Chalk may deteriorate under the action of water or frost and a protective layer of concrete must be laid as soon as the trench is bottomed up. Also swallow holes are liable to develop in chalk or limestone. Cavities in the rock are formed by underground water dissolving the rock away and the overburden collapses into the cavity, forming a swallow hole. As water movement causes these holes, soakaways should be kept well away from buildings.

Gravel has a high compressive resisting quality. There may be a loss of fine particles in water-bearing ground, and for this reason foundations on gravel or sand should be kept above the water table. If this is not possible, water should be drained away from the excavations.

Dense beds of sand, provided they are confined, make a good subfoundation. Sand has a high shear strength and is only slightly compressible. Avoid water from adjacent high ground washing out fine particles, as this scouring action can considerably weaken the sand and make it less stable. In some cases de-watering may be needed.

In some soils water drains away quickly. Other soils, such as fine sand and silt, tend to retain water and in freezing conditions ice lenses may form. The resulting pressure may lift the ground surface and this is known as frost heave. The ground below a building is sheltered by it and this results in the setting up of unequal pressure arising from frost heave which can cause cracking and failure of foundations. In soils of this type foundations should be at a depth of not less than 600 mm below surface level.

The strength and stability of clays are affected by water content. Clays shrink on drying and swell again when wetted. Depth of foundations should therefore be below seasonal moisture movement in the clay, with a minimum depth of 900 mm. The drying out process may be accelerated by tree roots and extend deeper. A building on shallow foundations should not be closer to a single tree than 1½ times the mature height of the tree or twice the height when near a group of trees. When a site has been cleared of trees or well established hedges there is a risk of the clay swelling and lifting the building, and time should be allowed for the clay to regain moisture. This may take several years.

The National House Building Council offers advice in Practice Note 3 regarding the houses having deep strip foundations near trees. It suggests that where houses are built near poplars, elms or willows, certain depths at various distances from the centre of a tree should be considered. See Table 9.

Table **9**   Depth of foundation near poplars, elms or willows.

Distance from tree equal to	Minimum depth of foundation
1/4 mature height	2.8m
1/2 mature height	2.3m
3/4 mature height	1.9m
mature height	1.5m

When erecting a house near other species of tree the foundation depths shown in Table 10 are appropriate. However, in all cases when building near groups of trees or trees in rows, it may be necessary to make foundations up to 50% deeper.

Table **10**   Depth of building foundations near other trees.

Distance from tree equal to	Minimum depth of foundations
1/4 mature height	2.4m
1/2 mature height	1.5m
3/4 mature height	1.2m
mature height	1.0m

Simple short bored pile foundations are frequently utilized to small buildings. See Fig. A44.

The greater care must be exercised if made-up ground is suspected. Movement and settlement is likely and in general it is advisable to use deep foundations passing through the fill, or possibly a raft foundation.

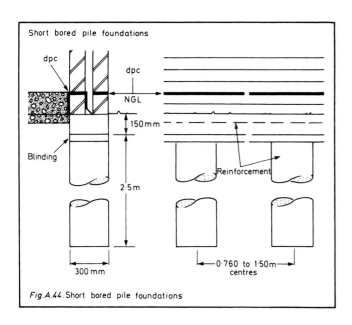

Fig.A.44. Short bored pile foundations

Adequate foundations are one of the most important aspects of building construction. A construction defect in the foundations of a structure is extremely difficult to remedy at a later date and therefore particular attention must be given to the building regulations which relate to foundations. It is significant that these regulations rely very largely upon Codes of Practice for foundations other than strip foundations.

Unless suitable precautions are taken there is frequently danger to the stability of adjoining structures. Particular attention must therefore be given to systematic planking and strutting as work proceeds.

The action of sulphate-bearing clays and ground water upon concrete is examined in detail in BRE Digest No. 31 (new series). It explains the conditions under which concrete may suffer deterioration and offers recommendations regarding measures which may be taken to safeguard foundations.

The various salts may be found in London clay and also in keuper marl and lower lias but not necessarily everywhere these soils are to be found. Oxford and Kimmeridge soils may also be affected.

The digest describes a method of taking soil samples when sulphates are thought to be present.

Trench fill foundations have been used for many years in clay subsoils but recent popularity of this method results from the labour intensive content of traditional foundations. The trench fill method uses more basic material but reduces considerably the labour content of foundation work (Fig. A45).

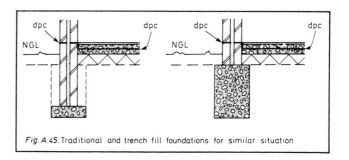

Fig.A.45. Traditional and trench fill foundations for similar situation

Exceptionally dry summers bring about many foundation problems in existing properties and emphasizes the care which should be exercised in the design of foundations, particularly on clay. The Building Research Station Geotechnics Division pioneered research on foundations for shrinkable clay over forty years ago and now has a detailed understanding of the mechanics of ground behaviour. BRE recommendations for foundation design have been set out in various papers, and in a series of digests.

Also a number of reviews of geotechnical conditions have been sponsored by the Minerals and Land Reclamation Division of the Department of the Environment. Four reviews have been prepared relating to

- Research into landsliding;
- Mining instability;
- Natural underground cavities, and
- Foundation conditions

in Great Britain.

These reviews assess the general state of knowledge regarding the various forms of land instability, giving a general picture of the scale and nature of the problems and how they may be overcome. Information regarding availability may be obtained from DPS/2, Department of the Environment, Transport and the Regions.

Design requirements

Requirements for foundation dimensions are shown in Fig. A46.

When one is satisfied regarding nature of the subsoil and other design provisions including the calculated load at the base of walls, the widths of strip foundations shown in Fig. A47 may be used.

BS 8103: Part 1: 1995 is a code of practice for stability, site investigations and ground floor slabs for housing. It contains tables enabling a 'load category' to be determined and further tables enabling fountain widths to be ascertained.

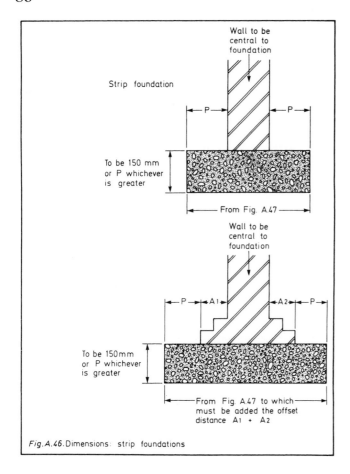

Fig.A.46. Dimensions: strip foundations

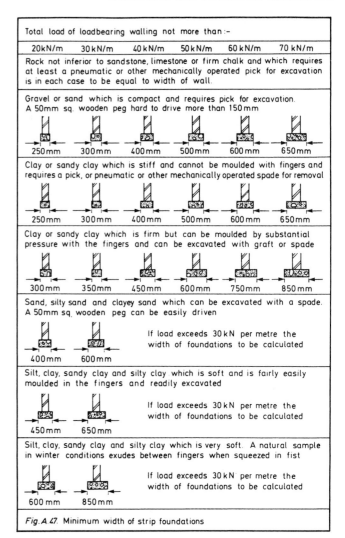

Fig.A.47. Minimum width of strip foundations

Concrete for foundations

Guidance is given for non-aggressive and also for chemically aggressive soils.

For chemically aggressive soils follow the provisions in BS 5328: Part 1: 1990, 'Guide to specifying concrete'. When working in non-aggressive soils concrete is to be composed of cement complying with BS 12 : 1989 together with fine and coarse aggregate conforming to BS 882: 1983.

- BS 12: 1989 specifies requirements for the composition, manufacture, together with the chemical and physical properties of ordinary and rapid hardening Portland cement.
- BS 882: 1983 deals with naturally occurring materials, crushed and uncrushed, used for making concrete for normal structural purposes. It contains requirements for sampling and testing the quality of aggregates and mechanical properties of coarse and fine aggregates.

The developer has a choice of concrete mixes.

- Firstly, the approved document says that the following is acceptable:

    50 kg cement
    0.1 m³ fine aggregate
    0.2 m³ coarse aggregate

- Secondly, by using grade ST1 concrete as specified in BS 5328: Part 2: 1991.

BS 5328: Part 2: 1991 describes methods of specifying mixes of concrete, both for site mixing and for supply as ready mixed. Known as a prescribed mix the developer may specify the mix proportions and is responsible for checking that the mix will give the performance required.

Grade ST1 is the recommended mix and is as follows:

Nominal maximum size of aggregate	40 mm	40 mm
(a) slump	75 mm	125 mm
cement	180 kg	200 kg
total aggregate	2010 kg	1950 kg

Nominal maximum size of aggregate	20 mm	20 mm
(b) slump	75 mm	125 mm
cement	210 kg	230 kg
total aggregate	1940 kg	1880 kg

- Thirdly, a developer may follow the very detailed advice in BS 8110: Part 1: 1985 which deals with the structural use of concrete in buildings and structures. It contains design objectives, detailing, reinforced concrete, prestressed concrete, precast and composite concrete. Concrete materials specification, construction and workmanship are included.

Section 6 is concerned with materials specification and construction.

6.1 contains recommendations on the choice and approval of materials, including:

- cement, ground granulated blast furnace slags and pulverized-fuel ashes;
- aggregates;
- water – much attention is given to proportions of water in the various types of concrete;
- admixtures.

6.2 is concerned with durability including:

- design;
- exposure conditions;
- mix proportions.

6.3 is about concrete mix specification and says that concrete should be specified, produced and tested in accordance with BS 5328: 1991. Recommendations are included for:

- compressive strength grade;
- mix parameters for durability;
- specification of constituent-materials;
- concrete to meet special circumstances, i.e.:
  - resistance to aggressive chemicals
  - resistance to thawing and freezing
  - minimum density
  - maximum density
  - very high strength
  - improved fire resistance
  - wear resistance
  - resistance to early thermal cracking
  - surface finishes
  - lightweight aggregate concrete.

A new edition – BS 8110: Part 1: 1997 – is now available.

Exposure conditions

Mild exposure conditions apply where surfaces of concrete are protected against weather or aggressive conditions.

Moderate exposure is related to concrete in contact with non-aggressive soil.

Severe exposure applies where a concrete is exposed to severe rain, alternative wetting and drying, occasional freezing or severe condensation.

Stepped foundations

When it is necessary to provide steps in foundations they should be within the dimensions shown in Fig. A48.

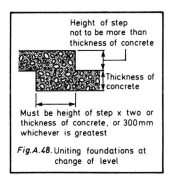

Height of step not to be more than thickness of concrete

Thickness of concrete

Must be height of step x two or thickness of concrete, or 300 mm whichever is greatest

Fig.A.48. Uniting foundations at change of level

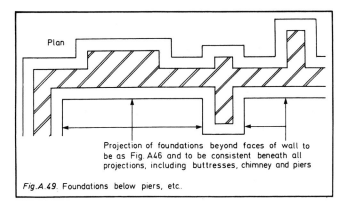

Plan

Projection of foundations beyond faces of wall to be as Fig. A46 and to be consistent beneath all projections, including buttresses, chimney and piers

Fig.A.49. Foundations below piers, etc.

Pier foundations

Foundations are to project beyond faces of piers, etc., the same distance as they project from the supported wall (Fig. A49).

Foundations near neighbour's building

The Party Wall, etc. Act 1996 requires that a notice be given to an adjoining owner when:

- the excavations are within 3 m of the neighbouring building and go deeper than the neighbour's foundations, or
- the excavations are within 6 m of the neighbouring building and will cut a line downwards at 45° from the bottom of the neighbour's foundations. (see Annexe C)

SECTION 2. EXTERNAL WALL CLADDING

This section of Part A gives guidance on the support and fixing of external wall cladding and arises from a number of failures in recent years. The actual requirements are dependent on the type and location of the cladding, and are directed at cladding which, because of its weight or height, could be a danger to people in the vicinity should it fall away from the wall. Cladding of stone or concrete is mentioned in the approved document, but some of the recommendations can be applied to curtain walling. Details are given of recommended fixings, but reliance is placed on Part C for the weather resistance of cladding.

The requirements will be met if wall cladding will

- safely support and transmit all loadings; dead; imposed and wind to the supporting structure;
- be securely fixed to the supporting structure to give both vertical support and lateral restraint;
- include provision to accommodate differential movement;
- be of durable material as will be the fixings; life of the fixings not to be less than the anticipated life of the cladding.

High standards of materials and workmanship are in particular required for the fixings, especially where they are not readily accessible for inspection and maintenance.

Loading

Wind loading should be obtained using CP3: Chapter V: Part 2: 1972, 'Wind loads'. A class A building size should be used to calculate ground roughness factor S_2 and in no case is factor S_3 to be less than 1.

Consideration is also to be given to the forces which may be applied to cladding by ladders or access cradles used when maintenance of the building is being undertaken. Likewise loads and forces which may arise from the fixing of sign-boards, antennae or overhead lines should be provided for.

CP3: Chapter V: Part 2: 1972 is now obsolete, and will eventually be withdrawn, but it may continue to be used. A new BS 6399: Part 2: 1997 is a code of practice for wind loads and contains a more complicated procedure for fixing the wind load of a building. Differences in the two codes are outlined in BRE Digest 406 – Wind actions on buildings and structures.

If cladding is to function as pedestrian guarding where the drop is 600 mm or more, the extra imposed loading required in Part K must be accommodated. This applies to stairs, ramps and vehicle barriers.

Where wall cladding is to be used on sports stadia a safety certificate may be required and the lateral pressures from crowds should be calculated on recommendations contained in the Home Office *Guide to Safety at Sports Grounds*, published in 1990.

Fixings

Strength of fixings should be ascertained

- by tests using samples of material to be anchored;

- by ascertaining inherent weaknesses likely to lower the strength of fixings;
- by checking for cracks, voids, shrinkage, etc., in masonry or concrete construction;
- by basing tests on the following:

  (i) BS 5080: Part 1: 1993, 'Structural fixings in concrete and masonry'. This specifies a method for conducting tests under axial tensile forces on structural fixings in concrete and masonry. Contents of the BS cover scope; references; definitions; base materials; installation and fixing; apparatus; procedures; presentation of results and test reports. There are three figures.

  (ii) BS 5080: Part 2: 1986, 'Structural fixings – determination of resistance to loading in shear'. This BS describes a method for conducting tests under shear force on structural fixings installed in concrete and masonry. It contains recommendations covering scope; definitions; principle; apparatus; base materials; installation and fixing; procedure; presentation of results and calculations; and test reports. There are four figures.

  (iii) Agrément Board MOAT No. 19: 1981, 'the assessment of torque expanded anchor bolts when used in dense aggregate concrete'.

- expanding bolt type fixings must have a working shear and tensile strength factor not less than

  (a) mean shear or tensile failure test load minus three times standard deviation obtained in tests $\times \frac{1}{3}$;
  (b) mean of loads causing

    (i) under direct tension a displacement of 0.1 mm;
    (ii) under direct shear a displacement of 1.0 mm.

- resin-bonded fixings must have a working shear and tensile strength factor not more than the mean shear and tensile failure test load minus three times standard deviation obtained in tests $\times \frac{1}{3}$;
- when considering resin-bonded fixings, some have a rapid loss of strength at temperatures above 50°C;
- parts of mechanical fixings and components must be lockable or otherwise secured to stop accidental slippage;
- allow for eccentricity of load on fixings to design against load spalling of anchoring material; there should be an allowance equal to 0.5 × diameter of the fixing.

Further information on fixings is available in BS 8200: 1985, 'Code of practice for design of non-loadbearing vertical enclosure buildings'. The purpose is to provide a systematic framework within which an enclosure can be designed and constructed. Contents cover general; performance criteria; design; production of components; site procedure; inspection and maintenance. There are eight appendices; 28 tables and 21 figures.

Clause 38 is referred to in the approved document. It gives details on 'attachment' with recommendations for

38.1 Support
38.2 Strength
38.3 Adjustment
38.4 Durability
38.5 Lightweight construction where there is fire requirement
38.6 Simplicity
38.7 Materials and methods.

BS 8298: 1989 is also helpful. It is a code of practice for the design and installation of natural stone cladding and lining, containing recommendations for the design, installation and maintenance of mechanically fixed facing units of natural stone as a cladding and as a lining.

Sections deal with materials and components; design; workmanship on site and maintenance. There are two appendices; 12 tables and 10 figures.

There are two clauses which are particularly appropriate.

Clause 6 relates to fixings and gives the various grades of stainless steel used for restraint and loadbearing fixings made from sheet, strip, plate and bar steel. It also gives various grades of copper and copper alloy fixings including aluminium bronze and silicon aluminium bronze, phosphor bronze and copper.

Clause 20 sets out methods of attachment and support including

20.1 Basic considerations
20.2 Loadbearing fixing
20.3 Restraint fixing
20.4 Combined loadbearing and restraint fixing
20.5 Face fixing
20.6 Soffit fixing
20.7 Block liners and other reinforcement.

Movement

Guidance on differential movement between wall cladding and the supporting structure is contained in BS 8200: 1985, 'Code of practice for design of non-loadbearing vertical enclosure buildings', and also BS 5268: Part 3: 1985, 'Code of practice for trussed rafter roofs'.

SECTION 3. RECOVERING EXISTING ROOFS

The replacement of existing roof coverings by new materials being heavier or lighter than the materials previously in situ has given rise to problems. New materials have been too heavy for the existing roof structure, and others have not been heavy enough to remain secure.

There is now a requirement, and the recovering of a roof falls within the definition of a material alteration which must meet the relevant requirement in Part A (1991 Regs, 3) and is within the definition of 'building work'.

Guidance is given on how to compare the existing and proposed loading:

- Firstly, to establish the load imposed by the new roof covering, an allowance must be made for the possible increase in load arising from water absorption. Dry plain clay tiles and concrete tiles should have 10.5% added and dry slates 0.3% to the dry weight.
- The existing roof structure should be subjected to a thorough inspection to establish

  (a) if the existing structure is able to support the new covering;
  (b) that vertical restraints are sufficient for the anticipated wind uplift which may occur if a light covering or underlay are used.

- Strengthen the roof if found to be necessary

  (a) by replacing any defective member, nails or other fixings, and vertical restraints;
  (b) provide extra trusses, rafters, braces, purlins, props; hangers, etc., as are necessary to provide adequate support for the new covering;
  (c) provide restraining straps, extra ties and fixings to walls to provide adequate resistance to wind uplift.

SECTION 4. CODES AND STANDARDS

This section refers to all building types and lists a number of British Standards and British Standard Codes of Practice which when followed will meet the requirements of A1 and A2 of Schedule 1 to the 1991 Regulations.

If alternative Codes and Standards are listed the whole of the design for the same material should normally be based on one of the codes only.

Loading

- Dead and imposed loads used in calculations are to be those contained in BS 6399: Part 1: 1984, 'Code of Practice for dead and imposed loads'.

This code was revised in 1996 and alterations relate to occupancy loadings, partition loadings and floor boards in multi-storey buildings.

Contents of the various sections include:

Dead loads
Imposed floor and ceiling loads
Reductions in total imposed loads
Imposed roof loads
Crane gantry girders
Dynamic loading (excluding wind)
Parapets and balustrades
Vehicle barriers for car parks.

For the purposes of the code a number of occupancy classes are listed and those are supported by numerous tables containing usages and loads.

The classes are:

Class	Typical structure
Residential (1)	Self-contained dwellings
Residential (2)	Apartment houses, boarding-houses, guest houses, hotels, lodging houses, residential clubs and communal areas in blocks of flats
Residential (3)	Hotels and motels
Institutional and educational	Colleges, hospitals, schools, prisons
Public assembly	Halls, auditoria, restaurants, museums, libraries, non-residential clubs, theatres, broadcasting, studios, grandstands
Offices	Offices and banks
Retail	Shops, department stores, supermarkets
Industrial	Workshops and factories
Storage	Warehouses
Vehicular	Garages, car parks, vehicle access ramps

Attention is directed to the footnote on page 3 of the code when it advises that when the snow is not uniform because of sliding, melting or the shape of the roof, the resulting load may be increased locally. BS 6399: Part 3: 1988, 'Code of practice for imposed roof loads'. Contents include scope; definitions; symbols and minimum imposed roof loads. There is advice regarding snow loads on roofs, on the ground, sliding down roofs and snow load coefficients. There are numerous diagrams. In exceptional circumstances the actual load may be greater than shown in BS 6399 and in that event the actual load is to be used. Guidance is also available in BRE Digest 290.

- Wind loading is to be calculated using the information given in CP3: Chapter V: Part 2: 1972, but in no case is the factor S_3 to be taken as less than 1.

  The code does not apply to buildings or structures that are of unusual shape or location for which special investigations may be necessary.

  S_3 is a statistical factor and takes account of the degree of security required and the period of time in years during which there will be exposure to wind.

  Calculation examples are included in Appendix C to the code. Complete contents of the code embrace procedures for calculating wind loads on structures, design wind speed, dynamic pressure of wind, pressure coefficients and force coefficients, and force coefficients for unclad structure. There are eight appendices together with many tables and figures.

  CP3: Chapter V: Part 2: 1972 is now obsolete, and will eventually be withdrawn, but it may continue to be used. A new BS 6399: Part 2: 1997 is a code of practice for wind loads and contains a more complicated procedure for fixing the wind load of a building. Differences in the two codes are outlined in BRE Digest 406 – Wind actions on buildings and structures.

Effect of wind (Figs A50 and A51)

Wind blowing against the face of a building is slowed down and exerts a pressure on that face. It is deflected around the sides and

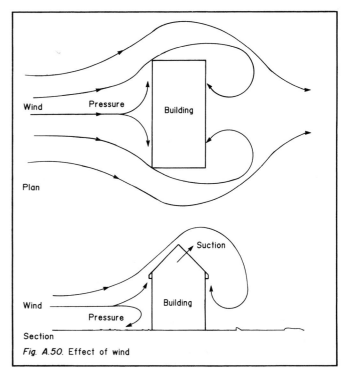

Fig. A.50. Effect of wind

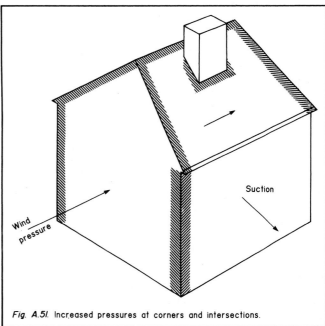

Fig. A.51. Increased pressures at corners and intersections.

over the roof. Pressure is reduced on these surfaces; an eddy is formed on the lee of the building creating suction in this area.

This explains why, when very strong gusts are experienced, glazing on one side of the building may be blown in, and on the opposite side, sucked out.

There are many fine details, such as the likelihood of negative pressure on the roofs up to 30° pitch even on the windward slope whereas roofs over 30° may develop a positive pressure on the windward side.

Design wind speed

This may be determined by using the following formula. When

V_s is design wind speed
V is basic wind speed for location of building taken from map in CP3

S_1 is topography factor
S_2 is a factor for ground roughness, building size and height above ground
S_3 is a probability factor to be not less than 1

$$V_s = V \times S_1 \times S_2 \times S_3$$

Topography (S_1)

Wind speed is not affected by height above sea level, but the following do have an influence.

(a) exposed hills which are high above surrounding ground
(b) valleys shaped in such a way that wind is funnelled.

The code recommends a factor of 1 for all normal sites.

Ground roughness (S_2)

Strong winds near to the ground are affected by the roughness of the terrain and there are four categories.

1. Open country with no shelter. (Examples are airfields, farmland without hedges, fens, moorland and coastal areas.)
2. Open country with scattered wind breaks. (This includes most farmland but does exclude well wooded parts.) The mean level of obstruction is taken as 2 m.
3. An area where there are many wind breaks. (Examples would be a small town and suburbs, wooded parkland and forest areas.) The top level of obstruction is assumed to be about 10 m.
4. Areas with large and numerous obstructions as are to be found in the centres of cities and towns. Roof height is taken to be 25 m or more.

Probability (S_3):

This relates to the probability that the basic wind speed will be exceeded in any period. Basic speeds shown on the map in the code are maximum 3 second gust speeds which are likely to be exceeded only once in 50 years.

The code gives a simple chart stating values for the factor S_3 using four probability levels and exposure periods from 2 years to 200 years. It also says that the S_3 factor should be taken as 1, except in the case of structures having either a short or an exceptionally long life.

Internal pressures

Total wind force on walls and roof depend upon the difference of pressure between inside and outside faces.

For example, open doors on the windward side wall will increase upward pressure on the roof and outward pressure on the lee faces, but the opening of doors or windows on the lee side will reduce such pressure (see Fig. A52).

The code provides information, tables and graphs covering pressure coefficients, total loads on whole structures, circular structures, frictional drag, loads during construction and a simplified method of calculation for simple structural shapes.

BRE Digests 119 and 346 are also essential reading.

If in exceptional circumstances the actual load is greater than BS 6399: Part 1: 1996 recommendations, the actual load should be used for design purposes.

Foundations

Foundations are to be in accordance with BS 8004: 1986. This is a large and very comprehensive code relating to foundations for all types of buildings, but does not cover special structures.

It is divided into the following sections:

Design of foundations
Shallow foundations
Deep foundations
Cofferdams and caissons
Dewatering
Pile foundations

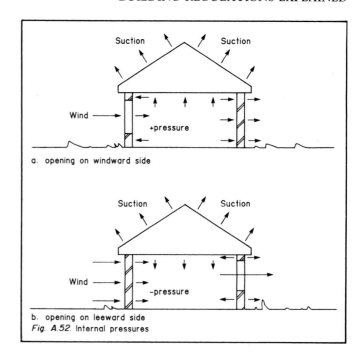

a. opening on windward side

b. opening on leeward side
Fig. A.52. Internal pressures

Excavation, demolition, shoring and underpinning
Durability of timber, metal and concrete structures.

The document is illustrated by many charts, tables and figures.

Structural work of reinforced, prestressed and plain concrete

BS 8110: Part 1: 1985 is a code of practice for design and construction of buildings and structures excluding bridges.

The aim of design is the achievement of an acceptable probability that the structure will perform satisfactorily during its projected life. All normal loads and deformations are to be sustained and the structure is to have adequate resistance to the effects of misuse and fire. There are numerous tables and illustrations.

BS 8110: Part 2: 1985 is a code of practice for special circumstances and gives guidance on ultimate limit state calculations and the derivation of partial factors of safety, with emphasis on deflections under loading and on cracking. Movement joints are considered and recommendations are made for both the provision and design of such joints.

BS 8110: Part 3: 1985 consists of design charts for singly reinforced beams, doubly reinforced beams, and rectangular columns.

The charts cannot be used to get the complete detailed design of any member but they may be used as an aid when analysing the cross section of a member at the ultimate limit state.

Structural work of composite steel and concrete construction

Compliance with CP 117: Part 1: 1965 is required. This refers to simply supported beams in buildings and refers to beams composed of rolled or built up structural steel sections. They may be with or without concrete encasement acting in conjunction with in-situ concrete slabs.

The code contains extensive information regarding the amalgamation of two structural elements of steel and reinforced concrete.

Structural work of steel

The designer is referred to several British Standards for compliance.

BS 5950: Part 1: 1985 is a code of practice for design in simple and continuous construction; hot-rolled sections.

The code embraces limit state design, properties of materials, design of structural elements, continuous construction, connections and loading tests. There are eight appendices, 39 tables and 32 figures.

BS 449: Part 2: 1969 is about the structural use of steel in buildings. It is a well-known design guide and was the basis of deemed to satisfy requirements.

Materials are covered, as are loading, design and details of construction, fabrication, erection and tests for the effectiveness of welding.

There are numerous tables. The original standard dealt with hot-rolled sections, plates and tubular shapes in several grades of steel. The addendum issued in 1975 covers the use of cold-formed steel sections from steel plate, sheet and strip not more than 6 mm thick.

BS 5950: Part 2: 1992 is a specification of hot-rolled sections. It covers materials, fabrication, assembly, erection, supports and foundations, and tolerances together with a number of tables and figures.

BS 5950: Part 4: 1982 is a code of practice for the design of floors with profiled steel sheeting. It may be used where the steel sheet and concrete act compositely in supporting loads and also where the steel sheet is used as a permanent shutter to support the concrete. This BS was updated in 1994 and included new recommendations on depth to span ratios.

BS 5950: Part 5: 1987, 'Code of practice for cold-formed sections'. Contents include limit state design properties of materials, local buckling; the design of members subject to bending; members in compression; members in tension; connections, simplified rules for commonly used members and testing. There are a number of diagrams and tables.

Structural work of aluminium

CP 118: 1969 is to be used. It deals with the use of alloys H30, N8 and H9 and also several other alloys. It contains recommendations for materials loading, design, testing, fabrication and erection. It also contains recommendations regarding protection of the material from corrosion. The design methods do not include limit state principles.

The approved document says that one of the principal or supplementary aluminium alloys designated in section 1.1 may be used. They are:

Principal alloys   H30, N8 and H9
Supplementary alloys   H20, H15, N3, N4, N5 and H17

Section 5: 3 relates to the fatigue acceptance test which applies to structures or parts subject to fluctuating loads of such size and frequencies as to render fatigue a possibility.

For the purposes of the code the approved document requires that the structure is to be classed as a safe life structure. A safe life design is one in which the structure is designed to have a fatigue life greater than its estimated service life.

Structural work of masonry

Two British Standards are listed:

BS 5628: Part 1: 1978, 'Unreinforced masonry';
BS 5628: Part 3: 1985, 'Materials and components – design and workmanship'.

BS 5628: Part 1: 1978. CP makes use of the limit state philosophy in arriving at design methods. Its recommendations cover the structural design of unreinforced masonry built in brick, block or natural stone, and also random rubble. It is clear from this code that wall thicknesses arrived at from strength considerations may not satisfy other standards for fire resistance or thermal insulation.

A new version of this BS was issued in 1992, and embraces the design of cantilever masonry walls when subject to wind loading.

BS 5628: Part 3: 1985 deals with materials and components – design and workmanship. It contains general recommendations for the design and construction of brick and block masonry. It does not cover natural stone masonry.

There are 21 appendices to the standard together with numerous figures and tables, including a map of wind zones.

Structural work of timber

There are two codes of practice applicable.

BS 5268: Part 2: 1991, which is a code of practice for permissible stress design, materials and workmanship, was updated in 1996 to reflect Eurocode DD ENV 1995-1-1.

BS 5268: Part 3: 1985 is a code of practice for trussed rafter roofs. Trussed rafters are assumed to be lightweight triangulated frameworks spaced at intervals not greater than 0.6 m. Timber members are to be fastened together on one plane by metal fasteners or plywood gussets. Ridge boards are not required but adequate bracing and satisfactory connections to the supporting structure are essential.

Contents of the code include materials, functional requirements, design, fabrication, handling, storage and erection. There are also sections on load testing, permissible spans together with a number of appendices, tables and figures.

SECTION 5. DISPROPORTIONATE COLLAPSE

A3 stipulates that a building is to be so constructed that in consequence of an accident the structure must not fail or collapse disproportionately to the cause of the damage.

This requirement extends to a building having five or more storeys, excluding a storey in the roof space when the roof slope does not exceed 70°. When the roof slope is in excess of 70° the storey is included.

Basement storeys are counted as storeys for the purpose of this requirement.

Regulations relating to disproportionate collapse first appeared in 1970 in consequence of the damage suffered by Ronan Point following a gas explosion. The explosion caused the dislodgement of a structural member and this set off a progressive collapse. The structure was required by the then amendment to be designed so that if any single member was dislodged the load it carried would be transferred to adjacent members which were to be able to sustain the additional loads.

The 1992 approved document offers advice regarding the requirements and sets a performance standard. A suitable choice of measures are taken to

• reduce, or avoid the dangers to which the building may be exposed;
• to reduce the sensitivity of the building to disproportionate collapse following an accident;
• ensure measures are taken to reduce the sensitivity of a building to disproportionate collapse in the event of local roof structure failure.

To show compliance, the following recommendations are to be adopted:

• Provide adequate horizontal and vertical ties. Several BS codes and standards contain recommendations which meet the requirements. In following recommendations given in the various codes of practice listed it is essential that advice given on ties, and on the effect of misuse or accident should be closely followed.

BS 5628: Part 1: 1978, 'Structural use of unreinforced masonry'. Clause 37.

The BS was replaced by BS 5628: Part 1: 1992, where section 4 deals with design and detailed considerations. The following clauses are of interest.

28 Consideration of slenderness of walls and columns
29 Special types of wall
30 Eccentricity in the plane of the wall
31 Eccentricity at right angles to the wall
32 Walls and columns subject to vertical loading

33 Walls subjected to shear forces
34 Concentrated loads – stresses under and close to a bearing
35 Composite action between walls and their supporting beams
36 Walls subject to lateral load and, in section 5,
37 Design for accidental damage.

BS 5950: Part 1: 1990, 'Code of practice for design in simple and continuous construction; hot-rolled sections'. The accidental loading clause 2.4.5.5 is to be chosen. Key elements are to be capable of accepting a load of not less than 34 kN/m² from any direction.

Accidental loads should be applied to members from appropriate directions together with reactions from other building components attached to the members which are subject to the same loading, but limited to the ultimate strength of these components or their connections.

BS 8110: Part 1: 1985, 'Code of practice for the design and construction of reinforced, prestressed or plain concrete'. Clause 2.2.2.2.

This clause is concerned with robustness. Structures should be planned to be not unreasonably susceptible to the effects of accidents. Situations where small areas of structure when damaged may lead to collapse of major parts of the structure are to be avoided.

(a) All buildings are to be capable of resisting notional horizontal design ultimate load applied to each floor or roof level simultaneously.
(b) All buildings are to be provided with effective horizontal ties

1. around the periphery
2. internally
3. to columns and walls.

A new edition – BS 8110: Part 1: 1997 – is now available.

BS 8110: Part 2: 1985, 'Reinforced, prestressed or plain concrete; code of practice for special circumstances'. Clause 2.6.

This clause is also concerned with robustness and contains the following paragraphs.

2.6.2.1   Design of key elements
2.6.2.2   Loads on key elements
2.6.2.3   Key elements supporting attached building components
2.6.3   Design of bridging elements
2.6.3.1   General
2.6.3.2   Walls. Length considered lost and lateral support.

• Where effective horizontal tying is provided but it is not possible to provide satisfactory vertical tying, each untied member should be considered in turn to ascertain if its position may be bridged if it were to be removed.
• Where effective vertical and horizontal tying are not possible it is necessary to consider the removal of each support member, one at a time in each storey, to establish the risk of collapse within the storey and the immediately adjacent storeys (Fig. A53). This

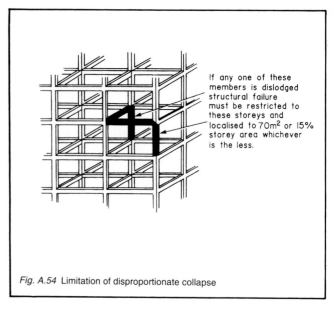

Fig. A.54 Limitation of disproportionate collapse

degree of collapse must not exceed 15% of the area of the storey or 70 m² whichever is the less Fig. (A54).
• Where the removal of a member suggests an excessive risk it is to be designed as a protected element (or key element) following recommendations contained in the appropriate BS or code mentioned above.

The 1991 Regs introduced A4 into Schedule 1. This required provision to prevent disproportionate collapse following roof failure in wide span buildings, and approved document advice became available.

However, following investigations it was found that most building failures occurred under 'normal' loading and were initiated by either loss of material strength or by construction deficiency, such as inadequate bearing or fixing of roof beams onto walls.

The D of E has said that if the buildings in question had been designed and constructed in accordance with appropriate British Standards then the reported material, construction or robustness deficiencies would not have resulted.

There has been considerable enhancement in the safety of buildings erected during the last 20 years, and requirement A4, Schedule 1, 1991 Regs has been removed by the 1994 Regs.

RELEVANT READING

British Standards and Codes of Practice

BS 12: 1996. Specification for Portland cement
BS 187: 1978. Specification for calcium silicate bricks
     Amendment 1: AMD 5427
BS 449: Part 2: 1969. The use of structural steel in building
     Amendment 1: AMD 416
                 2: AMD 523
                 3: AMD 661
                 4: AMD 1135
                 5: AMD 1787
                 6: AMD 4576
                 7: AMD 5698
                 8: AMD 6255
                 9: AMD 8859
BS 812: 1975–96. Testing aggregates
BS 882: 1992. Specification for aggregates from natural sources for concrete
BS 890: 1994. Building limes
BS 1142: 1989. Fibre building boards
BS 1199: 1976 and BS 1200: 1976. Building sands from natural sources

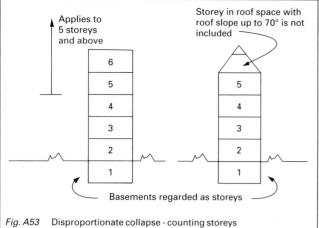

Fig. A53   Disproportionate collapse - counting storeys

BS 1243: 1978. Specification for metal ties for cavity wall construction
  Amendment 1: AMD 3651
         2: AMD 4024
BS 1297: 1987. Specification for tongue and groove softwood flooring
BS 3921: 1995. Clay bricks and blocks
BS 4471: 1987. Specification for the sizes of drawn and processed softwood
  Amendment 1: AMD 9434
BS 4721: 1981 (1986). Ready mixed building mortars
BS 4729: 1990. Shapes and dimensions of special bricks
BS 4887: Part 2: 1987. Mortar – specification for set retarding admixtures
BS 4978: 1996. Visual strength grading of softwood
BS 5224: 1995. Masonry cement
BS 5238: Part 1: 1997. Guide to specifying concrete
BS 5238: Part 2: 1997. Methods for specifying concrete
BS 5268: Part 2: 1996. Code of practice for permissible stress design, materials and workmanship
BS 5268: Part 3: 1985. Code of practice for trussed rafter roofs
  Amendment 1: AMD 5391
         2: AMD 8101
BS 5268: Part 5: 1989. Code of practice – preservation of timber
BS 5268: Part 6: Section 6.1: 1996. Dwellings not exceeding three storeys
BS 5268: Part 7: Section 7.2: 1989. Joists for flat roofs
BS 5268: Part 7: Section 7.5: 1990. Calculated basis span tables – timber
BS 5268: Part 7: Section 7.7: 1990. Purlins supporting sheeting or decking
BS 5390: 1999. Code of practice for site investigations
BS 5502: Part 22: 1987. Buildings and structures for agriculture
BS 5628: Part 1: 1992. Unreinforced masonry
BS 5628: Part 2: 1995. Structural use of reinforced and prestressed masonry
BS 5628: Part 3: 1985. Materials and components, design and workmanship
  Amendment 1: AMD 4974
BS 5642: 1978–83. Sills and copings
BS 5669: 1989. Specification for wood chipboard
BS 5950: Part 1: 1996. Code of practice for design in simple and continuous construction: hot rolled sections
BS 5950: Part 2: 1992. Specification for materials, fabrication and erection: hot rolled sections
BS 5950: Part 3: 1990. Code of practice for design of simple continuous composite beams
BS 5950: Part 4: 1994. Code of practice for design of floors with profiled steel sheeting
BS 5950: Part 5: 1987. Structural steel. Code of practice for design of cold-formed sections
  Amendment 1: AMD 5957
BS 5950: Part 8: 1990. Structural steel fire-resistant design
BS 5950: Part 9: 1994. Structural use of steelwork in buildings. Code of practice for stressed skin design.
BS 5975: 1975. Code of practice for false work
BS 6073: Part 1: 1981. Specification for precast concrete masonry units
  Amendment 1: AMD 3944
         2: AMD 4462
BS 6189: Part 1: 1990. Joist hangers for building into masonry walls
BS 6399: Part 1: 1996. Code of practice for dead and imposed loads
BS 6399: Part 2: 1997. Code of practice for wind loads
BS 6399: Part 3: 1988. Code of practice for imposed roof loads
  Amendment 1: AMD 6033
         2: AMD 9452
BS 6649: 1985. Clay and calcium silicate modular bricks
BS 6699: 1992. Specification for ground granulated blast furnace slag for use with Portland cement
BS 6750: 1986. Code of practice for modular construction in building

BS 7457: 1994. Specification for polyurethane (PU) foam system for masonry walls.
BS 8004: 1986. Code of practice for foundations
BS 8103: Part 1: 1997. Code of practice for stability, site investigation, foundations and ground floor slabs for housing
BS 8110: Part 1: 1997. Structural use of concrete – code of practice for design, materials and workmanship
BS 8110: Part 2: 1985. Structural use of concrete – code recommendations for use in special circumstances
  Amendment 1: AMD 5914
BS 8110: Part 3: 1985. Design charts for singly-reinforced beams, doubly-reinforced beams and rectangular columns
  Amendment 1: AMD 5918
BS 8118: Parts 1 and 2: 1991. Structural use of aluminium
BS 8200: 1985. Code of practice: non-loadbearing vertical enclosures for buildings
BS 8201: 1987. Code of practice for flooring: timber products and wood-based panel products
BS 8203: 1996. Code of practice for the installation of resilient floor coverings
BS 8204: Part 1: 1987. Code of practice for concrete bases and screeds to receive in situ flooring
BS 8298: 1994. Code of practice for design and installation of natural stone cladding
CP3: Chapter V: Part 2: 1972. Wind loads
  Amendment 1: AMD 4952
         2: AMD 5152
         3: AMD 5343

BRE Digests

12   Structural design in architecture
63   Soils and foundations (1)
64   Soils and foundations (2)
67   Soils and foundations (3)
141   Wind environment around tall buildings
157   Calcium silicate brickwork
161   Reinforced plastics cladding panels
163   Drying out buildings
194   The use of elm timber
196   External rendered finishes
223   Wall cladding – minimize defects
227   Estimation of thermal moisture movements and stresses. Part 1
228   Estimation of thermal moisture movements and stresses. Part 2
229   Estimation of thermal moisture movements and stresses. Part 3
235   Fixing concrete cladding panels
240   Low rise buildings on shrinkable clay (1)
241   Low rise buildings on shrinkable clay (2)
242   Low rise buildings on shrinkable clay (3)
246   Strength of brickwork and blockwork walls
251   Assessment of damage in low-rise buildings
263   Durability of steel in concrete. Part 1
264   Durability of steel in concrete. Part 2
265   Durability of steel in concrete. Part 3
268   Common defects in low-rise housing
269   Selection of natural building stones
273   Perforated clay bricks
274   Fill, Part 1: Classification of load carrying characteristics
275   Fill, Part 2: Site investigation, ground improvement and foundation design
276   Hardcore
281   Safety of large masonry walls
282   Structural appraisal of buildings with long span roofs
284   Wind loads on canopy roofs
295   Stability under wind load loose laid external roof insulation
298   Influence of trees on house foundations in clay soils
301   Corrosion of metals in wood
305   Zinc-coated steel
311   Wind scour of gravel ballast on roofs

BRE defects action sheets

   5   Trussed rafters – site storage
  18   External masonry walls, vertical joints
  25   External and separating walls – lateral restraint
  26   External and separating walls – lateral restraint
  27   External and separating walls – lateral restraint at pitched roof level
  28   External and separating walls – lateral restraint at pitched roof level
  32   Chipboard flooring – storage and installation
  44   Trussed rafter roofs – tank supports
  46   Masonry walls – chasing
  57   Suspended timber floors – joist hangers
  58   Suspended timber floors – joist hangers
  64   External walls – bricklaying in bad weather
  65   Oversites when weather bad
  74   Suspended timber ground floors
  75   External walls – brick cladding to timber frame – the need to design for differential movement
  76   External walls – brick cladding to timber frame – how to allow for movement
  81   Plasterboard ceilings – nogging and fixing – specification
  82   Plasterboard ceilings – nogging and fixing – site work
  83   Dual pitched roofs: trussed rafters, bracing and binders (Design)
  84   Dual pitched roofs: trussed rafters, bracing and binders (Site)
  96   Foundations on shrinkable clay: avoiding damage due to trees
  99   Suspended timber floors: notching and drilling of joists
 102   External masonry walls: assessing whether cracks indicate progressive movement (Design)
 110   Dual pitched roofs: trussed rafters – specification of remedial bracing (Design)
 111   Dual pitched roofs: trussed rafters – installation of remedial bracing (Site)
 112   Dual pitched roofs: trussed rafters – specification of remedial gussets
 115   External masonry cavity walls: wall ties – selection and specification
 116   External masonry cavity walls: wall ties – installation
 128   Brickwork: prevention of sulphate attack
 129   Freestanding masonry boundary walls: stability and movement
 142   Slate clad roofs: fixing of slates and battens (Design)

BRE good building guides

  20   Removing internal loadbearing walls in older buildings
  21   Joist hangers
  27   Building brickwork or blockwork retaining walls
  28   Part 1. Domestic floors: construction, insulation and damp-proofing
  29   Connecting walls and floors

BRE books and reports

Floor Loading in Warehouses. (J. S. Armitage and C. J. Judge) Ref. No. BR 109

Recommendations for the Procurement of Ground Investigation. (J. F. Uff and C. R. I. Clayton) BRE 1986. BR 94

Overcladding External Walls of Large Panel System Dwellings. (H. W. Harrison, J. H. Hunt and J. Thomson)

Structural adequacy and durability of large panel system dwellings:
   Part 1: Investigation of construction
   Part 2: Guidance on appraisal

BRE Report 1987 – Ref. BR 106. Design of normal concrete mixes

BRE Report 1987 – Ref. BR 104. A review of routine foundation practice

BRE Desk Study – Site investigation for low-rise building

BRE Report – Ref. EP2. Workmanship in masonry construction

Load tests on full scale cold formed steel roofs:
   Part 1: Sigma purlin system, Ref. No. BR 122
   Part 2: Zed purlin system, Ref. No. BR 123
   Part 3: Zeta purlin system, Ref. No. BR 124

Durability Tests for Building Stone (Ross and Butlin). Ref. BR 141

A Review of Routine Foundation Design Practice (G. H. Roscoe and R. Driscoll). Ref. BRE 1987 – BR 104

Bridges Foundations and Substructures (E. C. Hambley). BRE 1979 – SO 12

Roofs and roofing (H. W. Harrison) BRE 1996–302

Superplasticizers in concrete (A. Ali) BRE 1996–313

Review of the effect of fly ash and slag on alkali aggregate reaction in concrete (M. D. A. Thomas) BRE 1996–314

Factors affecting service life predictions of buildings (K. Bourke and H. Davis) BRE 1997–320

Timber drying manual (G. H. Pratt, K. W. Maun and A. E. Coday) BRE 1997–321

Fibre reinforced plastics for concrete reinforcement – position document BRE 1997–322

Avoiding deterioration in cement based building materials BRE 1997–324

BRE Laboratory reports

Performance of p.f.a. concrete in agressive conditions

  1.   Sulphate resistance, ref. BR 294
  2.   Marine conditions, ref. BR 295
  3.   Acidic ground waters, ref. BR 296
  4.   Carbonation, ref. BR 297

Monographs

The Designer's Guide to Wind Loading of Building Structures, Part 1. (N. J. Cook). The first of three part volume written to guide the designer of building structures. BRE

The Designer's Guide to Wind Loading of Building Structures (Supplement No. 1) (N. J. Cook) BRE.

Determination of Design Stresses in Plywood (W. T. Curry)

CHAPTER 11

Fire safety

(Approved document B)

Fire safety occupies an important and large part of building regulation requirements. Even so nothing is required to be done which is not directed at attaining a reasonable standard of health and safety for people in or about the building.

The aim can be embraced in five broad requirements:

- the structure is to be protected from collapse for a period of time;
- large buildings are to be subdivided into areas of manageable risk;
- there is to be adequate means of escape and facilities for firefighters;
- fire separation between adjacent or adjoining buildings is to be provided;
- the properties of materials and surfaces of buildings are controlled.

Approved document B 2000 edition describes processes to achieve the aims of the fire requirements in Schedule 1 of the Building Regulations 1991 (as amended).

B1 Means of warning and escape
B2 Internal fire spread on linings.
B3 Internal fire spread in structures.
B4 External fire spread.
B5 Access and facilities for the fire service.

The approved document is extensive and contains 19 sections supported by 6 appendices. The sections are as follows:

- Fire alarm and fire detection systems
- Dwellinghouses
- Flats and maisonettes
- Horizontal escape – buildings other than dwellings
- Vertical escape – buildings other than dwellings
- Provisions common to buildings other than dwellinghouses
- Wall and ceiling linings
- Loadbearing elements of structure
- Compartmentation
- Concealed cavities
- Protection of openings and fire stopping
- Car parks and shopping complexes
- External walls
- Space separation
- Roof coverings
- Fire mains
- Vehicle access
- Access for firefighters
- Venting heat and smoke from basements.

Appendices:

- Performance of materials and structures
- Fire doors
- Methods of measurement
- Purpose groups
- Definitions.
- Fire behaviour of insulating core panels
- Standards referred to

There is no definite obligation to adopt the solutions set out in the approved document, if the relevant requirement can be achieved in any other way.

PURPOSE GROUPS

The fire hazard present in a building depends largely upon the use to which a building is put, and in consequence buildings or compartments are grouped according to the purpose for which they are to be used.

There will be many instances where the designation of a building within a purpose group will not be clear cut. One building may have a number of different uses, each divided from the other by compartment walls or floors. Further, if a building or compartment is used for more than one purpose the main use will determine the group into which it is placed.

However, there may be a situation where the different use should be given a purpose group in its own right when the ancillary use is

- a flat, or a maisonette, or
- embracing more than one-fifth of the total floor area of a compartment or building, if the compartment or building exceeds $280\,m^2$, or
- storage occupying more than one-third of the total floor area of a building or compartment in purpose group 4 (shop and commercial) if the compartment or building exceeds $280\,m^2$.

Another circumstance which requires consideration is where a building may contain two or more uses that are not ancillary to each other. The approved document cites offices above shops, from which they are independent, and in this example each of the uses should be regarded as having a purpose group in its own right.

There are other examples of large buildings having a number of uses and, in these circumstances, it is necessary to consider the hazard that one part of the complex may have on another. There will be occasions when special measures must be taken to ensure that the risk is minimized.

LIST OF PURPOSE GROUPS

Residential – flats, maisonettes and dwellinghouses

In 1(a), 1(b) and 1(c), surgeries, consulting rooms, offices and other accommodation not exceeding $50\,m^2$ being part of the dwelling, and used by the occupant in a professional or business capacity, is permissible.

1(a) Flat or maisonette.
1(b) Dwellinghouse, containing a habitable storey having a floor level above 4.5 m from ground level.
1(c) Dwellinghouse not having a habitable storey with a floor level more than 4.5 m from ground level.

In 1(c) a detached garage not exceeding 40 m² in area is included in the group. A detached open carport not exceeding 40 m² is also included. A detached building consisting of a garage and open carport where neither the garage or carport exceeds 40 m².

Residential – institutional

2(a) Home for old people
Home for children
Hospital
Nursing home
Place of detention; where people sleep on the premises
School, or similar establishment having living accommodation
Treatment, care or maintenance of people who are ill, or have a mental or physical disability, or a handicap.

Residential – other than institutional

2(b) Boarding house
Hall of residence
Hostel
Hotel
Residential college
Any other residential premises not listed above.

Office

3. Offices or premises used for:
Administration
Banking
Bookkeeping
Building Society
Clerical work
Communications
Duplicating
Drawing
Editorial preparation of matter for publication
Filing
Handling money
Machine calculating
Police and fire service work
Postal, telegraph and radio
Sorting papers
Television, film or video recording including performance, not open to the public, and their control
Typing
Writing.

Shop and commercial

4. Shops or premises used for retail trade or business:
Barber or hairdresser
Food and drink for immediate consumption
Lending books or periodicals for gain
Premises where public may deliver or collect goods which may be hired, repaired, or receive other treatment (not including repair of a motor vehicle) where public may undertake repairs or other treatment
Retail by auction
Self-selection and over the counter wholesale trading.

Assembly and recreation

5. Place of assembly, entertainment or recreation including:
Amusement arcades
Art galleries
Bingo halls
Broadcasting
Casinos
Churches
Cinemas
Clinics
Concert halls
Conference centres
Crematoria
Dance halls
Dancing schools
Educational establishments
Exhibition centres
Entertainment centres
Funfairs
Gymnasium
Health centres
Law courts
Libraries open to public
Leisure centres
Menageries
Museums
Non-residential clubs
Non-residential day schools
Places of worship
Passenger stations
Public toilets
Recording and film studios when open to the public
Riding schools
Skating rinks
Sports pavilions
Sports stadia
Surgeries
Swimming pool building
Terminal for air, rail, road and sea travel
Theatres
Zoos.

Industrial

6. Factories and other premises used for:
Adapting
Altering
Breaking up
Cleaning
Generating power
Processing any article
Repairing
Slaughtering livestock
Washing.

Storage and other non-residential

7(a) Place for storing or depositing goods or materials (not including 7(b))
Any building not within purpose groups 1 to 6.
7(b) Car parks for:
Cars
Motor cycles
Passenger or light goods vehicles not exceeding 2500 kg gross.

FIRE SAFETY ENGINEERING

Fire safety engineering is considering the building as a complex system, with fire safety as one of the many aspects which can be attained by utilizing a number of equivalent methods. This alternative to the traditional approach of building regulations is described as the holistic approach. It entails designing a building to achieve acceptable standards of safety and involves an understanding of the fundamentals of fire safety. Of course, it must be proved that any particular solution achieves the overall standards which are set in Part B, but there will be circumstances when this is the only approach to achieve satisfactory fire safety.

It is inevitable that some of the rules outlined in Part B will be difficult to apply to existing buildings. This will be most apparent when work is undertaken on a building listed as being of architectural or historic interest, if some feature which would be damaged or removed is the very feature leading to the historical or architectural importance of the building. Fire safety engineering may be able, overall, to achieve the desired standards.

Important factors to be considered are:

- the calculated risk of a fire occurring and the most likely sources;
- the calculated degree of severity should a fire occur;
- the ability of the structure to withstand and resist extension of a fire from source, and the spread of smoke;
- the degree of danger to people in the building.

This entails consideration of an extensive variety of measures for inclusion in a building and the degree of importance of each measure, leading to the vital balance which may prove to be adequate in design and in fire.

- The measures taken to prevent fire. Fire arises from natural phenomena (as from lightning), carelessness, failure of electrical wiring and appliances and deliberate fire raising. Matches, smokers' materials and cooking appliances figure large in the sources of fire. Building materials and finishes are vital in measures taken to prevent fire spread, and the approved document gives guidance in this respect.
- Early warning. This is preferably by an automatic detector and warning system. They can be ionization or optical detector systems, heat detector or infrared sensors.
- The standard of means of escape must not be less than that described in the approved document.
- The control of smoke and preventing its movement about the building using doors, pressurization, venting or curtains, as some of the techniques available.
- Control of the speed a fire develops means that the availability of 'fuel' – the contents of a building, which are readily combustible – must be examined.
- The structure must be sufficiently resistant to the effects of fire.
- Steps should be taken to ensure that the fire is contained in the smallest possible part of the building by protecting structural elements, by devising compartmentation, and envelope protection of external walls and roof.
- Fire separation involving parts of buildings and between buildings achieved by using space and restrictions on openings in walls.
- Measures to secure control or extinguishment of a fire. In fire engineering this involves the use of auto suppression systems, of which sprinklers are well known, and they may be divided into wet, dry, alternating, recycling or pre-action methods. In specialist situations drenchers or deluge systems may justify consideration. Carbon dioxide or halons may also be considered. Compartmentation is designed to control the spread of fire within predetermined limits.
- Provisions to assist the fire service. For the first time an approved document sets out some requirements to assist the fire services and it is most important that the firefighters can get the appliances in the right place to enable them to perform adequately, and enter the building through safe bridgeheads.
- Are there any powers in legislation and statutory instruments requiring staff training and actions to be taken in the prevention of fire, and also if a fire should occur? In appropriate premises the Fire Precautions Act 1971 may be available, or in other premises may require licensing or registration by various bodies.
- If there are available any measures for ensuring continuing control of fire protection measures, these are also to be heeded in deciding which of the above should be incorporated, and receive more or less attention in a scheme for fire protection.

Additionally, extensive advice on fire safety is in BS 5588: Part 10: 1991, 'Code of practice for shopping complexes'.

Insurance

There will be occasions when the insurance industry, whose approach is probably wider than building control, will require higher standards and link those higher standards to the value of premiums, or even cover. The industry publishes a design guide available from the Loss Prevention Council.

Material alteration

There is currently no provision for continuing control of regulatory requirements, but when alterations are undertaken on an existing building they must not cause it to be less satisfactory in relation to Part B than it was before the work was carried out.

In this respect, reference may be made to

- Regulation 3 which defines building work, and
- Regulation 4 which sets out requirements for building work.

Performance

Satisfactory performance of components, structure, systems and products is dependent upon skilled site installation, testing and maintenance.

Independent schemes are in operation for the certification of installers and maintenance firms providing recognized standards of workmanship. Accredited product conformity certification schemes also provide for the supply of products or structures which satisfy the performance standards.

MEANS OF WARNING AND ESCAPE

This requirement is to enable provision to be made for people to escape to a place of safety by an adequate number of escape routes in suitable locations. During the process of escape people are to be protected from the effects of fire, smoke and the gases of combustion.

B1 requires the building to be designed and constructed with means of early warning of fire and means of escape from fire, to a place of safety outside the building.

The escape routes are to be capable of use in a safe manner at all times.

This requirement extends to all buildings, except any prison provided under section 33 of the Prison Act 1952.

Performance

The Secretary of State takes the view that the performance requirements of B1 can be met if the following standards are achieved:

- there are an adequate number of escape routes
- the escape routes are wide enough to accommodate those likely to use them
- the escape routes are suitably located
- where necessary the routes are to be protected from fire
- the routes are to be illuminated
- the exits are to be signed
- there are proper facilities to limit access of smoke to the escape route, or measures to limit a fire and remove smoke.
- the provision of early warning of fire.

In making a decision regarding the extent of these measures, regard may be had to the use of the building, its size and its height.

Other legislation

Probably the most well-known in building control circles is the Fire Precautions Act 1971, but there is quite a volume of legislation relating to numerous activities involving buildings. For example: animal boarding kennels; caravan sites; children's homes; storage of explosives; firework factories; houses in multiple occupation; homes for the elderly; homes for disabled people; nurseries; storage for petroleum; pipelines; schools and sports grounds.

- The Fire Precautions Act 1971, has been amended several times but remains an Act to make provision for the protection of persons from fire risks.

The Act applies to buildings which have been designated by Order. The latest designations include factories, shops, offices and railway premises. There are exemptions, and provisions dealing with instances where a building is divided into two or more parts and there are separate ownerships.

The Fire Precautions (Workplace) Regulations 1997 adds a 'workplace' as a designated use.

The owner of a designated building must obtain a fire certificate from the fire authority, and ensure that:

1. the means of escape in case of fire can be used safely and effectively at all times;
2. firefighting equipment is maintained in efficient working order;
3. persons employed in the premises receive instruction and training in what to do in case of fire.

- The Fire Precautions (Factories, Offices, Shops and Railway Premises) Order 1989 provides for exemption from the requirement to have a certificate when not more than 20 people will be employed and in which not more than 10 people will be employed above the ground floor. The exemption does not apply when two or more of the designated uses exist in the same building, and the aggregate number of employees is in excess of the above limits. It also does not apply if explosives or highly inflammable materials are stored, or used. The fire authority is given powers to exempt other premises limited to basements, ground floors and first floors.
- The Fire Precautions (Hotels and Boarding Houses) Order 1972 requires a fire certificate for all hotels and boarding houses when sleeping accommodation is provided for more than six people.
- The Housing (Means of Escape from Fire in Houses in Multiple Occupation) Order 1981 was made under the Housing Act 1980. It requires houses of three or more storeys (including a basement) in which the aggregate area of all floors exceeds $500\,m^2$. The area of stairways and basements are included in the measurements. Adequate means of escape must be provided.
- The Registration Homes Act 1984 and The Nursing Homes and Mental Homes Regulations 1984 require registration and adequate fire precautions, including facilities for the excavation of patients and staff in the event of a fire. The Residential Care Homes Regulations 1984 and 1986 require consultation with the fire authority regarding fire precautions and means of escape.
- The Education (School Premises) Regulations 1981 has requirements for means of escape in case of fire in day schools maintained by local authorities.
- The Safety of Sports Grounds Act 1971, as amended by the Fire Safety of Places of Sport Act 1987, says that designated sports grounds for more than 10 000 spectators are to have a safety certificate; application being made to the local authority. When the grounds are not designated a safety certificate must be obtained for every stand providing covered accommodation for more than 500 spectators.
- The Health and Safety at Work, etc., Act 1974 makes particular reference to premises where hazardous substances are stored and where the manufacturing process involves fire risks. The employers must provide training and supervision; keep the place of work safe, and provide safe means of ingress and egress.
- The Licensing Act 1964 enables the fire service to oppose the issue of a licence to sell intoxicating liquor if the means of escape is inadequate.
- The Gaming Act 1968 enables a fire authority to object to a licence being issued to casino or bingo halls if the premises are unsuitable.
- The Theatre Act 1968 and Cinematograph Act 1985 require satisfactory means of escape in case of fire before a licence can be issued.

Assessing the problem

It is unlikely that a fire will start in two different parts of a building at the same time. More than one third of fires are limited to the first item to be ignited and nine out of ten fires do not get beyond the room of origin. Such encouraging facts must not lead to complacency when assessing fire hazard problems.

Another factor is that only a small proportion of deaths arise directly or indirectly from contact with the flames. The majority are due to inhalation of smoke and its accompanying gases.

The combustible content of areas where there is the greatest possibility of fire occurring is vitally important but regulations do not provide control of what is the most frequent source of fire. Regulations limit themselves to the structure and its finishes.

Although most fires are limited to the area of source, the flames and smoke may spread to larger areas along cavities and along circulation routes and possibly limit vision on the approaches to escape routes and exits. Provision should therefore be made to limit the rapid spread of fire, smoke and gases.

In designing means of escape, the following are important factors:

- form of the building
- activities of the occupants
- likelihood of fire
- possible sources of fire
- possible routes of spread throughout the building.

Fire safety objectives

When designing a means of escape in case of fire it is necessary to have regard to several criteria:

- There should be an alternative way of escape from most situations where people may move away with backs to the fire.
- If direct escape is not available, people must be able to reach a place of relative safety, as, for example, a protected stairway giving a route to a final exit.

Thus the means of escape may be in three distinct parts:

1. in unprotected areas, which may be workrooms or circulation areas, and
2. in protected corridors, but more often in protected stairways, and
3. exit to the open air to get clear of the building.

In a number of large and complex modern buildings it may be possible to provide a place of reasonable safety within the building itself, but such areas require detailed evaluation and protection measures.

The approved document lists a number of aids which are not acceptable for building regulation purposes:

- lifts, other than those especially designed for use by firefighters and for removing disabled people;
- portable ladders and throwout ladders;
- fold-down ladders and similar apparatus.

Escalators are not regarded as having protected stairs capacity but mechanized walkways could be accepted as a route when the walkway is not moving.

Alternative means of escape

The aim in the provision of facilities to enable escape from fire, is to ensure the safety from being overcome by smoke or heat. This means that the speed of escapees' movement must be greater than it will take the fire to spread.

The approved document, while emphasizing the need for a quick escape, and the possibilities of a single escape route becoming impassable on account of smoke or fumes, does declare that in certain conditions a single direction escape may be acceptable.

The single route must present reasonable safety to the users. Called a 'dead end', the acceptance or otherwise will depend upon

- the number of people accommodated in the dead end, and
- the length of the dead end, and
- the use of the building, and
- the associated fire risk, and
- the size and height of the building.

Unprotected escape routes

An unprotected escape route is one which must be travelled before reaching the safety of a protected escape route, as, for example, from a desk across a room to a protected corridor.

Unprotected routes are to be limited in length so that people do not have to move excessive distances still exposed to the immediate hazard from fire and smoke.

Protected escape routes

The length of protected escape routes should be limited because the structure will not offer indefinite protection. A protected escape route can lead to the open air, or a protected stairway. The aim is to provide in a protected stairway an area considered safe from immediate danger from fire or smoke – an area virtually free from fire.

To achieve this, flames, smoke and gases are to be excluded from the routes to the greatest extent possible by

- a fire-resisting structure, or
- a smoke control system, or
- a combination of a fire-resisting structure and a smoke control system.

Stairs which are unprotected may be used on a daily basis by people engaged in a building, but such stairs have a very limited part to play in the overall provision of means of escape in case of fire.

Security of means of escape

To be effective, a means of escape must not present hazards or obstructions and the need for the security of a building may conflict with the needs for ready escape. Not only can security measures hinder escapees but also slow the entry of firefighters into a building.

It is therefore essential to ensure, at the design stage, that any problems in this direction are identified and overcome.

Disabled people

In Chapter 21, Part M of the regulations regarding requirements for access by disabled people are described and the approved document stresses that in the case of disabled persons management arrangements are a priority.

Guidance is available in BS 5588: Part 8: 1988, 'Code of practice for means of escape for disabled people'. This code gives details of the concept of 'refuges', and the use of an evacuation lift. Contents of the Code are as follows:

Section 1. General
　　1.　Scope
　　2.　Definitions
　　3.　Use of code
Section 2. Horizontal escape
　　4.　General
　　5.　Horizontal escape routes
　　6.　Refuges
Section 3. Vertical escape
　　7.　General
　　8.　Stairways
　　9.　Ramps
　　10.　Lifts
Section 4. Construction and fire-warning systems
　　11.　Refuges and evacuation lift enclosure
　　12.　Fire-warning systems
Appendices:
　　Advice on management
　　Application of code to existing buildings.

Health care premises

These premises, where people may be in varying degrees of ill-health and mobility, or bedridden, do present especial problems, and it may be that the total evacuation of a building in the event of fire can be avoided.

The specialized nature of health care premises does therefore require a different approach from the means of escape provisions set out in the approved document.

A number of guidance documents have been prepared by the Department of Health under an overall title of 'Firecode'.

- For new hospitals, follow 'Nucleus fire precautions recommendations', or Health Technical Memorandum 81.
- For work to existing hospitals relating to means of escape, follow the guidance in Home Office Draft 'Guide to fire precautions in hospitals'. There is also a draft 'Guide to fire precautions in existing residential care premises'.

In circumstances when a one- or two-storey house is converted into an unsupervised group house for not more than seven mentally handicapped or mentally ill people, for means of escape purposes it should be regarded as a 1(c) building – when the means of escape fulfil the requirements of Health Technical Memorandum 88.

Shopping complexes

Although guidance in the approved document can be applied to individual shops, it may be necessary to refer to BS 5588: Part 10: 1991 – Code of practice for shopping complexes, to resolve escape problems associated with this type of development.

Assembly buildings

In types of assembly buildings containing fixed seating such as theatres, lecture halls, conference centres or sports arenas, reference can be made to BS 5588: Part 6: 1991 – Code of practice for places of assembly, for advice on means of escape.

Schools

Advice in the approved document is appropriate for schools and other educational buildings.

Fire safety in Residential Care Premises

This is the title of a good practice guide to fire safety in residential care premises in England and Wales.

The guide embraces a full range of issues, including:

- number of residents;
- fire hazards within the premises;
- management policies;
- staff training in fire safety;
- fire precautions.

The document may be purchased from the Institute of Building Control.

MEASUREMENT

The approved document contains a number of methods of measurement which are applicable to B1 (Means of escape) and are not to be confused with the advice which relates to the whole of Part B in general.

Occupancy and building type

It is usual to make an estimate of the numbers of persons using a building and requiring escape facilities, and this is known as the occupancy load factor, or floor space factor. The factors may be related to a room or storey or extended to a whole building.

When calculating the area to be considered, stair enclosures, lifts and sanitary accommodation are ignored. The total for a whole building is arrived at by adding the area of each individual area together.

Table 1 Occupancy factors

1.	Standing spectator areas Bars without seating	0.3 m² per person
2.	Amusement arcade Assembly hall – includes general purpose place of assembly Bingo hall Club Crush hall Dance floor or hall Venue for pop concert or like events	0.5 m² per person

3.	Concourse
	Queuing area
	Shopping mall
	   Reference may be made to BS 5588:
	   Part 10: 1991, where detailed guidance
	   is available for common public areas
	   in shopping complexes.
	   (Refer to section 4)
		0.70 m² per person

4.	Committee room
	Common room
	Conference room
	Dining room
	Licensed betting office – public area
	Lounge bar – not including 1 above
	Meeting room
	Reading room
	Restaurant
	Staff room
	Waiting room
	   As an alternative the occupant capacity
	   may be taken as the number of fixed
	   seats provided when the occupants are
	   to be seated.
		1.00 m² per person

5.	Exhibition hall Studio – film, radio, TV, recording	1.5 m² per person
6.	Skating rink	2.0 m² per person
7.	Shop sales area, including main sales areas; supermarkets and department stores; shops giving personal services such as hairdressing; shop for the delivery or alteration of goods for cleaning, repair, or other treatment: premises for members of the public to carry out cleaning, repair or other treatment. (Exclude the shops in section 10 below.)	2.0 m² per person
8.	Art gallery Dormitory Factory production area Museum Workshop	5.0 m² per person
9.	Office	6.0 m² per person
10.	Shop sales area. Premises trading predominantly in furniture, floor coverings, cycles, prams, large domestic appliances or other bulky goods or cash and carry. (Exclude those in covered shopping complexes but include department stores.)	7.0 m² per person
11.	Kitchen Library	7.0 m² per person
12.	Bedroom Study bedroom	8.0 m² per person
13.	Bed sitting room Billiards or snooker room or hall	10.0 m² per person
14.	Storage and warehousing	30.0 m² per person
15.	Car park	two persons per parking space

Building types posing the greatest risks are hotels and boarding houses, residential institutions and recreation buildings. There are very special risks associated with high buildings and deep basements.

In applying the following occupancy factors it may be that a particular type of occupation is not directly applicable to the description given; in which case a reasonable value based upon a similar use may be employed.

When there is a mixed use, the most onerous factor is to be used.

The 2000 edition of the approved document increased the number of occupancy factors listed to 15 (Table 1).

TRAVEL DISTANCE

There are generally four stages to consider when evaluating travel distances.

1. Getting away from the room or area of fire origin.
2. Escape from the compartment of origin to a protected stair or area.
3. Escape to the ground level.
4. Escape into the open air.

Travel distances should be ascertained by measuring along the shortest route. There may be desks, fixed seating or other obstructions to the most direct route of escape, and therefore measurement is to be along the centreline of travel (Fig. B1.1), and along the pitch line on stairs (Fig. B1.2).

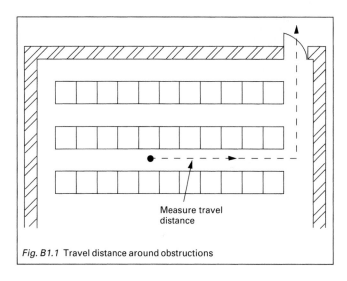

Fig. B1.1 Travel distance around obstructions

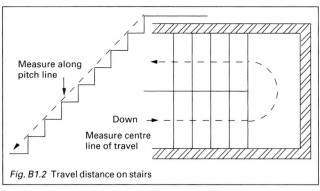

Fig. B1.2 Travel distance on stairs

There are also methods of measuring the width of doors, escape routes and stairways for the purpose of B1 (Figs B1.3 to B1.5).

Guiding rails for stairlifts may be ignored when measuring the width of stairs. Nevertheless the chair or carriage is to be parked in a position not causing obstruction on the stair or the landing.

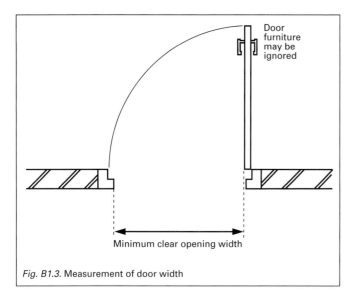

Fig. B1.3. Measurement of door width

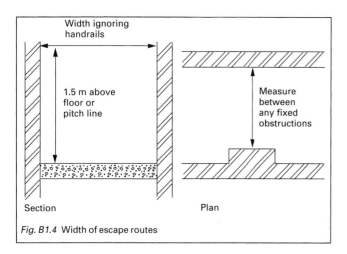

Fig. B1.4 Width of escape routes

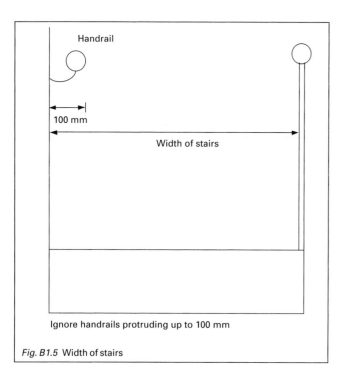

Fig. B1.5 Width of stairs

SECTION 1. FIRE ALARMS

The 1999 Regs inserted a requirement into B1 to the effect that there should be provided an early warning of fire in buildings. Approved document B 2000 in Section 1 sets out advice regarding the suitable provision of fire alarm and fire detection systems in dwellings and also in buildings other than dwellings. Suitable smoke alarms and fire detection systems may increase the level of safety by giving the occupants an early warning of fire.

DWELLINGS

Automatic fire detection and alarm systems are covered by the following British Standards:

- BS 5839: Part 1: 1988 which is a code of practice for system design, installation, and servicing. The installation should be at least L3 standard.

    Ceiling height has an effect upon the speed of reaction of detectors and the BS contains recommendations.

    Type L detectors are installed primarily for life safety, and a type L3 system is to be so designed that in any fire a warning is given at a sufficiently early stage to allow time for the escape routes to be used before they are choked with smoke.

- BS 5839: Part 6: 1995 being a code of practice for the design and installation of fire detection and alarm systems in dwellings. In this event the installation should be at least Grade E type LD3 standard.

Failing the installation of these systems, a suitable number of smoke systems should be provided. The alarms are to be mains operated and meet the following British Standard:

- BS 5446: Part 1: 1990 which is a specification for self-contained smoke alarms and point type smoke detectors. The detectors should be mains operated but may have a secondary power supply provided by rechargeable or replaceable batteries or a capacitor.

This BS covers smoke detectors based on ionization chamber smoke detectors and optical smoke alarms. These detectors have a different response to smouldering and fast flaming fires.

It is suggested that ionization chamber-based smoke alarms should be used in living rooms or dining rooms and that optical smoke alarms be used in circulation areas such as hallways or landings.

Large houses

For the purposes of B1 a house may be regarded as large if any of its storeys is greater than 200 m² in area.

A large house of more than three storeys (including a basement) should be fitted with an L2 system as specified in BS 5839: Part 1: 1988, and have a standby supply which need not fulfil the requirements of clause 16.5 regarding duration of the standby supply.

But in unsupervised systems the supply must be able to automatically maintain the system for 72 hours. Sufficient capacity should still remain to provide for the maximum alarm load for not less than 15 minutes.

A large house which does not have more than three storeys, including a basement, may be provided with an automatic fire detection and alarm system specified as Grade B type LD3 in BS 5839: Part 1: 1995.

Loft conversions

Habitable accommodation following a loft conversion in a one- or two-storey house is to have an automatic smoke detector and alarm system which is based on linked smoke alarms.

Flats and maisonettes

Follow the guidance given for houses, but the provisions are not

- to be applied to the common parts of blocks of flats;
- to include interconnection between installations in separate flats.

A maisonette is to be treated as a house with more than one storey.

Provision for alarms should be provided in student residential accommodation:

- an automatic detection system should be provided in a flat shared by a group of students which has its own entrance door;
- where a block of student flats have been compartmented the automatic detection system must give a warning in the flat or fire origin;
- in buildings where general evacuation is required the automatic alarm systems should be in accordance with BS 5839: Part 1: 1988.

Sheltered housing

These requirements do not apply to the common parts of sheltered housing or to sheltered accommodation in institutional or other residential groups.

- Detection units are to be connected to a central monitoring point.
- The warden, or other person in charge, must be able to identify the dwelling where the fire has been detected.
- The system may be connected to an alarm relay station.

Installation of smoke alarms

Smoke alarms are normally placed in circulation areas:

- near to a place where fires are likely to start;
- near to bedroom doors to give adequate warning to sleeping occupants;
- placed not more than 7.5 m from the doors of every habitable room;
- when more than one smoke alarm is fitted they are all to be interconnected in accordance with the manufacturer's instructions;
- there is to be at least one alarm on each storey;
- where door does not separate kitchen from circulation space or stairway an interlinked heat detector is to be fitted in kitchen;
- each alarm is to be fixed to the ceiling
  (a) not less than 300 mm from any wall,
  (b) not less than 300 mm from any light;
- sensors in ceiling-mounted devices
  (a) for smoke alarms 25 mm to 600 mm below ceiling,
  (b) for heat detectors 25 mm to 150 mm below ceiling;
- each alarm intended for wall mounting is to be fixed above level of door opening;
- alarms should not be fixed over stairs or other openings in floors, and should be readily accessible for maintenance;
- smoke alarms should not be fixed where they may be subjected to fumes, steam or condensation and give false alarms, as, for example,
  (a) near radiator, heater or air-conditioning outlet,
  (b) in a bathroom, shower, kitchen or garage,
  (c) in a boiler room or similar hot areas,
  (d) in an unheated porch, or similar very cold areas,
  (e) on surfaces normally warmer or colder than the space.

BS 5839: Parts 1 and 6 recommend that the manufacturer's instructions regarding operation and maintenance should be given to the occupier and this advice should be followed.

Power supplies

- Smoke alarms are to be permanently wired to a separate fused circuit on the distribution board or consumer unit.
- When installation does not include a stand-by power supply no other electrical equipment is to be connected to this circuit other than a dedicated monitoring device.

- Where a standby electrical supply is provided the smoke alarm may be connected to a lighting circuit.
- Devices for monitoring the main supply may be
  (a) audible or visible signals on each unit, or
  (b) audible or visible signals on a mains monitor.
  The position of any monitor should enable a warning to be very apparent to the occupants. It should be possible to silence an audible warning.
- It is preferred that a smoke alarm circuit is not protected by a residual current device – an rcd.
- Where an rcd is necessary because of electrical safety, the rcd serving the smoke alarm is not to be applied to any other current.
- Or, the rcd should operate independently of any rcd protection for socket outlets or portable equipment.
- Mindful of IEE regulations, cable suitable for domestic wiring, suitably colour coded, may be used.
- Any radio links interconnecting smoke alarms must not reduce the life of a standby electrical supply.
- Power to smoke alarms may also be supplied at low voltage via a control unit including a stand-by trickle charged battery.

BUILDINGS OTHER THAN DWELLINGS

In deciding upon the correct fire alarm or detection system appropriate to a particular building, it is necessary to take into account the type of occupancy and the method of escape to be provided. This may be simultaneous, phased or progressive horizontal evacuation.

Threat from fire is probably greatest in premises where people are sleeping and this is an important factor in deciding upon the mode of alarm and evacuation.

Where simultaneous evacuation is planned, a manual call point or fire detector will give an almost instantaneous warning, whereas when phased evacuation is to take place the alarm system will operate in two or more stages, giving appropriate warnings to the occupants of each area to be evacuated.

A fire detection system may involve any type of automatic sensor installation, sensitive to smoke, heat or radiation. Fire alarm systems will be activated and, where installed, automatic sprinkler systems brought into use.

Some buildings are also controlled under other legislation and there should be compatibility within the provision required. Consultation is always appropriate with Building Control and the Fire Authority.

Alarm systems

Arrangements for detecting fire should be provided in all buildings.

In small buildings it may be enough for a person detecting fire to call a warning that can be heard and understood by others present in the building. Elsewhere installation of a warning system that can be heard throughout the premises will be necessary. A simple arrangement is using manually-operated gongs or handbells, or by electrically-activated bells.

When these manually-operated methods are not adequate there should be installed in the building an electrically-operated fire warning system with sufficient sounders for the warning to be heard throughout the building. Manual call points should be placed near to exit doors.

Electrically-operated fire alarm systems are to meet the requirements of BS 5839: Part 1: 1988 Code of practice for system design, installation and servicing. Call points are to be in accordance with BS 5839: Part 2: 1983 Specification for manual call points.

BS 5389: Part 1: 1988 specifies types of system, some of which are sub-divided.

- Type L for the protection of life.
  Type L1 – systems installed throughout a building.
  Type L2 – systems installed in defined parts of a building.
  Type L3 – systems installed only for protection of escape routes.

- Type M – manual alarm systems.
- Type P for property protection.
  - Type P1 – systems installed throughout building.
  - Type P2 – systems installed in a defined part of a building.
- Type X – for multi-occupancy buildings.

Where there is any possibility that people may not react quickly to a fire warning and where people may be unfamiliar with the fire warning, a voice alarm system may be considered. This could be part of a public address system giving a distinct fire warning signal followed by clear verbal instructions to vacate the premises.

Voice alarm signals are to be installed in accordance with BS 5839: Part 8: 1998 Code of practice for design, installation and servicing of voice alarm systems.

A general fire alarm may be undesirable in premises such as large shops and places of assembly where a considerable number of people may be present. In such places staff must be trained in procedures leading to the safe evacuation of those present. Discreet alarms can alert the staff prior to instructions being given over the public address system when staff are in place to guide the public.

Staff alarm systems are to comply with BS 5839: Part 1: 1988 Code of practice for system design, installation and servicing.

Automatic systems

- Institutional and other residential premises are to have automatic fire detection and alarms in accordance with BS 5839: Part 1: 1988 Code of practice for system design, installation and maintenance.
- Office, shop, commercial, assembly, recreation, industrial and storage together with other non-residential occupancies may not normally require an automatic fire detection system.

Nevertheless there may be instances when a fire detection system is required:

- where a departure from guidance elsewhere in the approved document has been made, or
- where pressure differential systems or automatic door releases are installed, or
- where escape from an occupied part of a building could be placed in jeopardy, should a fire occur in a storage area, basement or part of a building temporarily not in use.

The recommendations contained in BS 5588: Part 7: 1997 Code of practice for precautions in atria buildings is to be followed when atria buildings are erected.

It is vital that fire detection and fire warning systems are correctly designed, installed and maintained, and an installation and commissioning certificate should be obtained.

SECTION 2. DWELLINGHOUSES

It is an unfortunate fact that people regularly suffer death or injury arising from fires within dwellings. Our local television screens all too frequently show smoke-stained façades of two- and three-storey dwellings where children or adults have perished. About 60% of fires in occupied buildings happen in dwellings; and they give rise to 90% of fatal cases and 80% of non-fatal injuries. In total this represents an average of about two fatalities and 28 non-fatal injuries each day.

Fuel for fires is concentrated in dwellings. Fats, oils, heat sources, electrical and gas appliances are in kitchens, and elsewhere in the premises furniture, upholstery, papers, television and other electrical equipment with wiring often overloaded live cheek by jowl. The slightest carelessness with a match or cigarette can result in a fire that develops at astounding speed.

Fire can start

- in any habitable room, or in a kitchen;
- in a room, whether it is occupied or not;
- when the occupants are awake and alert;
- when the occupants are asleep.
  Rooms with high hazard are:
  Living rooms
  Dining rooms
  Kitchens
  Bedrooms
  Dressing rooms.
  Rooms with low hazard are:
  Bath rooms
  Waterclosets
  Utility rooms
  Storage rooms.
  Areas protected;
  Entrance halls
  Stairways
  Landings.

The approved document makes the point that fire protection measures cannot significantly interfere with the day-to-day convenience of the occupants, or they would become unacceptable.

The measures proposed for one- and two-storey houses are simple to provide, requiring early warning of fire and the availability of a suitable exit from each storey.

In higher buildings exit via windows can be hazardous and it is therefore necessary to add protection to internal stairways. If there is a floor above 7.5 m above ground level there is a risk of the stairway becoming impassable before all the occupants can escape and therefore an alternative route is to be made elsewhere.

Houses in multiple occupation.

These are defined as 'house(s) which is (are) occupied by persons who do not form a single household' (Housing Act 1985, section 345).

Guidance in this section relates to houses in multiple occupation if there are no more than 6 residents. Additional precautions are necessary where the number of occupants exceeds 6. The registration authority should always be consulted in matters relating to houses in multiple occupation.

There are several official circulars which offer technical guidance:

- DOE circular 12/92
  Houses in multiple occupation. Guidance to local authorities on standards of fitness.
- Welsh Office circular 25/92
  Houses in multiple occupation. Standards of fitness.
- DOE circular 12/93
  Houses in multiple occupation. Guidance to local housing authorities.
- Welsh Office circular 55/93
  Houses in multiple occupation. Guidance on management strategies.

Inner rooms

There are guidelines on the use of inner rooms which have a route of escape through an access room (Fig. B1.6)

This arrangement is only acceptable when the inner room is one of the following:

- bathroom or shower room;
- dressing room;
- kitchen;
- laundry or utility room;
- WC;
- a room which is not more than 4.5 m above ground level served by one stair is to have a window or external door (Fig. B1.7). If a basement, there is to be a door or window in a protected stairway leading to a final exit.

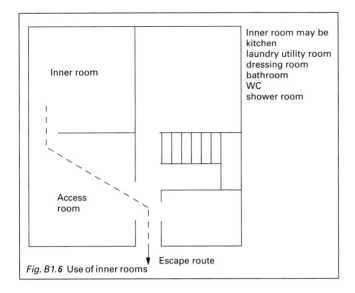

Fig. B1.6 Use of inner rooms

Inner room may be
kitchen
laundry utility room
dressing room
bathroom
WC
shower room

Inner room

Access
room

Escape route

- a sleeping gallery, not more than 4.5 m above ground level and where the distance between the foot of the access stairs to the gallery and the door to the room does not exceed 3 m. If the distance between head of access stair and any point on the gallery is more than 7.5 m, an alternative access is to be provided (Fig. B1.7). Any cooking facilities are to be enclosed in fire resisting construction and so placed that they do not interfere with escape from the gallery.

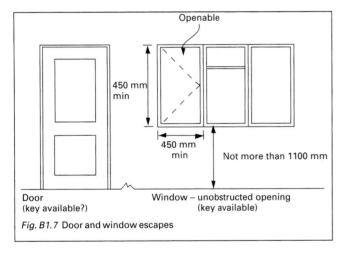

Openable

450 mm
min

450 mm
min

Not more than 1100 mm

Door
(key available?)

Window – unobstructed opening
(key available)

Fig. B1.7 Door and window escapes

Balconies and flat roofs

Where a roof forms part of a means of escape it is to meet the following requirements:

- the roof is to be part of the same building;
- the way from the roof is either a storey exit or an internal escape route;
- the roof and supporting structure, and any opening within 3 m of the escape route is to have 30 minutes' fire resistance;
- guarding may be necessary, and this is provided for in Part K, Chapter 19.

Escape from floors not above 4.5 m

All habitable rooms in upper storeys served by one stair are to have a window suitable for escape, or a door. This does not apply to kitchens. Two rooms may be served by one window if they have a communicating door, and each has direct access to the stairs.

All habitable rooms in the ground storey are

- to open directly into a hall leading to an exit, or
- have a suitable escape window or door (see Fig. B1.7).

This does not apply to kitchens.

In dwellings where a sleeping gallery is provided:

- the gallery is not to be in excess of 4.5 m above ground level;
- there should not be more than 3 m between the door to the room and the foot of the access stair to the gallery;
- if the distance from the top of the access stair and any point in the gallery is more than 7.5 m an alternative access or an escape window is to be provided, and
- any cooking facilities in a room containing a gallery are to be enclosed with fire resisting construction, or be placed remotely from the stairs to the gallery, in order not to cause a hazard on the escape route.

Basements

Basements present escape problems as smoke may impede the use of an internal stair. Should a basement contain a habitable room, there is to be an external door or escape window, or a protected stairway leading to a final exit.

Escape through doors and windows

We have seen that doors and windows may be used as a means of escape from basements, ground and first floors of dwellings. Such doors and windows must comply with the following requirements:

- have an unobstructed openable area not less than $0.33\,m^2$;
- have an unobstructed opening not less than 450 mm high and 450 mm wide;
- bottom of window opening is not to be more than 1100 mm above the floor.

Whether the position of a window is allowable for escape purposes is a matter for consideration in each case. Escape from first-floor windows may be by way of a ladder placed by a neighbour or the fire service; but escape may be achieved by simply dropping from the window to the ground below. Such an escape would be impeded by a conservatory or other structure having a fragile roof.

Escape from a ground storey or basement may be into an enclosed open space (Fig. B1.8) but only if the depth of the space is not less than the height of the building from which the escape is being made.

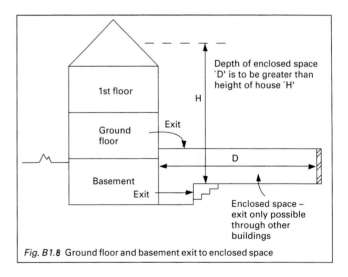

1st floor

Ground
floor

Basement

Exit

Exit

Depth of enclosed space
'D' is to be greater than
height of house 'H'

H

D

Enclosed space –
exit only possible
through other
buildings

Fig. B1.8 Ground floor and basement exit to enclosed space

A roof window or dormer intended for escape must be placed as shown in Fig. B1.9. As an alternative, the window may be in the end wall of a house.

One floor more than 4.5 m above ground level

Where a house has storeys more than 4.5 m above ground level (Fig. B1.10) there are additional provisions requiring

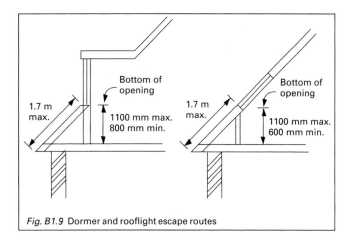

Fig. B1.9 Dormer and rooflight escape routes

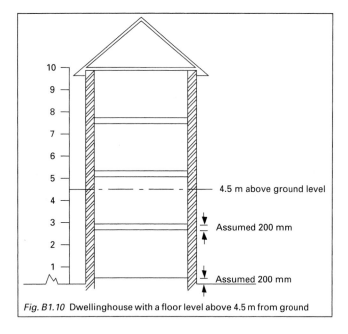

Fig. B1.10 Dwellinghouse with a floor level above 4.5 m from ground

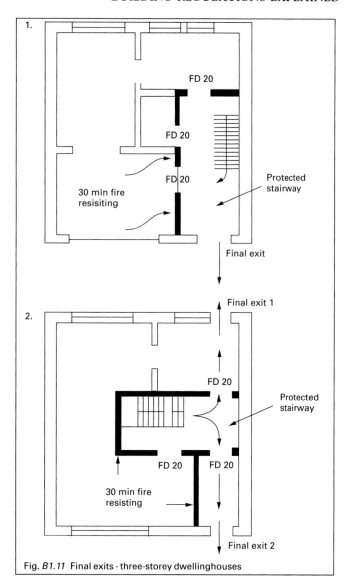

Fig. B1.11 Final exits - three-storey dwellinghouses

- a protected stairway, or
- top floor can be separated by fire resisting construction and provided with an alternative escape route.

If a protected stairway is to be provided, it should be

- protected to the final exit, or
- have access to at least two escape routes at ground level, separated by 30 minute fire resisting construction or self-closing fire doors.

The top storey is to be separated from lower storeys by fire resisting construction, and there is also to be an alternative escape route to its own final exit.

Self-closing fire doors in dwellings may be hung on rising butts.

More than one floor above 4.5 m

If a house has two or more storeys more than 4.5 m above ground level, it must, in addition to meeting the above requirements for a house with one floor above 4.5 m, have an alternative escape route provided from each storey or level which is situated at 7.5 m or more above ground level.

The access provided to the alternative escape route is

- along a protected stairway to an upper storey, or
- along a landing within a protected stairway enclosure to an escape route on the same storey,

the protected stairway at or about 7.5 m above ground level is to be separated from the lower storeys by fire resisting construction.

Air circulation systems

Precautions are essential when air circulation systems for heating, energy conservation or condensation control are installed in dwellings with a floor more than 4.5 m above ground level.

Guidance regarding steps to be taken to prevent the system conveying smoke or fire into a protected stairway are contained in the approved document.

- Transfer grilles are not to be fitted in walls, floors or ceilings enclosing a protected stairway.
- Ductwork passing through an enclosure to a protected stairway is to have all joints between ductwork and enclosure fire stopped.
- Ductwork carrying air into a protected stairway through its enclosure should have return air directed back to the plant.
- Air and return grilles should be positioned not more than 450 mm above the floor level.
- Room thermostats should be mounted in the living room between 1370 mm and 1830 mm above the floor and have a maximum setting not exceeding 27°C.
- Mechanical air circulation systems are to comply with BS 5588: Part 9: 1989 Code of practice for ventilation and air conditioning ductwork.

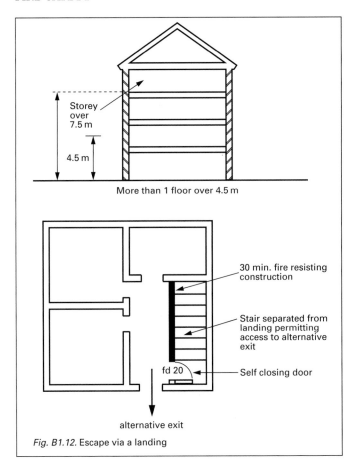

Fig. B1.12. Escape via a landing

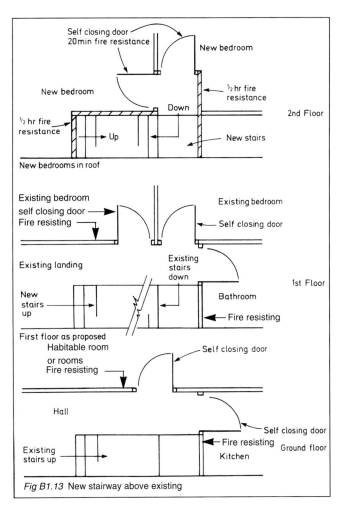

Fig B1.13 New stairway above existing

Passenger lifts

When a passenger lift in a house serves a floor which is more than 4.5 m above ground level, it is to be

- in an enclosure to a protected stairway, or
- in a fire resisting lift shaft.

LOFT CONVERSIONS

These provisions are an alternative to the requirement relating to houses having a floor more than 4.5 m above ground level but they are not suitable if

(a) the new storey has an area in excess of 50 m², or
(b) the new storey has more than two habitable rooms.

The provisions are (see Figs B1.13 and B1.14):

- existing stairs in ground and first floors are to be enclosed in fire resisting construction;
- enclosure to lead to final exit or give access to two escape routes at ground level, separated from each other by fire resisting construction and self-closing doors;
- doors to habitable rooms within the enclosure are to be self-closing, and rising butt hinges are acceptable;
- new doors to habitable rooms must be fire doors but existing doors need only be made self-closing;
- new and existing glazing in the enclosure to existing stair should be fire resisting, with the exception of bathroom and WC doors;
- glass should be fixed using suitable glazing system and beads;
- the new stairs may be a continuation up the existing stair enclosure and this can be by way of an alternating tread stair or fixed ladder with a fire door (FD20) at the top. Otherwise it can be in

an enclosure which is separated from the existing stairway and also from ground and first floor accommodation but opening onto an existing stairway at first floor level;

- the new floor to be 30 minute standard of fire resistance;
- measures are to be taken to prevent smoke and fire in the stairway from entering the new storey;
- escape windows are to be provided for each habitable room;
- a two-room conversion where each room has direct access to the stairs, and there is a communicating door between the two rooms, may rely upon one escape window (Fig. B1.15);
- space is required to place a ladder below each escape window.

The use of windows in a three-storey domestic loft conversion is acceptable for emergency escape. Rooms in the new storey are to have dormer window or roof light.

- that have a clear opening not less than 450 mm high by 450 mm wide;
- that are not more than 1.7 m from the eaves when measured along the roof surface (see Fig. B1.9).

The window may be placed in the end wall of a house if this is appropriate and escape may be over a suitable fire resisting roof of a ground storey extension

- smoke alarms are to be fitted, as described earlier.

Loft conversion guide

TRADA Technology Ltd and the Institute of Building Control have co-operated in the preparation of a 'Loft conversion guide'. This was sponsored by the Velux Company Ltd. It is a very useful booklet, which may well be developed into an approved document.

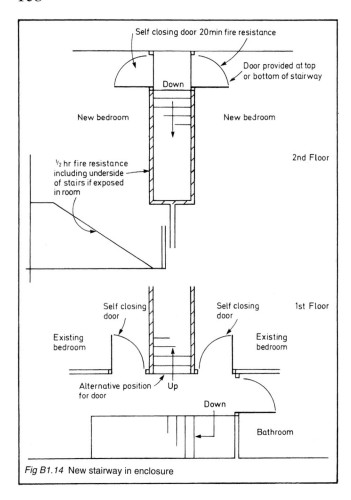

Fig B1.14 New stairway in enclosure

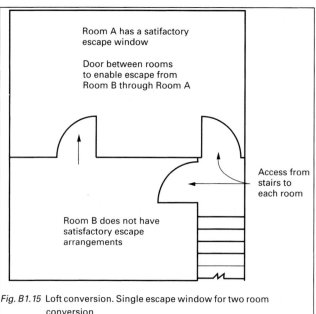

Fig. B1.15 Loft conversion. Single escape window for two room conversion

The booklet (some 54 pages) is divided into three parts, and includes many tables and diagrams.

Part 1 is concerned with assessing the feasibility of a loft conversion, and includes advice relating to the following:

- suitability of the roof space for conversion.
  Includes roof pitch, ceiling heights, access and fire escape;
- suitability of roof structure for conversion.

Discusses traditional and trussed rafter construction;
- suitability of existing construction for conversion.
  Covers timber sizes and placing of storage tanks, roof coverings and insulation, supporting walls and separating walls;
- there is a checklist.
- legislation involved.
  Relates to planning permission, building control with a list of regulations relating to roof conversions together with a legislation checklist.

Part 2 is devoted to technical information for designers and contains much useful guidance.

- To secure compliance with building regulations there is advice on structural stability, means of escape, fire resistance, spread of flame, sound insulation, ventilation of rooms and roofspace, vapour control and access (with diagrams indicating four stair arrangements), thermal insulation, plumbing, drainage and heating, together with a comprehensive checklist.
- The structural design of a conversion covering a number of specification solutions for each element of construction. Included are floor beams, floor decks, boarding, plywood, chipboard, roof beams, beam supports, rafters, and calculations.
- Roof windows and dormers are a vital part of any roof conversion and a number of diagrams are included. There is provision for ventilation, the installation of windows and dormers, carpentry, vapour control checks and operation. Again there is a comprehensive checklist.

Part 3 describes the procedures for carrying out a loft conversion which are best adhered to when undertaking the work. There are also notes on access for materials, electrics, plumbing, wall lining and decoration.

'The Loft Conversion Guide' is the copyright of the Velux Group and copies may be obtained from the Technical Services Department, The Velux Co Ltd, Woodside Way, Glenrothes, East Fife, Scotland KY7 4ND (01 592 77 82 50).

SECTION 3. FLATS AND MAISONETTES

There is guidance relating to the planning of flats and the means of escape requirements (Fig. B1.16). They lean heavily on the current BS 5588. Above two storeys, means of escape requirements have become more involved and automatic vents and even pressurization now have mention in the approved document.

Few provisions are outlined for escape from basement, ground and first floors in flats and maisonettes, but there are requirements relating to these floors:

- that each habitable room must open directly into a hallway or stair leading to an exit, or
- that there is a door or window through which an escape may be made, and
- that a fire alarm has been installed.

With increasing height, the hazards increase and the escape requirements become more involved.

Guidance in the approved document is restricted to common arrangements for flat or maisonette design. The code of practice embraces other design arrangements and guidance is given in clause 9 and 10 of BS 5588: Part 1: 1990.

Clause 3.9 relates to the internal planning of flats and says that a flat may be a single habitable room to a large family dwelling with many rooms. But there are two basic principles:

- ensure that the internal planning is such that a fire anywhere in the flat will not trap its occupants; and
- ensure that escape from the flat is possible.

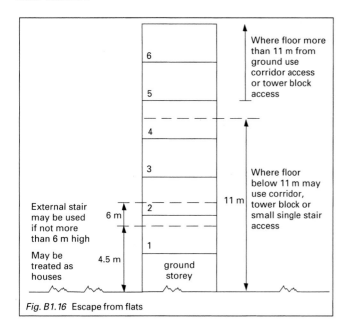

Fig. B1.16 Escape from flats

In open plan dwellings cooking facilities should be so located that people may move away and escape.

There are recommendations for flats with external entrances; basement flats; flats not more than 4.5 m above ground level and flats more than 4.5 m above ground. There are also recommendations relating to flats with galleries.

Clause 3.10 is concerned with the planning of maisonettes and follows the same line of thought. Maisonettes may occupy more than two storeys and may have separate rooms, or be open plan. In open plan occupants must be able to escape from the area where cooking facilities are located.

There are recommendations for maisonettes with no storey situated more than 4.5 m above ground, and also for maisonettes with one or more storeys in excess of 4.5 m above ground level.

In formulating the provisions contained in the approved document there have been several assumptions:

- Fire is usually in a dwelling.
- One must not rely on external rescue.
- Compartmentation offers protection against rapid fire spread.
- Where fires do occur in communal areas, materials and construction used should limit spread.

Important factors are the need for occupants to escape from their dwelling and reach the final exit from the building safely.

The guidance also relates to flats and maisonettes which may fall within the definition of 'house in multiple occupation'. These are defined as a 'house which is occupied by persons who do not form a single household' (Housing Act 1985, section 345). Advice on the implementation of this very wide definition is contained in official circulars:

DOE circular 12/92
Welsh Office circular 25/92
BRE circular 12/93
Welsh Office circular 55/93.

Sheltered housing

Many of the requirements for flats and maisonettes are appropriate for sheltered housing, but additional fire protection measures may be necessary. One- or two-storey bungalows or flats may be treated in the way described for small dwellinghouses, but where additional provisions are necessary guidance is available in clause 17 of BS 5588: Part 1: 1990.

This clause regards sheltered housing as a specialized form of accommodation consisting of self-contained dwellings with assis-

tance available at all times. The code deals with internal planning and escape routes.

Corridors should be subdivided to ensure a travel distance from flat to protected stairway or lobby not in excess of 7.5 m. Handrails are to be fixed on each side of stairs. There are also recommendations for communal areas and regarding the provision of furniture in common corridors (Fig. B1.17).

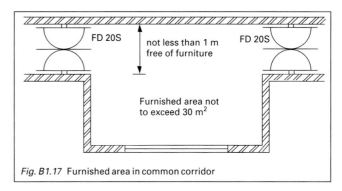

Fig. B1.17 Furnished area in common corridor

Inner rooms, basements, balconies and flat roofs

The guidance given in Section 1 – dwellings with regard to inner rooms, basements, balconies and flat roofs, applies equally to flats and maisonettes.

MEANS OF ESCAPE WITHIN FLATS AND MAISONETTES

- Automatic smoke detection and alarm systems should be fitted and the recommendations given in the section on dwellinghouses is to be followed.

In this connection they are not intended to be applied to common areas in flats and there is no requirement for connection between systems in individual flats.

In the case of a maisonette the installation should follow the recommendations for two-storey dwellinghouses.

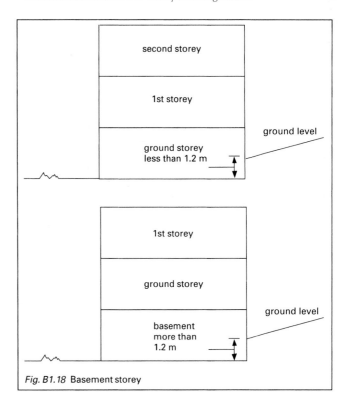

Fig. B1.18 Basement storey

- The risks associated with inner rooms are similar to the risks in dwellinghouses and the advice in that section should be followed.
- Likewise, the advice in the section on dwellinghouses regarding escape from basements is appropriate and should be followed (Fig. B1.18).
- The guidance in the section on dwellinghouses regarding escape to balconies and flat roofs is also appropriate. When a balcony is outside an alternative exit which is more than 4.5 m from ground level it must be a common balcony.
- An alternative exit must be remote from the principal entrance door to the dwelling and lead to a final exit or common stair. This can be by way of
  - (a) a door leading to an access corridor or common balcony, or
  - (b) an internal private stair which leads to an access corridor or common balcony which is at another level, or
  - (c) a door onto an external stairs, or
  - (d) a door onto a flat roof which is an escape route.
- No flat where the floor is not more than 4.5 m above ground level is to have an inner room unless it has a door or window as described in Section 1 – Dwellinghouses (Fig. B1.19).

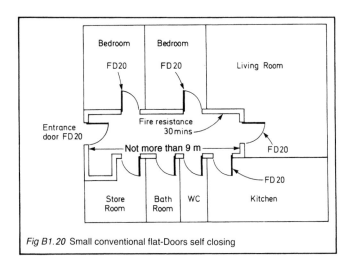

Fig B1.20 Small conventional flat-Doors self closing

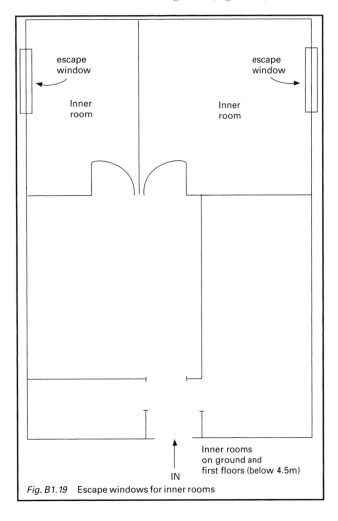

Fig. B1.19   Escape windows for inner rooms

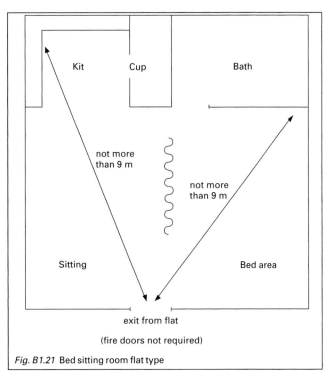

Fig. B1.21 Bed sitting room flat type

- An arrangement requiring cooking facilities to be as far as possible from the escape exit, combined with a maximum travel distance of 9 m from habitable areas to exit door (Fig. B1.21). A similar travel distance operates in a small flat having a separate bedroom (Fig. B1.22).
- Where there are habitable rooms with no direct access to the flat entrance door an alternative exit is to be provided in the part of the flat containing bedrooms (Fig. B1.23). The bedroom area must be separated from the living area by 30 minute fire-resisting construction and self-closing doors.

Planning maisonettes

A maisonette which has its entrance at ground level is to be treated as a dwellinghouse and escape measures should be arranged accordingly, being dependent on whether the height of the top storey is above or below 4.5 m from the ground.

Where a maisonette does not have its own external entrance at ground level and has a floor which is in excess of 4.5 m above ground level two alternative arrangements of internal planning are illustrated (Figs B1.24 and B1.25).

Flats with floor level greater than 4.5 m above ground

While still having regard to the requirements for inner rooms, the following arrangement for flats having a floor which is more than 4.5 m above ground level is acceptable.

- Provide a protected entrance hall/corridor with a maximum travel distance of 9 m from any habitable room to the exit door (Fig. B1.20).

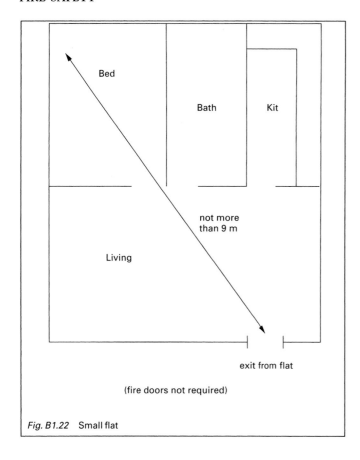

Fig. B1.22   Small flat

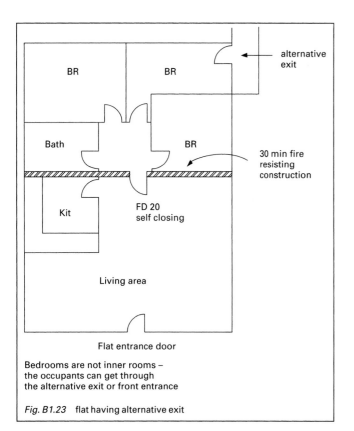

Bedrooms are not inner rooms –
the occupants can get through
the alternative exit or front entrance

Fig. B1.23   flat having alternative exit

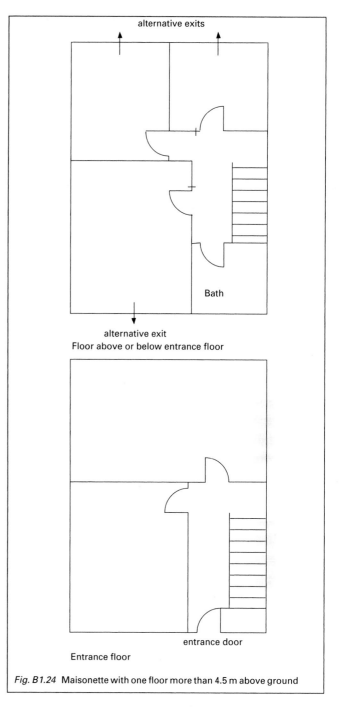

Fig. B1.24   Maisonette with one floor more than 4.5 m above ground

- Make an alternative exit available for each habitable room which is above or below the entrance floor of the maisonette.
- Form a protected entrance hall and landing and make all doors opening into these areas FD 20 fire doors; in addition to one alternative exit from each of the floors above or below the entrance floor.

An alternative exit from a flat or maisonette should meet the following requirements:

- be remote from the main entrance door
- lead to a final exit or common stair via a door onto an access lobby or common balcony; an internal stair leading to an access lobby or common balcony at another level, or a door to the external air, a door to an escape route over a flat roof, or a door to a common stair.

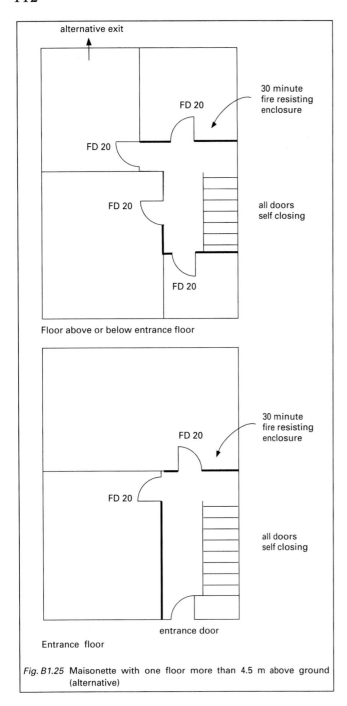

Fig. B1.25 Maisonette with one floor more than 4.5 m above ground (alternative)

Air circulation systems

Precautions are necessary to ensure that an air circulation system for heating, energy conservation or condensation control in a flat or maisonette which has a floor more than 4.5 m above ground level does not permit smoke or fire to spread to a protected landing or entrance hall.

- Transfer grilles are not to be fitted in walls, floors or ceilings enclosing a protected entrance hall, or a protected stairway or landing of a maisonette.
- Ductwork passing through the enclosure to a protected hall or stairway is to be adequately fire stopped.
- Returned air from the protected area is to be ducted back to the plant.
- Warm air, and return grilles are not to be more than 450 mm above floor level.
- A room thermostat is to be mounted between 1370 mm and 1830 mm above the floor and have a maximum setting of 27°C.

- In the event of a fire a mechanical system should close down or direct air away from protected escape routes.

Number of escape routes

Stage 1 of an escape route is to get out of the dwelling and stage 2 is to achieve the safety of a protected lobby or stair. We have seen that most dwellings must have an alternative escape route available, but a single escape route from the flat or maisonette entrance (see Figs B1.26 to B1.28) may be accepted

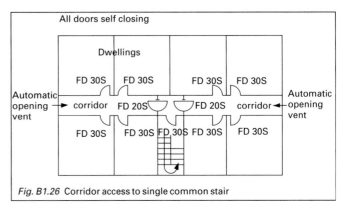

Fig. B1.26 Corridor access to single common stair

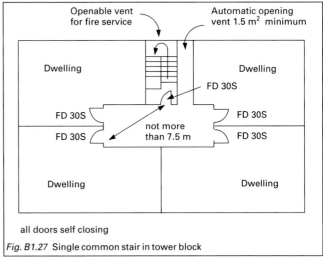

Fig. B1.27 Single common stair in tower block

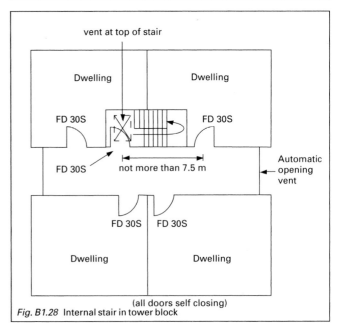

Fig. B1.28 Internal stair in tower block

- if there is a single common stair, and
- all the dwellings are separated from the common stair by a common corridor or protected lobby, and
- travel distances are limited, where escape is in only one direction, to 7.5 m.

Where a dwelling is situated at the dead end of a corridor which is served by two or more stairs (Fig. B1.29), the maximum travel distance from the dead end is 7.5 m; elsewhere the maximum travel distance is 30 m. Should the maximum travel distance not exceed 15 m the central corridor door may be omitted.

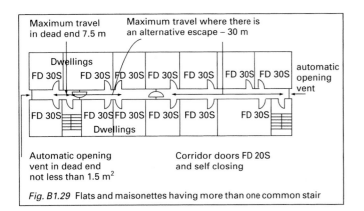

Fig. B1.29 Flats and maisonettes having more than one common stair

The above arrangements for a single common stair may be modified in the following circumstances – the top floor of the building is not to be more than 11 m from the ground, and the maximum number of storeys above the ground level storey is not to exceed 3. Additionally the stair must not connect with a car park which is not open sided (Figs B1.30 and B1.31).

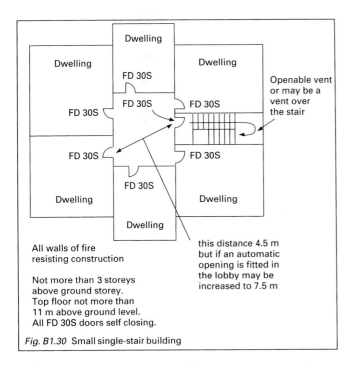

Fig. B1.30 Small single-stair building

The stairs are not to serve any ancillary accommodation unless

- there are no dwellings on the floor containing ancillary accommodation, and
- a protected lobby separates ancillary accommodation from the stairs or protected corridor. The lobby or corridor is to have at least 0.4 m² area of permanent ventilation or be protected by a mechanical smoke control system.

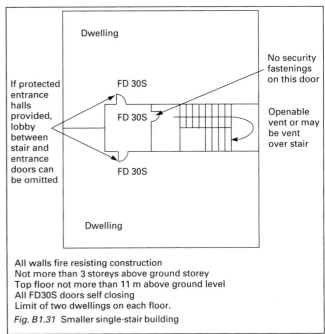

All walls fire resisting construction
Not more than 3 storeys above ground storey
Top floor not more than 11 m above ground level
All FD30S doors self closing
Limit of two dwellings on each floor.

Fig. B1.31 Smaller single-stair building

Where a balcony or deck access is provided guidance in clause 13 of BS 5588: Part 1: 1990 may be used. This says that unacceptable hazards may arise if opportunity is taken to use a balcony or deck approach for the erection of stores and other fire risks.

The soffit above a deck or balcony with a width of not less than 2 m should be designed with downstands placed 90° to the face of the building. Downstands are intended to hinder smoke rolling easily along the length of a balcony and to encourage it to flow to the outside edge of the soffit and disperse away from the face of the wall. Downstands should project 0.3 m to 0.2 m below any beam parallel to the face of the building (Fig. B1.32).

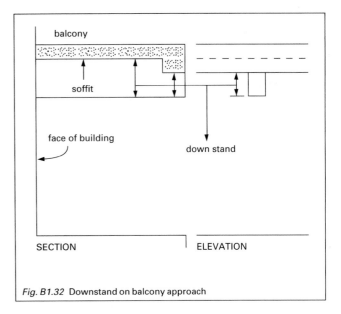

Fig. B1.32 Downstand on balcony approach

COMMON ESCAPE ROUTES

Escape routes in common areas should be planned to ensure that stipulated travel distances are not exceeded:

(a) The maximum travel distance from a flat or maisonette entrance door to the entrance door of a common stair, or the door to a lobby in single-stair flats, must not be more than 30 m when escape is possible in more than one direction.

(b) The maximum travel distance when escape is only possible in one direction is 7.5 m, and this may be reduced to 4.5 m in a small single-stair building which does not have an opening vent in the lobby (see Fig. B1.28).

If the dwellings in a storey have an independent alternative means of escape, maximum travel distances do not apply.

- It should not be necessary to pass through one stairway enclosure to reach another, but it is permissible to pass through a protected lobby to reach another.
- Common corridors should be protected corridors and the wall between each dwelling and the corridor is to be a compartment wall.
- It is always possible that some smoke may gain access to a lobby or common corridor – an entrance door could be left ajar in haste to escape. Some ventilation of common escape routes should be provided to assist smoke to disperse.

  (a) In single-stair buildings, except ones indicated in Figs B1.29 and B1.30, and in dead ends of a building with more than one stair (Fig. B1.28), there should be an automatic opening ventilator controlled by an automatic smoke detector fitted in the space to be ventilated. The ventilator is to have a free area not less than 1.5 m² and have a manual override.

  (b) Where there is more than one stair, common corridors are to extend to the outer walls. There are to be openable ventilators, operating automatically for use by firefighters. The free area of the ventilator at each end of the corridor is not to be less than 1.0 m².

- A common corridor between two or more storey exits is to be sub-divided by self-closing fire doors (Fig. B1.26) if necessary, in a fire-resisting screen. Location of doors should be so that smoke will not impede access to more than one stairway. Likewise, a dead end should have a self-closing fire door and any necessary fire-resisting screens to separate the area from the rest of the corridor.

- Where pressurization is employed as a smoke control system in escape corridors or stairways the cross corridor fire doors and the openable and automatically opening vents are to be omitted.

Pressurization systems are described in BS 5588: Part 4: 1998, 'Code of practice for smoke control in protected escape routes using pressurization'.

- Ancillary accommodation, such as stores, cupboards and drying rooms, etc., represent a potential fire hazard and should not open directly onto a protected lobby or a protected corridor which is part of a common escape route.
- Where more than one escape route is available, one may be over a flat roof if:

  (a) the roof is part of the same building;
  (b) it leads to a storey exit;
  (c) the escape route and its supporting structure is fire resisting, and any opening within 3 m of the route is fire resisting, and
  (d) the route is clearly marked and guarded by walls or other suitable protective barriers.

  (Part K; Chapter 19.)

Common stairs

The width of a common stair is not stipulated. When it is also a fire-fighting stair the width is to be not less than 1.1 m.

Protection

A common stair should be an area of relative safety and therefore it is necessary to ensure a satisfactory standard of fire protection.

- All common stairs are to be within a fire-resisting enclosure, and the level of fire resistance required is given in Appendix A, Tables A1 and A2 to the approved document (page 160 and 130).
- All protected stairways should take people directly to a final exit, or along a protected exit passageway to a final exit.

- If two protected stairways, or exit passageways, adjoin they are to be separated by an imperforate enclosure.
- A protected stairway is not to be used for any other purpose, with the possible exception of a lift well, or electricity meters. Guidance on the installation of electricity meters is contained in BS 5588: Part 1: 1990 Code of practice for residential buildings.
- There are situations when a protected stairway with an external wall of little fire resistance could be at risk from fire spreading from the adjoining structure (Fig. B1.40). This could occur where a protected stairway projects beyond, or is recessed from, or is in an internal angle of, the building. In these circumstances there is not to be any unprotected area of the building within 1.8 m of the stairway enclosure.

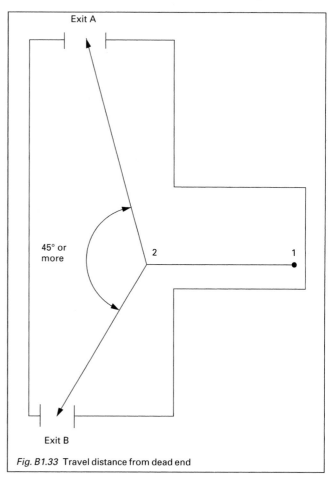

Fig. B1.33 Travel distance from dead end

- Gas pipes and meters are not to be placed in a protected stairway unless they comply with

  (a) Pipelines Safety Regulations 1996, and
  (b) Gas Safety (Installation and Use) Regulations 1998.

Basement stairs

Measures are necessary to prevent smoke from basements getting up into the escape routes from upper floors.

- Single escape stairs should not continue down to serve any basement storey. A separate stair should be provided for basements.
- Where there is more than one escape stair from an upper storey, one may terminate at ground level. Other stairs may go down to the basement if there is a ventilated protected lobby, or a ventilated protected corridor separating the stairs and accommodation at each basement level.

Ancillary accommodation

A common stair, which is part of the only escape route from a dwelling, is not to serve

- a covered car park;
- a boiler room;
- a fuel storage space;
- ancillary accommodation of similar fire risk, on the same storey as the dwelling.

A common stair which is not the only escape route may serve ancillary accommodation where it is separated by a protected corridor or lobby.

A stair serving an enclosed car park or a special fire hazard is to have

- at least 0.4 m² permanent ventilation, or
- be protected by a mechanical smoke extraction system.

External fire escape

A building may be served by a single access stair if it serves a floor not exceeding 6 m above ground level.

Where more than one escape route is provided, one may be an external stair if it meets the following requirements:

- six metres is the maximum height above ground level, a roof or a podium which has an independent protected stairway.

In each case the stairs are to meet the requirements of B1 section 6, described later.

MIXED USE BUILDINGS

In buildings limited to three storeys above the ground storey, stairs may be used by dwellings and non-residential occupancies. This requires protected lobbies at all levels to separate each occupancy from the stairs.

Where a building has more than three storeys above the ground level storey the following precautions are necessary.

- All stairs serving flats are not to communicate with other parts of the building except when the flat is ancillary to the main use of the building and has an independent alternative escape route.
- When a flat is ancillary to the main use, a stair serving the flat may be used by other parts of the building if
  - (a) it is separated on lower storeys by protected lobbies, and
  - (b) if an automatic fire detection and alarm system is fitted in the main part of the building, it must also embrace the flat.
  - (c) Any security measures in operation must not prevent use of the escape routes at all times.

If fuels such as petrol or LPG are present in the building, additional measures such as increased fire resistance between the storage area and the dwelling may be necessary.

Guidance is available in BS 5588: Part 0: 1996 Guide to fire safety codes of practice for particular premises/applications.

SECTION 4. HORIZONTAL ESCAPE: BUILDINGS OTHER THAN DWELLINGS

The horizontal part of a means of escape relates to the way available from any point to the storey exit on the floor being considered, and this section relates to all types of floor except dwellinghouses, flats and maisonettes.

Attention must be given to the following interrelated factors:

- the number of people involved
- the maximum travel distance to an exit
- the number of escape routes to be provided
- alternative routes to separate exits
- width of corridors and exits

- protection of corridors
- escape over available flat roofs.

Any decisions arrived at concerning horizontal escape provisions will naturally have regard to the facilities available for vertical escape, which are considered in the next section.

The guidance in this section is aimed at smaller types of buildings. Where more complex or specialized structures are concerned information is available in the range of British Standards numbered BS 5588.

Guidance is contained in Clause 10 of BS 5588: Part 11: 1997 Code of practice for shops, offices, industrial storage and similar buildings, in connection with industrial, office, small shops, storage or similar premises when no storey is in excess of 280 m² and there are no more than two storeys and a basement.

ESCAPE ROUTE DESIGN

In order to determine the number of escape routes and the number of exits to be provided it is first necessary to know the number of people in the room, area or storey being considered. The limits of travel distance to the nearest exit must be known and they are given in Table 2 compiled from the approved document.

When a multi-storey building is involved more than one stair may be needed, so that all parts of each storey will have access to more than one stair.

It is possible for dead end areas to be allowed if an alternative stair is available.

If a building contains a number of purpose groups the means of escape from

- purpose groups 1 and 2 – residential, or
- purpose group 5 – assembly and recreation;

is to be separate from means of escape provided for other parts of the building.

The aim is always to provide alternative routes from all areas in a building, but in the following circumstances a single route is permissible:

- when the storey exit is no further away than the limit for travel distance in one direction given in Table 2.

For this to be acceptable any room is not to have an occupant capacity in excess of 60 persons, unless it is an institutional building in purpose group 2(a), when the maximum number is 30 people in one room.

Occupant capacity calculations are described in section 1 (page 101).

- a storey having an occupant capacity not in excess of 60 persons when the limits of travel in one direction only are met (not to be applied to in-patient care in hospital).

Examples will arise where there is no alternative at the commencement of an escape route. There may be one exit from a room or corridor, and then escape routes are available in two directions.

To be acceptable the overall distance to the nearest storey exit must be within the limits when there is an alternative route. The first part of the escape – one direction only – is not to exceed the limit of travel when an alternative is not available.

In the example shown in Fig. B1.33 there is a dead end situation in a large open space. There are two exits available – A and B – and the routes from point 2 are at an angle of 45° plus 2.5° for each metre travelled from point 1. The lesser of the routes 1–2–B or 1–2–A is not to be more than the travel distance given in Table 2 for alternative routes and 1–2 is not to be longer than the travel distance where no alternative route is available.

- The number of occupants expected to be in a particular space may be specified as a basis for design. When the number of people likely to use a room, tier or storey is not known it can be

Table 2 Limitations on travel distance

Purpose group	Use of the premises or part of the premises	Maximum travel distance (1) where travel is possible in:	
		one direction only (m)	more than one direction (m)
2(a)	Institutional (2)	9	18
2(b)	Other residential		
	a. in bedrooms (3)	9	18
	b. in bedroom corridors	9	35
	c. elsewhere	18	35
3	Office	18	45
4	Shop and Commercial (4)	18 (5)	45
5	Assembly and Recreation		
	a. buildings primarily for disabled people except schools	9	18
	b. schools	18	45
	c. areas with seating in rows	15	32
	d. elsewhere	18	45
6	Industrial (6)	25	45
7	Storage and other non-residential (6)	25	45
2–7	Place of special fire hazard (7)	9 (8)	18 (8)
2–7	Plant room or rooftop plant:		
	a. distance within the room	9	35
	b. escape route not in open air (overall travel distance)	18	45
	c. escape route in open air (overall travel distance)	60	100

Notes:
1. The dimensions in the Table are travel distances. If the internal layout of partitions, fittings, etc. is not known when plans are deposited, direct distances may be used for assessment. The direct distance is taken as 2/3rds of the travel distance.
2. If provision for means of escape is being made in a hospital or other health care building by following the detailed guidance in the relevant part of the Department of Health 'Firecode', the recommendations about travel distances in the appropriate 'Firecode' document should be followed.
3. Maximum part of travel distance within the room. (This limit applies within the bedroom (and any associated dressing room, bathroom or sitting room, etc.) and is measured to the door to the protected corridor serving the room or suite. Sub-item (b) applies from that point along the bedroom corridor to a storey exit.)
4. Maximum travel distances within shopping malls are given in BS 5588: Part 10. Guidance on associated smoke control measures is given in a BRE report, *Design methodologies for smoke and heat exhaust ventilation* (BR 368).
5. BS 5588: Part 10 applies more restrictive provisions to units with only one exit in covered shopping complexes.
6. In industrial and storage buildings the appropriate travel distance depends on the level of fire risk associated with the processes and materials being used. Control over the use of industrial buildings is exercised through the Fire Precautions Act. Attention is drawn to the guidance issued by the Home Office *Guide to fire precautions in existing places of work that require a fire certificate Factories Offices Shops and Railway Premises*. The dimensions given above assume that the premises will be of 'normal' fire risk, as described in the Home Office guidance. If the building is high risk, as assessed against the criteria in the Home Office guidance, then lesser distances of 12 m in one direction and 25 m in more than one direction, would apply.
7. Places of special fire hazard are listed in the definitions in Appendix E.
8. Maximum part of travel distance within the room/area. Travel distance outside the room/area to comply with the limits for the purpose group of the building or part.

calculated using floor space factors. An example is an open plan shopping area where 5 m² is allowed per person (see section 1, page 10).

There are requirements for the minimum number of escape routes and exits from a room, tier or storey and it is likely that the following numbers will be increased when travel distances are taken into account:

- 1 escape route can serve a maximum of 60 people;
- 2 escape routes can serve a maximum of 600 people;
- 3 escape routes can serve in excess of 600 people.
- More than one escape route will be useless if the arrangement is such that they are all likely to be impassable simultaneously. The following criteria (Fig. B1.34) must therefore be followed:

  (a) the directions of the routes are 45° or more apart, or
  (b) when the directions are less than 45° apart they are separated by fire-resisting construction.

In this example alternative routes are available from point 1 as the angle A–1–B is in excess of 45°. The route 1–A or 1–B should have not more than the maximum travel distance permitted by Table 2.

The angle A–2–B is less than 45° and in this event alternative routes are not available, and the distance 2–B is not to be more than the maximum permitted distance for travel in one direction.

Only one exit – point B – is available for point 3 which should also not have to travel more than the maximum for one direction only.

Inner rooms

An inner room is one from which escape is only possible by passing through another room; called the access room, and there is therefore a serious hazard if a fire should occur in the access room.

Inner rooms are therefore only acceptable when the following requirements are met:

- there should not be more than 60 people in the inner room; but
- when the purpose group is 2(a) – institutional – the number of persons is limited to 30;
- the inner room is not to be a bedroom;
- the inner room is to be entered directly from an access room;
- escape from the inner room should not involve passing through more than one access room;
- the travel distance limits in one direction only in Table 2 are to be applied from any point in the inner room to an exit from the access room;
- the access room is to be in control of the same occupier and it is not to be a place of special fire risk, i.e. oil-filled transformer and

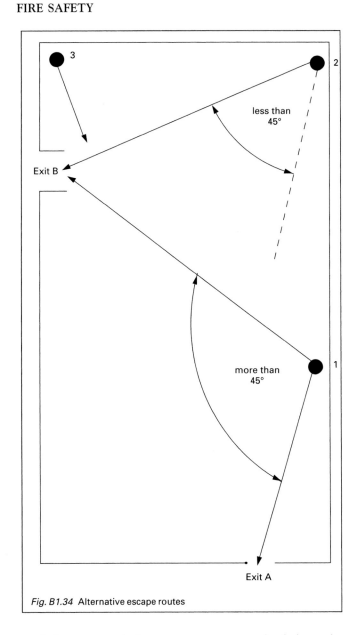

Fig. B1.34 Alternative escape routes

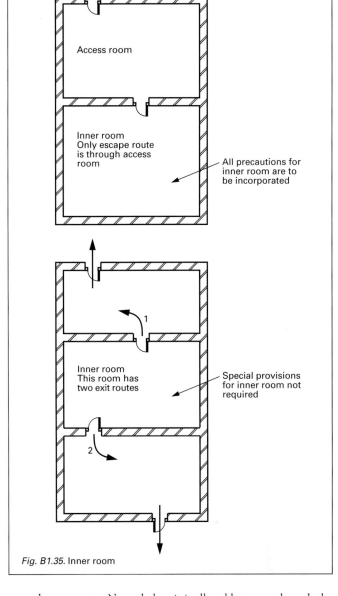

Fig. B1.35. Inner room

switchgear rooms, boiler rooms, storage space for fuel or other highly inflammable substances, and rooms housing a fixed internal combustion engine;

- (a) walls and partitions of the inner room are to be stopped not less than 500 mm below ceiling level, or
- (b) there should be a visual panel which need not be more than 0.1 m² in the door or wall of the inner room, or
- (c) an automatic fire detection and alarm should be fitted in the access room.

Requirements (b) and (c) are to enable the occupants of the inner room to see or have warning of any fire in the access room.

Central cores

When a building is planned with a central core, and it incorporates more than one exit, it is to be so arranged that the storey exits are remote from each other. No two exits are to be approached from the same lift hall, a common lobby, or undivided corridor, or linked by any of these.

Access to storey exits

A storey having more than one escape stair is to be arranged so that a person is not obliged to pass through one stairway to reach a second escape route. Nevertheless, it is allowable to pass through the protected lobby of one stairway to reach another stair.

Separation of circulation routes

If people are constantly passing through a protected stairway on what is a primary circulation route at the same level, self-closing fire doors are likely to become ineffective or even wedged open. This arrangement is therefore not acceptable and a protected stairway is not to be part of a primary circulation route between different areas of a building on the same level, unless the doors to the protected stairway and any passageway are fitted with an automatic release mechanism.

Different uses and occupancies

An area of a storey ancillary to the main use of a building and which is used for the consumption of food and drink is to have at least two escape routes provided. These routes should lead to a storey exit without entering any kitchen or similar area of fire hazard on the way.

The requirements for inner rooms apply.

In instances where storeys are split into different occupancies

- the escape route from one occupancy is not to pass through another occupancy, and

• if a common corridor or circulation space is included in the means of escape it is to be a protected corridor, or an automatic fire detection and alarm system is to be installed throughout the storey.

Dimensions of escape routes

All escape routes are to have a clear headroom of 2 m except when they pass through doorways. The width of escape routes depends upon the maximum number of persons they may be called upon to evacuate:

• for up to 50 persons the minimum width of route and exit is 750 mm which may be reduced to 530 mm for gangways between fixed storage racking, other than public areas in purpose group 4 (shop and commercial);
• for 51 to 110 persons the minimum width of route and exit is 850 mm;
• for 111 to 220 persons the minimum width of route and exit is 1050 mm;
• in excess of 220 persons, 5 mm is to be added for each person catered for;
• the minimum width of corridor in the pupil areas of schools is to be 1050 mm, increased to 1600 mm in dead ends.

Interpolation is not to take place when a width is less than 1050 mm.

When the actual number of people to be designed for is not known, a number is to be calculated using occupant capacity factors.

Where two or more storey exits are provided, it must be assumed that a fire may render one of them unusable. The other exits should be wide enough to permit all to leave quickly. When deciding on total width of exits needed, the largest exit is to be discounted. Stairways should be as wide as any storey exit discharging into them.

When two or more exits are in use, the total number of people which can escape is obtained by adding together the number of persons able to use each exit. This means that when three exits are available, each being 850 mm wide, 330 persons can be accommodated (3 × 110).

It is not correct to calculate 850 mm × 3 = 2550 mm giving a total of 510 persons.

The spacing of seating in auditoria is covered by BS 5588: Part 6: 1991, 'Code of practice for assembly buildings'.

Protected corridors

• A corridor has to be a protected corridor when

  (a) it serves bedrooms;
  (b) it is a dead end corridor (not including recesses and extensions described in Figs 10 and 11 of BS 5588; Part 11; 1997, Code of practice for shops, offices, industrial storage and other similar buildings; and
  (c) it is common to two or more different occupancies.

• Open planning does nothing to prevent the spread of smoke, but the danger can be seen and this enables the occupants to take appropriate action. There are occasions when a corridor is enclosed by partitions, but it is not a protected corridor. In the early stages of fire, these partitions do offer defence against the spread of smoke.

  In this situation the partitions are to be carried up to the soffit of the structural floor, or to a suspended ceiling. Openings into rooms are to have doors, but it is not necessary for them to be fire doors.

• It is necessary, when a corridor which provides access to alternative escape routes, and is more than 12 m long, to be subdivided by self-closing doors. This is to reduce the risk of smoke spread, and will ensure that the provision of doors and any associated screens midway between two storey exits will safeguard the route from smoke.

• Where a dead end part of a corridor gives access to a place where alternative escape routes are available, there is a possibility that smoke could affect the availability of both routes before people from the dead end have escaped.

  This may be avoided by pressurizing escape stairways and corridors with a system in accordance with BS 5588: Part 4: 1978. All dead end corridors more than 4.5 m long are to be separated by self-closing fire doors, from any part of the corridor which

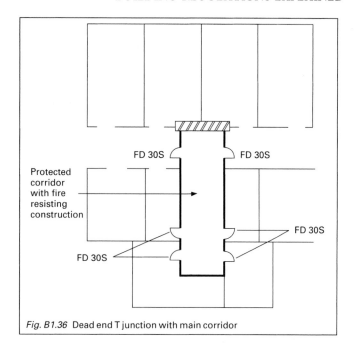

Fig. B1.36 Dead end T junction with main corridor

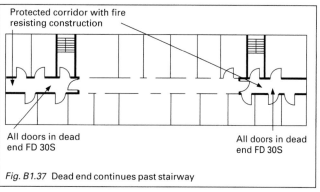

Fig. B1.37 Dead end continues past stairway

(a) gives two directions of escape (Fig. B1.36)
(b) goes past one storey exit to another (Fig. B1.37).

• Additional measures to prevent the escape of smoke are related to cavity barriers, described later in B3.
• Where an external escape route is within 1.8 m of an external wall, that part of the wall up to a height of 1.1 m above the paving level of the route is to be of fire-resisting construction.

Escape over flat roofs

If more than one escape route is provided, one route may be by way of a flat roof, provided that

• the route does not serve an institutional building;
• the route does not serve part of a building for use by the public;
• the roof is part of the same building from which people are escaping;
• the route leads to a storey exit or external escape;
• the roof and its supporting structure is of fire-resisting construction, and any opening within 3 m of an escape route is to be fire resisting;
• the route is adequately marked and guarded (Part K, Chapter 19).

PURPOSE GROUP 2(A). HOSPITALS AND OTHER RESIDENTIAL CARE PREMISES

• Guidance in Nucleus Fire Precautions recommendations; Health Technical Memorandum 81; Home Office Draft Guide to Fire Precautions in Hospitals, and the Draft Guide to Fire Precautions in

Existing Residential Care Premises, are appropriate for non-NHS premises.

- Progressive horizontal evacuation is advocated. In an effort to find a place of relative safety in a short distance, patients may be evacuated into adjoining compartments or subdivisions in the areas used for in-patient care. The arrangement enables patients to be relatively safe, but available for further evacuation should that prove to be necessary.
- The approved document advises that the principle of progressive horizontal evacuation may have use in some other residential buildings (Fig. B1.38).

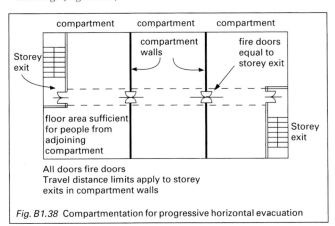

Fig. B1.38 Compartmentation for progressive horizontal evacuation

- In dealing with buildings where the 'Firecode' advice is not applicable, the following conditions should be observed:

  (a) Compartments should have sufficient area to accommodate the people normally there, and also people from an adjoining compartment. Numbers may be calculated using the design occupancy factors.
  (b) Each compartment is to have at least two escape routes. The second route may be by way of the third compartment if its exit is independent of the exits from the other compartments.

SECTION 5. VERTICAL ESCAPE: BUILDINGS OTHER THAN DWELLINGS

Vertical escape is the third stage in reaching safety at ground level. It links stage two where the occupants have reached a position protected from fire and should be able to leave the building without having to negotiate any further fire hazard. Vertical escape is usually by way of stairs, and lifts should not be used except in controlled circumstances when independent electrical supplies are installed and where operation is in the hands of management or the fire service.

The vital factors for vertical escape are the adequacy of stairs, both in numbers and dimensions.

ESCAPE STAIRS

- They must link with horizontal escape routes.
- Mixed occupancy buildings may require independent stairs.
- Is a single stair appropriate?
- Width is to be determined having in mind that a stair may have to be discounted if affected by smoke or fire.
- Large buildings may require the provision of stairs to serve as fire-fighting stairs for use by the fire service and this may affect the number of stairways required.
- Some buildings contain storeys, or parts of storeys, embracing occupancies in different purpose groups. When this is the case and when one of the purpose groups is

  (a) assembly or recreation, or
  (b) residential,

  there is to be means of escape which is independent from the other uses.

Where independent escape routes are not required because of the involvement of different purpose groups, there are some situations where it may be necessary to provide a single escape stairs:

- from a basement which is permitted to have a single escape route and from a building where no storey has a floor level above 11 m from the ground and where every storey is allowed to have a single escape route.

This applies when a storey is not used for in-patient care in a hospital; it has not more than 60 occupants per storey; and on each storey meets the requirements for one direction travel distances from Table 2.

In schools, floors above first floor should only be occupied by adults.

- When small premises are involved a single escape stairs is permissible when Clause 10 of BS 5588: Part 11: 1997, 'Code of practice for shops, offices and other similar buildings' is complied with.

MINIMUM WIDTH OF ESCAPE STAIRS

The width of a stairway must have regard to the number of people who will be using it in the event of fire. This is dependent upon the number of persons occupying each storey and how many storeys will be making a simultaneous discharge of people onto the stairway in a fire's early stages. The recommendations regarding widths of stairs are:

- a stair must not be less wide than any exit or exits which discharge into it;
- minimum widths are to be taken from Table 3;

Table 3 Minimum widths of escape stairs

Situation of stair	Maximum number of of people served (1)	Minimum stair width (mm) (2)
1a. In an institutional building (unless the stair will only be used by staff)	150	1000
1b. In an assembly building and serving an area used for assembly purposes (unless the area is less than 100 m²)	220	1100
1c. In any other building and serving an area with an occupancy of more than 50	over 220	see Note (3)
2. Any stair not described above	50	800 (4)

Notes:
1. Assessed as likely to use the stair in a fire emergency.
2. BS 5588: Part 5 recommends that firefighting stairs should be at least 1100 mm wide.
3. See Table 4 for sizing stairs for simultaneous evacuation, and Fig. B1.39 for phased evacuation.
4. In order to comply with the guidance in the Approved Document to Part M on minimum widths for areas accessible to disabled people, this may need to be increased to 1000 mm.

- to be no more than 1400 mm wide if vertical extent exceeds 30 m; may be wider if there is a central handrail;
- not become narrower at any point between entry and the final exit;
- where the stairs width is in excess of 1800 mm, there must be a central handrail, but the stair width on either side of the handrail is used in assessing the capacity of the stair (see Part K, Chapter 19);
- should an exit route from a stair also be an escape route from a ground or basement storey, the width must be increased to cater for all users.

In the approved document the maximum width of 1400 mm has been chosen for stairs in tall buildings following a research programme. It was found that people when making a long descent prefer to keep in reach of a handrail. Therefore, on wider stairs the central portion could be unused. This may give rise, on occasions, for additional stairs, and consideration should be given to using phased evacuation in buildings exceeding 30 m above ground.

Calculating minimum stair width

Width of stairs will arise from consideration of the number of persons to be evacuated, and whether total or part evacuation, or phased evacuation of the building is to be provided – and the number of stairs.

If the actual number of people involved is not known at the time of design, the figure is to be calculated using occupancy, or floor space factors from Table 1.

When two or more stairs are being provided it is also necessary to assume that one may be unusable for some reason. Each stair is to be discounted in turn in order to establish that the remaining stair or stairs are adequate for the number of persons leaving the building.

There are exceptions to this requirement:

- if on each floor, except the top storey, access to the escape stairs is through a protected lobby, it is not necessary to discount a stair;
- if the stairways are protected by a smoke control system complying with BSS 5588: Part 4: 1978, 'Code of practice for smoke control in protected escape routes using pressurization'.

The rule for discounting stairs does apply to buildings fitted with a sprinkler system unless the stairs are entered through lobbies or are protected by a smoke control system.

Simultaneous evacuation

When considering total evacuation the escape stairs (in association with the other escape provision) are to have a capacity sufficient to permit all the floors to be evacuated simultaneously and also to cater for those people temporarily on stairways during the evacuation.

Total evacuation arrangements are to be provided for:

- all stairs which serve basements;
- all stairs which serve buildings with open spatial planning; and
- all stairs which serve other residential or assembly and recreation buildings.

Designs based on simultaneous evacuation are included in BS 5588: Part 7: 1997, 'Code of practice for the incorporation of atria in buildings'.

Table 4 gives the capacity of stair widths from 1000 mm to 1800 mm which serve up to ten storeys, when total evacuation is to be used.

As an alternative and when designing for taller buildings, the following formula may be used for calculation purposes:

$$P = 200 w + 50 (w - 0.3) (n - 1), \text{ or}$$

$$w = \frac{P + 15n - 15}{150 + 50n}$$

where P = the number of people that can be served;
w = the width of stairs in metres, and
n = the number of storeys to be served.

- separate calculations are made for stairs or flights serving basement storeys and also upper storeys;
- P should be divided by the number of available stairs;
- the formula should be used to calculate widths of stairs where occupants are not distributed evenly within the building or a storey.

Phased evacuation

Phased evacuation may be used for any buildings and it requires careful planning, management and supervision. It works in the following way:

- The first people to be evacuated are those people of reduced mobility.
- People are evacuated from the floor of fire origin and the floor above.

As necessary, there may be a need to evacuate more people and this is done two floors at a time.

This method permits narrower stairs than would otherwise be required, and there is a practical advantage that, in large buildings, there is less disturbance.

Phased evacuation should not be used for basements, open plan buildings, and the purpose group's other residential or assembly and recreational buildings.

There are several conditions which must be satisfied before a building or part of a building is to have its escape arrangements designed for phased evacuation.

Table 4   Capacity of a stair for basements and for simultaneous evacuation of the building

No. of floors served	Maximum number of persons served by a stair width:								
	1000 mm	1100 mm	1200 mm	1300 mm	1400 mm	1500 mm	1600 mm	1700 mm	1800 mm
1.	150	220	240	260	280	300	320	340	360
2.	190	260	285	310	335	360	385	410	435
3.	230	300	330	360	390	420	450	480	510
4.	270	340	375	410	445	480	515	550	585
5.	310	380	420	460	500	540	580	620	660
6.	350	420	465	510	555	600	645	690	735
7.	390	460	510	560	610	660	710	760	810
8.	430	500	555	610	665	720	775	830	885
9.	470	540	600	660	720	780	840	900	960
10.	510	580	645	710	775	840	905	970	1035

Notes:
1. The capacity of stairs serving more than 10 storeys may be obtained by using linear extrapolation.
2. The capacity of stairs not less than 1100 mm wide may also be obtained by using the formula.
3. Stairs with a rise of more than 30 mm should not be wider than 1400 mm unless provided with a central handrail.
4. Stairs wider than 1800 mm should be provided with a central handrail.

- Approach to stairway on each storey is to be through a protected corridor or protected lobby. This does not apply to a top storey.
- Lifts are to be approached via a protected lobby at each storey.
- All floors are to be compartment floors.
- Where there is a storey with a floor more than 30 m above ground level an automatic sprinkler system is to be installed to protect the building.

  The relevant recommendations are contained in BS 5306: Part 2:1990, 'Specification for sprinkler systems'. This would have regard to the relevant occupancy rating and additional requirements for life safety. The provision does not apply to purpose group 1(a) – flats – when part of a mixed building.
- If the above conditions can be achieved, buildings which are less than 30 m high may be designed on the basis of phased evacuation.
- A fire-warning system in accordance with at least L3 standard given in BS 5839: Part 1: 1988, 'Code of practice for system design installation and servicing'.
- A telephone or intercom system or other suitable internal speech communication system is to be provided to give contact between a control point at the service access and a fire warden on every floor.

  Where it is appropriate to install a voice alarm, it should be in accordance with BS 5839: Part 8: 1998, 'Code of practice for the design, installation and servicing of voice alarm systems'.
- The minimum aggregate width of stairs designed for phased evacuation is shown in Fig. B1.39. There is an assumption that evacuation will be limited to two floors at a time. Assuming a minimum width of at least 1000, an alternative to using the table is to apply the formula $[(P \times 10) - 100]$ mm, where P is the number of occupants in the most heavily populated storey.

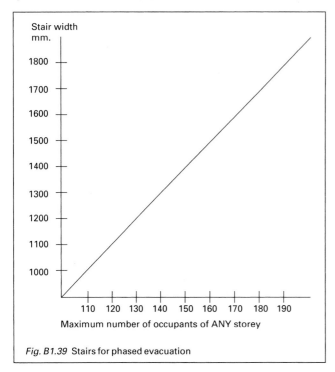

Fig. B1.39 Stairs for phased evacuation

Where stairs rise to more than 30 m there is to be a central handrail for stairs wider than 1400 mm.

PROTECTION OF ESCAPE STAIRS

To provide a satisfactory escape route in relative safety, stairways are to have a standard of fire protection and every internal escape stair is to be within a fire-resisting enclosure. There may also be additional measures when a stairway penetrates one or more compartment floors or is a firefighting shaft; discussed later.

But an unprotected stair, or accommodation stair, may be part of an internal route to a storey exit or a final exit. In these circumstances the number of people involved will be very small and the distance of travel is also limited. An example is in the requirements for small premises covered by Clause 10 of BS 5588: Part 11: 1997, 'Code of practice for shops, industrial storage and other similar buildings and raised storage areas'.

Sometimes an escape stair is to have the protection of a protected lobby, or a protected corridor. Therefore, when a stair

- is the only one for a part of the building, or a building having more than one storey above or below the ground storey; or
- is to any storey at a height which is more than 18 m; or
- where phased evacuation is proposed; or
- is in a building having a sprinkler system and the stair width has not been arrived at following the discounting of one stairway; or
- where the stair is a firefighting stair; or
- a smoke control system may be a considered alternative.

Protected corridors or protected lobbies are to be provided on all storeys including a basement, but are not required on the top storey.

A protected lobby is to be provided between an escape stairway and any place of special fire risk. The lobby is to have at least 0.4 m² of permanent ventilation, or have a mechanical smoke control system. Places of special fire risk are oil-filled transformer and switchgear rooms, boiler rooms, storage space for fuel or other highly flammable substances and rooms housing fixed internal combustion engines.

All protected stairways are to discharge directly to a final exit, or through a protected exit passageway to a final exit. A final exit is one which gives direct access to a street, passageway, walkway or open space in a situation which will allow the rapid dispersal of people from the building; smoke and fire no longer being a danger. Exits from protected stairways must be as wide, or wider than the stair leading to it.

- If two protected stairways are adjacent they, and any associated protected passageways to final exits, are to be separated by an imperforate enclosure.

Space within protected stairways

In order to maintain a protected stairway free of potential fire sources there is control of the space, and the following only are permitted:

- Sanitary accommodation and washrooms, but not cloakrooms. The only gas appliances that may be installed are water heaters and sanitary towel incinerators.
- A lift well; but not if it is a firefighting stair.
- A reception desk or enquiry office at ground or access level, and the area used for this purpose is not to exceed 10 m² This is not permissible if it is the only stair serving a part of, or the building.
- Cupboards enclosed in fire-resisting construction. But not permissible if it is the only stair serving a part of, or the building.
- External walls of protected stairways are to be protected (see Fig. B1.40).

Gas service pipes and associated meters are not to be incorporated within a protected stairway unless the installation is carried out in accordance with the Pipelines Safety Regulations 1996, and the Gas Safety (Installation and Use) Regulations 1998.

Measures are necessary to prevent smoke from basements getting up into the escape routes from upper floors.

- Single escape stairs should not continue down to serve any basement storey. A separate stair should be provided for basements.
- Where there is more than one escape stair from an upper storey, one may terminate at ground level. Other stairs may go down to the basement if there is a ventilated protected lobby, or a ventilated protected corridor separating the stairs and accommodation at each basement level.

External escape stairs

Where more than one escape route is available from a storey or part of a building, one may be an external stair, if

- for assembly and recreation buildings the escape is not to be used by members of the public; or
- for an institutional building the escape is to be used only by office or resident staff.

Where an external stair may be provided it is to be protected from the weather and also protected from fire within the building. Reference should be made to the advice regarding flats and maisonettes in section 6, and Figs B1.44 and B1.45, except that the stair may rise more than 6 m if protected from snow and ice.

The actual siting of a stair may offer some protection from the weather and it is not implied in the approved document that full enclosure is always necessary.

SECTION 6. PROVISIONS COMMON TO BUILDINGS OTHER THAN DWELLINGHOUSES

This section is to be read in conjunction with sections 3, 4 and 5 and contains some guidance on matters associated with escape routes generally. It brings together such matters as fire resistance, limitation of uninsulated glazing, suitability of doors and fastenings, the construction of escape stairs and materials of limited combustibility.

PROTECTION OF ESCAPE ROUTES

Information on the fire resistance test criteria and standard of performance are contained in a lengthy Appendix A to the approved document.

A 30 minute standard of fire resistance is suitable for the protection of most means of escape. There will be times when greater fire resistance is required and these will arise when considering the requirements of B3 (internal fire spread) and B5 (facilities for fire service).

- Walls, partitions and other enclosures that are to be fire resisting (including roofs that form means of escape) are to be not less than the minimum required by Appendix A, Table A1 – Specific provisions of test for fire resistance of elements of structure, and Table A2 – Minimum periods of fire resistance.

Doors

Test criteria are contained in Appendix B to the approved document and all doors which are to be fire resisting are to have an appropriate performance set out in Table B1 – Provisions for fire doors.

Glazed elements

If glazed elements in fire-resisting enclosures and doors are only able to fulfil requirements in terms of integrity, the use of glass is limited and is set down in Table A4 – Limitations on the use of uninsulated glazed elements on escape routes.

Where relevant performance can be met for both integrity and insulation there is no restriction on the use or area of glass. There are restrictions on the use of glass in firefighting stairs and lobbies contained in clause 9 of BS 5588: Part 5:1991, 'Code of practice for firefighting stairs and lifts'.

There is also guidance on safety glazing in Part N – Glazing materials and protection (Chapter 22).

Escape route doors

Doors on escape routes must be easy to open – delay can be a critical factor in a successful escape.

- Doors on escape routes should not be fitted with lock, latch or bolt or the fastenings must be simple and easy to operate from the side of approach.
- Fastenings should not require a key and it should not be necessary to operate more than one mechanism.
- Doors requiring security on final exits should use panic bolts.
- Door closing and hold open devices are described in Appendix B to the approved document.
- Where reasonably practical, doors are to be hung to open in the direction of escape. Where the number of persons expected to use the door is more than 60 then it must open in the direction of escape. Where very high fire risk is involved the number may be less than 60.
- (a) A door on an escape route must open for not less than 90°.
  - (b) The swing is to be clear of any change of floor level except a threshold or a single step on line of doorway.
  - (c) A door is not to reduce the effective width of an escape route across a landing.
  - (d) A door opening towards a corridor or a stairway is to be recessed and when fully open not project into the stairway or corridor.
- Vision panels are required in doors

  (a) when subdividing corridors;
  (b) where doors are hung to swing both ways.

  Refer to Part M – Access and facilities for disabled (Chapter 21) and Part N – Glazing (Chapter 22).
- The passage of people escaping can be obstructed by revolving doors, automatic doors and turnstiles. The doors are not to be placed on escape routes unless they are readily openable in an emergency.

  In particular, they are not to be placed across an escape route unless of the required width – fail safely to outward opening from any position of opening – have a monitored fail-safe system in the event of power failure and will fail safely to the open position should the power supply fail.

  An non-automatic swing door is to be installed alongside the automatic door.

Stairs

Escape stairs and associated landings are to be constructed of materials of limited combustibility. Materials of limited combustibility include composite products, of which plasterboard is an example, and when exposed as linings meet the appropriate flame spread rating. These materials may be used in the following locations:

- when it is the only stair serving the building, or part of a building (does not apply to a building of two or three storeys and in purpose group 1(a) or purpose group 3);
- when it is within a basement storey (does not apply to a private stair in a maisonette);
- when it serves a storey with a floor level in excess of 18 m above ground level or access level, or
- when it is an external stair (does not apply when the stair connects ground floor or paving level to a flat roof or floor which is not more than 6 m above ground level);
- when it is a firefighting stair.

It is permissible to attach combustible materials to the upper surface of stairs. This does not apply to firefighting stairs.

Firefighting stairs are discussed in section 18.

- Single steps should be avoided, but when absolutely necessary should be clearly marked.

Spiral stairs

Spiral stairs, helical stairs and fixed ladders may be incorporated into an escape route when complying with the following:

- Helical stairs and spiral stairs are to be in accordance with BS 5395: Part 2: 1984, 'Code of practice for the design of helical and spiral stairs'. Type E is intended for use by members of the public.

A type E stair is intended to be used by large numbers of people at one time; e.g. in a place of assembly. Typical outside diameter is between 2500 mm and 3500 mm.

Rise:	150 mm to 190 mm
Inner going:	150 mm
Minimum centre going:	250 mm
Maximum outer going:	450 mm
Twice rise + going:	min. 480 mm; max. 800 mm
Clear width:	1000 mm

- Fixed ladders are only suitable as an access to plant rooms, and are not to be used as a means of escape by members of the public. Fixed ladders should be made of non-combustible materials.

Guidance on the safety and guarding of stairs is contained in Part K (Chapter 19).

External walls of protected stairways

- There are situations when a protected stairway with an external wall of little fire resistance could be at risk from fire spreading from the adjoining structure (Fig. B1.40). This could occur where a protected stairway projects beyond, or is recessed from, or is in an internal angle of, the building. In these circumstances there is not to be any unprotected area of the building within 1.8 m of the stairway enclosure.

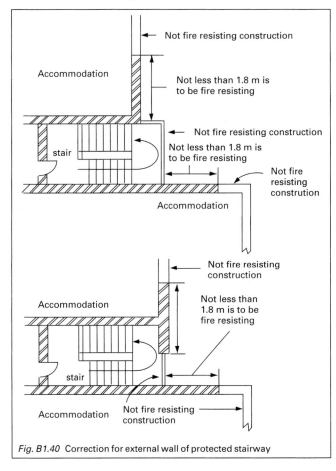

Fig. B1.40 Correction for external wall of protected stairway

External escape stairs

Any external escape provided is to meet the following requirements:

- All doors opening onto the stairs are to be fire resisting and self-closing. This does not apply to the door at the top of the stairs if there is only one exit leading onto the landing.
- Adjoining external walls are to be fire resisting

  (a) within 1.8 m of the flight and landings;
  (b) within 9 m vertically below flights and landings.

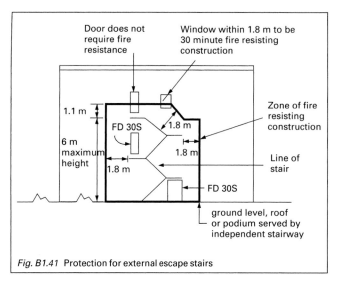

Fig. B1.41 Protection for external escape stairs

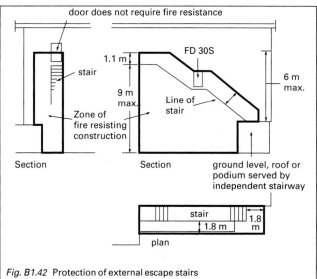

Fig. B1.42 Protection of external escape stairs

Above the top level of the stair only 1.1 m is required to be fire resisting unless it is a stair rising from the basement to ground level.

- Any part of the building within 1.8 m from the stairs to safety is to be fire-resisting construction. This does not apply if there is a choice of route at the foot of the stair, and one offers freedom from the effects of fire in the adjoining building.
- The maximum height above ground level, a roof or podium, for a stair without weather protection is 6 m. The amount of protection required will depend upon location and the degree of protection offered by the building.
- All glazing in fire-resisting construction is to be fire-resisting satisfying integrity standards. It is not to be openable.

General

- Escape routes are to have a clear headroom of not less than 2 m, and the only projection allowed below this height is for door frames.
- Foor surfaces on escape routes should minimize slipperiness when wet.
- Ramps on escape routes are not to be steeper than 1:12.
- A sloping floor or tier is not to have a pitch in excess of 35° from the horizontal.

Final exits

Final exits are to be not less in width than the escape routes they serve, and comply with the following:

- be sited to allow rapid dispersal of people from the building;
- have direct access to a street, passageway, walkway or open space;
- route is to be well defined and guarded where necessary;
- final exits are to be clearly marked, particularly when the exit is from a stair that continues down, or up, from the point of final exit;
- are to be clear of any risk of fire or smoke in a basement. This can arise when outlets to basement smoke vents are badly sited.
- are to be clear of openings to transformer chambers, refuse chambers, boiler rooms and similar hazards.

Escape over flat roofs

Any escape route provided over a flat roof is to meet the following requirements:

- the roof is to be part of the same building;
- route is to lead to a storey or external exit;
- the escape route, supporting structure and any opening within 3 m is to be fire resisting;
- route is to be well defined and protected by a wall or adequate barriers (see Part K – Chapter 19).

LIGHTING OF ESCAPE ROUTES

All escape routes are to have adequate lighting and the areas and routes shown in Table 5 are also to be provided with escape lighting which will function if the mains supply fails. Escape stairs are to have lighting which is on a separate protected circuit from that supplying any other part of the escape route.

Escape lighting requirements are given in BS 5266: Part 1: 1988, 'Code of practice for the emergency lighting of premises other than cinemas and certain other premises used for entertainment'. Also relevant is CP 1007: 1955, 'Maintained lighting for cinemas'.

Exit signs are to mark every doorway or other exit giving access to a means of escape. Letters and graphics are to be of adequate size and be in accordance with Health and Safety (Safety Signs and Signals) Regulations 1996, and BS 5499: Part 1: 1990, 'Specification for fire safety signs'. Other legislation may require additional signs in some buildings.

This does not apply to dwellings, or to those exits which are in ordinary use.

Protected circuits are to be used when they must function during a fire, and should consist of cable meeting the classification CWZ specified by BS 6387: 1983, 'Specification for cables required to maintain circuit integrity under fire conditions'.

The cable is to be routed through parts of the building where fire risk is negligible, and is to be separate from any other circuit.

LIFTS

It is generally unwise to use lifts when there is a fire in the building. The lift may become immobilized and the occupants trapped. There are, however, occasions when the overall fire management plan for a building may make use of a lift for the evacuation of disabled people.

When this option is to be used the lift installation has to be carefully sited and protected. A number of safety features must be provided in order to ensure that the lift remains usable during a fire. Guidance on the precautions to be taken are contained in BS 5588: Part 8: 1991, 'Code of practice for means of escape of disabled people'.

Lifts, by their very nature, connect the various floors of a building and constitute a risk of fire spread and danger to users of escape routes, unless adequate precautions are taken.

Shopping malls and atriums frequently feature wall climber lifts as a visual attraction. These lifts do not, of course, have a conventional well and the occupants may be at risk if they should pass through a smoke reservoir. The integrity of the smoke reservoir should be maintained to ensure the safety of people in the lift.

- A lift well should be within the enclosure of a protected stairway, or
- enclosed for the full height of the shaft by fire-resisting construction if it is placed where it could affect the means of escape, and
- a lift well passing through compartments should be constructed as a protected shaft.

Where escape is planned using phased or progressive horizontal evacuation and the lift well is not within the enclosure of a protected stairway, the lift is to be entered on every storey via a protected lobby.

Table 5 Provisions for escape lighting

Purpose group of the building or part of the building	Areas requiring escape lighting
1. Residential	All common escape routes (1), except in 2-storey flats
2. Office, Shop and Commercial (2) Industrial, Storage, Other non-residential	a. Underground or windowless accommodation b. Stairways in a central core or serving storey(s) more than 18 m above ground level c. Internal corridors more than 30 m long d. Open-plan areas of more than 60 m²
3. Shop and Commercial (3) and car parks to which the public are admitted	All escape routes (1) (except in shops of 3 or fewer storeys with no sales floor more than 280 m² provided that the shop is not a restaurant or bar)
4. Assembly and Recreation	All escape routes (1), and accommodation except for: a. accommodation open on one side to view sport or entertainment during normal daylight hours b. parts of school buildings with natural light and used only during normal school hours
5. Any purpose group	a. Windowless toilet accommodation with a floor area not more than 8 m² b. All toilet accommodation with a floor area over 8 m² c. Electricity and generator rooms d. Switch room/battery room for emergency lighting system e. Emergency control room

Notes:
1. Including external escape routes.
2. Those parts of the premises to which the public are not admitted.
3. Those parts of the premises to which the public are admitted.

Where a lift serves a basement and enclosed car park, access should be through a protected corridor or a protected lobby unless the lift is in a protected stairway.

Where there is a high fire risk, as for example in kitchens, living rooms and stores and lifts pass through these high-risk areas and deliver directly into corridors serving bedrooms, the lift should be approached by a protected corridor, or lobby, or be in the enclosure of a protected stairway.

If there is only one escape stair in a building, a lift is not to be taken down to basement level. Likewise a lift should not be taken to a basement level if it is within the enclosure to an escape stair which ends at the ground storey.

Whenever possible lift rooms are to be placed above the lift well. When the lift well is in a protected stairway which is the only stairway in the building, precautions must be taken to prevent the spread of fire and smoke into the stairway and this requires the machine room to be above the lift well, or if this is not possible, to be placed outside the stairway.

VENTILATION AND AIR CONDITIONING

Mechanical ventilation systems are to be designed to make sure that air movement is directed away from protected escape routes and their exits. Provision may also be made for a partial or complete shutdown in the event of fire.

Recirculating air distribution systems are to be installed in accordance with BS 5588: Part 9: 1989, 'Code of practice for ventilation and air conditioning ductwork', so far as it relates to operation under fire conditions.

In places of assembly, mechanical ventilation arrangements should follow guidance in BS 5588: Part 6: 1991, 'Code of practice for assembly buildings'.

• Ventilation and air-conditioning plant should be compatible with any system of pressurization when operating under fire conditions.

Mechanical ventilation and air-conditioning plant should be designed and installed according to the guidance in BS 5720: 1979, 'Code of practice for mechanical ventilation and air conditioning in buildings'.

REFUSE CHUTES AND STORAGE

The siting and construction of refuse storage chambers, refuse chutes and hoppers should follow the recommendations in BS 5906: 1980, 'Code of practice for storage and onsite treatment of solid waste from buildings'.

Refuse chutes and places provided for refuse storage should

• have fire-resisting construction separating them from other parts of the building, and
• not be placed in protected lobbies or protected stairways.

Access from the open air, or through a protected lobby, should be provided to rooms for storage of refuse, or containing refuse chutes. The lobby is to have not less than $0.2\,m^2$ of permanent ventilation.

Access to refuse storage containers is not to be placed near to windows of dwellings, an escape route or a final exit.

Shop storerooms

Fully enclosed walk-in storerooms in shops which are so sited as to prejudice the means of escape are

• to be separated from retail areas by fire-resisting construction, or
• provided with automatic fire detection and alarm system, or
• to be fitted with sprinklers.

Special fire hazard

Places of special fire hazard include oil-filled transformer and switchgear rooms, boiler rooms, storage space for fuel or other highly-inflammable substances, and rooms housing a fixed internal combustion engine.

They are to be enclosed with fire-resisting construction, and each side of elements of structure is to satisfy 30 minutes' exposure (load-bearing, integrity and insulation).

SECTION 7. INTERNAL FIRE SPREAD: LININGS

Internal spread of fire on surfaces may be limited by

• low rate of spread of flame on surfaces;
• provisions for walls, ceilings and rooflights which vary with use of a building and the location of the room or space;
• fire protection of suspended ceilings and other enclosures.

Therefore, the choice of materials for walls and ceilings can have a marked effect on the spread of fire and the rate at which it will grow.

Circulation areas should have careful attention as unsuitable linings may lead to a rapid spread of fire and create hazard for people along the route to an exit.

Regard must be given to the fact that some lining materials react more quickly or violently to intense radiant heating, and once ignited give off more heat when burning.

The requirement

B2. In order to limit the spread of fire inside a building, materials forming the internal linings must:

• **give adequate resistance to the spread of flame over the surface, and**
• **if ignited shall have a reasonable rate of heat release.**

For the purpose of B2, the term 'internal linings' includes materials lining any partition, wall, ceiling or other internal structure. The requirements do not extend to the upper surfaces of stairs or floors, or to fittings and furniture.

The performance of lining materials will be met if the spread of flame over the surface is restricted by making provision for them to have

• low rate of spread of flame; and
• low rate of heat release.

The degree of restriction necessary is related to the position of the lining, and the approved document gives guidance on these matters, but not on the extent to which the materials emit smoke and fumes.

There is an extensive Appendix A to Part B, and the different classes of performance are related to the methods of testing employed, and set out in the following BSs:

BS 476: Part 4: 1970 (1984), Non combustibility tests for materials
BS 476: Part 6: 1989, Method of test for fire propagation
BS 476: Part 7: 1971, Method of test to determine classification of the surface of spread of flame products.
BS 476: Part 11: 1982, Method of assessing heat emission from building products.

Where tests for thermoplastic materials are not appropriate to BS 476: Part 7: 1971, other tests are available and the material carries classifications TP(a) rigid, TP(a) flexible and TP(b).

CLASSIFICATION OF LININGS

There may be variations and specific provisions, but generally the surface linings of walls and ceilings are to satisfy the following classifications:

- Class 3
  To be used in small rooms, where the area is not more than

  (a) 4 m^2 in residential buildings, and
  (b) 30 m^2 in non-residential buildings.

  Domestic garages not more than 40 m^2 in area.
- Class 1
  To be used in other rooms (including garages, and also in circulation spaces within dwellings).
- Class 0
  To be used in other circulation spaces, including the common areas of flats and maisonettes.

In this context a room is an enclosed space within a building which is not used solely as circulation space. It includes conventional rooms and such spaces as warehouses and auditoria, and cupboards that are not fittings. 'Voids' are not included.

Materials having a resistance to the spread of flame control the spread of fire over wall and ceiling surfaces. Two criteria are used.

- Surface spread of flame.
  BS 476: Part 7: 1971 or 1987 relate to the surface spread of flame test for materials and determines the tendency of materials to support the spread of flame across surfaces and classifies them in relation to exposed surfaces of walls and ceilings.

  The various classes of tests that must be satisfied are as follows:

Class 1 spread of flame at 1½ minutes:	165 mm
Class 2 spread of flame at 1½ minutes:	215 mm
Class 3 spread of flame at 1½ minutes:	265 mm
Class 4 exceeds the limits for Class 3	

There is a small tolerance in respect of one specimen in a sample.

Class 4 is not acceptable to the building regulations.

Class 0 is an additional class which imposes greater control of both the spread of the flame and the rate at which the material releases heat. Class 0 is used generally as a designation for the walls and ceilings of escape routes. There are three basic requirements which must be met to achieve Class 0.

1. The material must have a surface spread of flame classification of Class 1.
2. The fire propagation index should not exceed 12.
3. The sub-index figure for i_1 should not exceed 6.

When a product is composed entirely of limited combustibility materials, or non-combustible materials it can be included in Class 0.

BS 476: Part 4: 1970 describes a test for determining whether building materials are non-combustible. Materials used in the construction and finishing of buildings are classified 'non-combustible' or 'combustible' according to their behaviour in the non-combustibility test. The test is intended for building materials, whether coated or not, and does not apply to the coating alone.

- Fire propagation
  BS 476: Part 6: 1981 gives a test for fire propagation for products. The result of the test is expressed as a fire propagation index, and provides a comparative measure of the contribution to the growth of fire made by an essentially flat material, composite or assembly. It is used as an assessment of the performance of wall and ceiling linings.

  The result 'L' ranges from 0 to 100+. When the 'L' figure is low it indicates that the sample tested did not contribute much to the heat of the furnace when it was being tested. Three sub-indices relate to the heat contributed at different stages of the test. They are i_1, i_2 and i_3. When a sample burns and releases heat quickly the sub-index will be high. Sub-index figure i_1 is derived from the first three minutes of test.
- Thermoplastic material
  When thermoplastic materials are used they must be bonded to a substrate (which is not a thermoplastic) material or used as linings to a Class 0 non-thermoplastic surface.

Walls

Wall linings include the surface of glazing, except the glazing in doors and a ceiling which slopes at an angle of more than 70° to the horizontal.

A wall does not include doors or door frames; window frames and frames in which glazing is fitted; architraves, cover moulds, picture rails, skirting and any similar narrow features; fireplace surrounds, mantelshelves and fitted furniture.

Parts of a wall in a room may have a lower class than those given above but in no case is it to be lower than Class 3. To achieve this reduction the total area of those parts is not to be more than 50% of the floor area of the room. There is, however, a maximum area which, for residential buildings, is 20 m^2 and for non-residential buildings, 60 m^2.

Ceilings

- Ceilings include the surface of glazing and any part of a wall sloping at an angle of 70°or less to the horizontal.
- A ceiling does not include trap doors and their frames, frames of windows and rooflights and frames in which glazing is fitted; architraves, cover moulds, picture rails and similar narrow features.

Suspended ceilings can contribute to overall fire resistance of a ceiling/floor assembly. The above lining classification is to be followed, and assembly must also comply with Table A3 of Appendix A.

Some concealed roof and floor spaces require cavity barriers, and requirements are set out in section 10. This may not be necessary if a fire-resisting ceiling is placed below the cavity as in Fig. B2.1. The ceiling is to have a 30 minute fire resistance and be imperforate, although certain pipes and cables are allowed; it should extend throughout the building or compartment and not be demountable.

Table A3   Limitations on fire-protecting suspended ceilings

Height of building or separated part (m)	Type of floor	Provision for fire resistance or floor (minutes)	Description of suspended ceiling
less than 18	not compartment	60 or less	Type A, B, C or D
	compartment	less than 60	
		60	Type B, C or D
18 or more	any	60 or less	Type C or D
no limit	any	more than 60	Type D

Notes:
Ceiling type and description
A. Surface of ceiling exposed to the cavity should be Class 0 or Class 1.
B. Surface of ceiling exposed to the cavity should be Class 0.
C. Surface of ceiling exposed to the cavity should be Class 0. Ceiling should not contain easily openable access panels.
D. Ceiling should be of a material of limited combustibility and not contain easily openable access panels. Any insulation above the ceiling should be of a material of limited combustibility.

Any access panels provided in fire protecting suspended ceilings of type C or D should be secured in position by releasing devices or screw fixings, and they should be shown to have been tested in the ceiling assembly in which they are incorporated.

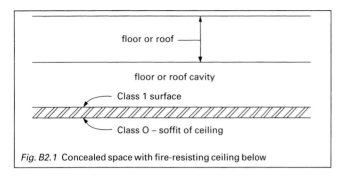

Fig. B2.1 Concealed space with fire-resisting ceiling below

Rooflights are also required to meet the relevant ceiling classification, but some plastic rooflights having at least a Class 3 rating may be used when a higher standard is required if the limitations in Table 6 and also Table 11 in section 15 are observed.

Special applications

Guidance is available in three publications regarding special applications.

- For supported structures, refer to BS 6661: 1986, Guide to the design, construction and maintenance of single skin air supported structures.
- When a flexible membrane is covering a structure which is not air supported, recommendations are contained in BS 7157: 1989, Method of test for ignitability of fabrics used in the construction of large tented stuctures.
- For information regarding the use of PTFE-based materials for tension membrane roofs and structures, refer to BRE Report 274, Fire safety of PTFE-based materials used in buildings (1994).

THERMOPLASTIC MATERIALS

A thermoplastic material is a synthetic polymeric material having a softening point below 200°C when tested to BS 2782: Part 1: Method 120A: 1976, 'Determination of the Vicat softening temperature of thermoplastics'.

Where thermoplastic materials cannot meet the standard required for linings, they can nevertheless be used in rooflights and lighting diffusers in suspended ceilings, but they must satisfy the following requirements.

- External windows to rooms may be glazed with thermoplastic materials if the material is a TP(a) rigid product. This does not apply in circulation spaces. Internal glazing is to be in accordance with the appropriate classification for linings; remembering that a wall does not include glazing in a door.
- Thermoplastic material may be used in rooflights in rooms and circulation spaces when

  (a) the lower surface is classified TP(a) rigid or TP(b);
  (b) the size and placing of the rooflights is in accordance with Table 6 and to the limitations on boundary distance in Table 11, section 15.

Thermoplastic rooflights are not to be used in protected stairways.

- Lighting diffusers (Fig. B2.2) which form part of a ceiling are to be translucent or open-structured elements allowing light to pass through. They may be part of a luminaire or used below rooflights or other sources of light.

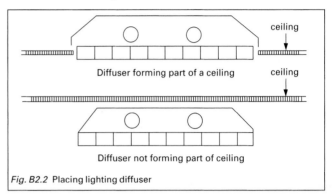

Fig. B2.2 Placing lighting diffuser

- Unless they have been tested, as part of the ceiling system being used to give fire protection. Thermoplastic lighting diffusers are not to be used in fire-protecting or fire-resisting ceilings.
- Nevertheless, ceilings to rooms and circulation spaces may incorporate thermoplastic lighting diffusers in accordance with the following:

  (a) Wall and ceiling surfaces within the cavity above a suspended ceiling are to comply with the appropriate classification of linings relating to the type of space below the suspended ceiling. (Does not relate to the upper surface of thermoplastic panels.)
  (b) There is no restriction of TP(a) rigid diffusers.
  (c) TP(b) are to comply with Table 6 and Fig. B2.3 which shows layout restrictions on Class 3 plastic rooflights, TP(b) rooflights and TP(b) lighting diffusers.

Table 6 Limitations applied to thermoplastic rooflights and lighting diffusers in suspended ceilings and Class 3 plastic rooflights

Minimum classification of lower surface	Use of space below the diffusers or rooflight	Maximum area of each diffuser panel or rooflight (1) (m²)	Maximum total area of diffuser panels and rooflights as percentage of floor area of the space in which the ceiling is located (%)	Minimum separation distance between diffuser panels or rooflights (1) (m)
TP(a)	any except protected stairway	No limit (2)	No limit	No limit
Class 3 (3) or TP(b)	rooms	5	50 (4)(5)	3 (5)
	circulation spaces except protected stairways	5	15 (4)	3

Notes:
1. Smaller panels can be grouped together provided that the overall size of the group and the space between one group and any others satisfies the dimensions shown in Fig. B2.3.
2. Lighting diffusers of TP(a) flexible rating should be restricted to panels of nor more than 5 m² each.
3. There are no limits on Class 3 material in small rooms.
4. The minimum 3 m separation specified in Fig. B2.3 between each 5 m² must be maintained. Therefore, in some cases it may not also be possible to use the maximum percentage quoted.
5. Class 3 rooflights to rooms in industrial and other non-residential purpose groups may be spaced 1800 mm apart provided the rooflights are evenly distributed and do not exceed 20% of the area of the room.

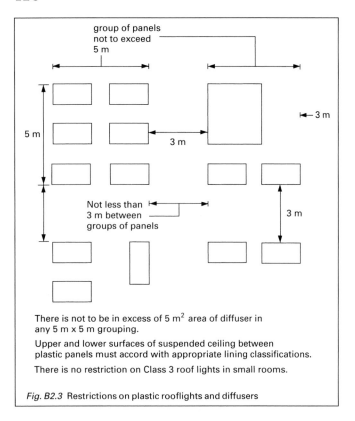

There is not to be in excess of 5 m² area of diffuser in any 5 m x 5 m grouping.

Upper and lower surfaces of suspended ceiling between plastic panels must accord with appropriate lining classifications.

There is no restriction on Class 3 roof lights in small rooms.

Fig. B2.3 Restrictions on plastic rooflights and diffusers

For the purpose of B2, thermoplastic materials are to have classifications 0 to 3, or TP(a) rigid, TP(a) flexible or TP(b) in accordance with the following:

TP(a) rigid:

(a) a rigid solid PVC sheet;
(b) solid polycarbonate sheet not less than 3 mm thick; (not double or multi-skin)
(c) multi-skinned rigid sheet made from unplasticized PVC, or polycarbonate which has a Class 1 rating;
(d) any other rigid thermoplastic product which has been tested by method 508A and is satisfactory.

TP(a) flexible: Flexible products not exceeding 1 mm thick fulfilling BS tests.

TP(b):

(a) rigid solid polycarbonate sheet material less than 3 mm thick, or multiple skin polycarbonate sheets not meeting TP(a) standards, or
(b) other products which have satisfied a test under method 508A.

Stretched skin ceilings

A room ceiling may be formed as either suspended or stretched skin membrane from panels of thermoplastic material which is to be TP(a) flexible. It is not to be a fire-resisting ceiling and each panel is not to be in excess of 5 m². Each panel is to be supported around its perimeter.

SECTION 8. INTERNAL FIRE SPREAD: STRUCTURE

Premature failure of a structure in the event of fire may be limited by

- elements of structure to have minimum periods of fire resistance;
- there is to be resistance to collapse of loadbearing elements;
- resistance to fire penetration of separating elements;
- resistance to excessive heat transfer through fire-resisting elements;

- considering, under fire conditions, the effects of dead, imposed and wind loads, the effect of heat on structural elements and structural integrity to limit the effect of fire-induced movement.
- the provision of cavity barriers.

The 2000 approved document sets out the requirements which are substantially the same as before, but there are some significant changes in the guidance offered. Fire-resistance standards have been altered, and are now given in minutes rather than hours.

Section 8 sets out to relate minimum periods of fire resistance aimed at minimizing the risk to

- people occupying a building;
- people in the vicinity of a building;
- firefighters who must enter the building should the need arise.

The definition of an element of structure remains very much the same as previously

(a) a member which is part of the structural frame, or any other beam or column;
(b) a loadbearing wall including part of a loadbearing wall;
(c) a floor;
(d) a gallery;
(e) an external wall;
(f) a compartment wall, including a wall separating two or more buildings;

Unless it serves the function of a floor, as an example a car park or escape route, a roof is not treated as an element of structure; or the structure is essential for the stability of an external wall which must have fire resistance.

Also the lowest floor of a building or a platform floor are not elements of structure for the purposes of Part B.

The requirement

B3. Internal fire spread – structure. Buildings are to be designed and constructed in such a way that in the event of fire

- **stability will be maintained for a reasonable period;**
- **walls separating two or more buildings are to resist the spread of fire;**
- **the spread of fire within a building is, where necessary, to be limited by compartmentation;**
- **concealed spaces within the structure are to be sub-divided and fire stopped to limit the spread of unseen fire and smoke.**

The requirements extend to buildings as extended and also to the walls separating semi-detached or terraced houses (Fig. B3.1). The subdividing of a building into compartments does not apply to material alterations to any prison under section 33 of the Prisons Act 1952.

The performance requirements of B3 will be met if:

- loadbearing elements will withstand fire for a given period without losing stability;
- using fire-resisting construction a building is divided into compartments;
- openings in fire-separating elements are protected;
- hidden voids are sealed and subdivided to prevent the spread of fire and smoke.

Factors which determine the extent of these various requirements are size and use of building and on the location of the element of construction.

INTRODUCTION

Fire resistance is measured by the ability of an element of construction to withstand the effects of fire by

- in the case of loadbearing elements, its resistance to collapse;
- in the case of fire-separating elements, its ability to resist penetration;

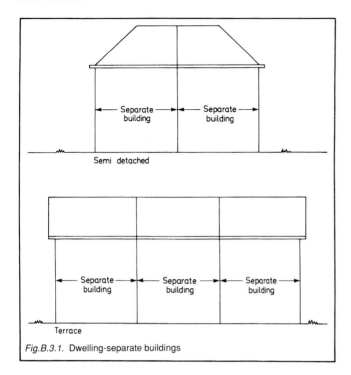

Fig.B.3.1. Dwelling-separate buildings

- also in the case of fire-separating elements, its resistance to the transfer of excessive heat.

Thus the various elements of construction are required to maintain loadbearing capacity, integrity and provide insulation from high temperatures.

LOADBEARING ELEMENTS OF STRUCTURE

Premature failure of a structure can and should be prevented by giving at least a minimum standard of fire resistance, and three benefits arise:

- reduces risk to occupants while they remain within the building;
- reduces the risk to firefighters entering the building;
- reduces danger to people nearby, who could suffer injury from falling materials should collapse occur.

Fire-resistance standards are set out in Appendix A to Part B, and include provisions:

- an element supporting another element of structure does not have less fire resistance than the supported element;
- elements common to more than one building or compartment are to meet the greater of relevant provisions;
- relating to elements of structure in single-storey buildings.

There are some concessions regarding elements of structure in basements where at least one side of the basement is open at ground level.

Further guidance

Guidance regarding other factors in loadbearing walls are contained in various sections of the approved document:

- compartment wall and walls common to two buildings – section 9;
- wall between house and domestic garage – section 9;
- protecting means of escape – sections 2–6:
- external walls – sections 13 and 14;
- firefighting shafts – section 18;
- compartment floors – section 9;
- wall enclosing special fire hazard – section 9.

DOMESTIC LOFT CONVERSIONS

Where existing two-storey family houses are altered to provide additional storeys, the floors, both old and new are to have a 30 minute standard of fire resistance.

However, if only one storey is being added, and the new storey does not contain more than two habitable rooms, and the total area of the new storey does not exceed 50 m², the existing floor construction, if it has a minimum of modified 30 minute standard of fire resistance, may be accepted if

- the floor does not separate circulation spaces, and the other requirements set out in section 2 are fulfilled.

When an existing floor does not achieve the full 30 minute standard it may be upgraded by fixing hardboard on the upper surface (Fig. B3.2).

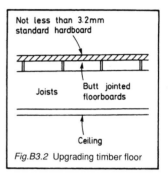

Fig.B3.2 Upgrading timber floor

RAISED STORAGE AREAS

Raised free-standing floors, often supported by racking, may be erected in single-storey industrial buildings. Whether considered a gallery or being large enough to be regarded as a floor, fire-resistance provisions do apply and Table A1, Appendix A, applies to these structures.

Reference is now made in the approved document to automated storage systems where people do not normally go onto the raised areas. In these instances it may not be necessary for full fire-resistance requirements to be applied to ensure safety of the occupants.

It may also be possible to reduce levels of fire resistance to storage tiers where people do go, if the following conditions are satisfied:

- there is only one tier and it is not used for any other purpose than storage;
- the number of persons likely to be on the storage area is low and does not include members of the public;
- only small number of persons on floor;
- floor is limited to not more than 10 m wide or long, and is not to occupy more than 50% of space in which it is sited;
- space above and below floor is open;
- means of escape to meet requirements of sections 4, 5 and 6.

The floor size may be increased to not more than 20 m wide or long if the lower level is fitted with an automatic detection and alarm system in accordance with BS 5839: Part 1: 1998, 'Code of practice for system design, installation and servicing'.

There are no limits on the size of the floor where a building is fitted throughout with an automatic sprinkler system in accordance with BS 5306: Part 2: 1990, Specification for sprinkler systems.

FLAT CONVERSIONS

If an existing house or other building is to be converted into flats, this is a material change of use, and should the existing building have timber floors there may be fire-resistance problems.

Table A2   Minimum periods of fire resistance

Purpose group of building	Minimum periods (minutes) for elements of structure in a:					
	Basement storey ($) including floor over		Ground or upper storey			
	Depth (m) of a lowest basement		Height (m) of top floor above ground, in a building or separated part of a building			
	more than 10	not more than 10	not more than 5	not more than 18	not more than 30	more than 30
1. Residential (domestic):						
a. flats and maisonettes	90	60	30*	60**†	90**	120**
b. and c. dwellinghouses	not relevant	30*	30*	60@	not relevant	not relevant
2. Residential:						
a. Institutional æ	90	60	30*	60	90	120#
b. Other residential	90	60	30*	60	90	120#
3. Office:						
– not sprinklered	90	60	30*	60	90	not permitted
– sprinklered (2)	60	60	30*	30*	60	120#
4. Shop and commercial:						
– not sprinklered	90	60	60	60	90	not permitted
– sprinklered (2)	60	60	30*	60	60	120#
5. Assembly and recreation:						
– not sprinklered	90	60	60	60	90	not permitted
– sprinklered (2)	60	60	30*	60	60	120#
6. Industrial:						
– not sprinklered	120	90	60	90	120	not permitted
– sprinklered (2)	90	60	30*	60	90	120#
7. Storage and other non-residential:						
a. any building or part not described elsewhere:						
– not sprinklered	120	90	60	90	120	not permitted
– sprinklered (2)	90	60	30*	60	90	120#
b. car park for light vehicles:						
i. open sided car park (3)	not applicable	not applicable	15*+	15*+	15*+	60
ii. any other car park	90	60	30*	60	90	120#

Single storey buildings are subject to the periods under the heading 'not more than 5'. If they have basements, the basement storeys are subject to the period appropriate to their depth.

Modifications referred to in Table A2: [for application of the table see next page]
$ The floor over a basement (or if there is more than 1 basement, the floor over the topmost basement) should meet the provisions for the ground and upper storeys if that period is higher.
*   Increased to a minimum of 60 minutes for compartment walls separating buildings.
**   Reduced to 30 minutes for any floor within a maisonette, but not if the floor contributes to the support of the building.
æ   Multi-storey hospitals designed in accordance with the NHS Firecode documents should have a minimum 60 minutes standard.
Reduced to 90 minutes for elements not forming part of the structural frame.
+   Increased to 30 minutes for elements not forming part of the structural frame.
†   Refer to paragraph 8.10 regarding the acceptability of 30 minutes in flat conversions.
@   30 minutes in the case of 3 storey dwellinghouses, increased to 60 minutes minimum for compartment walls separating buildings.

Notes:
1. Refer to Table A1 for the specific provisions of test.
2. 'Sprinklered' means that the building is fitted throughout with an automatic sprinkler system meeting the relevant recommendations of BS 5306 *Fire extinguishing installations and equipment on premises*, Part 2 *Specification for sprinkler systems*; i.e. the relevant occupancy rating together with the additional requirements for life safety.
3. The car park should comply with the relevant provisions in the guidance on requirement B3, Section 12.

However, if the building does not have more than three storeys and the means of escape comply with section 2, a 30 minute standard of fire resistance may be accepted for the elements of structure.

If there are four or more storeys in the altered building the full requirements for fire resistance are normally necessary.

Minimum periods of fire resistance

Table A2 gives the minimum periods of fire resistance, and the following notes apply:

- 'sprinklered' means that an automatic sprinkler system is fitted throughout the building;
- a car park is to comply with provisions in section 12.

FIRE-RESISTANCE STANDARDS

If one element of structure supports, carries or gives stability to another the fire resistance of the supporting element is not to be less than the fire resistance for the other element.

This principle may be varied when the supporting structure is in the open air, and unlikely to be affected by fire within the building, or the supporting structure is not in the same compartment and has a fire-separating element between the supported and the supporting structure.

An element of structure may form part of more than one building or compartment; in which event that element should be of the greater of the relevant provisions.

If one side of a basement is open at ground level, giving access for firefighting and facility for smoke venting the standard of fire resistance for the aboveground structure may be used.

Some elements of structure in a single-storey building may not require fire resistance but this must be provided if the element is part of a support for

(a) an external wall and B4 limits the extent of openings and other unprotected areas in the wall, or
(b) a compartment wall, including a separating wall, or a wall between a garage and a dwellinghouse, or
(c) a gallery.

In the above paragraph a ground storey of a building having one or more basement storeys may be regarded as a single-storey building. Fire resistance of the basement storey is to be that appropriate for basements.

Single-storey buildings are covered by the heading 'not more than 5' in Table A2. If there are basements then the standards appropriate to their depth are to be used.

SECTION 9. COMPARTMENTATION

The division of a building into a number of areas that are fire and smoke tight is called compartmentation. The aim is to contain fire spread in the area where it arises, and allow time for the occupants to make their escape.

As one of the more complex parts of the regulations the requirements of Part B continue to require provision for the separation of buildings or parts of buildings, not only vertically, but also horizontally in order to provide compartments. A high standard of fire resistance of walls and floors is also required in order to contain fire within compartments and prevent a whole building becoming engulfed.

Aimed to provide greater safety within a large building it is also possible that only property within a compartment of origin will suffer fire damage, and the remainder of the building may be virtually undamaged and in use again within the shortest possible time following a fire, with considerable financial savings.

It is emphasized that a building must be designed, constructed and located in relation to adjoining buildings so that they will not present a hazard to their neighbours.

Compartments are designed by subdividing a building by means of compartment walls, and compartment floors. The amount of subdivision depends upon

- use
- height, to the floor of the top storey
- area
- the availability of a sprinkler system.

Volume of a building is now no longer related to compartment sizes.

Single-storey buildings are likely to require compartmentation only when there is a real risk of people sleeping there. Most multi-storey buildings are affected by the requirements.

A simple example of protecting a single dwelling is the provision of a separating wall between it and an attached integral garage.

Junctions between different elements enclosing a compartment must be protected against fire penetration, and any openings between compartments are to be protected. Protected shafts are also to be provided where there are spaces connecting compartments in order to restrict fire spread.

Atria in buildings

When buildings contain one or more atria, advice is available in BS 5588: Part 7: 1997, 'Code of practice for the incorporation of atria in buildings'. So far as the approved document is concerned, the BS is only applicable where an atria breaches any compartmentation.

PROVISION OF COMPARTMENTATION

In circumstances requiring the provision of compartment floors this requirement does not extend to the lowest floor in a building. Under this heading there is guidance on compartmentation in different building types, the construction of floors and walls.

All purpose groups

The long-standing term 'separating wall' has been abandoned in the 1992 edition of the approved document and has been replaced by the description of 'a wall common to two or more buildings'.

These and walls provided to separate different occupancies are to be constructed as compartment walls. Floors too, in these circumstances are to be compartment floors.

However, if one of the different purposes is ancillary to the other the requirement does not apply. But in certain circumstances it is necessary to treat the different use as belonging to a purpose group in its own right (see Fig. B3.3):

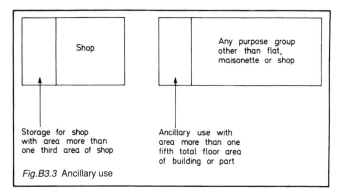

Fig.B3.3 Ancillary use

- where the ancillary use is a flat or maisonette, or
- the ancillary use relates to an area exceeding one-fifth of the floor area of the building or compartment which is more than $280\,m^2$ in area, or
- it relates to storage in a building or compartment of shop or commercial premises (purpose group 4) if storage exceeds more than one-third of the total floor area of the building or compartment which is more than $280\,m^2$ in area.

Buildings, where two or more main uses that are not ancillary, are to be considered as having a purpose group in their own right. As an example, offices over shops, from which they are independent.

In very large buildings with a complex mix of uses it may be necessary to consider special measures to avoid risks that one part may extend to its neighbours.

Walls and floors dividing a building into separate occupancies are to be made as compartment walls or compartment floors.

Special fire hazard

All places of special fire hazard are to be enclosed by fire-resisting construction.

Houses

Walls which are separating semi-detached houses or houses in terraces are to be constructed as a compartment wall and the houses regarded as separate buildings.

Attached and integral garages are to be given protection from the spread of fire, including the possibility of spilled fuel flowing from the garage into the house (Fig. B.3.4).

Flats

Certain walls and floors are to be compartment walls and floors. They are

- any floor except when it is in a maisonette, and

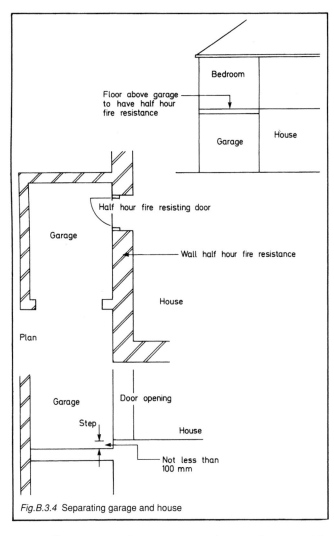

Fig.B.3.4 Separating garage and house

- any wall separating a flat or maisonette from another part of the building, and
- any wall enclosing a refuse chamber.

Institutional buildings

Certain walls and floors in institutional buildings, including health care, are to be compartment walls and floors:

- any floor;
- any wall dividing a building into compartments with a floor space not exceeding

    (a) 3000 m² for a single-storey building, or
    (b) 2000 m² for a multi-storey building.

- Upper storeys used for in-patients are to be divided into not less than two compartments. The arrangement must permit progressive horizontal evacuation of each compartment.

Other residential

All floors in purpose group other residential (group 2(b)) are to be constructed as compartment floors.

Non-residential

This requirement affects non-residential purpose groups:

Group 3 – Office
　　　　4 – Shop and commercial
　　　　5 – Assembly and recreation
　　　　6 – Industrial
　　　　7 – Storage and non-residential.

Walls and floors are to be constructed as compartment walls and floors in the following situations:

- any wall required to subdivide a building to achieve the size limits in Table 7;
- any floor of the building, or separated part, has a storey with a floor in excess of 30 m above ground level;
- if the building has one or more basements, the floor of the ground storey (excluding small premises);
- any basement floor when the building or separated part has a basement more than 10 m below ground level;
- if the building is part of a shopping complex, any wall or floor described in section 5 of BS 5588: Part 10: 1991, 'Code of practice for shopping complexes to be a compartment wall or floor', and;
- in shop or commercial, industrial or storage premises every wall or floor dividing a building into separate occupancies is to be a compartment wall or floor.

CONSTRUCTION OF COMPARTMENT WALLS AND FLOORS

All compartment walls and floors are to be constructed to form a complete barrier to fire between the compartments they separate, and to have the degree of fire resistance appropriate in Tables A1 and A2 in the appendix to Part B.

When there is a two-storey building in the shop, commercial and industrial purpose groups, and the upper storey is ancillary to the use below, it is possible to treat the ground storey as a single-storey building if

- the upper storey is not more than 500 m² in area or 20% of the ground storey area, whichever is less;
- the upper storey is compartmented from the ground storey;
- there is a separate means of escape from the upper storey.

When timber beams, joists, purlins and rafters are built into or carried through a compartment wall, openings are to be as small as possible and then carefully firestopped. Where trussed rafters bridge a wall, provision must be made to ensure that fire causing failure of any part of the truss in one compartment does not cause failure in any part of the truss in another compartment.

Between buildings

When compartment walls are common to two or more buildings they are to rise to the full height of the building in an uninterrupted vertical plane, which means that adjoining buildings are only to be separated by walls, not floors.

This does not apply to a compartment wall dividing a building into separate occupancies.

Separated parts

Where a compartment wall forms a separated part of a building it should rise to the full height of the building in an uninterrupted vertical plane. The two separated parts may have different standards of fire resistance.

Wall junctions

A junction of a compartment wall or compartment floor meeting another compartment wall, or an external wall, is to maintain fire resistance of the compartmentation.

This may be achieved by correctly bonding the construction or by the introduction of fire stopping.

Wall and roof junctions

If adequate protection is not given at the junction of a compartment wall and the the roof, fire can spread from one compartment to the next at an alarming rate. One method of protection is to carry the

Table 7 Maximum dimensions of building or compartment (non-residential buildings)

Purpose Group of building or part	Height of floor of top storey above ground level (m)	Floor area of any one storey in the building or any one storey in a compartment (m²)	
		in multi-storey buildings	in single storey buildings
Office	no limit	no limit	no limit
Assembly & Recreation Shop & Commercial:			
a. schools	no limit	800	800
b. shops – not sprinklered	no limit	2000	2000
shops – sprinklered (1)	no limit	4000	no limit
c. elsewhere – not sprinklered	no limit	2000	no limit
elsewhere – sprinklered (1)	no limit	4000	no limit
Industrial (2)			
not sprinklered	not more than 18	7000	no limit
	more than 18	2000 (3)	no limit
sprinklered (1)	not more than 18	14000	no limit
	more than 18	4000 (3)	no limit
	Height of floor of top storey above ground level (m)	Maximum compartment volume (m²)	
		in multi storey buildings	in single storey buildings
Storage (2) & Other non-residential:			
a. car park for light vehicles	no limit	no limit	no limit
b. any other building or part:			
not sprinklered	not more than 18	20000	no limit
	more than 18	4000 (3)	no limit
sprinklered (1)	not more than 18	40000	no limit
	more than 18	8000 (3)	no limit

Notes:
1. 'Sprinklered' means that the building is fitted throughout with an automatic sprinkler system meeting the relevant recommendations of BS 5306: Part 2, i.e. the relevant occupancy rating together with the additional requirements for life safety.
2. There may be additional limitations on floor area and/or sprinkler provisions in certain industrial and storage uses under other legislation, for example in respect of storage of LPG and certain chemicals.
3. This reduced limit applies only to storeys that are more than 18 m above ground level. Below this height the higher limit applies.

wall above the roof (Fig. B3.5) or alternatively for any building or compartment roof covering AA, AB or AC to extend at least 1500 mm on either side of the wall (Fig. B3.6). Similar treatment can also be used for a dwellinghouse and buildings or compartments in residential, office or assembly use (but not institutional) (Fig. B3.7).

Care must be taken to ensure that there is adequate stopping between the top of a wall separating buildings and the underside of roof coverings. Battens and tiles should be bedded as in Fig. B3.8.

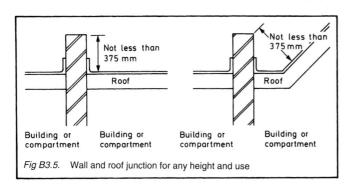

Fig B3.5. Wall and roof junction for any height and use

Hospitals

Compartment walls and floors in hospitals which have been designed on the basis of the Department of Health Care documents under the general title 'Firecode' are to be constructed of material of limited combustibility if they have a fire resistance of 60 minutes or more (unless fitted throughout with sprinklers).

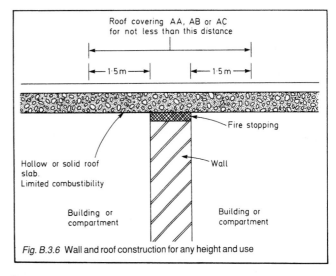

Fig. B.3.6 Wall and roof construction for any height and use

Openings

The only openings permitted in a compartment wall are when they are provided for a door which is a means of escape in case of fire; and to accommodate a pipe (Fig. B3.9).

Openings in other compartment walls or floors

The only openings permitted in a compartment wall or floor are:

• a door with the appropriate fire resistance, and correctly fitted (Appendix B);

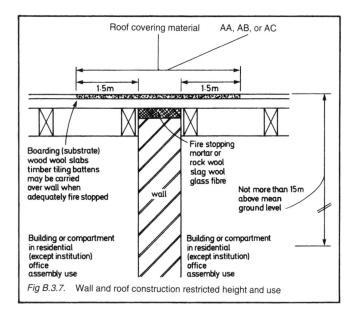

Fig B.3.7. Wall and roof construction restricted height and use

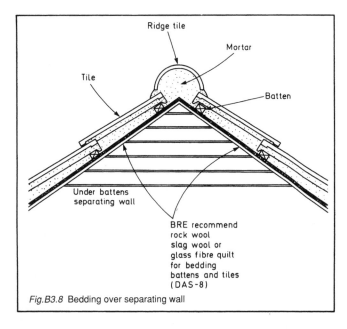

Fig.B3.8 Bedding over separating wall

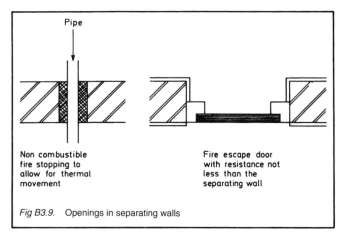

Fig B3.9. Openings in separating walls

- a pipe
  ventilation duct
  chimney
  appliance ventilation duct
  duct encasing flues (section 10);
- non-combustible refuse chute;
- a protected shaft;
- atria designed in accordance with BS 588: Part 7: 1997.

PROTECTED SHAFTS

Protected shafts (see Figs B3.10 to B3.12) allow the movement of people; the passage of goods; and services such as pipes and cables from one compartment to another. The elements enclosing a protected shaft may be a mixture of compartment and external walls, or compartment floors, and the protection offered is to delay the passage of fire and smoke between compartments.

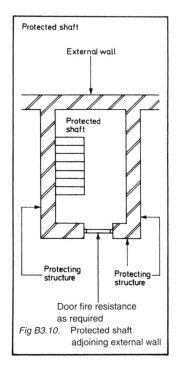

Fig B3.10. Protected shaft adjoining external wall

Sanitary accommodation and washrooms may be placed in a protected shaft, but otherwise the use should be limited to lifts, stairs, escalators, chutes, ducts and pipes.

The construction of a protected shaft should

- be a complete barrier between compartments connected by the shaft;
- have appropriate fire resistance (Table A1) with the exception of glazed screens;
- be provided with high and low level ventilation if piped inflammable gas is conveyed through the shaft.

There is provision for the introduction of glazed screens separating protected shafts from a lobby or corridor (see Figs B3.13 and B3.14), where the protected shaft contains a stairway.

The requirements are:

- a fire resistance of stair enclosure of not less than 60 minutes and
- the protected shaft is not a firefighting shaft;
- the glazed screen has not less than 30 minutes' fire-resisting integrity;
- the lobby or corridor enclosure has at least 30 minutes' fire resistance.

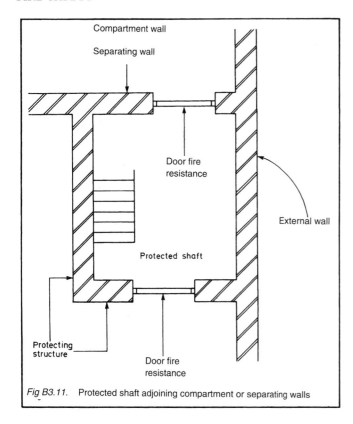

Fig B3.11.   Protected shaft adjoining compartment or separating walls

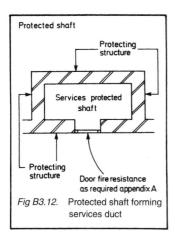

Fig B3.12.   Protected shaft forming services duct

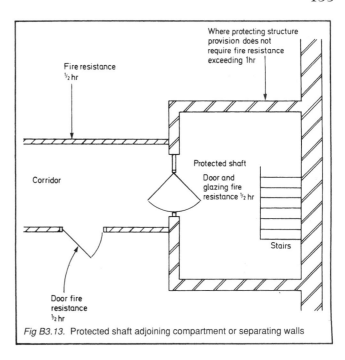

Fig B3.13.   Protected shaft adjoining compartment or separating walls

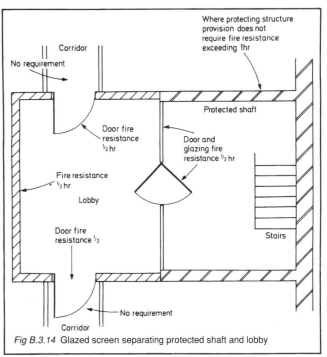

Fig B.3.14 Glazed screen separating protected shaft and lobby

Where these measures are not provided, the limits of glazing prescribed in Table A4 apply.

Where a protected shaft contains a lift or stair it should not also contain a pipe carrying oil or a ventilating duct. However, the mechanism for a hydraulic lift, and ducts for pressurizing the stairway to remove smoke, are allowed. A pipe completely separated from the protected shaft by fire-resisting construction, is not regarded as being in the shaft.

In order to include gas pipes of screwed steel or all welded construction in a shaft, they are to be installed in accordance with the Pipelines Safety Regulations 1996 and the Gas Safety (Installation and Use) Regulations 1998.

Adequate ventilation to the external air is to be provided in any protected shaft conveying piped flammable gas. Ventilation openings are to be at high and low level in the shaft.

The projection of a storey floor into a protected shaft must not limit the free flow of air throughout the shaft. Guidance is available in BS 8313: 1989, 'Code of practice for accommodation of building services in ducts'.

Openings

The external wall of a protected shaft is not normally required to have fire resistance, but there are provisions regarding the fire resistance of external walls of firefighting shafts in section 2 of BS 5588: Part 5: 1991, 'Code of practice for firefighting stairs and lifts'.

Openings in other parts of the enclosure of a protected shaft are not permitted, with the exception of

• in a wall common to two or more buildings, the following openings:

  (a) a door with the same fire resistance as the wall fitted with a self-closing device (Appendix B)
  (b) a pipe is allowed (section 11);

• other parts of the enclosure, not being an external wall may have the following openings:

(a) a door with appropriate fire resistance (Table B1) and fitted with a self-closing device (Appendix B);
(b) a pipe is allowable (section 11);
(c) inlets or outlets to ventilating ducts and openings for the ducts (section 11);
(d) where there is a lift, openings for cables, or a lift motor room at the bottom where openings should be of minimal size.

SECTION 10. CONCEALED SPACES

Following a spate of fires in schools and other buildings where fire spread rapidly through voids, it was inevitable that regulations provide for restriction in the movement of smoke and flame in the event of a fire in a building containing concealed voids.

The provision of firestopping became extended to embrace the blocking off of a cavity by a 'cavity barrier' in the same plane as the structural element.

A concealed space or cavity has the meaning demonstrated in Fig. B3.15 and to restrict the spread of smoke and flame, concealed spaces are to be protected as recommended in the approved document.

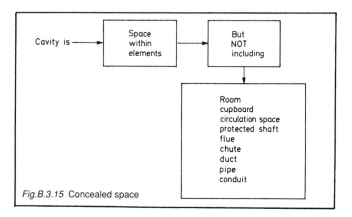

Fig.B.3.15 Concealed space

CAVITY BARRIERS

Requirements for cavity barriers have been much simplified by reference to Table 8.

There are additional provisions for some institutional and other industrial buildings – fewer barriers in timber frame construction and over large individual rooms. Also included are requirements for cavities over bedroom partitions, above non-protected corridors and within rainscreen cladding – the drained and ventilated cavities which are behind the outer cladding in some external wall construction.

Basically cavity barriers should close the edges of cavities:

- around openings through a wall, floor or other construction containing a cavity, and
- to separate a cavity in a wall, floor or other construction from another cavity.

Table 8 relates to the provision of cavity barriers in various locations and for the purpose group 1(b) to 7.

Where there are extensive concealed spaces all are to be subdivided in non-domestic buildings. Maximum distances between cavity barriers are specified and should not exceed those given in Fig. B3.16, and refer to purpose groups 2 to 7.

The requirement does not extend to a cavity which is in an external wall shown in Fig. B3.17 or is in a cavity between oversite concrete and the floor above not exceeding 1 m high as in Fig. B3.18.

Also excluded are the cavities between roof, roof sheet covering

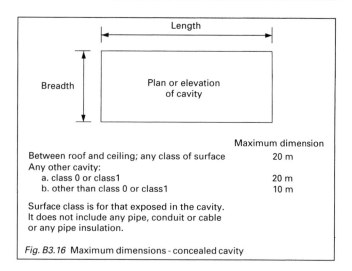

Fig. B3.16 Maximum dimensions - concealed cavity

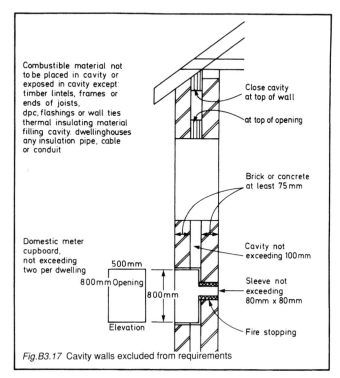

Fig.B3.17 Cavity walls excluded from requirements

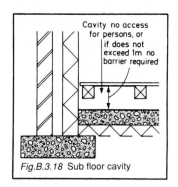

Fig.B.3.18 Sub floor cavity

and insulation below as in Fig. B3.19. Cavity barriers are necessary when there is a space between insulation and sheeting (Fig. B3.20).

Any room under a ceiling cavity exceeding the above dimensions only requires a cavity barrier on line of enclosing walls or partitions, but the cavity barriers are not to be more than 40 m apart.

Cavity barriers are to be placed at the intersection of fire-resisting

Table 8 Provision of cavity barriers

Cavity barriers to be provided:	Purpose group to which the provision applies (1)			
	1b & c Dwelling houses	1a Flat or maisonette	2 Other residential and Institutional	3–7 Office, Shop & Commercial, Assembly & Recreation, Industrial, Storage & Other non-residential
1. At the junction between an external cavity wall and a compartment wall that separates buildings; and at the top of such an external cavity wall. (2)	●	●	●	●
2. Above the enclosures to a protected stairway in a house with a floor more than 4.5 m above ground level (see Fig. B3.24). (3)	●	○	○	○
3. At the junction between an external cavity wall and every compartment floor and compartment wall. (2)	○	●	●	●
4. At the junction between a cavity wall and every compartment floor, compartment wall, or other wall or door assembly which forms a fire-resisting barrier. (2)	○	●	●	●
5. In a protected escape route, above and below any fire-resisting construction which is not carried full storey height, or (in the case of a top storey) to the underside of the roof covering. (3)	○	●	●	●
6. Where the corridor should be sub-divided to prevent fire or smoke affecting two alternative escape routes simultaneously (see Fig. B3.23) above any such corridor enclosures which are not carried full storey height, or (in the case of the top storey) to the underside of the roof covering. (4)	○	○	●	●
7. Above any bedroom partitions which are not carried full storey height, or (in the case of the top storey) to the underside of the roof covering. (3)	○	○	●	○
8. To sub-divide any cavity (including any roof space but excluding any underfloor service void) so that the distance between cavity barriers does not exceed the dimensions given in Fig. B3.16.	○	○	●	●
9. Within the void behind the external face of rainscreen cladding at every floor level, and on the line of compartment walls abutting the external wall, of buildings which have a floor 18 m or more above ground level.	○	●	●	○
10. At the edges of cavities (including around openings).	●	●	●	●

Key:
● provision applies
○ provision does not apply

Notes:
1. The classification of purpose groups is set out in Appendix D, Table D1.
2. The provisions in items 1, 3 and 4 do not apply where the cavity wall complies with Fig. B3.17.
3. The provisions in items 2, 5 and 7 do not apply where the cavity is enclosed on the lower side by a fire-resisting ceiling (as shown in Fig. B3.25) which extends throughout the building, compartment or separated part.
4. The provision of item 6 does not apply where the storey is sub-divided by fire-resisting construction carried full storey height and passing through the line of sub-division of the corridor (see Fig. B3.23), or where the cavity is enclosed on the lower side as described in Note 3.

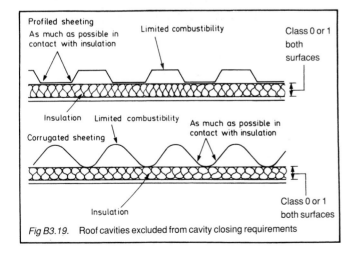

Fig B3.19. Roof cavities excluded from cavity closing requirements

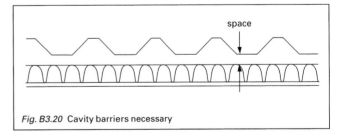

Fig. B3.20 Cavity barriers necessary

construction and elements containing a concealed space (Figs B3.21 and B3.22) in order to ensure that where a space around a wall, floor or other route there is a barrier to fire and smoke.

Compartment walls are carried up full storey height to a compartment floor or a roof, and it is important to continue a compartment wall through a cavity in order to keep the standard of fire resistance of the compartment wall.

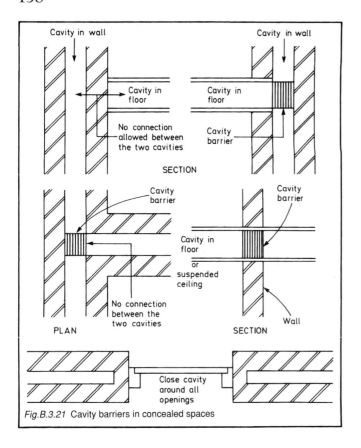

Fig.B.3.21 Cavity barriers in concealed spaces

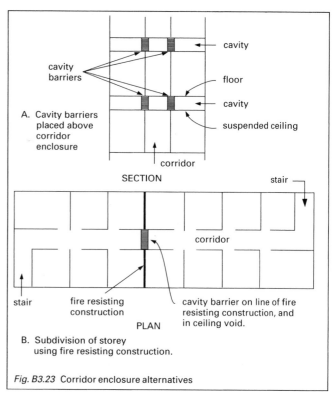

Fig. B3.23 Corridor enclosure alternatives

In purpose groups 3 to 7 provision for cavity barriers should be made above corridor enclosures which do not extend to full storey height – Table 8, item 7 and Fig. B3.23.

Three-storey houses

Provision should be made to limit the access of fire to the roof space of a three-storey house (Fig. B3.24) by either forming a fire-resisting ceiling or erecting a cavity barrier above the stairway.

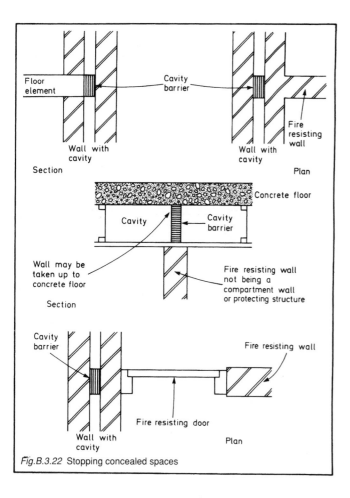

Fig.B.3.22 Stopping concealed spaces

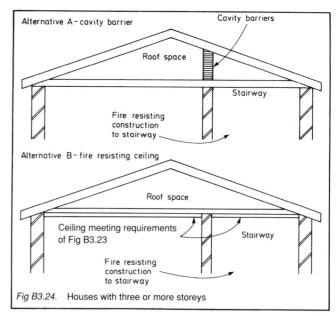

Fig B3.24.   Houses with three or more storeys

CONSTRUCTION AND FIXINGS

Any construction provided for another purpose which fulfils the provisions for cavity barriers may be accepted. At least 30 minutes' fire resistance is required when tested for integrity and 15 minutes for insulation; and the material is to satisfy this exposure from both sides (item 16, Table A1).

However, a cavity barrier placed in a stud wall or partition may comprise

(a) steel not less than 0.5 mm thick, or
(b) timber not less than 38 mm thick, or
(c) polythene sleeved mineral wool or mineral wool slab; under compression when installed, or
(d) plasterboard not less than 12.0 mm thick.

Where cavity barriers cannot be fixed to rigid construction, the connecting point should be fire stopped.

Fixing of cavity barriers must be such as to limit the possibility of failure from

• any movement of the building arising from subsidence, shrinkage, temperature change, and wind;
• services penetrating the cavity being affected by fire;
• fixings failing in fire;
• failure of abutting construction.

Roof members to which cavity barriers may be attached are not expected to have the same fire resistance.

MAXIMUM DIMENSIONS

The requirements regarding minimum dimensions of cavities in non-residential buildings do not apply to the following:

• in a wall which is fire resisting only because it is loadbearing;
• in a masonry or concrete external cavity wall as in Fig. B3.17;
• in a floor or roof cavity above a fire-resisting ceiling, and where maximum dimension does not exceed 30 m (see Fig. B.3.25);

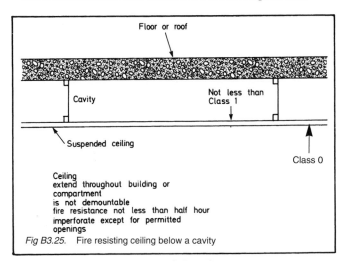

Fig B3.25. Fire resisting ceiling below a cavity

• below a floor next to the ground less than 1 m high and not provided with access for persons; unless there are openings in the floor permitting rubbish to gather in the cavity;
• behind the outside skin in rainscreen external wall construction, or by overcladding an existing external masonry or concrete external wall, if the cavity does not contain combustible insulation when requirements of Table 8, item 9, are followed;
• between double-skinned corrugated or profiled insulated roof sheeting when sheeting is of limited combustibility and each surface of insulation has a surface spread of flame of Class 0 or 1, and is in contact with both skins of cladding (Fig. B3.19);
• Within an underfloor service void.

If there is a room under a ceiling cavity which exceeds the sizes given in Fig. B3.16, cavity barriers are only to be provided on the line of the enclosing walls or partitions if the cavity barriers do not exceed 40 m apart and the exposed surfaces in the cavity are Class 0 or Class 1.

There is no limit to the size of a cavity when a concealed space is over an undivided area more than 40 m in both directions on plan, if

• the room and cavity are in one compartment;
• there is an automatic fire detection system in accordance with BS 5839: Part 1: 1988, 'Code of practice for system design, installation and servicing'. Detectors are not required in the cavity.
• the cavity is used as a plenum, and advice regarding recirculating air distribution in BS 5588: Part 9: 1989, 'Code of practice for ventilation and air conditioning ductwork', should be followed;
• exposed ceiling surface in cavity is Class 0 and supports and fixings are non-combustible;
• any pipe insulation has a flame spread rating of Class 1;
• electrical wiring in the void is in metal trays or conduit; and
• all other materials in the cavity are of limited combustibility.

OPENINGS

Openings in cavity barriers are limited to those illustrated in Fig. B3.26.

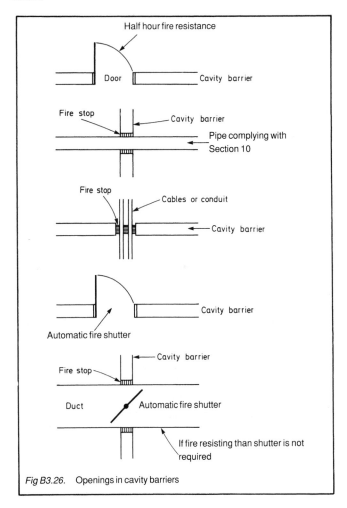

Fig B3.26. Openings in cavity barriers

SECTION 11. FIRE STOPPING

There are recommendations regarding the fire stopping of openings in elements of structure, or cavity barriers for pipes, conduits or cables. There are provisions to prevent the displacement of fire-stopping materials, and a list of recommended stopping materials.

An element of structure intended to give fire separation may have joints and openings for services to pass through. All openings and gaps are to be fire stopped in order that the fire resistance of the element does not suffer. Although intended to prevent the passage of fire there is also the benefit that smoke spread is inhibited.

Provisions for fire doors are contained in Appendix B.

OPENINGS FOR PIPES

Pipes which pass through
compartment walls
compartment floors   } unless pipe is in a
cavity barriers   protected shaft

must fulfil the requirements in the approved document. The requirements are

*   sealing the pipe with a proprietary system which has been tested to demonstrate that it will remain fire resistant; or
*   maximum nominal diameter of the pipe is not to be in excess of the dimensions given in Table 9. Openings are to be as small as possible, and fire stopped.

Table 9 Maximum nominal internal diameter of pipes passing through a compartment wall/floor

Situation	Pipe material and maximum nominal internal diameter (mm)		
	(a) Non-combustible material[1]	(b) Lead, aluminium aluminium alloy, PVC[2] fibre-cement	(c) Any other material
1. Structure (but not a wall separating buildings) enclosing a protected shaft which is not a stairway or a lift shaft.	160	110	40
2. Wall separating dwellinghouses, or compartment wall or compartment floor between flats.	160	160 (stack pipe)[3] 110 (branch pipe)[3]	40
3. Any other situation	160	40	40

Notes:
1. A non-combustible material (such as cast iron or steel) which if exposed to a temperature of 800°C, will not soften or fracture to the extent that flame or hot gas will pass through the wall of the pipe.
2. PVC pipes complying with BS 4514:1983 and PVC pipes complying with BS 5255:1989.
3. These diameters are only in relation to pipes forming part of an aboveground drainage system and enclosed as shown in Figs. B3.27 and B3.28. In other cases the maximum diameters against situation 3 apply.

There is provision to ensure that openings made for the passage of a pipe through the side of an enclosure or through a floor at the base or top of an enclosure is as small as is reasonably practicable and is fire stopped around the pipes. This means that in addition to any opening for a branch pipe or stack pipe, any opening for a water pipe must also be fire stopped.

There are also requirements for enclosures for drainage and water pipes (Fig. B3.26A).

When lead, aluminium or aluminium alloy, PVC or fibre cement pipes are used in

(a) a wall separating dwellinghouses, or
(b) a compartment wall or compartment floor between flats, use is limited to aboveground drainage system in an enclosure depicted in Figs B3.27 and B3.28.

Where not part of an aboveground drainage system, smaller pipe diameters shown for any other situation in Table 9 should be used.

As an alternative to the above methods a pipe of lead, aluminium or aluminium alloy, fibre cement or PVC with a nominal diameter not in excess of 160 mm may be protected by a non-combustible sleeve where it passes through the structure (Fig. B3.29).

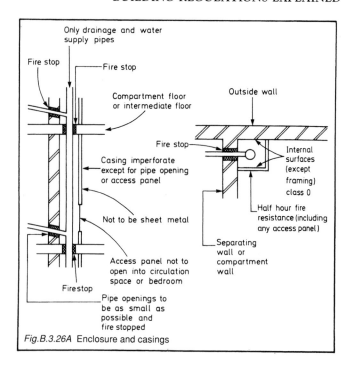

Fig.B.3.26A Enclosure and casings

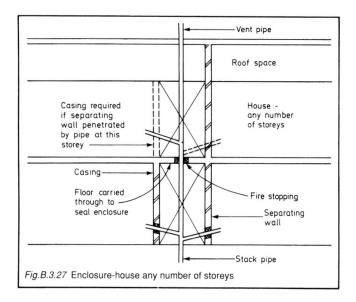

Fig.B.3.27 Enclosure-house any number of storeys

VENTILATING DUCTS

The form and degree of fire protection to be provided for ventilating ducts which pass from one compartment to another is dictated by its location.

*   Compartment walls and floors. A duct passing through a compartment wall (Fig. B3.30) should be fitted with an automatic fire shutter.
*   Protected shafts. When a ventilating duct forms or is included in a protected shaft the following precautions are advised:

(a) Automatic fire shutters fitted at the inlets and outlets to the shaft. Other arrangements including a shunt duct may be used to reduce as far as practical the possibility of fire spreading from one compartment to another.
(b) Materials used in construction should not encourage the spread of fire.

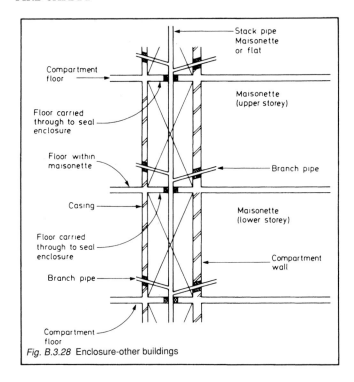

Fig. B.3.28 Enclosure-other buildings

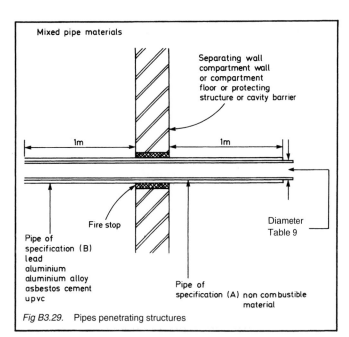

Fig B3.29.   Pipes penetrating structures

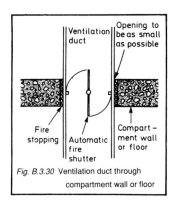

Fig. B.3.30 Ventilation duct through compartment wall or floor

(c) When a ventilating duct is within a protected shaft (Fig. B3.31) it should be placed within an enclosure (Fig. B3.32).

Alternative ways in which the integrity of compartments may be ensured when air-conditioning or ventilation ducts are installed, are contained in BS 5588: Part 9: 1989, 'Code for ventilating and air-conditioning ductwork'.

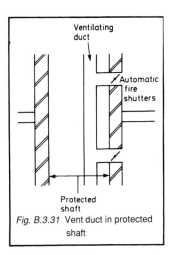

Fig. B.3.31 Vent duct in protected shaft

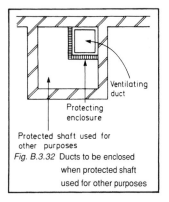

Fig. B.3.32 Ducts to be enclosed when protected shaft used for other purposes

CHIMNEYS AND FLUES

Chimneys, flues and ducts passing through or forming part of a compartment wall or floor are required to be encased in a structure having a fire resistance not less than half that required for the wall or floor (Fig. B3.33).

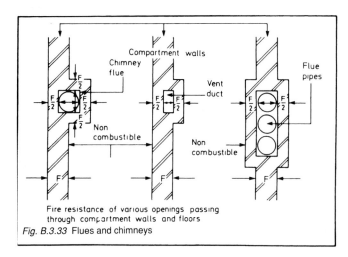

Fig. B.3.33 Flues and chimneys

FIRE STOPPING

Further recommendations regarding fire stopping of openings in elements, or cavity barriers, for pipes, conduits or cables are given (Fig. B3.34). The number of openings is to be the absolute minimum, kept as small as possible and fire stopped. There are provisions to prevent the displacement of fire-stopping material, and a list of recommended fire-stopping materials.

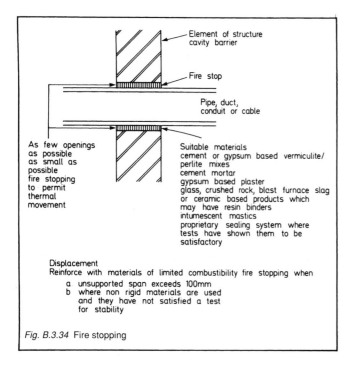

Fig. B.3.34 Fire stopping

SECTION 12. CAR PARKS AND SHOPPING COMPLEXES

The requirements for escape and construction are spaced among the previous sections where requirements relate to purpose group 7(b) which is concerned with car parks designed to admit and accommodate only cars, motor cycles, and passenger or light goods vehicles weighing no more than 2500 lb gross. The guarding of car parks is covered in Part K (Chapter 19).

The maximum travel distances are set out in Table 2, at 18 m in one direction only and 45 m when escape is available in two directions. There is also direct reference to Table A2 giving the minimum periods of fire resistance.

This section is therefore concerned with additional considerations to be borne in mind in the design and construction of car parks and shopping complexes.

CAR PARKS

In numerous respects buildings to be used as car parks are unlike other large buildings. Materials used will almost certainly be of limited combustibility and most are substantially open at the sides.

- The fire load is not particularly high.
- Fire spread from one vehicle to another is not a likely event.
- With a high standard of ventilation fire is not likely to spread from storey to storey.

Ventilation is therefore important and more concessions can be made for an open-sided car park than can be made for one which is not open.

The approved document looks at the several methods of ventilation and makes appropriate recommendations.

There are provisions in B1 to B5 which apply to car parks and all materials to be used should be non-combustible, except for

- surface finishes applied to floor or roof, or within adjoining buildings, compartment or separate part should be in accord with the requirements of B2 and B4;
- fire doors; and
- attendant's kiosk with an area of not more than 15 m²;
- any shop mobility facility.

Open-sided car park

The building may be regarded as an open-sided car park for the purposes of Table A2 in the following circumstances:

- there are no basement storeys;
- permanent ventilation at each parking level equal to at least 1/20th the floor area at that level; half is to be in each of two opposing walls;
- the car park is to be separated from any other use in the building.

Car parks not open sided

If the above standard of ventilation is not provided the car park is not open sided, in which event a different standard of fire resistance is required – see Table A2.

Ventilation is still essential; and this may be by natural or mechanical means.

- Where car parks are not open sided, natural ventilation is to be provided by permanent openings at each level with an aggregate of at least 1/40th of the floor area at that level. Half is to be in each of two opposing walls. Ceiling level smoke vents may be used as an alternative and they also are to have an aggregate permanent opening of at least 1/40th of floor areas. They are to be so arranged as to provide a through draught.
- Natural ventilation may not be possible in basement or enclosed car parks. In this event mechanical ventilation is to be provided.

  (a) It is to be independent of other ventilating systems; operate on 10 air changes per hour in the event of fire.
  (b) The system is to be in two parts, each able to extract 50% of the required rates. Each part is to be capable of single or simultaneous operation.
  (c) An independent power system is to be connected to each part.
  (d) Half the outlets are to be at low level and half at a high level.
  (e) Fans must be able to run at 300°C for not less than 60 minutes. Ductwork and fixings are to have a melting point of at least 800°C.

Information regarding the removal of hot smoke is available in BS 7346: Part 2: 1990, Specification for powered smoke and heat exhaust systems.

The BRE report No. 368 (1999), 'Design methodologies for smoke and heat exhaustion', describes an alternative method of exhausting smoke from enclosed car parks.

SHOPPING COMPLEXES

The various provisions in the approved document, while suitable for single shops, may not always be appropriate to the problems associated with a shopping complex. Problems are likely to be associated with fire resistance, surfaces, boundary distances and compartmentation.

A satisfactory standard of fire safety must be achieved and sections 5 and 6 of BS 5588: Part 10: 1991, 'Code of practice for shopping complexes', contains alternative recommendations and compensatory features which may be incorporated in the fire provisions of shopping complexes.

SECTION 13. EXTERNAL FIRE SPREAD

The threat of fire extending to adjoining properties and to people nearby is to be inhibited, as is the possibility of ignition by a fire in a neighbouring property. It is the roof and external walls which require attention. Once well alight a roof is a danger from burning brands being carried by convection currents to nearby buildings. Heat radiated through walls may ignite adjoining buildings and therefore great care is necessary to limit the amount of openings in external walls if buildings are close together.

The requirement

B.4. The requirement is that external walls of buildings are to resist the spread of fire:

- **over the walls, and**
- **from the building to another.**

Roofs are also required to resist the spread of fire:

- **over the roof, and**
- **from the building to another building.**

In planning to satisfy these requirements the following are to be taken into consideration:

- height of the building;
- use of the building;
- the siting of the building.

FIRE RESISTANCE

Performance

To satisfy the requirements of B4:

- External walls are to resist fire from an external source.
- Fire spread over wall surfaces is to be limited by providing construction with low rates of heat release.
- Thermal radiation is limited by control of unprotected areas, having taken into account distance from the boundary.
- Roof is constructed to resist spread of flame.
- Roof to resist fire penetration from an external source.

In making decisions regarding compliance with the performance required it is necessary to take into account distance from the boundary and height, when appropriate.

Walls

The construction of external walls must therefore be aimed to prevent fire spread. In the following requirements, construction is considered in conjunction with distance from the boundary.

Fire spread across open spaces depends upon:

- the extent of the fire in the building involved;
- the amount of risk it affords people in another building;
- the gaps between buildings;
- the protection of the facing walls

Section 14 is concerned with openings and other unprotected areas in external walls.

Construction

Because of distance from a relevant boundary it may be that some walls do not require fire resistance other than that which arises from the parts that are loadbearing. The relevant periods of fire resistance is dependent upon the mix of several factors: the use to which the building is to be put, its height and its area.

A reduced standard of fire resistance is available for a wall which is 1 m or more from the relevant boundary, and the wall may only require fire resistance from the inside.

The combustibility of external walls is restricted when

- less than 1 m from the relevant boundary;
- on high buildings regardless of distance from boundary;
- assembly and recreation buildings, also regardless of distance from a boundary,

the aim being to inhibit the possibility of ignition from an external source, and the spread of fire up the face of the external wall.

Standard

The appropriate fire-resistance standard is in Appendix A, Table A1.

PORTAL FRAMES

Steel portal frames are often used in situations where there may be no requirement for fire resistance of the steel structure.

When a building is near a site boundary the approved document has provisions for external walls to have sufficient fire resistance for it to remain standing and form a heat shield to prevent the spread of fire to an adjoining site. In cases when the external wall of a building cannot be wholly unprotected the rafters and columns of a portal frame may need to have fire protection.

Requirements for a single-storey building are shown in Fig. B4.1.

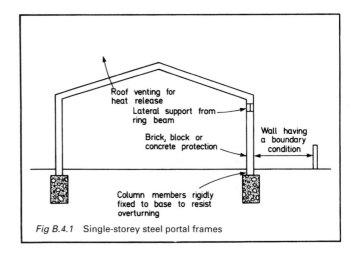

Fig B.4.1   Single-storey steel portal frames

The design method for foundations and connection of the portal frame, enabling an external wall to continue to perform its structural function during a fire is contained in *The Behaviour of Steel Portal Frames in Boundary Conditions* (2nd edition, 1990) obtainable from The Steel Construction Institute, Silwood Park, Ascot, Berks SL5 7QN. Guidance on other matters relating to portal frames is extensive.

Portal frames of reinforced concrete designed normally to have the necessary fire resistance can support external walls, having the same degree of fire resistance without specific requirements at the base to prevent overturning.

EXTERNAL SURFACES

External surfaces of walls are to be in line with the provisions illustrated in the following figures. In practice the total amount of combustible material may be limited by the requirements of section 14.

Where rainscreen construction is employed the surface of the outer cladding facing the wall is also to satisfy the provisions illustrated in Figs B4.2 to B4.4.

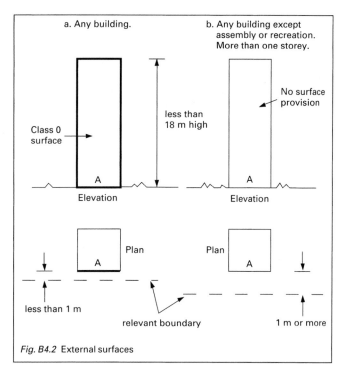

Fig. B4.2 External surfaces

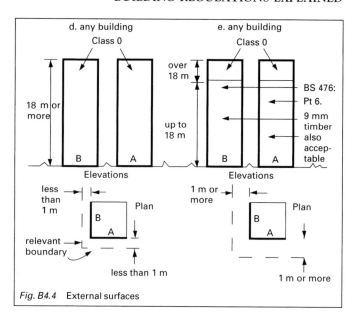

Fig. B4.4 External surfaces

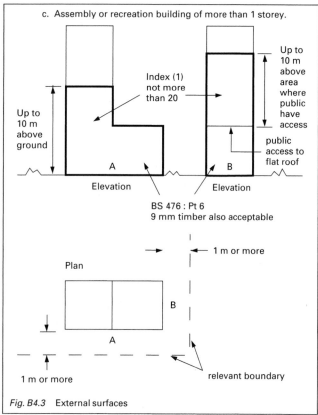

Fig. B4.3 External surfaces

Certain walls 1 m or more from the relevant boundary may have a surface classification (1) of not more than 20 when tested in accordance with BS 476: Part 6: 1981, and a short comment follows.

- Fire propagation

BS 476: Part 6: 1981 gives a test for fire propagation for products. The result of the test is expressed as a fire propagation index, and provides a comparative measure of the contribution to the growth of fire made by an essentially flat material, composite or assembly. It is used as an assessment of the performance of wall and ceiling linings.

The result 'L' ranges from 0 to 100+. When the 'L' figure is low it indicates that the sample tested did not contribute much to the heat of the furnace when it was being tested. Three sub-indices relate to the heat contributed at different stages of the test. They are i_1, i_2 and i_3. When a sample burns and releases heat quickly the sub-index will be high. Sub-index figure i_1 is derived from the first three minutes of test.

External envelope

Even when the provisions illustrated in Figs B4.2–B4.4 regarding external wall surfaces have been met a risk remains. This involves the use of combustible materials for cladding framework, or the use of combustible thermal insulation as an overcladding, or in ventilated cavities.

This risk in the external envelope should be avoided and in a building with a storey in excess of 20 m above ground level the only insulation material used in external wall construction is to be of limited combustibility. This does not apply to a wall as depicted in Fig. B3.17.

Advice is available in BRE report 135 – 'Fire performance of external thermal insulation for walls of multi-storey buildings' (1988).

SECTION 14. SPACE SEPARATION

The purpose of space separation between buildings is to reduce the risk that a fire may spread from one building to another by heat radiation. The proximity of a building to a boundary therefore influences the amount of unprotected area there is on the wall.

In reaching the provisions contained in the approved document, a number of aspects have been assumed:

- compartmentation is important in limiting the size of a fire;
- fire may affect a complete compartment but will not spread further;
- the purpose group of a building will influence the risk and intensity of a fire;
- a sprinkler system will offer control of a fire;
- residential (purpose group 1); assembly and recreation (purpose group 5) present a greater risk than the other uses;
- spread of fire between other buildings on the same site is a low risk to life;

- that the buildings on either side of the boundary have similar elevations, and are a similar distance from the boundary;
- where a wall is of fire-resistant construction the passage of radiation is discounted.

There will be occasions when a lower separation distance is appropriate, or an increased amount of unprotected area is necessary. In order to accommodate either of these design features may involve the formation of smaller compartments.

BOUNDARIES

By using the boundary, rather than another building, as the datum for space measurement enables calculation of permissible unprotected areas, even though there may not be a building on the adjoining site, or the area of unprotected wall it may have.

Relevant boundary

Measurements to boundaries are taken to a 'relevant boundary'. It means that if there is a legal boundary along a side, that is taken to be the relevant boundary. However, if the boundary is at an angle exceeding 80° to the face of the wall, that is not regarded as a relevant boundary. When a wall faces a street, stream or canal, measurement is taken to the centre-line, as shown in Fig. B4.5.

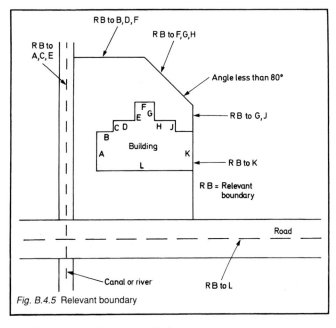

Fig. B.4.5 Relevant boundary

A relevant boundary is one which

- coincides with, or
- is parallel to, or
- is at angle of not more than 80° to the side of the building.

Notional boundary

Sometimes the degree of isolation necessary for buildings within the same site must be ascertained and for this purpose a 'notional boundary' is established, which then becomes the relevant boundary.

Establishing a notional boundary (Fig. B4.6) applies to residential, assembly or recreational buildings on the same site as another building.

UNPROTECTED AREAS

In relation to an external wall, an unprotected area (Fig. B4.7) means a part having less fire resistance than given in Table 2, and can be

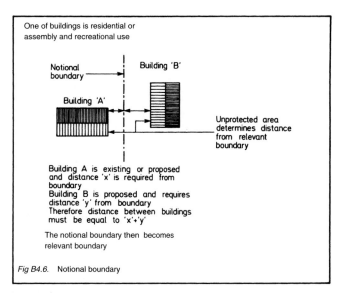

Fig B4.6. Notional boundary

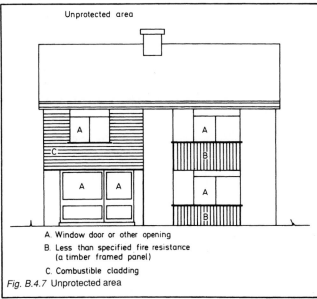

A. Window door or other opening
B. Less than specified fire resistance (a timber framed panel)
C. Combustible cladding

Fig. B.4.7 Unprotected area

- a window, door or other opening and measurements include the whole frame, i.e. the dimensions of the opening;
- any part of the external wall which has a fire resistance less than specified;
- any part of an external wall which has combustible material exceeding 1 mm thick attached or applied to its external face.

Unprotected walls of protected shafts are excluded when measuring unprotected areas, as also are parts of an external wall in an uncompartmented building more than 30 m above ground level (Fig. B4.8).

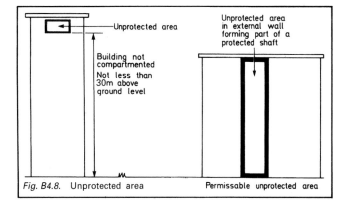

Fig. B4.8. Unprotected area

An external wall having the appropriate fire resistance but which is faced with combustible material more than 1 mm thick is regarded as having an unprotected area which is half the actual area of the combustible material (Fig. B4.9).

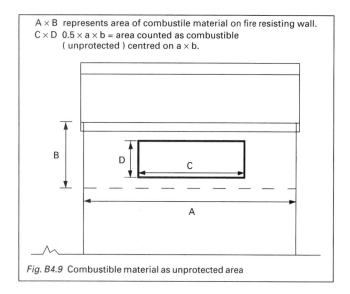

Fig. B4.9 Combustible material as unprotected area

There are a number of limited openings which may be placed in a protected wall and because of the small risk of fire spread may be ignored, and are illustrated in Fig. B4.10. There are no limitations on horizontal spacing of unprotected areas on either side of an internal compartment wall.

Reference may be made to section 2 of BS 5588: Part 5: 1991, 'Code of practice for firefighting stairs', for information on external walls for protected stairways and their relationship to unprotected areas elsewhere in the building.

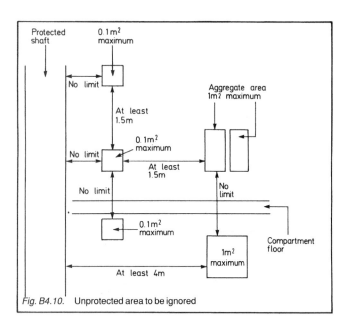

Fig. B4.10.   Unprotected area to be ignored

Canopies

Some of these structures are exempt from building regulations because they are within Class VI or Class VII of Schedule 2, but any which do not meet the criteria for size and distance from boundaries would not be exempt.

The approved document agrees that because of the need for canopies to be open sided to enable vehicles to pass limits about unprotected areas are unrealistic.

There is an extremely high degree of ventilation beneath a canopy, and provided it is at least 2 m away from the relevant boundary, separation distance is to be taken from the wall rather than the edge of the canopy.

Freestanding canopies above a limited risk or controlled hazard, and provided the canopy is at least 1 m from the relevant boundary, the requirements for space separation may be disregarded. Includes freestanding canopies over petrol pumps.

External walls

When a wall is within 1 m of the relevant boundary, including walls on the boundary, space separation requirements are met if

• unprotected areas are as shown in Fig. B4.10, and
• the remainder of the wall is fire resisting, inside and outside.

If a wall is 1 m or more from the relevant boundary, space separation requirements are met if

• the unprotected area is not in excess of that arrived at by calculation using method 1 or method 2;
• the remainder of the wall is fire resisting.

A BRE report 187 (1991), 'External fire spread – building separation and boundary distances' may be used in place of methods 1 and 2. Included in the report are the enclosing rectangle and aggregate notional area methods and they may be used instead of methods 1 and 2.

METHODS OF CALCULATION

The approved document gives two methods of calculating an acceptable amount of unprotected area in an external wall which is at least 1 m from the relevant boundary.

• For small residential buildings Method 1 may be used. It is not to be used if institutional buildings of group 2(a) are being considered.
• Most other buildings are appropriate for the use of Method 2.

The basis of Methods 1 and 2 is contained in the Fire Research Technical Paper No. 5, 1963, which is reprinted in the above report.

The desire is to achieve separation of a building from its boundary by not less than half the distance at which the total thermal radiation intensity from all unprotected areas would be $12.6\,\text{kW/m}^2$ in still air and if the radiation intensity from each unprotected area is

• for residential, office, assembly and recreational groups, $84\,\text{kW/m}^2$, and
• for shop and commercial, industrial, storage or other non-residential groups, $168\,\text{kW/m}^2$.

Where a sprinkler system is installed the boundary distance may be half of that for an unsprinklered building, subject of a minimum of 1 m.

Effect of sprinkler systems

A building having a sprinkler system in accordance with BS 5306: Part 2: 1990, Specification for sprinkler systems, may have reduced boundary distances

• half that required for a similar unsprinklered building, and not less than 1 m from the boundary, or
• the amount of unprotected area may be doubled and the boundary distance remain.

Atria

Clause 28.2 of BS 5588: Part 7: 1997, 'Code of practice for the incorporation of atria in buildings' should be adhered to.

Method 1: Small residential

This method is appropriate when used for dwellinghouses, flats and other residential purposes, but not including institutional. The building is to be 1 m or more from the relevant boundary,

- not to exceed 3 storeys in height;
- not to be more than 24 m in length;
- each side to meet requirements if

    (a) distance of side from relevant boundary, and
    (b) extent of unprotected areas are within limits, and
    (c) parts of building in excess of maximum unprotected areas are to be fire resisting.

Any unprotected area within the limits of Fig. B4.10 may be disregarded.

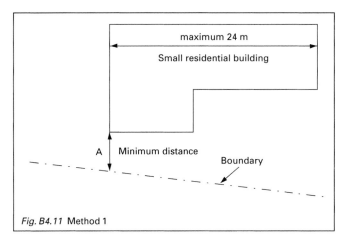

Fig. B4.11 Method 1

By reference to Fig. B4.11 the maximum amount of unprotected areas should fall within the following limits:

Maximum area of unprotected wall	Minimum distance (A) from side of building to relevant boundary
(m²)	(m)
5.6	1
12	2
18	3
24	4
30	5
No limit	6

Method 2: Other buildings or compartments

This method applies to any use where the building is at least 1 m from any point on the relevant boundary:

- building or compartment not to exceed 10 m in height, but

    (a) does not include open-sided car park – purpose group 7(b);
    (b) buildings or compartments in excess of 10 m high refer to BRE report 'External fire spread – building separation and boundary distances';

- each side will meet provisions if

    (a) the distance of side from relevant boundary, and
    (b) extent of unprotected area is within limits, and
    (c) parts of building in excess of maximum unprotected area are to be fire resisting.

Any unprotected area within the limits of Fig. B4.10 may be disregarded.

Minimum distance side of building to relevant boundary	Maximum total percentage unprotected area

Residential; office; assembly; recreation; open-sided car parks purpose group 7(b).

1 m	8%
2.5 m	20%
5 m	40%
7.5 m	60%
10 m	80%
12.5 m	100%

Shop and commercial, industrial; storage and other; non-residential.

1 m	4%
2 m	8%
5 m	20%
10 m	40%
15 m	60%
20 m	80%
25 m	100%

- Intermediate values may be calculated by interpolation.
- Sprinklered buildings – distance may be halved subject to a minimum of 1 m.

The original Enclosing Rectangle and Aggregate Notional Area methods contained in the 1985 approved document have been omitted from the 2000 edition, but are now contained in the BRE report 'External fire spread – building separation and boundary distances'.

However, these two methods may continue to be used, and the appropriate pages from the previous edition of this book now follow. Figure numbers remain as in the previous edition. The enclosing rectangle was previously described as Method 2 and the Aggregate Notional Area Method 3.

Enclosing rectangle

The enclosing rectangle method of determining distance from a relevant boundary is described in Method 2 and the approved document is supported by a number of tables. For each enclosing rectangle percentage table there are two sets of figures. One for separated parts or compartments with a shop, industrial or other non-residential use, the next figures relate to all other uses.

The enclosing rectangle shown in Fig. B102 is 4.5 m high × 5.5 m wide, but when using the tables it is necessary to find the size of the rectangle which coincides with the figure or the smallest size in which the illustration can be accommodated. In this example this will be 6 m high × 6 m wide, giving distances that range from 20% unprotected percentage at 1 m to 100% at 4 m.

In the approved document there are twelve stages to follow in order to fully establish the maximum area of unprotected area at a given distance from the relevant boundary.

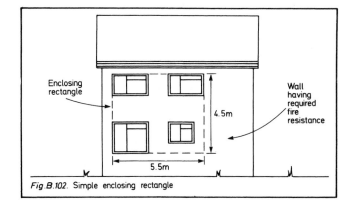

Fig.B.102. Simple enclosing rectangle

• Stage 1 is to ascertain the plane of reference. When a wall is parallel to the boundary the matter is simple as in Fig. B103(A).

The plane of reference should be in contact with the wall or some part of it (Fig. B103(B)). It may pass through projections such as a balcony or a coping but it must not cross the relevant boundary.

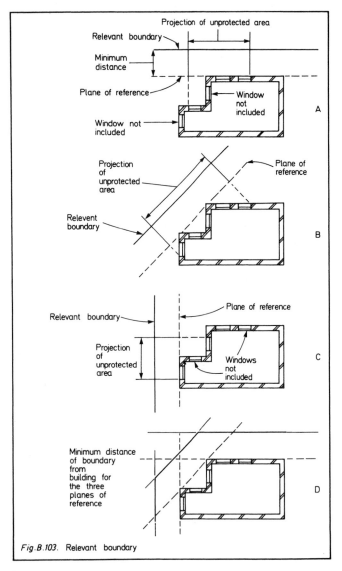

Fig.B.103. Relevant boundary

• Stage 2 is to mark on the plane of reference project lines relating to unprotected areas which are 80° or less to the plane of reference.
• Stage 3 involves construction of a rectangle enclosing all the unprotected areas which have been extended to the plane of reference. A simple example is in Fig. B102 but when there are other combustible areas on the wall, they must also be included.
• Stage 4. Reference to the tables in the approved document will identify the dimensions which will accept the rectangle drawn on the wall. The tables, which are reproduced in the following pages relate to enclosing rectangles between 3 m and 27 m high and with widths between 3 m to no limit.
• Stage 5 requires calculation of the total unprotected area and a determination of this area as a percentage of the enclosing rectangle, giving the unprotected percentage.
• Stage 6 passes on to finding the maximum unprotected area for a given boundary position or to find the nearest boundary for a building design, already demonstrated for a simple dwellinghouse.
    Maximum unprotected area for a given boundary position.
• Stage 7. Take the use of the building or compartment and with

the size of the enclosing rectangle refer to the tables. Read off the unprotected percentage allowed from the relevant boundary. Remember to use the plane of reference where this is appropriate.
• Stage 8. If it is found that the unprotected percentage is too great for the distance the amount of unprotected area or the extent of the enclosing rectangle must be modified.
• Stage 9 requires the designer to relate to all sides of the building having a relevant boundary.

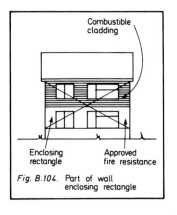

Fig. B.104. Part of wall enclosing rectangle

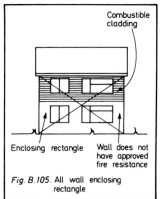

Fig. B.105. All wall enclosing rectangle

Nearest portion on boundary for desired building

• Stage 10. Refer to the tables and with the use of the building and size of the enclosing rectangle discover the minimum distance for the unprotected percentage.
• Stage 11. Repeat for all walls.
• Stage 12. These minimum distances when placed on the plans, establish an area around a building on which a boundary must not encroach.

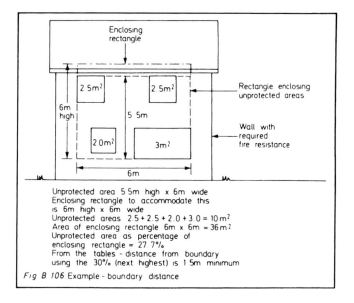

Unprotected area 5.5 m high x 6 m wide
Enclosing rectangle to accommodate this
is 6 m high x 6 m wide
Unprotected areas 2.5 + 2.5 + 2.0 + 3.0 = 10 m²
Area of enclosing rectangle 6 m x 6 m = 36 m²
Unprotected area as percentage of
enclosing rectangle = 27.7%
From the tables - distance from boundary
using the 30% (next highest) is 1.5 m minimum

Fig B 106 Example - boundary distance

Examples for larger buildings

• Four-storey uncompartmented building in Fig. B107
Rectangle enclosing unprotected areas 11.75 m × 20 m
Enclosing rectangle 12 m × 21 m = 252 m²
Unprotected areas 126 m²
Unprotected areas as percentage of enclosing rectangle
(100 × 126) 252 = 50%
Distance from boundary from tables 6.5 m minimum
(Residential office or assembly)

Table **J2** **Permitted unprotected percentages in relation to enclosing rectangles**

Width of enclosing rectangle [m]	20%	30%	40%	50%	60%	70%	80%	90%	100%
	Distance from relevant boundary for unprotected percentage not exceeding — **Minimum boundary distance** [m] Figures in brackets are for residential, office or assembly								
Enclosing rectangle 3 m high									
3	1.0 (1.0)	1.5 (1.0)	2.0 (1.0)	2.0 (1.5)	2.5 (1.5)	2.5 (1.5)	2.5 (2.0)	3.0 (2.0)	3.0 (2.0)
6	1.5 (1.0)	2.0 (1.0)	2.5 (1.5)	3.0 (2.0)	3.0 (2.0)	3.5 (2.0)	3.5 (2.5)	4.0 (2.5)	4.0 (3.0)
9	1.5 (1.0)	2.5 (1.0)	3.0 (1.5)	3.5 (2.0)	4.0 (2.5)	4.0 (2.5)	4.5 (3.0)	5.0 (3.0)	5.0 (3.5)
12	2.0 (1.0)	2.5 (1.5)	3.0 (2.0)	3.5 (2.0)	4.0 (2.5)	4.5 (3.0)	5.0 (3.0)	5.5 (3.5)	5.5 (3.5)
15	2.0 (1.0)	2.5 (1.5)	3.5 (2.0)	4.0 (2.5)	4.5 (2.5)	5.0 (3.0)	5.5 (3.5)	6.0 (3.5)	6.0 (4.0)
18	2.0 (1.0)	2.5 (1.5)	3.5 (2.0)	4.0 (2.5)	5.0 (2.5)	5.0 (3.0)	6.0 (3.5)	6.5 (4.0)	6.5 (4.0)
21	2.0 (1.0)	3.0 (1.5)	3.5 (2.0)	4.5 (2.5)	5.0 (3.0)	5.5 (3.0)	6.0 (3.5)	6.5 (4.0)	7.0 (4.5)
24	2.0 (1.0)	3.0 (1.5)	3.5 (2.0)	4.5 (2.5)	5.0(3.0)	5.5 (3.5)	6.0 (3.5)	7.0 (4.0)	7.5 (4.5)
27	2.0 (1.0)	3.0 (1.5)	4.0 (2.0)	4.5 (2.5)	5.5 (3.0)	6.0 (3.5)	6.5 (4.0)	7.0 (4.0)	7.5 (4.5)
30	2.0 (1.0)	3.0 (1.5)	4.0 (2.0)	4.5 (2.5)	5.5 (3.0)	6.0 (3.5)	6.5 (4.0)	7.5 (4.0)	8.0 (4.5)
40	2.0 (1.0)	3.0 (1.5)	4.0 (2.0)	5.0 (2.5)	5.5 (3.0)	6.5 (3.5)	7.0 (4.0)	8.0 (4.0)	8.5 (5.0)
50	2.0 (1.0)	3.0 (1.5)	4.0 (2.0)	5.0 (2.5)	6.0 (3.0)	6.5 (3.5)	7.5 (4.0)	8.0 (4.0)	9.0 (5.0)
60	2.0 (1.0)	3.0 (1.5)	4.0 (2.0)	5.0 (2.5)	6.0 (3.0)	7.0 (3.5)	7.5 (4.0)	8.5 (4.0)	9.5 (5.0)
80	2.0 (1.0)	3.0 (1.5)	4.0 (2.0)	5.0 (2.5)	6.0 (3.0)	7.0 (3.5)	8.0 (4.0)	9.0 (4.0)	9.5 (5.0)
no limit	2.0 (1.0)	3.0 (1.5)	4.0 (2.0)	5.0 (2.5)	6.0 (3.0)	7.0 (3.5)	8.0 (4.0)	9.0 (4.0)	10.0 (5.0)
Enclosing rectangle 6 m high									
3	1.5 (1.0)	2.0 (1.0)	2.5 (1.5)	3.0 (2.0)	3.0 (2.0)	3.5 (2.0)	3.5 (2.5)	4.0 (2.5)	4.0 (3.0)
6	2.0 (1.0)	3.0 (1.5)	3.5 (2.0)	4.0 (2.5)	4.5 (3.0)	5.0 (3.0)	5.5 (3.5)	5.5 (4.0)	6.0 (4.0)
9	2.5 (1.0)	3.5 (2.0)	4.5 (2.5)	5.0 (3.0)	5.5 (3.5)	6.0 (4.0)	6.0 (4.5)	7.0 (4.5)	7.0 (5.0)
12	3.0 (1.5)	4.0 (2.5)	5.0 (3.0)	5.5 (3.5)	6.5 (4.0)	7.0 (4.5)	7.5 (5.0)	8.0 (5.0)	8.5 (5.5)
15	3.0 (1.5)	4.5 (2.5)	5.5 (3.0)	6.0 (4.0)	7.0 (4.5)	7.5 (5.0)	8.0 (5.5)	9.0 (5.5)	9.0 (6.0)
18	3.5 (1.5)	4.5 (2.5)	5.5 (3.5)	6.5 (4.0)	7.5 (4.5)	8.0 (5.0)	9.0 (5.5)	9.5 (6.0)	10.0 (6.5)
21	3.5 (1.5)	5.0 (2.5)	6.0 (3.5)	7.0 (4.0)	8.0 (5.0)	9.0 (5.5)	9.5 (6.0)	10.0 (6.5)	10.5 (7.0)
24	3.5 (1.5)	5.0 (2.5)	6.0 (3.5)	7.0 (4.5)	8.5 (5.0)	9.5 (5.5)	10.0 (6.0)	10.5 (7.0)	11.0 (7.0)
27	3.5 (1.5)	5.0 (2.5)	6.5 (3.5)	7.5 (4.5)	8.5 (5.0)	9.5 (6.0)	10.5 (6.5)	11.0 (7.0)	12.0 (7.5)
30	3.5 (1.5)	5.0 (2.5)	6.5 (3.5)	8.0 (4.5)	9.0 (5.0)	10.0 (6.0)	11.0 (6.5)	12.0 (7.0)	12.5 (8.0)
40	3.5 (1.5)	5.5 (2.5)	7.0 (3.5)	8.5 (4.5)	10.0 (5.5)	11.0 (6.5)	12.0 (7.0)	13.0 (8.0)	14.0 (8.5)
50	3.5 (1.5)	5.5 (2.5)	7.5 (3.5)	9.0 (4.5)	10.5 (5.5)	11.5 (6.5)	13.0 (7.5)	14.0 (8.0)	15.0 (9.0)
60	3.5 (1.5)	5.5 (2.5)	7.5 (3.5)	9.5 (5.0)	11.0 (5.5)	12.0 (6.5)	13.5 (7.5)	15.0 (8.5)	16.0 (9.5)
80	3.5 (1.5)	6.0 (2.5)	7.5 (3.5)	9.5 (5.0)	11.5 (6.0)	13.0 (7.0)	14.5 (7.5)	16.0 (8.5)	17.5 (9.5)
100	3.5 (1.5)	6.0 (2.5)	8.0 (3.5)	10.0 (5.0)	12.0 (6.0)	13.5 (7.0)	15.0 (8.0)	16.5 (8.5)	18.0 (10.0)
120	3.5 (1.5)	6.0 (2.5)	8.0 (3.5)	10.0 (5.0)	12.0 (6.0)	14.0 (7.0)	15.5 (8.0)	17.0 (8.5)	19.0 (10.0)
no limit	3.5 (1.5)	6.0 (2.5)	8.0 (3.5)	10.0 (5.0)	12.0 (6.0)	14.0 (7.0)	16.0 (8.0)	18.0 (8.5)	19.0 (10.0)
Enclosing rectangle 9 m high									
3	1.5 (1.0)	2.5 (1.0)	3.0 (1.5)	3.5 (2.0)	4.0 (2.5)	4.0 (2.5)	4.5 (3.0)	5.0 (3.0)	5.0 (3.5)
6	2.5 (1.0)	3.5 (2.0)	4.5 (2.5)	5.0 (3.0)	5.5 (3.5)	6.0 (4.0)	6.5 (4.5)	7.0 (4.5)	7.0 (5.0)
9	3.5 (1.5)	4.5 (2.5)	5.5 (3.5)	6.0 (4.0)	6.5 (4.5)	7.5 (5.0)	8.0 (5.5)	8.5 (5.5)	9.0 (6.0)
12	3.5 (1.5)	5.0 (3.0)	6.0 (3.5)	7.0 (4.5)	7.5 (5.0)	8.5 (5.5)	9.0 (6.0)	9.5 (6.5)	10.5 (7.0)
15	4.0 (2.0)	5.5 (3.0)	6.5 (4.0)	7.5 (5.0)	8.5 (5.5)	9.5 (6.0)	10.0 (6.5)	11.0 (7.0)	11.5 (7.5)
18	4.5 (2.0)	6.0 (3.5)	7.0 (4.5)	8.5 (5.0)	9.5 (6.0)	10.0 (6.5)	11.0 (7.0)	12.0 (8.0)	12.5 (8.5)
21	4.5 (2.0)	6.5 (3.5)	7.5 (4.5)	9.0 (5.5)	10.0 (6.5)	11.0 (7.0)	12.0 (7.5)	13.0 (8.5)	13.5 (9.0)
24	5.0 (2.0)	6.5 (3.5)	8.0 (5.0)	9.5 (5.5)	11.0 (6.5)	12.0 (7.5)	13.0 (8.0)	13.5 (9.0)	14.5 (9.5)
27	5.0 (2.0)	7.0 (3.5)	8.5 (5.0)	10.0 (6.0)	11.5 (7.0)	12.5 (7.5)	13.5 (8.5)	14.5 (9.5)	15.0 (10.0)
30	5.0 (2.0)	7.0 (3.5)	9.0 (5.0)	10.5 (6.0)	12.0 (7.0)	13.0 (8.0)	14.0 (9.0)	15.0 (9.5)	16.0 (10.5)
40	5.5 (2.0)	7.5 (3.5)	9.5 (5.5)	11.5 (6.5)	13.0 (7.5)	14.5 (8.5)	15.5 (9.5)	17.0 (10.5)	17.5 (11.5)
50	5.5 (2.0)	8.0 (4.0)	10.0 (5.5)	12.5 (6.5)	14.0 (8.0)	15.5 (9.0)	17.0 (10.0)	18.5 (11.5)	19.5 (12.5)
60	5.5 (2.0)	8.0 (4.0)	11.0 (5.5)	13.0 (7.0)	15.0 (8.0)	16.5 (9.5)	18.0 (11.0)	19.5 (11.5)	21.0 (13.0)
80	5.5 (2.0)	8.5 (4.0)	11.5 (5.5)	13.5 (7.0)	16.0 (8.5)	17.5 (10.0)	19.5 (11.5)	21.5 (12.5)	23.0 (13.5)
100	5.5 (3.0)	8.5 (4.0)	11.5 (5.5)	14.5 (7.0)	16.5 (8.5)	18.5 (10.0)	21.0 (11.5)	22.5 (12.5)	24.5 (14.5)
120	5.5 (2.0)	8.5 (4.0)	11.5 (5.5)	14.5 (7.0)	17.0 (8.5)	19.5 (10.0)	21.5 (11.5)	23.5 (12.5)	26.0 (14.5)
no limit	5.5 (2.0)	8.5 (4.0)	11.5 (5.5)	15.0 (7.0)	17.5 (8.5)	20.0 (10.5)	22.5 (12.0)	24.5 (12.5)	27.0 (15.0)
Enclosing rectangle 12 m high									
3	2.0 (1.0)	2.5 (1.5)	3.0 (2.0)	3.5 (2.0)	4.0 (2.5)	4.5 (3.0)	5.0 (3.0)	5.5 (3.5)	5.5 (3.5)
6	3.0 (1.5)	4.0 (2.5)	5.0 (3.0)	5.5 (3.5)	6.5 (4.0)	7.0 (4.5)	7.5 (5.0)	8.0 (5.0)	8.5 (5.5)
9	3.5 (1.5)	5.0 (3.0)	6.0 (3.5)	7.0 (4.5)	7.5 (5.0)	8.5 (5.5)	9.0 (6.0)	9.5 (6.5)	10.5 (7.0)
12	4.5 (1.5)	6.0 (3.5)	7.0 (4.5)	8.0 (5.0)	9.0 (6.0)	9.5 (6.5)	11.0 (7.0)	11.5 (7.5)	12.0 (8.0)
15	5.0 (2.0)	6.5 (3.5)	8.0 (5.0)	9.0 (5.5)	10.0 (6.5)	11.0 (7.0)	12.0 (8.0)	13.0 (8.5)	13.5 (9.0)
18	5.0 (2.5)	7.0 (4.0)	8.5 (5.0)	10.0 (6.0)	11.0 (7.0)	12.0 (7.5)	13.0 (8.5)	14.0 (9.0)	14.5 (10.0)
21	5.5 (2.5)	7.5 (4.0)	9.0 (5.5)	10.5 (6.5)	12.0 (7.5)	13.0 (8.5)	14.0 (9.0)	15.0 (10.0)	16.0 (10.5)
24	6.0 (2.5)	8.0 (4.5)	9.5 (6.0)	11.5 (7.0)	12.5 (8.0)	14.0 (8.5)	15.0 (9.5)	16.0 (10.5)	16.5 (11.5)
27	6.0 (2.5)	8.0 (4.5)	10.5 (6.0)	12.0 (7.0)	13.5 (8.0)	14.5 (9.0)	16.0 (10.5)	17.0 (11.0)	17.5 (12.0)
30	6.5 (2.5)	8.5 (4.5)	10.5 (6.5)	12.5 (7.5)	14.0 (8.5)	15.0 (9.5)	16.5 (10.5)	17.5 (11.5)	18.5 (12.5)
40	6.5 (2.5)	9.5 (5.0)	12.0 (6.5)	14.0 (8.0)	15.5 (9.5)	17.5 (10.5)	18.5 (12.0)	20.0 (13.0)	21.0 (14.0)
50	7.0 (2.5)	10.0 (5.0)	13.0 (7.0)	15.0 (8.5)	17.0 (10.0)	19.0 (11.0)	20.5 (13.0)	23.0 (14.0)	23.0 (15.0)
60	7.0 (2.5)	10.5 (5.0)	13.5 (7.0)	16.0 (9.0)	18.0 (10.5)	20.0 (12.0)	21.5 (13.5)	23.5 (14.5)	25.0 (16.0)
80	7.0 (2.5)	11.0 (5.0)	14.5 (7.0)	17.0 (9.0)	19.5 (11.0)	21.5 (13.0)	23.5 (14.5)	26.0 (16.0)	27.5 (17.0)
100	7.5 (2.5)	11.5 (5.0)	15.0 (7.5)	18.0 (9.5)	21.0 (11.5)	23.0 (13.5)	25.5 (15.0)	28.0 (16.5)	30.0 (18.0)
120	7.5 (2.5)	11.5 (5.0)	15.0 (7.5)	18.5 (9.5)	22.0 (11.5)	24.0 (13.5)	27.0 (15.0)	29.5 (17.0)	31.5 (18.5)
no limit	7.5 (2.5)	12.0 (5.0)	15.5 (7.5)	19.0 (9.5)	22.5 (12.0)	25.0 (14.0)	28.0 (15.5)	30.5 (17.0)	34.0 (19.0)

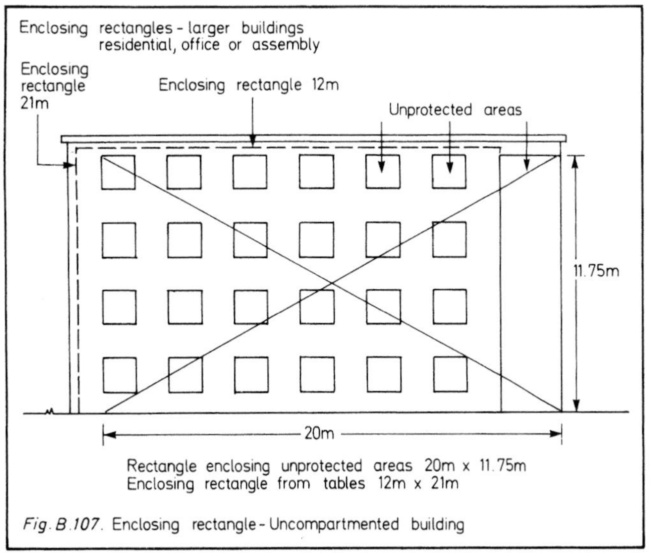

Rectangle enclosing unprotected areas 20m x 11.75m
Enclosing rectangle from tables 12m x 21m

Fig. B.107. Enclosing rectangle - Uncompartmented building

Table **J2** *continued*

Width of enclosing rectangle [m]	20%	30%	40%	50%	60%	70%	80%	90%	100%
	Distance from relevant boundary for unprotected percentage not exceeding — **Minimum boundary distance** [m] Figures in brackets are for residential, office or assembly								
Enclosing rectangle 15 m high									
3	2.0 (1.0)	2.5 (1.5)	3.5 (2.0)	4.0 (2.5)	4.5 (2.5)	5.0 (3.0)	5.5 (3.5)	6.0 (3.5)	6.0 (4.0)
6	3.0 (1.5)	4.5 (2.5)	5.5 (3.0)	6.0 (4.0)	7.0 (4.5)	7.5 (5.0)	8.0 (5.5)	9.0 (5.5)	9.0 (6.0)
9	4.0 (2.0)	5.5 (3.0)	6.5 (4.0)	7.5 (5.0)	8.5 (5.5)	9.5 (6.0)	10.0 (6.5)	11.0 (7.0)	11.5 (7.5)
12	5.0 (2.0)	6.5 (3.5)	8.0 (5.0)	9.0 (5.5)	10.0 (6.5)	11.0 (7.0)	12.0 (8.0)	13.0 (8.5)	13.5 (9.0)
15	5.5 (2.0)	7.0 (4.0)	9.0 (5.5)	10.0 (6.5)	11.5 (7.0)	12.5 (8.0)	13.5 (9.0)	14.5 (9.5)	15.0 (10.0)
18	6.0 (2.5)	8.0 (4.5)	9.5 (6.0)	11.0 (7.0)	12.5 (8.0)	13.5 (8.5)	14.5 (9.5)	15.5 (10.5)	16.5 (11.0)
21	6.5 (2.5)	8.5 (5.0)	10.5 (6.5)	12.0 (7.5)	13.5 (8.5)	14.5 (9.5)	16.0 (10.5)	16.5 (11.0)	17.5 (12.0)
24	6.5 (3.0)	9.0 (5.0)	11.0 (6.5)	13.0 (8.0)	14.5 (9.0)	15.5 (10.0)	17.0 (11.0)	18.0 (12.0)	19.0 (13.0)
27	7.0 (3.0)	9.5 (5.5)	11.5 (7.0)	13.5 (8.5)	15.0 (9.5)	16.5 (10.5)	18.0 (11.5)	19.0 (12.5)	20.0 (13.5)
30	7.5 (3.0)	10.0 (5.5)	12.0 (7.5)	14.0 (8.5)	16.0 (10.0)	17.0 (11.0)	18.5 (12.0)	20.0 (13.5)	21.0 (14.0)
40	8.0 (3.0)	11.0 (6.0)	13.5 (8.0)	16.0 (9.5)	18.0 (11.0)	19.5 (12.5)	21.0 (13.5)	22.5 (15.0)	23.5 (16.0)
50	8.5 (3.5)	12.0 (6.0)	15.0 (8.5)	17.5 (10.0)	19.5 (12.0)	21.5 (13.5)	23.0 (15.0)	25.0 (16.5)	26.0 (17.5)
60	8.5 (3.5)	12.5 (6.5)	15.5 (8.5)	18.0 (10.5)	21.0 (12.5)	23.5 (14.0)	25.0 (15.5)	27.0 (17.0)	28.0 (18.0)
80	9.0 (3.5)	13.5 (6.5)	17.0 (9.0)	20.0 (11.0)	23.0 (13.5)	25.5 (15.0)	28.0 (17.0)	30.0 (18.5)	31.5 (20.0)
100	9.0 (3.5)	14.0 (6.5)	18.0 (9.0)	21.5 (11.5)	24.5 (14.0)	27.5 (16.0)	30.0 (18.0)	32.5 (19.5)	34.5 (21.5)
120	9.0 (3.5)	14.0 (6.5)	18.5 (9.0)	22.5 (11.5)	25.5 (14.0)	28.5 (16.5)	31.5 (18.5)	34.5 (20.5)	37.0 (22.5)
no limit	9.0 (3.5)	14.5 (6.5)	19.0 (9.0)	23.0 (12.0)	27.0 (14.5)	30.0 (17.0)	34.0 (19.0)	36.0 (21.0)	39.0 (23.0)
Enclosing rectangle 18 m high									
3	2.0 (1.0)	2.5 (1.5)	3.5 (2.0)	4.0 (2.5)	5.0 (2.5)	5.0 (3.0)	6.0 (3.5)	6.5 (4.0)	6.5 (4.0)
6	3.5 (1.5)	4.5 (2.5)	5.5 (3.5)	6.5 (4.0)	7.5 (4.5)	8.0 (5.0)	9.0 (5.5)	9.5 (6.0)	10.0 (6.5)
9	4.5 (2.0)	6.0 (3.5)	7.0 (4.5)	8.5 (5.0)	9.5 (6.0)	10.0 (6.5)	11.0 (7.0)	12.0 (8.0)	12.5 (8.5)
12	5.0 (2.5)	7.0 (4.0)	8.5 (5.0)	10.0 (6.0)	11.0 (7.0)	12.0 (7.5)	13.0 (8.5)	14.0 (9.0)	14.5 (10.0)
15	6.0 (2.5)	8.0 (4.5)	9.5 (6.0)	11.0 (7.0)	12.5 (8.0)	13.5 (8.5)	14.5 (9.5)	15.5 (10.5)	16.5 (11.0)
18	6.5 (2.5)	8.5 (5.0)	11.0 (6.5)	12.0 (7.5)	13.5 (8.5)	14.5 (9.5)	16.0 (11.0)	17.0 (11.5)	18.0 (13.0)
21	7.0 (3.0)	9.5 (5.5)	11.5 (7.0)	13.0 (8.0)	14.5 (9.5)	16.0 (10.5)	17.0 (11.5)	18.0 (12.5)	19.5 (13.0)
24	7.5 (3.0)	10.0 (5.5)	12.0 (7.5)	14.0 (8.5)	15.5 (10.0)	16.5 (11.0)	18.5 (12.0)	19.5 (13.0)	20.5 (14.0)
27	8.0 (3.5)	10.5 (6.0)	12.5 (8.0)	14.5 (9.0)	16.5 (10.5)	17.5 (11.5)	19.5 (12.5)	20.5 (13.5)	21.5 (14.5)
30	8.0 (3.5)	11.0 (6.5)	13.5 (8.0)	15.5 (9.5)	17.0 (11.0)	18.5 (12.0)	20.5 (13.5)	21.5 (14.5)	22.5 (15.5)
40	9.0 (4.0)	12.0 (7.0)	15.0 (9.0)	17.5 (11.0)	19.5 (12.0)	21.5 (13.5)	23.5 (15.0)	25.0 (16.5)	26.0 (17.5)
50	9.5 (4.0)	13.0 (7.0)	16.5 (9.5)	19.0 (11.5)	21.5 (13.0)	23.5 (15.0)	26.0 (16.5)	27.5 (18.0)	29.0 (19.0)
60	10.0 (4.0)	14.0 (7.5)	17.5 (10.0)	20.5 (12.0)	23.0 (14.0)	26.0 (16.0)	27.5 (17.5)	29.5 (19.5)	31.0 (20.5)
80	10.0 (4.0)	15.0 (7.5)	19.0 (10.0)	22.5 (13.0)	26.0 (15.0)	28.5 (17.0)	31.0 (19.0)	33.5 (21.0)	35.0 (22.5)
100	10.0 (4.0)	16.0 (7.5)	20.5 (10.0)	24.0 (13.5)	28.0 (16.0)	31.0 (18.0)	33.5 (20.5)	36.0 (22.5)	38.5 (24.0)
120	10.0 (4.0)	16.5 (7.5)	21.0 (10.0)	25.5 (14.0)	29.5 (16.5)	32.5 (19.0)	35.5 (21.0)	39.0 (23.5)	41.5 (25.5)
no limit	10.0 (4.0)	17.0 (8.0)	22.0 (10.0)	26.5 (14.0)	30.5 (17.0)	34.0 (19.5)	37.0 (22.0)	41.0 (24.0)	43.5 (26.5)
Enclosing rectangle 21 m high									
3	2.0 (1.0)	3.0 (1.5)	3.5 (2.0)	4.5 (2.5)	5.0 (3.0)	5.5 (3.0)	6.0 (3.5)	6.5 (4.0)	7.0 (4.5)
6	3.5 (1.5)	5.0 (2.5)	6.0 (3.5)	7.0 (4.0)	8.0 (5.0)	9.0 (5.5)	9.5 (6.0)	10.0 (6.5)	10.5 (7.0)
9	4.5 (2.0)	6.5 (3.5)	7.5 (4.5)	9.0 (5.5)	10.0 (6.5)	11.0 (7.0)	12.0 (7.5)	13.0 (8.5)	13.5 (9.0)
12	5.5 (2.5)	7.5 (4.0)	9.0 (5.5)	10.5 (6.5)	12.0 (7.5)	13.0 (8.5)	14.0 (9.0)	15.0 (10.0)	16.0 (10.5)
15	6.5 (2.5)	8.5 (5.0)	10.5 (6.5)	12.0 (7.5)	13.5 (8.5)	14.5 (9.5)	16.0 (10.5)	16.5 (11.0)	17.5 (12.0)
18	7.0 (3.0)	9.5 (5.5)	11.5 (7.0)	13.0 (8.0)	14.5 (9.5)	16.0 (10.5)	17.0 (11.5)	18.0 (12.5)	19.5 (13.0)
21	7.5 (3.0)	10.0 (6.0)	12.5 (7.5)	14.0 (9.0)	15.5 (10.0)	17.0 (11.0)	18.5 (12.5)	20.0 (13.5)	21.0 (14.0)
24	8.0 (3.5)	10.5 (6.0)	13.0 (8.0)	15.0 (9.5)	16.5 (10.5)	18.0 (12.0)	20.0 (13.0)	21.0 (14.0)	22.0 (15.0)
27	8.5 (3.5)	11.5 (6.5)	14.0 (8.5)	16.0 (10.0)	18.0 (11.5)	19.0 (13.0)	21.0 (14.0)	22.5 (15.0)	23.5 (16.0)
30	9.0 (4.0)	12.0 (7.0)	14.5 (9.0)	16.5 (10.5)	18.5 (12.0)	20.5 (13.0)	22.0 (14.5)	23.5 (16.0)	25.0 (16.5)
40	10.0 (4.5)	13.5 (7.5)	16.5 (10.0)	19.0 (12.0)	21.5 (13.5)	23.0 (15.0)	25.5 (16.5)	27.0 (18.0)	28.5 (19.0)
50	11.0 (4.5)	14.5 (8.0)	18.0 (11.0)	21.0 (13.0)	23.5 (14.5)	25.5 (16.5)	28.0 (18.0)	30.0 (20.0)	31.5 (21.0)
60	11.5 (4.5)	15.5 (8.5)	19.5 (11.5)	22.5 (13.5)	25.5 (15.5)	28.0 (17.5)	30.5 (19.5)	32.5 (21.0)	33.5 (22.5)
80	12.0 (4.5)	17.0 (8.5)	21.0 (12.0)	25.0 (14.5)	28.5 (17.0)	31.5 (19.0)	34.0 (21.0)	36.5 (23.5)	38.5 (25.0)
100	12.0 (4.5)	18.0 (9.0)	22.5 (12.0)	27.0 (15.5)	31.0 (18.0)	34.5 (20.5)	37.0 (22.5)	40.0 (25.0)	42.0 (27.0)
120	12.0 (4.5)	18.5 (9.0)	23.5 (12.0)	28.5 (16.0)	32.5 (18.5)	36.5 (21.5)	39.5 (23.5)	43.0 (26.5)	45.5 (28.5)
no limit	12.0 (4.5)	19.0 (9.0)	25.0 (12.0)	29.5 (16.0)	34.5 (19.0)	38.0 (22.0)	41.5 (25.0)	45.5 (26.5)	48.0 (29.5)
Enclosing rectangle 24 m high									
3	2.0 (1.0)	3.0 (1.5)	3.5 (2.0)	4.5 (2.5)	5.0 (3.0)	5.5 (3.5)	6.0 (3.5)	7.0 (4.0)	7.5 (4.5)
6	3.5 (1.5)	5.0 (2.5)	6.0 (3.5)	7.0 (4.5)	8.5 (5.0)	9.5 (5.5)	10.0 (6.0)	10.5 (7.0)	11.0 (7.0)
9	5.0 (2.0)	6.5 (3.5)	8.0 (5.0)	9.5 (5.5)	11.0 (6.5)	12.0 (7.5)	13.0 (8.0)	13.5 (9.0)	14.5 (9.5)
12	6.0 (2.5)	8.0 (4.5)	9.5 (6.0)	11.5 (7.0)	12.5 (8.0)	14.0 (8.5)	15.0 (9.5)	16.0 (10.5)	16.5 (11.5)
15	6.5 (3.0)	9.0 (5.0)	11.0 (6.5)	13.0 (8.0)	14.5 (9.0)	15.5 (10.0)	17.0 (11.0)	18.0 (12.0)	19.0 (13.0)
18	7.5 (3.0)	10.0 (5.5)	12.0 (7.5)	14.0 (8.5)	15.5 (10.0)	16.5 (11.0)	18.5 (12.0)	19.5 (13.0)	20.5 (14.0)
21	8.0 (3.5)	10.5 (6.0)	13.0 (8.0)	15.0 (9.5)	16.5 (10.5)	18.0 (12.0)	20.0 (13.0)	21.0 (14.0)	22.0 (15.0)
24	8.5 (3.5)	11.5 (6.5)	14.0 (8.5)	16.0 (10.0)	18.0 (11.5)	19.5 (12.5)	21.0 (14.0)	22.5 (15.0)	24.0 (16.0)
27	9.0 (4.0)	12.5 (7.0)	15.0 (9.0)	17.0 (11.0)	19.0 (12.5)	20.5 (13.5)	22.5 (15.0)	24.0 (16.0)	25.5 (17.0)
30	9.5 (4.0)	13.0 (7.5)	15.5 (9.5)	18.0 (11.5)	20.0 (13.0)	21.5 (14.0)	23.5 (15.5)	25.0 (17.0)	26.5 (18.0)
40	11.0 (4.5)	14.5 (8.5)	18.0 (11.0)	20.5 (13.0)	23.0 (14.5)	25.0 (16.0)	27.5 (18.0)	29.0 (19.0)	30.5 (20.5)
50	12.0 (5.0)	16.0 (9.0)	19.5 (12.0)	22.5 (14.0)	25.5 (16.0)	27.5 (17.5)	30.0 (19.5)	32.0 (21.0)	33.5 (22.5)
60	12.5 (5.0)	17.0 (9.5)	21.0 (12.5)	24.5 (15.0)	27.5 (17.0)	30.0 (19.0)	32.5 (21.0)	35.0 (23.0)	36.5 (24.5)
80	13.5 (5.0)	18.5 (10.0)	23.5 (13.5)	27.5 (16.5)	31.0 (18.5)	34.5 (21.0)	37.0 (23.5)	39.5 (25.5)	41.5 (27.5)
100	13.5 (5.0)	20.0 (10.0)	25.0 (13.5)	29.5 (17.0)	33.5 (20.0)	37.0 (22.5)	40.0 (25.0)	43.0 (27.5])	45.5 (29.5)
120	13.5 (5.5)	20.5 (10.0)	26.5 (13.5)	31.0 (17.5)	36.0 (20.5)	39.5 (23.5)	43.0 (26.5)	46.5 (29.0)	49.0 (31.0)
no limit	13.5 (5.5)	21.0 (10.0)	27.5 (13.5)	32.5 (18.0)	37.5 (21.0)	42.0 (24.0)	45.5 (27.5])	49.5 (30.0)	52.0 (32.5)
Enclosing rectangle 27 m high									
3	2.0 (1.0)	3.0 (1.5)	4.0 (2.0)	4.5 (2.5)	5.5 (3.0)	6.0 (3.5)	6.5 (4.0)	7.0 (4.0)	7.5 (4.5)
6	3.5 (1.5)	5.0 (2.5)	6.5 (3.5)	7.5 (4.5)	8.5 (5.0)	9.5 (6.0)	10.5 (6.5)	11.0 (7.0)	12.0 (7.5)
9	5.0 (2.0)	7.0 (3.5)	8.5 (5.0)	10.0 (6.0)	11.5 (7.0)	12.5 (7.5)	13.5 (8.5)	14.5 (9.5)	15.0 (10.0)
12	6.0 (2.5)	8.0 (4.5)	10.5 (6.0)	12.0 (7.0)	13.5 (8.0)	14.5 (9.0)	16.0 (10.5)	17.0 (11.0)	17.5 (12.0)
15	7.0 (3.0)	9.5 (5.5)	11.5 (7.0)	13.5 (8.5)	15.0 (9.5)	16.5 (10.5)	18.0 (11.5)	19.0 (12.5)	20.0 (13.5)
18	8.0 (3.5)	10.5 (6.0)	12.5 (8.0)	14.5 (9.0)	16.5 (10.5)	17.5 (11.5)	19.5 (12.5)	20.5 (13.5)	21.5 (14.5)
21	8.5 (3.5)	11.5 (6.5)	14.0 (8.5)	16.0 (10.0)	18.0 (11.5)	19.0 (13.0)	21.0 (14.0)	22.5 (15.0)	23.5 (16.0)
24	9.0 (3.5)	12.5 (7.0)	15.0 (9.0)	17.0 (11.0)	19.0 (12.5)	20.5 (13.5)	22.5 (15.0)	24.0 (16.0)	25.5 (17.0)
27	10.0 (4.0)	13.0 (7.5)	16.0 (10.0)	18.0 (11.5)	20.0 (13.0)	22.0 (14.0)	24.0 (16.0)	25.5 (17.0)	27.0 (18.0)
30	10.0 (4.0)	13.5 (8.0)	17.0 (10.0)	19.0 (12.0)	21.0 (13.5)	23.0 (15.0)	25.0 (17.0)	26.5 (18.0)	28.0 (19.0)
40	11.5 (5.0)	15.5 (9.0)	19.0 (11.5)	22.0 (14.0)	24.5 (15.5)	26.5 (17.5)	29.0 (19.0)	30.5 (20.5)	32.5 (22.0)
50	12.5 (5.5)	17.0 (9.5)	21.0 (12.5)	24.0 (15.0)	27.0 (17.0)	29.5 (19.0)	32.0 (21.0)	34.5 (22.5)	36.0 (24.0)
60	13.5 (5.5)	18.5 (10.5)	22.5 (13.5)	26.5 (16.0)	29.5 (18.5)	32.0 (20.5)	35.0 (22.5)	37.0 (24.5)	39.0 (26.5)
80	14.5 (6.0)	20.5 (11.0)	25.0 (14.5)	29.5 (17.5)	33.0 (20.5)	36.5 (22.5)	39.5 (25.0)	42.0 (27.5)	44.0 (29.5)
100	15.5 (6.0)	21.5 (11.0)	27.0 (15.5)	32.0 (19.0)	36.5 (21.5)	40.5 (24.5)	43.0 (27.0)	46.5 (30.0)	48.5 (32.0)
120	15.5 (6.0)	22.5 (11.5)	28.5 (15.5)	34.0 (19.5)	39.0 (22.5)	43.0 (26.0)	46.5 (28.5)	50.5 (32.0)	53.0 (34.0)
no limit	15.5 (6.0)	23.5 (11.5)	29.5 (15.5)	35.0 (20.0)	40.5 (23.5)	44.5 (27.0)	48.5 (29.5)	52.0 (33.0)	55.5 (35.0)

- Four-storey compartmented building in Fig. B108
  Rectangle enclosing unprotected areas 5.5 m × 16 m
  Enclosing rectangle 6 m × 18 m = 108 m²
  Unprotected areas 39 m²
  Unprotected areas as percentage of enclosing rectangle
  (100 × 39) 108 = 36%
  Distance from boundary from tables (40%) = 3.5 m minimum
  (Residential, office or assembly)
- Four-storey compartmented building in Fig. B109
  Rectangle enclosing unprotected areas 5.5 m × 7 m
  Enclosing rectangle 6 m × 9 m = 54 m²
  Unprotected areas 19.5 m²
  Unprotected areas as percentage of enclosing rectangle
  (100 × 19.5) 54 = 36.1%
  Distance from boundary from tables (40%) = 2.5 m minimum
  (Residential, office or assembly)

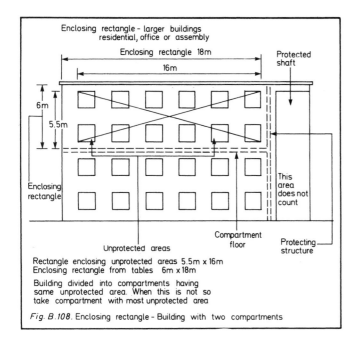

Rectangle enclosing unprotected areas 5.5m x 16m
Enclosing rectangle from tables 6m x 18m

Building divided into compartments having
same unprotected area. When this is not so
take compartment with most unprotected area

Fig. B.108. Enclosing rectangle – Building with two compartments

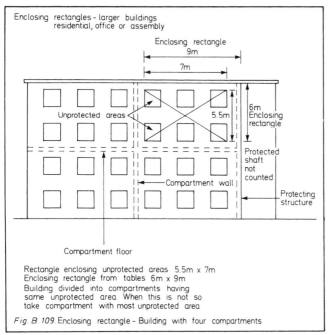

Rectangle enclosing unprotected areas 5.5m x 7m
Enclosing rectangle from tables 6m x 9m
Building divided into compartments having
same unprotected area. When this is not so
take compartment with most unprotected area

Fig. B.109. Enclosing rectangle – Building with four compartments

• Four-storey uncompartmented building in Fig. B110
  Rectangle enclosing unprotected areas 11.75 m × 20 m
  Enclosing rectangle 12 m × 21 m = 252 m²
  Unprotected areas 160 m²
  Unprotected areas as percentage of enclosing rectangle
  (100 × 160) 252 = 63.49%
  Distance from boundary from tables (70%) = 8.5 m minimum
  (Residential, office or assembly)

Aggregate notional areas

This is an alternative method of calculation which gives a more
accurate result and is generally more advantageous to a developer. It
is suitable for use when dealing with large irregular shaped buildings.
The method is time consuming as instead of making one calculation
upon a plane of reference, measurements are made at 3 m intervals
along the boundary.

A point is selected on the boundary referred to as the 'vertical

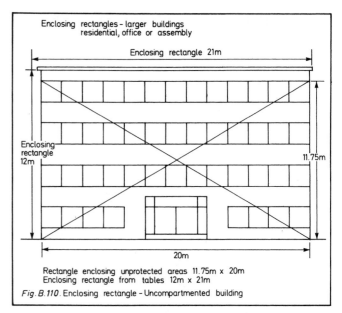

Rectangle enclosing unprotected areas 11.75m x 20m
Enclosing rectangle from tables 12m x 21m

Fig. B.110. Enclosing rectangle – Uncompartmented building

datum' and this is defined as a vertical line of unlimited height at any
point on the relevant boundary.

In order to have any effect upon the vertical datum, a window or
other unprotected area must not be screened from it and must be
within some limiting distance from it, and therefore such areas are
not taken into account, if they are:

• screened from it by any part of an external wall which is not an
  unprotected area:
• outside an arc of radius 50 m extending 90° on either side of the
  datum line, or
• facing away from it at an angle of less than 10° to it.

Fire issuing from any given point of an unprotected area will have
less effect in proportion to the distance from the unprotected area to
this boundary. There is a table of factors in the approved document
which quotes the figures to be used in calculations regarding this
method of calculating the aggregate notional area. They are the same
as in the 1976 Regulations.

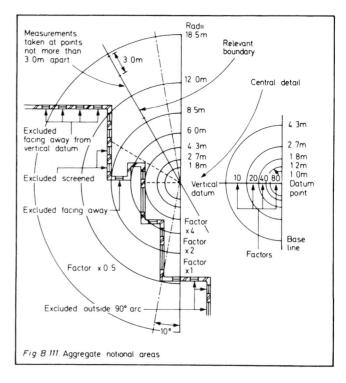

Fig B.111. Aggregate notional areas

For each vertical datum on the boundary an aggregate notional area is calculated, and this is made up of the area of each unprotected part in the side of a building or compartment multiplied by a factor based upon its distance from the vertical datum.

When the aggregate notional area of a particular vertical datum has been obtained, the provisions give a limit of $210\,m^2$ for buildings or compartments having a residential, assembly or office use. When the building, or compartment has a shop, industrial or other non-residential use the aggregate notional area is not to exceed $90\,m^2$.

The simplest method of using this technique is to make a protractor on tracing film or transparent plastic to the same scale as the building design being examined. The protractor must be marked with the distances and the factors (Figs B111 and B112).

The protractor is then applied at 3 m intervals along the relevant boundary and indicates quickly a way of finding out whether or not the requirements will be satisfied. If the lines are marked on it at an angle of 10° this will be useful in determining the areas which are within that angle to the vertical datum and therefore are not counted in the calculations.

Figure B111 shows a protractor with radii up to 18.5 m. The model illustrated in the approved document includes radii at 27.5 m and 50 m. Additional factors are 0.25 up to 27.5 m; 0.1 up to 50 m. Above 50 m there is no multiplication factor.

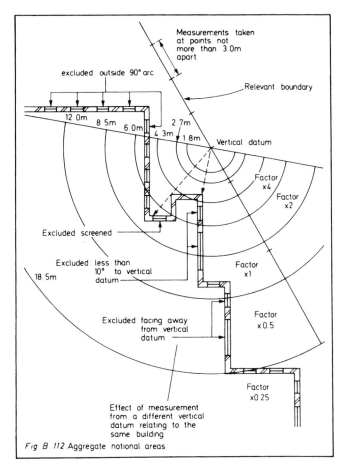

Fig B 112. Aggregate notional areas

SECTION 15. ROOF COVERINGS

The aim of this section is to limit the spread of fire across the roof covering where it has been exposed to fire from the outside.

BS 476: Part 3: 1958 is used to establish performance for the external fire exposure of roofs. This enables construction to be designated by 2 letters in the range A to D.

The construction of roofs is vital when one is mindful as to their vulnerability to attack by fire from adjoining buildings and flying brands. A very careful assessment of this risk is involved in the application of standards, relating to fire resistance and resistance to the spread of flame of materials used on the roofs of buildings.

Roofs are not required to resist fire from within the building, but a roof must not contribute to the growth of a fire by allowing rapid spread across its surface. The roof should be a barrier to fire from a nearby building due to radiation or flying brands.

The first letter in a roof designation relates to resistance to the penetration of fire as follows:

A   Roofs which can withstand penetration for one hour.
B   Roofs which can be penetrated between half hour and one hour.
C   Roofs which can be penetrated in less than half hour.
D   Roofs which are penetrated in the preliminary test.

The second letter relates to the spread of flame.

A   Roofs where there is no spread of flame.
B   Roofs where there is not more than 533.4 mm spread of flame.
C   Roofs where there is more than 533.4 mm spread of flame.
D   Roofs which continue to burn for 5 minutes after withdrawal of the flame, a spread more than 381 mm across the region of burning during the preliminary test.

It follows that an AA designation is the highest classification.

SEPARATION DISTANCES

The minimum distance from roofs of various classifications to the relevant boundary (which may be the notional boundary) are shown in Table 10. A wall separating a pair of semi-detached houses is not regarded as a relevant boundary for the purposes of section 14. There are no restrictions on roof coverings that are designated AA, AB or AC.

PLASTIC ROOFLIGHTS

There are limitations of size and distance from a relevant boundary for plastic rooflights (Fig. B4.12) which are to have at least a Class 3 lower surface. Alternatively, they may be of thermoplastic materials having a TP(a) rigid or TP(b) classification (see Tables 18 and 19).

Class 3 rooflights may be spaced 1800 mm apart in rooms in industrial and other non-residential buildings. The rooflights are to be evenly distributed and not be more than 20% of the area of the room.

A rooflight is not to be nearer than 1800 mm to the junction of a compartment wall and a roof, unless the wall is carried up above the roof not less than 375 mm.

A rigid thermoplastic sheet product manufactured from polycarbonate or unplasticized PVC which has a Class 1 rating for surface flame spread, is regarded as an AA designation when used in rooflights.

To achieve a TP(a) rigid classification a thermoplastic material must

• be a rigid solid PVC sheet; or
• a single polycarbonate sheet at least 3 mm thick, or
• multi-skinned polycarbonate rigid sheet made from unplasticized PVC or polycarbonate with a Class 1 rating, or
• other rigid thermoplastic product which satisfies method 508A test.

The TP(b) classification requires:

• a rigid solid polycarbonate sheet less than 3 mm thick; or multiple skin polycarbonate sheet products not meeting TP(a) test, or
• other products between 1.5 and 3.0 mm thick tested by method 508A.

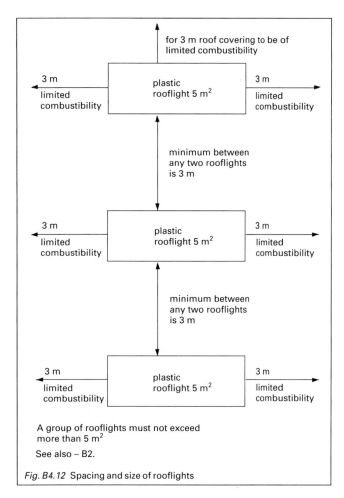

for 3 m roof covering to be of limited combustibility

3 m
limited combustibility | plastic rooflight 5 m² | 3 m
limited combustibility

minimum between any two rooflights is 3 m

3 m
limited combustibility | plastic rooflight 5 m² | 3 m
limited combustibility

minimum between any two rooflights is 3 m

3 m
limited combustibility | plastic rooflight 5 m² | 3 m
limited combustibility

A group of rooflights must not exceed more than 5 m²

See also – B2.

Fig. B4.12 Spacing and size of rooflights

When used in rooflights, a rigid thermoplastic sheet manufactured from polycarbonate or unplasticized PVC and achieves a Class 1 rating it can be given an AA designation.

To achieve this grading the material has to be tested to BS 476: Part 7: 1971 (1987 tor 1997), Surface spread of flame tests for materials.

Unwired glass

Unwired glass in rooflights must be at least 4 mm thick and is then regarded as having an AA designation.

Thatch and wood shingles

Thatch and wood shingles have an AA/BD/CD designation if a grading under BS 476: Part 3: 1958, External fire exposure roof tests, cannot be obtained.

However, thatched roofs may be placed closer to the boundary than indicated in Table 10 if

• rafters are overdrawn by construction with not less than 30 minutes' fire resistance;
• guidance given in Part J (Chapter 18) is followed; and
• a smoke alarm system is fitted in the roofspace (see Table 10).

ACCESS AND FACILITIES FOR FIRE SERVICE

The regulations require that there are such facilities provided to enable firefighters to get near enough to a building to use its appliances. There must also be provision for the personnel to get into and move about within a building on fire, with as much safety as can be provided.

Access to the outside of the building must be provided for at least three types of appliance: pumps, turntables and hydraulic platforms. Inside the building access for personnel depends upon the provision of safe routes and suitably placed supplies of water and foam with which to attack a fire. There are therefore requirements for stairs, lifts, wet and dry risers. The height and area of a building influences the provision to be supplied.

Requirements

B5. There are to be facilities to assist firefighters in the protection of life. Access is to be provided for fire appliances to gain access to the building.

The regulation is quite clear – it is not requiring any provision to protect goods or buildings; everything required by BS is directed at the protection of life.

Table 10   Limitations on roof coverings*

Designation† of covering of roof or part of roof	Minimum distance from any point on relevant boundary			
	Less than 6 m	At least 6 m	At least 12 m	At least 20 m
AA, AB or AC	●	●	●	●
BA, BB or BC	○	●	●	●
CA, CB or CC	○	● (1)(2)	● (1)	●
AD, BD or CD (1)	○	● (2)	●	●
DA, DB, DC or DD (1)	○	○	○	● (2)

Notes:

*   There are limitations on glass, thatch, wood shingles and plastics rooflights.
†   The designation of external roof surfaces is explained in Appendix A. (See Table A5, for notional designations of roof coverings.)

Separation distances do not apply to the boundary between roofs of a pair of semi-detached houses and to enclosed/covered walkways. However, see Fig. B3.6 if the roof passes over the top of a compartment wall.
Polycarbonate and PVC rooflights which achieve a Class 1 rating by test, may be regarded as having an AA designation.

● Acceptable.
○ Not acceptable.

1. Not acceptable on any of the following buildings:
   a. Houses in terraces of three or more houses;
   b. Industrial, Storage or Other non-residential purpose group buildings of any size;
   c. Any other buildings with a cubic capacity of more than 1500 m³.
2. Acceptable on buildings not listed in Note 1, if part of the roof is no more than 3 m² in area and is at least 1500 mm from any similar part, with the roof between the parts covered with a material of limited combustibility.

Table 11 Class 3 plastics rooflights: limitations on use and boundary distance

Minimum classification on lower surface (1)	Space which rooflight can serve	Minimum distance from any point on relevant boundary to rooflight with an external designation† of:	
		AD BD CA CB CC CD	**DA DB DC DD**
Class 3	a. balcony, verandah, carport, covered way or loading bay, which has at least one longer side wholly or permanently open.	6 m	20 m
	b. detached swimming pool.		
	c. conservatory, garage or outbuilding, with a maximum floor area of 40 m².		
	d. circulation space (2) (except a protected stairway).	6 m (3)	20 m (3)
	e. room (2).		

Notes:
† The designation of external roof surfaces is explained in Appendix A.
None of the above designations are suitable for protected stairways –
Polycarbonate and PVC rooflights which achieve a Class 1 rating by test, may be regarded as having an AA designation.
Where B3.6 & B3.7 applies, rooflights should be at least 1.5 m from the compartment wall.
Products may have upper & lower surfaces with different properties if they have double skins or are laminates of different materials. In which case the more onerous distance applies.

1. See also the guidance to B2.
2. Single skin rooflight only, in the case of non-thermoplastic material.
3. The rooflight should also meet the provisions of Fig. B4.12.

Table 12 TP(a) and TP(b) plastics rooflights: limitations on use and boundary distance

Minimum classification on lower surface (1)	Space which rooflight can serve	Minimum distance from any point on relevant boundary to rooflight with an external surface classification (1) of:	
		TP(a)	**TP(b)**
1. TP(a) rigid	any space except a protected stairway	6 m (2)	not applicable
2. TP(b)	a. balcony, verandah, carport, covered way or loading bay, which has at least one longer side wholly or permanently open.	not applicable	6 m
	b. detached swimming pool.		
	c. conservatory, garage or outbuilding, with a maximum floor area of 40 m².		
	d. circulation space (3) (except a protected stairway).	not applicable	6 m (4)
	e. room (3).		

Notes:
None of the above designations are suitable for protected stairways –
Polycarbonate and PVC rooflights which achieve a Class 1 rating by test, may be regarded as having an AA designation.
Where B3.6 & B3.7 applies, rooflights should be at least 1.5 m from the compartment wall.
Products may have upper & lower surfaces with different properties if they have double skins or are laminates of different materials. In which case the more onerous distance applies.

1. See also the guidance to B2.
2. No limit in the case of any space described in 2a, b & c.
3. Single skin rooflight only, in the case of non-thermoplastic material.
4. The rooflight should also meet the provisions of Fig. B4.12.

Performance

There are four targets to achieve in order to meet the performance standard required:

- access for appliances to get near the building;
- access into the building, and inside the building to enable firefighters to rescue people and fight fires;
- sufficient internal fire mains and other facilities to be provided;
- there is to be provision for venting heat and smoke from basements.

Guidance in the approved document, concerned only with the health and safety of people, relates to the provision of facilities to achieve the performance standard, and the amount of provision depends largely upon the size of the building. Some, or all of the following facilities maybe necessary:

- vehicle access;
- internal access for firefighting personnel;
- internal fire mains;
- venting of heat and smoke from basements.

Generally speaking, firefighting is carried out inside a building and the following considerations apply in high buildings or deep basements:

- special access provisions are required (section 17);
- fire mains are necessary (section 15);
- fire appliances must be able to get near to fire mains (section 16).

Table A5 Notional designations of roof coverings

Part i: Pitched roofs covered with slates or tiles

Covering material	Supporting structure	Designation
1. Natural slates 2. Fibre reinforced cement slates 3. Clay tiles 4. Concrete tiles	timber rafters with or without underfelt, sarking, boarding, woodwool slabs, compressed straw slabs, plywood, wood chipboard, or fibre insulating board	AA

Note: Although the Table does not include guidance for roofs covered with bitumen felt, it should be noted that there is a wide range of materials on the market and information on specific products is readily available from manufacturers.

Part ii: Pitched roofs covered with self-supporting sheet

Roof covering material	Construction	Supporting structure	Designation
1. Profiled sheet of galvanised steel, aluminium, fibre reinforced cement, or pre-painted (coil coated) steel or aluminium with a pvc or pvf2 coating	single skin without underlay, or with underlay or plasterboard, fibre insulating board, or woodwool slab	structure of timber, steel or concrete	AA
2. Profiled sheet of galvanised steel, aluminium, fibre reinforced cement, or pre-painted (coil coated) steel or aluminium with a pvc or pvf2 coating	double skin without interlayer, or with interlayer of resin bonded glass fibre, mineral wool slab, polystyrene, or polyurethane	structure of timber, steel or concrete	AA

Part iii. Flat roofs covered with bitumen felt

A flat roof comprising of bitumen felt should (irrespective of the felt specification) be deemed to be of designation AA if the felt is laid on a deck constructed of 6mm plywood, 12.5mm wood chipboard, 16 mm (finished) plain edged timber boarding, compressed straw slab, screeded woodwool slab, profiled fibre reinforced cement or steel deck (single or double skin) with or without fibre insulating board overlay, profiled aluminium deck (single or double skin) with or without fibre insulating board overlay, or concrete or clay pot slab (insitu or pre cast), and has a surface finish of:

a. bitumen-bedded stone chippings covering the whole surface to a depth of at least 12.5mm;
b. bitumen-bedded tiles of a non-combustible material;
c. sand and cement screed; or
d. macadam.

Part iv. Pitched or flat roofs covered with fully supported material

Covering material	Supporting structure	Designation
1. Aluminium sheet 2. Copper sheet 3. Zinc sheet	timber joists and: tongued and grooved boarding, or plain edged boarding	AA*
4. Lead sheet 5. Mastic asphalt 6. Vitreous enamelled steel 7. Lead/tin alloy coated steel sheet 8. Zinc/aluminium alloy coated steel sheet	steel or timber joists with deck of: woodwool slabs, compressed straw slab, wood chipboard, fibre insulating board, or 9.5mm plywood	AA
9. Pre-painted (coil coated) steel sheet including liquid-applied pvc coatings	concrete or clay pot slab (insitu or pre-cast) or non-combustible deck of steel, aluminium, or fibre cement (with or without insulation)	AA

Notes:
* Lead sheet supported by timber joists and plain edged boarding should be regarded as having a BA designation.

In other buildings there is a combination of facilities to be available:

- normal means of escape;
- availability of ladders and appliances;
- vehicle access required to some or all of the perimeter;
- ensure that dwellings and other small buildings are close to a point accessible to the fire brigade;
- for taller blocks of flats, access facilities within the building are required for firefighters. Compartmentation may simplify facilities required.

Smoke and gases from basements can escape up stairways from basements, making access difficult, and vents can reduce this problem. Venting reduces temperature, and improves visibility which assists search, rescue and firefighting in the basement.

SECTION 16. FIRE MAINS

Once in a building the firefighters require fire mains at strategic positions. These are vertical pipes with a fire brigade connection or a booster pump at the lower end and there are discharge points at different levels within the building. There may be falling mains to serve storeys below ground level.

There are two types of rising main. Dry risers are kept empty of water which is supplied through a hose from the fire service pumps. Wet risers are normally installed in taller buildings and the approved document requires them to be provided where the building has a floor 60m above the ground or access level. Wet risers are kept permanently charged with water provided from tanks and pumps within the building. There must be provision for the water supply to be topped up if necessary.

Buildings which do not have floors at more than 60 m above ground or access level may be fitted with either wet or dry risers.

As part of the firefighters' 'bridgehead' into high buildings there will be one or more firefighting shafts (section 18) and there is to be a fire main in every firefighting shaft. Outlets from the mains are to be in each firefighting lobby attached to the firefighting shaft.

Design and construction

Guidance in the design and construction of fire mains is contained in sections 2 and 3 of BS 5306: Part 1: 1976, 'Hydraulic systems, hose reels and foam inlets'.

SECTION 17. VEHICLE ACCESS

The approved document is concerned with vehicle access positions near the outside of buildings to enable them to operate efficiently. High-reach appliances such as turntable ladders and hydraulic platforms must be correctly placed to enable rescue to take place; pumping appliances must be able to direct water to the sites required, and enable firefighting and rescue activities to proceed efficiently. Access requirements increase with the floor area and height of the building.

Where fire mains are provided, pumping appliances require access to positions near the mains, and there needs to be doors nearby to enable personnel to enter. Facility for the rapid connection of hose to a dry main are required.

The following are access requirements to buildings not fitted with fire mains.

- Buildings with height of floor of top storey up to 11 m above ground. The type of appliance provided for is a pump. In each case the total floor area is the aggregate of all floors and where purpose group 7(a) is involved, height is to be measured to mean roof level.

Total floor area	Vehicle access required
Up to 2000 m²	to within 45 m or 15% perimeter, whichever is less
2000–8000 m²	15% of perimeter
8000–16 000 m²	50% of perimeter
16 000–24 000 m²	75% of perimeter
Over 24 000 m²	100% of perimeter

- Buildings with height of floor of top storey more than 11 m above ground level. The type of appliance provided for is one with high-reach capacity. In each case the total floor area is the aggregate of all floors and where purpose group 7(a) is involved, height is to be measured to mean roof level.

Total floor area	Vehicle access required
Up to 2000 m²	15% of perimeter
2000–8000 m²	50% of perimeter
8000–16 000 m²	50% of perimeter
16 000–24 000 m²	75% of perimeter
Over 24 000 m²	100% of perimeter

Measurement of perimeter walling is arrived at as in Fig. B5.1

In making arrangements for external access it must be borne in mind that the regulations cannot enforce requirements outside the site of works shown on deposited plans, or building notice, or initial notice.

All elevations required to have access for a vehicle are to have an entrance door at least 750 mm wide, and where a building has a dry main there is to be access for a pump not less than 18 m from the main inlet connection point. The inlet should be easily seen from the appliance.

Where a wet mains is fitted, a pumping appliance must be able to get within 18 m of the access to the mains and the emergency inlet for refilling the water tank. All to be clearly visible from the access area.

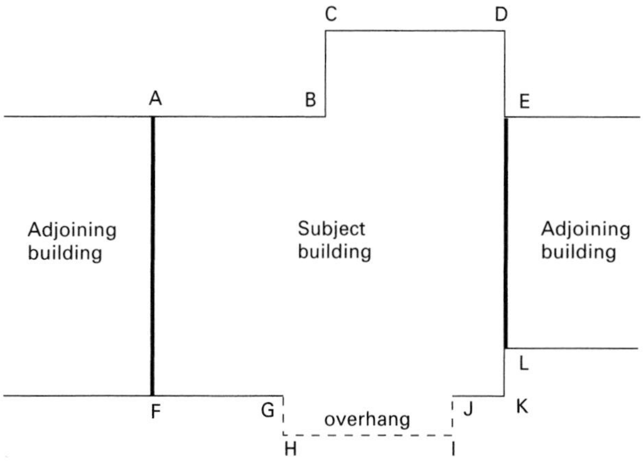

Common walls with adjoining buildings not included.

Perimeter A+B+C+D+E plus F+G+H+I+J+K+L

Adjoining building. If no adjoining buildings A+F and E+L are taken into measurement of perimeter

Fig. B5.1 Perimeter of building

Vehicle access is to be provided for a pump appliance to within 45 m of the dwelling enterance door to blocks of flats or maisonettes.

DESIGN

Access routes are also to meet the following dimensions and have in mind the weight of appliances particularly when designing the access surface and any manhole covers and the like which may be installed.

The weight of high-reach appliances is spread over a number of axles and therefore the access may be designed to accommodate 12.5 tonnes. Bridges and other structures over which appliances will pass should be designed to accept 17 tonnes.

Minimum widths	Appliances	
	Pump	High reach
Between kerbs	3.7 m	3.7 m
Width of gateways	3.1 m	3.1 m
Turning circle between kerbs	16.8 m	26.0 m
Turning circle between walls	19.2 m	29.0 m
Clearance height	3.7 m	4.0 m
Carrying capacity	12.5 tonnes	17.0 tonnes

Overhead obstructions are to be avoided (Fig. B5.2).

If there is a dead end access more than 20 m long it is to be provided with turning facilities, which may be a turning circle, or a hammerhead. It is not to be necessary for a fire appliance to reverse for more than 20 m

For buildings up to 11 m high there is to be access for a pump appliance near the building for the amount of perimeter specified. For buildings over 11 m, access routes are to be in accordance with Fig. B5.2 where the hardstanding is to be capable of supporting jacks with a load of 8.3 kg/cm²

SECTION 18. ACCESS FOR FIREFIGHTERS

For low rise buildings which do not have deep basements, access for firefighters may be satisfied by the facility of ladder access to upper

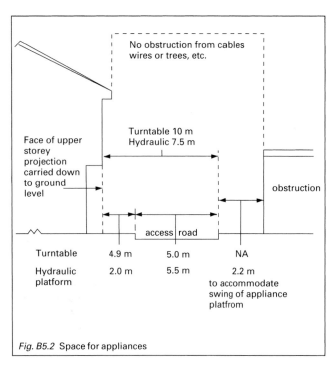

Fig. B5.2 Space for appliances

storeys combined with the normal provision for means of escape in case of fire.

However, in other buildings, where there are high floors or deep basements, the problems of getting to the source of a fire require that special provision is made for fire service personnel.

These facilities are to provide a secure operating base for the fire-fighters and enable them to take effective action.

Facilities required include firefighting stairs, firefighting lobbies and firefighting lifts, all of which may be grouped together and form a firefighting shaft (Fig. B5.3).

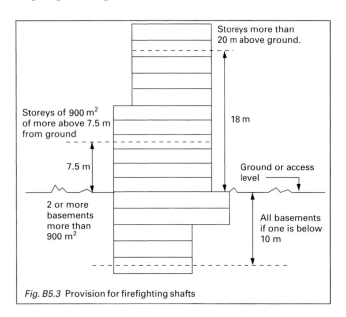

Fig. B5.3 Provision for firefighting shafts

FIREFIGHTING SHAFTS

Provision of firefighting shafts containing firefighting lifts is required

- in buildings where there is a floor more than 18 m above ground or fire service access level;
- in buildings where there is a basement more than 10 m below ground or fire service access level.

A firefighting shaft is required, but it is not necessary for it to contain a firefighting lift

- in a building with a storey 900 m² or more in area, if the floor is more than 7.5 m above ground or access level

Applies to the following purpose groups:
4.   Shop and commercial
5.   Industrial
7(a) Storage and other non-residential.

- in buildings with two or more basement storeys each in excess of 900 m².

A firefighting shaft provided for a basement is not required to serve upper floors unless a shaft is also required there because of the height and size of the building. Likewise, if a firefighting shaft is required to serve upper floors it is not required to serve a basement unless it is large enough or deep enough to require a shaft.

In shopping complexes firefighting shafts must be provided in accordance with the recommendations to be found in Section 3 of BSS 5588: Part 10: 1991, 'Code of practice for enclosed shopping complexes'.

Number and location

The number of firefighting shafts required depends upon whether a building is sprinklered or not.

- When there is a sprinkler system meeting BS 5306: Part 2: 1990, 'Specification for sprinkler systems', the minimum number of fire-fighting shafts required relate to the floor area of the largest storey over 18 m above ground level:

  (a) when it is less than 900 m²   1 shaft
  (b) between 900 m² and 2000 m²   2 shafts
  (c) over 2000 m²   2 + 1 for every further
                                         1500 m² or part thereof
- When a building is not fitted with sprinklers there must be at least one shaft for every 900 m² (or part thereof) of floor area of the largest floor above 18 m from ground level.
- When basements require firefighting shafts, the same 900 m² criteria should be used in calculating the number of shafts to be provided.

The placing of a firefighting shaft is related to the distance required to lay a hose from the shaft to the furthest part of a storey (Fig. B5.4).

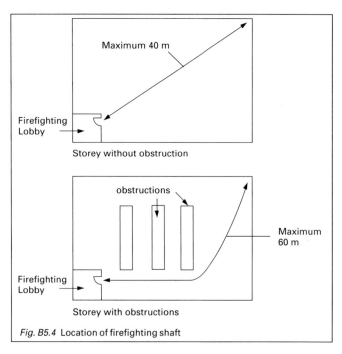

Fig. B5.4 Location of firefighting shaft

- The route around obstructions must not exceed 60 m, or
- if the internal layout is not known all parts of the storey should be within 40 m of the firefighting lobby.

DESIGN AND CONSTRUCTION

- All firefighting stairways and firefighting lifts are to be entered from the accommodation via a firefighting lobby (Figs B5.5 and B5.6). This does not apply to blocks of flats and maisonettes.

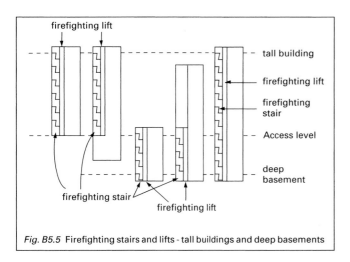

Fig. B5.5 Firefighting stairs and lifts - tall buildings and deep basements

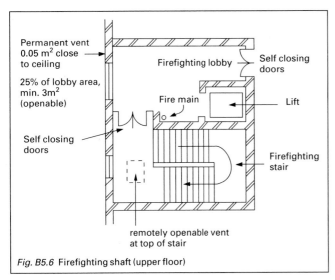

Fig. B5.6 Firefighting shaft (upper floor)

- There are to be fire main outlet connections and valves in every firefighting lobby.
- Firefighting shafts are to follow the recommendations in sections 2, 3, and 4 of BSS 5588: Part 5: 1991, 'Code of practice for firefighting stairs and lifts'.

Section 2 covers planning and construction,
Section 3 describes firefighting lift installations, and
Section 4 is concerned with electrical services.

The BS also provides guidance regarding ventilation measures required in a firefighting shaft.

Design of means of escape and compartmentation in blocks of flats and maisonettes should be in accordance with section 3 (flats and maisonettes) and section 9 (compartmentation); it is not necessary to place a firefighting lobby between the firefighting stairs and a protected corridor or lobby, which has been included for means of escape. Also a firefighting lift can open into the protected corridor or

lobby. However, the firefighting lift landing doors are not to exceed 7.5 m from the door of the firefighting stairs.

When roller shutters have been placed in compartment walls, it must be possible for them to be opened or closed manually by the firefighters.

SECTION 19. VENTING BASEMENTS, SMOKE OUTLETS

Should a fire occur in a basement area it is very natural for smoke, heat and gases to escape up stairways, and of course to fill the basement area, making search and rescue difficult.

Provision should therefore be made for smoke to escape to the outside. When this is practicable at least one or more outlets are to be provided, or mechanical extraction systems may be installed.

The requirement is therefore that suitably located smoke outlets are to be connected to the external air from every basement storey.

There are exclusions from this requirement:

- basements of single family dwellinghouses, purpose group 1(b) dwellinghouses with habitable storey more than 4.5 m above ground, and purpose group 1(c) dwellinghouse which does not have a habitable storey more than 4.5 m above ground level; or
- basement storey when floor area does not exceed 200 m²; and
- basement storey where floor is not more than 3 m below ground level;
- basement storey strong rooms.

If a basement has an external door or window the compartment containing the rooms with these doors and windows do not require smoke outlets.

Also, basements may be open to the air on at least one elevation, arising from

- different ground levels around the building,
- housing where sunken areas are provided to the front and rear of the property. Doors or windows may be fitted.

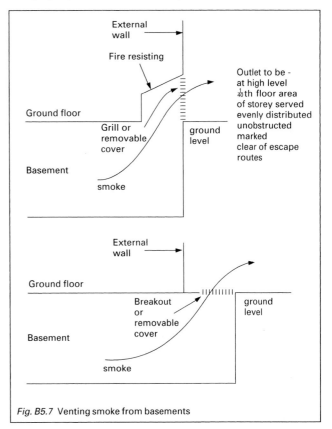

Fig. B5.7 Venting smoke from basements

In these circumstances smoke outlets are not required.

Natural smoke outlets should be placed at high level to induce cross ventilation and have a total clear area of not less than 1/40th of the basement floor area (Fig. B5.7).

- Places of special fire hazard should have separate smoke outlets;
- outlets to be covered with non-combustible grilles and kept unobstructed.

MECHANICAL EXTRACTION

If provided with a sprinkler system basement storeys may have a mechanical ventilation system to remove smoke, gases and heat. Sprinklers are not required to be placed elsewhere than in the basement to meet this requirement; and they should be installed in accordance with BS 5306: Part 2: 1990, 'Sprinkler systems'.

The mechanical system provided must ensure

- not less than 10 air changes per hour;
- that gas temperatures of up to 300°C can be extracted for at least 60 minutes;
- activation of a sprinkler system should automatically start extraction.

An automatic fire detection system conforming at least to L3 standard in BS 5389: Part 1: 1988, 'Code of practice for system design, installation and servicing' may be fitted.

Information on equipment for removing hot smoke is available in BS 7346: Part 2: 1990, 'Specification for powered smoke and heat exhaust ventilators'.

CONSTRUCTION

Non-combustible fire-resisting construction is to be used for outlet ducts, shafts, bulkheads, and the separation of ducts serving various basements.

BASEMENT CAR PARKS

The guidance offered in section 12 may be used in any basement car park.

PART B

APPENDIX A

Where in the approved document, performance is related to British Standards, European Standards or European Technical approvals may be appropriate if the product

- when tested, meets performance requirement, or
- when assessed by accredited laboratories meets performance requirement, or
- conforms to various tables in approved document, or
- if a fire-resisting element satisfies specification in Part 2, BRE report 'Guidance for the construction of fire resisting structural elements (1988)', or is in accordance with a relevant BS or Eurocode.

These tests may be undertaken by suitably qualified fire safety engineers, laboratories accredited to UKAS, the BRE, British Board of Agrément, and any other capable body notified to the UK government by a member state of the EU, as having the necessary expertise.

The use of product conformity certification schemes and registration of installers gives confidence in the level of performance of a material, product or structure.

Independant product conformity certification schemes provide a means of identifying materials, products or structures which have the required performance. They also provide confidence that the materials, products or structures are similar to those actually tested or assessed.

Likewise the independent accreditation and registration of installers, products or structures ensures that the installations are by knowledgeable contractors working to appropriate standards, giving confidence regarding behaviour in the event of fire.

Factors which have a bearing on fire resistance are:

- severity of the fire;
- height of the building; and
- nature of the occupancy of the building.

Standards of fire resistance have been recommended following assessment of the following:

- the fire load density;
- the height of top floor above ground level;
- the number of occupants;
- presence, or otherwise, of basements;
- if a single-storey building.

Fire-resistance performance is related to:

- BS 476: Parts 20–24: 1987, or
- BS 476: Part 8: 1972

regarding resistance to collapse, fire penetration and the transfer of excessive heat.

The approved document lists three organizations which have available publications regarding the testing of fire-resisting elements.

1. *Fire Tests on Building Products: Fire Resistance* (1983), Fire Protection Association, Melrose Avenue, Borehamwood, Herts WD6 2BJ.
2. *Fire Protection for Structural Steel in Buildings* (1988), Association of Specialist Fire Protection, Association House, 235 Ash Road, Aldershot, Hants GU12 4DD.
3. *Rules for the Construction and Installation of Firebreak Doors and Shutters* (1988), Fire Protection Association, Melrose Avenue, Borehamwood, Herts WD6 2BJ.

Internal linings

- Fire propagation indices are determined by reference to BS 476: Part 6: 1981 or 1989.
- Surface spread of flame is tested by reference to BS 476: Part 7: 1971 or 1987, when material may be classed 1, 2, 3 and 4. Class 1 is the highest.

Highest classification for lining materials is Class 0, when the material or the surface of a composite product is:

- composed of material of limited combustibility, or
- Class 1 with fire propagation index (1) not exceeding 12 and subindex (i_1) not exceeding 6.

Materials having a resistance to the spread of flame control the spread of fire over wall and ceiling surfaces. Two criteria are used.

- Surface spread of flame.
  BS 476: Part 7: 1971 or 1987 relate to the surface spread of flame test for materials and determines the tendency of materials to support the spread of flame across surfaces and classifies them in relation to exposed surfaces of walls and ceilings. The various classes of tests that must be satisfied are as follows:

Class 1 spread of flame at 1½ minutes	– 165 mm
Class 2 spread of flame at 1½ minutes	– 215 mm
Class 3 spread of flame at 1½ minutes	– 265 mm
Class 4 exceeds the limits for Class 3.	

  There is a small tolerance in respect of one specimen in a sample.
  Class 4 is not acceptable to the building regulations.
  Class 0 is an additional class which imposes greater control of both the spread of the flame and the rate at which the material

releases heat. Class 0 is used generally as a designation for the walls and ceilings of escape routes. There are three basic requirements which must be met to achieve Class 0.

1. The material must have a surface spread of flame classification of Class 1.
2. The fire propagation index should not exceed 12.
3. The sub-index figure for i_1 should not exceed 6.

When a product is composed entirely of limited combustibility materials, or non-combustible materials it can be included in Class 0.

BS 476: Part 4: 1970 describes a test for determining whether building materials are non-combustible. Materials used in the construction and finishing of buildings are classified 'non-combustible' or 'combustible' according to their behaviour in the non-combustibility test. The test is intended for building materials, whether coated or not, and does not apply to the coating alone.

- Fire propagation
BS 476: Part 6: 1981 or 1989 gives a test for fire propagation for products. The result of the test is expressed as a fire propagation index, and provides a comparative measure of the contribution to the growth of fire made by an essentially flat material, composite or assembly. It is used as an assessment of the performance of wall and ceiling linings.

The result 'I' ranges from 0 to 100+. When the 'I' figure is low it indicates that the sample tested did not contribute much to the heat of the furnace when it was being tested. Three sub-indices relate to the heat contributed at different stages of the test. They are i_1, i_2 and i_3. When a sample burns and releases heat quickly the sub-index will be high. Sub-index figure i_1 is derived from the first 3 minutes of test.

Appendix A: Table A1

Performance requirements for the resistance of elements of structure, doors and other forms of construction must be determined by reference to BS 476: Parts 20 to 24: 1987. BS 476: Part 8: 1972 in respect of items tested or assessed prior to 1 January 1988.

This part sets out methods of test and the rules for determining fire resistance of the elements of building construction. They relate to the following:

- Walls and partitions (loadbearing and non-loadbearing)
- Floors
- Flat roofs
- Columns
- Beams
- Suspended ceilings
- Doors
- Glazing
- Ceiling membranes

The tests involve heating the elements in a furnace with temperature and time to follow a curve from 556°C, at 5 minutes to 1193°C, at 360 minutes. Results are related to stability, integrity and insulation.

- *Stability.* Non-loadbearing constructions fail when collapse of the specimen takes place;
  Loadbearing constructions must support the test load for the prescribed heating period plus 24 hours. There is a reload test, and failure during this results in a specimen having a stability 80% of the time to collapse. Failure of floors, flat roofs and beams occurs when the specimen deflects in excess of the limit specified.
- *Integrity.* There is a failure when cracks or other openings occur through which flame or hot gases can pass causing flaming on a cotton wool pad.
- *Insulation.* Failure occurs when the mean temperature of the exposed surface of the specimen increases by more than 140°C above the starting temperature. Failure also occurs if the temperature of an unexposed surface increases by more than 180°C above the starting temperature.

Throughout the documents emphasis is placed upon the need for careful construction which should be checked to ensure that small details are not overlooked. Inappropriate fixings and joints can materially affect a fire-resistance rating.

Appendix A: Table A5

Roofs

BS 476: Part 3: 1958 is used to establish performance for the external fire exposure of roofs. This enables construction to be designated by 2 letters in the range A to D.

The construction of roofs is vital when one is mindful as to their vulnerability to attack by fire from adjoining buildings and flying brands. A very careful assessment of this risk is involved in the application of standards, relating to fire resistance and resistance to the spread of flame of materials used on the roofs of buildings.

Roofs are not required to resist fire from within the building, but a roof must not contribute to the growth of a fire by allowing rapid spread across its surface. The roof should be a barrier to fire from a nearby building due to radiation or flying brands.

The first letter in a roof designation relates to resistance to the penetration of fire as follows:

A   Roofs which can withstand penetration for one hour.
B   Roofs which can be penetrated between half hour and one hour.
C   Roofs which can be penetrated in less than half hour.
D   Roofs which are penetrated in the preliminary test.

The second letter relates to the spread of flame.

A   Roofs where there is no spread of flame.
B   Roofs where there is not more than 533.4 mm spread of flame.
C   Roofs where there is more than 533.4 mm spread of flame.
D   Roofs which continue to burn for 5 minutes after withdrawal of the flame, a spread more than 381 mm across the region of burning during the preliminary test.

It follows that an AA designation is the highest classification.

Notional designation of roof coverings are set out in Table A4; see Figs B37 to B44.

Felts referred to in the figures are types designated in BS 747: 1977. This BS gives four classes of felts, each divided into several types. Class 1 relates to bitumen felts with a fibre base, and Class 2 are bitumen felts with an asbestos base. Type 2B is a fine granule surfaced bitumen asbestos felt and type E is mineral surfaced bitumen asbestos felt.

BS CP 144: Part 3: 1970 is concerned with built-up bitumen felt

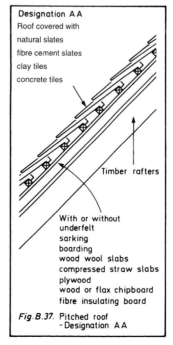

Designation A A
Roof covered with
natural slates
fibre cement slates
clay tiles
concrete tiles

Timber rafters

With or without
underfelt
sarking
boarding
wood wool slabs
compressed straw slabs
plywood
wood or flax chipboard
fibre insulating board

Fig. B.37. Pitched roof
- Designation A A

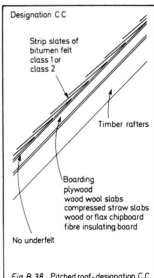

Designation C C

Strip slates of
bitumen felt
class 1 or
class 2

Timber rafters

Boarding
plywood
wood wool slabs
compressed straw slabs
wood or flax chipboard
fibre insulating board

No underfelt

Fig. B.38. Pitched roof-designation C C

Table A1 Specific provisions of test for fire resistance of elements of structure etc.

Part of building	Minimum provisions when tested to the relevant part of BS 476 (1) (minutes)			Method of exposure
	Loadbearing capacity (2)	Integrity	Insulation	
1. **Structural** frame, beam or column	see Table A2	not applicable	not applicable	exposed faces
2. **Loadbearing wall** (which is not also a wall described in any of the following items).	see Table A2	not applicable	not applicable	each side separately
3. **Floors** (3) a. in upper storey of 2-storey dwellinghouse (but not over garage or basement);	30	15	15	
b. between a shop and flat above;	60 or see Table A2 (whichever is greater)	60 or see Table A2 (whichever is greater)	60 or see Table A2 (whichever is greater)	from underside (4)
c. any other floor, including compartment floors.	see Table A2	see Table A2	see Table A2	
4. **Roofs** a. any part forming an escape route;	30	30	30	from
b. any roof that performs the function of a floor.	see Table A2	see Table A2	see Table A2	underside (4)
5. **External walls** a. any part less than 1000 mm from any point on the relevant boundary;	see Table A2	see Table A2	see Table A2	each side separately
b. any part 1000 mm or more from the relevant boundary (5);	see Table A2	see Table A2	15	from inside the building
c. any part adjacent to an external escape route (see Section 6, Fig. B1.44 and Fig. B1.44)	30	30	no provision (6)(7)	from inside the building
6. **Compartment walls** separating occupancies (see 9.20f)	60 or see Table A2 (whichever is less)	60 or see Table A2 (whichever is less)	60 or see Table A2 (whichever is less)	each side separately
7. **Compartment walls** (other than in item 6)	see Table A2	see Table A2	see Table A2	each side separately
8. **Protected shafts**, excluding any firefighting shaft a. any glazing described in Section 9, Fig. B3.13 and Fig. B3.14	not applicable	30	no provision (7)	
b. any other part between the shaft and a protected lobby/corridor described in Fig. B3.13 and Fig. B3.14	30	30	30	each side separately
c. any part not described in (a) or (b) above.	see Table A2	see Table A2	see Table A2	
9. **Enclosure** (which does not form part of a compartment wall or a protected shaft) to a: a. protected stairway;	30	30	30 (8)	each side
b. lift shaft.	30	30	30	separately
10. **Firefighting shafts** a. construction separating firefighting shaft from rest of building;	120	120	120	from side remote from shaft
	60	60	60	from shaft side
b. construction separating firefighting stair, firefighting lift shaft and firefighting lobby.	60	60	60	each side separately
11. **Enclosure** (which is not a compartment wall or described in item 8) to a: a. protected lobby;	30	30	30 (8)	each side
b. protected corridor.	30	30	30 (8)	separately
12. **Sub-division** of a corridor	30	30	30 (8)	each side separately
13. **Wall separating** an attached or integral garage from a dwellinghouse	30	30	30 (8)	from garage side
14. **Enclosure** in a flat or maisonette to a protected entrance hall, or to a protected landing	30	30	30 (8)	each side separately

Table A1 *Continued*

Part of building	Minimum provisions when tested to the relevant part of BS 476 (1) (minutes)			Method of exposure
	Loadbearing capacity (2)	Integrity	Insulation	
15. **Fire-resisting construction:**				
a. in dwellings not described elsewhere; each	30	30	30 (8)	
b. enclosing places of special fire hazard (see 9.12);	30	30	30	each side separately
c. between store rooms and sales area in shops (see 6.54).	30	30	30	
d. fire-resisting subdivision described in Section 10,	30	30	30	
16. **Cavity barrier**	not applicable	30	15	each side separately
17. **Ceiling** described in Section 10, Fig. B3.24 and Fig. B3.25.	not applicable	30	30	from underside
18. **Duct** described in paragraph 10.14e	not applicable	30	no provision	from outside
19. **Casing** around a drainage system described in Section 11, Fig. B3.27 and Fig. B3.28.	not applicable	30	no provision	from outside
20. **Flue walls** described in Section 11, Fig. B3.33.	not applicable	half the period specified in Table A2 for the compartment wall/floor	half the period specified in Table A2 for the compartment wall/floor	from outside
21. **Construction** described in Note (a) to paragraph 15.9.	not applicable	30	30	from underside
22. **Fire doors**	see Table B1			

Notes:
1. Part 21 for loadbearing elements, Part 22 for non-loadbearing elements, Part 23 for fire-protecting suspended ceilings, and Part 24 for ventilation ducts. BS 476: Part 8 results are acceptable for items tested or assessed before 1st January 1988.
2. Applies to loadbearing elements only (see B3.ii and Appendix E).
3. Guidance on increasing the fire resistance of existing timber floors is given in BRE Digest 208 *Increasing the fire resistance of existing timber floors* (BRE 1988).
4. A suspended ceiling should only be relied on to contribute to the fire resistance of the floor if the ceiling meets the appropriate provisions given in Table A3.
5. The guidance in Section 14 allows such walls to contain areas which need not be fire-resisting (unprotected areas).
6. Unless needed as part of a wall in item 5a or 5b.
7. Except for any limitations on glazed elements given in Table A4.
8. See Table A4 for permitted extent of uninsulated glazed elements.

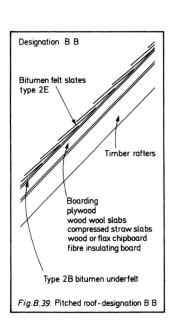

Fig.B.39. Pitched roof - designation B B

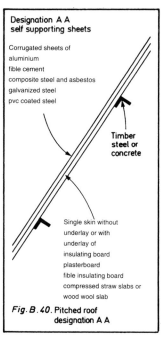

Fig.B.40. Pitched roof designation A A

and gives requirements for materials, design and construction of roofs and the preparation of surfaces upon which built-up bitumen felt roofing is to be placed, giving conditions under which satisfactory roof covering will be achieved.

Appendix A: Table A6

This table relates to the use of non-combustible materials. Non-combustible materials are those which

- do not flame or cause any rise in temperature of the specimen or furnace thermocouple when tested in accordance with BS 476: Part 11;
- are totally inorganic — as, for example, ceramics, concrete, fired clay, masonry, metals and plaster — with not more than 1% organic content;
- are concrete bricks or blocks to BS 6073: Part 1: 1981;
- are classified non-combustible by BS 474: Part 4: 1970.

These products may be used in the following situations:

- ladders – B1
- refuse chutes – B3
- suspended ceilings and supports – B3
- pipes – B3
- flue walls – B3
- car park construction – B3.

Table A4 Limitations on the use of uninsulated glazed elements on escape routes (These limitations do not apply to glazed elements which satisfy the relevant insulation criterion, see Table A1) (See BS 5588: Part 7 for glazing to atria; see BS 5588: Part 8 for glazing to refuges)

Position of glazed element	Maximum total glazed area in parts of a building with access to:			
	a single stairway		more than one stairway	
	walls	door leaf	walls	door leaf
Single family dwellinghouses				
1. a. Within the enclosure of:				
i. a protected stairway, or within fire-resisting separation shown in Section 2 Fig. B1.11 or	fixed fanlights only	unlimited	fixed fanlights only	unlimited
ii. an existing stair (see para 2.18).	unlimited	unlimited	unlimited	unlimited
b. Within fire-resisting separation:				
i. shown in Section 2 Fig. B1.12, or	unlimited above	unlimited above	unlimited above	unlimited above
ii. described in paras 2.13b & 2.20.	100 mm from floor	100 mm from floor	100 mm from floor	100 mm from floor
c. Existing window between an attached/integral garage and the house.	unlimited	not applicable	unlimited	not applicable
Flats and maisonettes				
2. Within the enclosures of a protected entrance hall or protected landing or within fire-resisting separation shown in Section 3 Fig B1.23.	fixed fanlights only	unlimited above 1100 mm from floor	fixed fanlights only	unlimited above 1100 mm from floor
General (except dwellinghouses)				
3. Between residential/sleeping accommodation and a common escape route (corridor, lobby or stair).	nil	nil	nil	nil
4. Between a protected stairway (1) and:				
a. the accommodation: or	nil	25% of door area	unlimited above 1100 mm (2)	50% of door area
b. a corridor which is not a protected corridor. Other than in item 3 above.				
5. Between:				
a. a protected stairway (1) and a protected lobby or protected corridor; or	unlimited above 1100 mm from floor	unlimited above 100 mm from floor	unlimited above 100 mm from floor	unlimited above 100 mm from floor
b. accommodation and a protected lobby. Other than in item 3 above.				
6. Between the accommodation and a protected corridor forming a dead end. Other than in item 3 above.	unlimited above 1100 mm from floor	unlimited above 100 mm from floor	unlimited above 1100 mm from floor	unlimited above 100 mm from floor
7. Between accommodation and any other corridor; or subdividing corridors. Other than in item 3 above.	not applicable	not applicable	unlimited above 100 mm from floor	unlimited above 100 mm from floor
8. Adjacent an external escape route described in para 4.27.	unlimited above 1100 mm from paving	unlimited above 1100 mm from paving	unlimited above 1100 mm from paving	unlimited above 1100 mm from paving
9. Adjacent an external escape stair (see Figs B1.41 and B1.42) or roof escape (see para 6.35).	unlimited	unlimited	unlimited	unlimited

Notes:
1. If the protected stairway is also a protected shaft (see paragraph 9.36) or a firefighting stair (see Section 18) there may be further restrictions on the uses of glazed elements.
2. Measured vertically from the landing floor level or the stair pitch line.
3. The 100 mm limit is intended to reduce the risk of fire spread from a floor covering.
4. Items 1c, 3 and 6 apply also to single storey buildings.

Appendix A: Table A7

This table is concerned with the use of materials of limited combustibility. They are generally materials in non-combustible cores with thin facings such as plasterboard.

Included are materials classified as non-combustible by BS 476: Part 4: 1970, but also materials having a specified performance under BS 476: Part 11: 1982. The test is appropriate for mixtures of materials whether manufactured or natural, providing they are reasonably homogeneous, although some non-homogeneous materials may be capable of being tested.

Results of tests on proprietary materials are contained in several publications listed in the approved document, and these are set out in the relevant reading section at the end of this chapter. The emphasis is on care in selecting a material and in its fixing.

For the purposes of the approved document, materials of limited combustibility are

- an acceptable non-combustible material;
- a material of density 300 kg/m³ or more which does not flame, and the rise in temperature of furnace thermocouple is not in excess of 20°C when tested to BS 476: Part 11;

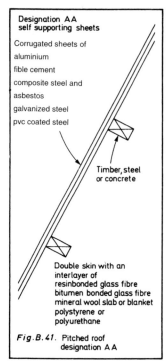

Fig.B.41. Pitched roof designation A A

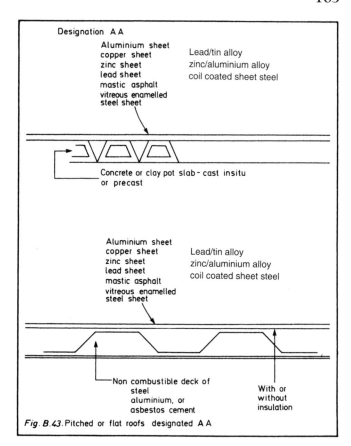

Fig. B.43. Pitched or flat roofs designated A A

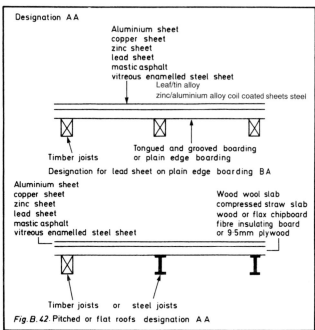

Fig. B. 42. Pitched or flat roofs designation A A

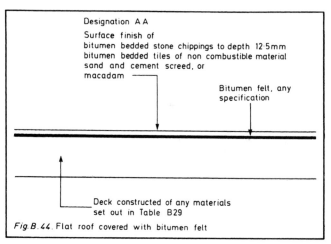

Fig.B.44. Flat roof covered with bitumen felt

Table 12 Pitched roofs covered with butumen felt. Notional designations.

Construction of deck	To CP 144: Part 3 1970: Layers			
	Upper Type; 1E	Upper Type 2E	Upper Type 2E	Upper Type 3E
	Lower Type 1B 1.3 kg/m²	Lower Type 1B 1.3 kg/m²	Lower Type 2B	Lower Type 3B
Concrete or clay pot slab Aluminium single skin or cavity deck with or without overlay of	AB	AB	AB	AB
fibre insulating board Asbestos-cement or single skin or cavity deck without overlay or with	AC	AB	AB	AB
fibre insulating board	AC	AB	AB	AB
screeded wood wool slab	AC	AB	AB	AB
Compressed straw slab	AC	AB	AB	AC
6 mm plywood, wood or flax chipboard (12.5 mm) 16 mm T & G boarding or 19 mm pe boarding	CC	BB	AB	AC

All butumen felt is to be BS 747: 1977.

- materials having a non-combustible core not less than 8 mm thick with combustible facings up to 0.5 mm on one or both sides. Any flame relating requirements are to be met.

These materials may be used in the following situations:

- stairs – B1
- above a suspended ceiling – B3
- reinforcement or support for fire stopping – B3
- roof coverings – B3 and B4
- roof deck – B3
- Class 0 in accordance with Appendix A
- ceiling tiles or panels – Table A3
- compartment walls and floors in hospitals
- insulation material in external walls
- insulation above fire-protecting suspended ceiling – Table A3

There is another material of limited combustibility listed in the approved document. It is a material having a density less than 300 kg/m^3 which does not flame for more than 10 seconds with a rise of temperature on the specimen is not in excess of 35°C and the furnace thermocouple is limited to 25°C when tested to BS 476: Part 11.

Materials meeting this specification may be used to insulate an external wall, or above any fire-protecting suspended ceiling (A3).

Appendix A: Table A8

The approved document contains in Table A8 some typical ratings of a number of generic materials or products. Plastic materials are not included as their range of performance is so dependent upon their formulation.

- Class 0
  (a) materials of limited combustibility cored products are not to have an index 'I' exceeding 12 or i_1 exceeding 6;
  (b) non-combustible material;
  (c) blockwork, brickwork, ceramic tiles and concrete;
  (d) plasterboard (plain or painted, or with PVC coating not in excess of 0.5 mm thick) with or without an air gap or fibrous or cellular insulating material behind;
  (e) woodwool cement slabs;
  (f) mineral fibre tiles or sheets with cement or resin (bonding).

The above materials also meet Class 1.

- Class 3
  (a) timber or plywood with a density exceeding 400 kg/m^3 plain or painted;
  (b) wood particle board or hardboard, plain or painted;
  (c) standard glass reinforced polyesters.

The approved document does not give any guidance on the treatment necessary to upgrade the timber products under Class 3 to Class 1, although this may be achieved by impregnating or coating with proprietary materials.

The following may achieve the stated ratings although the properties of products with the same generic description do vary, and the materials should be subject to tests.

- Class 0
  aluminium-faced fibre insulating board;
  flame-retardant decorative laminates on calcium silicate board;
  thick polycarbonate sheet;
  phenolic sheet;
  uPVC.
- Class 1
  phenolic or melamine laminates on a calcium silicate substrate;
  flame-retardant decorative laminates on a combustible substrate.

PART B

APPENDIX B

Protection of openings

The approved document sets down requirements in connection with openings allowed in elements of structure when protection against the spread of fire is required.

The first requirements apply to doors. Principles are that doors used for fire protection should normally be left closed, and a self-closing device fitted. There are a variety of hinges and other closing devices available, but when a door is activated by a fusible link it indicates that there is a necessity for the door to be kept open and a slight rise in temperature will allow the link to fuse, releasing the door, which will then close automatically. Electro-mechanical and electro-magnetic devices susceptible to smoke are also available and will hold open doors in stipulated circumstances. One of these

devices must ensure that on failure of the electricity supply the door will close when:

(a) smoke is detected automatically, or
(b) a manually operated switch is used, or
(c) the fire alarm system is brought into operation.

It may be necessary for doors to lift shafts to be kept open, possibly for long periods, as for example in a factory where lifts are closely integrated into the work floor and used to carry material from one floor to another, as part of a regular production programme.

Fire doors

Doors which must provide fire resistance should have the degree of protection required by Table B1. Where the suffix 'S' is given to the door rating it provides for restricted smoke leakage.

All fire doors are to be fitted with an automatic self-closing device, but they are not required for fire doors to cupboards and service ducts which are normally locked. There are situations where self-closing devices would be a hindrance to daily usage, and in these circumstances it is permissible to keep fire doors open using one of the following devices:

- fusible link; but not on a door used as a means of escape unless it meets the requirements for two doors, below;
- automatic release mechanism operated by an automatic fire detection and alarm system.

Two fire doors may be fitted in the same opening if

- the required fire resistance is the sum of the two individual ratings, and
- each door is capable of closing the opening;
- opening is a means of escape, both doors are to be self-closing although one may be fitted with an automatic self-closing device and held open by a fusible link; but only if the second door may be opened by hand and has a minimum of 30 minutes' fire resistance.

Fire doors provide only a limited insulation and, in consequence, the proportion of doorway openings in compartment walls is not to extend to more than 25% of the length of a compartment wall, unless the doors are made to provide both integrity and insulation to the required level as Table A2.

A heat sensor should be installed near a roller shutter placed across a means of escape to operate release in time of fire. This can be a fusible link or an electric heat detector. A smoke detector or fire alarm is not to be used to operate a door in these locations, unless the shutter is required to descend as part of the boundary of a smoke reservoir.

Rolling shutters are also to be made to function manually, in order that they may be opened or closed by firefighters.

Hinges for fire-resisting doors are to be manufactured from non-combustible materials with a melting point not less than 800°C. Guidance on door furniture is available in 'Code of practice for hardware essential to the optimum performance of fire resisting timber doorsets' (1993) published by the Association of Builders' Hardware Manufacturers.

All fire doors are to be marked on both sides with fire safety signs recommended by BS 5499: Part 1 identifying

- doors to be kept closed when not in use;
- doors kept locked; or
- doors held open by an automatic release mechanism.

Exceptions to this requirement are

- the door is in a dwellinghouse;
- doors to, and within flats or maisonettes;
- doors to bedrooms in 'other residential' purpose group 2(b) premises;
- doors to lifts.

BS 8214: 1990, 'Code of practice for fire door assemblies with non-metallic leaves', offers advice on design, installation and maintenance of such doors.

Table B1 Provisions for fire doors

Position of door	Minimum fire resistance of door in terms of integrity (minutes) (1)
1. In a compartment wall separating buildings	As for the wall in which the door is fitted, but a minimum of 60
2. In a compartment wall:	
a. if it separates a flat or maisonette from a space in common use;	FD 30S (2)
b. enclosing a protected shaft forming a stairway situated wholly or partly above the adjoining ground in a building used for Flats, Other Residential, Assembly and Recreation, or Office purposes;	FD 30S (2)
c. enclosing a protected shaft forming a stairway not described in (b) above;	Half the period of fire resistance of the wall in which it is fitted, but 30 minimum and with suffix S (2)
d. enclosing a protected shaft forming a lift or service shaft;	Half the period of fire resistance of the wall in which it is fitted, but 30 minimum
e. not described in (a), (b), (c) or (d) above.	As for the wall it is fitted in, but add S (2) if the door is used for progressive horizontal evacuation under the guidance to B1
3. In a compartment floor	As for the floor in which it is fitted
4. Forming part of the enclosures of:	
a. a protected stairway (except where described in item 9); or	FD 30S (2)
b. a lift shaft (see paragraph 6.42b);	FD 30
which does not form a protected shaft in 2(b), (c) or (d) above.	
5. Forming part of the enclosure of:	
a. a protected lobby approach (or protected corridor) to a stairway;	FD 30S (2)
b. any other protected corridor; or	FD 20S (2)
c. a protected lobby approach to a lift shaft (see paragraph 6.42).	FD 30S (2)
6. Affording access to an external escape route	FD 30
7. Sub-dividing:	
a. corridors connecting alternative exits;	FD 20S (2)
b. dead-end portions of corridors from the remainder of the corridor.	FD 20S (2)
8. Any door:	
a. within a cavity barrier;	FD 30
b. between a dwellinghouse and a garage;	FD 30
9. Any door:	
a. forming part of the enclosures to a protected stairway in a single family dwellinghouse;	FD 20
b. forming part of the enclosure to a protected entrance hall or protected landing in a flat or maisonette;	FD 20
c. within any other fire-resisting construction in a dwelling not described elsewhere in this table.	FD 20

Notes:

1. To BS 476: Part 22 (or BS 476: Part 8 subject to paragraph 5 in Appendix A).
2. Unless pressurization techniques complying with BS 5588: Part 4 *Fire precautions in the design, construction and use of buildings, Code of practice for smoke control using pressure differentials* are used, these doors should also have a leakage rate not exceeding 3 m³/m/hour (head and jambs only) when tested at 25 Pa under BS 476 *Fire tests on building materials and structures*, Section 31.1 *Methods for measuring smoke penetration through doorsets and shutter assemblies, Method of measurement under ambient temperature conditions.*

PART B

APPENDIX C

Methods of measurement

The whole of this appendix is devoted to defining the ways in which area, cubic capacity, height and the number of storeys have been calculated within Part B.

- Areas are indicated in Figs B113 and B114.
- Cubic capacity, see Fig. B115.
- Number of storeys, see Fig. B116.
- Galleries, see Fig. B117.
- Height of pitched and flat roofed buildings, see Fig. B118
- Height of mansard and monopitch roofed buildings, see Fig. B119.
- Height to top storey, see Fig. B120.

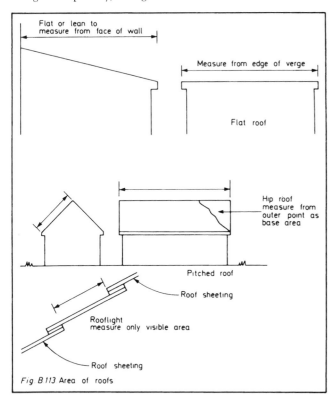

Fig. B113 Area of roofs

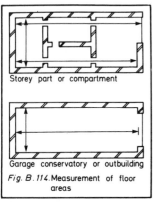

Fig. B.114. Measurement of floor areas

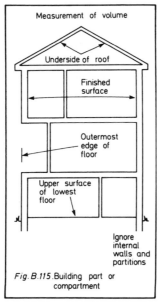

Fig. B.115. Building part or compartment

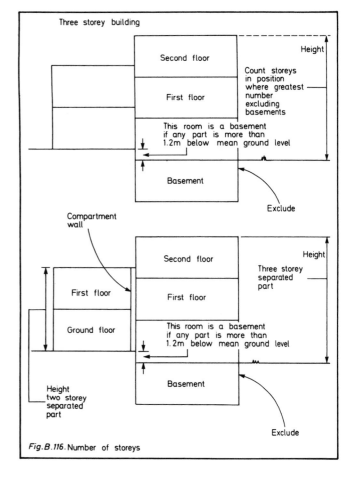

Fig. B.116. Number of storeys

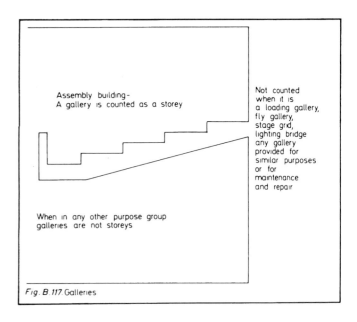

Fig. B.117. Galleries

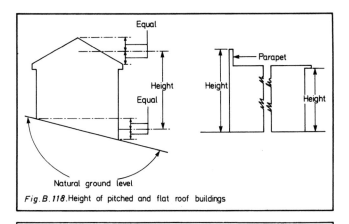

Fig.B.118. Height of pitched and flat roof buildings

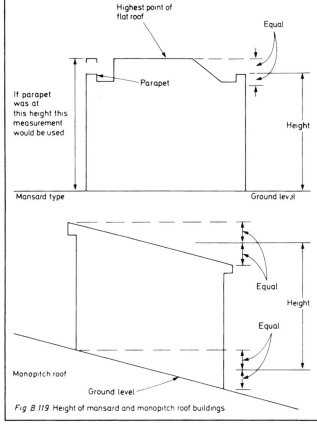

Fig B.119. Height of mansard and monopitch roof buildings

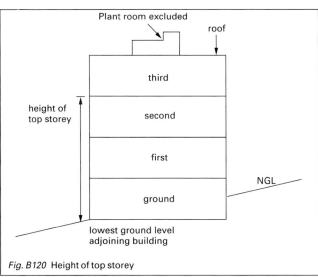

Fig. B120 Height of top storey

PART B

APPENDIX F, INSULATING CORE PANELS

Information was introduced first in the 2000 edition of the approved document to Part B. It relates particularly to the use of insulating core panels for internal structures, and the problems which may arise in the event of a fire.

The panels comprise an inner core of insulating material between and bonded to sheets of galvanized steel, or other material and possibly faced with PVC. They are used in positions where a high standard of hygiene is required. The most normal use is to create an enclosure where chilled or sub zero temperatures will be maintained in order to protect foodstuffs in storage or distribution.

Panels are normally attached to the structure by way of lightweight fixings and hangers.

Materials usually found in core panels are:

- expanded or extruded polystyrene
- polyurethane
- mineral fibre, or less frequently
- polyisocyanurate and modified phenolic.

In the event of fire these materials produce a large volume of smoke, although there are less problems associated with the use of mineral fibre than the other materials.

When subjected to high temperatures, the components of the panel will delaminate and the panel will lose its structural integrity. Fire can spread behind the panels and there is the likelihood that the panels may collapse.

This form of construction does therefore present firefighters with a number of problems:

- fire spread within the panels may be hidden and not readily detected;
- large quantities of toxic smoke may arise;
- fire may spread rapidly, leading to a flashover; and
- collapse of the system.

Design

It is recommended that an effort be made to identify the problems and then endeavour to provide a design solution, when consideration may be given to one or more of the following:

- remove the risk!
- provide distance between the risk and the panels;
- install a fire suppression system for the risk, or the enclosure;
- fix fire-resisting panels;
- use appropriate materials, fixing and jointing methods.

Specifying core panel materials

Care in selecting core panel materials will ensure that appropriate panels are used in specific areas

For example, mineral fibre core panels will be appropriate to use in bakeries, cooking areas, firebreaks in combustible panels, fire stop panels, general fire protection and hot areas.

All cores can be used in blast freezers, chill stores, clean rooms, cold stores and food factories.

The stability of a panel system may be improved in the event of fire by one or several of the following methods. Attention should be given to the use of combustible cores becoming exposed to fire.

- Endeavour to ensure that the panels and associated fixings will remain structurally stable.
- Protect building and panels from early collapse. This may involve the provision of supplementary support – perhaps involving fire protected heavier gauge steelwork fixings.
- Consideration should be given to the installation of non-combustible insulant cored panels when designated high-risk areas are involved.

- Ensure that combustible insulant is encased by non-combustible materials which will not become displaced in a fire.
- Prefinished and sealed areas should be incorporated to facilitate the penetration of services.

Machinery or other permanent loads should not be supported by a panel system. Any cavities between the system and other building elements are to have suitible cavity barriers.

Information about insulating core panels can be found in 'Design, construction, specification and fire management of insulated envelopes for temperature controlled environments' by the Association of Cold Storage Contractors (European Division), which is available from IACSE, Downmill Road, Bracknell, Berks RG12 1GH.

RELEVANT READING

British Standards and Codes of Practice

BS 476: Part 3: 1975. External fire exposure tests
BS 476: Part 4: 1970 (1984). Non-combustibility test for materials
    Amendment 1: AMD 2483
               2: AMD 4390
BS 476: Part 6: 1989. Fire propagation test for materials
BS 476: Part 7: 1997. Method of classification of the surface spread of flame
BS 476: Part 8: 1972. Criteria for fire resistance of elements of building construction
    Amendment 1: AMD 1873
               2: AMD 3816
               3: AMD 4822
BS 476: Part 10: 1983. Principles of application of fire testing
BS 476: Part 11: 1982. Methods of accessing the heat emission from building materials
BS 476: Part 13: 1987. Method of measuring the ignitability of products subject to thermal irradiance
BS 476: Part 20: 1987. Method of determination of fire resistance of elements of construction
    Amendment 1: AMD 6487
BS 476: Part 21: 1987. Methods for determination of fire resistance of loadbearing elements of construction
BS 476: Part 22: 1987. Methods for determination of the fire resistance of non-loadbearing elements of construction
BS 476: Part 23: 1987. Methods for determination of the construction of components to the fire resistance of a structure
BS 476: Part 24: 1987. Method for determination of the fire resistance of ventilation ducts
BS 746: Part 31.1: 1983. Measurements under ambient conditions
    Amendment 1: AMD 8366
BS 747: 1994. Specification for roofing felts
BS 1245: 1975 (1986). Specification for metal door frames
BS 2782: 1994. Methods of testing plastics (many parts)
BS 2782: Part 1: 1990. The vicat softening temperature of thermoplastics
BS 2782: Part 5: 1970. Miscellaneous methods of testing plastics, including Method 508A
BS 2782: 1970. Methods of testing plastics – Thermal properties: 120A to 120E:
BS 4422: Part 1: 1987. Glossary of terms
BS 4422: Part 2: 1990. Structural fire protection
BS 4422: Part 3: 1990. Fire detection and alarm
BS 4422: Part 4: 1994. Fire extinguishing equipment
BS 4422: Part 5: 1989. Smoke control
BS 4422: Part 6: 1988. Evacuation and means of escape
BS 4514: 1983. Specification for unplasticized PVC soil and ventilating pipes, fittings and accessories
    Amendment 1: AMD 4517
               2: AMD 5584
BS 5041: Part 1: 1987. Specification for landing valves for wet risers

BS 5041: Part 2: 1987. Specification for landing valves for dry risers
BS 5255: 1989. Plastic waste pipes and fittings
BS 5266: Part 1: 1988. Emergency lighting – other than cinemas and specified entertainment premises
BS 5268: Part 4: 1990 (section 4.2). Fire resistance of timber structures
BS 5306: Part 1: 1976 (1988). Hydrant systems, hose reels and foam inlets
    Amendment 1: AMD 4649
               2: AMD 5766
BS 5306: Part 2: 1990. Specification for sprinkler systems
BS 5395: Part 2: 1984. Design of helical and spiral stairs
    Amendment 1: AMD 6076
BS 5438: 1976. Tests for flammability of vertically oriented textile fabrics and fabric assemblies – Test 2: 1989
BS 5446: Part 1: 1990. Point type smoke detectors
BS 5499: Part 1: 1990 (1995). Specification for safety signs
BS 5516: 1977. Code of practice for patent glazing
BS 5588: Part 0: 1996. Guide to fire safety codes.
BS 5588: Part 1: 1990. Fire precautions in residential buildings
    Amendment 1: AMD 7840
BS 5588: Part 4: 1998. Code of practice for smoke control in protected routes using pressurization
BS 5588: Part 5: 1991. Code of practice for firefighting stairways and lifts
BS 5588: Part 6: 1991. Code of practice for assembly buildings
BS 5588: Part 7: 1997. Code of practice – atria in buildings
BS 5588: Part 8: 1999. Code of practice – means of escape for disabled people
BS 5588: Part 9: 1989. Ventilation and air-conditioning ductwork
BS 5588: Part 10: 1991. Code of practice for shopping complexes
BS 5588: Part 11: 1997. Shops, offices, industrial and other non-residential buildings
BS 5655: Part 10: 1995. Lifts and service lifts
BS 5720: 1979. Mechanical ventilation and air conditioning in buildings
BS 5803: Part 4: 1985. Flammability and resistance to smouldering
BS 5839: Part 1: 1988. Fire detection and alarm systems
BS 5839: Part 2: 1983. Specification for manual call points
BS 5839: Part 6: 1995. Design and installation of fire detection and alarm systems
BS 5839: Part 8: 1998. Design, installation and servicing of voice alarm systems
BS 5867: Part 2: 1980. Flammability requirements for curtains and drapes
    Amendment 1: AMD 4319
BS 5906: 1980. Storage and onsite treatment of solid waste
BS 5950: Part 8: 1990. Fire resistance – structural steel
    Amendment 1: AMD 8858
BS 6073: Part 1: 1981. Specification for precast masonry units
    Amendment 1: AMD 3944
               2: AMD 4462
BS 6203: 1991. Fire characteristics and performance of expanded polystyrene
BS 6206: Part 3: 1981 (1994). Fire performance of glass
BS 6262: 1994. Code of practice for glazing buildings
BS 6336: 1998. Development and presentation of fire tests and their use in hazard assessment
BS 6387: 1994. Cables required to maintain circuit integrity
BS 6661: 1986. Design, construction and maintenance of single-skin air-supported structures
BS 6750: 1986. Specification for modular construction in building
BS 7157: 1989. Test for ignitability of fabrics used in construction of large tented structures
BS 7346: Part 1: 1990. Specification for natural smoke and heat exhaust ventilators
BS 7346: Part 2: 1990. Specification for powered smoke and heat exhaust ventilators
BS 7801: 1995. Code of practice for safe working on escalators and passenger conveyors

BS 7807: 1995. Code of practice for designers of fire detection and alarm systems

BS 8202: Part 1: 1995. Code of practice for selection and installation of sprayed mineral coatings

BS 8214: 1990. Fire door assemblies with non-metallic leaves

BS 8313: 1989. Accommodation of building services and ducts

BSCP 1007: 1955. Maintained lighting for cinemas

BSCP 144: Part 3: 1970. Built up bitumen felt
    Amendment 1: AMD 2527
                2: AMD 5228

DD 240: 1997. Fire safety engineering

PD 6520: 1988. Fire test methods for building materials and elements of construction

BRE digests

161	Reinforced plastics cladding panels
208	Increasing fire resistance of existing timber floors
224	Cellular plastics for building
225	Fire terminology
230	Fire performance of walls and linings
260	Smoke control – design
285	Fires in furniture
288	Dust explosions
294	Fire risks from combustible cavity insulation
300	Toxic effects of fires
317	Fire-resistant steel structures: free-standing blockwork, filled columns and stanchions
320	Fire doors
388	Human behaviour in fire

BRE information papers

2/80 Roofs as barriers to fire

2/87 Fire and explosion hazards associated with redevelopment of contaminated land

3/80 Effects of a roof on a fire within a building

4/86 Ignition and growth of fire in a room

4/87 Fire safety considerations in the design of structural sandwich panels

5/88 Selection of sprinklers for high rack storage in warehouses

6/87 Fire behaviour of breather membranes

8/82 Increasing fire resistance of existing timber floors

9/85 Incidence and nature of fires in traditional and framed houses

11/83 Fire doors

17/89 Photoluminescent markings for escape routes

18/86 Passive fire precautions in LPS blocks of flats and maisonettes

18/87 Assessing the life hazard from burning sandwich panels

19/85 Simplified approach to smoke ventilation calculations

20/84 Important factors in real fires

20/85 Behaviour of people in fires

21/84 Fire spread in buildings

22/84 Smoke spread in buildings

23/82 Spontaneous combustion

25/80 Fire performance of loft insulation materials

16/91 Assisted means of escape of disabled people from fire in tall buildings

17/91 Statistical studies of fire

17/94 Emergency wayfinding lighting systems in smoke

BRE defects action sheets

7	Pitched roofs – preventing fire spread between dwellings
8	Pitched roofs – separating wall/roof junction. Preventing fire spread
29	External wall and separating walls – cavity barriers
30	Cavity barriers – installation
62	Electrical services – avoiding cable overheating – design

131 External walls: combustible external plastics insulation – horizontal fire barriers

132 External walls: combustible external plastics insulation – fixings

BRE books and reports

Fire safety in buildings (H. L. Malhotra)

Smoke control methods in enclosed shopping complexes (SO 17)

Interaction of sprinklers and roof venting in industrial premises (BR 57)

Factors determining life hazard in hotels and boarding houses (BR 82)

Factors determining life hazard in health care and residential health care buildings (BR 83)

Guidelines for the construction of fire-resisting structural elements (BR 128)

Fire and the Architect – the communication problem (BR 136)

Aspects of fire precautions in buildings (BR 225)

Design principles for smoke in enclosed shopping malls (BR 186)

External fire spread: building separation and boundary distances (BR 187)

Fire performance of external thermal insulation for walls of multi-storey buildings (BR 135)

Fire and disabled people in buildings (BR 231)

Fire safety of FTFE based materials used in building (Purser, Fardell and Scott) (BR 264, 1994)

Natural ventilation in Atria for environment and smoke control (BR 375)

Also relevant

Report on results of surface spread of flame on building products – 1976, from TSO

Report on fire test results on building products – 1976, from TSO

Results of fire resistance tests on elements of building construction – from TSO (two volumes)

Credibility and truth: a view of risk appraisal. Annual fire research lecture 1985

Factors determining life hazards from fire in group residential buildings:
    Part 1: Hotels and boarding houses
    Part 2: Health care residential premises and halls of residence

British statutes relating to fire (1425–1963)

Gale damage to buildings in the UK – an illustrated review by P. S. J. Butler

The overcladding of external walls on large panel system dwellings

Wake up – get a smoke alarm (Home Office 1990). TSO

Fire Code HTM–81. Fire precautions in new hospitals. TSO

Fire Code HTM–85. Fire precautions in existing hospitals. TSO

Fire Code HTM–88. Fire precautions in NHS housing in the community for handicapped or mentally ill people. TSO

Building Bulletin 7. Fire and the design of educational buildings. TSO

Rules for the construction and installation of firebreak doors and shutters. 1988. LPC

Guide to fire precautions in existing places of entertainment and like premises (1990). TSO

Safety signs and signals. Guidance on Health and Safety (Safety, Signs and Signals) Regulations 1996. (HSE. L64)

Fire and steel construction: Behaviour of steel portal frames in boundary constructions (1990.) (Available from Steel Construction Institute, Silwood Park, Ascot, Berks SL5 7QN)

Publications from the fire protection association

Fire test results on building products: fire resistance (1983)

CHAPTER 12

Site preparation and resistance to moisture

(Approved document C)

The 1992 edition of the approved document C gives recommendations for dealing with gaseous contaminants including radon, methane and other landfill gases. There are also further provisions relating to ventilation below suspended floors; the minimum residual cavity width when a wall cavity is partially filled with insulation is to be 50 mm and there is an increase in the thickness of polyethylene dpm.

Regarding radon, the areas where protection is required have been extended and provision included for basic protection to be applied universally, with greater provision in high-risk areas.

Regulation 4 requires that building work is to be carried out in accordance with the relevant paragraphs of Schedule 1 to the regulations.

Compliance with Part C is necessary when erecting any building, or when substantial alterations are made to a building.

Previous regulations showed that where wet processes are carried out in factories, the inclusion of moisture-resisting construction would not increase the protection of the health of employees were also exempt from Part C and this exemption continues.

Regulation C1. Preparation of site

The ground which is to be built upon must be cleared of all vegetable matter. This does not simply mean that grass and weeds should be cut down. More work is necessary to remove vegetable matter and pernicious weed roots and a minimum excavation of 150 mm is usual, but the vegetable matter should be removed to a depth which will prevent later growth (Fig. C1.1). This requirement does not apply to

- a building to be used for storing goods when the people employed in the building are only occupied in taking in, caring for, or removing the goods, or
- a purpose wherein the health and safety of employees in the building would not suffer.

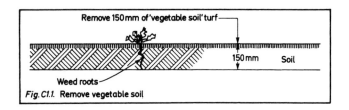

Fig. C1.1. **Remove vegetable soil**

Regulation C2. Dangerous and offensive substances

Steps are to be taken to avoid risk to health and safety which may be caused by any substances on or in the ground to be built upon.

When the Public Health Act was passed in 1936 there was concern that building might take place on ground filled with offensive material to the detriment of the health of persons living or working in the building.

So section 54 was introduced. This required a local authority to reject plans of a building to be erected on ground which had been filled up with material impregnated with faecal or offensive animal or vegetable matter. The alternative was to satisfy the local authority that the material had been removed or had become innocuous.

This requirement was transferred to section 29, Building Act 1984, and then promptly repealed by 1985 Regs, 18 to be replaced by C2.

C3 and C4 support this requirement and floors are to be constructed in a way that will resist moisture or vapour gaining access to the structure or into the building.

The increasing importance placed on the re-use of land following the decline of old industrial areas has revealed how crude was the requirement in the Public Health Act 1936 and how complex is the nature of contamination which may affect the soil in disused industrial areas.

This problem is acknowledged by government and there exists an inter-departmental Committee on Redevelopment of Contaminated Land (ICRCL for short!)

Contamination has not necessarily been deliberate. The accidental or careless leakage or spillage of materials used in industrial processes is a ready source of ground pollution, built up over a period of many years. Pollution may also arise from the settling of airborne particles, the storage of raw materials, wastes and residue or even the application of sewage sludge to the land.

These and many other sources may give rise to contamination to a degree that can restrict subsequent development. There can be damage to human health; to plant life and of course the actual new buildings or services which may come into contact with the polluted soil.

ICRCL has been studying this problem and issued a guidance document emphasizing that when it is proposed to purchase or re-use land previously used for industrial purposes or waste disposal the possibility of it being contaminated must be considered. Such matters must naturally be taken into account by planning authorities when devising their local plans and making land allocation maps.

Failure to carry out investigations may result in the value of the land being affected; emergency action becoming necessary during development to remedy contamination, with a consequent delay, and inevitably, adverse publicity.

Some sites which may present problems are old asbestos works, chemical works, railway sidings, petroleum storage and distribution sites, sewage works and farms, scrapyards, gas works and waste disposal sites.

One should also always be cautious when contemplating a site where there has been an industry making or using wood preservatives, metal mines, smelters, foundries, steel works and metal finishing works.

Any site where there has been nuclear installation demands caution as does munitions production and testing sites. Paper and printing works, and tannery works sites should be carefully investigated.

What are the hazards associated with a contaminated site? There is the possibility of pollutants being absorbed into food plants or surface contamination of food plants grown in gardens. Some contamination may render the soil incapable of growing plants.

Excavations of a site during construction work may lead to disturbance of contaminates releasing them into water courses and water supplies.

The possibility of chemical attack on building materials and services is real. Concrete may be affected by sulphates, metal corroded, and even plastics, rubber and other polymeric materials used in pipework or jointing in seals and protective coatings suffer damage. It is possible for some contaminants to penetrate through plastic pipework without actually causing structural collapse, but leading to pollution of water supplies.

It is therefore essential that hazards should be assessed for any site likely to create problems and in most instances expert advice must be obtained.

When acquiring a site a prospective purchaser must be satisfied that contamination from an earlier use does not preclude the new use he has in mind, or affect its value.

Quite clearly the early crude protection from rubbish, faecal and rotting matter contained in the 1936 Act did not visualize present-day problems.

Although the inadequacy of the old building regulations is apparent the government has taken first steps to remedy this situation. The new building regulations have this provision to cater for dangerous and offensive substances, requiring precautions to be taken to avoid danger to health. However, the new requirement only relates to contamination of the ground which is actually to be covered by new buildings.

A systematic approach is needed to obtain information about a site. It is first necessary to recognize that the site may be contaminated. Hazards must be identified, and their nature, concentration and distribution established.

All this information must be evaluated and it is then possible to decide if a site is suitable for the intended use.

It is possible to get some idea of the possible contaminants by walking carefully over a site using eyes and nose.

- Vegetation may be absent, poor in appearance or show unnatural growth – dwarf, deformed, lanky or scrubby are examples. The cause may arise from metals or metal compounds and if this should be so, the approved document suggests that no action other than careful normal site preparation is necessary.

  However, vegetation may suffer in consequence of the presence of organic compounds or gases. When this occurs then the relevant action is to arrange for removal of the offending material. This means that the contaminant, and also any contaminated ground, must be taken out to a depth of 1 m or as may be agreed with the local authority. Excavation depth agreed is to be below the level of the lowest floor. The material excavated must then be removed from the site and disposed of at a place agreed with the local authority.

- Surface materials may display unusual colours, or the contours may be unnatural and indicate the presence of wastes and residues.

  1. Possible contaminants are metals, metal compounds or fibres other than asbestos. When this is the case no action other than normal site preparation is necessary.
  2. However, should asbestos fibres be the contaminant there are two courses of action available – filling or sealing.

     Filling entails covering the site of proposed building with material that will not react adversely with the contaminant, and is suitable for making up levels. Actual depth of fill recommended is 1 m, or as may be agreed with the local authority. It is necessary to consider the types of filling to be used, and the design of the ground floor together.

     Alternatively one may resort to sealing the contaminant from the building. This entails laying a suitable imperforate barrier between the contaminant and the building. It must be sealed at any service entries, around the edges and at joints. If other mineral fibres are discovered no relevant action is recommended.
  3. Should the contaminant be oily or tarry waste, the relevant action may be removal, filling or sealing as agreed with the local authority.

     If the choice is to seal these contaminants from the building, polythene may not be suitable.

4. Organic compounds including phenols may be found to be the contaminant. In which case the methods to be adopted may be removal or filling as described above.
5. Coal or coke dust, or other potentially combustible material may also be the contaminant, and the remedy may be removal or filling with material completely non-combustible and unlikely to be affected by chemical reactions.
6. Refuse and waste, when found on the site of a building, should be removed as described above.

- Fumes and odours detected on site may indicate organic chemicals, possibly at quite low concentrations.

  1. If arising from corrosive liquids, the remedy may be by filling, sealing or removal. If removal is the appropriate action it may be that the nature of the liquid will require the work to be undertaken by specialists.
  2. Should the contaminants be of a faecal, animal or vegetable nature, it will be necessary to remove or resort to removal or filling.
  3. When flammable, explosive and asphyxiating gases are detected, including methane and carbon dioxide it will be necessary to arrange removal from the site.

     Additionally it is most important to ensure that the building is so constructed to ensure that there are no unventilated voids.

- Drums and containers, whether full or empty may be associated with a multitude of materials and when discovered should be removed together with all contaminated ground. The local authority may require this work to be undertaken by specialists.

Should there be any sign of possible contaminants the Environmental Health Officer should be notified, and if he confirms the presence of contaminants listed above the necessary precautions must be taken. There is an assumption in this recommendation that the site of the building will have at least 100 mm in-situ concrete laid over it.

From July 1999 new contaminated land provisions, part of the Environmental Protection Act 1990 became operative, with the object of encouraging the use of brownfield sites. These provisions require remedial measures to be taken if there are unacceptable actual or potential risks to health, or the environment. Local authorities will have records of sites they have investigated but they are not required to make public information relating to the sites examined.

Local authorities may issue notices requiring remedial work, and there is a right of appeal against these notices.

The BRE has established a service to give advice on assessing risks, and methods of protecting new buildings from landfill gases, fire, explosion and toxic chemicals.

ICRL Paper No. 59/83 describes methods of making a site investigation.

Removal, covering or sealing of contaminants are mentioned above, but there are also a number of more sophisticated techniques available.

- Bio-remediation, where micro-organisms are used to remove oily organic contaminants.
- Dilution. Clean materials are mixed with the contaminated soil to reduce its concentration.
- Incineration. Contaminated soil is incinerated on site.
- Stabilization. Using cement or lime to bond and prevent oily or metal contaminants escaping.
- Soil flushing or washing. Contaminants are flushed or washed from the soil. Ground water may be used for this purpose, but the contaminants removed will require safe disposal.
- Vacuum extraction. Liquids and gases may be withdrawn from soil by utilizing a vacuum.

Alternative approach

In the worst conditions only total removal of the material from the site of the building can provide a remedy. In other instances expert advice should be secured to reduce any risk to acceptable levels, and expert advice should also be obtained with regard to the handling of large quantities of contaminants.

There is a BS draft for development DD 175: 1988, 'Code of practice for the identification of potentially contaminated land and its investigation', which gives guidance on site investigations and BS 5930: 1981 is a code of practice for ground engineering and is relevant.

In outline, a site investigation will involve:

- Collecting information of past uses of the site from maps, rating records, local history societies and past employees, and any other available sources.
- Visual inspection of the site, looking for obvious evidence of tipping or spillage. Other signs are fumes and smells, unusual colours or surface materials and the lack of, or poor growth of vegetation in the area.
- Sampling the soil. Advice of an analyst is essential. Perhaps the best way is to sample on a grid system using a mechanical digger, taking samples near the surface, at an intermediate depth and at the base of the hole. Sometimes boreholes may be a suitable alternative.
- Having the sample analysed and assessing the findings. Information on 'trigger concentrations' is available in ICRCL 59/83. There are values of various contaminants below which a site could be regarded as uncontaminated. These vary with the proposed use of the site, and are intended only for guidance for those who interpret the findings of properly conducted site investigations.

RADON

Radon is a naturally-occurring, colourless, odourless gas which is radioactive and we have always received doses of radiation from it. Radon arises where uranium and radium are present and this occurs to some extent in all rocks and soils, although the most significant concentrations are to be found in granite, sandstone and phosphates in varying amounts throughout the country. Usually the gases dissipate harmlessly into the atmosphere.

Putting a building on a site can interrupt that dissipation and in consequence the gas may collect in spaces under and within dwellings leading to penetration, entering a dwelling through a number of possible routes such as:

- cavities in walls;
- construction joints;
- cracks in solid floors;
- cracks in walls below ground level;
- gaps around service pipes;
- gaps in suspended floors (concrete or timber).

If one is exposed to a high level of concentration within a dwelling for a long period, there is a risk of developing lung cancer.

Initially, building regulations required measures to be incorporated in new dwellings to protect occupiers from excessive exposure to radon in Cornwall, Devon, Somerset, Northamptonshire and Derbyshire. Further numerous surveys and measurements have been carried out by the National Radiological Protection Board (NRPB) and the results indicate that there are other areas outside these counties where a house, or an extension may require measures installed to prevent the ingress of radon (1999 Edition).

Guidance is given in the latest edition of a BRE report: No. 211, entitled 'Radon – guidance on protective measures for new dwellings'.

Affected areas are shown on a series of maps in 5 km grid squares, but current information can be obtained from local authority environmental health officers, building control officers and from approved inspectors. It is sensible to seek information when work is proposed in the districts of the following local authorities.

Requirements relate to a basic radon protection, supplemented by additional measures in higher risk areas.

Table 1. Local authorities affected

County	District
Cambridgeshire	Peterborough
Cornwall	Caradon
	Carrick
	Kerrier
	North Cornwall
	Penwith
	Restormel
Cumbria	South Lakeland
Derbyshire	Amber Valley
	Bolsover
	Chesterfield
	Derbyshire Dales
	High Peak
	North East Derbyshire
Devon	East Devon
	Mid Devon
	North Devon
	Plymouth
	South Hams
	Teignbridge
	Torbay
	Torridge
	West Devon
Dorset	North Dorset
	West Dorset
Gloucestershire	Cotswold
	Forest of Dean
	South Gloucestershire
	Stroud
Herefordshire	Leominster
Lancashire	Lancaster City
Leicestershire	Harborough
	Melton
Lincolnshire	Lincoln City
	North Kesteven
	South Kesteven
	West Lindsey
Northampton	Corby
	Daventry
	East Northamptonshire
	Kettering
	Northampton
	South Northamptonshire
	Wellingborough
Northumberland	Alnwick
	Berwick-on-Tweed
North Yorkshire	Craven
	Richmondshire
Nottingham	Ashfield
	Mansfield
Oxfordshire	Cherwell
	West Oxfordshire
Rutland	Rutland
Shropshire	Oswestry
	South Shropshire
Somerset	Bath and NE Somerset
	Mendip
	North Somerset
	Sedgemoor
	South Somerset
	Taunton Deane
	West Somerset
Staffordshire	Staffordshire Moorlands
Warwickshire	Stratford-on-Avon
Wiltshire	Salisbury

All new dwellings and additions are to have basic protection by provision of a damp-proof membrane in the floor, linked to a cavity tray. A good standard of workmanship is necessary, and all joints must be made gas tight. Whenever possible, service penetrations through the floor are to be avoided, but where necessary are to be carefully sealed. When sealed to a cavity tray this membrane should significantly reduce the opportunity for radon to enter the dwelling.

Where the proposed dwelling is in a higher risk area, the membrane and cavity tray provision is to be supplemented by subfloor ventilation or subfloor depressurization to reduce gas pressure on the membrane system.

Standards of radon protection are:

- basic – radon barrier embraces building;
- full – radon barrier, plus
    a. ventilated subfloor space through vents in outer walls,
    b. passive stack ventilation of subfloor space,
    c. fan-assisted stack ventilation of subfloor space.

Suitable materials for use in radon barriers are:

- 300 micrometre (1200 guage) polyethylene (Polythene) sheet;
- liquid coatings;
- flexible sheet roofing materials;
- self-adhesive bituminous-coated sheet products;
- asphalt.

When choosing a material it is important to consider the method of jointing and any difficulties which may be experienced in forming a gas-tight joint to membrane and cavity tray.

Figure C2.1 illustrates a basic method of protecting a dwelling and dissipating radon to the atmosphere.

There are also fan-assisted radon sump systems, and they may give rise to a noise problem in homes. BRE leaflet XL 9 describes how to design a sump system with a view to minimizing noise disturbance.

BRE research has established that quieter, less intrusive, low-speed mechanical extract fans are effective at reducing radon levels.

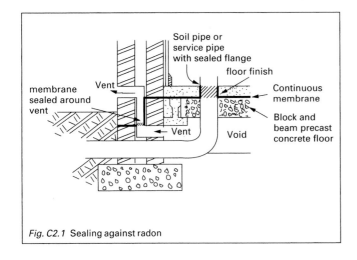

Fig. C2.1 Sealing against radon

LANDFILL GAS AND METHANE

Landfill gas is created by the action of anaerobic micro-organisms in biodegradable material and usually consists of methane and carbon dioxide. it can migrate through the ground from sources some distance away.

Similar gases may arise naturally and should be treated with the same caution as must be given to gas arising from old rubbish tips. The gas enters buildings through cracks and other openings in the structure from rhe ground beneath it.

Methane, when breathed, may cause unconsciousness and death; it will burn and can cause an explosion. There is a great variety of landfill gases; some of them being flammable and some toxic.

If a building is to be erected on a site that may give rise to gas production or is within 250 m of a site or suspected site of landfill or other contaminated ground, very careful investigation should be undertaken.

The BRE has published a report – 'The construction of new buildings on gas contaminated land' – and this offers guidance to the developer.

- When the level of methane in the ground is not likely to be greater than 1% by volume, the ground floor of a dwelling should be suspended concrete, and ventilated
- There should be a separate assessment of the carbon dioxide concentration and if this is in excess of $1\frac{1}{2}$ % by volume in the ground, some protection is to be considered. Where the volume rises to 5% then the design must incorporate specific measures against ingress of the gas.
- The installation of permanent continuous mechanical extraction systems to remove the gas from beneath a house is not recommended. Lack of maintenance could lead to failure of the system. A non-mechanical method will be acceptable on those sites where gas concentrations are low and may remain low.

Where gas concentrations seem to be higher or where non-domestic buildings are proposed it is advisable to seek advice from an expert. This may lead to a complete investigation into the source, and nature, of gases and also the possibility of increased gas generation in the future.

Other matters to have in mind are:

- the pressure of gas in the ground;
- a high gas concentration may have a limited impact if from a small natural deposit of silt or peat.

Extensive monitoring may be necessary to ascertain the present and estimate the future risks involved.

Expert advice should be secured to design protective measures to be incorporated in a building, and these measures should also have regard to the design of the structure.

The BRE report gives examples of construction techniques suitable for houses, industrial and commercial buildings. The various methods described are based upon practical experience.

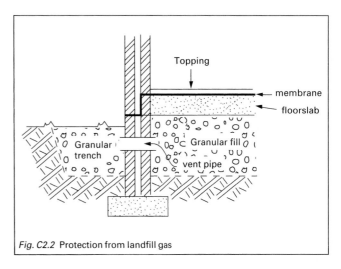

Fig. C2.2 Protection from landfill gas

Figure C2.2 illustrates a basic method of providing protection against landfill gas.

Regulation C3. Subsoil drainage

If necessary, because of damp ground conditions, subsoil drainage is to be provided to prevent moisture damaging the fabric of the building or gaining access to its interior.

On occasions it may be necessary to resort to tanking or the insertion of subsoil drains to divert water away from the building.

When the water table is high the site may be boggy or damp, even in dry weather, and when surrounded by higher land the site may have become a natural recipient of both surface and subsoil water from the surrounding area.

It may be necessary to insert new land drains where:

• surface water may gain access to the buildings or cause damage to the structure;
• ground water can rise to within 0.25 m of the lowest floor.

The outfall of a new land drainage system will depend upon the availability of a suitable ditch, an existing land drainage system or a public surface water sewer.

If, during construction, existing land drains are severed, they should be added to or relaid in such a way that water is diverted from the structure. (See Figs C3.1 to C3.6)

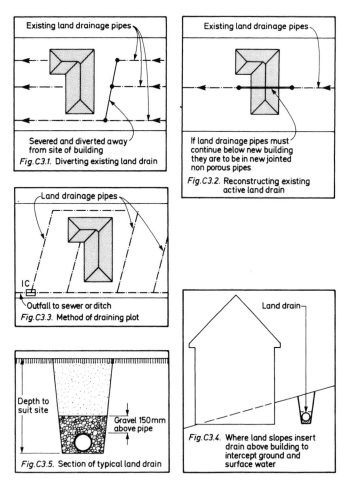

Fig. C3.1. Diverting existing land drain

Fig. C3.2. Reconstructing existing active land drain

Fig. C3.3. Method of draining plot

Fig. C3.4. Where land slopes insert drain above building to intercept ground and surface water

Fig. C3.5. Section of typical land drain

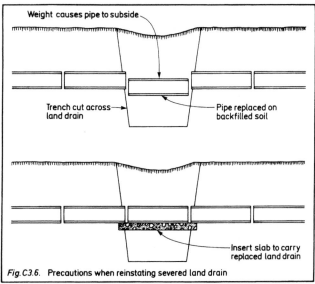

Fig. C3.6. Precautions when reinstating severed land drain

Regulation C4. Resistance to weather and ground moisture

Floors, walls and roof must resist the penetration of moisture to the interior of the building.

A leaking roof, window, or a door which allows rain to be driven in during severe gales is very annoying and occasionally does happen, even when the greatest care has been taken during construction. But a remedy is comparatively easy, and the owner's annoyance readily forgotten. A much more difficult fault to remedy is dampness rising up from the ground. Such moisture must not damage the floor or reach the upper surface of a floor. The utmost care must be taken in constructing those parts of a building which are next to the ground (see Fig. C4.1). It can be very expensive to remedy even the most simple fault when a building is occupied.

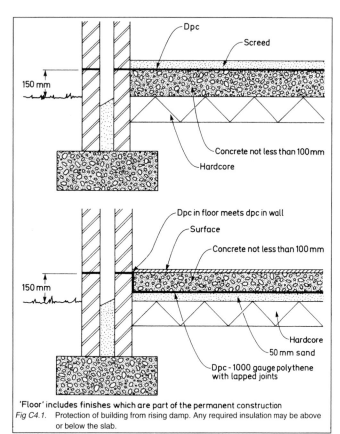

'Floor' includes finishes which are part of the permanent construction
Fig C4.1.   Protection of building from rising damp. Any required insulation may be above or below the slab.

(Fig. C4.1 does not include provision against radon gaining access up the external wall and is therefore not applicable to dwellings in radon affected areas.)

This provision does not apply when a building is to be used wholly for

• storing goods
• accommodating plant or machinery

if the people employed in the building are employed only in storing, caring, carrying or removing goods, plant or machinery, or

• any purpose where the requirements would not improve the protection of health and safety of people habitually employed therein,

and not be damaged by moisture rising from the ground.

Regulation C4 is most explicit and there are many ways of meeting the requirements which largely depend on the form of construction being used. In the case of a solid floor it requires the insertion of a damp-proof membrane within the floor. If placed below the concrete a suitable dpc would be 1200 gauge polythene with lapped joints, laid on a minimum of 50 mm of sand. 250 μm (1000 gauge) polyethylene may also be used if holding an Agrément certificate or is to PIFA standards. To lay directly on hardcore would cause damage to the polythene and a loss of its impervious requirement. Above the concrete a polythene sheet or three coats of cold bitumen should be protected by a screed, unless the dpm is pitchmastic or a similar material which may also serve as a floor finish.

Concrete should be composed of:

50 kg cement
0.11 m³ fine aggregate (maximum)
0.16 m³ coarse aggregate (maximum), or
BS 3238 mix ST 2.

The following is the BS mix ST 2:

Nominal maximum size of aggregate	40 mm	40 mm
(a) slump	75 mm	125 mm
cement	210 kg	230 kg
total aggregate	1980 kg	1920 kg
Nominal maximum size of aggregate	20 mm	20 mm
(b) slump	75 mm	125 mm
cement	240 kg	260 kg
total aggregate	1920 kg	1860 kg

Where there is embedded steel the concrete should be composed of

50 kg cement
0.08 m³ fine aggregate, and
0.13 m³ coarse aggregate,

or alternatively BS 3238 mix ST 4:

Nominal maximum size of aggregate	40 mm	40 mm
(c) slump	75 mm	125 mm
cement	280 kg	300 kg
total aggregate	1920 kg	1860 kg
Nominal maximum size of aggregate	20 mm	20 mm
(d) slump	75 mm	125 mm
cement	300 kg	330 kg
total aggregate	1860 kg	1800 kg

For more information, BS 3238 should be consulted.

BRE Digest 54 (second series) is entitled 'Damp proofing solid floors' and contains some very helpful notes regarding the types of dpc which are suitable for various floors. The digest also elaborates upon the effects of ground moisture on floor finishes.

There are a number of additives intended to be used as water proofing agents but it is doubtful if this form of construction can be regarded as adequate to meet the regulations.

If it is necessary to ensure that a floor itself is not affected by moisture as distinct from the necessity for keeping moisture from extending above the surface of the floor.

When solid floors are to be used incorporating timber (see Figs C4.2 to C4.5), it is necessary to prevent conditions under which decay will be encouraged, and BRE Digest 18 is concerned with the design of floors to prevent dry rot.

Timber embedded in concrete to form fillets must be treated with a preservative and BS 1282: 1975 is a guide to the choice, use and application of wood preservatives. Conditions under which wood is likely to deteriorate are set out and there are notes on preservative treatment including specifications and factors controlling effectiveness.

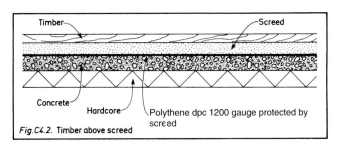

Fig. C4.2. Timber above screed

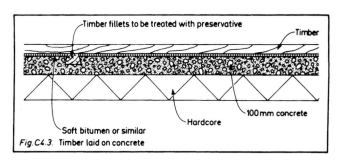

Fig. C4.3. Timber laid on concrete

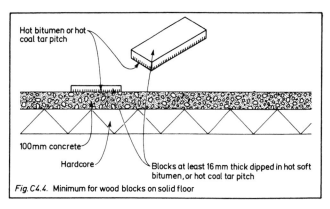

Fig. C4.4. Minimum for wood blocks on solid floor

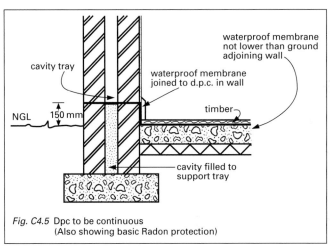

Fig. C4.5 Dpc to be continuous
(Also showing basic Radon protection)

Alternative approach

A satisfactory alternative is to follow recommendations given in BS 8102: 1990, 'Code of practice for protection of structures against water from the ground', includes advice regarding floors subject to water pressure.

BSCP 102: 1973, 'Protection of buildings against water from the ground', is also relevant, particularly clause 11.

HARDCORE

Great care must be exercised in the choice of all materials to be used in the erection of a building. This is particularly important when hardcore is being chosen and placed. Put in position when site conditions are usually at their worst, there are a number of hazards that can arise later. What then can be said about the choice and placing of hardcore?

Unsuitable hardcore can be the cause of chemical attack on concrete oversite slabs. Settlement or swelling can result in oversite concrete and partition walls cracking, and the remedy when these conditions arise is not simple.

No hardcore laid under a floor is to contain water-soluble

sulphates or other deleterious matter in such quantities as to be liable to cause damage to any part of the floor.

Regulation 7 takes this a step further by saying that proper materials are to be used and the work carried out in a workmanlike manner.

There are basically three difficulties which may arise from the failure to give careful attention to the selection and placing of hardcore: subsidence, heave and water-soluble sulphates. Hardcore may be placed at differing depths over a site, or the site may be slipping, having a little hardcore on one side and a much greater depth at the other. Also the internal portion of a wide and deep foundation trench may have been filled with hardcore. Inadequate overall compacting can lead to consolidation after completion of a building and in consequence a lack of support over part of its area; the slab may crack and gaps appear round the edges.

The National House Building Council requires a suspended floor to be constructed where the depth of filling would exceed 600 mm.

Heave can arise because of the chemical and volume changes in the hardcore. Some materials contained in hardcore that can give rise to this risk are: undehydrated lime, partially vitrified slags from old slag heaps and materials containing a significant proportion of clay. Placed in a dry condition and later becoming wet can cause swelling.

Many ground floor failures can be attributed to attack by water-soluble sulphates. Usually in damp conditions, the sulphates react with constituents of the Portland cement, producing a gradual breakdown of the concrete causing it to lift and expand. The surface of an affected floor will become cracked and distorted; in severe cases enclosing walls may be pushed outwards.

In these circumstances and in the light of requirements contained in the building regulations, what are the qualities of the various materials in daily use as hardcore?

- Blastfurnace slag arises from the making of iron. Often used as hardcore, it is a hard, strong material, and free draining. However, large volumes of slag should not be placed in wet stagnant conditions, as problems can arise from sulphur leaking into streams or underground water supplies. Also materials from old slag banks must be thoroughly examined and tested before use as they may be mixed with other industrial wastes. Slag from steel making may expand on wetting and is therefore not to be recommended.
- Brick rubble is often used. It is frequently readily available to the builder as his demolition works links with his need for hardcore elsewhere. Fine material is necessary to ensure adequate compaction. It is wise to place a water barrier between brick hardcore and a concrete floor, as many bricks contain sulphate, or old gypsum plaster may still be adhering. Bits of timber among the brick rubble may be the seat of a future outbreak of dry rot. Refractory bricks from chimneys and furnaces may swell when they become wet.
- Colliery spoil is also used for hardcore and the unburnt variety is to be preferred, as it usually has a lower sulphate content than burnt spoil or red shale. There have been many failures of ground floor slabs where burnt spoil was used for hardcore on wet sites.
- Chalk is widely used in some areas, but during prolonged periods of frost during construction there have been heave problems.
- Concrete rubble, as with brick rubble, if clean and properly graded is an acceptable hardcore material. Care is necessary to detect any mixture of other materials such as gypsum plaster, which in wet conditions could give rise to sulphate attack of the floor slab.
- Fly ash (pulverized fuel ash) arises from the precipitators of coal fired power stations, and is suitable for use as hardcore. Caution should be exercised when it is mixed with furnace bottom ash as the total sulphate content may be increased thereby.
- Gravel and crushed hard rock, suitably graded are excellent materials for use as hardcore.
- Quarry waste, if evenly graded, is a clean hard material which makes satisfactory hardcore.

There are many factors that may prompt a builder to use a particular hardcore on a particular site. Two important qualities are availability and cost. This yardstick should be applied against a material that is granular, will consolidate and drain readily. It should be chemically inert and not affected by water. Few of the materials available at reasonable cost meet these requirements completely.

Understanding of the qualities and weakness of each material is therefore a most important consideration in the decision-making process.

SUSPENDED CONCRETE GROUND FLOORS

To meet the requirements a solid suspended ground floor should be made of concrete composed of 300 kg cement for each cubic metre of concrete, at least 100 mm thick, or as spans and loads require, with 40 mm protection of reinforcement and a damp proof membrane (see Fig. C4.6 which includes basic radon protection).

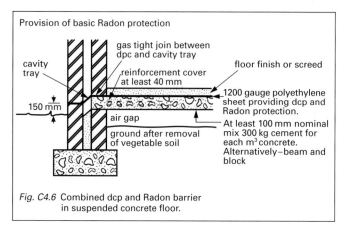

Fig. C4.6 Combined dcp and Radon barrier in suspended concrete floor.

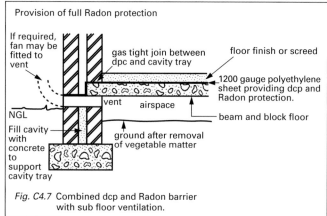

Fig. C4.7 Combined dcp and Radon barrier with sub floor ventilation.

Suitable damp proofing and full radon protection can be achieved by ventilating the subfloor space as in Fig. C4.7.

Other requirements are:

- If ground below floor has been excavated below level of adjoining ground a damp-proof membrane is to be incorporated; and
- Where there is a risk of an accumulation of gas there is to be a clear air space at least 150 mm from ground to underside of floor.
- If insulation is provided, 150 mm clear airspace is to be maintained.
- Ventilation openings are to be inserted in two opposing external walls so that air can move freely through the space. There are to be openings equal to 1500 mm² for each metre length of wall.

SUSPENDED TIMBER GROUND FLOORS

Suspended timber may only be used in proximity to the ground if there is a ground covering of concrete, composed of 50 kg cement to

0.13 m³ fine aggregate (maximum) and 0.18 m³ coarse aggregate (maximum). Alternatively BS 5328 mix ST 1 may be used. A ventilated space between the concrete and the timber is to be provided, and the timber protected by damp-proof courses.

The following is the BS ST 1 mix recommendation:

Nominal maximum size of aggregate	40 mm	40 mm
(a) slump	75 mm	125 mm
cement	180 kg	200 kg
total aggregate	2010 kg	1950 kg
Nominal maximum size of aggregate	20 mm	20 mm
(b) slump	75 mm	125 mm
cement	210 kg	230 kg
total aggregate	1940 kg	1880 kg

Concrete is to be 100 mm thick.

Alternatively the ground covering may be of concrete to the above mixes, or inert fine aggregate, not less than 50 mm thick placed on 300 μm (1200 gauge) polyethylene (polythene) having sealed joints and placed on a layer of sand or other material which will not damage the polythene. 250 μm (1000 gauge) polyethylene may also be used where there is an Agrément certificate or the material is up to PIFA standard.

Ventilated openings are necessary in outer walls and these should in total amount to no less than 1500 mm² for each metre run of outside wall (see Figs C4.8 and C4.9 which include suggested basic radon protection).

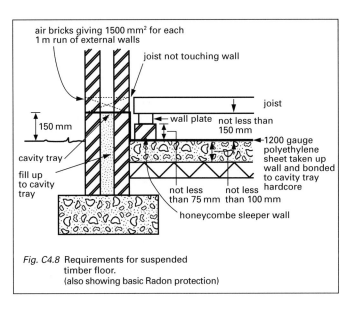

Fig. C4.8 Requirements for suspended timber floor.
(also showing basic Radon protection)

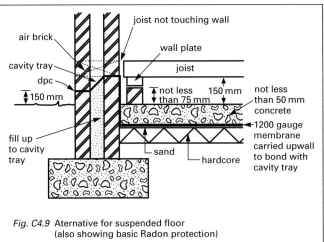

Fig. C4.9 Aternative for suspended floor
(also showing basic Radon protection)

Cross ventilation is vital and when a solid floor adjoins a suspended floor it is necessary to install pipes of at least 100 mm diameter through the solid floor in order to achieve through air movement (see Fig. C4.10). Provision must be made for protection against radon.

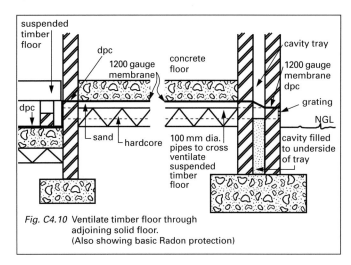

Fig. C4.10 Ventilate timber floor through adjoining solid floor.
(Also showing basic Radon protection)

More examples of radon protection are given in Figs C4.11, C4.12 and C4.13.

In Fig C4.11, a damp proof membrane is installed below the in situ concrete and must be sealed to the cavity tray with a gas-tight joint. It is vitally important that where an essential service penetrates the radon proof membrane that an air-tight seal is constructed around the service, and an example is given in Fig. C4.12.

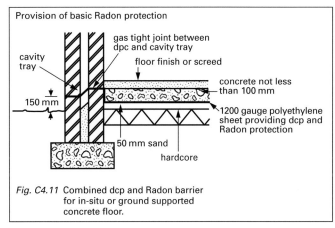

Fig. C4.11 Combined dcp and Radon barrier for in-situ or ground supported concrete floor.

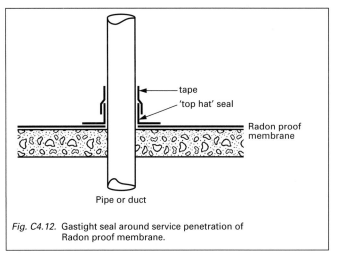

Fig. C4.12. Gastight seal around service penetration of Radon proof membrane.

When full radon protection is necessary, with passive or fan-assisted ventilation it may be necessary to construct a sump of loose laid bricks on edge, with ventilation ducts to external walls or up through the dwelling, as in Fig. C4.13.

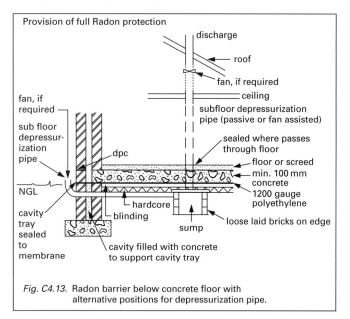

Fig. C4.13. Radon barrier below concrete floor with alternative positions for depressurization pipe.

Floor levels should be above the level of adjoining land, and ground covering should slope to an outlet in order that water will not lay on the surface covering. Alternatively a trench, wall or tanking should be employed to prevent egress of water (see Figs C4.14 and C4.15).

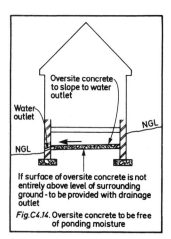

If surface of oversite concrete is not entirely above level of surrounding ground - to be provided with drainage outlet

Fig.C4.14. Oversite concrete to be free of ponding moisture

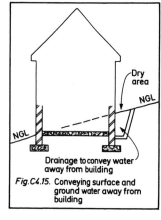

Drainage to convey water away from building

Fig.C4.15. Conveying surface and ground water away from building

Alternative approach

The performance required can also be achieved by following clause 11 of CP 102: 1973, Protection of buildings against water from the ground.

PROTECTION OF WALLS AGAINST MOISTURE

For the purpose of this requirement a wall embraces
    piers
    columns
    parapets
    chimneys (not free-standing).

Doors, windows and similar openings are not included in the definition.

Walls are required to resist the passage of moisture from the ground into the structure, or into the building. People working in buildings where wet processes are carried on are not likely to benefit from this protection, but regard must be taken of any possibility that the structure itself may be damaged by moisture from the ground.

Additionally the requirement does not apply to a building which is used only for storing goods, if people working in the building are employed only in storing, caring for or moving the goods.

Care must be taken when placing timber adjoining walls at ground level (Fig. C4.16).

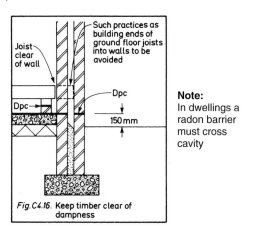

Fig.C4.16. Keep timber clear of dampness

Choice of walling materials below ground is also important. Very wet conditions or when there is the possibility of sulphate action or damage by frost all affect the choice of materials for use below dpc level.

On well-drained sites bricks that are available for other external parts of the building will probably give satisfactory service. When, however, there is a possibility of brickwork being very wet in winter, particular care should be taken in the choice of bricks or blocks. If sand lime bricks are used they should conform to BS 187: 1978, 'Specification for calcium silicate bricks'.

Any external or internal wall will be protected against moisture from the ground if a damp-proof course is inserted in a correct position. The requirement not only requires the insertion of a dpc but also prevents the use of materials which would allow the passage of moisture without damage to themselves.

Types of damp-proof course

There are three types of dpc – rigid, metal and flexible – having differing qualities and suitable for differing locations.

1. *Rigid.* Generally two courses of blue or engineering bricks, or natural slates laid break joint and bedded in mortar having all joints filled. These materials are suitable for heavy loads and for use in retaining walls. Not a suitable protection against moisture passing downwards through a wall.
2. *Metal.* The malleable materials used for this purpose are usually copper, lead or zinc. They are very suitable where the formation of trays or shaping is necessary.
3. *Flexible.* There are many dpcs which are flexible; the older types having a hessian base. They may also have a fibre base and be laminated with thin lead.

The newer types are polymer based and polythene and are covered by Agrément certificates. They are thinner, very resilient, but laps should be sealed with adhesive. They are easily punctured.

Suitable damp-proof courses are described in Table C4.1. When placed in an external wall a dpc must be not less than 150 mm above the finished level of the adjoining ground. (See Figs C4.17 and C4.18.)

An external cavity wall to a dwelling requires a cavity tray joined to the damp proof membrane to prevent gases from the ground entering the building.

Table **C4.1** **Suitable damp-proof course materials for situations shown**

Situation	Hessian bitumen	Asbestos bitumen	Lead bitumen	Aluminium bitumen	Pitch polymer	2 courses slate	2 courses Engineering bricks	Lead sheet	Copper sheet	Aluminium sheet
Base of cavity wall	Yes	Yes	Yes	Yes	Yes	Yes	Yes	Yes	Yes	Yes
Base of infill panel (apron)		Yes	Yes	Yes	Yes			Yes	Yes	Yes
Over lintel or arch			Yes	Yes	Yes			Yes	Yes	Yes
At jambs of opening (vertical dpc)			Yes	Yes	Yes				Yes	Yes
Base of parapet		Yes	Yes	Yes	Yes	Yes		Yes	Yes	Yes
Under coping	Yes	Yes	Yes	Yes	Yes			Yes	Yes	
Between panel and frame (vertical dpc)	Yes	Yes	Yes	Yes						
Between panel and frame Panel in front of frame (vertical dpc)	Yes	Yes	Yes	Yes						

Note: These recommendations do not necessarily apply to calculated loadbearing brickwork

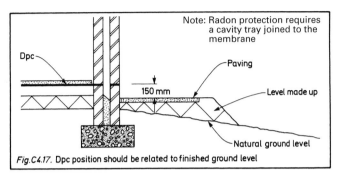

Note: Radon protection requires a cavity tray joined to the membrane

Dpc — Paving — 150 mm — Level made up — Natural ground level

Fig.C4.17. Dpc position should be related to finished ground level

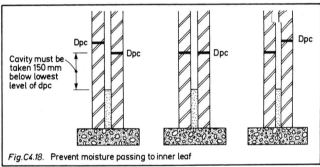

Dpc — Dpc — Dpc — Dpc

Cavity must be taken 150 mm below lowest level of dpc

Fig.C4.18. Prevent moisture passing to inner leaf

Damp-proof courses are the subject of BRE Digest 77 which lists materials in their groups.

Group A: Suitable for most situations and they alone should be used over openings and bridging centres.
Group B: Suitable for very thick walls and where there is high water pressure.
Group C: Materials that can withstand heavy loads but are more likely to be fractured by building movement than materials in Groups A and B.

Alternative approach

Provision is made for alternative methods of preventing dampness rising in walls and relate to the recommendations in clauses 4 and 5 of BS 8215: 1991, 'Code of practice for the design and installation of dpcs in masonry construction'.

Also BS 8102: 1990, 'Code of practice for protection of structures against water from the ground' includes recommendations for walls which must cope with ground water pressure.

WEATHER RESISTANCE OF WALLS

The penetration of rainwater into a building is a major cause of deterioration and the remedies may be time consuming and costly.

Methods which may be employed to meet requirements of the regulations in general terms vary from site to site. For instance a building to be erected on a sheltered site might utilize a form of construction which would be wholly unsuitable for an exposed position.

When making a decision in this connection one must be mindful that a spinney, or a range of buildings may at some time in the future be removed, resulting in a more severe exposure to the building.

A solid wall, even when rendered is suspect as moisture always penetrates to some extent; the greater the length of storm or force of wind which accompanies the rain, the greater the penetration. Therefore very great care is required in the specification for and the construction of a weather-resisting external wall. British Standards are helpful and BS 5262: 1976, 'Code of practice for external rendered finishes', deals extensively with techniques of application and repair of existing work.

The BRE booklet 'Protection from rain', contains many valuable hints and sketches to illustrate problems and recommended solutions.

New methods of construction bring new problems and very great care is necessary in the design and assembly of large concrete panel industrialized buildings. In particular, water may penetrate to internal surfaces or into the insulation in many different ways. Sealants and large block panels have long been a problem. There is so much that can go wrong with sealants, the wrong type used or applied under conditions which will cause cracking. There are also problems in the application of sealants. It must not be done too thickly; too thinly; too quickly or too slowly – all will produce adverse effects and demonstrate the need for sealant manufacturers to be consulted at the design stage. (See Fig. C4.19.)

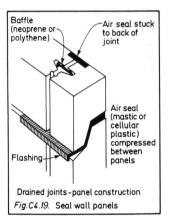

Baffle (neoprene or polythene) — Air seal stuck to back of joint — Air seal (mastic or cellular plastic) compressed between panels — Flashing

Drained joints – panel construction
Fig.C4.19. Seal wall panels

BS 6213: 1983, 'Selection of constructional sealants', has advice regarding the choice of a suitable sealant. It says the following should be considered:

- nature and cause of movement at the joints
- determine amount of movement
- select sealant with suitable dynamic properties with adequate movement accommodation
- seal shape to achieve satisfactory performance of the sealed joints
- remember to consider how maintenance and replacement may be affected.

Typical sealants may be acrylic, butyl, bituminous and rubber bitumins, epoxide, clear-resinous, polysulphide, polyurethane and silicone.

Life of sealants may be up to 10 or 20 years depending upon the type used and location. Application may be by gun, knife, trowel, poured or pretruded strips.

Parapets should contain a dpc at or near roof level which is lapped with its roof covering and a dpc below the coping (Fig. C4.20).

Chimney stacks are particularly liable to conduct rain into a building. Numerous ways are available to prevent this downward penetration of water through chimney construction. A damp-proof course of flexible or semi-rigid material may be provided and should extend over the whole area of the brickwork to form a moisture-resisting barrier in continuity with the apron flashing. An alternative method suggested in BSCP 131: 1974, 'Chimneys and flues for domestic appliances burning solid fuel', is that the dpc should be just above the underside of the roof timbers at the back of the chimney and laid in the form of a tray.

The position recommended by the BRE in 'Protection from rain', is shown in Fig. C4.21.

Some useful figures of unrestrained movement of framing and infilling materials are given in Table C4.2.

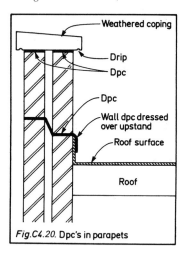

Fig.C4.20. Dpc's in parapets

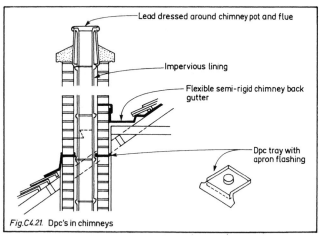

Fig.C4.21. Dpc's in chimneys

Table C4.2

Material	Thermal movement (for 50°C rise)	Reversible moisture movement (dry to saturated)
Steel	0 54–0 62	None
Aluminium	1 27	None
Copper	0 86	None
Glass	0 56	None
Bricks, fired clay	0 25–0 39	0 04–0 18
Bricks, calcium silicate	0 71	0 18–0 89
Dense concrete	0 49–0 71	0 36–1 07
Asbestos cement	0 42	0 36–1 43
Wood, along grain	–	0 02
Wood, across grain (radial)	–	30–50
Wood across grain (tangential)	–	50–150
Plywood	–	2–5
Chipboard	–	5–20
Rigid pvc	2 7	–
Glass reinforced polyester	0 8	–

SOLID EXTERNAL WALLS

When thinking of an external wall which must resist driving rain and snow, the first choice is almost automatically for some form of cavity construction.

However, if a solid wall is capable of absorbing moisture without it reaching the inside it may be permissible to use solid construction.

Thickness of the wall will be dependent upon choice of material, and its capacity to release the moisture following rain, by evaporation to the external air.

There is a procedure for evaluating the degree of severity of a wall to wind-driven rain set out in BSI Draft for Development, DD 93: 1984 called 'Method for assessing exposure to wind-driven rain'.

More advice is also available in BS 5628: Part 3: 1985, which is a code of practice for the use of masonry dealing with materials, components, design and workmanship. Reference may also be made to 'Thermal insulation – avoiding the risks' from BRE.

Technical solutions recommended are:

- when the wall is subject to very severe exposure it should be protected by an external cladding;
- when the exposure is severe the following are recommended:

  (a) brickwork not less than 328 mm thick;
  (b) dense aggregate concrete blockwork not less than 250 mm thick;

- lightweight aerated autoclaved concrete blockwork not less than 215 mm thick, and
- rendering the exposed face of the wall in two coats having a total thickness of not less than 20 mm and a scraped or textured finish. (See Fig. C4.22.)

Damp-proof courses are also required where there is a lean-to roof attached to a wall which in consequence becomes an internal wall below the junction (Fig. C4.23).

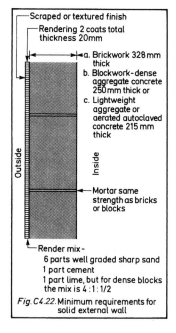

Fig. C4.22. Minimum requirements for solid external wall

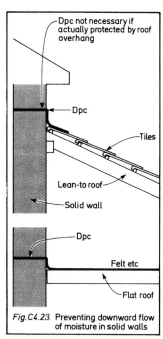

Fig. C4.23. Preventing downward flow of moisture in solid walls

Tops of unprotected walls must be protected by copings (Fig. C4.24).

DPCs are also necessary under openings where a sill is not a complete barrier, and also where an internal wall is carried up as an external wall.

A solid external wall may be insulated internally or externally.

When placed internally a cavity is to be allowed to break the path of moisture, and when placed on the outside of a wall it should have some protection from rain or snow (Fig. C4.25).

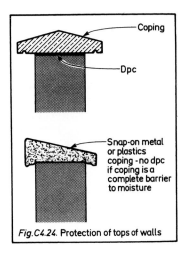

Fig.C4.24. Protection of tops of walls

with stone or cast stone or rubble-faced stone walls using facing stone or cast stone.

BS 6477: 1992, 'Silicone-based water repellants for masonry', describes three classes of repellants, defined according to the masonry surfaces for which they are suitable and contains advice on selection and use.

BS 5262: 1991, 'External rendered finishes', contains recommendations for a wide range of mixes according to the type of brick or block used, and the degree of exposure the wall must endure.

EXTERNAL CAVITY WALLS

Cavity walls to meet the requirements should have an outer leaf constructed of bricks, blocks, stone or cast stone, having a cavity at least 50 mm wide.

The internal skin may be constructed of masonry or a frame with lining.

A cavity may be bridged

- by wall tiles
- by damp-proof trays
- by insulating material
- when protected by a roof.

Any other opening or bridging must be protected by damp-proof materials to stop moisture penetrating to the inner leaf (see Figs C4.27 to C4.31).

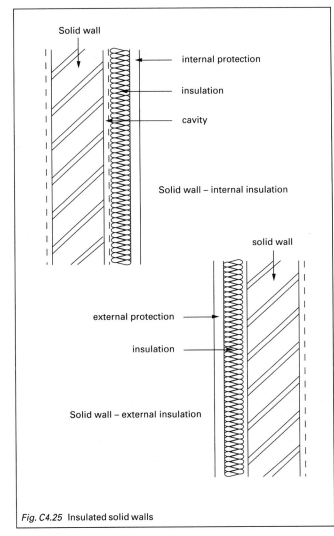

Fig. C4.25 Insulated solid walls

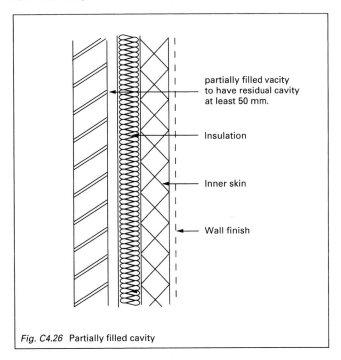

Fig. C4.26 Partially filled cavity

An alternative approach is available following recommendations in BS 5628: 1985, 'Code of practice – masonry, materials, design and workmanship', and BS 5390: 1976, 'Code of practice for stone masonry', section 3.

Methods described in BS documents will also meet requirements.

BSCP 121: Part 1: 1973, 'Brick and block masonry'. This code makes recommendations regarding type and quality of materials, mixing mortars, durability of jointing and exclusion of rain.

BS 5390: 1976, 'Code of practice for stone masonry', deals with the design, construction and selection of materials for walls ashlared

Alternative approach

Compliance with the relevant recommendations in BS 5628, 'Code of practice for masonry', Part 3: 1985; materials, components, design and workmanship relating to rain penetration of cavity walls is accepted.

CAVITY INSULATION

Subject to the following conditions, an insulating material may be inserted between the inner and outer leaves of a cavity wall.

- Rigid thermal insulating material to be subject of an Agrément certificate, or a European technical approval. To be installed in accordance with the requirements of the appropriate document.

- Urea–formaldehyde foam injected into cavity after wall built is to be in accordance with BS 5617: 1985, 'Specification of urea formaldehyde foam system suitable for thermal insulation of cavity walls'. The material should also be installed following BS 5618: 1985, 'Code of practice for thermal insulation of cavity walls'. Before foam filling is inserted into a wall it must be carefully examined to ascertain suitability. Anyone undertaking the survey must hold or operate under a current BSI Certificate of Registration of Assessed Capability or a similar authority issued by a competent body.
- Other types of insulating materials placed in the cavity after the wall has been built are to be in accordance with BS 6232, 'Thermal insulation of cavity walls by filling with blown man-made mineral fibre', Part 1: 1982, 'Specification for the performance of installation systems' and BS 5232 : Part 2, 'Code of practice for installation of blown man-made fibre in existing walls'.

When it is proposed to partially fill a cavity there is to be a residual cavity of at least 50 mm remaining (Fig. C4.26).

Prior to installation work being carried out the wall is to have been assessed using BS 8208, 'Guide to assessment of suitability of external cavity walls for filling with thermal insulants', Part 1: 1985, 'Existing traditional cavity construction'.

The person doing the work of installation should hold or operate under a current BSI Certificate of Registration of assessed capability or a similar document issued by a competent body.

Insulating material, the subject of an Agrément certificate or a European technical approval may be used and the work carried out as specified in the document by an approved installer, or operative directly employed by the holder of the document.

CLADDING FOR ROOFS AND WALLS

In addition to its decorative features cladding on walls can be used to protect a building from penetrating dampness arising from snow and rain. This is achieved by keeping water on the face of the cladding or preventing it from penetrating beyond the back of the cladding. It is important that the cladding itself should not be damaged by moisture.

Not all materials suitable for cladding a wall will be successful if applied to a roof.

Materials which deteriorate quickly without careful attention should only be used as part of weather-resisting construction when stringent conditions can be met. (See Table C4.3 and Figs C4.32 to C4.36.)

It is necessary for some sheet materials to be lapped at the edges. Provision should be made for structural and thermal movement when using jointless materials and sealed joints.

It is a requirement that every sheet, tile and portion of cladding is to be securely fixed.

Table **C4.3**   **Suitable cladding materials**

Quality	Nature
Impervious	Metal
	Plastic
	Glass
	Bituminous products
Weather resisting	Natural stone
	Natural slate
	Concrete tiles
	Concrete slates
	Clay tiles
	Wood
Moisture resisting	Bituminous products
	Plastic products

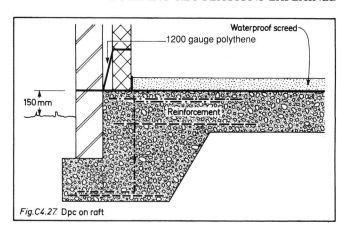

Fig.C4.27. Dpc on raft

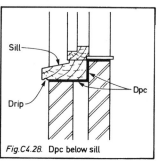

Fig.C4.28.   Dpc below sill

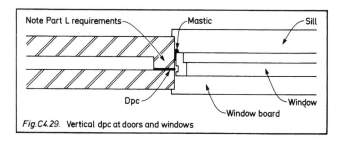

Fig.C4.29.   Vertical dpc at doors and windows

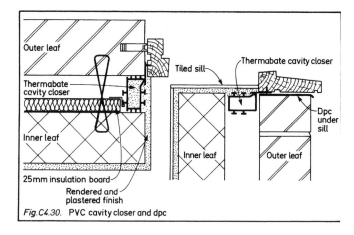

Fig.C4.30.   PVC cavity closer and dpc

Sarking felt underlay is an essential on pitched roofs. It must be carefully fitted and supported so that when necessary it can lead free water out of the roof space and clear of the inner leaf of the external walls. It is also essential that sarking felt is properly lapped around vent pipes or lapped and dressed over barge boards, as a protection against driving rain, and blown snow.

When dry joints are used water should not pass through them or the cladding should be so designed that water does not reach the back of the cladding or drain towards the face. The amount of exposure to wind and rain has an important bearing upon the design of cladding utilizing joints.

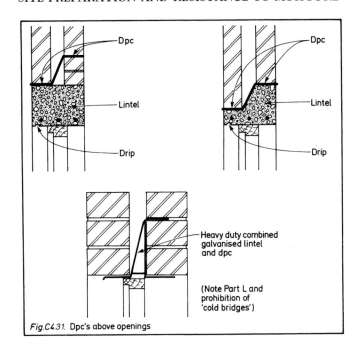

Fig.C4.31. Dpc's above openings

Heavy duty combined galvanised lintel and dpc

(Note Part L and prohibition of 'cold bridges')

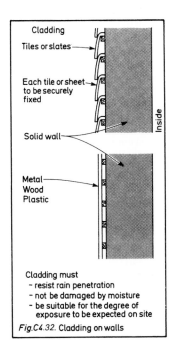

Cladding
Tiles or slates
Each tile or sheet to be securely fixed
Solid wall
Metal Wood Plastic
Inside

Cladding must
 - resist rain penetration
 - not be damaged by moisture
 - be suitable for the degree of exposure to be expected on site

Fig.C4.32. Cladding on walls

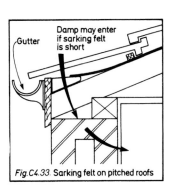

Gutter
Damp may enter if sarking felt is short

Fig.C4.33. Sarking felt on pitched roofs

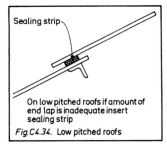

Sealing strip

On low pitched roofs if amount of end lap is inadequate insert sealing strip

Fig.C4.34. Low pitched roofs

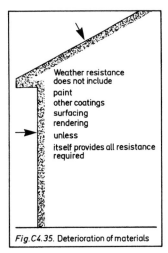

Weather resistance does not include
 paint
 other coatings
 surfacing
 rendering
 unless
 itself provides all resistance required

Fig.C4.35. Deterioration of materials

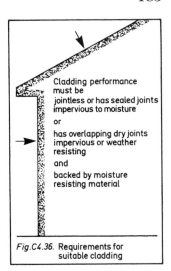

Cladding performance must be
 jointless or has sealed joints impervious to moisture
 or
 has overlapping dry joints impervious or weather resisting
 and
 backed by moisture resisting material

Fig.C4.36. Requirements for suitable cladding

Insulation can be incorporated in the construction if it is unaffected by moisture. Materials which are affected by moisture are to be protected.

Guidance is given regarding problems arising from cold bridges and condensation in Part F, Ventilation (Chapter 15), and in 'Thermal insulation – avoiding risks', published by BRE.

Cladding for roofs and walls may be designed and applied in accordance with the following British Standards.

BS 5534: 1990, 'Slating and tiling'
BSCP 143: 'Sheet roof and wall coverings'
    Part 1: 1958, 'Corrugated aluminium'
    Part 5: 1964, 'Zinc'
    Part 10: 1973, 'Galvanized corrugated steel'
    Part 12: 1970, 'Copper'
    Part 15: 1973, 'Aluminium'.
BS 6915: 1988, 'Specification for design and construction of fully supported lead sheet roof and wall covering'
BS 8200: 1985, 'Code of practice for the design of non-loadbearing external vertical enclosures of buildings'
BS 8298: 1989, 'Code of practice for design and installation of natural stone cladding and lining'
BSCP 297: 1972, 'Precast concrete cladding non-loadbearing'.

RELEVANT READING

British Standards and Codes of Practice

BS 187: 1978. Calcium silicate bricks
BS 743: 1970. Materials for damp-proof courses
BS 747: 1994. Specification for roofing felts
BS 882: 1992. Aggregate from natural sources
BS 1178: 1982. Milled lead sheet for buildings
BS 1521: 1972 (1994). Waterproof building papers
BS 3837: Part 1: 1985 (1996). Boards from expanded polyethylene beads
BS 3837: Part 2: 1990 (1996). Extruded polyethylene boards
BS 3921: 1985 (1995). Clay bricks and blocks
BS 4072: 1987. Wood preservation – water-borne copper/chrome/arsenic compositions
BS 4254: 1983 (1991). Two-part polysulphide sealants
BS 4255: Part 1: 1986. Specification for non-cellular gaskets
BS 5215: 1986. One-part gun grade polysulphide-based sealants
BS 5262: 1991. External rendered finishes
BS 5328: Part 1: 1990. Guide to specifying concrete
BS 5328: Part 2: 1990. Method for specifying concrete mixes
BS 5534: Part 2: 1986. Design charts for fixed roof slating and tiling
BS 5617: 1985. Specification for urea formaldehyde

BS 5618: 1985 (1996). Thermal insulation of cavity walls – urea formaldehyde

BS 5642: Part 1: 1978. Cills of precast concrete, cast stone, clayware, slate and natural stone

BS 6213: 1983 (1992). Selection of constructional sealants

BS 6232: Part 1: 1982. Performance of man-made blown fibre insulation systems
    Amendment 1: AMD 5428

BS 6262: 1982. Glazing for buildings

BS 6262: Part 4: 1994. Glazing – human impact

BS 6375: Part 1: 1983. Performance of windows – classification for weather tightness

BS 6398: 1983. Bitumen dpcs for masonry

BS 6492: 1984. Water repellants for masonry surfaces

BS 6576: 1985. Chemical dpcs

BS 6577: 1985. Mastic asphalt for roofing

BS 6676: Part 1: 1986. Specification for cavity batts

BS 6915: 1988. Specification for design and construction of fully supported lead sheet roof and wall coverings

CP 102: 1973. Protection of buildings against water from the ground
    Amendment 1: AMD 1151
              2: AMD 2196
              3: AMD 2470

CP 143: Sheet roof and wall coverings
    Part 1: 1968. Corrugated aluminium
    Part 5: 1964. Zinc
    Part 10: 1973. Galvanized corrugated steel
    Part 12: 1988. Copper
    Amendment 1: AMD 863
              2: AMD 5193
    Part 16: 1974. Semi-rigid asbestos bitumen

CP 144: Roof coverings
    Part 3: 1970. Built-up bitumen felt
    Part 4: 1970. Mastic asphalt

CP 153: Part 2: 1970. Durability of windows and rooflights

CP 297: 1972. Precast concrete cladding (non-loadbearing)

BS 5247: Part 14: 1975. Code of practice for sheet wall and roof covering
    Amendment 1: AMD 2821
              2: AMD 3502

BS 5250: 1989. Control of condensation

BS 5262: 1991. External rendered finishes

BS 5390: 1976 (1984). Code of practice for stone masonry
    Amendment 1: AMD 4272

BS 5534: Part 1: 1997. Design for slating and tiling

BS 5618: 1985. Code of practice for thermal insulation
    Amendment 1: AMD 6262
              2: AMD 7114

BS 5628: Part 3: 1985. Code of practice for masonry

BS 5930: 1981. Code of practice for ground engineering

BS 6093: 1993. Code of practice for design of joints and jointing in buildings

BS 6232: Part 2: 1982. Code of practice for blown fibre insulation

BS 6676: Part 2: 1986. Code for installation of batts

BS 8102: 1990. Code of practice for protection of structure against water from the ground.

BS 8200: 1985. Code of practice for non-loadbearing external vertical enclosures

BS 8208: Part 1: 1985. Thermal insulants for existing cavity construction

BS 8215: 1991. Code of practice for design and installation of dpcs in masonry

BS 8298: 1994. Code of practice for design of natural stone cladding and lining

BRE digests

54	Damp proofing solid floors
110	Condensation

144	Asphalt and built up felt roofings
145	Heat losses through ground floors
152	Repair and renovation of flood damaged buildings
163	Drying out buildings
180	Condensation in roofs
196	External rendered finishes
217	Wall cladding defects and their diagnosis
223	Wall cladding – designing to minimize defects
236	Cavity insulation
245	Rising damp in walls
268	Common defects in low rise traditional housing
270	Condensation in insulated domestic roofs
276	Hardcore
297	Surface condensation and mould growth
298	Influence of trees on house foundations in clay soils
312	Flat roof design – technical options
318	Site investigation for low rise buildings – desk studies
322	Site investigation for low rise buildings
324	Flat roof design: insulation
348	Site investigation for low rise buildings – a walk over survey
350	Part 1: General climate of UK
350	Part 2: Influence of microclimate
350	Part 3: Improving microclimate through design
364	Design of timbertloors to prevent decay
369	Interstitial condensation and fabric degeneration
380	Damp-proof courses
381	Site investigations for low rise buildings – trial pits
383	Site investigations for low rise buildings – soil description
410	Cementation renders for external walls
419	Flat roof design: bituminous waterproof membranes

BRE defects action sheets

9	Pitched roofs – sarking felt underlay – drainage from roof
10	Pitched roofs – sarking felt underlay – watertightness
12	Cavity trays in external cavity walls: preventing water penetration
16	Internal walls and ceilings: remedying mould growth
17	External masonry walls: insulated with mineral fibre cavity width batts – resisting rain penetration
22	Ground floors: replacing suspended timber with solid concrete – rising damp
33	Flat roofs: built up bitumen felt – remedying rain penetration
34	Flat roofs: built up bitumen felt – remedying rain penetration of abutments and upstands (design)
35	Substructure: dpcs and dpms – specification
36	Substructure: dpcs and dpms – installation
37	External walls: rendering – resisting rain penetration (design)
38	External walls: rendering – application (site)
59	Felted cold deck roofs
63	Flat and low pitched roofs – laying flexible membrane in bad weather
67	Inward opening external doors resistance to rain penetration
68	External walls – joints – detailing for sealants
69	External walls – application of sealants
70	External masonry walls: eroding mortars
71	External masonry walls: repointing
72	External masonry walls: repointing
73	Suspended timber ground floors – inadequate ventilation
75	External walls – designing for differential movement
76	External walls – allowing for differential movement
79	External masonry walls – partial cavity fill insulation resisting rain penetration
85	Brick walls: injected dpc
86	Brick walls: replastering
93	Chimney stacks: taking out of service (Design)
94	Masonry chimneys: DPCs and flashings – location (Design)
95	Masonry chimneys: DPCs and flashings – installation (Site)
97	Large concrete panel external walls: resealing butt joints

98　Windows: resisting rain penetration at perimeter joints (Design)
103　Wood floors: reducing risk of recurrent dry rot (Design)
106　Cavity parapets: avoiding rain penetration
107　Cavity parapets: installation of copings, dpcs, trays and flashings
114　Slated and tiled pitched roofs: flashings and cavity trays

BRE Good Building Guides

GBG 3　Damp-proofing basement rooms
GBG 18　Choosing external renderings
GBG 25　Buildings and radon
GBG 26　Minimizing noise from domestic fan systems and fan-assisted radon mitigation systems
GBG 33　Building damp-free cavity walls

BRE Information papers

2/87　Fire and explosion hazards association with redevelopment of contaminated land
2/88　Rain penetration of cavity walls: report of survey
5/84　Use of glass reinforced cement in cladding panels
7/83　Window to wall pointing
9/87　Joint primers and sealants: performance between porous claddings
25/82　Formaldehyde vapour from urea formaldehyde foam insulation
35/79　Moisture in timber based flat-roofs
8/86　Waterproof joints in large panel systems
14/86　Waterproof joints in large panel systems
15/85　Effect of rising water tables on settlement of open cast mining backfill
15/86　Waterproofing joints in large panel systems
7/95　Bituminous roofing membranes – performance in use

BRE leaflets

XL 9　Radon and buildings – minimizing noise from radon sumps
XL 10　Radon and buildings – 3. Protecting new extensions and conservatories

BRE Advisory Service publications

Protection from rain (1971).

BRE Reports (obtainable from TSO)

Driving Rain Index (R. E. Lacy) (1976)
Climate and Building in Britain (R. E. Lacy) (1977)
Moisture conditions in the walls of timber frame houses – effects of holes in vapour barriers (S. A. Covington and I. S. Mcintyre). BRE

Department of the Environment Interdepartmental Committee on the Redevelopment of Contaminated Land

ICRL 17/78. Notes on the redevelopment of landfill sites
ICRL 18/79. Notes on the redevelopment of gas works sites
ICRL 23/79. Notes on the redevelopment of sewage works and farms

ICRL 42/80. Notes on the redevelopment of scrapyards and similar sites
ICRL 59/83. Guidance on the assessment and redevelopment of contaminated land
ICRL 59/83. Guidance on the assessment and redevelopment of contaminated land. Second edition 1987
ICRL 61/84. Notes on the fire hazards of contaminated land Circular 17/89: Landfill sites – development control

All may be obtained from

Department of the Environment, Transport and the Regions, Publication Sales Centre, Unit 21, Goldthorpe Industrial Estate, Goldthorpe, Rotherham S63 9BL.

BRE books and reports

Measurement of Gas Emissions from Contaminated Land (D. Crowhurst)
Recommendations for the Procurement of Ground Investigation (J. F. Uff and C. R. I. Clayton)
Overcladding External Walls of Large Panel System Dwellings (H. W. Harrison, J. H. Hunt and J. Thomson)
BRE Report BR 100: 1987, Measurement of gas emissions from contaminated land
Radon sumps – guide to remedial measures (1992 BR 227)
Radon: guidance on protective measures for new dwellings (1999 BR 211)
Construction of new buildings on gas contaminated land (1991 BR 212)
BRE Desk Study, Site investigation for low rise building
Rain penetration through masonry walls: diagnosis and remedial measures. BRE Report 1988, No. BR 117
Measurement of Methane and other Gases from the Ground (D. Crowhurst and S. Manchester). Fire Station Ref. EP. 18
Performance of Building Materials in Contaminated Land (V. Paul). BRE Report Ref. BR 255: 1994
Polymeric anti-corrosion coatings for protection of materials in contaminated land. (1995 – BR 286)
Bibliography of case studies on contaminated land; investigation, remediation and redevelopment. (1995 – BR 291)
Radon in the workplace (1995 – BR 293)
Time-dependent modelling of soil gas movement: a literature review. (1995 – BR 298)
Review of the effect of fly ash and slag on alkali aggregate reaction in concrete. (1996 – BR 314)
HM Inspectorate of Pollution Waste Management: Paper 27 – *Control of Landfill Gas* (TSO)
Monitoring of Landfill Gas, published by Institute of Waste Management

CHAPTER 13

Toxic substances

(Approved document D)

There has been some evidence that fumes arising from urea–formaldehyde foam placed in the cavity walls of a dwelling could have an adverse effect on the occupants. This is a contentious issue, but even so this part of regulation requirements is intended to limit the extent that toxic fumes may penetrate into a building.

The approved document does not appear to preclude the insertion of urea–formaldehyde foam in walls where the inner leaf is not constructed of masonry. The only technical solution in the approved document relates to walls having an inner masonry leaf, but there are no limits of application set out in the third column of Schedule 1. It might be argued therefore that a sound vapour barrier behind plasterboard in a timber framed house would as far as practicable minimize the passage of fumes to occupied parts of the building.

CAVITY INSULATION

D1 requires that any insulating material inserted into a cavity wall must be accompanied by reasonable precautions to prevent toxic fumes from the material getting into any part of the building occupied by people.

An acceptable level of performance is given – fumes given off by formaldehyde foam must not get into occupied parts of a building in sufficient quantity to cause irritation and risk to the health of those people in the building.

UREA–FORMALDEHYDE FOAM

To fulfil the performance standard insulating materials which emit formaldehyde fumes may be used to insulate the cavity in a cavity wall. But there must be a continuous barrier to minimize as far as practicable the penetration of fumes into parts of a building occupied by people.

TECHNICAL SOLUTION

The sole technical solution offered by the approved document relates to a wall

- having an inner leaf built of bricks or blocks, and
- before foam is injected into the wall it has been carefully examined and found to be suitable.

This investigation must be carried out in accordance with BS 8208: Part 1: 1985, 'Guide to assessment of suitability of external walls for filling with thermal insulants'.

This code deals with existing traditional cavity construction. It gives guidance on factors to be considered when assessing the suitability of existing external cavity walls with masonry and/or concrete leaves for filling with thermal insulants.

The code applies when the height of the wall does not exceed 12 m, vertical members of structural frames do not bridge the cavity and there is no existing cavity insulation. Height of a wall is meas-

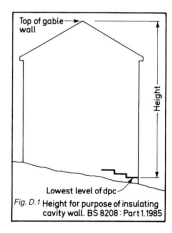

Fig. D.1 **Height for purpose of insulating cavity wall. BS 8208 : Part 1.1985**

ured in a way which is different from measurements specified in various parts of the regulations (see Fig. D1)

The code gives guidance on:

- *Form of construction and site conditions* and indicates that where there are cracks more than 1 mm wide, the cause should be established and specialist help obtained.
- *Age of the building.* Older buildings are likely to have more structural defects apparent, particularly if they have been empty for some time;
- *Condition of cavity.* There should be a continuous cavity width of not less than 40 mm. Mortar should not bridge cavities. Check for rat trap bond and record openings and trunking around vents;
- *Outer leaf.* The nature and condition of the outer leaf is to be carefully checked. Look for features which discharge water onto the wall, such as flush cills, faulty rainwater goods, wrongly installed overflow pipes and inadequate weatherings.
- *Inner leaf.* Dry lining may need to be sealed. The nature of the inner leaf should be checked, particularly where gaps may be associated with pipes, cables or built-in floor joists.
- *Services within cavity.* A check should be made to ascertain whether there are any pipes or cables within a cavity and whether flues cross a cavity.
- *Ventilation.* All essential vents should be trunked and made continuous through both leaves of the wall.

There are a number of illustrations and appendices.

Appendix A: Comparison of methods for assessing exposure to wind driven rain.

Appendix B: Estimation of cavity width and overall thickness.

The next requirement is that the installer must hold, or operate under, a BSI Registration Certificate enabling him to insert the foam.

MATERIALS

It is important that the material used is suitable, and it must comply with recommendations in BS 5617: 1985, 'Specification for urea–formaldehyde (UF) systems suitable for thermal insulation of cavity walls, masonry or concrete inner and outer leaves'.

This standard covers property requirements, the properties of the components and the production parameters of UF foam systems suitable for injection into masonry cavity walls to provide improved thermal insulation.

- Property requirements. Prescribed type tests and quality control tests.
- Processing requirements. Specifies foam system components and says that a foam system supplier must make a declaration that the system complies with eight parameters which are listed.
- Shelf-life limitation. Resins for use with UF foam systems are to remain usable for three months when stored at 20°C. The period commences with a quality control date to be stated on each container.

There area number of appendices:
Appendix A: Determination of effective density.
Appendix B: Determination of linear shrinkage.
Appendix C: Determination of water absorption.
Appendix D: Determination of wet density.
Appendix E: Determination of foam system gel time.
Appendix F: Determination of foam system stability.
Appendix G: Determination of foam system appearance.
Appendix H: Determination of solid resin yield.
Appendix J: Determination of free formaldehyde content of UF resin.
Appendix K: Determination of water tolerance of UF resin.
Appendix L: Determination of total acidity of foaming hardness.
Appendix M: Determination of resin viscosity by the break time method.

INSTALLATION

The next requirement hinges on the manner of installation, and this must be in accordance with BS 5618: 1985, 'Code of practice for thermal insulation of cavity walls (with masonry or concrete inner and outer leaves) by filling with urea–formaldehyde (UF) foam systems'.

This describes recommendations for the installation of urea–formaldehyde (UF) foam systems, which are dispensed on site to fill cavities of suitably constructed external walls to a height not exceeding 12 m. Walls are to have inner and outer leaves constructed of masonry or concrete. Random rubble walls are not covered by this code.

The code itself is brief, but is supported by extensive appendices:

- *Suitability of cavity walls for insulation.* It is essential that the installation contractor ascertains that the building is suitable or capable of being made suitable before an installation is carried out.

  Walls must be structurally sound, and to ensure that any gas which might evolve from the foam does not penetrate into the building, the inner leaf must be checked for continuity.

  Foam must not be installed between vapour barriers, because it would not dry out.
- *Filling process.* Materials, material preparation and quality checks are detailed. Post-installation activities include making good drill holes; all chimney vents and ducts should be checked and cleared if blocked or partially blocked. Any visible foam inside the property should be removed and the gap sealed.
- *Records.* The installation contractor is to keep records of all aspects of the installation.
- *Declaration.* The installation contractor should give a declaration that the work has been carried out in accordance with BS 5618 using foam complying with BS 5617.

Appendix A: Criteria for suitability of external cavity walls.
Appendix B: Natural stones used in the construction of masonry walls.

Appendix C: Calculation of exposure (tabular method).
Appendix D: Calculation of exposure (formula method).
Appendix E: Driving rain index.
Appendix F: Direction of prevailing wind.
Appendix G: Formaldehyde.
Appendix H: Flue checks.
Appendix K: Methods of determining cavity width.

The code of practice contains numerous tables and illustrations.

COMMENT

Approved document D sets out briefly via British Standards the standard of care required in construction; requiring reasonable precautions to prevent fumes from subsequently passing into any part of a building intended or to be used for occupation by people. When outer leaves of a cavity wall are more permeable than the inner, fumes have a better opportunity to go outside and dissipate.

Regulations only ask for a reasonable standard of care when constructing walls which are to have a foam infill. Indeed the same care ought to be employed in the construction of walls whether they are to be infilled or not. Similarly reasonable care must be taken when installing foam.

Installation is to be undertaken by people holding a current BSI Certificate of Registration of assessed capability, but what should a developer undertaking this work watch out for?

- A detailed pre-installation survey is necessary to assess the likelihood of fumes penetrating into the property.
- There should be a continuous barrier between the injected foam and the inside of the building. Defects found must be corrected.
- A check must be made that where suspended ceilings are installed, and the inner leaf above ceiling level has been carried up in brick or block construction.
- Gaps in the wall which can permit the passage of air across the cavity and into a building must be located and sealed with cement mortar.
- A search must be made for gaps under window sills, around untrunked vents and cavity vents.
- Air vents to occupied parts of the building must be sleeved.
- If any part of the cavity is open to an occupied part of the building it must be sealed.
- Any gaps in the inner leaf must be noted and sealed with cement mortar.
- Because masonry surfaces that are not plastered and are to have a decorative finish or dry lining are likely to be more permeable than a plastered wall it is preferable to inject foam during construction.
- Immediately following the installation an inspection should be made of the inside of the property. Any foam which has penetrated should be removed and the gap through which it passed, effectively sealed.
- The installer should leave a warning card with a responsible person on site.
- Arrangements must be made for the installer to respond immediately if he is notified of any fume emission.

TIMBER FRAME

The Building Research Establishment at its Scottish Laboratory has undertaken an investigation of houses which have received urea–formaldehyde foam cavity fill. None of the houses surveyed gave evidence of dampness arising from the treatment and formaldehyde fumes were not found to have caused problems.

Nevertheless BRE does not advise insulation that bridges the cavity in a timber-framed house, and the Scottish Building Standards Regulations prohibit foam cavity fill in timber-framed houses.

RELEVANT READING

British Standards and Codes of Practice

BS 5617: 1985 (1996). Specification for urea–formaldehyde foam systems

BS 5618: 1985 (1996). Code of practice for the thermal insulation of cavity walls
    Amendment 1: AMD 6262
                2: AMD 7114

BS 8208: Part 1: 1985. Assessment and suitability of external walls for filling with thermal insulants
    Amendment 1: AMD 4996

BRE information papers

7/84   Urea–formaldehyde foam cavity insulation: reducing formaldehyde vapour in dwellings
25/82   Formaldehyde vapour from urea–formaldehyde foam insulation
3/89   Subterranean fires in UK – the problem
23/89   CFCs and the building industry
8/93   Phase out CFCs – guidance air-conditioning systems

BRE digest

358   CFCs and buildings

BRE reports

Indoor air quality in homes
    Part 1: BR 299
    Part 2: BR 300

Department of the Environment

Householders guide to Radon (August 1988).
Circular 17/89: Landfill sites.

HM Inspectorate of Pollution

Management Paper 27– *Control of Landfill Gas*.

CHAPTER 14

Resistance to passage of sound

(Approved document E)

The building regulations have requirements aimed at reducing to acceptable levels the amount of noise which may be transmitted from one dwelling to its neighbour. Many dwellings have been found not to meet the performance standards set by earlier regulations; in fact the BRE in Digest 252 reported that half of the dwellings investigated during a survey failed to meet party wall requirements.

Put in its broadest context noise is a sound which is undesirable to the recipient. This simple description indicates quite adequately the cardinal fact that noise is personal to each individual.

It is possible to measure the energies and frequencies of simple and complex sounds, but it is impossible to foretell with any precision what a particular person's reaction to a specific noise may be.

Almost every activity is likely to produce noise of one kind or another. Control of noise at source is the best answer if it is possible, but failing this it may be controlled by absorption during transmission. Here we have very wide questions including building construction. Solutions are available, but require infinite care in application.

Materials containing lots of voids, which are generally very good thermal insulation materials, are also good absorbers of sound waves. Thus if they are used for lining wall and ceiling surfaces inside a room they reduce the level of sound within that room. But, on the other hand, these materials have a negligible effect upon the sound insulation properties of a dividing wall.

As an example, a 25 mm thick layer of mineral wool has an absorption of about 80% of sound waves. However, if the same material is used as a separating medium between two rooms the sound passing through is reduced only slightly (about three decibels).

It is rare to find good sound insulation and good thermal insulation together. A heavy structure is rather poor from the point of view of thermal insulation, yet excellent as insulation against airborne noises. Then again, if the cavity between two brick leaves is filled with thermal insulation material, the thermal insulation properties of the wall are improved enormously, while the acoustic insulation is only improved fractionally.

Sound consists of elastic waves which may be propagated through gases, liquids or solid materials. The maximum frequency range which can be detected by the human ear is limited.

The frequency of sound is the number of pulsations or waves that occur in a given period, and is expressed in Hertz. This used to be described as 'cycles' per second and Hertz (written Hz) is the metric equivalent. 1 Hz is equal to 1 cycle per second.

Sound pressure levels are measured on a scale of decibels (dB). It is a logarithmic scale representing the ability of the listener to recognize changes in pressure. The intervals on a decibel scale represent equal proportional changes and not equal absolute changes.

Thus when the intensity of sound waves is doubled the dB value is increased by approximately three. An increase of 10 dB means that the intensity of sound has increased 10-fold. When a wall has an insulation value of 30 dB, only about 0.1% of the sound intensity is transmitted through the wall.

It is possible to classify sounds according to their intensity or sound pressure; the human ear differs in its sensitivity to sounds of different frequencies. Sounds in the middle range, but possessing low intensities may appear as loud as sounds at low frequencies, but of high density. It is therefore necessary to give not only the number of dB of insulation of a wall, but also the range of frequencies, and were found in Part G of the early regulations. There are no similar tables included in the approved document to Part E of the 1991 Regulations. Frequently the expression dBA may be seen. This means that a correction factor has been included in particular circumstances.

Part E contains provisions devised to secure that only a limited amount of noise penetrates from one dwelling to its neighbour. Sound penetration of walls is mainly dependent upon the number and size of direct openings, and the fact that sound has the ability to outflank a wall or a floor.

When sound waves reach a porous surface such as carpets, curtains, plaster or rough cast stone work, some of the sound energy is changed into heat and dispersed. It is for this reason that an occupied dwelling is always less noisy than one which is bare.

The sound insulation standard of a given wall is only as good as its least effective part. This may be a door in the wall, penetration by pipes, inadequately filled mortar joints, or poor edge sealing.

Flanking sound is a problem which can easily be overlooked, and it is possible for noise travelling in this manner to act so as to almost neutralize the sound limitation value of a wall altogether. As an example, sound can travel into a cavity of a suspended ceiling or below floor boards, to produce a noise on the other side of a wall. This is achieved if there is inadequate party wall construction adjoining the cavities. Other examples where the sound insulation of walls and floors is put at risk is where there are common ducts or pipes penetrating the elements of structure. Noise will also travel through a window and enter another closely adjoining window in the same wall.

A very extensive source of annoyance in flats arises from impact noises on the floor above (Fig. E1). When something strikes the floor

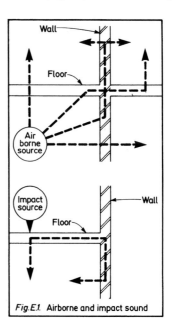

Fig.E1. Airborne and impact sound

some of the kinetic energy is converted into heat energy and some to sound. What fraction is converted into heat and what proportion into sound is entirely dependent upon the surface of the floor.

In general when there is impact with a hard non-resilient surface a much greater percentage of the kinetic energy is transformed into sound than when a blow falls on a soft surface, such as a carpet. Beyond that the amount of noise heard by the person on the other side of the surface is dependent upon the efficiency of noise transmission through the intervening solid or cavity.

It is unfortunate that when a wall or floor is an efficient insulator against airborne sounds, as for instance when it is solid and heavy, it may be poor as an insulator against impact noises.

The problems of voids have been mentioned, but except where they act as resonance chambers and are poorly sealed, voids are very good at preventing impact noises from passing through. Fibrous material is even better, as resonance cannot take place in it. One of the most troublesome noises in flats arises from people walking on the floor above, and while furnishings have an effect building design cannot be on the basis that the floors of flats will always have ankle deep carpets.

A solid concrete floor forming an ideal insulation against airborne sounds is a very poor insulator against impact noises. Thickness alone is quite useless in preventing the transmission of impact noise. If there is a solid connection between the source of the noise and the person hearing it, there will be little reduction in noise intensity.

Brick buildings are good insulators against impact noise, whereas concrete structures with steel skeletons are not nearly so efficient.

Impact noises can be muffled by adopting 'floating floors'. These are not attached rigidly to the building framework but allowed to rest on a layer of insulating material so that there is no direct connection between the floor and its supporting structure or the walls.

If a rigid contact does exist, then however carefully the floor has been constructed, failure is certain. Noise is so searching that failure will follow if the floating floor is in contact with the walls at its edges, or if nails have been driven through the fibrous bed of insulating material to the joist or slab below.

In blocks of flats an efficient type of flooring construction giving protection against both airborne and impact noises, is a heavy concrete ceiling, a layer of insulating material and another layer of heavy concrete forming the floor to the flat above. There must be no rigid connection with the walls or framing of the building. The floating slab must be carefully insulated against walls or frame.

When designing dwellings it is appropriate to keep in mind sound transmission problems which may arise (see Fig. E2) and help the building regulations by arranging rooms within each dwelling having the following points in mind.

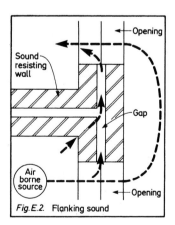

Fig. E.2. Flanking sound

- Noise is usually generated within living rooms, kitchens and circulation areas, from radio, television, etc., and some household appliances.
- If possible bedrooms should not be placed adjoining rooms in the neighbouring dwelling where noise is likely to be generated.

- Where living rooms adjoin party walls, better sound insulation will be achieved if one of the shorter dimensions utilizes the party wall.
- If dwellings can be arranged in stepped or staggered form, insulation from airborne sounds will be improved.
- Pipes likely to generate noise should not be fixed to a party wall.

Walls separating dwellings, or which separate a habitable room from another part of the same building not forming part of the dwelling, have to provide adequate resistance to the transmission of airborne sound.

There are some restrictions on external flanking walls in connection with the requirements for party floors. Where concrete floors are used they must extend through the inner leaf and be tied or bonded to every adjoining party wall and all other internal walls which give support.

As is evidenced when evaluating the sound insulation in old buildings, mass is the main factor for success with masonry walls. Laying bricks frog down greatly reduces the mass of a party wall, and in consequence greatly reduces the possibility of achieving the sound insulation requirements of building regulations.

Full mortar joints are also vital as direct air paths through or around a party wall will invite failure. It is also preferable to continue the same party wall construction unaltered up through the roof space, although not a requirement of the approved document.

Quite small gaps amounting to, in aggregate, as little as $100 \, mm^2$ in $10 \, m^2$ of a wall may seriously inhibit the sound insulation qualities of a wall. Flush-mounted fittings should not be mounted back to back on a party wall, but separated by at least 200 mm.

Plastered wall finishes usually give a better performance than dry lining.

Resilient thermal insulating materials fixed to a flanking wall and plastered, or wood wool slabs used as a permanent shuttering for concrete, may also have an adverse effect on the effectiveness of sound insulation.

The approved document E to the 1991 Regs includes some changes and extensions to previous recommendations.

There is an increase in the mass of some precast separating floors and concrete blocks separating walls, and where stairs form part of a separating element between dwellings, sound insulation is required in its construction.

Walls separating kitchens from a common circulation space are now subject to requirements and guidance is given regarding sound proofing to be installed when a material change of use takes place.

REGULATION REQUIREMENTS

Regulation 4 requires sound insulation work to be carried out in accordance with Part E of Schedule 1. When complying with any of the requirements care must be taken to ensure that it does not result in the failure to comply with any other requisite of the regulation.

Regulation 5 says that there is a material change of use when a building is used as a dwelling or contains a flat when previously it did not do so, and Regulation 6 extends the requirements in the case of a material change of use to include E1, E2, and E3. If a dwelling is formed adjoining or within a building of another type the requirements apply.

E1 requires that the following walls shall resist the transmission of airborne sound:

- a wall separating dwellings or separating a dwelling from another building;
- a wall separating a habitable room or a kitchen from another part of the building which is not used solely as part of the dwelling.

For the purpose of these requirements a kitchen is regarded as a habitable room (see Fig. E3). A wall resisting airborne sound is required between

- a dwelling and another dwelling or another building;

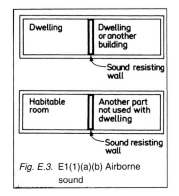

Fig. E.3. E1(1)(a)(b) Airborne sound

- a habitable room in a dwelling and

  (a) any part of the same building which is not part of the dwelling
  (b) a machinery or tank room
  (c) place for any other purpose except space used only for maintenance and repair;

- a habitable room or non-habitable room and a refuse chute in the same building.

There is an approved document setting down standards which fulfil the requirements of E1. These standards are given in constructional terms; nowhere are there requirements related to the actual dB reduction required in a particular situation.

Contents of the approved documents are as follows:

Section 1. Separating walls
            Refuse chutes
Section 2. Separating floors
Section 3. Similar construction
Section 4. Use of similar construction
Section 5. Conversions
Section 6. Test procedures

SECTION 1. SEPARATING WALLS

Refuse chutes

A wall which separates any habitable room or kitchen in a dwelling from any refuse chute should have an average mass of not less than $1320 \, kg/m^2$.

When a wall separates a refuse chute from a room which is not a habitable room in a dwelling, it is to have an average mass of not less than $220 \, kg/m^2$ (see Fig. E4).

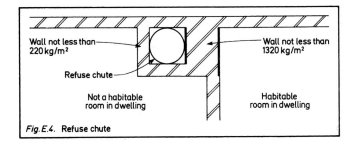

Fig. E.4. Refuse chute

Solid masonry walls

These are wall type 1 and rely upon mass for sound insulation qualities.

The approved document gives examples of four solid masonry walls which offer suitable insulation against noise transmission, emphasis being placed upon the need for

- filled mortar joints;

- sealing at junctions;
- precautions to prevent flanking transmission.

In the following examples the mass of a wall is shown as kg/m^2, and the density of materials used as kg/m^3.

Density of materials may be obtained from an appropriate Agrément certificate; a European technical approval or from the manufacturer. Building control officers may require confirmation of a manufacturer's figures and will look for a certificate from an approved independent body.

A. Brickwork plastered each side (Fig. E5):

- frogs upward;
- bond to include headers;
- weight including plaster not less than $375 \, kg/m^2$;
- 13 mm plaster on each face;
- coursing 75 mm;
- brick density $1610 \, kg/m^3$.

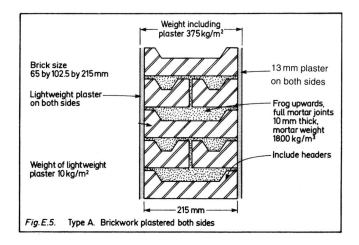

Fig. E.5.   Type A. Brickwork plastered both sides

B. Brickwork with plasterboard each side (Fig. E6):

- frogs upwards;
- bond to include headers;
- weight of masonry not less than $355 \, kg/m^2$;
- weight including plasterboard not less than $375 \, kg/m^2$;
- 12.5 mm plasterboard on each face;
- brickwork behind plasterboard should not be plastered as this may cause drumming;
- coursing 75 mm
- brick density $1610 \, kg/m^3$.

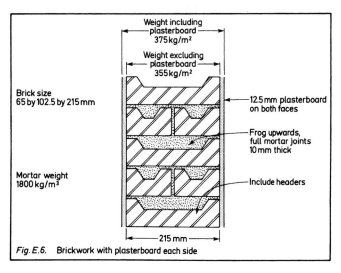

Fig. E.6.   Brickwork with plasterboard each side

C.   Concrete blockwork plastered each side (Fig. E7):

- blocks must extend full thickness of wall;
- weight including plaster not less than 415 kg/m²;
- at least 13 mm plaster on each side;
- coursing 110 mm;
- block density 1840 kg/m³.

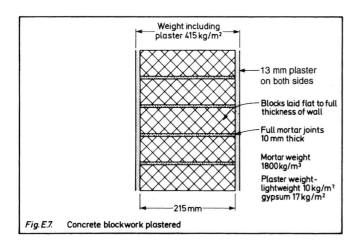

Fig. E.7.   Concrete blockwork plastered

D.   Concrete block plasterboard on each side (Fig. E7(A)):

- blocks must extend full thickness of wall;
- weight including plasterboard 415 kg/m²;
- at least 12.5 mm plasterboard on each side;
- coursing 150 mm;
- block density 1840 kg/m³.

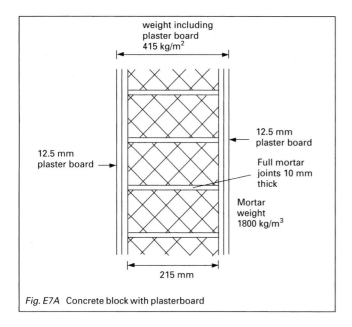

Fig. E7A   Concrete block with plasterboard

E.   Dense concrete (Fig. E8):

- can be in-situ or large panels;
- weight including plaster not less than 415 kg/m²;
- minimum density 1500 kg/m³;
- panel joints to be filled with mortar;
- where unplastered wall is 2200 kg/m³ a thickness of 190 mm is satisfactory.

With all these forms of construction precautions are to be taken to reduce transmission. These are necessary at junctions with the external wall, at floor levels, ceilings and roof spaces.

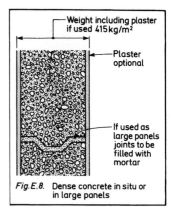

Fig. E.8.   Dense concrete in situ or in large panels

Junction with external wall

1. When the external wall is built of masonry, it can be either a solid wall or be the inner leaf of a cavity, there are the following requirements (Fig. E9):

- the sound-resisting wall should be bonded to the external wall, or;
- if butted to it, secured by wall ties at not more than 300 mm vertically;
- if weight is less than 120 kg/m² openings should be not less than 1 m high and not more than 700 mm from the face of the sound-resisting wall;
- seal cavity with mineral wool closer.

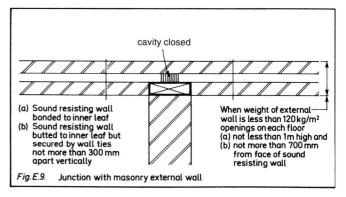

Fig. E.9.   Junction with masonry external wall

2. When the external wall is timber framed it is to be butted tightly against the sound resisting wall and fixed with ties spaced at not more than 300 mm intervals vertically (Fig. E10).

The joint between the sound resisting wall and the external wall lining is to be sealed with mastic or tape.

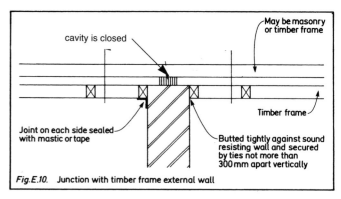

Fig. E.10.   Junction with timber frame external wall

Roof space

Sound-resisting walls in the roof space (Fig. E11) may be reduced to 150 kg/m². Rendering or painting required on one side when blocks of less than 1200 kg/m³ are used.

Joint between wall and roof is to be filled.

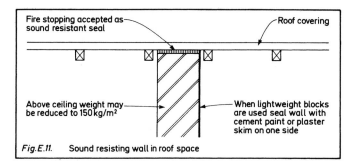

Fig.E.11. Sound resisting wall in roof space

Ceiling

Ceilings are to be 12.5 mm plasterboard or another material of not less than the same weight (Fig. E12).

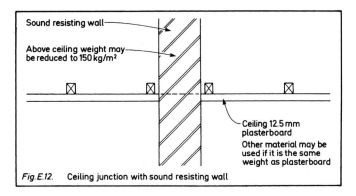

Fig.E.12. Ceiling junction with sound resisting wall

Floors

Bedroom or other intermediate floors adjoining a sound-resisting wall must not penetrate the wall, but there are no restrictions on the floor construction or the ceiling material (Fig. E13).

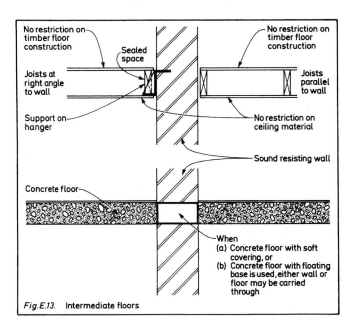

Fig.E.13 Intermediate floors

Careful workmanship is essential to ensure that there are no air-paths through the wall.

Concrete floors are discussed later.

Ground floors

There are no requirements regarding the type of ground floor that may be used or the way it can be joined to a sound-resisting wall.

Cavity masonry walls

These are wall type 2 and success arises from adequate mass and the amount of isolation gained.

The approved document gives examples of cavity walls which offer suitable insulation against noise transmission, emphasis being placed upon the need for

- filled mortar joints;
- sealing at junctions;
- precautions to prevent flanking transmission;
- leaves at least 50 mm apart;
- butterfly pattern wall ties at least 900 mm apart horizontally and at least 450 mm apart vertically;
- a laminate of plasterboard and mineral wool may be used where plaster or plasterboard are specified;
- cavity to rise to underside of roof;
- when external walls are filled with insulating which is not unbonded particles or fibre it must not be allowed to penetrate the cavity of the separating wall.

BS 5628: Part 3: 1985, 'Masonry materials, components, design and workmanship', places limitations upon the use of butterfly ties and requires a minimum masonry leaf thickness of 90 mm. It gives general recommendations for the design and construction of brick and block masonry, including materials and components.

A.   Two brick leaves plastered on room faces (Fig. E14):

- frogs upwards;
- weight including plaster not less than 415 kg/m²;
- 13 mm plaster on each room face;
- minimum cavity width 50 mm;
- mass including plaster 415 kg/m²;
- coursing 75 mm;
- brick density 1970 kg/m³.

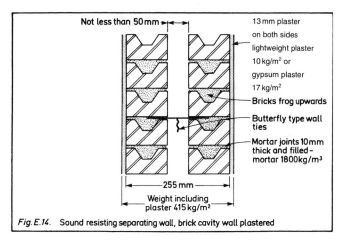

Fig.E.14. Sound resisting separating wall, brick cavity wall plastered

B.   Two concrete block leaves plastered on room faces (Fig. E15):

- weight including plaster not less than 415 kg/m²;
- 13 mm plaster on each room face;
- minimum cavity width 50 mm;
- coursing 225 mm;
- block density 1990 kg/m³.

C.   Two leaves lightweight concrete blocks (Fig. E16):

(a) plastered on both room faces; or
(b) drylined on both room faces.

- weight including plaster or drylining not less than 300 kg/m²;
- at least 13 mm plaster or drylining on each room face;
- minimum cavity width 75 mm;
- blockwork sealed with cement paint where passes through floor;
- coursing 225 mm;
- block density 1371 kg/m³.

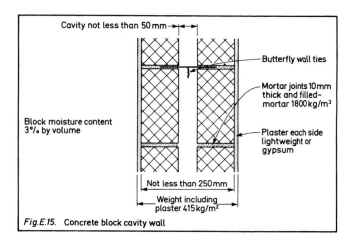

Fig. E.15. Concrete block cavity wall

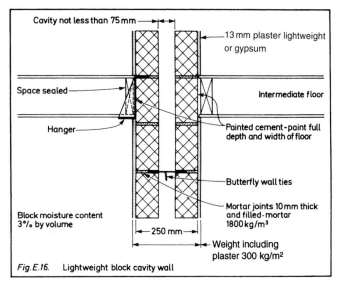

Fig. E.16. Lightweight block cavity wall

D. Two leaves concrete blocks having plasterboard on each side (Fig. E17):

- mass of masonry alone 415 kg/m²;
- at least 12.5 mm plasterboard on each face;
- coursing 225 mm;
- block density 1990 kg/m³.

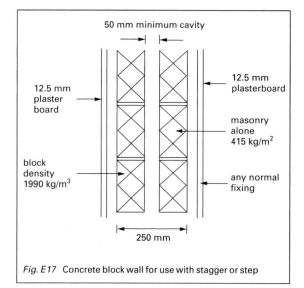

Fig. E17 Concrete block wall for use with stagger or step

E. Two lightweight aggregate concrete blocks (maximum density 1600 kg/m³) leaves plastered or drylined on each side (Fig. E17A):

- mass, including finish, 250 kg/m²;
- at least 13 mm plaster, or 12.5 mm plasterboard;
- coursing 225 mm;
- block density 1105 kg/m³.

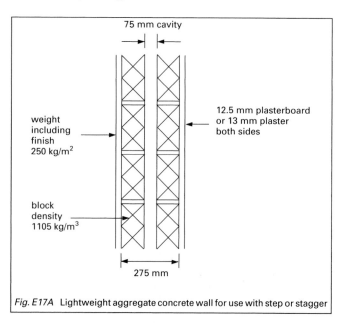

Fig. E17A Lightweight aggregate concrete wall for use with step or stagger

Types D and E are for use only in stepped or staggered situations where stepping or staggering can improve sound resistance.

Precautions must also be taken to reduce flanking transmission. Attention is therefore necessary at junctions with the external wall at floor levels, ceilings and roof spaces.

Junction with external wall

1. When the external wall is built of masonry it can be either a solid wall or be the inner leaf of a cavity, there are the following requirements (Fig. E18):

- the sound-resisting wall should be bonded to the external wall, or;
- if the external wall is butted to sound-resisting wall it must be secured by wall ties at not more than 300 mm intervals vertically;
- weight of inner leaf of external wall not less than 120 kg/m² unless cavity wall type B is used.

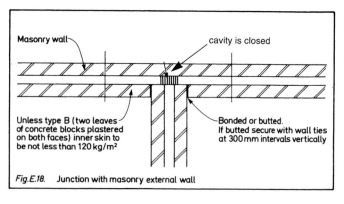

Fig. E.18. Junction with masonry external wall

2. When the external wall is timber framed it is to be butted tightly against the sound-resisting wall and fixed with ties spaced at not more than 300 mm intervals vertically (Fig. E19).

The joint between the sound-resisting wall and the external wall must be sealed with mastic or tape.

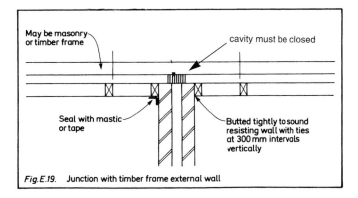

Fig. E.19. Junction with timber frame external wall

Any material which is used as a cavity closer must not rigidly connect leaves of separating walls.

There are no restrictions regarding junctions with partitions.

Roof space

Sound-resisting walls in the roof space (Fig. E20) may be reduced to 150 kg/m². When lightweight blocks are used one side must be sealed with plaster skim or cement paint.

Junction between wall and roof is to be sealed.

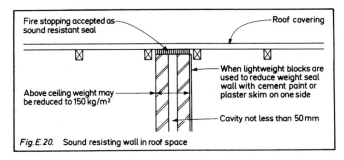

Fig. E.20. Sound resisting wall in roof space

Ceiling

Ceilings are to be not less than 12.5 mm plasterboard or another material of not less than the same weight (see Fig. E21).

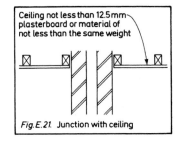

Fig. E.21. Junction with ceiling

Floors

Bedroom or other intermediate floors adjoining a sound-resisting wall must not penetrate the wall (Fig. E22), but there are no restrictions on the floor construction, or material for the ceiling.

Ensure that there are no airpaths through the wall.

Concrete floors are discussed later.

Ground floors

There are no requirements regarding the type of ground floor that may be used or the way it can be joined to a sound-resisting wall.

Masonry walls with lightweight panels

These are wall type 3. With this form of construction airborne sound resistance arises from the weight of the brick or concrete block core combined with panels and air spaces. In order to prevent gaps

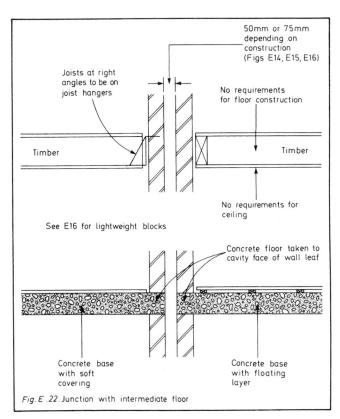

Fig. E.22. Junction with intermediate floor

leading to air paths this form of construction is only to be used when the ground floor is concrete. The lightweight panels must be isolated from the masonry core.

A/E. Brickwork and cellular core panels (Fig. E23):

- frogs upwards;
- weight of brickwork not less than 300 kg/m²;
- weight of panel including plaster finish, if used, to be not less than 18 kg/m²;
- panels not less than 25 mm from block core;
- all panel joints to be taped;

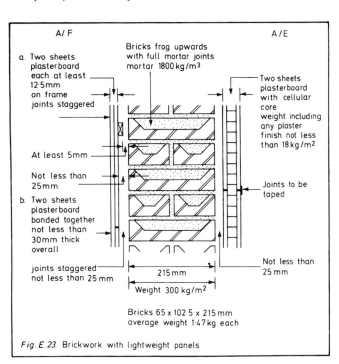

Fig. E.23. Brickwork with lightweight panels

- coursing 75 mm;
- brick density 1290 kg/m^3;
- panels fixed to floor and ceiling only.

A/F. Brickwork and two sheets plasterboard panels (Fig. E23):

- frogs upwards;
- weight of brickwork not less than 300 kg/m^2;
- when supported on a frame panel should comprise two sheets of plasterboard each not less than 12.5 mm thick;
- when plasterboard bonded without supporting frame, overall thickness to be not less than 30 mm;
- gap of 25 mm between sheets and core;
- gap of 5 mm between any frame and core;
- sheet joints to be staggered to obviate air paths;
- coursing 75 mm;
- brick density 1290 kg/m^3.

B/E. Concrete blockwork and cellular core panels (Fig. E24):

- weight of blockwork not less than 300 kg/m^2;
- density of block not less than 2200 kg/m^3;
- coursing 110 mm;
- weight of panel including plaster finish, if used, to be not less than 18 kg/m^2;
- panels not less than 25 mm from block core;
- all panel joints to be taped.
- fixings to be to floor and ceiling.

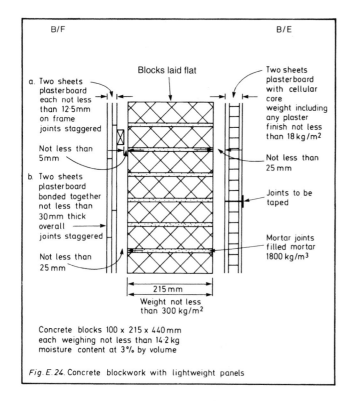

Fig. E.24. Concrete blockwork with lightweight panels

B/F. Concrete blockwork and two sheets plasterboard panels (Fig. E24):

- density of block not less than 2200 kg/m^3;
- weight of blockwork not less than 300 kg/m^2;
- when supported on a frame, panel to comprise two sheets plasterboard each not less than 12.5 mm thick;
- when plasterboard bonded without supporting frame, overall thickness to be not less than 30 mm;
- panels not less than 25 mm from block core;
- frame not less than 5 mm from block core;
- sheet joints to be staggered;
- coursing 110 mm.

C/E. Lightweight concrete blocks and cellular core panels (Fig. E25):

- block density less than 1600 kg/m^3;
- weight of blockwork not less than 160 kg/m^2;
- weight of panel including plaster finish, if used, to be not less than 18 kg/m^2;
- panels not less than 25 mm from block core;
- coursing 225 mm;
- panels fixed to ceiling and floor only;
- all panel joints to be taped.

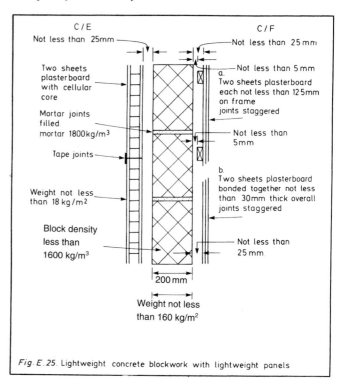

Fig. E.25. Lightweight concrete blockwork with lightweight panels

C/F. Lightweight concrete blocks and two sheets of plasterboard panels (Fig. E25):

- block density less than 1600 kg/m^3;
- weight of blockwork not less than 160 kg/m^2;
- when supported on a frame, panel to comprise two sheets of plasterboard each not less than 12.5 mm thick;
- when plasterboard bonded without supporting frame, overall thickness to be not less than 30 mm;
- panels not less than 25 mm from block core;
- frame not less than 5 mm from block core;
- sheet joints to be staggered.

D/E. Cavity brickwork or blockwork (Fig. E26):

- can be any mass;
- weight of panel including plaster finish to be not less than 18 kg/m^2;
- fix only to floor and ceiling;
- all panel joints to be taped;
- panels not less than 25 mm from brick core;
- all framing to be kept clear of core.

D/F. Cavity brickwork or blockwork (Fig. E26):

- can be any mass;
- where supported on frame, panel to comprise two sheets of plasterboard each not less than 12.5 mm thick;
- when plasterboard bonded without supporting frame, overall thickness to be not less than 30 mm;
- panels not less than 25 mm from core;
- all framing to be kept clear of core;
- sheet joints to be staggered.

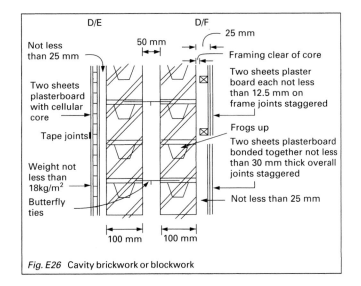

Fig. E26 Cavity brickwork or blockwork

With all these forms of construction precautions are to be taken to reduce the possibility of flanking transmission. These are necessary at junctions with partitions and external walls, at floor levels, ceilings and roof spaces.

Junction with external wall

A junction with external wall has the following requirements (Fig. E27):

- outer leaf of cavity wall may be of any construction;
- with a lightweight concrete block core the inner leaf of the cavity is to have an internal finish of isolated plasterboard or cellular core panels;
- if the separating wall has a brick core; a concrete block core or a lightweight concrete block core it may be plastered or drylined;
- joints are to be sealed with mastic or tape;
- insulation may be attached to a drylining finish and gaps of 10 mm or 25 mm are to be allowed between the lining and the inner leaf;
- if isolated panels are used for lining the inner leaf may be of any construction;

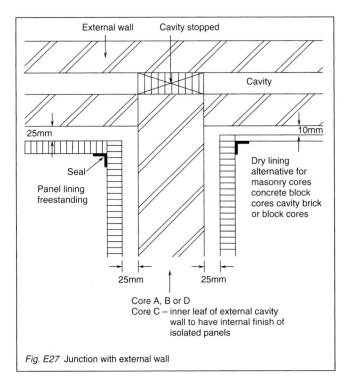

Fig. E27 Junction with external wall

- an inner leaf which is plastered or drylined is to have a total mass not less than 120 kg/m²;
- it is to be butt jointed to the separating wall core and have ties spaced not in excess of 300 mm vertical centres.

Junction with loadbearing partition

A junction with loadbearing partition has the following requirements (Fig. E28):

- partition not to be masonry;
- a continuous pad of mineral fibre is to be placed between the end of the partition and masonry core;
- joints between panels to be sealed with mastic or tape.

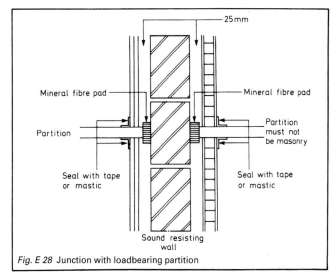

Fig. E 28 Junction with loadbearing partition

Junction with non-loadbearing partition

A junction with non-loadbearing partition has the following requirements (Fig. E29):

- partition not to be of masonry;
- partition to be butted tightly to panel of sound-resisting wall;
- joints to be sealed with mastic or tape.

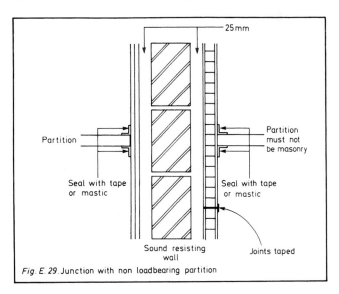

Fig. E. 29. Junction with non loadbearing partition

Roof space

If the sound resisting wall is of cavity construction, this must be taken to the underside of the roof (Fig. E30).

Sound-resisting walls in roof spaces may be reduced to 150 kg/m².

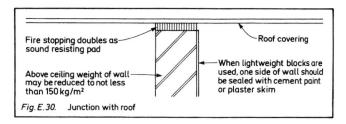

Fig. E.30.   Junction with roof

When lightweight blocks are used one side must be sealed with plaster skim or cement paint.

Junction between wall and roof is to be sealed.

Ceiling

Ceilings are to be not less than 12.5 mm plasterboard or another material or not less than the same weight (Fig. E31).

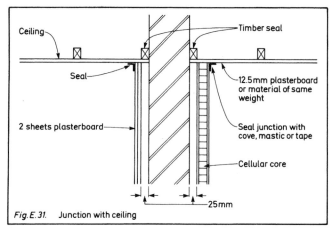

Fig. E. 31.   Junction with ceiling

Floors

Bedroom or other intermediate floors adjoining a sound-resisting wall must not penetrate the wall (Fig. E32):

- if floor joists are at right angles to masonry core they are to be supported on joist hangers.
- gaps between ends of joists to be closed with timber.
- lightweight panels to stand on battens.
- seal gaps with mastic, tape, or coving.

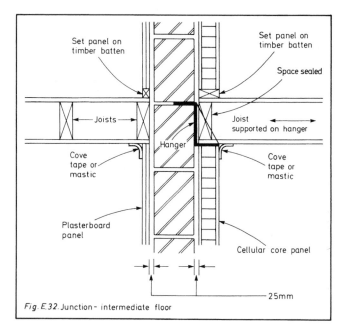

Fig. E.32. Junction - intermediate floor

A concrete intermediate floor may be used (Fig. E32(A)):

- the base only may be taken through the wall;
- it must have a mass of not less than 365 kg/m^2;
- if the core is a cavity wall the cavity must not be bridged;
- junction between ceiling and panels are to be sealed with mastic or tape.

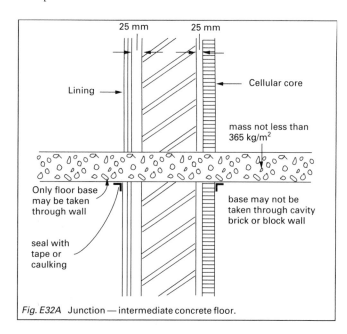

Fig. E32A   Junction — intermediate concrete floor.

Ground floor

A concrete ground floor may be used (Fig. E33):

- the ground floor may be a solid concrete slab;
- it may be suspended or laid on the ground;
- it may not be taken through the wall if it has a mass less than 365 kg/m^2;
- where the mass is less than 365 kg/m^2 the floor is not to bear on the core.

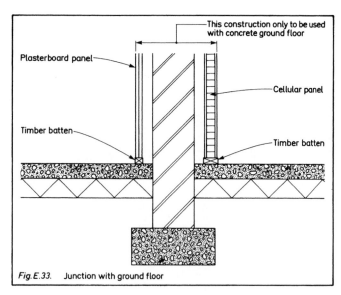

Fig. E.33.   Junction with ground floor

Timber-framed walls

These are shown as wall type 4 in the two examples of construction, including timber framed assembly that satisfy E1. To fulfil the standard of sound resistance required emphasis is placed upon the following:

- cladding, quilt and air space absorb sound;
- services should not penetrate the cladding;
- socket outlets should not be placed back to back in wall;
- if socket outlets are necessary in wall there should be 30 mm cladding placed behind the boxes;
- air paths should be avoided;
- any firestops preferably of flexible material and only fixed to one frame.

Two types of construction are illustrated. The first relies upon timber frames (Figs E34 and E35), and the second comprises two timber frames having a brickwork or blockwork wall (Fig. E36). There is no specification for the blockwork or brick wall in the approved document.

Requirements for cladding are

- must be not less than two sheets of plasterboard;
- may have plywood sheathing but not a requirement;
- cladding must not be less than 30 mm thick;
- joints between sheets are to be staggered;

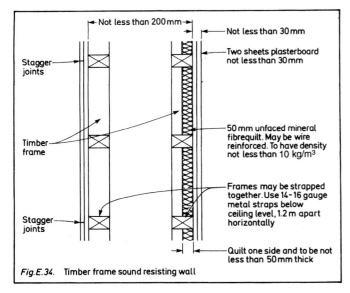

Fig.E.34. Timber frame sound resisting wall

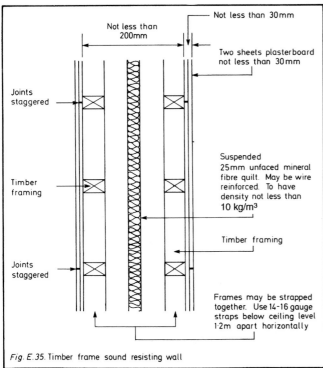

Fig. E.35. Timber frame sound resisting wall

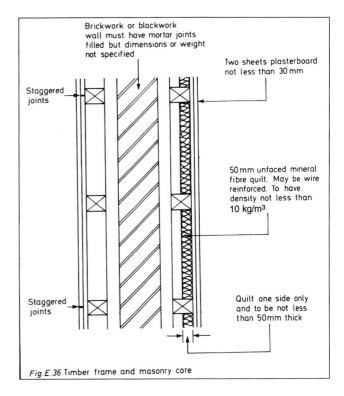

Fig.E.36. Timber frame and masonry core

Requirements for the absorbent curtain are

- unfaced mineral fibre quilt or batts;
- may be wire reinforced;
- density not less than $10 \, kg/m^3$;
- if suspended between frames not less than 25 mm thick;
- if fixed to one of the frames, not less than 50 mm thick;
- if fixed to each frame 25 mm for each quilt;
- may have plywood sheathing in cavity.

Precautions are necessary to reduce the possibility of flanking transmission. These are necessary at junctions with external walls, at floor levels, ceilings and roof spaces.

Junction with external wall

Requirements are (Fig. E37):

- no control if of external cladding to a timber-framed internal wall leaf;
- seal gaps between end of timber frame sound-resistant wall and any masonry outer skin of external wall;

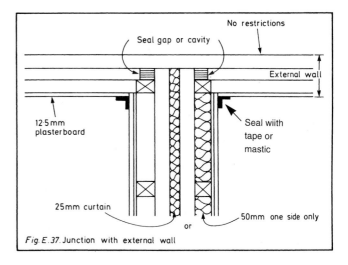

Fig. E.37. Junction with external wall

- line timber frame skin with plasterboard not less than 12.5 mm thick;
- no restrictions on internal partitions meeting a separating wall.

Roof space

Requirements are (Fig. E38):

- cladding in roof space may be reduced to not less than 25 mm, or;
- close cavity at ceiling level. Do not connect the two frames rigidly together, and use one frame with not less than 25 mm cladding on each side;
- seal connection between wall and underside of roof.

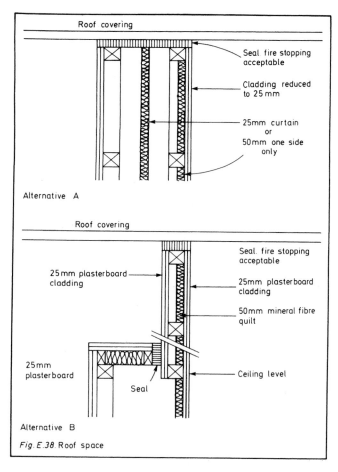

Fig. E.38. Roof space

Ceiling

Ceilings are to be not less than 12.5 mm plasterboard or another material of not less than the same weight (Fig. E39).

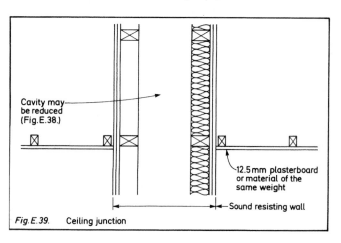

Fig. E.39.   Ceiling junction

Floors

There is a suggested method and freedom for the designer to devise alternative ways of forming junction with a bedroom or intermediate floor with sound-resisting wall (Fig. E40):

- stop air paths to cavity by taking cladding through floor;
- alternatively use solid timber edge to floor;
- seal spaces between joists at right angles to wall.

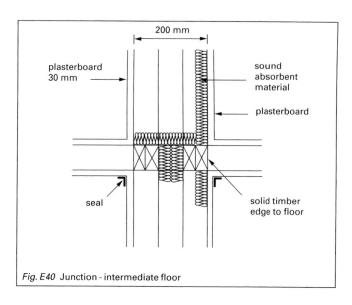

Fig. E40 Junction - intermediate floor

Ground floor

Requirements (Fig. E41):

- ground floor may be suspended concrete slab, or
- slab resting on ground;
- suspended floor to have mass not less than 365 kg/m².

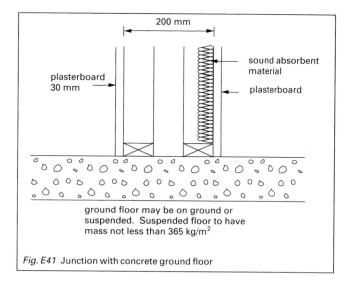

Fig. E41 Junction with concrete ground floor

SECTION 2. SEPARATING FLOORS

Sound insulation

The requirements make a distinction between floors above other rooms and floors overhead, and it is acknowledged that impact sounds from shoes, chairs, falling articles, etc. may cause a nuisance.

Floor finishes play a great part in the degree of impact noise in any situation. A hard floor surface gives poor insulation against impact noise, whereas a soft finish such as carpet with underfelt gives much better results. One cannot, however, rely upon people occupying a particular flat, laying a thick carpet on the floor.

It is important that structures associated with the floors are taken into account when determining the adequacy of sound resistance.

One of the principal reasons for poor results in the past has been the bridging or penetration of the resilient layer, giving a direct route for impact noise transmission. Additionally if a floating floor above a quilt is in contact with a wall at its edge, sound will be transmitted into the wall and thence to the room below.

Whereas separating walls are required to have resistance against airborne sound, floors must also have resistance to impact. The requirements are in E2 and E3 of Schedule 1.

E2. *Airborne sound (floors and stairs)*

A floor or a stair which separates a dwelling from:

- **another dwelling, or;**
- **another part of the same building not exclusive to the dwelling, must resist passage of airborne sound (Fig. E42).**

This does not apply to a floor separating the dwelling from space used only for inspection, maintenance or repair.

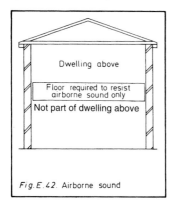

Fig. E.42. Airborne sound

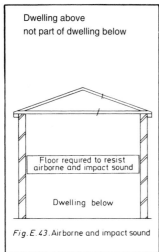

Fig. E.43. Airborne and impact sound

E3. *Impact sound (floors and stairs)*

A floor or stair above a dwelling which separates the dwelling from:

- **another dwelling, or;**
- **another part of the same building not exclusive to the dwelling, must resist the passage of impact sound (Fig. E43).**

The approved document gives examples of three types of floor, which are widely used:

- concrete base together with a soft covering;

- concrete base and a floating layer;
- timber base with a floating layer.

Type 1. Floor base with a soft covering

As with walls, weight offers a good resistance to airborne sound, and the soft covering adds protection against impact noise.

When it is required to insulate against airborne sound only, the soft covering will not be necessary. Any floor finish may be added.

Following the golden rule, pathways between elements on opposite sides of the sound-resisting floor should be avoided if flanking transmission is not to take place.

A.   Solid concrete slab (Fig. E44):

- weight of the base not to be less than 365 kg/m²; the weight may include floor screed and any ceiling bonded to the concrete;
- soft covering material must return to its original thickness after compression;
- give special attention to pipe and service penetrations;
- resilient material soft covering to have uncompressed thickness of 4.5 mm;
- alternatively a floor covering having a weighed impact sound improvement (ΔL_W) of at least 17 dB. Method of calculation is included in Annexe A to BS 5821: Part 2: 1984, 'Method for testing impact sound insulation'.

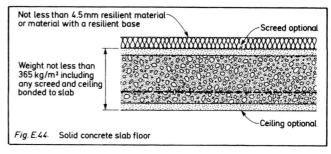

Fig. E.44. Solid concrete slab floor

This BS defines single-number quantities for the impact sound insulation of buildings and of floors, and gives rules for determining these quantities.

Contents:

1. Scope
2. References
3. Definitions
4. Procedure for evaluating single-number quantities
5. Statement of results.

There are two annexes relating to the evaluation of weighed impact sound improvement; five tables and one figure.

B.   Solid concrete slab retaining permanent shuttering (Fig. E45):

- weight of the base to be not less than 365 kg/m²; the weight may include floor screed and shuttering where it is solid concrete or metal, and any ceiling bonded to the shuttering;
- soft covering material must return to its original thickness after compression.

C.   Concrete beams with infilling blocks (Fig. E46):

- weight of the base to be not less than 365 kg/m²; the weight may include floor screed, blocks if they are clay or concrete and any ceiling bonded to the beams of blocks;
- all joints between beams and blocks are to be filled;
- soft covering material must return to its original thickness after compression.

D.   Concrete planks (Fig. E47):

- weight of the base to be not less than 365 kg/m²; the weight may include floor screed and ceiling when bonded to the beams;
- all joints between beams are to be filled;
- soft covering material must return to its original thickness after compression.

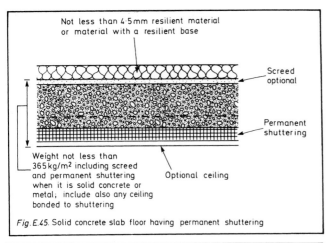

Fig.E.45. Solid concrete slab floor having permanent shuttering

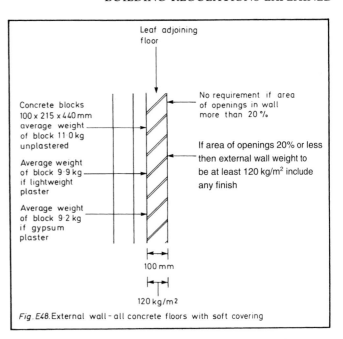

Fig.E.48. External wall – all concrete floors with soft covering

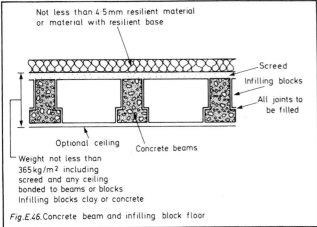

Fig.E.46. Concrete beam and infilling block floor

Avoiding flanking transmission

1. The approved document recommends minimum external wall construction which may be used in conjunction with all types of concrete floor having a soft covering (Fig. E48).
2. Junctions of floor and cavity walls must be carefully detailed. A concrete floor of any type with a soft covering should be built

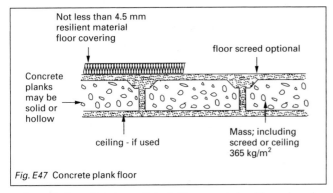

Fig. E47 Concrete plank floor

through the adjoining wall leaf. The screed should not be taken through, this applies whether or not floor beams are parallel or at right angles to the wall (Fig. E49).

3. Junctions of floor and a sound-resisting or solid internal wall (Fig. E50) should follow the recommendations:

   (a) if the wall weighs less than 375 kg/m² including any plaster finish, pass the floor base through the wall;

   (b) if the wall weighs more than 375 kg/m² including any plaster finish, the floor base may pass through the wall or be butted against it.

4. Great care must be taken when a pipe passes through a sound-resisting floor, to seal any gaps and limit the amount of noise that

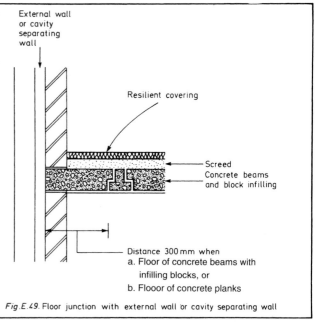

Fig.E.49. Floor junction with external wall or cavity separating wall

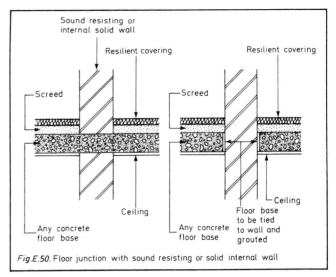

Fig.E.50. Floor junction with sound resisting or solid internal wall

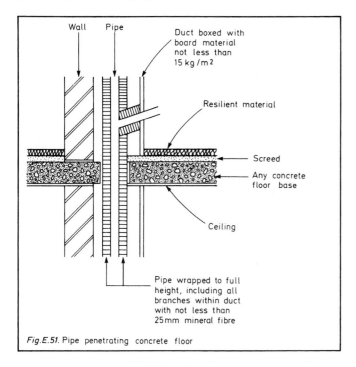

Fig.E.51. Pipe penetrating concrete floor

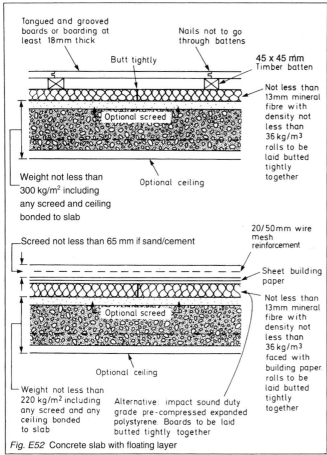

Fig. E52 Concrete slab with floating layer

may be conveyed along the pipe. Pipe should not be in contact with floor when passing through but be surrounded by 25 mm insulating material (Fig. E51).

• Refer to Part B (Chapter 11) for fire protection where pipes pass through separating floors.
• These requirements do not apply to gas pipes.
• Requirements in respect of gas pipes are contained in Gas Safety

(Installation and Use) Regulations 1998. Gas pipes may be placed in a separate vented duct or may remain exposed.

(There are requirements for ventilating gas pipe ducts at each floor.)

Type 2. Concrete base with floating layer

Success in limiting passage or airborne sound depends largely upon weight, both of the base and also the floating layer (Fig. E52). Impact sounds are diminished by the floating layer. As with all forms of sound-resisting construction meticulous care must be taken to avoid leaving gaps or pathways to encourage flanking transmission.

If resistance is only required for airborne sound the full construction should still be followed.

A. Solid concrete slab

Requirements (Fig. E52):

• weight of the base not to be less that 300 kg/m²; the weight may include screed and ceiling if bonded to the concrete;
• resilient layer 25 mm mineral fibre unless battens have integral closed cell resilient foam strip;
• 5 mm extruded closed cell polyethylene foam density 30–45 kg/m³ may be used only under a screed; it should be laid with lapped joints on a levelling screed and turned up at edges of floating screed.

B. Solid concrete slab with permanent shuttering

Requirements (Fig. E53):

• weight of the base not to be less than 300 kg/m²; the weight may include solid concrete or metal permanent shuttering, floor screed and ceiling finish.

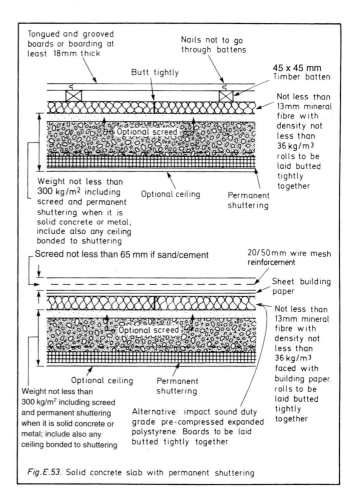

Fig.E.53. Solid concrete slab with permanent shuttering

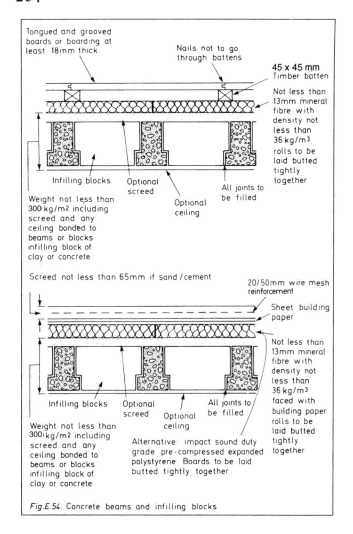

Fig.E.54. Concrete beams and infilling blocks

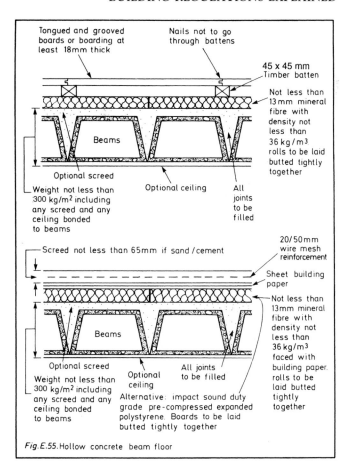

Fig.E.55. Hollow concrete beam floor

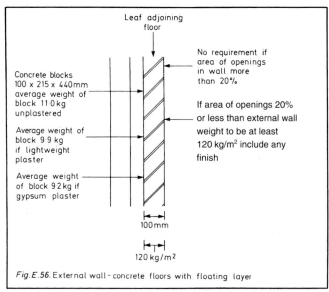

Fig.E.56. External wall – concrete floors with floating layer

C. Concrete beams with infilling blocks

Requirements (Fig. E54):

- weight of the base not to be less than 300 kg/m^2; the weight may include floor screed, blocks if they are clay or concrete and any ceiling bonded to beams or blocks;
- all joints between beams and blocks to be filled;
- maximum difference of 5 mm between surface of components;
- resilient layer 25 mm mineral fibre unless battens have integral closed cell resilient foam strip;
- 5 mm extruded closed cell polyethylene foam density 30-45 kg/m^3 may be used only under a screed; it should be laid with lapped joints on a levelling screed and turned up at edges of floating screed;
- if necessary, use a levelling screed.

D. Concrete planks

Requirements (Fig. E55):

- weight of the base must not to be less than 300 kg/m^2; the weight may include floor screed and ceiling finish if bonded to beams;
- all joints between beams are to be filled;
- maximum difference of 5 mm between surface of components;
- resilient layer 25 mm mineral fibre unless battens have integral closed cell resilient foam strip;
- 5 mm extruded closed cell polyethylene foam density 30–45 kg/m^3 may be used only under a screed; it should be laid with lapped joints on a levelling screed and turned up at edges of floating screed.

Avoiding flanking transmission

1. The approved document recommends a minimum external wall construction which may be used in conjunction with all types of concrete base floors having a floating layer (Fig. E56).

2. Junctions between a concrete floor having a floating layer and an external wall or a cavity separating wall must be carefully detailed. The concrete floor base should be taken through the adjoining wall leaf. Any screed should not be taken through. This applies whether or not floor beams are parallel or at right angles to the wall (Fig. E57).

3. Junctions between concrete floor with floating layer and a sound resisting or internal solid wall (Fig. E58), should follow the recommendations:

(a) if the wall weighs less than 375 kg/m^2 including any plaster finish pass the floor base through wall.

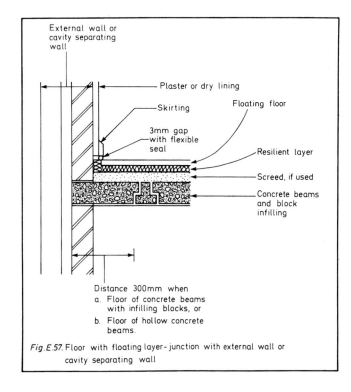

Fig.E.57. Floor with floating layer-junction with external wall or cavity separating wall

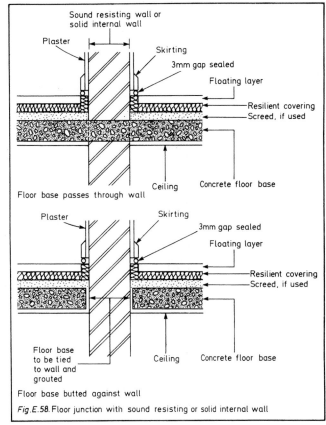

Floor base passes through wall

Floor base butted against wall

Fig.E.58. Floor junction with sound resisting or solid internal wall

(b) if the wall weighs more than 375 kg/m² including any plaster finish, the floor base may be passed through the wall or butted against it.

4. Great care must be taken when a pipe passes through a sound resisting floor to seal any gaps and limit the amount of noise that may be conveyed along the pipe. A pipe should not be in contact with the floor when passing through, but surrounded by 25 mm insulating material (Fig. E59).

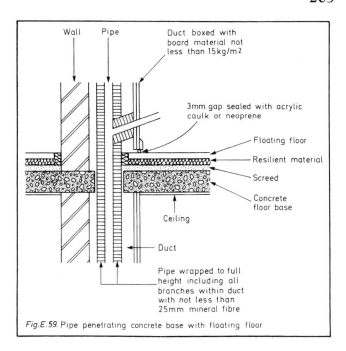

Fig.E.59. Pipe penetrating concrete base with floating floor

- Refer to Part B (Chapter 11) for fire protection where pipes pass through separating floors.
- These requirements do not apply to gas pipes.
- Requirements in respect of gas pipes are contained in Gas Safety (Installation and Use) Regulations 1998. Gas pipes may be placed in a separate vented duct or may remain exposed.

There are requirements for ventilating gas pipe ducts on each floor.

Type 3. Timber floor with floating layer

Timber radiates sound less efficiently than concrete floors and therefore weight is not so important. Airborne sound is reduced by the absorbent blanket (Fig. E60), or pugging and impact sound is reduced by the floating layer:

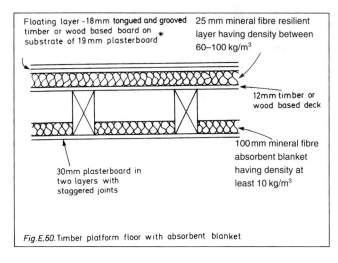

Fig.E.60. Timber platform floor with absorbent blanket

- any nails used on the floating layer should not penetrate through resilient layers or strip;
- plasterboard ceilings in two layers are to have staggered joints;
- when resistance to airborne sound only is necessary the full construction should still be followed.

A. Platform floor with absorbent blanket (Fig. E60)

- 18 mm t & g boards to have all joints glued
- cement-bonded particle board with staggered joints and in two

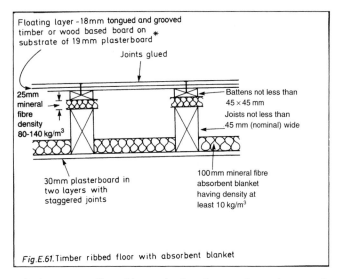

Fig.E.61. Timber ribbed floor with absorbent blanket

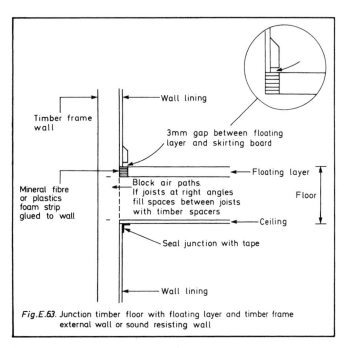

Fig.E.63. Junction timber floor with floating layer and timber frame external wall or sound resisting wall

thicknesses totalling 24 mm glued and screwed together may be used instead of 19 mm plasterboard.

- additional support round perimeter may be provided by a timber batten with foam strip along upper surface and attached to wall.

B.   Ribbed floor with absorbent blanket (Fig. E61)

C.   Ribbed floor with heavy pugging (Fig. E62)

- pugging may be chosen from the following:

  75 mm thick traditional ash
  60 mm thick 2–10 mm limestone chips
  60 mm thick 2–10 mm whin aggregate
  50 mm dry sand

  thicknesses given will achieve 80 kg/m²;

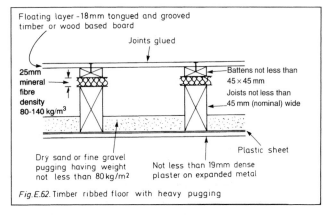

Fig.E.62. Timber ribbed floor with heavy pugging

- sand is not to be used in situations where it may become wet and overload the ceiling as for example in kitchens, water closets showers or bathrooms;
- alternative to 19 mm dense plaster ceiling is 6 mm plywood below joists with two layers staggered jointed plasterboard not less than 25 mm.

Avoiding flanking transmission

1.   The approved document emphasized the need to ensure that any junction between a floor having a floating layer and a timber-framed external wall, or sound-resisting wall must use construction which will block air paths (Fig. E63):

- normal methods of connecting floor to wall may be used;
- air paths are to be blocked;
- seal gap between floating layer and wall panel with resilient strip.

2.   Junction of a timber floor with floating layer with a solid masonry external wall or internal sound-resisting wall must be carefully

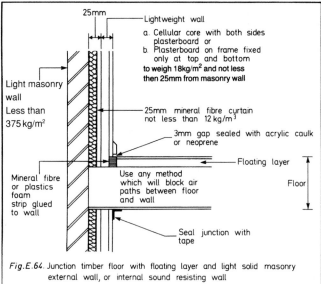

Fig.E.64. Junction timber floor with floating layer and light solid masonry external wall, or internal sound resisting wall

detailed. Light solid masonry walls weighing less than 375 kg/m² are shown in the figures (see Fig. E64).

3.   Junction between a timber floor having a floating layer with a solid heavy masonry external wall weighing 375 kg/m² or more and internal or separating walls are also referred to in the approved document (Fig. E65):

- there is to be a gap between wall and floating layer filled with resilient strip;
- there is to be a 3 mm gap between floating layer and skirting;
- any method of connecting floor to wall may be used;
- junction of ceiling and wall to be sealed with caulking or tape.

4.   Great care is necessary should a pipe pass through a sound-resisting floor (Fig. E66). All air gaps must be sealed to limit the amount of noise that may be conveyed along the pipe. A pipe should not be in contact with the floor when passing through but surrounded by 25 mm insulating material.

- Refer to Part B (Chapter 11) for fire protection where pipes pass through separating floors.
- These requirements do not apply to gas pipes.
- Requirements in respect of gas pipes are contained in Gas Safety

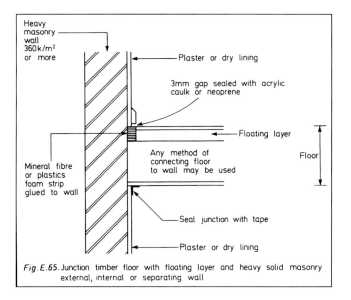

Fig. E.65. Junction timber floor with floating layer and heavy solid masonry external, internal or separating wall

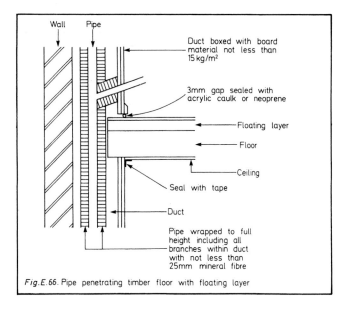

Fig. E.66. Pipe penetrating timber floor with floating layer

(Installation and Use) Regulations 1998. Gas pipes may be placed in a separate vented duct or may remain exposed.

There are requirements for ventilating gas pipe ducts on each floor.

SECTION 3. SIMILAR CONSTRUCTION

Section 3 in the approved document describes how requirements may be met by repeating a construction that has already been erected.

Before proceeding it is necessary to show that the sound-resisting qualities of the existing wall or floor are acceptable. It is also necessary to show that the new construction will have similar features to the old.

Tests are to be carried out on the existing construction following the methods given in BS.

- For airborne sound insulation: the weighed standardized level difference ($D'_{nT,w}$) can be ascertained by following BS 5821: Part 1: 1984, 'Method for rating the airborne sound insulation in buildings and interior building elements'.

This defines single-number quantities for the airborne sound insulation in buildings and of interior building elements such as walls, floors and doors, and gives rules to achieve requirements.

- For impact sound insulation: the weighed standardized impact sound pressure level ($L'_{nT,w}$) by following BS 5821: Part 2: 1984, 'Method for rating the impact sound insulation'.

Airborne sound

Minimum value:
Weighed standardized level difference, $D'_{nT,w}$

Test involving up to four pairs of rooms
Individual value (dB):   walls, 49   floors, 48
Mean value (dB):   walls, 53   floors, 52

Test involving at least eight pairs of rooms
Individual value (dB):   walls, 49   floors, 48
Mean value (dB):   walls, 52   floors, 51

Impact sound

Maximum value:

Weighed standardized sound pressure level, $L'_{nT,w}$

Test involving up to four pairs of rooms
Individual value (dB):   floors, 65
Mean value (dB):   floors, 61

Test involving at least eight pairs of rooms
Individual value (dB):   floors, 65
Mean value (dB):   floors, 62

Similar features

Sound insulation has a bearing not only upon construction but also on the size and shape of rooms. Important also is the position of doors and windows in the reduction of the possibility of flanking transmission.

Therefore the following features in the proposed building should relate to those in the existing building. They need not be the same:

- specification of sound-resisting walls and floors;
- wall mass not to be reduced;
- construction of walls and floors abutting sound-resisting walls and floors;
- arrangement of doors and windows;
- shape and size of rooms adjoining a sound-resisting wall;
- a step or stagger may be provided even though one does not appear in the existing building;
- the word 'floor' includes stairs.

Some differences are allowed for walls and floors

- construction of outer leaf of a masonry cavity wall;
- the inner leaf of a cavity wall but the weight must not be reduced and construction must be of the same type;
- material and thickness of floor with concrete base or a timber floor;
- small reduction in step or stagger;
- the type of timber floor if it is not a separating floor.

Test evidence may provide values to enable an existing construction to be used as the basis for a new construction. Failure of new construction to achieve these figures is not conclusive evidence of a failure to meet regulation requirements.

Test method

- Dwellings tested should be complete.
- Dwellings tested should be unfurnished.
- Doors and windows should be closed.
- Each wall or floor tested should be subjected to eight sets of measurements.

- If eight pairs of rooms are not available, use four pairs for testing, or as close to four as possible.
- Use pairs of large rooms.
- Only use pairs of a room and some other space if it is the only way to make up sets of four.
- One set of measurements to be taken between each pair.
- Airborne sound transmission measured should have sound source in a large room when spaces are unequal.
- Airborne sound transmission should be measured in the other space when being checked between room and other space.
- A test report is to be in a prescribed form.

Form of test report

1. Name, address and NAMAS accreditation number, if appropriate.
2. Name of person in charge of test.
3. Date of test.
4. Address of building tested.
5. Brief details of test equipment and test procedures.
6. Description of the building. This should include a sketch showing relationship and dimensions of rooms tested together with a description of external and separating walls, partitions and floors. Include details of materials, construction and finishes. There is to be an estimate of surface mass, kg/m^2 of external walls, partitions and floors, and also the dimensions of any step or stagger between rooms tested. Approximate sizes of windows and doors in external walls which are within 700 mm of a separating wall are to be included.
7. Results of the test are to be in tabular and graphical form giving the single number rating and the underlying data upon which the single number is based.

SECTION 4. USE OF SIMILAR CONSTRUCTION

Test chamber evaluation

Section 4 in the approved document describes how the functional requirements for new walls may be tested in a test chamber.

Details of test chamber construction are available from the BRE, Watford WD2 7JR.

Tests can show whether a separating wall and flanking construction can achieve a prescribed value.

Procedure

Insulation against airborne sound is measured between the lower pair of rooms and also between the upper pair of rooms as required in BS 2570: Part 4: 1980, 'Measurement of airborne sound insulation between rooms'. Measurements are to be standardized to a reverberation time of 0.5 second; made in the 1/3 octave bands and between 100 and 3150 Hz.

The purpose of BS 2750: Part 4: 1980 is to give procedures to measure the sound insulation between two rooms in buildings and to give field procedures to determine whether building elements have met specifications.

There are three annexes:

A   Example of a test procedure
B   Measurement of flanking transmission
C   Checking the loss factor.

There is also a table showing correction to sound pressure level readings.

Value required

Each measurement should give at least 55 dB as the modified weighed standardized difference value.

Room dimensions affect modified weighed standardized level difference values. Calculations are made by adding a constant K to the weighted standardized difference value. This constant arises from K = 10 log$_{10}$(3/L) + 1, where L (in metres) is the length of the room at right angles to the separating wall.

Test results

A form of test report is given in the approved document, and this gives evidence to show compliance with the regulations. The evidence is only valid for the type of construction tested.

However, the following features may be changed:

- dimensions of flank and separating walls;
- window and/or door openings may be placed in the flank wall;
- internal partitions may be attached to the flank wall.

Test chamber evaluation report

A report is to contain the following information:

1. Organization name, address and NAMAS accreditation number where appropriate.
2. If organization making acoustic measurements is not the above, details are to be given.
3. Date of test.
4. Test chamber to be described and also the method of attaching the test construction.
5. Details of test procedure and equipment.
6. Details of materials and test construction including separating wall; is junction between separating wall and flank wall bonded or tied; surface finish; mass in m^2 of walls; intermediate floor; separating wall in loft space; dimensions and any other special feature.
7. Results of test:
   (a)   standardized level difference (D$_{nT}$),
   (b)   weighted standardized level difference (D$_{nT,w}$)
   (c)   weighted apparent sound reduction index (R$'_w$)
   (d)   two modified weighted standardized level differences obtained by adding K to the weighed standardized level difference and a statement that the two values meet requirements.

SECTION 5. CONVERSIONS

None of the paragraphs E1, E2 or E3 in Schedule 1 which set out the requirements with regard to airborne and impact sound mentions conversions. Application is made via Regulations 5 and 6 to apply E1, E2 and E3 to a material change of use. The approved document in section 5 introduces requirements which extend to conversion work, remembering that the application is the same as that outlined in E1, E2 and E3 – construction separating dwellings from another dwelling or from another building.

Thus a floor, wall or stair in a building to be converted into separate dwellings may have to be upgraded to provide an acceptable standard of sound resistance. There may be instances when the existing construction is acceptable:

- when construction is generally similar or within 15% of the mass of construction already described for new separating walls and new separating floors;
- with regard to flanking construction attention is only required for floor penetration and sealing joints;
- laboratory tests for conversions are later described in section 6 and no remedial work will be required if the construction satisfies those test requirements.

Where it cannot be demonstrated that the existing construction may be accepted for sound insulating purposes there is one wall treatment, five floor treatments and one relating to stairs given in the approved document. Additional loads on existing structure are to be carefully evaluated and strengthened where necessary.

Wall treatment

The degree of resistance to airborne sound (Fig. E67) arises from

- the nature of the existing wall;
- the mass of an independent leaf;
- isolation of the independent leaf from the existing wall;
- the use of a suitable absorbent material.

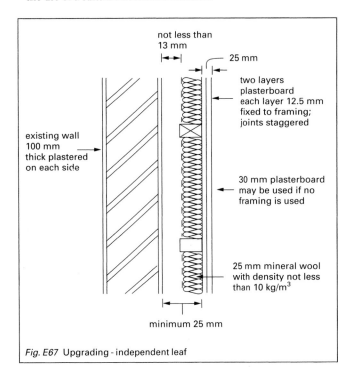

Fig. E67 Upgrading - independent leaf

The treatment may be limited to one side of a masonry wall which has a thickness of not less than 100 mm and is plastered on both faces. Walls not meeting this standard require the treatment on each side.

The perimeter of the new independent leaf is to be sealed with tape or mastic. The new leaf and its supporting frame is not to be in contact with the existing wall although absorbent material may bridge the cavity without being tightly compressed.

Care is required when making a junction of the new leaf and floors, and methods are illustrated in Figs E68 and E69.

First floor treatment

The first method of improving sound insulation in an existing floor involves adding an independent ceiling below the floor. Resistance to both airborne and impact noise transmission relies upon:

- the combined mass of the existing floor and the new independent ceiling;
- the absorbent material included in the new ceiling;
- lack of contact between floor and new ceiling;
- the lack of airways around the whole construction.

An existing ceiling made of lath and plaster may be left in-situ if fire resistance is also acceptable (Part B, Chapter 11) but when the existing ceiling is not lath and plaster it should be upgraded to 30 mm of plasterboard in two or three layers having staggered joints.

- Gaps in floorboarding are to be sealed with caulking or by applying a hardboard cover.
- There is to be a gap of not less than 25 mm between the underside of existing floor construction and top of new independant ceiling joists (Fig. E70).

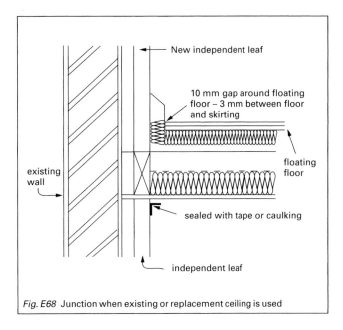

Fig. E68 Junction when existing or replacement ceiling is used

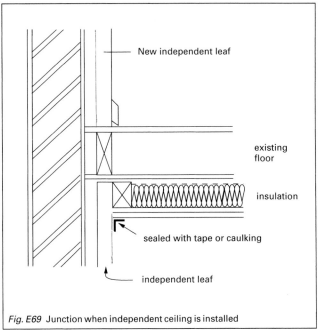

Fig. E69 Junction when independent ceiling is installed

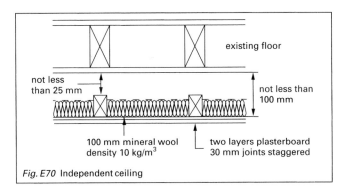

Fig. E70 Independent ceiling

- Perimeter of new ceiling is to be sealed with mastic or tape to prevent airways.
- In circumstances where the new ceiling is below the level of a window head (Fig. E71) it may be cut around the window to form a pelmet.

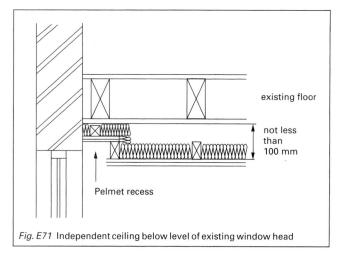

Fig. E71 Independent ceiling below level of existing window head

Second floor treatment

The second method of improving the sound insulation qualities of a floor involves placing a platform floor or floating floor on the surface of the existing floor (Fig. E72). The total sound insulation value then relies upon the mass of the floor and the effectiveness of the layer.

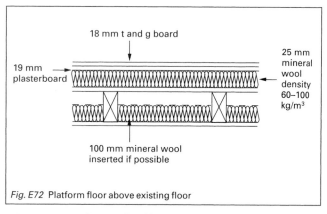

Fig. E72 Platform floor above existing floor

An existing ceiling made of lath and plaster may be left in-situ if its fire resistance is also acceptable (Part B, Chapter 11) but where the existing ceiling is not lath and plaster it should be upgraded to 30 mm of plasterboard in two or three layers having staggered joints.

- If possible, insert 100 mm absorbent layer of mineral wool between the joists (possible during ceiling replacement).
- If existing floor boards should be removed, replace by 12 mm thick boarding and place 100 mm absorbent layer of mineral wool between joists.

The new floating layer may be composed of

- 18 mm tongued and grooved timber or wood-based board, joints glued, and spot bonded to a substrate of 19 mm plasterboard; or
- material with a mass of not less than 25 kg/m² placed in single or double layers and having all joints glued;
- loadbearing resilient layer to be not less than 25 mm thick mineral wool which has a density of between 60 and 100 kg/m³;
- it is necessary to ensure that the layer of resilient material is of the correct density and will carry the expected load;
- there is to be a movement gap 10 mm wide surrounding the floating layer filled with resilient material;
- there is to be a 3 mm gap between floating layer and skirting board; the gap may be sealed with flexible material;
- if a new ceiling is installed its perimeter should be sealed with caulking or tape;
- additional support round the perimeter may be provided by a timber batten with a foam strip along the upper surface and attached to wall.

Third floor treatment

The third method is to develop a ribbed floor with absorbent material or heavy pugging (Fig. E73). The total sound insulation value depends upon the mass of the floor, the absorbent qualities and the material or pugging, and the resilient strips below the floating layer.

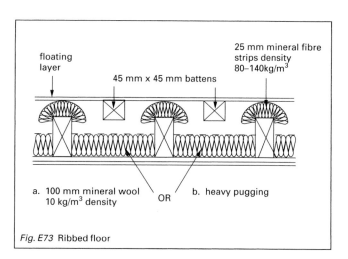

Fig. E73 Ribbed floor

An existing ceiling made of lath and plaster may be left in-situ if its fire resistance is also acceptable (Part B, Chapter 11) but where the existing ceiling is not lath and plaster it should be upgraded to 30 mm of plasterboard in two or three layers having staggered joints.

- If heavy pugging is to be used ensure that the ceiling is suitable.
- Existing joists are to be not less than 45 mm wide.
- After removal of floor boarding check that strutting is not required to ensure stability of the floor.
- In order to carry expected load, ensure that resilient layer is of the correct density.

There are two specifications for the floating layer contained in the approved document. The first gives two alternatives:

- 18 mm tongued and grooved timber or wood-based board all joints glued and spot bonded to a substrate of 19 mm plasterboard, or
- material with a mass of not less than 25 kg/m² placed in single or double layers and having all joints glued.

This method may be used if floor levels are critical:

- when plasterboard is incorporated in the floating layer the 45 mm × 45 mm battens should be placed above the joists;
- absorbent material placed between joists is to be 100 mm thick mineral wool of 10 kg/m³ density.

The second specification relates to a floating layer combined with pugging above the ceiling:

- 18 mm tongued and grooved timber or wood-based board, all joints glued, nailed or glued to 45 mm × 45 mm battens;
- battens are to be placed in the direction of the existing joists and may be sited directly above or between the joists;
- resilient strips on joists are to be 25 mm thick mineral fibre having a density between 80 and 140 kg/m³;
- pugging may be chosen from the following

  75 mm thickness traditional ash
  60 mm thickness 2–10 mm limestone chips
  60 mm thickness 2–10 mm whin aggregate
  50 mm thickness dry sand

  the thickness given will achieve 80 kg/m².

- sand is not to be used in a situation where it may become wet and overload the ceiling, as for example in kitchens, water closets, showers or bathrooms.

For both specifications there are the usual requirements regarding abutting construction:

* there should be a movement gap around the floating layer of 10 mm which should be filled with resilient material;
* there should be a 3 mm gap between floating layer and skirting; if a seal is used it is to be of flexible material;
* perimeter of a new ceiling is to be sealed with caulking or tape.

Fourth floor treatment

This method (Fig. E74) should only be used if there are strong technical reasons for not using the first, second or third methods. There is less reduction in ceiling height than occurs when using method 1, but sound insulation is lower.

* Existing floor should have gaps sealed by caulking or covering with hardboard.
* New ceiling to be not less than 30 mm thick formed from two or three layers of plasterboard with staggered joints.
* New ceiling supported on new ceiling joists, or
* New ceiling suspended from ceiling joists by 2 mm wire hangers or metal straps not exceeding 25 mm × 0.5 mm.
* There is not to be more than one fixing for each square metre.
* A minimum of 100 mm of mineral wool absorbent material having a density of not less than 10 kg/m³ is placed over the ceiling.
* Perimeter of the ceiling is to be sealed with tape or mastic.

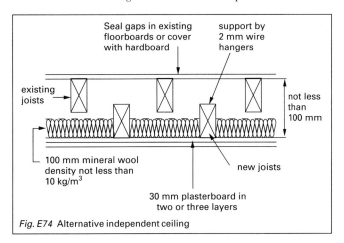

Fig. E74 Alternative independent ceiling

Fifth floor treatment

This method should only be used if there are strong technical reasons for not using the first, second or third methods. There is a lower floor level than results in following method 2.

An existing ceiling made of lath and plaster may be left in-situ if its fire resistance is also acceptable (Part B, Chapter 11) but when the existing ceiling is not lath and plaster it should be upgraded to 30 mm of plasterboard in two or three layers having staggered joints (Fig. E75). Such a new ceiling should be supported by resilient hangers, or fixed to new timber cross battens.

* If it is necessary to replace floorboards, 12 mm thick boarding should be used.
* 18 mm tongued and grooved boards or other boards with glued joints are to be used for the floating layer.
* 13 mm minimum thickness of wood fibre insulation board is to be used to form the resilient layer. The board is to comply with BS 1142: Part 3: 1989, 'Insulating board (softboard)'.

This specifies requirements for fibre building boards which are classified by international usage: (a) hardboard; (b) medium board; (c) medium density fibreboard (MDF) and (d) softboard.

There are four appendices, 10 tables and 17 figures.

This standard provides a co-ordinated statement for laboratories used for measurements of sound insulation of building elements.

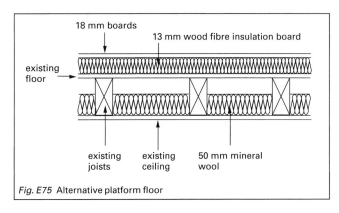

Fig. E75 Alternative platform floor

* 50 mm minimum thickness mineral wool to be used as absorbent material which need not be placed unless the ceiling or floorboards are removed for another purpose.
* There is to be a 10 mm gap around the floating layer filled with a resilient material.
* Perimeter of the ceiling is to be sealed with tape or mastic.

Treatment of stairs

In circumstances where stairs perform a separating function between a dwelling and another or from another part of the same building which is not used solely as part of the dwelling they are subject to sound insulation requirements. Fire-resisting requirements are dealt with in Part B (Chapter 11).

The sound insulation requirements for a stair (Figs E76 and E77) depend upon

* soft covering on the stair treads;
* mass of the stair;
* isolation and mass of independent ceiling;
* in a cupboard enclosure, the volume of absorbent material and airtightness;

Requirements are:

* not less than 6 mm of carpet or other soft covering should be placed on treads;

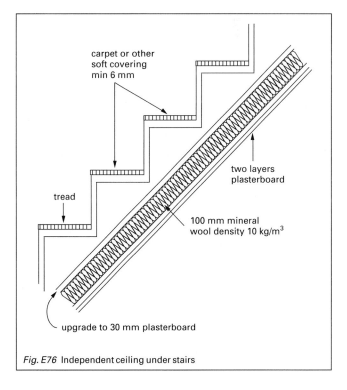

Fig. E76 Independent ceiling under stairs

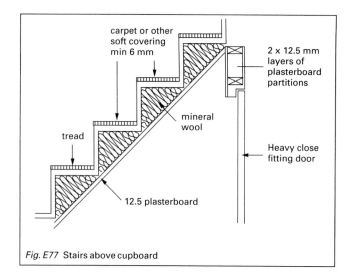

Fig. E77 Stairs above cupboard

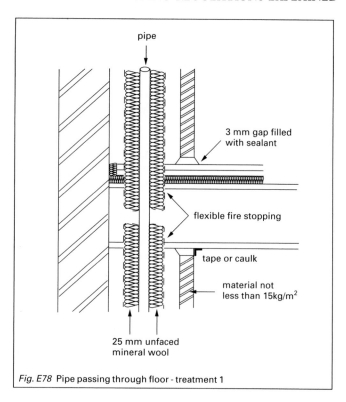

Fig. E78 Pipe passing through floor - treatment 1

- if there is no cupboard below the stairs an independent ceiling is to be constructed;
- form independent ceiling of two layers of plasterboard with a combined thickness of not less than 30 mm; joints are to be staggered and fixed to battens;
- independent ceiling to be not less than 100 mm below ceiling of stairs;
- place 100 mm mineral wool having a density of not less than 10 kg/m³;
- where ceiling abuts other construction it should be sealed with mastic or tape.

Where there is a cupboard below stairs the following is to take place:

- underside of stair is to be lined with not less that 12.5 mm plasterboard and mineral wool filled between risers and treads;
- cupboard walls formed of two layers of minimum 12.5 mm plasterboard or other material having an equal mass/m².
- use heavy close-fitting door.

Piped services

Great care is required, when a pipe or duct passes through a separating floor in a conversion, to seal any gaps and limit the amount of noise that may be conveyed along the pipe or into the surrounding construction. A pipe should not be in contact with the floor when passing through, but surrounded by 25 mm of insulating material (Fig. E78).

- Absorbent material should surround pipe to full height and be enclosed in a duct.
- Material forming duct to have a mass of not less than 15 kg/m².
- Enclosure to be lined or pipe wrapped with not less than 25 mm unfaced mineral wool.
- Gap of 3 mm to be left between enclosure and floating layer and seal with caulking or neoprene.
- When floor treatment 2 is used (Fig E72), the enclosure may extend down to floor base but must be isolated from the floating layer.
- Any penetration of separating floors by pipes or ducts should have flexible stopping and meet requirements of Part B (Chapter 11.)
- These requirements do not apply to gas pipes.
- Requirements regarding gas pipes are contained in Gas Safety (Installation and Use) Regulations 1998. There are requirements for ventilating gas pipes and ducts on each floor. Gas pipes may be placed in a separate vented duct or may remain exposed.

SECTION 6. TEST PROCEDURES

This section outlines ways which will meet the requirements for sound insulation in conversions by repeating a construction which has been built and tested in building or a laboratory.

Laboratories are to comply with BS 2750: Part 1: 1980, 'Measurement of sound insulation – recommendations for laboratories'. This standard provides a coordinated statement of requirements for laboratories used for measurements of sound insulation of building elements.

Contents:

1. Scope and field of apphication
2. References
3. Laboratories for airborne sound insulation measurements under diffuse conditions
4. Laboratories for impact sound insulation of floors and floor coverings.

- A better performance is required from laboratory tests than from field tests because of an adjustment for flanking transmittance.
- For conversion work it is assumed that a typical masonry wall comprises

  (a) half brick
  (b) plastered both sides
  (c) mass not more than 200 kg/m².

A typical timber floor comprises

  (d) timber floor of 22 mm plain edge boards
  (e) joists at 400 mm centres
  (f) ceiling of lath and plaster or not more than 30 mm plasterboard.

Validity

Results of tests are appropriate in situations that are similar to those which have been subjected to the test. If remedial treatment is required for walls or floors less than the above specification, then that form of construction must be tested.

Field tests

Field tests for separating walls:

- Method tested in conjunction with not less than two different half brick walls, plastered on each side.
- If of other construction, then that should be tested.
- Original walls are to be tested alone, followed by a test including the remedial treatment.
- Tests to be carried out in accordance with BS 2750: Part 4: 1980, 'Field measurements of airborne sound insulation between rooms' (page 208).
- Insulation values are to be established using BS 5821: Part 1: 1984, 'Method for rating the airborne sound insulation in buildings and interior building elements' (page 207).
- The weighed standardized level difference attained by both examples of the wall with remedial treatment are to be at least $D_{nT,w} = 49$ dB.

Field tests for separating floors

- Method tested with not less than two different types of remedial treatment.
- Where remedial treatment uses parts of original floor (except the joists), original floor is to be tested and then with the remedial treatment.
- Both results should be available for each type of original floor tested.
- If joists are retained but remedial treatment replaces the rest of the floor, two typical examples of remedial treatment are to be tested.
- Measurements should assess insulation to both airborne and impact/sound insulation.
- The report should state if improvement is only to one – impact or airborne sound insulation.
- Airborne tests are to be carried out in accordance with BS 2750: Part 4: 1980, 'Field measurement of airborne sound insulation between rooms' (page 208).
- Impact sound tests are to be carried out in accordance with BS 2750: Part 7: 1980, 'Field measurement of impact sound insulation of floors'. This gives a procedure to measure the impact sound insulation between two rooms in buildings, and to give a field procedure to determine whether building elements satisfy requirements.

    There are two annexes:

    A. Example of test procedure
    B. Measurement of flanking transmission.

    A table gives correction to sound pressure level readings.
- Insulation values of airborne insulation are to be calculated in accordance with BS 5821: Part 1: 1984, 'Method for rating the airborne sound insulation in buildings and interior building elements'. Values attained by floors with remedial work should not be less than $D_{nT,w} = 48$ dB.
- Insulation values for impact insulation are to be calculated in accordance with BS 5821: Part 2: 1984, 'Method of rating the impact sound insulation'. Results achieved by floors with remedial work is not to be less than $L'_{nT,w} = 65$.

Laboratory tests

- Separating walls should be tested alone and then with remedial work.
- Tests should be carried out in accordance with BS 2750: Part 3: 1980, 'Laboratory measurement of airborne sound insulation of building elements'. This BS specifies a laboratory method of measuring the airborne sound insulation of building elements such as walls, floors, doors, windows, façade elements and façades.

    There are three annexes:

    A. Measurement of flanking transmission
    B. Example of test procedure
    C. Checking the loss factor.

- Insulation values should be arrived at using BS 5821: Part 1: 1984, 'Method for rating airborne sound insulation in buildings and interior building elements' (page 207).
- The value reached by each combined test should not be less than $R_w = 53$ dB.
- Where parts of an original separating floor are to be used by remedial treatment a reproduction of the original floor is to be tested and then tested in conjunction with remedial treatment.
- Where remedial treatment replaces the original floor, other than the joists, then only the remedial treatment must be tested.
- Measurement of airborne and impact insulation should take place unless improvement is required to only one type of insulation.
- Airborne sound insulation of floors are tested in accordance with BS 2750: Part 3: 1980, 'Laboratory measurement of airborne sound insulation of building elements'.
- Impact sound insulation of floors are tested in accordance with BS 2750: Part 6: 1980, 'Laboratory measurement of impact sound measurement of floors'. This standard specifies a laboratory method of measuring impact noise transmissions through floors by using a standard tapping machine.

    Annexes:

    A. Example of test procedure
    B. Measurement of flanking transmission.

    Tables:

    1. Specification for rubber for hammer.
    2. Correction of sound pressure level readings.

- Weighed standardized level difference for airborne sound insulation is measured according to BS 5821: Part 1: 1984, 'Method for rating the airborne sound insulation in buildings and interior building elements' (page 207).
- Value attained by floor and remedial treatment is not to be less than $R_w = 52$ dB.
- Weighed impact sound pressure level is to be calculated with BS 5821: Part 2: 1984, 'Method for rating the impact sound insulation' (page 207).
- Value attained by floor and remedial treatment is not to be more than $L_{nT,w} = 65$.

Test reports

There is a prescribed form for giving results of tests of remedial treatments for separating walls, floors and stairs.

- Name and address of organization carrying out the test, and if appropriate, its NAMAS accreditation number.
- Name of person in charge of test.
- Date of test.
- Details of test equipment and test procedures.
- Description of treatment tested. This is to include a sketch showing position and dimensions of rooms tested; dimension of any step or stagger; description of existing construction; details of any openings within 700 mm of the separating element; the mass (m^2) of the existing construction; and an account of materials and methods used to upgrade existing construction.
- Results of tests of existing construction and upgraded construction in graphical and tabular form.

There is also an example of test report details relating to laboratory tests of remedial treatments.

- Name and address of organization carrying out the tests together with the NAMAS accreditation number where appropriate.
- Name of person in charge of test.
- Date.
- Details of test equipment and test procedures.
- Description of construction tested, giving description of base construction; mass (m^2) of base construction and a description together with a sketch of the upgraded construction.
- Results of tests on base and upgraded construction in tabular and graphical form. Single number ratings are to be shown as appropriate.

Calculating mass

Throughout the whole of the approved document to Part E the word 'weight' has been replaced by 'mass', and there is outlined a method for calculating mass.

- Mass throughout Part E has been expressed in kg/m^2.
- Mass may be obtained from actual figures given by manufacturers, or it may be calculated using a formula given in the approved document.
- The formula is linked to the coordinating height of masonry course and where this is not exactly as given in the formula the nearest height is to be used.
- The formula is based upon the following:

  M = the mass of $1\,m^2$ of leaf in kg/m^2
  T = the unplastered thickness of masonry in m
  D = the density of masonry units in kg/m^3 having 3% moisture content
  N = the number of finished faces
      if no finished faces N = 0
      if one finished face N = 1
      if two finished faces N = 2
  P = the mass of $1\,m^2$ of wall finish in kg/m^2.

- An assumed thickness of 13 mm plaster =

cement render	mass $29\,kg/m^2$
gypsum	mass $17\,kg/m^2$
lightweight plaster	mass $10\,kg/m^2$
plasterboard	mass $10\,kg/m^2$

- Coordinating height of masonry course. The formula

75 mm	M = T (0.79 D + 380) + NP
100 mm	M = T (0.86 D + 255) + NP
150 mm	M = T (0.92 D + 145) + NP
200 mm	M = T (0.93 D + 125) + NP

Where appropriate, densities of bricks and blocks may be obtained from Agrément certificates or ETA certificates. Manufacturers will also supply data, but building control staff may seek an assurance that the organization supplying the information is a properly accredited test house.

- Mortar density of $1800\,kg/m^3$ and a mortar joint of 10 mm are assumed values.
- If within 10% of the figures, the value is acceptable.

REMINDER

1. Sound-resisting walls

It is essential to

- lay bricks frog up;
- fill mortar joints;
- use joist hangers;
- seal wall surface where recommended.

Avoid

- back-to-back electrical sockets;
- plastered resilient surface layers;
- more than minimum connection between leaves of cavity walls;
- bridging cavities;
- air paths.

2. Sound-resisting floors

It is essential to

- use recommended grade and thickness of resilient layer;
- lay paper-faced resilient layer up under concrete;
- lay paper-faced resilient layer down under timber raft;
- turn resilient layers upon edges;
- ensure that structural base is level and smooth prior to laying resilient layer;

- leave narrow gap between skirting and floating layer;
- place rooms of similar function above each other where possible.

Avoid

- leaving any 'temporary' nails in floor;
- service pipes bridging resilient layer;
- gaps in resilient layer.

3. External flanking wall

It is essential to

- ensure good junction between masonry sound resisting walls and external walls;
- ensure adequate window and opening separation.

Avoid

- plastered resilient surface layers unless on expert acoustic advice.

RELEVANT READING

British Standards and Codes of Practice

BS 648: 1964. Schedule of weights of building materials
BS 1142: 1989. Specification of fibre boards
BS 2750: Part 3: 1995. Laboratory measurement of airborne sound insulation of building elements
BS 5228: 1975. Noise control on construction and demolition sites
BS 5228: Part 1: 1997. Code of practice for basic information for noise control
BS 5228: Part 2: 1997. Construction and demolition work
BS 5228: Part 4: 1992. Vibration control – piling
BS 5628: Part 3: 1985. Masonry materials, components, design and workmanship
BS 5821: Part 1: 1984. Rating airborne sound insulation in buildings and building elements
BS 5821: Part 2: 1984. Method of rating impact sound insulation
BS 5821: Part 3: 1984. Airborne sound insulation of façade elements
BS 7643: Part 3: 1993. Building construction: acoustical requirements
BS 8233: 1987. Code of practice for sound insulation and noise reduction for buildings
CP 153: Part 3: 1972. Sound insulation, windows and roof lights

BRE digests

293	Improving sound insulation – walls and floors
333	Sound insulation of separating walls and floors – Part 1: walls
334	Sound insulation of separating walls and floors – Part 2: floors
337	Sound insulation – basic principles
338	Insulation against external noise
347	Sound insulation of lightweight dwellings
379	Double glazing for heat and sound insulation

BRE information papers

24/82	Field measurements – plastered lightweight masonry
9/83	Ratings of sound insulation of post-1970 party walls
10/83	Ratings of sound insulation of party floors
2/81	Field measurement of sound insulation of timber joist party floors
8/81	Field measurement of the sound insulation of floors with floating screeds and hollow precast bases
6/88	Methods of improving the sound insulation between converted flats
9/88	Methods of reducing impact sounds in dwellings
12/89	Insulation of dwellings against external noise
13/82	Noise from neighbours – sound insulation of party walls in houses
19/92	Sound insulation and the 1992 Approved document E

21/93 The noise climate around houses
22/93 Effects of environmental noise on people at home
 6/94 Sound insulation provided by windows.

BRE defect action sheets

45   Sound insulation – intermediate floors in converted dwellings
46   Masonry walls – chasing
81   Plasterboard ceilings – noggings and fixings (Specification)
82   Plasterboard ceilings – noggings and fixings (Sitework)

104   Masonry separating walls – airborne sound insulation in new build housing (Design)
105   Masonry separating walls – improving airborne sound insulation between existing dwellings (Design)

BRE reports

Sound control for homes (BR 238, 1993)
Quiet homes: a guide to good practice and reducing the risk of poor sound insulation.

CHAPTER 15

Ventilation and condensation

(Approved document F)

Regulation 4 requires that building work is to be carried out in a way which meets the requirements of Schedule 1. When doing work complying with a requirement of Part F of Schedule 1, it must not lead to a failure to comply with any other requirement.

When a building is altered or extended no part of it is to be adversely affected by the work of alteration or extension.

There are only two requirements in Part F of Schedule 1. The first refers to means of ventilation and the second says that provision is to be made to prevent excessive condensation in roofs.

Building byelaws always sought to ensure that sufficient open space was left about dwellings to secure an adequate circulation of air. The need for this form of control arose from the congested conditions in Victorian working-class terraced houses and a minimum requirement for space at the front and rear was devised. It was found in practice that quite frequently doubt existed about the identity of the front or rear of some dwellings and, furthermore, as the byelaws made no reference to open spaces at the sides of dwellings it frequently happened that a window in a flank wall was only 1 m or less from the boundary. Nothing could be done if the adjoining owner erected a structure or a high fence that reduced the airspace to such a window. Reduction of daylight may have been another matter.

The 1965 Regs were claimed to cater for the vast changes in layout and design found in modern dwellings and housing estates. A zone or shaft of space open to the sky was required alongside a window to a habitable room. The requirements did not demand or make provision for daylighting as such. They were concerned with making adequate provision for the circulation of air in locations affording adequate ventilation to habitable rooms.

The basic concept behind those regulations, still contained in the 1976 Regs, was a relationship between an area of open space associated with the habitable room so that people living in the room could get more benefit from space provided.

Therefore a zone or shaft of space open to the sky of at least 3.6 m wide was required for windows serving habitable rooms. The method of defining the zones of open space admitted a considerable variety of arrangements, but in all cases it was mandatory to provide a zone unobstructed by rising ground or buildings above a specified sloping plane. The zone was also to be completely within site boundaries, except that half the width of an adjoining street, canal or river could be counted as part of a zone.

This requirement is missing from the current regulations as are also the following which were included within the 1976 Regs:

- shared land on housing estates (K2)
- preservation of zones of open space (K3)
- height of habitable rooms (K8).

Regulations do not now contain any requirement regarding the height of habitable rooms. One of the earliest requirements of building byelaws 90 years ago, the height required has been reduced from time to time until in the 1976 Regs it was 2.3 m. Now no height is specified, although the height above a landing must not be less than 2 m. Market forces are being left to dictate an acceptable minimum ceiling height.

The building regulations do not require that windows be installed although, if provided, they may have a ventilation function.

VENTILATION

Older readers will recall byelaw requirements for air bricks in habitable rooms, kitchens and bathrooms, but over the passage of time these provisions have been eroded and any ventilation problems arising compounded by fashionable draught-proofing.

Now we find that the main purpose of Part F is to raise the standard of ventilation, partly in response to concern about increased condensation risks and also to improve air quality.

Mechanical ventilation is now advocated in kitchens, bathrooms and shower rooms together with a new expression 'background ventilation' in addition to openable windows in habitable rooms.

Passive stack ventilation is an alternative option to mechanical extraction for domestic kitchens, bathrooms and sanitary accommodation.

The requirements for the ventilation of roof spaces apply to roofs as a whole and to all building types.

The requirements of F1 are that there should be 'adequate' means of ventilation, and this is directed towards the avoidance of condensation in addition to ensuring a supply of pure air for the occupants to breathe.

Three modes of ventilation are advocated:

- moisture extraction
- occasional rapid ventilation
- trickle ventilation.

A ventilation opening should have a smallest dimension of not less than 5 mm for slots or 8 mm for square or circular holes.

Performance

It is the Secretary of State's view that F1 will be satisfied if ventilation, as provided, will under normal conditions be capable 'if used' of

- restricting the accumulation of moisture which may give rise to mould growth;
- restricting the accumulation of pollutants arising within a building which would otherwise become a hazard to the health of people in the building.

Aim to achieve by provision of ventilation

- before it becomes generally widespread, water vapour arising in significant quantities should be extracted. Examples are from kitchens, utility rooms and bathrooms.
- before they become generally widespread, pollutants which are a hazard to health arising in significant quantities should be extracted. Examples are rooms where processes generate harmful contaminants, and rest rooms where smoking is allowed.
- when necessary dilute rapidly pollutants and water vapour arising in habitable rooms, occupiable rooms and sanitary accommodation.

- provide over long periods a minimum supply of fresh air for occupants. Where necessary disperse residual water vapour. This kind of ventilation should

  (a) not significantly affect comfort
  (b) be reasonably secure
  (c) provide protection against rain penetration.

Air conditioning or mechanical ventilation for non-domestic buildings would satisfy the above performance standards if they are

- designed, installed and commissioned to function in a manner which is not detrimental to the health of people in the building, and
- designed to ensure ease of maintenance.

OPEN-FLUED APPLIANCES

Where an open-flued appliance is to be provided, refer to Part J (Chapter 18).

OBJECTIVES

The approved document spells out three objectives: the importance of

- extracting moisture from rooms where it is produced in large quantities, for example in kitchens, bathrooms and shower rooms;
- occasional rapid ventilation for the dilution of pollutants and moisture which could give rise to condensation in habitable rooms and sanitary accommodation;
- background ventilation. The provision should not affect security and not significantly affect comfort in habitable rooms and kitchens. If it were to create draughts or other discomfort, then its use would be discouraged.

F1 in Schedule 1 requires that there is to be adequate means of ventilation provided for people in a building.

Requirements for non-domestic buildings were introduced by the 1994 Regs for the first time and reflect the relevant requirements of the Workplace (Health, Safety and Welfare) Regulations 1992, so that new building work should not require subsequent alteration.

The requirement for ventilation has become much more embracing than in the 1991 Regs and includes all buildings except a building, or space within a building

- into which people do not normally go; or
- which is used solely for storage; or
- which is a garage used solely in connection with a single dwelling.

If the requirements of approved document F are followed the work would, in accordance with section 23 of the Health and Safety at Work, etc ., Act 1974 avoid the service of a notice concerning the ventilation requirements in Regulation 6(1) of the Workplace (Health, Safety and Welfare) Regulations 1992. This regulation requires effective and suitable provision for the ventilation of an enclosed workplace by enough fresh or purified air.

The approved document contains a number of definitions:

- Bathroom includes a shower room and can include sanitary accommodation.
- Domestic buildings are those buildings used for dwelling purposes including dwellinghouses, flats, student accommodation and residential homes.
- Habitable room describes a room used for dwelling purposes but which is not solely a kitchen.
- Non-domestic buildings are all buildings not used as domestic buildings, including buildings where people temporarily reside, such as hotels.
- Occupiable room is a room in a non-domestic building such as a classroom, hotel bedroom, office. It does not include a bathroom,

sanitary accommodation, utility room or rooms or spaces used solely or chiefly for circulation, services plant and storage.

- Passive stack ventilation (PSV) comprises a ventilation system utilizing ducts from the ceiling of rooms to outlets on the roof. It operates by a combination of the natural stack effect arising from the movement of air due to the difference in temperature between inside and outside and also the effect of wind passing over the roof terminal.

  PSV, when used in conjunction with trickle vents, provides extract ventilation for kitchens, bathrooms and utility rooms. The passive stack outlet may be a ceiling mounted circular grill.

  Air movement up the stack must not be impeded and therefore the ductwork must offer minimal resistance to the air flow. Bends should be avoided but, when necessary, should be swept rather than sharp. Individual ducts should not be joined as this could provide cross contamination.

  Ductwork should be insulated and may be flexible or rigid with a diameter of 100–150 mm. Care is needed in the placing of terminals on the roof to avoid the possibility of downdraught.

- Sanitary accommodation (Fig. F1.1) relates to a space accommodating one or more closets or urinals. Sanitary accommodation wherein there are one or more cubicles (Fig. F1 .2) is regarded as a single space if there is free air circulation around the room or space and not to the cubicles separately.

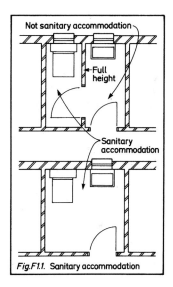

Fig.F1.1. Sanitary accommodation

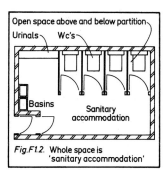

Fig.F1.2. Whole space is 'sanitary accommodation'

- Utility room is a room designed to be or likely to be used to contain clothes, washing and similar equipment such as

  (a) a sink
  (b) a washing machine
  (c) a tumble drier
  (d) other feature or equipment likely to produce significant amounts of water vapour.

- Ventilation opening includes any permanent or closable means of ventilation, providing it opens directly to the external air.

Ventilation openings comprise

1. openable parts of a window
2. a louvre
3. progressively openable ventilator
4. window trickle ventilator
5. a door opening directly to the external air

Ventilation openings are to have limited dimensions to minimize resistance to air flow. Ventilation openings may be protected by a screen, fascia or baffle to protect against insects.

Satisfactory ventilation by natural or mechanical means will be achieved if ventilation openings meet the following requirements (Fig. F1.3).

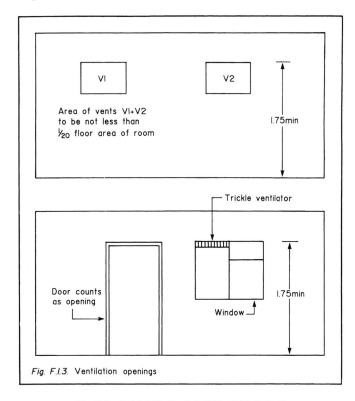

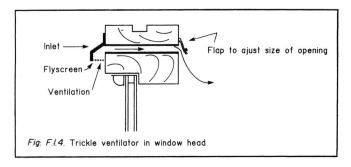

Fig. F.1.3. Ventilation openings

DOMESTIC BUILDINGS

Satisfactory ventilation may be achieved if provision is made as follows.

- Rapid ventilation: one or more ventilation openings, some part of it being not less than 1.75 m above floor level, and
- Background ventilation: one or more ventilation openings such as trickle ventilators, air bricks, 'hit and miss' ventilators or an appropriately designed opening window. The openings are to be controllable, secure and located with some part not less than 1.75 m above floor level in a position where draughts and the entry of rain will be avoided. Examples are shown in Figs F1.4, F1.5, F1.6, F1.7 and F1.8.

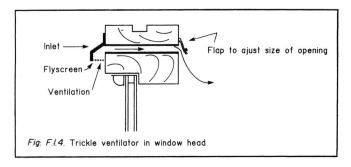

Fig. F.1.4. Trickle ventilator in window head

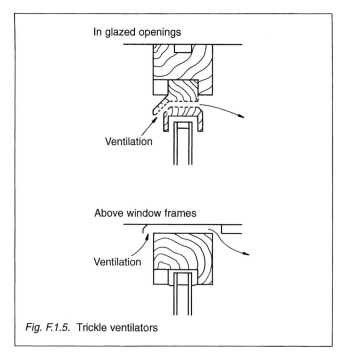

Fig. F.1.5. Trickle ventilators

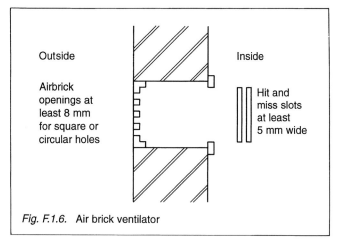

Fig. F.1.6. Air brick ventilator

Ventilation provision when fully open should have a minimum dimension of not less than 5 mm for slots or 8 mm for square or circular holes. Vertical sliding sash windows should have an adjustable opening to give the required area and be provided with a secure locking device. High-level top-hung window provision should be limited to use above ground storey level as a measure of security, and provided with a lock to hold in the required position, actuated by a removable key.

It should not be overlooked that some windows may be required as a means of escape, and must therefore meet the provisions of B1; and

- Extract ventilation may be provided by

  (a) mechanically operated manually and/or automatically by sensor or controller; or
  (b) PSV operated manually and/or automatically by sensor or controller. PSV provided in accordance with BRE IP 13/94, or BBA certification would be satisfactory, or
  (c) an appropriate open-flued heating appliance.

Open-flued appliances when being operated contribute to the ventilation of the room in which they are installed. Arrangements may be made for the provision of adequate extract ventilation when they are not firing. No additional extract ventilation will be required if

  (a) a solid fuel open-flued appliance is a primary source of heating, cooking or hot water production; or

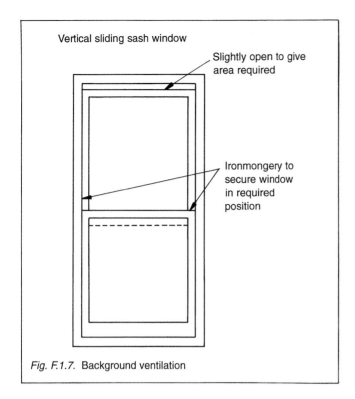

Fig. F.1.7. Background ventilation

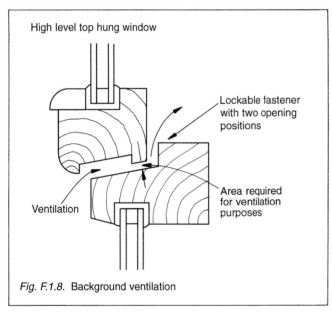

Fig. F.1.8. Background ventilation

(b) an open-flued appliance has a flue area at least equivalent to a 125 mm diameter duct, and the appliance combustion air inlet and dilution air inlet are permanently open. This means that no control dampers which could block the air flow are closed when the appliance is not in use.

The main changes to requirements are now outlined and they came into effect on 1 July 1995.

VENTILATION OF ROOMS

Habitable rooms

A habitable room is one which is used for dwelling purposes, but which is not solely a kitchen. It would therefore include a dining/kitchen.

Requirements are:

- opening windows equal to 1/20th floor area of the room;
- background ventilation equal to 8000 mm².

Kitchen

The requirement for the ventilation of kitchens (Fig. F1.9) in dwellings will be satisfied if there is an opening window and no minimum size is given in the approved document; and background ventilation totalling 4000 mm².

Also mechanical extract ventilation is required to be capable of extracting 30 l/s if it is adjacant to a hob, or 60 l/s if it is elsewhere; or PSV.

Mechanical extract ventilation should be rated

- not less than 30 l/s when incorporated within a cooker hood or located near the ceiling within 300 mm of the centre line of the space for the hob and under humidistat control; or
- at not less than 60 l/s when placed elsewhere.

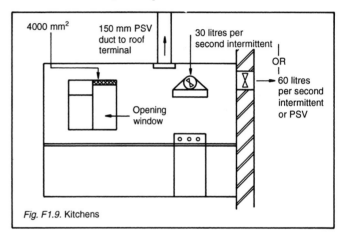

Fig. F1.9. Kitchens

Utility room

The 1994 Regs introduced a ventilation requirement for the first time and for the purposes of the building regulations requirements do not apply to a utility room which is accessible only from outside the building.

The requirements are

- an opening window, and no minimum size is given;
- background ventilation equal to 4000 mm²;
- mechanical extract ventilation capable of 30 l/s, or PSV.

Bathroom

A bathroom, which may or may not contain a WC, is to have

- an opening window, and no minimum size is given;
- background ventilation equal to 4000 mm²;
- mechanical extract ventilation capable of 15 l/s, or PSV. (Fig. F1.10)

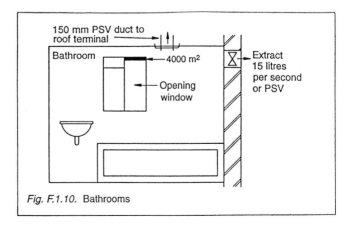

Fig. F.1.10. Bathrooms

Sanitary accommodation

Sanitary accommodation which is separate from a bathroom may achieve satisfactory ventilation if

- there is rapid ventilation via one or more ventilation openings having a total area of not less than 1/20th the floor area of the space, or a mechanical extractor capable of removing not less than 6 l/s;
- in addition, background ventilation equal to 4000 mm² is required.

As an alternative to the various requirements given for background ventilation, provision may be made by an average of 6000 mm² per room for the rooms listed with a minimum provision of 4000 mm² for each room.

Some rooms may have a combined function; for example, a kitchen diner and the individual provision for rapid, background or extract ventilation are not required to be duplicated if the greatest provision for the individual functions is made.

Non-habitable rooms

This requirement applies to kitchens, utility rooms, bathrooms and sanitary accommodation which are internal rooms and do not have openable windows.

- Mechanical extract ventilation rated as for rooms containing openable windows (page 219) the fan having a 15 minutes overrun and controlled automatically or manually. Where there is no natural light the fan may be controlled by the light switch; or
- PSV manually or automatically controlled by sensor or controller, or
- an open-flued heating appliance (see page 265).

In each case an air inlet is to be provided, which may be satisfied by a 10 mm gap under the door (Fig. F1.11).

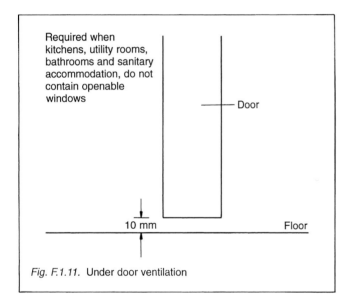

Fig. F.1.11. Under door ventilation

Ventilation through other rooms or spaces

- Two habitable rooms may be regarded as a single room. There is to be a permanent opening between the rooms equal to not less than 1/20th the combined floor area (Fig. F1.12).

Ventilation of a habitable room through an adjoining space may be satisfactory if

- the adjoining space is a conservatory or similar space; and
- there is an opening between the habitable room and the conservatory equal to 1/20th the combined floor areas (the opening may be closable);

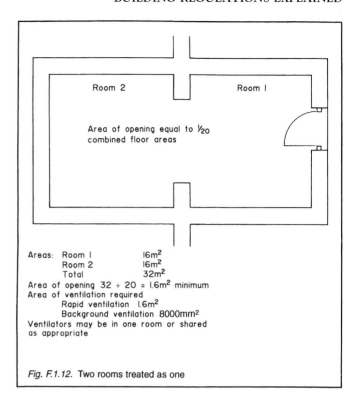

Fig. F.1.12. Two rooms treated as one

- there is an opening between the habitable room and the conservatory equal to 1/20th the combined floor areas (the opening may be closable);
- there is a ventilation opening, or openings in the space totalling 1/20th the combined floor areas;
- some part of the ventilation opening must be not less than 1.75 m above floor level.

To achieve background ventilation requirements there are to be ventilation openings to the space of at least 8000 mm². In addition there are also to be ventilation openings of at least 8000 mm² between the room and space. Openings are to be located in positions where undue draughts will be avoided (Fig. F1.13).

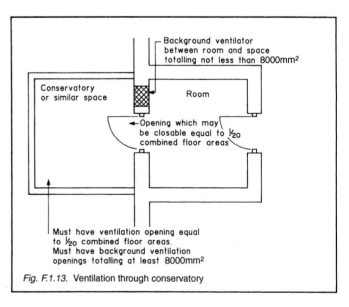

Fig. F.1.13. Ventilation through conservatory

MECHANICAL EXTRACT VENTILATION

Spillage of flue gases from open-flued combustion appliances can be caused whether or not extract fans, or extract air terminals and the appliances are located in the same room. The spillage of flue gases can be dangerous and it is necessary to ensure that the appliance will operate safely whether or not an extract fan is running.

- *Gas appliances.* Where appliance and fan are in the same room the rate of extraction should not exceed 20l/s. A spillage test as described in BS 5400: Part 1: 1990, clause 4.3.2.3 should be carried out whether or not the appliance or fan is in the same room.

This clause gives methods of test with the appliance connected. It first requires compliance with the free-flow test which is a visual check that smoke is leaving the correct terminal and there is no downdraught.

Then with the appliance connected, windows and doors are to be closed and any fans operated. Using manufacturer's instruction checks are to be made for 'spillage'. The appliance is not to be left connected to the gas supply if it has not satisfactorily passed the spillage test.

If a fan causes an appliance in another room to spill, a reduction in the extract rate may cure the problem, and there is further advice in BRE Information paper 21/92.

- *Oil-fired appliances.* These appliances should be installed in compliance with Technical Information Note T1/112 obtainable from Oil Firing Technical Association for the Petroleum Industry (OFTEC), Century House, 100 High Street, Banstead, Surrey SM7 2NN.
- *Solid fuel appliances.* Mechanical extract ventilation should not be installed in the same room. Further advice is available from HETAS – Heating Equipment Testing and Approved Scheme, PO Box 37, Bishop's Cleeve, Gloucestershire GL52 4TB.

ALTERNATIVE APPROACHES

The approved document allows the various requirements to be satisfied by adhering to the relevant recommendations in the following.

- BS 5925: 1991, 'Code of practice for ventilation principles and designing for natural ventilation'. The relevant clauses are 4.4, 4.5, 4.6.1, 4.6.2, 5.1, 6.1, 6.2, 7.2, 7.3, 12 and 13.
  Content of the various clauses is as follows:

4.4	The dilution and removal of airborne pollutants, including odour and tobacco smoke.
4.5	Sets out a method for the control of internal humidity.
4.6.1	Regarding an air supply to fuel burning appliances.
4.6.2	Requirements for an air supply for open-flued domestic heating appliances.
5.1	Sets out a method of designing and calculating air needs on a room-to-room basis.
6.1	Concerning the provision for ventilation and includes methods, infiltration, outlets and inlets, natural ventilation, mechanical ventilation and the effects of ventilation.
6.2	Regarding air supply and includes recommendations in connection with intermittent pollutant emissions, and ventilation effectiveness.
7.2	Is concerned with the limitations of natural ventilation and comments on quality, quantity and controllability.
7.3	Sets out the need for mechanical ventilation and its desirability.
12	Discusses the determination of natural ventilation rates, including wind only, temperature differences only, the combined effect of wind and temperature difference.
13	Describes the other mechanisms of natural ventilation including natural ventilation of spaces with openings on one side of a building, wind, temperature difference combined wind and temperature difference and pressure fluctuation.

- BS 5720: 1979, 'Code of practice for mechanical ventilation and air conditioning in buildings'. The relevant clauses are 2.3.2.1, 2.3.3.1, 2.5.2.9, 3.1.1.1, 3.1.1.3 and 3.2.6.

The code deals with the work involved in design, installation and maintenance of mechanical ventilation and air-conditioning systems. Recommendations made recognize the need to optimize the use of energy, reduce hazards and minimize effects detrimental to the environment.

Content of the clauses is as follows:

2.3.2.1	Says that a fresh air supply is required to maintain an acceptable non-odorous atmosphere, and gives minimum fresh air supply rates for mechanically ventilated spaces.
2.3.3.1	Covers air purity and filtration saying that ventilation should clean, freshen and condition the air within a space.
2.5.2.9	Gives recommendations regarding the installation of gas/oil-fired equipment.
3.1.1.1	In dealing with design considerations, says that the simplest form of extract system usually comprises one or more fans with a simple discharge to the open air.
3.1.1.3	Covers combined mechanical supply and extract systems which can provide better control of comfort conditions.
3.2.6	Ventilation and infiltration rates are discussed.

- BS 5250: 1989, 'Code of practice for the control of condensation in buildings'. The relevant clauses are 6, 7, 8, 9.1, 9.8, 9.9.1 to 9.9.3 and Appendix C.
  Content of the clauses is as follows:

6	Sets out design objectives; prevention of harmful surface or interstitial condensation; the prevention of mould growth and economical reduction of natural condensation.
7	Outlines design procedures.
8	Describes design principles considering occupant activity, heating and ventilating regime; building configuration; construction; heating and ventilating systems.
9.1	Deals with the general application of design principles.
9.8	Heating.
9.8.1	To minimize surface condensation the aim should be to maintain an air temperature at or above 10°C to 12°C in all parts of a building.
	Detailed calculations should take account of solar gains, ventilation areas, climatic conditions and the thermal response of the building.
9.8.2	If a warm air system will not be used for prolonged periods it is not recommended for high mass buildings, but it is ideally suited for low mass buildings with a fast response.
9.8.3	Hot water radiator systems are suitable for higher thermal mass buildings.
9.8.4	High-temperature radiant heaters include radiant gas and electric fires, but are liable to be used insufficiently to warm a heavy structure adequately. Gas fires must have a satisfactory air supply.
9.8.5	Electric storage heaters are particularly suitable for buildings occupied for long periods and for buildings of high mass.
9.8.6	Low-temperature radiant heaters usually have a low rate of heat output and are effective in controlling condensation.
9.8.7	Unflued gas appliances are to be discouraged because they release large quantities of water vapour into the room.
9.8.8	With open fires considerable quantities of air are required and there is much heat loss up the flue.
9.8.9	Controls should be provided, aimed at producing the temperatures required in each room.
9.9.1	Ventilation provision should aim to control the humidity of internal air between 40% and 70%. An ideal system of ventilation will provide freely controllable background ventilation, and extraction of moisture from kitchens and bathrooms.
9.9.2	Controllable slot ventilators should be provided in windows or walls. 4000 mm² background ventilation opening should be provided – more in large rooms.

9.9.3 Forced ventilation is preferable to natural ventilation because air changes can be controlled.

Appendix C. This sets out procedures to assess the risk of surface condensation.

- BRE Digest 398, 'Continuous mechanical ventilation in dwellings; design; installation and operation'. The information in this Digest may be used for the design of either

  1. continuous balanced (supply and extract) ventilation throughout a dwelling, or
  2. continuous mechanical extract ventilation for installation in kitchens, utility rooms, bathrooms and sanitary accommodation.

NON-DOMESTIC BUILDINGS

The 1994 Regs introduced for the first time requirements relating to the ventilation of non-domestic buildings. Provision is made for the mechanical ventilation of rooms; ventilation of common spaces; ventilation of specialist activities; car parks and the design of mechanical ventilation and air-conditioning plant.

Requirements to meet the performance standard will be satisfied if provision is made for

- Rapid ventilation. There should be one or more ventilation openings, such as an opening window, having some part at least 1.75 m above floor level; and
- Background ventilation. There should be an opening or openings such as trickle ventilators, air bricks, or suitable designed opening windows. Each should be adjustable and located 1.75 m above floor level, to prevent the ingress of cold draughts and rain; and
- Extract ventilation. There may be mechanical extract ventilation operated manually and/or automatically by sensor or controller.

VENTILATION OF ROOMS

Occupiable room

An occupiable room is one in a non-domestic building such as a hotel bedroom, or rooms in which people work. It does not include a bathroom or sanitary accommodation, circulation space, storage or plant rooms.

Requirements are

- Opening windows equal to 1/20th floor area of room;
- Background ventilation:

  (a) floor areas up to 10 m² = 4000 mm²,
  (b) floor areas more than 10 m² = at rate of 400 mm per m² of floor area.

There are more stringent requirements for specific rooms designed for heavy smoking, such as rest rooms.

Kitchen

This relates to the domestic type kitchen where the appliances and usage are of a domestic nature. The provision does not apply to commercial kitchens.

Requirements are:

- Opening windows; but no minimum size is given.
- Background ventilation – 4000 mm².
- Mechanical extract ventilation of 30 l/s near the hob or 60 l/s when located elsewhere in the room. PSV may be used as an alternative to extract ventilation, for domestic type facilities.

Bathrooms

This includes shower rooms. The requirements are:

- Opening window, but no minimum size is given.

- Background ventilation required is 4000 mm² for each bath and/or shower.
- Mechanical extract ventilation of 15 l/s for each bath or shower. PSV may be used as an alternative to extract ventilation.

Sanitary accommodation

These requirements also extend to washing facilities.

- Rapid ventilation requirements embrace opening windows equal to 1/20th the floor area, or mechanical ventilation capable of removing 6 l/s for each WC, or 3 air changes per hour.
- Background ventilation required is 4000 mm² for each WC.

Mechanical extraction associated with open-flued appliances may lead to the spillage of flue gases, and open-flued appliances must be able to operate safely, whether or not a fan is running.

MECHANICAL VENTILATION

The requirements are satisfied if

- In occupiable rooms mechanical ventilation is provided at not less than 8 l/s of fresh air per occupant, on the assumption that no smoking is permitted. When provision is made for light smoking there should be fresh air supplied at the rate of 16 l/s.
- In the case of kitchens, bathrooms and sanitary accommodation without windows where provision has been made for mechanical extract ventilation, and the fan has a 15 minutes overrun and is controlled either manually or automatically. Fans may be controlled by the light switch or occupant-detecting sensor. An air inlet must be provided, and may be a 10 mm gap below the door (Fig. F1.11).

COMMON SPACES

Common spaces in non-domestic buildings, where large numbers of people may gather, as in shopping malls and foyers, are to have provision for ventilation on the following basis.

- Natural ventilation. Appropriately sited ventilation openings having a total area not less than 1/50th the floor area of the common space; or
- Mechanical ventilation. Designed to provide a supply of fresh air of 1 l/s per 1 m² of floor area.

These requirements do not apply to common spaces used solely or chiefly for circulation purposes.

ALTERNATIVE APPROACHES

The approved document allows the various requirements for non-domestic buildings to be satisfied by following the relevant recommendations in the following.

- BS 5925: 1991, 'Code of practice for ventilation principles and designing for natural ventilation'. The relevant clauses are 5.1, 5.2, 6.1, 6.2, 7.3, 12 and 13.

  5.1 Sets out a method of designing and calculating air needs on a room-to-room basis.
  5.2 Is concerned with special applications including factories, garages, hospitals, large communal kitchens, underground rooms and internal common access lobbies.
  6.1 Discusses the provision for ventilation and includes methods, infiltration, inlets and outlets, natural ventilation, mechanical ventilation and the effects of ventilation.
  6.2 Is about air supply and includes recommendations regarding intermittent pollutant emissions.

7.3 Sets out the need for mechanical ventilation, and its desirability.
12 Discusses the determination of natural ventilation rates, including wind only, temperature differences only, and the combined effect of wind and temperature difference.
13 Describes the mechanisms of natural ventilation, including natural ventilation of spaces with openings on one side of a building and wind, temperature difference, combined wind and temperature difference and pressure fluctuation.

- By complying with the appropriate recommendations to be found in the CIBSE Guide A, 'Design data', and CIBSE Guide B, 'Installation and equipment data'. The appropriate sections are A4 – Air infiltration and natural ventilation, and B2 – Ventilation and air-conditioning requirements.

SPECIALIST ACTIVITIES

The approved document gives design guidance for ventilation in connection with a number of specialist activities.

- *Schools.* Ventilation for schools or other educational establishments may follow the foregoing requirements except for sanitary accommodation where 6 air changes per hour are required. It is also appropriate to follow requirements in the Education (School Premises) Regulations.

   Where noxious fumes may be generated in spaces, extra provision should be made for ventilation, including the use of fume cupboards designed in accordance with DFE Design Note 29.
- *Workplaces.* The recommendations for specific workplaces and work processes contained in HSE Guidance Note EH22, 'Ventilation in the workplace', are to be followed.
- *Hospitals.* Requirements differ depending on the functional uses, and these values may vary throughout the year. Requirements are to be found in the DHSS Activity Data Base with general guidance in Department of Health Building Notes appropriate to each specific departmental area.
- *Building services plant rooms.* There should be provision for emergency ventilation to control the dispersal of contaminating gases. Guidance is contained in HSE Guidance Note EH22, 'Ventilation in the workplace'. Other guidance is to be found in BS 4434: 1989, 'Specification for safety aspects in the design, construction and installation of refrigeration appliances and systems'.

   The contents are as follows.

1. The first part of this specification covers scope; definitions; refrigerant groups; categories of occupancy and the types of cooling systems.
2. Deals with refrigerant groups and cooling systems, and embraces the maximum charge of refrigerant, design and maximum working pressures, pressure vessels, heat exchangers, piping and fittings, as well as the protection against excessive pressure and instrumentation.
3. Covers installation generally, including the machinery room, safety provisions, refrigerant detection, electrical installation, the changing and discharging of refrigerant.
4. Is concerned with the testing, inspection, documentation, and marking of the installation.

   There are eight appendices; eight tables and two figures.
- *Rest rooms:* In 'mechanical ventilation' mention is made of extra provision for ventilation where smoking takes place in a room and this paragraph relates to 'rest rooms' where heavy smoking may take place.

   The workplace regulations stipulate that 'Rest rooms and rest areas shall include suitable arrangements to protect non-smokers from discomfort caused by tobacco smoke'.

   The approved document gives the following guidance to meet this requirement.

1. Natural ventilation. There should be an air supply to meet the requirement for an occupiable room, and also the removal of tobacco smoke particles through local extract ventilation.

2. Mechanical ventilation. The recirculation of air contaminated with tobacco smoke should be prevented. Smoke laden air is to be extracted to the external air at not less than 16 l/s per person.
- *Commercial kitchens:* CIBSE Guide B, Tables B.2.3 and B.2.11, should be followed.

CAR PARKS

The requirement for car parks below ground level; enclosed type car parks and multi-storey car parks will be satisfied if there is

- Naturally ventilated car parks. Well-distributed permanent natural ventilation is required and this means openings at each car parking level having an aggregate area not less than 1/20th of that floor level area. At least half is to be in two opposing walls.
- Mechanically ventilated car parks. There may be

1. Provision of permanent natural ventilation openings of not less than 1/40th the floor area, plus a mechanical system able to achieve not less than three air changes per hour; or
2. In the case of a basement car park the provision of mechanical ventilation capable of at least six air changes per hour; and
3. For exits and ramps, where cars queue inside the building, there is to be a local ventilation rate of not less than 10 air-changes per hour.

Alternative approach

An alternative approach provides for the requirement to be satisfied if the mean predicted pollutant levels are calculated. The ventilation system should then be designed to limit the concentration of carbon monoxide to not more than 50 parts per million, averaged over an eight-hour period and peak concentration, for example by ramps and exits, are not to exceed 100 parts per million for periods of not more than 15 minutes.

   There is further guidance in 'Code of practice for ground floor, multi-storey and underground car parks' published by the Association for Petroleum and Explosives Administration and in the CIBSE Guide B, Section B2–6 and Table B2–7.

DESIGN OF PLANT

Provision is to be made to protect fresh air supplies from contaminants injurious to health. Placing air inlets close to a flue; an exhaust ventilation system outlet; an evaporative cooling tower or from an area in which vehicles manoeuvre, or elsewhere where excessively contaminated air may be drawn into the building is to be avoided.

   There is guidance in steps to avoid legionella contamination to be found in HSE's 'The Control of legionellosis including legionnaire's disease'. There is also guidance on recirculated air in air-conditioning and mechanical ventilation systems in HSE's Workplace (Health, Safety and Welfare) Regulations 1992 approved code of practice and guidance L24.

Alternative approaches

The approved document allows the various requirements for mechanical ventilation/air-conditioning plant to be satisfied by following the relevant recommendations in the following.

- BS 5720: 1979, 'Code of practice for mechanical ventilation and air conditioning in buildings'. The relevant clauses are 2.3.2, 2.3.3, 2.4.2, 2.4.3, 2.5, 3.2.6, 3.2.8 and 5.5.6.

   Content of the various clauses is as follows.

2.3.2 This describes ventilation and air movement; fresh air supply; transfer of heat, moisture and air movement.
2.3.3 Concerned with air purity and filtration, covers the removal of particulate matter from air, and also the removal of fumes and smells from air.
2.4.2 Gives guidance for various applications – commercial

offices, hotel guest rooms, restaurants, cafeterias, bars, nightclubs, department stores and shops, theatres and auditoria and residential buildings.

2.4.3   Relates to special applications such as hospitals, operating theatres, computer rooms, 'clean' rooms, laboratories, libraries, museums, art galleries, industrial buildings.

2.5   Deals with statutory regulations and safety considerations, structural considerations, independent systems, fresh air quantities, ductwork, thermal and accoustic insulation, mechanical equipment, fans, refrigeration, electrical equipment, oil-fired heaters, gas-fired heaters, fire and smoke detection and smoke control.

3.2.6   Discusses ventilation rates and infiltration.

3.2.8   Is concerned with access and facilities. Essential that proper commissioning is carried out and that full facilities are provided for maintenance.

5.5.6   Covers the location and installation of air filters.

• by complying with relevant recommendations on CIBSE Guide B, 'installation and equipment data'. The appropriate sections are B2 – Ventilation and air-conditioning requirements, and B3 – Ventilation and air conditioning, systems, equipment and control.

ACCESS

There is to be reasonable provision for access for the purpose of replacing filters and also the provision of access points for cleaning ductwork.

Where no special provision is necessary in a plant room there is to be an access space 2 m high by 600 mm wide between plant and where space for routine cleaning is required it must be adequate to accommodate a kneeling person. Fig F1.14 shows the minimum necessary and additional space may be needed for access doors.

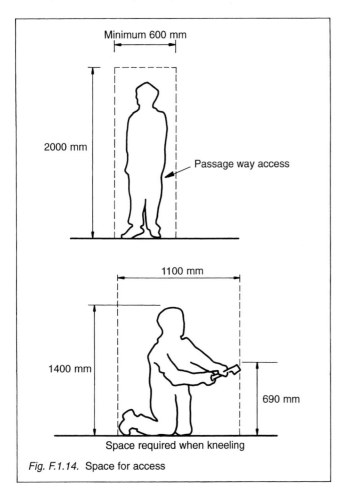

Fig. F.1.14. Space for access

For guidance on complex situations see BSRIA Technical Note 10/92, 'Space allowances for building services distribution systems'.

COMMISSIONING

The building control body is to be furnished with confirmation that systems installed to serve floor areas in excess of 200 m² comply with the above requirements, and are tested.

Compliance may be demonstrated by presenting test reports and commissioning certificates which certify that commissioning and testing have been carried out in accordance with CIBSE codes and that the systems perform in accordance with the specification.

CONDENSATION IN ROOFS

Current concern with energy saving involving dwellings with sealed off flues or no flues at all and draughtproofing has given rise to humid conditions. The moisture seeks locations to deposit itself and it is possible for much of the warm moist air to penetrate into roof voids.

The 1995 edition of approved document F will have, by its ventilation requirements, improved this situation.

The occupants of a dwelling produce a lot of water vapour in the course of a day. BS 5250 says that the total amount can be between 7.2 kg to 14.4 kg for a five-person family.

Warm moist air emanates from many sources. From our exhaled breath, from cooking, drying clothes near radiators, washing machines, dishwashers, tumble driers and any unflued gas or oil-burning appliances.

Broken down the figures given in BS 5250 for a five-person dwelling give the following moisture emission in kg:

5 persons sleeping for 8 hours	1.5 kg
2 persons active for 16 hours	1.7 kg
Cooking	3.0 kg
Bathing, dishwashing, etc.	1.0 kg
Washing clothes	0.5 kg
Drying clothes	5.0 kg
Paraffin heater	7.2 kg

Probably up to one-fifth of air ventilated from a house will find its way up through the roof space, and it is therefore essential to prevent large volumes of moist air getting into unheated roof voids (Fig. F2.1).

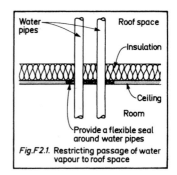

Fig.F2.1. Restricting passage of water vapour to roof space

Very simple steps will considerably limit large volumes of moist air entering the roof space and condensing causing dampness in insulation and in the structure:

• Seal any perforations through ceilings.
• Fix a compression seal to loft access hatches (Fig. F2.2).
• Ensure that loft access lids can be closed tightly.
• Include a vapour check above bathrooms and kitchens.

Requirements

• Those roofs where moisture from the building can permeate the insulation, i.e. cold deck roofs, must be ventilated.

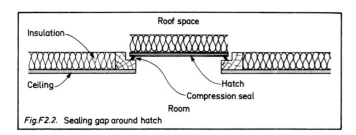
Fig.F2.2. Sealing gap around hatch

- Those roofs where moisture cannot permeate the insulation, i.e. warm deck roofs, do not require ventilation.
- The approved document says that for purposes of health and safety it may not always be necessary to ventilate small roofs such as those over porches and bay windows. It would seem desirable to ventilate over bays as a precaution.
- Pitched roofs require ventilation and where the ceiling of a room follows the roof pitch ventilation should be similar to that provided for a flat roof.
- A part of a roof having a pitch of 70° or more must be insulated as though it were a wall. However, the ventilation requirements of F2 extend to roofs of any pitch.
- Ventilation openings may be continuous or distributed along the length of eaves or side, provided the requisite area is available. They may be fitted with screens, fascias or baffles, as appropriate.

F2. Schedule 1, relates to roofs in all buildings. It says that adequate provision must be made to obviate excessive condensation in a roof or in a roof void which is above an insulated ceiling.

The aim is to limit condensation in roofs so that under normal conditions the thermal qualities of insulation and the structural performance of the roof will not be permanently diminished.

'Adequate' is defined in a dictionary as being sufficient for a particular purpose or need; able to satisfy a requirement or standard; suitable. Also having the necessary qualities to meet the demands of a situation.

This requirement has been in force since 1 April 1990 and was not amended by the 1994 Regs.

Approved documents set out provisions for meeting the performance required and they relate to cold roofs only. That is, roofs where the insulation is below the rain barrier (Figs F2.3 and F2.4).

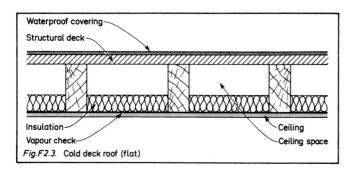
Fig.F2.3. Cold deck roof (flat)

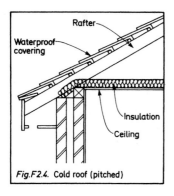
Fig.F2.4. Cold roof (pitched)

For insulation purposes a roof having a pitch of 70° or more is to be regarded as a wall, but for the purposes of F2 the recommendations apply to roofs of any pitch.

Roof pitch of 15° or more

Cross ventilation of the roof space is to be provided by having ventilation openings at eaves level. Area of the openings is to be equivalent to a continuous gap 10 mm wide running along the full length of the eaves (Fig. F2.5). Very great care must be taken to ensure that the ventilation gap is not impeded by insulation (Fig. 2.6). There are now a number of preformed slots available and when fixed it is not possible to block the opening with insulation.

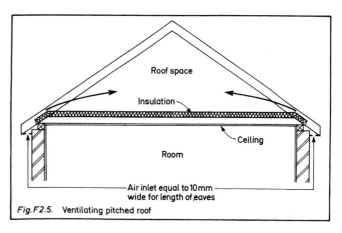
Fig.F2.5. Ventilating pitched roof

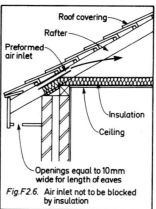
Fig.F2.6. Air inlet not to be blocked by insulation

A single slope roof when the upper part of the wall at high level is separating the roof space from a building will require the ventilation opening at the eaves, and also ventilation through the slope at high levels (Fig. F2.7).

Area of outlets at high level are to be equal to 5 mm wide continuous ventilation running the full length of the junction.

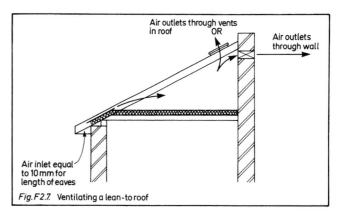
Fig.F2.7. Ventilating a lean-to roof

Another example of a single slope roof when the upper part of the wall at high level is exposed to the outside air will also require ventilation at eaves level, but the high level ventilation can be achieved by the correct area of openings in the wall.

An alternative approach is available and can be met by following BS 5250: 1989, 'Code of practice: control of condensation in buildings', clauses 9.1, 9.2 and 9.4.

BS 5250: 1989. Clause 9.1 provides guidance for condensation control and recommends an assessment of interstitial condensation risk. Calculation procedures are included in an appendix which gives tables to assist.

Recommendations are made with regard to the provision of unobstructed ventilation openings, including:

• external air outlets from moisture-producing appliances not to be sited near ventilation air inlets;
• when desired thermal performance can only be achieved by two or more layers of construction, careful attention is to be given to the balance of those insulants;
• ensure that construction, use, heating and ventilation all interrelate.

Clause 9.2 Vapour control layers.

A vapour control layer depends upon the material selected, workmanship and buildability. Holes or fixings penetrating the layer will downgrade its efficiency and should be sealed using an appropriate sealant. Vapour control layers should be placed on the 'warm' side of the construction and joints should be kept to a minimum.

To reduce the risk of degradation any polyethylene material used should be of virgin and not recycled origin.

Clause 9.4. Roofs.

9.4.1. General. A designer should take account of sources of dampness:

> weather
> interstitial condensation
> surface condensation
> water in wet construction.

Guidance is given as follows:

9.4.2	Pitched roofs
9.4.3	Flat roofs
9.4.3.1	Cold deck
9.4.3.2	Warm deck
9.4.4	Lightweight sheeted
9.4.5	Copper and similar roofs
9.4.6	Suspended ceilings under roof
9.4.7	Contains recommendations for ventilation together with illustrations
9.4.7.1	Tiled or slated: ceiling horizontal, insulation horizontal
9.4.7.2	Tiled or slated: ceiling inclined, insulation inclined
9.4.7.3	Tiled or slated: ceiling inclined, insulation inclined and of high vapour resistance
9.4.7.4	Pitched roof containing rooms: fully inclined ceilings
9.4.7.5	Pitched roof containing rooms: partially inclined ceilings
9.4.7.6	Pitched roof containing rooms: dormers
9.4.8	Timber flat roofs
9.4.8.1	Cold deck
9.4.8.2	Warm deck
9.4.8.3	Warm deck – inverted
9.4.9	Concrete flat roofs
9.4.9.1	Cold deck
9.4.9.2	Warm deck
9.4.9.3	Warm deck – inverted.

Roof pitch under 15°

When roofs are below a 15° pitch they are to have ventilation openings on two opposite sides to ensure cross ventilation. These same provisions apply to roofs with a pitch of 15° or more if the ceiling follows the pitch of the roof. More ventilation is required than for

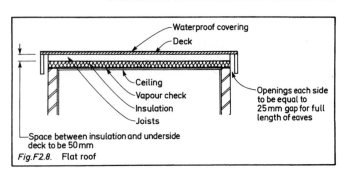

Fig.F2.8. Flat roof

roofs over 15° and the openings required must be equal to not less than a 25 mm gap running the full length of the eaves.

The ventilation gap must be at least 50 mm between insulation and the roof deck (Fig. F2.8).

Where for constructional reasons the joists are at right angles to the desired route for the flow of air, the required air space may be provided by fixing counter battens (Fig. F2.9).

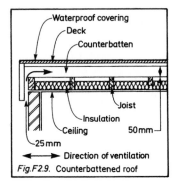

Fig.F2.9. Counterbattened roof

A requirement is that where roofs have a span of more than 10 m, or other than a simple rectangle on plan, more ventilation totalling 0.6% of the roof area may be required.

The approved document is quite emphatic that where the edge of a roof abuts walls or any other obstruction which prevents an adequate cross ventilation, some other methods of construction must be devised. It is also important that the movement of air outside ventilation openings is not restricted by obstructions.

The insertion of vapour checks help by reducing the amount of moisture reaching the roof space but they are not a substitute for adequate ventilation.

Vapour resistance of various membranes (MNs)/g

Aluminium foil	175 to 10,000
Bitumen impregnated paper	11
Kraft paper – single	0.2
Kraft paper – 5 ply	0.6
Tar infused sheathing paper	0.6 to 190
Polyethylene 60 μm	110 to 120
Polyethylene 100 μm	250 to 350
Polyethylene 150 μm	450
Emulsion paint (two coats)	0.2 to 0.6
Flat oil paint (two coats)	6 to 11
Gloss paint	7.5 to 40
Roofing felt	4.5 to 100
Vinyl-faced wallpaper	5 to 10
	(BS 5250)

(MNs)/g is the force needed to push a gramme of water vapour through the material.

When it is proposed to construct a roof having a pitch of 15° or more and the ceiling follows the slope of the roof the same requirements apply, namely a ventilation gap equal to 25 mm in width for the length of the eaves and an air space of 50 mm between roof covering and insulation (Figs F2.10 to F2.12). Outlets at a high level can be provided by one of several ridge tile systems now available.

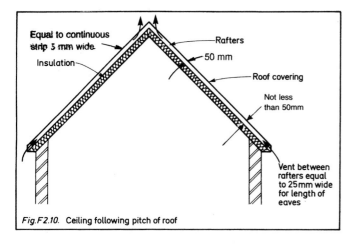

Fig.F2.10. Ceiling following pitch of roof

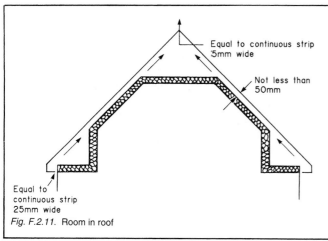

Fig. F.2.11. Room in roof

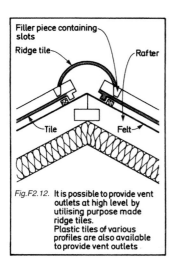

Fig.F2.12. It is possible to provide vent outlets at high level by utilising purpose made ridge tiles. Plastic tiles of various profiles are also available to provide vent outlets.

There is an alternative approach following the recommendations of clauses 9.1, 9.2 and 9.4 of BS 5250: 1989, 'Code of basic data for the design of buildings: the control of condensation in dwellings'.

These requirements have been outlined on previous pages.

RELEVANT READING

British Standards and Codes of Practice

BS 416: 1990. Ventilating pipes and fittings
BS 476: Part 24: 1987. Determination of fire resistance of ventilation ducts

BS 493: 1995. Air bricks and gratings for wall ventilation
BS 4016: 1972. Specification for building papers (breather type)
BS 5250: 1989 (1995). Code of practice: control of condensation in buildings
BS 5440: Part 1: 1990. Specification for installation of flues
BS 5925: 1991. Code of practice for design of buildings: ventilation principles
BS 7346: Part 1: 1990. Natural smoke and heat exhaust ventilators
    Amendment 1: AMD 7193
BS 7346: Part 2: 1990. Powered smoke and heat exhaust ventilators
BS 7346: Part 3: 1990. Smoke curtains

BRE digests

110 Condensation
170 Ventilation of internal bathrooms and WCs in dwellings
180 Condensation in roofs
270 Condensation in insulated domestic roofs
297 Surface condensation and mould growths
306 Domestic draughtproofing – ventilation considerations
309 Estimating daylight in buildings (1)
310 Estimating daylight in buildings (2)
312 Flat roof design – technical options
319 Domestic draughtproofing: materials, costs and benefits
336 Swimming pool roofs: minimizing condensation
398 Continuous mechanical ventilation in dwellings
399 Natural ventilation in non-domestic buildings

BRE defect action sheets

1   Slated or tiled pitched roofs: ventilation to outside air
3   Slated or tiled roofs, restricting entry of water vapour from the house
4   Pitched roofs – thermal insulation near the eaves
6   External walls – reducing risk from interstitial condensation
16   Internal walls and ceilings: remedying recurrent mould growth
59   Felted cold deck roofs: remedying condensation by converting to warm deck
73   Suspended timber floor – remedying dampness due to inadequate ventilation
91   Domestic gas appliances: air requirements
118 Slated or tiled pitched roofs: conversion to accommodate rooms, ventilation of voids
119 Slated or tiled pitched roofs: conversion to accommodate rooms, installing quilted insulation
136 Domestic draughtproofing – balancing ventilation needs against heat losses.

BRE information papers

13/87 Ventilating cold deck flat roofs
18/88 Domestic mechanical ventilation
21/89 Passive stack ventilation in dwellings
21/92 Spillage of gases from open-flued appliances
13/94 Passive stack ventilation systems
5/95   Performance of terminals for ventilation systems, chimneys and flues
6/95   Flow resistance and wind performance of some common ventilation terminals
13/95 The passive gas tracer method for monitoring ventilation rates in buildings.

BRE report

Thermal Insulation: avoiding risks (BR 262, 1994)
Background ventilation of dwellings. (Christine Uglow) (BR 162, 1989)
Minimizing air infiltration in office buildings (BR 265).

CHAPTER 16

Hygiene

(Approved document G)

Originally Part G contained requirements regarding the provision of a larder, or space to accommodate a larder. These have disappeared because of the argument that refrigerators are likely to be provided in dwellings, without the need for a requirement in the regulations.

Other requirements remain, although with some alteration or amendment.

- Buildings are to have sufficient sanitary conveniences for use by people living, visiting or working therein (G1)
- Dwellings are to have a bathroom with a fixed bath or shower and an adequate hot and cold water supply (G2)
- Unvented hot water systems of more than 15 litres capacity must have controls limiting temperature of the water to 100°C, and safety devices which operate to convey any discharge of hot water to a place where it will not cause danger to people (G3).

There is now confirmation that the only separation required between a WC and a kitchen is a door – a lobby is no longer necessary.

There are now references to European technical approvals.

G1. SANITARY CONVENIENCES AND WASHING FACILITIES

G1 (1) requires that sanitary conveniences must be provided which will be

- **in rooms provided for that purpose, or**
- **in bathrooms.**

Spaces containing sanitary conveniences are to be separated from rooms where food is prepared.

G1 (2) requires that adequate washbasins are to be provided in

- **rooms containing water closets, or**
- **rooms or spaces adjacent to rooms containing water closets.**

These rooms or spaces containing washbasins are to be separated from places where food is prepared.

G1 (3) requires that washbasins are to have a supply of hot and cold water.

G1 (4) requires sanitary conveniences and washbasins to be designed and installed so as to be easily and thoroughly cleansed.

Section 126 of the Building Act 1984 defines sanitary conveniences as meaning a closet or urinal. Closet includes a privy and the regulations are very silent regarding any guidelines for privies. However, a water closet is defined to mean a closet that has a separate fixed receptacle connected to a drainage system and separate provision for flushing from a supply of clean water either by the operation of mechanism or by automatic action.

The 1936 Act, section 51, places responsibility for seeing that the flushing apparatus is kept supplied with water and protected against frost upon the occupier of the premises.

The 1984 Act, section 26, required that new buildings should be provided with closets. However, this has been repealed by the 1985 Regs, 18 and replaced by G1.

The provision of sanitary conveniences in workplaces remains in the 1984 Act, section 65, which requires sufficient and satisfactory accommodation to be provided.

Performance

The approved document sets out an acceptable level of performance. There are to be adequate numbers of sanitary conveniences appropriate to the age and sex of the people provided for. Washbasins are also to be provided in or adjacent to rooms containing a water closet and the basins must have a supply of both hot and cold water.

It appears that G1 requires the provision of sanitary conveniences (closets and urinals) but does not require a washbasin where only urinals are installed.

All the fittings are also to be designed, sited and installed in such a way that they do not give rise to conditions which may be prejudicial to health.

Definitions

Two definitions are now introduced into the approved document.

- Sanitary conveniences means closets or urinals, thus repeating the definition in section 126 of the 1984 Act given above.
- Sanitary accommodation means a room containing closets or urinals. The space may also contain other sanitary fittings, as for example a bath, or bidet.

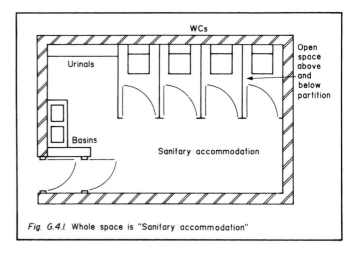

Fig. G.4.1. Whole space is "Sanitary accommodation"

Where the sanitary accommodation contains one or more cubicles it is nevertheless regarded as a single space if there is free circulation of air throughout the space.

Chemical closets

If there are no suitable facilities for disposing of foul water or no suitable water supply available, then closets or urinals utilizing chemical or other means of treatment may be installed. It is unlikely that such appliances will be used in dwellings which are required to have an adequate water supply and a proper drainage system. There are, however, in the approved document, no stated restrictions on their use or siting except that they must be separated from a place where food is prepared, and this includes a kitchen and any space in which washing up is done.

Workplace regulations

Regulations have been issued to implement provisions of the Workplace directive 89/654/EEC and they contain minimum health and safety requirements for the workplace.

They are the Workplace (Health, Safety and Welfare) Regulations 1992, and employers are under obligation to provide and maintain many facilities. As far as this chapter is concerned, there are two regulations.

There are to be suitable and sufficient sanitary conveniences in readily accessible places. The room containing them is to be kept in a clean and orderly condition, adequately lighted and ventilated. There are to be separate rooms containing conveniences for men and women except where the door of each separate room may be secured from the inside (Regulation 20).

There are to be suitable and sufficient washing facilities, including showers if necessary because of the nature of the work; washing facilities with hot and cold water are to be provided in the immediate vicinity of all sanitary conveniences (Regulation 21).

Numbers

All dwellings, be they houses, flats or maisonettes, are to have at least one water closet, and one washbasin.

There is also a requirement that a house in multioccupation must have the same provision of closets as a dwelling and they are to be accessible to all occupants.

A house in multiple occupation is one in which all the occupants are not part of a single household.

For buildings other than dwellings there are various requirements set out in other legislation and it will be necessary to provide the number of closets or urinals required, and fitted in accordance with the 1991 Regs.

The requirements for factories are set out in the Sanitary Accommodation Regulations 1938 and the basic requirement is for 1 WC for every 25 females and 1 WC for every 25 males.

However, where a factory employs more than 100 males and sufficient urinal accommodation is provided there may be 1 WC for each 25 up to the first 100, and thereafter 1 WC for every 40.

When calculating numbers of conveniences required, any odd number less than 25 or 40 as the case may be shall be reckoned as 25 or 40.

Sanitary conveniences are not to communicate with any workroom except through the open air or an intervening ventilated space.

Other regulations are concerned with food premises and these are the Food Hygiene (General) Regulations 1970. Numbers are not given, but in clause 16 every convenience is required to be

- kept clean and in efficient order, and
- so placed that no offensive odours therefrom can penetrate into a food room.

No food room which communicates directly with a room or other place containing a sanitary convenience is to be used for handling open food.

Wash handbasins are to be provided.

There are also regulations affecting offices, shops and railway premises known as the Sanitary Convenience Regulations 1964.

The basic requirement is for 1 WC where the number of persons does not exceed 5 at any one time.

Where greater number of persons are employed the following is to be provided separately for persons of each sex:

   (a) for females, and
   (b) for males where urinal accommodation is not provided.

Persons of each sex	No. of WCs
1–15	1
16–30	2
31–50	3
51–75	4
76–100	5
Exceeding 100	5 with the addition of 1 for every 25 persons exceeding 100

The following are for males when urinals are provided.

No. of males	No. of WCs	No. of urinals
1–15	1	–
16–20	1	1
21–30	2	1
31–45	2	2
46–60	3	2
61–75	3	3
76–90	4	3
91–100	4	4

with the addition of 1 sanitary convenience (WC or urinal) for every 25 persons by which the number of persons exceed 100, and of the additional number of sanitary conveniences not less than three-quarters shall be WCs.

The Education (School Premises) Regulations 1999 defines a sanitary fitting as a WC or urinal and in Article 3 requires every school to have sanitary fittings equal to the basic number, being the aggregate of:

- 10% of the number of pupils under 5 years of age, and
- 5% of the number of pupils who have attained 5 years of age, or
- 10% of the number of pupils in a special school.

When a building is being designed it is now necessary to refer to Part M – Facilities for disabled people – Chapter 21.

Siting

Limitation regarding siting and the provision of ventilated lobbies to prevent direct access from a closet compartment to habitable rooms have completely disappeared from provisions that meet the performance standard.

The space containing a closet or a urinal must be separated from a kitchen, or other food preparation room or a place where washing up takes place. That is the only restriction on siting.

The effectiveness of this change hinges on the word 'separated'. It appears that space accommodating a WC may now open directly into a kitchen. This relaxation stems from improved standards of hygiene and now that a washbasin and an extractor fan (Chapter 15, Part F) must be provided the approved document is relying upon separation to prevent offence or contamination in food preparation rooms and the latest approved document confirms that a WC need only be separated by a door from a food preparation room.

That is not to say that the older siting requirements for offices, shops, factories and food premises may be overlooked.

Washbasins are to be provided

- in the room containing the closet (Fig. G4.2), or
- in a room or space having direct access to the space containing the closet (Fig. G4.2), or
- in the case of a dwelling in a room adjacent to the room containing the closet (Fig. G4.3).

Washbasins provided to meet this requirement may not be in a room used for the preparation of food.

It would appear that in the case of a dwelling if a bathroom is adjacent to the water closet the requirement is met by a washbasin in the bathroom.

Design

Closets, urinals and washbasins are to be manufactured giving a

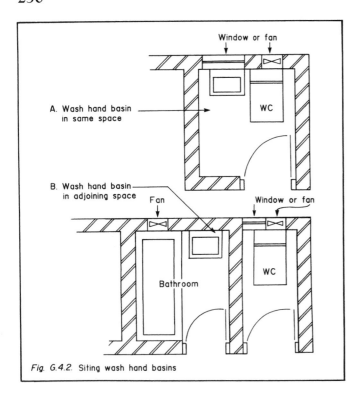

Fig. G.4.2. Siting wash hand basins

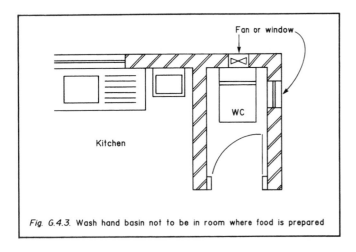

Fig. G.4.3. Wash hand basin not to be in room where food is prepared

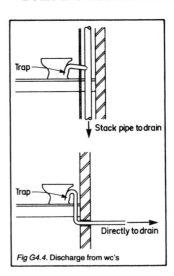

Fig G4.4. Discharge from wc's

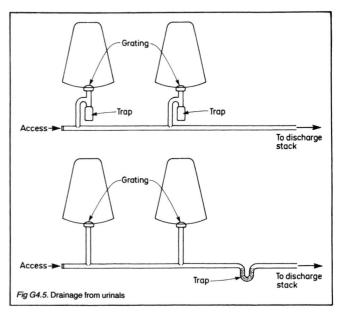

Fig G4.5. Drainage from urinals

surface which is smooth and non-absorbent. The appliance must be easily cleansed.

Any flushing apparatus for water closets or urinals is to be capable of cleansing the receptacle effectively. Connection is not to be made to any pipe other than a flush pipe or a branch discharge pipe.

The supply of hot water to a washbasin may be from a central source or from a unit water heater. A piped supply of cold water is also to be provided.

Installation

A water closet is to discharge through a trap and branch pipe into a discharge stack and drain (Fig. G4.4).

Urinals are to discharge through a grating to a trap via branch pipe to a discharge stack or drain (Fig. G4.5).

Automatic flushing systems for urinals

BS 1876: 1972 recommends the number of stalls which can be served by cisterns

Nominal size (litres)	Number of stalls, or equivalent width of slab served
4.5	1
9.0	2
13.5	3
18.0	4
22.5	5
27.0	6

Detailed provisions for drainage are contained in Part H, Chapter 17.

A washbasin is to discharge through a trap and a branch discharge pipe to a discharge stack. If located on the ground floor discharge may be into a gully or into a drain.

Small-bore systems

It is possible to obtain a water closet fitted with a macerator for shredding paper and solids which then enables all to be pumped through a small-bore pipe to a point of discharge in a drainage system or treatment plant via a discharge stack (Fig. G4.6).

If it is proposed to install one of these appliances the approved document requires:

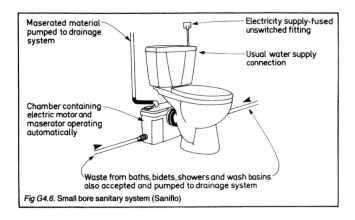

Fig G4.6. Small bore sanitary system (Saniflo)

- that there is also a closet discharging directly to a gravity system, and
- the system is the subject of a current Agrément certificate;
- it is installed and used in accordance with the conditions set out in the Agrément certificate.

The intention of the approved document is that there should be at least one conventional WC discharging into a gravity drainage system in each dwelling.

Use of a small-bore sanitary system is now possible and enables a WC to be located in positions that would otherwise present very difficult or expensive installation problems. As for example providing extra facilities for the elderly or disabled, or basement installations, a loft conversion, to mention just a few examples.

The approved document requires that such systems should be covered by an Agrément certificate.

In essence they comprise the use of a conventional WC and flushing apparatus. But discharge from the pan is into a small chamber where a macerator chops up paper and solids and a little pump forces the matter through a 22 mm dia. uPVC or copper discharge pipe to the house drainage system. Wastes from washbasins, bidets or baths may be discharged into and pumped to a drainage system by the same appliance.

Installation of small-bore systems require care and attention to manufacturers instructions as well as those contained in the Agrément certificate. For example, the small-bore discharge pipe should not contain any dips and rises in the final low gradient route towards its outlet. Bends, rather than elbows should be used at changes of direction, the discharge pipe carefully supported throughout its length, and, of course, allowance made in the total discharge length to compensate for resistance to flow at bends.

ALTERNATIVE APPROACH

The approved document also permits an alternative approach. Requirements can be met, providing it is in line with the other legislation, by following recommendations in BS 6465: Part 1: 1994, 'Code of practice of the scale of provision, selection and installation of sanitary appliances'. The relevant clauses are 3 to 8 inclusive.

Clause content is as follows:

Clause 3.	Definitions
Clause 4.	Exchange of information
Clause 5.	Design
Clause 6.	Toilets for disabled people
Clause 7.	Provision of appliances
Clause 8.	Workmanship.

Provision of appliances

Clause 7 containing recommended provision of appliances to various types of buildings is summarized. For more detail, reference should be made to BS 6465: Part 1: 1994.

The recommended number of WCs, urinals and washbasins should not be less than the following:

1. *Dwellings*

Bungalows and flats	1 WC and 1 washbasin. For 6 people and above: 2 WCs and 1 washbasin.
Houses and maisonettes	1 WC and washbasin. For 5 people and above: 2 WCs and 1 washbasin.
Small flats for 1 or 2 elderly people	1 WC and 1 washbasin.
Residential homes for the elderly	1 WC to every 4 people, plus 2 WCs for non-residential staff, and washbasins. 1 WC for visitors, and washbasin.

2. *Offices and shops, factories and non-domestic places of work*

| WCs for any group of staff | 1 to 5 people: 1 WC and 1 washing station. 6 to 25 people: 2 WCs and 2 washing stations. 26 to 50 people: 3 WCs and 3 washing stations. 51 to 75 people: 4 WCs and 4 washing stations. 76 to 100 people: 5 WCs and 5 washing stations. Over 100 people, add 1 WC and 1 washing station for each additional 25 people or part thereof. |
| WCs for males when urinals provided | 1 to 15 men: 1 WC and 1 urinal. 16 to 30 men: 2 WCs and 1 urinal. 31 to 45 men: 2 WCs and 2 urinals. 46 to 60 men: 3 WCs and 2 urinals. 61 to 75 men: 3 WCs and 3 urinals. 76 to 90 men: 4 WCs and 3 urinals. 91 to 100 men: 4 WCs and 4 urinals. Over 100 men, add 1 WC for each additional 50 men or part thereof, together with an equal amount of urinals. |

3. *Facilities for customers in shops and shopping malls*

Sales area 1000 m² to 2000 m²	Males: 1 WC, 1 urinal and 1 washbasin. Females: 2 WCs and 2 washbasins.
Sales area 2001 m² to 4000 m²	Males: 1 WC, 2 urinals and 2 washbasins. Females: 4 WCs and 4 washbasins.
Sales area in excess of 4000 m²	Facilities in proportion to the net sales area

In shopping malls the sum of the floor area of the shops is to be ascertained for use with the above requirements.

4. *Hotels*

| Ensuite accommodation | 1 WC and 1 washbasin for each guest bedroom. |
| Not ensuite accommodation | 1 WC per 9 guests. 1 bathroom per 9 guests containing washbasin and additional WC. |

5.   *Restaurants*

Male customers   1 WC per 100 up to 400.
                   Over 400, add at rate of 1 WC per 250 males or
                   part thereof.
                   1 urinal for every 50 males.
                   1 washbasin for every WC with 1 for every 5
                   urinals or part thereof.
Female customers   2 WCs for every 50 up to 200 females.
                   Over 200, add 1 WC for every 100 or part
                   thereof.
                   Washbasins at the rate of 1 per WC.
Additional toilet provision is to be made for staff.

6.   *Public houses and licensed bars*

Occupancy may be calculated at 4 persons per 3 m² of effective drinking area. In public houses it may be assumed that 75% are male and 25% female customers. In other premises a 50–50 mix may be assumed.

Male customers   1 WC for up to 150 males and add 1 WC for
                   every additional 150 or part thereof.
                   2 urinals for up to 75 males, plus 1 for every additional 75 or part thereof.
                   1 washbasin per WC plus 1 washbasin for each 5
                   urinals of part thereof.

Female customers   1 WC for up to 12 females and add 1 WC for
                   every 13 to 30 females, plus 1 WC for every additional 25 females or part thereof.
                   1 washbasin per 2 WCs.
Additional toilet provision is to be made for staff.

7.   *Concert halls, theatres, cinemas and similar buildings used for public entertainment*

Males   1 WC for up to 250 people plus 1 WC for every
                   additional 500 people or part thereof.
                   2 urinals for up to 100 people plus 1 for every
                   additional 80 males or part thereof.
                   Washbasins should be provided at the rate of 1
                   per WC, and in addition 1 per 5 urinals or part
                   thereof.
Females   2 WCs serve up to 40 females.
                   3 WCs for 41 to 70 females.
                   4 WCs for 71 to 100 females.
                   Plus 1 WC for every additional 40 females or part
                   thereof.
                   Washbasins should be provided at the rate of 1,
                   plus 1 per 2 WCs or part thereof.

8.   *Swimming pools*

Male bathers   2 WCs for up to 100 males with 1 extra WC for
                   every additional 100 bathers or part thereof.
                   1 urinal for every 20 males.
                   1 washbasin per WC and add 1 further washbasin
                   per 5 urinals or part thereof.
Female bathers   1 WC for every 5 females up to 50, and then add
                   1 WC for every additional 10 females or part
                   thereof.
                   1 washbasin, plus 1 extra for every 2 WCs or part
                   thereof.

In all situations additional toilet provision is to be made for staff, on the basis of recommendations for staff toilets in offices, shops and non-domestic premises used as places of work, as set out above.

G2. BATHROOMS

This requirement relates only to dwellings. A bathroom is to be provided containing a fixed bath or a shower. A suitable hot and cold water installation is also to be provided.

Section 33 of the 1961 Act allowed a local authority to reject plans of a dwelling or of a building being converted into a dwelling, if they did not show that a bathroom together with a hot and cold water system was to be provided. The requirement was not mandatory and was left to a local authority's discretion.

The same standard was taken up in the 1984 Act, section 27, and again it was not mandatory but by Regulation 18 of the 1985 Regulations, this section has been repealed and the requirement is incorporated into Schedule 1 as G2. However, the requirement is no longer discretionary and a bathroom must be provided in every new dwelling or conversion.

A bathroom to fulfil the G2 requirements must have:

- a fixed bath or shower and
- hot and cold water supply, and
- a connection to a foul drainage system.

The size of the bathroom is not stipulated but there is to be one in every house, flat or maisonette.

The approved document also says that where a house is occupied by people who do not form part of single household (in multi occupation) then at least the same provision is to be made and it must be accessible to all the occupants.

Each bath and shower is to have a supply of hot water which may be from a central source or from a unit hot water heater (Fig. G2.1).

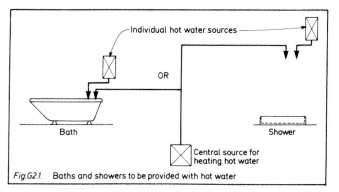

Fig.G2.1.   Baths and showers to be provided with hot water

Provision for drainage (Fig. G2.2) may be via a branch discharge pipe to a discharge stack or if on the ground floor straight into a gulley or a foul drain, and all is to be in accordance with the requirements of Part H of Schedule 1 (Chapter 17).

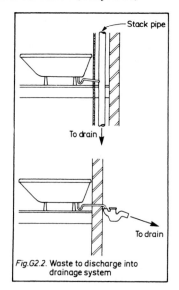

Fig.G2.2. Waste to discharge into drainage system

However, there may be a connection to a system incorporating a macerator and perhaps utilizing a small bore drainage system. Any such system must be covered by a current Agrément certificate and installed in accordance with the terms of the certificate.

Ventilation

Although Part G does not say anything about ventilating bathrooms, references must be made to F1 where it will be seen that mechanical ventilation should provide for the extraction of 15 litres per second (see Fig. G2.3).

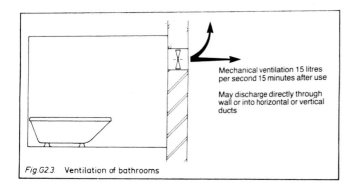

Mechanical ventilation 15 litres per second 15 minutes after use

May discharge directly through wall or into horizontal or vertical ducts

Fig.G2.3. Ventilation of bathrooms

G3. HOT WATER STORAGE

G3 makes provision for the installation of unvented water supply systems. When this was first mooted it created much alarm in the industry.

Our present vented system of hot water supply avoids the build up of high pressures and temperatures by permitting steam and boiling water to escape into the storage cistern. This cistern supplies water to the system by gravity. The only mechanical part is a ball valve in the cistern and even if that fails, water escapes through an overflow pipe to the open air.

Unvented hot water supply systems have been in use in Europe and America for some years, although when America went over to unvented systems in the early 1960s there were many reported explosions.

There has been considerable opposition to the introduction of unvented systems, which are acknowledged to have a number of advantages, including:

- no header tank or overflow in roof space
- better frost protection
- pressure feed gives better scope for siting the system
- balanced water pressure between hot and cold supply mixing is therefore easier and there is satisfactory pressure for showers
- pipework may be 15 mm (or less).

It is acknowledged that there is a need for high standards of both design and installation and it is in this climate that the approved document requires any system to be the subjected of an Agrément certificate and to be installed by an approved installer.

There appears to be a need for careful maintenance to ensure reliable operation at all times. If the mechanical controls are wrongly adjusted, jam, are tampered with or affected by water scale blockage, corrosion or age, circumstances can arise when high pressure will generate and the danger of an explosion arises.

Unvented systems in Europe and America are now reported as being used without trouble, and it is claimed that the servicing of appliances should not present any real problem.

One manufacturer has remarked that every gas boiler has to be looked at regularly. If not it can be just as lethal as an exploding water tank. The regulation requirements in connection with unvented hot water storage systems do not apply:

- when the storage capacity is 15 litres or less;
- when it is used in connection with a space heating system;
- when the system heats or stores water for industrial purposes.

With those exceptions, unvented hot water storage systems must have adequate precautions

- **to prevent stored water exceeding a temperature of 100°C at anytime, and**
- **to ensure that safety devices convey hot water to visible places where it will not give rise to danger to anyone in or about the building.**

Notices

The need to notify a local authority when it is proposed to install an unvented hot water storage system has been reinstated. This may be by way of a building notice or the deposit of full plans. Attention is directed to clauses 11, 12 and 13 of the 1991 B, Regs.

Regulation 11(1). A person intending to carry out building work or make a material change of use must serve a building notice on a local authority; or deposit full plans.

Regulations 12(4) is applicable. This requires service of a building notice accompanied by a statement giving

(a) the name and type of system to be installed;
(b) the name of the body which has approved or certified the system will meet the requirements of G3;
(c) the name of the body which issued a current registered operative identity card.

In Regulation 13(3) full plans consist of a description of the work and information required by Regulation 12(4). The appropriate water byelaws also apply to installations.

Definitions

The approved document describes criteria to be followed when installing an unvented hot water storage system which may be directly or indirectly heated. These have in mind that the possibility of explosion arises when temperature of the stored water is in excess of 100°C.

An unvented hot water storage system is an unvented vessel which either

(a) stores domestic hot water for subsequent use, or
(b) heats domestic hot water via a water jacketed tube or combi boiler, and
(c) is fitted with safety devices limiting temperatures to 100°C together with other safety devices.

Domestic hot water is water that has been heated for ablution, culinary or cleansing purposes.

They are related to systems which are the subject of a current Agrément certificate and are described as 'Units' or 'Packages'.

- A unit is a system incorporating safety devices, and other controls to stop primary flow, avoid backflow, govern working pressure, ease excess pressure and contain expansion. All the controls are to be fitted on the circuit by the manufacturer.
- A package is a system incorporating safety devices supplied in a kit form incorporating controls to stop primary flow; avoid backflow, govern working pressure and contain expansion, for fitting by the installer.

There are two sections dealing with these installations.

1. Hot water storage systems of not more than 500 litres capacity having a heat input of not more than 45 kW.
2. Hot water storage systems more than 500 litres capacity OR having a heat input more than 45 kW.

SMALL SYSTEMS

The system is to be in the form of a proprietory unit or a package, and is to be

- approved by a body which is a member of the European Organization for Technical Approvals (EOTA), i.e. The British Board of Agrément; or
- approved by a body having national Accreditation Council Certification Bodies (NACCB) accreditation as meeting the requirements of G3 having been tested in accordance with BS 7206: 1990, 'Specification for unvented hot water storage units and packages', or
- clearly demonstrated by independent assessment to meet the standards set by the above testing and accreditation bodies.

The first section sets out the provisions for an unvented hot water storage system having a storage vessel of not more than 500 litres. Heat input is not to exceed 45 kW and heating may be direct or indirect.

Direct heating

A unit or package directly heated is to have a minimum of two temperature-activated safety devices which operate in sequence:

- a non-self-resetting thermal cut out, and
- one or more temperature relief valves.

These controls are in addition to any thermostatic control fitted to maintain the temperature of stored water.

Non-self-resetting thermal cut outs are to be in accordance with BS 3955: 1986, 'Electrical controls for domestic appliances – general and specific requirements'.

The BS defines a thermal cut out as 'a device that during normal operation, limits the temperature of an appliance, or parts of it, by automatically opening the circuit or by reducing the current, and which is so constructed that its setting cannot be altered by the user'.

It also defines a non-self-resetting thermal cut out that requires resetting by hand in order to restore the current.

A non-self-resetting thermal cut out may also be in accordance with BS 4201: 1979 (1984) which is a specification covering thermostats supplied as independent units for incorporation in gas appliances for domestic, catering or commercial purposes.

Temperature relief valves are to be in accordance with BS 6283: Part 2: 1991, 'Safety devices for use in hot water systems – specification for temperature relief valves for use at pressures up to and including 10 bar', or BS 6283: Part 3: 1991, 'Specifications for combined temperature, and pressure relief valves for pressures up to and including 10 bar'.

Safety devices giving at least an equal degree of safety in preventing stored water exceeding 100°C at any time may be used if they are approved

- by members of EOTA (hold and Agrément certificate);
- by a body having NACCB (kitemarked to appropriate BS);
- following an independent assessment after meeting the standards set by the above testing and certification bodies.

Units and packages are to have the temperature relief valves located directly on the storage vessel. The valves are to be sized to give a discharge rating as in Appendix F of BS 6283: Part 2: 1991 or Appendix G of BS 6283: Part 3: 1991 which must equal the power input to the water.

Valves are not to be disconnected except for replacement or relocation in other devices or fittings. Valves are to discharge via a short length of metal pipe which has a diameter not less than the normal outlet size of the temperature relief valve. The discharge may be direct or by way of a manifold large enough to accept the total discharge from pipes connected to it. The discharge is to be over a tundish placed vertically, as near as possible to the valves, and not more than 500 mm distant.

Temperature relief valves are, according to BS 6283: Part 2 and Part 3: Temperature activated valves which open automatically at a specified temperature to discharge fluid. 'They are fitted to water heaters to prevent the temperature in the container from exceeding 100°C.'

BS 6283: Part 2 also defines a pressure relief valve as 'A pressure activated valve which opens automatically at a specified set pressure to discharge fluid. It is fitted to prevent the pressure in a water heater from exceeding a predetermined safe margin over the maximum working pressure when temperature controls fail to operate.'

The above devices are in addition to any thermostatic control which is fitted to maintain temperature of the stored water.

Indirect heating

- In indirectly heated units and packages the non-self-resetting thermal cut outs are to be wired to a motorized valve or some other device to shut off the flow to the primary heater

  (a) approved by the British Board of Agrément, or body having
  (b) a NACCB accreditation, or
  (c) an independent assessment after meeting standards set out by the above certification bodies.

- When the system is a unit incorporating a boiler the thermal cut out may be on the boiler.
- Temperature relief valve is to be located directly on the vessel and sized as required in Appendix F of BS 6283: Part 2: 1991 or Appendix G of BS 6283: Part 3: 1991 and be equal to the power input to the water.
- If an indirect unit or package has any alternative direct method of water heating fitted a non-self-resetting thermal cut out device is also required on the direct sources.
- The valves are to be sized to give a discharge rating not less than the heat input. Valves are not to be disconnected except for replacement or relocation in any other position or the connecting boss used to connect any other device or fitting.
- Each valve is to discharge via a short length of pipe with a diameter not less than the nominal outlet size of the temperature relief valve.
- Discharge is to be direct or by way of a manifold sized to accept the total discharge from the discharge pipes connected to it.
- Discharge is to be through an air break over a tundish located vertically as near as possible to the valves.

Installation

Whether or not the system is supplied as a unit or in package form, it must be installed by a competent person. This is someone holding a current Registered Operative Identity Card, issued by

- CITB, or
- Institute of Plumbing, or
- Association of Installers of Unvented Water Systems (Scotland and Northern Ireland), or
- a registered operative employed by company on the Agrément Board list of Approved Installers up to 31 December 1991, or
- an equivalent body.

The discharge pipe must be of suitable metal; is generally supplied by the manufacturer, and

- one pipe size larger than the discharge outlet size of the temperature relief valve;
- be via an air break to a tundish which should be set vertical;
- tundish should be placed in the same space as the unventilated hot water storage system, not more than 500 mm from the temperature relief valve;
- discharge pipe from tundish should not be more than 9 m long;
- when total hydraulic equivalent resistance is between 9 m and 18 m size should be twice that of the normal outlet size of the safety device;
- where total equivalent hydraulic resistance is between 18 m and 27 m, pipe size should be at least three sizes larger than the nominal outlet of the temperature relief valve;
- bends are to be taken into account when calculating flow resistance;

- there should be a vertical length of pipe not less than 300 mm long below the tundish before any bends or elbows;
- it should be fixed with a continuous fall;
- and finally, the discharge into the tundish should be visible.

Suitable discharges:

(a) Below a grating but above the water seal in a gully.
(b) Not more than 100 mm above an external surface. They must not be where children are likely to come into contact with the discharge. Surface could be grassed, hard standing or a car park. If children may have access there is to be a cage or other guard surrounding the discharge point.
(c) At a high level into a metal hopper or metal down pipe. End of discharge pipe to be visible. May be on to suitable roof surface and 3 m away from plastic guttering, and the tundish is to be visible.
(d) A single discharge pipe may serve up to six systems in blocks of flats. The common discharge pipe must be one pipe size larger than the largest individual discharge pipe connected.

When systems are installed in dwellings occupied by blind, infirm or disabled people the fitting of an electronically operated warning device should be considered.

As an alternative, pipe sizing may follow BS 6700: 1987, 'Specification for design installation and maintenance of services supplying water for domestic use within buildings and their curtilage'. (Refer to Appendix E, section E2 and Table 21.)

A method of calculating the size of a discharge pipe from tundish to outlet is illustrated in Fig. G3.1 and Table G1.

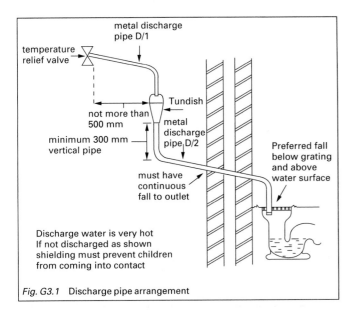

Fig. G3.1   Discharge pipe arrangement

Table G1   Determining the size of copper discharge pipes

Discharge pipe D/1 not less than (mm)	Valve outlet size	Discharge pipe D/2 not less than (mm)	Resistance, as length of straight pipe not more than (m)	Resistance given by each elbow or bend (m)
15	$G^{1/2}$	22	up to 9	0.8
		28	up to 18	1.0
		35	up to 27	1.4
22	$G^{3/4}$	28	up to 9	1.0
		35	up to 18	1.4
		42	up to 27	1.7
28	G^1	35	up to 9	1.4
		42	up to 18	1.7
		54	up to 27	2.3

Inspection

Site inspections of approved systems is unlikely to be necessary. In other cases building control officers or approved inspectors will use judgement in deciding whether to inspect the installation.

Here is an example of the calculations to determine whether a proposed discharge pipe from the tundish to the outlet (D/2) will be satisfactory.

Temperature relief valve:	$G^{1/2}$
Discharge pipe from tundish to outlet:	7 m long
Number of elbows on discharge pipe:	4

(Refer to Table G1)

22 mm straight copper pipe has maximum resistance of:	9 m
Deduct resistance of 4 × 22 mm elbows: 4 × 0.8 =	3.2 m
Maximum permitted length of 22 mm discharge pipe (D/2):	5.8 m

This is less than the actual length, is inadequate, and it is therefore necessary to calculate the next largest size:

28 mm straight copper pipe has maximum resistance of:	18 mm
Deduct resistance of 4 × 28 mm elbows: 4 × 1.0 =	4 m
Maximum permitted length of 28 mm discharge pipe (D/2):	14.0 m

The actual length of discharge pipe being 7 m means that a 28 mm dia. copper pipe is suitable.

When fitting the non-self-resetting thermal cut out it must be connected to the heat source in the way set out in the Regulations for Electrical Installations of the Institution of Electrical Engineers (Fig. G3.2).

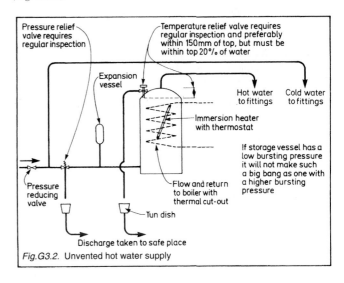

Fig. G3.2. Unvented hot water supply

LARGER SYSTEMS

This section sets out provisions for an unvented hot water storage system having a storage vessel of more than 500 litres, OR where the heat input exceeds 45 kW.

Such systems are not likely to be covered by an Agrément certificate. They will generally be individual designs for specific projects. Requirements are that the systems should

- be designed by an appropriately qualified engineer
- be designed to the same general safety requirements as in the smaller systems
- be installed by a competent person.

Safety devices

A system having a storage vessel exceeding 500 litres capacity and a heat input of not more than 45 kW is to have safety devices as recommended in BS 6700: 1987. This is for the design, installation, testing and maintenance of services supplying water for domestic use within buildings. The appropriate portion of the BS is section 2, clause 7.

7.1: Production of steam in a closed vessel or the heating of water under pressure to a temperature in excess of 100°C can be extremely dangerous. The standard deals only with low-temperature systems where the highest water temperature does not exceed 100°C at any time at any point in the system.

7.2: • energy supply to each heater is to be under effective thermostatic control
   • energy supply to each heater is to be fitted with a temperature manually reset energy cut out independent of thermostatic control
   • means of dissipating the heat input is to be provided in the form of either adequate venting or a temperature relief valve.

7.3: Pressure control. When necessary pressure reducing valves are to be used.

7.4: Maintenance of water level. Adequate means of supply to make up water is to be fitted. The hot water draw connection is to be located at the top of the hot water cylinder in conjunction with a suitably located vent.

Other equivalent practice specifications with suitable safety devices preventing water exceeding 100°C.

Temperature relief valves

An unvented hot water storage system having a heat input of more than 45 kW is to have the correct number of temperature relief valves

• either in accordance with BS 6283, Part 2 or 3 to give a combined discharge rating at least equivalent to the heat input, or
• an equally suitable temperature relief valve or valves marked with the set temperature in °C and discharge rating marked in kW.

The settings are to be measured in accordance with BS 6283: Part 2: 1991, Appendix F or BS 6283: Part 3: 1991, Appendix G and at least equivalent to the heat input and certified by the Agrément Board, or other recognized testing body, such as the Associated Offices Technical Committee.

The valves are to be factory fitted to the storage vessels and be within 150 mm of the top and within the top 20% of the volume of water.

Non-self-resetting thermal cut outs appropriate to the heat source are to be incorporated in the system.

Discharge pipes are also necessary and should be installed as already described.

British Board of Agrément

Each Agrément certificate sets out the position of the installation with regard to regulations, legislation and byelaws. If used in accordance with the provisions of a certificate, the installation will meet the relevant requirements.

• G3. Hot water storage
• L1. Heating system controls
• L1. Insulation of heating services
• Regulation 7. Materials and workmanship

The hot water storage systems satisfy requirements of Water Byelaws now being updated when the undertaker has obtained a blanket relaxation of Byelaw 46. Under new Model Byelaws now being implemented the systems are acceptable.

Each individual certificate deals in detail with the following:

• technical specification
• design data
• installation
• technical investigation

Figure G3.2 shows the approved document requirements in diagrammatic fashion. Figure G3.3 indicates how compact the arrangement of fittings may be when placed on top of the storage vessel (Megaflow).

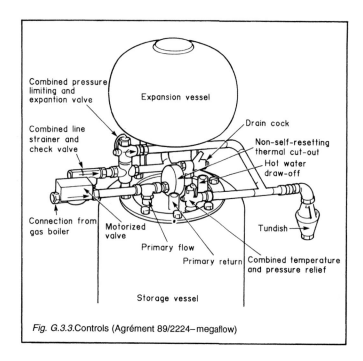

*Fig. G.3.3.*Controls (Agrément 89/2224–megaflow)

SUMMARY OF REQUIREMENTS

• Boiler to have a thermal cut out.
• Small-bore pipes are readily blocked by debris and therefore a strainer should be inserted.
• When mains pressure is high a pressure-reducing valve may be necessary.
• There must be no possibility of contaminating water in the mains and therefore a non-return valve should be installed to prevent backflow.
• An expansion vessel should be large enough to accommodate the volume of water as temperature rises.
• To prevent pressure in the system rising above the maximum designed for, there should be a pressure relief valve.
• Storage cylinder of special grade to accommodate the higher pressure.
• An automatic temperature relief valve to discharge fluid at a specified temperature, say 60°C.
• A thermostat.
• A non-self-resetting thermal cut out to cut out the boiler if temperature in storage cylinder rises to 90°C.
• A tundish, or dishes, to receive overflow and indicate to occupier that overflow is taking place.
• System must have an Agrément certificate or NACCB approval and be installed by a competent person.

RELEVANT READING

British Standards and Codes of Practice

BS 1125: 1987. Specification for WC flushing cisterns
BS 1188: 1974. Ceramic wash basins and pedestals
BS 1189: 1986. Cast iron baths for domestic purposes
BS 1254: 1981. Plastics WC seats
BS 1329: 1974. Metal hand rinse basins
BS 1390: 1990. Sheet metal baths

BS 4305: 1989. Baths made from cast acrylic sheet

BS 4814: 1990. Expansion vessels for sealed hot water systems

BS 4880: Part 1: 1973. Stainless steel slab urinals

BS 5503: Part 1: 1977. Vitreous china washdown pans connecting dimensions

BS 5503: Part 2: 1977. Materials, quality performance and dimensions

BS 5504: Part 1: 1977. Wall hung WC with close coupled cistern

BS 5504: Part 2: 1977. Wall hung WC with independent water supply

BS 5504: Part 3: 1977. Wall hung WC materials. Quality and functional dimensions

BS 5505: Part 1: 1977. Pedestal bidets, connecting dimensions

BS 5505: Part 2: 1977. Wall hung bidets, connecting dimensions

BS 5505: Part 3: 1977. Vitreous china bidets

BS 5506: Part 1: 1977. Connecting dimensions pedestal washbasins

BS 5506: Part 2: 1977. Connecting dimensions wall hung washbasins

BS 5506: Part 3: 1977. Washbasins materials, quality design and construction

BS 5520: 1977. Vitreous china bowl urinals

BS 5627: 1984. Plastics connectors for use with vitreous china WC pans

BS 6283: Part 1: 1991. Specifications for expansion valves for pressures up to and including 10 bar

BS 6283: Part 2: 1991. Specification for temperature relief valves for use at pressures up to and including 10 bar

BS 6283: Part 3: 1991. Specification for combined temperature and pressure relief valves for pressures up to and including 10 bar

BS 6283: Part 4: 1991. Specification for drop tight pressure reducing valves

BS 6340: Part 1: 1983. Guide on choice of shower units

BS 6340: Part 2: 1983. Specification for the installation of shower units

BS 6340: Part 5: 1983. Shower trays from acrylic materials

BS 6340: Part 6: 1983. Shower trays from porcelain enamelled cast iron

BS 6340: Part 7: 1983. Shower trays from vitreous enameled sheet steel

BS 6465: Part 1: 1994. Code of practice for scale of provision, selection and installation of sanitary appliances

BS 6700: 1987. Design, installation, testing and maintenance of services supplying water for domestic use

BS 7206: 1990. Specification for unvented hot water storage systems

BRE digests

170   Ventilation of internal bathrooms and WCs in dwellings.

308   Unvented domestic hot water systems

BRE information papers

5/85   Legionnaires' disease – Implications for design and use of hot water systems

9/90   Unvented hot water systems

12/83 Water economy with Skevington/DRE controlled flush valve for WCs

1/97   Unventilated hot water storage systems: microbial growth in expension vessels

BRE reports

Unvented domestic hot water systems. BRE Report 1988, No. BR 125.

Domestic heat pumps: performance and economics. BRE Report 1988, No. BR 126.

Potential health risks associated with expansion vessels in hot water installations. BRE Report 1996, No. BR 309.

BRE defect action sheets

109   Domestic hot water systems – protection against frost

139   Unvented domestic hot water storage systems – safety (design)

140   Unvented domestic hot water storage systems – installation and inspection

CHAPTER 17

Drainage and waste disposal

(Approved document H)

These requirements are related to the disposal of soil drainage and rainwater from buildings. Drainage of the site of a building is referred to in Part C3 (Chapter 12). Additionally cesspools and tanks are covered and there are provisions relating to the storage of solid waste awaiting removal from buildings.

There are several sections of the 1984 Act which relate to the drainage of new buildings and they are considered in some detail in Chapter 6.

Sections to be noted are:

- 1984 Act, section 21 – Provision of drainage
  Proper provision must be made for the drainage of new buildings, or an extended building. Plans are to be rejected if satisfactory provision is not shown.
- 1984 Act, section 22 – Drainage of buildings in combination
  Most housing sites are planned to be drained in combination by means of private sewers connected later to a public sewer. When this is not planned the local authority may require that the waste of any two or more buildings must be drained in combination.
- 1984 Act, section 60 – Use of ventilation and soil pipes
  A pipe for conveying rainwater from a roof is not to be used for taking soil or drainage from a sanitary convenience, nor as a ventilating pipe for soil drains.

Part H relates only to drainage systems up to the point of connection with a public sewer, or connection to a cesspool or septic tank where a public sewer is not available. The part deals with one of the basics of the protection of public health – the removal of offensive waste from a building to a point of satisfactory disposal.

Infiltration leads to surcharging of manholes in lower parts of a drainage system and this may cause danger to health. If a drain is liable to infiltration it is also equally capable of discharging foul contents into the surrounding ground.

Requirements for drainage and waste disposal are numbered as follows:

H1. Sanitary pipework and drainage
H2. Cesspools, septic tanks and settlement tanks
H3. Rainwater drainage
H4. Solid waste storage.

H1. (1) A drainage system conveying foul water from a building must be adequate, and
(2) 'foul water' is said to mean waste from:

- **sanitary conveniences;**
- **other soil appliances;**
- **water which has been used for cooking or washing.**

The statement that foul water does not include effluent arising from any trade or industry has been omitted from the approved document; reliance being placed upon legislation dealing with such waste. Some industrial processes require drains of special materials and construction to cope with the very damaging waste they have to carry. The Public Health (Drainage of Trade Premises) Act 1973 is concerned with premises from which trade effluents are discharged. The owner or occupier may not discharge trade effluent into sewers without

agreement from the water authority. The water authority may require plans and particulars of the trade process and effluent. Conditions may be attached to any consent to discharge which is issued. It may be necessary to pay for treatment of the trade effluent prior to discharge into the public sewers.

In order to minimize risks to health and safety of people the approved document sets the following level of performance for sanitary pipework and drainage:

- carry all the foul water to a foul outfall which may be a sewer, cesspool, septic tank or settlement tank
- not allow foul air from the drainage system to get into a building under working conditions
- be ventilated
- risk of blockage or leakage to be minimized
- have access for removing blockage should they occur.

Drainage systems are divided into two parts. Section 1 relating to those above ground and section 2 below ground.

Contents of each section are as follows:

Section 1. Sanitary pipework
  traps
  branch discharge pipes
  discharge stacks
  materials
  airtightness.

Section 2. Foul drainage
  layout
  pipe cover
  gradients and sizes
  materials
  bedding and backfilling
  clearance of blockages
  watertightness.

SECTION 1. SANITARY PIPEWORK

The capacity of any system of sanitary pipework or drainage must be adequate to carry anticipated flows from all the fittings connected to the system. Flow rates are given for a number of dwellings and separate appliances. Capacities are also related to the diameter and gradient of the pipes.

Flow rates from a number of appliances are given in Fig. H1.1.

It is very rare for all appliances in a building to discharge at the same time, and minimum stack and drain sizes are usually adequate.

Assuming a typical household having

1 WC
1 bath
1 or 2 wash handbasins
1 sink

the following flow rates are used in BS 5572: 1978:

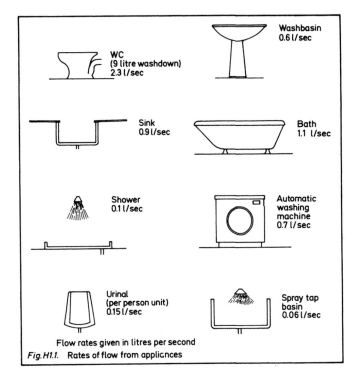

Fig. H1.1. Rates of flow from appliances

Flow rates given in litres per second

- WC (9 litre washdown) 2.3 l/sec
- Washbasin 0.6 l/sec
- Sink 0.9 l/sec
- Bath 1.1 l/sec
- Shower 0.1 l/sec
- Automatic washing machine 0.7 l/sec
- Urinal (per person unit) 0.15 l/sec
- Spray tap basin 0.06 l/sec

Table H1.1

Number of dwellings	1	5	10	15	20	25	30
Flow rate (litres per sec)	2.5	3.5	4.1	4.6	5.1	5.4	5.8

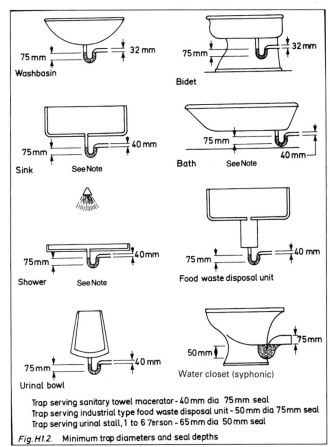

Trap serving sanitary towel macerator - 40 mm dia 75 mm seal
Trap serving industrial type food waste disposal unit - 50 mm dia 75 mm seal
Trap serving urinal stall, 1 to 6 person - 65 mm dia 50 mm seal

Fig. H1.2. Minimum trap diameters and seal depths

Pipe sizes

Pipe sizes given are nominal sizes used as a numerical designation in convenient round numbers. For individual pipe standards reference may be made to BS 5572: 1978 for sanitary pipework and BS 8301: 1985 for building drainage.

TRAPS

Appliances are required to have traps with a water seal adequate to ensure that in ordinary working conditions the seal is not destroyed. The requirement is quite fundamental and is intended to prevent the escape of foul sewer gases into positions where it could be harmful or dangerous.

This can arise by the loss of water seal emanating from fluctuation in pressure which may be brought about by back pressure, self-syphonage, induced syphonage or wind pressure.

Minimum recommended trap sizes and seal depths are shown in Fig. H1.2. Where a sink, bath or shower are installed on ground floors, and discharge to gully the depth of seal may be reduced to not less than 38 mm.

In order that there is adequate access for cleaning blockages

- where a trap forms part of an appliance the appliance should be removable
- all other traps should be removable or fitted with a cleaning eye.

Under working and test conditions the seal in a trap must not fall below 25 mm.

Self-sealing waste trap

A completely new approach to the prevention of foul air from waste discharge systems getting into living spaces has produced a self-sealing valve. Set to replace the water seal trap, the device utilizes a purpose designed membrane seal set within a tubular polypropylene body. The membrane opens when waste water passes through, but once the flow ceases, closes to create an airtight seal.

The fitting has been tested by WIMLAS who found that the valve provided a seal against foul air exceeding the protection expected from a 75 mm deep water seal trap. When correctly installed WIMLAS concluded that the valve could meet the requirements of H.1; and also 1991 Reg. 7 which provides for the use of tests and calculations to establish that a product or material is capable of performing the function for which it is to be used. (Hepworth Building Products – Hep$_v$0 plastics waste valve)

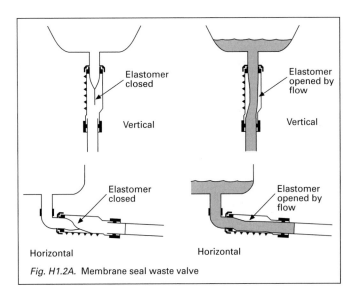

Fig. H1.2A. Membrane seal waste valve

BRANCH DISCHARGE PIPES

Branch discharge pipes above the ground floor may only connect to another branch pipe or a discharge stack (Fig. H1.3), and it must not be so arranged that it could cause crossflow into another branch discharge pipe (Fig. H1.4).

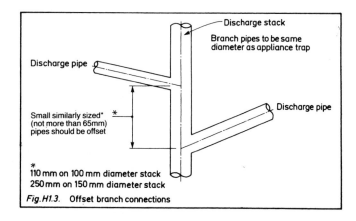

Fig.H1.3.   Offset branch connections

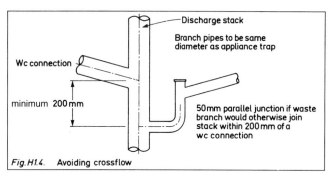

Fig.H1.4.   Avoiding crossflow

Branch discharge pipes on a ground floor are permitted to connect to a stub stack or discharge stack, to a drain, or where the pipe carries only waste water it may discharge into a gully (Fig. H1.5).

Fig.H1.5.   Ground floor waste pipe may enter gully

A branch discharge pipe serving a dwelling of up to 3 storeys is not to join a stack lower than 450 mm above the invert of the tail of the bend, at the foot of the stack (Fig. H1 .6).

For taller buildings the following apply:

- up to 5 storeys, branch waste not to discharge into stack less than 750 mm above invert;
- above 5 storeys if ground floor appliances do not discharge direct to a drain or gully it must connect to its own stack;
- for a building above 20 storeys the first floor wastes must also discharge into their own stack.

A discharge pipe from a ground floor WC may only discharge directly into a drain if the distance between the crown of the trap and invert of the drain is not more than 1.5 m (Fig. H1.7).

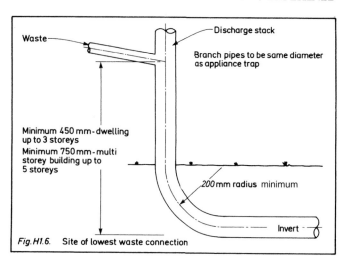

Fig.H1.6.   Site of lowest waste connection

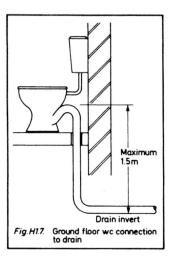

Fig.H1.7.   Ground floor wc connection to drain

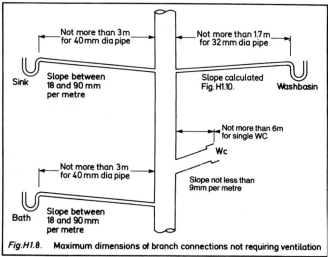

Fig.H1.8.   Maximum dimensions of branch connections not requiring ventilation

Unvented branch discharge pipes

When a branch waste pipe serves more than one appliance and is not ventilated, the length, size and gradient should be as follows (Fig. H 1.8). Bends should be avoided but when this is not possible the largest possible radius should be used. 65 mm pipes should have a centreline radius of not less than 75 mm.

- A minimum size 100 mm branch pipe may be used for a maximum of 8 WCs or a length not exceeding 15 m. Gradient should be between 9 mm and 90 mm fall per metre run.
- A minimum size 50 mm branch pipe may be used for a maximum

of 5 urinal bowls and the length should be as short as possible. Gradient should be between 18 mm and 90 mm fall per metre run.

- A minimum size 65 mm branch pipe may be used for a maximum of 7 urinal stalls and the length should be as short as possible. Gradient should be between 18 mm and 90 mm fall per metre run.
- A minimum size 50 mm branch pipe may be used for a maximum of 4 washbasins or a length not exceeding 4 m containing no bends. Gradient should be between 18 mm and 45 mm fall per metre run.

Junctions

Junctions on branch pipes are to be made with a sweep of at least 25 mm radius or at 45°.

Branch pipe connections of 75 mm diameter and above are to be made with a sweep of at least 50 mm radius or at 45°.

Ventilating branch pipes

If the length and slope of branch discharge pipes are not in excess of those shown in Fig. H1.8, separate ventilation to prevent water seals in traps being destroyed is not required. Fig. H1.9 shows maximum lengths when larger diameter waste pipes are used. Where lengths exceed these figures there should be a branch ventilating pipe to the outside air, a ventilating stack or a modified stack system.

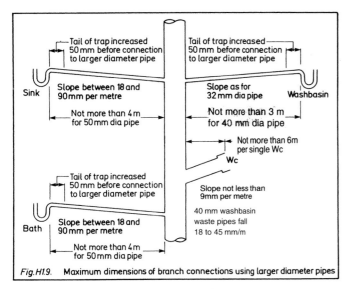

Fig. H1.9. Maximum dimensions of branch connections using larger diameter pipes

Branch vent pipes are required to meet the restrictions illustrated in Fig. H1.11.

When branch ventilating pipes are taken direct to the outside of a building they must not terminate lower than 900 mm above any opening into the building within 3 m (Fig. H1.12).

Separate ventilating stacks are unlikely to be required unless there are large numbers of ventilating pipes and they are some distance from the ventilating stack.

A separate dry stack may be fixed to provide ventilation for branch ventilating pipes instead of carrying them to a ventilated discharge stack or to the outside air (Fig. H1.13).

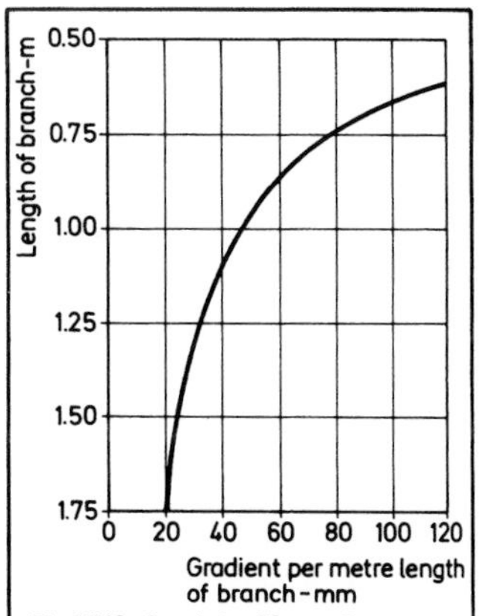

Where there are not more than 10 storeys and the building contains only dwellings, the ventilation stack should be not less than 32 mm diameter. Other buildings are covered by the alternative approach in BS 8301: 1985, 'Code of practice for building drainage'.

The lower end of a ventilating stack may be connected to a bend

Fig. H1.10. Graph for 32 mm diameter washbasin waste pipes

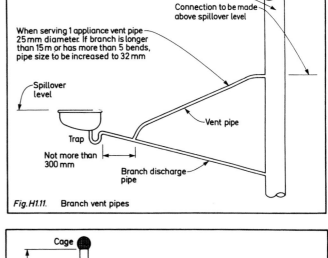

Fig. H1.11.   Branch vent pipes

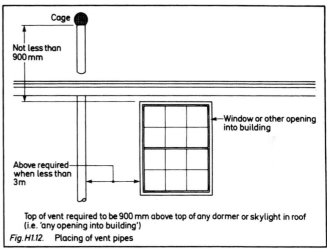

Fig. H1.12.   Placing of vent pipes

or connected to a ventilated drainage stack. When this is done the connection must be below the lowest branch discharge pipe.

The upper end of all ventilating stacks of ventilated discharge stacks must be taken to the outside air and finished as in Fig. H1.12. Any connection to a ventilated discharge stack is to be above the spill over level of the highest appliance.

DISCHARGE STACKS

All stacks should be connected to a drain via the largest radius bend possible but in no case less than 200 mm. There are to be no offsets in the wet part of a discharge stack. Where they are unavoidable in a building of up to three storeys then there is to be no branch connection within 750 mm of the offset. In a building of more than three storeys a ventilation stack may be necessary with connections above and below the offset. Where there are more than three storeys a stack must be placed within the building.

Stacks serving syphonic WCs are to have a minimum diameter of 75 mm and those for urinals 50 mm. Stacks serving washdown closets to be not less than 100 mm diameter. Maximum capacities for various sizes of discharge stacks are shown in Fig. H1.13.

Stub stacks

The rule is that discharge stacks must be ventilated to preserve trap seals of appliances connected to them (Fig. H1.14). But in certain circumstances stub stacks which are short unvented discharge stacks may be used. There may, however, still be other parts of the system requiring ventilation.

- A stub stack must be connected to a ventilated stack or a drain;

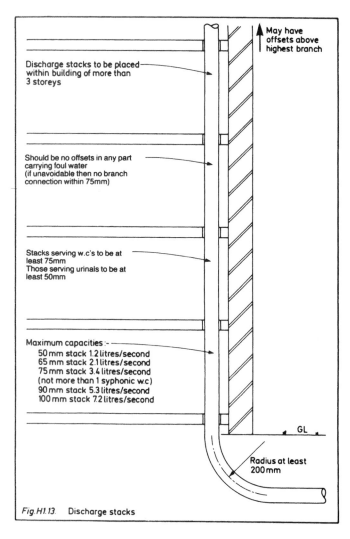

Fig.H1.13.   Discharge stacks

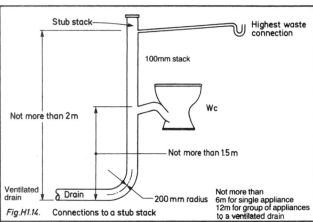

Fig.H1.14.   Connections to a stub stack

- A stub stack must have no branch joining it more than 2 m above the connection to a ventilated stack or drain;
- The crown of the lowest trap to a WC must be not more than 1.5 m above the drain invert.

Stack ventilation pipes

The dry part of a ventilation stack when it is above the highest branch on a discharge stack, may be reduced in one- or two-storey houses. The reduced diameter is not to be less than 75 mm.

When fitted with air admittance valves, discharge stacks are permitted to terminate inside a building. The only air admittance valves

that may be installed are those which are the subject of a current Agrément certificate. They must be installed and used in accordance with the terms of the appropriate certificate.

It is necessary to ensure that essential ventilation to below ground drainage is provided.

Advantages claimed for air admittance valves (see Figs H1.15 and H1.16) are:

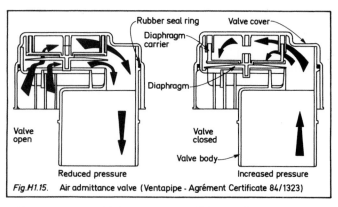

Fig.H1.15.   Air admittance valve (Ventapipe - Agrément Certificate 84/1323)

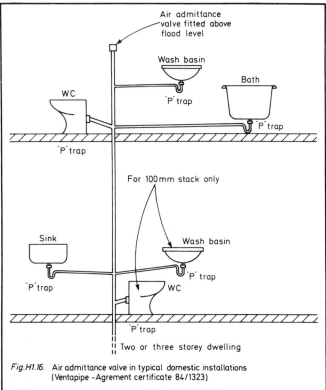

Fig.H1.16.   Air admittance valve in typical domestic installations (Ventapipe -Agrement certificate 84/1323)

- ventilating pipes may be terminated inside a building, thereby avoiding roof penetration;
- greater flexibility of design is available;
- suitability for use with either plastic or metal pipes.

An air admittance valve will prevent the release of foul air, and admit air under conditions of reduced pressure in the discharge pipes or stack.

When installed in up to 4 dwellings of no more than three storeys each and connected to the same drainage system, additional drain venting is not required (Fig. H1.17). When a drain serves more than four dwellings each with an air admittance valve, the drain should be vented through a vent discharge stack.

Agrément certificates have recommendations relating to additional dwellings, on drainage systems and also multi-storey domestic buildings.

Air admittance valves contain a diaphragm which lifts and allows air to be drawn into the system when it is subjected to negative pressure. When negative pressure ends the diaphragm returns to the

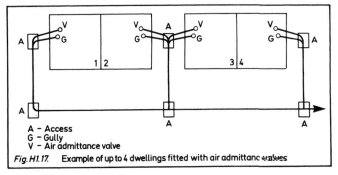

A – Access
G – Gully
V – Air admittance valve
Fig. H1.17.   **Example of up to 4 dwellings fitted with air admittance valves**

closed position thereby preventing escape of foul air into the building.

If a valve is to be installed in close proximity to a habitable space, when noise of operation may be disturbing, consideration should be given to the use of a suitable form of sound insulation.

Correctly installed the valve will open spontaneously when required, thereby allowing a supply of air to adequately ventilate the system.

CLEARING BLOCKAGES

Facilities and/or rodding points are to be provided for all traps, discharge pipes and discharge stacks. For the latter the top of a ventilating pipe may provide access if it is reasonably accessible.

AIRTIGHTNESS

A sanitary pipework installation is to be capable of satisfying an air or smoke test of positive pressure equal to not less than 38 mm water gauge for not less than three minutes.

All traps must maintain a water seal of not less than 25 mm for the three minutes. Smoke testing is not recommended for uPVC pipes.

MATERIALS

The approved document says that the following materials may be used for sanitary pipework:

- cast iron
  to be in accordance with BS 416: 1973 which specifies requirements for cast iron spigot and socket soil, waste and ventilating pipes with either type A or type B sockets.

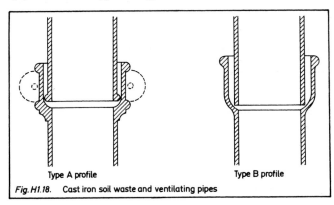

Type A profile   Type B profile
Fig. H1.18.   **Cast iron soil waste and ventilating pipes**

- copper
  BS 864 gives requirements for capillary and compression copper tube fittings.
  BS 864: Part 2: 1983, 'Specification for capillary and compression tube fittings'.
  BS 2871: Part 1: 1971 specifies requirements for copper tubes for the purpose of water and sanitation.

- galvanized steel
  BS 3868: 1973 sets down details of drainage stack units in galvanized steel.
- plastics, including
  ABS
  MUPVC
  polyethylene
  polypropylene
  BS 5255: 1976 relating to plastics waste pipes and fittings.
- uPVC
  BS 4514: 1983 together with amendment slip 1: AMD 4517 gives specifications for ventilating pipes, fittings and accessories.
- polypropylene
  BS 5254: 1976 with amendment slips 1: AMD 3588 and 2: AMD 4438 relates to polypropylene waste pipes and fittings.

The following materials are mentioned in the approved document in connection with traps:

- copper
  BS 1184: 1976 describes copper and copper alloy traps for use with baths, basins, sinks and washtubs, tubs and sinks.
- plastics
  BS 3943: 1979 details plastic waste traps giving depth of seal, inlet and outlet connections, access and angle of outlets. Included are tests for leakage, water seal and flow of water.

The approved document reminds users that when different metals are used they should be separated by a non-metallic material to prevent electrolytic corrosion.

It is essential that pipes be given adequate support but that this does not in any way inhibit thermal movement.

The approved document now reminds us that fire stopping provisions must not be overlooked.

ALTERNATIVE APPROACH

The performance required by section 1 of the approved document can also be achieved by following the relevant recommendations contained in clauses 3, 4 and 7 to 12 of BS 5572: 1978, 'Code of practice for sanitary pipework'.

The various clauses are summarized below, but the BS should be available to all concerned with the installation of sanitary pipework on an extensive scale. The document contains numerous informative tables and diagrams: particularly useful being those giving data and procedures used for pipe sizing.

Clause 3 gives definitions:

- access cover – a removable cover on pipes and fittings;
- branch discharge pipe – a waste pipe connecting a sanitary appliance to a discharge stack;
- branch ventilating pipe – a vent pipe connected to a branch discharge pipe;
- crown of trap – topmost point inside a trap outlet;
- depth of water seal – depth of water which would have to be taken out before air could flow freely through the trap;
- discharge pipe – a pipe carrying discharges from a sanitary appliance;
- size – nominal internal diameter of pipes;
- stack – a main vertical discharge or ventilating pipe;
- trap – fitting or part of an appliance or pipe to retain water so as to prevent the passage of foul air;
- ventilating pipe – a pipe installed to limit pressure fluctuations within the discharge pipe system.

Clause 4

Exchange of information

General. Emphasizes importance of consultation between clients, architects and engineers at all stages of the design of buildings.

Details of sewers and implications regarding surcharging, and a local authority's requirements should be ascertained.

Clause 7

7:1 General. System should comprise the minimum pipework necessary to carry away discharges quickly, quietly and without risk to health.

Typical discharge rates are given.

Appliance	Capacity (litres)	Discharge flow (litres/s)
Washdown WC	9	2.3
Urinal	4.5	0.15
Wash basin (32 mm outlet)	6	0.6
Sink (40 mm outlet)	23	0.9
Bath (40 mm outlet)	80	1.1
Automatic Washing Machine	180	0.7
Shower	–	0.1
Spray tap basin	–	0.06

Clause 7 also continues by giving recommendations regarding water seals in traps, limiting noise, resistance to blockage, durability, access for maintenance, replacement and accessibility for testing.

Also contained is information regarding the hydraulics and pneumatics of discharge systems, including descriptions:

* ventilated system used where there are large numbers of sanitary appliances in ranges or where they have to be widely dispersed;
* ventilated stack system is used in situations when close grouping of appliances makes it practicable to provide branch discharge ventilating pipes without the need for branch discharge pipes;
* single stack system is used in situations where close grouping makes it practicable, but only when the discharge stack is large enough to limit pressure fluctuations without the need for a ventilating stack.

Clause 8

Design

Typical features of domestic installations are single appliances connected to or closely grouped around a discharge stack.

Public buildings, offices, factories and schools frequently have appliances arranged so that they cannot be grouped around a stack as in domestic buildings.

This clause contains design details for traps; depth of seals, diameter, bottle traps, resealing traps and floor drainage.

Design factors are also included for discharge pipes and stacks, including sizing, gradients of branch discharge pipes, lengths, bends and junctions.

Discharge stack information is included. Recommendations for diameters, bends, branches, offsets, surcharging drains, intercepting tanks, termination of discharge stacks, discharge from urinals, and sinks are included.

Clause 9

Commonly used pipework arrangements, layout and sizing data

This clause contains data on the sizing of discharge and ventilating pipework and shows commonly used pipework arrangements.

It contains commonly used arrangements of branch discharge pipes including water closets and urinals, washbasins, bidets, sinks, baths, kitchen food waste and sanitary towel incinerators in single and multiple arrangement.

Commonly used arrangements of stack pipes and branches are also described and illustrated in detail.

Clause 10

Pipe sizing using discharge unit method

This method can be used for very tall or large buildings. In this method numerical values are used to express the load producinq

properties of various sanitary fittings, The method is based entirely on hydraulic loadings, but the clause contains a table giving approximate ventilating pipe and stack sizes to be used with the data.

Clause 11

Work on site

Contains recommendations regarding the jointing of pipes, the support and fixing of pipes, protection during building and construction.

Clause 12

Inspection and testing of completed installations

This clause describes inspections and tests which should be made during the installation of a discharge system as the work proceeds, and emphasizes that upon completion the work should be meticulously inspected. No cement droppings, pebbles or other objects are to be left in the pipes, and no jointing material is to be left projecting into the pipe bore.

Tests described include air tests, smoke, soap solutions, water and performance tests including induced syphonage and back pressure.

SECTION 2. FOUL DRAINAGE

A drain is not defined in the regulations but according to the 1984 Act, section 126, a drain is:

* used for the drainage of one building, or of buildings or yards appurtenant to buildings, within the same curtilage, and includes any manholes, ventilating shafts, pumps or other accessories belonging to the drains.

Likewise the term 'public sewer' is not defined in the regulations or the 1984 Act. One must refer to section 102, the Water Industry Act 1991, for a long discourse setting down the circumstances in which sewers and drains may become public sewers, but briefly they are:

* sewers and drains that have been vested in a local authority.

Another expression, regarding the status of sewers is that of 'private sewer' and this means a sewer which is not vested in a local authority. Officially the 1984 Act, section 126, says that:

* a private sewer means a sewer that is not a public sewer.

Depending upon a great variety of factors there are three systems of public sewerage which may be found in cities, towns and villages, and it is essential that new drainage systems are designed to make the correct connections to the various public sewers. This emphasizes the need to consult with a local authority and inspect the sewer maps prior to preparing plans for the new development:

* The 'combined' system comprises one pipe which receives all the foul water and surface water from an area. Whilst it is common in parts of the country, particularly in older areas, it is not popular as it requires numerous overflow arrangements to cater for large flows in time of storm in order to protect sewage treatment works and pumping stations from overloading.

The approved document reminds designers that when combined systems are in operation there may be a need to examine pipe sizes in order to accommodate peak flows.

* The 'separate' system comprises two sewers. One is there to receive foul water and the second pipe is for surface water. This is a very common arrangement and the authorities will not permit surface water to be put in a foul sewer, or foul water in the surface water sewer. In consequence the development requires separate drains for foul and for surface water.

- The 'partially separate' system will also be found. This comprises two sewers. One takes some surface water only as for example the highway drainage and surface water from the front of buildings. Foul water is taken to a separate sewer which also is permitted to receive a limited amount of surface water, possibly from rear roof slopes and back yards which are allowed to discharge into the rear-foul water gulley.

Particular care is to be exercised in the bedding and jointing of drains in a way which is appropriate to the material and location of each drain.

Much work in this connection has been undertaken in recent years. The then Minister of Housing and Local Government formed a working party in 1964. This was charged with the task of investigating the practical application of structural theory to the design and construction of buried pipe sewers with particular regard to structural stability, watertightness and maintenance costs.

Much of current practice stems from that beginning which emphasized the role of building regulations. This is to require the joints of drains and private sewers to be formed in such a manner that they remain watertight under all working conditions. This includes any differential movement between the pipe and the ground in any structure through or under which it passes.

LAYOUT

Drainage system layouts (Figs H1.19 and H1.20) should:

- be kept as simple as possible;
- have minimum changes of direction and gradient as practicable;
- change of direction and gradient should be as easy as practicable;
- be provided with access points only if they are necessary to enable any blockages to be cleared;
- connections of drains to other drains or sewers to be made obliquely, or in the direction of flow;
- be ventilated and provided with a vent pipe at the head of each main drain;
- have a ventilating pipe serving any branch more than 6 m serving a single appliance;
- have a ventilating pipe serving any branch more than 12 m serving a group of appliances;
- all drains having intercepting trap to have ventilating pipe;
- be kept in straight lines where possible but easy curves may be used if these do not inhibit clearance of blockages;

- limit the siting of bends to positions in or near manholes or inspection chambers, or at foot of discharge or ventilating stacks;
- have any bends provided to the largest practicable radius.

SETTLEMENT

Precautions are necessary to accommodate settlement:

- where pipes run near buildings
- where pipes run under buildings
- where pipes run on piles or beams
- where pipes are in unstable ground.

A drain is allowed to run under a building if it is protected by at least 100 mm of granular or other flexible material around a pipe (Fig. H1.21).

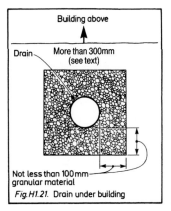

Fig. H1.21. Drain under building

Where excessive subsidence is possible:

- additional flexible joints, or
- suspension of drainage should be considered. If the crown of the pipe is less than 300 mm of the underside of the slab, surround with concrete made integral with the slab.

Unless care is taken differential settlement will result in shear fractures in pipes where they pass through walls or foundations. Settlement of a wall has this effect on a pipe rigidly fixed in wall as in order to accompany it on its downward path something must yield and the evidence is a vertical crack through the pipe.

To meet this situation there should be an opening in the wall to allow not less than 50 mm clear all around the pipe. This preserves the wall cavity and prevents the ingress of vermin (Fig. H1.21.1).

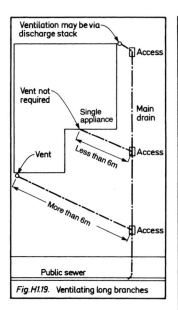

Fig. H1.19. Ventilating long branches

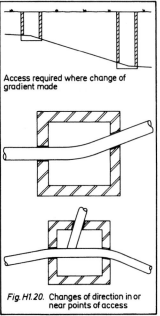

Fig. H1.20. Changes of direction in or near points of access

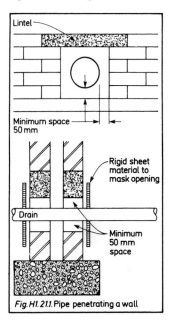

Fig. H1.21.1. Pipe penetrating a wall

The alternative is to build a short length of pipe into the wall with its joints not more than 150 mm away from the faces of the wall. This is connected on each side to rocker pipes each no longer than 600 mm and having flexible joints (Fig. H1.22).

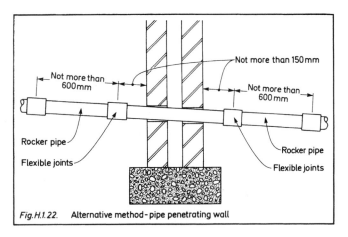

Fig.H.1.22. Alternative method - pipe penetrating wall

Where any drain or private sewer is constructed adjacent to a building it is necessary to take precautions to ensure that the drain trench in no way impairs stability of the building. Precautions required by the approved document are illustrated in Fig. H1.23.

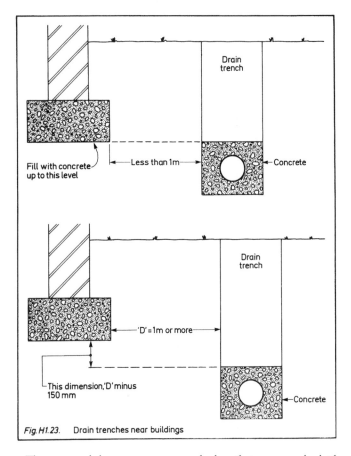

Fig. H1.23. Drain trenches near buildings

The approved document recommends that if pipes are to be laid on piles or beams, or in a common trench, or in unstable ground, or where there is a high water table the proposals should be discussed with the local authority.

Surcharging

Where there is a possibility that a drain may be surcharged, measures are to be taken to protect the building.

Drains which will be unaffected by surcharge should bypass any protection measures taken, and discharge by gravity wherever possible to a surcharge free outlet. If, however, this is not possible, discharge may be into the surcharged part of the system.

Measures available will depend upon particular circumstances, and the matter should be discussed with the local authority who will have information on the extent of possible surcharging, frequency, and measures found to be effective.

Protective measures are described in BS 8301: 1985, 'Code of Practice for building drainage'.

If it is necessary to install an anti-flood device, additional ventilation may be necessary in order to maintain trap seals (Fig. H1.23A).

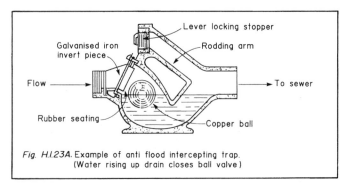

Fig. H.I.23A. Example of anti flood intercepting trap.
(Water rising up drain closes ball valve)

Rodent control

Local authorities have been waging a war against rodents in sewerage systems on a scientific and organized basis for very many years, but in some areas rodents are still a problem and it is possible for rats to gain access to buildings via sewers and drains.

The approved document draws attention to the problem.

- Local authorities can provide information on locations where infestation of drains and sewers is a problem,
- Local authorities can provide information regarding locally effective preventive measures.
- It may be necessary to seal drainage by adding sealing plates to pipework in inspection chambers instead of having the usual open channel.
- It may be necessary to resort to the installation of an intercepting trap. Such fittings were compulsory in many areas prior to the coming into force of building regulations; but they are liable to blockage unless they are regularly maintained – which is most unlikely (Fig. H1.23B).
- It may be necessary to combine a sealed drainage system with an intercepting trap in heavily infested areas.

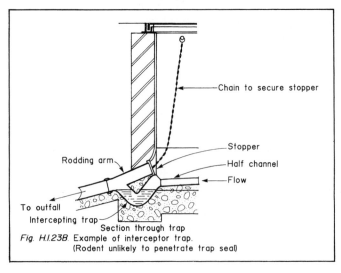

Fig. H.I.23B. Example of interceptor trap.
(Rodent unlikely to penetrate trap seal)

Depth of cover

The amount of cover to a drain or private sewer will usually depend upon:

- levels of the connections
- gradients
- ground levels.

Pipes require adequate cover to be protected from danger arising as a result of point loads which can occur when too little cover is provided. There is also a need for the pipes to be protected against the weight of backfilling when they are in deep trenches.

Where rigid pipes have less than the cover recommended in Figs H1.28 to H1.35 they are, where necessary, to be protected by being surrounded by not less than 100 mm concrete. There is to be a movement joint at each socket or sleeve face joint (Fig. H1.24)

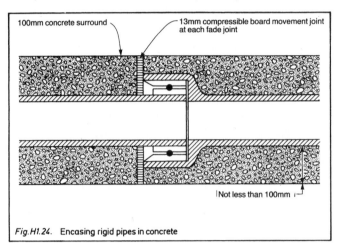

Fig.H1.24. Encasing rigid pipes in concrete

Care must be taken to protect flexible pipes. Those not under a road and having less than 0.6 m cover should be protected as shown in Fig. H1.25. Alternatively, the pipes should have concrete paving slabs placed above not less than 75 mm of granular or similar flexible filling.

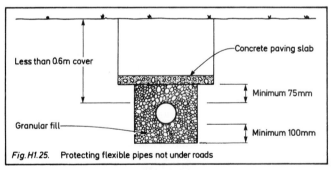

Fig. H1.25. Protecting flexible pipes not under roads

When flexible pipes are placed under a road and have less than 0.9 m cover they are to be protected by a reinforced concrete bridge as in Fig. H1.26.

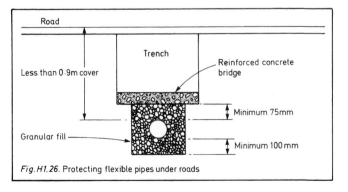

Fig.H1.26. Protecting flexible pipes under roads

GRADIENTS AND SIZES OF PIPES

The capacity of a drain depends upon two facts which are readily variable within a designer's wishes:

- diameter of the pipes
- gradient at which pipes are laid.

Sometimes there is difficulty in securing a gradient satisfactory for the chosen pipe size.

There are minimum pipe diameters:

- 75 mm diameter for drains carrying only waste water;
- 100 mm diameter for drains carrying soil water.

Recommended minimum gradients for foul drains are given in the approved document.

Peak flow (litre/s)	Pipe size (mm)	Minimum gradient	Maximum capacity (litre/s)
<1	75 mm dia.	1:40	4.1
<1	100 mm dia.	1:40	9.2
>1	75 mm dia.	1:80	2.8
>1	100 mm dia.	1:80 (minimum 1 WC)	6.3
>1	150 mm dia.	1:150 (minimum 5 WCs)	15.0

Long used figures for foul water from housing layouts are repeated from earlier editions of the book.

Drain receiving flow from	Pipe diameter	Minimum gradient
5 to 20 housing units	100 mm	1 in 80 (A)
20 to 150 housing units	150 mm	1 in 150 (B)
Smaller flows	100 mm	1 in 40

It is possible for (A) to be laid to a fall as flat as 1 in 130 and (B) 1 in 200. The flatter gradients are possible only where satisfactory joints and a high standard of supervision and workmanship can be guaranteed.

This is emphasized by the approved document which offers information regarding the discharge capacities of foul water drains running 0.75 proportional depth. It is one of the quirks of hydraulics that a sewer will have its greatest velocity when flowing at about 0.80 of its proportional depth and give its greatest discharge when operating at 0.94 of its proportional depth. (Fig H1.27)

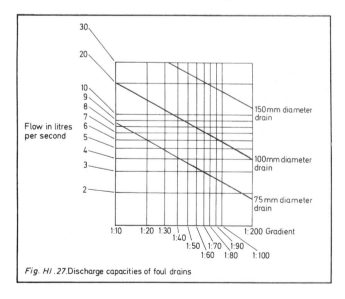

Fig. H1.27. Discharge capacities of foul drains

MATERIALS

Drain pipes are manufactured in asbestos cement, concrete, grey iron and vitrified clay. These pipes are rigid and will fracture before any significant deformation occurs. They can, however, be fitted with flexible joints. Plastic pipes are regarded as being flexible and will deform to a significant amount before collapsing. There are appropriate British Standards.

Any of the following materials may be used and will meet these standards:

- rigid pipes should have flexible joints;
- joints are to remain watertight under working conditions;
- nothing in pipes, joints or fittings should project into the pipeline;
- no obstructions are to be permitted to enter the pipeline;
- when different metals are used they must be separated by non-metallic materials to prevent electrolytic corrosion.

The approved document says that the following materials may be used for below ground gravity drainage:

Rigid pipes

- asbestos
  BS 3656: 1981 deals with asbestos – cement pipes, joints and fittings for sewerage and drainage. It contains details of the composition, geometrical characteristics, physical characteristics and classification of pipes by means of crushing strength related to diameter.

  Joint materials involving rubber rings and asbestos cement are specified. Requirements are given for crushing strength and chemical resistance.
- concrete
  BS 5911: Part 1: 1981 gives a specification for pipes and fittings with flexible joints. Details are given for precast cylindrical pipes, bends and junctions with flexible joints and also manholes.
- grey iron
  BS 437: 1978 is about cast iron spigot and socket drain pipes. Pipes and fittings may be manufactured by the sand blast or spun process. The pipes have spigot and socket joints suitable for caulking. Details of dimensions and protective coatings are included.
- vitrified clay
  BS 65: 1981 relates to vitrified clay fittings, pipes and joints. It sets out requirements for pipes with or without sockets for the construction of drains and sewers operating as gravity pipe lines. Normal pipes are suitable for all drains and sewers and the extra chemically resistant is suitable for drains and sewers in acid soil when extra resistance is required. Also BSEN 295: Parts 1, 2 and 3: 1991.

Flexible pipes

- uPVC
  BS 4660: 1973 is concerned with unplasticized underground drain pipes and fittings. Matters covered are material, colour, defects and dimensions of pipes and fittings. Information is given on sealing components together with recommendations for storage handling and installation. Pipes covered by this BS are suitable for laying under gardens, fields, driveways, yards and roads other than main roads. Diameters range from 110 mm to 160 mm.

  BS 5481: 1977 relates to unplasticized PVC pipes and fittings for gravity sewers. It covers a nominal diameter range of from 200 mm to 630 mm together with necessary joints and accessories. Materials, dimensions, construction, fittings, marking, testing and sealing components are all detailed.

When trade effluents are to be conveyed in a drain the manufacturer should be consulted regarding the type of pipe suitable for the effluent involved.

BEDDING AND BACKFILLING

Drains mainly depend for their strength upon the support and protection received from the bedding material, especially from that support at the sides. Uniform support should be provided on the prepared bed and good buttress support is necessary from the sides of the trench by well compacted side filling of a suitable quality.

The approved document gives three grades of filling which are to be used with both rigid and also flexible pipes.

1. Selected fill. This must be free from stones which are larger than 40 mm; lumps of clay more than 100 mm, bits of timber, any frozen material and vegetable matter.
2. Granular material which should comply with BS 882: 1983, Table 4 or BS 8301: 1985, Appendix D. Compaction fraction > 0.3 for Class N; > 0.2 for Class F and B. BS 882: 1983 relates to coarse and fine aggregates from natural sources and specifies requirements for naturally occurring materials crushed or uncrushed. Table 4 is included below.

Table 4.	Coarse aggregate — BS 882: 1983						
Sieve size (mm)	Percentage by mass passing BS sieves for nominal sizes						
	Graded aggregate (mm)			Single-sized aggregate (mm)			
	40 to 5	20 to 5	14 to 5	40	20	14	10
50.0	100	–	–	100	–	–	–
37.5	90–100	100	–	85–100	100	–	–
20.0	35–70	90–100	100	0–25	85–100	100	–
14.0	–	–	90–100	–	–	85–100	100
10.0	10–40	30–60	50–85	0–5	0–25	0–50	85–100
5.0	0–5	0–10	0–10	–	0–5	0–10	0–25
2.36	–	–	–	–	–	–	0–5

BS 8301: 1985 is a code of practice for building drainage and Appendix D details the compaction fraction test for the suitability of bedding material. Apparatus required is an open ended cylinder 250 mm long and 150 mm diameter; a metal rammer, and a rule.

The method is to put a representative sample of material into the cylinder loosely and without tamping. Strike off the top surface level with the top of the cylinder and remove all surplus spilled material. Lift the cylinder up clear of its contents and place on a fresh area of flat surface. Place about one quarter of the material back in the cylinder and tamp vigorously until no further compaction can be obtained. Repeat with the second quarter, tamp; add the third quarter, tamp; add the fourth quarter, tamping the final surface as level as possible. Measure down from the top of the cylinder to the surface of compacted material. Distance in millimetres, divided by height of the cylinder (250 mm) is regarded as the compaction fraction.

Compaction fraction	Suitability of material
0.15 or less	Suitable
Between 0.15 and 0.3	Suitable but requires extra care in compaction. Not suitable if the pipe is subject to waterlogged conditions after laying
Over 0.3	Unsuitable

3. Selected fill or granular material which is free from stones larger than 40mm.

These are the three bedding materials which may be used, and firstly the manner of protecting rigid pipes is illustrated. The illustrations also give minimum and maximum depths of cover for each type. Pipes which are laid with less cover require special protection as shown in Fig. H1.24.

The approved document gives four alternatives when considering the bedding of rigid pipes and the limits of cover for standard strengths of pipe in any width trench. Where site conditions do not fall within the limits described it will be necessary to undertake calculations and design protection against those conditions.

Bedding factors depend upon the nature and circumferential extent of the support provided by the bedding on which the pipe rests.

- Class D. The simplest and least strong of the classes described. However, the trench bottom must be trimmed by hand and a high standard of workmanship is necessary. It is not recommended to be used in compacted sand or gravel, very soft silt, clay, sandy clay. Pipes must not rest on bricks or similar hard packing.
- Class N. This class has the same bedding factor as Class D and may be used when it is not possible to ensure accurate hand-trimming of the trench bottom by shovel. Ground water must not be allowed to flow along the trench bottom. Clay or lean mix barriers may be necessary.
- Class F. Requires imported granular material and gives a stronger bedding factor than D or N. This class is generally suitable in all soil conditions but flow or ground water through the trench should be prevented.
- Class B. Highest bedding factor of the methods contained in the

approved document. It may be used when uniform support of the pipeline is required, and also when ease and speed of laying is important. This method affords support by granular fill to the side of each pipe.

100 mm diameter rigid pipes using bedding types D, N, F and B are illustrated in Figs H1.28 to H1.31.

150 mm diameter rigid pipes using bedding types D, N, F and B are illustrated in Figs H1.32 to H1.35.

Flexible pipes

Flexible pipes are pipes which deform to a significant amount before collapsing, and to meet building regulation requirements require support to limit deformation to 5% of the pipe diameter.

Bedding and backfilling for a narrow and open trench are shown

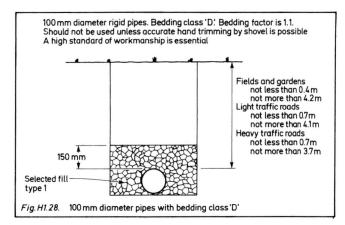

Fig.H1.28.   100 mm diameter pipes with bedding class 'D'

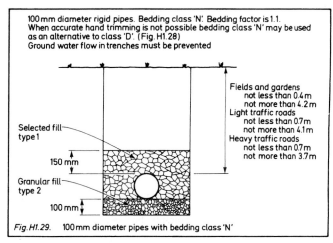

Fig.H1.29.   100 mm diameter pipes with bedding class 'N'

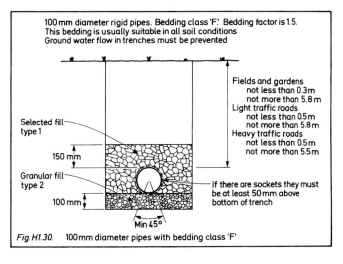

Fig.H1.30.   100 mm diameter pipes with bedding class 'F'

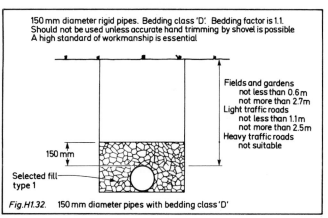

Fig.H1.31.   100 mm diameter pipes with bedding class 'B'

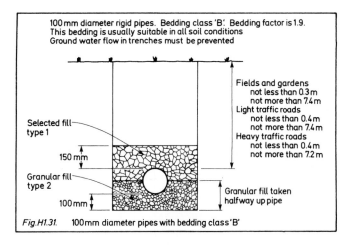

Fig.H1.32.   150 mm diameter pipes with bedding class 'D'

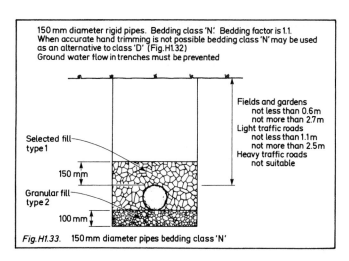

Fig.H1.33.   150 mm diameter pipes bedding class 'N'

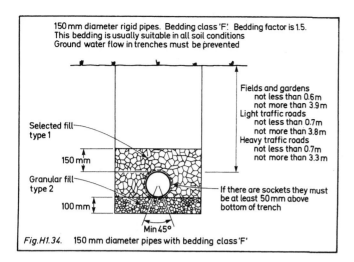

Fig.H1.34.   150 mm diameter pipes with bedding class 'F'

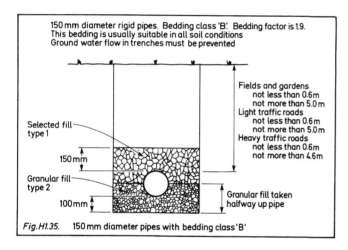

Fig.H1.35.   150 mm diameter pipes with bedding class 'B'

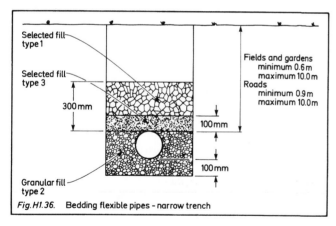

Fig.H1.36.   Bedding flexible pipes - narrow trench

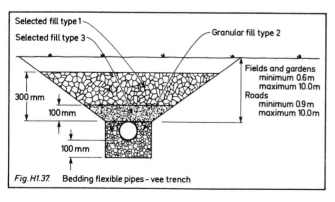

Fig.H1.37.   Bedding flexible pipes - vee trench

in Figs H1.36 and H1.37. In fields and gardens the minimum depth should not be less than 0.6 m; under roads the minimum depth allowed is 0.9 m and nowhere should the maximum depth exceed 10 m. Water flow in trenches should be avoided.

In circumstances where special protection is necessary, refer to Figs H1.24, H1.25 and H1.26.

PROVISION FOR CLEARING BLOCKAGES

The first principle is to do the utmost by way of good design and workmanship to prevent blockages arising. Without doubt badly made joints are a frequent cause of blockages. But however carefully a drain is constructed many factors, not least the carelessness of the occupiers of a building may give rise to a blockage. It is essential that clearance may be undertaken quickly, and without creating a nuisance, or the need to excavate to find a pipe.

Manholes are required on sewers and drains as a means of access for inspection and testing, and for the clearance. In present practice manholes are constructed at summits, changes in diameter, direction and gradient, and where sewers join.

Manholes have additional useful purposes as they are used for rodent control, television surveys, flow measurement and tracing illegal discharges.

Except for shallow drains or sewers, all manholes should be of adequate dimensions for entry and for rodding purposes. More modest access arrangements are acceptable for shallow installations.

The maximum spacing of access arrangements is contained in the approved document. Considerable difficulty is experienced in rodding long distances and these measurements should not be exceeded.

The sizing of manholes is governed by the need to provide for a man to be able to work within the chamber and insert stoppers without difficulty.

There are a number of plastics access fittings now available and many of them have Agrément certificates. They can be used with the various types of drain pipes now in use, although some are limited to be used within a system devised by the manufacturer.

It is particularly important that a drainage system within a building shall be absolutely watertight under all working conditions and remain so in the event of becoming choked. This may be achieved by a bolted inspection plate on a cast iron drain, or by a gas tight cover to an inspection chamber fitted with corrosion-resistant bolts to prevent lifting under pressure.

Overall the siting, spacing and type of access arrangement depends upon the layout, size and depth of the particular drainage system.

Provisions contained in the approved document are for normal methods of rodding, remembering that in many circumstances blockages may be cleared by rodding against the flow.

Types of access

Four types of access are in common use, and minimum dimensions are given for each arrangement. These dimensions should be increased at junctions where they do not allow enough space for branches. Suitable allowances are:

Length
Allow 300 mm for a
100 mm branch
Allow 380 mm for a
150 mm branch plus
adequate allowances on
the downstream side for
the angle of entry.

Width
Allow 300 mm for each side with
branches, plus 150 mm on the
diameter of the main drain
whichever is the greater

Types of access acceptable are:

- Rodding eye (Fig. H1.38). This is a very simple arrangement which may be used at heads of drains or on shallow drains for rodding in the direction of the flow.

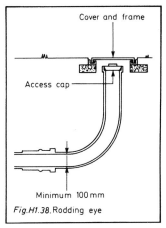

Fig.H1.38. Rodding eye

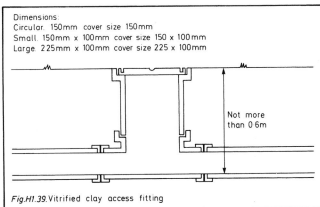

Dimensions:
Circular. 150mm cover size 150mm
Small. 150mm x 100mm cover size 150 x 100mm
Large. 225mm x 100mm cover size 225 x 100mm

Fig.H1.39. Vitrified clay access fitting

- Access fittings (Fig. H1.39). These are small chambers usually a component of the pipe system to facilitate inspection and rodding in both directions of shallow drains up to 600 mm deep. They are inexpensive and perform well near the head of the drain or adjacent to a terminal fitting.
- Inspection chambers (Figs H1.40 and H1.41). Larger chambers having an open channel with provision for junctions and may be performed in plastics or concrete or built on site in masonry. They may be oblong or circular on plan and do not have working space at drain level. Provision in the approved document is for inspection chambers of up to 1 m deep.

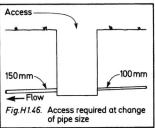

Cover 450mm x 450mm
Depth up to 1m
450mm x 450mm

Fig.H1.40. Inspection chamber

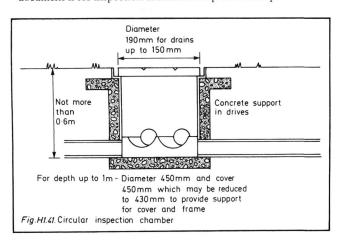

Diameter 190mm for drains up to 150mm
Not more than 0.6m
Concrete support in drives

For depth up to 1m - Diameter 450mm and cover 450mm which may be reduced to 430mm to provide support for cover and frame

Fig.H1.41. Circular inspection chamber

- Manholes (Figs H 1.42 to H1.44). These are large chambers with working space at drains level. These may be constructed of masonry, in situ concrete or prefabricated concrete components. Minimum dimensions are shown for shallow and deep manholes which may be circular on plan.

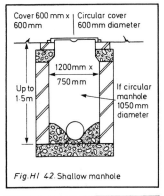

Cover 600 mm x 600mm | Circular cover 600mm diameter
Up to 1.5m
1200mm x 750mm
If circular manhole 1050mm diameter

Fig.H1.42. Shallow manhole

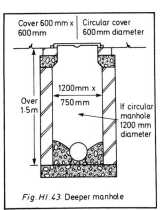

Cover 600 mm x 600mm | Circular cover 600mm diameter
Over 1.5m
1200mm x 750mm
If circular manhole 1200 mm diameter

Fig.H1.43. Deeper manhole

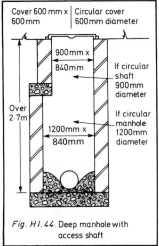

Cover 600 mm x 600mm | Circular cover 600mm diameter
Over 2.7m
900mm x 840mm
If circular shaft 900mm diameter
1200mm x 840mm
If circular manhole 1200mm diameter

Fig. H1.44. Deep manhole with access shaft

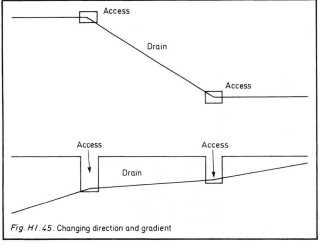

Access
Drain
Access

Access
Access
Drain

Fig. H1.45. Changing direction and gradient

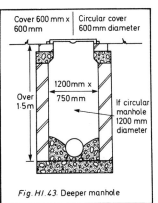

Access
150mm
100mm
Flow

Fig.H1.46. Access required at change of pipe size

Siting of access points

Access is to be provided at the following locations:

- at or near the head of each drain run;
- at a change of direction and at a change of gradient (Fig. H1.45);
- at a change of pipe size (Fig. H1.46);

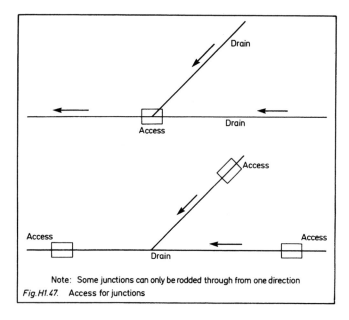

Note: Some junctions can only be rodded through from one direction

Fig. H1.47. Access for junctions

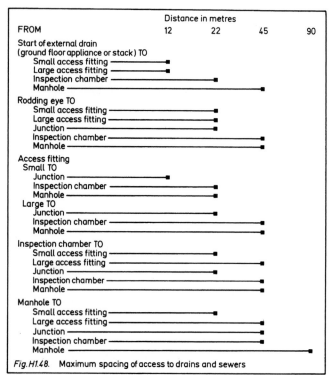

Fig. H1.48. Maximum spacing of access to drains and sewers

Manholes and inspection chambers

- Clay bricks and blocks
  BS 3921: 1985 gives the standard for brick and blocks for walling including engineering bricks which have a dense semi-vitreous composition suitable for manhole construction.
- Vitrified clay
  BS 65: 1981 sets out requirements for vitrified clay fittings in the construction of gravity sewers.
- Precast concrete
  BS 5911: Part 1: 1981 contains specifications for concrete cylindrical manholes both reinforced and unreinforced.
  BS 5911: Part 2: 1982 is a specification for precast concrete inspection chambers.
- In-situ concrete
  BS 8110: Part 1: 1985 deals with the design construction of structures in reinforced concrete.
- Plastics
  BS 7158: 1989 Specification for plastics inspection chambers.
  Plastics access fittings having a British Board of Agrément certificate may be used.
  Rodding eyes and access fittings (not including frames and covers). Includes the various materials acceptable in the construction of underground gravity drains and sewers.
- Asbestos
  BS 3656: 1981 deals with asbestos fittings.
- Grey iron
  BS 437: 1978 gives standards for cast iron spigot and socket drain fittings.
- uPVC
  BS 4660: 1973 sets out requirements for unplasticized underground drain fittings.
  BS 5481: 1981 also relates to unplasticized PVC pipes and fittings ranging from 200 mm to 630 mm diameter.

Rodding eyes and access fittings having a British Board of Agrément certificate may be used.

Other requirements for inspection chambers and manholes

- Have durable removable non-ventilating covers. These are usually cast iron, cast or pressed steel or precast concrete.
- Have mechanically fixed airtight covers in a building.
- Have metal step irons or fixed ladders when more than 1.0 m deep.
- Effluent should be conveyed through chamber in half round channels.
- Branches should discharge into channel at or above level of horizontal diameter.
- When angle of branch is more than 45° a three-quarter section bend is to be used.
- Channel and branches are to be benched (Fig. H1.49).

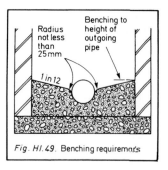

Fig. H1.49. Benching requirements

WATERTIGHTNESS

Gravity drains and private sewers should be water tested and achieve the following standard. Although it should be applied after laying and before backfilling or concreting the test may also be made imme-

- at a junction. The exception is where each run can be cleared from an access point (see Fig. H1.47).

Access arrangements must also be provided on long straight runs of drains or sewers. The maximum distances indicated on Fig. H1.48 are appropriate for pipe sizes up to and including 300 mm diameter.

Construction of access points

All points of access including inspection chambers and manholes should be watertight, preventing foul water under working conditions from escaping and groundwater from gaining access.

Various materials may be used, but compliance with these requirements is largely dependent upon the standard of workmanship employed.

The approved document says that the following materials may be used for the construction of access points.

diately prior to occupation. This means that great care must be taken during building operations carried out after drains have been back-filled, as it is an extremely expensive exercise to locate and remedy defects after backfilling.

Testing of systems up to 300 mm diameter

Water

- Test to pressure equal to 1.5 m of head of water measured above invert at head of drains, When testing, a section of the system should be filled with water and left to stand for 2 hours and topped up. Leakage over 30 minutes should then be measured and the approved document gives this requirement.
- Leakage should not be more than 0.05 litres per metre of a 100 mm diameter pipe. This equals a drop of 6.4 mm per metre.
- Leakage should not be more than 0.08 litres per metre of a 150 mm diameter pipe. This equals a drop of 4.5 mm per metre.

Long branches should be tested separately and, in order to avoid damage to the drain, the head of water at the lower end must not be more than 4 m (Fig. H1.50).

Alternatively, an air test may be used.

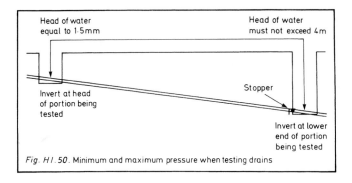

Fig. H1.50. Minimum and maximum pressure when testing drains

Air test

Test to ensure maximum loss in head on a manometer of 25 mm in 5 minutes for 100 mm gauge or 12 mm for a 50 mm gauge.

ALTERNATIVE APPROACH

This is a topic where the approved document says that the required performance can be met by following some of the recommendations in BS 8301: 1985, which is the code of practice for building drainage. The relevant clauses are:

Section 1
Section 2
Section 3 (except clause 10)
Section 4 (except clause 23)
Section 5 (clause 25 only)
Appendices

The code is comprehensive and describes the discharge and methods of determining pipe sizes.

This code sets out recommendations for the design, layout, construction, testing and maintenance of foul, surface water and groundwater drainage systems.

Section 1

General, contains definitions

- 'blinding' is material to fill interstices and irregularities in trench bottoms;
- 'branch drain' is a line of pipes installed to discharge into a junction or an access point on another line of pipes;
- 'branch vents' are connected to one or more branch drains;

- 'foul water' is any water contaminated by soil, waste or trade effluent;
- 'junction' is a fitting designed to receive discharges from a branch drain;
- 'inspection chamber' is a covered chamber constructed on a drain to give access thereto;
- 'manhole' is a working chamber on a drain or sewer;
- 'storm water' is rainwater discharged from a catchment area as the result of a storm.

The code advocates full exchange of information between all parties and close liaison should be maintained between all appropriate authorities.

Information available from the following various bodies is listed:

local authority
water authority
highway authority
police
local mining authority
electricity, gas and water supply authorities
building owner.

Listed is also the information a designer should make available to the various authorities, including the contractor.

The various enactments having an effect on drainage are also set out in the document.

Section 2

Materials and components

This section is concerned with the selection of materials, including cement, aggregates, concrete, cement mortar, pipes, joints and fittings.

Miscellaneous materials include manhole covers, step irons and ladders, gullies and gratings, bricks, block fill material and granular material for bedding.

Section 3

This emphasizes that a drainage system should be designed, installed and maintained so as to function without causing a nuisance or danger to health, arising from leakage, blockage or surcharge.

Advice is offered with regard to layout, pipe sizes and gradients, pipelines under and through buildings, differential movement and ventilation.

Foul drainage details are also given including hydraulic design principles, trade effluent, connections to sewers, cesspools and septic tanks.

Surface water drainage

Surface water drains are to discharge into a surface water sewer, or a combined sewer. If neither is available, to a soakaway, a watercourse or to a storage container.

Methods of determining the rate of flow, or run off are outlined.

Combined drain must discharge into a combined sewer.

Structure design of drainage

Contains information regarding the basis for design, with recommendations for pipes at shallow depths, special site conditions, rigid and flexible pipes.

There are also recommendations concerning the provision of gullies, interceptors and anti-flood devices. These are related to the drainage of garage areas and vehicle parks, grease traps, surcharge and protective arrangements.

Access arrangements for drains are set out in detail with recommendations regarding rodding eyes and access points, inspection chambers and manholes.

Ground movement in some areas is a severe problem. Movement may occur in superficial deposits or may result from underground mining operations. Advice is given regarding precautions that may be necessary.

Pumping

In connection with sewage and surface water lifting installations recommendations are given for the siting of the equipment, the investigations that are necessary to establish suitable subsoil and groundwater conditions.

Several types of sewage lifting apparatus are described including pneumatic ejectors, centrifugal pumps and the availability of packaged units.

Sewage lifting apparatus should be selected for its suitability for a particular application, its durability and economy in operation.

In choosing a system it is necessary to take into account the following:

- nature of the accommodation served
- space or site available
- sewage inflow rate
- required discharge head
- consequences of failure
- availability of maintenance facilities

Section 4

Work on site

This section commences by recommending how the work should be planned and organized.

Setting out should be done accurately and methods of trenching, excavation and timbering are given.

Information is also given concerning drains in heading, thrust boring, pipe jacking, auger boring and the cut and drive method of construction.

Pipe bedding and laying is detailed including clauses offering advice on laying pipes on granular beds, or bedded and haunched with concrete.

Various pipe-jointing techniques are also described.

Precautions to prevent displacement of drains during testing and the placing and compaction of sidefills are given.

Before pipe laying is commenced, the position and level of the drain or sewer to which it is proposed to make a connection should be confirmed.

Section 5

Inspection and testing

First stage and final testing are described, including pre-test procedures and cleansing. Details are given of the water test and air test.

Appendix   A. Bibliography
            B. Preparation of plans
            C. Background information on the discharge unit method for estimating drain flow rates
            D. Compaction fraction test for suitability of bedding material.

CESSPOOLS AND TANKS

A cesspool is a form of 'conservancy' sanitation. It really means that waste is contained in a tank for ultimate removal and disposal at some distant place. Cesspools are to have a capacity of $18\,m^3$ and for the cost of constructing a tank of this capacity it would be possible to install a small macerator pump to lift sewage quite a distance through a small-diameter polythene pipe to a sewer, should one be available.

When it is imperative that some form of on-site sewage disposal must be utilized then all the precautions suggested are most vital to prevent pollution of the ground, water supplies and buildings in an endeavour to protect the health of people living nearby.

In the event of surface or underground water becoming polluted, people consuming the water who are living considerable distances away may be affected.

The siting of a cesspool is most important. In byelaw days it varied from 15 m to 18 m away from a dwelling. This requirement was removed by the 1965 Regs, but it must not be thought that this removal of those distances from the regulations means that it is safe to construct a cesspool near a dwelling. Indeed in certain circumstances it may be necessary to be much further away.

Not covered by the regulations, but obviously an essential part of a septic tank installation is the form of secondary treatment involved whether by mechanical aeration or by irrigation.

It is necessary to approach the water authority serving the area to obtain consent before:

- surface water is taken through a new outfall into a water course;
- effluent from a sewage treatment plant is discharged into a watercourse;
- a sewage effluent is irrigated below the surface of the ground.

The discharges are affected by provisions contained in the Water Resources Act 1963 and Control of Pollution Act 1974.

H2 requires that cesspools, septic tanks and settlement tanks are to be:

- **of adequate capacity;**
- **impermeable to liquids;**
- **adequately ventilated;**
- **sited and constructed in such a way that it**

    1. **is not prejudicial to health of anyone**
    2. **will not contaminate any underground water or water supply**
    3. **will have adequate means of access for emptying.**

Performance

The approved document sets an acceptable level of performance.

- Cesspools are to be large enough to hold foul water from a building until such time as they can be emptied.
- Septic tanks and settlement tanks are to be large enough to enable breakdown and settlement of solid matter in the foul water.
- Cesspools, septic tanks and settlement tanks are to be so constructed that they do not leak contents into the surrounding soil or allow groundwater to enter the tank. They are to be adequately ventilated.
- Cesspools, septic tanks and settlement tanks must not be constructed in a position which will give rise to conditions prejudicial to health, or contaminate water supplies.
- There is to be satisfactory access for emptying.

Guidance in the approved document is brief, and relates only to general principles. It emphasizes that specialist knowledge is desirable and refers designers and builders to BS 6297: 1983, which is a code of practice for design and installation of small sewage works and cesspools (see page 256).

CESSPOOLS

Cesspools must have a capacity below the inlet level of at least 18 000 litres.

A tank of this minimum size would need to be, for example, about $3\,m \times 2.44\,m \times 2.44\,m$ below the top water level. It should be noted that $18\,m^3$ is the minimum capacity, and relates to the smaller dwelling. When the number of people involved is greater it is reasonable to expect a larger tank to be necessary. The great disadvantage of a cesspool is the recurring cost of emptying, and the fact that once full, any further foul water entering the tank will overflow and create very undesirable conditions. Quite prudent use of water by a four person family would fill the tank in about five weeks, so very frequent emptying is necessary.

This would normally be undertaken by a tanker vehicle and to facilitate access a cesspool must be so sited that the vehicle can get to within 30 m of the tank. The contents are not to be taken through a dwelling or place of work but access may be through an open or covered space.

Cesspools (Fig. H2.1)

- must be watertight and not permit water to escape or subsoil water to penetrate;
- may be constructed of brickwork, concrete, glass reinforced concrete, glass reinforced plastic or steel. When brickwork is used engineering bricks set in 1:3 cement mortar is to form walls not less than 220 mm thick. In situ concrete walls are to be not less than 150 mm thick;
- are to be covered and ventilated;
- are to have access for removal of contents, having dimensions of not less than 600 mm;
- are to have an access for inspection purposes at the inlet;
- are not have any openings except for inlet, access and ventilation.

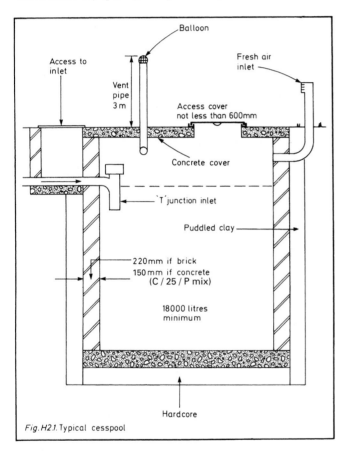

Fig. H2.1. Typical cesspool

In order to avoid undue disruption to tank contents velocity at point of entry should be limited. This may be achieved with drains up to 150 mm diameter, by laying the last 12 m of the incoming pipe at a gradient of 1:50 or flatter. The alternative is to install a dip pipe.

Reinforced plastics cesspools are available from several manufacturers, and should have an Agrément certificate.

SEPTIC TANKS AND SETTLEMENT TANKS

The approved document says 'septic tanks and settlement tanks' require a minimum capacity of 2700 litres (2.7 m³). In this phrase the two types of tank are linked together, but their functions are completely different.

A settlement tank is an appliance to receive sewage, and hold it quiescent for sufficient time to enable the mechanical settlement of solids in suspension to take place. The liquid then flows forward out of the tank for further treatment. Sludge which congregates in the base of the tank must be removed before it becomes septic. This means at weekly intervals, and involves a lot of labour for the de-sludging process, and will be found quite costly. The effluent which leaves a settling tank is quite different from that emanating from a septic tank.

The purpose of a septic tank is to receive sewage, permit its mechanical settlement, and then by the encouragement of anaerobic bacteria for the sludge to commence to digest.

In practice a large capacity is therefore required, and as part of the digestion process a thick crust will build up on top of the liquid.

At six-monthly intervals almost all the sludge should be removed. A little should remain to 'seed' the fresh sewage when the tank is brought back into operation.

As with the sizing of cesspools the capacity mentioned must be a minimum size, and the writer has for many years emphasized the need for a minimum of 3400 litres for a three-bedroom house.

Anyone who has been in the past concerned with the thankless task of designing and installing a small septic tank installation will welcome the range of factory prefabricated tanks now available.

As with cesspools provision must be made for a tanker vehicle to get within 30 m for emptying purposes. Contents are not to be taken through a dwelling or place of work, but access may be through an open or covered space.

Septic tanks and settlement tanks (Figs H2.2 to H2.5)

- must be watertight and not permit foul water to escape or subsoil water to penetrate;

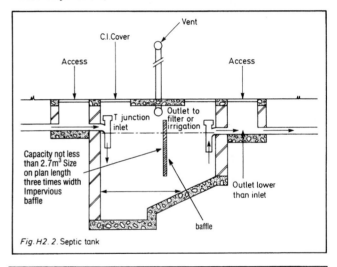

Fig. H2.2. Septic tank

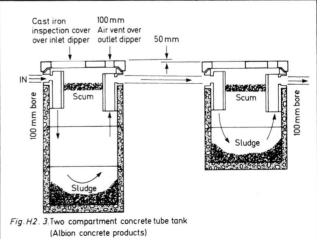

Fig. H2.3. Two compartment concrete tube tank (Albion concrete products)

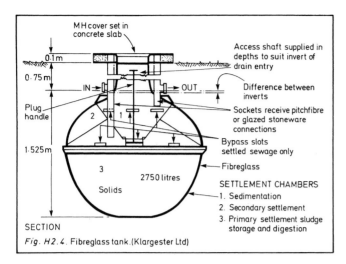

Fig. H2.4. Fibreglass tank.(Klargester Ltd)

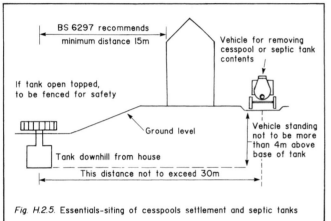

Fig. H.2.5. Essentials-siting of cesspools settlement and septic tanks

- may be constructed of brickwork, concrete, glass reinforced plastic or steel. Brickwork tanks should be built of engineering bricks set in 1:3 cement mortar. Brick walls must be not less than 220 mm thick and in-situ concrete walls at least 150 mm thick; concrete mix is to be C/25/P specification in BS 5328: Part 2: 1991 which is about the methods of specifying concrete mixes. C/25 has a characteristic compressive strength at 28 days of 25 N/mm^2.

  Contents of the BS include designated mixes, prescribed mixes and standard mixes – ST 5 gives a compressive strength of 25 N/mm^2. There is also a section related to designated mixes.

- may be open at the top or covered. If covered they must be ventilated and if open, protected by a fence surrounding the tank;
- if covered are to have access for removal of the contents, having dimensions not less than 600 mm;
- are to have access for inspection at the inlet and also the outlet.

It is more important to limit velocity at the point of entry than with cesspools, and similar arrangements may be made. With drains up to 150 mm diameter, the last 12 m of the incoming pipe may be laid at a gradient of 1:50 or flatter. The alternative is to fit a dip pipe to the inlet. In practice the flat gradient and the dip pipe will be found advantageous as flow directly across the tank will be avoided.

In septic tanks where the tank is less than 1200 mm wide from the inlet a dip pipe is to be installed.

Factory made tanks must be the subject of an Agrément certificate.

ALTERNATIVE APPROACH

The performance can also be met by complying with the relevant recommendations of BS 6297: 1983, 'Code of practice for the design and installation of small sewage treatment works and cesspools'.

The relevant parts are:

Section 1
Section 2
Section 3 (clauses 6 to 11)
Section 4
Appendices.

BS 6297: 1983 deals with installations suitable for domestic discharge ranging from single households up to 1000 population. General guidance is given on good design and installation practice.

Section 1

This section outlines the scope of the BS including also a number of definitions including:

- cesspool
  Covered watertight tank used for receiving and storing sewage from premises which cannot be connected to a public sewer and where ground conditions prevent the use of a small sewage treatment works including a septic tank;
- septic tank
  A type of settlement tank in which the sludge is retained for a sufficient time for the organic matter to undergo anaerobic decomposition.

A checklist of information required before designing a small sewage treatment works is included.

Section 2

This section is concerned with materials, including aggregates, mortar, cement, glass reinforced cement, concrete, glass reinforced plastics, steel tanks and clay concrete pipes and fittings.

Section 3

This section is concerned with design giving climatic considerations in addition to general design requirements. It is emphasized that in choosing the type of treatment the designer should compare costs of maintenance and operation as well as the initial capital cost of the works.

Sewage treatment works should be as far from habitable buildings as is economically practicable and direction of the prevailing wind should be considered in relation to any properties when siting a works. A small works serving more than one premises should not be less than 25 m from any dwelling and this should be progressively increased for larger treatment works.

General requirements for tanks

A cesspool must be watertight and there should be local emptying facilities available. The code states that a household of three people will produce 7 m^3 of waste for removal every three weeks. This is the capacity of a typical tanker.

The desirable minimum distance of a cesspool from any dwelling is 15 m.

There follows information about capacity, arrangement, shape, drain connection, ventilation and entry from confined spaces.

When a cesspool is abandoned it should be emptied and backfilled with hardcore or similar non-compressible material, as a safety measure.

Septic tank

Size of a septic tank is related to the number of persons who may be served and the following formula is given

$$C = (180P \times 2000)$$

where C = capacity of the tank in litres with a minimum value of 2720 litres, and
       P = the design population with a minimum value of 4.

Much information is given relating to design and sludge storage capacity, arrangement, inlets and outlets, and further treatment.

Recommendations are made regarding the prevention of blockages from rags and floating debris and the design of primary and secondary settlement tanks.

Section 4

This section is concerned with installation and says that workmanship should be of a good standard and wherever possible materials should be to the appropriate BS.

Upon completion, and before testing, all pipework should be thoroughly cleaned out. Before being put into use, tanks and other structures should be tested to hold liquid:

* soak for a period of 7 days;
* fall in water level over a further period of 7 days should not exceed 10 mm.

Appendices

Appendix A lists relevant British Standards.
Appendix B gives relevant sections of statutes applicable, including: Rivers (Prevention of Pollution) Act, which prohibits pollution of watercourses; Water Resources Act 1963, section 72, which controls discharges into underground strata; Control of Pollution Act 1974, sections 31 and 32, which control pollution of and discharges into rivers, and section 46, which enables water authorities to remedy or forestall pollution of water.

Other references have been overtaken by the 1984 Act and the Regs.

BS 6297 contains numerous illustrations.

RAINWATER DRAINAGE

Rainwater falling on the roof of a building must be taken away to a suitable point of discharge before it can do damage to the building or create damp conditions that may be injurious to the health of persons in the building.

Discharge from a rainwater pipe must not be connected into a foul water system unless the design of the sewerage system is intended to accept the rainwater run off. Even if there is no separate public sewer available the local authority may require separate drains up to the point of connection to an existing public sewer, so that at some future date the public system may have a separate surface water sewer.

There are many technical reasons why engineers do not now design systems as 'combined' and rigorously exclude water from the foul sewer.

When soakaways are to be used for rainwater disposal one must ensure that they are so sited that no dampness or damage will arise within the building itself.

H3. The requirement is that any system carrying rainwater from the roof of a building must be adequate for the purpose.

The approved document sets a level of performance, in order that danger to the health and safety of people in the building will not be at risk.

A rainwater drainage system must:

* convey the flow of rainwater to an outfall;
* minimize the possibility of blockage or leakage;
* have facility for clearing blockages should they arise.

An outfall includes a surface water or combined sewer, a private sewer, a soakaway or a watercourse.

Contents of the approved document relating to rainwater drainage are as follows:

Section 1. Gutters and rainwater pipes
  rainwater pipes
  materials

Section 2. Rainwater drainage
  combined systems
  layout
  depth
  gradients and sizes
  materials
  bedding and backfilling
  clearing blockages
  watertightness.

Provisions in the approved document do not apply for draining areas of 6 m² or less, including small roofs and balconies unless they receive water from a rainwater pipe or from paved or other hard surfaces.

DRAINAGE CAPACITY

* Assume that rainfall intensity is 75 mm per hour.
* Drainage system must be adequate to take expected flows at any point in the system.
* Size and gradient of gutters and pipes determine the capacity of a system.
* Rainwater or surface water must not be taken to a cesspool, septic or settlement tank.

Section 1: Gutters and rainwater pipes

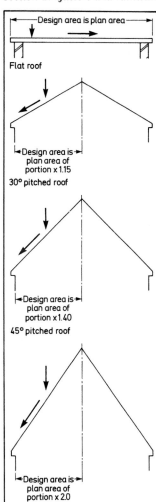

Fig.H3.1. Rainwater flow calculated on roof area to each gutter

In addition to the intensity of rainfall the flow of water from a roof depends upon the slope and a simple calculation may be made using a factor derived from the slope of roof to be drained (Fig. H3.1; see also Fig. H3.4).

Some roofs of 70° and over are regarded as walls (Fig. H3.2).

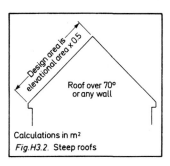

Fig.H3.2. Steep roofs

		Maximum roof area - m²				
		6	18	37	53	65	103
Gutter (mm/dia)		-	75	100	115	125	150
Outlet (mm/dia)		-	50	63	63	75	89
Flow capacity litres/second		-	0·38	0·78	1·11	1·37	2·16

When round edged outlets are used, smaller
downpipe sizes are permitted.

Fig. H3.3. Gutter and outlet sizes

The following are to be taken into account when calculating gutter and outlet sizes (see Fig. H3.3):

* largest effective areas which may be drained to the given gutter sizes;
* they relate to the gutter laid level and where distance from stop end to outlet is not more than 50 times the water depth;
* gutter is half round with sharp edged outlet at only one end;
* least size of outlet is shown;
* when outlet is not at end the gutter must be sized for the largest area draining into it;
* if there are two end outlets they may be not more than 100 times depth of flow apart.

Calculations are for gutters laid level, but they should be fixed with a fall in the direction of the nearest outlet. BS 6367: 1983 contains advice regarding the possibility of reducing size of gutter and pipe when gutters laid to a fall.

Gutters should be so placed that any overflow arising from abnormal rainfall will be discharged clear of the building.

May be prudent to provide a gutter and downpipe to a porch roof as shown in Fig. H3.4 (BRE defect action sheet 55/84).

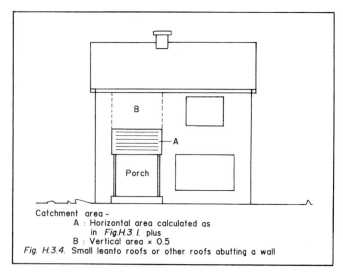

Catchment area -
   A : Horizontal area calculated as
        in *Fig.H.3.1.* plus
   B : Vertical area × 0.5
Fig. H.3.4. Small leanto roofs or other roofs abutting a wall

Rainwater pipes

* size must not be less than the gutter outlet;
* downpipe serving more than one gutter is to have an area not less than the combined areas of the outlets;
* should discharge into a drain or gulley;
* may discharge into another gutter;
* may discharge onto another surface if it is drained;
* rainwater pipes discharging into a combined drainage system must do so via a trap;
* gutters and rainwater pipes are to be firmly supported. Thermal movement is not be restricted;
* joints are to remain watertight under working conditions;
* rainwater pipes placed under a building must meet the same standards of material and construction as are applicable to a waste or a soil pipe and must also be capable of withstanding a water test.

Materials

Materials suitable for use as rainwater gutters and pipes are listed. When different metals are used they are all to be separated by non-metallic material to prevent electrolytic corrosion.

* Aluminium
  BS 2997: 1958, 'Aluminium rainwater goods' gives detailed information on the dimensions of rainwater gutters and pipes.
* Cast iron
  BS 416: 1973, 'Cast iron spigot and socket soil pipes and fittings'. The pipes are designed for use above ground for use in all types of buildings.
  BS 460: 1964, 'Cast iron rainwater goods' gives detailed information on the dimensions of rainwater pipes and fittings.
* Copper
  BS 1431: 1960, 'Wrot copper and wrot zinc rainwater goods'. There is the usual information regarding dimensions, quality and fittings.
* Galvanized steel
  BS 5493: 1977, 'Code of practice for protecting iron and steel against corrosion'. Guidance on how to ensure correct application and how to maintain it.
* Lead
  BS 1178: 1969, 'Milled lead and strip for building purposes', gives dimensions, thickness and tolerance on thickness marking.
* Low carbon steel
  BS 5493: 1977, 'Code of practice for protecting iron and steel against corrosion'. Guidance on how to ensure correct application and how to maintain it.
* Pressed steel
  BS 1091: 1963 is about pressed steel gutters, rainwater pipes, fittings and accessories.
* Unplasticized PVC
  BS 4514: 1983, 'Unplasticized soil and ventilation pipes' contains materials, dimensions, jointing and fixing details.
  BS 4576: Part 1: 1970, 'Unplasticized PVC rainwater goods' gives requirements for materials and dimensions of gutters, pipes and fittings.
* Zinc
  BS 1431: 1960, 'Wrot copper and wrot zinc rainwater goods' contains information regarding materials, dimensions, quality and fittings.

Alternative approach

The approved document says that the required performance can be met by following some of the recommendations in BS 6367: 1983, 'Code of practice for drainage of roofs and paved areas'.

The relevant clauses are in:

Section 1
Section 2
Section 3 (except clause 9)
Section 4
Section 5 (except clause 18)
Appendices

BS 6367: 1983, 'Drainage of roofs and paved areas' deals with the drainage of surface water from roofs, walls and paved areas and recommends methods of designing gutters, gutter outlets, rainwater pipes and inlets to gullies and to discharge stacks.

Section 1

This section outlines the scope, gives sources for definitions and gives letter symbols used in formulae.

Section 2

This section lists materials and components for rainwater goods. Materials should be used in sections that enable the system to withstand the maximum hydraulic head that could occur should a blockage take place at the lowest point.

Section 3

Design should endeavour to achieve a balance between the cost of the system and the frequency and consequences of flooding. Included are:

- Meteorological aspects of design. A design rate of rainfall of 75 mm per hour is generally satisfactory for roof gutters. Provision should be made for wind, snowfall and thermal movement.
- Mining subsidence. Rainwater systems may be displaced because of mining subsidence, and the Area Surveyor of the Coal Board may be able to predict movement and offer advice.
- Run off. Much depends upon the angle of descent of the rain and advice is offered in connection with flat roofs, paved areas, sloping roofs and vertical surfaces.
- Hydraulic design of roof drainage. The code offers design advice based upon three component parts:

  (i) gutter
  (ii) outlet
  (iii) pipework.

Details are given of design method, calculation of flows in gutters, gutter location and capacity, the effect of angles, valley, parapet and boundary wall gutters, gutter outlets, overflow weirs, rainwater pipes and outlets.

Drainage of flat roofs is described giving rate of run off, depth of water permissible, which should not exceed 30 mm when confined to relatively small areas adjoining outlets.

Procedures are given for determining the required dimension of a sump.

Section 4

This section covers work on site

- Types and spacing of fixings. Recommended fall is 1:350. Too steep a fall should be avoided. Fixing is detailed of eaves gutters, other types of gutter and rainwater pipes.
- Jointing. The three variables – gutter material, jointing material and type of fixing – are discussed, emphasis being upon the correct combination of these variables. Cast iron and steel gutters are covered together with asbestos-cement, aluminium, uPVC, copper and zinc gutters.
- Rainwater pipes. Joints are usually left unsealed when pipes are fixed externally. Information is given regarding horizontal runs of pipe, flat roofs, metal pipes, asbestos-cement and uPVC.
- Access to pipes. Cleaning eyes for access and rodding should be provided at the foot of each stack and at changes of direction.
- Encased pipes. Rainwater pipes should not be encased in concrete columns or structural walls.
- Warning pipes. Internal rainwater pipes must be able to with-stand the head of water likely to occur from a blockage. However, warning pipes should be fitted not more than 6 m above points where a blockage may occur. Warning pipes should discharge in a sensible position and be not less than 20 mm diameter.

Section 5

- Inspection. During construction all work should be visually inspected. Upon completion rainwater pipes should be inspected internally using mirrors and lights.
- Testing. All concealed work should be tested before it is enclosed. Internal rainwater pipes should be air tested. Gutters should be tested for leakage.

Appendices

Appendix A. Design rates of windfall and categories of risk.
Appendix B. Supplementary design for roof damage.

The Code of Practice is copiously illustrated by diagrams, tables and formulae to determine run off, flows and fitting sizes.

Section 2: Rainwater drainage

Almost all the requirements for rainwater drainage are in accord with those previously outlined for foul drainage. As with foul drainage there is no guidance given in the approved document regarding special arrangements needed for pumped systems. If a sewer or a private sewer, operated as a combined system has insufficient capacity the advice is that rainwater should be run in a separate system or to its own outfall.

Pipe gradient and sizes

Drains are to have sufficient capacity to carry the outflow. Flow is dependent upon the area being drained which may include paved and other hard areas. Size of pipe and gradient determines capacity of a surface water drain.

Surface water drains are to have a minimum of 75 mm diameter.

Discharge capacities of rainwater drains are similar to those for foul drainage and reference should be made to Fig. H1.27.

Rainwater drainage is to be constructed in the same way as foul drainage.

Soakaways

H3 now says that rainwater may be discharged into a soakaway. This is not, of course, for preference but is acceptable should there not be a more suitable outfall available. No guidance is given in the approved document: but soakaways must be adequate.

When a soakaway is being considered it is necessary to investigate the nature of the subsoil. Clay is almost impervious and it is most difficult to secure an adequate soakaway in a clay soil. To be effective a soakaway must be in permeable soil and constructed above the top of the water table.

Soakaways should be placed in land which is lower than the building being drained, and not placed near a building; certainly not closer than 5 m, otherwise the stability of foundations may be affected.

The use of soakaways in outlying rural areas has been common practice for many years. More recently, however, soakaways have been used in urban fully sewered areas to limit the impact of new upstream building works to avoid the cost of sewer upgrading outside the development (BRE Digest 365).

The function of a soakaway is to store the immediate stormwater run off and allow for its efficient infiltration into the soil. It is essential that the stored water soaks away through the chosen system sufficiently quickly to provide capacity necessary for any subsequent storm. It is also important to ensure that boggy areas do not arise near to or some distance away from the soakaway system.

The BRE digest contains advice upon the following factors:

size and slope; soil infiltration characteristics; calculating design rainfall; soil infiltration rates; construction details and includes several worked design examples.

It is necessary to provide for maintenance of a soakaway system and to that end all soakaways should include some form of inspection access so that the point of discharge of the drain to the soakaway can been seen (see Figs H3.5 and H3.6).

Alternative approach

The approved document says that the performance requirements can be met by following the relevant recommendations of BS 8301: 1985, 'Code of practice for building drainage'. The relevant clauses are:

Section 1
Section 2
Section 3 (except clauses 7 and 10)
Section 4 (except clause 23)
Section 5 (clause 25)
Appendices

As we have seen from the section on foul drainage (page 253) this code sets out recommendations for the design, layout, construction,

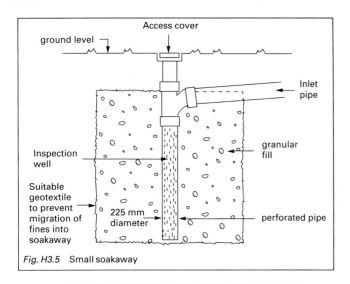

Fig. H3.5   Small soakaway

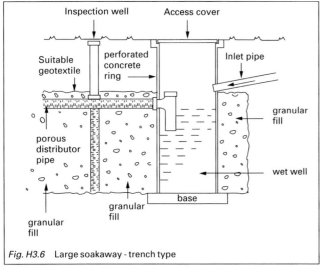

Fig. H3.6   Large soakaway - trench type

testing and maintenance of foul, surface water, and groundwater drainage systems.

The alternative approaches for both foul and rainwater are virtually identical. The only differences are:

• In section 3, foul water excludes clause 10 whereas rainwater excludes 7 and 10.

Clause 7 is concerned with foul drainage. Clause 10 covers groundwater drainage.

Advice is given regarding location, subsoil and groundwater conditions, and surface flooding. Also discussed are the types of apparatus available and factors to be evaluated in selecting apparatus.

Ejectors, air compressors and pumps should preferably be installed in duplicate to allow for standby duty and maintenance.

Wet well and dry well installations are discussed, together with methods of ventilation.

SOLID WASTE STORAGE

People are generally generating more and more rubbish, so any system of refuse storage now being installed must have regard for the future.

The requirements in Schedule 1 do not limit application in any way, and the approved document makes brief recommendations in connection with non-domestic developments.

The use of chutes and containers is the method most used in blocks of flats. The chute, which is a vertical pipe, runs from top to bottom of the building and has an inlet hopper on each floor. The hoppers must close off the chute when they are opened to receive refuse, otherwise people on the lower floors may find themselves covered with rubbish when they try to put their own into the chute.

There may be more than one chute in a building, and these discharge into storage containers in specially constructed chambers.

The 1984 Act, section 23, contained provisions regarding facilities which must be provided for refuse storage and removal in new buildings. However, regulation 18 of 1985 Regs repealed the first two clauses of section 23, and these are replaced by the two paragraphs of H4, in Schedule 1 of the regulations. The 1984 Act, section 23(3) and (4), remains in force and states that consent of the local authority is required before an access provided for the removal of refuse from a building is obstructed. Failure to secure approval before obstructing an access may lead to legal action by a local authority.

H4. There is to be:

• **adequate means of storing solid waste;**
• **adequate means of access for people in the building to the place of storage;**
• **adequate means of access to a street from the place of storage to enable solid waste to be removed.**

A level of performance is prescribed to ensure that the health of persons in and around buildings is not put at risk. Storage facilities must:

• provide sufficient capacity having in mind the quantity of waste and frequency of removal;
• be designed and sited in such a way that conditions prejudicial to health cannot arise;
• be in a position which is accessible for filling and for emptying.

The approved document gives provisions which will meet the above performance requirements.

DOMESTIC DEVELOPMENTS

Capacity

On the assumption that there is a weekly refuse collection service, each dwelling should be provided with a movable container of not more than $0.12\,m^3$ capacity, or a communal container of between $0.75\,m^3$ and $1.0\,m^3$ capacity. This is based on the assumption that each dwelling accumulates $0.09\,m^3$ refuse per week, and that there is a weekly collection. If a weekly collection is not available, larger containers, or more containers must be provided.

In blocks of flats each dwelling up to the fourth floor may have its own container, or may share a container. Above the fourth floor the occupants should be provided with a chute discharging into a container on the ground floor (Figs H4.1 and H4.2).

If the siting or operation of a chute is impracticable it will be necessary to make a satisfactory management arrangement for removing refuse from each dwelling.

Design

Containers not fed by a chute must have close fitting lids. Chutes are to have smooth non-absorbent surfaces and close fitting access doors, and be ventilated at the top and bottom.

It is not necessary for containers to be closed. If they are, the chamber should allow space for filling and removal. There is to be a clear space of 150 mm between and around the containers and be a minimum of 2 m high for communal containers. Openings for ventilation should be provided at top and bottom.

Siting

Containers and chutes are to be sited:

• so that householders do not carry refuse more than 30 m;
• within 25 m of a vehicular access.

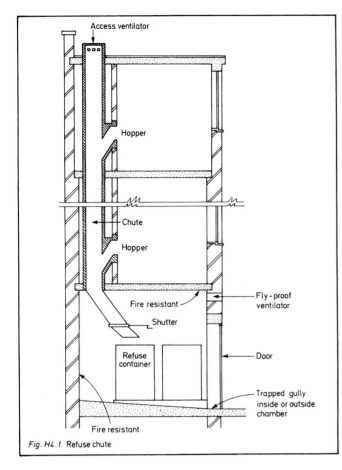

Fig. H4.1. Refuse chute

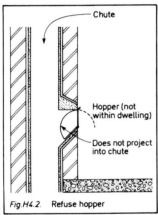

Fig.H4.2. Refuse hopper

Containers are to be sited in a position where it is not necessary for them to be taken through a building when being collected. However, it is permissible to remove them through a garage, carport or other covered space.

The provision applies only to new buildings.

NON-DOMESTIC DEVELOPMENTS

Non-domestic developments are now embraced by the approved document, although no constructional detail is given designers are advised to consult with the collecting authority, regarding:

- the volume and nature of the waste related to storage capacity, frequency of collection, and the size and type of containers which may be necessary;
- the method of storage and whether any on site treatment is proposed;

- the location of storage facilities and any treatment processes;
- hygiene arrangements;
- fire hazards and protection measures.

ALTERNATIVE APPROACH

The performance can be met by complying with the relevant clauses of BS 5906: 1980 (1987), 'Code of practice for storage and on site treatment of solid waste from buildings'.

This is concerned with the problem of storage and treatment of waste from large residential buildings, large and small commercial establishments and hospitals. It recommends good practice by collection and good hygiene. Information is given on the amount of refuse generated by residential buildings and commercial premises. Methods of storage are covered, ranging from dustbins to chutes and storage chambers. On site treatment such as compaction and shredding are outlined together with maceration and pipeline collection.

The clauses which are applicable to the performance required are:

Clause 3. Definitions
          4. Consultations
          5. Materials and components
          6. General principles for design and facilities
          7. Systems
          8. Choice of method of storage and collection
          9. Chutes
        10. Waste storage chambers
        12. Storage containers
        13. Roads
        14. Transportation
        15. Hygiene
              Appendix 'A'

Clause 3 gives a number of definitions, including:

- bulk waste container – movable container with a capacity of between $1\,m^3$ and $30\,m^3$;
- chute – ventilated vertical pipe passing from floor to floor of a building, terminating at its lower end at the roof of a waste storage chamber;
- communal waste storage container – movable receptacle $0.75\,m^3$ to $1\,m^3$ situated within a storage chamber.

Clause 4 emphasizes that at an early stage of design agreement should be reached between designers and appropriate authorities regarding:

- method of storage, on site treatment and method of collection;
- location of storage areas and access arrangements;
- capacity to be provided;
- responsibility for cleaning and maintaining storage facilities;
- environmental aspects;
- discharge of any liquid wastes;
- fire-resisting and firefighting arrangements.

Clause 5 lists materials, components, etc., all of which should be in accordance with British Standards.

Clause 6 gives general principles of the design of facilities. When designing for household waste it is recommended that the best available estimate of output per dwelling is 12 ± 1.2 kg per week at bulk density of 133 ± 13.3 kg/m^3.

For commercial waste it is recommended that detailed enquiries should be made of similar situations in the neighbourhood of the new development and for discussions with local authorities and waste contractors.

Clause 7 deals with systems of waste storage, on site treatment and collection. Waste may be stored in a variety of container types, or subjected to some form of on-site treatment. Pneumatic and waterborne pipeline systems are mentioned.

Clause 8 gives advice regarding the choice of method of storage and collection of waste in various types of building:

- Houses and bungalows. Individual storage containers for each dwelling and easily accessible for removal.
- Dwellings in blocks of not more than 4 storeys. Waste storage containers for communal use with chutes. Occupier should not have to carry refuse more than 30 m to chute hopper.
- Dwellings in blocks of more than 4 storeys. Communal chests are recommended.
- Commercial buildings. The range and size of commercial buildings, and the different activities that take place within them, means a wide range of waste-handling methods must be considered in order to select a system appropriate to the building and the nature of the waste.

Clause 9 considers chute systems. Chutes should not be less than 450 mm in diameter. A chute is to be ventilated. Hoppers should be in freely ventilated positions and not within any stairway enclosure, unless installed in accordance with detailed requirements set out in the BS.

Clause 10. Waste storage chambers. Waste storage chambers should be positioned at vehicle access level, away from the main entrance to a building. It should not be necessary to remove containers through a building.

Advice is given regarding dimensions of waste storage chambers, which will generally be determined by the number, size and type of containers to be accommodated, and its mode of removal.

Walls and roof are to be formed of non-combustible and impervious material.

Ventilation must be provided, and artificial lighting installed by means of sealed bulkhead fittings.

A tap should be provided to facilitate cleansing of the chamber.

Clause 12. Storage containers. Containers should preferably comply with the requirements of BS 1136 or BS 3495. When containers for which there is no BS are proposed to be used, they should be approved by the appropriate local authority.

Clause 13 contains details regarding roads and approaches to buildings. Access roads for solid waste removal should be constructed to carry an axle loading of not less than 11 tonnes.

Clause 14 discusses the transportation of containers and says that manual handling should be minimized.

Clause 15 examines the hygienic aspects of solid waste storage. House and trade waste decomposes rapidly, and all receptacles should be adequately cleaned; household waste wrapped; chutes cleaned and hoppers regularly cleansed.

Appendix A sets out some on-site treatment systems:

- volume reduction by compaction;
- shredding and crushing;
- baling;
- incineration;
- pipeline collection – airborne and waterborne.

RELEVANT READING

British Standards and Codes of Practice

BS 65: 1991. Specification for vitrified clay pipes and fittings and joints
  Amendment slips 1: AMD 4328
                            2: AMD 4394
BS EN295: 1991. Vitrified clay pipes
BS 416: Part 1: 1990. Spigot and socket systems of discharge and ventilating pipes
BS 416: Part 2: 1990. Socketless system of discharge and ventilating pipes
BS 437: 1978. Specification for cast iron spigot and socket drain pipes and fittings
  Amendment 1: AMD 5097
                    2: AMD 5631
BS 460: 1964. Cast iron rainwater goods

BS 864: Part 2: 1983. Specification for capillary and compression fittings for copper tubes
BS 882: 1992. Specification for aggregates from natural sources for concrete
BS 1184: 1976 (1981). Copper and alloy traps
BS 1091: 1963 (1980). Pressed steel gutters, rainwater pipes fittings and accessories
BS 1178: 1982. Specification for milled lead sheet for building purposes
BS 1427: Part 2: 1990. Plastics encapsulated manhole steps
BS 1431: 1960 (1980). Wrought copper and wrought zinc rainwater goods
BS 2997: 1976 (1980). Aluminium rainwater goods
BS 3656: 1981 (1990). Specification for asbestos cement pipes joints and fittings for sewerage and drainage
  Amendment 1: AMD 5531
BS 3868: 1995. Prefabricated drainage stack units: galvanized steel
BS 3921: 1985 (1995). Clay bricks
BS 3943: 1985. Specification for plastics waste traps
  Amendment 1: AMD 3206
                    2: AMD 4191
                    3: AMD 4692
BS 4514: 1983. Specification for unplasticized PVC soil and ventilating pipes, fittings and accessories
  Amendment 1: AMD 4517
                    2: AMD 5584
BS 4576: 1989. Unplasticized rainwater goods
BS 4660: 1989. Unplasticized PVC underground drain pipes and fittings nominal sizes 110 and 160 mm.
BS 5254: 1976. Polypropylene waste pipe and fittings
  Amendment 1: AMD 3588
                    2: AMD 4438
BS 5255: 1989. Plastics waste pipes and fittings
  Amendment 1: AMD 3565
                    2: AMD 5255
                    3: AMD 4472
BS 5328: Part 1: 1991. Guide to specifying concrete
BS 5328: Part 2: 1981. Methods for specifying concrete
BS 5328: Part 3: 1990. Specification for producing and transporting concrete
  Amendment 1: AMD 6927
BS 5328: Part 4: 1990. Sampling, testing and assessing concrete
  Amendment 1: AMD 6928
BS 5480: 1990. GRP pipes for water supply or sewerage
BS 5481: 1977. Specification for unplasticized PVC pipe and fittings for gravity sewers
  Amendment 1: AMD 3631
                    2: AMD 4436
BS 5572: 1994. Code of Practice for sanitary pipework
BS 5627: 1984. Plastics connectors
BS 5911: Part 1: 1988. Specification for concrete cylindrical pipes, bends, junctions and manholes. Unreinforced or reinforced with steel cages or hoops
  Amendment 1: AMD 6269
BS 5911: Part 2: 1982. Specification for inspection chambers and gullies
  Amendment 1: AMD 5146
BS 5911: Part 100: 1988. Unreinforced and reinforced pipes and fittings
BS 5911: Part 101: 1988. Glass composite concrete
BS 5911: Part 110: 1992. Specification for ogee pipes and fittings including perforated
BS 5911: Part 114: 1992. Specification for porous pipes
BS 5911: Part 120: 1989. Specification for reinforced jacking pipes and flexible joints
BS 5911: Part 200: 1994. Reinforced and unreinforced manholes and soakaways
BS 5906: 1980 (1987). Code of practice for storage and on site treatment of solid waste from buildings

BS 6087: 1990. Flexible joints for cast-iron drain pipes and fittings
    Amendment 1: AMD 6357
BS 6297: 1983. Code of practice for design and installation of small
    sewage treatment works and cesspools
BS 6367: 1983. Code of practice for drainage of roofs and paved areas
    Amendment 1: AMD 4444
BS 7158: 1989. Specification for plastics inspection chambers for
    drains
BS 8005: Part 0: 1987. Sewerage – introduction and guide to data
    sources
BS 8005: Part 1: 1987. Guide to new sewerage construction
BS 8005: Part 2: 1987. Guide to pumping stations and pumping
    mains
BS 8005: Part 4: 1987. Guide to design and construction of outfalls
BS 8005: Part 5: 1990. Guide to rehabilitation of sewers
BS 8301: 1985. Code of practice for building drainage
    Amendment 1: AMD 5904
             2: AMD 6580
    *Note: BSEN standards are European Standards, and the following
    were published in 1991:
BSEN 295: Part 1: 1991. Vitrified clay pipes and fittings. Require-
    ments.
BSEN 295: Part 2: 1991. Quality control and sampling.
BSEN 295: Part 3: 1991. Test Methods.

According to the foreword the UK is one of 18 countries 'bound to
    implement this European Standard'. It is a British standard pub-
    lished by BSI.

BRE digests

81   Hospital sanitary services – design and maintenance
83   Plumbing with stainless steel
170   Ventilation of internal bathrooms
248   Sanitary pipework Part 1 – design basis
249   Sanitary pipework Part 2 – design of pipework
292   Access to domestic underground systems
365   Soakaway design

BRE information papers

1/84   BREVAC Mechanical method of emptying sanitation cham-
       bers
21/89 Passive stack ventilation
6/90   Rats in drains

BRE defect action sheets

9   Pitched roofs – sarking felt underlay – damage from roof
39   Laying plastics drainage pipes
40   Plastics drainage pipes – storage and handling
41   Plastics sanitary pipework – jointing and support specification
42   Plastics sanitary pipework – jointing and support installation
49   Storing and handling clay and drainage pipes
50   Jointing and backfilling flexibly jointed clayware drainage
      pipes
55   Roofs, eaves, gutters and downpipe installation
56   Roofs, eaves, gutters and downpipe installation
89   Domestic foul drainage systems – avoiding blockages – specifi-
      cation
90   Domestic foul drainage systems – avoiding blockages – instal-
      lation
101   Plastics sanitary pipework – specifying for outdoor use – design
141   Water storage cisterns – warning pipes – design
143   Drainage stacks – avoiding roof penetration

CHAPTER 18

Heat-producing appliances

(Approved document J)

This part of Schedule 1 sets out to ensure that appliances producing heat are designed and installed to work efficiently and safely without giving rise to nuisances and also without harmful effect on the structure.

Electrical equipment is not affected by these requirements.

Requirements of Schedule 1 are asfollows:

J1. Heat-producing appliances which burn solid fuel, oil or gas are to be installed so that an adequate supply of air is available for combustion and for the efficient working of the flue or chimney.

The requirements also apply to incinerators.

J2. Adequate provision is to be made for the discharge of the products of combustion from heating appliances to the outside air.

J3. Heat-producing appliances, flue pipes and chimneys are to be installed and/or constructed so that the risk of a building catching fire from these sources is reduced to a reasonable level.

Contents of the approved documents have been simplified and are arranged by fuel type.

Section 1 contains provisions which refer to all fuel types.

Section 2. Here are the provisions for solid fuel appliances having a rated output up to 45 kW.

Section 3 contains provisions for individually flued non-fan-assisted gas-burning appliances having a rated input up to 60 kW. It also provides for an air supply for cooking appliances.

Section 4 deals with oil-burning appliances with a rated output not exceeding 45 kW.

PERFORMANCE

The approved document sets out an acceptable level of performance. Heat-producing appliances will meet the requirements in J1, J2 and J3 if they are so installed that they:

- get sufficient air for the correct combustion of the fuel and operation of the flue;
- are capable of being operated normally without the products of combustion becoming a danger to health;
- are capable of normal operation without giving rise to damage by heat or fire to the building.

The word 'fire' includes a chimney fire.

'Non-combustible' means capable of being classed as non-combustible when tested in the way described in BS 476: Part 4: 1970 (1984), 'Non-combustibility test for materials'.

Materials used in the construction and finishing of buildings may be classified as 'combustible' or 'non-combustible' depending on how they behave when subjected to a non-combustibility test. The test is intended for building materials whether coated or not, but it is not intended to apply to the coating alone.

A material is deemed to be non-combustible if, during the test, none of the three specimens either:

- cause temperature reading to rise by 50°C or more above the initial furnace temperature; or
- flame continuously for 10 seconds or more inside the furnace.

SECTION 1. GENERAL

Requirements for providing air to appliances.

The performance requirement for an adequate supply of air will be met in the following circumstances:

- it is a room sealed appliance (Fig. J1), or
- the room or space in which the appliance is fitted, has permanent air openings.

When a ventilation opening is made through an internal wall or partition to adjacent spaces or rooms, further openings of similar size are required to the outside air (Figs J2 and J3).

Ventilation openings should not be in fire-resisting walls.

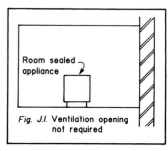

Fig. J.l. Ventilation opening not required

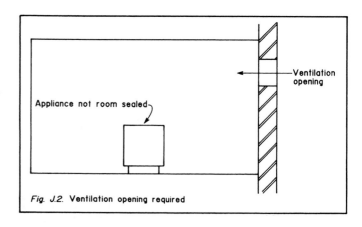

Fig. J.2. Ventilation opening required

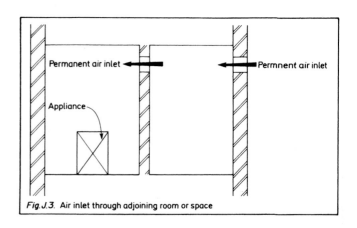

Fig.J.3. Air inlet through adjoining room or space

EXTRACT FANS

Any heat-producing appliance in a building where an air extract fan is installed must be able to operate effectively whether or not the fan is in use. Part F limits use of extract fans in rooms where there is an open flued appliance (see page 221).

FLUE PIPES AND CHIMNEYS

All heating appliances are to have a balanced or low level flue, or a chimney discharging to the external air (Fig. J4). The exception is those appliances designed to operate without discharging products of combustion to the outside air.

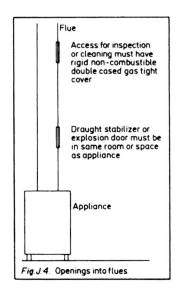

Fig. J.4. Openings into flues

OPENINGS INTO FLUES

Provision must be made for flue inspection and cleaning. Only the following openings into flues are allowed:

• for inspection and cleaning;
• a draught stabilizer or draught diverter.

COMMUNICATION BETWEEN FLUES

The only circumstance when a flue may have an opening into more than one room is when it is provided for inspection of or cleaning the flue (Fig. J5). The opening is to be fitted with a rigid non-combustible double-cased gas tight cover (Fig. J4).

A flue may serve more than one appliance in the same room.

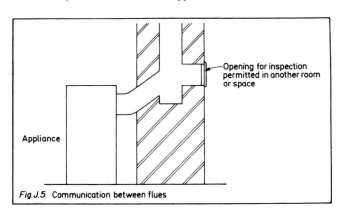

Fig. J.5. Communication between flues

FORMER CONTROL

When making alterations or additions to buildings having a chimney constructed prior to 1 February 1966, it is possible that it may not comply in all respects with current requirements.

In these circumstances, and if there is no obvious evidence that the chimney is unsatisfactory, it may be considered as satisfying the requirement.

It is suggested that great care should be exercised before deciding that there is no obvious evidence that a chimney is unsatisfactory.

SECTION 2. SOLID FUEL

These are the requirements for solid fuel burning appliances with a rated output up to 45 kW.

AIR SUPPLY

A room or space containing an open solid fuel burning appliance (Fig. J6) is to have available a permanent air entry opening equal to a total free area of not less than 50% of the appliance throat opening area.

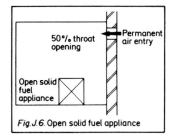

Fig. J.6. Open solid fuel appliance

BS 8303: 1986 which is a code of practice for the installation of domestic heating and cooking appliances burning solid mineral fuels makes reference to BS 1846: Part 1: 1968 for a definition of 'throat': 'It is that part of a flue, if contracted, which lies between the fireplace and the main flue.'

Any other solid fuel appliance is to be in a room or space containing an air entry or openings having a total free area of not less than 500 mm^2 per kW of rated output above 5 kW. If a flue draught stabilizer is used the total free area is to be increased by 300 mm^2 for each kW of rated output (Fig. J7).

These requirements emphasize the need to ensure an adequate supply of air for combustion and for ventilation.

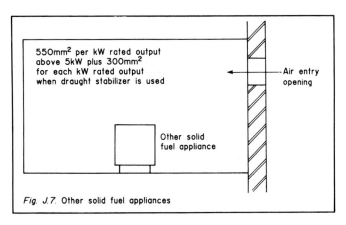

Fig. J.7. Other solid fuel appliances

SIZES OF FLUES

Flue pipe sizes must

- for flues, be at least equal to that of the flue outlet of the appliance;
- for chimneys, be not less than the dimensions given in Table J1. Irrespective of Table J1 they are never to be less than the size of the appliance flue outlet, or that recommended by the manufacturers.

The approved document gives diameter dimensions 'or square section of equivalent area'. Table J1 includes the approximate square section dimensions.

Wherever it is necessary to have an offset in a flue then the flue size is to be increased. Add 25 mm to the diameter in the case of a circular flue, or 25 mm to each side of a square flue.

Table **J1   Sizes of flues**

Type of installation	Minimum flue sizes
Fireplace with opening up to 500 mm × 550 mm	200 mm diameter 178 mm square
Fireplace recess with opening greater than 500 x 500 mm	free area 15% of the opening
Closed appliance up to 20 kW rated output burning bituminous coal	150 mm diameter 133 mm square
Closed appliance up to 20 kW rated output	125 mm diameter 111 mm square
Closed appliance between 20 kW and 30 kW rated output	150 mm diameter 133 mm square
Closed appliance between 30 kW and 40 kW rated output	175 mm diameter 155 mm square

OUTLETS FROM FLUES

The basis for calculating the height of flues above roof intersection is now more in accord with British Standards. The whole matter has been simplified and is shown diagrammatically in Fig. J8. Shaded areas are associated with roof slopes of less than 10°. On these latter roofs there are positions where the height of a flue is determined by a horizontal measurement, replacing the old general figure of 1 m.

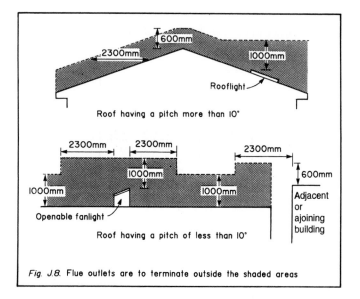

Fig. J.8. Flue outlets are to terminate outside the shaded areas

NEW BUILDINGS OVERREACHING ADJACENT CHIMNEYS

Anyone erecting or raising a building to a greater height than an adjoining building may be required to raise all chimneys within 6 ft of the new work (1984 Act, section 73).

DIRECTION OF FLUES

Flues must, wherever possible, be vertical, and horizontal runs avoided. It is permissible, however, to have a back outlet to an appliance horizontal for not more than 150 mm (Fig. J9).

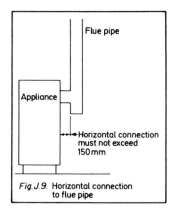

Fig.J.9. Horizontal connection to flue pipe

BENDS IN FLUES

Bends in flues should be avoided wherever possible. When it is necessary to introduce an offset it should be made an angle with the vertical of not more than 30° (Fig. J10). This is recommended in order to facilitate cleaning, to maintain good upward draught, to avoid collection of solid deposits on the slope and to prevent overheating to the upper surface of the slope.

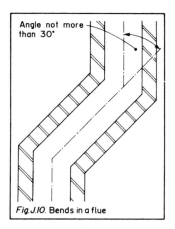

Fig.J.10. Bends in a flue

FLUE PIPES

Requirements with regard to flue pipes are that

- they should be used only to connect an appliance to a chimney;
- they do not pass through any roof space;
- if they have spigot and socket joints they are to be fitted with sockets uppermost.

Flue pipes may be manufactured from any of the following materials:

- Cast iron in accordance with BS 41: 1973 (1981), 'Cast iron spigot and socket flue or smoke pipes and fittings'. This covers flue or smoke pipes and bends 100 mm to 300 mm nominal bore and offsets of 100 mm to 150 mm and provides alternative socket profile to allow interchangeability between pipes manufactured by the sandcast or by the spun process.
- Mild steel having a wall thickness of not less than 3 mm.
- Stainless steel having a wall thickness of not less than 1 mm in accordance with BS 1449: Steel plate, sheet and strip, Part 2:

1983, 'Specification for stainless and heat resisting steel plate sheet and strip for Grade 316S11; 316S13; 316S31 and 316S33 or equivalent Euronorm 88–71 designation'.

The BS specifies requirements for flat-rolled stainless steel and heat-resisting steel products including plate 3 mm thick and above.

• Vitreous enamelled steel complying with BS 6999: 1989 which is a specification for vitreous enamelled low carbon steel flue pipes, other components and accessories for solid fuel burning appliances with a maximum rated output of 45 kW.

These flue pipes are intended for internal freely exposed use only. They are also suitable for gas- and oil-fired appliances. The standard contains many definitions and paragraphs dealing with:

3. materials
4. vitreous enamelling
5. construction. One requirement is that where access for inspection and cleaning is required, this must be carried out without disturbing the joints.
6. resistance to high temperature
7. instructions for installation.

Included in appendices are methods of test for heat resistance of vitreous enamel coating and methods of test for joints.

When a flue pipe is placed next to combustible material there must be an air space separating the pipe from this material. Distance across the air space depends upon whether or not a non-combustible shield is used. Minimum dimensions are indicated in Fig. J11. The approved document contains a recommendation regarding the width of a non-combustible shield, as shown in Fig. J12.

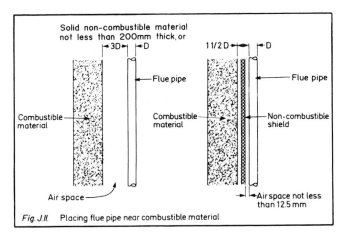

Fig. J.II. Placing flue pipe near combustible material

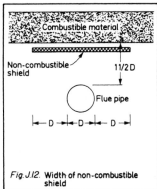

Fig. J.I2. Width of non-combustible shield

CHIMNEYS

When used for solid fuel appliances chimneys are to be capable of withstanding a temperature of 1100°C without undergoing any structural change affecting the stability or performance of the chimney.

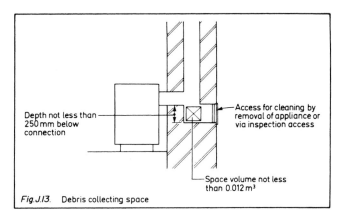

Fig. J.I3. Debris collecting space

If a chimney is not directly over an appliance, a debris collecting space should be provided which is accessible for emptying (Fig. J13).

Blockwork chimneys may be of refractory material; a combination of high alumina cement or pumice aggregates or lined in the same way as brick chimneys.

Minimum thickness of the walls of brick and blockwork excluding separate linings are specified for walls separating flues; another part of the same building; separate fire compartments and dwellings (Fig. J14).

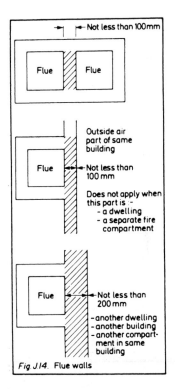

Fig. J.I4. Flue walls

CHIMNEY LININGS

Chimney linings that may be used are:

• clay flue linings having rebated or socketed joints in accordance with BS 1181: 1989, 'Clay flue linings and flue terminals'. This gives details of materials, workmanship, design, construction, dimensions, tests for clay flue linings, bends, flue terminals for heating appliances.

However, the 1989 edition of this standard has been withdrawn, to be replaced by BS 1181: 1999, 'Specification for clay flue terminals'. Clay flue liners are now covered by BS EN 1457: 1999, 'Chimneys – clay/ceramuic flue liners – requirements and test measures'. The performance of 22 classes of liners are specified in the standard.

It is considered that clay flue liners used in connection with solid fuel appliances and made in accordance with BS EN 1457: 1999 which achieve at least the classification A1 N2 are satisfactory.

- imperforate clay pipes with socketed joints as covered by BS 65: 1981, 'Vitrified clay pipes, fittings and joints'. It sets out the requirements for vitrified clay pipes and fittings with or without sockets, including materials, manufacture, dimensions and performance requirements;
- high alumina cement and kiln burnt or pumice aggregate pipes having rebated or socketed joints or steel collars surrounding the joints.

Linings are to be fitted with sockets or rebates uppermost and jointed with fireproof mortar. Space between liners and brickwork is to be filled with weak mortar or insulating concrete (Fig. J15).

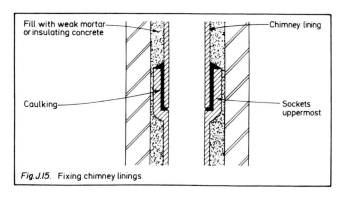

Fig.J.15. Fixing chimney linings

COMBUSTIBLE MATERIAL

Combustible material is not to be placed nearer than

- 200 m from a flue, or
- 40 mm from the outer surface of a masonry chimney or fireplace recess.

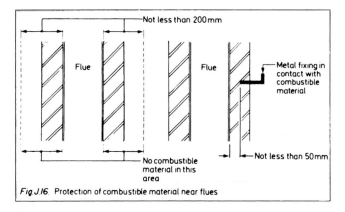

Fig.J.16. Protection of combustible material near flues

The following are excluded from the requirement:

    architrave
    dado
    floorboard
    mantelpiece
    picture rail
    skirting.

Metal fixings in contact with combustible material must be not less than 50 mm from a flue (Fig. J16).

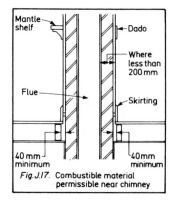

Fig.J.17. Combustible material permissible near chimney

FACTORY MADE INSULATED CHIMNEYS

A factory made insulated chimney which is serving a solid fuel appliance is to be constructed and tested to satisfy the relevant requirements given in:

- BS 4543: Part 1: 1990, 'Methods of test for factory made insulated chimneys', and
- BS 4543: Part 2: 1990, 'Specification for chimneys with stainless steel flue linings for solid fuel fired appliances'.

BS 4543: Part 1: 1990 relates to chimneys that are made in a factory and do not require any fabrication on site. The various components must be assembled in accordance with the manufacturers' instructions. Tests include joint test, draught test, thermal shock test, thermal insulation test, and strength test.

BS 4543: Part 2: 1990 refers to chimneys made from components to which Part 1 refers. The chimneys are to be used with solid fuel appliances. The standard covers materials, erection, finish, firestops and spaces, support assemblies, wind resistance, terminal assembly and rain cap, joints, cleaning traps and relative dimensions.

Factory made insulated chimneys must also be installed in accordance with the manufacturers' instructions or the relevant recommendations in:

- BS 6461: Part 2: 1984, which is a code of practice for factory made insulated chimneys for internal applications. This part gives recommendations for the installation of new factory made insulated chimneys for internal applications the operation of which depends on natural draught. It deals with chimneys of nominal internal diameters of 100 mm, 125 mm, 150 mm and 200 mm. Contents of various paragraphs are:

2. Definitions
3. Exchange of information
4. General – design – condensation – size of chimney – bends – openings – height and position of chimneys – terminals – stability and site testing.
5. Installation
6. Connecting flue pipes
7. Site testing.

Access for maintenance or replacement

Factory manufactured insulated chimneys are to be installed strictly in accordance with the manufacturers' instructions with particular reference to:

- the need for the chimney to be readily accessible throughout its length for maintenance or replacement;
- the requirement that joints should not be made within a wall. When this cannot be avoided the joint should be placed in a sleeve in order to isolate it from the wall.

Accessibility for maintenance and repair may be difficult if placed in flats or maisonettes. If it is found that proper access cannot be provided, factory made insulated chimneys should not be used.

Protection

An insulated metal chimney (Fig. J18) is not to:

- go through part of a building forming a separate compartment unless it is encased in non-combustible material having not less than half the fire resistance of the compartment wall or floor. (Also, see B3, Chapter 11);
- the gap between the outside of an insulated metal chimney and combustible material is not to be less than 'X' (see Fig. J20);
- go through a cupboard, storage space or roof space except when it is cased in a non-combustible material leaving a gap not less than 'X' from the outside of the chimney.

The approved document says that the gap 'X' is to be found by test in

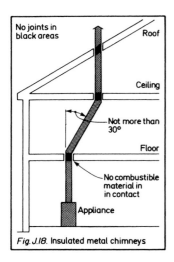

Fig. J.18. Insulated metal chimneys

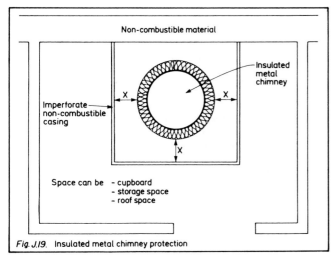

Fig. J.19. Insulated metal chimney protection

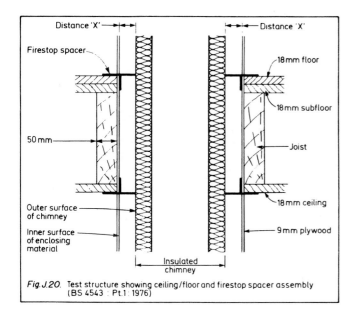

Fig. J.20. Test structure showing ceiling/floor and firestop spacer assembly (BS 4543 : Pt.1 : 1976)

accordance with BS 4543: Part 1: 1976, 'Methods of test for factory made insulated chimneys'.

Contents of this BS are as follows: Scope, references, definitions, chimney support test, structure and chimney joint test, draught test, thermal shock test, thermal insulation test, and strength test. This is supported by a table giving flue gas generator inputs and a number of illustrations.

Chimneys are tested on the basis of a gap of 40 mm to the enclosure or the manufacturer's specified clearance measured between the outer surface of the chimney and the interior surface of the enclosing material.

Surface clearance is designated 'X' on figures J19 and J20 and is normally 40 mm.

The casing used for enclosing a chimney is to be closed at each floor and ceiling level by a firestop or firestop spacer assembly.

CONSTRUCTIONAL HEARTHS

When an appliance is to be installed a constructional hearth is to be provided. It is to be solid, of non-combustible material to dimensions not less than those shown in Fig. J21. The thickness shown may include any solid non-combustible floor under the hearth.

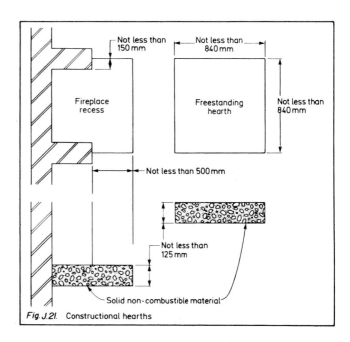

Fig. J.21. Constructional hearths

Combustible material may only be placed under constructional hearths in the following circumstances (Fig. J22):

- it is to support the edges of the hearth, or
- air space is provided.

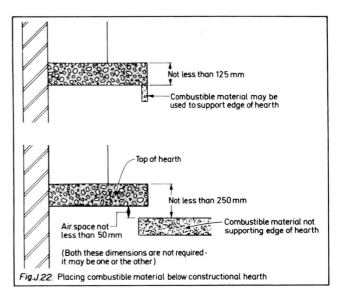

Fig. J.22. Placing combustible material below constructional hearth

FIREPLACE RECESSES

Fireplace recesses must be constructed of solid non-combustible material not less than the minimum dimensions given in Fig. J23.

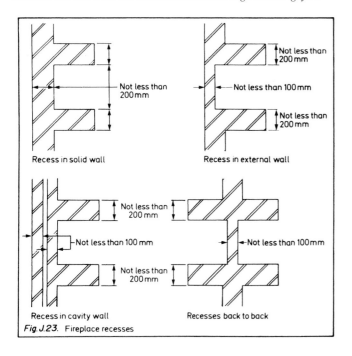

Fig. J.23. Fireplace recesses

WALLS

Walls not forming part of a fireplace recess, but near to a hearth, are to be protected by the introduction of a shield of solid non-combustible material (Fig. J24).

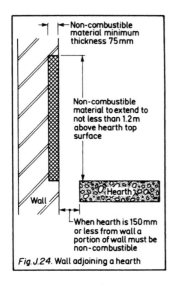

Fig. J.24. Wall adjoining a hearth

PLACING APPLIANCES

- An appliance must not be placed near the edge of a constructional hearth, or near to any combustible material that may be placed upon it (Fig. J25)
- An appliance is to be isolated from a wall containing combustible materials (Fig. J26).

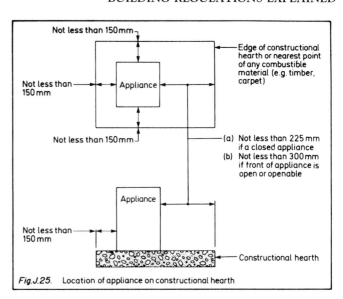

Fig. J.25. Location of appliance on constructional hearth

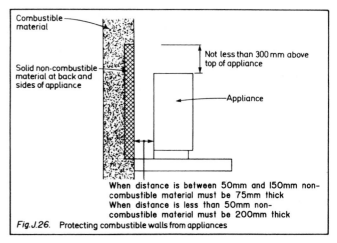

Fig. J.26. Protecting combustible walls from appliances

ALTERNATIVE APPROACH

The approved document provides that J1, J2 and J3, so far as they relate to solid fuel burning appliances with a rated output up to 45 kW, may be met by following the relevant recommendations in BS 8303: 1986, 'Code of practice for the installation of domestic heating and cooking appliances burning solid mineral fuels'.

The appliances are considered in the following four main categories.

1. Room heaters without convection.
2. Room heaters.
3. Independent boilers and freestanding cookers.
4. Warm air heating appliances.

Included in the recommendations are paragraphs relating to work on site; materials components and installation; structural and system design considerations, together with numerous illustrations.

SECTION 3. GAS

These are the requirements for individually flued gas-burning appliances which are not fan assisted. Rated output is up to 60 kW. Also included in this section is the provision of an air supply to cooking appliances.

The approved document describes provisions relating to:

- balanced flue appliances
- cooking appliances
- decorative appliances (solid fuel or log effect)
- natural draught and open flued appliances.

Installation must be in accordance with the Gas Safety (Installation and Use) Regulations 1998. Additional information is available in the Health and Safety Commission Code of Practice, 'Safety in the installation and use of gas systems and appliances'.

SOLID FUEL EFFECT APPLIANCES

These appliances simulate the burning of coal or wood with a live flame, and installation should follow the manufacturers' instructions if their appliances have been tested by an approved authority. Alternatively, installation may be in accordance with:

- the requirements for solid fuel set out in section 2, or
- the relevant recommendations in the following:
  (a) BS 5871: Part 1: 1991, 'Gas fires, convector heaters and fire/back boilers (1st, 2nd and 3rd family gases)'. This BS specifies the mode of installation of flued and flueless space-heating appliances.
  (b) BS 5871: Part 2: 1991, 'Inset live fuel effect gas fires of heat input not exceeding 15 kW (2nd and 3rd family gases)'.
  (c) BS 5871: Part 3: 1991, 'Decorative fuel effect gas appliances of heat input not exceeding 15 kW (2nd and 3rd family gases)'.

Each of the above British Standards has contents embracing:

1. Scope
2. Definitions
3. Exchange of information
4. Appliances
5. Materials and components
6. Location
7. Appliance sizing
8. Ventilation
9. Flueing
10. Appliance fixing
11. Fire precautions
12. Gas connections
13. Electricity supplies
14. Commissioning
15. Fireguards

Each BS also contains a number of appendices, tables and figures.

The expression 'family gases' is quite intriguing and is not clarified in the BS, but the explanation is simple. 1st family gases are the old-fashioned basic gases arising from treating coal in a retort, and known as towns gas. 2nd family gas is natural gas and 3rd family gas is liquid gas (LPG).

BATHROOMS AND GARAGES

Any appliance installed in a bathroom, shower room or a private garage must be of the room sealed type. This is also a requirement of the Gas Safety (Installation and Use) Regulations 1998.

AIR SUPPLY

In respect of a room or space containing a cooker (Fig. J27) there is to be:

- an openable window or other means of providing ventilation;
- where the room or space is less than 10 m³ there is additionally, to be a permanent ventilation opening of not less than 5000 mm².

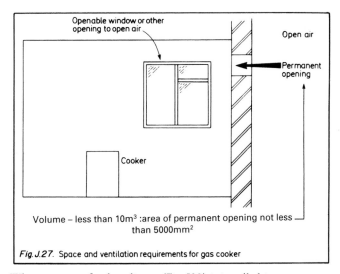

Volume – less than 10m³ :area of permanent opening not less than 5000mm²

Fig. J.27. Space and ventilation requirements for gas cooker

Where an open-flued appliance (Fig. J28) is installed in any room or space there is to be;

- a permanent ventilation opening of not less than 450 mm² for each kW of appliance input rating in excess of 7 kW.

These requirements for an air supply do not refer to balanced flue, or solid fuel effect appliances.

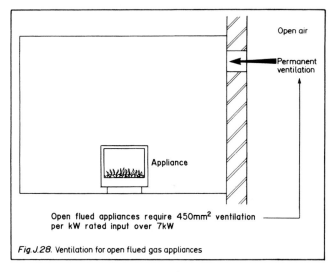

Open flued appliances require 450mm² ventilation per kW rated input over 7kW

Fig. J.28. Ventilation for open flued gas appliances

SIZES OF FLUES

The size of a flue is not to be less than:

- for a gas fire a cross sectional area of 12 000 mm² if the flue is round or 16 500 mm² for a rectangular flue, and the minimum dimension is not to be less than 90 mm (Fig. J29).

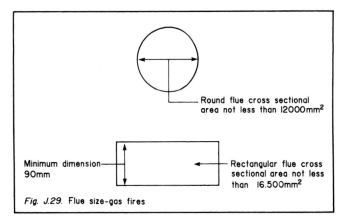

Fig. J.29. Flue size-gas fires

- for other appliances a cross sectional area equal to that of the appliance outlet.

These requirements for the size of flues do not refer to balanced flues or solid fuel effect appliances.

DIRECTION OF FLUES

Horizontal runs are to be avoided, and if a bend is necessary it should not be at a greater angle than 45° with the vertical (Fig. J30).

This requirement regarding the direction of flues does not refer to balanced flues.

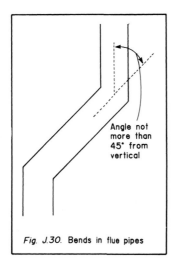

Fig. J.30. Bends in flue pipes

OUTLETS FROM FLUES

The approved document requires that a balanced flue outlet (Fig. J31) should be:

- placed in a position externally allowing a free intake of air and also the dispersal of the products of combustion, and
- not less than 300 mm from an opening into the building which is wholly or partly above the terminal, and
- if the terminal is in a position where persons come into contact with it, protection from damage is to be provided, and

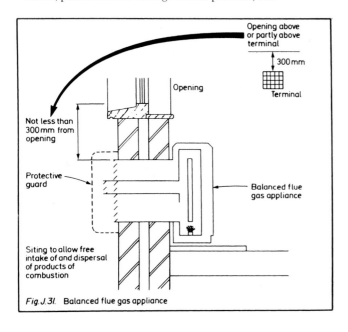

Fig. J.31. Balanced flue gas appliance

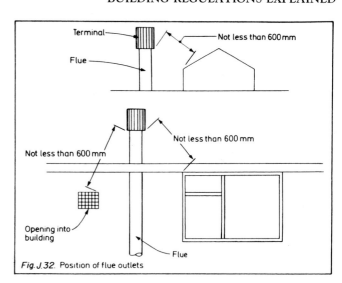

Fig. J.32. Position of flue outlets

- designed in such a way that will prevent the entry of any matter which might restrict the flue.

Provisions are also included regarding the outlets from any other appliances (Fig. J32), and they should be:

- placed in a position at roof level where air may pass freely across at all times, and
- not less than 600 mm from an opening into the building, and
- except in the case of a gas fire, flue outlets are to be fitted with terminals where any dimension across the axis is less than 175 mm.

MATERIALS FOR FLUE PIPES

Flue pipes may be of any of the following materials:

- Cast iron in accordance with BS 41:1973 (1981), 'Cast iron spigot and socket flue or smoke pipes or fittings'. This covers flue or smoke pipes and bends 100 mm to 300 mm nominal bore and offsets of 100 mm to 150 mm, and provides alternative socket profile to allow interchangeability between pipes manufactured by sandcast or by the spun process.
- Mild steel having a wall thickness not less than 3 mm.
- Stainless steel having a wall thickness of not less than 1 mm in accordance with BS 1449: Steel plate, sheet and strip: Part 2: 1983. This is a specification for stainless steel and heat-resisting steel plate and strip for Grade 316S11; 316S13; 316S31 and 316S33 or the equivalent Euronorm 88–71 designation.

The BS specifies requirements for flat-rolled stainless steel, and also heat-resisting steel products including plate 3 mm and above.

- Vitreous enamelled steel meeting the requirements of BS 6999: 1989 which is a specification for vitreous enamelled low carbon steel flue pipes; other components and accessories for solid fuel burning appliances with a maximum rated output of 45 kW.
- Sheet metal as described in BS 715: 1989, which is a specification for metal flue pipes, fittings, terminals or accessories for gas-fired appliances with a rated input not exceeding 60 kW.
- Asbestos cement in accordance with BS 567: 1989. This is a specification for asbestos cement flue pipes and fittings, light quality.
- Asbestos cement as described in BS 835: 1989, 'Specification for asbestos cement flue pipes and fittings, heavy quality'.
- Any other material fit for its intended purpose. Flue pipes which have spigot and socket joints are to be fitted with the sockets uppermost.

PROTECTION OF COMBUSTIBLE MATERIAL

- When combustible material is near to a flue pipe it must be protected by having a 25 mm wide separation (Fig. J33).
- In the event of a flue pipe passing through a wall, floor or roof it is to be encased by a non-combustible sleeve enclosing an air space not less than 25 mm around the flue pipe (Fig. J34).
- Where a flue pipe passes through a compartment wall or a compartment floor it is to be encased with non-combustible material not less than half the fire resistance required in the wall or floor. When measuring air space the 25 mm measurement may be taken from the outside of the inner pipe where a double wall flue pipe is being used (Fig. J35).

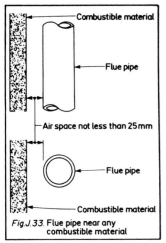

Fig.J.33. Flue pipe near any combustible material

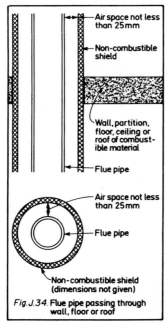

Fig.J.34. Flue pipe passing through wall, floor or roof

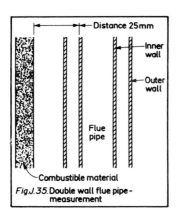

Fig.J.35. Double wall flue pipe - measurement

LINING BRICK CHIMNEYS

Chimney linings that may be used are:

- Clay flue linings having rebated or socketed joints in accordance with BS 1181: 1989, 'Clay flue linings and flue terminals'. This gives details of materials, workmanship, design, construction, dimensions, tests for clay flue linings, bends and flue terminals for heating appliances.
  BS 1181: 1989 has been replaced by BS 1181: 1999, 'Specification for clay flue terminals'. Also flue liners are now covered by BS

EN 1457: 1999, 'Chimneys – clay/ceramic flue liners – requirements and test methods'. 22 classes of liners are specified.
- Imperforate clay pipes with socketed joints as covered by BS 65: 1991, 'Vitrified clay pipes, fittings and joints'. It sets out the requirements for vitrified clay pipes and fittings with or without sockets, including materials, manufacture, dimensions and performance requirements.
- High alumina cement and kiln burnt or pumice aggregate pipes having rebated or socketed joints or steel collars surrounding the joints.

Linings are to be installed with the sockets or rebates uppermost. Joints between liners should be filled with weak mortar or insulating concrete (see Fig. J15).

CHIMNEY BLOCKS

Flue blocks used in the construction of masonry chimney are to be as described in BS 1289: Part 1: 1986, which is a specification for precast concrete flue blocks and terminals, and BS 1289: Part 2: 1989, a specification for clay flue blocks and terminals.

BS 1289: Part 1: 1986 specifies requirements for precast concrete blocks intended for use in bonded flues or in non-bonded flues, and precast concrete terminals.

The standard contains definitions and advice on materials, dimensions, dimension deviations, joints, compressive strength and flow resistance.

There are three appendices and numerous illustrations showing typical flue block installations.

BS 1289: Part 2: 1989 specifies requirements for clay flue blocks intended for use in bonded or non-bonded flues and clay terminals. They are intended for use with gas appliances.

The standard contains definitions and recommendations for materials and workmanship, dimensions, dimensional deviations, joints, compressive strength, flow resistance and sampling.

There are four appendices, including Appendix D which contains recommendations for the design and construction of clay flue block systems. There are numerous illustrations of components and installation methods.

FLEXIBLE FLUE LINERS

The circumstances in which a flexible flue liner may be used in a chimney are if:

- the liner is in accordance with BS 715: 1989, and
- the chimney was built before 1 February 1966, or it is already lined, or constructed of flue blocks in accordance with the approved document.

BS 715: 1989 applies to metal flue pipes, fittings, terminals and accessories for gas-fired appliances with rated input not exceeding 60 kW. Flue pipes are to have welded or folded seams and the requirements are based on the use of sheet carbon steel or aluminium.

DEBRIS COLLECTING SPACE

Where a chimney is not lined or is not constructed of flue blocks, a debris collection space is to be provided at the bottom of the chimney (see Fig. J13).

CHIMNEYS

The minimum thickness of a brick or blockwork chimney is not to be less than 25 mm.

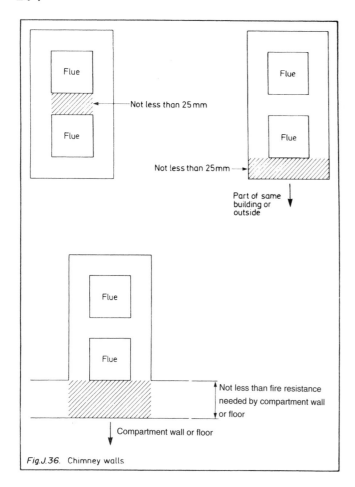

Not less than 25mm

Not less than 25mm

Part of same building or outside

Flue

Flue

Not less than fire resistance needed by compartment wall or floor

Compartment wall or floor

Fig.J.36. Chimney walls

Where chimneys pass through compartment walls or floors they must have walls with fire resistance not less than the fire resistance required for the compartment separating wall or floor (Fig. J36; see also B3, Chapter 11).

FACTORY MADE INSULATED CHIMNEYS

A factory made insulated chimney which is serving a gas appliance is to be constructed and tested to satisfy the relevant requirements given in:

- BS 4543: Part 1: 1990, 'Methods of test for factory made insulated chimneys', and
- BS 4543: Part 2: 1990, 'Specification for chimneys for solid fuel appliances', and
- BS 4543: Part 3: 1990, 'Chimneys for oil fuel appliances'.

It will be noted that factory made chimneys manufactured for solid fuel and oil fuel appliances may be used for gas installations.

BS 4543: Part 1: 1990 relates to chimneys that are made in a factory and do not require any fabrication on site. The various components must be assembled in accordance with the manufacturers' instructions. Tests include joint test, draught test, thermal shock test, thermal insulation test and strength test.

There are two tables and nine figures.

BS 4543: Part 2: 1990 refers to chimneys made from components to which Part 1 refers. The chimneys are to be used with solid fuel appliances. The standard covers materials, erection, finish, firestops and spaces, support assemblies, wind resistance, terminal assembly and rain cap, joints, cleaning traps and relative dimensions.

Included is an appendix, three tables and three figures.

BS 4543: Part 3: 1990 refers to chimneys designed for a rated

output up to 45 kW but with a flue gas temperature at the appliance exit not to exceed 450°C after any permanent draught break or before any stabilizer.

This standard covers materials, fire stops and spacers, chimney support tolerances and the testing sequences provided for in Part 1. There is an appendix, three tables and three figures.

Factory made insulated chimneys must also be installed in accordance with the manufacturers' instructions or the relevant recommendations in

- BS 6461: Part 2: 1984, which is a code of practice for factory made insulated chimneys for internal applications.

This part gives recommendations for the installation of new factory made insulated chimneys for internal applications, the operation of which depends upon natural draught. It deals with chimneys of nominal internal diameters of 100 mm, 125 mm, 150 mm and 200 mm.

The contents contain recommendations under the following headings – definitions; exchange of information; design; condensation; size of chimney; bends; openings; height and position of chimneys; terminals; stability; site testing; installation; connecting flue pipes.

There is an appendix, a table and four figures.

Access for maintenance or replacement

Factory manufactured insulated chimneys are to be installed strictly in accordance with manufacturers' instructions with particular reference to:

- the need for the chimney to be readily accessible throughout its length for maintenance or replacement;
- the requirement that joints should not be made within a wall. When this cannot be avoided the joint should be placed in a sleeve in order to isolate it from the wall.

Accessibility for maintenance or repair may be difficult if placed in flats or maisonettes. If it is found that proper access cannot be provided, factory made insulated chimneys should not be used.

Protection

An insulated metal chimney is not to:

- go through part of a building forming a separate compartment unless it is encased in non-combustible material having not less than half the fire resistance of the compartment wall or floor. (Also see B3, Chapter 11.)
- have a gap between the outside of an insulated metal chimney and combustible material that is less than 'X' (see Fig. J20).
- go through a cupboard, storage space or roof space except when it is cased in a non-combustible material leaving a gap not less than 'X' from the outside of the chimney.

The approved document says that the gap 'X' is to be found by test in accordance with BS 4543: Part 1: 1990, 'Methods of test for factory made insulated chimneys'.

Contents of this BS are as follows:

Scope; references; definitions; chimney support test; structure and chimney joint test, draught test; thermal shock test; thermal insulation test and strength test. This is supported by a table giving flue gas generator inputs and a number of illustrations.

Chimneys are tested on the basis of a gap 40 mm to the enclosure or the manufacturers' specified clearance measured between the outer surface of the chimney and the interior surface of the enclosing material.

Such a clearance is designated 'X' on Figs J19 and J20 and is normally 40 mm.

The casing used for enclosing a chimney is to be closed at each floor and ceiling level by a firestop or firestop spacer assembly (see Figs J18 to J20).

HEARTHS

A hearth is always to be provided for an appliance unless:

- any flame or incandescent material within the appliance is not less than 255 mm above the floor, or
- the appliance is in accordance with BS 5258, 'Safety of domestic gas appliances' or BS 5386, 'Specification for gas burning appliances'.

BS 5258 is split into a number of parts each of which apply to one type of appliance. Much of the information is directed towards the manufacturer but some details are included on installation.

The relevant parts are:

BS 5258: Part 1: 1986 Central heating boilers
Part 4: 1987 Ducted air heaters
Part 5: 1989 Gas fires
Part 7: 1977 Storage water heaters
Part 8: 1989 Combined appliances
Part 10: 1980 Flueless space heaters
Part 11: 1980 Combustion heaters

BS 5386: Part 1: 1976 describes gas burning appliances for instantaneous water heating. It includes the main constructional and performance characteristics of gas burning appliances.
Part 2: 1981 (1986) describes mini water heaters.

A back boiler, whether fitted by itself or in conjunction with another appliance is to have a hearth of solid non-combustible material to the dimensions shown in Fig. J38.

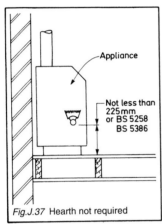

Fig.J.37 Hearth not required

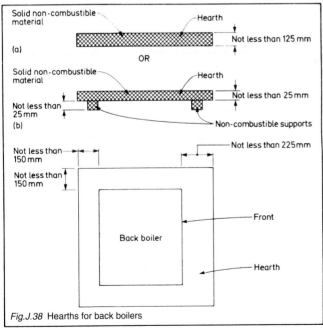

Fig.J.38 Hearths for back boilers

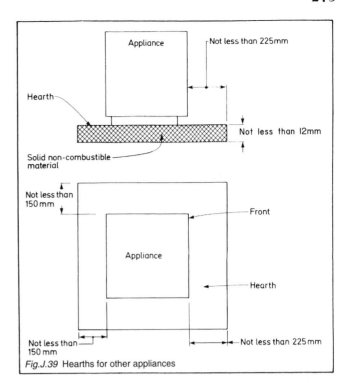

Fig.J.39 Hearths for other appliances

For all other appliances the hearth may be constructed of solid non-combustible material not less than 12 mm thick as in Fig. J39.

These requirements do not extend to solid fuel effect appliances.

SHIELDING APPLIANCES

When appliances do not comply with the appropriate parts of BS 5258 or BS 5386 the backs, tops, sides and any draught diverters are to be isolated from combustible material as shown in Fig. J40. Alternatively, a non-combustible shield may be used as in Fig. J41.

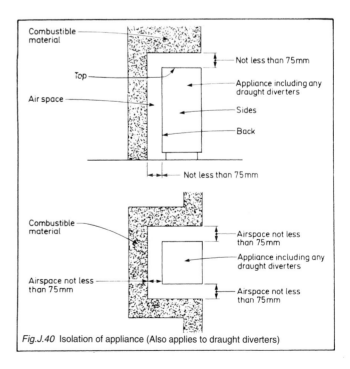

Fig.J.40 Isolation of appliance (Also applies to draught diverters)

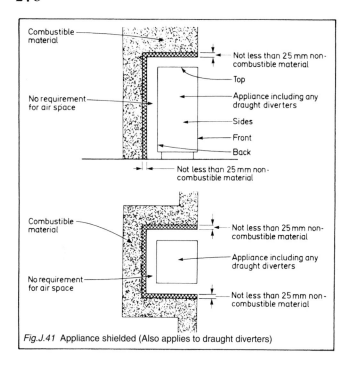

Fig.J.41 Appliance shielded (Also applies to draught diverters)

ALTERNATIVE APPROACH

The requirements of the approved document may also be satisfied by following the relevant recommendations in a number of British Standards.

The first is BS 5440 which is in two parts and relates to flues and ventilation of gas appliances of rated input not exceeding 60 kW (1st, 2nd and 3rd family gases).

- BS 5440: Part 1: 1990, 'Specification for the installation of flues, and deals with the complete flue and equipment from the appliance connection to discharge to the external air'.

Advice is given regarding the preliminary design considerations, natural and fanned draught for individual open fires, shared open flues, balanced flues, shared room sealed flues, natural draught and special categories of flue systems. There are three appendices, 14 tables and 25 figures.

- BS 5440: Part 2: 1989, 'Specification for the installation of ventilation for gas appliances'.

This covers materials for air vents and ducts, the size and position of air vents and the effect of extract fans. There is an appendix, three tables and three figures.

- BS 5546: 1990, 'Specification for installation of gas hot water supplies for domestic purposes (2nd family gases)'.

This BS contains recommendations for the selection and installation of appliances burning natural gas. The appliances considered supply hot water for domestic purposes in individual dwellings and other premises where a similar service is required. Many of the recommendations are obligatory under current gas supply regulations.

Advice is also given on system selection, installation, and planning together with details of water, gas and electricity supplies. Tables are given and there are numerous diagrams illustrating typical instantaneous systems and also storage systems and systems using circulators.

- BS 5864: 1989, 'Specification for the installation of gas-fired ducted air heaters of rated output not exceeding 60 kW (natural gas)'.

This deals with the installation of gas-fired ducted air heaters having

a rated input of not more than 60 kW for heating one room, two or more rooms or internal spaces simultaneously in domestic premises.

The BS relates mainly with the type of heater that incorporates a fan to circulate warm air by convection currents. It also covers combined air heater/water heaters.

Advice is given regarding the selection of suitable appliances and for safe and satisfactory installation ensuring that air supply and flue arrangements are satisfactory; that the appliance is accessible for maintenance and is safe in operation. Most of the practices recommended are obligatory under gas installation regulations.

- BS 5871: 1991 in three parts, 'Code of practice for the installation of gas fires, convectors and fire/backboilers (natural gas)'.

This covers the installation of space heating appliances burning gas, principally intended for heating single rooms in domestic and other premises. It also includes information on the installation of gas fires when combined with back boilers.

Four basic types of space heaters are described: radiant gas fires; radiant convector gas fires; gas convertors and fire/backboilers.

Any appliances described in the code may be used to heat a room or internal space to a desired temperature, but personal preference and knowledge of types of appliance available are the main factors in choosing an appliance.

- BS 6172: 1990, 'Specification for installation of domestic gas cooking appliances (natural gas)'.

This contains recommendations for the selection and installation of cookers or separate ovens, hot plates and grills.

- BS 6173: 1990, 'Specification for installation of gas catering appliances (2nd family gases)'.

The recommendations made in this BS cover all catering establishments with the exception of fish and chip shops.

- BS 6798: 1987, 'Specification for the installation of gas-fired hot water boilers or rated input not exceeding 60 kW'.

This specifies the selection and installation of gas-fired central heating boilers not exceeding 60 kW total rated input. It applies to boilers designed to operate in the condensing or non-condensing mode for the heating of domestic or commercial premises by the circulation of heated water in open or closed systems.

There are detailed recommendations and commentary concerning the selection and installation of boilers, fluing and the placing of the appliances. There are also recommendations regarding boiler compartments, airing cupboard installations, under stairs cupboard installations and fireplace installations. Provision is made for roof installation when great care must be taken including a non-combustible base of at least 12 mm thickness under the boiler. It should be possible to shut down the boiler without having to enter the roof space.

SECTION 4. OIL

These are the requirements for oil-burning appliances with a rated output up to 45 kW.

AIR SUPPLY

When an oil-burning appliance is installed in any room or space (Fig. J42) there is to be a permanent ventilation opening having an area of not less than 550 mm^2 for each kW of rated output above 5 kW.

The requirement does not apply when balanced flue appliances are fitted.

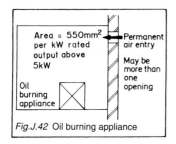

Fig.J.42 Oil burning appliance

SIZES OF FLUES

A flue size is to be at least

- in the case of a flue pipe, the same as the flue outlet from the appliance.
- in the case of a chimney, for a circular flue as shown in Table J2.

Table J2

Rated	Circular flue(diameter)	Equivalent square flue
up to 20 kW	100 mm	89 mm
20 kW to 32 kW	125 mm	111 mm
32 kW to 45 kW	150 mm	133 mm

Where a flue is of square section it is to have a cross sectional area not less than the area of a corresponding circular chimney and suggested dimensions for equivalent square flues are also given in Table J2. These requirements do not extend to balanced and low level flues.

DIRECTION OF FLUES

Horizontal runs are to be avoided and if a bend is necessary it should be at a greater angle than 45° from the vertical (see Fig. J30).

This requirement regarding the direction of flues does not refer to balanced flues.

OUTLETS FROM FLUES

The approved document requires that the outlet from a balanced, or low level discharge appliance (Fig. J43) should be:

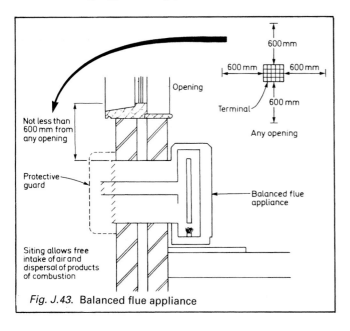

Fig. J.43. Balanced flue appliance

- placed in a position externally allowing the dispersal of the products of combustion and in the case of a balanced flue appliance, the free intake of air; and
- not less than 600 mm from an opening into the building; and
- if the terminal is in a position where persons could come into contact with it, protection from damage is to be provided;
- designed in such a way that will prevent the entry of any matter which might restrict the flue.

The outlet from a flue serving a pressure jet appliance may be terminated anywhere above the roof line.

Outlets which serve any other appliance must terminate above the roof line in a position indicated in Fig. J8 relating to solid fuel flue outlets.

CHIMNEY AND FLUE PIPES

The provision of flue pipes and brick and blockwork chimneys is dependent upon the temperature of the flue gases under the worst operating conditions.

Therefore, if the temperature is likely to exceed 260°C the following provisions should be applied where relevant.

1. A flue pipe should be used only to connect an appliance to a chimney.
2. A flue pipe should not pass through any roof space.
3. Flue pipes with spigot and socket joints are to be fitted with the socket uppermost.
4. Flue pipes may be manufactured from any of the following materials.

- Cast iron in accordance with B541: 1973 (1981), 'Cast iron spigot and socket flue or smoke pipes and fittings'. This covers flue or smoke pipes and bends 100 mm to 300 mm nominal bore and offsets of 100 mm to 150 mm and provides alternative socket profile to allow interchangeability between pipes manufactured by the sandcast or by the spun process.
- Mild steel having a wall thickness of not less than 3 mm.
- Stainless steel having a wall thickness of not less than 1 mm in accordance with BS 1449: Steel plate, sheet and strip, Part 2: 1983, 'Specification for stainless and heat-resisting steel plate sheet and strip Grade 316S11; 316S13; 316S31 and 316S33 or equivalent Euronorm 88-71 designation'.

The BS specifies requirements for flat-rolled stainless steel and heat-resisting steel products including plate 3 mm thick and above.

- Vitreous enamelled steel complying with BS 6999: 1989, which is a specification for vitreous enamelled low carbon steel flue pipes, other components and accessories for solid fuel burning appliances with a maximum rated output of 45 kW. These flue pipes are intended for internal freely exposed use only. They are also suitable for gas- and oil-fired appliances.

The standard contains many definitions and paragraphs dealing with

3. materials
4. vitreous enamelling
5. construction. One requirement is that where access for inspection and cleaning is required, this must be carried out without disturbing the joints
6. resistance to high temperature
7. instructions for installation.

Included in appendices are methods of test for heat resistance of vitreous enamel coating and methods of test for joints.

When a flue pipe is placed next to combustible material there must be an air space separating the pipe from this material. Distance across the air space depends upon whether or not a non-combustible shield is used. Minimum dimensions are indicated in Fig. J11. The approved document now contains a recommendation regarding the width of a non-combustible shield as shown in Fig. J12.

When used for oil fuel appliances chimneys are to be capable of withstanding a temperature of 1100°C without undergoing any structural change affecting the stability or performance of the chimney.

If a chimney is not directly over an appliance, a debris collecting space should be provided which is accessible for emptying (see Fig. J13).

Blockwork chimneys may be of refractory material; a combination of high alumina cement or pumice aggregates or lined in the same way as brick chimneys.

Minimum thickness of the walls of brick and blockwork excluding separate linings are specified for walls separating flues; another part of the building; separate fire compartments and dwellings.

The distances are 100 mm between one flue and another; or between a flue and the outside air or between a flue and another part of the same building (excluding a part which is a dwelling or is constructed as a separate fire compartment), and 200 mm between a flue and another compartment in the same building, another building or another dwelling (see Fig. J14).

CHIMNEY LININGS

(Where temperature likely to exceed 260°C.)

Chimney linings that may be used are:

- Clay flue linings having rebated or socketed joints in accordance with BS 1181: 1989, 'Clay flue linings and flue terminals'. This gives details of materials, workmanship, design, construction, dimensions, tests for clay flue linings, bends, flue terminals for heating appliances.
   BS 1181: 1989 has been replaced by BS 1181: 1999, 'Specification for clay flue terminals'. Also flue liners are now covered by BS EN 1457: 1999, 'Chimneys – clay/ceramic flue liners – requirements and test methods'. 22 classes of liners are specified.
- Imperforate clay pipes with socketed joints as covered by BS 65: 1981, 'Vitrified clay pipes, fittings and joints'. It sets out the requirements for vitrified clay pipes and fittings with or without sockets, including materials, manufacture, dimensions and performance requirements.
- High alumina cement and kiln burnt aggregate pipes having rebated or socketed joints or steel collars surrounding the joints.

Linings are to be fitted with sockets or rebates uppermost and jointed with fire proof mortar. Space between liners and brickwork is to be filled with weak mortar or insulating concrete (see Fig. J15).

MATERIALS FOR FLUE PIPES

If flue gas temperatures are unlikely to exceed 260°C the following provisions should be applied where relevant.

Flue pipes may be of any of the following materials:

- Cast iron in accordance with BS 41: 1973 (1981), 'Cast iron spigot and socket flue or smoke pipes and fittings'. This covers flue or smoke pipes and bends 100 mm to 300 mm nominal bore and offsets of 100 mm to 150 mm, and provides alternative socket profile to allow interchangeability between pipes manufactured by sandcast or by the spun process.
- Mild steel having a wall thickness not less than 3 mm.
- Stainless steel having a wall thickness of not less than 1 mm in accordance with BS 1449: Steel plate, sheet and strip: Part 2: 1983. This is a specification for stainless steel and heat-resisting steel plate and strip Grade 316S11; 316S13; 316S31 and 316S33 or the equivalent Euronorm 88–71 designation. The BS specifies requirements for flat-rolled stainless steel, and also heat-resisting steel products including plate 3 mm and above.
- Vitreous enamelled steel meeting the requirements of BS 6999: 1989, which is a specification for vitreous enamelled low carbon

steel flue pipes; other components and accessories for solid fuel burning appliances with a maximum rated output of 45 kW.
- Sheet metal as described in BS 715: 1986, which is a specification for metal flue pipes, fittings, terminals or accessories for gas-fired appliances with a rated input not exceeding 60 kW.
- Asbestos cement in accordance with BS 567: 1989. This is a specification for asbestos cement flue pipes and fittings, light quality.
- Asbestos cement as described in BS 835: 1973 (1984), 'Specification for asbestos cement flue pipes and fittings, heavy quality'.
- Any other material fit for its intended purpose. Flue pipes which have spigot and socket joints are to be fitted with the sockets uppermost.

PROTECTION OF COMBUSTIBLE MATERIAL

- When combustible material is near to a flue pipe it must be protected: by having a 25 mm wide separation (see Fig. J33).
- In the event of a flue pipe passing through wall, floor or roof it is to be encased by a non-combustible sleeve enclosing an air space not less than 25 mm around the flue pipe (see Fig. J34).
- Where a flue pipe passes through a compartment wall or a compartment floor, it is to be encased with non-combustible material not less than half the fire resistance required in the wall or floor.

When measuring air space the 25 mm measurement may be taken from the outside of the inner pipe where a double wall flue pipe is being used (see Fig. J35).

LINING BRICK CHIMNEYS

Chimney linings that may be used are:

- Clay flue linings having rebated or socketed joints in accordance with BS 1181: 1989, 'Clay flue linings and flue terminals'. This gives details of materials, workmanship, design, construction, dimensions, tests for clay flue linings, bends and flue terminals for heating appliances.
   BS 1181: 1989 has been replaced by BS 1181: 1999, 'Specification for clay flue terminals'. Also, flue liners are now covered by BS EN 1457: 1999, 'Chimneys – clay/ceramic flue liners – requirements and test methods'. 22 classes of liners are specified.
- Imperforate clay pipes with socketed joints as covered by BS 65: 1981, 'Vitrified clay pipes, fittings and joints'. It sets out the requirements for vitrified clay pipes and fittings with or without sockets, including materials, manufacture, dimensions and performance requirements.
- High alumina cement and kiln burnt or pumice aggregate pipes having rebated or socketed joints or steel collars surrounding the joints.

Linings are to be installed with the sockets or rebates uppermost. Joints between liners should be filled with weak mortar or insulating concrete (see Fig. J15).

CHIMNEY BLOCKS

Flue blocks used in the construction of masonry chimneys are to be as described in BS 1289: Part 1: 1986, which is a specification for precast concrete flue blocks and terminals, and BS 1289: Part 2: 1989, a specification for clay flue blocks and terminals.

BS 1289: Part 1: 1986 specifies requirements for precast concrete flue blocks intended for use in bonded flues or in non-bonded flues, and precast concrete terminals.

The standard contains definitions and advice on materials, dimensions, dimension deviations, joints, compressive strength and flow resistance.

There are three appendices and numerous illustrations showing typical flue block installations.

BS 1289: Part 2: 1989 specified requirements for clay flue blocks intended for use in bonded or non-bonded flues and clay terminals. They are intended for use with gas appliances.

The standard contains definitions and recommendations for materials and workmanship, dimensions, dimensional deviations, joints, compressive strength, flow resistance and sampling.

There are four appendices, including Appendix D which contains recommendations for the design and construction of clay flue block systems. There are numerous illustrations of components and installation methods.

FLEXIBLE FLUE LINERS

The circumstances in which a flexible flue liner may be used in a chimney are if:

- the liner is in accordance with BS 715: 1989, and
- the chimney was built before 1 February 1966, or it is already lined, or constructed of flue blocks in accordance with the approved document.

BS 715: 1989 applies to metal flue pipes, fittings, terminals and accessories for gas-fired appliances with rated input not exceeding 60 kW. Flue pipes are to have welded or folded seams and requirements are based on the use of sheet carbon steel or aluminium.

DEBRIS COLLECTING SPACE

Where a chimney is not lined or is not constructed of flue blocks, a debris collection space is to be provided at the bottom of the chimney (see Fig. J13).

THICKNESS OF WALLS

The minimum thickness of a brick or blockwork chimney is not to be less than 25 mm.

Chimney walls between a flue and

- a compartment of the same building
- another building, or
- another dwelling

are to have not less than the fire resistance required for the compartment or separating wall (Fig. J36; see also B3, Chapter 11).

When a chimney passes through a compartment floor or wall, it should have not less than half the fire resistance of the compartment wall or floor. A masonry wall or compartment floor may also form a chimney wall.

FACTORY MADE INSULATED CHIMNEYS

A factory made insulated chimney which is serving an oil appliance is to be constructed and tested to satisfy the relevant requirements given in:

- BS 4543: Part 1: 1990, 'Methods of test for factory made insulated chimneys', and
- BS 4543: Part 3: 1990, 'Chimneys for oil fired appliances'.

BS 4543: Part 1: 1990 relates to chimneys that are made in a factory and do not require fabrication on site. The various components must be assembled in accordance with the manufacturers' instructions. Tests include joint test, draught test, thermal shock test, thermal insulation test and strength test.

BS 4543: Part 3: 1990 refers to chimneys designed for a rated output up to 45 kW but with a flue gas temperature at the appliance exit not to exceed 450°C after any permanent draught break or before any stabilizer.

BS 6461: Part 2: 1984, 'Code of practice for factory made insulated chimneys for internal applications'. Recommends methods of installing chimneys dependent on natural draught.

Protection

An insulated metal chimney is not to:

- go through part of a building forming a separate compartment unless it is encased in non-combustible material having not less than half the fire resistance of the compartment wall or floor (also see B3, Chapter 11);
- have a gap between the outside of an insulated metal chimney and combustible material that is less than 'X' (see Fig. J20);
- go through a cupboard, storage space or roof space except when it is cased in a non-combustible material leaving a gap not less than 'X' from the outside of the chimney.

The approval document says that the gap 'X' is to be found by test in accordance with BS 4543: Part 1: 1976, 'Methods of test for factory made insulated chimneys'.

Contents of this BS are as follows:

Scope; references; definitions; chimney support test; structure and chimney joint test; draught test; thermal shock test; thermal insulation test and strength test. This is supported by a table giving flue gas generator inputs and a number of illustrations.

Chimneys are tested on the basis of a gap of 40 mm to the enclosure or the manufacturers' specified clearance measured between the outer surface of the chimney and the interior surface of the enclosing material.

Such a clearance is designated 'X' on Figs J19 and J20 and is normally 40 mm.

The casing used for enclosing a chimney is to be closed at each floor and ceiling level by a firestop or firestop spacer assembly (see Figs J18, J19 and J20).

Factory made insulated chimneys must also be installed in accordance with the manufacturers' instructions or the relevant recommendations in

- BS 6461: Part 2: 1984, which is a code of practice for factory made insulated chimneys for internal applications.

This part gives recommendations for the installation of new factory made insulated chimneys for internal applications, the operation of which depends upon natural draught. It deals with chimneys of nominal internal diameters of 100 mm, 125 mm, 150 mm and 200 mm.

The contents contain recommendations under the following headings: definitions; exchange of information; design; condensation; size of chimney; bends; openings; height and position of chimneys; terminals; stability; site testing; installation; connecting flue pipes.

Access for maintenance or replacement

Factory manufactured insulated chimneys are to be installed strictly in accordance with manufacturers' instructions with particular reference to:

- the need for the chimney to be readily accessible throughout its length for maintenance or replacement;
- the requirement that joints should not be made within a wall.

When this cannot be avoided the joint should be placed in a sleeve in order to isolate it from the wall.

Accessibility for maintenance or repair may be difficult if placed in flats or maisonettes. If it is found that proper access cannot be provided, factory made insulated chimneys should not be used.

HEARTHS

- If the surface temperature of the floor below an appliance is likely to exceed 100°C then a constructional hearth is to be provided.

It is to be solid, of non-combustible material, to dimensions not less than those shown in Fig. J21. The thickness shown may include any solid non-combustible floor under the hearth.

Combustible material may only be placed under constructional hearths in the following circumstances:

(a) it is to support the edges of the hearth, or
(b) an air space is provided (see Fig. J22).

- If the surface temperature of the floor below an appliance is unlikely to exceed 100°C the appliance may be placed on a rigid imperforate sheet of non-combustible material and no constructional hearth is required (Fig. J44).

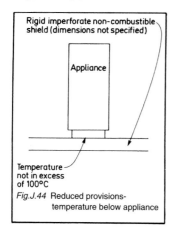

Fig.J.44 Reduced provisions- temperature below appliance

SHIELDING APPLIANCES

Where the surface temperature of the sides and back of an appliance is likely to exceed 100°C the appliance is to be separated from any combustible material by:

- non-combustible material not less than 25 mm thick, or
- an air space not less than 75 mm. (See Figs J40 and J41.)

ALTERNATIVE APPROACH

There is an alternative approach and the requirements may be met by following the relevant recommendations in BS 5410: Part 1: 1977, 'Installations up to 44 kW output for space heating and hot water supply purposes'.

Recommendations are given for oil-fired appliances and burners, oil systems from storage tank to burner, accommodation for boilers and air heaters, and for oil storage tanks for family dwellings; flues and chimneys, commissioning, maintenance and legal requirements.

There is an appendix, five tables and 11 figures.

RELEVANT READING

British Standards and Codes of Practice

BS 41: 1973 (1981). Cast iron spigot and socket flue or smoke pipes and fittings
BS 65: 1991. Specification for vitrified clay pipes, fittings and joints
    Amendment   1: AMD 8622
BS 476: Part 4: 1970 (1984). Non-combustibility test for materials
    Amendment 1: AMD 2483
                2: AMD 4390

BS 476: Part 11: 1982 (1988). Assessing heat transmission from building materials
BS 567: 1989. Asbestos cement flue pipes and fittings: light quality
    Amendment 1: AMD 5963
BS 715: 1993. Sheet metal flue pipes and accessories for fired appliances up to 60 kW
    Amendment 1: AMD 8413
BS 799: Part 5: 1987. Specification for oil storage tanks
BS 835: 1973 (1989). Specification for asbestos cement flue pipes and fittings (heavy quality)
    Amendment 1: AMD 5964
BS 1181: 1989 (1996). Clay flue linings and flue terminals
BS 1181: 1999. Specification for clay flue terminals
BS 1251: 1987. Open fireplace components
    Amendment 1: AMD 8454
BS 1289: Part 1: 1986. Specification for precast concrete flue blocks and terminals
BS 1289: Part 2: 1989. Specification for clay flue blocks and terminals
BS 1449: Part 2: 1989 (1996). Specification for stainless and heat resisting steel plate, sheet and strips
BS 1846: Part 1: 1994. Glossary of terms. Domestic appliances
BS 2869: Part 2: 1988. Specification for fuel oil for agricultural and industrial engines
BS 4543: Part 1: 1990 (1996). Factory-made insulated chimneys. Methods of test
BS 4543: Part 2: 1990 (1996). Factory-made insulated chimneys. Specification for chimneys for solid fuel appliances
BS 4543: Part 3: 1990 (1996). Factory-made insulated chimneys. Specification for chimneys for oil-fired appliances
BS 4876: 1984. Performance requirements for domestic flued oil-burning appliances
BS 5258: Part 1: 1986. Safety of central heating boilers and circulators
BS 5258: Part 4: 1987. Specification for fanned circulation ducted air heaters
BS 5258: Part 5: 1989. Gas fires
BS 5258: Part 7: 1977. Storage water heaters
BS 5258: Part 8: 1980. Combined appliances: gas fire/back boilers
BS 5258: Part 9: 1989. Specification for combined appliances
BS 5258: Part 12. 1990. Decorative log and other fuel effect fire appliances
    Amendment 1: AMD 7507
BS 5258: Part 13: 1986. Specification for convector heaters
BS 5258: Part 15: 1991. Specification for combination boilers
BS 5258: Part 16: 1991. Specification for live fuel effect gas fires
BS 5386: Part 1: 1976. Gas-burning appliances for instantaneous production of hot water for domestic use
    Amendment 1: AMD 2990
                2: AMD 5832
BS 5386: Part 2: 1981 (1986). Mini water heaters
BS 5386: Part 3: 1980. Domestic cooking appliances burning gas
    Amendment 1: AMD 4162
                2: AMD 4405
                3: AMD 4878
                4: AMD 5220
                5: AMD 6642
                6: AMD 6883
BS 5386: Part 4: 1991. Built-in domestic cooking appliances
    Amendment 1: AMD 6940
BS 5386: Part 5: 1988. Gas-burning instantaneous water heaters
BS 5386: Part 6: 1991. Domestic gas cooking appliances with forced convection ovens
    Amendment 1: AMD 6926
BS 5410: Part 1: 1977. Oil-fired installations up to 44 kW
    Amendment 1: AMD 3637
BS 5410: Part 2: 1978. Oil-fired installations of 44 kW or above
    Amendment 1: AMD 3638
BS 5440: Part 1: 1990. Code of practice, flues for appliances with rated input up to 60 kW
    Amendment 1: AMD 8819

BS 5440: Part 2: 1989. Specification: air supply for appliances with rated input up to 60 kW
Amendment 1: AMD 8128

BS 5546: 1990. Specification: installation of gas hot water supplies for domestic purposes
Amendment 1: AMD 6656
2: AMD 8129

BS 5864: 1989. Gas-fired ducted air heaters output up to 60 kW
Amendment 1: AMD 8130

BS 5871: Part 1: 1991. Gas fires, convectors and fire/back boilers

BS 5871: Part 2: 1991. Inset live fuel-effect fires input not exceeding 15 kW

BS 5871: Part 3: 1991. Decorative fuel-effect gas appliances input not exceeding 15 kW
Amendment 1: AMD 7033

BS 5955: Part 6: 1980. Unplasticized PVC pipework for gravity drains and sewers

BS 5955: Part 8: 1990. Thermoplastic pipes in cold water services and heating systems

BS 6073: Part 1: 1981. Precast concrete masonry units
Amendment 1: AMD 3944
2: AMD 4462

BS 6172: 1990. Installation of domestic gas cooking appliances

BS 6173: 1990. Installation of gas catering appliances

BS 6461: Part 1: 1984. Code of practice for masonry, chimneys and flue pipes
Amendment 1: AMD 5649

BS 6461: Part 2: 1984. Code of practice for factory-made insulated chimneys

BS 6798: 1987. Installation of gas-fired hot water boilers up to 60 kW

BS 6999: 1989 (1996). Vitreous enamelled low carbon steel flue pipes
Amendment 1: AMD 8949

BS 7455: Part 1: 1991. Light quality fibre cement pipes, fittings and terminals

BS 7566: Part 1: 1992. Factory-made chimneys specifying installation design information

BS 7566: Part 2: 1992. Factory-made chimneys: installation and design

BS 7566: Part 3: 1992. Factory-made chimneys – site installation

BS 7566: Part 4: 1992. Factory-made chimneys. Installation design, and installation

BS 8303: Part 1: 1994. Installation of domestic heating and cooking appliances burning solid fuel

BS 8303: Part 2: 1994. Domestic heating and cooking appliances. Installing and commissioning on site

BS 8303: Part 3: 1994. Domestic heating and cooking appliances. Design and site installation

BS EN 1457: 1999. Chimneys – clay/ceramic flue liners – requirements and test methods.

BRE digest

339 Condensing boilers

BRE information papers

3/83 Optimum start controls – low energy buildings
2/85 Performance of PSA low energy management system
6/85 Building energy management systems
10/88 Performance of gas-fired condensing boilers in family housing
11/89 Space and hot water heating: energy efficiency in large gas-fired systems
13/89 Energy efficient factories: design and performance
19/89 Condensing boilers: reducing space heating costs in large residential buildings
21/92 Spillage of flue gases from open flued combustion appliances
7/94 Spillage of flue gases from solid fuel combustion appliances
19/94 Condensing boilers: a review
16/95 Safety and environmental requirements of new refrigerants

BRE defect action sheets

91 Domestic appliances – air requirements (Design)
92 Balance flue terminals – location and guarding (Design)
109 Hot and cold water systems – protection against frost
126 Domestic chimneys: solid fuel design
127 Domestic chimneys: solid fuel installation
138 Domestic chimneys: rebuilding or lining existing chimneys

BRE reports

Unventilated Domestic Hot Water Systems (E. F. Ball). Ref. BR 125.
Construction of new buildings on gas contaminated land (1991)
Radon: guidance on protective measures for new dwellings (1991).

CHAPTER 19

Protection from falling, collision and impact

(Approved document K)

Regulation 4 requires that building work is to be carried out in accordance with the relevant paragraphs of Schedule 1 to the regulations.

When complying with any requirement the work must be undertaken in such a way that it does not result in the contravention of other regulations.

Compliance with Part K is necessary when erecting any building, or when material alterations are made to a building.

There are requirements for the pitch, headroom and clearance on stairs and ramps in various places and in different types of building. Stairs and ramps must also be properly guarded and in appropriate places fitted with handrails. There are also provisions regarding the guarding of roofs, rooflights and balconies to which people have access for purposes other than normal maintenance and repair.

Revised regulations are contained in the Building (Amendment) Regulations 1997, and set out to ensure that a building which complies with the building regulations will automatically comply with the corresponding health and safety regulations. The amended regulations came into force on 1 January 1998 and add requirements regarding the protection from collision with open windows and protection against impact from trapping by doors.

In addition, matters relating to stairs, ladders and ramps have been extended to cover access to areas used solely for maintenance; there is provision for the guarding of areas of lightwells and other sunken areas and the safety aspects of the design of loading bays.

These matters have an affinity with the requirements of the Workplace (Health, Safety and Welfare) Regulations 1992 which came into operation on 1 January 1993, and in consequence compliance with Part K requirements will prevent the service of an improvement notice under section 23(3) of the Health and Safety at Work Act 1974.

Compliance with building regulation	Prevents service of improvement notice under requirements of Workplace (Health, Safety and Welfare) Regulations 1992.
K1 (also M2)	Reg. 17
K2	Reg. 13
K3	Reg. 17 (loading bays)
K4	Reg. 15 (2)
K5	Reg. 18

All work subject to the requirements of the building regulations is to be carried out using proper materials and workmanship. Requirements are in regulation 7 amended by the 1999 Regs, and guidance given in the approved document to regulation 7 (Chapter 9).

When a stairway or ramp forms part of a means of escape in case of fire reference should be made to Part B discussed in Chapter 11. Also reference should be made to Part M – access and facilities for disabled people outlined in Chapter 21.

STAIRS, LADDERS AND RAMPS

K1. Stairs, ladders and ramps are to be designed, constructed and installed in such a way that they provide safe passage for people moving about in a building, or between different levels in the building.

The requirements under this heading are only applicable to stairs, ramps and ladders if they form part of a building. This includes in addition to stairs and ramps inside a building, those outside if they are part of a building. For example, steps leading to an entrance. Requirements have been extended to include stairs having a pitch in excess of 42° and also to provide control of ladders which form part of a building.

Where access is required solely for the purpose of maintenance, the approved document suggests that greater care can be expected from those using the access and therefore less demanding provisions may be acceptable.

PERFORMANCE

K1 will be met, in the Secretary of State's view, if stairs, ladders and ramps afford reasonable safety between levels in

- dwellings when the difference in level is in excess of 600 mm;
- other buildings when the change of level is two or more risers;
- other buildings where the change in level is at least 380 mm if it is not part of a stair.

    The requirements of Part K apply to stairs and ramps forming part of the means of access in an assembly building, including arenas, cinemas, sports stadia, theatres, etc. When steps form part of gangways serving areas for spectators, special consideration is to be given and reference should be made to guidance in the following documents:
- When it is a new assembly building refer to BS 5588: Part 6: 1991, 'Fire precautions – code of practice for places of assembly'.

    This gives guidance on measures that, in the event of fire, safeguard the lives of public and staff.

    Part 3.7 of Section 3 is concerned with stairs and final exits.
- 7.1   Protected stairways
- 7.2   Basement stairs
- 7.3   Access lobbies and corridors to protected stairways. There is a table giving capacities of stairs related to the number of floors served. It contains recommendations for capacity; width; use of space in a protected stairway and separation.

    Part 8 relates to seating and gangways, and includes limitation on travel distances; exits; placing of gangways; number of seats in a row; seating widths; slope of a tier of seating and the guarding of balconies. There is a table setting out the number of seats in a row related to seat width and the number of gangways.
- Where it is work to an existing assembly building consult the 'Guide to fire precautions in existing places of entertainment and like premises' (Home Office, 1990).
- When it relates to a stand on a sports ground consult the 'Guide to safety at sports grounds' (Home Office, 1997).

SECTION 1. STAIRS AND LADDERS

However, whenever stairs, ladders and ramps are constructed they must be in accordance with these requirements.

The following terms are used (see Figs K1.1 to K1.5):

Term	Definition
Stair	Includes landings and flights
Private stair	Serves only one dwelling
Alternating tread stair	The tread is paddle shaped and has the wide portion alternating from one side to the other, on consecutive treads
Containment	A barrier preventing people from falling from a floor to the storey below
Helical stair	Describes a helix around a central void
Ladder	Series of rungs or narrow treads. Person normally ascends and descends facing the ladder
Ramp	Slope steeper than 1:20
Spiral stair	Describes a helix around a central column
Tapered tread	Nosing is not parallel to the nosing above it
Going	Width of tread less any overlap with next tread
Rise	Height of step including thickness of tread
Flight	Part of a stair or ramp that has consecutive steps or a continuous slope
Pitch	Line connecting nosings of all treads on a flight
Nosing	The front edge of a tread.

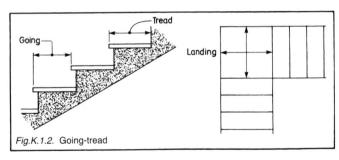

Fig.K.1.1. Stairs

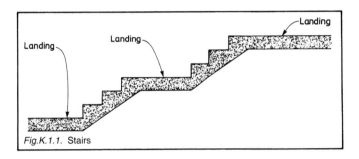

Fig.K.1.2. Going-tread

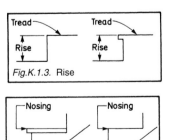

Fig.K.1.3. Rise

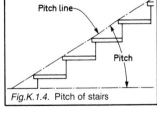

Fig.K.1.5. Nosing

Fig.K.1.4. Pitch of stairs

RISE AND GOING

Requirements regarding the steepness of stairs will be satisfied if all the steps have the same rise and going within the practical limits set for private stairs, institution and assembly stairs and other stairs.

Rise and going is calculated to ensure that twice the rise plus the going (2R + G) should fall within 550 mm and 700 mm.

- Private stairs to be used only for one dwelling.
- Institutional and assembly stairs serve a place where a substantial number of people may gather.
- Other stairs relate to all other buildings.

Thus the requirements which previously related to common stairways no longer appear and so the category 'other stairs' is appropriate.

- *Private stair.* The practical limits for rise is between 155 mm and 220 mm in conjunction with any going between 245 and 260 mm or, alternatively, any rise between 165 mm and 200 mm with any going between 223 mm and 300 mm. However, the approved document sets the maximum rise for a private stair at 220 mm and the minimum going at 220 mm. The maximum pitch for a private stair is 42° (see Fig. K1.5A)

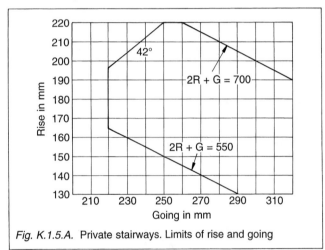

Fig. K.1.5.A. Private stairways. Limits of rise and going

- *Institutional and assembly stair.* The practical limits for rise is between 135 mm and 180 mm in conjunction with any going between 280 mm and 340 mm. The approved document sets the maximum rise for an institutional assembly stair at 180 mm and the minimum going of 280 mm. When the floor area of the building is less than 100 m² the going may be reduced to 250 mm (see Fig. K1.5B)

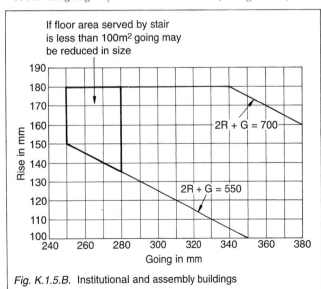

Fig. K.1.5.B. Institutional and assembly buildings

When access for disabled people is involved reference should be made to Part M, Chapter 21.

- *Other stairs.* Practical limits for rise is between 150 mm and 190 mm with any going between 250 mm and 320 mm (see Fig. K1.5C).

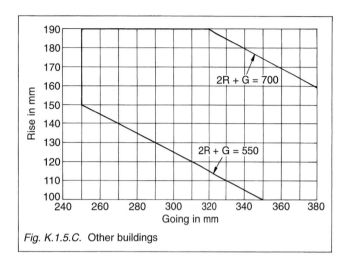

Fig. K.1.5.C. Other buildings

The approved document sets the maximum rise for other stairs at 190 mm with a minimum going of 250 mm.

When access for disabled people is involved reference should be made to Part M, Chapter 21 (see Fig. K.1.5C).

Pitch

Maximum pitch for a private stair is 42°, and gangways for seated spectators is set at 35°. Gangways may be required to be at different pitches in order to maintain sightline for spectators, and this may affect the main stairs.

The approved document gives a wide range of options with regard to rise and going, but BS 5395: Part 1: 1977 adds advice regarding optimum dimensions.

| Type of stairs | Optimum dimensions | |
	rise	going
A private stair used by a limited number of people who are generally very familiar with the stair, as for example in a dwelling	175 mm	250 mm
A semi-public stair used by larger numbers of people, some of whom may be unfamiliar with the stairs. This could apply to factories, offices, shops and also common stairs serving two or more dwellings.	165 mm	275 mm
A public stair used by a large number of people at one time. Examples are places of assembly or stairs used by people having ambulatory difficulties, or in old people's and children's homes	150 mm	300 mm

HEADROOM

There is to be a clear headroom of not less than 2 m measured from the pitch line and from floor level on landings.

In loft conversions if there is not enough space at the top of a stair to achieve a height of 2 m, headroom will be acceptable if it measures 1.9 m at the centre of the stair, reducing to 1.8 m on one side of the stair (Fig. K1.6).

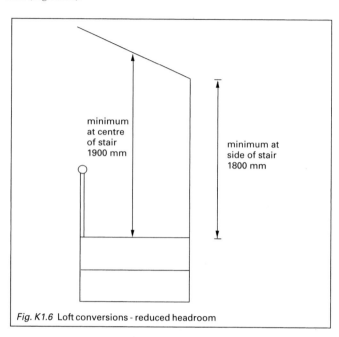

Fig. K1.6 Loft conversions - reduced headroom

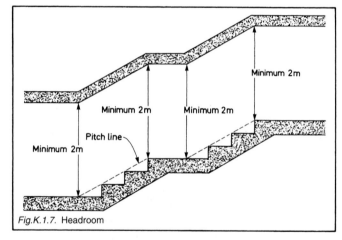

Fig.K.1.7. Headroom

OPEN RISERS

Open risers are permitted but there must be an overlap of 16 mm, on the treads (Fig. K1.8).

In dwellings, and other buildings likely to be used by children under 5, the space between each tread must be restricted so that a 100 mm sphere will not pass through (Fig. K1.9).

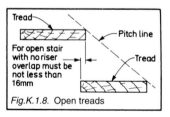

Fig.K.1.8. Open treads

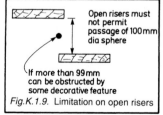

Fig.K.1.9. Limitation on open risers

TREADS

Treads must be level.

Sloping landings

Landings at the top or bottom of a flight may slope (Fig. K1.10), but the midpoint height of the adjoining step must be the same as the rise for all other steps in the flight.

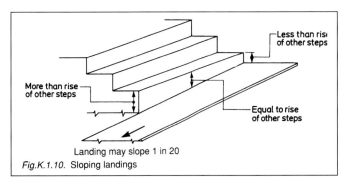

Fig.K.1.10. Sloping landings

Tapered treads

Consecutive tapered treads must have the rise and going within the dimensions allowed for straight flights and should have a minimum width of 50 mm at the narrow end. All tapered nosings must stem from the same point of origin, and have the same taper (see Figs K1.11 to K1.14).

Measurement of rise and going is made as follows.

- if flight is less than 1 m wide measure in the middle;
- if flight is more than 1 m wide measure 270 mm from each side;
- 2R + G must in all cases be within the permitted dimensions for the type of building;
- going of tread to measure not less than 50 mm at the narrow end;
- if consecutive tapered treads are used then a uniform going is to be maintained;

- if a stair consists of straight treads and tapered treads the going of tapered treads is not to be less than the going on the straight flight.

Stairs may be designed following recommendations in BS 585, 'Wood stairs': Part 1: 1989, 'Specification for stairs with closed risers for domestic use, including straight and winder flights and quarter or half landings'.

This BS specifies materials and construction requirements for stairs constructed of timber which may include plywood risers, or glue laminated wood components.

There is an appendix containing recommendations for site fixing of stairs and a table setting out minimum finished thickness of members. There are five figures.

WIDTH OF STAIRS

No recommendations for minimum stair widths are now given (Fig. K1.15), but the circumstances when forming a means of escape should be referred to Part B, Chapter 11, and if providing access for disabled people, Part M, Chapter 21.

Where, in a public building, a stair is wider than 1800 mm it is to be divided, and no flight is to be wider than 1800 mm.

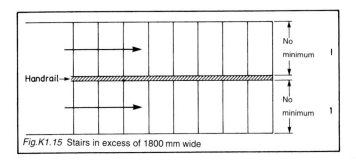

Fig.K1.15 Stairs in excess of 1800 mm wide

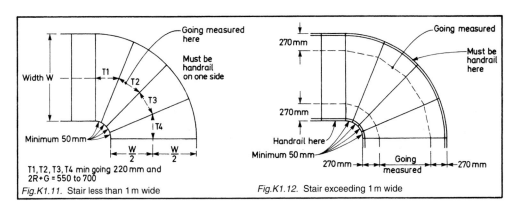

Fig.K1.11. Stair less than 1 m wide

T1, T2, T3, T4 min going 220 mm and 2R+G = 550 to 700

Fig.K1.12. Stair exceeding 1 m wide

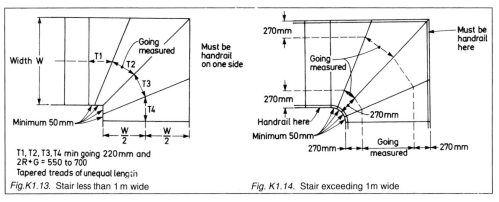

Fig.K1.13. Stair less than 1 m wide

T1, T2, T3, T4 min going 220 mm and 2R+G = 550 to 700
Tapered treads of unequal length

Fig. K1.14. Stair exceeding 1m wide

SPIRAL AND HELICAL STAIRS

The guidance given for rise and going is to be used, and stairs designed in accordance with BS 5395, 'Stairs, ladders and walkways': Part 2: 1984, 'Code of practice for the design of helical and spiral stairs'.

Stairs which will have going less than shown in the BS may be considered when undertaking conversion work where space is limited, and not more than one habitable room is served.

ALTERNATING TREAD STAIRS

This type of stair has been slowly gaining ground as it is designed to save space and has found favour when loft conversions are being undertaken. Such stairs may present a problem for a stranger but they can be used in reasonable safety by someone who is familiar with the alternate handed paddle-shaped treads (see Fig. K1.16).

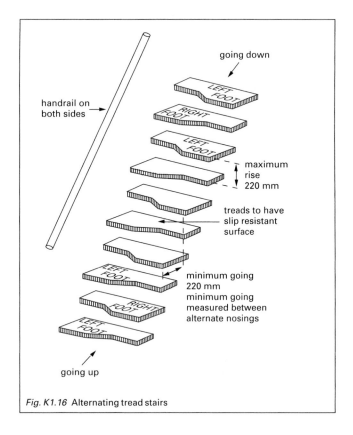

Fig. K1.16 Alternating tread stairs

Alternating tread stairs should only be used in the following circumstances:

- in one or more straight flights when undertaking loft conversions, but only if there is insufficient space for a stair meeting the standards already outlined;
- it should provide access to only one habitable room plus a bathroom or WC;
- when it provides access to a WC, there is to be another WC available in the dwelling;
- steps are to be uniform with parallel nosings;
- treads are to have slip-resistant surfaces;
- there is to be a handrail on each side of the stairs;
- tread sizes should be as for stairs with a minimum going of 220 mm and a maximum rise of 220 mm;
- if the stair is likely to be used by children under 5 years of age, it is to be constructed so that a 100 mm diameter sphere cannot be passed through the open risers.

FIXED LADDERS

There are now requirements to give some control over fixed ladders which form part of a building.

- A fixed ladder should only be used in loft conversions, and it is to have a handrail on each side. Even so it may only be used when there is not enough space without altering the existing space to install a stair meeting the requirements already outlined.

   The ladder may be used to provide access to only one habitable room. Retractable ladders may not be installed as a means of escape (see Part B, Chapter 11).
- Where stairs, ladders and walkways are to be installed in industrial buildings they are to be designed and constructed as recommended in

   (a) BS 5395, 'Stairs and ladders and walkways': Part 3: 1985, 'Code of practice for the design of industrial stairs, permanent ladders and walkways', or
   (b) BS 4211: 1987, 'Specification for ladders for permanent access to chimneys, other high structures, silos and bins'.

   BS 5395: Part 3: 1985 contains recommendations for the design of industrial type stairs, walkways, platforms, fixed ladders and companion way ladders.

   BS 4211: 1987 includes requirements for steel ladders with single bar rungs for permanent fixing. Also included are requirements for ladders to be used on silos, bins and stationary agricultural plant.

   There are three tables and six figures.

LENGTH OF FLIGHTS

If there are more than 36 risers in consecutive flights, there is to be at least one change of direction between flights (Fig. K1.17). This must be at an angle of at least 30°.

When a stairway is in a shop or a building used for shop or assembly purposes there must not be more than 16 risers in a flight.

Where gangways are of shallow pitch in assembly buildings the following should be consulted:

   (a) BS 5588: Part 6: 1991, 'Fire precautions – code of practice for places of assembly' (see page 282).
   (b) Home Office 'Guide to safety in sports grounds'.
   (c) Home Office 'Guide to precautions in existing places of entertainment and like premises'.

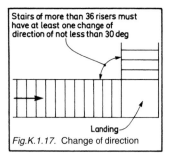

Fig.K.1.17. Change of direction

HANDRAILS

Flights 1 m wide and wider must have a handrail on each side. Less than 1 m wide requires only one handrail (Fig. K1.18).

Handrails are not required for the two bottom steps of a stairway (Fig. K1.19), except in the case of a public building or on access routes for disabled people when handrails should be provided beside the two bottom steps.

Handrails must be between 900 mm and 1000 mm above the pitch line of the stair to the top of the handrail.

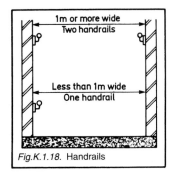

Fig.K.1.18. Handrails

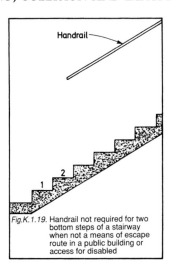

Fig.K.1.19. Handrail not required for two bottom steps of a stairway when not a means of escape route in a public building or access for disabled

LANDINGS

Landings are required at top and bottom of every flight. The width and length of every landing is not to be less than the least width of the flight.

Floor area can count as landing (Fig. K1.20).

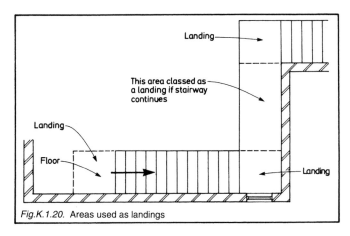

Fig.K.1.20. Areas used as landings

Landings are to be level. Where they are formed by ground at the top or bottom of a flight there may be a slope not exceeding 1:20 if the ground is paved or otherwise made firm.

Landings are to be clear of any permanent obstruction. However, a door may swing across a landing at the bottom of a flight if there remains a clear space not less than 400 mm across the full width of the flight (Figs K1.21 and K1.22).

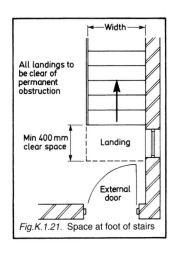

Fig.K.1.21. Space at foot of stairs

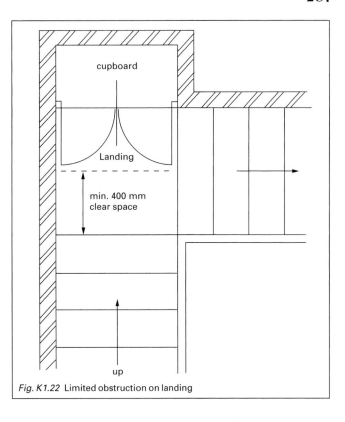

Fig. K1.22 Limited obstruction on landing

GUARDING OF FLIGHTS AND LANDINGS

- Dwellings. Flights and landings are to be guarded unless the drop at the side or sides is less than 600 mm (Fig. K1.23).

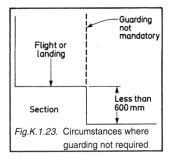

Fig.K.1.23. Circumstances where guarding not required

- Other buildings. Flights and landings are to be guarded when there are two or more risers.

   Suitable guarding includes:

   a wall
   a screen
   a railing
   a balustrade

and it must be so constructed that it is not easy to climb over it. Where the stairway is private, institutional or other used by children under 5, the design must ensure that any of the openings are too small for a child to fall through.

   Openings which will not permit a 100 mm diameter sphere to pass through would be safe (Fig. K1.24).

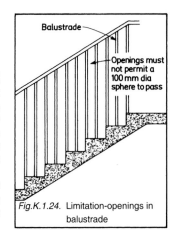

Fig.K.1.24. Limitation-openings in balustrade

It should be difficult for children to climb, and horizontal rails are not to be used.

Such a requirement would not be appropriate in a building constructed to accommodate old persons.

ACCESS FOR MAINTENANCE

If access is required for maintenance purposes at least once a month, the requirements for private stairs in domestic dwellings are appropriate or regard may be had to BS 5295: Part 3: 1985, 'Design of industrial type stairs, permanent ladders and walkways'.

In situations where access is required less frequently, portable ladders may be used.

The safe use of temporary means of access are referred to in the Construction (Design and Management) Regulations 1994.

SECTION 2. REQUIREMENTS FOR RAMPS

SLOPE

1:12 is the maximum slope permitted for a ramp.

HEADROOM

Ramps and landings must have a clear headroom of not less than 2 m (Fig. K1.25).

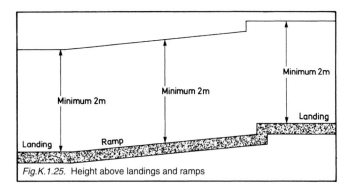

Fig.K.1.25. Height above landings and ramps

WIDTH OF RAMPS

No recommendation is now made for the width of ramps (Fig. K1.26), except for means of escape (Part B, Chapter 11) and access for disabled (Part M, Chapter 21).

Ramps are to be clear of any permanent obstructions.

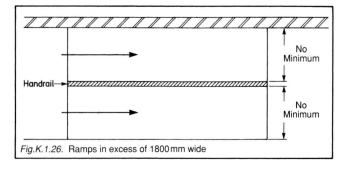

Fig.K.1.26. Ramps in excess of 1800 mm wide

HANDRAILS AND GUARDING

Ramps which are 1 m and more wide must have a handrail on each side. Less than 1 m wide requires only one handrail.

A handrail is not required if the rise of a ramp is less than 600 mm (Fig. K1.27).

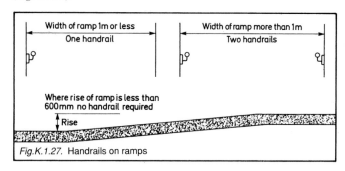

Fig.K.1.27. Handrails on ramps

Handrails are to be placed between 900 mm and 1000 mm above the ramp. Handrails must give firm support and a firm grip.

Where a ramp provides access for disabled people, reference should be made to Part M, Chapter 21.

Ramps and landings are to be guarded in the same manner as stairs.

Suitable guarding includes

a wall
a screen
a railing
a balustrade

and a handrail can form the top of any guarding if the height does not exceed 1000 mm.

LANDINGS

Length of a landing is to be not less than the width of the ramp.

Suitable barriers are to be provided in dangerous situations, to prevent people from falling.

PROTECTION FROM FALLING

K.2(a)   Users of the following are to be protected from the risk of falling:

- **stairs**
- **ramps**
- **floors**
- **balconies**
- **roofs to which people have access.**

This requirement relates only to stairs and ramps which form part of a building.

(b) Suitable barriers are to be erected to prevent people falling into any lightwell, basement area or similar sunken area connected to a building.

The standard of guarding requirements may vary with the use of a building. Buildings where large numbers of people gather, and they are unfamiliar with the premises, will require a higher standard of provision than a dwelling which will be familiar to the occupants.

Where access is required only for the purpose of maintenance it is to be expected that those gaining access will exercise care, and less demanding provisions may satisfy the requirement.

Guarding is unnecessary on such places as loading bays, where it would impede normal activities.

Protection is to be by way of suitable barriers adequate to prevent people from falling and these requirements supplement those already outlined under K.1 (see Fig. K2.1).

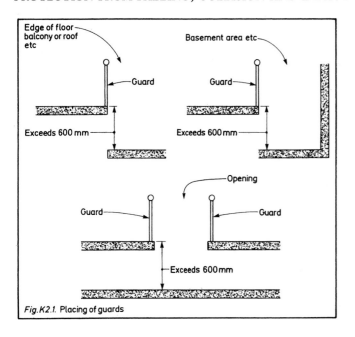

Fig. K2.1. Placing of guards

DESIGN

The 1998 approved document does not contain a stated strength requirement for guarding in particular circumstances. It requires guarding to be capable of resisting the horizontal forces given in BS 6399: Part 1: 1996, 'Code of practice for dead and imposed loads'. Barriers are to be capable of resisting impact from the side equal to the static forces given in the BS, and illustrated in Figs K2.4 and K2.5.

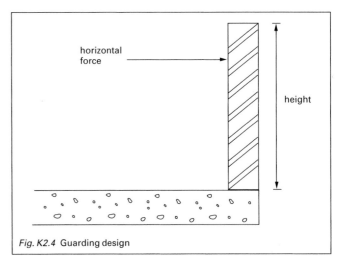

Fig. K2.4 Guarding design

Guarding is necessary for safety

- at the edges of any part of a raised floor, an opening window, gallery, balcony, roof, lightwell, basement area or similar sunken area adjoining a building;
- to rooflights and other openings (Fig. K2.2);
- vehicle parks, but not on ramps used only for vehicular access.

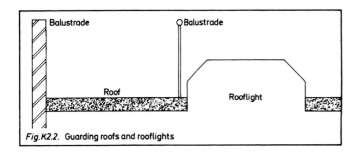

Fig. K2.2. Guarding roofs and rooflights

Requirements will be met if the following performance standards are achieved:

- pedestrian guarding is fixed in dwellings which will prevent people falling from a height exceeding 600 mm;
- pedestrian guarding is fixed in other buildings which will prevent people from falling more than the height of two risers; or 380 mm if the area being guarded is not part of a stair (Fig. K2.3).

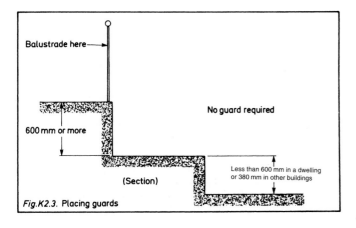

Fig. K2.3. Placing guards

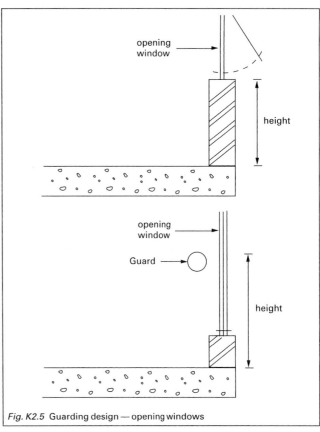

Fig. K2.5 Guarding design — opening windows

Guarding may be provided by a wall, parapet, balustrade or other construction which will give protection from falling.

Height of barriers

Barriers which are close to fixed seating may be as low as 800 mm in theatre and other assembly buildings in order to improve sight lines. The height requirements in various situations follow.

- Single family dwelling:
  stairs, landings, ramps, edges of internal floors
  height   900 mm for all elements
- Single family dwelling:
  external balconies and edges of roof
  height   1100 mm
- Factories and warehouses:
  subject to light traffic – stairs and ramps
  height   900 mm
- Factories and warehouses:
  subject to light traffic – landings and edges of floor
  height   1100 mm
- Residential, institutional, educational, office, and public buildings:
  all locations
  height   900 mm for flights; elsewhere 1100 mm
- Assembly:
  530 mm in front of fixed seating
  height   800 mm
- Assembly:
  in all other locations
  height   900 mm for flights; elsewhere 1100 mm
- Retail:
  all locations
  height   900 mm for flights; elsewhere 1100 mm
- All buildings:
  at opening windows (except roof windows in loft extensions)
  height   800 mm
- All buildings:
  at glazing to change of level to provide containment (Fig. K2.6)
  height   below 800 mm

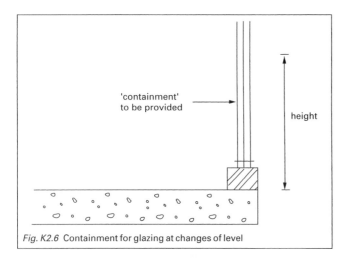

Fig. K2.6 Containment for glazing at changes of level

BS 6180: 1995, 'Code of practice for protective barriers in and about buildings' contains information regarding loads on infill panels.

When glazing is included in the barrier or guarding it should be glass blocks, toughened or laminated safety glass. Wired glass is not to be used.

Reference should also be made to Part N – Glazing: safety in relation to impact and cleaning (Chapter 22).

In buildings likely to be used by children under 5, precautions must be taken to prevent children falling through any guarding. This will be achieved if all openings are small enough to prevent a 100 mm diameter sphere from passing through.

Also guarding provision should be designed so that children will not find it easy to climb, and horizontal rails are to be avoided.

BS 6180: 1995, 'Code of practice for protective barriers in and about buildings', contains recommendations for barriers in and about buildings and places of assembly to protect persons from various hazards and to restrict or control movements of persons and/or vehicles.

Guarding maintenance areas

If access is required for maintenance purposes at least once a month, the requirements for private stairs is domestic dwellings are appropriate.

In situations where access for maintenance is required less frequently, temporary types of guarding or warning notices may be used. Such measures are not covered by building regulations and reference may be made to the Construction (Design and Management) Regulations 1994. The Health and Safety (Signs and Signals) Regulations 1996 may also be referred to.

Vehicle barriers are to be provided in a number of situations and be capable of resisting or deflecting the impact of a vehicle.

VEHICLE BARRIERS AND LOADING BAYS

K3.(1). The following are to be guarded with barriers where necessary to provide protection for persons in or about the building:

- **vehicle ramps forming part of a building; and**
- **any levels in a building to which vehicles have access.**

(2). People in a loading bay are to be protected from collision with vehicles using the bay.

The performance standard requires provision of a vehicle barrier which must be capable of resisting or deflecting the impact of vehicles. Barriers must be provided on the edge of any floor, roof or ramp which is level with or above the floor or ground, or on any route for vehicles (Fig. K3.1).

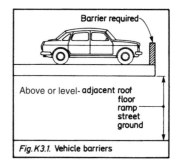

Fig. K3.1. Vehicle barriers

A wall, parapet, balustrade or similar construction may be used to form a barrier, which must be not less than 375 mm when placed on a floor or roof edge, and not less than 610 mm on a ramp edge (Figs K3.2 and K3.3).

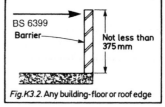

Fig.K3.2. Any building-floor or roof edge

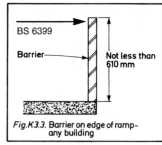

Fig.K3.3. Barrier on edge of ramp-any building

These barriers must be capable of resisting forces set out in BS 6399: Part 1: 1996, 'Code of practice for dead and imposed loads'.

This BS gives dead and imposed loads for use in the designing of buildings, including:

(a) Parapets and balustrades, which includes a table setting out horizontal loads on parapets and balustrades for all occupancy classes and also public assembly buildings.
(b) Vehicle barriers in car parks, which gives formula for determining horizontal force on 1.5 m of barriers.

LOADING BAYS

Loading bays are required to have enough exits or refuges from the lower level to avoid people being struck or crushed by vehicles using the bay.

The amount of provision to be made depends upon the size of the loading bay. The approved document recommends at least one exit point, but where more vehicles can be accommodated, then there should be at least two exits; one at each side of the bay.

An exit point may be omitted if a suitable refuge is constructed, to enable a person to seek protection at bay ground level and avoid being struck or crushed by a vehicle (see Fig. K3.4).

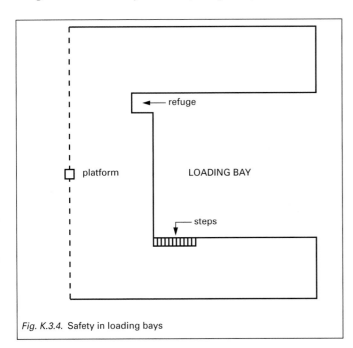

Fig. K.3.4. Safety in loading bays

There follows a provision which came into force on 1 January 1998, aiming to reduce the risk of collision with open windows, skylights and ventilators; and it also accords with regulation 15(2) of the Workplace (Health, Safety and Welfare) Regulations 1992.

PROTECTION FROM COLLISION

K4. Provision is to be made to avoid persons in the building from colliding with open windows, skylights or ventilators.

This requirement does not apply to dwellings.

Attention is directed to the requirements of Part B relating to the unobstructed dimensions of fire escape routes (Chapter 11) and also to Part M regarding access routes for disabled people (Chapter 21).

PERFORMANCE

The requirements of K4 can be met if windows, skylights and ventilators can be left open without the risk of people being injured by bumping into them.

Measures to be taken should aim to:

- install windows, skylights or ventilators so that projecting parts are kept away from people walking about the building, or
- the installation of features to guide people away from any hazard caused by an open window, skylight or ventilator.

It is reasonable to provide less demanding provisions in special cases as for example, maintenance spaces where people gaining access may be exercising great care.

The approved document gives some ways of meeting the requirements, and where parts of windows, skylights or ventilators project more than 100 mm either internally or externally into spaces used by people moving about, it should be:

- at least 2 m above the floor or ground;
- protected by a feature such as a barrier or rail 1100 mm high to prevent people colliding with the overhang;
- externally provide a roughened surface such as cobbles between a path and the wall (see also Part M, Chapter 21).

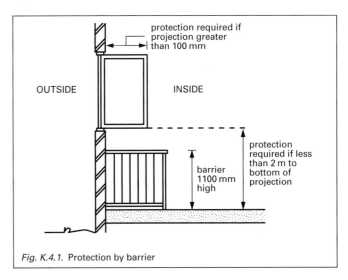

Fig. K.4.1. Protection by barrier

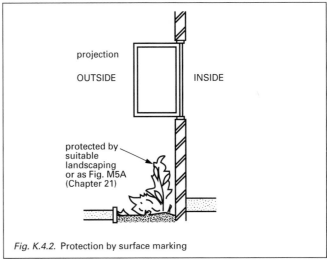

Fig. K.4.2. Protection by surface marking

In spaces used solely for maintenance the provision of a clear marking of the projecting part, making it easy to see, will fulfil the requirement .

There follows a requirement to prevent injury to people arising from being trapped by closing doors. This requirement is similar to Regulation 18 of the Workplace (Health, Safety and Welfare) Regulation 1992.

IMPACT FROM TRAPPING BY DOORS

K5. Ensure that no door or gate

- **if it slides or opens upwards, cannot fall onto any person; and**
- **if it is powered, cannot trap any person.**

It is to be so arranged that powered doors and gates are opened in the event of power failure. There is to be a clear view of the space on either side of a swing door or gate.

These requirements do not apply to dwellings, or to any door or gate which is part of a lift.

Attention is directed to Part B where there is guidance regarding doors on escape routes (Chapter 11) and Part M where there is advice relating to the design of internal and external doors (Chapter 21).

Performance

Steps are to be taken to prevent the opening of doors and gates creating a hazard to safety.

The approved document sets out some ways to avoid hazards, including the provision of features such as the following:

• Doors and gates which can be pushed open from either side, or are on main traffic routes are to have vision panels (see Fig. K5.1), unless they are low enough to see over. See also Part M (Chapter 21).

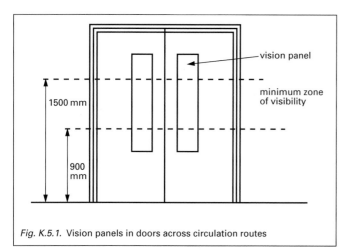

Fig. K.5.1. Vision panels in doors across circulation routes

• There should be a stop, or equally effective means of preventing a sliding door or gate coming off the end of the track. Additionally, a retaining rail should be fixed to stop the door or gate falling should the rollers leave the track or the suspension system fail.
• Upward opening doors are to be fitted with a mechanism to prevent them falling.
• Power operated doors and gates should have safety features to prevent injury to persons who are stuck or trapped. A pressure sensitive door edge operating the power switch is a suitable fitment. There should be a readily identifiable and accessible stop switch, and arrangements for manual or automatic opening in the event of a power failure.

TRANSITIONAL PROVISIONS

The 1991 Regs and 1992 edition of the approved document continue to apply to work where, before 1 January 1998 a building notice, initial notice or amendment notice was given to a local authority, or full plans had been deposited. If an amendment notice was given after 1 January 1998 relating to an earlier initial notice, the work added to the initial notice must comply with the current requirements of this part.

RELEVANT READING

British Standards and Codes of Practice

BS 585: Part 1: 1989. Wood stairs. Straight flights and quarter or half landings
    Amendment 1: AMD 6510
BS 585: Part 2: 1985. Performance required for domestic stairs
BS 4211: 1994. Ladders for permanent access to chimneys, other high structures, silos and bins
BS 4592: Part 1: 1995. Specification for open bar gratings
BS 4592: Part 2: 1987 (1994). Specification for expanded metal grating panels
BS 4592: Part 3: 1987. Specification for cold-formed planks
    Amendment 1: AMD 8735
BS 4592: Part 4: 1992. Glass-reinforced plastics open bar gratings
BS 4592: Part 5: 1995. Solid plates in steel, aluminium and glass-reinforced plastics
BS 5395: Part 1: 1997. Code of practice for stairs
    Amendment 1: AMD 3355
                2: AMD 4450
BS 5395: Part 2: 1984. Design of helical spiral stairs
    Amendment 1: AMD 6076
BS 5395: Part 3: 1985. Design of industrial stairs, permanent ladders and walkways
BS 5578: Part 1: 1978. Stairs – vocabulary
BS 5578: Part 2: 1995. Stairs – modular co-ordination
BS 5588: Part 6: 1991. Fire precautions – places of assembly
BS 6180: 1994. Code of practice for protective barriers in and about buildings
BS 6399: Part 1: 1996. Code of practice for dead and imposed loads

BRE defect action sheets

53   Stairs, safety of users. Design
54   Stairs, safety of users. Installation

BRE reports

Safety aspects of handrail design: a review (R. J. Feeney and G. M. B. Webber). (BR 260)
Alternating stair treads (G. Webber and R. J. Feeney). (BR 308)

Home Office publications

Guide to fire precautions in existing places of entertainment and like premises (1990).
Guide to safety at sports grounds (1997). (From the Stationery Office)

CHAPTER 20

Conservation of fuel and power

(Approved document L)

There have been insulation standards for many years. They were first introduced into the model building byelaws in 1952, although the requirements were very modest and have been improved in various amendments that have appeared from time to time.

These changes have been brought about by a re-think of the following factors in dwellings, and the need to conserve energy.

- Changes in living habits – houses unoccupied during the day as both husband and wife follow their own careers.
- Changes in methods of heating.
- Reduction of ventilation within the dwelling as double glazing and draughtproofing measures are installed and chimneys are no longer provided to each room.
- Condensation – possibly in consequence of one or all of the previous factors.

The minimum resistance to the transmission of heat which the various elements of walls, roofs and floors have to achieve are specified and this has been done by prescribing the maximum permitted thermal coefficient (U values).

There are unquestionable benefits of living in a thermally insulated building, although the requirements of the 1985 Regs were regarded as inadequate. Better standards lead to further saving of fuel, or alternatively a building will be warmer for the amount of fuel consumed.

Early regulations related to roof, walls and exposed floors, there being no concept of insulating the building as a whole; windows and rooflights were ignored.

This emphasizes the many facets which affect thermal insulation of a building. Walls and windows are composite, and each operation can affect the thermal qualities of the element. The thermal quality must vary greatly throughout each 24 hour period. A good example is the simple act of drawing curtains across windows; also the greater insulating value that thick lined curtains have compared with curtains of thin material.

Also within the building the temperature will vary from room to room and the influence of ventilation through the opening of doors cannot be gauged. Gaps around doors and windows are also an unknown factor as the space can vary so much from what might be regarded as a good fit to a big gap.

But perhaps the greatest influence on heat transfer is the direction and strength of the wind. As this varies, so different parts of the building may become warmer or cooler. Daytime radiation upon the external surface of a building increases the surface temperature and then come nightfall there is a cooling off. In consequence the external surface temperature becomes lower than the outside ambient temperature.

It is desirable always to use materials with the lowest possible thermal conductivity value. But there is another important factor, and that is the nature and occupancy of a building. For buildings which are fully occupied, such as flats or institutional buildings, one should use materials where the thermal storage quality is high. These are materials where the warm up and cool down period takes a long time.

For buildings occupied only part of the day – for instance offices which are used for about 8 or 9 hours a day – lightweight curtain walling of suitable insulation quality may be more appropriate. Such materials heat up quickly and there will be comparatively little heat required to warm up the structure the following day.

Regulations cannot cope with all the intricacies that these problems create and are therefore expected to give broad general approximations to achieve a better insulated building than in the past.

There were only three, but somewhat detailed regulations in Part L of the 1985 Regs, compared with the eight regulations and a lengthy schedule set out in the 1976 Regs, Parts F and FF.

However, the detailed requirements in the earlier regulations have been replaced by a new requirement in functional form that reasonable provision is to be made for the conservation of fuel and power in buildings. It applies to

- dwellings and
- other buildings which have a floor area in excess of $30\,m^2$.

This was brought about by The Building Regulations (Amendment) Regulations 1989 and the 1990 edition of approved document L. The 1991 Regs did not change Part L.

The 1994 Regs and the 1995 approved document have introduced a completely new look to Part L.

- Part L of Schedule 1 to the regulations is expanded to clarify the measures which must be taken and there is a requirement regarding lighting installations intended to serve more than $100\,m^2$ floor area in buildings other than dwellings.
- Part L now applies when a material change of use takes place.
- Energy ratings calculated by the standard assessment procedure (SAP) are taken into account in deciding upon the extent of insulation required.
- There are improved standards for fabric insulation, taking into account thermal bridges including mortar joints, timber joists and studs.
- Standard windows are now based on double glazing.
- The window area allowance includes windows, doors and rooflights.
- There are requirements around window and door openings and for reducing air leakage via windows and doors.
- Improved standards of thermal performance have been introduced in connection with hot water vessels and pipework.
- There are three methods of demonstrating that a dwelling fulfils the requirements:

  1. The elemental method has been kept but modified.
  2. Target U value method replaces calculation procedure.
  3. Energy rating method replaces the energy use approach.

- There are requirements to ensure that space temperatures and the provision of hot water have more flexible control.
- There are provisions where conservatories form an integral part of a new dwelling.
- Guidance is also given regarding appropriate provision when undertaking material alterations and change of use.

All this is expected to achieve a saving of 25 to 35% in energy requirements for space and water heating in new houses compared with dwellings built to recent standards, and similar improvements in other buildings to which the regulations relate. The actual saving may be a lot less.

L1. Conservation of fuel and power

- **Reasonable provision should be made for the conservation of fuel and power in buildings. It is to be achieved by limiting heat loss through the fabric of the building; controlling space heating and hot water systems; limiting heat loss from hot water vessels and hot water service pipes, and limiting heat loss from hot water pipes, and hot air ducts when used for space heating.**

   This requirement applies to:

   (a) dwellings; and
   (b) other buildings having a floor area greater than 30 m².

- **Artificial lighting systems are to be designed and constructed to use no more fuel than is reasonably necessary, including control systems.**

   This requirement only applies within buildings where more than 100 m² of floor area is to be illuminated. It does not apply within dwellings.

Other changes to the regulations are:

- L1 (Conservation of fuel and power) must be applied in cases of a material change of use (see Chapter 2).
- There is also a regulation, 14A, which provides for the calculation of an energy rating when a new dwelling is created by building work, or by a material change of use in connection with which building work is carried out.

MATERIALS

Typical insulation materials used in buildings are

- cellular glass
- expanded polystyrene insulation board
- fibre building board
- insulating board
- impregnated insulating board
- mineral fibre quilt or loose fill
- mineral fibre slab or rigid mat
- perlite granules
- phenolic foam
- polyurethane core to laminated board
- urea–formaldehyde foam cavity fill
- wood wool slab

PERFORMANCE

Many factors have a bearing upon the performance of a selected material for insulation.

One thing is certain; the improved standards of insulation are virtually untried in our climate and in consequence the utmost care is required in both specification and placing.

A number of factors have a bearing upon the performance of insulating materials:

- Combustibility. This can be an important aspect when considering plastics insulation and care is needed when plenty of combustion air is available. When plastics granules or partial fill boards are placed in cavity walls the top of the cavity should be sealed.
- Compatibility. Problems may arise when incompatible plastics are placed in contact. An example is the effect polystyrene can have on the insulation sheath of electric cables, and when foam plastics are used in conjunction with other plastics components. Any possible lack of compatibility should be investigated.
- Melting point. Glass fibre melts at 580°C and rock fibre above

1000°C whereas some plastics melt at 180°C. Even overheated flues, or contact with hot asphalt may cause damage to insulation having a low melting point.

- Other properties. Insulation materials may give off mildly poisonous gases, or have an objectionable smell. Very light insulation in exposed roofspaces may even move or form drifts like snow. There may be other useful properties, such as sound absorption.
- Size and shape. Materials are available in many sizes, shapes, thicknesses and rigidity. If the appropriate use is not set out on packaging, a check should be made with the manufacturer.
- Resistivity. Resistivity links conductivity to thickness of a material and is expressed in m² K/W.
- Thermal conductivity. Performance characteristics of insulation material is usually in the range 0.020–0.050 W/m K.

Table L1 Thermal conductivity of some building materials

Material or insulation	k value
Aerated concrete	0.16
Brickwork	1.45–0.72
Expanded polystyrene board	0.033–0.034
Granite	2.72
Glass fibre cavity batts	0.034
Plasterboard	0.16
Perlite polyurethane board	0.050–0.023
Polyisocyanurate board	0.020
Timber	0.14
Urethane spray coating	0.022
Wood wool	0.15

- Thermal stability. Some plastics have a high coefficient of thermal expansion, in which case they should not be used in a base for waterproofing.
- Vapour resistance. Vapour resistance of a material is important in the calculations to establish vapour diffusion when assessing the risk of interstitial condensation.
- Water absorbance. If a material is likely to become saturated, one that absorbs moisture is to be avoided as it may become ineffective. Closed cell plastics foam cannot become saturated and there are water resistant mineral fibre products. Water repellant additives are also available.
- Weather resistance. Insulation materials are not often exposed to the weather permanently, but facings to provide temporary weather resistance during construction may be necessary. UV light may damage plastics foams, so when exposed protection should be provided.

Manufactured sizes

Calculations may indicate a thickness of insulating material to be satisfactory in a given position, but the material may not be manufactured in that dimension. In those instances the next size thickness greater than the calculated dimension should be chosen.

Approved document

The approved document sets out an acceptable level of performance, stating that the requirements of L1 will be met by the provision of energy efficient measures which

- limit the heat loss and where appropriate maximize heat gains through the fabric of the building; more effective heating systems to be taken into account;
- control, as appropriate, the output of space heating and hot water systems;
- limit the heat loss from hot water storage vessels, pipes and ducts where heat does not contribute to space heating;
- use energy-efficient artificial lighting;
- use lighting systems which can be effectively controlled by manual or automatic switching.

SMALL EXTENSIONS TO DWELLINGS

The approved document makes it clear that if the construction of a small extension not exceeding 10 m² floor area is similar to the existing house, then reasonable provision is deemed to have been made.

LOW HEATING LEVELS

Because of the nature of the intended use of some commercial, industrial and storage buildings only low levels of heating, or even no heating at all, may be required, and in these circumstances it is not necessary to insulate the building fabric.

Low level of heating (Fig. L1) means

- for an industrial or storage building the output of the space heating system is not in excess of 50 watts per m² of the floor area, or
- for any other non-domestic building the output of the space heating system is not in excess of 25 watts per m² of the floor area.

There will be occasions when, because the use of a building is not known at the time of its construction, the levels of heating necessary cannot be established. However, when this occurs provision for the insulation and sealing of the building fabric must be undertaken.

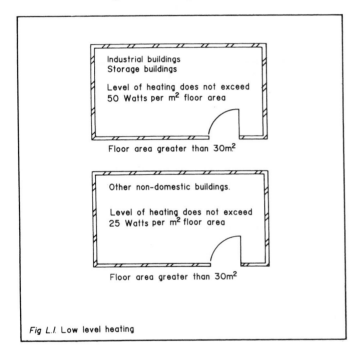

Fig L.I. Low level heating

RISKS

There are risks associated with improved thermal insulation. Aspects of construction which were not important in lightly insulated or uninsulated buildings may become significant when higher standards of insulation are installed. Some parts of the construction may remain colder and give rise to a greater risk of interstitial condensation. Also changes to traditional and well tried forms of construction aimed at improving insulation may give rise to damp penetration.

To avoid the risk of such building defects arising

- anticipate the effects
- comprehend the physical principles involved.

Traditional building often contains a considerable margin of safety inherent in the construction. Increased levels of insulation may erode this margin of safety, and in consequence attention must be given to

- correct design and specification
- good workmanship
- adequate supervision.

Summarized, the technical risks are

- surface condensation
- condensation in vent ducts that go through cold roof voids
- interstitial condensation in structures
- thermal stress
- fire risks
- overheating electric cables
- rain penetration.

Double and treble glazing can also introduce difficulties when escaping from fire. Overheating of electric cables muffed by insulation is a possibility and may dictate the down rating or re-routeing of cables likely to be affected. PVC cables should not be in contact with polystyrene insulation – polymer migration may be the consequence of the two being in contact.

The BRE publication 'Thermal insulation – avoiding risks' gives guidance on avoiding technical risks from the application of insulation. Also refer to Part F (Chapter 15) for guidance on the provision of ventilation to reduce the risks of condensation. Mechanical ventilation is now recommended in kitchens, bathrooms and shower rooms, together with background ventilation (in addition to openable windows) in habitable rooms. In addition the requirements for the ventilation of roofs apply to the roof as a whole – not simply to roof voids above an insulated ceiling – and to all building types.

Guidance is also available in the NHBC publication, *Thermal Insulation: Good Practice Guide*.

BRE Report: 'Thermal insulation – avoiding risks' (1994)

Although not an approved document this report from the BRE is to be regarded as inseparable from the implementation of Part L, and is essential reading.

Copiously illustrated, potential risks are explained, and advice on how to avoid each problem is given.

Walls

Rain penetration through cavity and solid walls
Masonry cavity walls
Masonry walls with internal insulation
Masonry walls with external insulation
Walls of timber-framed construction
Quality control checks for walls.

Roofs

Pitched roofs with ventilated roof space
Pitched roofs with sarking insulation
Pitched roofs of profiled sheet
Flat roofs with sandwich warm deck
Flat roofs with inverted warm deck
Flat roofs with cold deck
Quality control checks for roofs.

Floors

Concrete ground floors insulated below the structure
Concrete ground floors insulated above the structure
Concrete ground floors insulated at the edge
Suspended timber ground floors
Concrete and timber upper floors
Quality control checks for floors.

Windows

Window to wall junctions
Double-glazed windows
Quality control checks for windows.

Sizes of cables enclosed within thermal insulation
Insulation thickness for water supply pipes.

EXPOSED AND SEMI-EXPOSED ELEMENTS

There are two important definitions.

* Exposed element, which means an element exposed to the outside air. It is an element of construction which must be insulated to a recommended level.
* Semi-exposed element. This is an element separating a heated space from an unheated space which has one or more exposed elements that are not insulated to the recommended level.

(See Figs L2 to L7.)

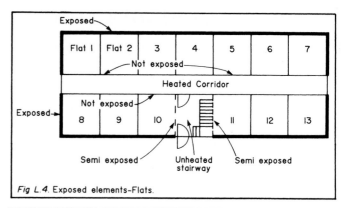

Fig L.4. Exposed elements-Flats.

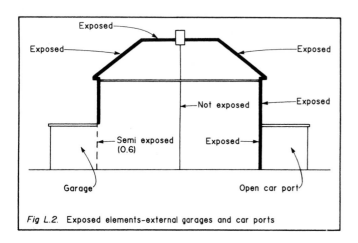

Fig L.2. Exposed elements-external garages and car ports

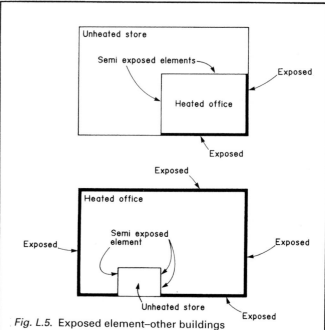

Fig. L.5. Exposed element–other buildings

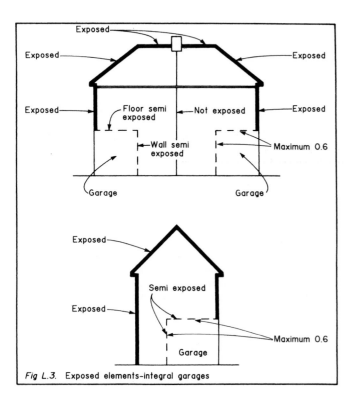

Fig L.3. Exposed elements-integral garages

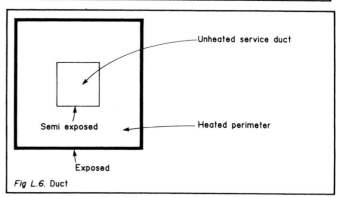

Fig L.6. Duct

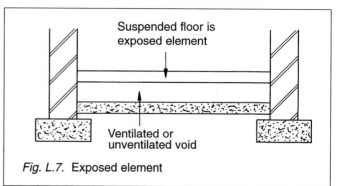

Fig. L.7. Exposed element

CALCULATION OF U VALUE

'U' value indicates thermal transmittance coefficient. That is the rate of heat transfer in watts through 1 m² of a structure when the temperature on each side differs by 1°C. This is expressed in W/m² K and can be calculated by using section A3 of the Chartered Institute of Building Service Engineers (CIBSE) Guide. The approved document says that calculations to two places of decimals are acceptable.

Thermal conductivity (the λ value)

Dimensions of insulation depend upon the thermal conductivity of materials used.

Heat passes through solid materials by means of conduction and the speed at which this occurs is dependent upon the thermal conductivity of the material. It is therefore a useful way of comparing materials.

Materials where the amount of solid matter is small and containing a large proportion of voids have the lowest thermal conductivity. Such materials include mineral wool, plastics foams and glass wool.

Moisture increases the thermal conductivity of a material and when materials containing cement are considered they are usually assumed to have a moisture content no greater than 3% by volume.

The thermal conductivity of a material is the quantity of heat which flows across a cube m² and 1 m thick, if the difference in temperature between the two faces is 1°C. Thermal conductivity of insulating materials is shown as W/m K.

Manufacturers or suppliers will give the thermal conductivity of their products but an authoritative publication giving the values of many materials is the CIBSE Guide, section A3.

There are tables in the approved document for windows, doors and rooflights. Appendix A to Part L contains tables which give a simple way to calculate the amount of insulation to achieve a given U value, for a number of varied constructions. The values shown take account of typical thermal bridging where appropriate.

Calculation of U value

Appendix B of Part L gives procedures for calculating U values and the following are generally taken into account.

- Thermal bridging effects of timber joists, structural and other framing, normal mortar bedding and window frames.

However, thermal bridging may be disregarded when the difference in thermal resistance between bridging material is less than 0.1 m² K/W, which means that *normal* mortar joints do not have to be taken into account when making brickwork calculations.

CALCULATED AREAS

Areas of

- walls
- roofs
- floors

are between finished internal surfaces including projecting bays of a building (Fig. L8).

Floor areas include non-usable space such as ducts and stair wells.

Areas of roofs are to be calculated in the plane of the insulation (Fig. L9).

Large complex buildings

It is appropriate to consider the needs of each part of the building, and thereby arrive at the measures applicable to each part.

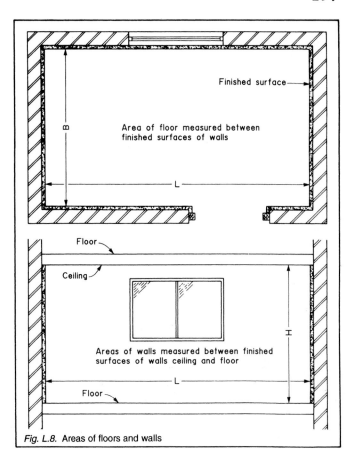

Fig. L.8. Areas of floors and walls

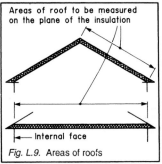

Fig. L.9. Areas of roofs

ENERGY RATING FOR DWELLINGS

The inclusion of the standard assessment procedure (SAP) into the approved document requirements went against the advice of the Building Regulations Advisory Committee and is based on the calculated energy cost for space and water heating expressed in a scale of 1 to 100 (100 being the best). Appendix G to the approved document gives reference data and an example of a calculation worksheet.

There is no obligation to obtain a particular SAP rating, but if a dwelling should have a rating of 60 or less, higher levels of insulation must be provided to meet certain U levels. A rating between 80 and 85 will demonstrate compliance with Part L, dependent upon the floor area of the dwelling.

SAP energy ratings should be determined at an early stage in design.

The rating obtained following SAP depends upon a number of factors.

- thermal insulation of the building fabric;
- efficiency and control of the heating systems;
- how the building is to be ventilated;
- solar gain;
- the price of fuel used for space and water heating.

The rating is not affected by characteristics such as

* the number and composition of residents;
* the ownership and efficiency of various domestic electrical appliances;
* individual heating patterns and temperatures.

Geographical location plays no part in SAP calculations. The inclusion of fuel costs was a contentious item, and the rates shown in the approved document are expected to be subject to amendment when fuel costs make this necessary.

Calculations

* SAP calculations should be submitted to the building control body.
* The competence of the person making calculations should be settled with the building control body prior to submission of the calculations.
* The building control body remains responsible for enforcement.
* Persons or bodies authorized by the Secretary of State (approved persons) as assessors for SAP calculations will be accepted as competent persons by the building control body.

DWELLINGS

There are three methods which may be used to show compliance with the requirements to limit heat loss through the fabric of dwellings. They are:

* an elemental method
* a target U value method
* an energy rating method.

ELEMENTAL METHOD

The elemental method is simpler than the other two and it requires elements to have a minimum level of insulation. Table 1 shows the U values for the elements depending upon whether the SAP energy rating is over 60 or below 60.

When the SAP rating is 60 or below, the U values shown in column (a) should be used, and column (b) applied when the SAP rating is in excess of 60. There is a possibility that the U values shown for windows, doors and rooflights may be modified, as will be seen later.

There are tables in Appendix A which, when used, will indicate an appropriate thickness of insulation to meet the required U values.

Windows, doors and rooflights

There is a basic allowance and the requirements will be met if the average U value for windows, doors and rooflights does not exceed

Table 1 Standard U values (W/m² K) for dwellings

Element	For SAP energy ratings of:	
	60 or less (a)	over 60 (b)
Roofs[1]	0.2	0.25[2]
Exposed walls	0.45	0.45
Exposed floors and ground floors	0.35	0.45
Semi-exposed walls and floors	0.6	0.6
Windows, doors and rooflights	3.0	3.3

Notes
1. Any part of a roof having a pitch of 70° of more may have the same U value as a wall.
2. For a flat roof or the sloping parts of a room-in-the-roof construction it will be acceptable if a U value of 0.35 W/m² K is achieved.

3.0 when SAP is 60 or less, and 3.3 when SAP is over 60. The area of the windows, doors and rooflights in these circumstances is not to be more than 22.5% of the total floor area.

To achieve this standard, windows may have sealed double-glazed units, or other means, such as secondary glazing using two or more panes of glass or other glazing materials such as acrylic, with a space between.

When an extension to an existing dwelling is to be undertaken the average U value of windows, doors and rooflights is not to exceed 3.3 W/m² K. To calculate the appropriate area for windows, doors and rooflights when designing extensions, the basic allowance may be applied to either

* the floor area of the extension; or
* the combined floor area of the existing dwelling and the extension.

Indicative U values for windows, doors and rooflights have been extended and provide for windows with wood, metal, thermal break or PVC-U frames, and these are shown in Table 2.

The design of doors can contain various panel arrangements but the values shown in Table 2 are generally acceptable. Single-glazed panels may be allowable in external doors providing the average U value for windows, doors and rooflights is not raised above the limit related to the area as given in Table 3.

Windows and doors with single-glazed panels which are protected by enclosed, unheated and draughtproof porches or conservatories are assumed to have a U value of 3.3 W/m² K.

When selecting and installing double-glazed windows care must be taken to ensure that condensation does not arise between the panes. Advice on this topic, and the many aspects of double-glazing installation, are given in BRE Report, 'Thermal insulation – avoiding risks' (1994).

The basic allowance for windows, doors and rooflights may be modified. The percentage area allowance is based on the average U values

Table 2 Indicative U values (W/m² K) for windows, doors and rooflights

Item	Type of frame							
	Wood		Metal		Thermal break		PVC-U	
Air gap in sealed unit (mm)	6	12	6	12	6	12	6	12
Window, double-glazed	3.3	3.0	4.2	3.8	3.6	3.3	3.3	3.0
Window, double-glazed, low-E	2.9	2.4	3.7	3.2	3.1	2.6	2.9	2.4
Window, double-glazed, Argon fill	3.1	2.9	4.0	3.7	3.4	3.2	3.1	2.9
Window, double-glazed, low-E, Argon fill	2.6	2.2	3.4	2.9	2.8	2.4	2.6	2.2
Window, triple-glazed	2.6	2.4	3.4	3.2	2.9	2.6	2.6	2.4
Door, half double-glazed	3.1	3.0	3.6	3.4	3.3	3.2	3.1	3.0
Door, fully double-glazed	3.3	3.0	4.2	3.8	3.6	3.3	3.3	3.0
Rooflights, double-glazed at less than 70° from horizontal	3.6	3.4	4.6	4.4	4.0	3.8	3.6	3.4
Windows and doors, single-glazed	4.7		5.8		5.3		4.7	
Door, solid timber panel or similar	3.0		—		—		—	
Door, half single-glazed, half timber or similar	3.7		—		—		—	

Table 3 Permitted variation in the area of windows and doors for dwellings

Average U value (W/m² K)	Maximum permitted area of windows and doors as a percentage of floor area for SAP energy ratings of:	
	60 or less	over 60
2.0	37.0%	41.5%
2.1	35.0%	39.0%
2.2	33.0%	36.5%
2.3	31.0%	34.5%
2.4	29.5%	33.0%
2.5	28.0%	31.5%
2.6	26.5%	30.0%
2.7	25.5%	28.5%
2.8	24.5%	27.5%
2.9	23.5%	26.0%
3.0	**22.5%**	25.0%
3.1	21.5%	24.0%
3.2	21.0%	23.5%
3.3	20.0%	**22.5%**
3.4	19.5%	21.5%
3.5	19.0%	21.0%
3.6	18.0%	20.5%
3.7	17.5%	19.5%
3.8	17.0%	19.0%
3.9	16.5%	18.5%
4.0	16.0%	18.0%
4.1	15.5%	17.5%
4.2	15.5%	17.0%

Note: The data in this table is derived assuming a constant heat loss through the elevations amounting to the loss when the basic allowance for openings of 22.5% of floor area is provided and the standard U values given in Table 1 apply. It is also assumed for the purposes of this table there are no rooflights.

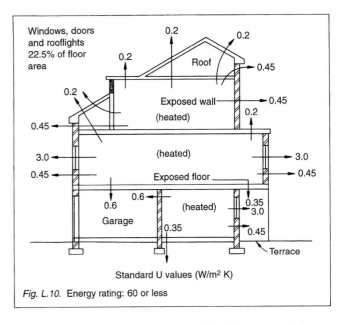

Fig. L.10. Energy rating: 60 or less

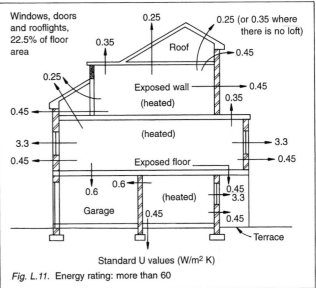

Fig. L.11. Energy rating: more than 60

given in Table 1, and the average will arise from the individual U values of the components proposed, related to their proportion of the total area of openings. This is demonstrated in Appendix E, which sets out to determine the permissible area of windows, doors and rooflights in the elemental method, and worked examples are given.

Whenever possible manufacturer's data should be used, but if properly certified manufacturer's figures are not available, the U value for components in Table 2 should be relied upon.

Areas of windows, doors and rooflights larger than those arising in the basic allowance may be used, but only if there is a compensating improvement in the average U value. It will be seen from Table 3 that there is a permitted variation in the area of windows and doors for dwellings, which is dependent upon the average U value of windows and doors in a dwelling.

Figures L10 and L11 summarize the fabric insulation standards and allowances for windows, doors and rooflights.

TARGET U VALUE METHOD

This, the second of the options available, allows more flexibility than the elemental method, and it also introduces optional allowances where high-efficiency heating systems are to be installed. Credit is also given where good passive solar designs are employed.

As with the elemental method, there are different criteria for dwellings with a SAP more, or less, than 60.

To satisfy this method, the calculated average U values are not to be more than the following targets.

• Where the SAP energy rating is 60 or less

$$\text{Target U value} = \frac{\text{Total floor area} \times 0.57}{\text{Total area of exposed elements}} + 0.36$$

• Where the SAP energy rating is more than 60

$$\text{Target U value} = \frac{\text{Total floor area} \times 0.64}{\text{Total area of exposed elements}} + 0.4$$

Total area of exposed elements will comprise the areas of walls and roof including windows, doors and rooflights which are exposed externally to the outside air. It also includes the area of the ground floors measured between finished internal surfaces of the exposed elements, including any projecting bays.

Where the SAP rating is less or more than 60, semi-exposed elements are to be insulated to achieve 0.6 W/m² K and omitted from the calculation of the average U value. Examples of the calculations are contained in Appendix F.

Solar gains

The target U value method relies upon a calculation which assumes an equal distribution of glazing on the north and south elevations of the dwelling. If the area of glazing on the south elevation is greater (±30°) than that facing north (±30°) then the benefit of solar gain may be taken into account. It can be taken as the actual window area less 40% of difference in area of windows facing north and south.

As an example of this option it is assumed that the total window area is $24\,m^2$ of which $14\,m^2$ faces south ($\pm30°$) and $9.5\,m^2$ faces north ($\pm30°$); the remaining $0.5\,m^2$ faces east. Therefore, the total area used in the calculations of U value can be reduced by 40% of $14 - 9.5 = 4.5 = 1.8\,m^2$.

High-efficiency heating systems

The method is based on a calculation assuming gas- or oil-fired hot water central heating with a seasonal efficiency of not less than 72%. More efficient heating systems permit a proportion of the benefits to be taken into account and the target U value may be increased by as much as 10%.

If, for example, a conventional gas boiler were to be replaced by a condensing boiler the seasonal efficiency would be increased to as much as 85% and in consequence the target U value could be increased by the maximum 10% permitted.

A high-efficiency electrical heating system involving a heat pump with a seasonal coefficient of performance of 2.5 with a mechanical ventilation system and heat recovery, could also lead to the target U value being increased by as much as 10%.

Any other heating system giving an improvement in seasonal efficiency between 72% and 85% would enable smaller increases in the target U value to be made.

ENERGY-RATING METHOD

The energy-rating method involves calculations using the Government's standard assessment procedure (SAP), which gives a scale of 1 to 100 and allows the use of any energy conservation measures.

It takes account of

- ventilation rates
- fabric losses
- water-heating requirements
- internal heat gains
- solar gains

and if the score is high enough, then this indicates compliance without further checks having to be made.

The SAP figure must relate to a dwelling, including each dwelling in a block of flats or a converted dwelling.

The appropriate ratings are 80 to 85, and the rating required is dependent upon the floor area of the dwelling, as shown in Table 4.

Table 4   SAP energy ratings to demonstrate compliance

Dwelling floor area (m²)	SAP energy rating
80 or less	80
more than 80 up to 90	81
more than 90 up to 100	82
more than 100 up to 110	83
more than 110 up to 120	84
more than 120	85

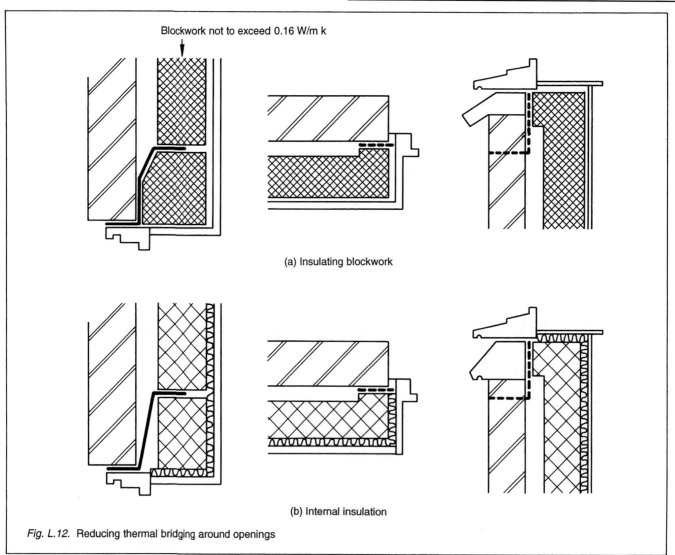

Blockwork not to exceed 0.16 W/m k

(a) Insulating blockwork

(b) Internal insulation

Fig. L.12. Reducing thermal bridging around openings

Calculation procedures

Whenever possible the U values of exposed walls and exposed floors are not to be more than $0.7\,W/m^2\,K$ and in respect of roofs, $0.35\,W/m^2\,K$.

THERMAL BRIDGING

Steps are to be taken to limit thermal bridging around windows, doors and other wall openings in order to avoid excessive heat losses, and also the possibility of condensation. To this end lintel, jamb and cill designs should be similar to those shown in Figs L12 and L13. If these designs are followed, heat losses because of thermal bridging can be ignored.

Referring to the figures, thermal conductivity of blockwork should not be more than $0.16\,W/m\,K$ (this would be autoclaved aerated concrete) and the frame should overlap the blockwork by not less than 30 mm for dry lining or 55 mm when lightweight plaster is used. At least 15 mm lightweight plaster should be applied to the internal faces of metal lintels, or alternatively they may be drylined.

It is also possible to demonstrate compliance by means of calculation, and Appendix D shows how this may be done around the edge of openings

INFILTRATION

Measures are to be taken to limit the infiltration of cold outside air into a dwelling through unintended air paths (Figs L14–L16), and as far as practicable the following precautions should be undertaken:

- Seal gaps between dry lining and masonry walls at the edges of openings and junctions with walls, floor and ceilings. This can be achieved by applying continuous bands of fixing plaster.
- Ensure that vapour control membranes in timber construction are adequately sealed.
- Fit draught stripping to the frames of windows, doors and rooflights.
- Insert a draught seal around access hatches to the loft space.
- Seal boxing for concealed services to the floor and ceiling.
- Seal piped services where they penetrate or project into voids or hollow constructions.

Further guidance is given in BRE Report 262, 'Thermal insulation – avoiding risks' (1994) and the NHBC 'Guide to thermal insulation and ventilation'.

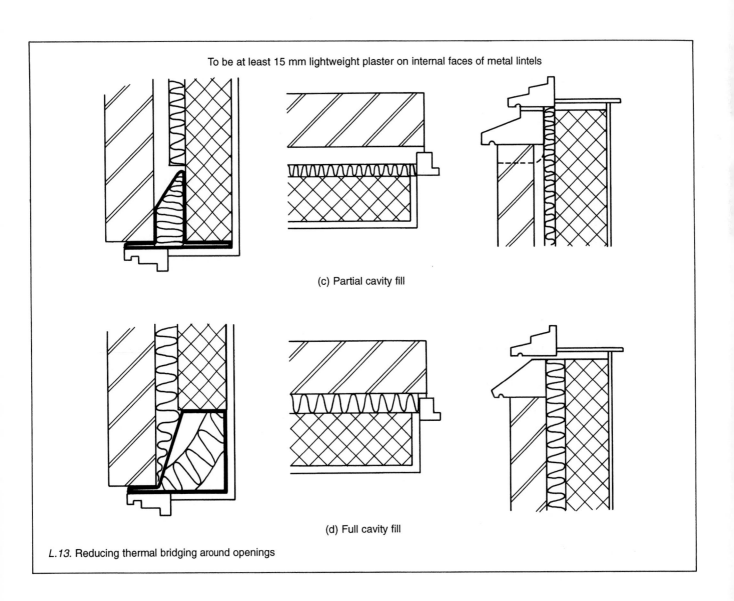

To be at least 15 mm lightweight plaster on internal faces of metal lintels

(c) Partial cavity fill

(d) Full cavity fill

L.13. Reducing thermal bridging around openings

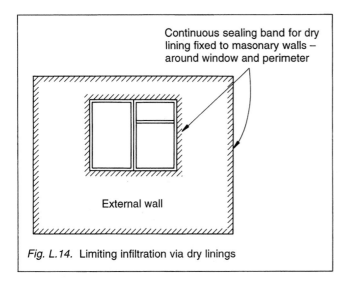

Fig. L.14. Limiting infiltration via dry linings

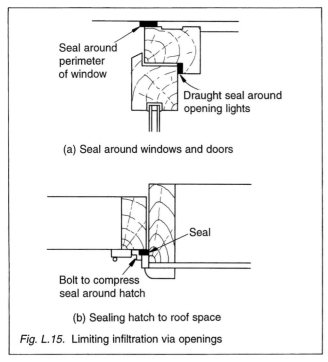

(a) Seal around windows and doors

(b) Sealing hatch to roof space

Fig. L.15. Limiting infiltration via openings

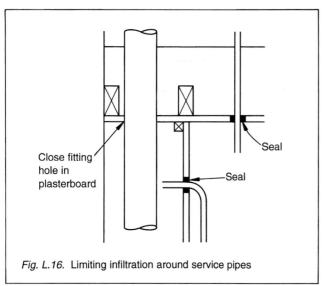

Fig. L.16. Limiting infiltration around service pipes

SPACE-HEATING CONTROLS

The requirements are that individual solid fuel, gas and electric fires or room heaters are to have integral controls embracing

- zone controls, and
- time controls, and
- boiler control interlocks.

Zone controls

These are appropriate for hot water heating systems, fan-controlled electric storage heaters and electric panel heaters to control temperatures in rooms which require different temperatures, such as bedrooms and living rooms. Control may be by room thermostats and/or thermostatic radiator valves. Other suitable temperature-sensing devices may be used.

In most dwellings two zones may be appropriate, but in single-storey open plan flats or bedsits the heating system could rely upon a single zone control.

Ducted warm air systems and flap-controlled electric storage heaters are not suitable for zone control, but they should be fitted with thermostats.

Timing controls

Heating systems require a timing device to control the periods when the system is in operation. The provision should be made for

- gas- and oil-fired systems, and
- solid fuel boiler systems where forced draught fans operate when heat is required.

Timing systems are not appropriate for use with solid fuel boilers operating only by natural draught.

Boiler control interlocks

Gas- and oil-fired water central-heating systems are to have controls to switch off the boiler when no heat is required and this can be by room thermostat or thermostatic radiator valves.

- When thermostats are used the system should only fire when a space heating or cylinder thermostat indicates heat is required.
- Systems utilizing thermostatic radiator valves are to have flow control or other devices to avert unnecessary boiler operation.

HOT WATER STORAGE CONTROLS

As with the heating controls, requirements stipulate efficient storage or hot water and controls on heating systems to limit the wastage of energy; and where the system is not one heated by a solid fuel boiler the following are necessary.

- The heat exchanger in the storage vessel is to have sufficient heating capacity for effective control. Hot water storage cylinders meeting the requirements of BS 1566: 1984 (1990), 'Copper indirect cylinders for domestic purposes', and BS 3198: 1981 (with amendments to June 1994), 'Specification for copper hot water combination units for domestic purposes'.

BS 1566: 1984 (1990) recommends sizes and grades, tolerances and materials, manufacture, connections, surface area and diameter of connections, optional features, testing and marking. Factory insulated cylinders are to have a heat loss not in excess of 1 W/litre when tested in accordance with the BS.

BS 3198: 1981 relates to combination hot water storage units for domestic purposes and includes materials, primary heaters, minimum areas of heating surface, manufacture, feed cistern, interconnection, valves, warning pipes and water level marking. Boss pipe connections are specified and performance regulations included. When factory insulated units are used the standing heat loss is not to be more than 1 W/litre of capacity. Also there is to be insulation between the hot water vessel and the cold feed cistern.

There are six appendices, one table and three figures.

- A thermostat which shuts off the supply of heat when the storage temperature is achieved. When used on a hot water central-heating system it is to be interconnected with the room thermostat to ensure that the boiler is turned off when no heat is required; and
- A timer which is installed as part of the central-heating system or as a local device enabling the supply of heat to be closed down during the periods when water heating is not required.

A system using a solid fuel boiler where the cylinder does not provide the slumber load, may satisfy the requirements by incorporating a thermostatically controlled valve.

Alternative approaches

The relevant recommendations, including zoning, timing and anti-cycling control requirements, will if adopted be satisfactory if these BS are followed.

1. BS 5449: 1990, 'Specification for forced circulation hot water central-heating systems for domestic premises'. The contents of this BS includes scope, definitions, design information, materials and components. Also included are design considerations including domestic hot water requirements, chimneys and flues, velocity and pressure loss in circuits, feed cisterns and expansion pipes. There are specifications relating to sealed systems, venting, heat emitters, insulation, water storage and system controls covering timing controls and temperature controls. There is advice regarding installation work on site, balancing and handing over.

    There are two appendices, 16 tables and four figures.
2. BS 5864: 1989, 'Specification for installation in domestic premises of gas fired ducted warm air heaters of rated output not exceeding 60 kW'. This specification contains scope, definitions and preliminary planning with materials and components used in installation. There is advice regarding the selection of appliances and sizing of the heater. Recommendations regarding installation relate to various types of buildings and there is also advice regarding air supply and ducting requirements. Thermostats are to be supplied with the appliance. Inspection and testing is detailed.

    There are four figures.
3. Other authoritative design specifications recognized by the heating fuel supply company may be acceptable.

INSULATION OF VESSELS, PIPES AND DUCTS

Hot water vessels

A 450 mm diameter × 900 mm high vessel of 120 litres capacity will satisfy the requirements if it has been given factory applied insulation which restricts standing heat losses to 1 W/litre or less.

Tests are to be in accordance with Appendix B4 of BS 1566: Part 1: 1984, 'Copper indirect cylinders for domestic purposes'.

Appendix B4 gives a method of tests for cylinders with side entry immersion heaters and the test procedure is intended to determine heat losses over successive 24 hour test periods with a water temperature of $65 \pm 2°C$ at the thermocouple position. Heat loss may be calculated in W/litre using the following formula:

$$\text{Heat loss} = \left(\frac{100E}{24C}\right)\left(\frac{(45)}{T_w - T_A}\right)$$

where　C　= cylinder capacity in litres
　　　　E　= energy in kW/h consumed in 24 hour period
　　　　T_w　= mean water temperature in °C over 24 hour period
　　　　T_A　= mean ambient temperature in °C over 24 hour period.

Other vessel sizes may satisfy the requirement by having the same insulation as a 120 litre vessel, or the equivalent. One way is to provide a 35 mm thick factory applied coating of PU foam, which has a minimum density of 30 kg/m and zero ozone depletion.

Additional insulation should be provided for unvented hot water systems, but it must not impede safe operation and the visibility of warning discharges (Chapter 16).

Pipes and ducts

In dwellings, thermal insulation is not necessary when the heat loss from a pipe or duct contributes to the heating of the building.

- Otherwise insulation is to have a thermal conductivity of not more than 0.045 W/m K and to be at least equal to the outside diameter of the pipe subject to a maximum thickness of 40 mm (see Fig. L17); or
- Insulation to the pipe may meet the requirements of BS 5422: 1990, 'The use of thermal insulating materials'.

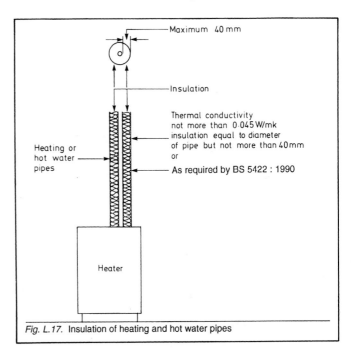

Fig. L.17. Insulation of heating and hot water pipes

This BS is comprehensive, and contains six sections.

1. Information regarding scope, definitions and details for correct assessments.
2. Refers to thermal conductivity, thickness, uniformity and bulk density.
3. Refers to thermal insulation of pipe ducts, tanks and vessels.
4. Is concerned with insulation of pipes, ducts, tanks and vessels.
5. Deals with thermal insulation for central heating, air conditioning, mechanical ventilation and hot water systems.
6. Outlines the use of thermal insulating materials on process pipework and equipment.

    There are tables of recommended thickness at three conductivity values: 0.040, 0.055 and 0.070.

The following are to be insulated for at least 1 m from the cylinder, or to a point where they become concealed:

- hot pipes
- vent pipes
- primary flow and return to heat exchanger.

Insulation is to be 15 mm of a material having a thermal conductivity of not more than 0.045 W/m K, or other material applied in a thickness giving the equivalent insulation.

Central heating and hot water pipework in unheated areas as

below the ground floor and in the roofspace may require an increased thickness of insulation as a precaution against frost damage.

Guidance is available in BRE report 'Thermal insulation – avoiding risks'.

CONSERVATORIES

There is a definition of conservatory in Part B, but for the purpose of insulation requirements, there is a separate definition.

A conservatory

* has not less than 75% of the roof area covered with translucent material; and
* has not less than 50% of the area of the external walls made of translucent material.

The requirements apply to a conservatory which is attached to and built as part of a new dwelling, and the actual requirements depend upon whether the conservatory is separated or not from the dwelling.

* If there is no separation between the conservatory and the dwelling, the conservatory is to be treated as an integral part of the dwelling.
* If there is separation between the conservatory and the dwelling, energy savings may be made if the conservatory is not heated. If fixed heating appliances are to be installed they are to have their own temperature and on/off controls.

Separation between a conservatory and a dwelling means

* separating walls and floors which are insulated to the same standard as semi-exposed walls and floors;
* separating windows and doors are to have the same U value and draught stripping as exposed windows and doors have elsewhere in the dwelling.

MATERIAL ALTERATIONS

A material alteration occurs if work, or any part of it, would adversely affect the existing building, if

* the building or a controlled service or fitting did not comply with a relevant requirement when previously it did so;
* the building or controlled service or fitting did not comply with a relevant requirement prior to the work, but as a result of the work it is more unsatisfactory.

A relevant requirement means any of the following:

Part A Structure
      B1 Means of escape
      B3 Internal fire spread – structure
      B4 External fire spread
      B5 Access for the fire service
      M   Access facilities for disabled people.

When contemplating alterations that are a material alteration it is necessary to consider insulation requirements and the extent of provision necessary will depend upon the circumstances of each case.

* Roof insulation. If the roof structure is substantially replaced it is necessary to provide insulation to achieve the U value for new dwellings.
* Floor insulation. If the structure of ground floors is to be substantially replaced, insulation is to be provided in heated rooms to the standard required for new dwellings.
* Wall insulation. If it is proposed to substantially replace complete external walls, a reasonable thickness of insulation is to be provided and measures to limit infiltration incorporated to the standard of new dwellings.
* Space heating and hot water systems. When carrying out work on these services, controls and insulation are to be provided as are required for new installations.

MATERIAL CHANGE OF USE

A material change of use occurs when a building is used as a dwelling when previously it was not used for that purpose; and also when a building contains flats when previously it was not used for that purpose.

A material change of use is most likely to involve insulation, although the extent of provision required will depend upon the circumstances of each case.

* Upgrading insulation in accessible lofts. Additional insulation is likely to be required to achieve a U value of not more than $0.35\,\text{W/m}^2\,\text{K}$ if the existing insulation gives a U value that is worse than $0.45\,\text{W/m}^2\,\text{K}$.
* Roof insulation. Where the roof structure is substantially replaced, insulation must be provided as for a new dwelling.
* Floor insulation. If the structure of ground floors is to be substantially replaced, insulation must be provided to the standard of new dwellings in heated rooms.
* Wall insulation. The substantial replacement of complete exposed walls will require a reasonable thickness of insulation and also means to limit infiltration to the standard of a new dwelling.
* Wall insulation. If substantial areas of internal wall surface is to be renovated, insulation of exposed and semi-exposed walls is to be undertaken. A reasonable thickness of insulating dry lining sealed with the means to limit infiltration is acceptable.
* Windows. When windows are replaced, the new windows are to be draught stripped and have an average U value not more than $3.3\,\text{W/m}^2\,\text{K}$.

  When dealing with listed buildings or buildings in conservation areas, and the existing window design needs to be retained, it is wise to discuss proposals with planning control before proceeding.
* Space heating and hot water systems. When carrying out work on these services, controls and insulation are to be provided as are required in new installations.

BUILDINGS OTHER THAN DWELLINGS

The requirement is that reasonable provision is to be made for the conservation of fuel and power by limiting heat loss through the fabric of the building; controlling space heating; limiting heat loss from water vessels and hot water services and limiting heat loss from hot water pipes and hot air ducts in buildings where the floor area is in excess of $30\,\text{m}^2$.

The approved document gives a requirement improving the U value of roofs, and also a new U value standard for commercial vehicle access and similar large doors.

Also for the first time, in buildings where more than $100\,\text{m}^2$ floor area is to be provided with artificial lighting, there is guidance on achieving lighting efficiency and control ability as an energy-saving measure.

As with dwellings, the approved document describes three ways which demonstrate how heat loss through the fabric of other buildings should be limited. They are

* an elemental method
* a calculation method
* an energy use method.

ELEMENTAL METHOD

Standard U values

The elemental method is simpler than the other two options and it requires elements to have a minimum level of insulation. The thermal performance of the elements shown in Table 5 must be achieved, and one way of doing this is by fixing insulation of a thickness which may be estimated by using the tables in Appendix A of

Table 5 Standard U values (W/m² K) for buildings other than dwellings

Element	U value
Roofs [1]	0.25 [2]
Exposed walls	0.45
Exposed floors and ground floors	0.45
Semi-exposed walls and floors	0.6
Windows, personnel doors and rooflights	3.3
Vehicle access and similar large doors	0.7

Notes
1. Any part of a roof having a pitch of 70° or more may have the same U value as a wall.
2. For a flat roof or insulated sloping roof with no loft space it will be acceptable if a U value of 0.35 W/m² K is achieved for residential buildings or 0.45 W/m² K for other buildings.

the approved document. Appendix C gives an alternative procedure for making satisfactory provision for floors.

Windows, doors and rooflights

There is a basic allowance and the requirement will be met when the average U value of windows, personnel doors and rooflights are not more than the basic allowance given in Table 5, and the areas of windows, personnel doors and rooflights is not more than the percentage given in Table 6.

Table 6 Basic allowance for windows, doors and rooflights for buildings other than dwellings

Building type	Windows and doors(1)	Rooflights
Residential buildings [2]	30%	
Places of assembly, offices and shops	40%	20% of roof area
Industrial and storage buildings	15%	
Vehicle access doors (all building types)	As required	

Notes
1. Percentage of exposed wall area.
2. Residential buildings (other than dwellings) means buildings in which people temporarily or permanently reside: for example, institutions, hotels and boarding houses.

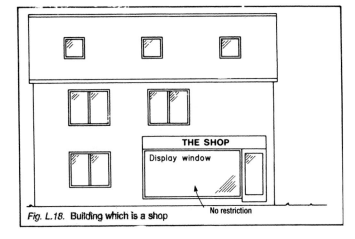

Fig. L.18. Building which is a shop

For the purpose of these calculations, display windows, shop entrance doors and similar glazing may be disregarded (see Fig. L18).

Windows and doors are shown as a percentage of exposed wall areas (see Fig. L19).

The basic allowance for windows, personnel doors and rooflights is not to exceed 3.3 W/m² K and this can be obtained by windows formed of double-glazed units or by other systems, including secondary glazing, incorporating two or more panes of glass or other glazing material, such as acrylic with a space between.

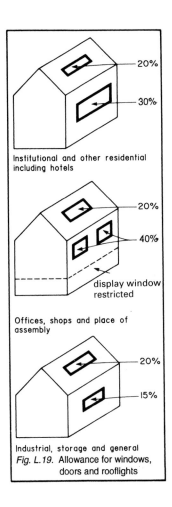

Fig. L.19. Allowance for windows, doors and rooflights

When an extension is to be made to an existing building the average U value of windows, personnel doors and rooflights is not to be greater than 3.3 W/m² K. To calculate the appropriate allowance given in Table 6, the basic allowance may be related to

- the wall area of the extension; or
- the combined wall area of the existing building and the extension.

Indicative U values for windows, doors and rooflights are given in Table 7 and they vary with the type of frame, whether it is wood, metal, thermal break or PVC-U.

The design of doors can contain various panel arrangements but the values shown in Table 7 are generally acceptable. Single-glazed panels may be allowable in external personnel doors if thereby the average U value for windows, personnel doors and rooflights is not raised above the limit related to the area given in Table 8.

Windows and personnel doors with single-glazed panels which are protected by enclosed, unheated and draught-proofed lobbies are assumed to have a U value of 3.3 W/m² K.

When selecting and installing double-glazed windows, care must be taken to ensure that condensation does not arise between the panes. Advice on this topic and the many aspects of double-glazing installation are given in BRE Report 262, 'Thermal insulation – avoiding risks'.

The basic allowance for windows, personnel doors and rooflights may be modified. The percentage area allowance is based on the average U values given in Table 6, and the average will arise from the individual U values of the components proposed, related to their proportion of the total area of openings. This is demonstrated in Appendix E which sets out to determine the permissible area of windows, doors and rooflights in the elemental method, and worked examples are given.

Whenever possible manufacturer's data should be used, but if

Table 7 Indicative U values (W/m² K) for windows, doors and rooflights

Item	Type of frame							
	Wood		Metal		Thermal break		PVC-U	
Air gap in sealed unit (mm)	6	12	6	12	6	12	6	12
Window, double-glazed	3.3	3.0	4.2	3.8	3.6	3.3	3.3	3.0
Window, double-glazed, low-E	2.9	2.4	3.7	3.2	3.1	2.6	2.9	2.4
Window, double-glazed, Argon fill	3.1	2.9	4.0	3.7	3.4	3.2	3.1	2.9
Window, double-glazed, low-E, Argon fill	2.6	2.2	3.4	2.9	2.8	2.4	2.6	2.2
Window, triple-glazed	2.6	2.4	3.4	3.2	2.9	2.6	2.6	2.4
Door, half double-glazed	3.1	3.0	3.6	3.4	3.3	3.2	3.1	3.0
Door, fully double-glazed	3.3	3.0	4.2	3.8	3.6	3.3	3.3	3.0
Rooflights, double-glazed at less than 70° from horizontal	3.6	3.4	4.6	4.4	4.0	3.8	3.6	3.4
Windows and doors, single-glazed	4.7		5.8		5.3		4.7	
Door, solid timber panel or similar	3.0		—		—		—	
Door, half single-glazed, half timber panel or similar	3.7		—		—		—	

Table 8 Permitted variation in the areas of windows, doors and rooflights for buildings other than dwellings

Average U value	Residential buildings percentage of wall area	Places of assembly, offices and shops percentage of wall area	Industrial and storage buildings percentage of wall value	Rooflights (all) percentage of roof area
2.0	55	74	28	37
2.1	52	69	26	35
2.2	49	65	24	33
2.3	46	62	23	31
2.4	44	58	22	29
2.5	42	56	21	28
2.6	40	53	20	27
2.7	38	51	19	25
2.8	36	49	18	24
2.9	35	47	17	23
3.0	34	45	17	22
3.1	32	43	16	22
3.2	31	41	16	21
3.3	**30**	**40**	**15**	**20**
3.4	29	39	14	19
3.5	28	37	14	19
3.6	27	36	14	18
3.7	26	35	13	18
3.8	26	34	13	17
3.9	25	33	12	17
4.0	24	32	12	16
4.1	23	31	12	16
4.2	23	30	11	15
4.3	22	30	11	15
4.4	22	29	11	14
4.5	21	28	11	14
4.6	21	27	10	14
4.7	20	27	10	13
4.8	20	26	10	13
4.9	19	26	10	13
5.0	19	25	9	13

Note
The data in this table is derived assuming a constant heat loss through the exposed wall or roof area as appropriate. The constant heat loss amounts to the loss through the wall or roof component plus the loss through the basic area allowance of windows, personnel doors or rooflights respectively as calculated using the U values in Table 5.

properly certified manufacturer's figures are not available the U value for components in Table 7 should be relied upon.

Areas of windows, personnel doors and rooflights larger than those arising in the basic allowance may be used, but only if there is a compensating improvement in the average U value. It will be seen from Table 8 that there is a permitted variation in the area of windows, doors and rooflights which is dependent upon the average U value of windows, doors and rooflights.

Figure L20 summarizes the fabric insulation standards and allowances for windows, personnel doors and rooflights.

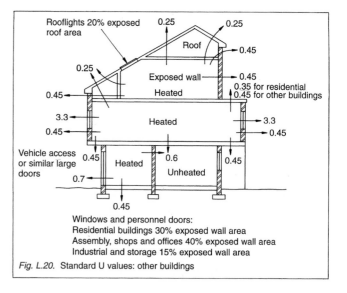

Fig. L.20. Standard U values: other buildings

CALCULATION METHOD

This method is regarded as being more flexible than the elemental method in selecting the areas of windows, personnel doors and rooflights; and also the levels of insulation required for the various elements.

It involves calculations to show that the rate of heat loss through the envelope of the proposed building is not more than from a notional building of the same size and shape that has been designed to meet the requirements of the elemental method.

It is possible to have larger areas of openings than are shown in Table 6, but if the proposed areas are smaller, then it is the smaller area which has to be assumed in the notional building.

Where the U value of the floor in the proposed building is better than 0.45 W/m² K and there is no added insulation, the better value is to be used in the notional building.

The calculation method is demonstrated in Appendix H.

ENERGY USE METHOD

This is also a calculation method and it permits freedom in the design of buildings where any valid energy conservation measure is used, and credit is given for useful solar and internal heat gains.

To meet the requirement the calculated annual energy use of the proposed building is to be less than that for a similar building designed in accordance with the procedure for the elemental method.

Buildings which are to be naturally ventilated should follow the recommendations in the CIBSE publication 'Building Energy Code 1981', Part 2(a) (worksheets 1a to 1e).

Using the calculation procedure

Using the calculation method or the energy use method it may be possible to obtain satisfactory solutions even when the U values of some elements are not as good as those set out in Table 5.

Nevertheless, the U values should generally be limited to ensure that

* in residential buildings the roof U value is not more than $0.45 \, \text{W/m}^2 \, \text{K}$; and
* in residential buildings exposed wall and floor values are not more than $0.7 \, \text{W/m}^2 \, \text{K}$; and
* in non-residential buildings the values of exposed roofs, walls and floors is not to be more than $0.7 \, \text{W/m}^2 \, \text{K}$.

THERMAL BRIDGING

Steps are to be taken to limit thermal bridging around windows, doors and other wall openings in order to avoid excessive heat losses, and also the possibility of condensation. To this end lintel, jamb and cill designs should be similar to those shown in Figs L12 and L13. If these designs are followed, heat losses because of thermal bridging can be ignored.

Referring to the figures, thermal conductivity of blockwork should not be more than $0.16 \, \text{W/m} \, \text{K}$ (this would be autoclaved aerated concrete) and the frame should overlap the blockwork by not less than 30 mm for dry lining or 55 mm when lightweight plaster is used. At least 15 mm lightweight plaster is to be applied to the internal faces of metal lintels, or alternatively they may be dry lined.

It is also possible to demonstrate compliance by means of calculation, and Appendix D shows how this may be done around the edge of openings.

INFILTRATION

Measures are to be taken to limit infiltration of cold outside air into a building through unintended air paths and as far as practicable the following precautions should be undertaken.

* Seal gaps between dry lining and masonry walls at the edges of openings and junctions with walls, floor and ceiling. This can be achieved by applying continuous bands of fixing plaster.
* Ensure that vapour control membranes in timber construction are adequately sealed.
* Fit draught stripping to the frames of windows, doors and rooflights.
* Insert a draught seal around floor and ceiling hatches.
* Seal boxing for concealed services to the floor and ceiling.
* Seal piped services where they penetrate or project into voids or hollow construction (see Figs L14, L15 and L16).

Further advice is given in BRE Report 265, 'Minimizing air infiltration in office buildings'.

SPACE-HEATING CONTROLS

The recommendations are relating to controls for space-heating systems and are not intended to apply to systems for commercial or industrial processes.

Requirements embrace

* temperature controls
* time controls
* boiler sequence controls.

Temperature controls

Room temperature controls are required to limit heat output of space-heating systems and the following provision should be made for each part of a space-heating system designed to be separately controlled:

* thermostats; or
* thermostatic radiator valves; and
* other equal forms of temperature sensing.

Where a space-heating system utilizes hot water, the temperature of the water flowing in the heating circuit is to be controlled by a temperature-sensing device (weather-compensating control) fixed to the outside of the building (see Fig. L21).

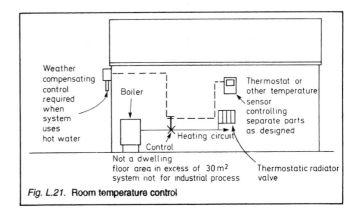

Fig. L.21. Room temperature control

Time controls

Not all buildings require continuous heating and when this is the case automatic controls are to be installed to ensure that temperature conditions within the building do not exceed those required during various periods (see Fig. L22). The following provision should be made.

* When the output rating of a system is 100 kW or less, it is only necessary to provide a clock-controlled automatic switch for shutting down and starting up the system as required. It may be set manually to give start and stop times.

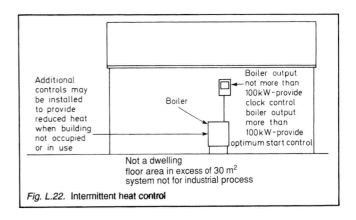

Fig. L.22. Intermittent heat control

• When the output rating of a system is more than 100 kW a more sophisticated control is required. It is necessary to install air optimum start control which will give start times for the heating system based on the rate at which the building will react when the heating is shut off and restarted.

This requirement does not prevent the fitting of controls to allow enough heating to prevent damage to the building, its services or contents. Such controls may be aimed at protection against frost, humidity or to prevent condensation by providing minimum heat when the building is not occupied or in use.

Boiler sequence controls

When a system is installed comprising interconnected boilers with an aggregate output exceeding 100 kW, automatic controls are to be fitted. Remembering that boilers operate most efficiently at or near full output, control is to be provided

• capable of detecting variations in the need for heat in the building, and
• by a method of sequence control, be capable of shutting down, starting and modulating boilers to ensure that only the minimum are operating to maintain normal temperature conditions and hot-water. (See Fig. L23.)

Design should be careful to ensure satisfactory hydraulics during sequence control, and ensure stable control.

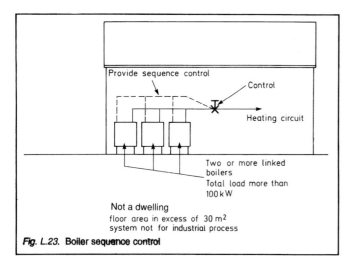

Fig. L.23. Boiler sequence control

HOT WATER STORAGE CONTROLS

Controls are necessary for hot water storage systems, and where a solid fuel boiler is not used, the requirement will be met if

• The heat exchanger in the storage vessel is to have sufficient heating capacity for effective control. Hot water storage cylinders meeting the requirements of BS 1566: 1984 (1990), 'Copper indirect cylinders for domestic purposes' and BS 3198: 1981 (with amendments to June 1994), 'Specification for copper hot water combination units for domestic purposes'.

Attention is to be given to the requirements in these standards for the surface areas and pipe diameters of heat exchangers.

• A thermostat which shuts off the supply of heat when the storage temperature is achieved. When used on a hot water central-heating system it is to be interconnected with the room thermostat to ensure that the boiler is turned off when no heat is required; and
• A timer which is installed as part of a central-heating system or as a local device enabling the supply of heat to be closed down during the periods when water heating is not required.

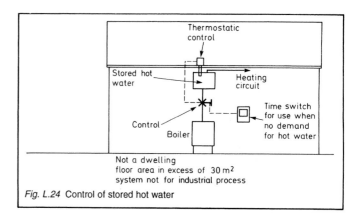

Fig. L.24 Control of stored hot water

A system using a solid fuel boiler where the cylinder does not provide the slumber load, may satisfy requirements by incorporating a thermostatically controlled valve (see Fig. L24).

Alternative approaches

Relevant recommendations in the following standards which achieve zoning, timing and boiler control performances equal to the approved document, may be adopted.

• BS 6880: 1988, 'Code of practice for low temperature hot water heating systems of output greater than 45 kW'. This standard is in three parts.
  Part 1 gives fundamental and design considerations. The principal contents are:
  Scope and definitions, fundamental considerations including comfort factors, application factors, structural and safety factors, energy conservation and energy management; quality and reliability.
  Design considerations include types of system, basis for design, energy conversion, subsystem designs, corrosion control, thermal insulation, automatic controls and general layout.
  There is one appendix, eight tables and 10 figures.
  Part 2 is concerned with the selection of equipment and covers scope and definitions. There is a section devoted to equipment groups and standards and another to utilization equipment. The section dealing with distribution equipment covers circulating pumps and pressurization equipment. Another section is devoted to energy conservation, and another to automatic controls including controller types and monitoring.
  There is one appendix, eight tables and 23 figures.
  Part 3 specifies the requirements for installation commissioning and maintenance. Installation covers safety factors, site facilities, storage, system installation, subsystems, energy conservation equipment and control equipment.
  There are recommendations for inspection and testing, commissioning, performance testing and a hand-over procedure. Another section relates to maintenance policy, safety considerations, personal records, and energy use.
  There is one appendix and two tables.
• CIBSE Applications Manual AM1: 1985, 'Automatic controls and the implementation for systems design'.

INSULATION OF VESSELS, PIPES AND DUCTS

The requirement is that provision is to be made to limit heat loss from hot water pipes and hot air ducts used for space heating in buildings having a floor area in excess of 30 m². It does not apply to vessels, piping and ducting systems which may form part of a commercial or industrial process.

Hot water vessels

A 450 mm diameter × 900 mm high vessel of 120 litres capacity will satisfy the requirements if it has been given factory applied insulation which restricts standing heat losses to 1 W/litre or less.

Other vessel sizes may satisfy the requirement by having the same insulation as a 120 litre vessel, or the equivalent. One way is to provide 35 mm thick factory applied coating of PU foam which has a minimum density of 30 kg/m and zero ozone depletion.

The above applies to vessels complying with BS 1566: 1984 (1990), 'Copper indirect cylinders for domestic purposes', or BS 3198: 1981 (with amendments to June 1991), 'Specification for copper combination hot water units'.

An equally satisfactory alternative treatment may be used.

When vessels are used which comply with BS 853: 1990, 'Specification for calorifiers and storage vessels for central heating with hot water supply', or the equivalent, the requirement will be achieved by installing 50 mm of material with a thermal conductivity of 0.045 W/m K. Other material providing the same measure of insulation may be used. In normal practice insulation will be protected by an outer metal casing.

Additional insulation should be provided for unvented hot water systems, but it must not impede the safe operation, and visibility of warning discharges (see Chapter 16).

Pipes and ducts

In other buildings, thermal insulation is not necessary for pipes and ducts when the heat loss contributes to the heating of a room or space on the building.

The requirement will be satisfied if

- insulation is to have a thermal conductivity of not more than 0.045 W/m K with a thickness at least equal to the outside diameter of the pipe, subject to a maximum thickness of 40 mm (see Fig. L17); or
- insulation for the pipe or warm air duct is in accordance with the requirements of BS 5422: 1990, 'Methods for specifying thermal insulating materials on pipes, ductwork and equipment'.

The following pipes are to have insulation:

1. Hot pipes connected to hot water storage vessels complying with BS 1566: 1984 or BS 3198: 1981, or an equivalent.
2. Vent pipes and primary flow and return to the heat exchanger where fitted.

Insulation is to extend for not less than one metre from their points of connection or to the point when they become concealed.

Insulation is to be 15 mm of a material having a thermal conductivity of not more than 0.045 W/m K, or other material applied in a thickness giving the same performance.

Hot pipes connected to hot water storage vessels meeting the requirements of BS 853: 1990, or the equivalent including the vent pipe and the primary and secondary flow and return pipes are also to be insulated as follows:

- insulation is to have a thermal conductivity of not more than 0.045 W/m K with a thickness of at least equal to the outside diameter of the pipe, subject to a maximum thickness of 40 mm; or
- insulation is in accordance with the requirements of BS 5422: 1990, 'Methods of specifying thermal insulation materials on pipes, ductwork and equipment'.

Central heating and hot water pipework in unheated areas, including spaces below ground floors and in the roofspace, may require an additional thickness of insulation as a precaution against frost damage.

Guidance is available in BRE Report 262, 'Thermal insulation – avoiding risks'.

LIGHTING

The requirement is that in buildings (excluding dwellings) where artificial lighting is to be provided for more than 100 m² floor area, the systems are to be designed and constructed to consume no more energy than is reasonable, and are to be provided with control systems.

There are a number of definitions in the approved document.

- Emergency escape lighting is that which provides illumination for the safety of people leaving an area, or attempting to stop a dangerous process before leaving an area.
- Display lighting is to highlight displays of exhibits or merchandise.
- Circuit watts is power consumed by lamps and associated controls and power factor correction equipment.
- Switch includes dimmer switches and switching includes dimming.

Display or emergency escape lighting is not affected by the requirements for efficacy and controllability of lighting.

The requirement for minimum efficacy of lamps will be met if at least 95% of the installed lighting capacity in circuit watts comprises lighting fittings utilizing lamps that are listed in Table 9.

Table 9 Types of high efficacy lamps

Light source	Types
High pressure Sodium Metal halide Induction lighting	All types and ratings
Tubular fluorescent	All 25 mm diameter (T8) lamps provided with low-loss or high frequency control gear
Compact fluorescent	All ratings above 11 W

Lighting controls

Lighting controls should aim to encourage the maximum use of daylight, and avoid unnecessary lighting during periods when spaces are not occupied. In addition, automatically switched lighting systems must not endanger the movement of persons within the building.

The provision of local switches should be in easily accessible locations within each working area or on the boundary between working areas, and on circulation routes.

Local switches may be:

- Switches requiring action by the occupant either manually or by remote control.
  Manual switches include rocker switches, press button and pull cords.
  Remote control switches may be infrared transmitters; sonic, ultra sonic or telephone handset controllers.
- Automatic switching systems, including controls which switch lighting off when they sense the absence of people.

Figure L25 demonstrates one way of satisfying the requirement and shows the distance on plan from any switch to the furthest fitting is not more than 8 m or 3 × height of fitting above the floor, if this distance is greater.

Alternative approaches

The requirement for the efficacy of lamps will be met if the installed lighting capacity has lighting fittings having lamps with an average initial (100 hour) efficacy of not less than 50 lumens per circuit watt.

Offices and storage buildings

The requirement will be met if the following are provided.

- Local switching for each working area, arranged to maximize the use of daylight.
- Controls such as time switches and photoelectric switches when they are appropriate.

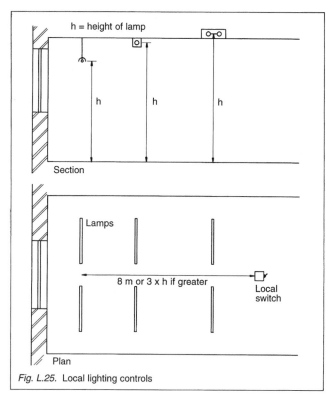

Fig. L.25. Local lighting controls

Other buildings

The requirement will be met if the following are provided and arranged to maximize the use of daylight:

• Local switching as described in previous paragraphs.
• Time switching which may be appropriate for large operational areas where occupation has clear working times.
• Photoelectric switching.

Guidance

If lighting controls are provided, designed in accordance with the recommendations in CIBSE publication, 'Code for interior lighting', the requirement will be satisfied.

Care must be taken to ensure that the alternative designs conserve the use of fuel and power as well as the measures given in the approved document.

MATERIAL ALTERATIONS

A material alteration occurs if work, or any part of it, would adversely affect the existing building, if

• the building or a controlled service or fitting did not comply with a relevant requirement when previously it did so;
• the building or controlled service or fitting did not comply with a relevant requirement prior to the work, but as a result of the work it is now more unsatisfactory.

A relevant requirement means any of the following:

Part   A   Structure
        B1   Means of escape
        B3   Internal fire spread – structure
        B4   External fire spread
        B5   Access for the fire service
        M   Access facilities for disabled people.

When contemplating alterations that are a material alteration it is necessary to consider insulation requirements and the extent of provision necessary will depend upon the circumstances of each case.

• Roof insulation. If the roof structure is substantially replaced it is necessary to provide insulation to achieve the U value for a new building.
• Floor insulation. If the structure of ground floors is to be substantially replaced, insulation is required to be provided in heated rooms to the standard required for new buildings.
• Wall insulation. If it is proposed to substantially replace complete external walls, a reasonable thickness of insulation is to be provided and measures to limit infiltration incorporated to the standard of new buildings.
• Space heating and hot water systems. When carrying out work on these services, controls and insulation are to be provided as are required for new installations.

MATERIAL CHANGE OF USE

A material change of use occurs when a building, after the change,

• is used as a hotel or boarding house, where previously it was not;
• is used as an institution, where previously it was not;
• is used as a public building, where previously it was not;
• the building is not an exempt building, where previously it was.

A material change of use is most likely to involve insulation, although the extent of provision required will depend upon the circumstances of each case.

• Upgrading insulation in accessible lofts. Additional insulation is likely to be required to achieve a U value of not more than $0.35 \, \text{W/m}^2 \, \text{K}$ if the existing insulation gives a U value that is worse than $0.45 \, \text{W/m}^2 \, \text{K}$.
• Roof insulation. Where the roof structure is substantially replaced, insulation must be provided as for a new building.
• Floor insulation. If the structure of ground floors is to be substantially replaced, insulation must be provided to the standard of new buildings, in heated rooms.
• Wall insulation. The substantial replacement of complete exposed walls will require a reasonable thickness of insulation and also means to limit infiltration to the standard of new buildings.
• Wall insulation. If substantial areas of internal wall surface is to be renovated, insulation of exposed and semi-exposed walls is to be undertaken. A reasonable thickness of insulating dry lining sealed with the means to limit infiltration is acceptable.
• Windows. When windows are replaced, the new windows are to be draught stripped and have an average U value of not more than $3.3 \, \text{W/m}^2 \, \text{K}$.

   Listed buildings or buildings in a conservation area may present a problem as planning requirements may not encourage windows of changed appearance, so the matter should be discussed with planning control.
• Space heating and hot water systems. When carrying out work on these services, controls and insulation are to be provided to the standard required in new installations.
• Lighting. Where lighting systems are to be substantially replaced, new lighting systems must be in accordance with the requirements outlined in this chapter.

APPENDIX A

Appendix A of the approved document to Part L contains 15 tables, and they relate to the calculation of the thickness of insulation required to obtain given U values.

These tables have been derived using the proportional area method and take into account the effects of thermal bridging in appropriate circumstances.

Intermediate values can be obtained from the tables by linear interpolation.

More accurate calculations may be made by utilizing the procedures that are set out in Appendices B and C.

Table A1 Base thickness of insulation between ceiling joists or rafters (Fig. L26)

| | Thermal conductivity of insulant (W/m K) | | | | | | |
	0.02	0.025	0.03	0.035	0.04	0.045	0.05
Design U value (W/m² K)	**Base thickness of insulating material (mm)**						
	A	B	C	D	E	F	G	H
1	0.20	167	209	251	293	335	376	418
2	0.25	114	142	170	199	227	256	284
3	0.30	86	107	129	150	171	193	214
4	0.35	69	86	103	120	137	154	172
5	0.40	57	71	86	100	114	128	143
6	0.45	49	61	73	85	97	110	122

Table A2 Base thickness of insulation between and over joists or rafters (Fig. L27)

| | Thermal conductivity of insulant (W/m K) | | | | | | |
	0.02	0.025	0.03	0.035	0.04	0.045	0.05
Design U value (W/m² K)	**Base thickness of insulating material (mm)**						
	A	B	C	D	E	F	G	H
1	0.20	126	145	166	187	209	232	254
2	0.25	106	120	136	152	169	187	204
3	0.30	86	104	116	129	143	157	171
4	0.35	69	86	102	112	124	135	147
5	0.40	57	71	86	100	109	119	129
6	0.45	49	61	73	85	97	107	115

In Tables A1 and A2 the proportion of timber has been taken at 8%, which corresponds to 48 mm wide timbers at 600 mm centres excluding any noggings. For constructions containing other proportions of timber, the U value may be calculated following the procedure in Appendix B.

ROOFS

The tables in this section enable calculations to be made to arrive at the amount of insulation required to achieve a given U value.

Example 1

In this first example it is necessary to find the thickness of insulation required to obtain a U value of 0.25 W/m² K when the insulation is placed between ceiling joists.

Fig. L.26. Insulation between ceiling joists and rafters

Choose an insulation with thermal conductivity value of 0.03 W/m K and refer to Table A1.

Column D, line 2, gives the base thickness of insulation   170 mm
There are several allowable deductions, dependent upon the form of construction used; for example the following are taken from Table A4.

19 mm roof tiles	(column D, line 13)	1 mm	
Roof space (pitched)	(column D, line 11)	5 mm	
10 mm plasterboard	(column D, line 7)	2 mm	8 mm
The base thickness less reductions is			162 mm

Therefore, 162 mm is the minimum thickness of the insulation layer between ceiling joists to obtain a pitched roof U value of 0.25 W/m² K.

Another example, using Fig. L26 and Table A1, uses insulation placed between rafters and seeks to obtain a U value of 0.35 W/m² K for the roof.

Choose an insulation with thermal conductivity value of 0.03 W/m K and refer to Table A1.

Column D, line 4, gives the base thickness of insulation   103 mm
Once again refer to Table A4 to determine the deductions in the base thickness that are allowed.

19 mm roof tiles	(column D, line 13)	1 mm	
10 mm plasterboard	(column D, line 7)	2 mm	3 mm
The base thickness less reductions is			100 mm

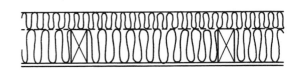

Fig. L.27. Insulation between and over joists and rafters

Example 2

In this example the thickness of insulation required to achieve a U value of 0.25 W/m² K for a pitched roof with insulation between, and over, ceiling joists is to be determined.

Insulation having a thermal conductivity of 0.035 W/m K is chosen (refer to Table A2).

Column E, line 2, gives a base thickness of   152 mm
This may be reduced by allowances in Table A4.

19 mm roof tiles	(column E, line 13)	1 mm	
Roofspace	(column E, line 11)	6 mm	
10 mm plasterboard	(column E, line 7)	2 mm	9 mm
The base thickness less reductions is			143 mm

This thickness of insulation over the joists is to be added to the 100 mm insulation between the joists in order to obtain a U value of 0.25 W/m² K; therefore base insulation thickness less the total reduction is   100 mm

   43 mm

Example 3

The thickness of an insulating layer required to give a U value of 0.45 W/m² K to an industrial roof (Fig. L29) with outer sheathing is required and it can be ascertained from Table A3 dealing with continuous insulation.

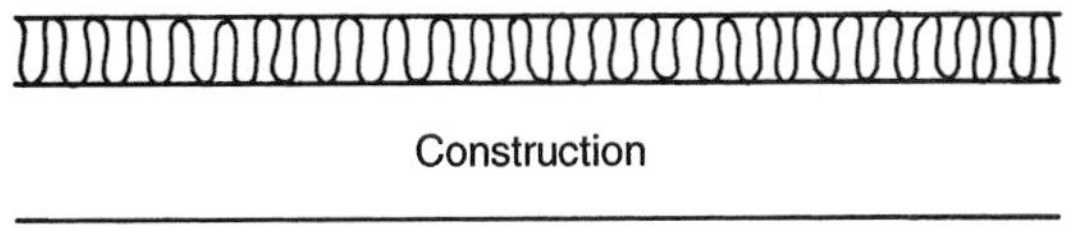

Fig. L.28. Continuous insulation

Using Table A3 and referring to column E, line 6, the base thickness insulation layer requirement is given as   73 mm

Construction of the roof may be taken into account to arrive at deductions, and from Table A4

Outer sheeting (no significant thermal resistance)	0
12 mm calcium silicate liner board (column E, line 10)	2 mm
The base thickness less reductions is	71 mm

Table A3 Base thickness for continuous insulation (Fig. L28)

		Thermal conductivity of insulant (W/m K)						
		0.02	0.025	0.03	0.035	0.04	0.045	0.05
Design U value (W/m² K)		Base thickness of insulating material (mm)						
		A	B	C	D	E	F	G	H
1	0.20	97	122	146	170	194	219	243
2	0.25	77	97	116	135	154	174	193
3	0.30	64	80	96	112	128	144	160
4	0.35	54	68	82	95	109	122	136
5	0.40	47	59	71	83	94	106	118
6	0.45	42	52	62	73	83	94	104

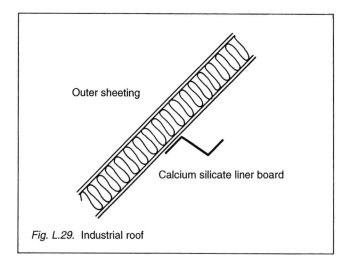

Fig. L.29. Industrial roof

Therefore, 71 mm is the minimum thickness of the insulation layer between the inner liner board and the outer sheeting to obtain a U value of 0.45 W/m² K.

Example 4

This refers to a concrete deck roof (Fig. L30) and requires the thickness of insulation to be ascertained in order to provide the roof with a U value of 0.45 W/m² K.

For this calculation Table A3 is used.

After choosing a thermal conductivity of 0.03 W/m K, the base thickness of the insulation is ascertained from Table A3, column D, line 6   62 mm

Allowances permitted by Table 4
3 layers felt   (column D, line 14)   1 mm
150 mm concrete   (column D, line 3)   14 mm   15 mm
(adjustment for thickness of concrete 1.5 × 9 = 14)

The base thickness less reductions is   47 mm

Therefore the minimum thickness of insulation necessary to achieve a U value for the concrete deck roof of 0.45 W/m² K is 47 mm.

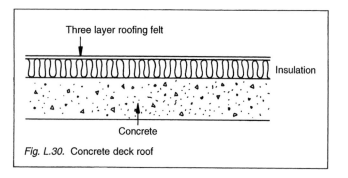

Fig. L.30. Concrete deck roof

Table A4 Allowable reductions in thickness for common roof components

		Thermal conductivity of insulant (W/m K)						
		0.02	0.025	0.03	0.035	0.04	0.045	0.05
Concrete slab density (kg/m)		Reduction in base thickness of insulating material (mm) for each 100 mm of concrete slab						
		A	B	C	D	E	F	G	H
1	600	11	13	16	18	21	24	26
2	800	9	11	13	15	17	20	22
3	1100	6	7	9	10	12	13	15
4	1300	5	6	7	8	9	10	11
5	1700	3	3	4	5	5	6	7
6	2100	2	2	2	3	3	4	4
Other materials and components		Reduction in base thickness of insulating material (mm)						
7	10 mm plasterboard	1	2	2	2	3	3	3
8	13 mm plasterboard	2	2	2	3	3	4	4
9	13 mm sarking board	2	2	3	3	4	4	5
10	12 mm Calcium Silicate liner board	1	2	2	2	3	3	4
11	Roofspace (pitched)	4	5	5	6	7	8	9
12	Roofspace (flat)	3	4	5	6	6	7	8
12	19 mm roof tiles	0	1	1	1	1	1	1
14	19 mm asphalt (or 3 layers of felt)	1	1	1	1	2	2	2
15	50 mm screed	2	3	4	4	5	5	6

WALLS

The four tables which follow enable calculations to be made to arrive at a required U value for walls. The tables give ready access to the base thickness layer, and provide for allowable reductions for common components, concrete components and timber framing in walls.

Example 5

The tables may be used to ascertain the thickness of insulation required to give a U value of 0.45 W/m² K for the following wall construction (see Fig. L31).
Outer leaf   102 mm thick
Cavity   50 mm
Inner leaf   150 mm concrete block density 600 kg/m³
Insulation   0.025 W/m K thermal conductivity
Lining   13 mm plasterboard.

Table A5 Base thickness of insulation layer

		Thermal conductivity of insulant (W/m K)						
		0.02	0.025	0.03	0.035	0.04	0.045	0.05
Design U value (W/m² K)		Base thickness of insulating material (mm)						
		A	B	C	D	E	F	G	H
1	0.30	63	79	95	110	126	142	158
2	0.35	54	67	80	94	107	120	134
3	0.40	46	58	70	81	93	104	116
4	0.45	41	51	61	71	82	92	102
5	0.60	30	37	45	52	59	67	74

Table A6 Allowable reductions in base thickness for common components

	Thermal conductivity of insulant (W/m K)						
	0.02	0.025	0.03	0.035	0.04	0.045	0.05
Component	Reduction in base thickness of insulating material (mm)						
A	B	C	D	E	F	G	H
1 Cavity (25 mm min.)	4	5	5	6	7	8	9
2 Outer leaf brick	2	3	4	4	5	6	6
3 13 mm plaster	1	1	1	1	1	1	1
4 13 mm lightweight plaster	2	2	2	3	3	4	4
5 10 mm plasterboard	1	2	2	2	3	3	3
6 13 mm plasterboard	2	2	2	3	3	4	4
7 Airspace behind plasterboard dry-lining	2	3	3	4	4	5	6
8 9 mm sheathing ply	1	2	2	2	3	3	3
9 20 mm cement render	1	1	1	1	2	2	2
10 13 mm tile hanging	0	0	0	1	1	1	1

Table A7 Allowable reduction in base thickness for concrete components

	Thermal conductivity of insulant (W/m K)						
	0.02	0.025	0.03	0.035	0.04	0.045	0.05
Design (kg/m)	Reduction in base thickness of insulation (mm) for each 100 mm of concrete						
A	B	C	D	E	F	G	H
Concrete inner leaf							
1 600	9	11	13	15	17	20	22
2 800	7	9	11	13	15	17	19
3 1000	6	8	9	11	12	14	15
4 1200	5	6	7	9	10	11	12
5 1400	4	5	6	7	8	9	9
6 1600	3	4	4	5	6	7	7
Concrete outer leaf or single leaf wall							
7 600	8	10	13	15	17	19	21
8 800	7	8	10	12	14	15	17
9 1000	6	7	8	10	11	12	14
10 1200	4	6	7	8	9	10	11
11 1400	3	4	5	6	7	8	9
12 1600	3	3	4	5	5	6	7
13 1800	2	3	3	4	4	5	5
14 2000	2	2	2	3	3	4	4
15 2400	1	1	2	2	2	2	3

Basic thickness of insulation may be acquired from Table A5, taking the figure from column C, line 4   51 mm

There are several allowable deductions which may be ascertained from Table A6.

Brick outer leaf	(column C, line 2)	3 mm
Cavity	(column C, line 1)	5 mm
Plasterboard	(column C, line 6)	2 mm

Deductions are also available in Table A7.
Concrete block (column C, line 1) and this is adjusted for 150 mm block thickness 1.5 × 11   17 mm   27 mm

Therefore the minimum thickness of insulation required to get a U value of 0.45 W/m² K is   24 mm

Example 6

This example also utilizes Tables A5, A6 and A7, and relates to a wall (Fig. L32) constructed as follows.

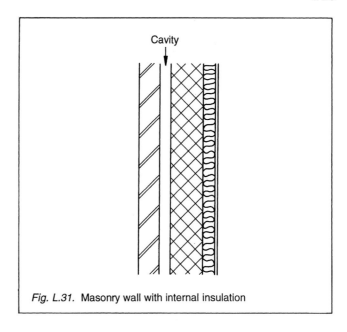

Fig. L.31. Masonry wall with internal insulation

Outer leaf	102 mm brick
Cavity	filled with insulation, thermal conductivity 0.035 W/m K
Inner leaf	100 mm concrete block density 1600 kg/m³
Lining	13 mm plasterboard on dabs

Base thickness of the insulation is obtained from
Table A1, column E, line 4   71 mm

This thickness may be reduced by taking account of the following reductions in Table A6.

Brick outer leaf	(column E, line 2)	4 mm
13 mm plasterboard	(column E, line 6)	3 mm
Airspace behind plasterboard	(column E, line 7)	4 mm

There is also a reduction available in Table A7
Concrete block   (column E, line 6)   5 min   16 mm   55 mm

In these circumstances the minimum thickness of insulation required to give a U value of 0.45 W/m² K is 55 mm.

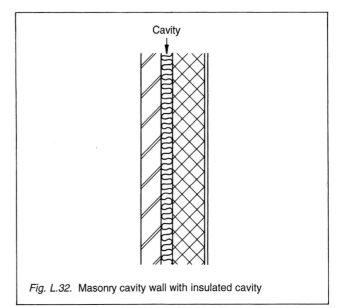

Fig. L.32. Masonry cavity wall with insulated cavity

Example 7

The thickness of the insulation necessary to achieve a U value of
0.45 W/m² K for the following wall construction (see Fig. L33) may
be obtainable from Tables A5, A6 and A7.

Outer leaf	102 mm brick
Cavity	50 mm
Insulation	Partial fill cavity batts having thermal conductivity 0.035 W/m K
Inner leaf	125 mm concrete block density 1000 kg/m³
Inner lining	13 mm plaster

The base thickness of insulation is
obtainable from Table A5 (column E, line 4)   71 mm

This figure may be reduced by giving credit
to the several construction components listed in
Table A6.

Brick outer leaf	(column E, line 2)	4 mm
Cavity	(column E, line 1)	6 mm
Plaster	(column E, line 3)	1 mm

Another allowance is available from Table A7.
Concrete block (column E, line 3) and this is
adjusted to 125 mm thickness (1.25 × 11)   14 mm   25 mm
                                                     ─────
                                                     46 mm

In these circumstances the minimum thickness of insulation to
achieve a U value of 0.45 W/m² K is 46 mm.

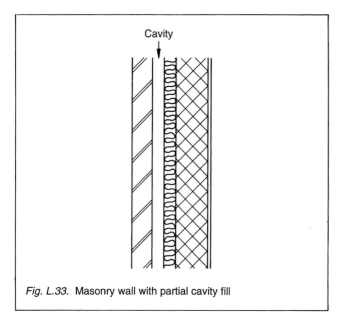

Fig. L.33. Masonry wall with partial cavity fill

Example 8

It is possible to discover the thickness of insulation required in a
semi-exposed wall (Fig. L34) taking information from Tables A5, A6
and A7. The wall has to achieve a U value of 0.6 W/m² K.
Details of construction are as follows.

Solid wall	215 mm aerated concrete blocks, density 600 kg/m³
Insulation	thermal conductivity 0.02 W/m K
Internal lining	13 mm plasterboard

Using Table A5 to obtain base thickness of
insulation (column B, line 5)   30 mm

This maybe reduced by taking account of Table A6.
   Plasterboard   (column B, line 6)   2 mm
and by referring to Table A7

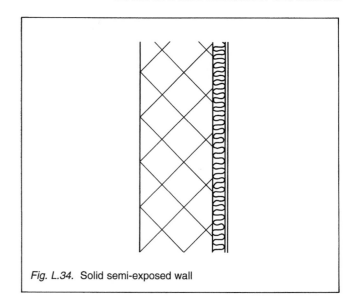

Fig. L.34. Solid semi-exposed wall

Concrete block   (column B, line 7)
adjusted to 215 mm thickness 2.15 × 8   17 mm   19 mm
The minimum thickness of insulation to
achieve 0.6 W/m² K is therefore   ─────
                                                   11 mm

Example 9

The thickness of insulation required to achieve a U value of
0.45 W/m² K in a timber-framed wall (Fig. L35) may be obtained by
using Tables A6, A7 and A8.

Construction of the wall is as follows.

Outer leaf	102 mm brick
Cavity	50 mm
Sheathing ply	9 mm
Inner leaf	90 mm timber frame filled with insulation of thermal conductivity 0.035 W/m K
Inner lining	13 mm plasterboard

Reference is made to Table A5 to obtain a
base thickness of insulation (column E, line 4)   71 mm

This base thickness may be reduced by making adjustments available
in Table A6.

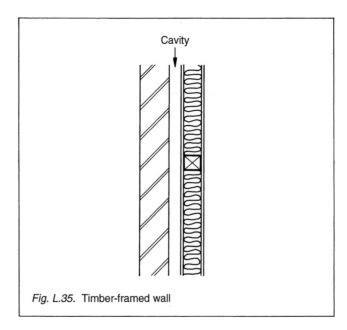

Fig. L.35. Timber-framed wall

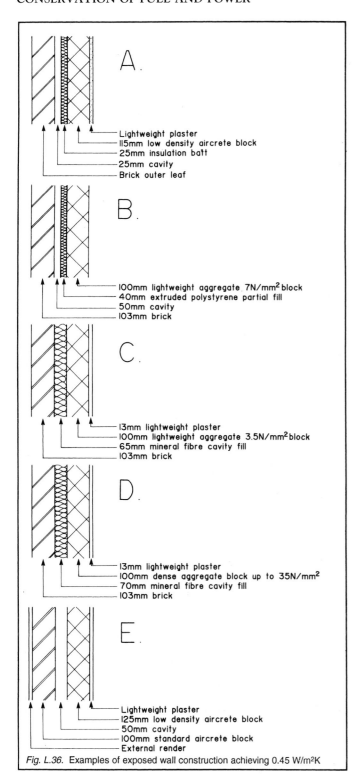

A.
— Lightweight plaster
— 115mm low density aircrete block
— 25mm insulation batt
— 25mm cavity
— Brick outer leaf

B.
— 100mm lightweight aggregate 7N/mm² block
— 40mm extruded polystyrene partial fill
— 50mm cavity
— 103mm brick

C.
— 13mm lightweight plaster
— 100mm lightweight aggregate 3.5N/mm² block
— 65mm mineral fibre cavity fill
— 103mm brick

D.
— 13mm lightweight plaster
— 100mm dense aggregate block up to 35N/mm²
— 70mm mineral fibre cavity fill
— 103mm brick

E.
— Lightweight plaster
— 125mm low density aircrete block
— 50mm cavity
— 100mm standard aircrete block
— External render

Fig. L.36. Examples of exposed wall construction achieving 0.45 W/m²K

Table A8 Allowable reduction in base thickness for insulated timber frame walls

	Thermal conductivity of insulant (W/m K)						
	0.02	0.025	0.03	0.035	0.04	0.045	0.05
Thermal conductivity of insulation within frame (W/m K)	Reduction in base thickness of insulating for each 100 mm of frame (mm)						
	A	B	C	D	E	F	G	H
1	0.035	42	53	63	74	84	95	105
2	0.040	38	48	58	67	77	87	96

Note: The table is derived for walls for which the proportion of timber is 12%, which corresponds to 48 mm wide studs at 400 mm centres. For other proportions of timber the U value can be calculated using the procedure in Appendix B.

Brick outer leaf	(column E, line 2)	4 mm
Cavity	(column E, line 1)	6 mm
Sheathing ply	(column E, line 8)	2 mm
Plasterboard	(column E, line 6)	3 mm

Also in Table A8 there is an allowance for timber frame which is adjusted for shallow members (0.9 × 74 mm)   67 mm   82 mm

The allowances have overcome the requirement for base thickness and therefore no additional insulation is required.

There are many alternatives available to the designer who may choose to use a cheaper insulant with a lower conductivity or he may wish to increase the standard of insulation by reducing the U value. These variations nay be explored by using the tables.

Further examples

Examples of wall achieving 0.45 W/m² K are shown in Fig. L36 and for a semi-exposed wall in Fig. L37.

FLOORS

The six tables in this section enable designers to determine the thickness of insulation required for solid floors in contact with the ground; suspended timber and concrete beam and block ground floors and also semi-exposed floors.

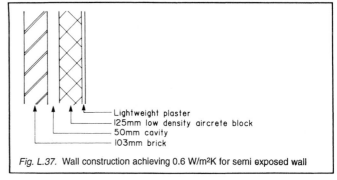

— Lightweight plaster
— 125mm low density aircrete block
— 50mm cavity
— 103mm brick

Fig. L.37. Wall construction achieving 0.6 W/m²K for semi exposed wall

Table A9 Insulation thickness for solid floors in contact with the ground (Fig. L39)

	P/A	0.02	0.025	0.03	0.035	0.04	0.045	0.05
	A	B	C	D	E	F	G	H

Insulation thickness (mm) for: U value of 0.25 W/m² K — Thermal conductivity of insulant (W/m K)

	P/A	0.02	0.025	0.03	0.035	0.04	0.045	0.05
1	1.00	62	77	93	108	124	139	155
2	0.90	61	76	91	107	122	137	152
3	0.80	60	75	90	105	119	134	149
4	0.70	58	73	87	102	116	131	145
5	0.60	56	70	84	98	111	125	139
6	0.50	52	66	79	92	105	118	131
7	0.40	47	59	71	83	95	107	119
8	0.30	39	49	59	69	79	88	98
9	0.20	24	30	36	42	48	54	60

U value of 0.35 W/m² K

	P/A	0.02	0.025	0.03	0.035	0.04	0.045	0.05
10	1.00	39	49	58	68	78	88	97
11	0.90	38	48	57	67	76	86	95
12	0.80	37	46	55	65	74	83	92
13	0.70	35	44	53	62	70	79	88
14	0.60	33	41	49	58	66	74	82
15	0.50	30	37	44	52	59	67	74
16	0.40	25	31	37	43	49	55	61
17	0.30	16	21	25	29	33	37	41
18	0.20	1	1	1	2	2	2	2

U value of 0.45 W/m² K

	P/A	0.02	0.025	0.03	0.035	0.04	0.045	0.05
19	1.00	26	33	39	46	53	59	66
20	0.90	25	32	38	44	51	57	63
21	0.80	24	30	36	42	48	54	60
22	0.70	22	28	34	39	45	51	56
23	0.60	20	25	30	35	40	45	50
24	0.50	17	21	25	30	34	38	42
25	0.40	12	15	18	21	24	27	30
26	0.30	4	5	6	6	7	8	9
27	<0.27	0	0	0	0	0	0	0

Table A10 Insulation thickness for suspended timber ground floors (Fig. L40)

Insulation thickness (mm) for: U value of 0.25 W/m² K — Thermal conductivity of insulant (W/m K)

	P/A	0.02	0.025	0.03	0.035	0.04	0.045	0.05
	A	B	C	D	E	F	G	H
1	1.00	95	110	126	140	155	170	184
2	0.90	93	109	124	138	153	167	181
3	0.80	91	106	121	136	150	164	178
4	0.70	88	103	118	132	145	159	173
5	0.60	85	99	113	126	139	153	166
6	0.50	79	92	106	118	131	143	156
7	0.40	71	83	95	107	118	129	140
8	0.30	57	68	78	88	97	106	116
9	0.20	33	39	46	52	58	64	69

U value of 0.35 W/m² K

	P/A	0.02	0.025	0.03	0.035	0.04	0.045	0.05
10	1.00	57	67	77	87	96	106	115
11	0.90	55	66	75	85	94	103	112
12	0.80	53	63	73	82	91	100	109
13	0.70	51	60	69	78	87	95	104
14	0.60	47	56	64	73	81	89	97
15	0.50	42	50	57	65	72	79	87
16	0.40	34	41	47	53	60	66	72
17	0.30	22	26	31	35	39	43	47
18	0.20	1	1	2	2	2	2	3

U value of 0.45 W/m² K

	P/A	0.02	0.025	0.03	0.035	0.04	0.045	0.05
19	1.00	37	44	51	57	64	70	77
20	0.90	35	42	49	55	62	68	74
21	0.80	33	40	46	53	59	65	70
22	0.70	31	37	43	49	54	60	65
23	0.60	27	33	38	43	49	54	58
24	0.50	22	27	32	36	40	44	49
25	0.40	15	18	22	25	28	31	34
26	0.30	4	5	6	7	8	9	10
27	<0.27	0	0	0	0	0	0	0

When calculating insulation thicknesses using Tables A9, A10 and A11 it is first necessary to arrive at a ratio P/A, where P is the floor perimeter length in metres and A is the floor area in square metres.

Referring to Fig. L38, which is the outline of a ground floor slab, to find the overall perimeter it is necessary to add the various dimensions

$$(7 + 6 + 6 + 1 + 4 + 4 + 4 + 2) = 34\,\text{m}.$$

The floor area of the slab is

$$(7 \times 6) + (4 \times 4) = 42 + 16 = 58\,\text{m}^2$$

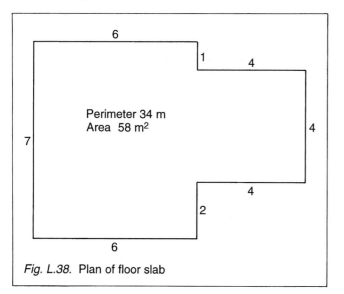

Fig. L.38. Plan of floor slab

Note: The table is derived for suspended timber floors for which the proportion of timber is 12%, which corresponds to 48 mm wide timbers at 400 mm centres.

For other proportions of timber the U value can be calculated using the procedure in Appendix B.

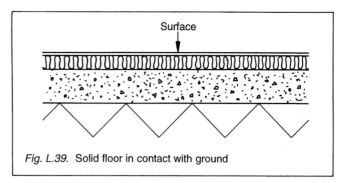

Fig. L.39. Solid floor in contact with ground

Therefore the P/A ratio is

Perimeter/Floor area = 34/58 = 0.58

The P/A figure is in column A of all three tables.

Example 10

By using the tables it is possible to determine the required thickness of an insulation layer, and if the required U value is 0.45 W/m² K for the floor slab shown in Fig. L38, using insulation with a thermal conductivity of 0.02 W/m K, the procedure is simple.

Take the P/A value of 0.58 and refer to Table A9, in which column B, line 23 indicates that 20 mm of insulation is necessary. The insulation may be placed above or below the slab.

Table A11 Insulation thickness for suspended concrete beam and block ground floors (Fig. L41)

		Insulation thickness (mm) for: U value of 0.25 W/m² K						
		Thermal conductivity of insulant (W/m K)						
	P/A	0.02	0.025	0.03	0.035	0.04	0.045	0.05
	A	B	C	D	E	F	G	H
1	1.00	60	75	90	104	119	134	149
2	0.90	59	73	88	103	118	132	147
3	0.80	58	72	86	101	115	130	144
4	0.70	56	70	84	98	112	126	140
5	0.60	54	67	80	94	107	121	134
6	0.50	50	63	75	88	101	113	126
7	0.40	45	57	68	79	91	102	113
8	0.30	37	46	56	65	74	84	93
9	0.20	22	27	33	38	43	49	54
		U value of 0.35 W/m² K						
10	1.00	37	46	55	64	74	83	92
11	0.90	36	45	54	63	72	81	90
12	0.80	35	43	52	61	69	78	87
13	0.70	33	41	50	58	66	74	83
14	0.60	31	38	46	54	61	69	77
15	0.50	27	34	41	48	55	62	69
16	0.40	22	28	34	39	45	50	56
17	0.30	14	18	21	25	29	32	36
18	0.20	0	0	0	0	0	0	0
		U value of 0.45 W/m² K						
19	1.00	24	30	36	42	48	54	60
20	0.90	23	29	35	41	46	52	58
21	0.80	22	28	33	39	44	50	55
22	0.70	20	25	31	36	41	46	51
23	0.60	18	23	27	32	36	41	45
24	0.50	15	18	22	26	29	33	37
25	0.40	10	12	15	17	19	22	24
26	0.30	2	2	2	3	3	4	4
27	<0.28	0	0	0	0	0	0	0

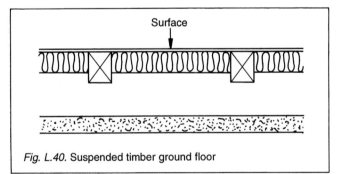

Fig. L.40. Suspended timber ground floor

Example 11

If the floor shown in Fig. L38 was a suspended timber floor, the P/A ratio would be the same and if it was desired to achieve a U value of 0.45 W/m² K using insulation with a conductivity of 0.040 W/m K. Table A10, column F, line 23, shows that 49 mm is required.

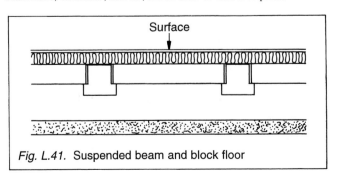

Fig. L.41. Suspended beam and block floor

Table A12 Exposed and semi-exposed upper floors of timber construction (Fig. L42)

		Thermal conductivity of insulant (W/m K)						
		0.02	0.025	0.03	0.035	0.04	0.045	0.05
Design U value (W/m² K)		Base thickness of insulation between joists to achieve design U values						
	A	B	C	D	E	F	G	H
Exposed floor								
1	0.35	61	76	92	107	122	146	162
2	0.45	42	53	63	74	84	95	106
Semi-exposed floor								
3	0.60	25	32	38	44	50	57	63

Note: Table A12 is derived for floors with the proportion of timber at 12% which corresponds to 48 mm wide timbers at 400 mm centres. For other proportions of timber the U value can be calculated using the procedure in Appendix E.

Example 12

The insulation thickness for a suspended concrete beam and block floor to achieve a U value of 0.45 W/m² K using insulation with a thermal conductivity of 0.35 W/m K by reference to Table A11, column E, line 23, gives a thickness of 32 mm.

There are no reduction allowances to be used in conjunction with Tables A9, A10 and A11.

EXPOSED AND SEMI-EXPOSED UPPER FLOORS

There are three small tables enabling rapid calculations to be made regarding the amount of insulation necessary for exposed and semi-exposed floors.

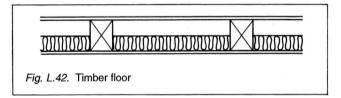

Fig. L.42. Timber floor

Example 13

It is proposed to install a timber semi-exposed floor using thermal material with a conductivity of 0.035 W/m K.

The base thickness from Table A12, column E, line 3		44 mm
There are allowable deductions which may be taken from Table A14.		
10 mm plasterboard (column E, line 1)	2 mm	
19 mm timber flooring (column E, line 2)	5 mm	7 mm
Therefore the thickness of insulation required to achieve 0.60 W/m² K is		31 mm

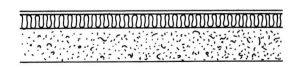

Fig. L.43. Concrete floor

Table A13 Exposed and semi-exposed upper floors of concrete construction (Fig. L43)

Design U value (W/m² K)		Thermal conductivity of insulant (W/m K)						
		0.02	0.025	0.03	0.035	0.04	0.045	0.05
		Base thickness of insulation to achieve design U values						
	A	B	C	D	E	F	G	H
Exposed floor								
1	0.35	52	65	78	91	104	117	130
2	0.45	39	49	59	69	79	89	98
Semi-exposed floor								
3	0.60	26	33	39	46	52	59	65

Example 14

It is proposed to install a concrete exposed floor using thermal material with a conductivity of 0.035 W/m K.

The base thickness from Table A13, column E, line 2	69 mm
There is one allowable deduction in Table A14 15 mm screed (column E, line 3)	4 mm
Therefore the thickness of insulation required to achieve 0.45 W/m² K is	65 mm

APPENDIX B

This appendix to the approved document Part L examines the proportional area method for determining the U value of structures containing repeating thermal bridges.

Full details of the calculation method are to be found in CIBSE Design Guide, section A3.

Two examples of this method of calculation are given in this appendix and they illustrate an application to frequently used designs.

The rules are as follows:

- If the element design does not include a continuous cavity then all the element layers have to be examined together.
- Where the design does include a continuous cavity the section is to be divided into two parts along the centre of the cavity, and these parts are then examined separately.
- 50% of the cavity resistance is to be allocated to each of the two parts.
- Thermal resistances are added together to get the total resistance of the element.

In making calculations, roof and ceiling joists should be restricted to the depth of insulation, ignoring the effects of any projection. Joists which are completely under insulation may also be ignored.

Table A14 Exposed and semi-exposed upper floors: allowable reductions in base thickness for common components

Component		Thermal conductivity of insulant (W/m K)						
		0.02	0.025	0.03	0.035	0.04	0.045	0.05
		Reduction in base thickness of insulating material (mm)						
	A	B	C	D	E	F	G	H
1	10 mm plasterboard	1	2	2	2	3	3	3
2	19 mm timber flooring	3	3	4	5	5	6	7
3	50 mm screed	2	3	4	4	5	5	6

Table A15 Thermal conductivity of some common building materials

Material	Density (kg/m³)	Thermal conductivity (W/m K)
Walls (external and internal)		
Brickwork (outer leaf)	1700	0.84
Brickwork (inner leaf)	1700	0.62
Cast concrete (dense)	2100	1.40
Cast concrete (lightweight)	1200	0.38
Concrete block (heavyweight)	2300	1.63
Concrete block (medium weight)	1400	0.51
Concrete block (lightweight)	600	0.19
Normal mortar	1750	0.80
Fibreboard	300	0.06
Plasterboard	950	0.16
Tile hanging	1900	0.84
Timber	650	0.14
Surface finishes		
External rendering	1300	0.50
Plaster (dense)	1300	0.60
Plaster (lightweight)	600	0.16
Calcium Silicate board	875	0.17
Roofs		
Aerated concrete slab	500	0.16
Asphalt	1700	0.50
Felt/bitumen layers	1700	0.50
Screed	1200	0.41
Stone chippings	1800	0.96
Tile	1900	0.84
Wood wool slab	500	0.10
Floors		
Cast concrete	2000	1.13
Metal tray	7800	50.00
Screed	1200	0.41
Timberflooring	650	0.14
Wood blocks	650	0.14
Insulation		
Expanded polystyrene (EPS) slab	25	0.035
Mineral wool quilt	12	0.040
Mineral wool slab	25	0.035
Phenolic foam board	30	0.020
Polyurethane board	30	0.025

Note: If available, certified test values should be used in preference to those in the table.

Example 1

It is necessary to ascertain the U value of a wall to the following specification.

External leaf	102 mm brick thermal conductivity 0.84 W/m K
Cavity	50 mm; thermal resistance 0.18 m² K/W
Sheathing	9 mm ply; thermal conductivity 0.14 W/m K
Timber studs	90 mm × 38 mm at 600 mm centres; thermal conductivity 0.14 W/m K
Insulation	thermal conductivity 0.04 W/m K between the studs
Plasterboard	13 mm; thermal conductivity 0.16 W/m K.

In examining the construction, thermal bridging of the insulation by timber studs is to be allowed for, and the wall must be considered as inner and outer leaves with the boundary down the centre of the cavity (Fig. L44).

Therefore, to ascertain the thermal resistance of the inner leaf it is first necessary to ascertain resistance through the section containing the timber studs, and then the section containing insulation, as follows.

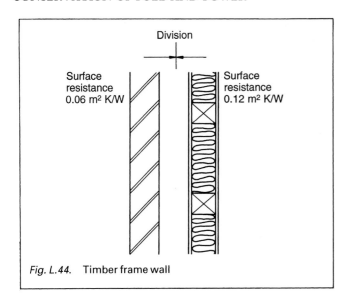

Fig. L.44. Timber frame wall

Resistance of inner leaf

Section containing timber stud:	m² K/W
Resistance of inside surface	0.12
Resistance of plasterboard (0.013 ÷ 0.16)	0.08
Resistance of timber stud (0.09 ÷ 0.14)	0.64
Resistance of sheathing ply (0.009 ÷ 0.14)	0.06
Half cavity resistance	0.09
Resistance of section through timber stud (R_t)	0.99

Section containing insulation: |
---|---
Resistance of inside surface | 0.12
Resistance of plasterboard (0.013 ÷ 0.16) | 0.08
Resistance of insulation (0.09 ÷ 0.04) | 2.25
Resistance of sheathing ply (0.009 ÷ 0.14) | 0.06
Half cavity resistance | 0.09
Resistance of section through insulation (R_{ins}) | 2.60

These resistances are combined:
Fractional area of timber stud:

$$F_t = \frac{\text{thickness of studs}}{\text{stud centres}} = \frac{38}{600} \; 0.063$$

Fractional area of insulation:
$F_{ins} = (1 - F_t) = (1 - 0.063) = 0.937$
The resistance of the inner leaf may then be obtained from:

$$R_{inner} = \frac{1}{\dfrac{F_t}{R_t} + \dfrac{F_{ins}}{R_{ins}}} = \frac{1}{\dfrac{0.063}{0.99} + \dfrac{0.937}{2.60}} = 2.36 \, \text{m}^2 \, \text{K/W}$$

Resistance of outer leaf

The difference between brick and mortar is less than 0.1 m² K/W and therefore the outer leaf is treated as an unbridged structure.

 | m² K/W
---|---
Half cavity resistance | 0.09
Resistance of brick (0.102 ÷ 0.84) | 0.12
External surface resistance | 0.06
Resistance of outer leaf (R_{outer}) | 0.27

Total resistance of wall

To obtain the total resistance of the wall add together the resistance of the inner and outer leaves.

$$R_{inner} + R_{outer} = 2.36 + 0.27 = 2.63$$

U value of the wall is given by:

$$U = \frac{1}{\text{Total resistance}}$$

Hence:

$$U = \frac{1}{2.63} = 0.38 \, \text{W/m}^2 \, \text{K}.$$

Example 2

There follows an example relating to a form of construction frequently used containing two thermally bridged layers.

1. blockwork by mortar joints, and
2. insulation by timber studs.

The wall is to be considered as inner and outer leaves with the boundary down the centre of the cavity (Fig. L45).

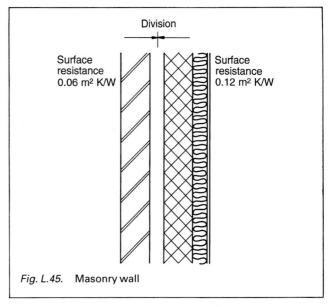

Fig. L.45. Masonry wall

Construction of the wall is as follows:

External leaf	102 mm, brick thermal conductivity 0.84 W/m K
Cavity	50 mm, thermal resistance 0.18 m² K/W
Inner leaf	125 mm × 440 mm × 215 mm block, thermal conductivity 0.11 W/m K; containing 10 mm wide mortar joints, thermal conductivity 0.8 W/m K
Insulation	30 mm, thermal conductivity 0.04 W/m K, between 50 mm × 30 mm timber battens at 600 mm centres, thermal conductivity 0.14 W/m K
Plasterboard	13 mm, thermal conductivity 0.16 W/m K

Resistance of inner leaf

This wall contains two bridged layers and the construction is such that it is difficult to confirm the precise location of the bridge in one layer with that of a bridge in the other layer.

There are four different combinations of paths through the blockwork and insulation and therefore an assumption has been made. As the pitch centres do not coincide it is assumed that heat flows through them in proportion to their relative areas.

The average resistance of non-bridged layers is as follows:

 | m² K/W
---|---
Resistance of half the cavity | 0.09
Resistance of plasterboard (0.013 ÷ 0.16) | 0.08
Resistance of inside surface | 0.12
Resistance of non-bridged layers (R_{nb}) | 0.29

In evaluating the resistance of bridged layers attention must be given to the four paths that are created:

1. block and insulation
2. block and timber
3. mortar and insulation
4. mortar and timber.

Resistances of the various materials are as follows:

Resistance of block (R_b = 0.125 ÷ 0.11)   1.14
Resistance of mortar (R_m = 0.125 ÷ 0.8)   0.16
Resistance of insulation (R_{ins} = 0.03 ÷ 0.04) 0.75
Resistance of timber (R_t = 0.03 ÷ 0.14)   0.21

The resistance of heat flow paths are arrived at as follows:

Resistance of block/insulation

$R_{b,ins} = R_b + R_{ins} + R_{nb}$
$= 1.14 + 0.75 + 0.29 = 2.18$

Resistance of block/timber

$R_{b,t} = R_b + R_t + R_{nb}$
$= 1.14 + 0.21 + 0.29 = 1.64$

Resistance of mortar/insulation

$R_{m,ins} = R_m + R_{ins} + R_{nb}$
$= 0.16 + 0.75 + 0.29 = 1.20$

Resistance of mortar/timber

$R_{m,t} = R_m + R_t + R_{nb}$
$= 0.16 + 0.21 + 0.29 = 0.66$

The following calculations show fractions of the face area of materials.

Blocks $F_b = \dfrac{440 \times 215}{450 \times 225} = 0.934$

(includes mortar joints)

Mortar $F_m = 1 - F_b$
$= 1 - 0.934 = 0.066$

Insulation $F_{ins} = \dfrac{550}{600} = 0.917$

(includes battens)

Timber $F_t = 1 - F_{ins}$
$= 1 - 0.75 = 0.083$

The following calculations show fractions of the face area of heat flow paths:

Block/insulation $F_{b,ins} = F_b \times F_{ins}$
$= 0.934 \times 0.917 = 0.856$

Block/timber $F_{b,t} = F_b \times F_t$
$= 0.934 \times 0.083 = 0.078$

Mortar insulation $F_{m,ins} = F_m \times F_{ins}$
$= 0.066 \times 0.917 = 0.061$

Mortar/timber $F_{m,t} = F_m \times F_t$
$= 0.066 \times 0.083 = 0.005$

The sum of resistances in parallel is shown by the following formula:

$$\frac{1}{R_{inner\ leaf}} = \frac{F_{b,ins}}{R_{b,ins}} + \frac{F_{b,t}}{R_{b,t}} + \frac{F_{m,ins}}{R_{m,ins}} + \frac{F_{m,t}}{R_{m,t}}$$

The figures in this case are:

$$\frac{1}{R_{inner\ leaf}} = \frac{0.856}{2.18} + \frac{0.078}{1.64} + \frac{0.061}{1.20} + \frac{0.005}{0.66} = 0.499$$

The resistance of the inner leaf is therefore:

$$R_{inner\ leaf} = \frac{1}{0.499} = 2.0\,\mathrm{m}^2\,\mathrm{K/W}$$

It is now necessary to calculate the resistance of the outer leaf, as follows.

	m² K/W
Resistance of external surface =	0.06
Resistance of the brick outer leaf (0.102 ÷ 0.84) =	0.12
Resistance of the half cavity =	0.09
Resistance of the outer leaf ($R_{outer\ leaf}$)	0.27

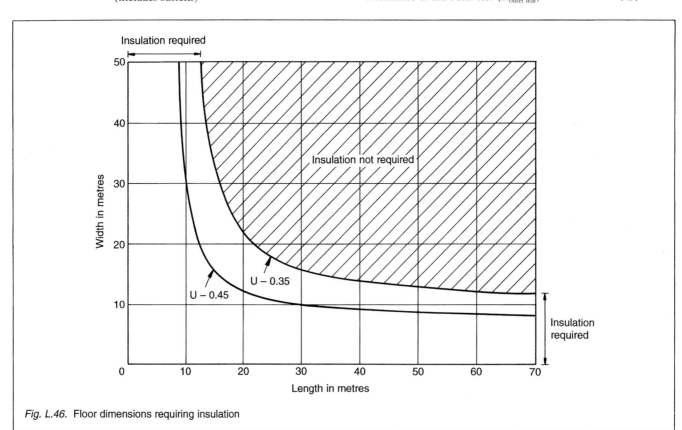

Fig. L.46. Floor dimensions requiring insulation

In these circumstances the total resistance of the wall is the sum of the inner and outer leaf resistances:

$$R_{total} = R_{inner\ leaf} + R_{outer\ leaf}$$
$$= 2.0 + 0.27 + 2.27\ m^2\ K/W$$

The U value of the wall is given by

$$U = \frac{1}{R_{total}} = \frac{1}{2.27} = 0.44\ W/m^2\ K$$

APPENDIX C

This appendix offers advice for meeting the requirement for U values of ground floors.

It is possible for a large ground floor to achieve a U value of 0.45 W/m² K or 0.35 W/m² K without additional insulation and Fig. L46 shows dimensions for a floor where insulation is not required.

Measurements are taken between the internal faces of external elements of the building, including any projecting bays.

When considering blocks of flats, semi-detached or terraced premises or similar, floor dimensions may be taken as for the premises themselves, or of the whole building.

Floor area when an extension is contemplated may be either the area of the extension or the complete building, including the extension.

U value of uninsulated floor

It is possible to arrive at the U value of an uninsulated floor from the ratio of its exposed perimeter to its area, using the following formula:

$$U_0 = 0.05 + 1.65\left(\frac{P}{A}\right) - 0.6\left(\frac{P}{A}\right)^2$$

where U_0 = U value of uninsulated floor
  P = length of exposed perimeter of floor in metres
  A = area of floor in square metres.

This equation applies to all types of uninsulated floors in contact with the ground, including

- concrete raft
- slab on ground
- suspended timber
- beam and block.

Garages, porches and similar unheated spaces attached to a building should be excluded when calculating P/A. However, the length of a wall adjoining the heated building and the unheated space is to be included when measuring the perimeter (see Fig. L47).

Table C1 indicates U values for uninsulated floors related to the P/A value, and the table has been formulated using the above equa-

tion. Reference to the table is sufficient to show compliance with the regulations, and linear interpolation may be used where necessary.

U value of insulated floor

The U value of an insulated floor may be obtained from

$$U_{ins} = \frac{1}{1/U_0 + R_{ins}}$$

where R_{ins} is the thermal resistance of the insulation. It should only include (a) the resistance of added insulation layers; and/or (b) any extra resistance of the structural deck over and above 0.2 m² K/W. U_{ins} includes the thermal insulation of the structural deck of suspended floors, and U_0 is a figure obtained from Table C1, or by calculation.

There are three useful BRE information papers:
IP 3/90   Floor U values
IP 7/93   Modification of floor U values by edge insulation
IP 14/94 Procedures for basements.

APPENDIX D

This appendix examines thermal bridging at the edges of openings and gives a procedure for ascertaining whether

- there is a high risk of condensation at the edges of openings; and/or
- there is a risk of significant heat losses at the edge of openings.

The minimum thermal resistance between inside and outside surfaces must be found. To do this the minimum thermal insulation paths are to be located and their thermal resistance calculated, having in mind the effect of thin layers such as metal lintels.

The minimum thermal resistance discovered must then be compared with satisfactory criteria and the need for any remedial action ascertained.

A minimum thermal resistance path via the junction of a window frame and reveal is shown in Fig. L48. This path which has the smallest thermal resistance is designated R_{min}, and it is from A to B on the figure.

Table C1   U values of uninsulated floors

Ratio P/A	U_o
0.1	0.21
0.2	0.36
0.3	0.49
0.4	0.61
0.5	0.73
0.6	0.82
0.7	0.91
0.8	0.99
0.9	1.05
1.0	1.10

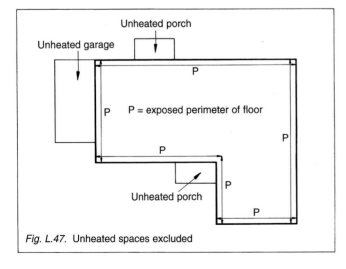

Fig. L.47. Unheated spaces excluded

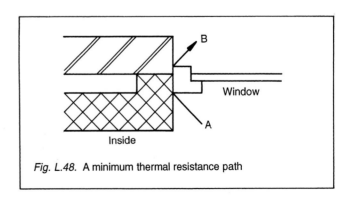

Fig. L.48. A minimum thermal resistance path

To arrive at R_{min} the total length from inside to outside is divided by the thermal conductivity of materials forming the joint.

Thin layers

Another calculation is necessary for thin layers such as metal lintels, not more than 4 mm thick. It is a modified calculation to establish minimum thermal resistance, wherein the effective thermal conductivity of the thin layer is taken as the largest of 0.1 W/m K or the thermal conductivity of materials that are on each side of it. This is R_{mod}.

Surface condensation

There is to be an assumption that the risk of surface condensation and mould growth at the edge of openings is negligible

*   when edges containing thin layers not more than 4 mm:
    R_{min} is not less than 0.10 m² K/W; and
    R_{mod} is not less than 0.45 m² K/W; or
*   in the case of other edge designs:
    R_{min} is not less than 0.20 m² K/W.

These figures are rounded to two decimal places, and the values do not apply when internal surface projections are used to avoid surface condensation

When an unacceptable risk is found, numerical calculation methods are to be used to analyse the problem and, where necessary, to improve the design.

Heat loss

Additional heat losses at the edge of openings may be disregarded when

*   edges containing thin layers are not more than 4 mm thick:
    R_{mod} is not to be less than 0.45 m² K/W; or
*   in the case of other edge designs:
    R_{min} is not to be less than 0.45 m² K/W.

In both instances the figures are rounded to two decimal places.

Compensating additional heat loss

Additional heat losses having been ascertained, and are such that they cannot be ignored, they can be taken into account in calculations.

*   in the case of dwellings the target U value method maybe adopted with the average U value increased using the following formula:

$$\frac{0.3 \times \text{Total length of relevant opening surrounds}}{\text{Total exposed surface area}} \ (W/m^2 \ K)$$

    The calculation procedure may be used for other buildings increasing the rate of heat loss by 0.3 × total length of relevant opening surrounds.
*   Compensating measures – for example, reducing the U value of the elements – should then be incorporated to ensure that
    (a) for dwellings the average U value is not in excess of the target value; or
    (b) for other buildings the rate of heat loss is not in excess of that for the notional buildings.

Example

A section of window jamb is shown in Fig. L49, where the inner leaf of blockwork is returned towards the outer leaf. The internal surface is plastered and it will be seen that a minimum resistance path exists between A, B and C.

Using the thermal conductivities shown in Fig. L49, R_{min} which is the sum of the resistance for each path segment, can be calculated.

Take the path segments A–B. The length 0.015 m, divided by conductivity 0.16 W/m K, gives resistance (R) of 0.094 m² K/W. Next, in the path segment B–C, the length 0.070 m is divided by conductivity

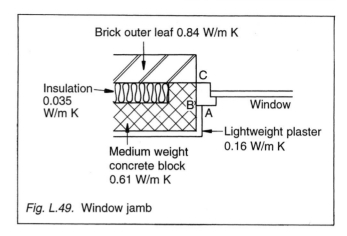

Fig. L.49. Window jamb

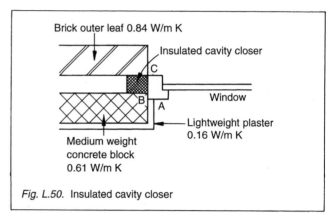

Fig. L.50. Insulated cavity closer

0.61 W/m K, giving a resistance (R) of 0.115 m² K/W. Minimum resistance, R_{min}, is therefore 0.94 + 0.115 = 0.209 m² K/W.

In this example the heat loss at the edge is not to be ignored as it is less than 0.45 m² K/W.

An attempt must be made to improve the situation, and this may be achieved by not returning the blockwork to the facing brickwork, but use an insulated cavity closer (see Fig. L50).

Take the path segment A–B in this revised construction. The length is 0.015 m, and this divided by conductivity 0.16 W/m K gives resistance (R) of 0.094 m² K/W. Next, the length of the path B–C is 0.070 m, and this is divided by 0.04 W/m K, the conductivity of the insulated cavity closer, gives a resistance (R) of 1.750 m² K/W, which when added to 0.094 m² K/W gives R_{min} = 1.844 m² K/W.

This being greater than 0.45 m² K/W, the additional heat loss can be ignored.

Alternative

The calculation procedure outlined in BRE IP 12/94, 'Assessing condensation risk and heat loss at thermal bridges around openings', maybe used as an alternative to the above.

APPENDIX E

This comprises a number of items being a remembrancer and guide through the elemental method applied to dwellings.

*   Check what the SAP rating is likely to be. Will it be more than 60, or less?

    Appendix G is devoted to estimation of a SAP rating including a worksheet. If the rating is likely to be 60 or less, the elemental method gives stringent U values for roofs, floors, windows, doors and rooflights, and they are to be found in Table 1.

*   What will the U value of the roof be?

It must be 0.25 if the SAP rating is more than 60 and 0.2 when the SAP is less than 60. Appendix A contains tables to assist in arriving at the correct thickness of insulation required, using certified manufacturer's data where possible.

- What will be the U value of exposed and semi-exposed walls?

    Exposed walls must achieve 0.45 and semi-exposed walls 0.6. Appendix A contains tables to assist in the calculation of correct insulation thickness.

- Is the ground floor U value satisfactory?

    When the SAP rating is more than 60 a U value of 0.45 is satisfactory, and where SAP is 60 or less 0.35 is necessary. Utilize Appendix C to calculate perimeter area and use the P/A tables in Appendix A to arrive at the thickness of insulation necessary.

- Are the U values of exposed or semi-exposed upper floors meeting requirements?

    The requirements are 0.6 for semi-exposed floors; 0.45 for exposed floors when the SAP rating is more than 60 and 0.35 for exposed floors when the SAP rating is 60 or less. The tables in Appendix A will assist in calculating the required thickness of insulation required.

- What is the average U value of windows, doors and rooflights?

    Refer to Table 2 for indicative U values for windows, doors and rooflights. There is an example which follows demonstrating how to calculate an average U value.

- The total area of windows, doors and rooflights is not to exceed the amount allowed.

    Ascertain the average U value and refer to Table 3 which gives the maximum permitted variation in the amount of windows and doors for dwellings. The permitted areas are given as a percentage of the floor area of the dwelling.

- If at any stage there has been an assumption that the SAP rating will be more than 60, this must now be confirmed.

    Complete, or get a 'competent' person to complete, the SAP worksheet which is to be found in Appendix G.

    There follows a method of calculating the permissible area of windows, doors and rooflights in the elemental method.

Example 1

It is assumed:

- Total floor area of dwelling, 80 m^2.
- It has double-glazed windows 15.4 m^2 with a U value of 2.9 W/m^2 K, giving a rate of heat loss per degree of 44.66 W/K.
- There are two half single-glazed doors having an area of 3.8 m^2, a U value of 3.7 W/m^2 K and a rate of heat loss per degree of 14.06 W/K.

This gives a total area of windows and doors of 19.2 m^2 and a total rate of heat loss per degree of 58.72 W/K.

To find the average U value for windows and doors, the total area of window and door openings is divided into the total rate of heat loss per degree. That is:

$$\frac{58.72}{19.2} = 3.1 \text{ W/m}^2 \text{ K}$$

Total area of openings as a percentage of floor area is:

$$\frac{19.2}{80} = 24\% \text{ of the floor area.}$$

Reference to Table 3 shows that for a U value of 3.1 W/m^2 K

(a) where the SAP rating is 60 or less, the maximum permitted area is 21.5%;
(b) where the SAP rating is over 60, the maximum permitted area is 24%.

The above example therefore only satisfies the requirements in a dwelling where the SAP rating is more than 60.

Included is an example of a single-storey assembly building which is to be erected having the following details.

Example 2

- Floor plan 16 m $\times$ 8 m = 128 m^2.
- Height to eaves = 4 m.
- Exposed wall area = 16 + 8 + 16 + 8 $\times$ 4 = 192 m^2.
- Roof open trussed and double pitched at 30° giving an area in the plane of insulation of

$$\frac{2 \times 4}{\cos 30°} \times 16 = 147.8 \text{ m}^2$$

- Windows area 60 m^2 with U value 3.6 W/m^2 K, giving rate of heat loss per degree of 216.0 W/K.
- Personnel door area 9.5 m^2 with U value of 3.0 W/m^2 K, giving rate of heat loss per degree of 28.5 W/K.
- Total for windows and personnel doors area is 60 + 9.5 = 69.5 m^2 and rate of heat loss per degree is 216.0 + 28.5 = 244.5 W/K.
- Rooflights total 25 m^2 with U value 3.8 W/m^2 K.

Using these figures the average U value for windows and doors arises from

$$\frac{\text{Total rate of heat loss per degree}}{\text{Total area of windows and personnel doors}}$$

$$= \frac{244.5}{69.5} = 3.5 \text{ W/m}^2 \text{ K.}$$

Area of windows and personnel doors as percentage of wall area:

$$\frac{69.5}{192} = 32.2\% \text{ of the gross wall area.}$$

Area of rooflights as percentage of roof area:

$$\frac{25}{147.8} = 16.9\% \text{ of roof area.}$$

Table 8 gives the permitted variation in the areas of windows, doors and rooflights for assembly buildings, and the permitted area for windows and doors with an average U value of 3.5 W/m^2 K is 37% of the wall area.

The permitted area of rooflights having a U value of 3.8 W/m^2 K is 17% of the roof area.

The proposed design meets these requirements on each count and the regulation is satisfied.

APPENDIX F

This contains a number of items, being a remembrancer and guide through the target U value method applied to dwellings and contains examples of the method applied to detached and semi-detached dwellings.

- Check what the SAP rating is likely to be. Will it be more than 60 or less?

    Appendix G is devoted to the estimation of a SAP rating including example dwellings included as a guide, and a worksheet. If the rating is likely to be 60 or less the target U value is lower.

- Surface area and U value are to be ascertained for the roof, exposed walls, ground and other exposed floors together with windows, doors and rooflights.

    The areas of semi-exposed elements are not included in the calculations. Find out the total exposed surface area of the dwelling as well as the total rate of heat loss per degree for the dwelling.

- Will any semi-exposed elements have a U value of 0.6 or less?

    The insulation standards for semi-exposed elements are contained in Table 1.

- Ascertain the average U value.

$$\frac{\text{Total rate of heat loss per degree}}{\text{Total external surface area}}$$

- Obtain the target U value.

Use the total floor area of the dwelling and total exposed surface area.

When SAP is 60 or less

$$\frac{\text{Total floor area} \times 0.57}{\text{Total area of exposed elements}} + 0.36$$

When SAP is more than 60

$$\frac{\text{Total floor area} \times 0.64}{\text{Total area of exposed elements}} + 0.4$$

- Compare the average U value with the target U value.

Where the average U value does not exceed the target U value the design meets requirements. If the average U value is more than the target U value, modifications are required; more efficient heating system, allowing for solar gains or altering design of fabric elements.

- An adjustment may be made to the average U value to be credited with solar gain.

Window area may be reduced if area of glazing facing south (±30°) is more than that facing north (±30°).

- An adjustment may be made to the target U value, giving credit for a more efficient heating system.

If a condensing boiler is used, the target U value may be improved by as much as 10%.

- If at any stage there has been an assumption that the SAP rating will be more than 60, this must now be confirmed.

Complete, or get a 'competent' person to complete, the SAP worksheet which is to be found in Appendix G.

Example 1

This illustrates the application of the target U value method for a detached dwelling of the dimensions shown in Fig. L51.

In this example the U values for walls and roof are not as good as would be required in the elemental method. The specification includes cavity walls, dry lined; metal-framed windows and doors with thermal breaks and sealed double glazing having 12 mm air space. The SAP rating is to be in excess of 60.

It is assumed that:

- Total/floor area of the dwelling is 56.2 m², with a U value of 0.45 W/m² K and a rate of heat loss per degree of 25.29 W/K (arrived at by multiplying area × U value).
- The double-glazed windows total 24.8 m² and have a U value of 3.3 W/m² K, giving a rate of heat loss per degree of 81.84 W/K.
- Area of doors is 3.8 m² with a U value of 3.3 W/m² K and the rate of heat loss per degree is 12.54 W/K.
- Wall area totals 121.4 m² and has a U value of 0.5 W/m² K; the rate of heat loss per degree being 60.7 W/K.
- Roof area totals 56.2 m² and has a U value of 0.3 W/m² K, which when multiplied together provides a rate of heat loss per degree of 16.86 W/K.
- Total exposed surface area is therefore 262.4 m² and the total rate of heat loss per degree is 197.23 W/K.

To calculate the target U value with a SAP rating in excess of 60 the following formula is used:

$$\frac{\text{Total floor area} \times 0.64}{\text{Total exposed surface area}} + 0.4$$

$$= \frac{112.4 \times 0.64}{262.4} = 0.67 \, \text{W/m}^2 \, \text{K}$$

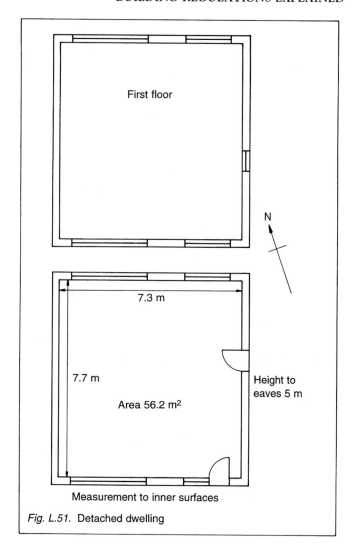

Fig. L.51. Detached dwelling

To obtain the average U value:

$$\frac{\text{Total rate of heat loss per degree}}{\text{Total external surface area}}$$

$$= \frac{197.23}{262.4} = 0.75 \, \text{W/m}^2 \, \text{K}.$$

This construction therefore does not meet the target U value and alterations in the specification must be contemplated.

Possibilities include:

- improve thermal resistance of exposed walls
- improve thermal resistance of doors and windows
- take account of solar gain
- provide more efficient space-heating system.

Walls: The wall specification could be varied to improve the U value from 0.5 W/m² K to 0.45 W/m² K.

Doors and windows: To improve the U value here a double-glazed or timber door having a U value of 3.0 W/m² K could be used and windows double glazed in wood frames provide a U value of 2.9 W/m² K.

Solar gain: Total window area is 24.8 m², and 15.0 m² faces south (±30°) with 9.3 m² facing north (±30°) (0.5 m² faces east). In these circumstances the average U value can be reduced by 40% of 15.0 – 9.3 = 2.28 m².

These adjustments give the following position:

- The exposed window area of 24.8 m² is reduced by 2.38 m² to

22.52 m² and has a U value of 2.9 W/m² K, giving the reduced rate of heat loss per degree of 65.31 W/K.

- Doors now have a U value of 3.0 W/m² K and the reduced rate of heat loss per degree is 11.4 W/K.
- Walls have the lower U value of 0.45 W/m² K, giving a reduced rate of heat loss of 54.63 W/K.
- In total, the exposed surface area is now 260.12 m² and the total heat loss per degree is 173.49 W/K.

The average U value is now

$$\frac{173.49}{260.12} = 0.67 \, \text{W/m}^2 \, \text{K}$$

and therefore complies with the target U value method.

A higher performance heating system could have been selected and if a condensing boiler were chosen a 10% relaxation of the target U value would be permitted, and this would mean that 0.67 + 10% would be increased to 0.74 W/m² K, which when associated with an improvement in only one of the elements would have given a satisfactory result.

Example 2

This is an example of a semi-detached house and when calculating requirements for these premises the party wall and the semi-exposed wall separating the garage from the dwelling are not included in measurements for the average U value or target U value (see Fig. L52).

The SAP energy rating is to be in excess of 60.

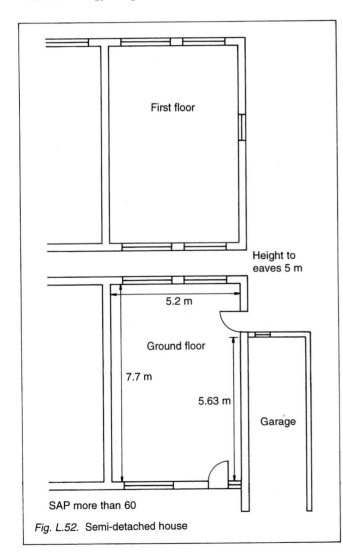

Fig. L.52. Semi-detached house

It is assumed that:

- The total exposed floor area of the dwelling is 40 m² with a U value of 0.45 W/m² K and a rate of heat loss per degree of 18.0 W/K.
- The double-glazed windows total 14.2 m² with a U value of 3.0 W/m² K, giving a rate of heat loss per degree of 42.6 W/K.
- Area of doors is 3.8 m² with a U value of 3.0 W/m² K; the rate of heat loss per degree is 11.4 W/K.
- Wall area totals 59 m² and has a U value of 0.55 W/m² K; the rate of heat loss per degree is 32.45 W/K.
- Roof area totals 40 m² and has a U value of 0.25 W/m² K which, when multiplied together, provides a rate of heat loss per degree of 10 W/K.
- The total exposed surface area is therefore 157.0 m² and the total rate of heat loss per degree is 114.45 W/K.

The target

For a dwelling with a SAP rating more than 60 the target U value is arrived at as follows.

$$\frac{\text{Total floor area} \times 0.64}{\text{Total exposed surface area}} + 0.4$$

$$= \frac{80 \times 0.64}{157} + 0.4 = 0.73 \, \text{W/m}^2 \, \text{K}$$

The proposal

Average U value for the semi-detached dwelling is calculated by applying the ratio of two values:

$$\frac{\text{The rate of heat loss per degree}}{\text{Total external surface area}}$$

$$= \frac{114.45}{157.0} = 0.73 \, \text{W/m}^2 \, \text{K}$$

which is precisely on the target without any further adjustment and therefore meets the regulations.

APPENDIX G

Regulation 14A requires that every new dwelling, and every dwelling created by a material change of use, is to have an energy rating. The rating is expressed as a score between 1 and 100 on the Government's standard assessment procedure (SAP). When a score is high enough it can be used to indicate compliance without further checks being made (see Table 4)

Anyone can undertake a SAP rating calculation, but unless prepared by an authorized assessor of the National House Energy Rating System or a 'competent' person, the calculations have to be submitted to the building control authority.

Appendix G contains a worksheet, guidance and tables to enable the calculations to be completed. The worksheet deals with overall dimensions, ventilation, heat losses, water energy requirements, internal gains, solar gains, internal temperatures, degree-days, space-heating requirements and fuel costs.

Finally one arrives at the SAP rating, which comprises an energy cost deflector which is currently set at 0.96, and this figure is expected to vary in the future, although not frequently.

Next, the energy cost factor (ECF) is determined by taking the total heating costs per year, multiplied by the energy cost factor minus 40 and then divided by the total floor area.

SAP rating is then obtained from Table 14 of Appendix G, and surprisingly the rating for 60, which figures so importantly in the requirements, does not appear. By interpolation the following would appear to be the SAP ratings between 59 and 63:

ECF-£/m²	SAP rating
3.3	63
3.37	62
3.45	61
3.52	60
3.6	59

The following SAP energy rating method for dwellings is reprinted with permission of HMSO.

THE SAP ENERGY RATING METHOD FOR DWELLINGS

The SAP energy rating and the building regulations

G1 This Appendix has been approved by the Secretary of State for the purpose of Regulation 14A of the Building Regulations 1991 as amended by the Building Regulations (Amendment) Regulations 1994.

G2 When calculating the SAP Energy Rating in compliance with Regulation 14A or for the purposes of following the guidance in Section 1 of Approved Document L the following considerations apply:

(a) The data used in calculations of SAP Energy Ratings should be obtained from the tables in this Appendix. The fuel cost data will be revised in future editions of the Approved Document.

(b) For each particular case the SAP Energy Rating should be calculated using the edition of this Appendix current at the date of giving notice or the deposit of plans in compliance with Regulation 11.

(c) When the final heating system is unknown, the SAP Energy Rating notified to the building control body in accordance with Regulation 14A should be calculated assuming a main system of electric room heaters and a secondary system of electric heaters, both systems using on-peak electricity.

(d) When undertaking SAP Energy Rating calculations for designs not intended for specific construction sites (e.g. type designs) the following assumptions should be made:

(i) two sides of the dwelling will be shaded;

(ii) the windows, doors and rooflights are all on the east and west elevations.

Example SAP energy ratings for different dwelling types

Example 1– Two bedroom mid-terrace house

Table G1 Data for the two bedroom mid-terrace house with electric storage heaters

Element	Description	Area	U value
Wall	Brick/cavity/dense block with 70 mm blown fibre cavity insulation	30.3	0.44
Roof	Pitched roof, 100 mm insulation between joists mm on top	27.3	0.25
Ground floor	Suspended timber, 25 mm insulation	27.3	0.37
Windows and doors	Double glazed (6 mm gap), wooden frame	11.7	3.3
Heating	Electric storage heaters (efficiency 100%)		
		SAP rating = 68	

Example 2 – Three bedroom semi-detached house

Table G2 Data for the three bedroom semi-detached house with gas boiler

Element	Description	Area	U value
Wall	Brick/cavity/dense block with 70 mm blown fibre cavity insulation	72.5	0.44
Roof	Pitched roof, 100 mm insulation between joists 50 mm on top	40	0.25
Ground floor	Solid concrete, 25 mm insulation	40	0.43
Windows and doors	Double glazed (6 mm gap), PVC-U frame	18.0	3.3
Heating	Central heating with gas boiler (efficiency 72%)		
		SAP rating = 79	

Example 3 – Three bedroom semi-detached house

Table G3 Data for the three bedroom semi-detached house with LPG boiler

Element	Description	Area	U value
Wall	Brick/cavity/aerated concrete block with insulated plasterboard	72.5	0.45
Roof	Pitched roof, 100 mm insulation between joists 100 mm on top	40	0.19
Ground floor	Concrete suspended beam and aerated concrete block, 40 mm insulation	40	0.35
Windows and doors	Double glazed (12 mm gap), wooden frame	18.0	3.0
Heating	Central heating with LPG boiler (efficiency 72%)		
		SAP rating = 58	

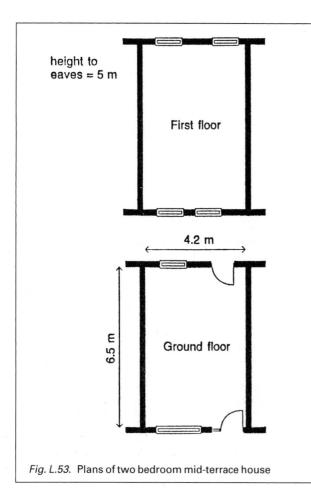

height to eaves = 5 m

First floor

4.2 m

6.5 m

Ground floor

Fig. L.53. Plans of two bedroom mid-terrace house

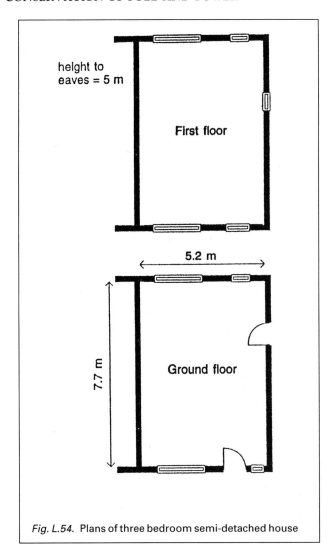

Fig. L.54. Plans of three bedroom semi-detached house

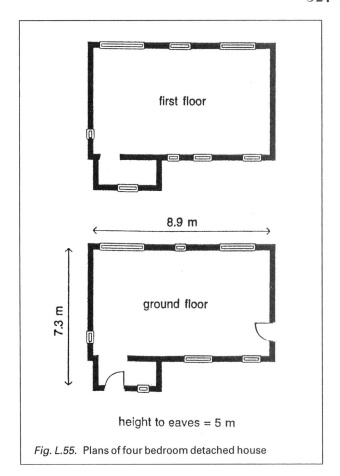

Fig. L.55. Plans of four bedroom detached house

Example 4– Four bedroom detached house

Table G4 Data for the four bedroom detached house with condensing boiler

Element	Description	Area	U value
Wall	Brick/partial cavity fill/ medium density block	116.5	0.45
Roof	Pitched roof, 100 mm insulation between joists 50 mm on top	50	0.25
Ground floor	Suspended timber, 35 mm insulation	50	0.45
Windows and doors	Double glazed (6 mm gap), wooden frame	24.9	3.3
Heating	Central heating with gas condensing boiler (efficiency 85%)		
		SAP rating = 85	

Example 5– Two bedroom bungalow

Table G5 Data for the two bedroom bungalow with gas boiler

Element	Description	Area	U value
Wall	Brick/cavity/aerated concrete block with insulated plasterboard	64.2	0.45
Roof	Pitched roof, 100 mm insulation between joists 50 mm on top	56.7	0.25
Ground floor	Concrete suspended beam and medium density concrete block, 25 mm insulation	56.7	0.45
Windows and doors	Double glazed (6 mm gap), PVC-U frame	13.4	3.3
Heating	Central heating with gas boiler (efficiency 72%)		
		SAP rating = 68	

Summary

This manual describes the Government's standard assessment procedure (SAP) for producing an energy rating for a dwelling, based on calculated annual energy cost for space and water heating. The calculation assumes a standard occupancy pattern, derived from the measured floor area of the dwelling, and a standard heating pattern. The rating is normalized for floor area so that the size of the dwelling does not strongly affect the result, which is expressed on a scale of 1–100: the higher the number the better the standard.

The method of calculating the rating is set out in the form of a worksheet, accompanied by a series of tables. A calculation may be carried out by completing, in sequence, the numbered boxes in the worksheet, using the data in the tables as indicated. Alternatively,

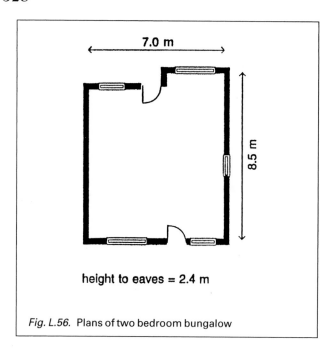

7.0 m

8.5 m

height to eaves = 2.4 m

Fig. L.56. Plans of two bedroom bungalow

a computer program approved for SAP calculations by the Building Research Establishment may be used.

Introduction

An energy rating aims to inform householders of the overall energy efficiency of a home in a way that is simple and easy to understand. Many different technical definitions for such a rating are possible, and several have been used to date in the United Kingdom. The standard assessment procedure (SAP) is the Government's recommended method for home energy rating.

The rating obtained from following the SAP depends upon a range of factors that contribute to energy efficiency:

- thermal insulation of the building fabric;
- efficiency and control of the heating system;
- ventilation characteristics of the dwelling;
- solar gain characteristics of the dwelling;
- the price of fuels used for space and water heating.

The rating is not affected by factors that depend on the individual characteristics of the household occupying the dwelling when the rating is calculated, for example:

- household size and composition;
- the ownership and efficiency of particular domestic electrical appliances;
- individual heating patterns and temperatures.

It is not affected either by the geographical location of the dwelling; a dwelling will have the same rating whether it is built in Caithness or Cornwall.

The procedure used to calculate the rating is based on the BRE Domestic Energy Model (BREDEM), which provides a framework for the calculation of energy use in dwellings. BREDEM exists in a number of standardized versions that differ in technical detail and in the precise requirements for data related to a particular application. Some versions are designed to be implemented as computer programs, while others are defined in such a way that they can be carried out without the use of a computer.

Worksheet calculations

The method of calculating the rating is set out in the form of a worksheet, accompanied by a series of tables. A calculation is carried out by completing the numbered entries in the worksheet sequentially. Some entries are obtained by calculation from entries already made or by carrying forward an earlier entry; highlighting is used to indicate where that applies. Other entries are obtained, using linear interpolation where appropriate, by reference to Tables 1 to 14.

APPROVED METHOD OF CALCULATING SAP ENERGY RATINGS

Experience has shown that there was a need for clarification of some issues concerned with the specification of inputs to calculations; as for example, to provide for new boiler types and district heating.

A second edition of the Government's guidance notes was issued in 1997 and this has been followed quickly by 'The Government's Standard Assessment Procedure (SAP) for energy rating of dwellings, 1998'. It gives a new method of calculating SAP ratings effective from 1 July 1999.

This document consolidates previous editions; the main changes being:

- a new optional way of calculating boiler seasonal efficiency;
- data for new heating control systems;
- an optional calculation procedure for CO_2 emissions.

Copies of 'The Government's Standard Assessment Procedure for energy rating of dwellings' may be obtained from the Enquiries Bureau, BRECSU, BRE, Garston, Watford WD2 7JR (01923 664258).

SAP WORKSHEET (VERSION – 9.53)

1. Overall dwelling dimensions

	Area (m²)		Av. Room height (m)		Volume (m³)	
Ground floor	_____ (1a)	×	_____	=	_____	(1b)
First floor	_____ (2a)	×	_____	=	_____	(2b)
Second floor	_____ (3a)	×	_____	=	_____	(3b)
Third and other floors	_____ (4a)	×	_____	=	_____	(4b)

Total floor area $\quad$ (1a) + (2a) + (3a) + (4a) = $\quad$ _____ (5)

Dwelling volume (m³) $\qquad$ (1b) + (2b) + (3b) + (4b) $\quad$ = $\quad$ _____ (6)

2. Ventilation rate

m³ per hour

Number of chimneys	_____	×	40	=	_____ (7)
Number of flues	_____	×	20	=	_____ (8)
Number of fans and passive vents	_____	×	10	=	_____ (9)

Air changes per hour

Infiltration due to chimneys, fans and flues $\qquad$ ((7) + (8) + (9)) ÷ (6) $\quad$ = $\quad$ _____ (10)
If a pressurisation test has been carried out, proceed to box (19)

Number of storeys $\qquad$ _____ (11)

Additional infiltration $\qquad$ ((11) − 1) × 0.1 $\quad$ = $\quad$ _____ (12)

Structural infiltration $\qquad$ 0.25 for timber frame $\qquad$ _____ (13)
0.35 for masonry construction

If suspended wooden floor, enter $\qquad$ 0.2 (unsealed) $\qquad$ _____ (14)
0.1 (sealed)

If no draught lobby, enter 0.05 $\qquad$ _____ (15)

Percentage of windows and doors draught stripped $\qquad$ _____ (16)
Enter 100 for new dwellings which are to comply with Building Regulations
Window infiltration $\qquad$ 0.25 − (0.2 × (16) ÷ 100) $\quad$ = $\quad$ _____ (17)

Infiltration rate $\qquad$ (10) + (12) + (13) + (14) + (15) +(17) $\quad$ = $\quad$ _____ (18)

If pressurisation test done $\qquad$ (measured L_{50} ÷ 20) + (10) $\quad$ = $\quad$ _____ (19)
$\quad$ else let (19) = (18)
Number of sides on which sheltered $\qquad$ _____ (20)
Enter 2 for new dwellings where location is not shown

Shelter factor $\qquad$ 1 − 0.075 × (20) = $\quad$ _____ (21)

If mechanical ventilation with heat recovery,
$\quad$ effective air change rate $\qquad$ ((19) × (21) + 0.17) $\quad$ = $\quad$ _____ (22)
(If no heat recovery, add 0.33 air changes per hour to value in box (22))

If natural ventilation, then air change rate $\qquad$ (19) × (21) $\quad$ = $\quad$ _____ (23)

If (23) ≥ 1, then (24) = (23) $\qquad$ _____ (24)

else (24) = 0.5 + ((23)² × 0.5)
Effective air change rate (enter (22) or (24)) $\qquad$ _____ (25)

3. Heat losses and heat loss parameter

ELEMENT		Area (m²)		U-value (W/m²K)		A × U (W/K)	
Doors		_____	×	_____	=	_____	(26)
Windows (single glazed)	0.9 ×	_____	×	_____	=	_____	(27)
Windows (double glazed)	0.9 ×	_____	×	_____	=	_____	(28)
Rooflights	0.9 ×	_____	×	_____	=	_____	(29)
Ground floor		_____	×	_____	=	_____	(30)

Walls (type 1)   [   ] × [   ] = [   ] (31)

Walls (type 2)   [   ] × [   ] = [   ] (32)

Roof (type 1)   [   ] × [   ] = [   ] (33)

Roof (type 2)   [   ] × [   ] = [   ] (34)

Other   [   ] × [   ] = [   ] (35)

Ventilation heat loss $(25) \times 0.33 \times (6)$ = [   ] (36)

Specific heat loss $(26) + (27) + ... (35) + (36)$ = [   ] (37)

Heat loss parameter (HLP) $(37) \div (5)$ = [   ] (38)

4. Water heating energy requirements

GJ/year

Energy content of heated water (Table 1, column (a))   [   ] (39)
  If instantaneous water heating at point of use, enter 0 in boxes (40), (43) and (48), and go to (49)

Distribution loss (Table 1, column (b))   [   ] (40)

  If no hot water tank, (i.e. combi or multipoint), go to (49)
Tank Volume   [   ] (41)

Tank loss factor (Table 2)   [   ] (42)

Energy lost from tank in GJ/year $(41) \times (42)$ = [   ] (43)

  If no solar panel, enter 0 in box (47) and go to (48)
Area of solar panel (m²)   [   ] (44)

  Solar energy available $1.3 \times (44)$ = [   ] (45)

  Load ratio $(39) \div (45)$ = [   ] (46)

  Solar input $(45) \times (46) \div (1 + (46))$ = [   ] (47)

Primary circuit loss (Table 3)   [   ] (48)

Output from water heater $(39) + (40) + (43) + (48) - (47)$ = [   ] (49)

Efficiency of water heater (Table 4)%   [   ] (50)

Energy required for water heating $(49) \div (50) \times 100$ = [   ] (51)

Heat gains from water heating $0.8 \times ((40) + (43) + (48)) + 0.25 \times (39)$ = [   ] (52)

5. Internal gains

Watts

Lights, appliances, cooking and metabolic (Table 5)   [   ] (53)

Water heating $31.7 \times (52)$ = [   ] (54)

Total internal gains $(53) + (54)$ = [   ] (55)

6. Solar gains

Enter the area of the whole window including frames and the value for solar flux obtained from Table 6. For Building Regulations assessment when orientations are not known, all vertical glazing should be entered as E/W.

Orientation	Area	Flux	Gains (W)
North	[ ]	× [ ]	= [ ] (56)
Northeast	[ ]	× [ ]	= [ ] (57)
East	[ ]	× [ ]	= [ ] (58)
Southeast	[ ]	× [ ]	= [ ] (59)
South	[ ]	× [ ]	= [ ] (60)
Southwest	[ ]	× [ ]	= [ ] (61)
West	[ ]	× [ ]	= [ ] (62)

Northwest		$\times$		$=$		(63)

Rooflights		$\times$		$=$		(64)

Solar access factor
Enter 1 for new dwellings where overshading is not known   (65)

Solar gains (UK average)	$((56)+57) + \dots + (64) \times (65)$	$=$		(66)

Total gains (W)	$(55) + (66)$	$=$		(67)

Gains/loss ratio (GLR)	$(67) \div (37)$	$=$		(68)

Utilisation factor (Table 7)   (69)

Useful gains (W)	$(67) \times (69)$	$=$		(70)

7. Mean internal temperature

°C

Mean internal temperature of the living area (Table 8)   (71)

Adjustment for gains	$0.2 \times R \times ((70) \div (37) - 0.4)$	$=$		(72)

R is obtained from the 'responsiveness' column of Table 4a

Adjusted living room temperature	$(71) + (72)$	$=$		(73)

Temperature difference between zones (Table 9)   (74)

Living area fraction (0 to 1.0)   (75)

Rest-of-house area fraction	$1.0 - (75)$	$=$		(76)

Mean internal temperature	$(73) - ((74) \times (76))$	$=$		(77)

8. Degree-days

Temperature rise from gains	$(70) \div (37)$	$=$		(78)

Base temperature	$(77) - (78)$	$=$		(79)

Degree-days (use (79) and Table 10 to adjust base)   (80)

9. Space heating requirement

GJ/year

Energy requirement (useful)	$0.000\ 0864 \times (80) \times (37)$	$=$		(81)

Fraction of heat from secondary system
Use value obtained from Table 11   (82)

Efficiency of main heating system (Table 4a)   (83)

Reduce by the amount shown in the 'efficiency' column of Table 4b where appropriate

Efficiency of secondary heating system (Table 4a)   (84)

Space heating fuel (main)	$(1.0 - (82)) \times (81) \times 100 \div (83)$	$=$		(85)

Space heating fuel (secondary)	$(82) \times (81) \times 100 \div (84)$	$=$		(86)

Electricity for pumps and fans
Enter 0.47 GJ for each central heating pump, 0.16 GJ for each boiler with a fan-assisted flue.
For warm air heating system fans, add 0.002 GJ $\times$ the volume of the dwelling (given in box (6)).
For dwellings with whole-house mechanical ventilation, add 0.004 GJ $\times$ the volume of the dwelling.   (87)

10. Fuel costs Fuel prices to be obtained from Table 12

	GJ/year	$\times$	Fuel price	$=$	£/year	
Space heating – main system	(85)	$\times$		$=$		(88)
– secondary system	(86)	$\times$		$=$		(89)

Water heating
If off-peak electric water heating:

				Fuel Price		£/year	
On-peak percentage (Table 13)		(90)					
Off-peak percentage	$100 - (90)$	$=$	(91)				
On-peak cost	$(51) \times ((90) \div 100)$	$\times$			$=$		(92)

Off-peak cost	$(51) \times ((91) \div 100) \times$ [ ]		$=$ [ ]	(93)

Else:

Water heating cost	$(51) \times$ [ ]		$=$ [ ]	(94)
Pump/fan energy cost	$(87) \times$ [ ]		$=$ [ ]	(95)
Additional standing charges (Table 12)			[ ]	(96)
Total heating	$(88) + (89) + (92) + (93) + (94) + (95) + (96)$		$=$ [ ]	(97)

11. SAP rating

Energy cost deflator (Table 12 footnote[2])		[ ]	(98)
Energy cost factor (ECF)	$((97) \times (98) - 40.0) \div (5)$	$=$ [ ]	(99)
SAP rating (Table 14)		[ ]	

Table 1   Hot water energy requirements

Floor area (m²)	(a) Hot water usage (GJ/year)	(b) Distribution loss (GJ/year)
30	4.13	0.73
40	4.66	0.82
50	5.16	0.91
60	5.68	1.00
70	6.17	1.09
80	6.65	1.17
90	7.11	1.26
100	7.57	1.34
110	8.01	1.41
120	8.44	1.49
130	8.86	1.56
140	9.26	1.63
150	9.65	1.70
160	10.03	1.77
170	10.40	1.84
180	10.75	1.90
190	11.10	1.96
200	11.43	2.02
210	11.74	2.07
220	12.05	2.13
230	12.34	2.18
240	12.62	2.23
250	12.89	2.27
260	13.15	2.32
270	13.39	2.36
280	13.62	2.40
290	13.84	2.44
300	14.04	2.48

Alternatively, requirements and gains may be calculated from the total floor area of the dwelling (TFA), using the following steps:

(a)   Calculate $N = 0.035 \times TFA - 0.000\ 038 \times TFA^2$, if TFA ≤ 420
     $= 8.0$ if TFA > 420

(b)   Hot water usage $= 0.85 \times (92 + 61 \times N) \div 31.71$

(c)   Distribution loss $= 0.15 \times (92 + 61 \times N) \div 31.71$

Table 2   Hot water cylinder loss factor (GJ/year/litre)
Multiply by cylinder volume in litres to get loss

Insulation thickness (mm)	Type	
	Foam	Jacket
None	0.0945	0.0945
12.5	0.0315	0.0725
25	0.0157	0.0504
38	0.0104	0.0331
50	0.0078	0.0252
80	0.0049	0.0157
100	0.0039	0.0126
150	0.0026	0.0084

Table 3   Primary circuit losses (GJ/year)

System loss	GJ/year
Electric immersion heater	0.0
Boiler with uninsulated primary pipework and no cylinder stat*	4.4
Boiler with insulated primary pipework and no cylinder stat*	2.2
Boiler with uninsulated primary pipework and with cylinder stat	2.2
Boiler with insulated primary pipework and with cylinder stat	1.3

A cylinder stat is required by the Building Regulations in new dwellings.

Table 4a   Heating system efficiency

This table shows space heating efficiency. Water heating efficiency is also obtained from this table when hot water is from a boiler system. For independent electric water heating use 100%. Use 70% for single-point gas water heaters. For multi-point gas water heaters and for heat exchangers built into gas warm air heaters, use 65%.

	Efficiency (%)	Heating type	Responsive-ness
CENTRAL HEATING SYSTEMS WITH RADIATORS			

1. 'Heating type' refers to the appropriate column in Table 8
2. Refer to Group 1 in Table 4b for control options
3. Check Table 4b for efficiency adjustment due to poor controls
4. Where two figures are given, the first is for space heating and the second is for water heating

	Efficiency (%)	Heating type	Responsive-ness
Gas boilers (including LPG) with fan-assisted flue			
1. Low thermal capacity	72	1	1.0
2. High or unknown thermal capacity	68	1	1.0
3. Condensing	85	1	1.0
4. Combi	71/69	1	1.0
5. Condensing combi	85/83	1	1.0
Gas boilers (including LPG) with balanced or open flue			
1. Wall mounted	65	1	1.0
2. Floor mounted, post-1979	65	1	1.0
3. Floor mounted, pre-1979	55	1	1.0
4. Combi	65	1	1.0
5. Room heater + back boiler	65	1	1.0
Oil boilers			
Standard oil boiler pre-1985	65	1	1.0
Standard oil boiler 1985 or later	70	1	1.0
Condensing boiler	85	1	1.0
Solid fuel boilers			
Manual feed (in heated space)	60	2	0.75
Manual feed (in unheated space)	55	2	0.75
Autofeed (in heated space)	65	2	0.75
Autofeed (in unheated space)	60	2	0.75
Open fire with back boiler to rads	55	3	0.50
Closed fire with back boiler to rads	65	3	0.50

Table 4a *continued*

	Efficiency (%)	Heating type	Responsive-ness
Electric boilers			
Dry core boiler in heated space	100	2	0.75
Dry core boiler in unheated space	85	2	0.75
Economy 7 boiler in heated space	100	2	0.75
Economy 7 boiler in unheated space	85	2	0.75
On-peak heat pump	250	1	1.0
24-hour heat pump	240	1	1.0
STORAGE RADIATOR SYSTEMS			
(Refer to Group 2 in Table 4b for control options)			
Old (large volume) storage heaters	100	5	0.0
Modern (slimline) storage heaters	100	4	0.25
Convector storage heaters	100	4	0.25
Fan-assisted storage heaters	100	3	0.5
Electric underfloor heating	100	5	0.0
WARM AIR SYSTEMS			
1. Refer to Group 3 in Table 4b for control options			
Gas-fired warm air with fan-assisted flue			
Ducted, with gas-air modulation	80	1	1.0
Room heater, with in-floor ducts	77	1	1.0
Gas-fired warm air with balanced or open flue			
Ducted (on/off control)	70	1	1.0
Ducted (modulating control)	72	1	1.0
Stub ducted	70	1	1.0
Ducted with flue heat recovery	85	1	1.0
Stub ducted with flue heat recovery	82	1	1.0
Condensing	94	1	1.0
Oil-fired warm air			
Ducted output (on/off control)	70	1	1.0
Ducted output (modulating control)	72	1	1.0
Stub duct systen	70	1	1.0
Electric warm air			
Electricairesystem	100	2	0.75
Air-to-air heat pump	250	1	1.0
ROOM HEATER SYSTEMS			
1. Refer to Group 4 in Table 4b for control options			
2. Check Table 4b for efficiency adjustment due to poor control			
Gas			
Old style gas fire (open front)	50	1	1.0
Modern gas fire (glass enclosed front)	60	1	1.0
Modern gas fire with balanced flue	70	1	1.0
Modern gas fire with back boiler (no rads)	65	1	1.0
Condensing gas fire (fan-assisted flue)	85	1	1.0
Gas fire or room heater with fan-assisted flue	79	1	1.0
Coal effect fire in fireplace	25	1	1.0
Coal effect fire in front of fireplace	60	1	1.0
Solid fuel			
Open fire in grate	32	3	0.5
Open fire in grate, with throat restrictor	42	3	0.5
Open fire with back boiler (no rads)	55	3	0.5
Closed room heater	60	3	0.5
Closed room heater with back boiler	65	3	0.5

Table 4a *continued*

	Efficiency (%)	Heating type	Responsive-ness
Electric (direct acting)			
Panel, convector or radiant heaters	100	1	1.0
Portable electric heaters	100	1	1.0
OTHER SYSTEMS			
Refer to Group 5 for control options			
Gas underfloor heating	70	4	0.25
Gas underfloor heating, condensing boiler	87	4	0.25
Electric ceiling heating	100	2	0.75

Table 4b Heating system controls

1. Use Table 4a to select appropriate Group in this table.
2. 'Control' refers to the appropriate column in Table 9.
3. 'Efficiency' is an adjustment that should be subtracted from the space heating efficiency obtained from Table 4a.
4. 'Temp' is an adjustment that should be added to the temperature obtained from Table 8.

Type of control	Control (%)	Efficiency (°C)	Temp
GROUP 1: BOILER SYSTEMS WITH RADIATORS			
No thermostatic control* of room temperature	1	5	0.3
Programmer + roomstat	1	0	0
Prog + roomstat* (no boiler off)	1	5	0
Prog + roomstat + TRVs	2	0	0
Prog + roomstat + TRVs* (no boiler off)	2	5	0
TRVs + prog + bypass*	2	5	0
TRVs + prog + flow switch	2	0	0
TRVs + prog + boiler energy manager	2	0	0
Zone control	3	0	0
Zone control (no boiler off)*	3	5	0
GROUP 2: STORAGE RADIATOR SYSTEMS			
Manual charge control	3	0	0.3
Automatic charge control	3	0	0
GROUP 3: WARM AIR SYSTEMS			
No thermostatic control* of room tenperature	1	0	0.3
Roomstat only*	1	0	0
Programmer + roomstat	1	0	0
Zone control	3	0	0
GROUP 4: ROOM HEATER SYSTEMS			
No thermostatic control	2	0	0.3
Appliance stat	3	0	0
Appliance stat + prog	3	0	0
Programmer + roomstat	3	0	0
Roomstat only	3	0	0
GROUP 5: OTHER SYSTEMS			
No thermostatic control of room temperature	1	0	0.3
Appliance stat only	1	0	0
Roomstat only	1	0	0
Programmer + roomstat	1	0	0
Programmer and zone control	3	0	0

* These systems would not be acceptable in new dwellings

Table 5 Lighting, appliance, cooking and metabolic gains

Floor area (m²)	Gains (W)	Floor area (m²)	Gains (W)
30	230	170	893
40	282	180	935
50	332	190	978
60	382	200	1020
70	431	210	1061
80	480	220	1102
90	528	230	1142
100	576	240	1181
110	623	250	1220
120	669	260	1259
130	715	270	1297
140	760	280	1334
150	805	290	1349
160	849	300	1358

Alternatively, gains may be calculated from the total floor area of the dwelling (TFA), using the following steps:

(a) Calculate $N = 0.035 \times TFA - 0.000\,038 \times TFA^2$, if TFA ≤ 420
   $= 8.0$ if TFA > 420
(b) Gains (W) $= 74 + 2.66 \times TFA + 75.5 \times N$, if TFA ≤ 282
   $= 824 + 75.5 \times N$ if TFA > 282

Note:
Gains from the following equipment should be added to the totals given above:
Central heating pump   : 10 W
Mechanical ventilation system   : 25 W

Table 6 Solar flux through glazing (W/m²)

	Horizontal	Vertical				
		North	NE/NW	E/W	SE/SW	South
Single glazed	31	10	14	18	24	30
Double glazed	26	8	12	15	21	26
Double glazed with low E coating	24	8	11	14	19	24
Triple glazed	22	7	10	13	17	22

Note:
1. For a rooflight in a pitched roof with a pitch of up to 70°, use the value under 'Northerly' for orientations within 30° of North and the value under 'Horizontal' for all other orientations. (If the pitch is greater than 70°, then treat as if it were a vertical window.)
2. For Building Regulations assessment when orientations are not known, all vertical glazing should be entered as E/W.

Table 7 Utilisation factor as a function of Gain/loss ratio (G/L)

G/L	Utilisation factor	G/L	Utilisation factor
1	1.00	16	0.68
2	1.00	17	0.65
3	1.00	18	0.63
4	0.99	19	0.61
5	0.97	20	0.59
6	0.95	21	0.58
7	0.92	22	0.56
8	0.89	23	0.54
9	0.86	24	0.53
10	0.83	25	0.51
11	0.81	30	0.45
12	0.78	35	0.40
13	0.75	40	0.36
14	0.72	45	0.33
15	0.70	50	0.30

Alternatively, the utilisation factor nay be calculated by the formula:

Utilisation factor $= 1 - \exp(-18 \div GLR)$,
where GLR = (total gains) ÷ (specific heat loss)

Table 8 Mean internal temperature of living area

Number in brackets is from the 'heating' column of Table 4a. HLP is item 38 on the worksheet

HLP	(1)	(2)	(3)	(4)	(5)
1.0 (or lower)	18.88	19.32	19.76	20.21	20.66
1.5	18.88	19.31	19.76	20.20	20.64
2.0	18.85	19.30	19.75	20.19	20.63
2.5	18.81	19.26	19.71	20.17	20.61
3.0	18.74	19.19	19.66	20.13	20.59
3.5	18.62	19.10	19.59	20.08	20.57
4.0	18.48	18.99	19.51	20.03	20.54
4.5	18.33	18.86	19.42	19.97	20.51
5.0	18.16	18.73	19.32	19.90	20.48
5.5	17.98	18.59	19.21	19.82	20.45
6.0 (or higher)	17.78	18.44	19.08	19.73	20.40

Table 9 Difference in temperatures between zones

Number in brackets is from the 'control' column of Table 4b. HLP is item 38 on the worksheet

HLP	(1)	(2)	(3)
1.0 (or lower)	0.40	1.41	1.75
1.5	0.60	1.49	1.92
2.0	0.79	1.57	2.08
2.5	0.97	1.65	2.22
3.0	1.15	1.72	2.35
3.5	1.32	1.79	2.48
4.0	1.48	1.85	2.61
4.5	1.63	1.90	2.72
5.0	1.76	1.94	2.83
5.5	1.89	1.97	2.92
6.0 (or higher)	2.00	2.00	3.00

Table 10 Degree-days as a function of base temperature

Base temperature (°C)	Degree-days	Base temperature (°C)	Degree-days
1.0	0	11.0	1140
1.5	30	11.5	1240
2.0	60	12.0	1345
2.5	95	12.5	1450
3.0	125	13.0	1560
3.5	150	13.5	1670
4.0	185	14.0	1780
4.5	220	14.5	1900
5.0	265	15.0	2015
5.5	310	15.5	2130
6.0	360	16.0	2250
6.5	420	16.5	2370
7.0	480	17.0	2490
7.5	550	17.5	2610
8.0	620	18.0	2730
8.5	695	18.5	2850
9.0	775	19.0	2970
9.5	860	19.5	3090
10.0	950	20.0	3210
10.5	1045	20.5	3330

Table 11 Fraction of heat supplied by secondary heating systems

Main heating system	Secondary system	Fraction
Central heating system	gas fires	0.15
with boiler and radiators,	coal fires	0.10
central warm-air system	electric heaters	0.05
or other gas fired systems		
Gas room heaters	gas fires	0.30
	coal fires	0.15
	electric heaters	0.10
Coal room heaters or	gas fires	0.20
electric room heaters	coal fires	0.20
	electric heaters	0.20
Electric storage heaters	gas fires	0.15
or other electric systems	coal fires	0.10
	electric heaters	0.10
Electric heat pump	gas fires	0.15
systems with heat storage or	coal fires	0.10
fan-assisted storage heaters	electric heaters	0.05

Table 12 Fuel prices and additional standing charges

	Additional standing charge (£)	Unit price (£/GJ)
Gas (mains)	38	4.26
Bulk LPG	51	7.11
Bottled gas – propane 47 kg cylinder		13.26
Heating oil		3.68
House coal		3.87
Smokeless fuel		6.54
Anthracite nuts		5.04
Anthracite grains		4.53
Wood		4.20
Electricity (on-peak)		22.38
Electricity (off-peak)	14	7.60
Electricity (standard tariff)		21.08
Electricity (24-hr heating tariff)	45	8.57

Notes
1. The standing charge given for electricity is extra amount for the off-peak tariff, over and above the amount for the standard domestic tariff, as it is assumed that the dwelling has a supply of electricity for reasons other than space and water heating. Standing charges for gas and for off-peak electricity are attributed to space and water heating costs where those fuels are used for heating.
2. The energy cost deflator term is currently set at 0.96. It will vary with the weighted average price of heating fuels in future, in such a way as to ensure that the SAP is not affected by the general rate of inflation. However, individual SAP ratings are affected by relative changes in the price of particular heating fuels.

Table 13 On-peak fraction for electric water heating

Dwelling floor area (m²)	Cylinder size (litres)		
	110	160	210
40 or less	12 (56)	6 (18)	1
60	14 (58)	7 (21)	3
80	17 (60)	9 (24)	4
100	19 (62)	10 (27)	5
120	21 (63)	12 (30)	6
140	23 (65)	13 (33)	6
160	25 (66)	15 (35)	7
180	27 (68)	16 (37)	8
200	29 (69)	17 (40)	9
220	30 (70)	18 (42)	9
240	32 (71)	19 (44)	10
260	33 (72)	20 (45)	11
280	34 (73)	21 (45)	11
300	36 (74)	21 (45)	12
320	36 (75)	22 (46)	12
340	37 (75)	23 (47)	12
360	38 (76)	23 (48)	13
380	39 (76)	24 (49)	13
400	39 (76)	24 (49)	13
420 or more	39 (77)	24 (50)	13

Table 13 shows percentage of electricity required at on-peak rates for cylinders with dual immersion heaters. The figures in brackets are for cylinders with single immersion heaters.

Table 14 SAP rating by energy cost factor

ECF (£/m²)	SAP rating	ECF (£/m²)	SAP rating
1.4	100	4	55
1.5	97	4.5	50
1.6	95	5	45
1.7	92	5.5	41
1.8	89	6	37
1.9	87	6.5	34
2	85	7	30
2.2	81	7.5	27
2.4	77	8	25
2.6	74	8.5	22
2.8	70	9	20
3	67	10	15
3.3	63	11	11
3.6	59	12	7
3.9	56	13	4
		14	1

The values in the above table may be obtained by using the formula:
$$\text{SAP Rating} = 115 - 100 \times \log_{10}(\text{ECF})$$

ANNEXE A. U VALUES FOR GLAZING

Table A1   Indicative U values (W/m² K) for windows, doors and rooflights

| | Type of frame | | | | | | | |
| | Wood | | Metal | | Thermal break | | PVC-U | |
Air gap in sealed unit (mm)	6	12	6	12	6	12	6	12
Window, double-glazed	3.3	3.0	4.2	3.8	3.6	3.3	3.3	3.0
Window, double-glazed, low-E	2.9	2.4	3.7	3.2	3.1	2.6	2.9	2.4
Window, double-glazed, Argon fill	3.1	2.9	4.0	3.7	3.4	3.2	3.1	2.9
Window, double-glazed, low-E, Argon fill	2.6	2.2	3.4	2.9	2.8	2.4	2.6	2.2
Window, triple-glazed	2.6	2.4	3.4	3.2	2.9	2.6	2.6	2.4
Door, half double-glazed	3.1	3.0	3.6	3.4	3.3	3.2	3.1	3.0
Door fully double-glazed	3.3	3.0	4.2	3.8	3.6	3.3	3.3	3.0
Rooflights, double-glazed at less than 70° from horizontal	3.6	3.4	4.6	4.4	4.0	3.8	3.6	3.4
Windows and doors, single-glazed	4.7		5.8		5.3		4.7	
Door, solid timber panel or similar	3.0		—		—		—	
Door, half single-glazed/half timber panel or similar	3.7		—		—		—	

Window U values should be obtained from Table A1, regardless of building age. The values apply to the entire area of the window opening, including both frame and glass, and take account of the proportion of the area occupied by the frame and the heat conducted through it.

ANNEXE B. DWELLINGS WITH MORE THAN ONE HEATING SYSTEM

General principles

The treatment multiple forms of heating in SAP is based on the following principles:

(1) the decision to include a secondary heating system should be based on the characteristics of the dwelling and the systems installed and not on the heating practices of the occupying household;

(2) secondary systems should only be included if they are based on 'fixed' appliances, unless portable appliances are necessary to achieve adequate heating.

To avoid excessive complexity and to reduce the extent to which surveyor judgement can influence the rating, the SAP considers only one secondary heating system per dwelling. Furthermore, secondary heating systems are governed by a set of rules that restrict the allowable combinations of heating systems and stipulate the proportion of heat supplied by the secondary system. Those restrictions will inevitably mean that the rating is based on assumptions about use that, in some cases, diverge considerably from what is actually practised by the occupying household. This is in line with the general principle that SAP is a rating for the dwelling and does not depend on who happens to be living in that dwelling. That does not preclude further estimates of energy consumption being made to take account of actual usage. Such estimates would not be part of SAP but could form the basis of advice given to the occupying household on how to make best use of the systems at their disposal.

Procedure for dealing with secondary heating systems

(1) Identify the main heating system. If there is a central system that provides both space and water heating and it is capable of heating at least 30% of the dwelling, select that system. If there is no system that provides both space and water heating, then select the system that has the capability of heating the greatest part of the dwelling.

(2) If there is still doubt about which system should be selected as the primary, select the system that supplies useful heat to the dwelling at lowest cost (obtained by dividing fuel cost by conversion efficiency).

(3) Use the responsiveness of the main heating system in calculating the mean internal temperature in Stage 7 of the SAP calculation.

(4) Decide whether a secondary heating system needs to be specified, bearing in mind that systems based on stored heat produced from electricity generally require a secondary system for their successful operation. Ignore all portable heaters, such as plug-in electrical radiators and fan heaters or free-standing butane and paraffin heaters. If no secondary system is to be specified, enter zero in box (82).

(5) If a secondary heating system is to be specified, use Table 11 to select the most appropriate description of the primary/secondary combination. Obtain the proportion of use for the secondary heating system from Table 11 and enter in box (82).

(6) Obtain the efficiency of the secondary heating system from Table 4 and enter in box (84).

(7) Calculate the space heating fuel requirements for both main and secondary heating systems as specified for entry in boxes (85) and (86).

Dwellings with inadequate heating systems

The SAP assumes that a good standard of heating will be achieved throughout the dwelling. For dwellings in which the heating system is not capable of providing that standard, it should be assumed that the additional heating is provided by electric heaters, using the fractions given in Table 11. For new dwellings that have no heating system specified, it should be assumed that all heat will be provided by electric heaters using electricity at the standard domestic tariff.

APPENDIX H

This appendix to Part 1 in the approved document completes 'Conservation of fuel and power'. It contains three illustrations of the way in which calculations may demonstrate compliance with requirements for insulation and lighting in buildings other than dwellings.

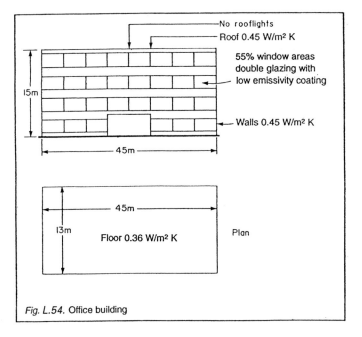

Fig. L.54. Office building

Labels in figure:
- No rooflights
- Roof 0.45 W/m² K
- 55% window areas double glazing with low emissivity coating
- Walls 0.45 W/m² K
- 15m
- 45m
- Floor 0.36 W/m² K
- 45m
- 13m
- Plan

Illustration 1

This relates to the fabric insulation of a new building.

A new four-storey office building is to be erected to the following design (see Fig. L54):

- 45 m × 13 m × 15 m high
- 55% window area. Double glazed in metal frames having thermal breaks; glass having low-E coating with 12 mm air gap.
- Exposed walls 0.6 W/m² K
- Roof 0.45 W/m² K
- Ground floor uninsulated. Reference to Fig. L46 shows that the dimensions of this floor are such that no insulation is required.

The following calculations are made to ascertain whether the building will satisfy the regulations regarding the conservation of fuel and power.

(a) Area of floor (45 m × 13 m)	585 m²
(b) Area of exposed wall and windows (45 m + 45 m + 13 m + 13 m) × 15 m	1740 m²
(c) Area of windows (55% of 1740 m²)	957 m²
(d) Area of personnel doors	14 m²
(e) Area of loading bay doors	27 m²
(f) Area of exposed walls (1740 m² − 957 m² − 14 m² − 27 m²)	742 m²
(g) Area of roof (45 m × 13 m)	585 m²

The rate of heat loss through walls, windows, doors and roof is:

- Floor (585 m² × 0.36 W/m² K) — 210.6 W/K
- Windows (957 m² × 2.6 W/m² K) — 2488.2 W/K
- Personnel doors (14 m² × 3.3 W/m² K) — 46.2 W/K
- Loading bay doors (27 m² × 0.7 W/m² K) — 18.9 W/K
- Exposed walls (742 m² × 0.6 W/m² K) — 445.2 W/K
- Roof (585 m² × 0.45 W/m² K) — 263.3 W/K

The total rate of heat loss as proposed and which must be compared with the rate of heat loss of a notional building — 3472.4 W/K

Notional building

The notional building may be given the basic allowance of 20% for rooflights even though there are not to be any in the proposal.

The area of openings in the proposed building is in excess of the basic allowance in Table 6, and in these circumstances calculations for the notional building must be based on the basic allowance of 40% for windows and doors.

It is first necessary to determine the P/A ratio using the perimeter measurement (45 m + 45 m + 13 m + 13 m), which is 116 m, and the area of the ground floor, which is 585 m². That is,

$$\frac{116}{585} = 0.2$$

Now refer to Table C1 which gives the U values for uninsulated floors, and in this instance the value is 0.36 W/m² K. Reference to Table 5 shows that this is less than the standard U value of 0.45 W/m² K and so 0.36 W/m² K is used in these notional calculations.

Areas of various elements to be taken into account are:

(a) Floor (45 m × 13 m)	585 m²
(b) Exposed wall and windows (45 m + 45 m + 13 m + 13 m) × 15 m	1740 m²
(c) Windows and personnel doors (40% of 1740 m²)	696 m²
(d) Loading bay doors	27 m²
(e) Exposed walls (1740 m² − 696 m² − 27 m²)	1017 m²
(f) Roof (45 m × 13 m − 117 m²)	468 m²
(g) Rooflights (20% of 585 m²)	117 m²

This is transmitted into the rate of heat loss for the notional building.

- Floor (585 m² × 0.36 W/m² K) — 210.6 W/K
- Windows and doors (696 m² × 3.3 W/m² K) — 2296.8 W/K
- Loading bay doors (27 m² × 0.7 W/m² K) — 18.9 W/K
- Exposed walls (1017 m² × 0.45 W/m² K) — 457.7 W/K
- Roof (468 m² × 0.45 W/m² K) — 210.6 W/K
- Rooflights (117 m² × 3.3 W/m² K) — 386.1 W/K

The total rate of heat loss from a notional building is therefore — 3580.7 W/K

As this is more than the proposal, the office building may be constructed as in the proposed design, as it meets the regulation requirement.

Illustration 2

This illustration relates to the lighting requirements for an extension to a community centre.

Table 9 lists types of high-efficiency lamps which, when used in 95% of installed lighting capacity, satisfies requirements. They are high-pressure sodium, metal halide, induction lighting, tubular fluorescent and compact fluorescent.

There will be occasions when a lighting scheme incorporates lamps that are not listed in Table 9 and in these circumstances some calculation is necessary to check that 95% of the lighting is as detailed in the table.

In this example a new extension comprising hall and changing rooms is to be added to an existing community centre. The lighting scheme incorporates lamps that are listed in Table 9 except that some low-voltage tungsten downlighters are to be placed in the entrance area, and have local controls.

The proposals are:

Main Hall
- 20 wall-mounted uplighters with 250 W high-pressure sodium lamps, mounted 7 m above the floor level. The furthest lamp will be 20.5 m from its switch, and this distance is less than three times the height of the lamp above the floor.
- There are to be 20 × 18 W compact fluorescent lamps to provide background lighting when required.

In the changing rooms, corridors and entrances there is to be:

- 10 × 58 W high-frequency fluorescent light fittings in changing rooms, each controlled by occupancy detectors.
- 6 × 58 W fluorescent light fittings placed in corridors and entrance areas, and switched locally.
- 6 × 50 W tungsten halogen downlighters in entrance area.

The schedule is as follows:

Main hall
(a) 20 × 250 W SON with 286 circuit watts per lamp.
    Total circuit watts   5720 W
(b) 20 × 18 W compact fluorescent with 23 circuit
    watts per lamp. Total circuit watts   460 W

Entrance, changing rooms and corridors
(c) 16 × 58 W HF fluorescent with 64 circuit watts per lamp.
    Total circuit watts   1024 W

Entrance
(d) 6 × 50 W low-voltage tungsten halogen with
    55 W circuit watts per lamp. Total circuit watts   330 W

Total   7534 W

The percentage of circuit watts consumed by fittings not listed in Table 9 is:

$$\frac{330 \times 100}{7534} = 4.4\%$$

In which case 95.6% of the proposed lighting, in circuit watts, is from lamps which are listed in Table 9, and therefore the requirements are satisfied.

Illustration 3

This illustration relates to evaluating the efficacy of a proposed lighting installation.

There is an alternative approach to the lighting requirements for buildings other than dwellings. Installed lighting is to comprise fittings incorporating lamps with an average initial (100 hours) efficacy of not less than 50 lumens per circuit watt.

A restaurant has been chosen to demonstrate how this requirement may be satisfied. The lighting specified:

- A mixture of concealed perimeter lighting using high-frequency fluorescent fittings and individual tungsten lamps over tables.
- Lights in dining area to be switched locally from behind the bar.
- Over-table lamps to have integral switches for customers' use.
- Lighting in kitchens and toilets to be switched locally.

The following enumerates the light sources proposed together with a calculation of the overall circuit efficacy:

(a) 20 × 60 W tungsten over-table lamps,
    have 60 circuit watts per lamp and
    lumen output of 710 per lamp
    Total circuit watts   1200 W
    Total lumen output   14 200 lm
(b) 24 × 32 W T8 fluorescent with
    high-frequency control concealed
    perimeter and bar lighting, with
    36 circuit watts per lamp, and a
    lumen output of 3300 each
    Total circuit watts   864 W
    Total lumen output   79 200 lm
(c) 6 × 18 W compact fluorescent with
    mains frequency control to toilets
    and circulation, with 23 circuit
    watts per lamp, and lumen output
    of 1200 per lamp
    Total circuit watts   138 W
    Total lumen output is   7200 lm
(d) 6 × 50 W T8 fluorescent with
    high-frequency control in kitchens
    with 56 circuit watts per lamp, and
    5200 lumen output per lamp
    Total circuit watts   336 W
    Total lumen output   31 200 lm

Totals   2538 W   131 800 lm

The average circuit efficacy is therefore

$$\frac{131\,800}{2538} = 51.9 \text{ lumens/watt}$$

This being better than the requirement, the scheme is satisfactory.

RELEVANT READING

British Standards and Codes of Practice

BS 699: 1984. Specification for copper direct cylinders for domestic purposes
    Amendment 1: AMD 5792
                            2: AMD 6600
BS 853: 1990. Calorifiers and storage vessels for central heating and hot water supply
BS 1566: Part 1: 1984 (1990). Double feed indirect cylinders
    Amendment 1: AMD 5790
                            2: AMD 6598
BS 1566: Part 2: 1984 (1990). Specification for single feed indirect cylinders
    Amendment 1: AMD 5791
                            2: AMD 6601
BS 3198: 1981. Specification for copper hot water storage combination units for domestic purposes
    Amendment 1: AMD 4372
                            2: AMD 6599
BS 3869: 1965. Rigid expanded polyvinyl chloride for thermal insulation purposes
BS 3958: Part 5: 1986. Specification for bonded man-made mineral fibre slabs
BS 3958: Part 2: 1982. Calcium silicate preformed insulation
BS 3958: Part 3: 1985. Metal mesh faced man-made mineral fibre mattresses
BS 3958: Part 4: 1982. Bonded preformed man-made mineral fibre pipe sections
BS 3958: Part 5: 1986. Bonded mineral wool slabs (temperatures over 50°C)
BS 3958: Part 6: 1972. Finishing materials, hard setting composition, self-setting cement and gypsum plaster
BS 4016: 1972. Specification for building papers (breather type)
BS 4046: 1991. Specification for compressed straw building slabs
BS 4841: Part 3: 1994. Specification for two types rigid polyurethane laminated roofboard
BS 4841: Part 2: 1975. Laminated board for use in wall and ceiling insulation
BS 5241: 1994. Rigid urethane foam when dispensed or sprayed on a construction site
BS 5250: 1989. Code of practice for control of condensation in buildings
BS 5284: 1993. Sampling and testing mastic asphalt and pitchmastic
BS 5422: 1990. Use of thermal insulating materials in temperature range −40 to +700°C
BS 5449: 1990. Forced circulation hot water central heating systems
BS 5608: 1978. Preformed rigid urethane foams for insulation of pipework
BS 5615: 1985. Specification for insulating jackets for domestic hot water storage cylinders
BS 5617: 1985 (1996). Specification for urea formaldehyde foam systems – cavity walls
BS 5618: 1985. Code of practice for thermal insulation of cavity walls
BS 5628: 1992. Use of masonry materials and components
BS 5803: Part 1: 1985. Specification for man-made mineral fibre thermal insulation mats
BS 5803: Part 2: 1985. Man-made mineral fibre thermal insulation in pelleted or granular form

BS 5803: Part 3: 1985. Specification for cellulose fibre thermal insulation

BS 5803: Part 4: 1985. Methods of test determining flammability and resistance to smouldering

BS 5803: Part 5: 1985. Specification for the installation of man-made mineral fibre and cellulose fibre insulation

BS 5864: 1990. Installation of gas fired ducted air heaters output up to 60 kW

BS 5925: 1991. Ventilation principles and designing for natural ventilation

BS 6203: 1991. Guide to fire characteristics and fire performance of expanded polystyrene used in building applications

BS 6232: Part 1: 1982. Specification for performance of blown man-made mineral wool

BS 6232: Part 2: 1982. Code of practice for installation of blown man-made mineral fibre in cavity walls with masonry and/or concrete leaves

BS 6676: Part 2: 1986 (1994). Code of practice for installation of batts filling a cavity

BS 7021: 1989. Thermal insulation of roofs internally by sprayed polyurethane or polyisocyanurate

BS 7456: 1991. Code of practice for stabilization and thermal insulation of cavity walls with polyurethane

BS 7457: 1991. Specification for polyurethane foam systems

BS 8207: 1986 (1995). Code of practice for energy efficiency

BS 8208: Part 1: 1985. Suitability of cavity walls for filling with thermal insulants
Amendment 1: AMD 4996

BS 8211: Part 1: 1988. Code of practice for energy efficient refurbishment of housing

BS 8216: 1991. Code of practice use of sprayed lightweight mineral coating for thermal and sound insulation

BRE digests

108	Standard U values
145	Heat losses through ground floors
190	Heat losses from dwellings
217	Wall cladding defects and their diagnosis
232	Energy conservation in artificial lighting
233	Fire hazard from insulating materials
236	Cavity insulation
251	Assessment of damage from progressive ground movement
254	Reliability and performance of solar collector systems
270	Condensation in insulated domestic roofs
277	Built-in cavity wall insulation for housing
294	Fire risk from combustible cavity insulation
295	Stability under wind load of loose laid external roof insulation boards
296	Stability under wind load – loose laid external roof insulation boards
297	Surface condensation and mould growth
306	Domestic draughtproofing
312	Flat roof design – the technical options
319	Domestic draught proofing: materials and benefits
324	Flat roof design – Insulation
342	Autoclaved aerated concrete
350	Part 1: General climate of UK
	Part 2: Influence of microclimate
	Part 3: Improving microclimate by design
369	Interstitial condensation and fabric degradation
379	Double glazing for heat and sound insulation

BRE information papers

3/81 Assessment of U values for insulated roofs

2/84 Movement of foamed plastics insulants in ward deck flat roofs

7/84 Urea–formaldehyde foam cavity insulation – reducing formaldehyde vapour in dwellings

1/87 Investing in energy efficiency – domestic hot water system

10/79 Field studies on the effect of increased thermal insulation in some electrically heated houses

22/86 Investing in energy efficiency – new housing

21/82 Moisture relations in timber framed walls

25/82 Formaldehyde vapour from urea–formaldehyde foam insulation

3/83 Optimum start controls in modern low energy buildings

16/85 The BREDEM domestic model

6/85 Selection of building energy management systems

7/85 Cost effectiveness of heat pumps in highly insulated dwellings

17/86 Investing in fuel efficiency (1)

20/86 Investing in fuel efficiency (2)

22/86 New housing – energy efficiency

1/87 Domestic hot water systems – energy efficiency

2/88 Rain penetration of cavity walls

13/88 Energy assessment for dwellings using BREDEM worksheets

22/88 Energy efficiency in housing stock

2/89 Thermal performance of lightweight inverted warm deck flat roofs

24/89 Opportunities for improving energy efficiency in housing

1/90 High tech mixed use buildings: attitudes to emergency efficiency

2/90 Greenhouse gas emissions and buildings in UK

3/90 U value of ground floors

4/91 Improving energy efficient performance of high rise housing

12/92 Energy audits and surveys

7/93 U value of solid ground floors with edge insulation

12/93 Heat losses through windows

12/94 Condensation risk and heat loss at thermal bridges around openings

14/94 U value for basements

15/94 Energy efficiency in new housing

16/94 Energy consumption in public and commercial buildings

4/95 A guide to the development of BREDEM

15/95 Potential carbon emission savings from energy efficiency in housing

1/98 Mortars for blockwork: thermal performance

4/98 Night ventilation for cooling office buildings

5/98 Metal cladding: assessing thermal performance

9/98 Energy efficient concrete walls using EPS

BRE monograph

Cavity insulation of masonry walls – dampness risks and how to minimize them

BRE defect action sheets

3	Slated or pitched roofs: restricting entry of water vapour from house
17	External masonry walls: insulated with mineral fibre cavity width batts
77	Cavity external walls – cold bridges
78	External walls – dry linings – avoiding cold bridges
79	External masonry walls partial cavity fill insulation – rain penetration

BRE good building guide

GB5   Choosing between cavity, internal and external insulation

BRE best practice programme

1   Energy efficiency in new buildings

BRE reports

Thermal insulation—avoiding risks. Ref. BR 262: 1994

Building regulations: conservation of fuel and power – the 'energy target' method of compliance for dwellings (B. R. Anderson). Ref. BR 150

An economic assessment of some energy conservation measures in housing and other buildings (John Plezzey)

Minimizing air infiltration in office buildings. Ref. BR 265: 1994

Double glazing units: a BRE guide to improved durability. Ref. BR 280: 1995

BREDAM–12: Model description. Ref. BR 313: 1996

Non-domestic building energy fact file. Ref. BR 339: 1998.

BRE pamphlet

Cost efficiency of thermal insulation

BSI

A designer's manual for the energy efficient refurbishment of housing

Department of Energy

Fuel efficiency booklet 8

The economic thickness of insulation for hot pipes

The following are obtainable from

Chartered Institution of Building Services, Delta House, 222 Balham High Road, London SW12 9BS

Building Energy Code: 1981: Part 2

Guide A: Design Data: Section A3: 1980

Code for internal lighting: 1994

Applications Manual AM1: 1985

Automatic controls and their implications for systems design

The future

It is anticipated that, shortly, new elemental U values of 0.35 for walls, 0.3 for floors and 0.2 for roofs will be embraced by Part L, to be followed later by even more stringent requirements of 0.3 for walls, 0.23 for floors and 0.16 for roofs.

CHAPTER 21

Access and facilities for disabled people

(Approved document M)

There has been legislation requiring that greater attention be given to the needs of disabled people in the design of buildings for a number of years.

Commencing with the Chronically Sick and Disabled Persons Act 1970, and the Town and Country Planning Act 1971 (replaced by the Town and Country Planning Act 1990), there have been several statutes requiring that buildings to which the public are admitted, including places where people work, and schools, should be accessible to disabled persons.

The Chronically Sick and Disabled Persons Act 1970 requires

- provision that is practicable and reasonable for disabled in the means of access to, and inside, a building or premises, and in the parking facilities and sanitary conveniences to be available;
- when a local authority serves notices requiring sanitary conveniences for persons visiting an inn, public house, beer house or place of public entertainment, provision must be made for the needs of disabled people.

The 1970 Act was amended by the Local Government (Miscellaneous Provisions) Act 1976 but it was not mandatory and there were no enforcement provisions. External arrangements for access fail to be dealt with under the 1971 Act, and as amended by the 1980 Act requires that planning authorities should draw attention to the need for adequate access arrangements in connection with appropriate buildings. All that happened in consequence of this legislation was that local authorities were able to make developers aware of their obligations by means of a note or letter when plans were received, or accompanying the permission when issued.

Designers and developers, with splendid exceptions, did not heed fully the legislation and it became necessary to issue building regulations as an amendment to the 1976 Regs, only three months before the 1985 Regs came into operation; such was the urgency of the matter.

The 1985 Regs, 4, required compliance with Schedule 2. This schedule was identical to Part T of the 1976 Regs and applied to all floors in new shops, offices, single-storey factories, educational buildings and other new single-storey buildings open to the public.

Reliance was placed upon British Standard 5810: 1979 and Department of Education and Science Design Note 18 as deemed to satisfy standards, but the recommendations in these documents were not mandatory.

It is noticeable that the requirements were directed at new buildings and they did not therefore relate to alterations and extensions to existing buildings. They did not refer to buildings subject to a material change of use (Regulation 6, para (1)).

Consultation documents, the forerunner of a recast of regulations, were issued in September 1986. The aim was to introduce a more flexible style echoing the rest of the building regulations.

The Building (Disabled People) Regulations 1987 stem from these consultations, and they came into force on 14 December 1987, and were slightly amended in 1988.

The principal change was the replacement of deemed to satisfy provisions by a functional requirement aiming for reasonable provision to be made for disabled people. As with the rest of regulations

this is to be achieved by use of an approved document. It also means that there is no provision for relaxation procedures as the new functional requirements cannot be relaxed although in certain circumstances it may be reasonable to dispense with the requirements. One cannot require that any provision be less than reasonable, which is the standard set in connection with all provision for disabled persons.

It will be seen that the current regulations differ from the old Schedule 2 in the following principles.

Part M of Schedule 1 to the 1991 Regs was recast involving a new definition of disabled people which includes people with impaired hearing or sight.

Additionally the requirements were extended to include non-domestic buildings which are newly erected or buildings which have been reconstructed and to building extensions which include a ground storey.

The Building Regulations (Amendment) Regulations 1998 amended the 1991 Regs by inserting a new Part M into Schedule 1. As amended, Part M has altered the limits of application so that, in general, the requirements regarding access and facilities for disabled people will apply to dwellings. Also M.3 has been amended to provide a new requirement regarding sanitary conveniences in dwellings.

The approved document to Part M sets out the Secretary of State's advice regarding application of the requirements to make it reasonably safe and convenient for disabled people to get into, and out of, and to move about within buildings.

The requirements are intended to ensure reasonable access and facilities for people who visit or work in a building.

There is now also a requirement that disabled people are able to enter new dwellings and can move about the principal storey. These provisions are not necessarily intended to enable fully-independent living for all disabled people.

M1. Interpretation

Part M commences by setting down definitions. The complex definition 'disabled people' contained in the National Assistance Act 1948 has been discarded for building regulation purposes and this much simpler description is used.

Disabled people are

- **people with physical impairment which limits their ability to walk, or**
- **people dependent upon a wheelchair for mobility, or**
- **people with impaired hearing or sight.**

In a number of appeal and determination decisions the Secretary of State has made it plain that he cannot determine what disabled people can or cannot do in a working environment. Much depends on individual circumstances, not only on the nature of the work, but also on the nature and degree of disability and the personal commitment of the individual concerned.

The term 'building' is set out in the approved document. It may include individual premises and

- **an office**
  This is a premises where the principal or sole use includes administration, clerical work, handling money and telephone or telegraph operating. Clerical work embraces writing, book keeping, sorting papers, typing, duplicating, machine calculating, drawing and editorial preparation of matter for publication. (1985 Regs, 2, not defined in 1991 Regs.)
- **a shop**
  This is a premises used for the carrying on of a retail trade or, business, and premises to which members of the public are invited for the purpose of delivering goods for repair, or other treatment of goods.
    Included are:

  (a) the sale of food and drink for immediate consumption;
  (b) retail sales by auction;
  (c) a barber or hairdresser. (1991 Regs, 2.)

- **a factory**
  A factory is any premises in which persons are employed in manual labour in any process for or incidental to:

  (a) the making of any article or part of any article, or
  (b) the altering, repairing, ornamenting, finishing, cleaning or washing or breaking up or demolition of any article, or
  (c) the adapting for sale of any article. (1984 Act, section 126 and Factories Act 1961, section 175.)

- **a warehouse**
  Warehouse is not defined in the regulations but the ordinary dictionary definition says it is a place where goods or merchandise are stored, or a large shop usually selling goods wholesale.
- **a school or other educational establishment**
  However, schools and other educational establishments erected in accordance with plans approved by the Secretary of State for Education and Science, or the Secretary of State for Wales are exempt from building regulations. (1984 Act, section 4(1)(a).)

Nevertheless the Secretary of State is unlikely to approve buildings which fail to meet his own requirements in Design Note 18.

The requirements therefore relate to those schools and educational buildings which do not require plans to be approved by the Secretary of State.

A definition of 'school' includes a Sunday school (1984 Act, section 126) but there is no definition in the Act, or regulations, of an educational establishment. Design Note 18 was intended to extend to special schools, further and higher education establishments and colleges, including all buildings for further and higher education.

- **student residential accommodation**
- **premises to which the public are admitted on payment or otherwise and is on the storey of the building containing the principal entrance to that building.**
- **an institution**
  This means an institution which may be described as a hospital, home, school or similar establishment used for living accommodation.
    It also includes a building used for the treatment, care or maintenance of people

  1. with disabilities due to illness or old age;
  2. with physical or mental incapacity;
  3. under 5 years of age who sleep on the premises.

There is an approved document setting out practical guidance for meeting the requirements, but there is no obligation to adopt any particular solution if the requisite standard can be achieved in some other way.

It is accepted in the foreword to the approved document that there may be circumstances when full provision of the requirements may not be reasonable. However, designers and building providers are encouraged whenever it is possible to go beyond the level of provision demanded by Part M. Reference to some of the literature listed in Relevant Reading at the end of the chapter will be helpful in these circumstances.

The approved document also defines a number of words or phrases when used in the document:

- 'access' means approach or entry;
- 'accessible' means that premises and their facilities are suitably designed to enable disabled persons to reach and use them;
- 'suitable' indicates that means of access or facilities are to be designed to enable a disabled person to use them;
- 'principal entrance storey' is the storey which contains the principal entrance to a building. When an alternative accessible entrance is to be provided the storey in which that entrance is placed is the principal entrance storey.

There are also definitions which apply only to terms used in that part of the approved document relating to dwellings:

- 'common' serves more than one dwelling;
- 'habitable room' in principal storey is a room used or intended to be used for dwelling purposes. It includes a kitchen but not a bathroom or utility room;
- 'maisonette' – a self-contained dwelling occupying more than one storey, but does not include a dwelling house
- 'point of access' – entry point where person normally alights from vehicle. May be within, or outside perimeter;
- 'principal entrance' – entrance one would normally expect a caller to approach. The common entrance to a block of flats;
- 'plot gradient' – slope between the point of access and the floor of the dwelling;
- 'steeply sloping plot' – where gradient is steeper than 1:15.

APPLICATION OF REQUIREMENTS

- Part M applies to new buildings and also to a building which has been demolished to such an extent that only the external walls remain.
- When reconstruction of a building other than a dwelling takes place, and it is not practical to make adjustments to levels of principal or other appropriate entrances to facilitate use by wheelchair users, or to provide a suitable new entrance, all other requirements of Part M will apply.
- When an existing building is extended or altered, one has to look at the 1991 Regs. 3(1) and 3(2) as amended, in deciding whether the work will adversely affect the existing building.

  With regard to an extension to a building other than a dwelling, Part M applies if the extension contains a ground storey, and will be met if arrangements for access and facilities in the existing building, including access from outside the building, are maintained. Thus, if a building should be extended at the existing point of access any new access is not to be less accessible and suitable. Also there is to be access from it to the existing buildings. A ramp should not be replaced by a stepped access.
- When an alteration to a building is proposed, it is important to recall that Part M is one of the requirements set out in the definition of a relevant requirement (1991 Regs, 3). Therefore, when a building is altered, access and facilities are to be maintained. This is not to say existing provisions must not be interfered with. They may be moved to other locations, but the general level of provision must be maintained.
- External features. It must be possible for sufficient facilities to be provided to enable a disabled person to get to the building from the edge of the site or from car parking arrangements provided on the site.

Limits of application

The requirements in Part M do not apply to:

- an extension to a dwelling;
- any other extension if it does not include a ground storey;
- a material alteration (a material alteration is referred to in 1991

Regs, 3, and Regulation 4 says that a material alteration must not result in an altered building having less satisfactory means of access and facilities for disabled people than it had prior to the alteration);
- any part of a building used only for inspection, maintenance or repair of the building or any service or fitting in the building.

INTRODUCING THE PROVISIONS

If Part M applies to a building, the following requirements are to be satisfied:

(a) Buildings other than dwellings

- Disabled people must be able to reach the principal entrance, or other appropriate entrances.
- Parts of a building are not to be a hazard to a person with defective sight.
- There must be access into and within a building, and to the various facilities provided.
- It must be possible for disabled people to use the facilities provided.
- Suitable sanitary conveniences must be provided.
- Suitable audience or spectator seating arrangements must be provided.
- People with hearing or sight defects are to have aids provided in auditoria, meeting rooms, reception areas and ticket offices.

The arrangements for access and facilities may be for the benefit of occasional visits, or employees, or both.

When premises in a storey are on more than one level, suitable means of access are to be provided between the levels. They are regarded as one storey.

Access for disabled persons is not required to those parts of a building used for the inspection, maintenance or repair of the building, or of its services or machinery.

The requirements of Part M do not extend to features outside a building unless they are required to provide access to a suitable entrance.

The requirements of M2, M3 and M4 in connection with schools or other educational establishments will be satisfied if provision is made in accordance with stipulated paragraphs of Design Note 18, 'Access for disabled persons to educational buildings'. The paragraphs are 2.1/2/4/6, 3.1, 4.1/2/4/6 and 5.1. Requirements for provision for people with defective sight and hearing may require inclusion as some features described as general design considerations.

(b) Dwellings

- Disabled people must be able to reach the principal, or a suitable alternative entrance from the point of access;
- there is to be access into and throughout the principal storey;
- there is to be sanitary accommodation in the principal storey.

These requirements relate to any purpose-built student living accommodation, but do not refer to traditional halls of residence which provide bedrooms and are not self-contained accommodation.

CONTENTS OF APPROVED DOCUMENT

The approved document 1999 edition, which became operative on 24 October 1999, contains guidance on dwellings as well as other buildings.

The advice is set out in ten sections.

The first five relate to buildings other than dwellings, and sections 6 to 10 are concerned with dwellings.

Section 1 – Means of access to and into buildings.
        2 – Means of access within buildings.
        3 – Use of buildings.

4 – Sanitary conveniences.
5 – Audience and spectator seating.
6 – Means of access: dwellings.
7 – Circulation within entrance storey: dwellings.
8 – Accessible switches and socket outlets.
9 – Lifts and common stairs in blocks of flats.
10 – WC provision: dwellings.

MEANS OF ACCESS

M2 requires that reasonable provision be made to enable disabled people to get into buildings and also to use the building.

One man's reasonableness is another man's onerous requirement and so one must look to the approved document to measure the standard of reasonableness expected.

Recommendations in the approved document cover the following:

- access to the building;
- access inside the principal entrance storey;
- access to other storeys;
- access within other storeys.

1. ACCESS TO BUILDINGS OTHER THAN DWELLINGS

An approach route into a building which is intended for use by disabled people should have areas of its surface formed in such a way that a disabled person is aware of the information being given. This may be achieved by the use of tactile paving slabs which should be placed where the access route crosses a carriageway, and also at the top of steps (Fig. M.1A).

Dropped kerbs should be provided for wheelchair users.

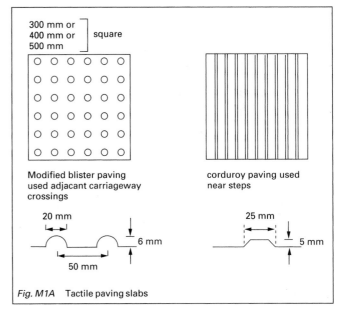

300 mm or
400 mm or   square
500 mm

Modified blister paving used adjacant carriageway crossings

corduroy paving used near steps

20 mm

6 mm

50 mm

25 mm

5 mm

Fig. M1A   Tactile paving slabs

The aim should be to provide access through the principal entrance to the building. Also when other entrances are provided for visitors, customers or staff each of these entrances should be both accessible and suitable.

An approach to an entrance should take account of the fact that changes of level do present difficulties for wheelchair users, ambulant disabled and people with impaired sight.

There may be exceptional instances when the space outside the principal entrance is just not suitable. This could be because the space is restricted, or could become congested or the site is on sloping ground. Should such circumstances arise an alternative entrance, if it is intended for general use may be accepted when it is accessible and suitable.

There should be suitable and accessible access from any car spaces provided, to the principal entrance. However, there may be situations when parking provision, although sited adjacent to the building, it is impracticable to provide suitable means of access to the principal entrance. In these circumstances an additional accessible entrance, if it is intended for general use, may be accepted where it is suitable.

If an access is close to a building it is important to ensure that projections from the building do not present a hazard to the disabled.

Many ambulant disabled find it easier to negotiate a stair than a ramp.

Design Note 18, 'Access for disabled persons to educational buildings' sets out acceptable provisions for schools and educational establishments as follows:

- approach (2.1);
- external ramps (2.2);
- external steps (2.4);
- handrails (2.6).

ENTRANCE APPROACH

The approved document contains information on 'level' approaches and emphasizes the need for adequate space for disabled people to pass. A level approach is one which has a gradient no steeper than 1:20.

A level or ramped entrance to a building is essential for wheel-chair users (Fig. M.1). The same arrangement is convenient for use by elderly people, and parents with prams. However, ramps are not the solution for all disabled people. For example, amputees and hemiplegics may find steps easier to negotiate than a long ramp.

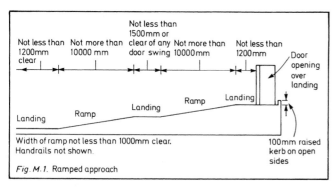

Fig. M.1. Ramped approach

The basic requirement is that the approach should not be steeper than 1:20 and it should have a width of at least 1200 mm. In circumstances where site constraints mean that a steeper ramp is the only way an approach may be achieved, then there are restrictions on the length of each slope and a requirement for landings.

Stepped approaches may be provided in addition to a ramped approach and they are to be designed in accordance with guidance given in the approved document.

- K1 required that stairways and ramps are to be constructed in such a way that they afford safe passage for people using them.
- K2 requires protection from falling for persons using staircases and ramps.

The requirements are illustrated in Chapter 19.

Ramped approach without associated stepped approach

The minimum requirements for these ramps are:

- non-slip surface;
- clear width not less than 1000 mm and surface width not less than 1200 mm;

- not steeper than 1:12; when ramp not longer than 5 m;
- for ramp not longer than 10 m to be not steeper than 1:15;
- have a landing at the top and the bottom the same width as the ramp and having a length clear of any door swing over it not less than 1200 mm;
- intermediate landings to be the same width as the ramp and not less than 1500 mm long clear of any door swing onto it;
- any open side is to be protected by a kerb not less than 100 mm high;
- guarding and handrails are to be in accordance with K1 and K2, and the requirements are illustrated in Chapter 19;
- irrespective of the requirements for guarding set out in K1 there is to be a suitable handrail on each side where length of ramp is more than 2 m (Figs M.2 and M.3).

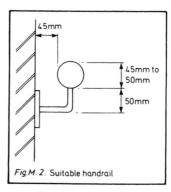

Fig.M.2. Suitable handrail

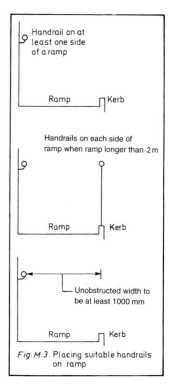

Fig.M.3. Placing suitable handrails on ramp

Stepped approach

When designing a stepped approach there are a number of factors which must be catered for – different disabilities present different problems for the person affected, for example:

- tripping and losing balance by persons with impaired sight;
- tripping by catching feet beneath nosings and treads is a risk borne by people wearing callipers or with stiff knee or hip joints.

Where a stepped approach is to be provided, it is regarded as suitable if it fulfils the following requirements:

- there is a tactile surface on top landings to give warnings (Fig. M.4A);
- the rise of flight between landings is not to exceed 1200 mm;
- intermediate landings are to have a depth not less than 1200 mm clear of door swing;
- there is to be a landing at the top and bottom of a stepped approach with length at least 1200 mm clear of door swing;
- step nosings are to be distinguished by contrasting brightness;
- risers are not to exceed 150 mm and are not to be open;
- the going should be uniform and at least 280 mm with any tapered step measured 270 mm from inside of the stair;
- suitable handrails are to be provided on each side of the flight of two or more risers;
- unobstructed width not less than 1000 mm;

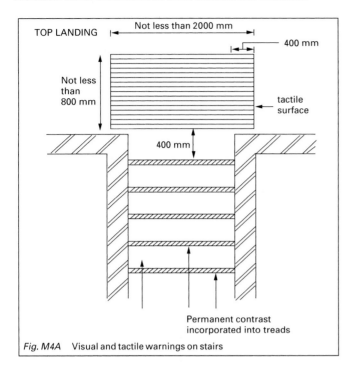

Fig. M4A   Visual and tactile warnings on stairs

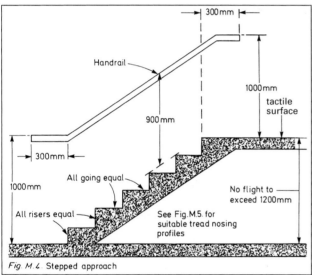

Fig. M.4. Stepped approach

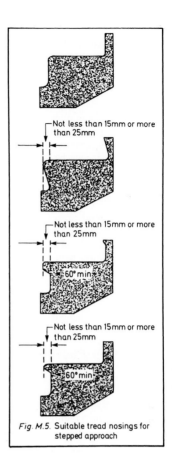

Fig. M.5. Suitable tread nosings for stepped approach

- in all other aspects the approach is to comply with K1, discussed in Chapter 19;
- comply with the dimensions shown in Figs M.2, M.4 and the tread details in Fig. M.5.

The design of a suitable handrail requires care and when provided in connection with ramp or stepped approach is to be in accordance with Fig. M.2 and as follows:

- not more than 900 mm above surface of a ramp or the line of nosings;
- not more than 1000 mm above surface of any landing;
- extend 300 mm beyond the top and bottom of a ramp or the top and bottom nosings of a stepped approach;
- terminate in a closed end which does not project into a route of travel;
- dimensions of suitable handrail as shown in Fig. M.2, and see also Fig. M.4.

ACCESS ROUTES

Parts of a building may occasionally obstruct a route and create a hazard to partially sighted and blind people. It is therefore particularly important that windows and doors which open outwards should not protrude over an adjacent path (Fig. M.5A).

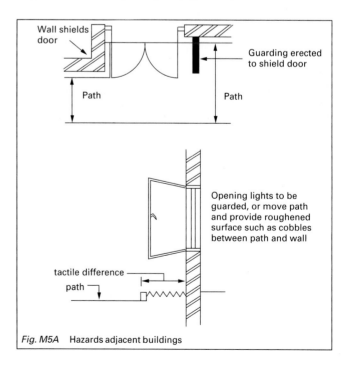

Fig. M5A   Hazards adjacent buildings

Access to a building must be convenient whatever way disabled persons may arrive, and the requirements of M2 will be satisfied if

- the principal entrance is accessible and suitable for customers and visitors alike;
- where space outside the principal entrance is severely limited, or the ground may be sloping, another alternative entrance giving adequate internal access may be provided;
- if access from car parking spaces to the principal access may be unsuitable, suitable alternative entrance may be provided;
- an entrance provided for staff may be used if it is accessible and suitable.

Schools and educational establishments

Design Note 18 makes recommendations for arrivals, departures and parking in connection with educational buildings. It is recommended that a setting down area which is level and suitably sized should be provided in a convenient position near the point of entry to the building.

For disabled people arriving in their own vehicles parking spaces should be provided:

- for ambulant disabled people 2.8 m wide;
- for wheelchair users 3.2 m wide.

Requirements stemming for the use of Design Note 18 are that approaches to the building are to be clearly defined, well lit and not less than 1200 mm wide. Ramps should not exceed 1:12 and have a raised kerb on the edge. If the length of a ramp is more than 3 m but less than 6 m the maximum gradient should not exceed 1:16. Over 6 m but not more than 10 m long a ramp should not have a gradient in excess of 1:20 (Fig. M.6).

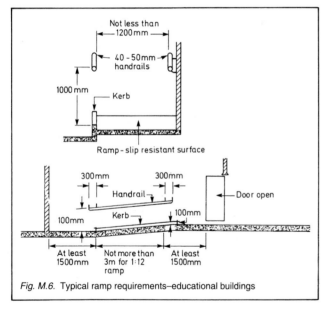

Fig. M.6. Typical ramp requirements–educational buildings

A level landing not less than 1200 mm × 1500 mm unobstructed and clear of any gate or door swing is necessary at the head and foot of each ramp.

Suitably designed handrails are to be provided on each side of ramps which exceed 1:20 gradient.

All surfaces should be firm, well drained and slip resistant. Additionally, all ramps and landings must be adequately lighted.

Stepped approaches are to have goings of not less than 280 mm with risers not more than 150 mm. The rise of a flight of steps must not exceed 1200 mm and all risers and goings must be uniform.

Suitably designed handrails are required on each side of steps. Surfaces must be well drained and slip resistant. All steps and associated landings are to be adequately lighted.

Handrails are to be continuous and unbroken at landings and they

must always extend horizontally at least 300 mm at the top and bottom of any ramp or staircase.

Handrails for ramps are to be fixed so that the top of the rail is not more than 1000 mm above the ramp surface.

For steps handrails are not to be more than 850 mm above line of the step nosings, or more than 1000 mm above finished landing or ground level (Fig. M.7).

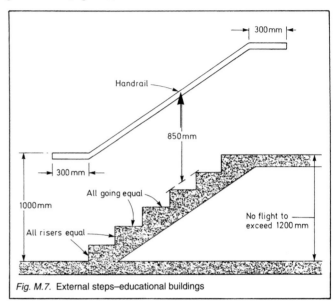

Fig. M.7. External steps–educational buildings

ENTRANCE DOORS

The approved document concentrates on the size of the opening in connection with entrance doors. It points out that disabled people cannot react quickly to avoid collisions, and therefore they should be able to see others approaching from the other side of a door, and should also be seen. It does not refer to other important considerations necessary in the provision of a satisfactory access for disabled persons:

- thresholds should be flush;
- single steps associated with doors to be avoided;
- glazed doors to be clearly identifiable;
- ironmongery easily usable;
- door closers to have minimum practical opening pressure;
- doors to have kicking plates.

The requirement for a suitable means of access is for a minimum clear opening width as provided by a 1000 single leaf external door set or by one leaf of an 1800 double leaf external doorset.

Reference is made to Table 2 to BS 4787 (850 mm and 810 mm respectively). In no case is it to be less than 800 mm.

There is to be an unobstructed space 300 mm wide on the side next to the leading edge of the door.

An entrance door is to be provided with a vision panel extending from a height of 900 mm to 1500 mm (Fig. M.7A).

BS 4787: Part 1: 1980 is entitled 'Dimensional requirements for door sets' and covers metric sizes for dimensionally coordinated wood doorsets, door leaves and frames.

All dimensions are given including doorstep width, rebate depth clearances, permissible deviations, position of latch/lock and hinges. In addition minimum margins around glazing are given.

REVOLVING DOORS

Revolving doors present problems for disabled persons and any entrance fitted with a revolving door is to be supplemented by

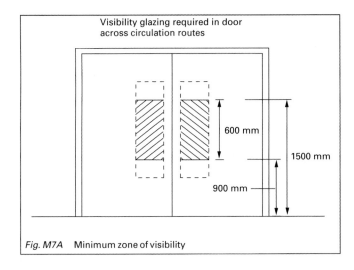

Visibility glazing required in door across circulation routes

Fig. M7A   Minimum zone of visibility

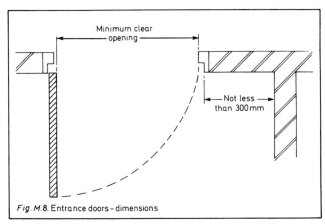

Fig. M.8. Entrance doors – dimensions

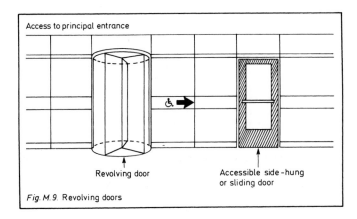

Access to principal entrance

Revolving door

Accessible side-hung or sliding door

Fig. M.9. Revolving doors

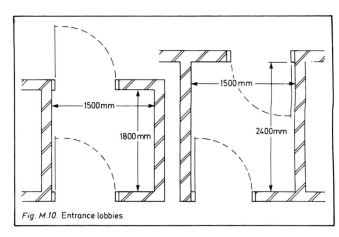

Fig. M.10. Entrance lobbies

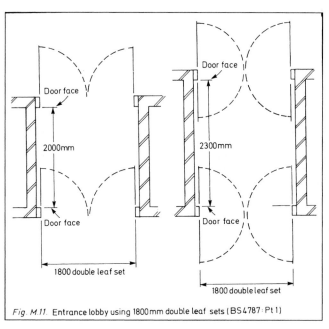

Fig. M.11. Entrance lobby using 1800mm double leaf sets (BS 4787 : Pt 1)

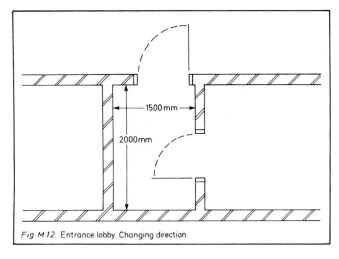

Fig. M.12. Entrance lobby. Changing direction

an accessible side hung or sliding door as described above in regular use.

There are some quite large revolving doors which move very slowly.

They are capable of accommodating several people and these doors are considered suitable.

ENTRANCE LOBBY

Entrance lobbies should be easily negotiable and allow for wheelchair manoeuvre. Doors should be placed for easy unobstructed approach and any mats should be close fitting with the floor surface, and firm. Minimum dimensions are given in Figs M.10, M.11 and M.12.

Schools and educational establishments

The clear opening for single leaf doors is not to be less than 800 mm for entrance doors leading to spaces intended for wheelchair users.

A clear wallspace of 300 mm is required on the door handle side to aid wheelchair users when opening the door.

Unobstructed circulation space is required within all lobbies accessible to wheelchair users (see Figs M.13 and M.14).

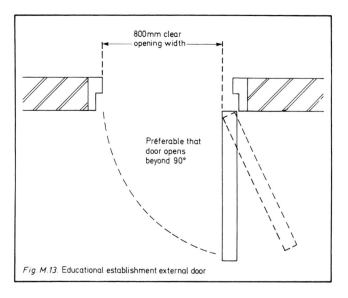

Fig. M.13. Educational establishment external door

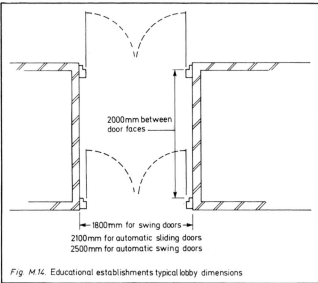

1800mm for swing doors
2100mm for automatic sliding doors
2500mm for automatic swing doors

Fig. M.14. Educational establishments typical lobby dimensions

2. ACCESS WITHIN BUILDINGS OTHER THAN DWELLINGS

The most suitable means of access from one storey to another is by a passenger lift, but this may not always be a reasonable requirement.

If a passenger lift is not installed a stair is necessary and this should meet the needs of ambulant disabled people. The stair must be designed to be suitable for use by persons having impaired sight.

Lifts

To be suitable a lift should accommodate a wheelchair and facilitate entry and exit of a wheelchair. Controls are to be within reach, and as an aid to people with sensory impairment a method of making the occupants aware of the floor reached by the lift should be provided. Controls to limit the possibility of a disabled person coming into contact with closing doors are also desirable.

Requirement M2 will be achieved if a passenger lift is installed which serves any storey above or below the principal entrance storey when it contains

- more than 280 m² nett floor area in a two-storey building;
- more than 200 m² nett floor area in a building of more than two storeys.

Suitable means of access must be provided from a lift to the remainder of the storey.

In order to determine the nett floor area of a storey, add together the areas of all parts of the storey which use the same street entrance or have access from an indoor mall. Areas may be in more than one part of the same storey, or they may be used for different purposes.

The following are excluded:

- areas of vertical circulation;
- sanitary accommodation;
- maintenance areas.

To meet the requirements for a lift (see Fig. M.15):

- There should be a clear landing not less than 1500 mm wide and not less than 1500 mm deep in front of the lift doors.
- The lift must be accessible.
- Lift doors should have a clear opening width of not less than 800 mm.
- Internal dimensions of the lift should be not less than 1100 mm wide by at least 1400 mm deep.
- There should be controls on the landing and inside the lift that are not less than 900 mm and not more than 1200 mm above the floor, placed not less than 400 mm from the front wall.
- There should be tactile indication near the lift call button to identify the storey.
- If more than three floors are served there is to be a tactile indication by the lift buttons within the car.
- If more than three floors are served the lift is to have visual and voice indication of the floor reached.
- The lift should be provided with a signalling system giving

    (a) five seconds' indication that the lift is answering a landing call;
    (b) five seconds' dwell time before doors begin to close.

- Doors may be operated by a re-activating device relying on a photo eye or infrared beam. Door edge pressure methods are not permissible. Minimum time for a door to remain fully open is three seconds.

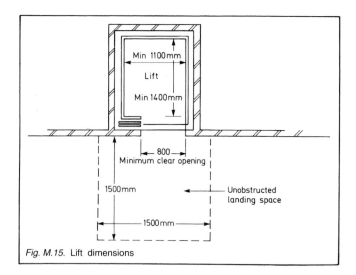

Fig. M.15. Lift dimensions

Wheelchair stairlifts

In a building containing small areas of limited use it may not be practical to install a lift but nevertheless it will be reasonable to expect access for wheelchair users.

Such small areas may have what the approved document calls a 'unique function or facility'. Unique functions such as the following may justify the installation of a wheelchair stairlift:

- a small library gallery
- staff rest room
- training room, etc.

Where a storey having a nett floor area exceeding 100 m² contains a unique facility and is not large enough to justify a passenger lift, it should nevertheless be accessible to wheelchair users and a suitable location for a wheelchair lift will be found.

Platform lift

There are occasions when the provision of a ramp within a building may be very difficult or impossible, and in these circumstances it may be possible to install a platform lift. A platform lift is not to be installed as a substitute for a stair for use by ambulant people.

Thus a platform lift may complement a stair access.

Reference is made to the following British Standards where details of some of the above requirements are detailed:

- BS 5655: Part 1: 1986, 'Safety rules for the construction and installation of electric lifts'.
- BS 5655: Part 2: 1988, 'Safety rules for the construction and installation of hydraulic lifts'.
- BS 5655: Part 5: 1989, 'Specifications for dimensions for standard lift arrangements'.
- BS 5655: Part 7: 1983, 'Specification for manual control devices, indicators and additional fittings'.
- BS 5776: 1979, 'Specification for powered stairlifts'.
- BS 6440: 1983, 'Powered lifting platforms for use by disabled people'.

ACCESS WITHIN OTHER STOREYS

An access is to be provided from any lift, ramp or stairway to:

- all relevant premises within the storey;
- any sanitary or other accommodation supplied in the storey to meet requirements in Part M.

The approved document sets out minimum requirements for access within a building covering:

- internal doors;
- corridors and passageways;
- internal lobbies;
- hotel and motel bedrooms;
- lifts;
- stairways;
- internal ramps.

Internal doors

The requirement for a suitable means of internal access is for a minimum clear opening width as provided by a 900 mm single leaf internal doorset or by one leaf of an 1800 mm double leaf internal doorset.

Reference is made to Table 1 to BS 4787 (770 mm and 820 mm respectively). In no case is it to be less than 750 mm (Fig. M.16).

There is to be an unobstructed space 300 mm wide on the side next to the leading edge of the door.

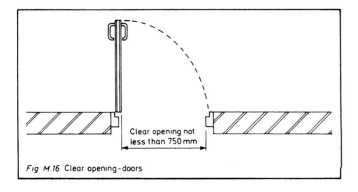

Fig. M.16. Clear opening – doors

This is not necessary if the door is opened by automatic control or it is sited in a position where it may be reasonable to expect assistance. Glazed doors across circulation routes are to be provided with visibility glazing as in Fig. M.7A

Schools and educational establishments

The clear opening width per single leaf is not to be less than 750 mm where access is given for wheelchair users. A clear wallspace not less than 300 mm is required to the side of the door by the door handle. There is to be unobstructed circulation space within all lobbies for use by people in wheelchairs (Fig. M.17).

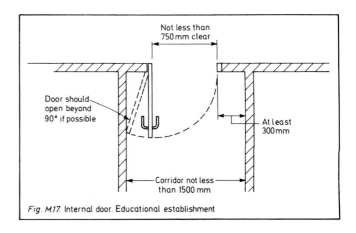

Fig. M.17. Internal door. Educational establishment

Corridors and passageways

The approved document simply says that to be suitable corridors and passageways should have a clear width of not less than 1200 mm. A corridor or passageway which is accessible by stairway alone, or is in an extension where access is gained through an existing building, is to have a clear width not less than 1000 mm. There should be no obstruction caused by radiators, fire extinguishers or other appliances.

Additionally it is desirable:

- to splay or round off corner;
- avoid unexpected changes in level;
- avoid deep pile carpets.

Figure M.18 shows some recommendations given in the design guidance notes of the Access Committee for England.

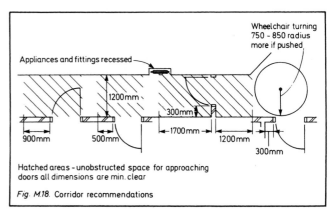

Hatched areas - unobstructed space for approaching doors all dimensions are min. clear

Fig. M.18. Corridor recommendations

Internal lobbies

All internal lobbies should be large enough to permit wheelchair manoeuvre (Fig. M.19), but it is possible for internal lobbies to be a little smaller than those required for principal entrance lobbies.

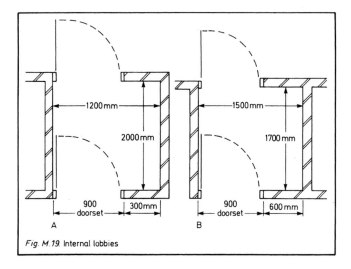

Fig. M.19. Internal lobbies

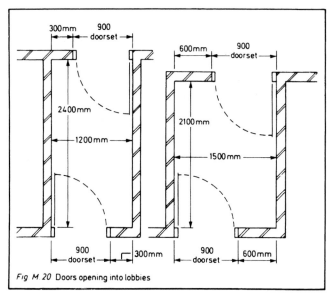

Fig M.20 Doors opening into lobbies

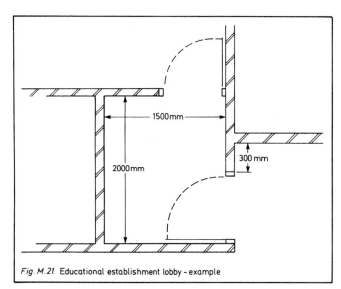

Fig. M.21. Educational establishment lobby - example

Schools and educational establishments

There is to be unobstructed circulation space within the lobbies. Recommended dimensions are as in Fig. M.19(B) and Fig. M.20. (See also Fig. M.21.)

Lifts

Schools and educational establishments

Minimum requirements for the design of lifts is a clear internal floor plan size of 1400 mm depth and 1100 mm wide. Doors are not to be less than 800 mm wide, although a preferred dimension is 900 mm to reduce the risk of damage from wheelchairs.

The landing clear space is somewhat larger than that required by the regulations. It isto be not less than 1800 mm deep and 1500 mm wide (Fig. M.22).

Design Note 18 makes the point that suitably delayed door closing action will aid entry and exit, also minimize the possibility of damage to door edges by wheelchairs.

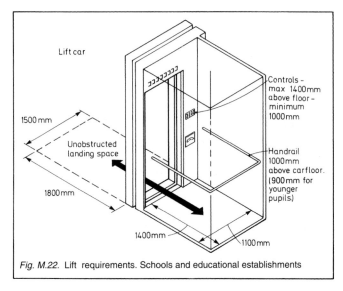

Fig. M.22. Lift requirements. Schools and educational establishments

Stair

When an internal stair is to be provided it is regarded as suitable if it fulfils the following requirements:

- step nosings to be distinguished by contrasting brightness.
- the rise of a flight between landings is not to exceed 1800 mm;
- intermediate landings are to have a width and depth not less than the width of the flight; and not less than 1200 mm long clear of door swing;
- there is to be a landing at the top and bottom of a stairway; and not less than 1200 mm clear of door swing;
- risers are to be uniform and not to exceed 170 mm;
- goings should be uniform and at least 250 mm for tapered treads measure 270 mm from inside of stair;
- suitable handrails are to be provided on each side of a stairway (see Fig. M.2); if stair comprises two or more risers; risers are not to be open;
- unobstructed width not less than 1000 mm;
- in all other aspects the stairway is to comply with K1 discussed in Chapter 19;
- comply with the dimensions shown in Fig. M.23;
- handrails should not be more than 900 mm above surface of line of nosings and be not more than 1000 mm above the surface of any landing;
- handrails should extend 300 mm beyond the top and bottom nosings of a stairway;
- handrails should terminate in a closed end which does not project into a route of travel.

Exceptionally, the provision for the rise of a flight may differ, where particular storey heights or the need to gain access beneath an intermediate landing dictates. Stairs are to be suitable for people with walking difficulties or impaired sight, but it may not be reasonable to require tactile warning for internal stairs.

Suitable tread nosing profiles are show in Fig. M.5.

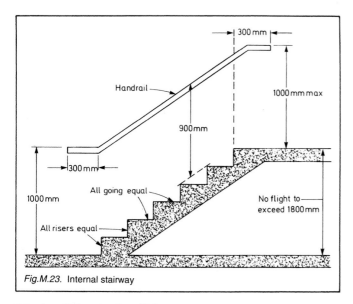

Fig.M.23. Internal stairway

Schools and educational establishments

Design Note 18 sets out the minimum requirements for internal staircases (Fig. M.24). Stair goings are to be not less than 250 mm with risers not exceeding 163 mm. Total rise of each flight is not to exceed 1800 mm. Handrails are required on each side of stairs. Treads and landing surfaces are to be slip resistant and must be adequately illuminated.

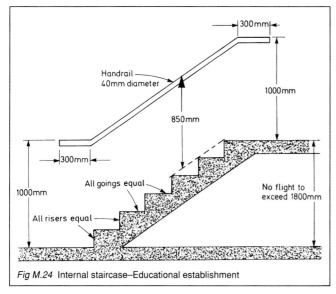

Fig M.24 Internal staircase—Educational establishment

Internal ramps

To be suitable internal ramps should be constructed as for external use.

The minimum requirements for these ramps are:

- non-slip surface;
- clear width not less than 1000 mm; surface width not less than 1200 mm;
- not steeper than 1:12; when ramp is not longer than 5 m;
- for ramp not longer than 10 m to be not steeper than 1:15;
- have a landing at the top and the bottom, the same width as the ramp and whose length clear of any door swing over it is not less than 1200 mm;
- intermediate landings to be same width as the ramp and not less than 1500 mm long clear of any door swing onto it;
- any open side is to be protected by a kerb not less than 100 mm high;

- guarding and handrails are to be in accordance with K1 and K2. K1 requires that stairways and ramps are to be constructed in such a way that they afford safe passage for people using them. K2 requires protection from falling for persons using stairways and ramps. The requirements are illustrated in Chapter 19;
- irrespective of requirements of K1 there is to be a suitable handrail on each side where length of ramp is more than 2 m.

Schools and educational establishments

When internal ramps are required they are to meet the following minimum requirements (Fig. M.25):

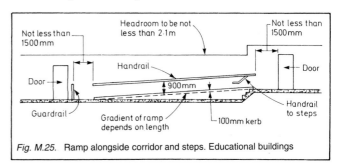

Fig. M.25. Ramp alongside corridor and steps. Educational buildings

- Gradient. When length not exceeding 3 m gradient not to exceed 1:12. Length more than 3 m but not exceeding 6 m, gradient not to exceed 1:16.
- Width. A ramp width of 1500 mm is preferred but the width is not to be less than 1200 mm.
- Landings. A level landing not less than 1200 mm × 1500 mm which must be unobstructed and clear of any door swing is required at the head and foot of each ramp. Where necessary the same dimensions apply to any intermediate landings.
- Suitably designed handrails are to be provided on each side of ramps.
- Kerbs. Kerbs at least 100 mm high are to be provided to all exposed sides of ramps unless safety balustrades are fitted.
- Surfaces. All surfaces are to be slip resistant and adequately illuminated.

3. THE USE OF BUILDINGS OTHER THAN DWELLINGS

Efforts are to be made when designing buildings to ensure that disabled people have access to the facilities provided, and to use them.

The scale and range of building types is very wide. However, it is possible to give guidance on wheelchair space and sanitary facilities in a limited number of buildings. Additionally some guidance is given to enable people with hearing difficulties to take part in such activities as conferences, committees, seminars and similar activities.

Common facilities should be sited in a storey having access for wheelchair users, for example:

- canteens;
- cloakrooms;
- doctors' and dentists' surgeries;
- other health facilities.

Inside the building

Requirements set out in the approved document stipulate that access should be provided:

- to facilitate movement in a building;
- to provide sufficient space for wheelchair users;
- to provide a convenient way of vertical circulation;
- to provide features enabling people with defective hearing and sight to find their way.

Restaurants and bars

Disabled people should be able to visit alone or with friends. If self-service and waiter service are available then it is reasonable for a disabled person to have access to both. Difficulties may be caused by changes in floor level, but all areas should be accessible to ambulant disabled people.

- access must be provided to the full range of services offered;
- there is to be wheelchair access to bars and self-service counters, and not less than half the seating area;
- if services are divided into different areas or different storeys, not less than half of each area is to be accessible to wheelchair users.

Hotels and motels

When guest bedrooms are available there are the following requirements:

- The entrance door to each guest bedroom is to be accessible to disabled people.
- One guest bedroom out of every 20 should be suitable by reason of size and layout for use by a person in a wheelchair.
- The entrance doors of other guest bedrooms are to have an opening width of 750 mm but the 300 mm space alongside the door may be omitted.
- WC and basin are to be arranged and equipped with rails as in Fig. M.26A.

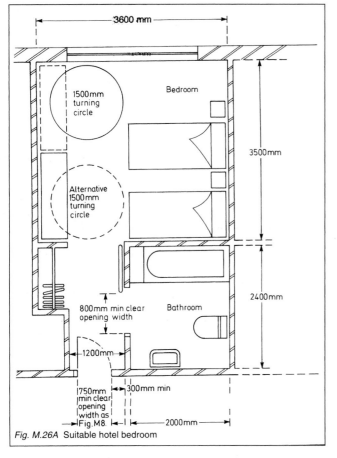

Fig. M.26A Suitable hotel bedroom

Guest bedrooms	Suitable for wheelchairs
up to 20	1
21 to 40	2
41 to 60	3

The requirements for bedroom entrance doors to be accessible is for a minimum clear opening width as Fig. M.26A. Reference is made in the approved document to Table 1 of BS 4787: Part 1.

CHANGING FACILITIES

Introduced into the 1991 approved document M are requirements for dressing cubicles and shower compartments. For a disabled person the absence of adequate facilities of this nature at swimming pools and in other recreational buildings can present very real difficulties.

Therefore the following are all critical issues:

- manoeuvring space for a wheelchair;
- space for transfer to a seat;
- seats, taps, shower heads and clothes hooks fixed at suitable heights.

Shower compartments and dressing cubicles dimensioned as in Figs M.26B, M.26C and M.26D will satisfy the requirements of M.2.

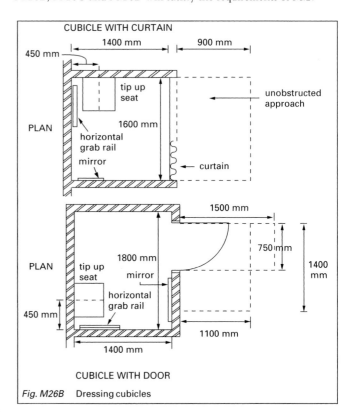

Fig. M26B   Dressing cubicles

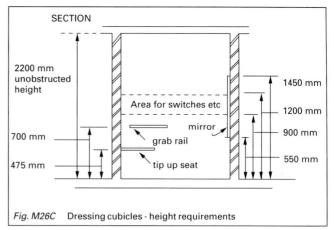

Fig. M26C   Dressing cubicles - height requirements

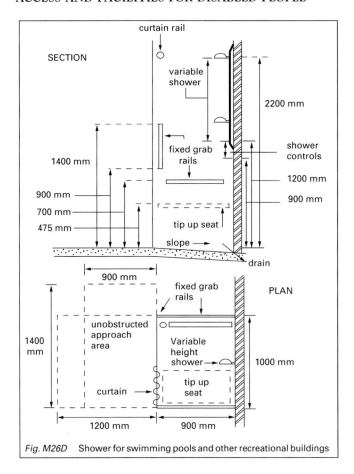

Fig. M26D Shower for swimming pools and other recreational buildings

AIDS TO COMMUNICATION

The aim is to enable a person with impaired hearing to enjoy a public performance and to play a proper part in discussions. This means that a signal some 20 dB above the 'normal' level must be available for the disabled person, and there are two systems which may provide this facility.

The induction system depends on a signal from a microphone being passed to an amplifier which directs a current through a loop surrounding the relevant space. A magnetic field is created and this is picked up by a hearing aid and converted for the wearer to hear.

Sound may spill beyond the boundary of the loop and in these circumstances if confidentiality is required it is difficult to achieve.

An infrared system radiates invisible light and this is picked up by a personal receiver, demodulated and converted for the wearer to hear.

The possibility of spill beyond the boundary is less likely with the infrared system, but the receiver must wear a stethoscope to receive the signals.

The system selected must be able to suppress reverberation, internal and external noises.

M2 is satisfied if aids to communication are available at booking and ticket offices, if the customer is separated by a glazed screen.

Aids should also be available in large reception areas in auditoria and meeting rooms of more than 100 m² in area.

Systems provided should include features ensuring the recipient hears sound without loss or distortion because of bad acoustics or extraneous noise.

The choice of system remains with the building owner.

4. SANITARY CONVENIENCES IN BUILDINGS OTHER THAN DWELLINGS

Special toilet facilities for disabled people should be readily available. Unisex toilets for disabled people help to minimize costs and space requirements; and they also allow any necessary assistance for disabled people to be given by helpers of the opposite sex.

Sanitary conveniences provided must be suitable in number, designed for use by disabled people and given a suitable means of access.

The regulations do not limit requirements to provision for visitors or customers. Where in any of the buildings, sanitary conveniences are provided for employees, provision must also be made for disabled employees. With suitable siting and access arrangements the same conveniences may serve both employees and visitors.

The requirements regarding sanitary conveniences were extended in the 1998 Regs in order to embrace dwellings, but with regard to buildings that are not dwellings the requirement remains as follows.

M3(3) requires that where sanitary conveniences are provided in a building which is not a dwelling, reasonable provision is to be made for disabled people.

Sanitary conveniences in a building should be just as available for disabled people as they are for others, having in mind the nature and scale of the building. Travel distances should have regard to the fact that some disabled people may need to get to a WC quickly.

The number and siting of WCs for the disabled will relate to the size of the building. Consideration must be given to ease of access, and a person in a wheelchair should not have to move through more than one storey to reach a WC.

A unisex WC has an access separate from other sanitary accommodation, and it has a number of advantages:

- easily identifiable;
- likely to be available when required;
- the user may be assisted by a person of either sex;
- less demanding of space than an integral arrangement.

An integral WC, as its name suggests, is included within the usual separate provision for each sex:

- it requires duplication to secure the same measure of provision for both sexes;
- assistance by a member of the opposite sex is unlikely.

To be flexible, the approved document suggests that if the building is large there can be a mixture of unisex and integral facilities.

WC compartments designed for use by people using wheelchairs will be similar in layout whether or not they are unisex or integral, and they must provide

- space for wheelchair manoeuvre;
- space for backward, diagonal, frontal and lateral movement onto the WC;
- hand-washing and hand-drying facilities which may be reached from the WC before transfer back to the wheelchair;
- sufficient space to permit someone to assist in the transfer.

If sanitary accommodation is to be installed in upper or lower storeys and no lift is available there must be reasonable provision to assist people having walking difficulties, are unsteady on their feet or who need some assistance to stand up or sit down.

Provision for ambulant disabled people must be by at least one WC compartment in each range of WC compartments placed in storeys which are not designed to be available to wheelchair users. This is additional to the WCs provided for wheelchair users in the principal entrance storey.

Suitable dimensions are shown in Figs M.27, M.28 and M.28A.

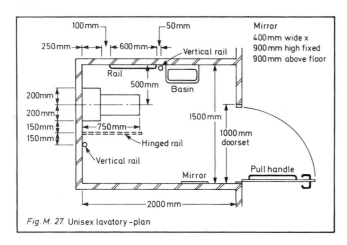

Fig. M. 27. Unisex lavatory - plan

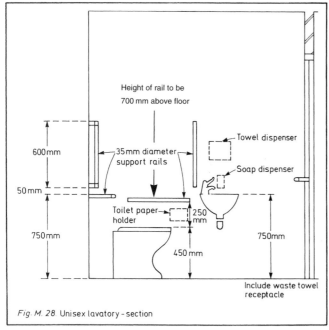

Fig. M. 28. Unisex lavatory - section

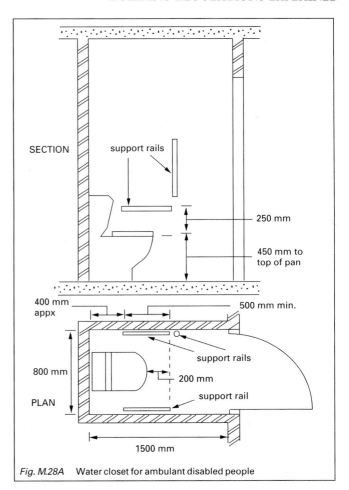

Fig. M.28A   Water closet for ambulant disabled people

Wheelchair users

● Staff

(a) WCs provided may be of the unisex type or integral within the male and female facilities.

(b) Provision for wheelchair users of both sexes may be on alternate floors if cumulative horizontal travel distance from work station to WC does not exceed 40 m.

(c) Where lift access is provided, the general provision for sanitary conveniences are located where everyone using the building has unrestricted access.

(d) Where there is only stair access in a building wheelchair users' sanitary provision is to be provided in the principal entrance storey. The exception to this requirement is where that storey contains only the principal entrance and vertical circulation areas.

(e) If there is more than one sanitary provision the layouts should be handed.

● Visitors and customers

M3 will be satisfied if unisex compartments are provided for use by visitors and customers.

● Hotel and motel guest bedrooms

In connection with hotel and motel bedrooms where en-suite accommodation is the norm, then similar suitable arrangements should be included in those bedrooms allocated to disabled people. Where en-suite facilities are not provided in guest bedrooms then unisex provision is to be provided within easy reach.

School and educational establishments

Design Note 18 requires that there should be at least one WC compartment accessible to a person in a wheelchair. The compartment should have dimensions not less than 1750 mm × 1400 mm (Fig. M.29).

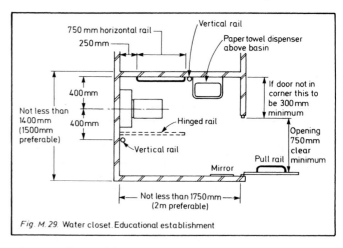

Fig. M. 29. Water closet. Educational establishment

Fittings, all accessible to someone in a wheelchair, ought to include:

toilet roll holder;
hand towel dispenser;
towel disposal bin;
soap dispenser;
chemical sanitary disposal container or incinerator mirror;
small shelf above lavatory basin;
coat hook.

The provision of an alarm call may also be considered.

5. AUDIENCE OR SPECTATOR SEATING

When an audience or spectators sit in seating which is fixed or arranged in tiers, wheelchair spaces are to be provided.

There must be a suitable means of access to enable disabled people to get to them.

M4 says that if a building contains audience or spectator seating reasonable provision is to be made to accommodate disabled people.

Wheelchair spaces are to be provided in theatres, cinemas, concert halls, sports stadia and similar locations.

A wheelchair space is a clear space not less than 900 mm wide and 1400 mm in depth. The space should be accessible to the user and afford a clear view of the event taking place.

The space may be kept permanently clear for use by a disabled person or may be one from which seating can easily be removed for the occasion.

The approved document recommends that wheelchair spaces should not all be together, but dispersed throughout the audience in convenient and accessible places.

A minimum of 6 or 100th of the total seats available to the public are to be made available as 'wheelchair spaces'. This means that for up to 600 seats provision must be made for 6 wheelchair spaces.

Total seating	Provision for wheelchair spaces
up to 600	6
601 to 700	7
701 to 800	8
801 to 900	9
901 to 1000	10
etc.	

In a large stadium it is reasonable to provide a smaller proportion of wheelchair places.

In a theatre or similar auditorium provision should be made for any mobile companion of a disabled person to be able to sit alongside the wheelchair. There are many arrangements where this desirable feature may be met (see, for example, Fig. M.30). A suggested notional disposition of wheelchair spaces in a theatre is shown in the approved document.

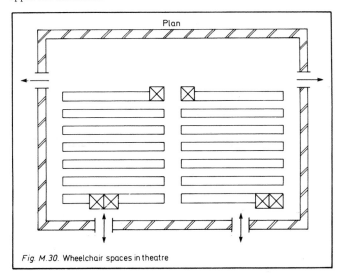

Fig. M.30. Wheelchair spaces in theatre

In a stadium or arena it is frequently arranged that wheelchair spaces are placed in front of the crowd at the lowest level. However, in the approved document there is a diagram showing how a platform may be formed to accommodate wheelchairs within a seated audience (Fig. M.31).

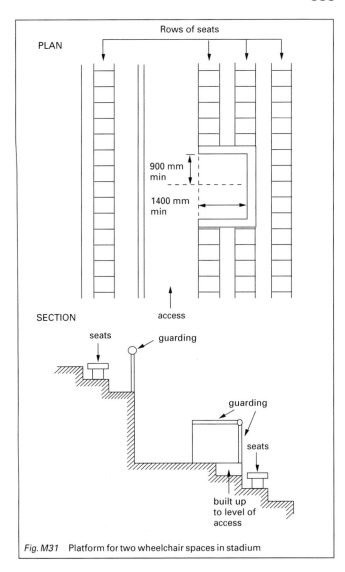

Fig. M31 Platform for two wheelchair spaces in stadium

6. ACCESS TO AND INTO DWELLINGS

The objective of this requirement is to make reasonable provision to enable disabled people to approach a dwelling and gain access. The facility provided should be suitable for use by wheelchair users and also those dependent on sticks or crutches.

Authority is in **M2 which requires reasonable provision for disabled people to gain access to and use a dwelling.**

The requirements do not apply to:

• a material alteration, or
• an extension to a dwelling.

A level plot is the most desirable when considering the provision of access, but this ideal will not always be present. Therefore, on steeply sloping plots, it may be necessary to consider the provision of ramps or even a stepped approach.

In normal circumstances the access facility should be provided to the principal entrance. Where in a particular situation this is not possible, access may be provided via a suitable alternative entrance.

It will be noted that the requirements relate to the principal entrance, or to a suitable alternative. It is not intended that the provisions be used for other entrances, as for example an entrance from the rear garden or terrace.

A driveway may double up as a level or a ramped approach or as the commencement of a suitable final approach to the principal entrance. In the absence of a driveway, the commencement of the access to be provided will be on the boundary with the street.

It is important that the whole length of the access provided should be suitable for use by wheelchair users, as well as those people who rely upon sticks or crutches.

Essential features are:

- solid enough to support user and wheelchair;
- smooth surface to facilitate movement of wheelchair, and ambulant users;
- loose-laid materials, such as gravel or shingle, should be avoided;
- not to have a crossfall steeper than 1:40.

When using a drive, or a car parking area, the width of the access should be additional to the space required by a parked vehicle.

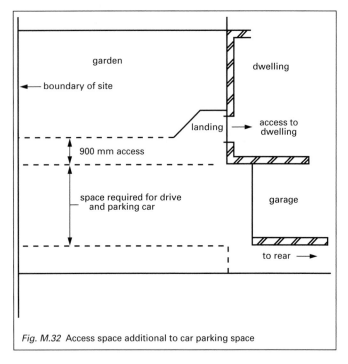

Fig. M.32 Access space additional to car parking space

A 'level' approach is desirable and would meet the requirement if:

- the gradient is not steeper than 1:20 and
- the width is not less than 900 mm, and
- the surface is firm and even.

Ramped approach

If the gradient from the point of access towards the principal entrance is more than 1:20 but less than 1:15 then a ramped approach may be provided.

Requirements for a ramp are:

- individual flights not to be more than 10 m long, where gradient is not steeper than 1:15, or
- individual flights not to exceed 5 m for gradients which are not steeper than 1:12, and
- there are to be top and bottom landings, each at least 1.2 m long, exclusive of the space occupied by the swing of any door or gate opening onto it;
- unobstructed width is to be not less than 900 mm;
- the ramp has to have a firm and even surface.

Stepped approach

When the topography of the site is such that from the point of access to the entrance, the gradient exceeds 1:15, then a stepped approach

may be provided. In this case the steps are to be designed to meet the needs of ambulant disabled people.

- the rise of a flight between landings is not to exceed 1.8 m;
- there are to be top and bottom, and if necessary, intermediate landings;
- the length of any landing is to be at least 900 mm;
- the unobstructed width is to be at least 900 mm;
- steps are to have suitable nosing profiles (see Fig. M.5);
- each step is to have a uniform rise of between 75 mm and 150 mm;
- each step is to have a going of at least 280 mm;
- when tapered treads are incorporated, the measurement is to be at a point 270 mm from the inside of the tread;
- a flight of three or more risers is to have a continuous handrail on one side; and
- the handrail is to have a grippable profile, be between 850 mm and 1000 mm above the pitch line, and extend 300 mm beyond the top and bottom nosings.

Access into the dwelling

When access to the principal entrance consists of a level or ramped approach it is necessary to provide an 'accessible' threshold. A level or near-level threshold is required. This involves a radical re-think in the design of the threshold, not forgetting damp proofing (Part C – Chapter 12), insulation (Part L – Chapter 20) and protection against the ingress of methane or radon gas (Part C – Chapter 12) (see Fig. M.33).

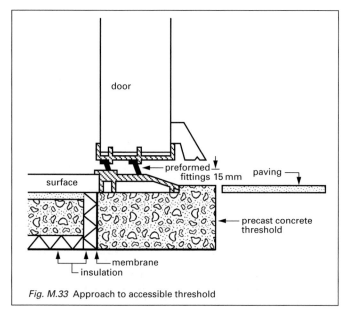

Fig. M.33 Approach to accessible threshold

An accessible threshold is also to be provided to entrance-level flats.

When the approach to the entrance is by way of a stepped ramp it is still necessary to provide an accessible threshold. In the event that the provision of a step into the dwelling is unavoidable, the rise is not to exceed 150 mm which is also the maximum rise allowed in a stepped approach.

Entrance doors

The minimum clear opening for an entrance door for use by disabled people is not to be less than 775 mm.

Thresholds

A document offering advice for house builders and designers in making the thresholds of dwellings accessible to wheelchair users and those people of limited mobility is available from TSO. It is not an approved document, but provides official support to the requirement of the approved document.

Entitled 'Accessible thresholds in new housing' it sets out to offer design solutions to secure a 'level' access and at the same time minimize the risk of water getting into the dwelling.

The guide acknowledges that the design solutions offered are not exhaustive and that the principles set out may be achieved by other design approaches. However, the solutions that are offered are based upon the experience of builders and others concerned with disability issues.

In considering threshold design, regard must be given to:

- the external landing and its drainage.
  This should be level and large enough to enable disabled people to approach and face the door. There should be provision to avoid standing water and to limit the quantity of surface water likely to reach the threshold.
- the threshold, sill and junction with the external landing.
  The sill and threshold profile, whilst allowing access to disabled ambulant and wheelchair users, must minimize the opportunity of surface water getting into the dwelling.
- the internal junction of the threshold and the floor finish.
  This should provide for differences in level between the threshold unit and the internal floor surface to accommodate the occupant's choice of type and thickness of floor covering.

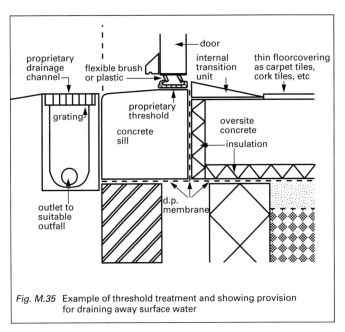

Fig. M.35 Example of threshold treatment and showing provision for draining away surface water

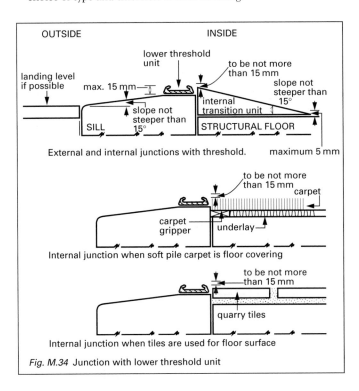

Fig. M.34 Junction with lower threshold unit

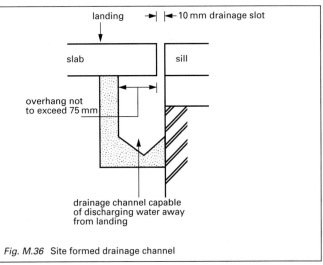

Fig. M.36 Site formed drainage channel

The limits within which the threshold is to be constructed are given in Fig. M.34

Working within these principles, an example of preformed threshold is shown in Fig. M.33, and a further example embracing the need for a drainage channel to convey surface water from the access and landing is illustrated in Fig. M.35. As an alternative to the proprietary drainage channels an on-site drainage slot 10 mm wide discharging to a channel below may be formed to convey surface water away from the doorway. It should be possible to rod the discharge channel for the purpose of clearing debris.

If the water is not likely to seep away into the adjoining soil, provision should be made to take it to a suitable outfall.

A porch or projecting canopy above the front door helps to prevent driving rain reaching the threshold.

7. CIRCULATION IN ENTRANCE STOREY OF DWELLING

The aim is to assist access in the entrance or principal storey of a dwelling, into habitable rooms, and WCs which may be in a bathroom at that level.

Corridors and passageways in these situations must be wide enough to facilitate use by a wheelchair user, and obstruction by radiators and other fixtures should not be overlooked.

In circumstances when an alternative entrance has to be used, it may be necessary to pass through a room to get to the remainder of the entrance storey. It is therefore necessary to ensure that the route may be negotiated by a disabled person.

The provision of internal doors of adequate width is important, and it should be noted that width may depend upon whether a wheelchair user can approach the door head on, or must turn into the door opening.

The requirements for corridors, passageways and internal doors are:

- when serving a habitable room or a space containing a WC:

1. Doorway clear opening not to be less than 750 mm, and corridor width at least 900 mm when approach to the doorway is head on.

2. With a doorway opening of 750 mm, the corridor width must be at least 1200 mm when the approach is not head on.
3. With a doorway opening of 775 mm, the corridor width must be at least 1050 mm when approach is not head on.
4. Should the doorway opening be 800 mm, then the corridor width may not be less than 900 mm when the approach is not head on.

Thus a corridor width of 900 mm will serve a door opening of 750 mm approached head on; or a door opening of 800 mm when the approach is not head on.

Obstructions

- Radiators and other permanent obstructions may be placed in a corridor, providing they do not exceed 2 m in length, and do not reduce the corridor width at that point to less than 750 mm.
- Permanent obstructions are not to be fitted directly opposite a door, because this may prevent a wheelchair user turning into or out of the room (see Fig. M.37).

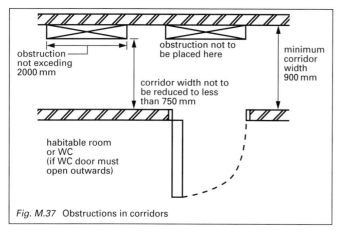

Fig. M.37 Obstructions in corridors

Vertical circulation in entrance storey

When building on a steeply-sloping site, a stepped change of level may be neccessary within the entrance storey. In these circumstances provision should be made for ambulant disabled people to move about within the storey.

Therefore any stair provided should follow requirements for rise and going in Part K (Chapter 19), and be provided with handrails:

- clear width of flight to be not less than 900 mm; and
- when rise of flight comprises three or more risers there are to be handrails on each side, including any intermediate landings.

8. SWITCHES AND SOCKET OUTLETS IN DWELLINGS

The requirement is to provide for those people who have limited reach and to assist them by putting wall-mounted switches and socket outlets at suitable heights.

Thus socket outlets and other equipment should be placed at appropriate heights between 450 mm and 1200 mm above finished floor level.

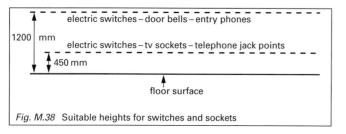

Fig. M.38 Suitable heights for switches and sockets

9. ACCESS IN BLOCKS OF FLATS

This section of the approved document is devoted to the provision of passenger lifts and common stairs in blocks of flats. The objective is that reasonable provision should be made for disabled people to visit residents on any storey.

The provision of a lift is the ideal solution, but there is no requirement that a lift should be installed.

Common stairs

In those circumstances, the design of stairs should meet the needs of ambulant disabled people, and in particular be suitable for people with impaired sight.

There should be:

- step nosings emphasized by using a bright colour contrasting with the rest of the tread;
- landings at top and bottom, meeting the requirements of K1 (Chapter 19);
- tread nosing profiles are to be in accordance with Fig M.5;
- a rise not exceeding 170 mm to each step;
- uniform going for each step not less that 250 mm;
- going on tapered treads to be measured 270 mm from the inside of the tread;
- no open risers;
- a handrail on each side of flights and landings when the stair consists of two or more risers.

Passenger lifts

When a lift is provided, it must be appropriate for use by an unaccompanied person in a wheelchair and provide enough time for a disabled person to enter without being trapped by closing doors.

Any lift provided must have:

- a minimum load capacity of 400 kg;
- a clear landing not less than 1500 mm × 1500 mm in front of its entrance;
- doors which give a clear opening not less than 800 mm;
- a car not less than 1250 mm long and 900 mm wide. Consideration may be given to other sizes if tests or experience show suitability for use by an unaccompanied person in a wheelchair;
- controls on the landings and in the car not less than 900 mm or exceeding 1200 mm above the floor surface and not less than 400 mm from the front wall;
- tactile indication on landings and also near the lift call buttons identifying the storey;
- tactile indication on or near the car lift buttons to identify the chosen floor;
- a signalling system giving visual evidence that the lift is responding to a call;
- a dwell time of 5 seconds before fully-open doors commence to close. An over-riding reactivating system may be incorporated, but a door edge pressure system is not to be used, and the minimum time for the door to remain fully open is 3 seconds;
- visual and audible indication of floor reached in installations serving more than 3 storeys.

10. PROVISION OF WC

The objective is to provide a WC in the entrance storey of a dwelling. It should be sited in a position where it is not necessary to use a stair to gain access to it from the habitable rooms in the entrance storey.

In dwellings where there are no habitable rooms in the entrance storey, the WC may be sited in either the principal storey, or the entrance storey. The WC may be placed in a bathroom.

The approved document mentions that it may not always be prac-

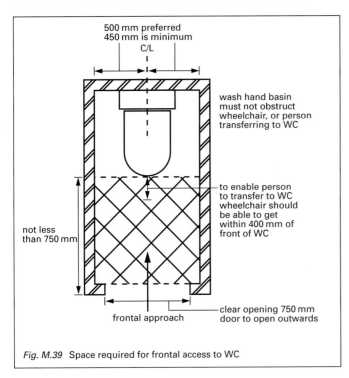

Fig. M.39 Space required for frontal access to WC

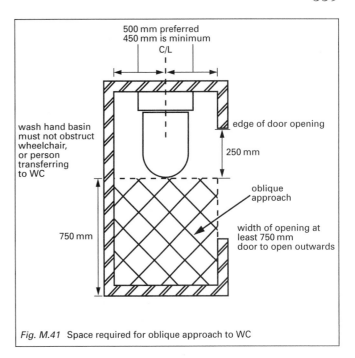

Fig. M.41 Space required for oblique approach to WC

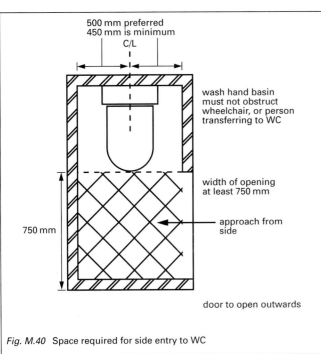

Fig. M.40 Space required for side entry to WC

tical for a wheelchair to be fully accommodated in the WC compartment.

It is required that:

- the door to the WC compartment is to open outwards;
- the door is to be fitted to facilitate access by wheelchair users and have a clear opening width not less than 750 mm when the approach is head on, or between 750 mm and 800 mm when the approach is not head on. Wider door openings do facilitate use by a person in a wheelchair;
- the WC compartment to provide a clear space for wheelchair users to transfer to the WC;
- the washbasin to be fitted where it will not impede the movement of a disabled person.

FEES

Fees are not to be charged for dealing with plans deposited under the building regulations when a local authority is satisfied:

- that the whole of the building work in question consists of an alteration; and
- that the building work is solely for the purpose of providing means of access to enable disabled people to get into any part of a building, or providing facilities intended to secure their greater health, safety, welfare or convenience; and
- that the building work is to be carried out in relation to

  (a) an existing building to which members of the public are admitted; or
  (b) an existing dwelling which is, or is to be, occupied by a disabled person.

- fees are not to be levied in connection with the provision or extension of a room in a dwelling when the sole use of the room will be:

  (a) undertaking medical treatment of a disabled person; or
  (b) the storage of medical equipment to be used by a disabled person; or
  (c) providing necessary accommodation or facility by adapting or replacing existing provision for use by a disabled person.

This exemption is contained in the Building (Local Authority Charges) Regulations 1998, regulation 9, and includes all persons falling within the definition of 'disabled persons' in section 29(1) of the National Assistance Act 1948, as extended by the Mental Health Act 1959.

This definition is much wider than that contained in Part M, and embraces people who are blind, deaf or dumb. Other people who are substantially and permanently handicapped by congenital deformity, mental disorder, illness or injury are also included.

FIRE ESCAPE

The approved document does not contain any recommendations for special fire escape precautions necessary for disabled people. Reference should be made to the approved document to Part B discussed in Chapter 11 but the following comments may be found to be useful.

BS 5588 draws attention to legislation requiring buildings to be designed so that they are accessible to disabled persons. The common requirement is that areas used by the public or employees are to be accessible to disabled people and that includes those confined to wheelchairs.

The designer has two basic questions to satisfy:

- the inaudibility to deaf persons of fire alarms, and this is dealt with in BS 5839: Part 1: 1988.

This is a code of practice for the installation and servicing of fire detection and alarm systems. It covers the planning, installation and servicing of fire alarm systems in and around buildings. It covers the systems from simple installations with one or two manual call points to complex installations with automatic detectors, control and indicating equipment and connection to the public fire service. It also covers systems capable of providing signals to initiate, in the event of fire, the operation of ancillary services such as fixed fire extinguishing systems. Design considerations also include audible and visual alarms:

- the exit of non-ambulant people, especially those confined to wheelchairs.

Lifts are normally disregarded as a means of exit in an emergency but in a multi-storey office building it may be the only practicable way to get a chairbound person out of the building. It is imperative that suitable arrangements are made for capable people to assume responsibility for supervising the removal of a chairbound person via a suitable lift. The lift and the arrangements must be agreed with the appropriate fire officer.

It is not possible to predict how many disabled people may be in the office building at any time. Some disabled people may be expected to use the escape stairs, and emergency arrangements should provide for the able bodied to help disabled people down the stairs.

Although the proportion of disabled people in shops may be higher than in office buildings, when a shop is being planned it will not be possible to predict the amount of usage by disabled people.

However, it can be expected that some members of the staff and a number of customers will have a disability of some kind. As with offices it may also be expected that in the event of an emergency involving evacuation of the building the majority of those disabled will be able to use the escape stairs, assisted by able bodied members of the staff without impeding escape routes.

As with offices it may be possible for special arrangements to be made for a lift to be utilized when under the control of a responsible person, such as a safety officer, or fireman.

Providing refuge

BS 5588: Part 8: 1988 is an important code of practice dealing with means of escape for disabled people.

The code is applicable to all buildings, except single family dwellings – houses, flats and maisonettes.

Management systems are essential and Appendix A to the code provides a management aid for making the best use of facilities recommended.

Definitions include 'firefighting lift', which is one designated to have additional protection and controls that enable it to be used under direct control of the fire brigade.

There is a recommendation that a 'refuge' should be provided. This is an area that is both separated from a fire by fire-resisting construction and provided with a safe route to a storey exit. It provides a temporary safe space for disabled persons to await assistance for their evacuation.

There is also a definition of 'wheelchair stairlift'. This is an appliance for transporting a person or persons with a wheelchair between two or more levels by means of a guided carriage.

Principles and philosophy underlying the provision of means of escape are:

- planning and protecting escape routes;
- construction and finishing of escape routes using suitable materials;
- segregation of higher fire risk areas;
- means of giving warning.

Some disabled persons in wheelchairs, or of limited mobility, but not requiring a wheelchair, will not be able to use stairways without assistance, and for this reason it is necessary to provide refuges on all storeys, except in the case of small buildings of limited height.

A refuge is therefore an area separated from the fire by fire-resisting construction and provided with a safe route to a storey exit. This is a temporarily safe place for the disabled person to await assistance.

The space provided needs to be not less than 700 mm × 1200 mm, but ideally it should be larger – 900 mm × 1400 mm.

The following are listed as examples of temporary refuges (see Figs M.42 and M.43):

- compartment; protected lobby; protected corridor or a protected stairway;
- a flat roof; balcony; podium or similar place which is adequately protected, or remote from any fire risk;
- other arrangements which may give an equal measure of safety.

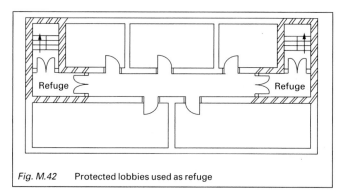

Fig. M.42   Protected lobbies used as refuge

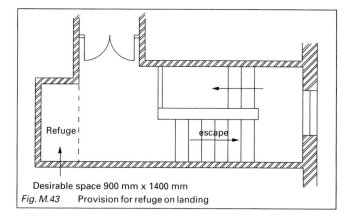

Desirable space 900 mm x 1400 mm

Fig. M.43   Provision for refuge on landing

The code gives a number of illustrations indicating how a refuge may be secured for buildings when an evacuation lift is provided and when such lifts are not available.

A refuge should be provided in connection with each protected stairway. The code also provides recommendations in connection with stairs, ramps, lifts, construction and fire-warning systems. Especially does it emphasize the need for an unimpeded escape route, free of all projections which may limit the width available.

Appendix A is devoted to advice to management and emphasizes that fire procedures should be preplanned by persons having control of a building. It is essential to identify the needs of disabled people and to make proper arrangements for their assistance. This means that the necessary aid is to be made available to assist a disabled person to get out of the building from each of the refuge points provided, and responsibility for so doing is made clear to those concerned. It also means that the refuge space must be kept completely clear at all times.

Techniques of evacuation are described, as is the management of evacuation lifts; fire plan strategies and alarm systems.

Schools and educational establishments

Design Note 18 offers advice in connection with educational establishments, and in principle it is appropriate for consideration in connection with all buildings to which the regulations relate.

- Normal escape distances and conditions may need to be modified to cater for slower movements.
- Dead end conditions should be avoided.
- Regularly used routes are preferable.
- Well signed routes.
- Particular attention is needed to escape doors so that they may be opened by a disabled person.
- In single-storey buildings window escapes are unsuitable for disabled people.
- In multi-storey buildings it may be possible to:

  (a) transfer on the same level to some other nearby fire protected compartment;
  (b) have link bridges to adjoining blocks.

DEVELOPMENT CONTROL

The Department of the Environment, Transport and the Regions publishes policy notes from time to time, including advice regarding access for the disabled.

Developers are reminded that suitable access for disabled people is a factor which may be taken into account when local authorities are considering the granting of planning permission. So we have building regulations concerned with the building, and planning control looking at the environs of a building.

There are a number of suggestions regarding aspects where planning departments may direct their attention when considering proposals for buildings to which disabled people must have provision for access:

- Special car parking and setting down points.
- Alterations to existing buildings should make provision wherever possible.
- Provision for buildings of architectural or historic interest may be difficult, but access provision should be provided if practical.
- Multi-level shopping centres should have suitable routes for disabled people.
- Pedestrian zones pose special problems for disabled people so there should be reserved parking areas in or convenient to a zone.
- Consideration of units for disabled people should be a facet in the design of private housing.
- There should be contact between developers and local disabled organizations.

The Access Committee for England is a charitable body, and able to supply many booklets and pamphlets at very small cost. The Access Design Guidance Notes for developers contains much advice which supplements both the regulation requirements and planning considerations, including the placing of fittings, portable ramps, seating, counters, enquiry hatches, telephones and car parking, to name but a few.

Their car parking suggestions are shown in Fig. M.44.

One of their publications indicates that provision for the disabled is not a luxury for a few. It says 'most of us are disabled at some time in our lives through accident or injury, or the infirmities of old age, by young children in prams or pushchairs, or by shopping or luggage. Accessible buildings are accessible to all.'

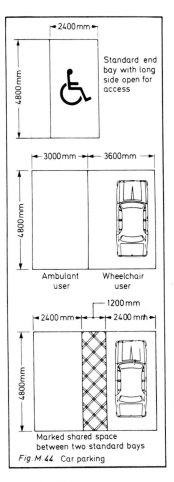

Fig. M.44. Car parking

Disability Discrimination Act 1995

This Act provides that employers are not to discriminate against disabled employees, and also provides for the adaptation of premises for individual employees.

The provision is to be reasonable for the circumstances of each case. There is no direct link between the Act and Part M of the 1991 Regs, although Part M could provide useful guidance.

Responsibility for administering these requirements rests with the Department of Education and Employment.

RELEVANT READING

British Standards and Codes of Practice

BS 4787: Part 1: 1980. Specification for dimensional requirements – door sets

BS 5588: Part 3: 1983. Code of practice – fire precautions in office buildings

BS 5588: Part 8: 1988. Code of practice – means of escape for disabled people

BS 5619: 1978. Code of practice for design of housing for disabled people

BS 5655: Part 1: 1986. Safety rules for the construction and installation of electric lifts
Amendment 1: AMD 5840

BS 5655: Part 2: 1988. Safety rules for the construction and installation of hydraulic lifts
Amendment 1: AMD 6220

BS 5655: Part 5: 1989. Specifications for dimensions for standard lift arrangements

BS 5655: Part 7: 1983. Specification for manual control devices, indicators and additional fittings
Amendment 1: AMD 4912

BS 5776: 1996. Specification for powered stairlifts

BS 5810: 1979. Code of practice for access for the disabled to buildings

BS 5839: Part 1: 1988. Code of practice for fire detection and alarm systems

BS 5839: Part 6: 1995. Code of practice for the design and installation of fire detection and alarm systems in dwellings

BS 6440: 1983. Powered lifting platforms for use by disabled people
Amendment 1: AMD 4027
               2: AMD 6523

BRE digest

301   Escape of disabled people from fire

Information paper

9/97   Emergency lighting and wayfinding systems for visually-impaired people

BRE Report

Fire and disabled people in buildings (T. J. Shields). (BR 231: 1993)

Department of the Environment, Transport and the Regions

Guidance on the use of tactile paving surfaces. Obtainable from the Mobility Unit, Zone 1/11, Great Minster House, 76 Marsham Street, London SW1P 4DR.

Department of Education and Science

Design Note 18, available from the department at Elizabeth House, York Road, London SE1 7PH

Access Committee for England

Design guidance notes for developers, available from 126 Albert Street, London NW1 7NF (sae please)

Disabled Living Foundation

Extensive information lists with lists of aid centres throughout the country from 380 to 384 Harrow Road, London W9 2HU (sae please)

Royal Town Planning Institute

Practice advice note 9 – Access for disabled people from 26 Portland Place, London W1N 4BE

Safety in sports grounds

1.   Guide to safety at sports grounds. Obtainable from TSO
2.   Designing for spectators with disabilities. The Football Stadium Advisory Design Council. Available from the Sports Council
3.   Guidance Notes – access for disabled people. Available from English Sports Council.

Other helpful addresses

Welsh Council for the Disabled,
    Caerbargdy Industrial Estate, Bedwas Road, Caerphilly, Mid Glamorgan C58 3FL
Royal Association for Disability and Rehabilitation
    25 Mortimer Street, London W1N 5AB
Centre on Environment for the Handicapped
    35 Great Smith Street, London SW1P 3BJ
John Groom's Association for the Disabled
    10 Gloucester Drive, London N4 2LP

CHAPTER 22

Glazing: safety in relation to impact, opening and cleaning

(Approved document N)

Part N was introduced by the Building Regulations 1991, and is the first occasion that glazing, against which people may stumble or fall or simply walk into, has been the subject of a requirement in building regulations.

Revised regulations are contained in the Building (Amendment) Regulations 1997, and set out to ensure that a building which complies with the building regulations will automatically comply with the corresponding health and safety regulations. The amended regulations came into force on 1 January 1998 and add requirements regarding the safe opening and closing of windows, and safe access for cleaning windows and reflecting requirements in the Workplace (Health, Safety and Welfare) Regulations 1992 which came into operation on 1 January 1993, and have three requirements having an affinity to Part N:

- protection against collision with glazing (Regulation 14); and
- dangers associated with the opening and closing of windows (Regulation 15); and
- the safe cleaning of windows (Regulation 16).

Therefore compliance with Part N requirements will in accordance with section 23(3) of the Health and Safety at Work, etc. Act 1974 prevent the service of an improvement notice made under regulations 15(1) and 16 of the Workplace (Health, Safety and Welfare) Regulations 1992.

Regulation 8 states that the requirements of Part N do not require things to be done if they are not for the purpose of securing reasonable standards of health and safety of people who may be in or about the building.

Building work is defined in Regulation 3 to mean

- the erection or extension of a building;
- the material alteration of a building;
- the provision or extension of a controlled service or fitting in or in connection with a building;
- the work when a material change of use is to take place.

Thus the requirements regarding glazing are applicable to new buildings and also to those portions where there was none previously in the building and to extensions or material alterations.

When a mixed-use development is taking place, it may be that a portion of the building could be used as a dwelling whereas another part has a non-domestic use. In these circumstances, and if the requirements for dwellings and non-domestic use differ, the standards for non-domestic use are to apply to those parts of the building which are shared.

The requirements do not extend to exempt buildings, except conservatories and porches (Chapter 5), or to replacement glazing, although there may be consumer protection legislation which is appropriate.

Guidance on fire-resisting glazing and the reaction of glass to fire is contained in Chapter 11. There are also requirements when glazing forms part of protection from falling from one level to another, and information on this aspect of glazing is referred to in Part K (Chapter 19).

MATERIALS

All work subject to the requirements of building regulations is to be carried out using proper materials and workmanship. Requirements in regulation 7 of the 1991 Regs, as amended by the 1999 Regs, and guidance is given in the approved document to regulation 7 (Chapter 9).

There are requirements regarding glazing which are additional to those contained in Part N; for example, the Consumer Protection Act 1987 which, in section 10, lays down a duty that suppliers and distributors are to ensure that a product is safe.

The General Product Safety Regulations 1994 apply to all EU Member States and came into operation in October 1994.

- Where a product conforms to specific rules there is a presumption that the product is safe.
- If no such rules exist, conformity is to be assessed using UK national standards, harmonized European standards or EOTA technical approvals.

When making an assessment it is necessary to take into account the measure of safety that consumers may reasonably expect.

PROTECTION AGAINST IMPACT

Part N contains two requirements which are to be followed in what are referred to in the approved document as 'critical locations'.

N1. Glazing which people are likely to come into contact while moving in or about the building must

- **when broken on impact, break in a way not likely to cause injury, or**
- **resist collision without breaking, or**
- **be guarded or protected from impact.**

PERFORMANCE

The aim of Part N is to limit the possibility of people being cut or pierced should they fall against glazing. The most likely places for such accidental impact are in doors and panes of glass alongside doors. Glazing at low level in walls and partitions also provides a risk of injury.

The greatest risk of injury in doors and door side panels occurs from the floor up to shoulder level and also when glazing is near to door handles and push plates if the door should stick.

Hands, wrists and arms are at risk and there is always the possibility that a collision between a person's waist and shoulder level may lead to a fall through the glazing and extensive injury.

Children are especially vulnerable when glazing is at low level.

N1 therefore requires that, in what are described as critical locations, the particles of any broken glass should be relatively harmless, or the glazing should be tough enough to limit the risk of breakage, or contact with the glazing should be prevented.

TYPES OF GLASS

There are numerous types of flat glass but the ones most frequently encountered are clear float (annealed) glass made by passing glass as it emerges from the furnace over a bath of molten tin. It is suitable for all classes of glazing and is available in thicknesses from 3 mm to 25 mm.

Toughened or tempered glass is produced by subjecting annealed glass to heating followed by rapid surface cooling, producing high compression in the surface of the glass. Toughened glass requires more force than annealed glass to effect breakage and when it breaks it fragments into small almost harmless pieces. It cannot be cut to size or drilled.

Laminated glass is formed by bonding two or more pieces of glass together and inserting a plastics interlayer between each pair of panes.

CRITICAL LOCATIONS

The places considered critical for the purpose of N1 are:

* from finished floor level to 800 mm above that level (Fig. N.1) in internal and external walls and partitions;
* from finished floor level to 1500 mm above that level in a door or glazing close to either edge of the door (Fig. N.2.)

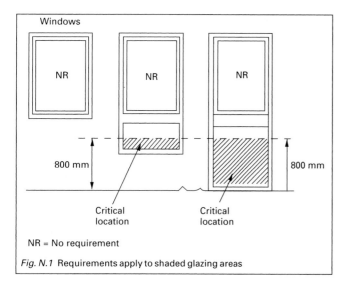

Fig. N.1 Requirements apply to shaded glazing areas

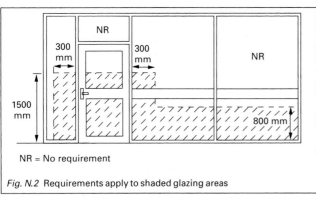

Fig. N.2 Requirements apply to shaded glazing areas

Reducing risk

The requirement gives three options for reducing the risks in critical locations:

* If it breaks, glazing should break safely.

In practice this will involve the use of laminated or toughened glass as defined in BS 6206: 1981, which is the specification for impact performance requirements for flat safety glass and safety plastics for use in buildings.

Clause 5.3 of the BS is the one which defines the standard to be achieved when test pieces must not break, or break safely in one of the following ways:

(a) Numerous cracks or fissures may appear but no shear or opening develops through which a 76 mm diameter sphere can be passed freely. There is also a limit on the weight of particles which become detached during testing.
(b) Disintegration occurs but the ten largest crack-free particles must not weigh more than a stated limit.
(c) Breakage resulting in several pieces which may or may not be retained in the frame. For those exposed after impact the perimeter must not be sharp, and there is a limit to the size of any pointed protrusion.

Glazing in a critical location is to meet class C test requirements when installed in a door. When a door side panel has a width of more than 900 mm it should meet class B test requirements.

BS 6206 also covers safety performance levels; safety breakage criteria; safety glass and safety plastics; and durability.

All panels are to be marked, giving identification, name or trademark; type of material; BS 6206 and classification.

* Glazing should be in small panes or be tough or robust.

Annealed glass increases in strength the thicker it is. Some annealed glass is suitable for glazing large areas as in factories, offices, showrooms, shops and public buildings. Acceptable dimensions for various thicknesses of annealed glass, as given in the approved document, are shown in Fig. N.3.

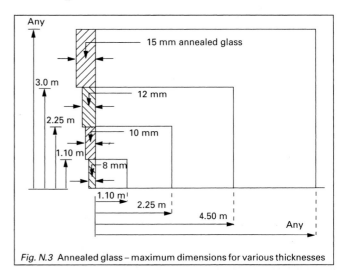

Fig. N.3 Annealed glass – maximum dimensions for various thicknesses

Polycarbonates or glass blocks are inherently strong and are not limited to the areas shown in Fig. N.3.

The approved document regards a small pane as an isolated pane or one of a number of panes contained within glazing bars (Fig. N.4). The smaller dimension is not to exceed 250 mm and area is not to be more than $0.5\,m^2$.

Small panes are measured between glazing beads or other fixings; and when annealed glass is used its thickness is not to be less than a nominal 6 mm.

In traditional leaded or copper lights when fire resistance is not a factor, 4 mm glass may be used.

* The glazing should be given permanent protection.

Glazing may be protected by a permanent screen (Fig. N.5) which must be tough and also prevent a 75 mm diameter sphere from hitting the glass. This means that all apertures in the screen should

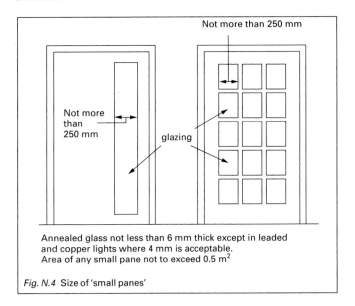

Annealed glass not less than 6 mm thick except in leaded and copper lights where 4 mm is acceptable.
Area of any small pane not to exceed 0.5 m²

Fig. N.4 Size of 'small panes'

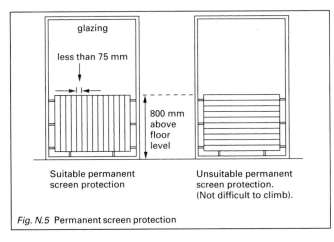

Fig. N.5 Permanent screen protection

have one dimension less than 75 mm. Additionally, the screen is to be not less than 800 mm high above finished floor level. A climbable screen does not meet the criteria set out in the approved document.

When glazing in a critical location is given satisfactory permanent screen protection, the glass need not necessarily comply with requirement N1.

MANIFESTATION OF GLAZING

There are on occasions large areas of glazing proposed within a building which may not be immediately apparent as a barrier and people may walk into it. Such areas of glazing must be highlighted to direct attention to the fact that a barrier is present.

N.2 says that transparent glazing with which people moving about the building are likely to collide is to incorporate features which can be seen.

This requirement applies to all buildings which are not dwellings.

CRITICAL LOCATIONS

Most danger of a person colliding with a transparent glazed screen will occur when adjoining parts of a building or its immediate surroundings are at or about the same level. In these circumstances a person may attempt to walk through the unseen screen and be involved in a collision.

All large areas of transparent glazing are regarded as critical loca-

tions and include glazed areas both internal and external of factories, public buildings, shops, showrooms offices and any other non-domestic building.

The word 'manifestation' is used to indicate measures taken to emphasize the presence of these areas of glazing.

Common design features such as the presence of door framing, large 'push' or 'pull' door handles, mullions or transoms may be of such dimensions that a permanent additional manifestation may not be necessary.

In the following circumstances 'manifestation' will not normally be required:

(a) when a single pane glazed door has a substantial frame (Fig. N.6);
(b) when a single pane glazed door is not framed, or has a very narrow framing, but large easily seen handles, push or pull plates are affixed (Fig. N.6);
(c) when door height glazing is not more than 400 mm wide (Fig. N.7);
(d) when door height glazing is provided with a rail which is not less than 600 mm or more than 1500 mm above finished floor level (Fig. N.7).

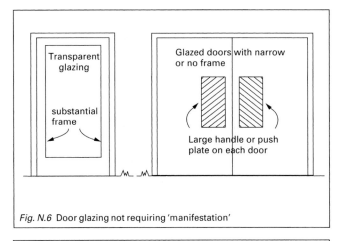

Fig. N.6 Door glazing not requiring 'manifestation'

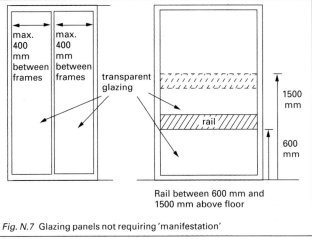

Fig. N.7 Glazing panels not requiring 'manifestation'

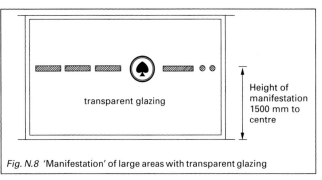

Fig. N.8 'Manifestation' of large areas with transparent glazing

Where none of the features shown in Fig. N.7 or similar design solutions are not used it is necessary to resort to the permanent manifestation of large uninterrupted glazed areas. This may take the form of solid or broken lines of sufficient dimensions to be clearly visible (Fig. N.8). Company logos may also be used; or the choice of features is almost endless.

Now follow the two requirements N3 and N4, which are concerned with the safe opening of doors and windows together with safe access for cleaning windows.

SAFE OPENING OF WINDOWS

The following requirement has been introduced in order to lessen the possibility of people being involved in an accident by over-reaching to open or close a window, or having to climb to stand on an unsafe surface in order to open or close a window.

N.3. Windows, skylights and ventilators which can be opened by people are to be constructed or equipped in such a way that they may be opened, closed or adjusted safely.

This requirement was added by the 1997 Regs, and does not apply to dwellings.

Window controls

Window controls are to be placed in a position where they can be operated safely, and where access to the window is not obstructed, a control may be placed up to 1.9 m above a firm and stable standing.

In instances where there is some obstruction in front of a window, the height of the control must be reduced, and in the example, an obstruction 600 mm deep, which would include any recess, requires the control not to be higher than 1.7 m when the obstruction does not exceed 900 mm above the floor.

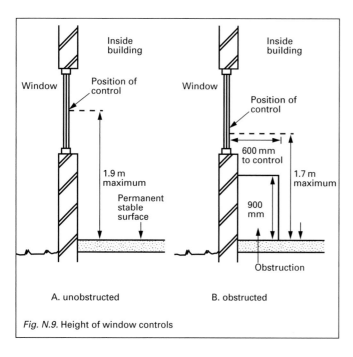

Fig. N.9. Height of window controls

There will be circumstances when the controls cannot be placed in a position which is safe for a person to attempt to reach. When this occurs, provision is to be made for a manual or electrically-operated remote control system to be installed.

Opening limitators are to be provided on any window above ground level where there is a danger of anyone falling through the window. The alternative is to provide guarding to the window (see Chapter 19).

SAFE CLEANING OF WINDOWS

The following requirements relate to the internal cleaning of windows, and also to external cleaning, including the use of ladders not more than 9 m long when extended.

N.4. All windows, skylights or any transparent or translucent walls, ceilings or roofs are to be safely reachable for cleaning.

This requirement was added by the 1997 Regs, and does not apply to:

- dwellings, or
- transparent or translucent surfaces which are not intended to be cleaned.

Performance

The requirement will be met if safe provision is made for cleaning both sides of glazed surfaces in positions where there is a danger of falling more than 2 m.

Access for cleaning windows, etc.

Glazed surfaces are required to be capable of being cleaned safely by someone standing on the ground, or on a floor or some other permanent solid surface. When this cannot be achieved, provision has to be made restricting the height of a window and the reach required to effect safe cleaning.

- Size and design should permit the external surface of a window to be cleaned from within the building.
- When a walkway is necessary, it has to be at least 400 mm wide, and provided with a guarding not less than 1100 mm high. Alternatively, there must be suitable anchorages for a sliding safety harness.
- Suspended cradles or travelling ladders, complete with attachments for safety harnesses, may be necessary.
- Suitable anchorage points for safety harnesses or abseiling hooks may provide the access required for cleaning the windows.
- When no other means can be found to clean the windows, space is to be available for scaffold towers in positions where access for cleaning the windows may be achieved.

BS 8213: Part 1: 1991 is a Code of practice for safety in use and during cleaning of windows and doors. It also includes guidance on cleaning materials and methods.

Critical dimensions are shown in Figs N.10 and N.11, demonstrating the dimensions given in the 1998 approved document and setting out the maximum downward, horizontal and angular reach permitted to satisfy the requirements of N.4.

- When a window may be reversed for cleaning, it is to be fitted with a mechanism which will hold it firmly in the reversed position.
- A firm level surface is to be provided in a safe place and large enough to accommodate a ladder set at an angle of 75° to the ground and reaching to the window to be cleaned.

    The approved document refers to ladders not more than 6 m long when extended and also to ladders having a maximum length of 9 m.

    When a 6 m ladder is used, normal soil will usually provide a suitable standing surface. However, when a ladder exceeds 6 m but is less than 9 m long, a firm level surface must be provided for it to stand upon. At the upper end there should be an eyebolt for using to secure the ladder to the building.

TRANSITIONAL PROVISIONS

The 1991 Regs and 1992 edition of the approved document continue to apply to work where before 1 January 1998 a building notice, initial notice or amendment notice was given to a local authority, or

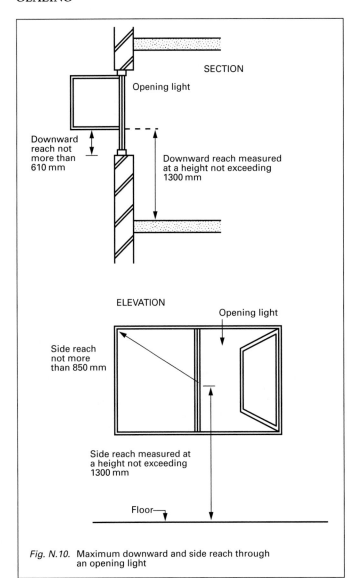

Fig. N.10. Maximum downward and side reach through an opening light

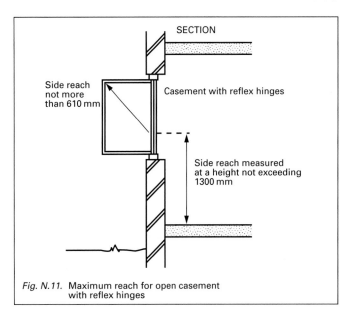

Fig. N.11. Maximum reach for open casement with reflex hinges

RELEVANT READING

British Standards and Codes of Practice

BS 952: Part 1: 1995. Glass for glazing – classification
BS 952: Part 2: 1980. Terminology for work on glass
BS 6206: 1981. Specification for impact performance requirements for flat safety glass and safety plastics for use in buildings
BS 6262: 1982. Code of practice for glazing for buildings
BS 6262: Part 4: 1994. Code of practice for safety – human impact
BS 8213: Part 1: 1991. Code of practice for safety in use and during cleaning of windows and doors
BS EN 572–2: 1995. Float glass
BS EN 572–3: 1995. Polished wired glass

BRE information paper
18/81 Accidents involving glass in domestic doors and windows, some implications for design

full plans had been deposited. If an amendment notice was given after 1 January 1998 relating to an earlier initial notice, the work added to the initial notice must comply with the current requirements of this part.

CHAPTER 23

Other approved documents

All approved documents discussed in the preceding chapters were approved by the Secretary of State and were prepared by the Department of the Environment after extensive consultation. Now this officially created series of approved documents has been supplemented by two prepared privately.

The Secretary of State is empowered to approve such a privately prepared document by section 6 of the Building Act 1984. This section enables the Secretary of State to issue any document which in his opinion gives suitable practical guidance to the building regulations. The document may come from any source.

1. TIMBER INTERMEDIATE FLOORS FOR DWELLINGS

The first of what is to be a series of approved documents prepared by bodies other than the Department of the Environment is about timber intermediate floors in dwellings but it does not relate to compartment floors. It has been prepared by the Timber Research and Development Association (TRADA) and as it is approved by the Secretary of State the guidance given has the same authority as in other approved documents.

What is so different about the TRADA document is that it applies to one element in a building. In this one approved document is gathered together all the recommendations to meet various requirements referring to that element. There is practical guidance on meeting paragraphs A1, B2, B3, J3 and L1 in Schedule 1 of the Building Regulations 1991 applying to timber joist floors which must have 30 minute or 60 minute fire resistance in dwellings.

In addition to the guidance given in meeting the regulation requirements, this approved document offers supplementary advice on 'good building practice' and is to be regarded as evidence supporting compliance with the requirements.

TRADA produced this document believing that the element approach is a practical one. The designer, contractor and controller having this multi-performance data available may save time and effort in researching through the appropriate official approved documents.

The TRADA document gives design guidance for intermediate timber joist floors carrying a domestic imposed load of $1.5\,\mathrm{kN/m^2}$ in buildings up to three storeys above ground level and used as single occupancy dwellings. The document also applies to floors within maisonettes where the building is not in excess of three storeys and has not more than four self-contained dwellings per floor accessible from one staircase.

STRUCTURE

All the guidance under this heading relates to compliance with the requirements of A1 of the 1991 Regs – Loading.

FLOOR JOISTS

Maximum clear spans have been calculated in accordance with BS 5268: Part 7: 1989, 'Recommendations for the calculation basis for span tables', section 7.1: Domestic floor joists.

Table 1 in the document shows spans for both SC3 and SC4 timber and they have been prepared on the basis of BS 5268: Part 2: 1991, 'Permissible stress design, materials and workmanship', and includes the deflection criterion of 0.0003 times span, or 14 mm, whichever is the lowest.

The span tables are very similar to those in the D of E approved document (Chapter 10) though not identical. There is no information for 50 mm wide joists, but this is included in the D of E document. TRADA says that the difference in span between 47 mm and 50 mm width joists is not significant. Guidance is given on accommodating non-loadbearing partitions by the provision of extra joists, or the adjustment of maximum span where partitions cross the joists.

Table 2 of the document lists the species and grade combinations which satisfy Classes SC3 and SC4 timber.

There are two good practice recommendations. In order to ensure that floors and ceilings are fixed level, joists should be regularized or be ALS or CLS as described in BS 4471: 1987, 'Specification for sizes of sawn and processed softwood'. Moisture content of the timber as fixed should not exceed 19%.

SUPPORT FOR TIMBER JOISTS

Three methods of support are described: built into cavity walls; on joist hangers or timber ledgers; and as platform floors in timber-framed walls.

Good practice emphasizes the need for good workmanship and the elimination of air paths through walls. Joists should only be built into single leaf masonry walls if cladding is separated from the wall by a drained and vented cavity and the wall is also protected by a breather membrane.

Joist hangers should meet the requirements of BS 6178: Part 1: 1990, 'Specification for joist hangers for building into masonry walls of domestic dwellings'. Joints in joists which are not made over load-bearing wall should be designed in accordance with BS 5268: Part 2: 1991, 'Code of practice for permissible stress design, materials and workmanship'.

LATERAL SUPPORT

Lateral support should be provided for masonry external and compartment walls more than 3 m long and also internal masonry load-bearing walls, using 30 mm × 5 mm galvanized mild steel or other durable straps.

Joist hangers and restraint straps should bear on the central point of a brick or block and be fitted before laying subsequent masonry courses.

Special care is necessary when an opening in a floor adjoins a supported wall and sufficient straps should be provided on each side of the opening equal to the number which would be necessary if there were no opening. Straps to wall are not to be more than 2 m apart.

Where joists

- have 90 mm bearing on supported wall, or
- have 75 mm bearing on timber wallplate at each end, or
- are carried on a restraint type hanger at not more than 2 m centres,

longitudinal straps are not required.

Straps are not required in conjunction with timber-framed walls and a platform floor.

STRUTTING

Strutting is required at midspan when the joist span is between 2.5 m and 4.5 m. Two rows of strutting are required, each at one-third span position where the joist span exceeds 4.5 m (Fig. 23.1).

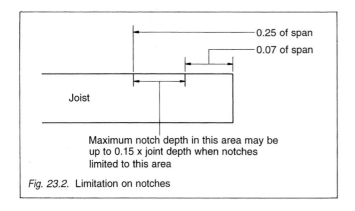

Fig. 23.1. Strutting to floor joists

NOTCHING AND DRILLING

Requirements for notching and drilling joists are similar to the D of E approved document, and the recommendations in BS 5268: Part 2: 1991, 'Code of practice for permissible stress design, materials and workmanship' are also acceptable (Fig. 23.2).

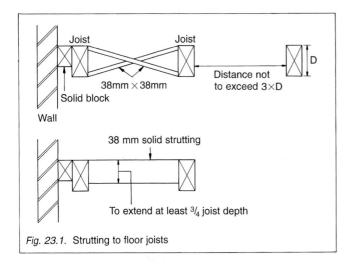

Fig. 23.2. Limitation on notches

ALLOWANCE FOR DIFFERENTIAL MOVEMENT

This is not a requirement, but it is good practice to provide for differential movement when timber joists bear on different materials.

TRIMMERS AND TRIMMING JOISTS

There are two tables relating to spans in the document. The first gives spans for trimmers and the other spans for trimming joists, and allowance is made for supporting lightweight non-loadbearing internal partitions. The design has been based upon BS 5268: Part 7: section 7.1: 1989, 'Recommendations for the calculation basis of span tables'.

The approved document for Part A (Chapter 10) does not contain the equivalent tables.

There is also a table giving requirements for fixing double member trimmers using nails, bolts or double-sided toothed plate connectors, and yet another lists fixings and spacings for double member trimming joists using wire nails, M12 bolts and 51 double-sided plate connectors. Advice is also given regarding loads which need to be supported from each end of a double member trimmer.

FLOOR DECKING

Softwood tongued and grooved

Boarding is to be 16 mm finished thickness where joist centres do not exceed 500 mm and not less than 19 mm thick when maximum joist centres are 600 mm. This assumes an imposed load of 1.5 kN/m² and the alternative concentrated load of 1.4 kN. All boarding is to satisfy BS 1297: 1987, 'Specification for tongued and grooved softwood flooring' (Fig. 23.3).

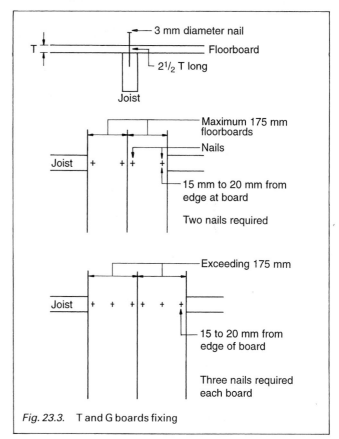

Fig. 23.3. T and G boards fixing

Plywood

Plywood is to be WBP or exterior bonded of a grade listed in BS 5268: Part 2: 1991, 'Code of practice for permissible stress design, materials and workmanship'. Other plywoods are to be independently certified as suitable for the purpose.

For the same loading as t and g boards at maximum joist centres nominal thickness of plywood required is 15mm for centres not exceeding 450mm and 18mm when joist centres are not in excess of 600mm if the following plywoods are used.

American construction and industrial plywood

C–D grade: exterior unsanded
C–C grade: exterior unsanded

Finnish birch faced plywood 1/1, 1/11, etc., sanded
Finnish conifer plywood 1/1, 1/11, etc., sanded.
Canadian douglas fir plywood and Canadian softwood plywood select tight face; select and sheathing grades unsanded are suitable for use at joist centres 450mm and 600mm when the nominal thickness is 15.5mm
Swedish softwood plywood P30 grade; unsanded may be 16mm thick when joist centres are 450mm and 19mm thick when joist centres do not exceed 600mm.

Plywood floor decking should be continuously supported along walls and other abutments (Fig. 23.4).

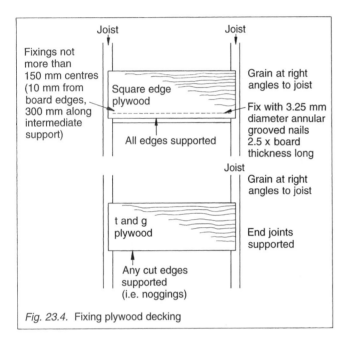

Fig. 23.4. Fixing plywood decking

Particle board

Chipboard floor decking is to be either Type C4 or C5 meeting BS 5669: Part 2: 1989, 'Specification for wood chipboard'. Type T2 cement-bonded particle board may also be used when manufactured in accordance with BS 5669: Part 4: 1989, 'Specification for cement-bonded particle board'.

When subject to an imposed load up to 1.5kN/m² and the alternative concentrated load of 1.4kN, chipboard Type C4 and cement-bonded particle board Type T2 should be a minimum 18mm thick for maximum joist spans of 450mm, and 22mm thick for maximum joist spans of 600mm.

If oriented strand board is to be used it must be independently assessed as suitable and thickness to spans should also be independently certified (Fig. 23.5).

For fixing, 3mm diameter annular grooved nails are to be used with length equal to 2½ times thickness of board, spaced at 300mm centres on continuously supported edges, and any intermediate supports. Nails or screws should be not less than 9mm from edges. Drill cement-bonded particle boards prior to nailing, and countersunk holes are required for all boards when screws are used.

A good practice note advises gluing all t and g joints to reduce creaking!

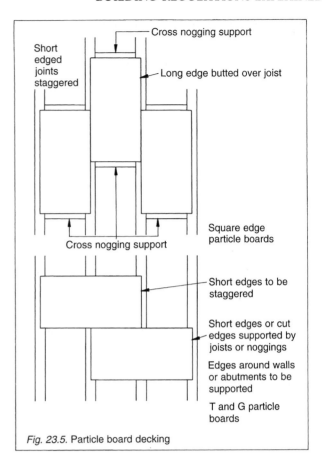

Fig. 23.5. Particle board decking

SUPPORT FOR NON-LOADBEARING PARTITIONS

The TRADA approved document offers guidance relating to the support of non-loadbearing partitions up to 81.5kg/m run. When partitions are at right angles to joists they should be secured to joists, having a span reduced by 10%.

Partitions aligned above floor joists or offset may require the addition of one or two joists to provide the support required.

When non-loadbearing partitions are in excess of 81kg/m joists are to be designed and sized to meet span and loading requirements.

FIRE

Requirements are limited to the spread of fire along surfaces and the floor structure.

Surfaces

The requirements relate only to ceilings below floors to which the approved document relates, and the minimum classification for ceilings is as follows:

- Spaces with floor area not exceeding 4m² (small rooms) Class 3
- Other rooms Class 1
- Circulation spaces Class 1
- Circulation spaces in common areas of blocks of flats and maisonettes Class 0

Classes 1 and 3 must satisfy the requirements of BS 476: Part 7: 1971 or 1987, 'Method for classification of surface spread of flame on products'.

Class 0 is either composed of materials of limited combustibility or a Class 1 material with a fire propagation index of not more than 12 and a sub-index (i₁) of not more than 6. BS 476: Part 6: 1981 or 1989 sets out the method of test for fire propagation products.

Examples of materials with a Class 3 rating include timber or plywood having a density exceeding 400 kg/m³ and wood particleboard or hardboard, whether painted or not.

Materials for Class 0 and 1 include plasterboard, which may be painted or not or have a pvc facing not exceeding 5 mm thick, mineral fibre tiles or sheets with cement or resin binding.

If it is intended to use other materials they should have been independently tested to show compliance with the requirements.

Properly tested impregnation and/or surface treatments may be used to upgrade timber-based materials from Class 3 to Class 1.

Structure

Maisonettes

- Ground or upper storey: Height to finished floor level of top storey not more than 5 m = 30 minutes; not more than 20 m = 60 minutes. The latter may be reduced to 30 minutes where floor does not provide structural or lateral support.
- Basement storeys, including floor above, and subject to a maximum depth of 10 m = 60 minutes (Figs 23.6 and 23.7).

Dwellinghouses (1 and 2 storey)

- Ground or upper storey (see Figs 23.8 and 23.9).
- Basement storeys, including floor above, and subject to a maximum depth of 10 m = 30 minutes (Fig. 23.6).

Dwellinghouses (3 storeys)

- Ground or upper storeys not more than 20 m above ground = 30 minutes (Fig. 23.6).

Other forms of floor construction may be used but they must have been tested.

Also floor construction certified by NAMAS or in accordance with BRE Report 1988, 'Guidance for the construction of fire resisting structural elements', or meeting BS 5268: Part 4: section 4.2: 1990, 'Recommendations for calculating fire resistance of timber stud walls and joists in floor construction'.

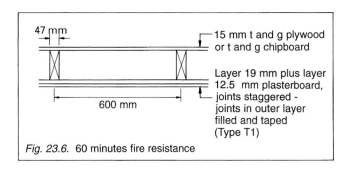

Fig. 23.6. 60 minutes fire resistance

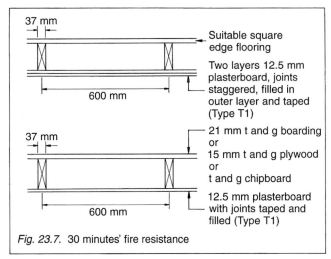

Fig. 23.7. 30 minutes' fire resistance

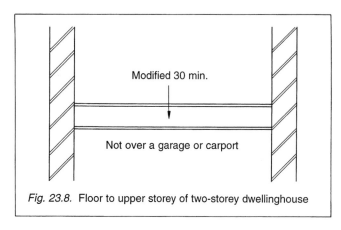

Fig. 23.8. Floor to upper storey of two-storey dwellinghouse

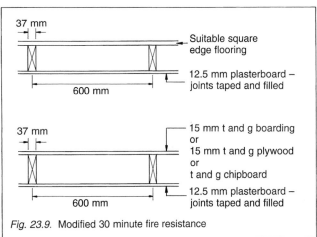

Fig. 23.9. Modified 30 minute fire resistance

Maximum centres for timber supporting plasterboard is 600 mm.

- For plasterboard thickness 12.5 mm use nails 40 mm long or screws 36 mm long.
- For two-layer 12.5 mm plasterboard use nails 50 mm long or screws 50 mm long.
- For 19.0 mm and 12.5 mm layered plasterboard use nails 65 mm long or screws 60 mm long.

Nails to have a shank diameter of 2.5 mm with minimum 7.0 mm head diameter.

Beams projecting below the ceiling are to have the same fire resistance as the floor construction. This can be achieved by including fire resistance into the beam design or by applying plasterboard or timber cladding. See BS 5268: Part 4: section 4.1: 1978, 'Method of calculating fire resistance of timber members'.

Attic room floors

When an attic room floor becomes part of the roof at the edges it should have the same degree of fire resistance as the floor.

CONSERVATION OF FUEL AND POWER

Floors between habitable rooms and unheated spaces must be insulated to limit heat loss. Guidance is given in Approved document L (see Chapter 20).

SOUND INSULATION

A good practice note included relates to a WC floor and the provision of insulation between it and a habitable part of a dwelling. 60 mm mineral wool should be packed lightly between joists.

INSULATING SERVICES

Unless heat loss from pipes and ducts in floors is intended to contribute to the heating of the room or space they should be enclosed in insulation material. Guidance is contained in BS 5422: 1990, 'Methods for specifying thermal materials on pipes, ductwork and equipment'.

Flue pipes passing through floors should meet the requirements for isolation and protection of J3 in Chapter 18.

(The approved document *Timber Intermediate Floors for Dwellings* is the copyright of and may be purchased from Timber Research and Development Association, Stocking Lane, Hugbenden Valley, High Wycombe, Bucks HP14 4ND. Telephone 0494 563091.)

RELEVANT READING

BS 476: Part 6: 1981 and 1987. Method of test for fire propagation for products
BS 476: Part 7: 1971 and 1987. Classification of the surface spread of flame
BS 476: Part 8: 1972. Fire resistance of elements of building construction
BS 476: Part 21: 1987. Fire resistance of loadbearing elements of construction
BS 1230: Part 1: 1985. Plasterboard
BS 1297: 1987. T & G softwood flooring
BS 4471: 1987. Sizes of sawn and processed softwood
BS 4543: Part 1: 1990. Methods of testing factory-made chimneys
BS 4978: 1988. Softwood grades for structural use
BS 5250: 1989. Condensation in buildings
BS 5268: Part 2: 1991. Timber permissible stress, design, materials and workmanship
BS 5268: Part 4: Section 4.1: 1978. Calculating fire resistance of timber members
BS 5268: Part 4: Section 4.2: 1990. Fire resistance of timber stud walls and joisted floor construction
BS 5268: Part 7: Section 7.1: 1989. Span tables – domestic floor joists
BS 5422: 1990. Specifying thermal insulating materials on pipes and ductwork
BS 5669: Part 2: 1989. Wood chipboard
BS 5669: Part 4: 1989. Cement bonded plasterboard
BS 6178: Part 1: 1990. Joist hangers
BS 6399: Part 1: 1984. Dead and imposed loads
BS 8212: 1988. Drylining and partitioning using gypsum plasterboard

2. BASEMENTS FOR DWELLINGS

The TRADA Approved Document was published in July 1992, and it was not until July 1997 that the second privately prepared document was issued by the British Cement Association.

The *Basements for Dwellings Approved Document* was produced under a Department of the Environment Research Contract, and the National House Building Council was the principal sponsor. There was a steering group which directed the preparation, comprising representatives from the British Cement Association; Department of the Environment; Institute of Building Control; and the National Housebuilding Council.

As with the TRADA document, concentration is upon one topic – the construction of basements in dwellings. Not many single dwellings now have basements, so the application will usually be to blocks of flats of up to a maximum height of 15 m. This one approved document relates to all aspects of the 1991 Regs to basements. These include Parts A; B; C; E; F; H; J; K; L; and N. This only leaves Parts D (toxic substances); G (bathrooms); and M (facilities for disabled) which are not applicable to basements.

Following the example of the TRADA document there is supplementary advice on 'good building practice' which may be regarded as evidence supporting compliance with the requirements.

This is a large Approved Document comprising some 82 pages with numerous illustrations and tables.

STRUCTURE

Requirements of the 1991 Regs regarding loading will be met in regard to basements in dwellings by following approved document A when elements are not required to resist the effects of lateral earth loads. However, when vertical and/or lateral earth loads, and vertical loads due to surcharge, have to be provided for, advice in the basement approved document should be followed.

The document applies to buildings having a maximum height from the lowest ground level to the ridge of 15 m, and there are limitations on the height of basement walls as shown in Fig. 23.10.

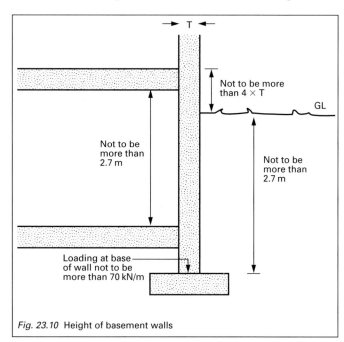

Fig. 23.10 Height of basement walls

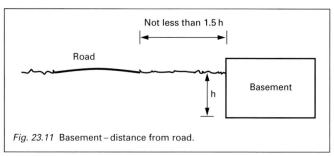

Fig. 23.11 Basement – distance from road.

A basement should not be too near a road, as shown in Fig. 23.11.

The space between road and dwelling may, however, be used as an access drive or parking in connection with the dwelling.

The sizes of openings in a basement retaining wall are slightly modified from those given in approved document A, and the advice with regard to chases is given in limiting depths and lengths.

- vertical chases are not to exceed 15 mm in depth;
- horizontal chases are not to exceed 15 mm in depth and are limited in length to 600 mm between vertical restraint walls;
- a chase must not impair the stability of a wall.

There are requirements for a wall designed as a propped cantilever. This is a wall, supporting a lateral load due to earth loads, which

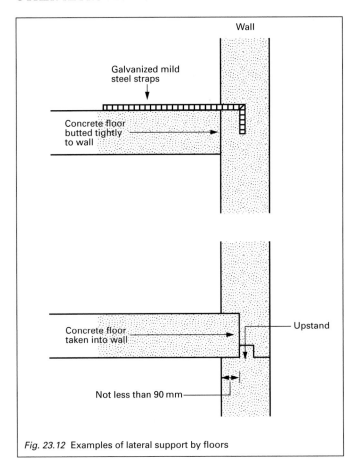

Fig. 23.12 Examples of lateral support by floors

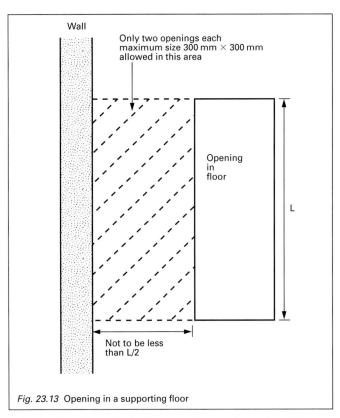

Fig. 23.13 Opening in a supporting floor

resists the movement at its base and is propped by an upper floor within the storey height. It must have sufficient strength and stiffness to transfer forces to the supporting floor.

Support for a propped cantilever wall should not be expected of a normal timber floor.

The floor is to be adequately secured to the supported wall, and several examples are given. The simplest method is to butt the floor tightly against the wall and provide fixing by means of galvanized mild steel straps.

When a basement is to be constructed in ground conditions likely to result in swelling or heave (as, for example, desiccated clays or where trees are to be removed), design must provide for any lateral pressures against the walls.

A floor used for lateral support is to be stiff, without undue movement or deflection. There are limitations on the number and size of openings in the floor as shown in Fig. 23.13.

MASONRY RETAINING WALLS

The compressive strengths of bricks and blocks should be:

- bricks – 20 N/mm²
- blocks – 7 N/mm² or 10 N/mm² as necessary.
- The minimum thickness of 100 mm should be used for the leaves of reinforced masonry grouted cavity walls, and the grouted cavity is to be at least 100 mm wide.

Materials and workmanship

Wall ties are to be in accordance with BS 1243: 1978, Specification for metal ties for cavity wall construction to be Type 1 or 2 of DD 140: Part 2: 1987, 'Recommendation for design of wall ties', and 6 mm diameter. They are to be galvanized in accordance with BS 729: 1971 (1994), 'Specification for hot dipped galvanized coatings, with a minimum mass of zinc coating of 940 kg/m²'.

In the low lift method of construction, wall ties should have a maximum horizontal spacing of 900 mm and a vertical separation of 450 mm. In high lift construction, the maximum spacings should be 900 mm and 300 mm.

Reinforcement should be grade 460 complying with BS 4449: 1988, Specification for carbon steel bars for reinforcement of concrete; BS 6744: 1986, 'Specification for austenitic stainless steel bars'; or BS 970: 1991, 'Requirements for carbon, carbon manganese, alloy and stainless steel', whichever is appropriate.

Where masonry is buried or continually submerged in fresh water, or where the exposure category is sheltered/moderate or moderate/severe, reinforcement should be of carbon steel.

A severe or very severe exposure category arises when masonry may be subjected to freezing when wet, or heavy condensation. This requires reinforcement of galvanized carbon steel. Austenitic stainless steel or carbon steel coated with 1 mm stainless steel reinforcement should be used when the wall may be exposed to salt water, moorland water or corrosive fumes.

Appropriate British Standards to be consulted are BS 5628: Part 3: 1985, 'Masonry components, design and workmanship', and BS 5628: Part 2: 1985, 'Structural use of reinforced concrete'.

There are Tables giving the minimum masonry and reinforcement requirements for propped walls retaining 2.7 m soil, the requirements for an unpropped wall retaining 2.1 m soil, and also for an unpropped wall retaining a maximum of 1.6 m soil.

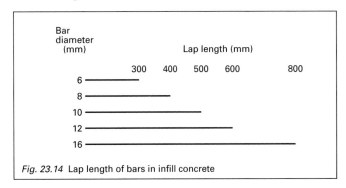

Fig. 23.14 Lap length of bars in infill concrete

Reinforcement in a grouted cavity wall is to have 40 mm cover from the inside face of the masonry leaves, and is to be adequately lapped.

Horizontal reinforcement should be continuous; not spaced more than 500 mm apart, and have adequate lapping which must be continuous at corners.

Mortar is to be class (i) or (ii) as in BS 5628: Part 2: 1995, or a cement, lime, sand mix of dry materials – 1:0 to $\frac{1}{4}$:3.

Concrete fill for reinforced masonry may be an RC30 mix or RC35 mix to BS 5328: Part 2: 1997, Methods for specifying concrete mixes; or a cement, lime, sand, 10 mm maximum size aggregate mix of 1:0 to $\frac{1}{4}$:3:2, by volume in dry materials.

Measures should be taken to prevent early age shrinkage in high lift masonry construction.

CONCRETE RETAINING WALLS

The guidance given with regard to in situ concrete retaining walls assumes the characteristic strength of concrete to be 35 N/mm² and walls are to comply with BS 8110: Part 1: 1997, 'Structural concrete: design and construction', with some exceptions regarding materials and workmanship.

There are three Tables setting out minimum reinforcement and thickness requirements for:

- a propped wall retaining up to 2.7 m soil;
- a cantilevered wall retaining up to 2.1 m soil; and
- a cantilevered wall retaining up to 1.6 m soil.

Materials and workmanship

Reinforcement is to meet the requirements of BS 4449: 1988, 'Steel fabric for reinforcement of concrete', and should be carbon steel in:

- buried concrete;
- concrete submerged in fresh water; or
- external concrete.

Specialist advice must be obtained in circumstances where concrete is exposed to sea water or flowing water with a pH at or below 4.5.

Reinforcement is to be provided having an area dependent upon the thickness of the wall; it is to be adequately supported and lapped.

There is to be continuity of horizontal reinforcement achieved by adequate lapping and made continuous at corners. A minimum lap length of 500 mm is acceptable and spacing may not exceed 750 mm.

The amount of horizontal reinforcement required in each face is:

- for 300 mm thick walls = 375 mm²/m;
- for 200 mm thick walls = 250 mm²/m.

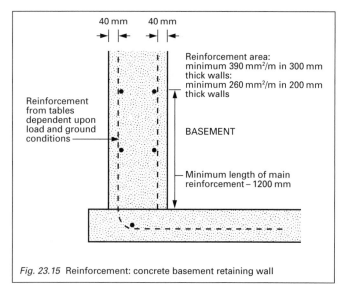

Fig. 23.15 Reinforcement: concrete basement retaining wall

Walls thicker than 300 mm should have a minimum area of steel equal to not less than 1.3 times the thickness of the wall.

Guidance is given for aggressive and non-aggressive soil conditions. For non-aggressive soil concrete is not to be less than RC35 in BS 5328: Part 2: 1997, 'Methods of specifying concrete mixes'. In aggressive soil conditions the provisions of BS 5328: Part 1: 1997, 'Guide to specifying concrete', are to be followed. A slump test should be not less than 75 mm.

FOUNDATIONS

There are requirements regarding foundations of both plain and reinforced concrete.

Ground conditions below the basement should be of a reasonably uniform type and condition in order that stability of the structure is not impaired.

Design provisions

The provisions of Approved Document A with regard to strip foundations may be applied to a basement in a dwelling, except in the case of retaining walls, when a number of modifications apply.

- Walls carrying vertical loads and/or lateral earth loads are to have minimum widths related to a much more extensive range of factors, including 'moment' than is given in Table 12 of Approved Document A.
- The projection of raft foundations beyond an external wall is prescribed; see Fig. 23.16.

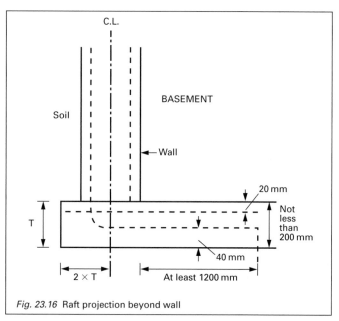

Fig. 23.16 Raft projection beyond wall

- A concrete blinding is to be provided, and thickness of concrete strip or raft foundations is not to be less than 200 mm.
- In chemically aggressive soils, follow BS 5328: Part 1: 1997, 'Guide to specifying concrete', and in non-aggressive conditions rely on concrete grade R35 to BS 5328: Part 2: 1997, 'Methods of specifying concrete mixes'.
- Area of steel in mm²/m should not be less than 1.3 times foundation thickness, when foundation is in excess of 200 mm.

Precautions are to be taken to prevent differential settlement when a basement is under only part of the total plan area of the building.

Provision is to be made for adequate support of an internal loadbearing wall resting on the foundation, as shown in Fig. 23.17.

There is a Table detailing bottom reinforcement requirements under loadbearing walls, and another with extensive details of top

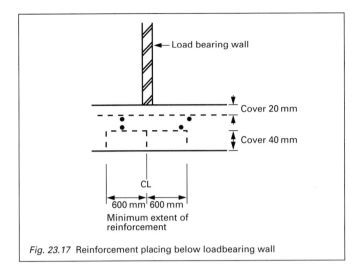

Fig. 23.17 Reinforcement placing below loadbearing wall

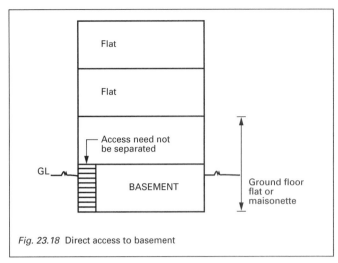

Fig. 23.18 Direct access to basement

reinforcement in raft foundations. This latter requirement caters for loads as a result of soil/structure interaction, and is required over the whole area of the foundation.

FIRE SAFETY

This part of the approved document sets out ways of complying with the 1991 Regs, Part B if a building includes a basement. Generally guidance in Approved Document B is to be followed, coupled with the following.

Means of escape

In a dwellinghouse where no storey has a floor in excess of 4.5 m above external ground level, it is not necessary to separate the basement from the rest of the dwelling. Separated means that the basement is divided from the rest of of the building by fire resisting construction and is maintained by preventing fire and smoke arising in the basement from gaining access to the stairway.

In circumstances where the basement is not separated, rooms on ground and upper storeys are regarded as 'inner rooms' (see Chapter 11). When the only escape route from a basement is through another room there is a risk from a fire in the outer room. This situation may only be accepted when the inner room is:

- a bathroom; or
- a dressing room, shower room or WC; or
- a kitchen; or
- a laundry; or
- a utility room; or
- another room in the basement which has an external door or openable window suitable as an escape window.

When a basement is part of a ground floor flat or maisonette and there is direct access from the flat or maisonette, it is not necessary for it to be separated (see Fig. 23.18).

Where it is necessary to have separation between the basement and a storey above, any vertical duct passing through the separation and upper storeys is to have the same fire resistance as the separating construction.

Alternative escape routes

When there is a bedroom in a basement which is separated from the rest of the dwelling, an alternative escape route is to be available. This may be achieved by providing an external door or window. If a door, the threshold height is not to be in excess of 150 mm. If a window, it is to have an unobstructed opening of not less than 850 mm × 500 mm which is not more than 1100 mm above the internal floor level (see Fig. 23.19).

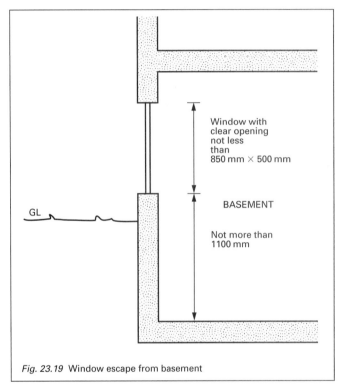

Fig. 23.19 Window escape from basement

Conditions which are acceptable for a basement storey exit to lead into an enclosed area reiterate those shown in Fig. B1.9 in Chapter 11.

The provisions regarding smoke alarms contained in Approved Document B are to be observed in basements.

Stairways

Basement stairs are to comply with Approved Document K. Careful measures are necessary to prevent a fire in a basement endangering the dwelling in storeys above.

- When a common stair in a block of flats is part of the only escape route from a dwelling, it is to be taken down to the basement. The basement is to have a separate stair with a separate final exit. Basement bedrooms or other habitable rooms are to have escape doors or windows (see Fig. 23.19).
- Where there is more than one common stair in a block of flats, and there is interconnection on each floor, one stair is to end on the ground floor. The other stair may go to the basement storey and be provided with a ventilated protected lobby in the

basement. Basement bedrooms or other habitable inner rooms are to have an escape door or window.

- A common stair forming the only escape route from a dwelling must not also serve spaces having a fire risk such as a boiler room, fuel storage space or car park.
- When separated by a protected and ventilated lobby, a common stair in a block of flats which is not the only escape route, may serve ancillary accommodation which has a fire risk.

Common stair construction

All escape stairs and associated landings from the basements of blocks of flats or maisonettes are to be constructed of materials of limited combustibility, although the upper tread surface may be of combustible material, except when it is a firefighting staircase or landing. This follows requirements of Approved Document B.

Fire resistance

Elements of structure in houses and flats are to follow the requirements of Approved Document B, but the minimum periods of fire resistance for a basement which is not more than 10 m below ground level are stated:

- for a dwellinghouse, the period is 30 minutes for elements of structure in a basement storey;
- for flats and maisonettes the period is 60 minutes.

Provision of fire doors

Fire resisting doors are to fulfil the requirements of BS 476: Part 22: 1987, 'Determination of the fire resistance of non-loading elements of construction'. The number indicates a door's integrity in terms of minutes and, when the suffix 'S' is added, it indicates that smoke leakage at ambient temperature is to be restricted. Fire doors should be fitted with automatic self-closing devices. The position of fire doors is as follows:

FD 30 S a. In compartment wall between a basement flat or maisonette and a common space.
       b. In a compartment wall between a basement area and a common space.
       c. Door at basement level between a ventilated lobby and staircase to upper floors in block of flats.
FD 30   Door between garage and basement.
FD 20   Door between basement and ground and upper storeys in a single family dwelling where there is an upper storey more than 4.5 m above ground.

SITE PREPARATION AND RESISTANCE TO MOISTURE

The guidance given in Approved Document C is to be followed (Chapter 12) and emphasis is made upon the extra care which should be taken when making preparation for a building which includes a basement.

Good practice requires an early site investigation involving:

- a desk study;
- ground conditions including subsoil;
- establishing water table and movement;
- location of existing drains and services;
- presence of contamination in the site;
- presence of radon in the site.

When difficulties are anticipated it is recommended that guidance given in BRE publications BR211, 'Radon: guidance on protective measures for new dwellings', and BR 212, 'Construction of new buildings on gas contaminated land', should be referred to. One is also reminded of the importance of 1991 Reg. 7 in considering

appropriate materials and workmanship suitable for a basement in the particular circumstances of the site (Chapter 9).

The water table is regarded as:

- high when it is above the underside of the lowest basement floor slab;
- low when it is permanently below the underside of the lowest basement floor slab;
- variable when it fluctuates between these two levels.

All existing underground drains and services discovered must be re-routed around the building, or dealt with in another satisfactory manner, with the agreement of the appropriate statutory authority. Active subsoil drains should be repaired, and rerouted where necessary as described in Chapter 12.

Movement of ground water coupled with the shape of a building can pose problems associated with hydrostatic pressure and good practice dictates:

- waterproofing basement to withstand the full hydrostatic head, or
- provision for the water to continue to drain away.

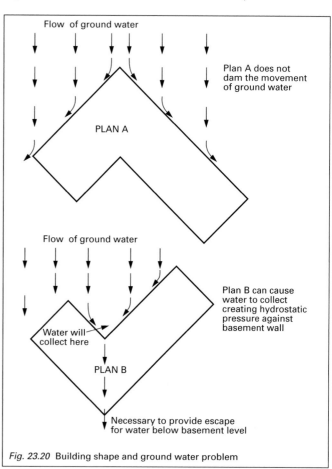

Fig. 23.20 Building shape and ground water problem

Walls and floors are to resist ground moisture reaching the internal surfaces of the basement and this calls for particular attention being given to wall and floor junctions.

BS 8102: 1990, 'Code of practice for the protection of structures against water from the ground', gives four levels of protection:

- Grade 1 permits some seepage and damp patches on internal surfaces;
- Grade 2 permits no water penetration but moisture vapour is tolerated;
- Grade 3 permits no water penetration and provides a dry environment maintained by heating and ventilation;
- Grade 4 requires a controlled environment including air conditioning.

Good practice recommends Grade 2 for garages, plant rooms and workshops and Grade 3 for habitable accommodation. It is advisable that all basements are ventilated.

The continuity of waterproofing is paramount, particularly in wall base details and laps in membranes.

An evaluation of the options should ensure:

• each item has been considered and careful assessment made of associated factors;
• the waterproofing system chosen meets the conditions set out in any appropriate Technical Approval;
• the design accords with the supplier's recommendations.

There are tables to assist in site evaluation, as follows:

• characteristics of soils which affect basement construction;
• requirements for continuity;
• suitability of primary waterproofing systems. This table lists the seven categories of waterproofing systems.

   1. Bonded sheet membranes.
   2. Cavity drain membranes.
   3. Bentonite clay active membranes.
   4. Liquid applied membranes.
   5. Mastic asphalt membranes.
   6. Cementitious crystallization active systems.
   7. Proprietary cementitious multi-coat renders, toppings and coatings.

• acceptability of construction types.

Tanked construction alone is not recommended when the water table is high.

There are figures illustrating the three types of basement construction:

• Type A – tanked protection;
• Type B – structurally integral protection; water resistant concrete;
• Type C – drained cavity, which is dependent upon an external structure to reduce the ingress of water and silt to be removed by the cavity system.

There are further figures demonstrating the effect of foundation designs on the continuity of waterproofing – linking with superstructure; step changes in construction; and the penetration of services through waterproofing (which is to be avoided wherever possible).

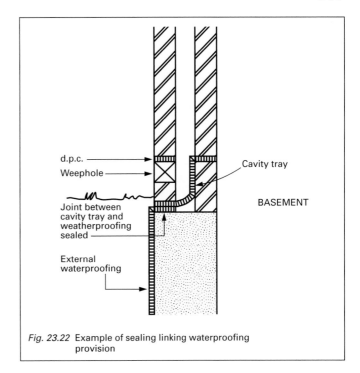

Fig. 23.22 Example of sealing linking waterproofing provision

RESISTANCE TO THE PASSAGE OF SOUND

There is additional and brief guidance in meeting the requirements of Part E when basement walls which separate one dwelling from another are to resist the passage of airborne sound.

Wall Types 1, 3a, 3b, and 3c described in Chapter 14 will meet the requirements for basement retaining walls and it is recommended that where a separating wall abuts an external wall, extra care is necessary (see Fig. 23.23).

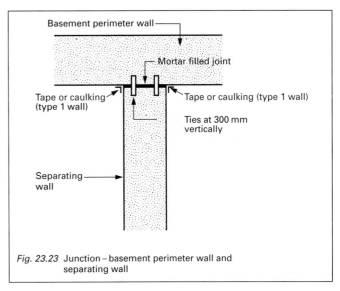

Fig. 23.23 Junction – basement perimeter wall and separating wall

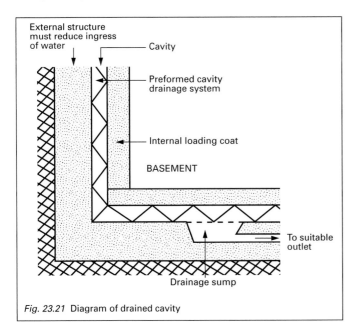

Fig. 23.21 Diagram of drained cavity

Airborne sound is to be resisted by a floor or stair between the basement and ground floor if it separates a dwelling from another part of the building which is not part of the dwelling.

Airborne and impact protection is necessary when a floor or stair between the basement and ground floor separates two dwellings. Appropriate guidance is given in Approved Document E (Chapter 14).

VENTILATION

All habitable rooms, bathrooms, kitchens, sanitary accommodation and utility rooms located in basements are to meet the ventilation requirements in Approved Document F (Chapter 15).

There are no additional requirements in the British Cement Association document, but it does contain several items of advice regarded as good practice:

- heated or unheated non-habitable rooms should be provided with ventilation required in Approved Document F for non-habitable rooms not containing windows;
- habitable rooms in basements should be provided with cross-flow background ventilation. Internal doors should have a 10 mm gap at the bottom;
- single-sided basements require ventilation in the exposed face, and background ventilation provided by a passive stack, ducts or extract fans.

DRAINAGE AND WASTE DISPOSAL

Drainage installed in a basement is to comply with the requirements of Approved Document H (Chapter 17) and BS 8301: 1985, 'Building drainage', and the following requirements are noted.

- If there is a possibility of surcharge which could lead to flooding of the basement, a flood valve should be installed;
- any pipe passing through the basement structure is to have provision made for differential movement, and the pipe sealed to the waterproofing system;
- where a gravity connection to a main is not possible refer to BS 8301: 1985 for guidance on sewage lifting equipment.

Surface and ground water

Surface water drainage should also comply with Approved Document 'H' and BS 8301: 1985.

When there is an access ramp falling to private garages, the drainage should incorporate an interceptor trap, or a trapped gully having a cleanable sump.

The outfall of ground water is to be made through a catchpit capable of intercepting and retaining soil and other suspended matter.

Lightwells and access areas at basement level are to be provided with surface water drainage, including:

- gullies which must resist blockage by debris, and
- if the site is sloping, arrange to direct surface water from lightwells.

Ground water drainage should discharge directly or through a catchpit to a watercourse, or through a catchpit into a soakaway. A soakaway must be not less than 5 m from the basement. When it is necessary to discharge ground water into a public sewer, arrangements must be made with the drainage authority.

HEAT PRODUCING APPLIANCES

The advice contained in Approved Document J is to be followed when installing heat producing appliances in a basement. The Gas Safety (Installation and Use) Regulations 1998 also apply and may limit the use of certain LPG appliances in basements.

Good practice advice is also given that LPG containers should not be placed in basements, and where LPG installations are fixed in buildings having a basement, an LPG detector should be fitted.

Products of combustion

Flue terminals serving room-sealed appliances should be not less than 300 mm above external ground level, and precautions taken to prevent them becoming ineffective by rubbish, leaves or snow. In addition, a balanced flue outlet should not be sited in a lightwell, as wind pressures may hold products of combustion within the well.

STAIRS, RAMPS AND GUARDS

Ramps and guards are to be provided as required by Approved Document K (Chapter 19).

External ground levels may be cut away to provide lightwells for basement windows, and the provision of guards, as in Fig. 23.24, is recommended.

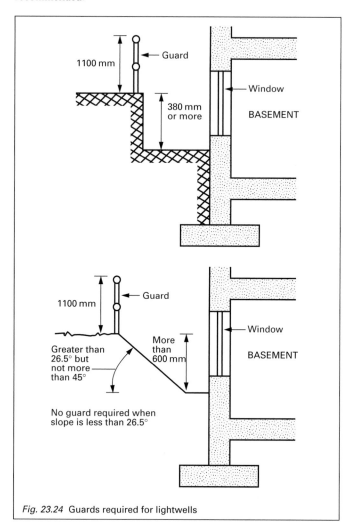

Fig. 23.24 Guards required for lightwells

CONSERVATION OF FUEL AND POWER

Generally the guidance in Approved Document L (Chapter 20) is to be followed in basements to dwellings, but there is much extra advice in the British Cement Association document.

Naturally the nature of soil surrounding a basement influences conditions in the building and there is information regarding thermal conductivity of the ground. Clay or silt is 1.5 W/mK; sand or gravel 2.0 W/mK; and homogeneous rock 3.5 W/mK.

It is recommended that a satisfactory limitation of heat loss when basements are a consideration may rely upon one of the methods described in Approved Document L.

Elemental method

This is intended to show the thermal performance of the construction elements when basement walls and floors are taken to be exposed elements. Continuous insulation of an appropriate thickness is to be provided and there are tables to be used to arrive at the base thickness of an insulation layer; and for floors in contact with the ground.

By way of an example, take a basement wall which is to have an internal insulation lining and achieve a U value of 0.45 W/m²K. The wall is assumed to be at a constant depth of 2.5 m in the surrounding ground and the construction is shown in Fig. 23.25. Using tables in the approved document, base thickness of an insulating layer with a thermal conductivity of 0.025 W/mK is given as 20 mm; this would be increased to 28 mm for insulation having a thermal conductivity of 0.035 W/mK.

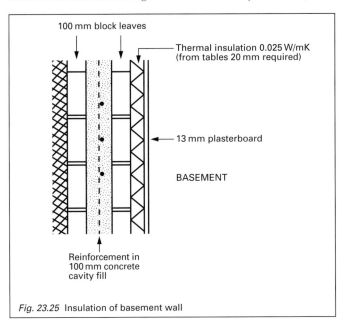

100 mm block leaves

Thermal insulation 0.025 W/mK (from tables 20 mm required)

13 mm plasterboard

BASEMENT

Reinforcement in 100 mm concrete cavity fill

Fig. 23.25 Insulation of basement wall

When the depth of soil around a basement is varied, then the average depth is determined before making calculations.

Included are tables to determine the insulation thickness for solid floors in contact with the ground for U 0.25 W/m²K; 0.35 W/m²K; and 0.45 W/m²K. The U value of a basement floor is arrived at using the ratio of the floor perimeter as in Part L and also the depth below ground level.

In order to determine the thickness of insulation (at 0.02 W/mK) required to achieve a U value of a basement floor, reliance is placed upon the formula

$$\frac{\text{perimeter length}}{\text{floor area}}$$

and the included tables.

For example, if the perimeter length is 32 m and the floor area 52 m², we have

$$\frac{32}{52} = 0.6$$

Assuming an average depth of 2 m, reference to tables indicate that 5 mm of insulation is required. Were the average depth to be 1.5 m, an increase to 8 mm of insulation would be necessary.

Insulation placed below the floor slab is to be sufficiently robust to maintain the load, be durable in that situation and keep its insulating properties. If insulation is placed above the slab, it should be capable of adequately supporting the flooring placed upon it.

Calculation methods

There are alternative procedures for determining the U value and insulation requirements for walls and floors of basements.

For evaluating the U value of basement floors, there are tables relating to the average depth below ground level related to the P/A area ratio and the thermal resistance needed to achieve indicated U values at various depths.

Included in the approved document is a formula to enable the average U value of a basement storey to be calculated:

$$U_{ave} = \frac{A_f U_{bf} + H_a P U_{bw} + A_w U_w}{A_f + A_w + H_a P}$$

Where:

A_f = area of basement floor (m²)
A_w = area of walling above ground level (m²)
H_a = average depth below ground level (m)
P = perimeter of floor (m)
U_{ave} = average value of basement (W/m²K)
U_{bf} = U value of basement floor (W/m²K)
U_{bw} = U value of walls below ground level (W/m²K)
U_w = U value of basement walls above ground level (W/m²K)

Target U value method

This method, detailed in Approved Document L, allows higher U values for some parts of the fabric to be offset against lower U values elsewhere, and may be used in conjunction with the calculation methods above.

Energy rating method

Energy rating for a dwelling including a basement may be calculated using the Standard Assessment procedure (Chapter 12). Adjustments must be made as follows:

• overall dwelling dimensions are to assume the basement floor is the ground floor when on a sloping site;
• when basement is completely below ground, the floor is regarded as an 'other floor';
• ventilation should be on the basis of 'structural infiltration';
• to ascertain heat losses and heat loss parameters, the basement floor is to be taken as the ground floor;
• windows in the basement (if any) should be entered according to their orientation. If not known they are to be shown E/W.

U values determined by PrEN 1190

The document includes a procedure and equations from PrEN 1190, used to determine the heat transfer from a basement and U values of floor and wall elements.

Thermal resistance of building products:

• when available, use certified test results;
• alternatively, use appropriate design values in a European Standard or European Technical approval;
• below ground, thermal resistance products are to reflect the moisture conditions;
• heat capacity of floor materials are to be neglected.

GLAZING

Glazing is to follow the guidance given in Approved Document N (Chapter 22) with the following additional requirements:

• glazing within 800 mm of (a) external ground or (b) floor level which is not given a protective barrier is to meet test requirements of Class C in BS 6206: 1981, 'Impact performance of flat safety glass and safety plastics in buildings'.
• glazing installed in a door side panel having a pane width more than 900 mm must have glass to meet the test requirements of Class B in BS 6206: 1981, if it is within 1500 mm of the ground or adjoining floor.

(The Approved Document 'Basements for Dwellings' is the copyright of, and may be purchased from, the British Cement Association, Century House, Telford Avenue, Crowthorne, Berkshire RG45 6YS Telephone 01344 762676.)

RELEVANT READING

British Standards

BS 187: 1978. Calcium silicate (sandlime and flintstone) bricks
    Amendment 1: AMD 5427
BS 476: Part 22: 1987. Fire resistance of non-loading elements of
construction
BS 729: 1971 (1994). Hot-dipped galvanized coatings on iron and
steel articles
BS 970: Part 1: 1991. Specification and testing carbon, carbon man-
ganese, alloy and stainless steels
    Amendment 1: AMD 7226
             2: AMD 8973
BS 1243: 1978. Metal ties for cavity wall construction
    Amendment 1: AMD 3651
             2: AMD 4024
BS 3921: 1985. Specification for clay bricks
    Amendment 1: AMD 8946
BS 4449: 1988. Carbon steel bars for reinforcement
BS 4483: 1985. Steel fabric for reinforcement
BS 5328: Part 1: 1997. Guide to specifying concrete
BS 5328: Part 2: 1997. Methods for specifying concrete mixes
BS 5628: Part 1: 1992. Structural use of unreinforced masonry
    Amendment 1: AMD 7745
BS 5628: Part 2: 1995. Structural use of reinforced and prestressed
masonry
BS 5628: Part 3: 1985. Materials and components, design and work-
manship (masonry)
    Amendment 1: AMD 4974
BS 6073: 1981. Specification for precast masonry units
    Amendment 1: AMD 4508
BS 6206: 1981. Impact performance requirements for flat safety glass
and safety plastics for use in buildings
    Amendment 1: AMD 4580
             2: AMD 5189
             3: AMD 7589
             4: AMD 8693

BS 6399: Part 1: 1996. Loadings for buildings – dead and imposed
loads
BS 6744: 1986. Austenitic stainless steel bars for reinforcement
BS 7348: 1990. Electrical apparatus for detection of combustible
gases in domestic premises
    Amendment 1: AMD 6869
BS 8002: 1994. Earth retaining structures
BS 8004: 1986. Foundations
BS 8007: 1987. Design of concrete structures for retaining aqueous
liquid
BS 8102: 1990. Protection of structures against water from the ground
BS 8110: Part 1: 1997. Concrete design and construction
BS 8110: 1985. Concrete code of practice for special circumstances
    Amendment 1: AMD 5914
BS 8110: Part 3: 1985. Design charts for singly-reinforced beams,
doubly-reinforced beams and rectangular columns
    Amendment 1: AMD 5918
BS 8301: 1985. Code of practice for building drainage
    Amendment 1: AMD 5904
             2: AMD 6580

BRE Reports

Radon: guidance on protective measures for new dwellings.
1992 – Ref BR 211.
Construction of new buildings on gas contaminated land.
1991 – Ref BR 212.

British Cement Association

Basement waterproofing: design guide.
1994 – Ref 48.058.
Basement waterproofing: site guide.
1994 – Ref 48.059.

Definitions

Here is a selection of definitions appropriate to building control. In each example a reference is made to the source.

1984 Act	Building Act 1984
1985 AI Regs	Building (Approved Inspectors) Regulations 1985
1989 Act	Water Act 1989
1990 Act	Town and Country Planning Act 1990
1990 LBC Act	Planning (Listed Buildings and Conservation Areas) Act 1990
1991 CP Regs	Construction Products Regulations 1991
1991 Regs	Building Regulations 1991
1996 PW Act	Party Wall, etc. Act 1996
AD	Approved Document

Access: approach or entry (AD–M)

Accessible: access is facilitated (AD–M)

Access room: room which is an escape route from inner room (AD–B)

Accommodation stair: provided for convenience of occupants (AD–B)

Acknowledged rule of technology: technical provision acknowledged by majority of representative experts (1991 CP Regs, 2)

Adjoining owner: owner of land, buildings, storeys or rooms adjoining those of a building owner (1996 PW Act, section 20)

Agriculture: includes horticulture, fruit growing, seed growing, fish farming and breeding and keeping livestock (1991 Regs, Schedule 2)

Alternative escape routes: separate escape routes (AD–B)

Alternative exits: separated exits (AD–B)

Alternative stair tread: a stair with paddle shaped treads (AD–K)

Appliance ventilation duct: conveys combustion air to a gas appliance (AD–B)

Appointed person: person appointed to consider representations about prohibition notice (1991 CP Regs, Schedule 4)

Approved inspector: a person appointed by the Secretary of State, or a designated body, to inspect plans and certify work under provisions in the building regulations (1984 Act, section 49)

Appropriate attestation procedure: procedure in relation to construction product in relevant technical specification (1991 CP Regs, 2)

Approved body: an approved laboratory, a certification body or an inspection body (1991 CP Regs, 2)

Approved laboratory: testing laboratory designated for purposes of EC Directive (1991 CP Regs, 2)

Approved person: a person approved by the Secretary of State, or a designated body, to give certificates to the effect that work carried out in accordance with the deposited plans will comply with the regulations specified in the certificate (1984 Act, sections 16(9) and 17(1))

Area: calculated to the internal finished faces of a building (1991 Regs, 2)

Atrium: vertically open space through three or more storeys enclosed at top (AD–B)

Authorized officer: an officer of a local authority authorized in writing to act in a matter of a specified kind. By virtue of his appointment for the purposes within his province, a proper officer, or a district surveyor (1984 Act, section 126)

Automatic self-closing device: not to include rising butts, except where used in a flat, or maisonette, cavity barrier or on the door from a dwelling into a garage (AD–B)

Automatic release mechanism: device allowing door to close automatically when hazard occurs (AD–B)

Basement: a storey where the floor is at some point more than 1.2 m below surface of adjoining ground (AD–B)

Bathroom: includes shower room (AD–F)

Boundary: includes up to centre of street, railway, canal or river (AD–B)

Building: any permanent or temporary building, but does not include any other kind of structure or erection; includes part of a building (1984 Act, section 121; 1991 Regs, 2; 1985 AI Regs, 2). Also includes any structure or erection and any part of a building, but not plant or machinery comprised in a building (1990 Act, section 336), shop, office, factory, warehouse, school, other educational building, student residential accommodation, institution, premises where public admitted (AD–M)

Building matter: is any building or other matter to which building regulations are applicable (1984 Act, sections 11(8) and 12(12))

Building operations: include rebuilding, structural alterations and additions to buildings and other operations normally undertaken by a person carrying on a business of a builder (1990 Act, section 336)

Building Regulations: made by Secretary of State:

(a) for securing health, safety and welfare and convenience of persons in or about buildings
(b) furthering conservation of fuel and power
(c) preventing waste, undue consumption, misuse or contamination of water.

(1984 Act, sections 1 and 122)

Building notice: a notice given by developers to council as an alternative to the submission of full plans (1991 Regs, 11(1))

Building owner: the owner of land wishing to exercise party wall rights (1996 PW Act, section 20)

Building work: means the erection of a building

- provision, alteration or extension of services or fittings, material alteration
- work in consequence of a change of use

(1991 Regs, 3(1))

Buttressing wall: wall offering lateral support to another wall (AD–A)

Cavity barrier: seals a cavity against passage of smoke and flame (AD–B)

Cavity width: distance between two leaves in a cavity wall (AD–A)

Ceiling: exposed overhead in a room, circulation space or protected shaft including a soffit and the actual rooflight (AD–B)

Certification body: body designated as a certification body for the purposes of EC Directive (1991 CP Regs, 2)

Cesspool: includes a settlement tank or other tank for the reception of foul matter from buildings (1984 Act, section 126)

Circulation space: access between room and protected shaft and an exit (AD–B)

Class 0: a classification indicating restriction of spread of flame more strict than Class 1 (AD–B)

Closet: includes a privy (1984 Act, section 126)

Common balcony: open air walkway used as escape route (AD–B)

Common stair: stairway used by two or more dwellings (AD–K)

Community: the European Economic Community, The European Union (1991 CP Regs, 2)

Compartment: portion of building constructed to prevent spread of fire (AD–B)

Compartment wall/floor: used to divide buildings into compartments (AD–B)

Common space: space used by one or more dwellings (AD–F)

Completion certificate: local authority may be required to provide (1991 Regs, 15)

Concealed space (cavity): an enclosed space within a building but not including a room, cupboard, circulation space, protected shaft or space in a flue duct, chute, pipe or conduit (AD–B)

Conservatory: one storey building having glazed walls and roof (AD–B)

Construct: carrying out operations designated in building regulations failing to be treated as construction or erection of a building. Conversion of a moveable object into a building (1984 Act, section 123)

Construction Products Directive: Directive 89/106/EEC (1991 Regs, 2)

Controlled service or fitting: bathrooms, hot water storage, sanitary conveniences, drainage, and waste disposal; heat-producing appliances, heating system controls and insulation of heating services (1991 Regs, 3)

Containment: barrier which prevents people falling (AD–K); noxious material in ground to be covered by building (AD–C)

Contravention: means failure to comply, and contravene has a corresponding meaning (1984 Act, section 126)

Corridor access: common horizontal internal access (AD–B)

Crown authority: means Crown Estate Commissioners, a Minister of the Crown, a government department, or any other person or body whose functions are performed on behalf of the Crown (not in Her Majesty's private capacity) or a person acting in right of the Duchy of Lancaster or the Duchy of Cornwall (1984 Act, section 44)

Crown building: is a building in which there is Crown interest or Duchy interest (1984 Act, section 44)

Crown interest: interest belonging to Her Majesty in right of the Crown, or belonging to a government department, or held in trust to Her Majesty for the purpose of a government department (1984 Act, section 44)

Customs officer: officer within meaning of Customs and Excise Management Act 1979 (1991 CP Regs, 19)

Dead end: escape in one direction only (AD–B)

Deadload: weight of walls, permanent partitions, floors, roofs, finishes, services, and any other permanent construction (AD–A)

Defective state: premises which are in such a state as to be prejudicial to health or a nuisance (1984 Act, section 76)

Deposit of plans: submission of plans to a local authority for purpose of securing approval of proposed work (1984 Act, sections 16 and 124)

Designated body: body designated by Secretary of State to approve inspectors (1985 AI Regs, 3)

Designated use: as designated under section 1, Fire Precautions Act 1971 (1985 AI Regs, 2)

Development: the carrying out of building, engineering, mining or other operations in, or under land or the making of a material change of use of any buildings or other land (1990 Act, section 55)

Directive: the Council Directive 89/106/EEC on the approximation of laws, regulations and administrative provisions relating to construction products (1991 CP Regs, 2)

Direct distance: shortest distance to protected stairway (AD–B)

Disabled persons: people with physical impairment limiting their ability to walk or makes them dependent on a wheelchair, or with impaired hearing or sight (AD–M)

Disposal main: means any outfall or other pipe which (a) is a pipe for conveyance of effluent to or from any sewage disposal works, whether of a sewage undertaker or of any other person, and (b) is not a public sewer (1989 Act, Schedule 19)

Domestic buildings: buildings used for dwelling purposes including dwellinghouses, flats, student accommodation and residential homes (AD–F)

Domestic purposes: except in relation to sewers means in relation to a supply of water to premises for drinking, washing, cooking, central heating and sanitary purposes (1989 Act, section 189)

Door: can have one or more leaves. Includes a shutter, cover or other form of protection to an opening in a wall, floor or protected shaft (AD–B)

Drain: includes a pipe and associated works provided in pursuance of sections 12(6), 14(5), 21(4), or 26 of Control of Pollution Act 1974, section 12(6): disposal authority may construct, lay and maintain pipes (1984 Act, section 18(5))

Drain: a drain used for the drainage of one building or of buildings and yards within the same curtilage. Includes manholes, ventilating shafts, pumps and other accessories belonging to the drain (1984 Act, section 126)

Drainage: includes the conveyance, by means of a sink or any other necessary appliance, of refuse water and the conveyance of rainwater from roofs (1984 Act, section 21(2))

Duchy interest: an interest belonging to Her Majesty in right of Duchy of Lancaster or belonging to the Duchy of Cornwall (1984 Act, section 44)

Dwelling: includes a dwellinghouse and a flat (1991 Regs, 2); unit of residential accommodation (AD–B)

Dwellinghouse: does not include a flat or a building containing flats (1991 Regs, 2)

Earth closet: is a closet which has a moveable receptacle for the reception of faecal matter and its deodorization by the use of earth, ashes or chemicals, or by other methods (1984 Act, section 126)

EC certificate of conformity: a certificate of conformity issued by a certification body (1991 CP Regs, 2)

EC declaration of conformity: a declaration of conformity made in accordance with regulations made under Directive (1991 CP Regs, 2)

EC mark: the symbol 'CE' (1991 CP Regs, 2)

Effluent: any liquid, including particles of matter and other substances in suspension in the liquid (1989 Act, section 189)

Electromagnetic: a device susceptible to smoke which will hold a door open but close it automatically when smoke detected, the electricity supply fails or the fire alarm system is operated (AD–B)

Element of structure: part of structural frame, load-bearing wall, separating wall, compartment wall, gallery and protecting structure (AD–B)

Emergency lighting: for use when public supply fails (AD–B)

Enforcement authority: the Secretary of State and any authority or council on whom functions are conferred (1991 CP Regs, 2)

Engineering operations: include the formation or laying out of means of access to highways (1990 Act, section 336)

Erect: carrying out operations designated in building regulations falling to be treated as construction or erection of a building. Conversion of a moveable object into a dwelling (1984 Act, section 123)

Escape lighting: illumination of escape route (AD–B)

Essential requirements: requirements applicable to works which may influence technical characteristics (1991 CP Regs, 2)

European technical approval: favourable technical assessment of the fitness for use of a construction product (1991 CP Regs, 2)

Evacuation lift: for use by disabled people in fire (AD–B)

Exempt body: building regulations may exempt

- a local authority
- a county council
- other bodies that act under an enactment for public purposes and not for its own profit.

Exemption relates to compliance with regulations that are not substantive (1984 Act, section 5)

Exposed:

(a) dwellings
exposed to atmosphere
separating dwelling from part of a building ventilated by openings or ducts equal to 5% of enclosing walls;

(b) other buildings
exposed to atmosphere
separating heated part of building from part not heated and which is exposed to outside air

(1985 Regs L1, Schedule 1)

External cladding: material fixed on the outside face of an external wall (AD–B)

External wall: includes a roof pitched at 70% or more from the horizontal when adjoining space where people have access (AD–B)

Factory: any premises in which persons are employed in manual labour in any process for or incidental to:

(a) the making of any article or part of any article, or

(b) the altering, repairing, ornamenting, finishing, cleaning or washing or breaking up or demolition of any article, or

(c) the adapting for sale of any article

(Factories Act 1961, section 175; 1984 Act, section 126)

Factory production control: permanent internal control of production exercised by the manufacturer (1991 CP Regs, 2)

Final certificate: a certificate furnished by an approved inspector indicating that he is satisfied that any work specified in an initial notice has been completed (1984 Act, section 51)

Final exit: end of an escape route (AD–B)

Fire authority: authority carrying out functions set by the Fire Services Act 1947 (Fire Precautions Act 1971, section 43(1); 1984 Act, section 126; 1985 AI Regs, 2)

Fire door: intended to resist passage of fire and smoke (AD–B)

Fire resistance: ability to satisfy BS 476 requirements (AD–B)

Fire stop: seal to close imperfections between elements and thereby restrict the passage of smoke and flame (AD–B)

Firefighting lift: can be used in fire under control of fire service (AD–B)

Firefighting lobby: protected lobby (AD–B)

Firefighting shaft: protected enclosure (AD–B)

Firefighting stair: protected stair (AD–B)

Flat: separate self-contained premises for residential use and forming part of a building from some other part of which it is divided horizontally (1991 Regs, 2; 1985 AI Regs, 2)

Flight: part of stair (AD–K)

Floor: lower surface of any space (AD–C)

Floor area: total area of every floor in a building, or extension, calculated between finished internal surfaces of the enclosing walls, or if at any point there is no wall, then the outermost edge of the floor (1991 Regs, 2)

Foul water: waste from sanitary conveniences and water from cooking and washing; excludes trade effluent (1991 Regs H1, Schedule 1)

Foul outfall: is a sewer, cesspool, septic tank or settlement tank (AD–H)

Full plans: plans of a proposal submitted by developer to local authority for approval (1991 Regs, 13(3))

Functions: includes power and duties (1984 Act, section 126)

Gallery: a floor which may include a raised storage area not more than one half the area of the space into which it projects (AD–B)

Going: depth of tread minus overlap with next tread (AD–K)

Habitable room: a room for dwelling purposes, not including a kitchen or scullery (AD–F); a room for dwelling purposes including kitchen, not including bathroom (AD–B)

Harmonized standard: established by European standards organization (1991 Regs, 2)

Height: of building measured from mean ground level to top of parapet or midway between eaves and top of gable whichever higher (1991 Regs, 2)

Helical stair: describes helix around central void (AD–K)

Highway authority: where the highway is repairable by the inhabitants at large, the council in whom the highway is vested (1984 Act, section 126)

House: a dwellinghouse whether it is a private dwellinghouse or not (1984 Act, section 126)

Imposed load: arises from occupancy; weight of moveable partitions, distributed, concentrated, impact inertia and snow loads; does not include wind loads (AD–A)

Initial notice: to be given by approved inspector to local authority indicating that he is an approved inspector for work proposed to be carried out (1984 Act, section 47)

Inner room: where escape is only possible by passing through another room (AD–B)

Inspection body: body designated as an inspection body for the purposes of the Directive (1991 CP Regs, 2)

Institution: a hospital, home, school, or similar place used as living accommodation, or for the treatment, care or maintenance of persons suffering from disabilities due to illness or old age, or other physical or mental disability, or under the age of five where such people sleep on the premises (1991 Regs, 2)

Ladder: access to another level (AD–K)

Limited combustibility: may be a material with a non-combustible core having combustible facings (AD–B)

Limits of supply: the area within which the statutory water authority is authorized by or under an Act to supply water (1984 Act, section 126)

Listed building: one which, for the time being, is included on a list compiled by the Secretary of State as being of architectural or historic interest (1990 LBC Act, section 1)

Local authority: a district council; a London borough council; common council, City of London; Sub-Treasurer of the Inner Temple or the Under-Treasurer of the Middle Temple. For sections 1 to 58 of the 1984 Act – the Council of the Isles of Scilly (1984 Act, section 126)

Maisonette: self-contained dwelling, but not a dwellinghouse, which occupies more than one storey (AD–M)

Map of sewers: a map of public sewers which must be kept by a local authority (1936 Act, section 32(1)); a map of pipes kept by an authority (Control of Pollution Act 1974, section 28(1); 1984 Act, section 18(5))

Material alteration:

(a) is alteration to a building which would 'adversely affect' the existing building unless other work were done

(b) putting insulation into cavity walls

(c) underpinning a building

(1991 Regs, 3)

Material change of use:

(a) building is used as a dwelling

(b) change to a hotel or institution

(c) conversion into flats

(d) exempt building changed to non-exempt use

(e) change to use as a public building

(f) change to institution

when previously the building was not used for these purposes (1991 Regs, 5)

Minor part product: construction product included in list of products which play a minor part in respect to health and safety drawn up by European Commission (1991 CP Regs, 2)

Minor work: alterations or extensions to dwellinghouses having not more than two storeys; provision, alteration or extension of services or fittings in or in connection with any building (1991 Regs, 9)

Modifications: additions, omissions and amendments (1984 Act, section 126)

Moisture: vapours as well as liquid (AD–C)

National technical specification: specification which member State regards as complying with essential requirements (1991 CP Regs, 2)

Non-domestic building: all buildings not domestic, including buildings where people temporarily reside, such as hostel (AD–F)

Notice to warn: warning relating to construction products which do not satisfy relevant requirements (1991 CP Regs, 9)

Notification: a notification in writing (1991 CP Regs, Schedule 4)

Notional boundary: an assumed boundary for the purposes of fire protection (AD–B)

Occupiable room: a room in a non-domestic building such as an office workroom, classroom or hotel bedroom. Does not include bathroom, sanitary accommodation, utility rooms, plant or storage rooms (AD–F)

Office: premises used for purposes of administration, clerical work, handling money or telephone and telegraph operating; includes such activities as writing, book-keeping, sorting papers, filing, typing, duplicating, machine calculating, drawing and editorial work (1991 Regs, 2)

Officer: person authorized in writing to assist authority in carrying out its function (1991 CP Regs, 2)

Outfall – rainwater: includes a surface water or combined sewer or a private sewer and a watercourse (AD–H)

Owner: a person for the time being receiving rackrent whether on his own account or as agent or trustee for another person, or who would receive it if the premises were let at a rackrent (1984 Act, section 126)

Party fence wall: wall which is not part of a building and which stands on land of different owners to separate the adjoining land (1996 PW Act, section 20)

Party structure: a party wall or floor partition separating buildings or parts of buildings having separate entrances or staircases (1996 PW Act, section 20)

Party wall: stands on the land of different owners to a greater extent than the projection of any artificially-formed support, and separates buildings belonging to different owners (1996 PW Act, section 20)

Passive stack ventilation (PSV): ventilation system using ducts from the ceiling of rooms to roof terminals (AD–F)

Plans – full: includes drawings of any description and also specifications or other information in any form (1991 Regs, 13; 1984 Act, section 126)

Plans certificate: a certificate given by an approved inspector indicating that he is satisfied plans are not deficient and that work carried out to those plans will not contravene Building Regulations (1984 Act, section 50)

Platform floor: an access floor supported by a structural floor over a concealed space (AD–B)

Plot gradient: measurement taken between finished floor level and point of access (AD–M)

Point of access: point at which person visiting a dwelling would normally alight from a vehicle (AD–M)

Prejudicial to health: injurious, or likely to cause injury to health (1984 Act, section 126)

Premises: includes buildings, land, easements and hereditaments of any tenure (1984 Act, section 126); includes any place and any ship, aircraft or vehicle (1991 CP Regs, 2)

Prescribed: stipulated in the building regulations (1984 Act, section 126)

Principal entrance: entrance one would normally expect to approach, or common entrance to block of flats (AD–M)

Principal regulations: the building regulations 1985 (1985 AI Regs)

Principal entrance storey: principal entrance to building (AD–M)

Private sewer: a sewer that is not a public sewer (1984 Act, section 126)

Private stairway: stairway serving one dwelling (AD–K)

Prohibition notice: prohibits person from supplying any construction products not satisfying requirements (1991 CP Regs, 9)

Proper materials: have EC mark; conform to harmonized standard or BS or have Agrément certificate

Protecting structure: encloses a protected shaft but does not include

- an external wall, separating wall or compartment wall
- a compartment floor or a floor on the ground
- a roof

(AD–B)

Public body: a local authority, a county council, or other body that act under an enactment for public purposes and not for its own profit (1984 Act, section 8(5))

Public body's final certificate: notice to be given to a local authority when a public body is satisfied that work has been completed in compliance with building regulations (1984 Act, section 58 and para 3, Schedule 4)

Public body's notice: when a public body intends to carry out building work, and is satisfied that it can be adequately supervised by its own servants or agents, it may serve a notice on the local authority to that effect (1984 Act, section 54(1))

Public body's plans certificate: certificate given to local authority indicating that plans have been inspected by a servant or agent competent to assess the plans, and that they indicate compliance with building regulations (1984 Act, section 58 and para 2, Schedule 4)

Public building: consisting of or containing

(a) theatre, public library, hall or other place of public resort
(b) school not exempt from building regulations
(c) place of public worship

(1991 Regs, 2)

Public sewer: sewers that are vested in a local authority (water authority) (1936 Act, section 20; 1984 Act, section 126)

Publicized information: information which has been disclosed in any civil or criminal proceedings or has been included in a notice to warn (1991 CP Regs, 25)

Purpose group: based upon the nature of occupation for fire protection purposes (AD–B)

Rackrent: not less than two-thirds of the rent at which the property might reasonably be expected to be let free of tenants' rates and taxes and less probable annual expenses (1984 Act, section 126)

Ramp: slope steeper than 1:20 (AD–K)

Records: includes any books or documents and any records in non-documentary form (1991 CP Regs, 2)

Relevant boundary: a boundary which has relation to a side or external wall of a building (AD–B)

Relevant building: a building where it is proposed to use materials unsuitable for permanent building (1984 Act, section 20(8))

Relevant national standard: national standard corresponding to harmonized standard (1991 CP Regs, 2)

Relevant period: for passing or rejection of plans – five weeks or an extended period of not more than two months agreed between council and person depositing plans. Where person has given notice that he intends to demolish a building the relevant period before commencing is six weeks (1984 Act, sections 16(12), 81(4) and 126)

Relevant person: means Minister of Crown, government department; weights and measures authority and any person having regulatory functions conferred by any enactment (1991 CP Regs, 25)

Relevant requirements: means Parts A, A1, B3, B4, B5 and M (1991 Regs, 3)

Relevant technical specification: European technical approval, a national technical specification, or a relevant national standard (1991 CP Regs, 2)

Rise: height of step including thickness of tread (AD–K)

Rooflight: includes skylights, lantern lights, dome lights or similar (AD–B)

Room: does not include an enclosed circulation space or protected shaft, can include cupboards that are not fittings and auditoria or similar large spaces (AD–B)

Sanitary accommodation: space containing one or more closets or urinals; if there is free circulation of air, room with cubicles counts as a single space (AD–F)

Sanitary conveniences: means a closet or urinal (1984 Act, section 126)

Section 36 notice: a notice given by local authority requiring owner to pull down or remove work, or allowing him to do such alterations as are necessary to make it comply with the regulations (1984 Act, section 36)

Separating wall: a wall separating two adjoining buildings (AD–B)

School: includes a Sunday school (1984 Act, section 126)

Sewer: does not include a drain used in connection with one building, but includes all sewers and drains used for draining buildings and yards; includes manholes, ventilating shafts, pumps and other accessories belonging to the sewer (1984 Act, section 126)

Shop: premises used for carrying on a retail trade or business. Premises to which members of the public are invited for the purpose of delivering goods for repair or other treatment, or themselves carrying out repairs to or other treatment of goods. Includes:

- sale of food or drink for immediate consumption
- retail sales by auction
- barber, or hairdresser

(1991 Regs, 2)

Signature: includes a facsimile of a signature by whatever process reproduced (1984 Act, section 93)

Spacing: distance between two timber members measured on plane of floor, ceiling or roof of which they form part (AD–A)

Span: distance between any centres of two bearings or supports; clear span is between faces of supports (AD–A)

Special foundations: involves an assemblage of beams or rods to distribute the load (1996 PW Act, section 20)

Spiral stair: describes helix round central column (AD–K)

Standard scale: a scale of fines (1984 Act, section 126; Criminal Justice Act 1982, section 75)

Stair: includes landings and flights (AD–K)

Statutory undertakers: authorities responsible for railways, canals, internal navigation, dock, harbour, tramway, gas, electricity, water or other public undertakings (1984 Act, section 126)

Statutory water undertakers: a body empowered by statute to supply water (1984 Act, section 126)

Steeply sloping plot: plot gradient in excess of 1:15 (AD–M)

Street: a highway, including a highway over a bridge, a road, lane, footway, square, court, alley, passage whether or not it is a thoroughfare (1984 Act, section 126)

Street: any highway, road, lane, footway, alley or passage; any square or court; any land laid out as a way whether it is for the time being formed as a way or not (1991 New Roads Act, section 48)

Street work: include breaking up or opening a street or any sewer, drain or tunnel under it. It does not include maintenance, traffic signs or crossings of verge and footway (1991 New Road Act, section 48)

Structure or erection: includes a vessel, hovercraft, aircraft or other moveable object of any kind (1984 Act, section 121(2))

Suitable: access and facilities for disabled people (AD–M)

Substantive requirements: means the requirements of building regulations concerning design and construction and the provision of services, fittings and equipment, as distinct from procedural requirements (1984 Act, section 126)

Supply: includes offering to supply, agreeing to supply, exposing for supply and possessing for supply (1991 CP Regs, 2)

Supplier: in relation to a prohibition notice means the person on whom the notice is or was served (1991 CP Regs, Schedule 4)

Supported wall: wall receiving lateral support from a wall, buttress, pier or chimney (AD–A)

Surface water: includes water from roofs (1984 Act, section 126)

Suspension notice: prohibits a person from supplying suspect construction products for a period not exceeding six months (1991 CP Regs, 10)

Tapered tread: nosings not parallel (AD–K)

Unprotected area: comprises window, door or other openings, or wall with less than the required fire resistance; or wall with combustible material exceeding 1 mm thick attached to its external face (AD–B)

Utility room: contains clothes washing and similar equipment such as sink, washing machine, tumble drier or other equipment producing significant volumes of water vapour (AD–F)

U value: thermal transmittance coefficient calculated in Watts (Joules per second) lost for each 1 m^2 of fabric, per degree Kelvin of difference between the external and internal temperatures of air (1991 Regs L1, Schedule 1)

Ventilation opening: opening to external air such as windows, louvres, airbricks or openable ventilators; includes a door which opens to external air (AD–F)

Wall: includes a roof having a pitch of 70° or more from the horizontal (1991 Regs L1, Schedule 1); vertical construction including piers, columns and parapets (AD–C)

Wall surface: includes glazing and ceiling which slopes at an angle of more than 70° from the horizontal (AD–B)

Water closet: a closet with a separate fixed receptacle connected to a drainage system; there must be separate provision for flushing from a supply of clean water (1984 Act, section 126)

Wind load: the effect of wind pressure or suction (AD–A)

Window: includes a door having more than 2 m^2 of glass; lintel sill and jamb may be regarded as window for purposes of Part L (1991 Regs L1, Schedule 1)

Working days: not to include Saturday, Sunday, bank holiday or public holiday (1991 Regs, 14; 1985 AI Regs, 2)

Workplace: does not include a factory, but does include any place where people are employed otherwise than in domestic service (1984 Act, section 126)

Works: construction works, including both buildings and civil engineering works (1991 CP Regs, 2)

Sources of information and research

It is not possible for an individual to absorb all the knowledge involved in the construction industry, and one must become aware of the constant changes which are taking place.

Technical journals play a vital part, but frequently only specialist information services can give expert advice quickly.

A list of some important sources of information is given in this appendix. Altogether there are about 550 organizations in the construction industry which to a greater or lesser degree, give information.

In the first issue of *Construction* some years ago the Department of the Environment gave the following advice to those who seek information:

- Ask personally; a question can be completely altered if it is passed through several people.
- Say who you are.
- Say why you need the information.
- State the urgency.
- Give as many relevant facts as possible.
- Be clear and don't jump to the answer.

IMPORTANT ADDRESSES

Department of the Environment, Transport and the Regions Building Regulations Division
3rd Floor
Eland House
Bressenden Place
London SW1 5DW
(0207 890 5742)

The Stationery Office

Under new ownerships, Her Majesty's Stationery Office is known as The Stationery Office. DoE and other official publications may be purchased from the following bookshops:

123 Kingsway, London WC2B 6PQ
   (Counter service only)
   (0207 242 6393)
68–69 Bull Street, Birmingham B4 6AD
   (0121 236 9696)
33 Wine Street, Bristol BS1 2BQ
   (0117 926 4306)
9–21 Princess Street, Manchester M60 8AS
   (0161 834 7201)
The Oriel Bookshop, The Friary, Cardiff CF1 4AA
   (029 2039 5548)
The Parliamentary Bookshop, 12 Bridge Street, Parliament Square, London SW1A 2JX
   (0207 219 3890)

Mail and telephone orders only to the following:

SO Publications Centre, PO Box 276, London SW8 5DT.
   (0207 873 0011)

There are Accredited Agents in most large towns – see the Yellow Pages.

British Standards

British Standards Institution, 389 Chiswick High Road, London W4 4AL
   (0208 996 9001)
Sales Department, 101 Pentonville Road, London N1 9ND
   (0207 837 6801)

Building Research Establishment

Documents emanating from the Building Research Establishment may be purchased from:
CRC Ltd, 151 Rosebery Avenue, London EC1R 4QX
   (0207 505 6622)

Agrément Certificates

British Board of Agrément, PO Box 195, Bucknalls Lane, Garston, Watford WD2 7NG
   (01923 665300)

Building Centre Enquiry Service

A permanent exhibition is on show, and there is a bookshop devoted to building.
26 Store Street, London WC1E 7BT
   (0207 637 1022)

A SELECTION OF ADDRESSES

Access Committee for England
12 City Forum, 250 City Road, London EC1V 8AF (0207 250 0008)

Aluminium Federation
Broadway House, Calthorpe Road, Birmingham B15 1TN (0121 456 1103)

Aluminium Window Association
11 Cleeve Cloud Lane, Prestbury, Cheltenham, Glos. GL52 5SE (01242 677 459)

Aluminium Rolled Products Manufacturers' Association
Broadway House, Calthorpe Road, Birmingham B15 1TN (0121 456 1103)

Architecture and Surveying Institute
St Mary House, St Mary Street, Chippenham, Wilts. SN15 3BR (91249 655398)

Architectural Cladding Association
60 Charles Street, Leicester LE1 1FB (0116 253 6161)

Asbestos Removal Contractors Association
Friars House, 6 Parkway, Chelmsford, Essex CM2 0NF (01245 259 744)

Association for Environmental Conscious Building
Windlake House, The Pump Field, Coaley, Gloucester GL11 5DX
(01453 890 757)

Association of British Roofing Felt Manufacturers
38 Bridlesmith Gate, Nottingham NG1 2GQ (0115 958 8209)

Association of British Solid Fuel Appliance Manufacturers
134 Renfrew Street, Glasgow G3 5TG (0141 332 0826)

Association of Builders' Hardware Manufacturers
42 Health Street, Tamworth, Staffs. B79 7JH (01827 52337)

Association of Building Component Manufacturers
Service House, 61–63 Rochester Road, Aylesford, Kent HE20 7BS
(01622 715577)

Association of Building Engineers
Jubilee House, Billingbrook Road, Weston Favel, Northampton NN3
8NW (01604 404 121)

Association of Corporate Approved Inspectors
c/o Waterlinks House, Richard Street, Aston, Birmingham, B7 4AA
(0121 333 2811)

Association of Insulation Manufacturers
38 Bridlesmith Gate, Nottingham NG1 2GQ (0115 941 5269)

Association of Specialist Fire Protection Contractors and Manufacturers
Association House, 235 Ash Road, Aldershot, Hants. GU12 4DD
(01252 21322)

Association of Municipal Engineers
1–7 Great George Street, London SW1P 3AA (0207 222 7722)

ASTA Certification Services
ASTA House, Chestnut Field, Rugby, Warwickshire CB21 2TL
(01788 578435)

Autoclaved Aerated Concrete Products Association
Cowley House, 9 Little College Street, London SW1P 3XS (0207
222 0666)

Association of Tank and Cistern Manufacturers
12 Appletons, Hadlow, Tonbridge, Kent TN11 0DT (01732 850744)

Automatic Door Suppliers Association
411 Limpsfield Road, Warlington, Surrey CR6 9HA (01883 624961)

Boiler and Radiator Manufacturers Association
Savoy Tower, 77 Renfrew Street, Glasgow G2 3BZ (0141 332 0826)

Box Culvert Association
60 Charles Street, Leicester LE1 1FB (0116 253 6161)

Brick Development Association
Woodside House, Winkfield, Windsor SL4 2DX (01344 885 651)

British Adhesives and Sealants Association
33 Fellowes Way, Stevenage, Herts. SG2 8BW (01438 385 514)

British Aggregate Construction Materials Industries
156 Buckingham Palace Road, London SW1W 9TR (0207 730
8194)

British Bathroom Council
Federation House, Stoke on Trent ST4 2RT (01782 74074)

British Blind and Shutter Association
42 Heath Street, Tamworth, Staffs. B79 7JH (01827 52337)

British Cement Association
Telford Avenue, Crowthorne, Berks. RG45 6YS (01344 762676)

British Ceramic Research
Queens Road, Penkhull, Stoke on Trent ST4 7LQ (01782 845431)

British Ceramic Tile Council
Federation House, Station Road, Stoke on Trent ST4 2RT (01782
747147)

British Combustion Equipment Manufacturers Association
The Fernery, Market Place, Midhurst, West Sussex GU29 9DP
(01730 812782)

British Constructional Steelwork Association
4 Whitehall Court, London SW1A 2ES (0207 839 8566)

British Concrete Masonry Association
Grove Crescent House, 18 Grove Place, Bedford MK40 3JJ (01234
353745)

British Concrete Pumping Association
28 Ecclestone Street, Victoria, London SW1W 9PY (0207 730
7117)

British Constructional Steelwork Association
4 Whitehall Court, Westminster, London SW1A 2ES (0207 839
8566)

British Equipment and Information Technology Association
8 Southampton Place, London WC1A 2EF (0207 405 6233)

British Fire Protection Systems Association
48A Eden Street, Kingston on Thames, Surrey KT1 1EE (0208 549
5855)

British Flat Roofing Council
177 Bagnall Road, Basford, Nottingham NG6 8FJ (0115 942 4200)

British Floorcovering Manufacturers Association
125 Queens Road, Brighton BN1 3YW (01273 29271)

British Flue and Chimney Manufacturers Association
Sterling House, 6 Furlong Road, Bourne End, Bucks. SL8 5DG
(01628 531186)

British Foundry Association
Bridge House, Smallbrook, Queensway, Birmingham B5 4JP (0121
643 3377)

British Glass Manufacturers Confederation
Northumberland Road, Sheffield S10 2UA (0114 268 6201)

British Hardware Association
Brook House, 4 The Lakes, Bedford Road, Northampton NN4 7YD
(01604 22023)

British Hardware Federation
225 Bristol Road, Edgbaston, Birmingham B5 7UB (0121 446 6688)

British Independent Steel Producers Association
5 Cromwell Road, London SW7 2HX (0207 581 0231)

British Industrial Fasteners Federation
Grove Hill House, 245 Grove Lane, Handsworth, Birmingham B20
2HB (0121 523 3496)

British Institute of Architectural Technologists
397 City Road, London EC1V 1NE (0207 278 2206)

British Iron and Steel Producers Association
5 Cromwell Road, London SW7 2HX (0207 581 0231)

British Laminated Plastics Fabrication Association
6 Bath Place, Rivington Street, London EC2A 3JE (0207 457 5000)

British Lock Manufacturers Association
42 Heath Street, Tamworth, Staffordshire B79 7JH (01827 52337)

British Metals Federation
16 High Street, Brampton, Huntingdon, Cambs. PE18 8TU (01480
455249)

British Non-Ferrous Metals Federation
18 Greenfield Crescent, Edgbaston, Birmingham B15 3AU (0121
456 3322)

British Plastics Federation
6 Bath Place, Rivington Street, London EC2A 3JE (0207 457 5000)

British Pre-Cast Concrete Federation
60 Charles Street, Leicester LE1 1FB (0116 253 6161)

British Ready-Mixed Concrete Association
The Bury, Church Street, Chesham, Bucks. HP5 1JE (01494 791050)

British Rubber Manufacturers Association
90–91 Tottenham Court Road, London W1P 0BR (0207 580 2794)

British Standards Institution
Quality Assurance Services, PO Box 375, Milton Keynes, Bucks. MK14 6LL (01908 220908)

British Standards Institution
Testing Services Division, Maylands Avenue, Hemel Hempstead, Herts. HP2 4SQ (01442 230442)

British Stainless Steel Association
8th Floor, Bridge House, Smallbrook, Queensway, Birmingham B5 4JP (0121 643 3377)

British Timber Merchants Association
Stocking Lane, Hughenden Valley, Bucks. HP14 4JZ (01494 563602)

British Urethane Foam Contractors Association
PO Box 23, Kenilworth, Warwickshire CV8 2YZ (01926 513187)

British Wood Preserving and Damp Proofing Association
Building No. 6, The Office Village, 4 Romford Road, Stratford, London E15 4EA (0208 519 2588)

British Woodworking Federation
82 New Cavendish Street, London W1M 8AD (0207 580 5588)

Builders Merchants Federation
15 Soho Square, London W1V 5FB (0207 439 1753)

Building Conservation Trust
Apartment 39, Hampton Court Palace, East Molesey, Surrey KT8 9BS (0208 943 2277)

Building Regulations Division
Department of the Environment, 3rd Floor, Eland House, Bressenden Place, London SW1 5DU (0207 890 5742)

Building Research Establishment
Bucknall Lane, Garston, Watford, Herts. WD2 7JR (01923 664664)

Building Services Research and Information Association
Old Bracknall Lane West, Bracknall, Berks. RG12 7AH (01344 426511)

Calcium Silicate Brick Association Ltd
Rectory Road, Grays, Essex RM17 (01375 393434)

Cavity Foam Bureau
PO Box 79, Oldbury, Warley, West Midlands B69 4PW (0121 544 4949)

Cement Admixtures Association
Harcourt, The Common, Kings Langley, Herts. WD4 8BL (01923 264314)

Chartered Institute of Building
Englemere, Kings Ride, Ascot, Berks. SL5 7TB (01334 630 700)

Chartered Institute of Building Services Engineers
Delta House, 222 Balham High Road, London SW12 9BS (0208 675 5211)

Clay Pipe Development Association
Copsham House, 53 Broad Street, Chesham, Bucks. HP5 3DX (01494 791456)

Clay Roofing Tile Council
Federation House, Station Road, Stoke on Trent ST4 2SA (0782 744631)

Concrete Block Association
60 Charles Street, Leicester LE1 1FB (0116 253 6161)

Concrete Brick Manufacturers Association
60 Charles Street, Leicester LE1 1FB (0116 253 6161)

Concrete Lintel Association
60 Charles Street, Leicester LE1 1FB (0116 253 6161)

Concrete Pipe Association
60 Charles Street, Leicester LE1 1FB (0116 253 6161)

Concrete Repair Association
235 Ash Road, Aldershot, Hants. GU12 4DD (01252 21303)

Concrete Society
3 Eatongate, 112 Windsor Road, Slough SL1 2JA (01753 693313)

Construction Industry Computing Association
1 Trust Court, Histon, Cambridge CB4 4PW (01223 236336)

Construction Industry Council
26 Store Street, London WC1E 7BT (0207 637 8692)

Contract Flooring Association
4c St Mary's Place, The Lace Market, Nottingham NG1 1PH (0115 941 1126)

Copper Development Association
Orchard House, Mutton Lane, Potters Bar, Herts. EN6 3AP (01707 650711)

Copper Tube Fittings Manufacturers Association
10 Greenfield Crescent, Edgbaston, Birmingham B15 3AU (0121 456 3322)

Council for Aluminium in Building
11 Cleeve Cloud Lane, Prestbury, Cheltenham, Glos. GL52 5SE (01242 677459)

Council for Energy Efficiency Development
PO Box 12, Hazlemere, Surrey GU27 3AH (01428 654011)

Council for Registered Gas Installers (CORGI)
4 Elmwood, Chineham Business Park, Crockford Lane, Basingstoke RG24 0WG (01256 708133)

Decorative Gas Fire Manufacturers Association
Bridge House, 10 Freemantle Business Centre, Milbrook Road East, Southampton SO15 1JR (01703 631593)

Decorative Paving and Walling Association
60 Charles Street, Leicester LE1 1FB (0116 253 6161)

Department of Energy
1 Palace Street, London SW1E 5HE (0207 238 3000)

Design Council
Haymarket House, Oxenden Street, London SW1Y 4EE (0207 208 2121)

Disabled Living Foundation
380–384 Harrow Road, London W9 2HU (0207 289 6111)

Domestic Heating Controls Group
TACMA, Leicester House, 8 Leicester Street, London WC2H (0207 437 0678)

Door and Shutter Manufacturers Association
42 Heath Street, Tamworth, Staffs. B79 7JH (01827 52337)

Draught Proofing Advisory Association
PO Box 12, Hazlemere, Surrey GU27 3AH (01428 654011)

Dry Lining and Partition Association
82 New Cavendish Street, London W1M 8AD (0207 580 5588)

Electrical Contractors Association
34 Palace Court, London W2 4HY (0207 229 1266)

Electronic Data Interchange
235 Ash Road, Aldershot, Hants. GU12 4DD (01252 336318)

Energy Device Advice Scheme
Camargue House, Wellington Road, Cheltenham, Glos. GL 52 2AG (01242 577277)

Energy Information Centre
PO Box 147, Grosvenor House, High Street, Newmarket CB8 9AL (01638 663030)

Energy Technology Support Unit
Harwell Laboratories, Didcot, Oxon. OX11 0RA (01235 821000)

EURISOL UK
Mineral Wool Association
39 High Street, Redbourne, Hants. AL3 7LW (01582 794624)

Environmental Services Association
154 Buckingham Palace Road, London SW1W 9TR (0207 824 8882)

Expanded Polystyrene Vacity Insulation Association
5 Belgrave Square, London SW1X 8PH (0207 235 9483)

External Wall Insulation Association
PO Box 12, Hazlemere, Surrey GU27 3AH (01428 654011)

Federation of Civil Engineering Contractors
Cowdray House, 6 Portugal Street, London WC1A 2HH (0207 404 4020)

Federation of Environmental Trade Associations
Stirling House, 6 Furlong Road, Bourne End, Bucks. SL8 5DG (016285 31186)

Federation for the Repair and Protection of Structures
Association House, 235 Ash Road, Aldershot, Hants. GU12 4DD (01252 342072)

Federation of Master Builders
Gordon Fisher House, 14–15 Great James Street, London WC1N 3DP (0207 242 7583)

Federation of Piling Specialists
39 Upper Elmers End Road, Bedrenham, Kent BR3 3QY (0181 663 0947)

Federation of Plastering and Drywall Contractors
18 Mansfield Street, London W1M 9FG (0207 580 5404)

Fencing Contractors Association
St John's House, 23 St John's Road, Watford WD1 1PY (01923 248895)

Fibre Cement Manufacturers Association
PO Box 117, Hexham, Northumberland NE46 3LQ (01434 601393)

Fire Extinguishing Trades Association
48A Eden Street, Kingston on Thames, Surrey KT1 1EE (0208 549 8839)

Fire Protection Association
Melrose Avenue, Borehamwood, Herts. WD6 2BJ (0208 207 2345)

Fire Research Station
Melrose Avenue, Borehamwood, Herts. WD6 2BL (0208 207 2345)

Flat Glass Manufacturers Association
Prescot Road, St Helens, Merseyside WA10 3TT (01744 288882)

Flat Roofing Contractors Advisory Board
Fields House, Gower Road, Haywards Heath, West Sussex RH16 4PL (01444 440027)

Galvanizers Association
6 Wrens Court, 56 Victoria Road, Sutton Coldfield, West Midlands B72 1SY (0121 355 8838)

Glass and Glazing Federation
44–48 Borough High Street, London SE1 1XB (0207 403 7177)

Glassfibre Reinforced Cement Association
The Wigan Investment Centre, Waterside Drive, Wigan, Lancs. WN3 5BA (01942 705550)

Glued Laminated Timber Association
Chiltern House, Stocking Lane, Hughenden Valley, High Wycombe, Bucks. HP14 4ND (01494 565180)

Guild of Master Craftsmen
166 High Street, Lewes, East Sussex BN7 1XU (01273 478449)

Guild of Surveyors
1 Alexandra Street, Queens Road, Oldham OL8 2AV (0161 727 2389)

Gypsum Products Development Association
c/o KPMG, 165 Queen Victoria Street, London EC4V 4DD (0207 311 2942)

Heating and Ventilating Contractors Association
ESCA House, 34 Palace Court, Bayswater, London W2 4JG (0207 229 2488)

House Builders Federation
82 New Cavendish Street, London W1M 8AD (0207 580 5588)

Institute of Architectural Ironmongers
8 Stepney Green, London E1 3JU (0207 790 3431)

Institute of Building Control
92–104 East Street, Epsom, Surrey KT17 1EB (01372 748282)

Institute of Ceramics
Shelton House, Stoke Road, Shelton, Stoke on Trent, Staffs. ST4 2DR (01782 202116)

Institute of Clerks of Works
27 Rosendale Close, Gossops Creek, Crawley, W. Sussex RH11 8NO (01293 562 959)

Institute of Concrete Technology
PO Box 255, Beaconsfield, Bucks. HP9 1JE (01494 674572)

Institute of Construction Management
8 Rookley, Nestley Abbey, Southampton SO31 5PH (023 8045 3680)

Institute of Energy
18 Devonshire Street, London W1N 2AU (0207 580 7124)

Institute of Fire Protection
Melrose Avenue, Borehamwood, Herts. WD6 2BJ (0208 207 2345)

Institute of Fire Safety
PO Box 687, Croydon, Surrey CR9 5DD (0208 654 2582)

Institute of Maintenance and Building Management
Keets House, 30 East Street, Farnham, Surrey GU9 7SW (01252 710 994)

Institute of Materials
1 Carlton House Terrace, London SW1Y 5DB (0207 839 4071)

Institute of Plumbing
64 Station Lane, Hornchurch, Essex RM12 6NB (01708 472791)

Institute of Quality Assurance
PO Box 712, 61 Southwark Street, London SE1 1SB (0207 401 7227)

Institute of Quarrying
7 Regent Street, Nottingham NG1 5BS (01159 411315)

Institute of Wastes Management
9 Saxon Court, St Peters Gardens, Northampton NN1 1SX (01604 20426)

Institution of British Engineers
Alne House, Damzey Green, Tamworth in Arden, Solihull B94 5BB (01564 742915)

Institution of Civil Engineering Surveyors
26 Market Street, Altrincham, Cheshire WA14 1PF (0161 928 8074)

Institution of Fire Engineers
148 New Walk, Leicester LE1 7QB (0116 255 3654)

Institution of Structural Engineers
11 Upper Belgrave Street, London SW1X 8BH (0207 235 4535)

International Concrete Brick Association
60 Charles Street, Leicester LE1 1FB (0116 253 6161)

International Waterproofing Association
38 Bridlesmith Gate, Nottingham NG1 2GQ (0115 950 4225)

Interpave – Precast Concrete Paving Association
60 Charles Street, Leicester LE1 1FB (0116 253 6161)

Intumescent Fire Seals Association
20 Park Street, Princes Risborough, Bucks. HP27 9AH (01844 275500)

Lattice Girder Floor Association
60 Charles Street, Leicester LE1 1FB (0116 253 6161)

Lead Development Association
42 Weymouth Street, London W1N 3LQ (0207 499 8422)

Lead Sheet Association
St John's Road, Tunbridge Wells, Kent TM4 9XA (01892 513351)

Lighting Industry Federation
Swan House, 297 Balham High Road, London SW17 7BQ (0208 675 5432)

Local Authority National Type Approval Confederation
35 Great Smith Street, Westminster, London SW1 3BJ (0207 664 3000)

Loss Prevention Council
Melrose Avenue, Borehamwood, Herts. WD6 2BJ (0208 207 2345)

Manufacturers Association of Domestic Unvented Water Supply Systems
12 Appletons, Haddon, Tonbridge, Kent TN11 0DT (01732 850744)

Mastic Asphalt Council
8 North Street, Ashford, Kent TN24 8JN (01233 634466)

Metal Roof Deck Association
Fields House, Gower Road, Haywards Heath, West Sussex RH16 4PL (01444 440027)

Metal Cladding and Roofing Manufacturers Association
18 Mere Farm Road, Noctorum, Birkenhead, Merseyside L43 9TT (0151 652 3846)

Meteorological Office
London Road, Bracknell, Berks. RG12 2SZ (01344 420242)

Mortar Products Association
74 Holly Walk, Leamington Spa, Warwickshire CV32 4JD (01926 38611)

National Association of Chimney Lining Engineers
St Mary's Chambers, 19 Station Road, Stone, Staffs. ST15 8JP (01785 811 732)

National Association of Lift Makers
33–34 Devonshire Street, London W1N 1RF (0207 935 3013)

National Association of Loft Insulation Contractors
PO Box 12, Haslemere, Surrey GU27 3AH (01428 654011)

National Association of Plumbing, Heating and Mechanical Services Contractors
Units 14–15, Ensign House, Ensign Business Centre, Westwood Way, Coventry CU4 8AJ (01203 470626)

National Association of Shopfitters
NAS House, 411 Limpsfield Road, The Green, Warlington CR6 9HA (01883 624961)

National Cavity Insulation Association
PO Box 12, Haslemere, Surrey GU27 3AH (01428 54011)

National Clayware Federation
7 Castle Street, Bridgwater, Somerset TA6 3DT (01278 458251)

National Council of Building Material Producers
26 Store Street, London WC1E 7BT (0207 323 3770)

National Federation of Builders
82 New Cavendish Street, London W1M 8AD (0207 580 5588)

National Federation of Clay Industries
Federation House, Station Road, Stoke on Trent, Staffs. ST4 2SA (01782 744631)

National Federation of Demolition Contractors
Resugaun House, 1A New Road, The Causeway, Staines, Middx. TW1 8DH (0207 456 799)

National Federation of Housing Associations
175 Grays Inn Road, London WC1X 8UF (0207 278 6571)

National Federation of Plastering, Dry Lining and Partition Contractors
82 New Cavendish Street, London W1M 8AD (0207 580 5588)

National Federation of Roofing Contractors
24 Weymouth Street, London W1N 4LX (0207 436 0387)

National Federation of Terrazzo-Mosaic Specialists
PO Box 50, Banstead, Surrey SM7 2RD (01737 360 673)

National Fireplace Association
Bridge House, Smallbrook, Queensway, Birmingham B5 4JP (0121 643 3377)

National GRP Construction Federation
18 Mansfield Street, London W1M 9FG (0207 580 5588)

National Heating Consultancy
Central Office, PO Box 370, London SE9 2RP (0208 274 2442)

National Home Improvement Council
26 Store Street, London WC1E 7BT (0207 636 2562)

National House Building Council
Chiltern Avenue, Amersham, Bucks. HP6 5AP (01494 434477)

National Prefabricated Building Association
PO Box 3, Kenilworth, Warwickshire CV8 1SD (01926 851165)

National Quality Assurance Ltd
Gainsborough House, Houghton Hall Park, Houghton Regis, Dunstable, Bedfordshire LU5 5ZX (01582 866766)

National Society for Clean Air
136 North Street, Brighton BN1 1RG (01273 263313)

National Society of Master Thatchers
High View, Little Streetly, Yardley Hastings, Northamptonshire (01672 870 225)

Natural Slate Quarries Association
26 Store Street, London WC1E 7BT (0207 323 3770)

Office of Fair Trading
Field House, 15–25 Breams Buildings, London EC4A 1PR (0207 242 2858)

OFTEC
Eaton House, Nashgrove Lane, Finchampstead, Berks. RG40 4HE
(01734 735 050)

Ordnance Survey
Romsey Road, Maybush, Southampton SO9 4DH (02380 792912)

Paint Research Association
8 Waldegrave Road, Teddington, Middx. TW11 8LD (0208 977 4427)

Paintmakers Association of Great Britain
Alembic House, 93 Albert Embankment, London SE1 7TY (0207 582 1185)

Partitioning and Interiors Association
692 Warwick Road, Solihull, West Midlands B91 3DX (0121 705 9270)

Plastic Pipe Manufacturers' Society
c/o Wenham Major, 89 Cornwall Street, Birmingham B3 3BY (0121 236 1866)

Plastic and Rubber Advisory Service
6 Bath Place, Rivington Street, London EC2A 3JE (0207 908070)

Plasterers Craft Guild
56 Burton Road, Kingston on Thames, Surrey KT2 5TF (0208 546 1470)

Polyethylene Foam Insulation Association
235 Ash Road, Aldershot, Hants. GU12 4DD (01252 336 318)

Power Fastenings Association
42 Heath Street, Tamworth, Staffs. B79 7SH (01827 52337)

Precast Concrete Frame Association
60 Charles Street, Leicester LE1 1FB (0116 253 6161)

Precast Flooring Association
60 Charles Street, Leicester LE1 1FB (0116 253 6161)

Prestressed Concrete Association
60 Charles Street, Leicester LE1 1FB (0116 253 6161)

Roofing Tile Association
60 Charles Street, Leicester LE1 1FB (0116 253 6161)

Royal Commission on Historical Monuments in England
Fortress House, 23 Savile Row, London W1X 2JQ (0207 973 3500)

Royal Commission on Ancient and Historical Monuments in Wales
55 Blandford Street, London W1H 3AF (0207 208 8200)

Royal Institute of British Architects
66 Portland Place, London W1N 4AD (0207 580 5533)

Royal Institution of Chartered Surveyors
12 Great George Street, Parliament Square, London SW1P 3AD (0207 222 7000)

Royal Town Planning Institute
26 Portland Place, London W1N 4BE (0207 636 9107)

Sand and Gravel Association
1 Bamber Court, 2 Bamber Road, London W14 9PB (0207 381 8778)

Single Ply Roofing Association
38 Bridlesmith Gate, Nottingham NG1 2GQ (0115 924 0499)

Society for the Protection of Ancient Buildings
37 Spital Square, London E1 6DY (0207 377 1644)

Solid Fuel Advisory Service
Hobart House, Grosvenor Place, London SW1X 7AE (0207 235 2020)

Specialist Ceilings Association
Holly House, Queensway, Leamington Spa, Warwick CV31 3LT (01926 435533)

Sprayed Concrete Association
235 Ash Road, Aldershot, Hants. GU12 4DD (01252 21302)

Stainless Steel Advisory Centre
Shepcote Lane, PO Box 161, Sheffield S9 1TR (0114 244 0060)

Stainless Steel Fabrication Association
Renfrew Street, Glasgow G2 3BZ (0141 332 0826)

Steel Castings Research and Trade Association
7 East Bank Road, Sheffield S2 3PT (0114 272 8647)

Steel Construction Institute
Selwood Park, Ascot, Berks. SL5 7QN (01990 23345)

Steel Windows Association
The Building Centre, 26 Store Street, London WC1E 7BT (0207 637 3571)

Stone Federation
82 New Cavendish Street, London W1M 8AD (0207 580 5404)

Structural Concrete Consortium
60 Charles Street, Leicester LE1 1FB (0116 253 6161)

Structural Precast Association
60 Charles Street, Leicester LE1 1FB (0116 253 6161)

Suspended Access Equipment Manufacturers Association
82 New Cavendish Street, London W1M 8AD (0207 580 5404)

Thermal Insulation Contractors Association
Charter House, 450 High Road, Ilford, Essex IG1 1WF (0208 514 2120)

Thermal Insulation Manufacturers and Suppliers Association
235 Ash Road, Aldershot GU12 4DD (01252 336318)

Timber and Brick Information Council
Gable House, 40 High Street, Rickmansworth, Herts. WD3 1ES (01923 778 136)

Timber Research and Development Association
Stocking Lane, Hughenden Valley, High Wycombe HP14 4ND (01494 563091)

Timber Trade Federation
26–27 Oxenden Street, London SW1Y 4EL (0207 839 1891)

United Kingdom Accreditation Service (UKAS)
21–47 High Street, Feltham, Middx. TW3 4UN (0208 917 8400)

Warrington Fire Research Centre
Holmesfield Road, Warrington, Cheshire W1A 2DS (0192 565 5116)

Waterheater Manufacturers Association
Melverton Lodge, Hanson Road, Victoria Park, Manchester M14 5BZ (0161 224 0977)

WIMLAS Ltd,
St Peter's House, 6–8 High Street, Iver, Bucks. SL0 9NG. (01753 737744)

Wood Wool Slab Manufacturers Association
26 Store Street, London WC1E 7BT (0207 323 3770)

Zinc Development Association
42 Weymouth Street, London W1N 3LQ (0207 499 6636)

Relevant legislation

LEGISLATION

Much of the legislation affecting day-to-day construction has been gathered together in the Building Act 1984.

Of this, a considerable amount is mentioned in various chapters throughout the book. However, there are many more requirements which may affect building works, and some of the Acts are listed in this Annexe, together with a brief explanation. The appropriate Act must be consulted before any decision is taken.

Prescription Act 1832

Person who has access to light to a window for 20 years may have acquired a right to continue to receive the light (see Right of Light Act 1959).

Town Improvement Clauses Act 1847

Naming and numbering streets

This ancient piece of legislation, still operative, says that it is the duty of local authorities to name streets and to number houses, as they consider it necessary (section 64).

The local authority is given power to require occupiers of buildings to display numbers as directed and this must be done within two weeks of receiving a written notice to do so (section 65; see also Public Health Act 1925).

Town Police Clauses Act 1847

Tall people will appreciate it when there is compliance with the above Act. It requires blinds, shades, coverings, awnings and other projections along or over a footpath to be not less than 8 feet above the ground (section 28).

Public Health Act 1875

Protection from liability

Commencing with the Public Health Act 1875, an officer of a local authority properly carrying out his functions cannot be subjected to any sort of action or liability claim against him/her personally. Should a claim be brought the local authority concerned is to meet the expenses. This has been brought up to date by section 305 of the Public Health Act 1936, section 1(4) of the Public Health Act 1961 and section 39 of the Local Government (Miscellaneous Provisions) Act 1976.

Public Health Acts Amendment Act 1890

Platforms for use on public occasions

Any roof, platform, balcony or other structure to be used by spectators for any show, procession, open-air meeting, etc. must be safely constructed to the satisfaction of the surveyor to the council.

Ancient Monuments Consolidation and Amendment Act 1913
Ancient Monuments Act 1931
Historic Buildings and Ancient Monuments Act 1953
Field Monuments Act 1972

This legislation provides for protection of ancient monuments, such as buildings, structures and other work whose preservation is deemed to be of national importance. Three months' notice of any alteration or work affecting them must be given to the Department of the Environment.

Celluloid and Cinematograph Act 1922

Large quantities of raw celluloid or celluloid film may not be stored in a building:

- unless the local authority has been notified;
- unless the premises are provided with an adequate means of escape in case of fire;
- if it is below premises used for residential purposes;
- if a fire occurred and it would interfere with any means of escape;
- if there is a risk of fire spread to another part of the building;
- unless the regulations concerning the storage of celluloid are observed.

An appeal may be made against an adverse decision to a magistrates' court (sections 1 and 2)

Public Health Act 1925

Before naming a street, a developer must send the proposed name to a local authority. The council then has one month to object to the name and should they do so the developer has three weeks in which to make an appeal to a petty sessional court (section 17).

A council may by order alter the name of a street (section 18).

Street name plates must be placed by the local authority in a conspicuous position (section 19).

Anyone who defaces, removes or obscures a street nameplate may be liable to a fine (section 21) (see also Town Improvement Clauses Act 1847).

Public Health Act 1936

Care of closets

The occupier is to ensure that every watercloset is kept supplied with enough water for flushing. He is also obliged to protect the closet against frost. Anyone failing to do so may be liable to a fine (section 51).

Notices and forms

Every notice given to a local authority must be in writing and all notices, refusals or consents given by a local authority must also be in writing (section 283).

All notices and other documents issued by a local authority are to be signed by someone specially authorized by the local authority for that purpose. Facsimile signatures may be used (section 284).

A council may serve a notice by any of the following methods:

- by delivering it to the addressee;
- by leaving it or sending it to the person's usual or last known residence;
- if a company is involved, by sending or delivering it to their secretary at the registered or principal office;
- when the name of the owner or occupier cannot reasonably be discovered, or if the premises are unoccupied, by sending it addressed to the owner or the occupier at the address of the premises to which the communication relates;
- if there is no one on the premises to receive the notice it can be affixed in a conspicuous place, e.g. the main entrance door (section 285).

Culverting of watercourses

When someone proposes to build on open land the local authority have power to require him to fill in any watercourse or ditch and culvert it where it crosses or abuts the site. The work has to be done before or during building operations. Disputes may be referred to a magistrates' court, and a fine may be imposed for non-compliance (section 262).

A developer must not culvert a ditch or watercourse without first obtaining approval to his proposal from the local authority. Six weeks are allowed for the council to make up its mind and any dispute may be settled in the magistrates' court.

If, for any reason the local authority says that the culvert is to be capable of carrying more water than it did previously, they must bear the extra cost (section 263).

Clean Air Acts 1956 and 1993

Impose controls on buildings in connection with the height of chimneys, installation of grit and dust arrestment plant and types of furnaces installed. A number of regulations have been made under the terms of these Acts.

Right of Light Act 1959

Enables an owner to register a notice with a local authority aimed at legally preventing a neighbour from acquiring a right of light over his property (see Prescription Act 1832).

Public Health Act 1961

Conditions attached to a consent to discharge trade effluent into a sewer may be made under the following heads:

- periods when trading effluent may be discharged;
- the exclusion of condensing water;
- removal of any particular deleterious constituent;
- temperature and pH of effluent;
- the making of payments;
- provision of sampling;
- provision and maintenance of flow meters and other equipment;
- keeping of records;
- requiring returns to be made to the water authority (section 59).

Farm effluent

Effluent from premises used for agriculture or horticultural purposes or for scientific research is regarded as a trade waste (section 63).

Factories Act 1961

This Act lays down a number of important duties on persons undertaking building or civil engineering construction. In particular, persons undertaking such work must, within seven days of starting work, serve a notice on the Factories Inspector:

- containing the name and address of the person undertaking the work;
- stating the place and nature of the operations or works;
- indicating whether mechanical plant is to be used.

This requirement does not apply if the operations can be completed in less than six weeks (section 127).

For the purpose of this Act a place where building or civil engineering operations is being carried out is regarded as a 'factory'.

Non-compliance with the Act is actionable in the courts.

Offices, Shops and Railway Premises Act 1963

The Act makes provision for securing the health, safety and welfare of persons employed in offices and railway premises. There are sections dealing with heating, ventilation, lighting, sanitary conveniences, washing facilities, drinking water, floors, passages and stairs. Fire precautions, including means of escape are also included.

Fire Precautions Act 1971

Fire certificates

The occupiers of a number of premises are required to obtain fire certificates before the buildings can be used. The Secretary of State is empowered to make Orders saying which buildings require certificates, and they are limited to the following uses:

- sleeping accommodation;
- institution providing treatment and care;
- entertainment and recreation;
- teaching, training and research;
- when public enter on payment or otherwise;
- a place of work.

The following designating orders are appropriate:

- Hotels and Boarding Houses Order 1972;
- Fire Precautions (Non-Certified Factory, Office, Shop and Railway Premises) Order 1976;
- Factories, Offices, Shops and Railway Premises Order 1976.

Failure to be in possession of a certificate may result in a fine or imprisonment, or both (section 1).

Fire certificates are not required in connection with

- premises used for public worship;
- private dwelling houses (section 2).

Application for fire certificates must be made on a special form requiring all the uses on the premises to be stated. Plans of premises are usually required.

Having received an application the fire authority must inspect the premises and issue a certificate when they are satisfied that:

- adequate means of escape in case of fire will be provided;
- the escape route will be safeguarded;
- adequate provision has been made for fire fighting;
- the fire alarm system is adequate.

If following an inspection the fire authority is not satisfied with the proposals they may give notice to the owner setting out the matters that must be improved, altered or provided before they may issue a certificate (section 5).

A fire certificate will specify the use to which a building may be put, and in addition, state the measures necessary to:

- ensure all escape routes are maintained and kept free from obstruction;
- ensuring that any equipment provided is well maintained;
- instruct employees regarding procedures to be followed in case of fire.

Failure to comply with a fire certificate may result in a fine or imprisonment, or both (section 7).

A fire authority may inspect the premises from time to time, and may require amendment of a certificate. The occupier is obliged to tell the fire authority if he intends to make a material alteration to the premises; he must also notify the authority if explosives or highly inflammable materials are to be stored (section 8).

Consultations

When plans are deposited under building regulations, relating to alterations or extensions, or new buildings, to be used as:

- boarding houses
- factory
- hotel
- office
- railway premises
- shop.

The local authority is obliged to consult the fire authority before approving the plans. This also affects an approved inspector (section 16).

On those occasions when a fire authority propose to require alterations to a factory they must consult the local authority before serving any notice on the owner or occupier (section 17).

Defective Premises Act 1972

This legislation imposes a duty on anyone erecting a dwelling to ensure that the work is done in a workmanlike manner so that the completed building is fit for occupation.

Responsibility is not limited to builders, as architects, quantity surveyors, site managers, etc., who are involved in the construction may be affected and can be liable for any breach.

It also requires an owner to ensure that anyone entering the premises does not suffer injury in consequence of lack of maintenance or repair.

Water Act 1973

All water authorities have a duty to provide sewers and to dispose of effluent discharged into them (section 14).

Local authorities operate under an agency arrangement with water authorities (section 15) but this does not apply to the treatment of sewage.

Charges may be made for the use of a sewer provided by a water authority (section 16).

Adoption of sewers

A water authority may by agreement take over or 'adopt' a sewer being constructed or to be constructed in the future (section 18).

Control of Pollution Act 1974

A local authority is required to keep maps showing pipes used:

- for water collection or disposal systems:
- for conveying heat by air, steam or water;
- as outfall pipes from sewage treatment works.

The maps are to be available for inspection without charge (section 28).

A local authority may serve a notice imposing requirements regarding the way works are to be carried out. The notice may specify plant or machinery, hours of working and noise levels (section 60).

Persons intending to carry out work on a construction site may ask the council for prior approval at the time application is made for building regulation approval.

When application is made for prior approval it must contain particulars of the works, methods to be employed and steps which will be taken to minimize noise. The local authority must give its decision within 28 days (section 61).

Town and Country Amenities Act 1974

This Act extends a local authority's powers in conservation areas. The Act provides that the demolition of all buildings within a conservation area are brought under control. However, regulations do permit the removal of certain small buildings.

To be really sure a developer should always consult the local authority prior to undertaking any work in a conservation area; even the demolition of a garden wall, or replacement of a window, as there may be a special order in force.

Additionally anyone intending to cut down, top, lop, etc., a tree within a conservation area must give the local authority six weeks' advance notice of his intention to do so.

There are heavy fines for failure to meet these requirements.

Health and Safety at Work Act 1974

The purpose of this Act is to secure the health, safety and welfare of people at work, and also people whose health or safety may be affected by work activities.

There are several detailed regulations which are linked to EU Directives (see Chapter 9).

Safety in Sports Grounds Act 1975

When any sports stadium has accommodation for more than 10 000 spectators, the Secretary of State may require that it should have a safety certificate.

Certificates will be issued by county councils and are of two kinds:

- a general safety certificate will be issued when the stadium is to be used on a regular basis;
- a special safety certificate is appropriate for a specific activity or occasion (section 1).

Contents of certificates

Safety certificates are to contain requirements to secure the safety of the stadium, including:

- number of spectators permitted;
- number allowed in various parts of the stadium;
- size and situation of entrances and exits;
- proper maintenance of all entrances and exits;
- the numbers, locations and strength of crush barriers.

Plans of the premises are to be included with the appropriate certificate (sections 2 and 4).

Upon receipt of an application for a certificate, the local authority must send a copy to the police chief and the district council, and consult them regarding the contents of any certificate issued (section 3).

There is provision for an appeal to the Secretary of State, against anything omitted from a certificate, or the refusal of a local authority to amend it (section 5).

The Secretary of State may make regulations:

- setting out procedures and detailing information to be provided;
- authorizing the charging of fees;
- setting out procedures for appeals;
- detailing provision of safety requirements in sports grounds;
- requiring that records be kept of attendance figures.

The Safety in Sports Ground Regulations 1976 have been made under this authority (section 6).

Local Government Miscellaneous Provisions Act 1976

Dangerous trees

A council may initiate action in connection with a tree which is dangerous. Complaint to the local authority may be made by the owner or occupier of the land on which the tree stands, or land which is affected by the tree (sections 23 and 24).

Dangerous excavations

Some authorities are empowered to provide protection for dangerous excavations on private land to which the public have access. Any work is to be at the local authority's expense (sections 25 and 26).

Highways Act 1980

Many facets of the Highways Act 1980 have a relation to construction taking place on adjoining land.

Generally a highway extends from hedge to hedge, fence to fence or wall to wall. All footpaths and grass verges within this width are regarded as 'highways'.

A highway is a strip of land over which the general public may pass and repass.

There are several categories of highway in addition to the obvious metalled road running alongside a building site.

On such highways one may travel on foot, on a horse or by vehicle. There are also bridleways where one may walk, ride a horse or a bicycle, or public footpaths where one may walk.

Sometimes there are motorways to contend with, or walkways, or cycle paths, and there are many facets of the Highways Act 1980 which affect builders.

Adoption

All new roads, streets or footpaths – highways – must be adopted or taken over by the appropriate highway authority before they may be maintained at public expense.

A person wishing a highway to be adopted may serve a notice on the local authority giving three months' notice of his wish for the council to take over the highway (section 37).

If the council is not satisfied that the highway is of a suitable standard to justify it being maintained at public expense, it may go to a magistrates' court and the dispute must be argued.

Another method is to arrange for the adoption of new highways by entering into an agreement with the highway authority.

Such an agreement will detail the standards of construction required by the council together with a bond or other surety to meet the expense of completing the road if the developer should for any reason not complete the work (section 38).

A type of highway called a 'walkway' may also be the subject of an agreement between an owner and a council. Such an arrangement enables a walkway to be provided through, over or under parts of a building, and provides for maintenance and cleaning.

Building and improvement lines

A highway authority may prescribe a frontage line for buildings, called a 'building line' (section 74). When a building line has been prescribed no building or permanent excavation may be made nearer to the centre of the highway than the building line. This does not preclude boundary walls and fences. Application for consent to erect buildings or other permanent excavations in front of a prescribed building line must be made to the highway authority.

When a highway authority considers:

- that a street is narrow or inconvenient or without a sufficiently regular boundary line, or
- it is necessary or desirable that a street be widened it may prescribe an 'improvement line' (section 73).

Damage

There are numerous instances of actions which may cause damage to a highway and a penalty may be incurred, if without lawful authority or excuse (section 131):

- a ditch or excavation is made in a highway;
- soil or turf is removed from any part of a highway;
- anything whatsoever is deposited so as to cause damage to the highway, or
- to light a fire or discharge a firearm or firework within 50 ft, from the centre of a highway, and in consequence thereof the highway is damaged.

The magistrates' court may inflict a fine upon anyone found guilty. Upon a second conviction the fine may be doubled.

Unauthorized marks made on highways may also invoke the wrath of the council (section 132). If anyone without consent or reasonable excuse permits or otherwise inscribes or affixes:

- any picture
- letter, or
- sign, or
- other mark

on the surface of the highway, or upon any tree, structure or works on or in a highway he may upon conviction be fined.

If a footway is damaged by or in consequence of any excavation or other work on land adjoining a street, the highway authority may make good the damage and recoup the cost from the person responsible (section 133).

Footpaths

When a footpath is no longer needed for public use the highway authority may by order extinguish the public right of way over the path (section 118). This may not be done until the action is confirmed by the Secretary of State or in the event of there being no objections, it is confirmed as an unopposed order. These are referred to as 'public path extinguishment orders'.

There are circumstances where bridleways and public footpaths cross a site and if they were to remain could impede development. Where the owner, lessee or occupier of land which is crossed by a bridleway or footpath is able to convince the local authority that to secure the efficient use of the land or provide a shorter or more commodious path the council may make a 'public footpath diversion order' (section 119). The order must be confirmed by the Secretary of State unless unopposed.

When planning permission has been given to enable development to be carried out in accordance with the approved plans the council may make an order to stop up or divert a footpath (section 210). An alternative route is usually required, and the costs will be borne by the developer. The decision in Ashby v Secretary of State in 1980 enables the order to be made retrospectively where the development has already been carried out.

Private access

If a highway authority considers that a private means of access from the highway to any premises is likely to cause a danger or to interfere unnecessarily with traffic, an order may be made to stop up the access (section 124).

As a condition of the order it must be shown that no access to the premises from the highway is required or there is other access.

Section 125 contains further powers for the stopping up of a private means of access to premises adjoining or adjacent to the route of the road. A new means of access must be provided.

An agreement may be made between the occupier of premises and the highway authority (section 127). Such an agreement may make provision for compensation.

A person using an access which has been stopped up is liable upon conviction to a fine (section 128).

When the occupier of any premises adjoining or having access to a highway maintainable at the public expense habitually takes or permits to be taken a mechanically propelled vehicle across a kerbed footway or a verge the highway authority may (section 184):

- construct a vehicle crossing over the footway or verge and recover the cost, or
- impose conditions on the footway or verge as a crossing.

Obstruction

There are many ways in which a highway may become obstructed deliberately or by lack of care. When carrying out work on a site, if a developer creates a situation endangering users of the highway, then that may be a public nuisance.

Examples of public nuisance have been:

- vehicle standing on highway for a long period and causing obstruction (Dymond v Pearce 1972);
- excavations adjoining the highway which were unguarded (Jacobs v LCC 1950);
- unguarded excavations in highway (Haley v LEB 1965)

Section 137 also makes obstruction of a highway a criminal offence.

A highway authority may serve notice on the owner or occupier of land directing him to alter any wall, fence, hoarding, paling, shrub, tree or other vegetation which is regarded as causing an obstruction at corners (section 79).

It may also serve notices restraining owners and occupiers from causing or permitting any building, wall, fence, boarding, paling, tree, shrub or other vegetation to be erected or planted on the land.

It is an offence without lawful authority or excuse to deposit anything whatsoever on a highway to the interruption of any user of the highway. A person found guilty may be fined (section 148).

When something has been deposited on a highway and it constitutes a nuisance the person responsible may be required to remove it forthwith (section 149).

A builder's skip may not be deposited on a highway without permission from the highway authority (section 139).

Any permission given may contain conditions relating to:

- the siting of the skip;
- its dimensions;
- the manner in which it is to be coated with paint and other material for the purpose of making it immediately visible to the oncoming traffic;
- the care and disposal of its contents;
- the manner in which it is to be lighted or guarded;
- its removal at the end of the period of permission.

If a builder's skip is deposited on a highway without permission the owner is liable to a fine.

A builder's skip must:

- be properly lighted during the hours of darkness;
- be marked with the owner's name, address and telephone number;
- be removed as soon as practicable after it has been filled.

Failure to comply with these requirements can also result in a fine.

A 'builder's skip' is defined to mean a container designed to be carried on a road vehicle and to be placed on a highway or other land for the storage of builders' material, or for the removal and disposal of builders' rubble, waste, household and other rubbish or earth.

'Owner' in relation to a builder's skip subject to a hiring agreement is also defined. If the hiring agreement is for not less than one month, or a hire purchase agreement, the owner is the person in possession of the skip under that agreement.

The highway authority or a constable in uniform may require the owner of a skip to remove or reposition it (section 140). Failure to comply can result in a fine.

Anyone who mixes or deposits on a highway any mortar or cement or any other substance which is likely to stick to the highway surface, or which if it enters drains or sewers is likely to solidify, may be fined (section 170).

It is permissible to do the mixing in a receptacle or on a plate which prevents the substance from coming into contact with the highway, and from entering any drains or sewers connected with the highway.

Consent of the highway authority is required to temporarily deposit building materials, rubbish or other things in a street or to make temporary excavation in it (section 171).

Any obstruction or excavation must be properly fenced and lighted.

Trees and shrubs may cause both physical and visual obstruction on a highway. No tree or shrub is to be planted in a made up carriageway, or within 15 ft from the centre of a made up carriageway

(section 141). If a tree or shrub is so planted the highway authority may require its removal. Once again failure to comply may lead to a fine and a continuing daily penalty.

A highway authority may give a licence to the occupier or owner of any premises adjoining a highway to plant and maintain trees, shrubs, plants or grass in such parts of the highway as may be specified in the licence (section 142).

The owner of land must not allow soil or refuse from his land to fall, or be washed or carried onto a street (section 151). Also the material must not be allowed to enter a sewer or gully in such quantities as to choke the sewer or gully.

Another section of the Act (section 152) enables a highway authority to require an owner to remove or alter any porch, shed, projecting window step, cellar, cellar door, cellar window, sign, signpost, sign iron, showboard, window, shutter, wallgate, fence, or other obstruction or projection which has been erected or placed against or is in front of a building and is an obstruction or projection which has been erected or placed against or is in front of a building and is an obstruction to the safe or convenient passage along a street. Offenders may be fined.

A door, bar or gate must not open outwards onto a street, except in the case of a public building, and only if the highway authority has given permission (section 153).

An unauthorized door, bar or gate may attract a notice for its removal. Failure to comply may result in a fine.

The occupier of premises adjoining a highway may be required to construct and maintain channels, gutters or downpipes necessary to prevent (section 163):

- water from the roof or any other part of the premises falling upon persons using the highway, or
- so far as is reasonably practicable, surface water from the premises flowing onto, a footway of the highway.

Failure to comply may mean a fine and a continuing daily penalty.

No scaffolding or other structure which obstructs a highway may be erected unless it is authorized by a licence in writing from the highway authority (section 169).

A highway authority may require the corner of a building intended to be erected at the corner of two streets to be rounded or splayed off to the height of the building if necessary (section 286).

If a person without consent or excuse erects a building or fence, or plants a hedge in a highway he is liable to a fine (section 138).

Hoardings

Anyone intending to erect or take down a building in a street or court or to alter or repair the outside of a building in a street or court, must erect a close boarded hoarding to separate the building from the street or court (section 172).

Hoardings are to be securely fixed to the satisfaction of the council (section 173).

Walkways

An agreement may be made between owner and a council enabling walkways to be provided through, over or under parts of a building and for the maintenance and cleaning of the walkway (section 35).

Power of entry

A person, duly authorized in writing, has the authority to enter any land at any reasonable time for the purpose of surveying that or any other land in connection with the exercise of a highway authority's functions (section 289).

Land adjoining streets

If in or on any land adjoining a street there is an unfenced or inadequately fenced source of danger to persons using the street, the owner or occupier may be required to undertake such works of repair, protection, removal or enclosure as will obviate the danger (section 165).

The occupier of land adjoining a highway may be required to remove barbed wire which is a nuisance to the highway (section 164).

A local authority may require the forecourt of a premises abutting a street which is regarded as a source of danger, obstruction or inconvenience to be fenced (section 166).

Any retaining wall exceeding 4 ft 6 in high and within 4 yards of a street may only be erected in accordance with plans, sections and elevations that have been approved by the local authority (section 167). There is a fine for failure to comply.

Anyone carrying out building work in or near a street must take all reasonable precautions to ensure that the work is carried out so as to avoid causing damage to persons in the street (section 168).

Over or under highways

No one may erect a building over any part of a highway, or alter buildings so constructed, without a licence granted by the highway authority (section 177).

Contravention may lead to a fine with a continuing daily penalty.

It is not lawful to fix or place any overhead beam, rail, pipe, cable, wire or other similar apparatus, over, along or across a highway without the consent of the highway authority (section 178).

Failure to comply may lead, upon conviction, to a fine and a continuing daily penalty.

A licence is required from the highway authority to construct a bridge over a highway (section 176). Failure to secure a licence or comply with its conditions may lead to a fine and a daily penalty.

One may not construct a vault, arch or cellar under a street without consent from the highway authority (section 179).

Anyone who does so may upon conviction be fined and subjected to a daily penalty. The cellar may also have to be removed.

No person may make an opening in the footway of a street to form an entrance to a cellar or vault without consent from the council (section 180). There is the possibility of a fine upon conviction.

Anyone without lawful authority who places any apparatus in or under a highway, or breaks upon a highway for the purpose of maintaining, repairing or reinstating any apparatus is guilty of an offence and liable to a fine (section 181).

Highway authorities may give licences for such work, subject to conditions.

More about adoption of streets

As with sewers, streets may be adopted or unadopted. Adopted streets imply that they are highways maintainable by the public – they are streets which the local highway authority is obliged to maintain and repair.

Unadopted streets are also known as private streets and the highway authority is not under any duty to maintain or repair them – this must be done by the owners or frontagers.

A local highway authority may carry out private street works to reconstruct unadopted streets to their specification, but this will only be done at the expense of the owners of premises having a frontage to the street. When the work is completed the street will be adopted and thereby become maintainable by the public at large. The authority is unlikely to adopt a substandard street (sections 205 to 218).

Adoption by agreement

A local highway authority may, by agreement with someone who intends to make a new street, undertake to adopt it upon completion. The street must be constructed to the authority's specification and be the subject of a maintenance period prior to adoption (section 38).

Advance payments

Provision is made for the operation of an advance payments code. It operates when plans are deposited in accordance with the requirements of building regulations. It applies in connection with a new building or a site which fronts an unadopted street when private street works may take place in the future. Construction must not commence until the local highway authority is given a sum calculated to cover the cost of street works relating to the frontage.

There are exceptions:

- where a building is to be on land exempted from private street works charges;
- where a building is to be within the curtilage of an existing building;
- when the plans relate to an area in which the advance payments code is not operative;
- where the developer of the street has entered into a section 38 agreement (see above);
- where it is unlikely that the street will be made up in a reasonable time;
- when the street may not be joined to an adopted street within a reasonable time;
- when the street was more than half constructed in 1951 and is less than 10 yards long;
- when the street was substantially completed in 1951;
- when the land is in the ownership of a county council or the British Transport Commission;
- where the street has been laid out at government expense on an industrial estate;
- where the street is substantially built up and frontages may be for industrial premises or because private street works may not be carried out for some time (section 219).

Charges are fixed by the local highway authority and notice must be given to the developer within six weeks of the plans being passed. One month is given for the developer to make an appeal to the Secretary of State.

When land is to be subsequently sub-divided the charge is to be apportioned in relation to the frontage of each property (section 220).

Payments which are made to meet private street works liabilities are based on the estimate, and it may be necessary to pay more later; and there is the unlikely possibility of refund (section 222).

When plans have been deposited and approved for a period of three years the local authority may declare them to be of no effect, meaning that they are cancelled (1984 Act, section 32). In these circumstances any payments made under the advance payments code must be refunded (section 223).

These matters must be entered in the Land Charges Register by the local authority (section 224).

Housing Act 1980

Lodging houses

When a house or part of a house is let in lodgings, or occupied by members of more than one family it must be provided with adequate means of escape in case of fire (section 12).

Local Government Miscellaneous Provisions Act 1982

This legislation contains a provision which becomes operative when a building is unoccupied, or the occupier is temporarily absent. Should it become necessary the local authority may undertake works required to prevent unauthorized entry or to stop it becoming a danger to public health.

Notice must be sent to the owner and occupier 48 hours before work is to be commenced. There may, however, be occasions where the owner or occupier cannot be located, or immediate action by the council is necessary.

Local authorities are given wide powers to enter the building, and if necessary, the adjoining land. Costs may be recovered and where appropriate a magistrates' court may share the costs among third parties (section 29).

Building Act 1984

Many requirements are outlined in the various chapters, and the following are also of interest to all concerned with the construction and maintenance of buildings.

Defective and dangerous buildings

Various items of legislation relating to defective or dangerous buildings have been brought together in the Building Act 1984.

For example, if a local authority is satisfied that a building or structure, or any part of it is in such a condition, or is carrying such loads as to be dangerous they may apply to a magistrates' court. The justices, upon considering the evidence, may take the following action:

- if the danger is because of the condition of the building they may require repair or demolition, and the removal of all rubbish arising from the work;
- if the danger is because of overloading they may restrict use of the building until remedial work has been undertaken.

Failure to comply with a magistrates' order in the time allowed may lead to a fine and the local authority entering to do the work at the owner's expense (section 77). However, in the case of listed buildings or subject to preservation orders or buildings in conservation areas, the local authority must be cautious. Before taking action under section 77 the local authority must, as required by the 1990 LBC Act, section 56, consider whether they should use powers to carry out urgent preservation works or otherwise to acquire the building in need of repair.

There are occasions when a building or structure, or a part of it, is in such a dangerous state that action must be taken at once. In these circumstances the local authority may take such steps as they consider necessary and whenever possible they are obliged to notify the owner of their action without delay.

The local authority may recover the expenses from the owner, but they cannot do so if the owner can show that immediate action was not necessary (section 78).

In practice the local authority will have delegated this responsibility to one of its senior authorized officers who may act without reference to the council or committee.

There is also provision in the Act for dealing with ruinous or dilapidated structures which may not be dangerous within the meaning of the above paragraph, but simply an eyesore.

So if a building is considered to be so ruinous or dilapidated that it is seriously detrimental to the amenities of the area, the council may require the owner to repair or restore it. However, the owner has an option and he may, if he wishes, demolish the building, or part of it and remove the rubbish (section 79).

This option is not available if the building should be listed, or in a conservation area, when planning consent to demolition would have to be obtained.

Sometimes a local authority may conclude that premises are in such a state as to be prejudicial to health or a nuisance and that unreasonable delay would occur if the procedure set out in the Housing Act 1936 were to be followed. In these circumstances the local authority may decide to do any necessary work themselves and send a notice to the owner giving details of the work proposed.

The owner may within seven days tell the council that he will remedy the defective state himself. If he fails to do so after nine days the council can carry out the work and recover the cost (section 76).

Demolition

Before commencing a demolition, notice must be given to the local authority which may then set out their requirements in connection with the proposed demolition.

As with all notices, it must be in writing, and a copy is to be sent to the occupier of the building, the electricity supplier and British Gas (section 80). A person failing to give this notice is liable on summary conviction to a fine.

Some work is exempt from this requirement:

- internal work when the building will continue to be used;
- buildings not more than 1750 ft^3 measured externally;
- greenhouses, conservatories, sheds or prefabricated garages;
- agricultural buildings.

The power to serve notices on the person making or about to make a demolition is in section 81, and it may also be used when the authority are themselves requiring the building to be demolished.

The local authority response must be within six weeks of receiving the notice of intended demolition, or within seven days if they themselves are making a demolition order.

A copy of the council's notice is to be sent to:

- the owner and occupier of any adjacent building;
- British Gas;
- electricity supplier;
- fire brigade if there is to be burning on the site;
- the Health and Safety Executive where hazardous premises are affected (section 81).

The person undertaking demolition may be required to:

- shore up any adjacent buildings;
- weatherproof walls or adjacent buildings that are exposed;
- repair and make good any damage to adjacent buildings;
- remove rubbish from the site;
- disconnect and seal off sewers and drains;
- remove drains and sewers;
- make good drain and sewer trenches;
- ensure disconnection of gas, electricity and water supplies as required by the undertakings;
- ensure any burning on site is in accord with fire brigade requirements;
- protect the public and minimize harm to amenity.

An appeal to a magistrates' court may be made:

- if the notice requires an adjoining building to be shored, and the owner of that building is not entitled to that support, and should pay towards the cost;
- if there is a requirement that an adjoining building be weatherproofed and the owner may not be entitled to have that provision without making a contribution to the cost (section 83).

Repair of drain

No one may,

- unless in a case of emergency, repair, alter or reconstruct a drain or sewer, or
- having carried out such work, cover over the drain or sewer

without giving the local authority at least 24 hours' notice. The section relates to drains which communicate with a sewer, a cesspool or other receptacle for drainage. If the requirements of this section are not observed, the local authority may take proceedings for a fine (1984 Act, section 61).

Disconnection of drain

Anyone who carries out work resulting in a drain being taken out of use must disconnect and seal up any unused drain or portion of a drain (1984 Act, section 62).

Improper repairs

If a watercloset or drain or soil pipe is so constructed or repaired as to be defective, the person undertaking the work is liable to a fine on summary conviction (1984 Act, section 63).

Provision of closets

A local authority is enabled to require the owner of a building to provide sufficient appliances when the existing arrangements are inadequate (1984 Act, section 64).

Type approval

The Secretary of State may approve any particular type of building or any other matter to which building regulations apply. He may confine such an approval to a particular regulation requirement.

A type approval may be made following an application made to him, or the Secretary of State may make one on his own initiative.

The Secretary of State will issue a certificate saying:

* the type of building matter to which the certificate relates;
* the requirements of the building regulations;
* where applicable, the class or classes of case to which the certificate applies.

The effect is that if a particular building, or building matter, is covered by a type approval certificate it is deemed to comply with the regulations to the extent stated in the certificate.

There is provision for charging fees to applicants. The certificate may have a time limitation.

The power to make a type approval may be delegated to any person or body, as the Secretary of State sees fit (sections 12 and 13).

Means of escape from fire

A local authority is required to ensure provision of means of escape from fire from buildings of more than two storeys if any upper storey is more than twenty feet above the ground.

The requirement applies where a building:

* is let in flats or tenements; or
* is used as a boarding house, boarding school, children's home, hospital, hotel, inn, nursing home or similar institution; or
* is used as a restaurant, shop, store or warehouse and has sleeping accommodation for employees on any upper floor.

Where in an existing building provision is inadequate, the local authority must serve notice on the owner setting out what must be done to meet the requirements.

The local authority may undertake the work in default and a penalty may be imposed by the courts (section 72).

Raising a chimney

If anyone erects, or increases the height of a building to a greater height than an adjoining building he must also, if necessary increase the height of any chimney of flue on the adjoining building if it is within 6 feet of the taller building.

The adjoining owner is given 14 days in which to give notice of his intention to do the work himself and recover the cost from the person erecting the higher building.

If neither party raises a chimney covered by this section, the council may carry out the work and recover its expenses (section 73).

Power to enter

A local authority has the power to authorize any of its officers to enter premises for a variety of purposes and this power is contained in a number of Statutes. For the purposes of the 1984 Act and the Regulations an authorized officer has the power to enter premises to ensure that a local authority's responsibilities are enforced.

The officer having this authorization will usually be an engineer, surveyor or building control officer. The authorized person must carry an identity card authenticated by the local authority stating the nature of the power of entry delegated, and the document must be produced upon request.

The authorized officer may not demand access to any premises except a factory or workplace unless 24 hours' written notice has been given.

Should admission be refused, or if the premises are unoccupied, or if the need for an inspection is very urgent, the authorized officer may obtain a warrant allowing entry, from a Justice of the Peace (section 95).

An authorized officer may be accompanied by anyone necessary to assist in the inspection, and a warrant remains in force until its purpose has been achieved.

It is an offence for anyone who enters a factory or workplace in the course of official duties to disclose any trade secret (section 96).

Anyone who wilfully obstructs an authorized officer in the course of his duties is committing an offence and upon conviction may be fined (section 112).

Notices to be in writing

Every notice given to a local authority must be in writing and all notices, refusals or consents given by a local authority must be in writing. There is special exemption in connection with 'stage' notices.

The Secretary of State may prescribe the form in which notices or documents relating to building control must be formulated (section 92).

All notices and other documents issued by a local authority are to be signed by a 'proper officer', that is someone specially authorized by the local authority for that purpose. Facsimile signatures may be used (section 93).

A council may serve a notice by one of a number of ways depending upon the circumstances:

* by delivering it to that person;
* by leaving it or sending it to that person's usual or last known address;
* if a company is involved, by sending or delivering it to its secretary at the registered or principal office;
* when the name of the owner or occupier cannot reasonably be found, by sending it addressed to the owner or occupier at the address of the premises to which the communication relates;
* if there is no one at the premises to receive the notice, it can be affixed to the premises in a conspicuous place (section 94).

Enforcement notices

Any notice requiring work to be carried out must:

* detail the work which must be undertaken;
* state the time allowed before the work must be completed.

There is a right of appeal, but if this right is not exercised and the work is not carried out in the time allowed the local authority may enter and carry out the works. Expenses are recoverable from the person to whom the notice was addressed, and additionally he may be liable for a fine for defaulting (sections 99 and 102).

If a local authority do enter and do the work, they are allowed to sell surplus materials if they are not claimed by the owner and taken away within three days (section 100).

Breaking open streets

The local authority or any person properly and legally undertaking work in connection with sewers, drains, water mains and services is empowered to break open streets for that purpose (section 101).

Local Government Access for Information Act 1985

The purpose of this Act is to provide greater access for the public to council and committee meetings. More importantly it provides for access to reports and documentation in addition to requiring local authorities to publish certain information.

Agendas and reports for committees and sub-committees have to be made available to the public three clear days before a meeting takes place.

When officers write their reports for a meeting they must list any background papers from which they have obtained information, or used to support their judgement. Background papers may include:

* inspection records;
* practice notes or other documents of an advisory or interpretative nature;

- articles, technical reports or learned papers on the subject;
- trade literature;
- letters received following any consultation procedure;
- letters received offering support or objection to the matter under report.

Copies of the documents are to be available for inspection by the public for a period of four years. Copies may also be requested, but they must be paid for.

Housing Act 1985

Housing authorities may require work to be carried out to make a dwelling fit for occupation and also relates to houses in multiple occupation. Examples of provision which may be required are:

- facilities for preparation and cooking of food;
- means of escape in case of fire and other fire precautions;
- storage accommodation;
- baths, washbasins, showers and water closets, together with hot and cold running water supplies.

Latent Damage Act 1986

The Limitation Act 1980 said that an action founded on a breach of legal duty giving rise to damage or injury could not be brought later than six years from which the cause of action accrued.

This six-years' period has been extended by the Latent Damage Act 1986 to allow the claimant three years from the date the damage is discovered, or ought to have been discovered. Also anyone who purchases already damaged property, but the damage is not known or could not be known, has a right to initiate proceedings.

There is protection also for builders, architects and surveyors who may have been faced with proceedings after a long period alleging negligence in being responsible for or not detecting a defect.

Proceedings cannot be brought alleging a lack of duty or care after a period of 15 years from the breach of duty.

Town and Country Planning Act 1990

In addition to obtaining approval for development under the Building Regulations there may also be the need for planning permission, and one approval does not automatically imply the other is agreeable.

There is, therefore, a need for planning permission for much that is done to property. Development is the carrying out of building, engineering, mining or other operations in, on or under land, or the making of material change in the use of any buildings or other land. Expressly included as development is the use of a single dwellinghouse for two or more separate dwellings, the deposit of refuse, or waste material on an existing dump thereby extending its superficial area or its height above the level of adjoining land, and the display of advertisements.

Some operations are expressly excluded:

- Works of maintenance or improvement or the alteration of buildings affecting only the interior of the building, or which do not materially effect the external appearance of the building. The exception does not relate to any alteration providing additional space below ground level. Work on the interior of a listed building may require listed building consent from the local authority.
- The use of a building or land within the curtilage of a dwellinghouse for any purpose incidental to the enjoyment of the house as such.
- The use of land or any building occupied therewith for agriculture or forestry.
- Changes in use within a class stated in an Order made by the Secretary of State.

The current Order is the Town and Country Planning (Use Classes) Order 1987 which specifies 16 classes. A change which involves a switch from one to another class will constitute development requiring consent (section 55).

Planning (Listed Buildings and Conservation Areas) Act 1990

Prior to proceedings under the 1984 Act, sections 77 or 79 in respect of dangerous or dilapidated buildings, a local authority must consider whether to carry out urgent repair work themselves, or issue a repair notice, or compulsorily acquire the building (section 56).

Environmental Protection Act 1990, *Waste from building and demolition site*

It is an offence to treat, keep or dispose of controlled waste in such a way as to cause pollution of the environment or harm human health (section 33).

There is a duty of care and all reasonable measures must be taken to prevent illegal disposal (section 34).

A code of practice on waste management and the duty of care may be purchased from TSO.

Planning and Compensation Act 1991

This Act states that planning permission is required for certain works of demolition (section 13).

The main concern is relating to the demolition of dwellinghouses and the Secretary of State is to exclude from planning control the demolition of any building that does not contain a dwellinghouse or one or more flats occupied for residential use independently of the main use of the building.

Thus, planning permission will not be required to demolish offices, shops or factories except where the subject building contains a dwellinghouse or flat. Domestic garages, garden sheds and other small domestic structures are excluded from the requirement.

There is a need for cooperation between planning officers and building control staff concerned with the administration of the 1984 Act, section 80

Regard must be taken of the limitations affecting alterations and demolition of a listed building or building in a conservation area.

Water Resources Act 1991

Every local authority is required to keep a map showing public sewers. It must be available for inspection (section 199).

Anyone constructing a sewer may be required by the water authority to lay it in such a way that it forms part of the general system of sewers. This might involve putting the line of sewer in a different place, using a larger diameter pipe, or putting the sewer at a different depth or to another outfall.

In this connection the water authority must make a financial contribution towards the extra costs involved. In case of disagreement an appeal may be made to the Secretary of State, a magistrates' court or the matter may be settled by arbitration (section 112).

Right to connect

The drains from any property may be connected to a public sewer, but may only discharge domestic sewage without prior approval. Consent is required for the discharge of trade effluent.

When separate sewers are provided it is not permitted to discharge:

- foul water into a surface water sewer;
- surface water into a foul sewer except when the water authority are agreeable.

No connection may be made to a storm overflow pipe.

The water authority must be given 21 days' notice of an intention to make a connection (section 106).

When a water authority receive a notice that drains are to be connected to a public sewer, they may decide that they wish to do the work themselves. If this is the case, the developer must be told within 14 days, and he will be required to meet the cost of work undertaken by the authority (section 107).

Forbidden discharges

It is forbidden to allow the following to enter a public sewer:

- matter likely to injure the sewer
- matter which will interfere with the flow in a sewer
- matter which will prejudicially affect the treatment or disposal of the contents of a sewer
- any chemicals, steam or hot liquid which alone, or in combination with other contents of the sewer, will be a danger, a health risk or a nuisance
- petrol or carbide of calcium.

For the purpose of this requirement, 'petrol' means crude petrol, oil made from petroleum or bituminous materials and any product or mixture of petrol.

Upon conviction one may be fined together with a daily penalty for a continuing offence (section 111).

Existing drains

There may be occasions when premises are drained to a sewer or a cesspool, and although the arrangement is effective it is not adapted to the general system in the area. When this is so the water authority may alter the drainage system at their own expense, including filling in any old drains or cesspools.

There is a right of appeal to a magistrates' court (section 42).

If a local authority believe that any drain is defective and admits subsoil water or is prejudicial to health or a nuisance, they may test the drain and if necessary, open up the ground. A water test under pressure is not allowed, and if the drains are found to be satisfactory, the ground must be reinstated and any damage made good (section 114).

Trade effluent

The occupier of any trade premises may discharge trade effluent into public sewers, but first he must obtain consent from the water authority.

A 'trade effluent notice' must be delivered to the water authority and is to contain the following information:

- nature and composition of the effluent
- maximum daily rate of discharge
- highest rate of discharge.

When a water authority give consent to the discharge of trade waste they may impose a number of conditions:

- specifying the sewer into which the discharge may be made
- limiting the nature and composition of the discharge
- limiting the maximum daily discharge
- limiting the highest rate of discharge (some trade effluents are exempt)
- there is no increase in the daily maximum discharge rate
- when in that same period the effluent discharged was subject to an agreement and charges were paid (sections 1 and 2).

Agreements

A water authority may enter into an agreement with the owner or occupier of trade premises to receive and dispose of trade effluents. They may be also agree to carry out work at the owner's expense.

The owner or occupier will be required to produce:

- plans of drains or sewers;
- information regarding the drains or sewers (sections 1 and 2).

Water Industry Act 1991

An owner or occupier may require water undertakers to provide a water main to give a supply to domestic premises (section 41).

By agreement the cost must be repaid to the undertaker over a period of 12 years (section 43).

Owner or occupier may be required by water undertaker to carry out repairs to prevent leakages or excessive consumption (section 75).

A sewerage undertaker may adopt sewers after receiving a request from the owner; this is achieved by a vesting declaration (section 102).

New Roads and Streetworks Act 1991

Part III of this Act replaces the Public Utility Street Works Act 1950 and some sections of the Highways Act 1980. The aim is to provide improved conditions relating to the conduct and coordination of streetworks. The Act applies to highways maintained at public expense and also to private streets, although the latter are exempt from some requirements.

A street is any

- highway, road, lane, footway, alley or passage;
- square or court;
- land laid out as a way whether it is for the time being formed as a way or not.

Streetworks include breaking up or opening a street or any sewer, drain or tunnel under it. It does not include maintenance, traffic signs, or a footway and verge crossing (section 48).

It is an offence to place apparatus in a street or to break up or open a street without a statutory right or a streetworks licence (section 50).

Undertakers having a statutory right must give 7 working days' notice to the street authority (section 54).

Street authorities are required to keep a register of street works. It must be available for inspection free of charge at all reasonable hours (section 53).

Street authorities are to use their best endeavours to coordinate streetwork and other works for road purposes (section 59).

Access to Neighbouring Land Act 1992

Any intrusion onto your neighbour's land without his consent is a trespass, but this Act provides machinery whereby basic preservation works may be undertaken by having access to a neighbour's property.

If your neighbour is not agreeable to giving reasonable access to his property then one may go to court for permission. The court may make an order allowing access and setting conditions including a payment to your neighbour.

The proposed work must be reasonably necessary for the preservation of the whole or any part of the property and cannot be carried out or would be substantially more difficult to carry out without entry onto the neighbour's land.

The court may state how long the access is permissible, the hours of work and what precautions are to be taken to protect your neighbour's property.

If an access order is made then, subject to its conditions, the neighbour is obliged to permit access.

Home Energy Conservation Act 1995

This makes it the duty of local authorities who are housing authorities to prepare a report drawing up energy conservation proposals in relation to residential accommodation. Measures are to be practicable, cost effective and designed to result in a significant improvement in the energy efficiency of residential accommodation.

The Act does not confer any powers to make grants, or loans; or give any power of entry to carry out works; or require any person to carry out works.

Housing Grants, Construction and Regeneration Act 1996

Sets out four main types of grant payable by local housing authorities.

1. Renovation grant relates to the improvement or repair of a dwelling, or the provision of dwellings for letting by conversion.

2. A common parts grant is intended for the improvement or repair of the common parts of a building containing flats.
3. An HMO grant is available for the improvement or repair of houses in multiple occupation, and also for converting buildings to be houses in multiple occupation.
4. A disabled facilities grant is for adapting or providing facilities in the home of a disabled person.

The amount of grant payable is linked to the financial resources of the applicant.

Party Wall, etc. Act 1996

This Act extends to England and Wales similar requirements relating to party walls as are contained in the London Building Acts (Amendment) Act 1939 and was effective from 1 July 1997. It sets out the procedures to be followed together with the rights and obligations when building work affects the property of an adjoining owner.

Various notices must be exchanged between the owners, but the building control officer or an approved inspector are not involved except so far as building regulations apply to the proposed work.

There are a number of definitions and they are important for interpretation of the requirements:

- 'adjoining owner' and 'adjoining occupier' mean any owner or occupier of land, buildings, storeys or rooms adjoining those of the building owner;
- 'building owner' is the owner of land wishing to exercise rights under the Act;
- 'party fence wall' is a wall which is not part of a building and which stands on land of different owners to separate the adjoining land (Fig. 1.);

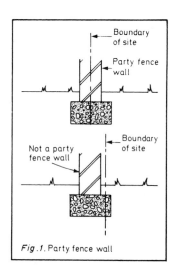

Fig. 1. Party fence wall

- 'party structure' is a party wall; a floor or floor partition separating buildings or parts of buildings having separate entrances or staircases;
- 'party wall' means: (i) a wall which forms part of a building and stands on lands of different owners to a greater extent than the projection of an artificially formed support on which the wall rests (Fig. 2), and (ii) so much of a wall not being a wall referred to in (i) as separate buildings belonging to different owners (Fig. 3).
- 'special foundations' involves an assemblage of beams or rods to distribute the load.

The Act relates to the land of different owners when either owner wants to build on the line of junction. A building owner wishing to build a party wall or fence on the line of junction is to give the adjoining owner a month's notice of his intention.

The adjoining owner may be agreeable and then give notice indicating his consent to the building of a party wall or a party fence wall:

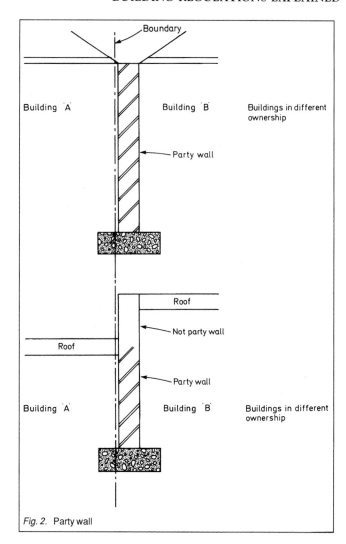

Fig. 2. Party wall

- the wall is to be built half on the land of each of the two owners; or in such other position as may be agreed; and
- the expense of building the wall may be shared between the two owners in proportions to be agreed.

When the adjoining owner does not give written consent within fourteen days, the building owner may only build the wall at his own expense and completely on his own land.

If a building owner wishes to build on the line of junction, and wholly on his own land, he must give the adjoining owner one month's notice, and he may place projecting footings and foundations below the level of the land of the adjoining owner. The work will be completely at the building owner's expense and he must compensate the adjoining owner and occupier for any damage to their property (section 1).

Repairs

A building owner has a wide range of rights in connection with the repair of a party wall or party fence wall. He may

- underpin, thicken or raise the wall;
- make good, repair, or demolish and rebuild;
- demolish a partition not conforming to statutory requirements and replace it with a satisfactory construction;
- demolish a party structure of insufficient strength or height and rebuild;
- cut into for any purpose, as for example the insertion of a dpc or flashings;
- undertake work of underpinning;

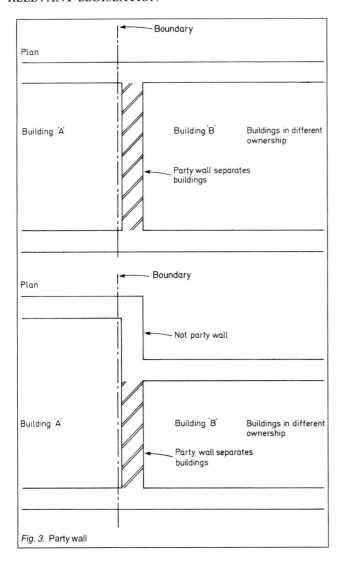

Fig. 3. Party wall

- expose a party wall previously enclosed, but subject to the provision of adequate weathering.

All damage incidental is to be made good (section 2).

Notices

Before exercising his rights, a building owner must serve on the adjoining owner a 'party structure notice' containing the following information:

- name and address of the building owner;
- nature and particulars of the proposed work;
- plans, sections, details and loads of any special foundations;
- date the proposed work will begin.

The notice is to be served not less than two months before work begins and will cease to have effect if the work is not commenced within twelve months, or not carried out with due diligence (section 3).

Counter notices

Upon receipt of a party structure notice the adjoining owner may respond by sending the building owner a 'counter notice', requiring work to be undertaken which may reasonably be required for his convenience. When special foundations are involved, the adjoining owner may require them to be placed at a greater depth or constructed of sufficient strength to be carried by columns of any intended building of the adjoining owner.

A counter notice must specify the works required and include plans, sections and particulars of those works. It is to be served within a month of receiving the party structure notice.

A building owner must comply with those requirements unless the works would:

- be injurious to him; or
- cause unnecessary inconvenience; or
- cause unnecessary delay.

A dispute arises if there is no response to these notices within fourteen days of them being served (section 4).

Adjacent excavation and construction

When a building owner intends to excavate for and erect a building or structure within 3 m from a building of an adjoining owner, and the excavation is to a lower level than the level of the existing foundations, he may be required to underpin the foundations of the adjoining building (Fig. 4).

Likewise, if excavations for a new building or structure are to be within 6 m from the building of an adjoining owner, foundations of the existing building may have to be underpinned in the circumstances shown in Fig. 5.

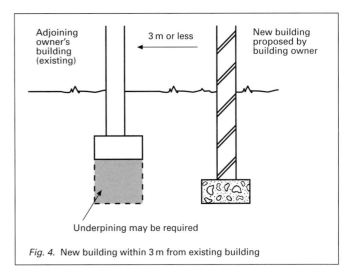

Fig. 4. New building within 3 m from existing building

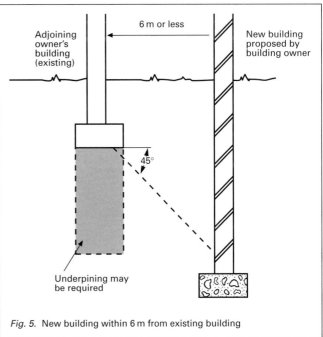

Fig. 5. New building within 6 m from existing building

There is provision for the exchange of notices between the two owners (section 6).

Rights

A building owner must:

- not cause unnecessary inconvenience to any adjoining owner or adjoining occupier;
- compensate any adjoining owner or adjoining occupier for any loss or damage resulting from the work;
- make and maintain hoardings, shoring and temporary construction to protect adjoining land and its security;
- not deviate from his plans unless agreed between the owners (section 7).

Rights of entry

A building owner, his servants, agents and workmen is given a right of entry on any land or premises for the purpose of executing work, and may if necessary make a forcible entry if accompanied by a constable, or other police officer (section 8).

Resolution of disputes

When a dispute arises between a building owner and an adjoining owner, both parties may agree to appoint a surveyor to settle the problem. An alternative is for each party to appoint a surveyor and the two surveyors select a third surveyor to adjudicate on the dispute.

There are detailed requirements regarding the duties, responsibilities and the costs involved in the appointment of the surveyors, and the award:

- the right to execute;
- time and manner of execution, including plans and schedules;
- the costs in making the award. May include or exclude making inspections and any other matter arising out of the dispute.

Fourteen days are allowed in which an appeal may be made to the County Court against an award (section 10).

Other requirements

Further sections of the Act refer to:

- Expenses (11)
- Security of expenses (12)
- Account for work carried out (13)
- Settlement of account (14)
- Method of serving notices (16)
- Recovery of sums (17)
- Crown land (19)

Exception

The Act does not apply to land situated in Inner London and in which there is an interest belonging to the Inner Temple, Middle Temple, Lincoln's Inn or Gray's Inn.

Local authorities

Local authorities and their property are not exempt from the 1996 PW Act.

In certain circumstances there is provision for an 'appointing officer' to select a third surveyor. This is the only aspect of administration involving a local authority, which is to name an appointing officer; most likely a building control officer. The local authority does not have any part to play in the resolution of disputes.

LOCAL LEGISLATION

In addition to the national statutes there are a number of local Acts which give the local authorities special powers, many of them relating to building control.

Local Acts are operative in the following areas:

Year received Royal Assent	Area in which Act is operative
1856	Middlesex
1957	East Ham
1976	South Glamorgan
1980	Cheshire
	Isle of Wight
	Merseyside
	South Yorkshire
	West Midlands
	West Yorkshire
1981	Avon
	Derbyshire
	East Sussex
	Greater Manchester
	Kent
1982	Cumbria
	Humberside
1983	Hampshire
	Staffordshire
1984	Lancashire
	Cornwall
1985	Bournemouth
	Clywyd
	Leicestershire
	Nottinghamshire
	Poole
	Surrey
	Tyne and Wear
1986	Berkshire
	Poole
	West Glamorgan
1987	Cleveland
	Dyfed
	Essex
	Mid Glamorgan
	West Glamorgan
	York

Requirements in each local Act vary; some are more onerous than others; but the following is a list of the principal matters covered in many of the Acts:

Access for fire brigade
Approval of plans – void after three years
Buildings used for storage of flammable substances
Byelaws for temporary structures
Dust from building operations
Entertainment clubs
Fire precautions for high buildings
Fire precautions in large storage buildings
Fireman's switches
Fire and safety precautions in public and other buildings
Means of escape from fire in certain buildings
Night cafés and take away cafés
Oil-burning equipment
Power to order alteration of chimneys
Storage of flammable materials
Safety requirements for parking places
Safety of stands
Separate drainage
Signs on buildings regarding hazards
Traffic signs on buildings.

Those local authorities which have a local Act containing provisions which control the construction or location of buildings must keep a copy available for free inspection at their offices during normal working hours (1984 Act, section 90).

Repeal

The Party Wall, etc. Act 1966 (Repeal of Local Enactments) Order 1997 repealed sections XXVII to XXXII of the Bristol Improvement Act 1847.

Appeals and determinations

An appeal may be made in respect of a decision made by a local authority not to relax or dispense with a particular requirement in the building regulations. Appeals may be made to the Secretary of State (see Chapter 4) under the 1984 Act, section 39.

The Secretary of State may make a determination under the 1984 Act, section 16(10) (see Chapter 4) on a matter arising under the building regulations.

A number of appeal decisions and determinations are reproduced in this Annexe. They demonstrate the open approach which must be made in interpreting the regulations and the approved documents. There is no room for prejudice or bias, or to those seeking to make regulations embrace more, or less, than intended and which is apparent from the clear logical assessment of each problem.

The Secretary of State emphasizes that each appeal or determination is, of course, only valid in the context of the particular case, but it is nevertheless thought that each may be of general interest. The details of the applicant and the local authority have been omitted and it is not possible to reproduce the drawings. The text of the appeal or determination brings out the salient facts.

It will be noticed that Part B continues to give rise to most of the decisions.

Each of the decisions is Crown copyright, and is reproduced by permission.

The decisions are preceded by the author's comments in italics, and this is followed by the D of E reference number and the date of issue, ready for identification.

APPEALS

PART A: STRUCTURE

Listed building – upgrading fire resistance (A.1)

The local authority refused to dispense with a requirement in respect of the loading of upper floors consequent upon upgrading of fire resistance of the ceilings. The decision contains a paragraph outlining the reasoning that A.1 should be applied in this particular case.

Ref.: 15.3.103   Date: 10 October 1996

The appeal

The building work to which this appeal relates is to effect fire precaution works to a mid-eighteenth century Grade II listed building of three storeys comprising two former buildings which abut each other on a corner. Following earlier use as a public house, the building was subsequently used as a commercial hotel; each bedroom contained a wash hand basin and was let individually, with bathrooms being shared. It is understood that the building was subsequently converted into flats by your client which are contained on the first and second floors, and that the building was subsequently confirmed as a House in Multiple Occupation (HMO). The ground floor comprises a shop to the front with kitchen and store to the rear.

When the use of the building was first confirmed as an HMO, Building Regulations approval was given under the 1985 Building Regulations for fire precaution work which was carried out by your client in 1988. It is understood that, as a result of a subsequent court case and order, the District Council's Environmental Health Department required your client to carry out improvements to the fire safety of the building. A full plans application for this work was made to the District Council.

The work involved provision of 60 minutes' fire resistance to the ground floor ceiling over the shop premises, and the addition of a 12.5 mm plasterboard and skim to bring the fire resistance of the ceilings to the first and second floors up to 30 minutes. An L2 AFD system was to be installed. However, the Building Control Department of the District Council raised the question as to whether the joists to the upper floors were adequate to bear the additional load of the plasterboard and skim. As a result your client employed a structural engineer. It was agreed that the joists to the upper floors needed strengthening to support the additional loading to achieve the 30 minutes' fire resistance. A full plans application to effect this work was then approved.

Your client subsequently decided that the District Council's request that the upper floors be strengthened was excessive and you then, on your client's behalf, requested a dispensation from Part A of the Building Regulations 1991. You did not specify which of the three requirements within Part A you were seeking a dispensation from but for the purposes of this appeal the Secretary of State has assumed this was Requirement A1 ('Loading') with respect to the upper floors. This application was refused by the District Council on the grounds that the strengthening of the floors was necessary to carry the additional weight imposed by the provisions required to achieve the improved fire resistance. It is against that refusal that you appealed to the Secretary of State.

The Appellant's case

On behalf of your client you contend that the requirement that there should be structural upgrading of the upper floors is excessive. You have contrasted the position in 1988 when there was no request from the District Council for calculations or justification of the structure when fire precaution work was carried out, with that of your 1995 application for additional fire precaution work when such matters were questioned by the District Council.

In addition you have referred to the fact that there is a general assumption that a 'reasonable' attitude should be taken by local authorities when processing Building Regulations applications relating to listed buildings. You also contend that the 'reasonable' approach is confirmed, particularly in respect of HMOs, in Circular 12/92. You have enclosed with your appeal various extracts from Approved Document 'B' ('Fire Safety'); DOE Circular 12/92 'Houses in Multiple Occupation'; and BRE Digest 366 'Structural appraisal of existing buildings for change of use'.

In addition you have argued that, in any event, Regulation 6 of the Building Regulations 1991 (as amended) excludes the need for compliance with Part A in the particular circumstances of this case.

The District Council's case

The District Council contend that the upper floors should be strengthened in order to sustain the additional loading imposed by the 12.5 mm plasterboard and skim, which they consider to be essential for bringing the fire rating of the upper ceilings up to 30 minutes. They believe that the strengthening of the floors is essential to enable them to sustain the loading during the initial stages of fire. In the absence of this strengthening, they state that the floors could collapse earlier than the 30 minute fire exposure, thereby jeopardizing the safety of occupants of the building.

The District Council also rehearse the fact that their consulting structural engineer was a party to the discussions with your client's structural engineer which led to agreement between all parties on the need to strengthen the upper floors if they were to be up-graded for fire resistance.

The District Council also state that they were concerned about the ability of the upper floors to sustain the additional load because they were old and 'live'. The District Council have also explained that DOE Circular 12/92 requires 60 minutes' fire resistance to HMOs but that this can be reduced to 30 minutes where an L2 AFD system is in place. In this particular case the District Council required 60 minutes' fire resistance to the ceiling of the ground floor because this contained shop premises but required only an up-grading to 30 minutes' fire resistance to the upper floor ceilings because an L2 AFD system was to be installed.

The Department's view

The Department fully accepts and would encourage a reasonable approach to the application of Building Regulations to listed buildings. However, whilst it is clear that the up-grading of the fire resistance is to sustain a listed building in an apparently appropriate use, you have provided no evidence to demonstrate in what way the listed building would be harmed or endangered by the proposed strengthening work.

In so far as the District Council has taken account of the installation of an L2 AFD system and reduced the fire resisting requirements to 30 minutes to the two upper floors, the Department believes that the District Council has acted reasonably and within the spirit of Circular 12/92. However, the fact remains that the upper floors would need to be strengthened even to sustain the loading of this lower specification; and the work required to effect this strengthening formed a part of the Building Regulations full plans approval given in April 1996.

You have challenged the right of the District Council to apply the requirements of Requirement A1 to this particular case on the grounds that Regulation 6 of the Building Regulations 1991 (as amended) cannot require, in this particular instance, compliance with Part A. However, your proposal to omit the strengthening work to the upper floors whilst upgrading the fire resistance of the ceilings would, on the evidence provided, result in non-compliance of the work in respect of Requirement A1. The work would then fall within the definition of Regulation 3(2) of the Building Regulations 1991 for a material alteration. As a material alteration the work would constitute 'Building work' within the meaning of the regulations and therefore be required to accord with Regulation 4 which requires building work to comply with the relevant requirements contained in Schedule 1 to the regulations. The Department is therefore satisfied that your proposed work would be required to comply with Requirement A1.

The decision

The Secretary of State has given careful consideration to the circumstances of this particular case. He is concerned to ensure that the application of the Building Regulations to listed buildings is done with reasonableness and balance. However, in this particular case it appears that at one stage all parties agreed to the need for the specified strengthening work without objection being raised on grounds of

harm or unacceptable change being done to this listed building. Against this the Secretary of State must balance the fact that the work is required as part of the need to secure and sustain essential fire safety precautions in a residential context.

The Secretary of State has noted that the District Council has accepted a lower fire resistance specification to the upper floors because of the installation of the L2 AFD system. He considers it important that this lower specification is achieved in full and accepts that on the evidence provided the upper floors will need strengthening to effect this work. He has therefore concluded that the District Council came to the correct decision in refusing to dispense with Requirement A1. Given the nature of the structural issue involved it would also not have been appropriate to consider relaxing Requirement A1. Accordingly, the Secretary of State dismisses your appeal.

PART B: FIRE SAFETY

Four storey office building – means of escape (B.1)

This relates to office premises involving two tenancies. An alternative escape route passed through a lobby area which could contain smoke by the time detectors were activated. In consequence the appeal was dismissed.

Ref.: 45.3.48   Date: 21 March 1995

Careful consideration has been given to the facts of the case, and to your representations and those of the Borough.

The building work involves alterations to office accommodation with an overall floor area (including lobby) of 455 square metres to form a separate tenancy of about 50 square metres on the fourth floor of an office block. The block has a protected main stairway at one end and an external escape stair at the other. The single entrance to the new tenancy is from the main common stair and lift lobby area. Alternative escape from this new tenancy could only be achieved through this lobby into the main tenant's accommodation, and then via the external stair at the far end of the floor.

You have stated that the doors from the lobby to the main tenant's accommodation area will be free of fastenings and outside normal office hours will be capable of being opened from the lobby by means of a break glass facility. Direct travel distance within the small tenancy to the exit door to the lobby is about 8 metres, with a further 3 metres through the lobby to the main tenant's doors if alternative escape is necessitated. Smoke detectors are to be located in the common area and in the main tenant's accommodation. There are no proposals for smoke detectors elsewhere.

The Council have rejected the proposals on the basis that the smaller tenancy has only a single stair for escape which they state is unacceptable for a building with a floor at this height. They quote paragraph 4.2 ('Number of escape stairs') of Approved Document 'B' ('Fire Safety') in support of this which in turn refers to paragraph 4.5 ('Single escape stairs'). This suggests that a single stair is only acceptable from a building, or part of a building, where the top floor is not more than 11 metres above ground level.

The Council have discounted the alternative escape route through the main tenancy because there is no control over the access, and also because the route to this main tenancy area is still through the common stair and lift lobby area which may be smoke logged. They acknowledge the smoke detection system but point out that smoke will already be in the common area when the detectors are activated.

They also consider that fire exit signs in the common lobby area will confuse persons escaping from the main tenant's area.

Alternative escape through the main tenant's accommodation would have been acceptable to the Council if a second exit door had been provided so that the route was independent of the common stair and lift lobby area.

You consider that there would be adequate safeguard against the main tenant's entrance door into the main lobby area being obstructed because the two tenancies are associated companies. With regard to a second, direct exit into the main tenancy area you consider this would restrict the usable floor area of the smaller tenancy and necessitate moving computer equipment. You also point out that sub-letting parts of office buildings into different tenancies is common practice.

The Department's analysis of the case is as follows. It notes that the floor area of the smaller tenancy is relatively small (50 square metres) and that the drawing shows that it is only going to be occupied by two people. It recognizes that there could of course be more people than this, but it is unlikely that there will be many.

It also notes that the doors providing alternative escape through the main tenancy form the main entrance for that unit. It is unlikely therefore that the doors will be obstructed during normal office hours because the entrance is needed for main access. This is all the more likely to be so given that the two tenancies are associated companies. However, the Department considers that it would be important to ensure that the suggested break glass facility was installed to cater for persons working outside the normal hours. Given that people in the smaller tenancy would be familiar with that route of travel, the Department does not consider that any additional escape signs located in the common area would cause confusion.

The Department notes that your proposals provide for only one escape route from the small tenancy via the common stair and lift lobby area and that you do not wish to provide a second direct exit door to the main tenancy. On the assumption that the lobby area is a protected route to the ground floor it is the Department's view that the proposed single escape route should be relatively risk free. Travel distances are well within the permitted limits given in Approved Document 'B' for travel in one direction and the smoke detection system will give early warning. However, the Department fully acknowledges the Council's point that by the time the smoke detector is activated smoke will already be in the common space.

In these circumstances the Secretary of State is of the opinion that the means of escape from the smaller tenancy is not capable of being safely and effectively used at all material times and accordingly dismisses your appeal.

It is not the function of the Department to advise on solutions to particular questions not raised between appellants and local authorities; this is a matter for the parties concerned to resolve. However, it may be helpful if the Department observes that, in the circumstances of this particular case, it would have been satisfied that the means of escape complied with the Statutory Requirement 'B1' of Schedule 1 to the Building Regulations 1991 if the smoke detection system had been extended in the escape route to at least the floor below. In the Department's opinion this would give sufficient early warning for the occupants of the smaller tenancy to escape in safety, via the alternative route.

Converted three storey dwelling – means of escape (B.1)

This proposal involved open stair arrangements up to the second floor and it was not intended for the dwelling to be occupied on a permanent basis. The decision indicates that the design cannot be based upon occasional occupancy.

Ref.: 45.3.49   Date: 20 March 1995

Careful consideration has been given to the facts of the case and to your representations and those of the Borough.

The proposal is to retain the external walls of the existing three storey property whilst demolishing the remainder of the structure and re-erecting the dwelling. The ground floor living accommodation will consist of a kitchen/dining room; the rest of the space will be occupied by an integral garage which is fire separated from the living accommodation. The first floor will contain two bedrooms, a study and bathrooms and the second floor consists of a living room. All floors are to be constructed in concrete, as is the spiral stair that provides access to all floors.

The spiral stair is to be fully fire protected at first floor level with all doors opening onto the stairway being half hour fire resisting and self-closing. The intention is that at ground and second floor levels the stair will be open to the accommodation. Alternative escape from the first floor will be via windows that have the minimum opening size recommended in Approved Document 'B' ('Fire safety') for an inner room situation. On the second floor there is to be an external balcony for the full width of the house which you state gives access to neighbouring properties. Located above the stairway will be a sliding domed rooflight that can be used for escape onto the roof. A type L1 comprehensive smoke detection system is to be installed throughout the dwelling to British Standard BS 5839: Part 1: 1988 ('Fire detection and alarm systems for buildings: Code of practice for system design, installation and servicing').

The Borough Council have stated that there should be a fully protected route at all levels in a new three storey dwelling to comply with the guidance given in Approved Document 'B'. They were prepared to accept the open plan arrangement at ground floor level providing the stair was fully protected at first and second floor levels and the smoke detection system installed. However, because you have since stated that you intend to have the stair open to the second floor living accommodation they have rejected the proposals. They consider that the removal of the passive protection at this level will prejudice the occupants if a fire occurs on the ground floor.

You consider that your proposals are reasonable because the roof and balcony form alternative escape routes from the second floor and the smoke detection system will provide early warning. In addition you have offered to provide manipulative chain ladders. Although you appear to doubt that there could be hot gases you have offered to have the rooflight linked to the smoke detection system to give automatic smoke venting if this is considered necessary. You have further pointed out in support of your case, that the property will not be occupied on a permanent basis but will be the London residence of the owner whose permanent home is elsewhere. However, the Borough Council do not consider a manipulative type chain ladder to be an acceptable method of escape and have cited paragraph 0.29 ('Criteria for means of escape') of the Approved Document in support of this view. They have also rejected your argument about occasional occupancy on the grounds that it is not relevant to consideration of means of escape.

It is the Department's view that what has to be considered in this case is whether the proposals will provide safe escape for the occupants of the second floor in a fire situation. Clearly the worst scenario would be where a fire occurs in the kitchen/dining room on the ground floor because it would put the stair out of action. The Department accepts that the balcony is available as an alternative route but this assumes that persons on the second floor are fit enough to use it, given its height above ground. In the case of elderly or very young persons they might need to be rescued. Furthermore, if the fire is in the kitchen/dining room rescue is not likely to be possible from the internal stair and they might therefore have to wait for external rescue. Persons in this situation are entitled to a reasonable degree of safety while so waiting and this can sometimes only be achieved by the provision of passive fire protection. Manipulative ladders do not provide such protection and in the Department's view the Borough Council's judgement was correct in dismissing them as a viable option.

The applicant has suggested the installation of an automatic opening vent; however, firm proposals have not been submitted with the application to the Department either with regard to the provision of such vent or to details of its capability for keeping the escape routes clear of smoke. Also it is doubtful in a domestic situation if the degree of maintenance would be guaranteed for such a sophisticated device.

As has been pointed out by the Borough Council, it is normal for a fully protected route to be provided in a new three storey dwelling. The Department considers that the Borough Council have shown flexibility in accepting the open plan arrangement at ground floor level. They have clearly taken account of the other factors in this case and the Department supports the judgement they have arrived at. Equally, the Department supports their argument that a further reduction in means of escape standard will prejudice safe escape and considers that the stair should have fire separation from the living accommodation at second floor level.

The Department also supports the Council's view that in considering the means of escape in case of fire the design of the building cannot be based upon the occasional occupancy of the building. What has also to be considered is the continuing use in the future by other occupants to whom the house could be sold at any time.

In these circumstances, therefore, the Secretary of State is of the opinion that the means of escape are not capable of being safely and effectively used at all material times and accordingly dismisses your appeal.

Alterations to flat – means of escape (B.1)

The local authority refused to dispense with the requirements of B.1 in connection with the internal alterations to an upper ground floor flat. The alterations will create an inner room and escape from this room is in dispute. Escape into an enclosed lightwell would not be acceptable.

Ref.: 45.3.51   Date: 23 March 1995

Careful consideration has been given to the facts of the case and to your representations and those of the Borough.

This appeal relates to the internal alteration to an upper ground floor flat consisting of the removal of a plasterboard wall between the entrance hall and the reception room to create an enlarged reception room which would have the effect of converting the bedroom at the end of the existing hall into an inner room. It is compliance of the means of escape from this inner room that is at issue.

There are two windows in the bedroom which have openable sizes that conform with the recommendations given in Approved Document 'B' ('Fire safety') for an inner room situation. These windows both open onto a lightwell. The floor of the flat is about 3 metres above the floor of the lightwell. No dimensions have been given for this well area but it is understood to be totally enclosed.

The Borough Council take the view that achievement of satisfactory means of escape would involve, in part, provision of an external escape stair. You have stated that the erection of such a stair would not be permitted by the housing association. You have therefore offered to provide safe escape by fixing a vertical ladder to the outside wall leading down to the floor of the lightwell; alternatively by providing a ladder to give upward escape to the roof which provides access to the other roofs of the flats and is protected by a parapet. You also state that you have been given to understand that all similar flats have made identical alterations.

The Borough have rejected the proposals on the grounds that safe escape from the inner room (bedroom) has not been provided. They do not consider that escape via a ladder to the roof to be an acceptable solution since ladders do not provide a safe means of escape. They also state that the roof escape is not adequately lit or protected in accordance with Approved Document 'K' ('Stairs, ramps and guards').

With regard to escape into the lightwell the Borough make reference to sub-paragraph 1.18(b) ('Windows and external doors for escape') of Approved Document 'B' which they interpret to mean that an external escape stair must be provided. This sub-paragraph also refers to diagram 1 ('Conditions for it to be acceptable for a ground or basement storey exit to lead to an enclosed space') of Approved Document 'B' which suggests that for escape into an enclosed space to be acceptable then the depth of space from the building to the rear wall should exceed the height of the building. The Borough Council state that this is not the case since the depth of lightwell is less than the height of building. They also consider that sub-paragraph 1.18(c) of Approved Document 'B' is applicable which gives guidance on the positioning of roof or dormer windows for escape purposes. In their view escape from the windows in question will rely upon rescue; and because the lightwell onto which they open is enclosed such rescue would be unavailable.

In considering cases of this type the Secretary of State is conscious of the large number of fires that occur in dwellings. In looking at this particular case he cannot take into account other neighbouring flat alterations which may have been carried out under different circumstances or previous legislation.

With regard to the proposal for escape via the roof the Department agrees with the Borough's view that this is unsatisfactory and likely to be unsafe for the reasons they have stated.

With regard to escape into the lightwell the Department does not accept the Borough's interpretation that Paragraph 1.18(b) suggests that an external escape stair is necessary from an inner room window of this height. The guidance does not state that stairs are required and sub-paragraph 1.16(e) ('Inner rooms') makes it clear that such windows are considered suitable for escape and rescue up to first floor level. Paragraph 2.11 ('Planning in flats where the floor is not more than 4.5 metres above ground level') of the Approved Document refers also to paragraph 1.18 where inner rooms are provided in flats that do not have a floor level in excess of 4.5 metres above ground level. In this case the floor is about 3 metres above ground level and therefore the window should be acceptable for emergency escape or rescue without an external escape stair.

The Department also does not accept the Borough's interpretation of sub-paragraph 1.18(c) in this case. The sub-paragraph refers to diagram 4 ('Position of dormer window or rooflight suitable for escape or rescue purposes from a loft conversion') which gives recommendations for the location of windows in a pitched roof for escape purposes. The flat is shown in the marked drawing No. 4 to have a flat roof and the windows in question are on an external wall. In the circumstances of this case the Department does not therefore consider this sub-paragraph to be applicable. Therefore with regard to the question of whether an external escape stair is necessary the Department supports your view and considers that Requirement 'B1' would be satisfied without such provision.

However, there still remains the issue of whether the lightwell is suitable for escape into and will constitute a place of safety. As has been pointed out by the Council, Diagram 1 of Approved Document 'B' suggests that an enclosed space may be suitable if the depth exceeds the height of the building. This is so that persons may await rescue at a safe distance from the burning building. In this case no dimensions of the lightwell have been provided but the Council have indicated in the drawing marked No. 1 that the lightwell is totally enclosed by walls. They have further stated that the lightwell would not appear to comply with the requirements of Diagram 1 of the Approved Document for enclosed spaces. The Department accepts these points but would observe that if the depth of the lightwell had equalled the height from ground to the roof of the building containing the flat, excluding the parapet, or if there had been a permanent exit route from the lightwell, the Secretary of State would have taken the view that compliance with Requirement 'B1' would be achieved. However, this is not the case.

In the circumstances, the Secretary of State is of the opinion that whilst there is satisfactory means of escape from the window of the bedroom to the lightwell without the need to install an external escape stair or the provision of fixed ladders, the enclosed lightwell does not provide a place of safety in accordance with Requirement 'B1' of the Building Regulations 1991, and accordingly dismisses your appeal.

Thatched house – external fire spread (B.4)

Using 'Thatchbatts' and fire retardent treatment to the thatch, the owner wanted to place an extension within 800 mm of a boundary. The approved document looks for a distance of 12 m whilst in this instance the local authority might have accepted 6 m. However, the appeal was lost because 600 mm is inadequate separation. Separation distances should not only rely upon whether thatch has received fire retardent treatment.

Ref.: 45.3.64   Date: 19 April 1995

Careful consideration has been given to the facts of the case, and to your representations and those of the District Council.

The building work involves the two storey extension to a thatched detached dwelling house. The extension is to have a thatched roof to match the existing roof. The new roof has a plan area of about 22 square metres and represents about 29% of the complete house roof as extended. The existing house is within 1 metre of the north side boundary with the adjoining residential property close to this boundary. On the south side the proposed extension will be located 600 millimetres from the boundary although the thatch overhang could decrease this distance. The adjoining residential property on this boundary side, house number 67, also has planning permission for an extension, although the thatched extension in question will only be adjacent to the driveway of this neighbouring property.

The construction details for the new extension show the placement of 'Thatchbatts' between rafters on the underside of the roof and the application of a fire retardant treatment to the thatch reeds.

The District Council rejected your plans on two grounds – one of which was non-compliance of your proposals to thatch the extension with Requirement 'B4' of Schedule 1 to the Building Regulations 1991. Your clients subsequently applied for a relaxation of Regulation 'B4' which was refused by the District Council. It is against this refusal to relax that you have appealed to the Secretary of State.

Within the documents submitted for this appeal you have included examples given by your thatch consultant of previous thatched roof appeals submitted to the Secretary of State where both underdrawing the roof and the impregnation of the thatch with a fire retardant have been used. You have also submitted the Loss Prevention Council's assessment report of the external fire exposure roof performance of a traditional thatched roof assembly which included the use of 'Thatchbatts' and the same proposed fire retardant treatment. This report concluded that the proposed thatch roof assembly would achieve an 'AA' rating if it were to be tested in accordance with British Standard BS 476: Part 3: 1958, 'Fire tests on building materials and structures: external fire exposure roof tests'.

With regard to the efficiency of the proposed fire retardant treatment, you refer to a test report on a ten year specimen which shows that such leaching out is limited. You have also referred to the thatching of the Globe Theatre in London and to examples of internal uses of the fire retardant in buildings.

Your clients understand the fire risk problem associated with thatch but have stated that any other form of roof cladding would be completely out of keeping, against planning conditions, and would have a detrimental effect on the visual appearance of the property.

Your clients have agreed to remove one chimney and would be prepared to consider removing the second chimney in order that no fire or sparks from open fires would affect the property. You consider that the chance of fire occurring is minimal and the thatched extension does not constitute a risk because there is only the neighbouring driveway adjacent to the proposed extension on the south side of the boundary.

The District Council has observed that the two storey thatched extension will be approximately 600 millimetres from the boundary when completed as opposed to the 6 metres recommended distance given in Table 17 ('Limitations on roof coverings') of Approved Document 'B' ('Fire Safety'). In their view the provisions made for the fire protection to the new thatched roof do not comply with Requirement B4(2) in that adequate resistance to fire spread over the roof, and from one building to another, has not been demonstrated. This is particularly so when Regulation 7 ('Materials and Workmanship'): Section 1 – 'Materials', which advises on ways of establishing the fitness of materials, is also taken into account.

The District Council have expressed doubt about the durability of the fire retardant treatment to be applied to the thatch, stating that in their view there is a tendency for this to leach out. They consider that any leaching out would make the top surface of the thatch susceptible to fire spread. They state that there is no fire test available on a leached sample. Without evidence to the contrary, they do not consider material with such a limited performance suitable in this instance.

The District Council have pointed out that they have, with suitable fire precautions, relaxed/dispensed Requirement B4 in the past where the distance to the boundary was marginally less than that recommended in Approved Document 'B'. However, they consider that the construction of a thatch roof at a distance of 600 millimetres to a boundary – possibly closer if the thatch overhang is taken into consideration – constitutes an unacceptable risk.

The precise content of your application to the District Council for a relaxation was in respect of Requirement B4, Table 17 of Approved Document 'B'. In considering your appeal against refusal of your application by the Council, the Department is bound to point out that it is the Requirement itself and not the guidance in the Approved Document which has to be considered. Consequently, whilst Table 17 of Approved Document 'B' can be taken into account as guidance when considering the case, it would not be appropriate for the Department to consider relaxing the Table itself whatever the merits of the case might be.

In your submission you have referred to a report from the Thatching Advisory Services Limited which instances previous appeal applications made to the Department. However, these cannot be taken into consideration since each case has to be looked at individually and on its merits. The company has also cited the treatment for the thatching of the Globe Theatre in London as an example similar to that referred to in this appeal case but the Department agrees with the District Council's view that this is not particularly relevant to this case.

With regard to adjoining properties, the Department notes that planning permission has been granted to erect an extension to house number 67. The current owners have expressed concern about fire spread to this property because of the proposed thatched roof extension of house number 65 which is the subject of this appeal. The Department considers that it is right that this concern is given proper consideration.

Although the Local Fire Authority have no legislative jurisdiction over dwellings, they were consulted in accordance with Section 15 (consultation with fire authority) of the Building Act 1984 by the District Council and consider the thatch roof extension to be unsatisfactory. The Department has taken note of this opinion.

Whilst the District Council appear to have been prepared to accept the new thatched roof at a distance of 6 metres to the boundary, the Department notes that the recommended distance given in the guidance in Approved Document 'B' (Table 17) for thatch in this situation is 12 metres. Notwithstanding that the 12 metres distance may be considered onerous in this case because of the proposed fire precautions, thatch does need to be located at a reasonable distance from boundaries to avoid risk of fire spread. The Department accepts that at present the adjacent property may not be exactly in line with the proposed thatched extension but it would be unfair to dictate conditions on any development that the adjoining owners may wish to make in the future.

In the Department's view the main purpose for setting limits on the minimum distances to boundaries is to prevent major conflagration between buildings. The Department notes that the existing house is already close to the boundary on both sides and concludes that the thatched extension will make the boundary situation worse.

The Department has taken into account the proposed fire precautions. With regard to the application of fire retardants to the thatch, the Department considers that there is no continuing control over domestic dwellings to ensure that the thatch is re-treated when necessary. The Department therefore supports the District Council's view that to locate a thatch roof within 600 millimetres of a boundary could prejudice fire safety.

In the circumstances, therefore, the Secretary of State has concluded that the proposed thatch roof will not resist the spread of fire over the roof and from one building to another, having regard to the use and position of the building. He accordingly dismisses your appeal.

Four storey nursing home – external escape (B.1)

The appeal is against a local authority refusal to accept unenclosed external stairs as a satisfactory means of escape. The Secretary of State held the same opinion as the local authority and dismissed the appeal.

45.3.66   Date: 25 May 1995

Careful consideration has been given to the facts of the case, and to the representations made by you and the Borough.

The appeal relates to a new four storey purpose built nursing home consisting of a lower ground floor and three upper floors; all floors having bedroom accommodation for the residents. A fully-protected internal stair is provided for normal access and escape. A lift is also provided for access but this has not been designed as an escape lift.

To provide alternative means of escape, an unenclosed external stair has been provided at the two ends of the building and it is the exposure of these stairs to the elements that is at issue. Openings from windows and doors where necessary have been up-graded to ensure that the use of the external stair cannot be affected by smoke and flames.

Other aspects of fire safety and means of escape comply with the recommendations given in Approved Document 'B' ('Fire Safety') for a building in purpose group 2a ('Institutional'). A sophisticated fire alarm system has been installed and corridors have additional fire doors provided to give increased compartmentation.

Conditional approval to your full plans application for this nursing home was given on the 5 August 1993. Your plans indicated that the external staircases would be clad and the second condition in the approval required that full constructional details of the glazed enclosures to the stairs should be submitted and approved by the Borough. You subsequently applied to the Borough for relaxation to Requirement B1 insofar as you no longer wished to cover the two staircases. Having consulted the Fire Authority and the Area Health Authority, the Borough Council refused your application on the 25 August 1994. It is against that decision that you appealed to the Secretary of State on the 27 September 1994. At that time the two staircases had been installed without any enclosures.

The Borough Council consider the enclosure of the stairs to be an essential feature of the means of escape in case of fire, and is necessary to meet the requirements of Part B. In support of their rejection of the proposals they make reference to paragraph 4.35 of Approved Document 'B' which advises that the use of external stairs in institutional buildings is only appropriate where the route serves only office or staff accommodation. In their view paragraph 4.36 of the Approved Document, which deals with the partial protection of external stairs and to which you have made reference, is not applicable because of the advice given in paragraph 4.35.

In the Borough's view the stairs in question, situated at the two ends of the building, are essential to the escape strategy of the building and would be used by residents in the case of emergency to reach a place of safety. In view of the level of care intended, they consider that it must be assumed that the majority of residents would require assistance during the escape process. The Borough Council consider that for this reason it is essential that the stairways are afforded the necessary protection, particularly in inclement weather. The lift is not an escape lift and has therefore been discounted for escape purposes.

The Borough Council have stated that the Fire Authority did not respond to their consultation on the relaxation application. However, at the time of submission the Fire Authority agreed with their interpretation of Part B of the Regulations, which included the enclosure of the stairs.

You consider that adequate means of escape to a place of safety outside the building, which is capable of being safely and effectively used at all material times, has been provided. You also contend that the Fire Authority stated at a meeting that the enclosure of the stairs was not necessary.

In support of your claim as to why enclosure of the two external stairs is unnecessary you have given the following reasons:

(i) The external stairs from the building are to be used in emergency situations only and are not for access by either staff, residents, or public visitors.

(ii) Residents will in most cases not be able to use the stairs by themselves and in an emergency all residents are likely to require assistance, thereby creating a much safer exit on these stairs. You consider that an enclosure around the stair could therefore actually hinder evacuation.

(iii) The stairs have chequer plate treads with raised safety grip surfaces and if required additional grip can be provided to the surface of the stairs by the application of an additional coating.

(iv) All steelwork is decorated on a regular basis to ensure weather protection is maintained.

(v) Paragraph 4.36 of Approved Document 'B' suggests that sufficient protection from the weather may be required although full enclosure of the stair may not be necessary.

(vi) The building itself provides a good degree of protection due to the positioning of the stairs in relation to the building and two out of the four floors have emergency exits direct on to solid ground, reducing the need to use the stairs.

(vii) The upgrading of the access lift to an escape lift has been offered in lieu of providing stair enclosures if this is considered necessary.

(viii) A previous determination case submitted to the Secretary of State for the Environment under Requirement 'K1' ('Stairs and ramps') is considered relevant to this case. In this instance the Secretary of State took the view that the council's demand for the enclosure of an external stair at a rest home went beyond Requirement 'K1'.

There appears to be some disagreement between yourself and the Borough Council as to whether the Fire Authority were in agreement or not with the two external stairs remaining unenclosed. Clearly the Fire Authority have been consulted but in the absence of any written statement from them the Department is unable to take account of either your or the Borough Council's opinion with regard to the views of the Fire Authority.

You have also suggested that the lift could be upgraded to become an escape lift. However, the Department can only consider the proposals as presented and in this case the appeal relates to whether two unenclosed stairs at the two ends of a nursing home are considered acceptable. It would be for you to reach agreement with the Borough Council as to whether these two external stairs could be dispensed with in favour of an escape lift.

Finally you have referred to a previous determination by the Secretary of State relating to whether an unenclosed stair at a rest home complied with Requirement 'K1' of the Regulations. Each appeal to the Secretary of State is considered on its own merits and in respect of the specific circumstances of the case. The Department does not consider the decision on this previous determination to be relevant to this particular case – particularly since it related to a 'K1' and not a 'B1' case.

You have appealed to the Secretary of State against the Borough Council's refusal to relax the functional requirements of Requirement B1 ('Means of escape') of Schedule 1 to the Building Regulations 1991. You have, however, specified in what respect you require a relaxation by referring to Approved Document 'B' and the guidance therein concerned with when it is, and is not, appropriate to have an unenclosed external staircase as a means of escape. The Department has therefore needed to consider whether Requirement B1 would be met without the enclosure of the two external stairs, thus leaving them unprotected from the elements.

The Department takes the view that it is important that the likely mental and physical condition of the occupants of the home is given full consideration. It was with these types of occupant in mind that the guidance given in paragraph 5.35 of Approved Document 'B' was formulated. This paragraph clearly recommends against the use of unenclosed external stairs for residents of institutional buildings. It follows that paragraph 4.36 is irrelevant to this case because it relates to buildings where external stairs are considered acceptable. The Department has also taken account of the location of the stairs in relation to the building. In all the circumstances, it is of the view that it could be hazardous to make residents of a nursing home escape down an open external stair.

Having given careful consideration to this case, the Secretary of State is of the opinion that adequate escape has not been provided and that the Borough Council's judgement in refusing your application to relax Requirement B1 was correct in the circumstances. Accordingly, he dismisses your appeal.

Air supported structure – means of escape (B.1)

The number of emergency exits and travel distances from an air supported structure covering tennis courts gave rise to this appeal. The decision accepted that in the circumstances the two emergency exits proposed satisfied the requirements of B.1.

Ref.: 45.3.69   Date: 18 October 1995

The appeal concerns the erection of an air supported structure which includes ground beam and pumped drainage. The intention is that the dome will contain either two full-sized tennis courts or eight short tennis courts. In general the design of the structure follows the guidance given in BS 6661: 1986 ('Design, construction and maintenance of single skin air supported structures').

The overall dimensions of the structure are shown on the drawings to be 34.2 metres by 38 metres. Drawing LCTC 4 revision 'B' shows two emergency exits on opposite sides of the structure and an entry airlock lobby adjacent to the exit giving access to the clubhouse. The emergency exits are provided with signage. It is the number of emergency exits and travel distances to them that is at issue.

You have stated that the number of persons (including squad training numbers and limited numbers of spectators) inside the air dome could be as many as 40 to 50 although this could be as low as two to ten persons at most times. You also state that the structure falls within class 2A of BS 6661 which permits up to 99 persons to be accommodated within this type of enclosure.

Your original full plans application was rejected by the Borough Council. One of the reasons for rejection was non-compliance with Requirement B1 ('Means of escape') of the Building Regulations 1991. Your proposed measures to ensure a satisfactory means of escape were considered inadequate. You then applied to the Borough Council under section 8 of the Building Act 1984 for a relaxation of Requirement B1. This was rejected on the grounds that the travel distances to the two emergency exits of 39 and 42 metres respectively would be excessive. It is against that decision that you appealed to the Secretary of State on 13 October 1994.

The Borough Council have stated that the structure is classified as being within purpose group 5 ('Assembly and recreation') of Table D1 of Approved Document 'B' ('Fire safety'). For this purpose group Table 3 of the Approved Document suggests that the travel distance where escape is available in more than one direction should not exceed 32 metres. The Borough Council also point out that whilst BS 6661 recommends that the distance between exits should not exceed 30 metres and that the maximum travel distance should also not exceed 30 metres, the travel distances in this case, even allowing for a 2 metres side wall standing clearance, are 39 metres in one direction and 42 metres in another.

The Borough Council state that the occupation levels of the enclosure are in the order of 40 to 50 persons with a possible 64 persons for short tennis, and that no other features such as managerial or staffing criteria have been put forward to compensate for the extended travel distances. They also consider that there is a possibility of the area being used for special or other non-tennis club related activities – in the event of which occupancy numbers could considerably increase.

The local Fire Authority have stated that they cannot support any application for a relaxation that would excessively extend travel distances in a collapsible structure.

You consider that a structure of this size and intended use (two tennis courts or eight short tennis courts) is not likely to have more than 50 persons in it at any one time. You consider two emergency exits to be adequate and have given examples of other similar structures erected elsewhere. You had also agreed to obtain a letter from the Tennis Club stating the maximum intended occupancy of the structure; however, such a letter has not been provided as part of this submission.

You state that the principal purpose of your appeal is to establish whether or not BS 6661 intended for there to be no more than 30 metres between emergency exit doors or whether no part of the air structure should be more than 30 metres from an emergency exit door.

The Department's analysis of this case is as follows. Table D1 of Approved Document 'B' suggests that the type of establishment concerned in this case should be classified as 'Assembly and recreation'. Table 3 of the Approved Document suggests that for this purpose group travel distances should not exceed 32 metres where escape is available in more than one direction. In this case the Borough Council have stated that the travel distances are significantly greater than this.

The Department has therefore needed to consider the possible threat to life safety that could be caused by the increase in travel distances. The structure is primarily intended for playing tennis although the Department accepts that there will be variations in the layout of the courts. There would appear to be some disagreement between the parties as to how many persons will be in the enclosure at any one time. The occupancy figures range between 2 and 99 persons. However, the Department takes the view that the occupancy figures you have put forward of between 40 and 50 persons are realistic, and it is prepared to accept these.

The recommendations given in the Approved Document for travel distances in assembly buildings have to take account of the varying physical abilities of large numbers of people who may be unfamiliar with escape routes. In this case it can be expected that a high proportion of the occupants will be tennis players who are not likely to cause a problem for escape by being concentrated in any one particular area of the courts. The drawings you submitted indicate that the most likely position for spectators to be located is at the two opposite ends of the structure nearest the two emergency exits, which in the Department's view have adequate fire exit signs.

In view of the situation in this case the Department is of the opinion that the two emergency exits, as shown on drawing LCTC 4 revision 'B', satisfy Requirement B1 of the Building Regulations 1991.

In all the circumstances, therefore, the Secretary of State is satisfied that the building work in question (i.e. that which is in accordance with drawing LCTC 4 revision 'B') complies with Requirement B1 of Schedule 1 to the Building Regulations 1991. Accordingly, he declines to uphold your appeal on the grounds that it is not necessary to consider the case for relaxing the requirement in question.

You should note that appeals are considered on their individual merit and in the light of the specific circumstances of the case. This appeal decision should not therefore be regarded as establishing a precedent.

As part of your appeal submission you asked for clarification of the guidance given in the British Standard relating to travel distances to emergency exits. However, it is not for the Secretary of State to comment on British Standards within the context of an appeal submission relating to the requirements of the Building Regulations. If you wish to pursue the question of clarity of the guidance given in a British Standard then you may wish to write to the British Standards Institution.

Four storey house – door to the kitchen (B.1)

Means of escape from a four storey house, where the top storey has recently been formed by building a mansard extension is in dispute in this appeal. The Council wanted a self-closing FD20 fire door to the lower ground floor kitchen, and the Secretary of State agreed that the local authority was right to refuse a dispensation.

Ref.: 45.3.71   Date: 18 October 1995

The appeal relates to a four storey terraced house in which the top storey has recently been formed by the construction of a mansard extension. The house now consists of three above ground storeys and a lower ground storey.

With the exception of the kitchen door at lower ground level, all other existing doors have been replaced with self-closing FD20 fire doors. A linked mains operated smoke detection system has also been installed. At present the kitchen is fitted with a sliding door because of the lack of space. The door is not fitted with a self-closing device, although you state that it has been fire protected.

The work was progressed on the basis of a Building Notice together with various plans, etc. You were advised by the Council that 'The room to the new storey and all habitable rooms and the kitchen opening onto the stairway shall be provided with minimum FD20 fire resisting and self-closing doors, as indicated on the drawings'. You resisted complying with the Council's requirement in respect of fire resisting the kitchen door and making it self-closing.

You were then served with a section 36 Notice under the Building Act 1984 on 21 June 1994 requiring you to carry out the specified work to the kitchen door. You responded to the Notice by requesting a dispensation from Requirement B1 which was refused by the Council. After a subsequent exchange of correspondence with this Department you formally appealed to the Secretary of State against the Council's decision not to dispense with Requirement B1.

In support of their decision to reject your application for a dispensation, the Borough Council quote paragraph 1.20(a)(i) and (ii) of Approved Document 'B' ('Fire safety'). They point out that because of the scope of the works carried out, the existing doors have been changed to self-closing FD20 doors. In their view the kitchen door should also be fitted with a self-closing device because kitchens constitute a higher fire risk.

The Borough Council state that they would have no objection to a sliding door being retained provided that the existing door is adapted so that it achieves not less than 20 minutes' fire resistance, and that an adequate self-closing device is fitted which would render the door self-closing in such a manner that the door is effectively smoke stopped against the opening. They clearly state that they are not prepared to accept any alternative other than an FD20 self-closing fire door to the kitchen. However, they have suggested other ways of achieving this specification such as by fitting double doors or relocating the door.

Your application for a dispensation and this subsequent appeal was made on the grounds that you object to the use of any form of self-closing device on the lower ground floor kitchen door. You consider the fitting of such a device as an imposition. In support of your case you state that the kitchen door is currently a sliding door type which has been fitted in order to minimize the use of space in a small room and that the fitting of a hinged door in this limited space would be impracticable. Furthermore, you consider the kitchen door to be superfluous because it is never closed for personal and domestic reasons. You also suggest that other houses in the same street have had the doorway completely removed to create an open plan effect.

You state that the kitchen door has been lined to give fire protection and that mains operated smoke detectors have been fitted on all floors. You also state that while all the other doors in the house have been changed to fire check doors fitted with self-closing devices, their operation has proved to be impractical.

In considering cases of this type, the Secretary of State takes the view that it is important to provide for the safe escape of occupants. He is very conscious of the fact that a large number of fires occur in dwellings. The Department has therefore needed to consider the safety of the occupants of the house who may have to use the single stair to escape in a fire situation.

The Department notes that the house has three above ground storeys plus the lower ground storey which contains the kitchen. Paragraph 1.20 of Approved Document 'B' suggests that in a house of this height the stair should be a protected stair. The Department also notes that the Borough Council have pointed out to you the dangers of not fitting a self-closing device to the kitchen door and why reliance cannot be placed on a smoke detection system alone.

You have raised the point that other similar properties in the same street have had the kitchen door removed. However, the Secretary of State is not in a position to take account of other circumstances. Each case put to him on appeal must be considered on the basis of its individual merit and the particular circumstances of that case.

The Department understands the need to minimize the use of space in your kitchen by installing an appropriately constructed door. However, it is the safety of the occupants and the need to ensure that the single stair is protected which is the paramount consideration. It is therefore of the opinion that the omission of the self-closing device from the kitchen door would prejudice safe escape in the event of fire.

Having given careful consideration to this case, the Secretary of State is satisfied that the Borough Council came to the correct decision in refusing your application for a dispensation. Accordingly, the Secretary of State dismisses your appeal.

Basement storage area – means of escape (B.1)

The reason for this appeal was the provision being made for means of escape from the basement storage area newly enlarged as part of works of alteration to a retail store. The Council would not accept extended travel distances being created and the Secretary of State supported this stance in his decision.

Ref.: 45.3.68   Date: 8 December 1995

The building work to which this appeal relates has involved works of refurbishment to, and the bringing into greater use of, the existing basement area of a retail store.

The basement area comprises two approximately equal but separate longitudinal areas with one access mid-way joining the one to the other. The previous occupant used only a small separated part of the basement but the whole area has now been fully refurbished so that the two basement areas are now in full use as a retail area and as a stockroom. The latter is accessed from the retail area via a fire resisting lobby which leads to a further fire protected lobby area before the stockroom is entered. The storey exit is from this protected area to a lightwell where stairs are available for escape to the ground floor. Work is now complete and the store is in operation.

Apart from giving access to the retail area and toilets the protected area also gives access to an office and the staff rest room/canteen. Each of these rooms is stated by the appellant to be provided with FD30S fire doors. He also states that he is prepared to install smoke detectors in both these rooms and in the retail area outside the lobby. The installation of the detectors is, however, dependent on the outcome of the appeal.

The Council rejected your proposals on the grounds that they do not show compliance with Requirement B1 ('Means of escape') of the Building Regulations 1991, because of the extended travel distances from the stockroom to the storey exit. You then applied to the Council for a relaxation to Requirement B1 which they refused. The Council have not sought the comments of the Fire Authority on your application for a relaxation on the grounds that the work shown on the plans was considered to be in contravention of B1.

The Council have offered the following points in support of their decision to reject your application for a relaxation:

(i) one of the doors opening onto the escape corridor/lobby is from the staff canteen/rest room. The Council consider this room to be at equal or greater fire risk than the retail area;

(ii) the two doors in question may not in fact be properly referred to as FD30S since they have been modified on site as a result of the removal of air transfer grilles;

(iii) during each visit made to the store by the Council, the door to the staff canteen has been found to be wedged open. This is in spite of the assurances that management have given that good practice will be adhered to;

(iv)   the escape corridor in question continues to be used as a staff cloakroom and locker room; and

(v)   the travel distance from the basement exceeds the maximum referred to in Approved Document 'B' ('Fire Safety'). Therefore a safe protected corridor/lobby arrangement was proposed by the Council as a compensating feature. To complete the protection of this lobby the Council proposed that the two doors, one serving an office and one serving a staff canteen, should either be re-located or provided with double door protection. The second option would rely on the provision of linked smoke detectors in the rooms concerned.

These precautions were considered necessary because any person working in the storeroom would have no alternative means of escape, and rescue was considered difficult because this is a basement area.

In support of your appeal against the Council's decision to refuse to relax Requirement B1 you make the following points:

(i)   your request for a relaxation is made in respect of that part of the recommendation in Approved Document 'B' which deals with dead-end travel distances. The issue relates to the dead-end condition from the basement stock room;

(ii)   the 18 metre actual travel distance limit given in Table 3 of Approved Document 'B' is satisfied but only up to the point where the occupants of the stockroom would enter the fire resisting corridor/lobby;

(iii)   you consider that the major risk to this dead-end condition is from the basement retail area which is separated from the lobby by two sets of fire resisting doors and screens;

(iv)   it is considered that the standard of fire resistance and proposed early warning system is sufficient compensation to offset the additional few metres' travel within the lobby;

(v)   the Council's view that the doors opening into the corridor/lobby could be wedged open should not be taken into account. You have stated that both management and staff have been acquainted with the need to keep them shut, and that the manager has been informed that the practice of wedging the staff room door open must be stopped; and

(vi)   you are confident that the scheme provides reasonable means of escape and is in compliance with Requirement B1.

The Department observes that refurbishment of the basement area in this case has resulted in extended travel distances from the stockroom to the storey exit. The storey exit being the exit to the lightwell where stairs are available for safe escape to the ground floor. Actual travel distances have not been quoted by either yourselves or the Council. However, the Department estimates that the maximum travel distance from the stockroom to the storey exit is approximately 24 metres. Table 3 ('Limitations on travel distance') of Approved Document 'B' recommends a maximum travel distance of 18 metres where escape is available in a single direction only.

To overcome the travel distance problem the Council had proposed that a protected corridor/lobby be provided between the storey exit and the stockroom. The intention was that this should be imperforate except for the door opening to the toilets and the separate lobby approach to the basement sales area. As an alternative they were prepared to accept the doors from the staff room and office opening into the enclosure, provided lobby protection was provided to both rooms. The installation of linked smoke detectors was also a condition for this second option. You have offered single door protection to each room and a smoke detection system.

You have lodged an appeal with the Secretary of State against the refusal of the Council to relax the functional requirements of Requirement B1 insofar as they affect the single direction travel from the basement. The precise content of your application to the Council was in respect of the recommendation in Approved Document 'B' on travel distances. However, in considering your appeal against refusal of your application by the Council, the Department is bound to point out that it is the Requirement itself and not the supporting guidance in any of the Approved Documents which has to be considered. In this particular case, although Table 3 of Approved Document 'B' can be taken into account as relevant, it is nevertheless only guidance and as such it would be neither logical for the Department to consider relaxing the Table nor indeed is there any statutory procedure for doing so.

Regulation 8 of the Building Regulations has the effect of constraining the requirements of Requirement B1 to the purpose of securing reasonable standards of health and safety for persons in and about buildings. It is the Department's view that it is therefore very difficult to substantiate a case for relaxing the purpose of such a requirement.

In this particular case the area in question is in a basement where rescue and smoke dispersal may be more difficult to achieve than in an 'above ground' situation. If the proposed additional provision of smoke detectors and single door protection could be judged to be an acceptable alternative to the extended travel distance, then your proposals would comply with Requirement B1 and there would be no call for the Secretary of State to consider your appeal other than by declining to uphold it on the grounds that it was not necessary for him to address the issue of relaxation. However, in this particular case it is the Secretary of State's view that your additional proposals do not constitute adequate provision to achieve compliance. Having full regard to this, and the fact that Requirement B1 is concerned only with health and safety, and is a life safety issue, he is therefore of the view that the Borough Council came to the correct decision in refusing your application to relax Requirement B1. He accordingly dismisses your appeal.

You should note that the Secretary of State's decision on this appeal is based on your design proposals and not the management procedures of the store. The latter are matters for you to consider under continuing control procedures as appropriate.

Seven storey building – means of escape (B.1)

This appeal relates to the provision of means of escape from an existing seven storey building to be divided into two separate parts, both of which will be converted into self-contained flats. It was decided that a fire near the stairs in a lower storey could be a hazard to escape, and therefore a ventilated lobby between each flat and the single stairs was necessary.

Ref.: 45.3.78   Date: 18 March 1996

The building to which this appeal relates is an existing six storey building plus a part-basement storey. The property was previously a hostel with one internal staircase. Means of escape was by this staircase together with an external one.

The proposed building work comprises the division of the building into two and conversion of both parts into self-contained flats. Both parts are to rely on a single stair for access as well as escape. Both stair enclosures also contain a lift for access purposes, which are not the subject of this appeal. The proposal shows one flat on the basement and ground floor and three flats on each of floors one to five. The remaining space on the basement and ground floors is intended for use as garage parking. The existing internal stair at one end will serve two of the flats on each of the upper floors. A new internal stair is proposed at the other end which will serve the third flat on each floor. The division of the building into two parts will mean that there is no means of access between the two flats at one end and the third flat at the other end of each of the upper floors. Consequently the two stairs in the building will be single escape stairs.

The proposal is to pressurize the stairs and provide internal lobbies within each flat. The proposal does not incorporate provision of a common ventilated lobby between the escape stair and the entrance doors to the flats.

You gave a Building Notice to the Borough Council in respect of the proposed work. They required you to provide a common ventilated lobby between the escape stair and the entrance doors to the flats in accordance with British Standard BS 5588 Part 4: 1978,

'Code of practice for smoke control in protected escape routes using pressurization', in order to comply with Requirement B1 ('Means of escape') of the Building Regulations 1991 (as amended). Under the particular circumstances of this case you considered this neither necessary nor readily achievable. You therefore sought a relaxation of the Requirement B1 from the Borough Council which they refused. It is in respect of that refusal that you have appealed to the Secretary of State.

The Department notes that you are formally requesting the Secretary of State to relax the need for common ventilated lobbies between the pressurized stair and the entrance door to the flats. However, you should note that whilst the Secretary of State must clearly take into account the question of the provision of these lobbies, he can only give formal consideration to the relaxation of Requirement B1, 'Means of Escape', itself which makes no specific reference to common ventilated lobbies.

In support of your appeal you have questioned the need to provide a common lobby to the staircase in view of the fact that each flat has an internal lobby and the staircase will be pressurized. In addition, you have questioned the need to provide such a lobby with automatic opening vents. In this respect you have asked the Department to note that if a common lobby is provided the existing staircase shaft does not have an external wall and therefore has no windows – for this reason it is proposed to pressurize the stairs.

You have also pointed to the fact that existing buildings often have physical restrictions which do not allow every regulation to be met and, in your view, relaxation ought to be granted if it does not pose a danger to the occupants. In these circumstances you consider that a relaxation should be granted.

The Borough Council accepts that the building does not lend itself to the normal requirement of ventilated lobby separation between the horizontal and vertical escape routes. In their view the only solution was to pressurize the staircase in accordance with BS 5588 Part 4: 1978.

They point out that this British Standard states that pressurization is in lieu of common lobby ventilation requirements only and that other fire protection arrangements are still necessary. They do not feel that they can support the omission of common lobbies themselves and have refused to relax this design requirement.

In this case the Secretary of State takes the view that he is being asked to relax Requirement B1 only insofar as it is the Borough Council's view that common lobbies (albeit unventilated) are required in addition to a pressurized stair in order to meet this Requirement.

The Department's analysis of this case is as follows. This building is a six storey building (plus a part basement storey) and the height of the top floor is approximately 16 metres above ground level. This is above the height that a fire authority could reasonably be expected to facilitate external rescue. The Department takes the view that a building of this height should not be permitted the same concessions given to smaller single-stair buildings. Diagram 10 ('Flats or maisonettes served by one common stair') of Approved Document 'B' ('Fire Safety') sets out the basic recommendations for a single-stair building where the height of the top floor is in excess of 11 metres above ground level.

In a building of this height the normal recommendation is for an automatically ventilated common lobby between the stair and the flats. The ideal solution would be for alternative means of escape to be provided but the Department accepts that this is not possible in this case. However, in view of the fact that only single direction escape is available the Department considers it reasonable for lobby protection to be provided to each single stair.

In some situations pressurization may be considered as an alternative to lobby ventilation and the Borough Council appears to have accepted this. However, the Department agrees with the Borough Council in that pressurization should not be regarded as an alternative to providing a common lobby between the single stairs and the entrance doors to the flats. In a building of this height it is important that adequate protection is provided to the single stair. The Depart-

ment, therefore, does not support your view that this can be achieved by the provision of internal lobbies in the individual flats rather than a common lobby, and that your proposals would be sufficient not to pose a danger to the occupants.

In all these particular circumstances, the Secretary of State has therefore concluded that the Borough Council came to the correct decision in refusing your application for a relaxation of Requirement B1 ('Means of escape') of the Building Regulations 1991 (as amended) and hereby dismisses your appeal.

It is understood that, since lodging your appeal to the Secretary of State, you have agreed with the Borough Council a solution to the design problem which formed the subject of this appeal.

Shop – fire escape after refurbishment (B.1 and B.3)

This appeal relates to the provision of means of escape from a shop created by alteration and refurbishment, from two shops involving three accommodation stairs in a five storey mid-terrace building.

The floor area is disputed and the decision emphasizes that it is necessary to include all the space including that occupied by ancillary rooms and display areas.

Ref.: 45.3.76   Date: 5 June 1996

The building work comprises alterations at basement and ground floor level to convert two small shops into one larger unit by forming large openings in the former separating wall. In addition to the retail areas on the basement and ground floor levels, new ancillary accommodation has also been formed within the existing basement area. This ancillary accommodation consists of an office, WC, display area, coats room, staff room and three stock rooms.

There are three accommodation stairs connecting the ground floor sales area with the basement which are provided for both access and escape purposes. Two of these stairs discharge on the ground floor near the rear of the shop and the third discharges approximately 3 metres from the shop entrance. Escape from the ground floor is only available from the front of the shop and is via two adjacent entrance doors.

The first full plans application relating to this case was rejected by the Borough on the ground of lack of information. A further full plans application showing the detailed information on shop fitting was rejected on the grounds that the united shops would not comply with the requirements of the Building Regulations 1991 where previously they had done so. A resubmission was rejected; but this was followed by the giving of a Commencement Notice. A letter from you requesting a 'waiver' on the matter of non-compliance of the front staircase, was taken by the Borough to be an application for dispensation of Requirement B1 ('Means of Escape') and Requirement B3 ('Internal fire spread (Structure)') of Schedule 1 to the Building Regulations. The Borough refused this application stating that all three staircases should be within protected shafts. It is against that refusal that you have appealed to the Secretary of State.

The Appellant's case

In support of your appeal you state that you have chosen to use the guidance given in BS 5588: Part 2: 1985 ('Code of practice for shops'), as an alternative approach to the relevant recommendations given in Approved Document 'B' ('Fire Safety'). Paragraph 9.2.4 ('Stairways') of BS 5588: Part 2 states:

(a) In a small shop comprising a bar or restaurant, all stairways should be protected stairways discharging to a final exit.

(b) In a small shop that is not a bar or restaurant, stairways should be protected stairways discharging to a final exit except that a stairway may be open if it does not connect more than two storeys, and delivers into the ground storey not more than 3 metres from the final exit, and either

(1) the storey is also served by a protected stairway; or
(2) it is a single stairway in a small shop with the floor area in any storey not exceeding 90 m² and, if the shop contains three

storeys, the stairway serving either the top or the bottom storey is enclosed with fire-resisting construction at the ground storey level and discharges to a final exit independent of the ground storey.

The main elements of your submission in support of your appeal are that:

(i)   The premises can be classified as a small shop.
(ii) The public areas of each storey are 90 square metres or less.
(iii) The means of escape are justified under Section 9 BS 5588: Part 2: 1985 ('Code of practice for shops').

In making your representations you particularly wish to bring to the Department's attention the following drawings:

(i)   MCH 1 – Plan of fixed permanent displays (the plan now marked No. 1).
(ii) MCH 2 – Area plan and schedule (the plan now marked No. 2).

You contend that both the ground and basement floors fall within the 90 square metre criterion given in the BS Code. You state that the floor area of the basement is 72.8 square metres and that of the ground floor 90.8 square metres. You contend that your measurements of the floor area reflect the true thickness of the walls which incorporates the permanent display area as indicated on the above drawings. You consider that it is the usable floor space which is relevant; and because of the permanent display areas the usable space has, in your view, been considerably reduced. The result of the store construction is that the floor area which can be occupied by customers on the ground and basement floors does not exceed 90 square metres.

Because in your view the premises constitute a small shop with limited floor capacity you believe that your provisions for escape are justified under Section 9 of BS 5588: Part 2. You point out that your premises comprise three stairs leading from the basement to the ground floor from which escape to the street is via two exits. All stairways and exits are clearly seen and easily accessible. You argue, moreover, that the actual travel distances comply with the recommendations given in BS 5588: Part 2 and are as follows:

(i)   Ground floor – 23 m (measured to a storey exit).
(ii) Basement – 14 m (measured to the foot of the open stair).

In particular you are therefore of the opinion that the open stairway discharging to the ground floor within 3 metres of a final exit should be permitted. Additionally in your view the three stairs are sited in such a way (i.e. two at the rear and one in the centre) as to afford alternative direction of travel from the basement.

The Royal Borough Council's case

The Borough have made the following main points in support of their decision to reject your application for dispensation:

(i)   The floor area should be measured to the inner surface of the enclosing walls. On this basis the floor area of the basement is 189 square metres and that of the ground floor 177 square metres. In their view therefore the premises are twice the area of a shop for which a reduced standard is allowed under paragraph 9.2.4(b)(2) of BS 5588: Part 2, and therefore a protected stair is required.
(ii) The three open stairs are not satisfactory for escape. A person escaping from the basement area is not presented with alternative safe routes of escape as these are within close proximity. If they should choose one of the rear staircases they would be presented with a travel distance of 18 metres at ground floor level.
(iii) The exits to the street at ground level are in close proximity and cannot be regarded as two exits for the purposes of escape. Consequently the travel distances quoted by the applicant are considered to be irrelevant.

(iv) Approved Document 'B' states that all floors over shop basements should be constructed as compartment floors and therefore any opening must be within a protected shaft. They consider this to be in conflict with the British Standard. However, they are prepared to accept open stairs providing the basement storey is served by one protected stairway discharging to a final exit as required by paragraph 9.2.4(b)(1) of the BS Code.

The Department notes that compartmentation had been raised as an issue by the Borough because Approved Document 'B' suggests that a floor above a basement should be a compartment floor. This would therefore mean that all stairs need to be enclosed in a protected shaft. However, the Borough are prepared to accept open stairs providing the basement storey is served by one stairway which is protected and which discharges to a final exit, as suggested in paragraph 9.2.4(b)(1) of the BS Code.

The Department's views

It is the Department's view that the main issue in this case centres on whether the premises can be considered to be a small shop in terms of paragraph 9.2.4 of BS 5588: Part 2, and how the floor area is measured. The Department accepts that the British Standard can be used as an alternative approach to the guidance given in Approved Document 'B'. However, the Department is not responsible for producing British Standards and it is therefore not for the Department to give interpretation of them. The Department can only consider whether your interpretation of the Code satisfies the functional requirements of the Building Regulations 1991, which in this case is Requirement B1 ('Means of escape').

You consider that three open stairs are acceptable for means of escape from the basement because the floor area is under 90 square metres. This measurement of floor area not only excludes all ancillary rooms and areas but also excludes the element of floor area occupied by permanent display units. The Department notes that the actual total floor area of the basement storey as a whole is nearly double this figure.

In excluding all ancillary rooms and areas from the floor area calculation, you are in effect ignoring the means of escape from such areas. The Department takes the view that means of escape from the ancillary areas of a basement cannot be ignored. If it is your wish that ancillary areas should not be included in the floor area, then in the Department's view such areas would need to be fully compartmented from the sales area and provided with independent escape routes.

The Department agrees with the Borough Council's view that your basis of measurement of the floor area of the shop is unacceptable and that the shop cannot be considered to be a small shop within the definition contained in BS 5588: Part 2. It follows that it is also unacceptable for all three stairs from the basement to be open to the accommodation.

The Secretary of State's decision

The Secretary of State has given careful consideration to all the facts of this case. In the particular circumstances of this case he has concluded that the Borough came to the correct decision in refusing to dispense with Requirements B1 and B3 of Schedule 1 to the Building Regulations 1991 (as amended) with respect to your proposed installation of three accommodation stairs. He accordingly dismisses your appeal.

In giving careful consideration to the technical detail of this case the Department has considered what design solution would achieve compliance with Requirements B1 and B3. It may therefore be helpful to you if the Department observes that in its view compliance with Requirements B1 and B3 would be achieved if one of the three stairs was to be protected, as suggested by the Borough.

Loft alteration – storage room (B.1 and B.3)

Here a storage area has been created in the roof of a two storey dwelling, with finishes and fittings which would make possible its use for habitable purposes. It could be used in the future as a playroom or study. In these circumstances means of escape and internal fire spread requirements apply.

Ref.: 45.3.83   Date: 15 August 1996

Your proposals involve the alteration of the roof space of a two storey semi-detached house to form what you state to be a storage room. The new loft room is insulated, lined with plasterboard, and has been provided with light fittings and power points. The floor of the new loft room has been strengthened and the Borough Council have confirmed that this upgrading satisfies Approved Document 'A' ('Structure') with regard to domestic loading.

The drawings submitted with the appeal show an existing loft window in the gable wall and also indicate that a new double glazed velux rooflight is to be installed. Access to the new room is via a pull down loft ladder. In addition, the Department estimates from the drawings provided that the floor area of the new room is approximately 19 square metres and the height at the centre of the roof pitch is approximately 1.9 metres.

It is understood from the Borough Council that the work was substantially completed prior to you submitting full plans for approval under the Building Regulations dated 10 July 1995. This was followed shortly by an application to the Borough Council to dispense with Requirements B1 and B3. The Borough Council rejected your full plans application on grounds of non-compliance with Requirements B1 and B3; and also rejected your application for a dispensation of both requirements. However, you hold the view that because the intention is to use the loft space as a store room these requirements in the regulations are not applicable. It is against the refusal by the Borough Council to dispense with Requirements B1 and B3 that you have appealed to the Secretary of State.

The Appellant's representations

In support of your appeal against the Borough Council's refusal to dispense with Requirements B1 and B3 of the Building Regulations 1991 (as amended), insofar as they relate to the conversion of the loft space to a room, you have stated that you do not consider that Requirements B1 and B3 are applicable in this case since the work that has been carried out was only intended to upgrade an existing facility. You add that the loft area will be used for exactly the same purpose as it was before the work was carried out – i.e. the storage of household items which are superfluous to everyday requirements.

You state that if at any time in the future it was decided to change the use of the loft space from storage to habitable accommodation, then the relevant requirements of the Regulations would be implemented by your clients. You also contend that there is insufficient head room to use the loft space as a bedroom or other habitable room.

You therefore consider that it is unjust for the Borough Council to have refused building regulation consent for what is not in fact a change of use but the continuation of existing usage in a safe and clean way.

The Borough Council's representations

The Borough Council have refused your application to dispense with Requirements B1 and B3 on the grounds that it would be inappropriate given the potential risk to life involved.

In considering the submitted plans and dispensation application, the Borough Council took account of the room size and standard of finish. They were of the opinion that although headroom is restricted for adults, the room could be used for habitable purposes, particularly for children.

In support of their refusal to allow your application for dispensation, the Borough Council also offered the following observations:

(i) The new room has plastered walls and ceiling and is provided with electric lighting and power points.

(ii) The room is accessed by a retractable loft ladder.

(iii) No provision has been incorporated within the first and second floors to provide the 30 minutes' fire resistance suggested in Approved Document 'B', in order to satisfy Requirement B3.

(iv) No provision has been incorporated to provide a safe means of escape from the second floor room to an exit at ground floor level in order to satisfy Requirement B1.

The Borough Council also consulted with, and enclosed the comments from the Fire Prevention Officer. He had no comments to make in respect of the area being used for storage purposes only, but states that the proposed alterations would not be satisfactory for a habitable room of any sort.

The Borough Council are therefore of the opinion that it would be inappropriate to dispense with such fundamental requirements as means of escape and internal fire spread, considering the risk to life.

The Department's views

In considering cases of this type the Department must consider the safety of the occupants of the house who may have to escape from the loft space as altered in a fire situation and what the implications would be of relaxing or dispensing with Requirements B1 and B3.

You have stated in this case that the alterations to the loft space are intended to cater for improved storage facilities. What the Department has therefore needed to consider is whether it would be reasonable and safe in these particular circumstances to dispense with Requirements B1 ('Means of escape') and B3 ('Internal fire spread – structure').

The Department notes that the standard to which the roof space is being altered is high, given that the stated intention is to use it for storage only. You have also stated that should the use change from storage to habitable accommodation then alterations would be carried out so that the building regulation requirements are satisfied for the new purpose. However, it is important to recognize that in this instance there will be no on-going control and that even if your clients as the present occupants of the property were willing to upgrade the new second floor accommodation, future occupants might not be.

Whilst the Department accepts that there is some limitation on head room in the new loft room, it considers that the standard of accommodation is such that the new room could be put to frequent use, such as for a children's play room or study. If this were to be the case then it would be reasonable for occupants of the new room to expect that there should be provision of a safe and permanent access and means of escape, at least to the standard suggested for loft conversions in paragraphs 1.23–1.31 of Approved Document 'B' ('Fire Safety').

With regard to Requirement B3, again the Department is of the opinion that it would be reasonable for any occupants of the new room to be provided with structural fire protection at least to the standard suggested in the Approved Document for loft conversions.

Given the particular circumstances of this case, the Department considers that it would not be appropriate to dispense with Requirements B1 and B3. It also observes that it accepts the Borough Council's view that the alterations to the loft space do not show compliance with the Building Regulations in respect of Requirements B1 and B3.

The decision

You have appealed to the Secretary of State against the decision by the Borough Council not to dispense with Requirements B1 and B3 of the Building Regulations. He regards both as life safety matters and would not therefore normally consider it appropriate to relax or dispense with these requirements. Additionally he is very conscious of the fact that a large number of fires occur in dwellings.

The Secretary of State has given careful consideration to all the facts of this case and has concluded that they do not justify either the relaxation or dispensation of Requirements B1 and B3. He has therefore concluded that the Borough Council came to the correct decision in refusing your application to dispense with requirements B1 and B3 and accordingly dismisses your appeal.

Retail arcade – means of escape (B.1)

Alterations and extensions are to create a retail arcade and cinema development. There is a dispute regarding the means of escape from the gallery/mezzanine level. The developer wished to rely upon accommodation stairs rather than a totally protected stairs and this the Council did not accept, and suggested an adequate smoke extraction/ventilation system.

Ref.: 45.4.12   Date: 27 June 1996

Section 13 of the Hampshire Act 1983 ('Fire precautions in certain large buildings') relates to either: the erection of a building of the warehouse class or which is intended for trade or manufacturing use and which exceeds 7000 cubic metres; or to the erection of a building so used or intended to be used which as extended would exceed 7000 cubic metres.

Section 13(2)(a) of the Hampshire Act 1983 provides the discretionary power that a district council may reject plans and particulars unless it is shown to their satisfaction that the building which is the subject of the operation will be provided with: '(i) fire alarms (whether automatic or otherwise) and a fire extinguishing system, or either such alarms or such system; and (ii) effective means of removing smoke in case of fire'.

Section 13(5) of the 1983 Act provides that a person who is aggrieved by the action of a district council which has rejected plans, or imposed conditions, may appeal to the Secretary of State for the Environment.

The proposal

The appellants' proposal is to construct a single storey retail store as part of the third phase of a shopping development. They state that the floor area of the retail area is 3310 square metres and that there is a storage and ancillary accommodation area at the rear occupying 546 square metres. The total area is therefore stated by the appellants to be 3856 square metres and the compartment size is stated by the Fire Authority to be in excess of 17000 cubic metres. However, the appellants have since confirmed in their letter of the 13 December 1995 that the storage and ancillary area will be compartmented from the retail area. The Department therefore takes the view that the floor area in question is 3310 square metres.

The roof height is shown on the drawing to vary between approximately 5.5 m and 8.5 m and the proposal is to incorporate automatic fire detection. The proposal does not incorporate any provision for a sprinkler system. At an earlier stage the appellants had offered a nominal smoke venting system, but it is assumed for the purposes of this appeal from paragraph 8 of the appellants' letter that the proposal excludes any provision for smoke venting.

Because the single storey unit is in excess of 7000 cubic metres, section 13 of the Hampshire Act 1983 is applicable. The Borough Council have invoked this section and rejected the proposals on the grounds that they do not meet the fire precaution provisions of section 13(2)(a) as quoted in paragraph 4 above.

The Borough Council have stated that they require the proposals to incorporate a sprinkler system and smoke extraction system. The appellants do not consider that it is necessary to meet these requirements and have therefore appealed to the Secretary of State in respect of this grievance.

The Appellants' representations

The appellants have made the following points in support of their appeal:

(i) The building in question is essentially a single storey building and will be used for the sale of household goods which will not be displayed on racking. The maximum height of combustible material across the whole of the shop floor will not exceed 2 m and therefore the combustible content would not exceed 7000 cubic metres.

(ii) The storage and ancillary area at the rear will be compartmented from the retail area.

(iii) A building Regulation application for the fit-out has received approval, therefore other fire matters such as means of escape and compartmentation are considered to be satisfactory. In particular the applicant wishes to emphasize his claim that the proposals comply with Requirement B5 which requires adequate access and facilities for the Fire Service.

(iv) The reduction of potential fire damage should be a matter between the building occupier and his insurers who in this case are prepared to accept the building without the provision of sprinklers or a smoke extract system.

(v) The occupier trades from many similar outlets in the UK and has not been asked to make provision for these additional requirements elsewhere.

(vi) The occupier is prepared to provide some smoke detection within the unit which will be linked to a central receiving station in order that the Fire Service can respond quickly.

The appellant therefore suggests that the health and safety of persons in and around the building is satisfactory without the imposing of the additional requirements specified in section 13(2)(a) of the Hampshire Act 1983.

The Borough Council's representations

The Borough Council consulted the Fire Authority and in view of their strong recommendation for the installation of sprinklers and the requirement for smoke extraction, they considered they had no alternative but to support the Fire Authority. The Borough Council also point out that the nominal amount of smoke venting which the appellant originally said he was prepared to agree, could not practically be calculated and designed without sprinklers. In the view of the Borough Council such a smoke venting system could not be considered as a compensatory factor for a sprinkler system.

The Fire Authority have indicated that the products to be sold in this store would mean that an unsafe fire condition could develop within five minutes. They point out that should a fire develop unchecked beyond a certain size then it is unlikely that it could be controlled within the building. The Fire Authority state that until such time as there is a regulatory control for single compartment single storey buildings, they recommend section 13 of the Hampshire Act 1983 should be applied in full.

The Department's views

The Department takes the view that the Hampshire Act 1983 provides permissary powers of additional fire precautions regulation to district councils and that these powers should not unreasonably be curtailed on appeal simply because the proposals show compliance with the Building Regulations 1991 (as amended). The issue before the Secretary of State is whether it is reasonable in this particular case for the Borough Council to have used its discretionary powers of section 13(3)(b) of the Hampshire Act 1983 to require the installation of an effective means of removing smoke and a fire extinguishing system.

The main use of the building is retail and the appellant has stated that non-retail areas will be provided with fire separation. The floor area of the retail area is approximately 3310 square metres and it is noted that an automatic smoke detection system is to be installed and linked to the Fire Authority. Section 13 of the Hampshire Act 1983 makes provision, where appropriate, for the automatic protection of the larger type of warehouse and commercial premises. However, in this particular case the Department takes the view that it would be unreasonable to require the insallation of a sprinkler system in this size of retail development.

With regard to smoke extraction, the Department notes that the appellants were originally prepared to provide 'nominal' smoke extraction, although this has not been taken as forming a part of the present proposal. The Department accepts such a system would be difficult to design for an uncontrolled fire situation, such as could occur where sprinklers are not provided. However, given the fact that the proposal incorporates automatic smoke detection and a fire alarm system, which it is assumed will be installed to the Borough Council's requirements, the Department considers the proposals to be satisfactory.

The decision

The Secretary of State has given careful consideration to the exposition of the appellants' grievance and the facts of this case. He has come to the conclusion that the appellants' proposals are satisfactory and it is unreasonable in this particular case for the Borough Council to seek to impose the fire precaution provisions of section 13(2)(a)(i) and (ii) of the Hampshire Act 1983 relating to sprinkler and smoke extraction systems. Accordingly, he upholds the appeal.

Refurbishment of two storey shopping complex (B.1)

New suspended ceilings were to be installed and problems arose regarding the amount of timber within the ceiling void; and recessed lights. It was decided not to relax B.1, particularly as it is a life saving matter; but the last paragraph is important.

Ref.: 45.3.86   Date: 3 October 1996

The building work to which this appeal relates concerns the refurbishment of an existing two storey shopping complex which comprises 32 shop units, enveloping enclosed, semi-open and open market areas. Part of the refurbishment works has involved the replacement of the suspended ceilings above the shopping mall walkways. The new ceilings consist of plasterboard supported on softwood framing with recessed light fittings.

The retail units and the mall are provided with a full sprinkler system, an automatic smoke control system, and an automatic smoke detection and alarm system. In addition, the public walkways are provided with emergency lighting. The ceiling void has been provided with lateral cavity barriers at a maximum of 10 metre centres and with longitudinal barriers located either side of the service ducts.

In addition to the new ceilings, new openings were introduced between the lower and upper walkways; and new doors provided to the mall entrances which were previously open. The percentage of covered area remains unchanged although three new units were introduced which had the effect of reducing the amount of public walkway.

It is understood that the Borough Council wrote formally alleging that the mall ceiling did not comply with the Building Regulations. They were concerned that there was the possibility of fire breaking into the ceiling void via the lighting unit and igniting the softwood timber framing. You produced an assessment report for your clients which states that fire would not rapidly spread along the public walkways, if at all, and that Requirement B1 would not therefore be jeopardized. You then applied to the Borough Council for a relaxation of Requirement B1 which they refused. It was against that refusal that you appealed to the Secretary of State.

In support of your appeal, you state that although the suspended ceilings in the mall are softwood-framed, resulting in a considerable amount of timber within the void created between the ceiling and the soffit of the structural floor slab, the timber is largely protected by the plasterboard ceiling surfaces which have 30 minutes' fire resistance. You accept that this fire resistance has been interrupted by the installation of recessed lighting units. However, you point out that in most situations sprinkler heads have been provided immediately outside the shop units.

The Appellant's case

You consider that the work satisfies Requirement B1 for the following reasons:

(i) The recessed lights are of metal construction and are fitted tightly into the false ceiling.

(ii) Any flowing layer of gases resulting from a fire will move rapidly along the soffit of the ceiling, rise up through the voids, and be extracted at high level.

(iii) It is unlikely that the timber within the ceiling void would readily ignite because of the cooling effect of the sprinklers located both within the mall and the shop units.

(iv) The ceilings are effectively divided into small rectangular fire compartments, approximately 10 metres by 5 metres, because of the cavity barrier spacing.

(v) In the unlikely event that fire would be able to spread via ceiling voids, then there is a high degree of compartmentation throughout the complex which would have the effect of limiting fire spread. The market hall, which was previously uncompartmented, has now been separated by a compartment wall and fire shutters.

(vi) The existing sprinkler system has been extended to cover the walkways, although it is not the intention to display or exhibit in them.

The Metropolitan Borough Council's case

The Metropolitan Borough Council gave the following reasons for rejecting your application for a relaxation of Requirement B1:

(i) The malls have low ceiling heights which have been compounded by the type of ceiling construction used. The Metropolitan Borough Council had made the suggestion to your clients that use could be made of a non-combustible ceiling with a 25% grid system to allow the void behind the ceiling to be used as part of the smoke reservoir. This suggestion was not taken up and as a result the depth of smoke reservoir was reduced. This will have the effect of increasing the smoke layer temperature to the maximum permissible and greatly increase the required smoke fan extraction rates. Although you have stated that the smoke control system is extremely well designed, the Metropolitan Borough Council consider that this design was out of necessity given that your client declined to take up their suggested solution.

(ii) The Metropolitan Borough Council and the Fire Authority contend that the penetration of the 30 minutes' fire resisting ceiling by light fittings will allow the timber within the ceiling void to become involved in any fire. This in turn could affect the design consideration of the finely balanced smoke extraction system.

10. The local Fire Authority support the Metropolitan Borough Council's views, and state that since the primary danger associated with fire in its early stages is not flame but smoke, it is essential that safe means of escape are not compromised by breaches of the fire resistance of the public mall ceilings. The Fire Authority are also concerned about safe fire fighting operations.

The Department's view

The Department is of the opinion that should fire break through the light fittings into the ceiling void it would tend to be contained by the cavity barriers for sufficient duration to permit safe escape. Any fire tending to break out from these ceiling reservoirs would be controlled and cooled by the sprinkler system below the ceiling with hot gases being exhausted from the extraction system. The Department does not therefore support the Metropolitan Borough Council's view that the combination of low ceiling height and combustible ceiling supports will add significantly to the fire risk in terms of life safety. The Department notes that a sprinkler system, smoke extraction system, and early warning system have been provided.

Paragraph 5.3 of BS 5588: Part 10: 1991 ('Code of practice for shopping complexes') suggests that 'Materials of construction used in the malls should themselves not contribute unduly to fire growth'. This document also suggests that whether or not sprinklers need to be provided in malls will depend on the extent to which any significant fire load is introduced. In this context the document refers to trading stalls and displays. In this case a sprinkler system has been provided and fire starting in one of the shop units would have to break through the light fittings in the mall ceiling to involve the timber ceiling supports. The Department believes such a conflagration to be unlikely. The Department therefore takes the view that the softwood timber framing in the ceiling voids would not in itself contribute unduly to fire growth, nor would it readily ignite and prejudice safe escape in fire.

The decision

The Secretary of State takes the view that he is being asked to relax Requirement B1 of the Building Regulations 1991 (as amended) to the extent that combustion material be permitted in the void above the mall ceilings. However, Requirement B1 is functional and contains no specific requirement relating to combustible materials. Moreover, Requirement B1 is a life safety matter which the Secretary of State would not normally consider relaxing in any event. Having given careful consideration to this case the Secretary of State therefore dismisses your appeal against the refusal by the Metropolitan Borough Council to relax Requirement B1 of Schedule 1 to the Building Regulations 1991 (as amended). In making his decision he has also taken account of the issue of protection to firefighters, which was one of the concerns raised by the Fire Authority.

You will no doubt wish to note that as indicated the view formed by the Department in analysing this case was that the building work as executed appears to comply with Requirement B1.

Showroom and office – ventilation and means of escape (B.1 and F.1)

The proposal was to form a showroom and office in the ground floor of a workshop, together with a new first floor above it. The decision accepts the means of escape proposed (B.1) but found ventilation proposals to the small office unsatisfactory (F.1).

Ref.: 45.3.97   Date: 9 October 1996

The building work to which this appeal relates concerns internal alterations to an existing single storey industrial unit to form a ground floor showroom in one corner of the unit with a new first floor above it. The first floor will consist of two separate rooms; one being designated for office use and the other for light storage. Access and egress from the storage area is via the office which in turn exits onto a new protected stair which leads direct to a ground floor final exit.

The new first floor retains the existing sloping roof and the drawing submitted shows the headroom to vary from approximately 800 mm at the lowest end of the storage room to 1.8 m at the highest end of the office. The door to the protected stair is situated at the high end of the office and the combined floor area of the store and office is approximately 36 m².

It is proposed that there should be no windows serving the office in the external walls. The submitted drawing shows a 1200 × 900 mm top hung window (with a 8000 mm² trickle ventilator) in the longer of the two internal walls. The approximate floor area of the office is 13.3 m².

Your full plans application for this proposed work was rejected on various grounds by the District Council, including non-compliance with Requirements B1 and F1. The Council was concerned that the proposed headroom in the first floor was unacceptable for escape purposes and that it was not acceptable to ventilate the office area through the adjacent workshop. However, you considered that the occupancy of the showroom and office would be minimum and that

leasehold constraints on external alterations to form a window warranted a relaxation to the relevant requirements. You therefore applied to the District Council for relaxations in respect of Requirements B1 and F1 which were refused. It was against that refusal that you appealed to the Secretary of State.

The Appellant's case

Requirement B1   You are of the opinion that the District Council has designated the whole of the proposed first floor as forming part of the escape route. However, you point out that the most frequently used section would only be the high side of the office. In your view it would be impracticable to use any other part of the first floor on a regular basis because of the limited headroom available.

You further state that the available headroom within the usable section of the office varies between 1.8 m and 2.1 m, and that the maximum available headroom at 1 m from the office escape route wall is approximately 1.9 m to the underside of the roof cladding. You further argue that even this most frequently used section would only be used for approximately one hour per day by one person.

Requirement F1   You have given the following grounds to substantiate your case for a relaxation of Requirement F1. The building is occupied on a leasehold basis which prevents you from puncturing the external skin. It is therefore not possible to insert a window in the external walls to provide ventilation. Because the maximum usage of the office will only be for a cumulative period of one hour per day by one person, you consider the provision of the openable window with trickle ventilator in the internal wall will improve ventilation. You consider that the workshop is a relatively large open area, and therefore that the provision of the internal window should provide the necessary adequate ventilation for the very low level of usage of the office envisaged.

The District Council's case

Requirement B1   The District Council rejected your application for a relaxation of Requirement B1 on the grounds that the proposed headroom on the escape route was not acceptable. They considered that a reduction in headroom on a means of escape route was not within the spirit of Approved Document 'B' ('Fire Safety') which recommends a minimum headroom of 2 m.

The District Council also takes the view that although the appellant has stated that the office will only be occupied by one person for one hour per day, the potential use could be much greater.

The Council consulted the Fire Authority who were of the opinion that there is insufficient headroom in both the proposed office and storage area for either of these two locations to be considered as habitable rooms.

Requirement F1   The District Council have stated as their grounds for refusing to relax Requirement F1 that they consider ventilation of a working office via a factory production area would not meet the requirements of F1 because the air circulating within the room would be contaminated with smells and pollution from the processes involved.

The Department's view

Requirement B1   It is the means of escape from within the two new first floor rooms that is at issue in this case. It is for the Department to consider the issue in terms of the life safety aspects with respect to the Regulations. However, it is not for the Department to express an opinion on the suitability of the rooms for the proposed use.

Paragraph 5.24 of Approved Document 'B' suggests that all escape routes should have a clear headroom of not less than 2 m. Appendix E of the document defines an escape route as the 'route forming that part of the means of escape from any point in a building to a final exit'. The submitted drawing indicates that the new escape stair has a clear headroom of 2 m. What therefore needs to be considered is whether it is reasonable to apply the 2 m headroom criterion to this small office and storeroom in terms of life safety.

With regard to the storage room, the Department estimates from the drawing that its floor area is approximately 17 m², with the headroom being shown to vary from 0.8 m to 1.2 m. In addition, there are also floor loading restrictions. Therefore because of the storeroom's limited space and height restrictions the Department is of the opinion that occupation of the room will be minimal and it would be reasonable to ignore the 2 m headroom criterion for escape purposes.

With regard to the new office, this also has limited headroom which will restrict its use and the Department estimates its floor area to be approximately 19 m². Although you have stated that the room will only be used by one person for one hour per day, the Department supports the District Council's view that this could be exceeded.

However, the Department considers that occupancy of the new office will still be low and that occupants will be aware of the headroom restriction because it will form part of their normal working environment. The Department notes that escape from the office is direct to the new protected stair and is therefore of the opinion that it would also be unreasonable to apply the 2 m headroom criterion to this room for escape purposes.

Requirement F1 Requirement F1 requires that 'There should be adequate means of ventilation provided for people in the building'. The question to consider is therefore whether the proposed window and ventilator will satisfy this requirement for the occupant of the office.

Paragraph 2.2 and Table 2 on page 9 of Approved Document 'F' ('Ventilation') suggest that Requirement F1 can be met in occupiable rooms (e.g. offices) by providing an opening window and ventilator on an external wall with the following minimum sizes:

(a) for rapid ventilation – the window should have an open area of at least 1/20 of the floor area. In this case the floor area is about 13.3 m², so the required open area is 0.67 m²;
(b) for background ventilation – the ventilator should have an open area of 400 mm² per m² of floor area, which in this case is 5320 mm².

The ventilator proposed exceeds this minimum area, and there would be no difficulty in selecting a window having the required open area.

However, the window and ventilator open into the workshop – not the outside. The effects of this are that:

(i) ventilation rates will be reduced as there will be no wind; and
(ii) any airborne pollutants in the workshop which are a hazard to health will be able to enter the office.

Although it may be that with the present usage there are no significant pollutants in the workshop, this may change. There is also the possibility that the occupant of the office may spend longer in there. Such changes could take place without further control under the Building Regulations (although the Workplace Regulations would still apply). In the view of the Department, therefore, the proposal does not comply with Requirement F1 and there are no substantive grounds for relaxing that requirement.

The decision

The Secretary of State has given careful consideration to the facts of this case. Requirement B1 is a life safety matter which the Secretary of State would not normally consider appropriate to relax or dispense with. However, in this case he considers that your proposals do in fact demonstrate compliance with Requirement B1, but insofar as there is a difference of opinion regarding compliance between you and the District Council he hereby formally relaxes Requirement B1 to the extent necessary for you to carry out your proposals and accordingly upholds your appeal in this respect.

With regard to Requirement F1, the Secretary of State has concluded that your proposals do not comply with the requirement and that it would be necessary to relax or dispense with that requirement in order for you to proceed with the work as proposed. He has considered the reasons you have put forward for relaxing the requirement and concluded that they do not constitute substantive grounds for doing so. It follows that he considers that the District Council came to the correct decision in refusing your application to relax Requirement F1 and accordingly dismisses your appeal in this respect.

Single storey building – increase in height and reclad (B.3 and B.4)

This appeal relates to internal and external fire spread in respect of alterations to increase the height and re-clad a single storey building. The problem arose because of the modest distance between the building and the boundary. The decision was that B.3 and B.4 could not be relaxed.

Ref.: 45.3.87   Date: 10 October 1996

The appeal

The building work to which this appeal relates concerns alterations to an existing single storey building which it is understood is to be used for the storage of car parts. Details of the existing building have not been provided but drawing 3402–02 shows that the building is to be increased in height and re-clad. The increased height of the side wall is indicated on the drawing to be approximately 2 metres and is to be supported on existing brickwork construction.

The land on which the building stands, and that on which all adjacent properties stand, is understood to be in the ownership of the City Council. The location of the adjoining buildings is not shown on the drawings submitted but it is also understood that the distance to the boundary of your client's building from the north and south walls is 3 metres; and that the boundary distance of the adjacent building to the north is 5 metres and that of the adjacent building to the south is a minimum of 3 metres.

Your full plans application was for the removal of the 'existing roof and construction of new roof on existing walls' and was refused by the City Council on several grounds including non-compliance with Requirement B4 ('External fire spread'). It appears that work progressed and that the actual construction consisted of a new steel frame built off of the existing brickwork with a new cladding system. The Council requested details of the proposed cladding but details of its fire resistance and the fire protection to the supporting steelwork were not specified by you before work progressed. The City Council requested remedial work to provide appropriate fire resistance of the external walls of the north and south elevations and appropriate fire resistance of the supporting structure to the walls.

You did not proceed with the remedial work because you considered that given that all the land was owned by the City Council it was unreasonable to take the site boundary as the relevant one as opposed to taking the total distance between the adjacent buildings. You therefore applied for a relaxation/dispensation of Requirements B3 and B4, which was rejected by the City Council. It is against that refusal that you appealed to the Secretary of State.

The Appellant's case

The grounds for your appeal are based on the fact that the City Council own the land on which your client's building stands and that on which all the adjacent buildings stand. You contend that it would not be logical for the premises on each side of your client's building to be sold or extended to the boundaries because, firstly, planning permission would not be granted and, secondly, owners of the land would object.

In addition, you point to the fact that the adjacent building to the north has what you term a 'firebreak' in excess of 5 m and the building to the south a 'firebreak' of a minimum of 3 m.

The City Council's case

The City Council rejected your application to dispense with Requirements B3 and B4 because they considered that there was insufficient distance to the boundary of your client's building from the south and north elevations to justify the amount of unprotected area proposed and to justify unprotected supporting steelwork. They considered that the fire loading within the building would be increased due to the additional height of the walls and that no compensatory features such as sprinklers had been offered.

The City Council requested confirmation that the cladding and supporting structure would achieve 60 minutes' fire resistance. Such confirmation was not provided and therefore on the assumption that the external walls formed unprotected areas, and using the enclosing rectangle method, the City Council calculated that the distance to the boundary should be 7 m. The City Council state that in fact the distance to the boundary is only 3 m.

The City Council also argue that although the building is currently used for the storage of car spares, there is no continuing power of control over the type of storage and highly flammable materials could be stored in the future. It is also pointed out that whilst at the present time the lease of the properties would appear to be both restrictive and controllable by the Council, this situation could change should the Council sell any of the properties.

The City Council has consulted with the Fire Authority who were of the opinion that a relaxation of Requirements B3 and B4 would be unreasonable in the circumstances. The City Council also gave notice to adjacent occupiers under section 10(2) of the Building Act 1984.

The Department's view

The Department takes the view that the question before the Secretary of State is whether it would be appropriate to relax Requirements B3 and B4 of the Building Regulations 1991 in this particular case. It also takes the view that the need for such a relaxation arises because the external walls and the supporting structure on the north and south elevations do not achieve the degree of fire resistance recommended in Table A2 on page 98 of Approved Document 'B' ('Fire Safety'). For a building such as this the recommendation given in the table is for the provision of 60 minutes' fire resistance.

In order to consider a relaxation in this case it is necessary to assume that the boundary to the property concerned is not in dispute, and that the distance to the north and south boundaries is less than the distance that would be considered acceptable for wholly unprotected areas. The Department has effected its own boundary calculation and these are supportive of the City Council's argument that acceptable boundary distances would exceed those provided by the site boundaries.

You have argued that it is more appropriate in this case to consider the distance to other buildings rather than the distance to your client's own boundary and justified this on the basis that the City Council have control over your client's site and the sites of all the other buildings surrounding it. However, you have not produced evidence of any planning restrictions that exist in relation to the adjoining buildings, or that building works in the form of extensions could not be carried out under the terms of any of the leases for these other buildings. The Department takes the view that as you have yourself extended the height of your building presumably within the terms of your lease, then it is reasonable to assume that the adjoining buildings could likewise be extended.

The decision

The Secretary of State has taken the view that he is being asked in this appeal to relax Requirements B3 and B4 of Schedule 1 to the Building Regulations 1991 (as amended). He has given careful consideration to all aspects of this case and in particular to the arguments put forward by you with respect to the common ownership of the land by the City Council. However, he has concluded that for the purposes of considering the relevant boundary it would be inappropriate to consider any other boundary distance than that provided by the existing site boundaries and that the City Council came to the correct decision in refusing your application to relax Requirements B3 and B4. He accordingly dismisses your appeal.

Extension to factory – animal feed storage bins (B.3 and B.4)

A local authority would not accept that provisions for the prevention of internal and external spread of fire were adequate in respect of an extension to a factory complex handling and storing animal feed in bins. The applicant argued that the structure was a type of plant and the Fire Authority did not object. The appeal was upheld.

Ref.: 45.3.90   Date: 21 February 1997

The appeal

The building work to which this appeal relates is an extension to an existing facility and forms part of a factory complex for the manufacturing, storage and distribution of animal feeds. The extension, which will contain 18 double skinned steel bulk animal feed storage bins, is shown to be clad with plastic coated corrugated steel supported on a steel frame. There is no fire resistance protection provided to the external envelope of the extension.

The extension is stated by the District Council to be 18 m high and is shown on the sketch which forms part of their submission (marked 'Plan A') to be 11.5 m in length. The existing part of the animal feeds facility is shown on the same sketch to be 14 m in length. The existing and extended facility are single storey although there are some high level walkways provided to give access to the bins. There are a total of seven product transfer conveyors feeding material into both new and existing storage bins but these machines are situated above the 12 m level.

The main point at issue is the non-provision of fire resistance to the external wall on the southern boundary. This is shown to be located 8.5 m from the actual site boundary; and you have confirmed that the wall is situated approximately 19 m from the nearest process machinery within the existing production facilities.

The District Council state that the occupancy of the extension will be limited and that the means of escape is adequate.

These proposals were the subject of a full plans application which was rejected by the District Council on grounds of non-compliance with Requirement B4 and unsatisfactory fire safety within the premises. You then applied for a dispensation of fire protection to southern elevation of new structure. The Council believed there was a case for dispensing with the requirements for structural fire protection (internal fire spread) but that the risk of fire could not be eliminated and that its effects must be controlled to protect the occupiers of adjoining land and buildings. You resisted the additional fire precaution work involved on the grounds of cost and the Council therefore proposed an alternative precaution in the form of roof venting.

A re-submission of the full application was made by you, but you were unable to agree to the alternative proposals put forward by the Council. Your plans were therefore rejected; and subsequently your application for dispensation was refused. The letter of refusal of the application for dispensation from the Council does not specify the relevant requirements but refers only to 'fire resistance to elements of structure and to external fire spread'. The Department has therefore assumed that your subsequent appeal to the Secretary of State against that refusal is in respect of Requirements B3 and B4.

The Appellant's case

You state that you would not normally consider this type of structure as a building, but rather as a type of plant, and would not therefore expect that there would be a need to seek Building Regulation approval. However, as a matter of course you sought the view of the District Council. It became clear in this instance that you would be required to make an application. You state that you have been involved in this type of work for a number of years, and in very similar situations, and took the view that you would be required to satisfy Building Regulations in respect of the structure and fire escape requirements. You did not expect to come under scrutiny in respect of fire protection, but the District Council considered that due to the area and proximity of the new works relative to the neighbouring southern boundary, 90 minutes' fire protection to that elevation of the extension would be a requirement. However, your view is that the plant would function unclad but that this would create planning problems.

In support of your appeal you have also made the following points:

(a) The tunnel formed by the support steel through which the bulk lorries drive to load the products from the bins is naturally vented each end because of the full-size permanent openings necessary for the vehicles to pass.
(b) The bin wall area has its own in-built fire protection in that it has two layers of 3 mm mild steel plate 104 mm apart, in addition to the external corrugated sheet cladding.
(c) Since the rejection from District Council, you have further considered the matter in respect of compliance with Regulation 4(2)(a) of the Building Regulations 1991 which requires that any building which is extended must comply with the relevant requirement, and be no more unsatisfactory in relation to that requirement than before the work was carried out. You contend that the existing work did not comply and that the new work is no more unsatisfactory than it was before.
(d) In particular, you have drawn attention to the Fire Authority's comments (see below).

The District Council's case

The District Council state that your description of the works as 'Erection of 18 No Storage Bins' is a little misleading. It is the Council's view that the works consist of the extension of an existing building and existing facility, and comprises more than the provision of storage bins because they include milling facilities. The District Council point out that prior to rejecting the application for dispensation, the District Council offered an alternative approach involving the provision of roof vents to ventilate the products of combustion so as to reduce the risk of radiation across the site boundary.

The District Council state that they rejected your application for a relaxation of Requirement B3 and B4 of the Building Regulations 1991 because they did not consider that the building would prevent fire spread to and across the relevant boundary. In support of their decision they state that:

(a) The guidance given in Approved Document 'B' ('Fire Safety') suggests that the boundary elevation and supporting steelwork should be provided with 90 minutes' fire resistance. However, the Council accepts that 90 minutes' fire resistance may be excessive and has conceded, subject to certain conditions, that no added fire resistance would be required to either the steel frame or the external cladding.
(b) The new work itself does not comply with Requirement B4; and although the existing facility did not comply, the extension causes the existing building to be more unsatisfactory.
(c) If the south elevation were to be wholly unprotected then the distance to the boundary would need to be 20 m whereas it is in fact only 8.5 m.

The District Council accept that little information is available on the combustion properties of animal feed products but state that it contains grains, fats/oils and it cannot be considered as inert. They contend that there is the possibility that the extension may be put to some other industrial use in the future or another type of product stored in the bins.

The Fire Authority's and Health and Safety Executive's comments

The Fire Authority were consulted and they state that in this particular case it would not be unreasonable for the dispensation to be granted. They have made the following points:

(a) The majority of potential fire load will consist of feed products stored within the bulk silos/bins.
(b) It would be normal for this type of organic material to burn by smouldering rather than by flaming. This, combined with the good insulating properties of the product, would result in a deep seated type of fire within the storage silo.
(c) Such materials are usually difficult to ignite and would constitute a low fire risk. Also a fire in the product is likely to be confined to a single silo, its spread being prevented by the twin skin metal walls.

The Health and Safety Executive were consulted and state that they are not aware of any significant recent fires in bins storing such material. They also point out that smouldering fires in such bins may be possible but develop slowly and would not be expected to cause a significant risk of fire spread beyond the confines of a single bin.

The Department's view

You have appealed to the Secretary of State against the decision of the District Council to refuse to dispense with Requirement B3 ('Internal fire spread (structure)') insofar as it affects the fire resistance of the steel work supporting the new wall on the southern boundary, and Requirement B4 ('External fire spread') insofar as it affects the fire resistance of that wall. Part B requirements are life safety matters and the Secretary of State does not normally consider relaxing or dispensing with them.

The Department accepts the District Council's view that this appeal case should be considered as new building work and not considered as an item of plant. The new extension is fully clad and the risk of fire spread across an adjoining boundary, in this case the southern boundary, therefore needs to be addressed.

Requirement B4(1) states that:

The external walls of the building shall resist the spread of fire over the walls and from one building to another, having regard to the height, use and position of the building.

The Department accepts that the southern boundary wall of the new extension and hence to the combined new and existing facility as a whole, will not meet the recommendations given in Approved Document 'B' with regard to the extent of unprotected area and the distance to that boundary. However, this is a purpose-built building containing double skinned storage bins/silos for animal storage feeds. The Department considers that it is, therefore, reasonable to take account of the specialized nature of the building.

The assessment provided by the Health and Safety Executive is similar to that provided by the Fire Authority. The Department is also of the view that the risk of fire is low and that if a fire starts it will be a smouldering fire and will be confined to a small area.

The Department is therefore of the opinion that in this case Requirement B4 will have been satisfied in respect of the southern boundary without any further works being undertaken, either by way of the provision of fire resistance or ventilation. It follows that if the external wall on the southern boundary satisfies Requirement B4, then Requirement B3 is also satisfied in that it would be unnecessary to provide fire resistance to the structural frame supporting that wall.

The Secretary of State's decision

The Secretary of State has given careful consideration to the facts of this case and the points put forward by all parties. He takes the view that he is being asked to dispense with the Requirement B3 ('Internal fire spread (structure)'), insofar as it affects the fire resistance of the steelwork supporting the new wall on the southern boundary and Requirement B4 ('External fire spread') insofar as it affects the fire resistance of that wall.

Part B requirements in the Building Regulations are life safety matters and the Secretary of State would not normally consider it appropriate to dispense with or relax them. However, in the circumstances of this particular case, the Secretary of State takes the view that your proposals do in fact demonstrate compliance with Requirements B3 and B4 of Schedule 1 to the Building Regulations 1991 (as amended), but insofar as there is a difference of opinion regarding compliance between you and the District Council he hereby formally relaxes Requirements B3 and B4 to the extent necessary for you to carry out your proposals. Accordingly he upholds your appeal in this respect.

Public House – means of escape (B.1)

A City Council was concerned regarding the number of persons who may be on the first floor, and the need to upgrade the escape route. The appeal was dismissed.

Ref.: 45.3.102   Date: 29 May 1997

The appeal

The building work indicated on the full plans application form submitted to the City Council comprises new internal refurbishment, new ground floor bar servery and new shop front. This appeal relates to the means of escape from the first floor bar area and in particular the number of persons that will be permitted in the first floor bar area. The City Council have accepted seating accommodation for 30 persons on the first floor and the appellant wants 50 persons.

The building is existing and comprises four above ground storeys and a basement storey with the public house occupying the basement (which is not at issue), ground and first floors. The ground floor comprises bar area, bar servery and gent's toilet. The first floor comprises bar/restaurant area with full seating facilities, bar servery, small kitchen and ladies' toilet.

The first floor is served by a single protected stair which leads to a ground floor final exit and which also forms the alternative escape route from the rear of the ground floor bar. Lobby protection is provided between the ground floor bar and the stair enclosure. New smoke detectors have been provided in each ceiling bay of the ground and first floor bars in addition to emergency lighting.

The Appellant's case

You have appealed to the Secretary of State against the refusal of the City Council to relax Requirement B1 of the Building Regulations to the extent that would permit 50 persons to occupy the first floor bar.

You state that you have the agreement of the Fire Officer to increase the number of persons on the first floor to 50 subject to certain undertakings. However, the response from the Fire Officer on the consultation form states that there should be a maximum 30 persons on the first floor unless an additional protected route is provided.

You have provided a method statement from your client as to how a limit of 50 persons in the first floor bar will be controlled and they state that 48 seating spaces will be available. You also state that your client enjoys an established use of the first floor for the purpose which it is intended to continue, without alteration.

You state that you have investigated the possibility of the provision of a secondary means of escape and have come to the conclusion that this would not be feasible.

The City Council's case

The City Council conditionally approved the plans on the basis that the first floor plan showed seating accommodation for 30 persons. They state that they do not grant relaxations under the Building Regulations 1991 and therefore refused to relax Requirement B1 to permit 50 persons on the first floor.

The technical issues on which the City Council have based their decision are as follows:

(a) Paragraph 7.12 of BS 5588: Part 2: 1985 ('Code of practice for shops') suggests that two protected escape routes should be provided from a restaurant or bar. The premises are not a small shop but the provisions of paragraph 9.21 of the code which restricts the seated occupancy to 30 persons were considered reasonable.

(b) The premises have been extensively refurbished and at the design stage it would have been possible to provide a second escape route from the first floor bar which would have enabled an occupancy increase.

(c) The City Council regard the stair to the street as less than ideal and the means of escape through a single 800 mm door, down these stairs, through an inward opening door and past a non-fire resisting lobby, as inadequate for 50 persons in a licensed premises.

(d) The majority of measures proposed by the appellant address defects in exiting construction and do not offer any additional compensatory protection.

The Department's view

You have stated that your client enjoys an established use of the first floor for the purpose which it is intended to continue, without alteration. However, the Secretary of State is only able to base his decision on the basis that all parties accept that the Regulations are applicable.

The Department notes that the City Council have used the recommendations given in BS 5588: Part 2 as the basis for their decision and that they were prepared to accept 30 persons in the first floor bar area served by a single stair.

The Department does not see the occupancy figure of 30 as a finite limit and notes that for some situations Approved Document B suggests that a total storey occupancy of 50 persons may be acceptable. However, for an increase in numbers to be considered acceptable it may be necessary to upgrade the escape route to satisfy the concerns of the City Council.

You have not stipulated how many non-seated persons there will be on the first floor; however, the method statement provided by your client states that there will be 48 seating spaces available. The Department considers that it is probable that there will be more than two non-seated persons on the first floor and therefore takes the view that neither the guidance given in the British Standard nor that given in Approved Document B is being followed.

The Secretary of State's decision

The Secretary of State has taken the view that he is being asked to relax Requirement B1 of Schedule 1 to the Building Regulations 1991 (as amended). He has given careful consideration to all aspects of the case and the arguments put forward but he does not consider it appropriate to relax the requirement. He accordingly dismisses your appeal.

You may, however, wish to take account of the Department's view put forward above, in conjunction with the City Council.

Department store – roof extension (B.1)

In dispute was the necessity, or otherwise, of lobby protection to a Georgian staircase in a listed building. A number of compensating features were to be installed, and this proved to be an acceptable alternative.

Ref.: 45.3.88   Date: 30 May 1997

The appeal

Your proposals relate to Phase 6 of the roof development of the existing nine storey department store (of which eight are above ground), which is a listed building in respect of both the external façade and certain interior features. The existing roof level accommodation comprises three levels on the sixth, seventh and eighth floors. The sixth and seventh floors contain engineering offices, workshops, plant rooms and architects' offices, etc. The eighth floor contains a small conference area.

This phase of the roof works involves the removal of some existing roof accommodation and the construction of two levels of roof extension. The extended sixth floor (roof level 1) will mainly be for storage, and the extended seventh floor (roof level 2) will mainly be for office accommodation. The eighth floor (roof level 3) will remain unchanged.

The existing and proposed floor areas are as follows:

(i)   sixth floor – currently 2314 m²; to be extended by 806 m²;
(ii) seventh floor – currently 326 m²; to be extended by 1016 m²;
(iii) eighth floor – currently 265 m²; no extension.

You have stated that four stairs are available from the roof, two of which already benefit from lobby protection. Stair No. 5 is to be extended and upgraded, including the provision of lobbies. This means that three of the four stairs have (or will have) lobby protection but the fourth stair, which is an existing Georgian stair and which currently serves all floors up to the fifth floor level, will not benefit from lobby protection.

The existing Georgian stair is linked to roof level 1 by another existing accommodation stair which you propose to provide with lobby protection. However, the drawings submitted as part of the appeal show that full lobby protection is only provided at the sixth floor level (roof level 1). One of the escape routes from the seventh floor (roof level 2) is via a designated external route to the link stair at roof level 1.

The building is designed for phased evacuation and you state that the top storey, i.e. the eighth floor (roof level 3), is approximately 31.5 metres above street level.

These proposals were the subject of a full plans application which was rejected on several grounds of non-compliance, but primarily in respect of Requirement B1. You then re-submitted the full plans application, together with a request for relaxation to paragraphs 4.26b of Approved Document 'B' ('Fire Safety'). Approved documents are documents approved by the Secretary of State to provide practical guidance on meeting the requirements of Schedule 1 and Regulation 7 of the Building Regulations 1991. As guidance it is neither appropriate – nor is there any statutory procedure – to relax or dispense with matters in these approved documents. However, the Borough Council took your application as one 'to dispense with the requirements of B1 with regard to lobbying the "Georgian" stair area' and refused it.

You consider that the retention of the Georgian staircase as existing without lobbies will not make the building any less satisfactory in relation to Requirements B1, B3, B4 and B5 than it was before. You therefore appealed against the refusal of your application for a relaxation/dispensation to the Secretary of State.

The Appellant's case

You state that your application for a relaxation/dispensation relates only to paragraph 4.26 of Approved Document 'B' and not to Requirement B1 in its entirety. You make the following points in support of your application:

(i)   The appeal relates to an existing nine storey building where the existing Georgian staircase does not currently have the benefit of lobby protection. The proposed works do not extend this existing staircase.
(ii) You make reference to paragraphs 0.10–0.13, and 0.17 of Approved Document 'B' which mainly relate to existing buildings and in particular to buildings of special architectural or historic interest.
(iii) The proposals do not extend the height of the building and additional lobby protection has been provided on the fifth and sixth floors to the link staircase.
(iv) The store has an operational fire department which has facilities to assist the fire service. A fire safety strategy for the store has also been provided as part of your appeal submission.
(v)   Taking into account the additional stairs provided as part of the Phase 6 development, the capacities of the stairs are not exceeded by the current proposals.

(vi) The store is a listed building and the listing covers both the external façade and internal features. Also, the existing Georgian stair is a highly decorative timber staircase which is considered to be of architectural and historic merit.

In addition, you state that a number of fire engineering provisions have been included in the proposals. These comprise: a sprinkler installation to all areas; a fire alarm system to all areas; a smoke extract system to all areas; wet risers to staircase 5; the provision of two hose reels at each level; and a voice evacuation and public address system.

You consider that the above provisions should be taken into account when considering a variation of Approved Document 'B'. You also consider that the provision of the new facilities does not make the building less satisfactory than before, in relation to Requirement B1 of the Building Regulations 1991.

The Borough Council's case

The Borough Council contend that the existing staircase, serving the new roof extension which comprises two additional storeys, should be provided with lobby protection. They consider that lobbies are necessary to provide additional protection for the increased number of persons accommodated on the new roof extension. In reaching their decision the Borough Council have referred to paragraph 4.26 of Approved Document 'B' which indicates situations where an escape stair needs the added protection of a protected lobby. In particular, they consider the following sub-paragraphs of paragraph 4.26 of the approved document to be applicable to this case:

(b) where a stair serves any storey at a height greater than 20 m; or
(c) where the building is designed for phased evacuation; or
(d) in a sprinklered building in which the stair width has not been based on discounting one stairway.

The Borough Council consider that the measures you have proposed as compensatory features are measures which would be required for this type of building to be in accord with paragraph 4.22 of Approved Document 'B'; and they do not, therefore, consider them to be additional measures which negate the need for lobby protection.

The Borough Council do not consider the fire shutters, which you have also proposed as a compensatory feature, to be relevant on the basis that not all openings to the stair are so protected.

The Department's view

The Department considers that the Borough Council were correct in giving consideration to the need for lobby protection to the existing Georgian stair at all levels throughout the store. Notwithstanding that the building, as existing, has eight above ground floors, the sixth and seventh floors (roof levels 1 and 2) are being extended and the works therefore constitute 'building work' in terms of Regulation 3 and must comply with the requirements of Regulation 4 of the Building Regulations 1991. The guidance given in paragraph 4.26 of Approved Document 'B' with regard to the provision of lobby protection could therefore be considered to be appropriate for this case. However, you have particularly requested that the Department takes into consideration the fact that the building is listed and you have referred to guidance given in the approved document on such buildings.

The Department has therefore needed to consider the risk to the building occupants, and in particular to the occupants of the new extended floors on the roof, of not providing lobby protection to the Georgian stair.

The Borough Council have referred to stair widths and capacities but have submitted no figures to the Department to justify any cause for concern over this issue. A table of exit capacity figures relating to the zonal roof population densities has been submitted as part of the store's fire safety strategy and this has not been disputed by the Council.

The Department accepts that some of the 'fire engineering' features that you have proposed as compensatory features would normally be expected to be provided in a building of this height. However, having assessed the proposals and the need for lobby protection the Department considers the following points to be relevant as compensatory features:

(i) Extensive compartmentation provided throughout the building both horizontally and vertically.

(ii) Means of escape routes which are available both vertically and horizontally; horizontal movement being into evacuation zones with smoke fill times having been calculated for each zone.

(iii) A full smoke detection system to a property protection standard (P1) which has been installed throughout the building.

(iv) The provision of smoke extract systems to all areas.

(v) The provision of a public address and voice evacuation system.

(vi) The provision of a comprehensive 'Fire Safety Strategy' for the store which states that five fire officers must be on duty whilst members of the public are on the premises. The level of store fire officer training and expertise is such that they are able to provide a breathing apparatus team.

The Department has taken account of the above features and the alternative escape routes available. It has also borne in mind the fact that there are listed building constraints. It is the Department's opinion that Requirement B1 of the Building Regulations 1991 will be satisfied without the provision of lobby protection to the Georgian stair throughout its length.

The Secretary of State's decision

You have appealed to the Secretary of State against the decision of the Borough Council not to dispense with Requirement B1 in respect of the provision of lobby protection to the Georgian staircase. He has given careful consideration to the facts of this case and the points made by both parties. He has also paid particular attention to the fact that internal features of the building are listed.

Compliance with Requirement B1 is a life safety matter and the Secretary of State would not normally consider it appropriate to relax or dispense with the Requirement. In this particular case it is the safety of the public and employees which is at issue.

However, in this case he considers that your proposals do in fact demonstrate compliance with Requirement B1. Therefore insofar as there is a difference of opinion regarding such compliance between you and the Borough Council, he hereby formally relaxes Requirement B1 to the extent necessary for you to carry out your proposed building work and accordingly upholds your appeal in this respect.

Department store – firefighting shaft (B.5)

A firefighting shaft and lift were not to be provided in association with alterations to the upper floors. There were a number of compensating features and these led to the appeal being successful.

Ref.: 45.3.89   Date: 30 May 1997

The appeal

Your proposals relate to Phase 6 of the roof development of the existing nine storey department store (of which eight are above ground), which is a listed building in respect of both the external façade and certain interior features. The existing roof level accommodation comprises three levels on the sixth, seventh and eighth floors. The sixth and seventh floors contain engineering offices, workshops, plant rooms and architects' offices, etc. The eighth floor contains a small conference area.

This phase of the roof works involves the removal of some existing roof accommodation and the construction of two levels of roof extension. The extended sixth floor (roof level 1) will mainly be for storage, and the extended seventh floor (roof level 2) will mainly be for office accommodation. The eighth floor (roof level 3) will remain unchanged and you state will accommodate 15 people only.

The existing and proposed floor areas are as follows:

(i) sixth floor – currently $2314\,m^2$; to be extended by $806\,m^2$;

(ii) seventh floor – currently $326\,m^2$; to be extended by $1016\,m^2$;

(iii) eighth floor – currently $265\,m^2$; no extension.

You have stated that four stairs are available from the roof, two of which already benefit from lobby protection. One of these stairs (marked stair No. 5) is to be extended and upgraded, including the provision of lobbies. This means that three of the four stairs have (or will have) lobby protection but the fourth stair, which is an existing Georgian stair and which currently serves all floors up to the fifth floor level, will not benefit from lobby protection.

One of the escape routes from the seventh floor (roof level 2) is via a designated external route to the link stair at the sixth floor (roof level 1). This link stair connects with the existing unlobbied Georgian stair which is the subject of a separate Requirement B1 ('Means of escape') appeal to the Secretary of State in respect of the non-provision of lobby protection to this stair.

You state that protected shafts are provided with wet risers. However, your clients do not propose to provide any firefighting shafts incorporating firefighting lifts.

These proposals formed a full plans application which was rejected by the Borough Council on several grounds including non-compliance with Requirement B5 ('Access and facilities for the fire service') by virtue of the fact that your clients had made no provision for a firefighting shaft incorporating a firefighting lift. You were able to satisfy the Borough Council's schedule of defects with the exception of provision of the firefighting shaft and lift. You therefore made a re-submission of your full plans application together with a request to the Borough Council to relax section 17 of Approved Document 'B' ('Fire Safety') on the grounds that, given the listed building status of the building, the fire engineering strategy would provide an acceptable alternative.

Approved documents are documents approved by the Secretary of State to provide practical guidance on meeting the requirements of Schedule 1 and Regulation 7 of the Building Regulations 1991. As guidance it is neither appropriate – nor is there any statutory procedure – to relax or dispense with matters in these approved documents. However, the Borough Council took your application as one 'to dispense with the requirements of B5 with regard to the firefighting lift' and refused it.

You consider that on various grounds your proposals would justify the relaxation/dispensation of Requirement B5 in respect of the need to provide firefighting shafts, incorporating firefighting lifts, in buildings with floors in excess of $20\,m^2$ above ground level. You therefore appealed against the refusal of your application for a relaxation/dispensation to the Secretary of State.

The Appellant's case

You state that your application for a relaxation related only to section 17 of Approved Document 'B' and not to Requirement B5 in its entirety. You made the following points in support of your original application for a relaxation:

(i) The appeal relates to an existing nine storey building which currently is not provided with any firefighting lifts and the proposed work involves extending the accommodation on the 6th and 7th floors (roof levels 1 and 2) only.

(ii) Paragraphs 0.10–0.13 and 0.17 of Approved Document 'B' are relevant because they mainly relate to the application of Part B of the Building Regulations 1991 to existing buildings, and in particular to buildings of special architectural or historic interest.

(iii) The proposals do not extend the height of the building and additional lobby protection has been provided on the fifth and sixth floors to the link staircase.

(iv) The proposals do not involve the construction of additional floors in the building.

(v) The store has an operational fire department which has facilities to assist the fire service. In addition, a fire safety strategy for the store has been provided as part of the appeal submission.

(vi) The store is a listed building and the listing covers both the external façade and internal features.

In addition, you state that a number of fire engineering provisions have been included in the proposals. These comprise: a sprinkler installation to all areas; a fire alarm system to all areas; a smoke extract system to all areas; wet risers to Stair No. 5; provision of two hose reels at each level; provision of hand-held extinguishers at each level; and a voice evacuation and public address system.

You consider that the above provisions should be taken into account. You also take the view that the proposed alterations do not make the building less satisfactory than before in relation to Part B of the Building Regulations. Furthermore, you consider that fire protection and escape facilities in respect of the floors under consideration have been significantly improved. The building is designed for phased evacuation and you have stated that the top storey, i.e. the eighth floor (roof level 3), is approximately 31.5 metres above street level.

The Borough Council's case

In support of their decision to refuse your application to relax/dispense with Requirement B5, the Borough Council maintain that the firefighting lift is required to serve the new roof extension. They consider that the roof extension to the store, comprising two additional storeys on the roof, is subject to Requirement B5 by virtue of Regulation 4. They point out that under Requirement B5 the substantive requirement is that the building be designed and constructed so as to provide facilities to assist firefighters in the protection of life. For a building of this height a firefighting shaft with a firefighting lift is deemed to be necessary by the Borough Council. They point out that a firefighting lift enables firefighting personnel to reach a high-level fire quickly with their equipment.

The Borough Council recognize that firefighters already have the task of reaching the roof to tackle a fire at that level in the building as existing. However, the Council's view is that with the installation of a firefighting lift to the existing roof or even to the fifth floor, firefighters would be able to reach a fire on the seventh floor quickly without expending energy climbing stairs.

In reaching their decision, the Borough Council had regard to Section 17 of Approved Document 'B' which gives guidance on the provision of firefighting shafts incorporating lifts. They have also referred to BS 5588: Part 5: 1991 ('Code of practice for firefighting stairs and lifts').

The Borough Council consider that many of the 'fire engineering' provisions put forward by you as compensatory features are already necessary for this building in order to satisfy Requirements B1 and B3 of the Building Regulations 1991. They do not consider such features to be additional measures which negate the need for the provision of a firefighting lift.

The Borough Council accept that in-house firefighters may tackle a fire in its early stages but their view is that the local fire brigade may also still need to attend and be able to reach a fire on the upper storeys. They can therefore find no reason to dispense with Requirement B5 of the Building Regulations 1991.

The Department's view

The Department considers that the Borough Council were correct in giving consideration in this case to the need for the provision of firefighting shafts incorporating firefighting lifts. Notwithstanding that the building, as existing, has eight above ground floors, the sixth and seventh floors (roof levels 1 and 2) are being extended and the works therefore constitute 'building work' in terms of Regulation 3 and must comply with the requirements of Regulation 4 of the Building Regulations 1991. The guidance given in Section 17, 'Access to building for firefighting personnel' of Approved Document 'B', with regard to the provision of firefighting shafts and lifts, could therefore be considered to be appropriate to this case. However, you have particularly requested that the Department takes into consideration the fact that the building is listed and you have referred to general guidance given in Approved Document 'B' relating to the possible appropriateness of varying the provisions for such buildings.

Diagram 45 and paragraph 17.2 of Approved Document 'B' relate to the provision of firefighting shafts, and suggest that buildings with a floor at more than 20 metres above ground or fire service access level should be provided with firefighting shafts containing firefighting lifts. The Department has therefore needed to consider the risk to life safety, with particular respect to the new extended floors on the roof, of not providing a firefighting lift.

The Department accepts that some of the 'fire engineering' features that you have put forward as compensatory features would normally be expected to be provided in a building of this height. However, after assessing the proposals with regard to the need for the installation of a firefighting lift, the Department considers the following points to be relevant in this case:

(i) The extensive compartmentation provided throughout the building, both horizontally and vertically.

(ii) The sprinkler system provided to all areas.

(iii) The provision of a full smoke detection system to a property protection standard (P1), installed throughout the building.

(iv) The provision of smoke extract systems to all areas.

(v) The comprehensive 'Fire Safety Strategy' for the store which has been provided as part of your appeal submission and which states that five fire officers must be on duty whilst members of the public are on the premises. The level of store fire officer training and expertise is such that they are able to provide a breathing apparatus team.

(vi) The fact that the new roof extensions will be relatively small when compared with the building as a whole.

The Department has taken account of the above features, together with the listed building constraints and is of the opinion that Requirement B5 of the Building Regulations 1991 will be satisfied without the provision of a firefighting lift. In reaching this conclusion the Department has obtained expert advice with respect to firefighting procedures and operations.

The Secretary of State's decision

You have appealed to the Secretary of State against the decision of the Borough Council not to relax/dispense with Requirement B5 in respect of the provision of firefighting shafts, incorporating fire lifts. He has given careful consideration to the fact of this case and the points made by both parties. He has also paid particular attention to the fact that internal features of the building are listed.

Compliance with Requirement B5 is a life safety matter and the Secretary of State would not normally consider it appropriate to relax or dispense with the requirement. However, in this case he considers that your proposals do in fact demonstrate compliance with Requirement B5 of Schedule 1 to the Building Regulations 1991. Therefore, insofar as there is a difference of opinion regarding such compliance between you and the Borough Council, he hereby formally relaxes Requirement B5 to the extent necessary for you to carry out your proposed building work and accordingly upholds your appeal in this respect.

Listed building – conversion to prep school (B.1)

The developer and the local authority could not agree regarding the amount of protection to be afforded to two external stairs to be provided when the listed building was converted into a preparatory school.

Ref.: 45.3.100   Date: 26 June 1997

The appeal

The appeal relates to the change of use and conversion of a listed residential building, which has remained unoccupied and unmaintained for the last 20 years, to a non-residential educational establishment for 3–11 year old children.

The building is of brickwork construction with timber floors and consists of basement, ground, first and second storeys. The proposal is to provide two external stairs so that each floor, including the ground floor, has an alternative escape route. Both external stairs discharge at basement level.

One of the external stairs serves the second floor only and is shown on drawing GR 1612/93 rev B (sheet 4) to have a height of 8.9 m from basement ground level to the top landing. The stair has four straight flights and three intermediate landings. It is sited within an open well of the building, with various windows on all floors and in each of the three sides of the building facing onto it. The other stair provides alternative escape from the ground and first floor and is shown on drawing GR 1612/93 rev B (sheet 4) to have a height of 6.12 m from basement ground level to the top landing. This stair is sited within a right angle formed by the walls of the building at the rear. It comprises three straight flights with two intermediate landings. Secondary Georgian wire glass has been installed in the two windows adjacent to the route of this stair. It is understood that the approved plans propose canopies over the exit doors at the top of both stairs.

The normal access/egress route from the first and second floors is via a single internal split level stairway which is shown on the drawings to be provided with fire protection. This stairway discharges in the ground floor passageway leading to the front entrance.

A full alarm and detection system is to be provided and will be installed to an L1 standard in accordance with BS 5839: Part 1: 1988.

The proposals are based on a fire engineering approach to meeting the requirements of Part B; both parties having been concerned to respect the listed building status of the building. Your original full plans application for the alterations and the provision of the two external fire escapes was approved by the Borough Council on 8 September 1994. However, it appears that you subsequently requested a reduction in the level of weather and fire protection to these stairs as provided for in the approved full plans.

9. *In the case of the stair serving the second floor only*, the Borough Council have required the following:

(a) weather protection in the form of a simple canopy covering over the exit door at the top of the stair;
(b) fire resistance to two second floor windows opening onto the route of the fire escape marked 'A' in red on drawing GR1612/93 rev G (sheet 2) which the drawing shows as opening from a WC and a meeting room;
(c) fire resistance to six windows also facing the route of the fire escape marked 'A' in red on drawing GR1612/93 revision G (sheet 1). Three of these windows are on the first floor, two opening from a passageway and one from the room notated as the head's office. The other three windows are on the ground floor with two opening from a passageway and one from the secretary's room.

In the case of the second alternative escape stair serving the ground and first floor, the Borough Council has required as a minimum that weather protection in the form of a simple canopy covering should also be provided to the exit door at the top of this stair.

However, you believe that the Borough Council's requirement listed at paragraphs 9 and 10 above are not necessary for you to achieve compliance with Requirement B1. You therefore applied for a dispensation/relaxation of the provisions which was refused by the Borough Council. It is against that decision that you appealed to the Secretary of State.

The Appellant's case

Your grounds of appeal centre on the Borough Council's refusal to dispense with paragraphs 2.45 and 4.36 of Approved Document 'B' ('Fire Safety'). It is your contention that the building's existing fire

precautions secure reasonable standards of health and safety for persons in or about the building.

Your appeal relates to two specific issues:

(1) the provision of secondary fire resisting glazing to windows opening onto the external escape stair serving the second floor; and,
(2) the weather protection of the uppermost doors/landings giving access onto both escape stairs.

In respect of issue (1) above (secondary fire resisting glazing) you make the following points:

(i)   The fire alarm system will result in the speedy evacuation of persons from the building (approximately 90 seconds).
(ii) The risk of fire spread in the areas in question is minimal because they have a lower fire load than the classrooms, and in addition the rooms in question are separated from the rest of the building by fire resisting construction.
(iii) The windows in question are the only form of ventilation and it would be necessary to mechanically ventilate these rooms if the fire resisting glazing were installed. There would also be a tendency to prop the fire doors open.

In respect of issue (2) above (weather protection) you make the following points:

(iv) The proposed canopies will provide minimal protection to the areas in question and it will always be necessary to provide additional management procedures to keep the stairs free from snow and ice.
(v)   The stairs serving the second floor are protected on three sides by the building and the lower rear stair has some protection on two sides from the projecting box gutter.

The Borough Council's case

The Borough Council point out that when considering the fire safety design of the building, reference was made to Approved Document 'B' ('Fire Safety') which in turn refers to Building Bulletin 7 for educational establishments. However, due to the building layout and the fact that it is a listed building the Borough have accepted that compliance with Building Bulletin 7 is not possible. Meetings, which included the Fire Authority, were held to agree an acceptable fire engineering solution.

In response to your appeal submission the Borough Council have made the following points:

(i)   The secondary means of escape forms an integral part of the escape route design and needs to be protected at all times to ensure it is effective.
(ii) They dispute your contention that the building can be evacuated in 90 seconds in a real fire.
(iii) They do not accept that the fire loading in some of the rooms in question can be considered as low.
(iv) Originally a completely covered route was put forward for both stairs but consideration was given to the sheltered nature of the building, the site geography, and the fact that the building is listed. The Borough Council concluded that as a minimum some weather protection should be provided to the landings serving the uppermost doors opening onto the two external stairs.
(v)   A fire engineering solution consists of a number of integral parts and should not be jeopardized by the removal or dispensation of any of those parts.

The Department's views

You originally applied to the Borough Council for a relaxation/dispensation of paragraphs 2.45 and 4.36 of Approved Document 'B'. However, approved documents are documents approved by the Secretary of State to provide practical guidance on meeting the requirements of Schedule 1 to, and Regulation 7 of, the Building

Regulations 1991 (as amended). As guidance it is therefore neither appropriate – nor is there any statutory procedure – to relax or dispense with matters in these approved documents. The Borough Council therefore took the view that your application was for the dispensation of Requirement B1 and you state that you subsequently amended your application to that effect.

Paragraph 4.36 of Approved Document 'B' gives recommendations for protection of the external stair against inclement weather, particularly where the stair exceeds 6 m in height. Paragraph 2.45 of Approved Document 'B', although primarily applicable to flats and maisonettes, is referred to in paragraph 4.36 and gives recommendations with respect to the fire protection of external stairs.

The Department notes that the conversion of this building from residential to that of a non-residential school invokes Regulation 5 ('Meaning of material change of use') and, as a consequence, Regulation 6 ('Requirements relating to material change of use'). This is because the definition of 'public building' in Regulation 5(e) includes a school or other educational establishment. Appendix 'D' of Approved Document 'B' also suggests that educational establishments should be included in the 'Assembly and Recreation' purpose group.

Paragraph 4.35 of Approved Document 'B' suggests that an external stair in an assembly building should not be used by members of the public. The inference of this paragraph is that whoever the persons are who are using the stair they should be able to do so safely. In the case of a school this situation may not always be achievable, given the use to which schools can be put after normal school hours and the ages of the children during school hours.

The Department is of the opinion that the Borough Council have acted reasonably in permitting external stairs to be provided with the minimum of weather protection. Whilst the Department accepts this approach by the Borough Council, it can see no justification in permitting a further reduction to the degree of weather protection which the Borough were prepared to accept as a minimum.

With regard to the fire protection to the external stair serving the second floor, the Department considers that it could reasonably be argued that the two windows on the second floor (i.e. the windows to the WC and meeting room) could be considered to satisfy Requirement B1 if they remained openable and were not provided with fire resistance. This takes account of the early warning system and the fact that the rooms are on the top floor. However, the Department does not take the same view with regard to the windows on the lower floors. The Department has reached this view because it is important that account is taken of the ages of the children and the fact that they may not all be available for orderly escape when the fire alarm system is activated. The Department therefore is of the opinion that all the windows on the ground and first floors that open onto the external stair serving the second floor should be fixed shut and provided with fire resistance in terms of integrity.

The Secretary of State's decision

You have appealed to the Secretary of State against the Borough Council's decision to refuse to relax/dispense with Requirement B1 of the Building Regulations 1991. Requirement B1 is a life safety matter and as such the Secretary does not normally consider it appropriate to either relax or dispense with it.

The Secretary of State has given careful consideration to the facts of this case and the arguments put forward by both parties. Additionally, he has noted that the building is listed but has not identified any specific arguments in favour of relaxation relating to the fact that the building is listed. He has concluded that there are no extenuating circumstances which would justify consideration of a relaxation of Requirement B1 in this case; particularly giving that children would comprise the majority of persons at risk. He has therefore concluded that the Borough Council came to the correct decision in refusing to relax/dispense with Requirement B1 of Schedule 1 to the Building Regulations 1991 (as amended) and hereby dismisses your appeal.

PART E: RESISTANCE TO PASSAGE OF SOUND

Conversion of dwelling into eight flats for elderly people (E.1, E.2 and E.3)

The proposal is to convert an existing dwelling into eight flats for seven frail, elderly people and a house mother. The local authority refused to dispense with required standards for airborne sound for walls and ceilings and impact sound for floors.

The Secretary of State dispensed with the requirements but this was subject to a direction relating to any other future use; something a local authority could not do.

Ref.: 45.3.91   Date: 25 June 1996

The building to which this appeal relates is an existing, substantially built, Victorian single dwelling which is being converted into a *Home* of eight flats to provide sheltered accommodation for seven frail, elderly people and a house mother. The main accommodation will be on three floors comprising of three studio type self-contained flats (i.e. bedroom with en-suite bathroom and kitchenette), together with shared lounge and dining room, and main kitchen on the ground floor; four studio type flats on the first floor; and one larger flat on the second floor (for the house mother).

In addition to the alteration work required, the building work also involves the building of a two-storey extension at either side of the property after demolition of an existing garage and outrigger; a two storey external fire escape and a single-storey rear extension; together with a disabled access ramp at the front of the property.

The conversion has planning permission subject to conditions, one of which is: 'The building shall be used as a Home and for no other purpose, including any other purpose in Class 2 of the Schedule to the Town and Country Planning (Use Classes) Order 1987.'

Your original full plans for this work were approved on 5 September 1995 and were accepted by the Council as showing compliance with the requirements of Part E. However, whilst accepting that Part E was applicable to the building work, because of the costs of meeting these relatively new requirements, and the fact that your clients do not believe that noise will be a problem to the proposed elderly occupants, your client requested the Council to dispense with the requirements of Part E. The Council was particularly concerned that if it dispensed with the requirements of Part E it would not have any subsequent Building Regulation control locus to ensure reasonable provision for sound insulation if the building were to be sold as flats. The Council therefore refused to dispense with Requirements E1, E2 and E3. It was against that refusal that you appealed to the Secretary of State.

The Appellants' case

In support of your appeal you state that, in view of the proposed use as sheltered housing for the elderly, the noise levels generated in respect of both airborne and impact sound will be minimal, and certainly no greater than that found within a domestic property. You contend that such noise as will be generated will be adequately contained by the existing structure separating the individual bedrooms.

In your letter you accepted the Council's view that the proposals constitute the creation of flats (i.e. individual dwellings). However, you point out that the Planning Permission granted by the Council is subject to a condition requiring the use of the property as 'a Supportive Home and for no other purpose . . .' Your understanding is that should your clients, or future owners, wish to let or sell individual bedrooms as flats this would require a further application for Change of Use and a subsequent application for Building Regulation approval to meet the standards required for individual dwellings.

The Borough Council's case

The Council view the conversion of the property as a material change of use by virtue of the living accommodation falling within the definition of flats (i.e. separate self-contained premises). In so doing, they contend that all buildings converted into flats should achieve standards of sound insulation suitable for all types of domestic occupancy. They are concerned that if these standards are not achieved they would have no powers under Building Regulations to require up-grading work for sound insulation if the building were to be sold as flats at some future date.

The Council also state that they are mindful of the fact that requirements for sound insulation in flat conversions are relatively new and were introduced to address an existing problem. In their view the additional costs of such works is the inevitable result of higher standards being applied. With all this in mind they take the view that it would be inappropriate to dispense with the requirements.

The Department's view

The Department has noted that although the individual units afford a degree of privacy to the residents, they are not designed as fully self-contained residential units. There is no dispute that the units are not designed to include facilities for the keeping and preparation of main meals. In addition it is noted that the home will be run on the basis that residents will take their main meals in the communal dining room, and that there will be a common sitting room with television. There will also be a resident house mother and other nursing and domestic care is also provided as occasion requires. From these facts it would seem that the property will have a use amounting far more to a residential home than to a block of self-contained, serviced flats.

On one view, therefore, the home is not a block of self-contained flats. In that case the material change of use would be from a residential to an institutional use as defined in Regulation 5(d) of the Building Regulations 1991. If that were the case the requirements of Part E of Schedule 1 to the Regulations would not apply to that change of use by virtue of the effect of Regulation 6(1).

The Department, however, notes that both the Council and the appellant have acted on the view that the units in the Home may and should be regarded as self-contained flats. The building work would then fall to be defined as a material change of use under Regulation 5(b) and would be required to comply with Part E by virtue of Regulation 6(1)(e).

The Department notes that whichever view is taken it would appear that if, as postulated by the Council, the building were to be sold as flats sometime in the future, a material change of use should be triggered by either Regulation 5(a) or 5(g) respectively and the Council would have a Building Regulation control locus to apply the full compliance of Part E if that was appropriate.

Notwithstanding the above point, the Secretary of State is being asked to consider an appeal based on the view that the proposed units are to be self-contained flats. The Department is prepared to accept that view as the basis on which the appeal should be decided.

Appeal decision

The Secretary of State has considered the facts of this case. In particular, he has considered the concerns of your clients to minimize and avoid unnecessary costs in respect of noise insulation; and he has also considered the legitimate desire on the part of the Council to ensure that any future change of use can be properly controlled in respect of future compliance with Part E.

In the particular circumstances of this case, the Secretary of State has concluded that it is appropriate to dispense with Requirements E1, E2 and E3 of the Building Regulations 1991 (as amended) in their application to the proposed building work to be carried out at the premises, subject to the following direction:

that the dispensation shall cease at such time as the building is no longer in institutional use to house the frail and elderly in individual private units in combination with the provision – on a community basis within the building – of main meals, cooking and eating facilities, lounge and nursing or domestic care.

Change of use – sound insulation to stairs (E.2)

The proposal involved conversion into a mixed domestic and non-domestic use. Although a ground floor workshop did not currently cause noise problems this may not always be so.

Ref.: 45.1.101   Date: 10 October 1996

The appeal

The building to which this appeal relates is an end of terrace two storey house which has been previously converted to non-domestic use, and which has now been converted further by your client to form a mixed domestic and non-domestic use. Before the current conversion of the building the ground floor comprised a reception room, two offices, a storage room and a kitchen; and the first floor comprised three offices, a kitchen and a WC. The ground floor now comprises a studio flat converted from the office in the centre of the building and the kitchen, together with the remaining reception and adjoining storage room; and the first floor comprises a two bedroom flat which has been created out of all the previous accommodation on this floor. The reception and adjoining room (marked as storage on the drawing) on the ground floor are described in the correspondence as reception/office and workshop respectively.

The sound insulation of that part of the floor between the first floor flat and the studio flat below has been improved, but the rest of the floor over the ground floor reception and storage room has not been so improved.

Your original full plans application for this work was rejected by the City Council on the grounds of non-compliance with Requirements B3, E2, E3 and F1. You re-submitted plans together with a written application for the dispensation of Requirement E2 and E3. These plans were also rejected together with your dispensation application.

The City Council believes that a reasonable degree of sound insulation should be provided between the first floor flat and the workshop and storage area below. You do not consider this to be necessary and therefore appealed to the Secretary of State against the City Council's refusal to dispense with Requirement E2.

The Appellant's case

You argue on behalf of your client that application of Requirement E2 is excessive for the following reasons:

(i)   The business is only open from 0900–1730 on Mondays to Fridays so that outside these hours there will be less noise than is normal in a dwelling;

(ii) only one person is employed in the workshop below and there has not been an intensification of use;

(iii) the occupants of the first floor flat are in full-time employment and are therefore unaware of the activities within the premises below; and

(iv) the occupants of the flat have never complained about the noise, even during working hours.

Finally, you have confirmed that the need for additional sound insulation between the first floor flat and the studio flat on the ground floor is accepted and is not being contested.

The City Council's case

The City Council have rejected your plans because they consider that Requirement E2 is applicable, and that the floor separating the first floor flat from the reception and workshop rooms below should be upgraded to provide reasonable insulation against airborne sound, in the same way as provided for the floor between the studio flat and first floor flat.

Whilst the City Council does not dispute that there is no noise problem at present, they consider the reasons put forward by you for dispensation of Requirement E2 are reliant on the building's current use and the present occupier's circumstances and tolerances. They point out that should the circumstances of those persons residing in the first floor flat change to that of shift working, illness, or unemployment, or if they should leave and new persons take up residence, then the lack of adequate sound insulation could be detrimental to their health and result in legal action being taken under the Environmental Protection Act 1990. Similarly, the City Council argue that should the user of the workshop change, or additional persons be employed or hours of operation be extended (none of which can be controlled under the Building Regulations), excessive noise could result, again constituting a nuisance to the occupier of the flat above.

The Department's view

Requirement E2 is that:

A floor or a stair which separates a dwelling from another dwelling, or from another part of the same building which is not used exclusively as part of the dwelling, shall resist the transmission of airborne sound.

In the Department's view the latest alterations have resulted in a material change of use as defined under Regulation 5(b) of the Building Regulations 1991 (as amended). As a consequence Regulation 6(1)(e) applies and compliance is therefore required with Requirement E2. In this instance clearly Requirement E2 is applicable and although there is no noise problem at present the Department accepts the City Council's concern that the noise level in the workshop could increase, and that the occupants of the flat could spend more time at home, with the result that airborne noise could become a problem.

The decision

The Secretary of State has given careful consideration to the particular circumstances of this case. He is satisfied that Requirement E2 would normally be applicable in such circumstances, and has considered the arguments you have put forward for dispensing with Requirement E2. However, he has concluded that there are no special circumstances which would warrant relaxation or dispensation of Requirement E2 and that the City Council came to the correct decision in refusing your application to dispense with Requirement E2. He accordingly dismisses your appeal.

Maisonette – sound resistance (E.1, E.2 and E.3)

It was proposed to form a maisonette above an existing ground floor flat. The local authority refused permission because of the lack of adequate sound resistance between the maisonette and the flat. The appeal was dismissed.

Ref.: 45.3.108   Date: 7 May 1997

The appeal

The proposed work to which this appeal relates comprises the alteration of an existing three storey semi-detached dwelling which was converted to three self-contained flats (one on each floor) some time ago. As owner of the second (top) floor flat your client has now purchased the first floor flat and proposes to convert both flats into a maisonette.

It is not clear from the plans how many bedrooms there are in total in the first and second floor flats; but it appears that the first floor flat has one bedroom. The plans show that after conversion the maisonette will have three designated bedrooms – one on the first floor and two on the second. However, there will also be two lounges – one on each floor, and the one on the second floor could become an extra bedroom or be interchanged with the bedroom on the first floor, making another living room in which noise can be produced on the first floor.

These proposals were the subject of a full plans application which was rejected by the Borough Council on the grounds that they did not show compliance with requirements E1, E2 and E3.

However, your client considers that there was no justification to impose the requirements of Part E of the Building Regulations because the proposal merely combines the top two flats into one and that the separation between ground and first floor flats effectively remains unchanged. You therefore applied to the Borough Council for a dispensation from Requirements E1, E2 and E3 which was refused by the Borough's Development Control Sub-Committee. It was against that refusal that you then appealed to the Secretary of State.

Before considering your application for dispensation the Borough Council formally consulted the occupier of the ground floor flat under section 10(2) of the Building Act 1984 and the response is indicated in the Borough Council's enclosures.

The Appellant's case

In support of your original application to dispense with the requirements of Part E, you state that the room use and layout on the first floor are to remain unchanged, and that the only alterations on the first floor are at the entrance area in order to provide access to the upper floor within the maisonette. You consider that as the room uses and layout of the first floor are not changed there will be no difference to the occupation of this floor and that there is no reason to suspect that any noise problems will occur. You therefore consider that there does not seem to be any justification for additional sound insulation and that it is therefore not reasonable to impose the requirements of Part E to the alteration work.

The Borough Council's case

The Borough Council consider that construction of the maisonette constitutes a material change of use under Regulation 5(g) of the Building Regulations 1991 (as amended) and that the requirements of Part E therefore apply. The Borough have therefore asked your client to improve the sound insulation of: (a) the first floor structure; (b) the wall between the common area and the habitable rooms; and (c) the stairs from ground to first floor.

The Borough Council also consider that because the maisonette will have three bedrooms, future occupation by a family is more likely with the attendant possibility of greater noise generation.

The ground floor flat occupier's view

The occupant of the ground floor flat has commented that the only noise problem at present occurs in the kitchen when it is used by guests late at night or when it is used in the early hours and the washing machine may be operated. The kitchen is over the second bedroom in the ground floor flat. The occupant believes that the appellant is willing to soundproof this floor area but would not be happy with the installation of suspended ceilings if soundproofing were to be required overall.

The Department's view

You have appealed to the Secretary of State against the refusal of the Borough Council to dispense with all three requirements of Part E of the Building Regulations 1991. These requirements are:

E1   'Airborne sound (walls)'
E2   'Airborne sound (floors and stairs)'
E3   'Impact sound (floors and stairs)'

They exist to protect occupants of buildings from the harmful effects of the passage of sound which in some circumstances can affect health. The Secretary of State would therefore not consider it appropriate to dispense with, or relax, these requirements except in special circumstances.

In the Department's view the circumstances you have described do not constitute ones which would warrant dispensation or relaxation. In this particular case sleeping accommodation is involved, and the transmission of sound may result in sleep disturbance which can be injurious to the health of occupants. Moreover, the maisonette will have three bedrooms and therefore have the potential to house more occupants with the accompanying potential for noise nuisance to be greater than before. Although the present occupier of the ground floor flat may not be very concerned about the present noise levels, another occupier of the ground floor flat might be more sensitive to these noise levels. Moreover, those noise levels themselves might increase depending on the occupiers of the maisonette.

In considering this case the Department has formed a judgement as to what would be necessary to achieve compliance with all three requirements of Part E. The Department agrees with the Borough Council's judgement that the first floor should be upgraded to reduce noise transmission. It also would seem reasonable to require the upgrading of the wall on the ground floor that separates the common entrance from the main bedroom. However, because the stairs are over a cupboard with a hallway on one side and a kitchen at the end, the Department's judgement is that it should not be necessary to treat these areas.

The Secretary of State's decision

The Secretary of State has given careful consideration to the facts of this case and the points made by all parties.

Compliance with the requirements of Part E may be a matter affecting the health of occupants, particularly in a domestic situation where sleeping accommodation may be subject to noise. In the particular circumstances of this case the Secretary of State considers that it would not be appropriate to relax or dispense with the requirements of Part E, and has concluded that the Borough Council came to the correct decision in refusing to dispense with the requirements. Accordingly, he dismisses your appeal.

PART F: VENTILATION

House extension – means of ventilation (F.1)

The appeal related to a two storey extension to a detached house and included vertically sliding windows installed without trickle ventilation, as indicated on approved drawings. It was not appropriate to dispense with the requirements of F.1.

Ref.: 45.3.59   Date: 7 February 1995

Careful consideration has to be given to the facts of the case and the representations made by the District Council and yourself.

The building work in this case has comprised the construction of a two storey extension to a two storey detached house located in a conservation area. The extension contains primarily a kitchen annexe and a dining room on the ground floor, and two bedrooms and an en suite bathroom on the first floor. The work has been completed and includes vertically sliding windows which do not incorporate trickle vents for the provision of background ventilation.

Your plans were approved by the District Council in August 1993 and included specific provision for background ventilation to the windows throughout. The Council noted the absence of trickle ventilators after the work had been completed in March 1994. The District Council required that background ventilation be provided as indicated on the approved drawings. You then applied to the District Council for a dispensation of Requirement F1 which the Council refused. It is this refusal which is the subject of your appeal to the Secretary of State under section 39 of the Building Act 1984.

You have explained that the new windows were manufactured to match the existing ones. This resulted in an insufficient thickness in the top rail to install the trickle ventilators. You have also argued that remedial action to install background ventilation either through trickle ventilators in the bottom rails of the new windows, or ventilators in the walls, would be visually unacceptable because the house is in a conservation area.

The District Council has made the point that the approved plans indicated that trickle ventilators were to be installed in the windows

throughout and that they were not made aware of your reluctance to install them until the work had been completed. It is their stated view that some form of background ventilation could reasonably be provided.

The Department's analysis of this case is as follows. Although your proposals referred to the installation of trickle ventilators throughout the extension, it is the Department's view that the rooms which should be provided with background ventilation are the dining room, kitchen annexe, and bedrooms 2 and 4; although it is not clear whether ventilation has in fact been provided in the kitchen. Paragraph 0.7(c) of Approved Document F ('Ventilation') (1990 edition) states that part of the objective of Requirement F1 is to provide a means of 'achieving background ventilation which is adequately secure and does not significantly affect comfort in habitable rooms and kitchens, so as to encourage its use'. It is the Department's view that in order to satisfy the requirements of Part F1 of the Building Regulations 1991, background ventilation should be provided.

In the circumstances, the Secretary of State has concluded that it would not be appropriate to dispense with Requirement F1 of Schedule 1 to the Building Regulations 1991. Accordingly, he dismisses the appeal on the grounds that it is not appropriate to dispense with the requirement in question.

It is not the function of the Department to advise on specific ways of achieving compliance with a particular statutory requirement. However, you may wish to note that trickle ventilators are only one example of how background ventilation can be provided and alternative ways are permissible. In this particular case the vertical sliding windows could provide background ventilation by being opened a little at the top, without the installation of trickle vents. A solution along these lines would, however, need to be compatible with achieving adequate security so as to encourage the use of this method of background ventilation. Diagram 1 of the 1995 Approved Document 'F1' ('Means of ventilation') (which comes into force on 1 July 1995) illustrates such a solution.

Conversion of barn into dwellinghouse – ventilation (F.1)

This proposal involves the conversion of a Grade II listed building into a dwelling having two bathrooms. Ventilation of the bathrooms is in dispute. The local authority required mechanical extract fans and the owner appealed. The decision rejected the appeal and made suggestions, including the provision of passive stack ventilation.

Ref.: 45.3.57   Date: 21 April 1955

Careful consideration has been given to the facts of the case, and to your representations and those of the Borough Council.

This appeal relates to a material change of use involving work to convert a previously derelict timber-framed, two storey farm barn to a single, private house. The conversion incorporates three bedrooms, a guest room and two bathrooms. It is compliance of the ventilation of the two bathrooms that is at issue.

You submitted plans by way of a Building Notice to the Borough Council. Following an inspection on site, the Council wrote to you explaining that they considered that the converted building contravenes the Building Regulations 1991 in that mechanical extract fans are required to the bathrooms to comply with Requirement F1 of the Regulations but have not been provided. You then applied to the Council for a relaxation of this requirement which was refused. It is against this decision that you have appealed to the Secretary of State.

In support of your application for a relaxation of Requirement F1 you contend that:

(a) because the barn is a Grade 2 listed timber-framed barn on a very prominent site, extract grilles would significantly detract from the careful restoration achieved to date;

(b) the residence is now a private house with an individual owner–occupier; both bathrooms have large openable windows and the owner will not use any form of mechanical extraction, even if fitted;

(c) the owner is prepared to enter into a written agreement to install fans if the residence is sold.

In support of their decision to refuse your application for relaxation the Borough Council have made the following points in respect of yours at (a) to (c) in paragraph 6 above. These are:

(a) that after discussions with the Planning Department (both Development Control and Listed Buildings), it is their view that the installation of black matt finish extract grilles, similar to that already installed in the kitchen, should not present an aesthetic problem;

(b) that the converted barn is now a private residence to which the requirements of the Building Regulations must be applied as with any other dwelling and irrespective of individual occupier's preferences concerning the installation of mechanical extraction; and

(c) that any agreement would be unenforceable by the Council and that this point would appear to contradict your reason for requesting a dispensation in respect of point (a).

The Council have opined that ventilation is very necessary in bathrooms in order to overcome condensation problems and that any relaxation of the requirement needs to be given serious consideration.

Approved Document F ('Ventilation': 1990 Edition) states that 'the objective of the requirement is to provide a means of extracting moisture from areas where it is produced in significant quantities (kitchen, bathroom and shower-room)'. The guidance given in paragraph 4.1 of the Approved Document ('Ventilation of bathrooms') explains that 'the requirement will be satisfied by the provision of mechanical extract ventilation capable of extracting at a rate of not less than 15 litres per second . . .').

The Department notes that one of the bathrooms has a shower cubicle which increases the risk of condensation due to the higher levels of water vapour produced with showers. It must also take into account that you have provided no evidence that the large opening windows in the bathrooms will extract moisture to the outside effectively, particularly from the shower cubicle, which would reduce the risk of condensation.

Having given careful consideration to this case, the Department takes the view that the windows to the bathrooms do not provide ventilation that complies with Requirement F1 of the Building Regulations 1991, and that in the circumstances it is reasonable to expect the ventilation of the bathrooms to meet that requirement.

In considering this case the Secretary of State has been concerned to balance the over-riding needs of health and safety explicit in Requirement F1 with the constraints which conversion of this Grade 2 building may impose on full compliance with the requirement. In this particular case the Secretary of State has concluded that the provision of ventilation to the two bathrooms does not comply with Requirement F1. He also accepts the Borough Council's view that the appearance of the barn will not be unreasonably prejudiced by installation of the grilles. Accordingly, he dismisses your appeal.

It is not the function of the Department to advise on solutions to particular questions not raised between an appellant and a local authority; this is a matter for the parties concerned to resolve. However, it may be helpful if the Department offers the following informal observations in respect of the concern about the outlet terminals from the mechanical extraction being unsightly.

In the Department's view it should be possible for the outlet terminals to be installed discretely as suggested by the Borough Council (and as already done for the kitchen extract fan and other service outlets). Alternatively, the outlet terminals could be installed in special ridge tiles. Consideration could also be given as an alternative to mechanical extraction, to the installation of Passive Stack Ventilation. Guidance on this has recently been published by the Building Research Establishment in an Information Paper 13/94, 'Passive stack ventilation systems: design and installation'. These are only suggestions and it is accepted that there may be other ways of installing the outlets more suited to the property and the minimization of their aesthetic impact.

Extension of dwelling – ventilation of enlarged rooms (F.1)

Two bedrooms are extended on the first floor and existing windows reused. This resulted in an inadequate area for rapid ventilation and the appeal was dismissed. However, a solution to the problem suggested by the local authority was, in the Department's view, acceptable.

Ref.: 45.3.95 Date: 12 September 1996

The building to which this appeal relates is an existing detached three bedroomed two storey house with an existing extension at ground floor level. The new work which is the subject of this appeal comprises a first floor extension built over part of the existing ground floor extension. It contains two rooms each of which adjoin two bedrooms at the rear of the property. The new rooms communicate with the existing rooms through the original window openings which have been enlarged to form door openings, although doors have not been fitted. The new rooms are not regarded as 'new' but are taken to be extensions of the original bedrooms. Both extensions have been fitted with windows containing only top-hung fanlights.

Your original full plans for this work showed double glazed windows comprising side-hung sashes and trickle vents to each of the two extensions. It was on this basis that you received conditional approval for those plans. However, the Borough did not accept that the windows as installed were in compliance with Requirement F1 because there was inadequate provision for rapid ventilation and resolved to serve a Section 36 Notice if appropriate. You then applied to the Borough Council for a relaxation of Requirement F1 which was refused. It was against that refusal that you appealed to the Secretary of State.

The Appellant's case

In support of your appeal you state that because of limited finances your client elected to re-use the existing windows. On the basis that the bedrooms were double bedrooms and became enlarged double bedrooms, the assumption was that the occupancy would not increase and that therefore the demand for ventilation would not increase either. You state that there is clearly a contravention of the Regulations but in this instance your client feels that the opening windows provided are adequate for his needs. You also make the point that the house is situated on a very busy road with the result that because of the heavy vehicle noise the windows are rarely opened.

The Borough Council's case

The Borough Council state that the approved plans were endorsed to indicate that the new windows serving the bedrooms would provide 5% of the new floor area as opening windows, plus background or trickle ventilation to each room.

The Borough state that the floor area in bedroom 1 as extended is $27.06\,m^2$ which at 5% of the floor area would require $1.38\,m^2$ of opening windows. Bedroom 2 as extended has a floor area of $25.36\,m^2$ which at 5% of the floor area would require $1.28\,m^2$ of opening windows. In contrast, the Borough states that the two top-hung fanlights which have been installed in each bedroom provide an opening area of approximately $0.5\,m^2$ to each room, which they consider to be a substantial reduction on that recommended in Approved Document 'F' ('Ventilation').

The Borough also state that the lack of ventilation was discussed with your client following which alterations to the window frames were made to ensure the requisite background ventilation was provided. However, no alteration in the size of the opening windows was made to provide for rapid ventilation. The Borough Council then suggested to your client that to compensate for the loss of opening area, mechanical extraction at the opposite end of each bedroom might be acceptable, provided it was of adequate capacity.

Although not a recommendation included in Approved Document 'F' the Borough took the view that such an arrangement would effectively allow additional air movement/ventilation and meet the spirit of Requirement F1. However, your client was not prepared to accept this suggestion and submitted his application for a relaxation of Requirement F1.

The Borough conclude by stating that the requirements for natural ventilation of habitable rooms by the provision of opening windows equivalent to the minimum of 1/20 of the floor area have remained unchanged for many years and that the bedrooms, as extended, in this case are of considerable size having depths of 7.8 m and 6.8 m. They therefore feel it is essential that the minimum standards recommended within Approved Document 'F' for ventilation should be provided to ensure the health of those occupying the rooms.

The Department's view

Requirement F1 states that: 'There shall be adequate means of ventilation provided for people in the building'. The guidance in Approved Document 'F' covers both rapid ventilation to extract or dilute pollutants and moisture, and background ventilation to provide a minimum supply of fresh air. Relaxation of Requirement F1 has been sought in respect of rapid ventilation.

You contend that as the occupancy of the two bedrooms has not increased the ventilation requirements should remain the same as before the extensions. In this case the bedrooms have single-sided ventilation (i.e. the fresh air comes from one end of the room only and there is no design provision for cross flow). In the Department's view this is not an efficient method of ventilation and is not recommended for rooms having a length exceeding about 2.5 times their height (BRE Digest 399). The height of these bedrooms is not given but it is assumed that they are unlikely to be more than 2.4 m so for effective ventilation of the rooms their lengths should not exceed more than about 6 m. However, as the two rooms are 7.8 m and 6.8 m long respectively, it would not in the Department's view be prudent to reduce the ventilation area below the minimum recommended in Approved Document 'F'.

The Department also considers it appropriate to consider the possibility that a future owner might fit doors in the present door openings. In this event it would be more difficult to provide adequate ventilation to the original bedrooms.

The Department notes your clients' point concerning noise from the busy road and the fact that the windows are rarely opened. The implication appears to be that improved ventilation would be of no benefit to health because the windows would still remain closed. However, on the other hand, improved ventilation would offer the opportunity for rapid ventilation if required, thus allowing the closure or reduction of open window shortly thereafter.

Diagram 2 of Approved Document 'F' provides guidance for circumstances such as these. It provides that two rooms may be treated as a single room for ventilation purposes if there is:

(a) an area of permanent opening between them equal to at least 1/20 of the combined floor area, and
(b) provision for rapid ventilation by opening windows which have an open area equal to at least 1/20 of the combined floor area, and
(c) background ventilation provided by an opening of 8000 mm².

The windows fitted provide an open area of approximately 0.5 m² for each room. Bedroom 1 as extended has a floor area of 27.06 m² and bedroom 2 as extended has a floor area of 25.36 m². The opening areas are therefore about 1.8% and 2% respectively, instead of the 5% of the combined floor area as recommended in Approved Document 'F'.

The decision

The Secretary has given careful consideration to the facts of this case. He has concluded that it would not be appropriate to relax Requirement F1 so as to provide a considerably reduced capacity for

rapid ventilation and that the London Borough came to the correct decision in refusing your application for such a relaxation. The Secretary of State therefore dismisses your appeal.

You may wish to note that in analysing this case the Department considered the appropriateness of installing extractor fans on the external walls at the far end of each bedroom as an alternative to enlarging the window openings. In the Department's view this could be an acceptable way of complying with Requirement F1.

PART K: STAIRS, RAMPS AND GUARDS

Dwellinghouse – loft conversion (K)

The council had refused to relax requirements with regard to a spiral staircase with alternating treads giving access to two rooms and a WC, and the appeal was dismissed.

Ref.: 45.3.63   Date: 13 February 1995

Careful consideration has been given to the facts of the case, and to the representations made by you and by the Borough.

This appeal concerns a new spiral alternating-tread staircase in a bungalow which connects the ground floor with two rooms and a WC that have been built in a first floor loft conversion. The staircase has been installed and the building work completed. The ground floor comprises two bedrooms, a bathroom, a kitchen, and a study located between a dining room and lounge.

The full plans application was first rejected by the Borough on the grounds that it did not accord with the guidance in Approved Document 'K' ('Stairs, ramps and guards') of the Building Regulations 1991. You re-submitted revised details to the Borough and gave reasons why the spiral alternating tread type of staircase was necessary. The Borough considered that your proposal did not accord with the recommendations in Approved Document 'K' and refused to relax the requirements relating to Part 'K' of the Building Regulations 1991. You then appealed to the Secretary of State against that decision.

The Borough states that it refused to dispense with the requirements of K1 because a staircase of this type does not offer safety to users for the following reasons:

(a) the guidance for alternating treads is given in Approved Document 'K', paragraphs 1.22 to 1.24; and
(b) paragraph 1.24 gives further guidance on how to measure the rise and going of the treads to a straight flight of an alternating staircase and the parameters they have to fall within.

It is the Borough's view that paragraph 1.23 of Approved Document 'K' specifically excludes alternating treads in staircases of a spiral design.

You stated that the rooms on the first floor roof space were to replace the existing ground floor study which intrudes into your living accommodation; and that the alternative of extending the premises at ground floor level was not practical because of lack of land area. You pointed to the particular limitations of the existing space at ground floor level as the reason for proposing this particular design of spiral staircase and as the reason for requesting a relaxation of Requirement K1.

In your letter of 31 August 1994, you acknowledged that the guidance in the Approved Document 'K' refers to straight flights for alternating treads, but you did not agree with the Borough's view that this necessarily excludes the use of a spiral alternating tread stair. You refer to the stair as a 'Private' one and argue that the users of it are familiar with this type of stair and use it regularly. You also state that it was selected for its proven reliable construction, safety, and for the 'Private' use intended; and that its safety has been demonstrated in its regular use.

In giving careful consideration to this case, the Department has taken account of the views expressed by the Borough, and also noted that the stair is currently in use and the purpose which it serves. The Department considers it important to also take into account the safety of future occupants of the bungalow, and the safety of the present occupants in an emergency situation. The Department accepts that the advice in Approved Document 'K' is for general guidance, and proposals offering a lower standard of safety may in certain circumstances be acceptable. However, in this particular case it is considered that the staircase represents a safety hazard and does not comply with Requirement K1.

It follows that the Secretary of State is satisfied that in all the circumstances of this case the Borough came to the correct decision in refusing your application for relaxation. Accordingly, the Secretary of State hereby dismisses your appeal.

Three storey house – guarding low-level windows (K.2)

Replacement windows will have a cill height only 300 mm above floor level and the local authority would not accept the safety measures proposed by the developer. The decision in this appeal was that the measures provided did comply with K.2, but the appeal was dismissed.

Ref.: 45.3.50   Date: 3 May 1995

Careful consideration has been given to the facts of the case, and to the representations made by the Borough and by the appellant.

The appeal relates to structural work involving the insertion of new lintels and replacement windows to the front and rear of an existing three storey family dwelling. The windows concerned number eight and are vertical sliding sashes with a cill height of about 300 mm above the floor. The work constituted a material alteration and therefore was required to comply with the relevant requirements of the then Building Regulations 1985. The work was completed in December 1992.

The appellant states that the original plans approved by the Borough made no mention of barriers or safety glazing to the windows. He also states that it was originally agreed by the Borough that the windows were in such a location and of such a size as to be acceptable glazed in 4 + 4 double glazing. These points are contested by the Borough. Following a survey of the premises the Borough wrote to the appellant stating that guarding had not been provided to the windows and that they did not therefore comply with the Building Regulations 1985. The appellant subsequently had the lower vertical sashes limited to an opening of 100 mm.

The restriction to 100 mm was acceptable to the Borough but it required in addition the glazing to the lower sections of the windows to be of toughened glass, glass blocks, or laminated safety glass. The appellant then applied anti-shatter film to the relevant glazing and furnished the Borough with the manufacturer's details. However, the Borough did not find this acceptable in terms of compliance with respect to guarding Requirement K2. In the Borough's letter of the 9 September 1993 it was stated that due to the revisions of Approved Document 'K' ('Stairs, ramps and guards') (1992 Edition) it would be acceptable to use wired glass. The appellant rejected this suggestion on the grounds that the breakage of a sheet of wired glass could result in sharp edges which could be extremely hazardous.

The Borough's letter of the 3 November 1993 pursued the issue of containment and expressed concern that the test results on the anti-shatter film furnished by the appellant gave no indication that the glass would remain in place after fracture.

The appellant then applied for a dispensation in respect of Requirement 'K2' of the Building Regulations 1985. This application was refused by the Borough on the grounds that even with the limited opening of the sashes the meeting rail provided guarding only at a height of 1350 mm above floor level and that the glazing below the recommended height should be of glass blocks, toughened glass or laminated safety glass as recommended in Approved Document 'K' ('Stairs, ramps and guards') – paragraph 1.2 ('Design') of K2/K3; 1985 Edition). The Borough considered that the anti-shatter film applied by the appellant had insufficient evidence attaching to it to

show that it would comply with Requirement 'K2'. The Borough considered that to dispense with the Requirement would result in an unacceptable risk to the user, particularly young children. It was against this refusal to dispense with the requirement that you appealed to the Secretary of State.

The Department's analysis of this case is as follows.

When the appellant's work was executed the Building Regulations of 1985 applied and it is to this standard which the work must be judged. Requirement K2 of Schedule 1 to those Regulations required that:

Stairs, ramps, floors and any other roof to which people would normally have access, shall be guarded with barriers where they are necessary to protecting users from the risk of falling.

The windows at issue have a cill height of 300 mm above floor level. The 1985 edition of the Approved Document 'K' ('Stairs, ramps and guards') advises in respect of Requirement 'K2' that in this situation the edges of a floor should be guarded to a height of 900 mm above floor level and that glazing below this level should be of glass blocks, toughened glass or laminated safety glass; and that it should not have gaps which a 100 mm diameter sphere could pass through.

The Department has given careful consideration to the work as executed and it is its view that because of the design of the windows, the measures the appellant has taken to limit their opening, and the application of film to the glazing that it provides the users with reasonable protection from the risk of falling. It is therefore the conclusion of the Department that the work as executed complies with Requirement 'K2' of the Building Regulations 1985.

In reaching a decision on this appeal against refusal of an application for dispensation of Requirement K2 the Secretary of State is of the clear view that, in these particular circumstances, it would be inappropriate to uphold such an appeal because to do so would be tantamount to implying that it was not necessary to protect users from the risk of falling. Accordingly, the Secretary of State dismisses the appeal. However, he has also had regard to the technical aspects of this case and observes that in his view the work as executed and modified by the appellant complies with Requirement 'K2' of the Building Regulations 1985.

Loft conversion – headroom to new stairs (K.1)

In dispute was the design and construction of a staircase to connect the ground floor of an existing bungalow to a bedroom and bathroom in the roof. It involved the lack of adequate headroom and the fact that the applicant was of short stature was not an adequate reason for relaxing K.1.

Ref.: 45.3.85   Date: 19 April 1996

This appeal relates to an existing three bedroom bungalow with dining room and living room. The building work has comprised alterations to form a bedroom and bathroom within the roof space together with a new stair to provide access from the ground floor; and a laundry room and extended hallway on the ground floor.

You received full plans approval conditionally for the building work. However, you were notified by the Borough Council that the stairs to the new rooms in the roofspace space were in contravention of the Building Regulations, because, in their view insufficient headroom had been provided. However, you contend that it was unreasonable to cite an 'initial contravention' because this part of the structure was not complete. You were notified by the Borough Council that your subsequent remedial work to recess the ceiling had achieved a head clearance of only 1.8 metres above the pitch line of the stairs as well as the top step at the landing.

You responded to the Borough Council setting out six options for increasing the head height and set out the implications of each option. You concluded that the best answer to the problem was to do nothing (your option 7) because the other options were physically, structurally, or environmentally dangerous. The Borough Council indicated that they were not satisfied with that solution and you then applied for a relaxation or dispensation of Requirement K1. This was considered by the Borough Council as a request for a relaxation and refused by them. It is against this refusal that you have applied to the Secretary of State under section 39 of the 1984 Act.

Your original full plans made provision of 1900 mm head clearance in accordance with Approved Document 'K' ('Stairs, ramps and guards'). The actual headroom varied between a minimum of 1780 mm and 1820 mm at two regions on the centre line of the completed stair.

In support of your appeal against the refusal by the Borough Council to relax Requirement K1 you have explained in detail how the 1900 mm was calculated on the basis of the unopened structure but that on opening up an existing structural beam was found to very much larger than expected. This could not be safely cut away for structural reasons with the result that the stairs had been constructed slightly (about 2000 mm towards the eaves) which reduced the clear headroom. You contend that discovering unexpected sizes of existing timber members is a common problem in old timber-framed structures.

The Borough Council favoured two of your potential options to overcome the problem but you regarded all of these as structurally dangerous or dangerous to users. One of these options was to construct a rooflight over the pinch point on the stair to produce additional headroom. However, you have opposed this option on the grounds of the difficulty of adequately detailing the rooflight to ensure watertightness and the effect on the structural integrity of the surrounding roofing timbers.

You have described the stair as a standard and gentle one with a 200 mm rise and 225 going, and 820 mm width flights. In your view this, together with the half landing, means that the stair poses no risk to users, nor do they distract users' attention from the clearance. You further argue that because of the low ceiling height in the bedroom and bathroom in the new roof space, the risk of a person bumping their head in these areas is much greater than on the stair, although this is not found to be a problem in use by you and your wife. It follows that with a relatively low clearance in the bedroom, no one who was particularly tall would choose to live in the house thus further reducing the risk. You contend also that the head clearance is therefore much less of a risk to health and safety in this particular property than normal.

In earlier submissions to the Council you dealt at some length with the question of the characteristics of the user. You explained that you and your wife are relatively short and do not find the roof space ceiling height and the stairs an inconvenience. Because there are three bedrooms and the main bathroom on the ground floor, you explained that tall visitors could be accommodated downstairs rather than upstairs. With respect to use by subsequent purchasers, you argue that because the ceiling to the bedroom and bathroom in the roofspace are very low over much of the area, the house would not be of interest to tall purchasers.

The Council state that the staircase does not 'offer safety to users moving between levels of the building'. From their perspective the issue is that they do not consider a reduced ceiling height headroom of 1.78 m to be safe, given that this is significantly lower than the 1.9 m recommended in Approved Document 'K' for loft conversions. The Council acknowledge that your original drawings show the headroom of 1.9 m, although they state that this was achieved by compromising thermal insulation. The Council also argue that the works could have been classed as 'alterations' rather than forming a loft conversion and that their conditional approval sought a headroom of 2.0 m.

The Council also contend that the extent of the deficiency of headroom is more extensive than the individual points outlined by you. They consider that the headroom is not only inadequate at these points but extensively between these points and on the top landing for a distance of more than 500 mm from the top nosing.

With regard to the options you put forward, the Council state that only two were considered worthy of further investigation and that it was recognized that careful consideration would have to be given to the structural aspects of such solutions. The Council has also made it clear that they are not proposing or prepared to consider compensating the risk of structural failure for a risk of falling, or visa versa.

With regards to the comparison of safety risks you have made in respect of the new bedroom and bathroom, and the stairs which give access to them, the Council have stated that – given that they have no control over the headroom in the roofspace they are neither condoning nor accepting the low ceiling to the new bedroom and bathroom. However, they argue that 'banging one's head' on the bedroom ceiling would not result in a risk of falling several metres. They argue that there is statistical information confirming the significant risk of injury and death from falling on a staircase but they are not aware of such data of similar injury within habitable rooms. They conclude that the suggestion that the intended use of the bedroom and bathroom would result in 'less than normal' risk of falling in this instance is impossible to substantiate.

Finally the Council have expressed some surprise that in view of the history of the works carried out on the property, and your involvement with those works, that the larger beam was not known about at the time of your full plans application for the building of the roof in 1993/94.

The Department's analysis of this case is as follows. Approved Document 'K' recommends that the headroom over a stair should be 2 m, but for practical reasons in loft conversions it is reasonable to reduce this to 1.9 m on the centreline and 1.8 m at the side of the stair. In this case the headroom is not lower at the side than at the centre, but the headroom is below 1.9 m in two regions. The lowest points are 1.8 m about halfway up the stair and 1.78 m at the top of the stair.

You describe yourself and your wife as relatively short and have stated that you have no difficulty with the use of the stairs. However, the Building Regulations are concerned with the safety of people in and about a building – i.e. 'people' in general rather than people with specific characteristics, such as height. It follows that it would not be appropriate for the Secretary of State to relax a regulation based on the specific characteristics of the current occupants except perhaps in very exceptional circumstances.

The circumstances in this particular case emanate from the discovery of a structural constraint which was not appreciated until the work was opened up. The Council have favoured options (your options 3 and 4) to overcome the problem which they have acknowledged will require structural consideration but which they believe to be viable solutions worth pursuing. However, you have claimed that neither option is viable and that the only solution is to retain the stair as it is.

The Approved Documents provide only guidance. However, in this case the failure of the stair to accord with the recommended headroom measurements is 122 mm which is a very significant shortfall in this context. It follows that a person of above average height could straighten up after the first pinch point only to impact on the second. A fall, especially from the head of the stair, could result in serious injury. There would therefore need to be very exceptional circumstances for the Secretary of State to consider relaxing Requirement K1.

The particular circumstances are the constraint imposed by the beam. From the documentation the Department is not convinced that a structurally viable and sound solution cannot be devised which will afford greater safety to the stair. The Department also has to consider the implications of relaxing a standard on the basis that there may be a structural/design constraint which cannot be overcome for valid practical reasons or because of a reluctance to meet the additional cost of doing so. Very particular circumstances would have to prevail for the Secretary of State to consider relaxing the standards of safety so as not to create a general precedent which could be used by others on any occasion when approved plans could not be complied with because of subsequent difficulties discovered during construction.

The Secretary of State has given very careful consideration to the facts of this case and to the arguments of both parties. He has concluded that there is a safety risk attaching to the stair as constructed and that the circumstances of this particular case do not warrant a relaxation of Requirement K1. He has therefore concluded that the Borough Council came to the correct decision in refusing to relax Requirement K1. He accordingly dismisses your appeal.

Restaurant – handrail to stairs (K.1)

Handrails should be designed and placed to aid balance and provide support when required. In this case the Secretary of State supported the local authority in its refusal to dispense with the handrail requirement in K.1.

Ref.: 45.3.93   Date: 14 June 1996

The premises to which this appeal relates comprise a ground floor restaurant and a lower ground floor containing a bar with associated seating and toilet facilities. The building work has involved the refurbishment and alteration of these premises and has included the alteration of the staircase to the rear which connects the ground and lower ground floor facilities. There is also a second staircase connecting these two floors in the centre of the premises but this is not at issue.

From drawing No. 94/018/9 (now marked No. 14) it appears that the stairs are constructed in three flights; the upper flight being about a metre wide, while the other two flights appear to be about 1600 mm wide (i.e. centre flight) and 1150 mm wide (i.e. lower flight). The upper and centre flights of these stairs are duplicated to the left and right of the centre of the stairwell, whilst the lower flight is common. The guarding comprises a wall 125 mm wide with a convex top to form a 'handrail' which can be grasped.

Your full plans application for the refurbishment and alterations was rejected by the Borough on the grounds, inter alia, that no details regarding Part K had been furnished in respect of staircases or other balustrades. You re-submitted a full plans application which was rejected on grounds relating to Part B of the Building Regulations. In a letter the Borough reported that they had conducted a recent survey of the work completed by your clients. They reported that it did not comply with the Building Regulations in respect of the following:

- the appropriate handrails had not been provided to the rear staircase leading to the lower ground bar area;
- the appropriate handrails had not been provided to the central staircase leading to the staff area and toilets; and
- there was insufficient head height on the lower flight of the rear staircase.

It appears that the issue of compliance of the central stairs and the insufficient head height of the rear staircase were resolved leaving the question of appropriate handrails to the rear staircase unresolved.

Your clients wished to retain the guarding wall with convex top rather than provide the handrails considered appropriate by the Borough. You therefore applied to the Borough for dispensation from Requirement K1 which was refused. It is in respect of that refusal that you then appealed to the Secretary of State.

The Appellants' case

In support of your case for dispensation of Requirement K1 in respect of the provision of handrails to the rear stair, you state that you consider the height, strength, and all matters relating to the staircase, to be satisfactory with the exception of the thickness of the wall/actual handrail. You also state that the desire to retain the design of the stair is because it has been constructed to an authentic Southern European design giving wide open proportions with natural tiled balustrade and handrail. The interior design architects have commented that such a design is similar to thousands of staircases throughout Southern Europe. You consider that the proportions are an inherent part of the design and that although the stair is wide it is broken by intermediate landings giving a flight of eight risers from

basement to landing; three wide intermediate risers from the half landings; and only five risers to the ground floor level.

The Borough Council's case

The Borough Council states that it has refused to dispense with Requirement K1 on the grounds that:

(a) the balustrading enclosing the stairwell is of a design which is not suitable to be used as a handrail; and
(b) the intermediate flight is greater than 1 metre in width and thus requires a handrail on both sides.

With respect to ground (a) the Borough Council consider that it is appropriate in some cases to consider the top of the required guarding as an adequate handrail provided the height can be matched. However, in this instance they consider that the wall providing the guarding is of a satisfactory height but is approximately 125 mm wide, and because the upper surface is convex, it is extremely difficult to grip. The Borough Council have also referred to Approved Document 'K' ('Stairs, ramps and guards'). They state that whilst no specific guidance as to satisfactory designs for handrails is given there, Clause 12.3.5.4 of BS 5395: Part 1: 1977, Diagram 5 in Approved Document 'M' ('Access and facilities for disabled people'), and the general performance guide given on page 4 of Approved Document 'K' indicate that a higher standard of provision should be considered in public buildings where the occupants are unfamiliar with their surroundings.

Finally, the Borough Council state that, following tests made on site by their officers, they are of the opinion that the configuration of the balustrading is such that if someone did stumble whilst descending the stairs it would not be possible to grip the 'handrail' and thus prevent falling.

The Department's views

This appeal relates to the refusal by the Borough Council to dispense with Requirement K1 insofar as that Requirement requires the provision of an appropriately designed handrail. Requirement K1 states: 'Stairs, ladders and ramps shall offer safety to users moving between levels of the building.'

Paragraph 1.27 of Approved Document 'K' covers the location of handrails but does not give specific advice on their design. However, the guidance in Approved Documents is not intended to be exhaustive and it is reasonable to assume that the design of handrails contributes to the safety of users, and so implicitly are covered by Requirement K1.

It follows that it is important to be clear as to what the likely requirements for a safe handrail should be. In the Department's view a handrail should: (i) aid balance; and (ii) provide support if needed (e.g. to assist elderly users or to prevent a fall). The design requirements for a handrail to perform the first function are minimal. However, for a handrail to provide support it must be within easy reach, and be easy to grip tightly.

In the Department's view, BS 5395: Part 1: 1977, provides the most relevant advice. It states that the section of a handrail which will provide the most comfortable grip should be circular with a diameter between 45 mm and 50 mm. It goes on to state that: 'whatever the shape it should be capable of being readily gripped by the hand'. In the Department's view the convex top of the guarding which constitutes the 'handrail' clearly does not approximate to this BS recommendation and does not meet the criterion that it should be readily gripped by the hand.

In the Department's view the principle concern for complying with Requirement K1 in the context of this case must be that persons should be capable of preventing themselves from falling. Statistics demonstrate that significant personal injury can result from falls on staircases. The Borough Council has referred in particular to the possibility of someone stumbling during descent and it being impossible for them to grip the 'handrail' to prevent falling. The Department also has in mind the fact that because the stair gives access to a bar area, it is unlikely to be brightly lit and may be used by persons whose sense of balance is impaired by alcohol.

It is therefore the opinion of the Department that a safe handrail is essential in this particular situation and that the 'handrail' as constructed does not provide a sufficient level of safety. Requirement K1 has clear implications for safety which in certain circumstances may in fact be life safety matters. In the particular circumstances of this case – particularly given that the staircase is to be used by members of the public – the Department considers there are no grounds for dispensing with or relaxing Requirement K1.

In the context of this appeal, the Borough Council has also raised the question as to whether a handrail should be required on both sides of the intermediate flights of stairs because they are greater than 1 metre in width. Paragraph 1.27 of Approved Document 'K' recommends that: 'Stairs should have a handrail on at least one side if they are less than 1 m wide. They should have a handrail on both sides if they are wider.' Having regard to the fact that the Department considers there are no grounds for relaxing or dispensing with Requirement K1, it observes that in its view to comply with Requirement K1 handrails of a form suggested in BS 5395: Part 1: 1977, should be provided on both sides of the staircase on the middle and lower flights.

The Secretary of State's decision

The Secretary of State has given careful consideration to all the facts of this case and in the particular circumstances has concluded that the guarding 'handrail' as constructed does not provide a reasonable standard of safety, and that the Council therefore came to the correct decision in refusing to dispense with Requirement K1 with respect to the 'handrail'. In reaching this conclusion he has also considered whether it might have been more appropriate to have considered a relaxation to Requirement K1. However, in his view Requirement K1 in the context of this particular case is a matter of serious public safety and in extreme cases could be a matter of life safety. He has concluded that relaxation would not be appropriate either. He accordingly dismisses your appeal.

Offices – installation of spiral staircase (K.1)

The proposal to install a spiral staircase from ground to first floor in offices was rejected by the Council who considered the inadequate dimensions dangerous. The Secretary of State does not consider lightly the relaxing of K.1 and dismissed the appeal.

Ref.: 45.3.104   Date: 7 February 1997

The appeal

The building to which this appeal relates was constructed as a two storey dwelling in 1896, and converted into offices in 1983. There are now two suites of three offices on each floor (i.e. six offices on each floor). Access between the ground and first floors is via a commonly shared internal stair.

Recently your client, as occupier of suite 1 on the ground floor, took over suite 4 on the first floor which is directly above suite 1. To improve communications between the two suites he wishes to install a new stair to link them more directly than via the common internal stair.

The proposed new stair is to be of a small spiral type and it will be located in a corridor (detailed in drawing 9604.01a). In order not to restrict the corridor (and thereby obstruct the means of escape) the stair must be small. As it is a spiral stair, the overall diameter must fit in the corridor, and consequently the clear width of the stair, which is less than the radius, is very small. The following dimensions are proposed:

(a) Clear width of 470 mm
(b) 'Going' of 140 mm, 'rise' of 210 mm.

Your client submitted these proposals, together with provision for a fire screen, as a full plans application to the District Council which were rejected on the grounds of failure to comply with Requirement

K1. Your client accepted that the proposed spiral stair did not comply with Requirement K1 and therefore requested the District Council to dispense with or relax Requirement K1. This application was refused. It was against that refusal that you then appealed on his behalf to the Secretary of State.

The Appellant's case

In support of your client's appeal against the refusal for a dispensation/relaxation of K1, you state that:

(a) The alteration works should not disregard a number of very fine architectural features inside the building;
(b) The occupants of the building would be perfectly capable of recognizing the characteristics of the proposed staircase; would be aware of its narrowness and steepness; and would exercise due care and attention when using it;
(c) The alternative of a straight stair would be out of proportion to the place of its location and its length would be unacceptable in the circumstances;
(d) Currently there are only five members of staff employed in the two suites; and
(e) Without the proposed stair the door to the existing common stair would have to remain unlocked – allowing easy access for the general public to suite 1 on the ground floor.

The District Council's case

The District Council say that the clear width of the stair is below that recognized in Approved Document 'K' ('Stairs, ramps and guards') and other guidance on stairs for use in any type of premises. They state that the minimum dimension recognized is 600 mm, and that this is only recommended for use in special circumstances. They also say that the 'going' is not sufficient to provide a safe footing for the sole and heel of a shoed foot, and that it would be particularly dangerous for women in high-heeled shoes.

The District Council also believe that a conventional stair could be accommodated, although at great cost, which would be more in keeping with the character of the premises. They point out that the building is not listed and do not see an over-riding need to preserve the existing interior finishes at the expense of reduced safety.

The Department's view

You have appealed to the Secretary of State against the refusal of the District Council to relax or dispense with Requirement K1. Your arguments in support of this are based on the need to have an internal staircase for security reasons and the necessity of installing one which does not disturb the architectural features of the building or which will involve the loss of office space.

Requirement K1 is designed to protect the safety of occupants and users of a building, and there are times when this can be a matter of life safety. The Secretary of State would not, therefore, normally dispense with Requirement K1, and would not lightly relax it except in exceptional circumstances. In the Department's view the circumstances of this case are not uncommon and do not justify a relaxation of Requirement K1; and more specifically, would certainly not justify dispensation from Requirement K1. In the Department's view, the particular problem of security might be overcome by use of such devices as entry-phones, or push-button security combination locks and/or internal push-bars, which would inhibit the free movement of employees between the two suites to the minimum.

You will wish to note that in considering this case, the Department has considered whether in fact the proposed staircase could be considered to comply with Requirement K1 in the context of this office environment. Requirement K1 states that:

Stairs, ladders and ramps shall offer safety to users moving between levels of the building.

Approved Document 'K' does not contain guidance on the design of spiral stairs, but refers instead to BS 5395: 'Stairs, ladders and walkways: Part 2: Code of Practice, for the design of helical and spiral stairs'. This Standard recommends a clear width of 800 mm for situations like this. It suggests that 600 mm width will be acceptable to give access to a small room in an office. It does not refer to widths as low as 470 mm.

The Department would not normally regard the width of a stair as a factor which affects safety. However, with spiral stairs the 'rise', 'going', and width are inter-related and a narrow stair leads to a small 'going'. In the Department's opinion the combination of narrow width and small 'going' makes this particular stair unsafe for use in an office environment where staff may need to carry files, etc., and may wear unsuitable shoes. Currently up to five staff may use the stair. Under extenuating circumstances, a design such as this might be acceptable for a small domestic loft conversion, but it is not appropriate for regular use in a work place.

The Department has also noted that the proposed guarding at the top of the stair is 900 mm high. This compares with the recommendation in Approved Document 'K' of a height of 1100 mm.

Secretary of State's decision

The Secretary of State has given careful consideration to the facts of this case, and the argument put forward by both parties.

You have appealed to the Secretary of State against the decision of the District Council not to dispense with or relax Requirement K1 in respect of the installation of this spiral staircase. Compliance with Requirement K1 can be a matter of life safety and the Secretary of State would not normally consider it appropriate to dispense with the requirement, and would not lightly consider relaxing it except in exceptional circumstances. In this particular case, he considers the circumstances to be not uncommon, and does not consider that they justify a dispensation or, indeed, a relaxation of Requirement K1. The Secretary of State therefore considers that the District Council came to the correct decision in refusing your application for a dispensation or relaxation, and accordingly dismisses your appeal.

Dwelling house – unusual staircase (K.1)

The proposal was to install a staircase comprising steel treads supported on a central beam. Guarding was to be tensioned steel cable. The local authority did not accept that this provided sufficient protection for children, and the appeal was lost.

Ref.: 45.3.98   Date: 18 February 1997

The appeal

The building work to which this appeal relates comprises the installation of a new stair in an existing house which has been extensively altered and extended. The first floor comprises a study and living room.

The stair in question is the only one connecting the ground and first floors. It comprises steel treads supported on a central beam. The proposed guarding is 900 mm high and consists of tensioned stainless steel cables, appropriately spaced (sic), parallel to the pitch of the stair. The guarding to the landing is to be of a similar design but 1100 mm high.

A full plans application for the proposed alterations and extensions to the dwelling incorporating the installation of the stair was submitted to the Borough Council and approved by them, subject to an endorsement that the guarding to the stair would satisfy the requirements of Part K. Although you indicate that you in fact believe that the proposed staircase design complies with Requirement K1 you applied for a relaxation of Requirement K1. This application was refused by the Borough Council and it is against that decision that you appealed to the Secretary of State.

The Appellant's case

Your client is the owner and sole resident of the property, and the alterations and extensions are for his personal use. Great care has gone into the interior and exterior design of the property.

You contend on behalf of your client that although the guarding wires travel parallel to the staircase, their small section and flexibility render them a much greater obstacle to climbing than a conventional solid timber balustrade. You believe that this argument applies to the entire balustrade including the staircase landing surround, but most particularly to the staircase itself where the wires are close to a 45° angle.

You have cited three examples of very similarly designed staircases and it appears to be your assumption that dispensations are being regularly granted for the use of similar designs in other parts of the country. With regard to this specific staircase you contend that the proposed design is:

(i) inherently safe within the context of the relevant Building Regulation;
(ii) is effectively the same as ones which are receiving dispensation elsewhere in the country;
(iii) presents a virtually unclimbable balustrade detail on the staircase itself.

You conclude by stating that the flexibility of the balustrade presents an adequate obstacle on the stairwell surround but that if required your client would be prepared to consider substituting glass infill panels on the landing area, although these would not be compatible with the design of the staircase itself.

The Borough Council's case

The Borough Council points out that Requirement K1 states that:

Stairs, ladders and ramps shall offer safety to users moving between levels of the building.

Paragraph 1.29 of Approved Document K ('Stairs, ramps and guards') states that in order to satisfy the requirements for safety of children on stairways, the construction should be such that:

(a) a 100 mm sphere cannot pass through any opening in the guarding, and
(b) children will not readily be able to climb the guarding.

The plans approved by the Council contained an endorsement that the guarding would satisfy the provisions of Part K but it is the judgement of the Council that the proposed stair is of a construction which does not meet the standards stated on the approved plan.

In addition, the Borough Council state that the grounds for relaxation were based on the fact that no children were currently living on the premises and that the straining wires would not be climbable due to their steepness. However, the Council take the view that although children are not in residence at the present time that does not mean that children will never visit the premises or dwell there in the future. With regard to the straining wire, the Council take the view that there would be a degree of flexibility in the structure and as such this would allow movement when any bodyweight was applied. Such movement would not only allow purchase to be achieved but would also allow the space between the wires to extend beyond the recommended 100 mm spacing.

The Borough Council therefore concluded that the proposals did not meet the objectives for safety as set out in Approved Document 'K' and rejected your application for a relaxation of Requirement K1.

The Department's view

You have appealed to the Secretary of State against the refusal of the Borough Council to relax Requirement K1. Your arguments in support of this are based on your client's desire to incorporate this particular design of stair into the residence which is for his personal use; that the design of the stair is inherently safe; and is effectively the same as ones which you say are receiving dispensation elsewhere in the country.

Local Building Control Authorities have the power to dispense with or relax the requirements in the Building Regulations. Such decisions do not necessarily come to the attention of the Department. Any case referred to the Secretary of State on appeal under Section 39 is considered on its individual merit.

Requirement K1 is designed to protect the safety of occupants and users of the building, and there are times when this can be a matter of life safety. The requirements in the Building Regulations 1991 are designed to secure reasonable standards of health and safety for persons in or about buildings. It follows that no differentiation should be made in respect of 'persons' and that children must be incorporated within this definition. The Secretary of State would not lightly relax Requirement K1 except in exceptional circumstances. In the Department's view the circumstances of this case, judged in the context of a dwelling environment, do not justify the relaxation of Requirement K1 because to do so would clearly prejudice the safety of young children.

You will wish to note that in considering this case, the Department has analysed the degree to which the proposed staircase could be considered to comply with Requirement K1.

The relevant guidance in Approved Document 'K' is that:

(a) flights and landings in dwellings should be guarded at the sides where there is a drop of more than 600 mm (paragraph 1.28);
(b) guarding for stairs and landings in single family dwellings should be (at least) 900 mm high (diagram 11 on page 11);
(c) the handrail should be between 900 mm and 1000 mm high and the handrail can form the top of the guarding if the heights can be matched (paragraph 1.27); and that
(d) except on stairs not likely to be used by children under 5 years (for example in accommodation for old people) the guarding to a flight should prevent children being held fast by the guarding and should be such that children will not readily be able to climb it (paragraph 1.29). In addition, it should be noted that paragraph 1.9 says:

All stairs which have open risers and are likely to be used by children under 5 years should be constructed so that a 100 mm diameter sphere cannot pass through the open risers.

The Borough Council appear to accept that (a), (b) and (c) above would be satisfied and the centre of their concern is in respect of (d) and the safety of children.

In the Department's opinion although the wires of the stair guarding are assumed to be spaced at 100 mm intervals, it is very likely that the spacing will increase if the wires are forced by a child. It is difficult to judge by how much the wires will deflect although clearly this will be more at the centre than at the ends, and whether the increased opening would offer a greater risk of injury to the child. In the Department's view the risk would not be increased significantly, and the guarding in this particular case would appear to be reasonably safe in this respect.

It is also not easy to judge how difficult it would be to climb the stair guarding. As there are no intermediate vertical supports there would be nothing to stop a child's foot sliding except friction. However, if the wire deflected when a foot was placed on it, the angle of the wire to the horizontal would decrease thus slightly reducing the tendency to slip. However, on the assumption that the wire was under sufficient tension, and would be kept taut, it is likely that this reduction would be negligible. On this basis, it is the opinion of the Department that the guarding would be difficult to climb.

Although few details of the guarding to the landing are given – such as the height of the first wire above the floor – the fact remains that the wires would be horizontal which would make them easy to climb and unsuitable for young children. This problem could be overcome by using glass infill panels as suggested by you.

Notwithstanding the comments above, and although the Borough Council has not commented upon the fact, the Department views with concern the large opening in the guarding between the treads and the first guard wire, and also the fact that the treads are open. In the Department's view these openings present a high risk to young children.

The Secretary of State's decision

You have appealed to the Secretary of State against the decision of the Borough Council not to relax Requirement K1 in respect of the installation of this steel staircase. He has given careful consideration to the facts of this case, and the arguments put forward by both parties. All decisions are considered on the basis of their individual merits.

Compliance with Requirement K1 can be a matter of life safety and the Secretary of State would not lightly consider relaxing it except in exceptional circumstances. Whilst sympathizing with your client's desire to introduce innovative design into his dwelling, he has had to bear in mind that this is a domestic environment in which it must be assumed that children of any age may live or to which they may have access. The Secretary of State has therefore concluded that the circumstances do not justify a relaxation of Requirement K1 and that the Borough Council came to the correct decision in refusing your application for a relaxation. Accordingly, he dismisses your appeal.

Residential property – balcony and stairway (K.2)

A local authority refused to dispense with K.2 in respect of a replacement balcony and stairway at the rear of a residential property. It was found that a child could quite easily climb the guarding to be provided and the appeal was dismissed.

Ref.: 45.3.105   Date: 12 March 1997

The appeal

The building work to which this appeal relates comprises the replacement of an external balcony and stairway that gives access to the rear garden of the house. The cast iron balcony is attached to the side of the house and is accessed at ground floor level from the study at the rear of the house, and through the French windows at the rear of the main reception room located at the front of the house. The balcony runs to the rear corner of the house where the cast iron stairway gives access down to the garden. The drawings show a height above ground level for the balcony of approximately 2.7 m at the reception room end. At the time of your appeal the stairway and balcony had been installed.

The stairway is exposed on both sides and has guarding to each comprising pairs of parallel vertical members joined at top and bottom by a scrolled single short vertical section connecting these to the rail and treads. Because the balcony is against the house wall it has guarding on one side only. This comprises alternate 600 mm panels – one style of panel basically replicating those on the stair but with the addition of a 100 mm bar at the top with 100 mm horizontal oval holes; and the other comprising large decorative oval-shaped scroll work incorporating a horizontal bar at the bottom and top.

These proposals were the subject of a full plans application which was given conditional approval subject to modifications to the deposited plans being made. The Borough Council's concern was that the decorative scroll panels on the balcony appeared to be climbable and they required confirmation as to how this would be prevented to satisfy Requirement K2. You then wrote to the Borough Council explaining that your clients had redesigned the panels so that the decorative swirls and details were closer together and more tightly packed to prevent feet being able to pass through the panels.

Although in a subsequent letter to the Borough Council you stated that your clients were probably going to provide some form of mesh to make the scroll panels unclimbable you subsequently wrote stating that your clients felt that the requirement was unreasonable and the scroll panels were not inherently climbable. You then requested a dispensation from Requirement K2 which was refused by the Borough. It is against that refusal that you subsequently appealed to the Secretary of State under Section 39 of the Building Act 1984.

The Appellant's case

Referring to drawing D2670/08, which is assumed to show the re-designed scroll panels to the balustrade guarding, you state that these are not 'climbable' as the closeness of the scrolling details prevents the structure from being climbable for the purposes of paragraphs 3.3 of Approved Document 'K' ('Stairs, ramps and guards'). You argue that the closeness of the redesigned details now prevents feet being able to pass through the panels and that the structure is not climbable.

The Borough Council's case

The Borough Council refer to paragraph 3.3 of Approved Document 'K' which advises that where buildings are likely to be used by children under 5, the construction of the guarding should be such that:

(a) a 100 mm sphere cannot pass through any opening in the guarding; and
(b) children will not readily be able to climb it.

The Council considers that the redesigned scroll panels remain climbable and are contrary to the guidance given in paragraph 3.3.

The Borough Council state that they do not wish to discourage artistic flair and have suggested that the scroll panels should be covered by glass panels or fine mesh which they consider would not detract significantly from the original design. The Borough also note that although your clients first appeared to agree with this suggestion, they subsequently changed their mind.

The Department's view

You have appealed to the Secretary of State against the refusal of the Borough Council to dispense with Requirement K2 in respect of the large decorative scroll work proposed for the balcony guarding. Requirement K2 exists to protect the safety of occupants and users of a building, and there are times when this can be a matter of life safety, particularly where young children are likely to have access to that which is guarded. The Secretary of State would not normally therefore consider it appropriate to dispense with Requirement K2, and would not lightly relax it except in exceptional circumstances. In this case you have not sought to claim any exceptional circumstances; and, to the contrary, it appears clear that the proposal relates to an unexceptional domestic environment in which it must be assumed that young children may live or to which they may have access.

Requirement K2 states:

Stairs, ramps, floors and balconies, and any roof to which people normally have access, shall be guarded with barriers where they are necessary to protect users from the risk of falling.

As already stated at paragraph 8 above, Approved Document 'K' provides guidance at paragraph 3.3 which says that where a building is likely to be used by children under 5 the construction of the guarding should be such that children cannot be held fast by it and that children will not readily be able to climb it. Given that the property is a dwelling, the question as to whether the guarding could be readily climbed by young children needs to be addressed.

In the Department's view although the gaps may be rather tight for an adult foot, there is nonetheless room for a child's foot and a child would find it quite easy to climb the scroll panels. It is therefore the Department's view that the revised design of the scroll panels does not demonstrate compliance with Requirement K2.

The Secretary of State's decision

You have appealed to the Secretary of State against the decision of the Borough Council decision not to dispense with Requirement K2 in respect of the guarding to the proposed balcony. He has given careful consideration to the facts of this case and the points made by both parties.

Compliance with Requirement K2 can be a matter of life safety and the Secretary of State would therefore not normally consider it appropriate to dispense with the requirement, or lightly consider relaxing it except in exceptional circumstances. The proposals relate to a domestic environment in which it must be assumed that chil-

dren of any age may live or to which they may have access. In these circumstances the Secretary of State considers that it would not be appropriate to dispense with, or indeed relax, Requirement K2; and has concluded that the Borough Council came to the correct decision in refusing to dispense with Requirement K2. Accordingly, he dismisses your appeal.

Maisonette – guarding stair (K.1 and K.2)

The guarding in this example did not include features to make it safe for children, as the flat was unlikely to be used by children. This argument was not accepted and the appeal was lost.

Ref.: 45.3.107   Date: 8 May 1997

The appeal

This appeal relates to the guarding associated with a new stair which connects the basement and ground floor of the new maisonette. The stair has guarding on both sides, and there is similar guarding round the edges of the balconies it leads to on the ground floor. This guarding is 900 mm high and comprises a hand rail at the top and another rail about half way up, with slender vertical supports 1 m apart on the balcony, slightly closer on the stair.

Requirement K1 states that:

Stairs, ladders and ramps shall offer safety to users moving between levels of the building.

Requirement K2 states that:

Stairs, ramps, floors and balconies, and any roof to which people normally have access, shall be guarded with barriers where they are necessary to protect users from the risk of falling.

Approved Document K gives guidance on the design of safe stairs and guarding which will satisfy the requirements, and, in paragraphs 1.9, 1.29 and 3.3 it gives specific advice on guarding and open riser stairs which are *likely* to be used by children under five years old.

Your letter suggests that the staircase design does not comply with paragraphs 1.9 and 1.29 of the Approved Document to Part K. Paragraph 1.9 is relevant to staircases with open risers. However, the drawing WG/30 indicates that the staircase does not have open risers and this presumption is supported by the Borough Council's letter which only refers to guarding to the stairs and balconies. Thus, in consideration of this appeal, it has been presumed that the risers are not open, and that the issue to be considered for the staircase is the need for safety features as recommended in paragraph 1.29 which is concerned with guarding.

Although you do not refer to the guarding of the balconies in your letter, it is referred to in the Borough Council's letter. As it is of similar design to the stair guarding it has been assumed that you wish it to be considered in the appeal.

You do not consider it necessary to make special provision for children to use the stair, and so asked the Borough Council to dispense with the requirements. This request was refused.

The Appellant's case

You state that the flat has one bedroom and a box room and has been designed for a childless couple or single adult. You add that there is insufficient storage and poor access for anyone with children, and the owner will not let it to couples with children.

You agree that your design does not include features to make it safe for use by children, but you claim that in view of the above, it is unlikely that the flat will be used by children, and the stairs are *not likely* to be used by children under 5 years, and so such features are unnecessary.

The Borough Council's case

The Borough Council argues that the Building Regulations are intended to protect everyone who uses the building, not just the permanent occupants. They feel that the guidance in the Approved Document Part K should be applied, and so will not dispense with the requirements with regard to the guarding for the stairs and balcony.

The Department's view

The purpose of requirements K1 and K2 is to protect the safety of the users of stairs, ramps, floors and balconies, and this may be a matter of life safety.

In this case, you argue that children are not likely to use the stair and balconies because the maisonette is not large enough to accommodate a family with children. In the Department's view it is likely that, during the life of the building, children will visit and perhaps live in the maisonette, and it should be designed with their safety in mind by taking account of the guidance given in paragraphs 1.9, 1.29 and 3.3.

The Secretary of State's decision

You have appealed to the Secretary of State against a decision of the Borough Council not to dispense with the requirements of Part K in respect of the staircase and guarding. He has given careful consideration to the facts of this case and the points made by both parties. As the purpose of the requirements is to protect the safety of the users, the Secretary of State would not normally consider it appropriate to dispense with the requirements, and would only relax them in exceptional circumstances.

As this appeal relates to a domestic situation the Secretary of State feels it is reasonable to assume that young children may live in or visit the maisonette at some stage.

The Secretary of State feels that this circumstance is not exceptional and considers that it would not be appropriate to dispense with or relax the requirements, and has concluded that the Borough Council came to the correct decision in refusing to dispense with the requirements. Accordingly he dismisses your appeal.

PART M: ACCESS AND FACILITIES FOR DISABLED PEOPLE

Amusement arcade – provision of sanitary conveniences (M.3)

The provision of WC facilities for disabled persons in a conversion and extension of a shop to an amusement arcade was in dispute. It was established that reasonable provisions should be made for disabled persons and the Secretary of State suggested that it would be reasonable to provide suitable ambulant disabled toilets.

Ref.: 45.3.58   Date: 21 June 1995

Careful consideration has been given to the facts of the case, and to the representations made by you and the Borough Council.

The appeal relates to the conversion and extension of a baker's shop, premises, and living accommodation into an amusement arcade. The building comprised a three-storey front portion containing the shop fronting onto a highway. The rear portion of the building comprised part two-storey and part single-storey. The ground storey of the rear portion contained a bedroom, kitchen and a compartment housing a WC, a shower and a washbasin. The upper storey of the rear portion contained a kitchen, bathroom and WC, which formed part of the living accommodation at that level – the remainder being within the front portion of the building.

The building work involved to achieve the conversion from the original shop and premises into an amusement arcade has involved the demolition of the rear portion and its reconstruction, and the gutting of the ground storey of the front portion. Access into the arcade from the adjoining street is level. The arcade is served, at the rear, by an office and workshop. Within the workshop is a compartment containing a WC and wash-hand basin, intended for the use of members of staff. There is also an exit door at the rear of the building, providing an alternative means of escape. This exit door gives access to 11 steps leading up to a rear garden.

Initially the Borough Council was not satisfied with your proposals for WC provision, although the scheme was eventually approved on the understanding that a WC, suitable for a wheelchair user, was to be provided adjacent to the workshop. However, in the event the WC provision as built did not accord with the approved plans.

You subsequently applied to the Borough Council for a dispensation from Requirement M3 of the Building Regulations. However, they refused your application on 9 June 1994. It is against that decision that you appealed to the Secretary of State on the 30th June 1994.

The Borough Council state that the intention of Requirement M3 is to ensure that all premises where the regulation is relevant have arrangements in place, so as to place as little constraint as possible upon their use by disabled people. They add that this applies to disabled people's places of work or to places which are visited in other circumstances, and that where sanitary conveniences are required to be provided they should be suitable for use by those disabled people.

They contend that this case involves a building to which Requirement M3 applies, and that people will visit and work in the Amusement Arcade. When assessing the proposals it was considered by the Borough Council that adequate provisions would be made if a unisex facility were provided and it was accessible to the staff and those people visiting the premises whether they suffered from a disability or not.

Your application for a dispensation was made on several grounds. You stated that your client could not employ a wheelchair-bound person because the amusement arcade is frequently staffed by just one person who must be very mobile to deal with any disturbances, etc. In addition, you drew attention to the fact that because supervision is conducted from the rear of the premises, means of escape in the event of emergency from that part of the premises would necessitate negotiating a flight of steps up to the rear garden. Given that, you argue that employment of a wheelchair user would be impractical. You further argue that the Borough Council have been inconsistent in being content not to require a safe means of escape for disabled persons from the ground floor yet requiring a disabled person's WC.

You state that no provision for sanitary conveniences was made in the original premises, in which up to four people were employed. You add that the provision has been increased for the current building, for which there will be a maximum number of two staff. You also state that there is no requirement either for staff or public toilets within the building, and that the proposal to include one in the secure staff area was for convenience only. You make reference to paragraphs 0.4 and 0.5 of Approved Document M ('Access and facilities for disabled people') which describe the obligations with respect to different kinds of extension. Paragraph 0.5 recommends a standard of provision where extensions have independent access equivalent to that which would be expected in a new building, but you make the point that the new extension is approached through the existing building and not independently.

You suggest that the employment of a wheelchair user in such a work environment would be impractical, particularly when it is to be expected that a staff member will be frequently alone and could be called on to deal with disturbances, emergencies and security. Among your concerns in this respect is the ability of a wheelchair user, in the event of fire, to clear the arcade, deal with the fire, as well as to escape. Such escape could be from the rear of the building which would involve the negotiation of a storey-height flight of steps.

Before considering the nature of sanitary provision, the Department has needed to consider whether there is a statutory obligation to make such provision. You suggest that there is no such obligation, whereas the Borough Council refer to Requirement G1(1) of Part G ('Hygiene') of the Building Regulations which deals with the adequate provision of sanitary conveniences. In any event the requirements of the Workplace (Health, Safety and Welfare) Regulations 1992, and the approved code of practice which accompanies them, are applicable in these circumstances. They are retrospective in nature and the code of practice has no lower limit in terms of the number of workers for whom sanitary provision is to be made.

The Secretary of State has consistently taken the view that it is not for him to determine what sort of work can be undertaken by a disabled person. Much will depend on the type of disability, the abilities of the individual concerned, and the potential for enterprises to adapt working practices. In addition to this, the Secretary of State has to take into account the fact that the use of the building may change several times during its lifetime and that, once a decision has been made, there may be no means of controlling the situation in changing circumstances. That said, a number of considerations have to be taken into account with respect to this building.

The building is not large and the potential to accommodate a workforce with varying physical or sensory abilities is limited. You have mentioned the need to deal with disturbances, emergencies and security and, in the event of fire, clearing the arcade and dealing with the fire itself. However, these considerations should not be overemphasized. It is not, normally, the responsibility of workers to deal directly with such incidents, other than by ensuring that proper procedures are carried out and the appropriate bodies notified. Of significance, however, is the need in this instance to negotiate a flight of steps to achieve means of escape at the rear of the premises. It is the Department's view that this fact is sufficient to suggest that this workplace could only provide employment for a person who can manage stairs.

Requirement M3 states that: 'If sanitary conveniences are provided in the building, reasonable provision shall be made for disabled people.' You applied to the Borough Council for dispensation of this Requirement. Had the Borough Council granted your application there would have been no obligation upon you to comply in any way with Requirement M3 anywhere within the premises. In this instance the provision for sanitary conveniences for employees is required under the Workplace (Health, Safety and Welfare) Regulations 1992; and in any event you have already made some form of provision in this area. Given these circumstances there appears to be no good reason why Requirement M3 should not be complied with in some reasonable way.

It follows that the Secretary of State is satisfied that in all the circumstances of this case the Borough Council of Restormel came to the correct decision in refusing your application for dispensation. Accordingly, the Secretary of State dismisses your appeal.

In analysing the issues involved in this appeal the Department has considered what would constitute reasonable provision for compliance with Requirement M3 in this particular case. The following observation may therefore be helpful. In the Department's view it would be reasonable to limit the sanitary provision to that suitable for an ambulant disabled person. Diagram 17 of Approved Document M includes guidance on the design of a suitable WC compartment of this type. However, this example does not contain a wash-hand basin. Further guidance which may be of benefit can be found in the approved code of practice which accompanies the Workplace (Health, Safety and Welfare) Regulations 1992.

Petrol filling station – access for disabled (M.2)

In this appeal the Department of the Environment examined the building regulations to see if specific pedestrian routes were required from the curtilage to the buildings. It was concluded that there is no requirement for the provision of a designated route to give access to a building.

Ref.: 45.3.67   Date: 28 June 1995

The appeal relates to the erection of a petrol filling station on a triangular island site, bounded on one side by a railway track and by public roads on the other two sides. The development comprises a petrol filling forecourt and canopy with an associated building con-

taining a shop and car sales, together with a car wash to the rear. Further to the rear there is a building for 'rapid fit'.

There is a public footpath outside one of the boundaries of the site which allows pedestrians to skirt an adjacent roundabout. A flight of steps gives access up an embankment from this footpath to the forecourt. It is the need to provide a dedicated pedestrian route and associated tactile surfaces from the vehicular and public footpath access at the boundary which has been at issue.

Your first full plans application was rejected following a non-response to the District Council's request for further information regarding, among other matters, pedestrian access from the curtilage of the site to the entrance to the buildings, including the 'Rapid Fit'. Following re-submission of plans it was agreed that access from the curtilage of the site to the entrance of the shop and car sales building would be provided in order to comply with Requirement M2. The plans were, however, rejected on other grounds and remained as such at the date of this appeal being lodged. However, work was commenced and the agreed provision for access was not carried out. The District Council then instigated legal proceedings for this and other contraventions.

You state that you agreed reluctantly to the access provisions in the hope of receiving approval of your plans but that it was always intended that the question of pedestrian access would be the subject of further discussions with the Council. However, the matter could not be resolved by discussion and you applied to the District Council for a dispensation of Requirement M2. The District Council refused this application, citing 11 reasons for doing so. It was against this refusal that you appealed to the Secretary of State.

Your grounds for the application for dispensation were that, because the site fronts a busy road, it was unlikely that disabled people, other than those arriving by motor vehicle in order to purchase fuel, would visit it. In that context, you argue that disabled drivers should be able to get to the building without the need for designated routes. You consider that a disabled person, arriving on foot, can follow the same procedure as a disabled driver. You also claim that the dedicated route shown is disruptive to vehicle manoeuvre.

The District Council relies on statements in the Approved Document 'M' ('Access and facilities for disabled people') that Part M applies to features which are needed to provide access to a building from the edge of its site and that reasonable provision should be made so that disabled people can reach the principal entrance of the building from the edge of the site curtilage. The Council refers to the negotiated provision for a segregated route from the edge of the site curtilage to the shop, which it was proposed to mark by bollards and white lining. Tactile paving was to be incorporated where the route crosses the carriageway.

The District Council justifies its judgement by pointing out that the site is served by an adjacent pedestrian footpath with crossing points at road intersections with the roundabout and contends that the shop is frequented by local residents, many of whom are pedestrians. In support of its contention, the Council points to the construction of the stepped approach up the embankment and onto the forecourt. The authority states that the segregated route from the head of the steps and leading to the sales building has not been provided and that the area intended for the route is now used for displaying vehicles for sale, thereby creating an obstruction and forcing pedestrians to mix with vehicular traffic on the forecourt.

It would appear that you concede that local residents use the sales building on the site and that, because the more agile residents were taking a shortcut across the embankment on the edge of the site, it was decided to provide the stepped approach rather than allow the embankment to suffer continuing damage.

In giving careful consideration to this appeal the Department has asked whether, in this instance, the Building Regulations require specific pedestrian routes to give access to buildings; and, if they do, what the nature of the design of such a route should be.

Paragraph 1.11 of Approved Document 'M' refers to the provision of tactile warnings for the benefit of people with impaired sight on pedestrian routes which are intended to be used by disabled people.

However, that should not be taken to imply that the principle of providing a dedicated pedestrian route would be a condition of compliance with the Building Regulations. Moreover, even if in principle it was, much would depend – in practice – on the nature of the building and on its design and that of its immediate surroundings. The filling station environment is one in which pedestrian and vehicular movements are normally mixed. In such an environment, drivers with mobility difficulties would be expected, along with other drivers, to take account of vehicle movement in ensuring their own safety and that of others.

The Secretary of State is satisfied that there is no requirement, through the Building Regulations, for the provision of dedicated pedestrian routes to give access to buildings. In his view to require such a route, when one would not normally be provided – as well as insisting on the incorporation of facilities for people with impaired sight – is unreasonable. However, you have appealed against the refusal of the District Council to dispense with Requirement M2. If the District Council had granted your application there would have been no obligation upon you to comply with Requirement M2 anywhere within the site and buildings.

It follows that the Secretary of State is satisfied that in the circumstances of this case, the Wrekin District Council came to the correct decision in refusing your application for dispensation. Accordingly, the Secretary of State dismisses your appeal.

In analysing the issues involved in this appeal, the Department has considered what would contribute to, or constitute, reasonable provision for compliance with Requirement M2 in this particular case. The following observations may therefore be helpful.

In the Department's view it is arguable that there should be tactile paving at the top of the stepped approach at the embankment in order to give warning of the existence of the steps and of vehicle manoeuvres on the forecourt. A decision on whether there should be tactile paving at the bottom of the stepped approach is a matter for the highway authority, together with any proposals for footpath crossings adjacent to the roundabout. The Department would also observe that it is arguable whether the erection of bollards will necessarily provide a safe environment for people whose sight is sufficiently impaired to warrant tactile paving as a warning device. The bollards may be of value in ensuring that the drivers avoid the footpath but their existence might create an obstruction for people with impaired sight.

Betting shop – access for disabled persons (M.2)

A pharmacy is to be converted into a betting shop and as the work would not result in a building, which taken as a whole, would not be more unsatisfactory than its existing form, the proposal should not be regarded as a material alteration. In that case the work is not subject to the building regulations.

Ref.: 45.3.74   Date: 15 September 1995

The building work to which this appeal relates involves the conversion of an existing pharmacy shop located on one side of a market square, into a licensed betting shop. The existing shop has level entrance from the pavement. The level entrance floor area extends approximately half way into the depth of the shop. The remaining half of the floor area to the rear of the shop is raised and accessed by three steps which extend across the full width of the shop.

The construction work is to include the replacement of the existing shop front containing central double entrance doors, by what is described on the plans as a traditional one with an entrance door to one side recessed back from the pavement. The first section of floor (i.e. at the front of the shop) is to be raised in order to provide a single level throughout. As a consequence of the raising of this floor area the new entrance door will open onto an internal mat well and three steps beyond. The building work also comprises improved staff facilities, including a female/disabled WC.

Your full plans application was rejected on grounds of noncompliance with Requirement M2 of Schedule 1 to the Building

Regulations 1991. You then applied under section 8 of the Building Act 1984 for a dispensation from Requirement M2 which was refused by the District Council. It is against that decision that you have appealed to the Secretary of State under section 39 of the 1984 Act.

The District Council's view is that there is at present direct and unaided access to the building from the site boundary. They contend that this should be maintained and that if the internal arrangement is to be improved this should be done within the building, such as by a ramp and/or the introduction of a platform lift at the change in levels between floors, so as to enable disabled persons to gain access to the whole of the customer areas, including the sanitary accommodation.

In forming their judgement the District Council have consulted, and been supported by, two Access Committees. The Highway Authority was also consulted.

In support of your appeal for dispensation from Requirement M2 you have argued on behalf of your clients that there are both external and internal constraints on the provision for access which can be made. Externally, you state that the pavement between the premises and the carriageway is very narrow and is not wide enough to accommodate a wheelchair. In addition, the step up to pavement level from the road is in excess of 150 mm, thus making wheelchair access difficult.

You argue that internally, because of the size of the property, it would be impracticable to install an entrance ramp. Instead, your clients propose to follow the practice they adopt in many of their other shops throughout the country of providing assistance by means of a call button intercom system sited at the entrance door which will be connected directly to the staff area within the premises. Assistance will be given upon demand. You claim that this practice is recognized and accepted in many other of your clients' premises.

The District Council accepts that the pavement outside the premises is not wide enough to take a wheelchair and that there is a raised kerb. Although they acknowledge that these are matters outside the site boundary (and therefore outside the scope of the Building Regulations) they consider that 'the arrangement was material to the assessment because wheelchair users would have difficulty accessing a call button initially and if successful would have to "park" in the carriageway to wait, presenting a hazard to both pedestrians and vehicle users, before being manhandled into the building'. To enhance the direct access which the District Council consider should be available, they have suggested that a dropped kerb could be provided.

Your request for dispensation was in respect of Requirement M2 of Schedule 1 to the Building Regulations 1991. This Requirement, along with the other three Requirements in Part M of Schedule 1, is limited in its application and does not apply to four specific types of building work. The relevant exclusion of Requirement M2 in the context of this case is potentially '(b) a material alteration'. It therefore follows that irrespective of whether the proposed work could be said to be categorized as a 'material alteration' or not, Requirement M2 would not appear to apply to the *proposed* work.

Given that Requirement M2 is not applicable to the *proposed* work, the Secretary of State is of the view that your application for a dispensation was neither appropriate nor necessary. In these circumstances he therefore feels obliged to refuse to uphold your appeal on the grounds that to do otherwise would be to imply that Requirement M2 *does* apply to the *proposed* work.

You will wish to note that in giving careful consideration to this appeal the Department has considered whether, on completion, the work would result in a building which, *taken as a whole*, would not comply with a relevant Requirement (as defined in Regulation 3(3)) whereas previously it had; or in a building which before commencement of work did not comply with the same relevant Requirement and which, on completion of all of the works would be more unsatisfactory. To achieve this the Department has considered whether the building contains provisions which, were they to be measured against the Requirements of Part M, would demonstrate that the building already made reasonable provision for access and facilities for disabled people. It has concluded as follows.

To the extent that there exists at present a change in level requiring the negotiation of three steps, effectively dividing the floor of the shop into two, reasonable provision by current standards does not exist – notwithstanding the fact that there is level entry from the pavement to the front half of the premises. Given that the standard of provision of access and facilities for disabled people in the existing building would not adequately measure against that which would be sought were the same building to be constructed today, it is then necessary to ask whether it would measure any worse as a result of the proposals.

In the Department's view, this operation resolves itself into the issue as to whether the provision of a wholly level floor requiring steps at the entrance (where formerly there were none) is less satisfactory than the retention of the existing split level shop floor with the need to negotiate three steps in the centre of that floor. In either event, a wheelchair user would need some assistance to negotiate a change in level when visiting the proposed betting shop.

However, on balance the Department believes that it would be better for the interior of the premises to be at one level so that, for a wheelchair user, access to and from the counter and other facilities – which may be needed several times during the course of a visit – would be made easier. From a management point of view, assistance is going to be required at some stage. You have stated that your clients would be prepared to provide such assistance from the pavement into the premises. In the Department's view, this management commitment would be best called upon on arriving and leaving the premises, rather than when negotiating two levels of a shop floor during the course of a disabled person's visit.

It follows that the Department has concluded that, in this instance, those elements of the work which affect access and facilities for disabled people would not result in a building which *taken as a whole* would be more unsatisfactory than it is in its present form. It further follows that in the Department's view the particular work in question does not constitute a 'material alteration' as defined in Regulation 3(2)(b). The work cannot therefore constitute 'building work' and the corollary of this must be that the question of dispensation from the statutory obligation of compliance under Regulation 4(2)(a) should not arise.

Retail foodshop – lift for disabled persons (M.2)

This appeal hinges upon the suitability of a goods lift to provide satisfactory access to the first floor where a staff rest room and cloakroom were placed. This proved not to be acceptable, but solutions to the problem were offered.

Ref.: 45.3.73   Date: 16 February 1996

Your appeal concerns shop fitting works to an existing two storey retail food shop. The ground floor has a nett sales area of 420 square metres and a backup area of 95 square metres to the rear of the building. The latter area comprises a goods handling area, a goods lift, a cleaner's store, and a WC which is adapted for disabled persons' use. The second floor area is 348 square metres and comprises an administrative and cash office, female toilets, a staff rest room, a cloakroom, and a beer and wine store. It also contains a 5000 cubic feet storage room for freezer food, a chiller, and a dry goods storage and backup area.

There was an existing ambulant stair to the rear of the building and a goods handling lift with manual doors formed part of the original proposal. Your original proposals for the shop fitting of this building were rejected by the Borough Council on six grounds; No. 1 of which relates, in part, directly to this appeal – *viz:* 'The lift is not shown to be accessible as required in Part M of the Building Regulations . . .'

In subsequent correspondence the Borough elaborated on this ground for rejection and stated (with reference to Requirement M2) that 'The lift is not shown to be accessible to the disabled in that:

(i)   use should be possible without accompaniment;
(ii) manual operation of gates is not acceptable.'

You contended that given the circumstances and purpose for which the lift was primarily intended, compliance with Requirement M2 was unreasonable. You therefore applied to the Borough Council for a dispensation of Requirement M2 which was refused. It was against this refusal that you subsequently appealed to the Secretary of State.

A substantial amount of correspondence has been submitted relating to the further attempts by you to meet the problems identified by the Borough Council in respect of their contention that your proposals for the lift did not comply with Requirement M2. There have also been several Building Regulation applications which were rejected on the same grounds. The general thrust of both the Borough's and your arguments is as follows.

The Borough Council argues that it does not consider that to insist on compliance with Requirement M2 is unreasonable in this particular case for the following reasons:

(i)   The first floor area is in excess of $280\,m^2$ which is considered to be the minimum size below which it would be unreasonable to require a passenger lift.
(ii) There are offices situated on the first floor together with the staff room and cloakroom.
(iii) Although the major part of the first floor is for storage purposes disabled personnel involved in stock taking may require access.
(iv) The use of the first floor could change in the future, and it may become office accommodation throughout.
(v)   Facilities for disabled personnel and able-bodied personnel should be similar. The location of the staffroom and cloakroom at first floor level presents disabled personnel with a disadvantage.
(vi) There may be a requirement for personnel working with food to remove certain protective clothing when using the WC. There are no changing facilities adjacent to the sanitary accommodation for disabled personnel.

The Borough Council agreed to consider the potential use of the goods lift by a person who was a wheelchair user. The Council's disability policy adviser assessed its ease of use and found that the gates were difficult to open without assistance.

You state that because the lift is used for the transportation of heavy goods, it requires to be robust particularly in the case of the landing and car doors, otherwise they become damaged during their normal working life. In this regard you state that the lift fully complies with BS 5655: Part II: 1988.

You point out that a special handle for wheelchair users had been provided at the time the Council's disability policy adviser assessed the lift and that although he found it difficult to undo the manually-operated loading gate he did not find it impossible. You also contend that there is nothing in Requirement M2 which states that the building has to be used by the disabled person on his own, without assistance from able-bodied persons; and that there is signage at the lift entrance which states that disabled persons using the lift should call for assistance. You argue that such an arrangement is not contrary to Requirement M2 and is reasonable in the circumstances.

With respect to the assessment made by the Borough Council, you claim that this was subjective. You claim that this assessment in fact demonstrated that the gate could be opened (albeit with difficulty) by a wheelchair user using the additional handle. You conclude that depending upon the upper body strength of the wheelchair user, some people may find it difficult to open the gate, whilst others will not.

With regard to the lift itself, you submit that reasonable provision has been made to make it useable by a wheelchair user. Every effort has been made to ensure the doors run smoothly and lift buttons specifically designed for disabled persons have been incorporated, together with the special handle.

You contend that the alternative to these arrangements would be a lift with automatically powered doors. You believe that automatic doors would not be strong enough to withstand the type of usage

expected and would require constant replacement and costly maintenance. You also contend that the incorporation of automatic doors
would involve major works and would not be reasonable in terms of
the cost.

As part of your case, you have quoted various parts of the text contained in Approved Document 'M' ('Access and facilities for disabled
people'). In particular, you have quoted all of paragraph 2.9 which
concludes that it is not reasonable to require a lift to be provided in
every instance. You have also quoted the opening sentence of paragraph 2.13 which provides guidance on the nett floor areas in excess
of which the provision of a passenger lift will satisfy Requirement
M2. You argue that paragraph 2.13 does not state that Requirement
M2 can only be satisfied by meeting the requirement of paragraph
2.13 and that it follows that having the assistance of able-bodied
persons to open the manual doors for those who cannot open them
on their own, may be considered reasonable in certain circumstances.

You refer to the figure of 280 square metres of nett floor area
quoted for a two storey building in paragraph 2.13. You state that the
first floor area in this case is 348 square metres in total. However, this
area includes the cold store, chiller and storage handling areas where,
you argue, a disabled person would not be expected to work because
such work involves the handling and movement of heavy stock,
much of it frozen, with and without pallet trucks. You conclude that
as there are no other functions carried out in this area, the only area
which a disabled person is likely to require access to would be the
cash/administrative office and staffroom. You therefore contend that
only the offices and staffroom area should be taken into account in
calculating the nett area of this floor and that on this basis it would
fall below the 280 square metres criteria and there would be no
requirement at all to provide a lift and that ambulant stairs would
suffice.

The Department's analysis of this case is as follows. The first question for consideration is whether the extant guidance suggests that
wheelchair access should be provided to the first floor. The second
question is, if it is required, is the modified goods lift of adequate
design to comply with Requirement M2?

You have questioned the basis on which the nett floor area of the
first floor should be calculated. However, the last sentence of paragraph 2.13 provides clear guidance on what should be excluded in
assessing the nett area and in the Department's view it is clear that
the first floor in this case cannot be precluded from passenger lift
access on the basis of this argument. Rather, the pertinent questions
for the Department are: whether the tasks involved in working in the
storage area on the first floor could be undertaken by a wheelchair
user; and whether the first floor offers 'common areas' and 'unique
facilities'.

In respect of the potential for a wheelchair user performing the
tasks described on the first floor, you have argued that this is
unlikely. Against that, the Borough Council has asserted that stock
taking tasks could be undertaken by a wheelchair user; and that, in
any event, the function of the first floor accommodation could
change at some time in the future.

It is not for the Secretary of State for the Environment to determine what people can or cannot achieve in a work environment, as
this has as much to do with the abilities and commitment of an individual as with the nature of the work itself and how procedures can
be adjusted. However, there has to be some practical limit as to how
far that view can be sustained.

On balance, in this instance, the Department accepts that it seems
unlikely that the tasks involved in working in the storage area could
be undertaken by a wheelchair user. Nevertheless, the Department
acknowledges the argument put forward by the Borough Council
concerning the potential change in function of the accommodation
and considers that this should carry some weight.

The first floor of the premises contains a staff rest room and a
cloakroom. These should both be regarded as common facilities and
paragraph 3.3 of Approved Document 'M' sets objectives for
'common facilities, e.g. canteens and cloakrooms . . . and states that

they 'should be located in a storey to which wheelchair users have
access'. A staff rest room facility is also included among the list of
'unique facilities' contained in paragraph 2.15. Therefore, on grounds
of access to both 'common areas' and 'unique facilities' the guidance
indicates that access for wheelchair users should be provided to the
first floor.

In reaching a conclusion on this case the Secretary of State is con-
strained by the procedure by which the case has been put to him.
You have appealed to the Secretary of State against the Borough
Council's refusal to dispense with Requirement M2. To have dispensed with this Requirement would have been to do so in respect of
all access provision throughout the building. In this regard the
Department notes that, in any event, it appears that access issues –
other than vertical access – are not matters at issue. The Secretary of
State is therefore satisfied that the Borough Council came to the
correct decision in refusing your application to dispense with
Requirement M2. Accordingly, he dismisses your appeal.

In giving careful consideration to this case, you will wish to note
that the Department considered if, or how far, your proposals could
be considered to comply with Requirement M2. The following observations may therefore be helpful.

Even if it was to be concluded that the employment potential on
the first floor did not warrant access for wheelchair users – and the
Department gives no definitive view on this because this case is not a
determination – the fact remains in the Department's view that the
first floor contains a 'unique facility'. Paragraphs 2.15 and 2.16 of
Approved Document 'M' recommend that if a storey exceeds 100
metres square, but is not large enough to warrant a passenger lift, then
it should be made accessible to wheelchair users by use of a wheelchair stairlift. Equally, the 'common facility' of the rest room on the
first floor also requires access for wheelchair users by some form of lift.

Given that a lift is already installed which has the potential to
provide access for a wheelchair user, then that lift could be made
convenient for use by a wheelchair user. In the Department's view a
prerequisite for such convenient use should be that the door or doors
open automatically thus allowing independent use by a wheelchair
user, rather than an individual having to summon assistance as you
have proposed in this instance. It is the Department's view that the
provision of manual over-ride for the automatic operation of the
door or doors should facilitate the movement of goods and obviate
the potential damage which was one of your chief concerns.

Student residential accommodation – access and facilities for disabled people (M)

*The proposal is to erect a block to accommodate more than 500 stu-
dents. The appellants claim that the self-contained flats were for residential purposes only, and that Part M did not apply to dwellings.
Common areas in the building were designed to meet Part M, but there
was no provision within the flats.*

Ref.: 45.3.42   Date: 20 February 1995

Careful consideration has been given to the facts of this case, and
the representations made by the appellants and by the City Council.

The development in question consists of a student living accommodation block which is elliptical in shape and varies from two to
four storeys in height. It is to be built around an inner courtyard
within which there is to be a single storey building, incorporating
administrative offices and six dwellings for staff.

It is understood that the student living accommodation block is
designed to house 504 students. The building comprises, predomi-
nantly, groups of self-contained accommodation, each housing six
students. Each group is approached by a stairway and contains six
bedrooms, each with en-suite toilet and washing facilities, and a
shared dining kitchen. Each group has its own 'private' entry. The
building also contains, on one side of the ellipse, two rooms, vari-
ously described by the appellants as assembly or common rooms, and
two sets of laundry and WC facilities; the last-named including a
'unisex' wheelchair WC.

You contend that the requirements of Part M of Schedule 1 to the Building Regulations 1991 should apply to the development. The appellants have contended that the requirements of Part M should not apply because the living accommodation comprises self-contained flats. Although it was open to them to challenge this view by referring the question to the Secretary of State for his determination as a 'matter arising', in the context of section 16(10) of the Building Act 1984, they chose instead to make application to the City Council for a dispensation of the requirements of Part M. This was refused by you on 4 February 1994 and it was against this decision that they have appealed to the Secretary of State under section 39 of the Building Act 1984.

In support of their case the appellants claim that the student living accommodation in dispute comprises separate and self-contained flats; that the flats are constructed for residential purposes only; and that each flat forms part of a building from which it is divided horizontally. Under these circumstances they consider that the criteria for defining flats in Regulation 2 of the Building Regulations have been fully satisfied and that therefore Part M of Schedule 1 of the Building Regulations 1991 does not apply. In addition, the appellants have stated that the Assembly Rooms in the proposed building have not been provided for the exclusive use of the residents of the flats; and that the Hall Managers' accommodation is to provide self-contained accommodation for the managers to live and work in.

You, on the other hand, contend that the proposed student living accommodation represents a 'building' as defined in Approved Document 'M' ('Access and facilities for disabled people'), and that the proposal is therefore required to comply with the relevant sections of Part M: namely M2 ('Access and use') and M3 ('Sanitary conveniences'). You consider that the student bedrooms, living rooms and bathrooms have not been designed to provide suitable access or facilities for any wheelchair-disabled or other disabled persons. However, you state that the common areas of the development – i.e. the common room, laundry and reception area – are all designed to comply with Part M and are not a matter of dispute.

You further contend that the proposals appear to satisfy all the criteria put forward by the Department to define traditional halls of residence. In support of this you state that:

(a) the proposal contains common areas such as common rooms, reception areas and laundry;
(b) all managers are accommodated within the proposals;
(c) tutors are accommodated within the scheme;
(d) a student refectory is available within easy pedestrian access and is suitable for wheelchair-disabled persons;
(e) the proposal is on-campus.

In the light of the points in the above paragraph, and in the absence of what you consider to be reasonable grounds for such an application, you consider it inappropriate to dispense with the requirements of Part M with respect to the student living accommodation.

In considering this case, the Secretary of State has been conscious of the need to balance the sensible and realistic provision of access and facilities for the disabled with the departmental view (expressed in a letter of 25 February 1994 addressed to Chief Executives of building control authorities) on the application of Part M to some types of student living accommodation. The letter confirmed the Department's intention to encompass, within the scope of Part M, buildings covered by other relevant primary legislation: in this instance, section 8 of the Chronically Sick and Disabled Persons Act 1970 which is concerned with the provision, inter alia, of a 'hall of a university' and provisions for the needs of persons using the building who are disabled. The letter also referred to the fact that the viability and practicability of applying access requirements to new dwellings was being considered by the Department and that student accommodation in the form of 'dwelling' was being considered as part of that exercise. A consultation exercise is now underway on proposals to extend the application of Part M to new dwellings. The letter to

Chief Executives concluded by acknowledging the efforts of university and college authorities to make provision for disabled students in housing schemes and stated that the Department would continue to encourage this approach.

The Department's analysis of this case is as follows. It is the Department's view that some accommodation for students, particularly that in which the accommodation is self-contained and provides for self-catering, can be defined as a 'dwelling'. Dwellings are excluded from the requirements of Part M and it follows that its provisions would not apply to any student living accommodation which can be defined as a 'dwelling'. However, it is also the Department's opinion that Part M does apply to the traditional 'hall of residence' – the kind which would have been most common at the time the Chronically Sick and Disabled Persons Act 1970 came into force and which includes, within it, not only student bedrooms but also central catering and other amenities. Having carefully examined the plans for this particular development, it is the Department's view that the living arrangements would appropriately be described as a group of self-contained flats and, therefore, not subject to the requirements of Part M.

Having given careful consideration to this case, the Secretary of State has come to the following conclusion. In his view the proposed student living accommodation in the elliptical building constitutes 'dwellings' as defined by Regulation 2(1) of the Building Regulations 1991 and it would not be appropriate to apply Part M to this accommodation in view of the limits on application of the requirements of Part M. The Secretary of State accordingly upholds the appeal with respect to dwellings in the development.

With respect to the common parts of the development, the Secretary of State notes that these are separate from the dwellings. It is understood that they are not intended for the exclusive use of the dwellings but rather are to be used, for example, as assembly areas for the whole campus or let out to clubs or societies. The Secretary of State has therefore decided that all the common parts of the development are subject to the requirements of Part M and accordingly dismisses the appeal in respect of these parts. In so doing he notes that you have stated in respect of these common parts of the development that there is no dispute with the appellants regarding compliance.

Petrol filling station – sanitary convenience and access for disabled (M.2 and M.3)

The local authority wanted a ramped access and a suitable WC compartment for disabled people. There are arguments regarding duties attendant upon the owners by the Health and Safety at Work Act and the Petroleum Regulations. Although the appeal is dismissed, circumstances indicate compliance with M.2 and M.3.

Ref.: 45.3.41   Date: 25 April 1995

Careful consideration has been given to the facts of the case, and to the representations made by you on behalf of the appellant and to those of the Borough Council.

The building work involves the construction of a kiosk on a petrol filling station forecourt which forms part of a supermarket development. The accommodation in the kiosk comprises attendant, customer and retail areas; together with store cupboards and a staff toilet segregated from these areas at the rear.

A full plans application by you was rejected by the building control authority on the grounds of non-compliance of the plans for the kiosk in respect of Requirements M2 and M3 of Schedule 1 to the Building Regulations 1991. The Borough's requirement was that there should be provision for ramped access and that there should be a suitable WC compartment for disabled people.

You have resisted complying with these requirements on the grounds that the toilets in the kiosk are for the exclusive use of employees serving the filling station and that your clients cannot safely employ and train attendants in wheelchairs to carry out the duties placed upon them by the relevant Petroleum Regulations and the Health and Safety at Work Act 1974. You therefore sought a dispensation from Requirements M2 and M3 in respect of employees in the kiosk which was refused by the Borough Council. It is the refusal by the Borough to dispense with these requirements which forms the subject of your appeal to the Secretary of State.

In your submission to the Department you have described the responsibility for safety placed upon petrol filling station employees by guidance (HS(G)) as issued by the Health and Safety Executive. Given these duties you have argued that it is difficult to foresee a disabled person in a wheelchair being able to safely fulfil all these demands as well as the remainder of the guidance requirements; and even less likely in an emergency, particularly if a disabled customer in a motor vehicle was also to be involved. Your clients have concluded that there would be a potential risk to the safety and welfare of the employee and the customer which your clients would be unwilling to take.

You have concluded by stating that your clients have a positive policy towards employing people with disabilities throughout its establishment and would not discriminate against an applicant whose disability did not prevent them from carrying out the duties safely. Your clients would therefore consider a wide range of disabled applicants for employment within the petrol filling station.

The Borough Council consulted the local access group. The group have pointed out that the Building Regulations apply to the building and not the owners of the buildings, and have therefore requested that proper access and facilities be provided in the kiosk. They have expressed the hope that, in the future, your clients will receive an application for employment in the kiosk by a disabled person and will not have to refuse it because the building is unsuitable.

The Borough Council have not accepted your argument that the legislative requirements of the Petroleum Regulations and the Health and Safety at Work Act 1974 should take precedence over the requirements of the Building Regulations, particularly in the case of a new building. They argue that any subsequent alteration or variation in the petroleum legislation would tend to show that dispensation from Part M had pre-empted any subsequent decisions in respect of the employment of disabled people. They further argue that dispensation would mean that there would be no facilities for customers with disabilities.

However, having argued the case for not being able to dispense with Requirements M2 and M3, the Borough Council have concluded that the requirement for access within the staff side only of the kiosk could reasonably be limited to a non-wheelchair approach – i.e. one suitable for ambulant disabled persons and those with sight or hearing impairment. The Borough Council have further stated that the construction of the kiosk is in accordance with Plan No. 6109/9D (dated February 1992) and not in accordance with the subsequent drawing No. 6109/P2 (dated October 1993) relating to the application for relaxation/dispensation. The Borough have stated that they accept that the design is in accordance with the requirements of Part M and in particular with Requirement M3 ('Sanitary conveniences'). They consider that whilst the design does not provide for wheelchair access to the sanitary conveniences, the design has been based on access for ambulant disabled people. The Borough have also quoted the aim for judging compliance with M3 contained in paragraph 4.1 of the Approved Document 'M' ('Access and facilities for disabled people') which refers, inter alia, to bearing 'in mind the nature and scale of the building in which the provision is to be made'.

In considering your appeal the Department has noted in particular your argument that the legislative requirements of the Petroleum Regulations and the Health and Safety at Work Act 1974 prevent a disabled person in a wheelchair from being able to safely fulfil these demands and requirements. Responsibility for licensing petrol filling stations rests either with local fire brigades or with other departments of local authorities. In the Department's view it must be for these authorities to decide whether the fact that a disabled person can, or cannot, perform relevant duties is material to the granting of a licence; and, if so, whether the employment of a disabled person mil-

itates against the issuing of a licence. It should not, therefore, be for the Secretary of State to pre-empt such considerations and decisions by foreclosing on the opportunity for a disabled person to be employed in the kiosk by deciding in favour of a dispensation of Requirements M2 and M3.

In giving careful consideration to your appeal, and the particular points above, the Secretary of State has been anxious not to sustain any preconceptions about the abilities of disabled persons. He is also very conscious that to uphold your appeal for dispensation of Requirements M2 and M3 would imply that no provision would need to be made for access and WC facilities for disabled customers and employees within the kiosk. In the Secretary of State's view this would be unreasonable. Having regard to all the circumstances the Secretary of State therefore dismisses your appeal.

The Secretary of State notes that it is clear from the position taken by the Borough Council in their letter of the 2 March 1994 that they appear to be content that the kiosk as constructed provides access for wheelchair users, whether customers or employees, and therefore complies with Requirement M2; and that although the sanitary convenience proposed would not be suitable for use by a wheelchair user it would be for an ambulant disabled person, and in the circumstances, complies with Requirement M3. Your client's letter of 11 March 1994 has accepted this as a very practical resolution of the matter. However, it should be noted that the Secretary of State considers all cases on their individual merit, and solutions to compliance developed in one particular set of circumstances should not be taken as creating a precedent.

School sports centre – installation of stairlift (M.2)

The appeal revolves around the local authority's insistence that a wheelchair stairlift be installed to serve a lower ground floor. In reaching his decision the Secretary of State was 'conscious of the need to balance the general needs of disabled persons against the specific realities of particular circumstances'.

Ref.: 45.3.92   Date: 24 September 1996

The appeal

The school is an independent boarding school for boys aged 8–13 years and currently has an attendance of 255 boys. The building which is the subject of this appeal comprises a newly constructed purpose-built two storey sports centre. The ground floor is the entrance level and is accessible for wheelchair users. It contains a large viewing gallery, a rifle range, a fitness training room, a kitchen, and toilets – including one suitable for the disabled. The viewing gallery overlooks the lower ground floor which contains the main sports hall, several stores, two fives courts and three squash courts. There are three sets of stairs giving access to the main sports hall at three of its four corners together with access to the other facilities.

The City Council state that full plans approval was granted with conditions to the effect that a wheelchair lift should be installed (understood to be in staircase No. 3) in order to achieve compliance with Requirement M2 ('Access and use'). However, this lift was not installed and on discovering this the City Council took action under section 36 of the Building Act 1984 to secure its installation. As there are no disabled pupils at the school and the sports centre is restricted to use by pupils and staff only, your clients did not consider it reasonable to have to install the stairlift.

At the City Council's suggestion, the school then applied to the Council for a dispensation of Requirement M2 which was refused. It was against that refusal that you subsequently appealed on behalf of the school to the Secretary of State.

The Appellant's case

In support of your case for dispensing with Requirement M2 in respect of the provision of a stairlift between the two levels, you have argued on behalf of the school that the sports centre will not be open to the public and would only be used by current pupils and staff.

There are, moreover, no disabled pupils in the school. You have explained that the original planning permission for the sports centre restricts its use to the school only so that it cannot be used by the general public. On this basis you contend that since there are no disabled persons in the school, and that no one else is able to use the centre, it would be unreasonable to require the school, at this stage, to install a disabled person's stairlift. You do, however, comment that if the circumstances change, and if in particular there was someone disabled in the school, then it would undertake to install the stairlift as required by the regulations.

The City Council's case

The City Council states that the Building Regulations require reasonable provision to be made for disabled people to gain access to and to use the building and that this requirement applies to those who work in the building or who visit it. In this particular case where access is required to the lower ground main sports hall area, Approved Document 'M' ('Access and facilities for disabled people') indicates that compliance may be achieved by the installation of a full passenger lift. Following representations made to the City by you, and having regard to all the circumstances, it was agreed that in this instance a wheelchair stairlift would be an acceptable means of securing compliance. The City Council make the point that you now appear to consider the requirement too onerous or inappropriate but that you have not put forward any alternative proposals following construction of the building.

The City Council state that they have given careful consideration to your application for dispensation for Requirement M2 in respect of the wheelchair stairlift, but do not accept that because the school at this moment in time has no disabled pupils or employees that the requirement is too onerous or inappropriate. They argue that to provide no alternative solution clearly fails to satisfy the requirement for reasonable provision to be made. In addition they state that they understand from discussions with the Bursar that the school, for financial reasons, would like to open the centre to the public and that they would be reviewing this matter in the future. The Council state that they would have no continuing powers available under the Building Regulations to ensure the future installation of a new wheelchair stairlift in staircase No. 3.

The Department's views

The Department's views are as follows. Your original application and subsequent appeal was in respect of the unspecified dispensation or relaxation of Requirement M2 in respect of the new sports centre. In the Department's view it would be inappropriate to consider dispensation or relaxation of Requirement M2 in respect of the entire building; and in any event it is noted that wheelchair access has been provided for use of the ground floor facilities, including the viewing gallery. In the Department's view the question for the Secretary of State is whether it would be appropriate to relax Requirement M2 in respect of vertical access between the ground and lower ground floor levels so as to require no provision of a mechanical lift of any type.

In the Department's view the main consideration in this case must be whether the facilities on the lower ground level are likely to be needed to be accessed by the school's pupils or staff. The Department considers that the stores would not need to be accessed by wheelchair users and the squash courts and five courts, and the main sports hall, would only be likely to require access by those who will be participating in active sport. It also needs to be recognized that access for spectators, including wheelchair users, has been satisfactorily provided from the entrance level viewing gallery.

In addition, the Department notes that the use of the sports centre is restricted by condition 4 of the Planning Consent granted by the City Council in respect of the development. As a result of this condition, the centre cannot be used generally by the public and cannot therefore be viewed in the context of a leisure centre where it would be appropriate to provide access for wheelchair users for the playing of sports.

The decision

The Secretary of State has given careful consideration to the facts of this case and has been very conscious of the need to balance the general needs of disabled persons against the specific realities of particular circumstances.

In this particular case he has noted that the sports centre is restricted to the private use of the school by pupils and staff, none of whom are disabled. He considers that it is therefore appropriate to draw a clear distinction between this situation and that of a leisure centre to which the public would be admitted. In these circumstances, the Secretary of State has therefore decided to uphold your appeal by relaxing Requirement M2 ('Access and use') in respect of the provision of any form of mechanical lift between the two floor levels of the new sports centre. He directs that such a relaxation shall only be applicable for so long as the centre is restricted to the private use of the school.

Factory and office – installation of lift (M.2)

In dispute was the provision of a lift for the use of disabled people wishing to gain access to the first floor. It was argued that wheelchair users would not be using the first floor and a semi-ambulant stairs was provided. M.2 was relaxed and the appeal upheld.

Ref.: 45.3.82   Date: 20 January 1997

The appeal

The building work to which this appeal relates is an extension to an existing factory to provide a food processing area, loading and unloading bays, plant room. In addition there is a separate two storey office block approximately 32 × 12 metres attached at first floor level to the factory floor area. There are also new sewage treatment works.

Only half of the area on each floor of the new two storey block is in fact devoted to office use. On the ground floor there is a canteen and kitchen occupying approximately half the floor which serves both the factory and office workers. There is also a disabled persons' toilet. Half the first floor area provides staff offices and toilets. The other half of the first floor area, nearest to the factory floor area, is separated by doors and contains male and female changing rooms and toilets serving only the occupants/staff of the food processing factory. A stair links the ground and first floors. A lift shaft has been constructed but no gear or car installed.

Your clients submitted a full plans application to the District Council which were rejected. Among the items listed in the rejection notice was a requirement for a lift to be provided in order to satisfy Approved Document 'M' ('Access and facilities for disabled people').

However, your client took the view that because the first floor of the office block does not exceed 280 m², and because no disabled persons could be employed on the factory floor area because of the nature of the process, access for wheelchair users was not required to the first floor. He therefore appealed for a relaxation of Requirement M2 ('Access and use') which was rejected by the District Council. It is against that refusal that you subsequently appealed to the Secretary of State on behalf of your clients.

The Appellant's case

You state that the office block is ancillary to the food processing factory but is connected to it by a corridor. A lift was initially proposed but it was decided during construction that, in order to save unnecessary expense, the lift would be deleted from the proposals.

You state that access for disabled persons from the ground floor offices to the first floor offices and factory front was to be via a semi-ambulant stair, and that the details of the stair and its layout had been approved as fit for this purpose by Building Control. You contend that it is a provision of Part M of the Building Regulations, if an upper floor does not exceed 280 m² in extent, then a lift is not necessary for access by disabled persons. The area to which disabled persons are admitted on the first floor of the office building is limited to 160 m².

You also state that although the first floor of the office block is linked by a corridor to the welfare facilities for factory workers and the main food production facilities, the nature of the operations therein and the standard of hygiene which prevail preclude entry to disabled persons. You state that no disabled persons are employed in poultry killing, evisceration or processing sections of the factory because it would not be safe to do so. Floors are constantly wet and liable to be slippery and standards of hygiene are rigorous. All workers wear protective clothing including rubber boots. All areas are protected from contamination by hand wash facilities and foot-baths. It would not be possible to eliminate germs or bacteria from mobility aids, particularly wheelchairs, and consequently employment to disabled persons is restricted to the office premises.

You also make the point that stringent EU legislation governing hygiene standards in food production restrict access to areas designated for use in killing or processing of meat for food. Double doors with security locks control access and egress between the food production areas and the offices, and it is this physical barrier which determines the extent of floor space on the first floor to which disabled persons may require access.

In conclusion, you believe that a lift is not essential to achieve compliance with Part M and that this should be recognized by an acknowledgement that the absence of a lift does not preclude the District Council from granting approval under the Building Regulations 1991.

The District Council's case

The District Council state that although the work was described as an extension, it in fact is comprised of a new facility and therefore Part M of the Building Regulations applies. The Council state that the proposed plans had always shown the provision of a lift to the first floor levels and it was not until at a very late stage of the construction process that it became clear to the Council that, although a lift shaft had been constructed, the lift car and mechanisms were not to be installed. In a letter you confirmed that to be the case. Because it was assumed that a lift was to be installed, the Council state that although the stair may be acceptable for the purpose of complying with Part K ('Stairs and ramps') of the Building Regulations 1991, they did not consider it in relation to Part M.

The District Council noted comments about employment restrictions in connection with the provision of a lift for the benefit of disabled people and was unaware of any legislation which would preclude the employment of disabled people in the production area, even though some practical difficulties may arise as a result of such employment. The Council further commented that they were not in a position to judge what disabled people could or could not achieve in terms of employment. In this respect, they maintained that a lift should be installed as indicated originally on the full plans submission.

The District Council accept that the first floor area to the office accommodation is less than the floor area threshold of 280 m² as indicated in Approved Document 'M', but point out that this does not include the factory/processing area which leads on from the office.

The District Council contend that no segregation exists between these areas and that the total floor area at first floor level to the new two storey block exceeds the threshold suggested in Approved Document 'M' for the provision of a lift. The Council maintain that access to the production area for disabled people can only be gained via a lift.

The Department's view

The Department notes that the extension to the existing factory contains a ground floor entrance that is separately accessed and so the requirements of Part M of the Building Regulations 1991 rightly apply to this extension. The point at issue is the area to which it is reasonable for disabled people to gain access to and use the building on the first floor level. The District Council consider that access should be to the combined first floor area of the new office block and

its ancillary accommodation as well as to the factory floor area. This comprises the office and its ancillary accommodation together with the food production area. You contend that it is only the office accommodation on the first floor to which it is reasonable to consider the requirements of disabled people to gain access to and that because this area is below 280 m² the guidance in Approved Document 'M' indicates that a lift is not required.

The Department notes that at ground floor level staff who are wheelchair users are able to gain access to office areas for employment, and to the staff canteen which serves factory workers and office staff. There is also a disabled toilet provided at ground floor level, where satisfactory provision is required for disabled people, including wheelchair users.

It is not for the Secretary of State to determine what disabled people can or cannot achieve in a work environment, as this is as much to do with the abilities and commitment of an individual as with the nature of the work itself, and how procedures can be adjusted. However, there has to be some practical limit as to how far that view can be sustained. On balance, in this case the Department accepts that there are restrictions on the employment of disabled people – particularly those who are wheelchair users – in a food production area such as this.

It follows that access for wheelchair users is not required to the factory floor area and the ancillary accommodation contained on the first floor of the office block; and that the first floor area under consideration in terms of achieving compliance with Requirement M2 is therefore 180 m². This is well below the threshold suggested in Approved Document 'M' of 280 m² for which a passenger lift is recommended.

The Department also notes that the 180 m² of floor area on the first floor contains no unique facilities which anyone using the building might reasonably expect to use (such as a staff rest room – see paragraph 2.15 of Approved Document 'M') and which might warrant the provision of a wheelchair stairlift.

It follows that in the particular circumstances of this case, the Department considers that the provision as constructed complies with Requirement M2.

The Secretary of State's decision

The Secretary of State has given careful consideration to the the facts of this case, and the arguments put forward by both parties.

The Secretary of State accepts the arguments that you have put in respect of the restrictions on disabled persons – particularly those who are wheelchair users – being employed in the factory area. Given the size of the remaining first floor area contained within the office block and which is not associated with the factory process; the fact that there are no unique facilities contained therein; and on the assumption that the stair satisfies the needs of ambulant disabled persons and those with impaired vision; he has concluded that the work complies with Requirement M2. However, insofar as there is a difference of opinion regarding compliance between you and the District Council, and on the basis of the assumption above, he hereby formally relaxes Requirement M2 to the extent necessary for your work to comply, and accordingly upholds your appeal in this respect.

Beauty salon – sanitary conveniences (M.3)

In fitting out the premises, space intended to be used as a WC compartment for wheelchair users was provided with 'powder room' facilities, thus denying wheelchair access. The applicants claimed that there was no demand for a wheelchair WC in a beauty salon; but the appeal was dismissed.

Ref.: 45.3.96   Date: 31 January 1997

The appeal

The building work to which this appeal relates has involved the fitting-out of a new, ground floor, shop unit comprising a ladies' beauty salon. The unit had not been previously occupied and was not completed to an occupiable standard by the original developer – i.e. it was left as a blockwork shell with no shop front. The blockwork shell did, however, incorporate a WC compartment suitable in dimensions for wheelchair users. Part of your fitting-out work included changes to the positioning of the WC and basin within the existing compartment so as to accommodate 'powder room' facilities.

You state that you were unaware that Building Regulations approval would be required for this work. However, the District Council contacted you after work had commenced and requested that you apply for Building Regulations approval. You submitted a full plans application acknowledging that the final point to be resolved was a question of the disabled WC. The District Council took the view that the disabled WC compartment should be re-fitted as originally installed by the developer so as to provide access for and use by wheelchair users.

However, for operational reasons you did not wish to revert to the previous compartment layout and therefore applied to the District Council for a relaxation of Requirement M3.

In a letter the District Council took the view that in practice they could not consider a relaxation of Requirement M3 because 'the Regulations now refer to minimum requirements to achieve an adequate standard and clearly a relaxation of this standard would result in an inadequate standard'. The District Council therefore considered your application on the basis of a request for a dispensation and refused it. It was against that refusal that you applied to the Secretary of State under section 39 of the Building Act 1984.

The Appellant's case

In support of your application for a relaxation of Requirement M3 you stated that you had had no demand for a wheelchair WC and that your business was one which offered no services applicable to non-ambulant disabled persons on site. However, as a company, you did provide a home visit service, primarily for weddings, and this service would be available to non-ambulant disabled persons, although you had never had a request for this.

Your earlier representations justifying the adequacy of the ambulant disabled WC compartment had asked if the District Council could relax the immediate requirement for a wheelchair WC and accept a WC for ambulant disabled, as provided after the fitting-out work. You offered to ensure that, in conjunction with your landlords, you would provide an undertaking to re-install a wheelchair WC in the event that there was any change from the present use of the unit.

You argued that your business was one which, by its nature, could not accommodate seriously disabled customers. You further explained that this was purely due to the nature of the equipment you use and the treatments you offer. However, you could, and do accommodate and welcome, any disabled customers who, you state, are quite satisfied with the WC facilities you provide.

You also explain that it is important that you provide your customers with a relatively comfortable and well-equipped WC area, again reflecting the type of business you are involved in. You conclude that it is because of this situation that you cannot see any way in which you can revert the existing WC to one suitable for a wheelchair user without removing an important existing feature (i.e. the 'powder room' facilities) of the salon which would compromise the design of the salon by providing a facility for which you have no requirement or demand.

You have further argued that because you would not be prepared to reduce your standards and risk losing custom by making the specified changes to the WC compartment, this would result in the closure of the salon; which in turn would have the knock-on effect of having to lay off three full-time staff and removing a business from the area.

In further representations to the Council to try to resolve this issue you explained that from your cumulative experience of operating two salons over a period of four-and-a-half years that you never had a single occasion when a totally wheelchair-bound person had visited you. You have also explained that you have had customers who are wheelchair users but are semi-ambulant and they have no difficulty with the existing arrangements.

You conclude by explaining that the WC compartment for ambulant disabled people is already adequate in the context of the regulations given the particular circumstances of your business. You believe that the granting of a dispensation would be linked to the building's existing use and not be indefinite, so that if there were a change of use in the future requiring a wheelchair WC this could be provided.

The District Council's case

In support of the District Council's decision to refuse your application for dispensation, they state that the Building Regulations application was for the fit-out of a shop unit in a recently completed building. The proposed unit had not been previously occupied and was not completed to an occupiable standard by the original developer, i.e. it was left as a blockwork shell without a shop front. However, the disabled WC cubicle in question had been constructed with the correct spatial dimensions and a WC and basin positioned in accordance with diagram 16 of the Approved Document 'M' ('Access and facilities for disabled people'). The associated handrails and supports were not fitted by the original developer.

The District Council state that the subsequent amendments you made to this layout have restricted the use of the WC facilities by moving the WC and basin, and by providing a work-top and cupboards along the right hand side of the cubicle. It was because the spatial dimensions have been much reduced from those recommended in Approved Document 'M' that they found the provision unsatisfactory.

The District Council state that you have acknowledged that all new buildings of the type in question are required by the Building Regulations to have adequate toilet facilities. They also state that the specification, in line with that contained in Approved Document 'M', which provides guidance as to what is deemed adequate, was the standard they were looking for.

The WC compartment itself was, prior to the fitting-out of the shop, the right size and shape and the WC was in the correct position. All that was needed to provide facilities in accordance with the guidance in Approved Document 'M' was the installation of a few support rails. During the fitting-out works, you had moved the WC and placed floor cupboards along one side of the compartment making it impossible for the wheelchair-bound to use the facilities.

The District Council noted your arguments that the beauty salon services were not applicable to fully-disabled people (they assume you meant wheelchair users). They pointed out that the building was designed to be fully accessible by wheelchair users as required in Approved Document 'M'. They also dispute your arguments about some of the services you offer and propose that these should be equally applicable to the disabled as to the able bodied; for example, manicure, pedicure and waxing.

The District Council acknowledge your commitment, and that of your landlord, to bring the accommodation up to the standard of requirement of Approved Document 'M' when the building ceases to be used as a beauty salon, but they contend that such an arrangement would not be enforceable by the Council since the Building Regulations only require such WC accommodation in new buildings. The District Council, during their discussions with you had made several suggestions on how the WC compartment could be re-arranged to satisfy the requirements without compromising the 'powder room' facility which is incorporated in the compartment.

The District Council argue that the requirements relating to access and facilities for the disabled have been included in the Building Regulations to reduce the way in which the built environment discriminates against people with disabilities. They conclude they would be failing in their duty to enforce the regulations if they conceded to your request for a dispensation.

The Department's view

The Department accepts that it was appropriate for the District

Council to seek to apply the requirements of Part 'M' of the Building Regulations 1991 to the fitting-out works of this previously unoccupied shop unit.

The Department notes your lack of awareness when you took over this unoccupied shop unit that the Building Regulations would also apply to the fitting-out works. The works to the WC compartment involved the re-positioning of the WC and basin, and provision of additional cupboards and work surfaces along one side to provide the 'powder room' facility. This has resulted in a layout which does not enable the WC to be used by wheelchair users and, in the Department's view, has therefore compromised provision for the disabled in this new shop unit because the facilities offered by your business cannot be regarded as accessible and usable by wheelchair users.

The Department has also noted that the District Council considered that it was not obliged to consider your application to relax Requirement M3 because they considered the requirements in the Building Regulations 'refer to minimum requirements'. The requirements are written in functional terms and although there are some life safety requirements such as requirement B1 ('Means of escape') where the Secretary of State would not normally consider relaxation to be appropriate, there are other requirements (for example in Part M) where a relaxation may be justifiable. In any event, a local authority is statutorily obliged to consider applications for relaxations.

The Secretary of State's decision

The Secretary of State has given careful consideration to the facts of this case. You appealed to the Secretary of State against the refusal by the District Council of your application to dispense with Requirement M3 ('Sanitary provision'). In so applying you accepted that the requirement applied and that your work did not fully comply with that requirement. The Secretary of State takes the view that it is clear from the work as it now stands that you have compromised access to the WC compartment by wheelchair users and have therefore failed to make adequate provision to comply with Requirement M3 and compromised the use of the facilities you provide on your premises to wheelchair users.

The Secretary of State has noted that you consider that reverting the WC compartment to a specification suitable for wheelchair users would eliminate the 'powder room' provision currently in the compartment and which you consider is a necessary facility to conduct your business. However, he therefore cannot accept that this presents sufficient justification to consider dispensing with Requirement M3 and has concluded that the District Council came to the correct decision in refusing to dispense with that Requirement.

The Secretary of State has also considered whether relaxation of Requirement M3 – as opposed to a dispensation of M3 – might have been appropriate. However, he has concluded that in these particular circumstances it would not be appropriate to relax Requirement M3.

The Secretary of State accordingly dismisses your appeal.

Conversion of two shops into one – access for disabled (M.2)

This appeal demonstrates most aptly the need to check regulations; and then check again. Here the appellant had applied for a dispensation in respect of M.2, only to be refused by the Council. He then appealed to the Secretary of State, only to be told that the requirements of M.2 did not apply to a material alteration. The rule is to establish what is a material alteration, and then which regulations may be applied.

Ref.: 45.3.94   Date: 20 May 1997

The appeal

The building work involved relates to the uniting of two existing separate but adjacent shop units (Units C8 and C9) into a single shop unit by the removal of the dividing wall. Unit C8 has a floor area of 240 square metres and Unit C9 has a floor area of 45 square metres. Both existing shop units had level access from the shopping mall outside but there was a difference in floor levels between the

two units of about 420 mm – Unit C9 being at the lower level. The uniting has resulted in a single shop unit on two different floor levels linked by three steps designed to ambulant disabled standards. The level access doors of both units to the shopping mall have been retained. There is a covered walkway between both these doors.

The above work was carried out but the City Council took the view that because provision for access and use by wheelchair users had not been provided between the two different floor levels in the united shop, Requirement M2 had not been complied with. Although you accepted that you would normally be required to facilitate the passage of wheelchair users between the two floor levels by means of a ramp or lift arrangement, you rejected the technical options put forward by the City in order to achieve compliance for various reasons including the fact that you considered them to be too onerous. You then applied for dispensation of Requirement M2 in respect of the united premises. This was refused and it is against that refusal that you appealed to the Secretary of State.

The Appellant's case

You state that the Company acquired Unit C8 – which had not traded for some time, and that at the Landlord's suggestion also took the small Unit C9 which was being traded by a temporary tenant. The Landlord agreed to remove the short dividing wall to make the shopping frontage more attractive to customers. The Company agreed to the suggestion and acquired each unit on its own separate lease for five years only.

You explain that the small Unit C9 is only used for goods display and does not have a cash and wrap station. The main shop unit C8 has its own front doors from the shopping mall and full disabled access is available to this unit – the principal part of the shop where staff and tills are located. You explain that you had fully considered the requirements for access for the disabled and decided that in this case reasonable provision had been achieved by installing steps suitable for the ambulant disabled to link the two units internally, and by the retention of both the shop front doors which provide level access for wheelchair users from the mall. You consider that your building work has improved the facilities and not made them worse.

In support of your request for a dispensation you argue that Unit C9 is a small unit and that a ramp or lift would take up an inordinate amount of floor space in this lower unit; and that the levelling of the floor in C9 would mean the closing of the smaller unit's door which you state would not be permissible. You also argue that there is access between the two levels albeit via the two main entrances to each unit and the covered walkway which is no more than 12.5 metres in length. You further add that if changes had to be made, then the store would not be viable.

The City Council's case

The City Council have explained that their reasons for refusing to dispense with the requirements of M2 were as follows:

(i)   The Centre is a recently-constructed shopping centre, in which Units C8 and C9 were originally separate units each with its own access complying with Part M of the Building Regulations. The removal of the wall between the units has caused a contravention of Part M by creating one larger unit with different floor levels.

(ii) To gain access from one unit to the other wheelchair users would have to do so via the two front doors, involving a travel distance of 14 metres along the shopping mall which has little protection from the weather.

(iii) Wheelchair users wishing to purchase goods from Unit C9 would need to take the goods, unpaid for, out of the shop in order to gain access to the tills in Unit C8.

(iv) The local authority have consulted the Access Group who have supported their view.

(v)   The Council appreciate that the provision of a ramp may be onerous in this instance, but argue that there are other methods of providing satisfactory access such as the provision of a platform lift or the raising of the floor level – or a reasonable proportion of it – in Unit C9.

The Department's view

The City Council has taken the view that Part M is applicable to these works because the creation of one larger unit with different floor levels has caused a contravention of Part M. The Department considers that the removal of the dividing wall between the units brings the work within Regulation 3(1)(c) of the Building Regulations – i.e. the material alteration of a building. However, the 'limits on application' relating to Part M in Schedule 1 of the Building Regulations 1991 mean that such a material alteration is not required to comply with Requirement M2. In other words you were under no statutory obligation to make the provision for access and use that you have already carried out.

The Department notes that your provision for use of the premises comprises an ambulant staircase but also notes the issues raised in the appeal documents concerning the ability for wheelchair users to pay for goods without having to leave the premises through one door and re-enter through the other. Although the effect of this appeal is to require no action on your part, the Department observes there will clearly be merit if this problem could be addressed by you.

The Secretary of State's decision

The Secretary of State has given careful consideration to this case and concluded that Requirement M2 is not applicable to it. Accordingly, he upholds your appeal against the decision of the City Council.

Petrol filling kiosk – access for disabled (M.2)

The local authority refused to dispense with the Requirement M.2 in respect of the staff area of a petrol filling station. However, the appeal was allowed.

Ref.: 45.3.75   Date: 19 June 1997

The appeal

The petrol filling station to which this appeal relates is situated on the opposite side of an access road to an existing store and car parking area. It comprises six pump islands providing a total of 12 pumps. The building to which the appeal relates is to one side of the pump forecourt and comprises a sales area which is accessible to customers who are wheelchair users. Beyond this sales area an inner lobby provides access to a wheelchair WC which is available to customers, and to a store and electrical equipment room. A second lobby, immediately beyond the first, provides access to a consol area used by staff to serve customers and to an office and a staff room. The consol area is constructed over a raised floor which extends into the second lobby and forms a step up from the lobby floor. Attached to the sales building, but separately accessible, is another store and a room containing recycling plant for an adjacent single car wash. Adjacent to the car wash is a single-jet wash facility also.

It appears that during discussions about the full plans application, the Borough Council stated that for the building to comply with Requirement M2 of the Building Regulations, staff who are wheelchairs users should be able easily to obtain access to the consol and that the proposed plans did not demonstrate that this would be possible. You contended that an employee using a wheelchair would not be able to meet the various conditions of the relevant licence and that it was therefore unnecessary for the area of the building to be used by staff to comply with Requirement M2. You therefore requested from the Borough Council a dispensation/relaxation of Requirement M2 in respect of access by wheelchair users to the staff areas of the building. The Borough Council refused the application and it is against that decision that you appealed to the Secretary of State.

The Appellant's case

Your application for a dispensation/relaxation was supported by a statement that the requirements of the County Council's Trading Standards Office, and the guidance published by the Health and Safety Executive with respect to the construction and operation of petrol filling stations, each militated against the employment of a wheelchair user. However, the application conceded that it might be possible to employ an ambulant disabled person or persons.

The statement from the County Council Trading Standards Office contended that persons confined to wheelchairs would be unable to meet the requirements and conditions contained in a licence issued in respect of premises where petroleum spirit is stored. Those conditions include requirements about the competence and ability of employees, in the event of an emergency, to be able to operate any plant or equipment and to carry out adequate supervision of the event and any other steps that are absolutely necessary. Such actions would include the prompt cleaning of small leaks and petrol spillage, by the application of sand or other absorbent materials.

A letter to you from the Flammable Liquids Policy Section of the Health and Safety Executive referred to the Executive's publication HS(G)41, 'Petrol Filling Stations: Construction and Operation', as having the status of guidance only. In offering advice on training and experience of station attendants, it was said not to be intended to exclude people with disabilities from working in them. In the Executive's view, it was for an employer to decide on the competence of a prospective employee following an assessment of site circumstances and the abilities of the individual concerned. Competence would include the ability to deal not only with routine operations but also with foreseeable incidents such as fire or explosion or the spillage of petroleum spirit. These tasks should be able to be performed without the individual putting himself/herself or others at risk.

A statement by your clients has also been submitted describing their employment policy. They describe themselves as an 'equal opportunities employer', in accordance with which all decisions about people are based on an objective assessment of their suitability for a particular job. It is said to be guided by the relevant regulatory authorities: in this instance, the view of the local Trading Standards Office that the employment of a person confined to a wheelchair would not be appropriate.

In a letter, you have drawn attention to factors which you consider militate against the employment of a wheelchair user; namely, that the need to position petrol tanker vehicles alongside either pump islands or 'offset-fill' points for underground tanks would leave insufficient space for a wheelchair user to take action in an emergency.

Having regard to all these factors your clients contend that, in such circumstances, it is unnecessary to make provision for controls which are accessible by wheelchair users.

You have also submitted a copy of a letter, from the Petrol Retailers' Association. The letter describes the factors which in your view appear to militate against the employment of a disabled person. It includes a statement that 'the petroleum licence requires licensees to employ persons with enough practical and theoretical training, knowledge and actual experience to carry out a particular task safely and conveniently' and that 'they would also be required to take appropriate action following laid down emergency procedures in the event of a fire or spillage of petroleum spirit'.

Among the conditions of licence are said to be that:

(i) any leaks or spills shall be prevented from escaping from the licensed premises;

(ii) small leaks shall be cleared up promptly by the application of dry sand or other absorbent material.

You contend that an employee who is a wheelchair user would not be able to comply with these particular conditions and that disabled persons employed on a petrol filling station would have to be supervised at all times and have someone available to carry out necessary emergency action should it be required.

In your letter you have stated that two members of staff are scheduled to operate the petrol filling station at all times, but have re-affirmed your clients' concerns about employing wheelchair users in their petrol filling stations having regard to their safety and that of customers in the event of evacuation because of fire, or a major incident. You further state that your clients would consider employing a wheelchair user if the company's concerns could be suitably allayed.

The Borough Council's case

The Borough Council state that in their view provision should be made in new buildings to accommodate staff who are wheelchair users and that the proposals do not make such provision. In particular they have drawn attention to the fact that:

(i)   wheelchair-bound staff would not easily gain access to the consol due to the step at the end of the raised floor in the second lobby; and

(ii) doors are positioned too close to the walls to enable the convenient operation of door furniture.

The Borough Council contends that although some tasks in a filling station may be difficult for a wheelchair user to undertake, there remain others which would be manageable – for instance, serving at the consol and working in the office. Such a person, the Borough believe, could assist an able-bodied person quite adequately. They have since stated that in their view the information in your letter only enhances the need for provision to be made for at least one appropriately-trained wheelchair user.

The Disabled Persons Liaison Committee, a group based in the same area as the proposed development, has seen the proposals and submitted its views to the Council. These were taken into account by the Borough Council at the time the application for dispensation was being considered. The Committee considered the application for dispensation unreasonable and one which, had it been conceded, would have set a dangerous precedent.

The Department's views

The Department's views are as follows. Responsibility for licensing petrol filling stations rests either with local fire brigades, or with other departments of local authorities. It would be for those authorities to decide in a particular case whether a person with a particular disability could or could not perform all the relevant duties which are material to the granting of a licence and, if so, whether the employment of a particular disabled person would preclude the granting of a licence.

The issues for consideration centre on the Borough Council's belief that a wheelchair user could assist an able-bodied colleague *vis-à-vis* your client's view that the need to deal with emergencies outside the building would render the employment of a wheelchair user not feasible.

In the Department's view there is also the consequential and specific point as to whether in this particular case, given the scale of this particular service station and its operational requirements, it is reasonable to require access to, and use of, the consol as constructed by a wheelchair user. In the absence of agreement during the design and construction stage, the Borough Council are requiring adjustments to the positioning of doors and the provision of suitable access to the consol platform for a wheelchair user. In the Department's view the provision of access may be difficult to achieve for convenient use by a wheelchair user and the safety of other employees given the particular scale of the building and accommodation. Such difficulties as might arise in this respect could be construed as being a result of your proceeding in the absence of acceptance of compliance of your client's proposals with Requirement M2 by the Borough Council.

However, in considering this appeal it is also for the Secretary of State to look at the broader principles as to whether it would be reasonable to require suitable provision for wheelchair users to have access to, and use of, consol areas in self-service petrol stations. In the Department's view consols are invariably raised for operational reasons such as supervising of the forecourt and serving standing cus-

tomers. These design considerations, plus the uncertainty of the granting of a licence for the storage of petroleum where an employee is a wheelchair user, combine generally to make it unreasonable to expect access to be provided for a wheelchair user to use a consol area.

In the Department's view the above considerations do not necessarily imply that it would not be appropriate to require reasonable provision for an ambulant disabled person to gain access to, and to use, the consol area. Your client has already conceded that the employment of an ambulant disabled person might be possible and this was reflected in the original application for dispensation/relaxation.

The Secretary of State's decision

The Secretary of State has given careful consideration to this case and the points made by all parties. He considers that the combination of design constraints, and concern over potential licensing constraints, indicate that it would be unreasonable to insist on provision for wheelchair users to gain access to, and to use, a consol area. He does, however, consider that it would be reasonable to make provision for ambulant disabled employees to have access to, and to use, a consol area.

The Secretary of State has therefore decided to uphold your appeal by relaxing Requirement M2 in this particular case so as to require reasonable provision for access to, and use of, the staff area (including the consol area) of the petrol filling station by disabled persons except those who are wheelchair users.

For the purposes of this relaxation the staff area is defined as the lobby marked No. 2 on the plans and the office, staffroom and consol to which this gives access. Access via lobby No. 1 to the WC which is designed for use by wheelchair users is therefore not included in the relaxation.

Tourist information centre – sanitary conveniences (M)

A Grade I listed building presented serious difficulties in the provision of sanitary convenience for visitors, and in the circumstances it was not considered reasonable to make provision in a new extension.

Ref.: 45.3.113   Date: 6 August 1997

The appeal

The work involved a two storey extension to a Grade 1 listed building, built on the site of a demolished section of the original building, which contained storage rooms and ground floor WCs. The existing building is used as a museum and the ground floor of the extension is to accommodate a tourist information centre. The entrance to the new tourist information centre is through an existing door opening which used to serve the storerooms and the fire escape staircase. The floor level at the back of the centre, where the WCs are located, is lower than that at the entrance because it provides a link with the existing building and has to match its floor levels. Three steps link the two levels within the centre.

The Appellant's case

You explain that the existing building was totally unsuitable for disabled persons before any work was carried out, due to the uneven floors, level changes, indistinct steps, etc., and lack of a lift from ground to first floor, and this has not changed. You believe that no remedial measures can be implemented as the building is a Grade I listed building. You argue that because the new extension to form the tourist information centre utilizes an existing entrance and door which had previously been closed off to the public, you are therefore not dealing with a totally new building or a new entrance to an existing building.

You contest that the Building Regulations Part M require that an extension to an existing building should be *at least as accessible as the building being extended* and argue that you have complied with this.

You also note that *there is no obligation to carry out improvements within the existing building* as long as *the extension does not adversely affect the existing building*. You explained that as access from the original tourist information office (which was located in the existing building) was restricted, you are not making access arrangements from the new tourist information centre to the existing building any more difficult for disabled persons to negotiate.

With regard to the provision of new WCs in the Tourist Information Centre, you argue that these are not mandatory and the dispute has arisen because these are only being provided for staff and not for members of the public.

You have included evidence from the Town Council, who operate the museum, explaining that the nature of the building makes it extremely difficult for it to employ disabled persons and so questioning the value of a wheelchair accessible WC for staff. You also refer to European Law, whereby it is a requirement to state whether or not public buildings are suitable for disabled people, and explain that the building is not suitable and could not advertise as being such. As a consequence, you do not believe the District Council are correct to insist on the provision of a WC accessible to disabled people, and believe that it would be reasonable to relax the requirements of Part M in relation to sanitary conveniences in this case because exceptional circumstances apply.

As a concession to the District Council's insistence on providing a WC accessible by wheelchair users, you have offered that members of the public could gain access to the WCs provided for the staff with their supervision and have offered the provision of a disabled portable ramp and staff assistance to wheelchair users. As an alternative, you have created a new entrance at the rear of the centre to provide level access to the WCs via an external path.

You explain that the level changes in the new tourist information centre are due to the differences in the existing floor levels. Construction had started using your original design with ramps and WCs for staff and visitors, but this was ruled impractical by the experts from the Heart of England Tourist Board because the ramps would take up all the useful working area, making the new tourist information centre unworkable. This resulted in a redesign on site in which you provided steps suitable for ambulant disabled, rather than the two ramps in the original design, to access the rear of the centre and decided that toilets should only be provided for staff. You do not accept the District Council's proposals for a platform lift because this would be too costly within the tight budgetary constraints and would intrude on access to the information counter.

The District Council's case

In support of its refusal of your application for relaxation of Part M, the District Council explain that the extension is completely new with the exception of the front wall which contains the entrance door. They note that this door is accessible in both approach and width, and furthermore note that it is to be the principal entrance of the whole building which is accessed via the extension. They have taken the view that the requirements of Part M should apply to the extension since the proposal was not governed by any of the limits on application specified in Schedule 1 to the Regulations. They also note that the proposal has its own principal entrance and is independently approached and entered from the boundary of the site. They conclude it should therefore be treated the same as a new building.

They argue that where Part M applies reasonable provision should be made for sanitary accommodation for disabled people. Following an earlier rejection, amended plans were submitted showing a wheelchair disabled WC at the lower floor level which was accessible from the upper entrance level via ramps. They concluded that although not in strict accordance with the guidance this was deemed a reasonable compromise. They also consulted the Council's Conservation Architect regarding the internal proposals in this two storey extension to a Grade 1 listed building and he advised that there would be no objection to alterations within the extension since he was primarily concerned with the external appearance.

During the building work, the District Council report that you decided to depart from the approved plans with regard to the ramped area but that the disabled WC was to remain and steps were constructed down from the entrance level to the lower level at the rear where the WC was located. They rejected the solution offered of a portable ramp and assistance from staff arguing that disabled people should be allowed to be independent and not to rely on the help of others. They argue that this solution would have been impractical without assistance due to the steepness of the portable ramp and the lack of guarding.

They also rejected as inappropriate the alternative solution proposed to provide access to the rear of the extension at the same level as the WC because this would mean disabled people who could not manage steps having to travel via an external route around the building to the rear in order to use the WC. They saw this as unreasonable since a wheelchair user would have to leave the building, whatever the weather, follow a path around to a heavy iron gate to the rear boundary wall and negotiate a gravel path to the rear door of the extension. They further note that the rear entrance is inaccessible due to a ramp which does not comply with diagram 1 of the Approved Document and has a double-leaf door which does not comply with paragraph 1.35 (total width of the two leaves is only 1100 mm).

They conclude that WCs should be provided for both staff and visitors and have suggested that a platform lift would have been an acceptable method of satisfying the requirements, without detriment to the useful space in the extension.

The Department's view

The Department accepts that it was appropriate for the District Council to apply the requirements of Part M of the Building Regulations 1991 to this new extension since it contained a ground storey and was independently approached and entered from the boundary of the site. In such a case the guidance in the Approved Document states that the extension *should be treated in the same manner as a new building*. However, in this case it is important to consider what is reasonable, bearing in mind this is an extension to an existing Grade 1 listed building and has to accommodate differences between existing floor levels. The primary purpose of the new extension is to provide a tourist information centre and not to provide WCs for members of the public. The size of the centre also has a bearing on what is reasonable. The Department notes that access has been provided into the tourist information centre but inside there are necessary changes of level to match the existing floor levels within the Museum, which results in a lower level at the rear, where the WCs are located which would be inaccessible to wheelchair users.

In your final proposals you decided to provide WCs only for staff and not for members of the public who would be visitors to the centre. Requirement M3 states that *if sanitary conveniences are provided in a building, reasonable provision shall be made for disabled people,* in this case sanitary conveniences have only been provided for members of staff. Provision of staff WCs in this context is considered reasonable as the demolished sections of the building contained WCs. It has been argued that the staff cannot be wheelchair users because of the nature of the existing building. However, staff could be employed who are ambulant disabled and the Department notes that such people could access and use the sanitary conveniences provided. As a consequence the Department takes the view that with regard to the provision of sanitary conveniences for members of staff reasonable provision has been made as required by M3. The Department notes that no formal provision has been made for sanitary conveniences for visitors to the centre.

The Secretary of State's decision

The Secretary of State has given careful consideration to the facts of this case. You appealed to the Secretary of State against the refusal by the District Council of your application to relax Part M. As explained above, consideration has been limited to a relaxation of

Requirement M3, 'Sanitary conveniences'. The Secretary of State takes the view that the primary purpose of the new extension is to create a new tourist information centre with toilet facilities for staff and not provide sanitary conveniences for visitors. He has decided that reasonable provision of sanitary conveniences has been made for staff under requirement M3. Given the purpose of the new extension, and the inaccessibility of the building, he does not consider it reasonable to require sanitary accommodation for visitors.

The Secretary of State accordingly upholds your appeal and relaxes Part M to require only the provision of sanitary conveniences for members of staff.

Sanitary convenience in dental surgery (M.3)

Plans of a small dental surgery contained details of a suitable WC compartment for use by disabled people. However, the compartment was not constructed to those dimensions and it was not regarded as a suitable case for dispensation or relaxation of M.3.

Ref.: 45.3.117   Date: 9 February 1998

The appeal

The building work to which this appeal relates comprises a newly-constructed single storey dental surgery built onto one side of a large detached dwelling. The new building comprises a surgery and associated reception and storage facilities; together with a WC compartment contained in the lobby at the front of the building. The access to the lobby is ramped and provides independent access from any associated with the dwelling.

Your full plans application was accompanied by drawing 1119/1B (No. 1) which shows a disabled WC compartment contained within the reception area rather than the lobby area. Following a written request from the District Council concerning three points on provisions relating to Part M, the drawing was amended to show provision in that WC compartment for handrails and a mirror, etc. Following a further written request concerning provisions for requirements relating to Part A and Part B, full plans approval was given.

Construction commenced and following one inspection the District Council wrote to you itemizing a number of matters which did not appear to comply with the Building Regulations. Among these was a WC compartment which had been provided in the lobby, as opposed to the reception area, with dimensions smaller than those shown on the plan. The letter advised that the compartment should accord with Approved Document 'M' ('Access and facilities for the disabled') in respect of size, and fixtures and fittings, in order to comply with Requirement M3. This matter was still outstanding when the District Council wrote to you giving 21 days' notice for the work to be brought into compliance and warning that failure to do so would result in enforcement action.

You took the view that the WC facilities as provided were appropriate for the amount of use to which your client will be putting the surgery and applied on his behalf for a relaxation of Requirement M3. The District Council treated this application as one for a dispensation which was refused. It is against that refusal that you appealed to the Secretary of State under section 39 of the Building Act 1984.

The Appellant's case

You acknowledged that the WC compartment had not been constructed in accordance with the approved plans but consider that it is adequate for the scale and nature of the dental practice. Your client's main practice is in another town and his use of these premises is limited to 3 days a week of which one or two are part-time only. The new building replaced a similar-sized surgery elsewhere with similar facilities, but you believe that the new facilities are an improvement on those offered there.

In addition you make the point that the WC is contained in the entrance lobby and therefore is separate from the reception area. You explain that the receptionist will be able to help any disabled person from the lobby to the WC.

The District Council's case

In explaining its reasons for refusing the dispensation, the District Council comment that the building has not been built in accordance with the approved plans. They explain that if sanitary conveniences are provided within the building, reasonable provision should be made for people with disabilities. They also explain that sanitary conveniences should be no less available for people with disabilities than for able-bodied people. With regard to assistance being available from a receptionist, they take the view that custom precludes assistance from a member of the opposite sex.

In considering the size of the WC compartment for wheelchair users, the Council state that the design should:

(a) enable necessary wheelchair manoeuvre;
(b) allow for transfer to the WC;
(c) have facilities for hand washing and hand drying within easy reach from the WC, prior to transfer back to the wheelchair;
(d) have space to allow a helper to assist in the transfer;
(e) ensure that the cubicle door can open outwards in the event of an emergency.

However, in a note to the refusal notice the Council state that they are mindful of the small size of the building and the nature of its use, and would consider a smaller WC compartment having overall dimensions and facilities as indicated in BS 5810 (Figure 13).

The Department's view

Although physically joined to one side of the dwelling, for the purposes of the application of Part M of Schedule 1 to the Building Regulations 1991, the dental surgery stands to be defined as an independent building designed for a dental practice with its own access independently approached and entered from the boundary of the site.

You have appealed to the Secretary of State against refusal by the District Council to dispense with Requirement M3 ('Sanitary conveniences'). Requirement M3 states that:

If sanitary conveniences are provided in the building reasonable provision shall be made for disabled people.

The point at issue is not the different location but the size of the WC compartment as constructed, which is shown with dimensions 910 mm wide and 1730 mm long and in consequence is smaller than the WC compartment illustrated in Approved Document 'M' or the alternative illustrated in Figure 13 of BS 5810.

Paragraph 4.1 of the Approved Document 'M' states that the objectives of Requirement M3 is that sanitary conveniences should, in principle, be no less available to disabled people than for able-bodied people; and that the aim is to provide solutions which will most reasonably satisfy that principle, whilst bearing in mind the nature and scale of the building. In this context the Department notes the willingness of the District Council to accept a smaller compartment based on BS 5810. The Department also notes and accepts the District Council's suggested requirements of a WC compartment as expressed above.

In the Department's view the size of the WC compartment as constructed is such that it would will be too small to enable a wheelchair user to enter and close the compartment door before transferring onto the WC.

Except in the exceptional circumstances of a particular case the Secretary of State would not normally consider it appropriate to dispense with a requirement of Part M. Nor would he lightly consider relaxing such a requirement except in exceptional circumstances presented by the appellant.

In this case, it is clear that the premises may be used by any member of the public who happens to be one of your client's patients. It is reasonable to assume that some of those patients may well be disabled persons who will require access to sanitary facilities as specified under Requirement M3. The fact that the premises may only be used on a limited number of days does not appear to be material to what the needs of a disabled person may be when the premises are in operation. The Department does not therefore accept your case as being a substantive one for dispensing with, or indeed even relaxing, Requirement M3. The Department does, however, accept that irrespective of the number of days the premises may or may not be in use, the actual scale of those premises is such that it could be inappropriate to adopt the compartment illustrated in Diagram 16 of Approved Document 'M', always providing that reasonable provision has been made in accordance with an accepted standard, such as that illustrated in Figure 13 of BS 5810, and therefore is considered to be in compliance with Requirement M3.

The Secretary of State's decision

The Secretary of State has given careful consideration to the facts of this case; the particular circumstances that have led to this appeal becoming necessary; and the particular circumstances of your client's case. He has concluded that there is no case to support either the dispensation or the relaxation of Requirement M3 ('Sanitary conveniences') of Schedule 1 to the Building Regulations 1991 (as amended); and has concluded that the District Council came to the correct decision in refusing your application to dispense with Requirement M3. Accordingly he dismisses your appeal.

Sanitary conveniences in a small, single storey office (M.3)

Although large enough to accommodate a wheelchair user, the arrangement of fittings and the door precluded such use and the appeal against the District Council's rejection of the plans was refused.

Ref.: 45.3.116   Date: 11 February 1998

The appeal

The building work to which this appeal relates comprises the construction of a small, single storey office in the grounds of, and immediately to the rear of, your dwelling. The plan area is shown as being 8.5 m × 4.5 m and is to comprise a single office with lobby and WC facilities located at one end taking up approximately a quarter of the total floor area. At the time you lodged your appeal the work was under construction.

These proposals were the subject of a full plans application to which the District Council requested various amendments and clarifications including full access internally and externally in respect of Requirement M2 ('Access and use') and in respect of Requirement M3 ('Sanitary conveniences'). Your full plans application was rejected because of inadequate information under Regulation 13 and non-compliance with Requirement M3.

With respect to the latter requirement the District Council took the view that the provision of a full wheelchair accessible WC compartment would be more beneficial and cost effective given the nature of the use of the building and the fact that owners and users could change over time. However, you took the view that for the use and purpose for which the office was being constructed, ambulant toilet provision would be adequate. You therefore applied for a relaxation of Requirement M3 in order that you might provide an ambulant WC compartment rather than one suitable for full wheelchair access. The District Council refused your application. It is against that refusal that you appealed to the Secretary of State under Section 39 of the Building Act 1984.

The Appellant's case

In the case you put to the District Council for a relaxation of Requirement M3 you opined that the provision of a wheelchair WC compartment was unduly onerous and proposed to provide an ambulant disabled WC instead. You referred to a 1993 determination by the Secretary of State with respect to sanitary conveniences for the disabled in small office premises of 185 square metres where four people – including a husband and wife – were employed. You pointed out that your proposal was one-sixth of the size of this particular office.

In further correspondence you explained that the office is to accommodate your small company which is concerned with drainage management and site surveys, and the associated reconditioning/upgrading of computer systems and other business support services which go with this main function. The company comprises yourself and your wife as partners, two full-time staff, and one part-time member. There are no defined jobs functions which means that a member of staff may be involved with computers on one day, and drainage and sight surveys on another. On this basis you consider that a wheelchair user would be unable to meet the requirements of the posts. You do, however, accept that an ambulant disabled person might be suitable and for that reason a relaxation of Requirement M3 was requested in order to make such provision in compliance with the requirement.

The District Council's case

The District Council, and its access officer, have made several points in justification of their judgement to refuse your application for a relaxation:

(i) There appears to be no disadvantage in terms of space required between providing a wheelchair WC compartment and an ambulant WC compartment, and there should be no additional cost implications.

(ii) Wheelchair WC compartments as defined by BS 5810 often contain handrails, covering 98% of all disabilities, resulting in many ambulant disabled people gaining better assistance than from ambulant layouts.

(iii) The advantages of full wheelchair provision against ambulant ones are immense and must be preferred every time. The District Council's view is that ambulant provision is usually incorporated where either space demands it or to compliment existing wheelchair facilities. Neither circumstance applies in this instance.

(iv) Although the Disability Discrimination Act 1995 will not apply to companies employing less than 20 people, the District Council takes the view that such issues of 'non-appropriate jobs' are becoming less of a justification for not providing accessible buildings and appropriate levels of service. In this particular instance they consider that a wheelchair user or less ambulant employee could conceivably operate exclusively in the office, even in a small organization such as this.

(v) Not withstanding the circumstances of the present occupiers, the building may ultimately be owned and occupied by employers who may still work within the office use class. It would not be appropriate to restrict the facilities to exclude their use by wheelchair users, particularly as prospective purchasers might rely on the current minimum requirements expected of a building constructed in 1997.

The Department's view

It should be noted that all appeal cases are considered on the facts and their individual merits. Previous determinations and/or appeal decisions may not therefore necessarily be relevant.

You appealed to the Secretary of State against the decision of the Mid Beds District Council's decision not to relax Requirement M3 in respect of the provision of a WC compartment suitable for ambulant disabled people as opposed to wheelchair users. Requirement M3 states that:

If sanitary conveniences are provided in the building reasonable provision shall be made for disabled people.

Except in the exceptional circumstances of a particular case the Secretary of State would not normally consider it appropriate to dispense with a requirement of Part M. Nor would he lightly consider relaxing such a requirement except in exceptional circumstances presented by the appellant.

Paragraph 4.1 of the Approved Document 'M' ('Access and facilities for the disabled') states that the objective of Requirement M3 is that sanitary conveniences should, in principle, be no less available to disabled people than for able-bodied people; and that the aim is to provide solutions which will most reasonably satisfy that principle, whilst bearing in mind the nature and scale of the building. In this context the Department notes that the presumption in Approved Document 'M' is that reasonable provision normally relates to WC compartments for wheelchair users, and that those for ambulant disabled people relates to compartments installed on storeys to which the only access is by a stairway, or are in some other way inaccessible to wheelchair users. Since in this case the office facility is at ground level only, the Department considers it is appropriate to provide for wheelchair users.

The Department does, however, accept that the actual scale of these premises is such that it could be inappropriate to adopt the compartment illustrated in Diagram 16 of Approved Document 'M', always providing that reasonable provision has been made in accordance with an accepted standard for use by wheelchair users, such as that contained in BS 5810.

The Department also notes that in your submission you have not sought to argue your case for relaxation on the grounds of the lack of space to accommodate a WC compartment for wheelchair users. Moreover, the District Council have observed that the size of the proposed WC compartment is larger than that illustrated in Approved Document 'M' and that the proposed lobby separating the compartment from the office is of an equivalent size. The problem has arisen because the arrangement of the sanitary fittings and the door is such that wheelchair users would not be able to gain access to the WC or be able to use it. In particular, the Department notes that the compartment door opens inwards, which is not recommended for either wheelchair users or ambulant disabled people.

Given that the compartment is of sufficient size to accommodate wheelchair users, and the repositioning of the sanitary fittings and amending of the door arrangement would enable wheelchair users to gain access and use the WC, the Department concludes that compliance with Requirement M3 is readily achievable. This fact must be set against the fact that in the Department's view you have not put forward any substantive case to justify a relaxation of Requirement M3 such that provision should be made only for ambulant disabled persons as opposed to wheelchair users.

The Secretary of State's decision

The Secretary of State has given careful consideration to the facts of this case and your particular circumstances, and to the points made by both parties. He has concluded that there is no case to support either the dispensation or relaxation of Requirement M3 ('Sanitary conveniences') of Schedule 1 to the Building Regulations 1991 (as amended); and has therefore concluded that the District Council came to the correct decision in refusing your application to relax Requirement M3. Accordingly, he dismisses your appeal.

PART N: GLAZING

Shop front – manifestation to glazing (N.2)

The council wanted shopfront areas of glazing to incorporate manifestation as the windows were not intended for display but to enable people to see merchandise in the shop. The Secretary of State dismissed the appeal, at the same time saying that manifestation was unnecessary; the windows were sufficiently well indicated to comply with N.2.

Ref.: 45.3.65　Date: 3 May 1995

Careful consideration has been given to the facts of the case, and to the representations made on behalf of the appellant and by the Borough Council.

This appeal relates to building work to install a new shopfront and internal alterations to the same shop. The shop front comprises two sets of recessed double doors in the centre, with glass panes to the left and right of the doors. Each door is about 900 millimetres wide, is framed, and fitted with a large 'D'-shaped handle. The window panes are about 2160 millimetres wide and 2460 millimetres high and do not incorporate any manifestation above floor level. They are supported on stallrisers about 150 millimetres high. The bottom rails of the door frames are also about 150 millimetres in depth.

The shopfront is adjacent to the pavement, and there is a canopy covering at first floor level to protect shoppers from the rain, which extends about half way across the pavement. The shop does not normally have any conventional displays in order to encourage people to look directly into the shop and view the merchandise on sale.

The Borough Council gave conditional approval to the work on the 7 March 1994 subject to the condition that suitable and adequate permanent manifestation should be provided at a height of 1500 mm from the finished floor level to the shopfront window glazing. However, the appellant's client maintained the view that manifestation was not necessary and the appellant applied to the Borough Council on the 15 June 1994 for dispensation of Requirement N2. The application was refused on the 30 August 1994 and it is against that decision that the appellant appealed to the Secretary of State.

The Department's analysis of this case is as follows.

Requirement N2 requires that 'Transparent glazing, with which people are likely to collide while in passage in or about a building, shall incorporate features which make it apparent'. The Borough Council have asked for permanent manifestation to the shop window glazing at 1.5 metres above floor level in accordance with the guidance contained in Approved Document N ('Glazing – materials and protection') to comply with Requirement N2. They have given the following reasons for doing so:

(a) No control exists over the positioning of the free standing clothing racks.
(b) Shop units on either side of Unit 1 have manifestation, so Unit 1 will be out of step.
(c) The 170 mm upstand is not sufficient to mark the window, especially at dusk for people with impaired vision.

The appellant considers that manifestation to the shop windows is inappropriate in this case for the following reasons:

(a) The stallrisers fitted at the bottom of the glazing define its location.
(b) The area of glazing is not excessively large compared to that in many shopfronts – the largest pane being approximately 2050 millimetres wide.
(c) The glazing is not being installed in an area where there was none before – the Centre is an existing retail unit which must have had shopfronts before its refurbishment.
(d) The shop front doors are framed and feature a large white 'D' handle to each leaf.
(e) The appellants do not use window displays. They rely on potential customers viewing the whole shop, and they feel that lines and logos on the shopfront glazing would interfere with people doing this and would detract from the overall image of the shopfront.

The Department observes that in the case of conventional shop window displays, these usually make the presence of glazing obvious. What therefore has to be decided is whether, in the particular circumstances of this case, there are already sufficient cues to prevent the likelihood of people colliding with the glazing so as to make manifestation unnecessary.

In the Department's view the cues are as follows:

(i) Photographs of the shopfront submitted by the Borough Council appear to show that whilst the merchandise which can be seen through the window is not a window display, it may be positioned close enough to the shop window to make passage difficult.

(ii)   The stallrisers mark the foot of the glazing and define its extent.
(iii) The area of the windows is not large compared with the area of the doors which are clearly defined.
(iv)   The shop front faces onto an outdoor pavement and therefore shoppers will expect the merchandise to be protected from the weather by a window.
(v)   The shop is one of a row of shops.

Offsetting considerations which might make collision likely are:

(i)   The shop may not always have merchandise close to the windows.
(ii)   When it rains shoppers will walk close to the shop fronts to take advantage of the covering which may increase the likelihood of collision.

Having carefully considered these factors it is the Department's view that, on balance, the presence of the windows is already sufficiently well indicated to comply with the Requirement N2.

In reaching a decision on this appeal against refusal of an application for dispensation of Requirement N2 the Secretary of State has concluded that, in these particular circumstances, it would be inappropriate to uphold such an appeal because to do so would be tantamount to implying that it was not necessary to protect persons from the risk of colliding with transparent glazing. Accordingly, the Secretary of State dismisses the appeal. However, he has also had regard to the technical aspects of this case and observes that in his view the work as executed complies with Requirement N2 of Schedule 1 to the Building Regulations 1991.

DETERMINATIONS
PART A: STRUCTURE

Suitability of site (A.1 and A.2)

It was proposed to erect 20 dwellings on a site known to have geological problems. Engineering consultants were involved and several possible solutions examined. However, the Secretary of State concluded that there was a long-term risk of an indeterminate nature and that the proposals did not meet the requirements of A.1(1)(b) or A.2(b).

Ref.: 45.1.76   Date: 10 April 1995

The Secretary of State takes the view that he is being asked to determine under section 16(10) (a) whether plans of your proposed work are in conformity with the requirements of A1 ('Loading') and A2 ('Ground movement') of Schedule 1 to the Building Regulations 1991. In making his determination the Secretary of State has not considered whether the plans conform to any other relevant requirements.

The point at issue in this case is the suitability of the site for redevelopment, with particular reference to the suitability of the ground and the design of foundations. The area of the site and its contours are shown on your submitted drawing No. 91 1201/1A.

In August 1993 you made an application for Building Regulation approval (Step 1: Layout and Sub-strata) to the City Council for the development of this site. The proposed development comprised a mixture of semi-detached bungalows and terraced houses totalling 20 dwellings. Your application was rejected in respect of Requirement A1 ('Loading') and Requirement A2 ('Ground movement'). Following further discussions and consultancy assessment a further application was made for Building Regulation approval for the same development but with site treatment involving drainage wells to stabilize the site. This application was rejected on 19 July 1994 again in respect of Requirements A1 and A2. It is in respect of this second rejection and proposed treatment of the site that you have sought a determination.

The City Council has referred to the history of this site. They

state that during the winters of 1977/8 and 1978/9 a landslip developed on and adjacent to the site. As a result dwellings 37 to 47 (odd numbers) were demolished. The Council believes that the landslip was a result of regrading work carried out during construction of three properties immediately up slope, and also due to probable porewater pressure changes. Land drainage was installed in 1980 designed only to retard and eventually halt the movement of the slope in its undeveloped condition.

Because of the known geological problems, the City Council employed independent engineers to assess the site investigation reports which you submitted in respect of your 1993 application. A geotechnical engineer from these consultants commissioned the reports accompanying the City's statement. As a result of the examination and comments from these consultants, the City Council considered it appropriate to reject your application in 1993.

The City Council's consultants were asked to examine your subsequent application of 1994. The Council took the view that the scheme would rely upon the permanent effectiveness of the proposed drainage wells. They considered that maintenance of these wells would not be an appropriate condition to apply to an approval of your Building Regulations application and that because the effects of non-maintenance could be so severe your proposals were considered not to achieve compliance. The City Council took the view that the wells would result in the lowering of the groundwater and it was their view – as well as that of the City – that this could certainly affect the adjacent properties to such an extent as to be unacceptable in terms of stability. On the basis of its own judgements and the technical advice of its consultants, the City Council consider that they made the correct decisions.

The statement of your engineering consultant dated 11 August 1994 states that he was aware of what he describes as a shallow slip in the period 1977/9 which in his view was stabilized by the installation of the land drains. Your consultant was also aware of an earlier slip but considered this had no connection with the site. Although both movements were shallow (i.e. confined to the top few metres of ground) investigation by your consultant demonstrated that a deep slip along a pre-existing ancient slip surface was probable if reasonable assumptions were made about the residual strength of the ground and likely maximum values of the water table.

Your consultant took the view that as no deep-seated movement was known to have occurred on this, or on the adjacent site, within the historic period, it was sufficient to assess whether the proposed redevelopment of the site would have an adverse effect on any potential slip surfaces.

Your consultant then carried out a number of analyses and it was concluded that your proposed development would have no adverse effect on any deep-seated slip surface that might, or might not, exist below the site. The possibility of shallow slips was considered but stability was considered to have been achieved by the land drainage. Nevertheless, as a precaution it was decided to design the foundations of the proposed buildings in such a way that they could easily withstand any small movements that might possibly occur due to creep or slow mass movement. This formed the basis for the proposed work which was the subject of your 1993 application for Building Regulations approval.

Your consultant's statement refers to the fact that in rejecting your application in 1993, the City Council stated that a substantial increase in the factor of safety of any possible slip would be required before the development could be approved. Your consultant considers that because of the deep nature of these slips under consideration (up to 17 m at the position of the site) and the fact that the site only covers a small part, this is a difficult requirement to meet. However, a scheme was proposed involving deep drainage which would have the effect of draining the ground on the site below the level of any credible slip, into the underlying sand. This scheme was revised after considering comments from the parties involved and submitted for Building Regulations approval in June 1994. Your consultant's statement disputes the City Council's view that the drainage wells would require long-term maintenance and that the action of draining the area may affect the surrounding properties. Your consultants maintain that the original proposals without drainage complied with the Building Regulations and more specifically with Requirements A1(1)(b) and A2(b).

Your consultant's statement goes on to state that you are nonetheless proposing to drain the site to increase its stability and that the drawing down of the water table over a wide area will increase the stability of the ground under the surrounding properties as well as on the site itself. It is expected that there will be some settlement associated with this reduction in water table. Your consultant has calculated that this settlement will be fairly uniform across the width of any building (maximum 3–4 millimetres in 10 metres) and that therefore the stability of any adjacent building would not be impaired. Your consultant states that these probable settlements are less than those normally accepted with the growth or removal of trees; and that the calculations indicate that no damage will be caused to adjacent buildings and that therefore the City Council's comments in their minutes concerning the effect on surrounding properties, are incorrect. The statement also refers to the fact that monitoring of adjacent buildings will be offered.

Your consultant's statement refers finally to the maintenance of the drainage system. The proposed wells will be 300 millimetres in diameter, incorporate filters, and have an open slotted pipe in their centre and a sludge bucket at their base. It is intended to site all wells in an accessible position. They will be easily cleaned and monitored; and the efficacy of the drainage system will also be easily checked. Your consultant contends that the Building Regulations do not have any requirement which states that any part of the structure should be able to survive the design life of the building without maintenance. On this basis your consultant maintains that your proposals meet the requirements of the Building Regulations; and that because the proposals are capable of being monitored they are unlikely to fail, but that in the event of their doing so they are easily maintained or replaced without affecting the building in any way. The management company for the site would be responsible for ensuring the wells were so maintained.

Although the sequence of events is accepted as described by the City, your consultant maintains that the Council failed to consider the later calculations regarding the drainage and maintenance of the wells. The revised calculations had been made after the drainage system had been recast to take into account its effect on the surrounding properties and earlier comments from the City Council's consultants and a world-renowned expert on slope stability problems who gave specialist advice to your consultant. These were submitted to the Council in June 1994 but your consultant considers the City Council rejected them seemingly without any further consideration.

In considering cases of this type, the Secretary of State is concerned to assess the long-term risks for the safety of the works proposed as well as the ramifications for surrounding property which may, perforce, be affected by land which is rendered unstable by that development. The latter issue may well have direct implications by way of blight on existing property.

The Department has given careful consideration to this case. All documentation submitted has been considered, including your revised calculations referred to in paragraph 15 above. The Department's analysis and conclusions are as follows.

The Department takes the view that small creep movements leading to ground instability will occur even if the factor of safety is substantially greater than one. This may well require precautionary measures to be taken for safety of all new buildings. The 'stability calculations' enclosed with your letter of 17 August are insufficient for checking the adequacy of the design against local ground strain. The same ground strains may pose maintenance difficulties with the proposed deep-well drains.

In your consultant's statement of 11 August 1994, it is admitted that there is a possibility that settlement of neighbouring properties may be caused by the proposed drainage. In addition it is noted that the calculations assume laterally uniform ground conditions. In the Department's view if this assumption is valid, damage could result

due to differential settlement which may be larger than the calculations seem to predict.

With regard to the proposed well drainage system, the Department accepts the City Council's view that there can be no guarantee that it would be kept maintained over the life of the building and the site could then be exposed to the risk of slope instability over the longer term. Aside from the fact that the new dwellings would thereby be put at risk, failure to maintain the drainage system could well have an effect on the adjacent properties the owners of which would have no management control over the maintenance of the drainage system.

The Secretary of State has concluded that, in general, the proposal poses a long-term risk of an indeterminate nature. He therefore determines that your proposals do not comply with the requirements of A1(1)(b) on account of the impairment of the stability of the other buildings; and that they do not comply with the requirements of A2(b) on account of the lack of assurance that buildings will be safe from land-slip in the long term. In making this determination the Secretary of State also observes that the uncertainty of the success of your proposal could well cause blight on the property values in the surrounding area.

Garage and loft – loading and internal fire spread (A1 and B3)

Disagreement here hinged on the interpretation of 'garage' and 'storage and other non-residential use' but the determination found the proposal acceptable.

Ref.: 45.1.147   Date: 9 September 1997

In making the following determination, the Secretary of State has not considered whether the plans conform to any other relevant requirements.

The proposed work

The building to which the proposed building work relates is a new, detached two storey timber framed and clad outbuilding. The outbuilding will be ancillary to an existing dwelling house and will be sited approximately 20m from the house. The ground floor of the outbuilding is open plan with garaging space for three vehicles. The first floor, also open plan, is designated as a loft space to be used primarily for storage purposes, although you accept that it could be adapted for other uses. Access to the first floor is via an external timber stair. A single window is located in the opposite end of the building to the first floor entrance door. An interlinked smoke detection system is to be installed. The building is fully enclosed.

The ground floor and first floor areas are 48 m² and 48.5 m² respectively. The elements of structure will achieve 30 minutes' fire resistance and the structural design is therefore compatible with the loading associated with this degree of fire resistance. The floor construction incorporates exposed timber floor joists as a feature of the design, with two layers of flooring grade chipboard above. Your submission includes a performance report from your fire consultant stating that the timber floor construction will achieve 30 minutes' fire resistance.

These proposals were the subject of a full plans application which was rejected by the Borough Council on grounds of non-compliance with Requirements A1 ('Loading') and B3 ('Internal fire spread (structure)'). The Borough Council contend that the elements of structure should be designed to achieve 60 minutes' fire resistance instead of the proposed 30 minutes in order to comply with Requirement B3. They further contend that, as a consequence of increasing the level of fire resistance, the building structure should be redesigned to accommodate the increase in loading in order to comply with Requirement A1. However, you consider that the rejection of your proposals by the Borough Council is unreasonable and it is in respect of whether it is necessary for the elements of structure to achieve 60 minutes' fire resistance, and therefore whether a consequent redesign of the structure is needed, that you have applied for this determination.

The Applicant's case

You contend that the building need only achieve 30 minutes' fire resistance. The structure has been designed accordingly for a loading compatible with elements of structure that have 30 minutes' fire resistance. You make the following points in support of your application:

(i) The reason for the Borough Council's rejection in your view is that the floor area exceeds the permitted 40 m² floor limit for a detached garage as set out in Appendix D of Approved Document 'B'. You consider, however, that the guidance does not specifically exclude a garage with greater floor area.

(ii) You consider the assumption that a domestic garage with a floor area in excess of 40 m² may be put to a non-domestic use to be unreasonable and contend that such a change of use could be controlled under either the Building Regulations or planning legislation.

(iii) You point out that in the case of a dwelling with a floor area in excess of 40 m², the elements of structure are required to achieve only 30 minutes' fire resistance, even though in theory the whole of the ground floor could be given over to garaging. You contend that an occupied dwelling constitutes a greater fire risk than an unoccupied domestic outbuilding.

(iv) You state that your client has erected many other similar two storey outbuildings in other areas and has had the proposals accepted by the relevant building control authorities.

The Borough Council's case

The Borough Council accept that the new outbuilding is for domestic use but maintain that, because its floor area is greater than 40 m², it is classified under the 'storage and other non-residential' purpose group. Appendix A of Approved Document 'B' requires that the elements of structure of buildings in this purpose group should achieve 60 minutes' fire resistance. The Borough Council therefore have based their decision on the assumption that the increase in fire resistance is necessary because of a possible increase in fire load and risk. They also state that their acceptance of the timber staircase as the means of access to the upper level is reliant on the elements of structure achieving 60 minutes' fire resistance.

The Department's view

The Department takes the view that the question for determination is whether the proposal to provide elements of structure that are designed to achieve 30 minutes' fire resistance will show compliance with Requirement B3 of the Building Regulations.

You have made reference to other similar projects where the elements of structure have been designed to achieve 30 minutes' fire resistance and for which you have received approval from the relevant authorities. However, each case for determination must be considered on its individual merits.

The Department considers that it is the risk to the occupants of the first floor loft space which has to be taken into consideration in this case. The drawings provided with your submission do not incorporate the level of finishes and completion which would normally be provided for a habitable space and the Department's view is that the use of the loft space appears to be more compatible with that of storage than of habitable use.

The Department accepts that the floor area of the three bay garage area exceeds by 8 m² the 40 m² guidance criterion given in Appendix D of Approved Document 'B'. The Department understands the concerns of the Borough Council and accepts that there is an argument for placing the new outbuilding in the 'storage and other residential' purpose group which would then suggest that 60 minutes' fire resistance would be appropriate.

In this case, however, the Department takes the view that the fire loading in the ground floor garage area will not constitute an unacceptably high risk to occupants of the first floor loft space. The Department therefore considers that elements of structure designed to achieve 30 minutes' fire resistance will be acceptable.

In reaching this conclusion the Department has taken account of the size of the building, its location, its proposed use and the provision of an early warning system. Although the proposals have not been rejected for failure to comply to Requirement B1, in the Department's view the external timber staircase would be acceptable in the circumstances.

With regard to Requirement A1, the Department notes that all parties appear to consider that the structural design is adequate for elements of structure incorporating 30 minutes' fire resistance. The Department therefore considers that compliance with Requirement A1, in respect of elements of structure incorporating a fire resistance greater than 30 minutes, to be no longer an issue.

The determination

The Secretary of State has given careful consideration to the particular circumstances of this case and determines that your proposals incorporating 30 minutes' fire resistance demonstrate compliance with Requirement B3. Because your plans achieve compliance without the need for additional structural loading, the Secretary of State also determines that they demonstrate compliance with Requirement A1.

PART B: FIRE SAFETY

Thatched dwellinghouse – external fire spread (B.4)

A two storey thatched extension to a thatched dwelling was to be erected. Thatch is to be treated with a fire retardent solution. The point at issue was the width of separation between the dwelling and the boundary. It was found that as precautions were to be taken the overall fire risk was not to be made worse.

Ref.: 45.1.94   Date: 15 September 1995

Your letter requested the Secretary of State to determine under section 16(10)(a) whether plans of the proposed work were in conformity with Requirement B4 ('External fire spread') of Schedule 1 to the Building Regulations 1991. In making his determination he has not considered whether the plans conform to any other relevant requirements.

Your proposals show the erection of a two storey rear extension to a dwelling house to provide an additional bedroom, dining room and adjacent conservatory; the two storey extension will be thatched to match a new replacement thatch roof covering proposed for the existing house. The new extension has a plan area of approximately 20 square metres which represents an increase of about 30% over the existing house whose plan area is approximately 60 square metres.

You state that the original house was designed and built by a noted British architect, about 80 years ago. It is sited at an angle to the boundary with another property such that the boundary distance varies between 6.5 metres and 2.5 metres. The new extension increases this side of the house by about 4.5 metres and means that the new extended thatched roof will only be about 2 metres from the boundary.

Your proposal is to remove the thatched roof from the existing house which has no fire resistance. Both the existing house and the extension will then be provided with new thatch to maintain consistency in colour and depth. All the thatch will be soaked in a fire retardant solution. The thatch will be laid on an aluminium foil fire barrier under which 100 mm thatch batts will be laid between joists.

The extension is clad in elm boarding to match the existing house, but this has been accepted by the District Council in respect of Requirement B4. The only point at issue is the separation from the boundary of the thatched roof extension.

You contend that because of the building orientation the new thatch will be no closer to the adjoining properties than the existing thatch roof is at present. Furthermore, the other property which has the adjacent boundary, has the advantage of being 19.8 metres away in linear distance from your property which is reduced at worst by only 700 mm following the erection of the extension.

The owner of the other adjoining property has written via you to the Department in support of allowing the thatch extension to your client's house. He considers that this will preserve the visual impact of this property not only for his but for others in the village.

You maintain that the fire precautions for the new thatch roof (i.e. the fire retardant solution and the thatch batts) will provide a greater degree of fire resistance than the existing roof has at present. This will stop fire spreading not only from the inside of the dwelling through the roof but also from an external fire to the inside.

You have drawn attention to the planning aspect that the use of any other roofing material would cause concern since re-tiling the existing house and extension would not be in keeping with the inherent character. In addition the roof timbers would require strengthening.

You also state that planning permission for the extension has been granted on condition that it is constructed in materials that match to ensure harmonization with the existing building. You consider that the provision of the thatch roof will safeguard the integrity of the property and contribute to the environment in which it resides.

The District Council have rejected the proposal for a thatched extension to the house on the basis that it did not show compliance with Requirement B4 of the Building Regulations 1991.

In this case the proposed extension is shown to be constructed using elm cladding and a reed thatched roof. The District Council accepted the elm boarding on the basis that the new unprotected wall area was not excessive but they were not prepared to accept the thatch.

The District Council point out that the thatch roof of the existing house does not accord with the guidance given in Approved Document 'B' ('Fire Safety') which recommends that thatch roofing should be sited at least 12 metres from the boundary. This is to ensure that firstly, if a fire occurs in this building, the intensity of the fire will not spread to an adjoining property; and secondly, if a fire over an adjoining boundary occurs, even if it is only a garden bonfire, the required isolation from the boundary gives this property a recognized degree of protection. Even if the extension is looked at in isolation, the District Council consider that fire risk potential is too great to be acceptable.

In analysing this case it is therefore for the Department to consider the overall fire safety aspect of the proposals in terms of Requirement B4 ('External fire spread').

The Department notes that the existing thatched house does not comply with the present recommendations given in Approved Document 'B' with regard to the distance to two boundaries and that no fire precautions have been provided. The existing roof is only about 3 metres from the boundary with one property. However, the worst situation is created along the boundary line with the other property which is adjacent to the new extension. Here the existing roof is only about 2.5 metres from the boundary. The proposed thatched extension will decrease the distance to this boundary from between 2.5 metres to 2 metres over a length of boundary line of about 4.5 metres.

The Department would not consider this to be acceptable on its own but what needs to be considered is the influence that the proposed fire precautions will have on the overall situation. These precautions will comprise removal of the thatch roof from the existing house and incorporation of the same fire precautions as for the extension. It can therefore be argued that the overall fire performance of the roof will have been improved when consideration is given to the distance that the existing house is sited from both boundaries.

The other aspect to be considered is that this is a site of architectural importance and is clearly under strict planning control. Also the houses on each side of the boundary are spaced well away from your client's house with regard to distance and orientation. The Department is therefore of the opinion that conflagration between buildings will be extremely unlikely.

The Department therefore takes the view that the overall fire risk will not be made worse by the proposed thatched extension.

In these circumstances, therefore, the Secretary of State determines that the proposals for the erection of the extension with thatch roof covering show compliance with Requirement B4 of Schedule 1 to the Building Regulations 1991.

It should be noted that the Secretary of State's determination of this case is based on the clear understanding that the full fire precautions as proposed – including those for the existing roof – are incorporated. It should also be noted that his decision relates only to the circumstances of this particular case and should not be considered to be a precedent for any other similar case involving thatch roof extensions.

Conversion of offices to flats – means of escape (B.1)

This relates to a five storey building of self-contained flats and having a single stairway. Each flat was to have an internal entrance lobby having an FD30S door and this proved in this example to be acceptable in place of a ventilated common lobby.

Ref.: 45.1.109   Date: 5 December 1995

Your letter requested the Secretary of State to determine under section 16(10)(a) whether plans of your proposed work were in conformity with the requirements of B1 ('Means of escape') of Schedule 1 to the Building Regulations 1991. In making his determination the Secretary of State has not considered whether the plans conform to any other relevant requirements.

Your request for a determination concerns a five storey building and the proposed conversion of offices to flats, with one flat on each of floors 1 to 4. The ground floor of the building contains a retail unit which is fire separated from the rest of the building and is not therefore at issue. It is the non-provision of a common ventilated lobby on each floor containing a flat that is at issue.

Your proposal is for the flats to be self-contained and that the building be considered as a small single stair building, as described in paragraph 2.19 of Approved Document 'B' ('Fire safety'). The height of the top floor above ground level is approximately 14 metres. The single stair, as shown on your plans, serves all four floors above ground floor level although the first and second floor flats were dealt with by the Borough Council under a separate submission. The third floor is shown as an existing office unit although you have stated that it is your intention to provide a flat on this floor also.

Your proposals show the entrance door to each flat opening directly into the stair enclosure. Each flat is provided with a protected internal entrance lobby. All doors to habitable rooms opening onto this lobby are FD30S fire doors as is the front entrance door to each flat. It is proposed to install an addressable central fire alarm system to an L2 standard in accordance with BS 5839: Part 1: 1988 ('Fire detection and alarm systems in buildings'), to give early warning of fire.

The Borough Council have rejected your proposals on the grounds that they consider the proposed work would not show compliance with Requirement B1 ('Means of escape'). They point out that sub-paragraph 2.18(a) of Approved Document 'B' recommends the provision of a ventilated lobby separating the single staircase from the dwelling with a single entrance door and that this has not been indicated on your submitted plans.

The Borough Council acknowledge that a smoke vent at the head of the stairs is now to be provided. However, they consider that since the top floor is approximately 14 metres above ground level the modified provision contained within paragraph 2.19 of Approved Document 'B' (small single stair buildings) should not, in these circumstances, be applied.

Your fire consultant accepts that the building exceeds the 11 metres threshold given in paragraph 2.19 of Approved Document 'B' for a small single stair building. However, you state that it would not be practicable to incorporate a ventilated lobby between the accommodation and the common stair at each level as recommended in paragraphs 2.18 and 2.23 of Approved Document 'B'.

You propose that the floors containing flats will be provided with a fire alarm and automatic detection system meeting the L2 standard as described in BS 5839: Part 1: 1988. In addition, smoke detectors will be provided in the common stairway and in each of the flats' internal entrance lobby. An openable vent having an area of 1.5 square metres will also be provided at the head of the common stairway for use by the fire service. You also consider that the provision of self-closing FD30S doors in the enclosures of the entrance halls of the flats will be an enhancement over the normal arrangements which would only require self-closing FD20 doors.

In considering this case the Department has assumed that the third floor will not remain as offices and that floors one to four will each contain a single flat. On this basis the Department's analysis of this case is as follows. Sub-paragraph 2.18(a)(i) of Approved Document 'B' suggests that every dwelling should be separated from the common stair by a protected lobby and paragraph 2.23 suggests that this lobby should be ventilated. The alternative to this arrangement is where the conditions relating to a small single stair building, as set out in paragraph 2.19 of the Approved Document, are complied with. Sub-paragraph 2.19(a) suggests a height limitation of 11 metres (for the top floor above ground level); in this case the height of the top floor is approximately 14 metres.

You suggest that it would be difficult to provide a common ventilated lobby on each of the proposed residential floors of this existing building because of the small plan area. However, this in itself is not sufficient reason for not providing a ventilated lobby and what the Department has needed to consider is the safety of the occupants of the building as a whole.

The Department notes that there is only one flat per floor and that the stairway itself is provided with ventilation at the top. Instead of a common lobby there is an internal entrance lobby provided to each flat and each door opening onto this internal lobby is an FD30S door. This means that when considered in conjunction with the flat entrance door, there is 60 minutes' fire resistance between the stairway and each habitable room and two sets of smoke seals. In addition an addressable automatic detection system is to be installed in accordance with BS 5839, L2 standard. The Department therefore considers that in the circumstances the omission of a ventilated common lobby will not present an undue risk to the occupants of this building.

In all these particular circumstances, therefore, the Secretary of State determines that your proposals, which exclude provision for a ventilated common lobby, show compliance with Requirement B1 of Schedule 1 to the Building Regulations 1991 in respect of means of escape.

You should note that the Secretary of State considers each case on its individual merits. This decision should not therefore be taken as creating a precedent for the omission of ventilated common lobbies on the assumption that this will necessarily secure compliance with the Building Regulations 1991 in any other case.

Loft conversion – means of escape (B.1)

In this conversion the stairs were to be open to the ground floor living accommodation and a mains operated interlinked smoke detection system installed. This was not acceptable to the local authority.

Ref.: 45.1.104   Date: 6 December 1995

The Secretary of State takes the view that he is being asked to determine under section 16(10)(a) whether plans of your proposed work are in conformity with the requirements of Part B1 ('Means of escape') of Schedule 1 to the Building Regulations 1991. In making his determination the Secretary of State has not considered whether the plans conform to any other relevant requirements.

Your request for a determination concerns your proposals to convert the roof space of your clients' two storey terraced house to form a bedroom. The existing arrangement consists of two ground floor living rooms and two first floor bedrooms. The staircase is centrally located between the two ground floor living rooms and dis-

charges directly into the dining room against the party wall and opposite the main exit.

The new loft room is to be fitted with a 30 minutes' self-closing fire door and a fire escape velux window. The new stair is to be protected to the first floor landing. The proposed escape route is then through a first floor bedroom to a window which has the correct opening size recommended in paragraph 1.18(a) of Approved Document 'B' ('Fire Safety') for an inner room. The door to this bedroom is to be replaced with a 30 minutes' self-closing fire door and a similar fire door is to be provided on the first floor landing to separate the ground floor from the upper floors. Self-closing mechanisms are also to be fitted to all doors to the other habitable rooms.

It is your proposal to leave the stairs open to the ground floor living accommodation rather than enclosing them. A mains operated interlinked smoke detection system is to be fitted as a compensatory feature, with smoke detectors located in every habitable room and a heat detector in the kitchen.

The Borough Council have rejected your proposals on the grounds that they do not show compliance with Requirement B1 of the Building Regulations 1991 by virtue of the fact that the existing staircase is not shown to be enclosed at ground floor level. It is in respect of this issue that you have sought this determination.

The Borough Council accept that smoke alarms provide a measure of safety but point out that the intention of your proposals is still to provide a means of escape route via a first floor bedroom window. They also point out that the recommendation for loft conversions contained within paragraph 1.24 of Approved Document 'B' is that a staircase should be enclosed within its entirety and discharge to a final exit.

It is the opinion of the Borough Council that since the stair is not enclosed the occupants of the upper floor, particularly those who may suffer from any disablement, will be denied a safe passage to a final exit because of smoke. They also consider this view to be consistent with the Secretary of State's decision on previous similar determination cases.

You fully accept that the most recent fire safety guidance suggests that there should be a protected primary means of escape route to a front or rear door. However, you consider that this would be impossible to achieve in this case without reducing the dining room to an unacceptably small size. You have therefore offered the following design features which you consider more than compensate for the omission of a protected stair.

(a) A fire escape roof window to the front of the property in accordance with normal guidelines.
(b) A fully protected stair to first floor level, with fire doors leading to the landing and front bedroom. The window of this bedroom is suitable for escape and has an 850 mm by 500 mm clear opening.
(c) The introduction of a mains operated interlinked smoke alarm with battery back-up to BS 5446 ('Specification for components of automatic fire alarm systems for residential purposes') to all habitable rooms, including a heat detector in the kitchen.
(d) Self-closing devices to be fitted to the existing doors of all other habitable rooms.
(e) A fire door to the loft room itself.

You consider that the level of protection offered is at least equal to the current guidelines and leaves the occupier well protected in the event of fire and indeed better protected than they are at present. You further consider that by allowing a 30 minutes' fire resisting safe passage from the loft room to the first floor bedroom you are effectively giving the occupants of the loft room the same degree of safety as any first floor occupant.

In addition, you point out that there is the added safety of double fire doors between the ground floor and the new loft room – i.e. the door at the top of the ground to first floor stairs and the door to the new loft room itself – as well as the early warning system.

In considering cases of this type, the Secretary of State takes the view that it is important to provide for the safe escape of occupants.

He is very conscious of the fact that a large number of fires occur in dwellings. The Department has therefore needed to consider the safety of the occupants of the new second floor unit in the event of a fire situation.

The Department's analysis of the case is as follows. In this case you are claiming that by providing a protected route of travel to the first floor, the proposals show at least the same degree of safety that is available for the occupants of any other open plan house where there are only two storeys. However, the Department does not accept this argument because risk increases with height.

The Department accepts that an open plan arrangement is acceptable in a two storey house providing the windows of any inner rooms have unobstructed opening of the size recommended in paragraph 1.18 of Approved Document 'B'. However, the persons needing to escape in such a house are already on the same storey as the escape window and probably in the same room. However, in this particular case a third storey is being proposed and the persons needing to escape would have to descend down a flight of stairs and select the correct bedroom to enter in order to reach the escape window. This would also assume that the bedroom floor, which in this case is above the lounge, had sufficient fire resistance to allow time for such escape.

The Department notes that you have made reference, as part of your submission, to the provision of an escape window from the new loft room. However, the provision of such a window is a recommendation given in paragraph 1.29 of Approved Document 'B' and as such cannot therefore be considered as a compensatory measure.

You have stated, as part of your submission, that you consider it unlikely that a disabled person would be on the top floor. However, it must be recognized that there is no on-going control over who may use the new loft room and it would, of course, be quite possible for a young child to occupy it. Given that there can be no control over the users, the Department considers that it would be unreasonable to dispense with the primary protected route of travel to a ground floor final exit either for self-escape or for rescue.

In all these particular circumstances, therefore, the Secretary of State determines that your proposals incorporating an existing staircase, which is shown not to be enclosed at ground floor level, do not show compliance with Part B1 of Schedule 1 to the Building Regulations 1991 in respect of means of escape.

Five storey building – means of escape (B.1)

It is proposed to erect a five storey reinforced concrete building comprising flats on the upper floors; a retail unit on the ground floor and a basement car park.

The issue is the safety of people on the upper floors should a fire take place in the basement, and to achieve the required standard will require an additional lobby to protect the lift in the basement.

Ref.: 45.1.107   Date: 15 December 1995

Your letter requested the Secretary of State to determine under section 16(10)(a) whether plans of the proposed work were in conformity with Requirement B1 ('Means of escape') of Schedule 1 to the Building Regulations 1991. In making his determination he has not considered whether the plans conform to any other relevant requirements.

Your proposals show work to erect a five above ground storey and a basement storey reinforced concrete building containing 35 flats (23 × 1 bedroom and 12 × 2 bedrooms) on the first, second, third and fourth floors. The ground floor is to contain a retail unit and is intended to be let as a fast food restaurant. The development is on an island site and car parking facilities for the flats is provided in a double height basement car park containing 48 spaces and 1 reserved space for disabled persons.

The flats are accessed via a self-contained ground floor entrance containing the lift and a single staircase. The single staircase gives access to all upper floors but not to the basement. It is, however, proposed to extend lift access to the basement level. The fire escape

lobbies to the upper floors are mechanically vented and there is to be a smoke extraction system in the basement. The retail unit is completely fire separated from the flats and the basement, as well as from the stairs and lift. It does, however, incorporate a refuse chute which discharges into the basement refuse store.

Vehicular access to the basement is via a wide ramp at the rear of the building. There is also a concrete stair at basement level which discharges direct to the street. This is completely fire separated from the single stairs which give access from ground level to the floors above.

The basement lift lobby will be separated from the concrete stair to the street by a 60 minutes' fire resisting steel roller shutter operated by a fusible link and which is connected to the fire alarm system. The intention is that in the event of a fire this shutter would close, thus retaining the direct access from the basement to the street and would also isolate the upper parts of the building from the basement.

The basement lift lobby also gives access to a small refuse store designed to take chuted waste from the kitchen of the ground floor restaurant unit, a lift motor room, and an electrical intake room. All these areas will be separated from the basement car park by the steel fire shutter.

The proposal to continue the lift into the basement was considered unsatisfactory by the Borough Council on the grounds of non-compliance with Requirement B1 ('Means of escape'). They were also concerned that the lift at basement level would open onto a protected lobby which also would give access to a refuse area and lift motor room. The Council therefore rejected your plans and it is in respect of these issues that you have requested a determination.

Your letter states that your understanding of Part M ('Access and facilities for disabled people') of Schedule 1 to the Building Regulations 1991 is that the building should be accessible to disabled persons. In your statement attached to your letter you state that the proposal was developed in conjunction with the Council's Planning Department and in particular the Council's officer concerned with disabled access. A car parking space has been allocated in the basement for disabled persons with a dedicated aisle for wheelchair access. You consider lift access from the basement to the upper floors, for the benefit of disabled persons, to be particularly important.

You have made reference to a similar determination case which was upheld by the Secretary of State. You claim that the decision in that case accepted that a car fire does not present a serious threat where the parking area is effectively open to the external air via the vehicle access ramp and where mechanical ventilation will be incorporated. In addition you consider that the high basement ceiling, which is a consequence of the double height car park stacking system, will be beneficial. You also point out that the building will be managed by a professional company and that all systems will be maintained and tested regularly.

You state that the refuse room, lift motor room, and electrical intake room are separated from the lift lobby by fire doors. However, a subsidiary lobby could be formed if necessary and this additional proposal is indicated on drawing number 753 SK 74X (the marked plan No. 6) which you have included along with your other plans.

The Council rejected your original proposals on the grounds that the building has a single staircase and a lift giving access between the basement car park and all the upper levels. They have noted that at basement level the lift and basement stair to the street terminate within a fire resisting enclosure separated from each other by a protected lobby incorporating a fire shutter with a fusible link. They have also noted that a refuse area and lift room open into the protected lobby.

The Council consider this situation to be unsatisfactory for the following reasons:

(i)   in the event of a fire in the basement, occupants of the lift and those in the lobby area would be trapped if the fire shutter operated and the lift defaulted to the basement level and they would not therefore be able to travel from the building to a place of safety – the basic requirement for means of escape in case of fire;

(ii) there is a danger that, in the event of a fire in the refuse area, smoke would be drawn up the lift shaft causing possible danger to occupants on the upper levels; and

(iii) disabled persons can access the upper levels of the building via the external entrance and lift at ground floor level without the need to enter the lift at basement level.

In support of their rejection the Council also refer to paragraphs 5.41 ('Fire protection of lift installations') and 5.48 ('Refuse chutes and storage') of Approved Document 'B' ('Fire Safety'). Paragraph 5.41 suggests that a lift should not be continued down to serve a basement storey if it is in a building served by only one escape stair. Paragraph 5.48 suggests that refuse storage rooms and chutes should not be located within protected lobbies.

In your application for this determination you have particularly asked that the Secretary of State give a determination on the balance of probabilities of a fire occurring which would affect the lift shaft. However, the statutory provisions of the Building Act 1984 only enable the Secretary of State to determine whether plans show compliance with the relevant requirements of Schedule 1 to the Building Regulations 1991. In this case the relevant Requirement is B1, 'Means of escape in case of fire'.

You have also referred to previous determination cases which you consider to be similar to this and which have been upheld by the Secretary of State. However, each case is considered on its individual merit and the particular circumstances pertaining.

In the Department's view the primary consideration in this case is the safety of the occupants of the upper floors if a fire starts in the basement area. The building is a single stair building but it is only the lift shaft which it is proposed to continue down into the basement car park and you state that this is being proposed for the benefit of disabled persons. This proposal to provide access for disabled persons is taken as your perceived need to comply with Part M of Schedule 1 of the Building Regulations 1991 referred to in your letter of 6 July 1995. However, the limits on application of Part M specifically exclude dwellings and the common parts of dwellings. Provision for disabled persons in this context cannot therefore, and does not, form a material consideration in this determination which relates solely to the question of compliance with Requirement B1.

The Department agrees with the Council that to continue the lift down into the basement in a single stair building would not accord with the recommendations given in Approved Document 'B'. However, the Department notes that the drawings show double fire door protection between the basement car park and lift lobby. In addition the proposal incorporates a 60 minutes' fire resisting steel shutter on a fusible link and linked to the fire alarm system. The intention of this is to provide additional isolation to the lift shaft from the basement.

The Council have raised the issue of persons being trapped in the lift lobby area at basement level if a fire occurred and activated the shutter. The Department considers that the probability of this occurring is remote and that persons could be trapped by fire and smoke in any situation where a lift continued down into the basement. In the unlikely event of such an occurrence, persons would be in a protected zone until they could be rescued or the lift wound back up to the ground floor. The Department therefore considers this arrangement to be acceptable in that adequate precautions would have been taken to ensure that smoke from a fire in the basement car park does not prejudice escape from the upper levels of the building.

The other issue raised by the Council is the risk of fire spread into the lift shaft from the refuse area, lift motor room, or electrical intake room. All of these are higher fire risk areas and a fire starting in one of these three areas could be of sufficient severity to prejudice the lift shaft. The original proposals showed only single fire door protection between these areas and the lift lobby. The Department considers that the Council were correct to raise this as an issue. However, your later latest proposal submitted by you (i.e. drawing number 753 SK

74X) shows additional fire separation between these areas and the lift lobby. The Department considers that if this additional separation is incorporated then adequate precautions will have been taken to prevent fire spread into the lift shaft from these areas.

The Secretary of State has given careful consideration to this case and in the circumstances considers that if compliance is to be achieved there should be an additional lobby to provide separation between the lift shaft and the refuse, lift motor, and electrical intake rooms – as proposed and shown on your drawing number 753 SK 74X (the marked plan No. 6). Subject to the assumption that this additional lobby is provided, the Secretary of State determines that the proposals for continuing the lift into the basement show compliance with Requirement B1 of the Building Regulations 1991.

You should note that each case is considered on its individual merit and the particular circumstances pertaining. It follows that this case should not be taken as creating a precedent whereby any proposal for a lift shaft in a single stair building to be extended to basement level can be assumed to achieve compliance with Requirement B1. Any such proposals must be judged carefully in respect of the technical design and precautions incorporated in them to address any potential hazard.

Loft conversion – means of escape (B.1)

The roof space of a two storey chalet type house is to be converted to one bedroom with a window escape provision some 5 m up a 55° slope, from the eaves. This is not acceptable to the local authority, or the Secretary of State.

Ref.: 45.1.110   Date: 20 December 1995

The Secretary of State takes the view that he is being asked to determine under section 16(10)(a) whether plans of your proposed work are in conformity with the requirements of Requirement B1 ('Means of escape') of Schedule 1 to the Building Regulations 1991. In making his determination the Secretary of State has not considered whether the plans conform to any other relevant requirements.

Your request for a determination concerns your proposals to convert the roof space of a two storey house to form a bedroom.

The pitch of the roof to the front elevation, which is to contain an escape rooflight, is 55 degrees. The new escape rooflight is to be located about 5 metres from the eaves, up the pitch of the roof. The eaves themselves are approximately 2.4 metres vertically above ground level. Two dormer windows to the first floor rooms are also on this elevation, below the new escape rooflight.

You suggest that safe rescue from the new second floor room via the escape rooflight would be possible by laying a ladder along the slope of this front elevation. However, the District Council have rejected your proposals on the grounds that they do not show compliance with Requirement B1 of the Building Regulations 1991 by virtue of the fact that the new second floor escape rooflight does not provide for adequate means of escape. It is in respect of this rejection that you have sought a determination as to whether this proposed escape rooflight is sited in a position that makes it suitable for escape or rescue.

The District Council point out that Approved Document 'B' suggests the following two ways in which Requirement B1 can be met for a two storey house with a loft conversion:

(i) enclosure of the existing stairway with fire resisting construction (paragraphs 1.24 and 1.25), or

(ii) provision of a suitable window located to allow rescue from the ground (paragraphs 1.29 to 1.31).

The District Council state that the escape rooflight does not accord with the guidance given in diagram 4 of Approved Document 'B' ('Position of dormer window or rooflight suitable for escape or rescue purposes from a loft conversion'), in that the bottom of the window opening is substantially more than the suggested 1.7 metres maximum dimension above the eaves – the actual dimensions are approximately 5 metres, with the height of the eaves above ground being about 2.4 metres.

The District Council do not consider that it would be possible for a ladder to be rested against the eaves to facilitate a rescue from the escape window 5 metres away. The ladder would have to be laid along the 55 degree roof slope which would make rescue difficult for the following reasons:

(i) The ladder would be liable to slip backwards because of the angle.

(ii) The ladder would be difficult to use considering that it would be resting on the sloping section of the roof such that feet rests and hand grips would be unsatisfactory and considered unsafe.

(iii) There are two dormer eaves that are directly below the escape window that would inhibit the laying of a ladder below the window.

In conclusion, the District Council consider that the proposed window is not suitable for means of escape purposes because the location of this window would make rescue by ladder extremely difficult in the event of fire. This view is supported by the local fire authority.

You state that the District Council have rejected your proposals because they do not show compliance with Approved Document 'B' and in particular diagram 4 of that document. However, you point out that approved documents are for guidance only and are not mandatory.

You state that the roof pitch to the front elevation which contains the escape rooflight is 55 degrees and that you consider that a ladder could safely be laid along its slope for escape purposes. You do not consider that the Approved Document caters for the roof of a chalet-type dwelling such as in this particular case. You also state that the escape rooflight is still no higher than that provided in a conventional roof containing a loft conversion. You consider that this is an exceptional case and still complies with escape requirements; in your opinion it may even be safer than escape from a conventional roof. You also state that you have obtained approval in other areas for similar type roofs.

In considering cases of this type, the Secretary of State takes the view that it is important to provide for the safe escape of occupants. He is very conscious of the fact that a large number of fires occur in dwellings. The Department has therefore needed to consider the safety of the occupants of the new second floor unit in the event of a fire situation and whether adequate provision for means of escape for such occupants has been made.

In support of your case, you have made reference to previous similar cases where local authority approval has been granted. However, the Secretary of State is not in a position to take account of other circumstances. Each case put to him must be considered on the basis of its individual merit and the particular circumstances of that case.

The Department's analysis of this case is as follows. The Department accepts, as you have pointed out, that the guidance given in approved documents is not mandatory. However, what needs to be shown is that adequate means of escape has been made. In this case it is quite clear from the front elevation shown on drawing 95/117/2 that the existing dormer windows, which are sited below and directly to either side of the space a ladder would have to occupy, cause obstruction and would make rescue from the new loft rooflight extremely hazardous. The Department has consulted experts in fire rescue and this is also the view reached by them.

Given therefore the difficulty that would be experienced in accessing the rooflight the Department considers that the questions as to whether it would be acceptable for the ladder to be laid along the roof slope, or whether the ladder would slip, do not need to be, and have not been, addressed in this particular determination.

In all these particular circumstances, therefore, the Secretary of State determines that your proposals for siting the new loft escape rooflight do not show compliance with Requirement B1 of Schedule 1 to the Building Regulations 1991 in respect of means of escape.

Block of flats – means of escape (B.1)

A new block of flats with six above ground storeys is to be erected, and the local authority want sprinklers to be provided in each flat. The applicant offered a fire engineering solution and the determination accepted the proposals offered.

In this instance the Secretary of State emphasized that the case should not be taken as a precedent.

Ref.: 45.1.101   Date: 22 December 1995

The Secretary of State takes the view that he is being asked to determine under section 16(10)(a) whether plans of your proposed work are in conformity with the Requirement B1 ('Means of escape') of Schedule 1 to the Building Regulations 1991. In making his determination the Secretary of State has not considered whether the plans conform to any other relevant requirements.

Your request for a determination concerns the erection of a new block of flats with six above ground storeys and a basement plant area. The basement is fire separated from the above ground storeys and is not at issue.

Seven flats are contained on each of the ground, first, second and third floors; five flats on the fourth floor; three flats on the fifth floor and one flat on the sixth floor. All 37 flats are set around an atrium which provides access via open balcony decks to the lift, single staircase, refuse chute and service ducts. Means of escape is direct to the single staircase which is to be pressurized and discharges to ground floor level.

The City Council have rejected your proposals on the grounds that they do not show compliance with Requirement B1 of the Building Regulations 1991. It is in respect of this rejection that you have sought a determination to resolve the issue of the effect that the open atrium will have on means of escape via the balcony decks and the acceptability of the fire engineering solution which you have proposed.

In support of their judgement the City Council point out that the situation where the horizontal route of escape is via an open atrium is not covered by any current UK code of practice. They therefore consider that the recommendations for a protected corridor given in paragraph 2.22 of Approved Document 'B' ('Fire Safety') should be adhered to. They also make reference to the latest draft code for atria buildings BS 5588: Part 7: 'Fire precautions in the design, construction and use of buildings: code of practice for atria buildings' which would require sprinklers to be provided throughout the building.

The City Council consider that the low design fire size of 0.5 MW can only be achieved with the aid of sprinklers installed within the risk areas and not with the aid of the proposed drencher system over the entrance door to each flat. They state that the reduction in fire size is critical because of the sizing of the slot extract system and ductwork. The City Council also consider that the design could be seriously affected by wind pressures on the building.

Given that the lobbies between the front doors of the flats and the single staircase are unprotected, the City Council are of the opinion that the provision of sprinklers in each individual flat is the only option.

Your submission for this determination is based on a fire engineering solution as an alternative to prescribed recommendations given in other documents such as the draft British Standard on atria design. You consider your proposals to be less restrictive and that the arrangement is very efficient in terms of planning. You state that fire safety has been the designers' paramount concern. You have therefore offered the following enhancements as an alternative to prescriptive recommendations:

(a) Sixty minutes' fire separation horizontally at each floor level and vertically between the residential accommodation and the balcony decks open to the atrium. Sixty minutes' fire resisting doors provided to flat entrances, refuse room lobby, and service duct doors; and two 30 minutes' fire resisting doors provided to the staircase.

(b) Direct means of escape distances less than 7.5 metres from the flat entrances to the staircase.

(c) Local smoke exhaust ductwork over each flat entrance door. This can be activated by smoke detection on the outside of the flat entrance door and the smoke detectors within the dwelling's rooms, or a flow switch within the drencher system.

(d) Drencher heads located over flat entrance doors and a sprinkler head located in the refuse lobby.

(e) Smoke detection in the dwellings comprising detectors in every room and hallway linked to the fire alarm which will activate the stair pressurization system and the smoke exhaust system. These detectors will operate on a double knock principle (i.e. the room of fire origin and the hallway will operate the fire alarm). The intention of this level of smoke detection is to avoid false alarms.

(f) Smoke detection in the general circulation space to be of L2 classification to BS 5839: Part 1: 1088 ('Fire detection and alarm systems for buildings: Code of practice for system design, installation and servicing').

(g) A fully addressable fire alarm system with manual call points adjacent to the stairwell. The fire alarm will be linked direct to the local fire service station and to the management suite.

(h) Pressurization of the stairway to be in accordance with BS 5588: Part 4: 1978 ('Fire precautions in the design, construction and use of buildings: Code of practice for smoke control in protected escape routes using pressurization').

(i) High level ventilation provided by automatic opening vents at the top of the atrium to act as inlet and be used as required by the fire service for smoke clearance.

(j) Dry riser installation to the stair well.

By offering the above enhancements you consider that the level of protection for the open atrium is in excess of what would be provided if the scheme had full compartmentation on each floor.

You also dispute the City Council's contention that the design fire is based on 0.5 MW, since this is the value of heat flux released at the flat door. You state that based on a typical room size of 20 square metres the actual fire loading is 3 MW.

In considering cases of this type, the Secretary of State takes the view that it is very important to provide for the safe escape of occupants. He is very conscious of the fact that a large number of fires occur in dwellings.

The Department's analysis of this case is as follows. The Department accepts that unless adequate precautions are taken where an open atrium is provided throughout the height of the building a situation could arise where smoke from a fire at a lower level could spread through to the upper floors. The Department has therefore needed to consider whether sufficient measures have been proposed to ensure that persons are able to escape safely.

The installation of an adequate smoke control system should ensure that smoke and other products of combustion are extracted, so that more time is available for escape. The efficiency of this system also depends on the height at which smoke/gas layers can be maintained. In this case you have attempted to show that a smoke layer can be maintained by positioning the smoke extract above each flat entrance door. You estimate that the time to 'flashover' in any one flat will be approximately 3.6 minutes and the assumption is that the hot smoke will exit into the circulation area via the flat entrance door. The drencher system will reduce the heat flux of the gases and the Department assumes that the intention is that the cooling produced by the drenchers will stop the combustion of unburnt fuel gases outside the door.

The Department takes the view that if the drencher system operates satisfactorily then the heat release calculations you have provided are acceptable. However, in reaching this view the Department has considered it necessary to make two assumptions with regard to the satisfactory operation of the drencher system, as follows:

(a) The drenchers would be designed so as not be activated until smoke temperatures reach between 150 and 200 degrees C so that down-drag of smoke by the drenchers in the early stages of fire will not affect means of escape.

(b) The drenchers would be of a type that produces a spray instead of sparge pipe type droplets.

The Department would normally support the City Council's arguments for a full sprinkler system. However, in this case the Department accepts that the design philosophy, although unusual, is basically sound. In the circumstances and on the assumptions set out in 15 above, the Department does not therefore consider that a fully sprinklered building, with sprinklers being provided throughout all the residential accommodation, is justified.

Therefore, in all these particular circumstances and based on the assumptions set out in paragraph 15 above, the Secretary of State determines that your proposals – which exclude the provision of sprinklers in the individual flats – comply with Requirement B1 of Schedule 1 to the Building Regulations 1991 in respect of means of escape.

You should note that this case should not be taken as a precedent whereby it can be assumed that omission of a sprinkler system in the individual flats of an atrium access, single stair designed residential building will necessarily achieve compliance with Requirement B1. Each case must be considered on its individual merits having regard to the design, the alternative solution proposed, and the particular circumstances of the case.

Two storey nursing home – means of escape (B.1)

In this case the local authority is insisting upon the provision of self-closing devices on bedroom doors. The Secretary of State finds that the submitted scheme is acceptable on the ground floor, but does not comply with the Building Regulations on the first floor. He does not require the self-closing devices but suggests how the problem may be overcome.

Ref.: 45.1.102   Date: 11 January 1996

The Secretary of State takes the view that he is being asked to determine under section 16(10)(a) whether plans of your proposed work are in conformity with the requirements of Part B1 ('Means of escape') of Schedule 1 to the Building Regulations 1991. In making his determination the Secretary of State has not considered whether the plans conform to any other relevant requirements.

Your request for a determination concerns the erection of a new 61-place, two storey nursing home for the elderly; constructed using timber frame with masonry outer leaf. The bedroom wings contain mainly single rooms, although there are some double ones. On both the ground and first floors there are two day rooms which give access onto the bedroom corridors.

You have stated that the bedroom wings have been designed so that each floor is its own fire compartment. Each compartment has been divided into two smoke containment areas/sub-compartments containing about six or seven bedrooms – and in some instances a day room – from which means of escape is available in two directions.

Smoke detectors are to be provided in each bedroom. An automatic smoke detection system is to be provided throughout the building and this will be fully addressable so that if smoke is detected the precise location will be identified. It is not your intention to provide self-closing devices on the bedroom doors although these doors will have 30 minutes' fire resistance. Bedroom corridors are sub-divided with fire doors held open by automatic release mechanisms that will be activated by the fire alarm system. The doors to the day rooms are also to be held open on automatic self-closing mechanisms.

The District Council have rejected your proposals on the grounds that they do not show compliance with Requirement B1 of the Building Regulations 1991 by virtue of the fact that none of the bedroom doors are to be fitted with self-closing devices. It is in respect of this issue that you have sought this determination.

The District Council consider it important that the building be constructed so that fundamental measures of safety from fire are incorporated. They are not prepared to take account of fire safety procedures and staff levels as these are part of the requirements for the registration of the home. They consider any reduction in escape route protection to be unacceptable and that any assumed inconvenience to any one resident must be acceptable to ensure safety for all.

The District Council state that evacuation will need to be of the highest standard because attendance time for the fire service could take up to 20 minutes. They consider the omission of door closers to be a direct contravention of the Building Regulations and have therefore rejected the plans on this basis. They also state that should an automatic self-closing device be a hindrance in any way, then doors can be held open by a choice of automatic release mechanisms. The District Council state that the Fire Authority have wholly endorsed the view that all bedroom doors must be fitted with self-closing devices.

You have stated that the home is intended for the use of elderly mentally infirm and frail elderly persons and that you have been advised that such patients become distressed when confronted with closed doors. You consider that any self-closing device fixed to a bedroom door will inhibit the ease of door operation and confine the occupant to the room.

You consider that automatic release mechanisms are also unacceptable for two reasons. Firstly, you consider that such devices do not allow privacy; a resident will often request that a door be left only slightly open. Secondly, you consider that when released owing to the fire alarm being activated, they would prevent or restrict independent egress from the bedroom. You are further of the opinion that closed doors everywhere would not reflect the right atmosphere of general wellbeing and confidence of movement throughout the home.

You consider that, in the event of fire occurring in one of the bedrooms, there is more advantage to be gained by allowing the resident to escape than to be trapped by a self-closing device. You have also made the point that bedroom door closers have not had to be fitted in other similar buildings and have made reference to similar appeal/determination cases upheld by the Secretary of State. You point out that bedroom door closers are not a requirement of the Home Office Draft Guide to fire precautions in existing residential care premises. You state that a fire safety policy document for the home has been submitted and that the home will have a minimum of seven staff available at night to respond to any emergency.

For the purposes of this determination the Department has considered this case in two parts – the ground and the first storey. This is because whilst additional escape routes via patio doors are available from the ground floor day rooms, no such alternative escape routes are available from the first floor day rooms. In the Department's view this makes the situation on the two floors quite different.

You have made reference to previous determination cases in support of this case. However, each case put to the Secretary of State on appeal must be considered on the basis of its individual merit and the particular circumstances of that case.

The Department's analysis of this case is as follows. It accepts that paragraph B2 of Appendix 'B' ('Fire doors') to Approved Document 'B' ('Fire safety') states that all fire doors should be fitted with automatic self-closing devices. However, reference is also made in Approved Document 'B', notably paragraphs 0.34 ('Health care premises') and 3.26 ('Hospitals and other residential care premises of Purpose Group 2a'), to other Health Care codes which do not always require such closers to be fitted.

The Department accepts your assertion that self-closing devices can be a hindrance to the normal day-to-day functioning of this type of residence. However, it also has to be accepted that if these self-closing devices are to be omitted, then compensatory features that give the occupants at least the same degree of safety, have to be provided. In this case a full smoke detection system and vertical fire separation dividing the building into six/seven bedroom units is proposed. Both of these proposals go beyond the recommendations given in Approved Document 'B'. In addition, there are no 'dead-end' conditions in the locations where it is proposed not to fit door closers.

However, the Department notes that some parts of the corridors containing the bedrooms without door closers also give direct access to day rooms. These day rooms could accommodate more than the bedroom occupancy within that corridor and therefore means of escape could be prejudiced if a fire occurred in a bedroom not fitted with a door closer. The Department considers it important that the day rooms on the first floor are separated from the corridors containing bedrooms without closers. The Department does not consider this to be important on the ground floor because alternative escape via patio doors is available from the day rooms.

The District Council have suggested that automatic release mechanisms can be fitted to the doors if self-closing devices are considered to be a hindrance. Although it is the intention to provide some doors with this system it could lead to onerous wiring and maintenance requirements and the Department is of the opinion that what has been proposed is a superior arrangement in terms of occupant safety.

In all these particular circumstances, therefore, the Secretary of State determines as follows. In respect of the ground floor he considers that your proposals to omit self-closing devices from bedroom doors comply with Requirement B1 of the Building Regulations 1991. In respect of the first floor he considers that your proposals to omit self-closing devices from those bedroom doors opening into corridors that give direct access to day rooms do not comply with Requirement B1 of the Building Regulations 1991.

In analysing the issues involved in this determination, the Department has considered what would constitute reasonable provision to comply with Requirement B1 in respect of the first floor proposals. The following observations may therefore be helpful. In the Department's view the non-provision of door closers to those bedrooms on the first floor opening into corridors that give direct access to day rooms would be acceptable if additional fire doors – which could be fitted with automatic release mechanisms – had been shown to be placed across the first floor bedroom corridors separating those corridors from the access into the day rooms.

Bank building – vertical division to provide two premises (B.1)

Plans have been rejected because the scheme does not provide adequate means of escape from a basement; and this was so determined.

Ref.: 45.1.115   Date: 8 February 1996

The Secretary of State takes the view that he is being asked to determine under section 16(10)(a) whether plans of your proposed work are in conformity with the Requirement B1 ('Means of escape') of Schedule 1 to the Building Regulations 1991 (as amended). In making his determination the Secretary of State has not considered whether the plans conform with any other relevant requirements.

Your request for a determination concerns existing bank premises comprising a basement (containing primarily two document stores, a safes room and male and female toilets); a ground floor containing a redundant public banking section and an open plan processing centre extending from the front of the building to the rear; and a first floor containing primarily an operations room, staff room and toilets. Two internal stairs give access from the ground floor to the basement.

The proposed building work involves the vertical division of the premises into two separate occupancies, with your clients maintaining control over one part of the divided premises. The building work will include division of the basement to retain two storage rooms for your clients – one designated as a document store and the other as storage.

As a result of the division only one stair will give access from the ground floor area to be held by your clients to their part of the divided basement. Single direction escape only is therefore available from the basement via these stairs. The proposals show both storage rooms open on to the single stair enclosure with the stairs discharging into the open plan ground floor processing centre. A final exit door is located at each end of this processing centre. The Department estimates from the drawings submitted that there is an actual travel distance of 21 metres from the furthest point in the document store to the point in the ground floor processing centre where alternative escape is available. A fully automatic fire detection system is to be installed throughout the building.

The Borough Council has rejected your proposals on the grounds that they do not consider the means of escape to be satisfactory in that part of the basement to be retained by your clients. It is in relation to this decision that you have applied for a determination in respect of the compliance of the work with Requirement B1.

The Borough Council consider the scope of works to materially affect the existing means of escape in that there were originally three exits. After sub-division of the building, escape from the basement is only possible in one direction and this is via an accommodation stair which discharges into the ground floor processing centre.

The Borough Council accepts that smoke detection will give early warning. However, they consider that alerting occupants of the basement to a fire on the ground floor would not prevent that fire making escape impossible.

The Borough Council refers to two paragraphs of Approved Document 'B' ('Fire safety'). Paragraph 4.25 suggests that every internal stairway should be a protected stairway, and paragraph 4.28 suggests that a protected stairway should either discharge to a final exit or by way of a protected exit passageway to a final exit. In this case the proposals show the single stair from the basement discharging into the ground floor accommodation. They also point out that the proposals do not comply with the recommendations given in BS 5588: Part 3: 1983 ('Code of practice for office buildings').

The Borough Council also state that they consider they have been helpful in suggesting alternative schemes that would provide a protected route to external air from the head of the basement stairs.

You state that the intention of dividing the property is to enable the bank to continue its office use in a reduced area, the public banking section of that branch having been abandoned. You accept that the work involved in the division of the basement, which results in single direction escape, will not accord with any particular solution contained in Approved Document 'B'. However, you suggest the functional requirement for means of escape will be satisfied by the introduction of engineered safety measures and make the following points in support of this:

(a) The building is an existing structure and rigid compliance with guidance given to the Building Regulations will be unduly restrictive.
(b) The basement rooms are used for storage of redundant files and cheques and will be visited infrequently by bank staff who will be familiar with fire procedures.
(c) The fire risk is low and adequate fire separation will have been provided between the basement area and the ground floor level.
(d) Automatic smoke detection and sounders will have been provided throughout the premises.
(e) There are very many similar situations in other banks and buildings and the possibility of persons becoming trapped is small.

In addition, you claim that the Local Fire Authority consider the proposals to be acceptable although written comments have not been received from that Authority.

The Department's analysis of this case is as follows. You imply that an accommodation stair, i.e. a stair that is open to the accommodation, should be acceptable in this case. The Department acknowledges that paragraph 4.25 of Approved Document 'B' suggests that such a stair may be considered acceptable provided that the distance of travel and the number of people involved is very limited. As an example the provisions permitted for small shops given in BS 5588: Part 2: 1985 ('Code of practice for shops') is quoted in the paragraph. In this case the number of people likely to use the basement rooms is low. However, the travel distance, even when taken to the point in the ground floor office where escape is available in alternative directions, is considered by the Department and the Borough Council to exceed the recommended maximum single direction travel distance given in Approved Document 'B'.

In addition, if the guidance for small shops is followed for an accommodation stair, then the recommendation given in BS 5588: Part 2 is that the stair should terminate within 3 m of a final exit. The proposals show that in one direction the distance of the head of the stair is approximately 6 m from a lobby with another 4 m of travel to a final exit. In the opposite direction there is approximately 9 m of travel to a final exit from the head of the stair.

The Department accepts that the smoke detection system will give early warning and that the numbers of persons using the basement rooms will be small. However, basements present more of a risk to safe escape than do above ground storeys. Moreover, as has been pointed out by the Borough Council, the escape route through the ground floor office accommodation could be prejudiced by the time the smoke detectors are activated. This could be particularly the case if, as the Department considers likely, staff remain in the basement to lock up before vacating it. In reaching this opinion the Department has taken account of the statement made by the bank confirming that access to the basement will only be required to retrieve/deposit files and to retrieve/deposit cash in the safe.

The Secretary of State has given careful consideration to all the facts in this case and has concluded that the proposals for means of escape from the basement are not satisfactory. The Secretary of State therefore determines that the proposals do not show compliance with Requirement B1 of Schedule 1 to the Building Regulations 1991 (as amended).

Alterations and extensions to existing shop – means of escape (B.1)

Here a local authority will not accept the escape provisions proposed, which include linking two stairs by means of a 22 m long corridor.

A comprehensive fire detection and alarm system together with adequate signings and emergency lighting were determined to show compliance with the Building Regulations.

Ref.: 45.1.111   Date: 16 February 1996

Your application has been made under section 30 of the Building Act 1984 which was repealed on 11 November 1985. However, the Secretary of State has taken the view that he is being asked to determine under section 16(10)(a) whether plans of your proposed work are in conformity with the Requirement B1 ('Means of escape') of Schedule 1 to the Building Regulations 1991 (as amended).

Your request for a determination relates to building work to refurbish existing shop premises fronting a High Street and the erection of an extensive longitudinal extension to the rear.

The existing shop premises comprise a basement in the form of vaults; a ground floor sales area; a first and second floor which are to be refurbished for staff offices and facilities respectively; and a third floor which is to be used for stores and archives. The new extension is to comprise a lower ground floor abutting but not connected to the existing vaults; a ground floor sales area which is to link in at the same level as the sales area of the existing shop to form one continuous floor area; and an upper sales area on the first floor which will be at the same level as the refurbished staff offices in the existing building.

The upper floors of the existing building are to be served by a single staircase terminating at first floor level. The new extension is to be served by two staircases. The first (marked 'stair A' on the plans) is located approximately one third along the length of the new extension from the front of the premises. It connects the upper sales area direct to a final exit at lower ground level. There is no access to this stair from the ground floor sales area. The second stair (marked 'stair B' on the plans) is located at the rear of the new extension and connects the upper sales area, ground sales area, and the lower ground level where there is a final exit.

Escape within the new extension is therefore available via both stairs A and B. To achieve escape from the second and first floors of the existing building a protected corridor at first floor level is proposed, linking the stair in the existing building to stair A.

An 'L1' fire alarm and detection system, as defined in BS 5839: Part 1: 1988, is proposed as a compensatory feature.

The Borough Council has rejected your proposals on the grounds that, in their view, they do not show compliance with Requirement B1 of Schedule 1 to the Building Regulations 1991 (as amended). You contend that they do meet the Requirement and have therefore sought a determination from the Secretary of State.

In rejecting the proposals the Borough Council refer to paragraph 4.28 of Approved Document 'B' ('Fire safety') which recommends that every protected stairway should discharge directly to 'a final exit or by way of a protected exit passageway to a final exit'. However, the proposals show that the stair from the existing building discharges into a protected corridor at first floor level. This corridor connects with a new stair (stair A) in the rear of the existing building with a travel distance of 22 m between stairways. In the Council's opinion this route does not accord with paragraph 4.28 of Approved Document 'B'.

They further argue that it is possible stair 'A' could be compromised in the event of fire at first floor level; people at the High Street end of the upper sales area and within the upper storeys could then become trapped at the front of the building. Having consulted with the local Fire Authority the Borough Council points out that they fully agree the proposals do not provide satisfactory means of escape in the event of fire.

Finally, the Borough Council take the view that a stairway connecting the first floor directly to the ground floor and exiting to the street from the existing building, could be provided, thereby achieving compliance.

You consider your proposals for means of escape to be satisfactory on the basis that the lobby between the first storey retail accommodation and 'stair A' serving the storeys above this level will not form part of the normal vertical circulation in the building and will function only as an emergency exit. Consequently there will not be the possibility of doors being routinely wedged or propped open.

You further argue that the escape corridor from the front of the (existing) building to stairway A is unlikely to be used by large numbers of the public since there will be two other exits from the first floor sales area, and staff who will be expected to use the escape corridor route will be familiar with it.

Whilst acknowledging that the structure will not give fire protection indefinitely, you contend that the automatic fire detection and alarm system will give the earliest possible warning of an outbreak of fire in the building.

Finally, you also refer to paragraphs 4.28 and 0.28 of Approved Document 'B', pointing out that paragraph 4.28 states that a protected stairway may discharge to 'a protected exit passageway' and that paragraph 0.28 refers to the use of a 'place of relative safety'.

The Department's analysis of this case is as follows. The proposal is to link the new stair at the front of the existing building to a new stair in the proposed extension with a protected corridor. The issue relates only to whether the 22 m long corridor is acceptable for means of escape purposes. The Department agrees with the Borough Council that this arrangement does not accord fully with paragraph 4.28 of Approved Document 'B' in that the protected passageway does not lead to a final exit but to another protected stairway.

What therefore needs to be considered in this case is the risk that the route of travel via a 22 m long protected corridor poses to the occupants of the building. The new single stair at the front of the (existing) building terminates at first floor level but serves only a limited amount of staff accommodation on the existing second and third floors. The link corridor is imperforate apart from the connection at each end to the stairways and lobby access from the front section of the first floor retail sales area. This provides alternative escape from that part of the retail area.

Clearly the worst scenario is if a fire occurred on the first floor

retail area. However, the Department notes that only those members of the public located at the front section of the retail area would have any need to use the link corridor as an escape route. The other occupants of the building having to use the corridor would all be members of staff from the second and third floors who should all be familiar with the building.

The Department accepts that the fire resistance of the corridor will not hold out indefinitely in a fire but the drawing shows that the link corridor has a one hour fire-resisting rating to the walls, floor and ceiling. The Department is therefore of the opinion that adequate protection is provided to persons needing to use the first floor link corridor for means of escape. This view takes account of the comprehensive fire detection and alarm system to be incorporated in the building and assumes that adequate fire signage and emergency lighting, in accordance with the relevant British Standard, is also to be provided.

The Secretary of State has given careful consideration to the particular circumstances of this case. On the assumption that signage and emergency lighting is to be provided as referred to in paragraph 20, he has concluded that your proposals do show compliance with Requirement B1 of Schedule 1 to the Building Regulations (as amended).

Two storey retail store – refurbishment (B.1 and B.3)

It is proposed to alter and refurbish a two storey retail store and the dispute concerns the provision being made for the means of escape, and internal fire spread.

The decision accepted that the measures being proposed by the applicant showed compliance with the building regulations.

Ref.: 45.1.114   Date: 26 February 1996

Your letter requested the Secretary of State to determine under section 16(10)(a) whether plans of the proposed work were in conformity with Requirements B1 ('Means of escape') and B3 ('Internal fire spread (Structure)') of Schedule 1 to the Building Regulations 1991. In making his determination he has not considered whether the plans conform to any other relevant requirements.

Your proposals show work to alter and refurbish a retail store which has two above ground storeys. They include the removal of existing stock rooms and the creation of a new sales area at first floor level. The retail store forms part of a shopping complex although the new first floor sales area does not connect with the mall.

The proposals included the removal of an existing sprinkler system installed in the first floor ceiling void as part of the refurbishment programme; however, a full sprinkler system is to be provided below ceiling level. The first floor ceiling void is shown on the drawings to have a depth of approximately 1600 mm.

The Borough Council have stated that all works comply with the Building Regulations except that the ceiling void in the first floor ceiling must be protected by the re-instatement of a sprinkler system within this void. They therefore passed your plans but on condition that you submit full details of the proposed sprinkler system to be installed, both above and below the new suspended ceiling, complying with British Standard BS 5306: Part 2: 1990 ('Specification for sprinkler systems'). You believe that your work complies without the re-instatement of the sprinkler system in the first floor ceiling void and it is in respect of this issue that you have requested a determination.

The Borough Council state that the need for sprinkler protection arises out of the following two Building Regulation requirements:

(i)   B1 – Since this retail store forms part of a shopping complex, the appropriate guidance document for assessing means of escape is British Standard BS 5588: Part 10: 1991 ('Code of practice for shopping complexes'). This document recommends that sprinklers be provided throughout the complex and states that the sprinkler system should be designed in accordance with British Standard BS 5306: Part 2: 1990 ('Specification for sprinkler systems').

(ii) B3 – Table 12 ('Maximum dimensions of building or compartment (multi-story non-residential buildings)') of Approved Document 'B' ('Fire safety') suggests that the floor area of any one storey or compartment in a shop should not exceed 2000 square metres. The Borough Council point out that the ground floor exceeds this area and is not separated from the first floor and in their view it follows that the shop will only comply with the compartment size limitations providing the complete building is given sprinkler protection. This means that the sprinkler system should comply with British Standard BS 5306: Part 2, since this is the reference document quoted for sprinkler installations in Approved Document 'B'.

The Borough Council have pointed to paragraph 26.6.1 of BS 5306: Part 2 which suggests that ceiling voids should be provided with sprinklers where the depth of void exceeds 800 mm and have noted that in this determination case the depth of void is indicated as being approximately 1600 mm. They further point out that it is not proposed to fit cavity barriers in the void and that the distance across the void will exceed the maximum recommendation of 40 metres given in paragraph 9.13 ('Maximum dimensions of concealed spaces') of Approved Document 'B' by an average of 6 metres.

As part of the negotiations over the point at issue, you submitted a formal appeal under sections 8 and 9 of the Building Act 1984, at the suggestion of the Borough Council, for a 'relaxation or dispensation' of Requirement B3. On 17 March 1995, the Borough Council issued a formal notice of refusal of your appeal for a relaxation/dispensation of Requirement B3 to permit the first floor ceiling void to be unsprinklered. They particularly wished to have it noted that had the void been protected by an automatic smoke detection system then it is likely that your application would have been viewed in a different light.

The Borough Council would also have been prepared to accept a cruciform arrangement of cavity barriers in the ceiling void as an alternative to sprinklers. They are concerned that two escape routes could be lost simultaneously if neither sprinklers nor cavity barriers are provided in the void.

The Borough Council have consulted the fire authority and it is that Authority's considered opinion that sprinkler protection within the first floor ceiling void should be retained under the criteria laid down in BS 5306: Part 2.

You contend that it is not necessary to provide sprinklers in the first floor ceiling void and refer to a similar appeal case that was upheld by the Secretary of State. In addition you have made the following points in support of your case:

(i) The first floor ceiling will have been upgraded to achieve a class 'O' (the highest product performance classification for lining materials) surface spread of flame rating and the materials used within the ceiling void will be of limited combustibility. In addition the suspended ceiling is continuous, largely imperforate, and although originally provided with sprinkler protection, will have been reduced in area by approximately 25%.

(ii) All new main structural members will be provided with two hours' fire protection and means of escape has been improved at first floor level.

(iii) Your clients have a pro-active approach towards fire and safety which ensures high standards.

(iv) The ceiling void above the existing ground floor sales area is approximately 1050 mm deep and does not have sprinkler protection. You consider that this sets a precedent for not providing sprinklers in the first floor ceiling void.

(v) The new first floor sales area is not directly linked to the shopping mall.

(vi) A full sprinkler protection system is being provided at ceiling level on the first floor and a smoke extraction facility via the supply air ducts, triggered by an in-duct smoke detector, is also being provided.

(vii) The building and shopping centre insurers have been

approached and have indicated that they consider it acceptable for the first floor ceiling void to remain unsprinklered.

The Department's analysis of this case is as follows. It notes that the Borough Council have expressed concern at the possibility of two escape routes being lost simultaneously, but the Department does not share this concern. Should the main escape stair within the shop become prejudiced through smoke then there are separate routes into the two staff areas. These are located at each end of the retail area with each of these areas having an escape stair. Once in the staff areas members of the public could wait in relative safety until they were able to evacuate the building and the ceiling cavity in question does not communicate with either area. It is also noted that the first floor sales area does not communicate directly with the shopping mall. The Department does not therefore consider that the unsprinklered ceiling void would prejudice escape from either the mall or the store itself.

With regard to Requirement B3 of the Regulations, the Department accepts that BS 5306: Part 2 recommends that ceiling cavities in excess of 800 mm should be provided with sprinkler protection. However, the Department notes from the drawings submitted that there are a number of downstand beams within the ceiling cavity that could partially restrict the spread of smoke in the early stages of a fire. The Department considers it likely therefore that the reservoirs formed by these downstands would provide localized hot areas sufficient to activate the ceiling mounted sprinklers. The Department does not therefore consider the absence of sprinklers in the ceiling void to be a threat to life safety but more an issue of property protection.

The Department therefore considers in this case that the proposal to omit sprinklers from the ceiling void to the first floor shows compliance with Requirements B1 and B3 of the Building Regulations 1991.

In these particular circumstances, therefore, the Secretary of State determines that the proposals for the alteration and refurbishment of the retail store show compliance with Requirements B1 and B3 of the Building Regulations 1991 (as amended).

Lighting on an escape route (B.1)

This determination relates to the amount of emergency lighting to be provided in the refurbishment of premises falling with purpose group 5 (assembly and recreation).

The decision allows that some relaxation may be made, but only if additional lighting is provided in three rooms.

Ref.: 45.1.105   Date: 18 April 1996

Your request for a determination relates to the extent of emergency lighting provision required in the refurbishment of the premises. In making his determination the Secretary of State has not considered whether the plans conform to any other relevant requirements.

The building is one of three existing buildings being refurbished and refitted. The building is a five storey building and the work consists basically of converting open plan offices to rooms appropriate for academic use. The building had previously been altered and extended to form office accommodation in 1990.

The building has been placed in purpose group 5 (assembly and recreation) as defined in Table D1 of Approved Document 'B' ('Fire safety') and will contain laboratories, fume cupboards and preparation areas containing chemicals, research and teaching equipment. You have submitted to the Department a list of chemicals to be used and stored in the building.

All rooms – with the exception of toilets and two basement rooms – have windows, and emergency lighting has been installed in corridors, lobbies, stairs and other rooms designated in the fire risk assessment submitted by the applicant. There are no proposals to provide emergency lighting in other rooms such as laboratories that are associated with the assembly purpose group.

The building contains a fully addressable fire alarm system consisting of manual call points and smoke detectors installed to BS 5839: Part 1.

Your full plans application for this building was rejected on grounds, among others, of non-compliance with six matters in respect of Part B of Schedule 1 to the Building Regulations 1991. One of the requirements cited under Part B was Requirement B1 ('Means of escape') on the grounds that the provision for emergency lighting in the premises were inadequate. It is in respect of this that you have applied for a determination. You have stated that the other issues relating to the remainder of Part B and all other Parts cited in the rejection will be resolved. However, in the view of the City Council to achieve compliance with Requirement B1 you should install emergency lighting in all laboratories, lecture theatres and seminar rooms used by students.

You contend that your proposals for emergency lighting do comply with Requirement B1. You state that in requiring this emergency lighting to be installed, the Council are relying on paragraph 5.33 and Table 9 of Approved Document 'B', which sets out the location of such lighting. It is because the building is stated by the Council to fall within purpose group 5 (assembly and recreation) that the requirement for emergency lighting is made.

You accept that the building falls within purpose group 5, but in his view paragraph 5.33 and Table 9 of Approved Document 'B' are not applicable. You take this view because paragraph 3.32 of the Approved Document states that the design of the means of escape in schools and other education buildings in purpose group 5, should be in accordance with Building Bulletin 7 (Fire Design in Educational Buildings) issued by the Department of Education and Science.

You refer, in particular, to paragraph 56 of Building Bulletin 7 which suggests that in the case of halls and large spaces without natural lighting, or which are likely to be used during the hours of darkness, emergency lighting should be considered. Paragraph 66 of the bulletin emphasizes that consideration should be given to the provision of emergency lighting to escape routes when the buildings may be used during the hours of darkness.

You consider that you have fully complied with the recommendations given in BS 5266 in that you have provided illumination to escape routes and ensured that firefighting equipment and fire alarm call points can be readily located. In addition you consider that you have complied with paragraph 9.3.5 of BS 5266 with regard to recommendations for emergency lighting in non-residential premises used for teaching, training and research.

You have reported that a lighting test was carried out at night in areas not provided with emergency lighting and that it was found that light spillage through vision panels and street lighting would enable persons to negotiate obstructions and make a safe passage to an exit door.

You therefore consider that the Council are incorrect in stating that you have failed to comply with the Building Regulations in relation to the provision of emergency lighting.

In support of their decision to reject your proposals on grounds of non-compliance with Requirement B1, the City Council make reference to Table 9 of Approved Document 'B' which gives recommendations for the provision of escape lighting. Table 9 indicates that for assembly buildings, such as the one in this case, escape lighting should be provided in all escape routes and accommodation. In view of the use of the building the Council requested that, in addition to the corridors and staircases, emergency lighting should also be provided in all other areas apart from small offices occupied by staff only.

The Council point out that the building will be used during the hours of darkness as well as in daytime. Because of its height, layout and relationship to minor roads, very little street light illumination would pass into the building to aid people in an emergency. In view of this the Council do not feel it is justified to provide a reduced standard of emergency lighting.

With regard to the Department of Education and Science's Building Bulletin 7, the Council do not consider this to be an appropriate document to use since the building is not an existing building approved by that Department.

The Council consulted the local Fire Authority who concluded that a compromise could be evolved which would be appropriate to the use of the premises. The Fire Authority have only recommended the installation of emergency lighting in the larger seminar rooms and laboratories.

The Department's analysis of this case is as follows. You have made reference to BS 5266 and, in particular, to paragraph 9.3.5 of that standard. However, this is not conclusive with regard to the siting of emergency lighting, it simply states the type of system that should be used. For this type of non-residential premises it recommends a non-maintained, one hour duration system.

Reference has also been made to Building Bulletin 7 published by the Department of Education and Science. You have pointed to paragraph 56 of this document which refers to emergency lighting only needing to be considered in halls and large spaces without natural lighting, or when they are intended for use during the hours of darkness. However, in this instance the Department of the Environment does not consider the issue as to whether or not Building Bulletin 7 is the appropriate document to use to be material in this case. The issue for consideration is the safety of the occupants in an emergency situation and whether the proposals satisfy the functional requirement for means of escape.

The Department considers that the Council are correct in placing the building in the assembly purpose group as defined in Table D1 of Approved Document 'B'. However, the Department notes that from the occupancy figures provided by you that none of the rooms, with the exception of the lecture theatre, will hold more than ten persons. It can be argued therefore that from a means of escape viewpoint the situation equates more to offices than to areas of assembly. It is also expected that the room occupants are likely to be familiar with the layout.

The Department has therefore taken account of the low occupancy and usage of the rooms, and the recommendations for escape lighting recommended in paragraph 5.33 of Approved Document 'B' (as defined in BS 5266: Part 1: 1988), and concluded that your proposals for escape lighting comply with Requirement B1 with the exception of:

Laboratory room 216
Laboratory room 115
Lecture room G14.

The Secretary of State has given careful consideration to this particular case and in the circumstances has concluded that your proposals for escape lighting do not comply with Requirement B1. However, it is his view that if escape lighting were to be provided additionally in the above three rooms then compliance with Requirement B1 would be achieved.

Means of escape – alterations to two terraced properties (B.1)

The buildings are Grade II listed and were originally constructed as three houses. This determination revealed that wherever possible, flexibility in achieving compliance with building regulations should be applied to listed buildings.

The high level of fire resistance of the floors and the fire detection system was accepted as compliance in respect of the four storey building, but not for the five storey block where the risk to life was greater.

Ref.: 45.1.130   Date: 21 August 1996

In making the following determination the Secretary of State has not considered whether the plans conform to any other relevant requirements.

The proposal

The proposed work comprises the alteration of two terrace properties to form separate single dwellinghouses by subdivision along the original party wall lines. The buildings are Grade II listed and were originally constructed as a terrace of three houses (numbers 74, 75 and 76) about 100 years ago. The complete terrace was converted to a hotel between the wars but this fell into disuse in the late 1980s and has remained vacant.

No. 76 has been converted to flats and does not form part of this determination. The rear of No. 74 is attached to a two storey cottage (74A) which is fire separated from No. 74 and is also not at issue.

No. 74 has five storeys above ground level. Your drawing number 911/03 shows proposals to alter No. 74 to provide the following accommodation:

Ground floor – kitchen, dining room;
1st floor – study, living room;
2nd floor – sitting room, WC, bedroom;
3rd floor – 2 bedrooms, bathroom;
4th floor – guest bedroom/sitting room, bathroom, kitchen.

No. 75 has four storeys above ground level. Your drawing number 911/02 shows proposals to alter No. 75 to provide the following accommodation:

Ground floor – utility room, kitchen, dining room;
1st floor – living room, study, bathroom;
2nd floor – 2 bedrooms, bathroom;
3rd floor – 2 bedrooms, bathroom.

A separate sketch indicates that there is also a basement with three storage areas to No. 75 and is shown to be fire separated from the rest of the house.

The proposals also show that No. 74 and No. 75 will each have a single protected stair. There are no proposals to provide an alternative means of escape route from the upper storeys of either house.

You submitted two separate full plans applications for the above alteration work which was rejected by the Borough Council on the grounds that they did not show compliance with Requirement B1 as both applications showed floors exceeding 4.5 m above ground level without satisfactory additional provision for means of escape. However, you contend that within the constraints of a listed building you have provided adequate compensatory measures and that your proposals do comply with Requirement B1 in respect of both proposed dwellings at No. 74 and No. 75. It is in respect of the issue of compliance in both properties that you applied for a determination.

The Applicant's case

You accept that the guidance given in paragraphs 1.20 and 1.21 of Approved Document 'B' ('Fire safety') suggests that an alternative escape route should be provided from the upper floors of dwellings with more than one floor over 4.5 m above ground level.

You point out, however, that these buildings were originally constructed as dwellings and served that purpose for many years; and that the Grade II listing presents severe restrictions on additions such as the provision of external fire escape stairs. You also point out that the provision of external fire escape stairs would not be possible because of restricted space.

In support of your contention that your proposals comply with Requirement B1 of the Building Regulations 1991 you refer to the following compensatory features incorporated in your proposals:

(a) The provision of an extensive smoke detection/alarm system throughout the building installed to BS 5839.
(b) The upgrading of floors to give a full 60 minutes' fire resistance.
(c) Full protection to the stairway.

You have also made reference to approvals given to other similar buildings, and in particular No. 76 which was converted to flats.

The Borough Council's case

The Borough Council considers that the recommendations of sub-paragraph 1.20(a) of Approved Document 'B' with regard to the provision of a protected stair have been met but sub-paragraph 1.20(b) has not. They point out that sub-paragraph 1.20(b) suggests that the

top storeys should be fire separated from the lower storeys and be provided with an alternative escape route.

In addition, the Borough Council point out that in the case of No. 74, paragraph 1.21 of Approved Document 'B' has not been satisfied. Paragraph 1.21 refers to clause 4.4 of BS 5588: Part 1: 1990 ('Code of practice for residential buildings') which in turn suggests that alternative escape routes should be provided from each storey situated 7.5 m or more above ground or access level.

The Borough Council state that the compensatory fire safety package put forward by you was carefully considered but was not accepted for the following reasons:

(a) There would be no continuing control over the effectiveness of the electrical warning system and it would not be subject to re-inspection. Also self-closing devices could be removed and, without continuing control, the effectiveness of the only means of escape could be compromised.
(b) The provision of an enclosed internal protected stairway is a requirement of the Regulations for this height of building and was not therefore considered to be an additional compensatory feature.
(c) Sixty minutes' fire separation is a requirement of the Regulations and was therefore also not considered to be an additional compensatory feature.

The Borough Council point to the large number of deaths that occur through fire in domestic houses and they suggest that the proposals carry considerable risk due to the heights involved and the single stair situation.

The Department's view

In addition to paragraph 1.21 of Approved Document 'B', the Borough Council have also made reference to sub-paragraphs 1.20(a) and 1.20(b) of the Approved Document. These sub-paragraphs give guidance for houses with one storey having a floor more than 4.5 m above ground level, i.e. a typical three storey house. The council state that compliance with sub-paragraph 1.20(a) has been achieved but compliance with sub-paragraph 1.20(b) has not. However, the introduction to paragraph 1.20 makes it clear that sub-paragraphs (a) and (b) are alternatives and need not both be complied with. In any event as both properties, No. 74 and No. 75 have *more than one floor over 4.5 m* above ground level (No. 74 has 5 storeys above ground level and house No. 75 has 4 storeys above ground level), the Department considers that paragraph 1.21 of the Approved Document provides the more appropriate guidance for both properties.

Paragraph 1.21 suggests that in such cases reference should be made to clause 4.4 of BS 5588: Part 1: 1990, which in turn suggests that alternative escape routes should be provided from each storey situated 7.5 m or more above ground or access level. To be in accord with this guidance No. 74 would therefore need to have alternative escape provided from the top two storeys and No. 75 would need to have alternative escape provided from the top storey.

The Department notes that both properties are Grade II listed and takes the view that wherever possible account should be taken of the constraints that this listing may impose on the alteration and conversion of such buildings. However, the Department must also give full consideration to the importance for the occupants of such buildings being able to escape, or be rescued safely, in a fire situation.

No. 75

On the basis of the following assumptions the Department takes the view that in order to achieve compliance with Requirement B1, a single escape stair would be acceptable for No. 75 (i.e. the four storey property) excluding the basement storey. The assumptions are that:

(a) in the absence of any specific information in your proposals, the basement storey is fire separated from the above ground storeys to a standard that is acceptable to the Council;

(b) in the absence of information about the specific type of installation you proposed a smoke detection and alarm system is provided throughout the building and installed in accordance with BS 5839: Part 1: 1988 ('Fire Detection and alarm systems for buildings') to an L1 standard;

(c) all floors, as proposed by you, are upgraded to achieve 60 minutes' fire resistance;

(d) in the absence of specific information in your proposals, the stairway forms a protected enclosure which achieves a minimum of 30 minutes' fire resistance with all doors opening onto the stairway being 30 minutes' fire resisting and self-closing.

The above view takes account of the concern that the Council had for the lack of on-going control.

No. 74

With regard to No. 74 which has 5 above ground storeys, the Department takes the view that the top floor (i.e. the fourth floor) should be provided with an alternative means of escape in order to comply with Requirement B1.

The determination

The Secretary of State is particularly concerned that wherever possible flexibility in achieving compliance with the Building Regulations should be applied to listed and other buildings of historic or architectural interest. He has given careful consideration to the facts of this case. In particular, he has taken account on the one hand of the constraints which listed building status may impose on achievement of compliance with the Building Regulations, and on the other of the fact that Requirement B1 is a life safety matter and of the large number of fires which occur in dwellings.

Taking all these factors into account the Secretary of State has concluded as follows. He determines that subject to the assumptions listed in sub-paragraphs 21(a)–(d) above, your proposals for No. 75, which incorporate a single protected stair, show compliance with Requirement B1 of Schedule 1 to the Building Regulations 1991 (as amended); and that your proposals for No. 74 do not show compliance with Requirement B1 of Schedule 1 to the Building Regulations 1991 (as amended).

Six storey office building – means of escape (B.1)

Sub-letting of a portion of the building will involve tenants sharing corridors with the main occupier remaining responsible for the building management. The Local Authority wants separate escape corridors.

The Secretary of State suggests that a legal agreement may provide an acceptable solution.

Ref.: 45.1.122   Date: 15 October 1996

The Secretary of State takes the view that he is being asked to determine under section 16(10)(a) whether plans of your proposed work are in conformity with Requirements of B1 ('Means of Escape') of Schedule 1 to the Building Regulations 1991 (as amended). In making his determination the Secretary of State has not considered whether the plans conform to any other relevant requirements.

The proposal

Your request for a determination concerns the need or otherwise, consequent upon sub-letting, for additional building work to form shared escape corridors on each floor of a headquarters building of a large publishing corporation. The building is rectangular in plan and consists of office accommodation on six floors in the west section of the building and on five floors, with a roof terrace above, in the east section. Three atria interconnect all floors.

The building is divided vertically into three fire compartments, each compartment being separated by two hours' fire resisting construction and each containing one of the atria. It was originally designed for single occupation but because of company restructuring,

one third of the building is to be sub-let. However, as the major occupier, the publishing company will be responsible for managing the building as a whole, including the fire safety strategy. The application stresses that the compartmentation is for fire separation only and is not necessarily intended as a demarcation of occupancies.

There are four firefighting shafts, located at the two ends of the two compartment walls, which are designed to serve the building as a whole. The building is designed for single stage evacuation. A smoke detection and alarm system to an L1 standard has been installed, as a single system throughout the building, to BS 5839: Part 1: 1988 ('Fire detection and alarm systems for buildings'). The three atria are used for smoke extraction in the event of fire. The building is considered as a complete unit for fire safety matters and not one of sub-tenanted areas.

In 1994 it was proposed to sub-let the southern third of the building. With the approval of the District Council some revisions were made to the building to accommodate this. However, the Council also required that new fire rated corridors be constructed at two positions adjacent to the escape stairs on each floor level (11 in total) to create a shared corridor route to the south east and south west escape stairs from the office areas in the centre third of the building and the sub-let accommodation in the southern third. You contested the need for such corridors, proposing instead to create shared circulation spaces leading to the stairs.

You then applied to the District Council for a relaxation of Requirement B1 which they refused. After subsequent correspondence with this Department you applied for a determination that your proposal to create a circulation space, rather than construct new shared corridors, would secure compliance with Requirement B1.

The Applicant's representations

You are basically in disagreement with the District Council over the interpretation of paragraph 3.13 of Approved Document 'B' ('Fire safety'). Paragraph 3.13 states:

Where any storey is divided into separate occupancies (i.e. where there are separate ownerships or tenancies of different organisations):

(a)   the means of escape from each occupancy should not pass through any other occupancy; and

(b)   if the means of escape include a common corridor or circulation space then either it should be a protected corridor or a suitable automatic fire detection and alarm system should be installed throughout the storey. See also paragraph 8.10 on compartmentation.

You do not agree with the Council's requirement that new corridors need to be constructed at two positions on each floor, thereby creating a corridor route shared by the main occupant and sub-tenant leading to the south east and south west escape stairs. You accept the wording of paragraph 3.13(a) of the Approved Document. However, your interpretation of paragraph 3.13(b) is that since the building has a single fire detection and alarm system giving full cover throughout the building, then the route to the escape stair can be a shared circulation space.

You also consider that because the main occupant is to be responsible for the management of the building as a whole, including fire safety, then it will be easier for them to police the common area to make sure it is kept clear.

You believe that paragraph 3.13 of Approved Document 'B' is open to interpretation and that the approach you seek recognizes the practicality of buildings in use. You therefore propose to create a shared circulation space on each floor leading to the protected stairs which will allow more visibility and will assist management.

The Council's representations

The Council have refused to accept your alternative proposal of creating shared circulation spaces once the southern third of the building is sub-let. In support of their decision they state that it will be necessary for occupants of the central section of the building to pass through the wall separating the tenancies in order to reach the protected escape stairs in the southern third of the building. If access was not available to the two protected stairs in the southern third the travel distance and width of the remaining stairs would not satisfy the guidance given in Approved Document 'B'.

The Council point out that paragraph 3.13(a) of Approved Document 'B' suggests that means of escape from each occupancy should not pass through any other occupancy. They also make reference to BS 5588: Part 3: 1983 ('Fire precautions in the design, construction and use of buildings') which gives similar guidance.

The Council accept that an automatic fire detection and alarm system has been installed throughout the building as originally designed. However, in their view, although this system means that the corridors need not be protected ones, this does not obviate the need to provide corridors to separate the occupancies.

Finally, the Council report that the local Fire Authority support their interpretation of Requirement B1 of the Building Regulations in this case.

The Department's view

The question for determination has been put to the Secretary of State in terms of a difference of interpretation of Approved Document 'B'. However, it is important to recognize that the approved documents provide only general guidance on how to comply with the statutory functional requirements of the Building Regulations – in this case Requirement B1. There is no obligation to adopt any particular solution in the approved documents and they do not therefore preclude other solutions being adopted provided they will achieve compliance.

The building in this case is divided vertically into three fire compartments and you have made it clear that the fire rated walls are not necessarily lines of tenancy demarcation. From this statement your question for determination might be construed as one of what constitutes an acceptable boundary between tenancies in an open office space. However, because sub-letting per se does not constitute a material alteration as defined in Regulation 3 to the Building Regulations, there would be no 'building work' to consider in terms of its compliance with the requirements contained in Schedule 1 of the regulations.

In order to reach a conclusion in this case the Department has therefore taken the view that the building was designed for the purpose of sub-letting the southern third, and that the demarcation of tenancies is coincidental with the compartment wall at the southern end of the building.

The District Council have expressed concern that access through the sub-let areas to the protected escape stairs on each floor could become obstructed and unavailable for escape. This is the scenario that gave rise to the guidance given in paragraph 3.13 of Approved Document 'B' where it is suggested that escape should not be through another occupancy. However, what needs to be looked at is the risk in any particular situation.

The Department notes that the sub-letting arrangements are for a number of floors and could possibly involve more than one sub-tenant. If one of the escape routes through the sub-tenancy were to be obstructed then persons needing to escape would have to move to the alternative exit on the floor. This could prejudice safe escape, particularly from the fire floor, and it is also possible that the risk could be increased because of smoke issuing from the atria.

It is therefore essential that if the proposals are to be considered acceptable then there must be a clear obligation on the part of the sub-tenant/tenants to keep the escape routes on their side of the tenancy clear. The Department accepts the Council's view that the provision of a common fire alarm system will not in itself keep escape routes clear. However, it is the Department's view that to not erect separate corridors leading to the escape stairs would be acceptable if

such a proposal was accompanied by a legal agreement making it an obligation on the sub-tenant to maintain clear escape routes. It would have to be a requirement of each sub-tenancy that such an agreement would have to be entered into.

The determination

The Secretary of State has given careful consideration to the particular circumstances of this case. He has concluded that your proposal to create shared circulation spaces but not to erect shared corridors leading to each of the relevant escape stairs on each floor, would only be able to show compliance if accompanied by a statement that it shall be a condition of each sub-tenancy that a legally binding agreement shall be entered into making it an obligation on the sub-tenant(s) to maintain clear escape routes. No such statement accompanied your application for determination. The Secretary of State therefore determines that your proposal does not comply with Requirement B1 of Schedule 1 to the Building Regulations 1991 (as amended).

Block of flats – conversion of roofspace (B.1)

The proposal is to convert the roofspace to form additional flats but the existing building constraints do not permit the making of fire lobbies. Improvements proposed will benefit occupiers of the existing flats, but do not provide adequate safety for the occupants of new flats on the fourth storey.

Ref.: 45.1.119   Date: 15 October 1996

In making the following determination the Secretary of State has not considered whether the plans conform to any other relevant requirements.

The proposed work

The proposed work comprises the conversion of the roof space of four separated blocks within a horseshoe shaped complex of flats, to construct a new self-contained flat on the new fourth storey of each block. The proposal is therefore to construct a total of four new flats which are indicated on the plans submitted with the application as flats A, B, C and D.

The proposed new accommodation is shown on the drawings to be as follows:

Flat A – bedroom, kitchen, bathroom and lounge;
Flat B – bedroom, bedroom 2/study, kitchen, bathroom, lounge and dining area;
Flat C – bedroom, bedroom 2/study, kitchen, bathroom, lounge and dining area; and
Flat D – bedroom, kitchen, bathroom and lounge.

Also included in the proposals is the upgrading of the existing flat entrance doors to doors of an FD30S standard and the provision of an integrated fire alarm system in the halls of all flats, both new and existing, together with panic buttons on communal landings.

Each new flat is served by a single stair which also serves up to two existing flats on each of the ground, first and second floors of each block. The new stairways are continuous with the existing stair and will be constructed from the existing top floor landing.

The Borough Council took the view that your proposals comprised a material alteration under Regulation 3. They rejected your full plans application on grounds of insufficient information and non-compliance of your proposals with Requirement B1 insofar as there was no provision for lobby protection to each of the single stairs. The Borough Council contend that the minimum standard of fire protection to achieve compliance with Requirement B1 cannot be provided. However, you contend that your proposed up-grading of entrance doors and alarm system would achieve compliance and would enhance fire safety for the existing flats. It is in respect of this question that you have sought a determination.

The Applicant's case

You accept that paragraph 2.18(I) of Approved Document 'B' ('Fire safety') suggests that each dwelling should be separated from the common stair by a protected lobby or common corridor. You point out that the size and layout of the existing stairways preclude the construction of a lobby outside each flat and practical problems prevent the existing corridors being upgraded to meet the guidance.

However, you take the view that adequate provisions with regard to Requirement B1 of the Building Regulations 1991 (as amended) will have been made and make the following points in support of this:

(a) the general construction of the flats complex is masonry walls and concrete floors;
(b) all existing flat entrance doors will be upgraded to an FD30S standard (those currently in place have little or no fire resistance);
(c) a linked fire alarm system will be installed to offset the lack of lobby protection and detectors will be provided in the halls of all flats, with panic buttons also being provided on communal landings; and
(d) at present there is no mechanism by which an occupier of an existing flat can raise any alarm, which could result in the outbreak of fire on any of the lower levels going unnoticed until it is too late.

You also accept that the existing building constraints will not permit the construction of fire lobbies either within the communal landings or within each existing flat. However, in your opinion the proposals will improve the current situation by the installation of new fire doors together with a mains operated early warning system to the relevant British Standard.

The Borough Council's case

The Borough Council consider paragraph 2.19 of Approved Document 'B' to be appropriate to this case. It permits the use of a single stair when protected in accordance with diagram 12 of the Approved Document. However, in this case, the existing stairway does not conform with the recommendations given in diagram 12 with regard to lobby protection. The Council state that diagram 12(b) of Approved Document 'B', which is appropriate to this case, suggests that a fire lobby should be provided between the flat entrance doors and the common stairway. The Council point out that it is also suggested in the notes to diagram 12(b) that if the dwellings have protected entrance halls then the lobby between the common stair and dwelling entrance is not essential.

The Council are of the opinion that the minimum standard of fire protection suggested in paragraph 2.19 of Approved Document 'B' cannot be provided. They point out that no fire lobby protection has been or can be provided within the confines of the existing structure and, alternatively, it is not possible to form protected entrance halls within each of the existing flats.

In conclusion, the Council take the view that the proposals to provide additional flats in the roof space, forming a new fourth storey, fail to provide the minimum protection required by the Building Regulations for the occupants of the building. They consider that any lessening of the standards suggested in Approved Document 'B' may well endanger life if fire should occur in the premises.

The Council state that they have consulted with the Fire Authority who support the Council's view that the proposals do not meet the necessary criteria.

The Department's view

In this case you are proposing the provision of a fire/smoke alarm system, linking all flats, as a compensatory feature to offset the recommendation given in diagram 12(b) of Approved Document 'B' for lobby protection to the common stair.

The upgrading of flat entrance doors is also being put forward by

you as a feature. However, Table B1 of Approved Document 'B' gives recommendations that such doors should be fire doors in any event. The Department therefore considers that these proposals cannot be regarded as compensatory features.

The Department accepts your point that the proposals will benefit the residents of the existing flats. However, what needs to be considered is the safety of the occupants in the new flats on the new fourth storey.

The main difference between escape from dwelling houses and from flats is that in flats people tend to rely on the safety that the compartmentation provides and should not need to attempt to escape unless the fire is in their flat. However, a situation could arise when they may still need to make their escape, or to be rescued. The risk to such a safe escape or rescue becomes greater from the higher storeys in a building. In this case the Department takes the view that Requirement B1 means that the occupants of the new fourth storey must be able to effect a safe escape, or to be rescued, should the need arise. However, this could only be achieved if the common stairway does not become smoke-logged, and in the Department's view the provision of a linked smoke detection and alarm system would not prevent this occurring.

The determination

The Secretary of State has given careful consideration to the particular circumstances of this case. He has concluded that the Borough Council are correct in their judgement that lobby protection to the common stair should be provided in accordance with the recommendations given in Approved Document 'B'. The Secretary of State therefore determines that your proposals are not in compliance with Requirement B1 of Schedule 1 to the Building Regulations 1991 (as amended).

Conversion of thatched house into two dwellings (B.4)

The work involves division of the existing house and the erection of a two storey addition at the rear.

The dispute hinges upon the fear of external fire spread because of the nearness to a new boundary, but relying upon precautions to be taken the proposal was found to comply with the Building Regulations.

Ref.: 45.1.100   Date: 8 December 1996

Your letter requested the Secretary of State to determine under section 16(10)(a) whether plans of the proposed work were in conformity with requirement B4 ('External fire spread') of Schedule 1 to the Building Regulations 1991. In making his determination he has not considered whether the plans conform to any other relevant requirements.

Your proposals show work to divide into two separate dwellings an existing thatched two storey dwellinghouse. The dwelling comprises a Grade II listed building of architectural and historic interest known as 'The Thatched Cottage' plus a twentieth-century single storey extension constructed of rendered blockwork and a roof of asbestos cement roof slates known as 'Pastoral'. Your proposals include two new extensions. One is to be a two storey extension at the rear of 'The Thatched Cottage' on the north elevation. It will contain a ground floor kitchen and a first floor bathroom, and is to have a thatched roof consisting of 300 mm thatch treated with a fire retardant combined with an underlining of thatch batts and a barrier foil. Planning considerations have indicated that the location of the extension at the rear of the building is the only practical one.

The second extension for which planning permission has been granted is to be 'Pastoral'. It will also be on the north side and will be single storey and constructed of similar material to 'Pastoral'. This means that the roof will also be constructed of asbestos cement slates.

As a result of the proposed division of the present dwelling a new boundary will be formed comprising a small length of wall at the north west corner of 'Pastoral' plus a newly established garden boundary.

A third dwelling in separate ownership exists to the south of 'Pastoral' but this building has no impact on the proposals.

Your proposals were rejected by the District Council on grounds which included non-compliance with Requirements B3 ('Internal fire spread (Structure)') and B4 ('External fire spread'). You submitted a further full plans application to the District Council which was again rejected on grounds, inter alia, of non-compliance with B4 on 20 December 1994. The District Council's concern was with the proximity of the thatched roof extension to 'The Thatched Cottage' to the new boundary formed by the proposed division of the dwelling. It was in respect of this second rejection on grounds of non-compliance with B3 that you applied to the Secretary of State for a determination.

The District Council have suggested to your client that the proposals might have been considered more favourably if the extension had been sited further away from the boundary. They consider that any structural problems relating to this could have been resolved.

The District Council have also expressed the following points of concern:

(i) in their view your statement that the boundary line is 3.7 metres from the extreme edge of the thatch is incorrect since the drawings show that the extension to 'The Thatched Cottage' is only 0.5 metres from the adjoining property ('Pastoral'). This distance is maintained for 1.5 metres after which it is sited 3.7 metres from the boundary;

(ii) the conflicting opinions given by experts on the question of the 'AA' rating given to the treated thatch and the long-term affect of the treatment on the thatch itself. It has also been suggested by thatch experts that a full fire test should be carried out on the thatch assembly to British Standard BS 476: Part 3: 1958 ('External fire exposure roof tests').

In view of these different opinions relating to the use of fire retardants and the close proximity of the adjacent property to the proposed thatched roof extension, the District Council does not consider it is able to approve the scheme under Requirement B4.

You state that the relocation of the proposed extension had been considered but this would require the removal of the two buttresses. The buttress is a large, heavy mass of masonry built to counteract movement in the wall of the cottage. It was concluded that if this means of support were removed the wall would start moving again unless extensive remedial work were to be undertaken to prevent it. This could involve the demolition and reconstruction of at least part of the wall making the construction of the extension in this location uneconomically expensive.

You point out that at present, with the two elements of the building in use as a single dwelling, the extension can be built as designed since there is no relevant boundary required between 'The Thatched Cottage' and 'Pastoral'. The point at issue has arisen as a result of the proposal to separate 'The Thatched Cottage' and 'Pastoral' into two separate dwellings.

You state that tests of materials used in the thatch roof assembly showed the following results:

(i) a test on 150 mm thick thatch impregnated with the fire retardant treatment achieved a 'CA' designation;

(ii) a test on untreated thatch protected with aluminium barrier foil and two layers of 'thatch batts' achieved an 'A' designation for the resistance to fire penetration;

(iii) a laboratory report which has been provided, showed that on a 10-year-old sample of thatch there was an average leaching of the fire retardant out of the end 3.5 centimetres of the reed. This was considered to be small when compared to the length of reed used which is between 1.4 metres and 2 metres long.

An appraisal by the Loss Prevention Council of these tests concludes that on the evidence of the tests conducted, the proposed roof construction would achieve an 'AA' designation.

You state that in this case the boundary line is 3.7 metres from the extreme edge of the thatch and because of the 'AA' designation accords with the guidance given in Approved Document 'B'.

In support of your case you point to previous appeal/determination cases relating to the distance of thatch to the boundary which have been upheld by the Secretary of State.

The Department's analysis of the case is as follows. What has to be considered is the level of risk associated with the extension to 'The Thatched Cottage' using a thatched roof construction in respect of its location to the new wall and garden.

The Department accepts that if the two dwellings, i.e. 'The Thatched Cottage' and 'Pastoral', had remained as a single dwelling house then the issue relating to a possible conflict with Requirement B4 would not have arisen.

Table 17 ('Limitations on roof coverings') of Approved Document 'B' ('Fire safety') recommends that thatch should be sited at least 12 metres from the relevant boundary if performance under British Standard 476: Part 3: 1958, cannot be established.

You have stated that the distance of the new thatch to the proposed garden boundary will be 3.7 metres, although the District Council state that for a distance of 1.5 metres along the wall boundary the new thatch is only 0.5 metres from 'Pastoral'. From the drawings submitted to the Department this would appear to be correct.

It also appears to the Department that there are likely to be planning constraints on any proposed development to either property. Planning considerations have resulted in the extension to 'Pastoral' having a roof constructed with asbestos cement slates. Therefore the risk of fire spread from the thatched extension to the extension to 'Pastoral' is low.

The Department accepts, however, that the risk of fire spread is greatest at the point where the new thatched roof extension is only 0.5 metres from 'Pastoral'. However, this risk is only applicable to 1.5 metres and in essence is no worse than if the two dwellings were looked at as a semi-detached property. The note to paragraph 14.4 ('Separation distances') of Approved Document 'B' suggests that the boundary formed by the wall separating a pair of semi-detached houses may be disregarded for the purpose of considering recommendations in the Approved Document for roof coverings in relation to Requirement B4.

The Department also acknowledges that the proposed fire precautions associated with the new thatch, which include a fire retardant, barrier foil, and the provision of thatch batts, will assist with a reduction in fire spread.

In these particular circumstances, therefore, the Secretary of State determines that the proposals for the erection of the extension to 'The Thatched Cottage' with the thatched roof covering as proposed, show compliance with Requirement B4 of the Building Regulations 1991.

Terrace building – change of use (B.1)

The problem is one of means of escape from a building to be changed from commercial to part residential use. The local authority wants a protected lobby between a ground floor shop and the common entrance hall. Space is at a premium, but compliance with B.1 may be achieved by providing a suitable sprinkler system.

Ref.: 45.1.123   Date: 15 October 1996

In making the following determination the Secretary of State has not considered whether the plans conform to any other relevant requirements.

The proposal

The proposed work concerns the change of use, from commercial to part residential, of an existing terrace building comprising three storeys plus a basement. It is proposed to convert the first and second floors from existing commercial use to form a maisonette, with the addition of a small new third floor.

The proposed accommodation shown on the drawings submitted with the application is as follows:

Basement – retail;
Ground floor – retail;
First floor – kitchen and living/dining area;
Second floor – bedroom 1 and bedroom 2; and
Third floor – bathroom and small cupboard.

The new maisonette is served by a single stair which terminates in a ground floor protected common entrance corridor/hall, which also has the shop entrance opening from it. The stair serving the maisonette is separated from this common hall by an FD30S fire door located at the bottom of the flight.

The shop and maisonette share the same front entrance door from the street which opens into the common entrance corridor/hall. Double leaf shop entrance doors also open directly into the shop from this corridor/hall and are designated as FD30S fire doors.

The distance between the common front entrance door and the fire door at the bottom of the stairs to the maisonette is shown on the drawing to be approximately 4 m.

The proposals show that an integrated fire alarm system incorporating smoke detectors will be installed at all levels of the building so the occupants of the maisonette will have early warning of a fire starting in the retail areas.

You submitted a full plans application for the above work which was rejected by the Borough Council on the grounds that it did not show the provision of a protected lobby between the ground floor shop and the common entrance hall, and therefore did not comply with Requirement B1. However, you consider that your proposals do show compliance with Requirement B1 and it is in respect of this question that you have sought a determination.

The Applicant's case

You make the following points in support of your application:

(a) the proposals involve the retention of the ground floor shop and conversion of the upper floors to a maisonette, which is the use preferred by the local planning department and the building's historic use. You point out that if the use of the upper floors had remained commercial then a lobby would not have been required;

(b) the shop premises are already extremely small, having a floor area of only 42 square metres, and the provision of a lobby would further reduce their lettability;

(c) the majority of adjoining shops in the street with different uses on the upper floors do not have lobbies provided on the ground floors;

(d) the building is for the first time to be provided with an integrated fire alarm system incorporating smoke detectors at all levels;

(e) the distance from the fire door separating the stairs to the street door is only 3 m (shown on the drawing as approximately 4 m);

(f) rescue would be available from the first floor balcony overlooking the street, and access is available onto the main roof via the bathroom window on the top floor and via this onto adjoining roofs;

(g) it would be possible to make it a condition of use that the front door would always be held open to the street when the ground floor premises were open for business. You consider that this would help with smoke dispersal; and

(h) as the building is to contain only one small two bedroom maisonette and one small shop, the normal occupancy of both is unlikely to exceed 4–5 people.

The Borough Council's case

The Borough Council have rejected the proposals on the grounds that they do not comply with Requirement B1 of the Building Regulations 1991. In reaching that decision they took account of paragraph 2.46 of Approved Document 'B' ('Fire safety'), which suggests

that stairs may serve both dwellings and non-residential occupancies providing that the stairs are separated from each occupancy by protected lobbies at all levels. In this case the Borough point out that the protected lobby to the shop entrance has not been provided.

The Borough Council discount your suggestion that alternative escape is available via the roofs of adjoining premises on the basis of the location and small size of the bathroom window. They also discount the use of the balcony for rescue purposes.

With regard to the lettable space remaining after the provision of a lobby, the Borough contend that this is not a building regulations issue.

In conclusion, the Borough Council welcome the installation of an automatic fire detection system but do not consider the installation of such a system to be adequate compensation for the omission of a protected lobby to the only escape route from the maisonette.

The Department's view

You have mentioned in your submission similar cases where lobby protection has not been provided to the common entrance hall. However, each case for determination must be considered on its merits and it would not be appropriate for the Department to take account of other situations or circumstances in other buildings not the subject of this determination.

In rejecting the proposals the Borough Council have taken account of the guidance given in paragraph 2.46 of Approved Document 'B' which applies to flats in mixed use buildings and suggests that lobby protection should be provided to stairs.

In this case, the upper floors are being converted to a maisonette and the ground floor entrance hall/corridor is common to both the maisonette and shop. The double leaf fire doors opening from this corridor/hall provide the only means of access to the shop.

You have suggested the following as compensatory features for the non-provision of lobby protection from the shop:

(i) the installation of an integrated fire alarm system incorporating smoke detectors at all levels of the building; and

(ii) the imposing of a condition on the occupiers that the front entrance door be left open when the shop is in use.

The Department is aware that the main difference between escape from dwelling houses and escape from flats/maisonettes is that in flats/maisonettes people tend to rely on the safety that the compartmentation provides and should not need to attempt to escape immediately unless the fire is in their dwelling. However, a situation could arise when they may still need to make their escape, or to be rescued. The risk to such a safe escape or rescue becomes greater from the higher storeys in a building. In this particular case the maisonette will occupy the three upper floors of the building.

The Department's view is that escape should be available via the designated protected route and not via a balcony that is shown on the drawing to be approximately 0.5 m deep. In addition, the Department does not consider it appropriate for the escape route to be by via a small third floor bathroom window on to adjoining roofs. In the circumstances, safe escape or rescue would therefore only be available provided the single escape route does not become smoke-logged. However, the Department does not consider that the provision of a linked smoke detection system would prevent smoke-logging of the single escape route.

With regard to imposing conditions about keeping the front entrance door open when the shop is in use, the Department does not consider it to be within its remit to do so, particularly in view of the potential problems of ensuring on-going control.

Finally, the Department notes that the Borough Council have been content to allow paragraph 2.46 of Approved Document 'B' to be applied to a mixed use development containing a maisonette. Additionally, although the maisonette does not have its own external entrance at ground level, the Borough have insisted on the provision of properly designed alternative escape routes from the upper floors of the maisonette, as suggested in paragraph 2.15 of Approved Document 'B'. In the Department's view the Borough have shown flexibility in the application of the guidance in Approved Document 'B'.

The determination

The Secretary of State has given careful consideration to the particular circumstances of this case. He considers that the Borough Council have shown flexibility in applying the guidance given in Approved Document 'B'. He also accepts the Borough Council's judgement that a lobby should be provided in the shop to protect the single escape route to the maisonette. Your proposals do not provide for such a lobby and in the circumstances the Secretary of State therefore determines that your proposals do not comply with requirement B1 of Schedule 1 to the Building Regulations 1991 (as amended).

You will wish to note that in analysing this case, the Department considered whether compliance with Requirement B1 could be achieved in any other way than by provision of a protected lobby. In the Department's view, given the small size of the shop, the provision of a basic sprinkler system instead of a protected lobby might secure compliance with Requirement B1. The adequacy of such a solution would be a matter for the Borough Council.

Three storey building – conversion to residential use (B.1)

The rooms above an ironmonger's shop are to be converted and there is no provision for lobby protection at the foot of the stairs. Resiting of the shop door would overcome the problem.

Ref.: 45.1.131   Date: 19 December 1996

The Secretary of State takes the view that he is being asked to determine under section 16(10)(a) whether plans of your proposed work are in conformity with Requirement B1 of Schedule 1 to the Building Regulations 1991 (as amended). In making his determination the Secretary of State has not considered whether the plans conform to any other relevant requirements.

The proposed building work

The proposed work concerns an existing three storey, terrace building that was constructed around 1890. The building fronts onto the village High Street which is a conservation area.

The first and second floors of the building were previously used for office accommodation. The proposal is to convert these two floors into a maisonette. Drawing H/HS/1 (Plan No. 2) shows the second floor designated as a bedroom and the first floor as a lounge/dining room or bed-sit. A bathroom and new kitchen is also located on the first floor.

The ground floor is as existing and is occupied by an ironmonger's lock-up shop on long lease. The stair serving the maisonette is located against a party wall and is protected with 30 minutes' fire resisting construction. It discharges to a small ground floor lobby which gives access to the High Street via an external door and an open porch. The shop entrance door opens directly into the lobby at the bottom of the stairs serving the maisonette – there is no provision for lobby protection.

Although there is a patio to the rear of the property this is in the ownership of your client's neighbours who also own land at the side of the premises and adjacent to the rear patio. Because of concerns over security these neighbours feel unable to grant your client permission to erect a fire escape or ladder.

The proposals therefore incorporate by way of compensatory measures a magnetic catch to the shop door linked to the fire alarm and smoke detection system, and a full alarm and smoke detection system on both upper floors. The District Council were not minded to accept these proposals on the grounds that they did not demonstrate compliance with Requirement B1 of the Building Regulations 1991.

The Applicant's case

You consider that the existing arrangement whereby access is gained to the ironmonger's shop from the lobby, at the bottom of the access stairs to the maisonette, is reasonable. You give the following reasons in support of this:

(i)   The door to the ironmonger's shop is already fire resisting and self-closing but the proposal is to provide a magnetic catch linked to the fire alarm and smoke detection system.

(ii) A full alarm and smoke detection system is offered on both upper floors to a standard that would be above the normal recommendations given in Approved Document 'B' ('Fire safety') for dwellings.

You state that an alternative solution would be to seal the ironmonger's door and reposition this in the angled side of the shop front which forms part of the open porch. The ironmonger rejects this proposal on the grounds of security, heat loss, and the narrow pavement. You state that planning permission would be required for this change and that you have discussed this with planning officers. Your view is that the old-fashioned architectural merit of the shop front would be lost by such an alteration.

You are therefore requesting the Department to determine that the alternative solution of a magnetic hold-open device on the fire door to the ironmonger's shop, and the fire alarm and smoke detection system, are appropriate.

The District Council's case

It is the District Council's view that the conversion of the first and second floor offices to a maisonette constitutes a change of use as defined in Regulations 5 and 6 of the Building Regulations 1991. In support of their judgement that the proposals do not demonstrate compliance with Requirement B1 the Council quote the guidance given in Approved Document 'B', and in particular paragraph 2.46 which states:

In buildings with not more than 3 storeys above the ground storey, stairs may serve both dwellings and non-residential occupancies, provided that the stairs are separated from each occupancy by protected lobbies at all levels.

Because lobby protection is not proposed between the shop and the stairway it is the District Council's opinion, after considering the compensatory proposals offered by you, that a satisfactory means of escape from the maisonette would not be achieved. They consider that a fire within the shop could render the sole means of escape from the maisonette unusable.

The Council state that the only viable option to achieve an acceptable standard of means of escape is for you to relocate the shop entrance door to the front elevation of the shop front (from which the Department assumes that elevation facing the pavement but excluding the angled side) and thus facilitate the provision of a completely separate fire protected staircase. They point out that your concern for heat loss and security could be addressed in the design of a new shop front and that the need to obtain planning permission is not a relevant building regulations consideration.

The Department's view

The point at issue in this case is the safe means of escape from the maisonette which occupies the first and second floors of the building. The Department accepts that the conversion of the two upper floors from office to residential accommodation constitutes a material change of use and that therefore Requirement B1 is a relevant requirement which must be complied with.

Paragraph 2.46 of Approved Document 'B' suggests that in buildings such as this where the stair serves both dwellings and non-residential accommodation then lobby protection needs to be provided to the stair. The Department notes that in this case the proposal is for only single door protection to be provided between the ironmonger's shop and the stairway serving the residential accommodation. Such a situation could mean that if a fire occurred in the shop then escape could be prejudiced from the residential accommodation because the complete stairway could become smoke-logged. The door in question is the only means of access into the shop and could therefore put residents at risk if it is does not function properly as a fire resisting door-set.

You have offered a magnetic hold-open device on the shop door which would be operated on the activation of the fire alarm. The Department does not consider this to be a satisfactory alternative solution to lobby protection since the occupants of the maisonette will have no control over the maintenance of such a system. In addition it is the Department's view that the level of early warning system proposed for the upper floors will not be of benefit if a fire occurs in the ground floor shop.

The District Council have suggested that the shop entrance door could be relocated to the front elevation of the shop. However, the Department notes from plan No. 2 that it appears that the shop entrance door could be relocated in the angled side section of the shop front, adjacent to the common entrance porch. The relocated shop door would then open onto this porch which has open access to the street. The Department would consider this arrangement acceptable because the shop entrance door, although still opening into the common entrance area, would effectively be in open air.

The Department notes your concern regarding the potential effect of relocating the front door on the conservation area of the High Street. However, against that must be weighed the fact that Requirement B1 is a 'life safety' matter and that, in any event, the option for achieving compliance by re-locating the front door either on the angled side or on the front elevation of the shop front appears not to have been the subject of a planning application and that you have not submitted any written evidence from the planning section of the District Council to support your view about the planning problems.

The determination

The Secretary of State has given careful consideration to the particular circumstances of this case. He considers that your current proposals which incorporate a magnetic catch on the shop entrance door and a full alarm and smoke system will not provide an acceptable standard of escape. Accordingly, the Secretary of State determines that your proposals do not comply with Requirement B1 of Schedule 1 to the Building Regulations 1991 (as amended).

Three storey dwelling – means of escape (B.1)

A new extension to a three storey dwelling caused the local authority to decide that when completed the extended building would be more unsatisfactory in escape provision than before the work was carried out. A smoke detector system helped secure approval for the proposal.

Ref.: 45.1.127   Date: 19 February 1997

In making the following determination the Secretary of State has not considered whether the plans conform to any other relevant requirements.

The proposal

The building to which this determination relates is a three storey detached house built in 1903. The ground floor comprises in part and at the rear an extension containing a kitchen and conservatory. The second floor contains two habitable rooms whose windows are in the gable ends which face to the front and rear of the property. There is little fire separation – all ceilings are lath and plaster, and all doors opening off the stairway are standard hardwood panel doors.

The proposal is to demolish the existing rear single storey extension and replace it with a new single storey extension which will be 1.15 m deeper than the existing. The ceiling over the new kitchen extension is to have 30 minutes' fire resistance. The Borough Council believe that at present the windows of the second floor

rooms could be used for escape. However, although the window in the front has unobstructed ladder access the rear windows can only be accessed with a ladder extending over the extension at what is understood to be a 55° pitch. The new extension would decrease the pitch of the ladder to 47°. The two second floor rooms are not interconnected and escape via the front window would involve crossing the landing at the top of the stairs.

Because of these potential implications for escape a mains operated automatic smoke detection system with battery back-up is proposed, with detectors being located in the hall/landing of all floors. Other than these proposals the property will remain as existing.

Your full plans application for this work was found to be defective by the Borough Council in respect of Requirement B1 ('Means of escape') and not in compliance with Regulation 4(2)(a) of the Building Regulations 1991. The Borough Council consider that the rear windows of the second floor room form an important means of emergency escape in case of fire. Because the new extension would decrease further the pitch of an escape ladder, they take the view that the building which does not comply at present, would be made more unsatisfactory in relation to Requirement B1 and would therefore fail to comply with Regulation 4(2)(a) of the Building Regulations 1991. However, you consider that your plans do show compliance and it is in respect of this question that you have sought a determination.

The Applicant's case

You consider that the Borough Council is being unreasonable in their application of the Building Regulations and make the following points in support of your application:

(i) The Borough Council appears to be applying the Building Regulations as if the proposal was for a loft conversion. However, the second floor rooms are existing and the windows were never designed for escape and any such use would be coincidental.

(ii) There is no protective refuge on the second floor where it would be safe to wait for rescue by ladder and this would make the provision of either a protected corridor between the two second floor rooms, or the provision of an escape rooflight, redundant. You also contend that it would be impractical to provide either in any event.

(iii) The provision of the smoke detection system will enhance the protection afforded to the whole building and mitigates against any perceived risk from the new extension.

(iv) Ladder rescue from the rear second floor room is still possible and is only slightly worse than that previously existing. The new extension will be provided with a 30 minutes' fire resisting ceiling to make ladder rescue safer.

The Borough Council's case

The Borough Council has rejected your proposals on the basis that they do not show compliance with Regulation 4(2)(a) of the Building Regulations 1991 which requires that building work shall be carried out so that after being completed any building which is extended complies with the relevant requirements of Schedule 1 or, where it did not comply with any such requirement, is no more unsatisfactory in relation to that requirement than before the work was carried out.

The Borough Council accepts that access to the rear second floor windows is difficult because of the existing extension but they are of the opinion that the new, deeper extension makes access worse. They therefore conclude that Regulation 4(2)(a) would be contravened because the building, which does not comply at present, would be made more unsatisfactory in relation to Requirement B1 ('Means of escape'). The Borough Council also make the following points in support of their judgement:

(i) The new extension would necessitate pitching the ladder at an even shallower angle than required with the existing extension. A ladder pitched at a shallow angle would make rescue difficult and it could break under the combined load of two persons.

(ii) It may not necessarily be the fire brigade that carries out the rescue.

(iii) The smoke detection system is not considered to be a substitute for the worsening access to the windows.

(iv) Both second floor rooms could be made interconnecting or, alternatively, an escape window could be provided in the roof slope.

The Department's view

The Department takes the view that the Secretary of State is being asked to determine whether your proposals would result in means of escape from the existing second floor of the dwelling being, at worst, no more unsatisfactory in terms of Requirement B1 of the Regulations, and therefore being in compliance with Regulation 4(2)(a).

In a situation where a *loft conversion* is carried out to an existing two storey house, fire separation is provided to enable persons to wait in a degree of safety for rescue via the window if necessary. However, in this case the second floor rooms are provided with little fire separation and it might therefore tend to develop a false sense of security to suggest that persons on that floor should wait in the room to be rescued. The same argument can be applied to whether it would be reasonable to expect the two rooms to be interlinked or whether a new escape window should be fitted in the roof slope. In the circumstances the Department does not consider that such suggestions would be reasonable.

The Department accepts that because of the proposed increase in the depth of the ground floor extension, rescue by ladder will be more difficult than before. However, given the lack of fire separation to the second floor, the Department considers that the new proposed smoke detection system will be more beneficial in that it will give early warning of fire. This will enable persons in the rear second floor room to move across the small landing to the front room which is provided with unobstructed window access, before the landing becomes smoked logged. The Department considers that this should result in a situation which, at worst, will be no more unsatisfactory than exists at present but which, in fact, is likely to be a safer arrangement than exists at present.

The determination

The Secretary of State has given careful consideration to the particular circumstances of this case. He accepts the Borough Council's judgement that rescue from the rear window will be made more difficult but considers that the proposed installation of a smoke detection system will at worst result in a situation where compliance with Requirement B1 will be no more unsatisfactory than previously. Accordingly, the Secretary of State determines that your proposals do not contravene Regulation 4(2)(a) of the Building Regulations.

Two storey dwelling – means of escape (B.1)

The dispute arose in connection with a proposal to create a sleeping gallery in a two storey extension. The council maintained that the galleried sleeping area constituted a floor more than 4.5 m above ground level. As such it did not comply with B.1 and the council attitude was sustained in this determination.

Ref.: 45.1.132   Date: 7 May 1997

In making the following determination the Secretary of State has not considered whether the plans conform to any other relevant requirements.

The alteration work

The building work relates to alterations to an existing detached 1930s house to form a two storey side extension and a single storey

front extension. The application for determination relates only to the two storey extension.

Drawing ML/256/1 (Plan No. 2) submitted with the application, shows that the first floor of the side extension will contain a master bedroom with en suite bathroom. Drawing ML/56/2 (Plan No. 1) shows that the first floor level of the extension is set at about 400 mm lower than the rest of the first floor.

To increase the usable size of the master bedroom the proposals show the provision of a sleeping gallery constructed into the roof space. The gallery is open to the bedroom along the side facing the window and access to the gallery will be via open stairs from within the bedroom which lead up from the window wall. The first floor area of the bedroom is 13.5 square metres and the floor area of the sleeping gallery is 8.7 square metres.

The gallery is set 3 m back from the front wall of the bedroom, with the underside of the roof being exposed to both the gallery and main bedroom areas. You state that doors to the stairway will be made self-closing and the drawing shows that smoke detectors are to be located: in the ground floor hall, on the first floor landing, and on the gallery floor.

Your original proposals were contained in a full plans application which was rejected by the Borough Council on several grounds, including non-compliance of the galleried sleeping area with Requirement B1. You then requested consideration by the Borough Council of a relaxation of Requirement B1 but subsequently decided to pursue a determination. This was based on your judgement that your original proposals, coupled with the additional compensatory proposals of making all doors off the hallway/escape route self-closing, did in fact comply with the basic functional requirements of B1. It is in respect of this question that you requested a determination.

The Applicant's case

In support of your case that the functional requirements of B1 can be met by your proposals, you make the following points:

(i) The property is a single family dwelling and the bedroom suite is for the personal accommodation of the owners.

(ii) The sleeping gallery area is of insufficient size to be used independently of the bedroom suite and is fully open to, and an integral part of, the suite. Any person on the sleeping gallery would be fully aware of what was happening below.

(iii) There is no obligation to adopt any particular solution given in approved documents. Account should therefore be taken of the fact that the hall/stairway doors are to be made self-closing and that smoke detectors are to be installed at all levels.

(iv) The bedroom has a suitable escape window and the bedroom door opening onto the stairway could be changed to a fire door.

The Borough Council's case

The Borough Council's rejection of the proposals with regard to the two storey extension was on the basis that the proposed galleried sleeping area constituted a floor which would be more than 4.5 m above ground level and should therefore comply with the recommendations given in paragraphs 1.20–1.22 of Approved Document 'B' ('Fire safety').

The Borough Council have made the following points in support of their rejection of the proposals:

(i) The proposals show that a galleried sleeping area is to be incorporated into part of the roof space. The floor of the gallery is in excess of 4.5 m above ground level and therefore a protected stairway should be provided to a final exit point.

(ii) The gallery is open along one side to the master bedroom below and the access to the gallery is via open stairs from within the bedroom.

(iii) The gallery is large enough to be used for sleeping accommodation, and means of escape would be via unprotected stairs to the bedroom below and then along the full length of the bedroom to the unprotected staircase.

(iv)   The provision of smoke detectors and self-closing doors is not a suitable alternative to a fully protected stairway from the second floor.

The Department's view

The issue is whether the means of escape from the sleeping gallery level above the first floor bedroom is in compliance with Requirement B1 of the Building Regulations 1991.

The intention behind the guidance given in Approved Document 'B' for loft conversions is that the means of escape should be via protected or partially protected stairs. Window escape is only considered to be a secondary route and is an alternative in loft conversions to the fully protected stair route that is normally required for new three storey houses.

In this case the proposals do not include either a protected or partially-protected escape route from the sleeping gallery. As a result the only route is across the full length of the first floor bedroom to the stairway.

The Department acknowledges that smoke detectors are being provided and that this is in excess of the recommendations given in Approved Document 'B' for loft conversions. However, for a detection system to be considered as suitable compensation for a protected route of travel, the assumption would have to be that persons on the sleeping gallery would be able to respond immediately the alarm was activated. The Department does not necessarily believe that this would always be the situation – for example if a very young or very elderly person were on the gallery.

In the Department's view, therefore, the addition of the gallery level is effectively the same as adding another storey and the Department does not consider that the proposals show adequate means of escape from this level.

The determination

The Secretary of State has given careful consideration to the particular circumstances of this case, and has also had particular regard to the fact that a large number of fires occur in residential accommodation. He has concluded that the Borough Council were correct in their judgement that your proposals do not show adequate means of escape from the gallery level. The Secretary of State therefore determines that your proposals are not in compliance with Requirement B1 of Schedule 1 to the Building Regulations 1991 (as amended).

Large retail building – means of escape (B.1)

The appellant wished to install sliding/folding security gates at the bottom of a stair. The council argued that the gates were not visible within the store and persons wishing to escape would find the route unexpectedly blocked. This was also the determination of the Secretary of State.

Ref.: 45.1.133   Date: 8 May 1997

In making the following determination the Secretary of State has not considered whether the plans conform to any other relevant requirements.

The proposed work

The building to which this determination relates is a large retail one comprising sales floor areas on the ground and first floors, with stockrooms on the second. Stair No. 16 is an escape one leading down from the first floor to exit G8 at ground level, as shown on drawing 1725/17. It is shown to have a 60 minutes' fire resisting enclosure and there is no access from this stair to the ground floor. The stair discharges at ground level to a new enclosed lobby which in turn exits onto an external service yard.

The proposal is to guard this exit to the service yard by the instal-

lation of sliding/folding security gates at the bottom of stair No. 16. Drawing 1725/17 also shows a security gate across exit G11 at ground level but the rejection notice issued by the City Council refers only to the security gates proposed at the bottom of stair No. 16 (exit G8).

These proposals were the subject of a full plans application which was rejected by the City Council on the grounds that they did not demonstrate compliance with Requirement B1. However, you contend that your client's proposals do comply and it is in respect of this question that you have sought a determination.

The Applicant's case

You state that the security of alcoves is a common problem in many of your client's retail stores and the usual solution is to install sliding security gates of the type being proposed in this case. You also make the following points in support of your judgement that your proposals comply with Requirement B1:

(i)   To maintain the means of escape, gates are unlocked and padlocked open as part of the opening up procedure and locked shut when a store closes. A strict management regime would be in place to ensure that this procedure is followed in this store.

(ii)   The Fire Officer has requested an audible alarm system which would operate when the gates were closed, but it would be difficult to establish responsibility for this.

(iii) The management procedure is a 'tried and tested' procedure that is open to verification at any time and does not rely on any equipment such as mechanical or electrical links.

The City Council's case

The City Council make the following points in support of their judgement to reject your proposals:

(i)   The shutter is not visible from within the store and persons making their escape would be unaware if it was shut, by which time they would have travelled a considerable distance.

(ii)   If management procedures are to be effective they should be augmented by the provision of an automatic alarm which will register in the store if the shutter is closed.

(iii) In addition to the safety of members of the public, concern is expressed over staff working late and cleaners/maintenance personnel who may also attempt to use this exit in an emergency.

(iv)   Approved Document 'B' ('Fire safety') recommends that doors on escape routes should be fitted with a simple fastening that can be readily openable.

The Department's view

The Department notes and is prepared to accept the proposed management procedures to be adopted by your client to ensure that the gate is open during the shop trading hours. However, the City Council have raised the issue of persons being in the shop after trading has finished. The Department considers that the City have a valid concern in this respect since the numbers of persons likely to be in the store after trading has finished is an unknown factor and should not be ignored. For example, large retail stores employ persons after trading hours for duties such as shelving and cleaning, etc. However, these issues have not been addressed by you.

In this case the Department considers that the worst scenario would be if a fire occurred on an upper floor, possibly after normal trading hours when the proposed security gate might be locked. Persons descending stair No. 16 to make their exit from the first floor would then have to go back up the stair towards the fire to make their escape via another exit. Therefore, in the Department's view the proposed siting of a lockable security gate at the bottom of stair No. 16 could prejudice safe escape from the first floor.

In analysing this case the Department has considered possible design solutions which might have made your proposals acceptable in terms of compliance with Requirement B1. In the Department's view there are possibly two solutions which you may consider appropriate to discuss with the City Council. The first could be a suitably fire resistant door leading into the ground floor area of the shop from the lobby at the base of stair No. 16. This would then provide an alternative ground floor escape if the security gate was found to be locked, rather than persons having to retreat back up the stair to the first floor where escape might be blocked by a fire. A second solution might be the provision of a warning device indicating that the gate was locked, as suggested by the City Council.

The determination

The Secretary of State has given careful consideration to the particular circumstances of this case. He has concluded from the information available that the City Council were correct in their judgement that your proposals do not show compliance with Requirement B1. The Secretary of State therefore determines that your proposals are not in compliance with Requirement B1 of Schedule 1 to the Building Regulations 1991 (as amended).

Terraced dwelling – alteration to upper floors (B.1)

The proposal was to create an additional space on a half landing level by stripping out a bathroom. It was determined that this would make an existing situation worse.

Ref.: 45.1.138   Date: 28 May 1997

In making the following determination the Secretary of State has not considered whether the plans conform to any other relevant requirements.

The proposal

The proposed building work concerns alterations to the upper floors of an existing terraced dwelling house which contains habitable accommodation on basement, ground, first, second and third floors. A single stair provides access to all levels. The subject of this determination is the creation of a space at the rear of the stairway enclosure which will be open to the stairway at half landing level between the first and second floors. The space is to be created from a fully-enclosed bathroom which is to be completely stripped, and the door and part of the partition wall removed, so that the space is fully open to and forms part of the stairway. The floor area of the space in question is shown on the drawing to be approximately 4.5 m². The space will be enclosed by the rear wall which has a window, and on two other sides.

These proposals were the subject of a full plans application which was rejected by the Borough Council on several grounds of non-compliance with the Building Regulations, including Requirement B1 ('Means of escape'). It is understood that you were able to subsequently meet the Borough Council's concerns in respect of all of these grounds with the exception of the proposed open space which the Borough Council considered had the potential to be used as a habitable area and as such would compromise the requirements of B1. However, you did not accept that the creation of an open space would constitute non-compliance with Requirement B1 and therefore applied for a determination in respect of this question.

At the date of your application the building work relating to the creation of the open space had not been commenced. However, in a subsequent letter, copied to you, the Borough Council informed the Department that work to create the open space had taken place.

The Applicant's case

You disagree with the Borough Council's interpretation that the half landing provided between the first and second floor is a room and that it endangers the integrity of the escape route. You contend that a room is an enclosed space, i.e. with walls on all sides. You state that

the space is to be opened up to be incorporated into the existing small landing and that it is not uncommon in domestic architecture for generous circulation space to be provided.

You have stated to the Borough Council that the space in question is not a habitable room and that it is being integrated with the stairway to allow light into this area. You also state that it will be part of the landing and that your client could limit its possible misuse by not providing power to the space.

The Borough Council's case

The Borough Council state that the proposed space should not be open to the landing but be enclosed as it will have the potential of being used as a habitable space which would introduce a fire risk into the stairway enclosure. The schedule to the Borough Council's notice of rejection required that an FD20 fire-resisting and self-closing door should be provided to the space.

The Borough Council point out that fire in a circulation space poses the most immediate and serious threat to the occupants and it is essential to ensure that the potential for fire is minimized. They contend that the space created could be used for a variety of purposes such as a library, study, or play area; or would have the potential to be so used in the future.

The Department's view

The question for determination is whether the space shown to be open to the stairway on the first floor plan is acceptable in terms of Requirement B1 of the Building Regulations 1991.

You contend that it is not uncommon in domestic architecture to have a generous circulation space. The Department accepts that this might be the case where the landing area clearly forms part of the circulation space and cannot be used for anything else because of the need to access rooms, etc. However, the Department does not consider this to be the situation in this case because the space in question is enclosed on three sides and in effect forms an annex to the landing.

The Department notes that this dwelling house has four above ground storeys served by a single stair and paragraph 1.21 of Approved Document 'B' ('Fire safety') suggests that in such a situation the guidance given in clause 4.4 of BS 5588: Part 1: 1990 ('Fire precautions in the design, construction and use of buildings') should be followed. This paragraph suggests that an alternative escape route should be provided from each storey situated 7.5 m above ground level. Such an alternative escape route is not shown to have been provided in this case.

The Department accepts that in this case the dwelling is existing. Nevertheless, the building work must comply with the requirements of Regulation 4 of the Building Regulations 1991 and should not make an existing situation worse.

The determination

The Secretary of State has given consideration to the particular circumstances of this case. In doing so he has been conscious of the fact that a large number of fires occur in residential accommodation. He has concluded, and hereby determines, that your proposals as forwarded in your application do not comply with Requirement B1 ('Means of escape') of Schedule 1 to the Building Regulations 1991 (as amended).

Residential care home – means of escape (B.1)

Fire separation at the base of the main stair was in dispute here, and there were no compensatory features balancing the fact that evacuating the occupants could be difficult and lengthy.

Ref.: 45.1.141   Date: 29 May 1997

In making the following determination the Secretary of State has not considered whether the plans conform to any other relevant requirements.

The proposal

The building to which this determination relates is a new, part-two storey and part-three storey residential care home for the elderly. The building consists of two wings with all floors shown to contain sleeping accommodation and residents' lounges, additionally the kitchen and laundry is located on the ground floor.

The second floor is served by two stairs and the first floor by three. Two of the stairs are located at each end of the wings of the building and discharge to protected final exit lobbies at ground floor level. The central main stair discharges to an entrance foyer which gives access to each wing of the building via fire doors held open on automatic hold-open devices. The foyer also contains a lift, a reception area with office and gives direct access to a residents' lounge.

The Borough Council have accepted the central stairs discharging into the ground floor entrance foyer on the basis that a screen and door will be provided at the bottom of the stairs so that the stairway itself has some protection from fire. You do not consider it to be necessary to provide this enclosure at the bottom of the stair and it is this that is at issue.

The Applicant's case

You acknowledge that your proposal does not fully comply with the recommendations given in Approved Document B. However you take the view that your proposal not to provide fire separation at the bottom of the main staircase gives a satisfactory and safe alternative. You are therefore requesting that the Secretary of State determines that the Borough Council's requirement for a separating screen at the base of the main staircase is not required in order to satisfy the requirements of the Building Regulations.

You make the following points in support of your application:

(i)　The staircase is adequately protected without the provision of a screen at its base because the entire foyer area can be considered to be part of the main stairway compartment.

(ii)　The provision of a screen at the base of the stair would only serve to complicate the means of escape from the building at ground floor level. It is also considered important that the everyday functioning of the building is not disturbed by unnecessary obstructions that may have a detrimental effect on the safety of the occupants.

(iii)　The stair is fully protected on the upper floors and alternative escape is available from all floors.

(iv)　It is accepted that Approved Document B suggests that a protected stairway should not form part of the primary circulation route between different parts of the building. The reason given for this is that self-closing fire doors could become ineffective. However, in this case the doors would be fitted with hold-open type overhead door closers linked to the fire alarm system which would close the doors in the event of fire or power failure.

(v)　In the event of fire the occupants of the ground floor would have been evacuated before the second floor occupants reach ground floor level and therefore the stairway would not be used as a circulation route at this stage.

The Borough Council's case

The Borough Council has rejected your proposals on the basis that because of the lack of protection at the base of the main stair at ground floor level, they do not show compliance with Requirement B1 of the Building Regulations.

The Borough Council makes the following points in support of its rejection:

(i)　Approved Document B suggests that every internal escape stair should be a protected stair and if the lobby at the bottom of the main stair is not provided then this would cause the stairway to become a circulation space between different parts of the building at the same level.

(ii)　The Approved Document also suggests that every protected stairway should discharge directly to a final exit. The Borough Council considers that it has already compromised on this issue since the stair does not discharge to a final exit but to a protected foyer.

(iii)　The Borough Council do not consider hold-open devices linked to the fire alarm system to be suitable for a stairway.

The Department's view

The department takes the view that the Secretary of State is being asked to determine whether your proposals to omit fire separation at the base of the main stair at ground floor level will show compliance with Requirement B1 of the Building Regulations.

The Borough Council have already accepted your proposal for the stair to discharge into the main entrance foyer and that this foyer gives access to a lounge and both wings of the building, one of which contains the laundry and kitchen. On this basis it would appear to the Department that in requiring a screen at the base of the stair the main concern of the Borough Council is to protect the stairway from the products of combustion of a fire occurring on the ground floor.

The Department notes that this case concerns the safe means of escape for the occupants of a residential care home with bedroom accommodation on all three floors. In the Department's view it is very likely that evacuation of the occupants in fire could be both difficult and lengthy.

The Department accepts the benefits that hold-open devices can provide both in terms of everyday access and safe escape. However, this is a nursing home where a failure of the device could leave the stairway exposed to either or both ground floor wings. Also given the probable length of time that it could take to evacuate the building the Department takes the view that in this case the stair should be given protection and that the ground floor screen should not be omitted.

The determination

The Secretary of State has given careful consideration to the particular circumstances of this case. He considers that your proposal to omit the protecting screen at the bottom of the stair on the ground floor will not provide an acceptable standard of escape. Accordingly the Secretary of State determines that your proposals do not comply with Requirement B1 of Schedule 1 to the Building Regulations 1991 (as amended).

New dwellings – means of escape (B.1)

New dwellings were to have windows with an opening restrictor mechanism, which in an emergency could be pushed hard to sever a shear pin. The Council did not like this, but the determination went in favour of the developer.

Ref.: 45.1.134　Date: 24 June 1997

In making the following determination the Secretary of State has not considered whether the plans conform with any other relevant requirements.

The proposal

The building work involves the development of 24 dwellings. The plan submitted to the Department as part of your submission shows a pair of two bedroom semi-detached houses. The stairs discharge into the ground floor living room and therefore the bedrooms are designated as inner rooms.

The windows to be installed in each bedroom comprise a pair of side-hung casements. According to the manufacture's literature these are designed for both children's safety and safe escape. A restrictor mechanism is fitted to limit the opening size and emergency opening is made possible by pushing hard on the window to break a shear pin. When this pin is broken the window can be opened fully giving an unrestricted opening of at least 500 mm wide by 850 mm high.

Your original proposals were conditionally approved subject to conditions relating to Parts A, B, K and N of the Building Regulations 1991. With respect to Part B the conditions were that the first floor bedrooms off the ground floor living room, 'require to have an escape window with an unobstructed opening that is at least 850 mm high and 500 mm wide, and the bottom of the window opening between 800 mm and 1100 mm above the floor.'

Because of the need to break the shear pin in the windows to establish the recommended minimum width for escape, the Borough Council has taken the view that the windows result in a non-compliance with Requirement B1. However, you believe that the windows are suitable for escape purposes and that the type of hinges fitted in them does satisfy Requirement B1. You therefore applied to the Secretary of State for a determination in respect of this question.

The Applicant's case

You accept that because the stairs open into a ground floor living room then the upstairs bedrooms are inner rooms. However, you point out that in terms of the opening size of the windows they conform in every respect to the recommendations given in paragraph 1.18 of Approved Document 'B' ('Fire safety'). The windows are at least 850 mm high and 500 mm wide, and the bottom of the opening is between 800 mm and 1100 mm above the floor.

The window fitted is of the type where a shear pin is inserted into the hinge so that during normal, everyday use the window has additional safety and easy-clean features. In an emergency situation an additional push on the casement will break the shear pin, thereby enabling the window to open fully.

The manufactures of the special hinges have confirmed to you that no other building control authority has questioned the suitability of this product when used in similar circumstances. You also state that you have discussed the situation with the local Fire Authority who do not see a problem with this type of window.

The Borough Council contend that additional force would be necessary to break the shear pin, but your view is that there is not a standard that quantifies the amount of force necessary to open windows and that therefore this is a subjective matter.

The Borough Council's case

The Borough Council have confirmed to the Department that in their view the provison for means of escape from the bedrooms is unsatisfactory. This is because the bedrooms are inner rooms and that, in their view, the windows do not comply with their conditional approval for a minimum opening size.

The Borough Council state that when fully open the windows only have an opening width of 385 mm. In order to fully open the window for escape in case of fire the occupant has to 'push hard on the window to break the shear pin'. The quote is taken from the manufacturer's literature. The Borough points out that this will require both a technical knowledge of how to break the obstruction (shear pin) and the physical strength to do it.

Finally the Borough Council state that they have not been able to find any other authority which has approved the use of this product for similar situations.

The Department's view

The question which has been submitted for determination is whether the bedroom windows as designed will be suitable for escape purposes should this become necessary and whether they therefore satisfy Requirement B1 of the Building Regulations 1991.

The Department notes that both parties accept that the bedrooms are inner rooms because the stairs discharge into a habitable ground floor room. The Department also notes that what is specifically at issue in this case is the openable width of the window. Paragraph 1.18(a) of Approved Document 'B' suggests that the unobstructed opening width should be at least 500 mm; whereas the Borough Council state that the opening width with the shear pin in place is only 385 mm.

The Department notes that the windows are specially designed for cleaning and child safety but with the provision of a frangible shear pin that when broken will permit the window to be opened fully for escape purposes. When the window is in this fully open position the unobstructed width is at least 500 mm.

The Department accepts that in an emergency situation some additional force will be necessary to break the shear pin so that the window can be fully opened for escape. The Department also accepts that some persons, such as the very young or frail elderly, may not be able to exert the force necessary to break the shear pin. However, the Department considers that it is likely that these type of persons would, in any event, need to be either provided with assistance for self-escape or would need to be rescued.

The Department does not accept the Borough Council's argument that a technical working knowledge would be required of how to break the shear pin. In the Department's view that is insufficient grounds for rejecting this type of window when the other safety features are taken into consideration. It could also be argued that a large person trying to exit through the restricted window opening would break the shear pin by default. Nonetheless, the Department does recognize it would not be immediately apparent to a visitor, or even an occupant, that the casements are retained in their restricted opening position by a shearable pin. It therefore strongly recommends that your client should consider in – conjunction with the manufacturers – the practicality of attaching to the window units in some manner a suitable and permanent notice to the effect that additional force is necessary to open the window fully for fire escape purposes.

The determination

The Secretary of State has given careful consideration to the particular circumstances of this case and the arguments put forward by both parties. He has concluded, and hereby determines, that in the circumstances of this particular case the proposed propriety windows with shear pins comply with Requirement B1 of Schedule 1 to the Building Regulations 1991 (as amended).

In issuing this determination the Secretary of State draws your attention to the Department's strong recommendation regarding the desirability of a permanent notice as suggested above.

Loft conversion – fixed ladder access (B.1)

A new single room in the roofspace was to have a vertical ladder access, and full fire separation between stairway and the new second floor was not to be provided.

Ref.: 45.1.137   Date: 24 June 1997

In making the following determination the Secretary of State has not considered whether the plans conform to any other relevant requirements.

The proposed work

The proposed work to which this determination relates is the alteration of a two storey semi-detached house to form a single room in the roof space designated on the drawing as a store/playroom. Access to the new second floor room is to be via a vertical ladder from the existing first floor landing. The proposals as submitted by you to the Department show that a hinged self-closing gate will be provided at the top of the ladder from which it is concluded that full fire separation between the new second floor room and the stairway is not intended.

An emergency escape rooflight in the rear elevation roof slope is proposed, which would be reached from the rear garden. However, access to the rear garden is only available through the garage or over the flat roof of that garage. A mains-operated smoke detection system is to be incorporated in the property, replacing the existing battery-operated smoke detectors.

Your original, re-submitted proposals were rejected by the City Council because of their non-compliance with the Building Regulations 1991 in respect of the need for mains-operated smoke alarms; the unsuitability of a fixed ladder for means of escape purposes; and the need for suitable guarding to be provided around the proposed access hatch. The City Council's letter of 26 April 1996 states that it was understood that you were prepared to incorporate mains linked smoke detectors to BS 5446 and guarding to the access; but that you believe that the specified fixed ladder satisfied the requirements of B1 and K1 ('Means of escape' and 'Stairs and ramps' respectively). It is in respect of the acceptability of the fixed ladder for compliance with Requirement B1 that you applied for this determination.

The Applicant's case

You contend that the existing configuration of the dwelling makes the provision of a stair complying with paragraphs 1.1 to 1.7 of Approved Document 'K' ('Stairs, ramps and guards') impossible without alteration to the existing space. You confirm that the ladder serves only one habitable room and contend that it is therefore permissible under paragraph 1.25 of Approved Document 'K'.

You also point out that you are to replace the existing battery-operated smoke detectors with a mains-operated system and will be providing a hinged, self-closing gate at the head of the ladder.

The City Council's case

The City Council has rejected your proposals on the basis that reasonable provision for means of escape has not been made. They base this judgement primarily on the limited fire separation between the new room and the rest of the dwelling; the difficulty in effecting external rescue from the rear garden; and the provision of a ladder which will therefore need to function as the means of escape. They make the following points in support of their judgement to reject your proposals:

(i) A protected stairway approach has not been achieved due to the omission of a fire resistant enclosure between the new room in the roof and the stairway. It is also unclear whether doors serving first floor airing and ground floor under-stair cupboards will be upgraded to achieve 30 minutes' fire resistance.

(ii) The proposed degree of smoke detection is acknowledged, but it is considered that this will not provide adequate compensation for the omission of an enclosed staircase.

(iii) The ladder to be provided has only a limited headroom of approximately 1.6 m over the new second floor landing area and it is considered that some persons such as the young, elderly or infirm may have difficulty in negotiating this ladder in an emergency.

(iv) An escape window has been indicated on the rear elevation roof slope. However, there are possible difficulties in readily gaining access to the rear garden. This is because access is only possible through the garage or over the garage flat roof if the garage should be locked.

(v) Your reference to Approved Document 'K' is considered to be misleading since a ladder would not generally be opposed if it were not also being relied on for means of escape.

The Department's view

Your application for a determination was in respect of the acceptability or otherwise of your proposal to use a fixed ladder for access to the new second floor room. However, the City Council appear to have rejected your proposals on a number of grounds although in response to the City's points you have only addressed the issue of the ladder. The Department's views on the main question put and on the additional grounds of rejection are set out below insofar as this is possible on the information provided:

(i) Guidance on fire protection to existing stairways in loft conversions is given in paragraphs 1.24 to 1.26 of Approved

Document 'B' ('Fire safety') and the Department considers that in this case nothing need be done beyond the measures recommended in those paragraphs. Paragraph 1.26 states that existing doors need only be made self-closing and in the Department's view this would not preclude the use of rising butt hinges provided they could be shown to work to the City Council's satisfaction.

(ii) Paragraph 1.28 of Approved Document 'B' gives guidance on the fire separation of a new storey and the Department considers that it is important that adequate fire separation is provided if a person is to wait in relative safety to be rescued from that new storey. The Department would therefore agree with the City Council in this case that the provision of a self-closing hinged gate at the top of the ladder will not provide adequate fire separation between the new second floor room and the stairway.

(iii) With regard to access to the rear garden and the escape window located in the rear elevation roof slope, the Department has taken note of the recommendations with respect to pedestrian access given in paragraph 1.31 of Approved Document 'B' and notes the City's concern that access could be obstructed by the garage being locked. However, the Department considers that any delay caused by having to cross the flat garage roof would be minimal in life safety terms and therefore considers that it would be unreasonable, in this particular case, to reject the proposals on grounds of uncertainty of access to the rear garden for the purposes of rescue.

(iv) With regard to the provision of ladder access to the new second floor room, paragraph 1.25 of Approved Document 'K' gives guidance on the acceptability of this in certain limited circumstances. The Department accepts your point that there is limited space and considers that in these circumstances it would be unreasonable to insist on the provision of a stair. However, the Department considers that all relevant aspects of the guidance given in Approved Document 'K' should be followed to enable escape to be made as safely as possible. The City Council has raised the issue of headroom and the Department accepts the City's judgement that a headroom of 1.6 m is unsafe for escape purposes and should be considered unacceptable. Paragraph 1.25 of Approved Document 'K' also suggests that a ladder should be provided with fixed handrails on both sides and it is unclear from the information submitted as to whether such handrails are proposed.

The determination

The Secretary of State has given careful consideration to the facts of this case. He is particularly conscious of the fact that a large number of fires occur in dwellings.

The specific question put to the Secretary of State for determination was whether a fixed ladder would comply with the Building Regulations. Because the fixed ladder is the only means of access to the second floor extension it must also be capable of acting as the primary means of escape. It is therefore important to demonstrate that the fixed ladder complies with Part K. However, in this instance the guidance in Approved Document 'K' ('Stairs, ramps and guards') has not been followed and therefore the Secretary of State does not consider the proposals to be acceptable in terms of Requirement B1. The Secretary of State accordingly determines that your proposals as submitted do not show compliance with Requirement B1 of Schedule 1 to the Building Regulations 1991 (as amended).

In making this determination the Secretary of State draws your attention to the Department's views expressed on other issues above, which may be of assistance to you.

Multi-storey commercial building – refurbishment (B.1 and B.5)

Fire doors providing separation between a reception area and the fire-fighting shaft were to be removed. The Local Authority thought that this would prejudice safe access for the fire service and this was supported by the Secretary of State.

Ref.: 45.1.140   Date: 4 July 1997

In making the following determination the Secretary of State has not considered whether the plans conform to any other relevant requirements.

The proposal

The building to which this determination relates is an existing multi-storey commercial building that is being refurbished on a floor by floor basis to suit new tenants. It is the refurbishment of the ground floor that is the subject of this appeal.

The City Council state that the building has 11 floors above ground floor and a basement car park. The building is provided with a single firefighting shaft comprising stairs and two lifts, one of which is designated as a firefighting lift. The shaft connects all floors including the basement car park. You state that alternative escape is available although you have not been specific about the number or location of available routes. The sketch plan of the ground floor, which forms part of your submission, shows an alternative exit from the lift lobby to the service yard. The plans also show that new smoke detectors will be installed in the reception area.

The ground floor of the building does not contain any accommodation other than a reception/waiting area that you state occupies approximately 50 square metres. The existing double doors which provided fire separation between the reception area and the firefighting shaft have been removed so that the reception area is now open to the shaft.

The Secretary of State is therefore being asked to determine whether it is necessary to provide new fire resisting doors between the reception area and the firefighting shaft in order to satisfy Requirements B1 and B5 of the Regulations.

The Applicant's case

You originally requested that the City Council relax/dispense with Requirement B1 and B5. However, notwithstanding this your view is that you consider the current situation, whereby no fire separation between the ground floor reception area and the firefighting shaft has been provided, complies with the Requirements. You do not consider that the removal of the doors will adversely affect the existing situation.

You state that your client feels very strongly that the existing fire doors separating the reception area from the firefighting shaft need to be removed in order to create a more attractive entrance area. You also point out that the firefighting shaft does not totally comply with the current guidance and that the shaft will be upgraded during the refurbishment.

You point out that paragraph 4.31 of Approved Document B permits a reception area of 10 square metres providing the stair is not the only stair serving the building. In your case you state that the furniture contained in the reception area will not exceed 6 square metres and you do not therefore consider that the removal of the doors will affect means of escape.

With regard to Requirement B5 you do not consider that paragraph 4.31 of the Approved Document excludes the provision of a reception area within a firefighting shaft. You do not consider that a fire on the upper floors will affect the reception area, even with the separating doors removed.

You accept that the Code of practice for firefighting stairs and lifts, BS 5588: Part 5: 1991, does in fact limit the maximum lobby size to 20 square metres. However you consider that by allowing controlled use of the space, abuse of a larger area would not occur and you have provided a list of low combustible furniture that would be used.

The City Council's case

The City Council, after consultation with the Fire Authority, consider that the proposal to remove the fire separation between the reception area and the firefighting stairs do not show compliance with Requirements B1 and B5.

The City Council have expressed considerable concern about the proposals and give the following points in support of their rejection:

(a) The firefighting stairway and firefighting lift would not be approached through a firefighting lobby in accordance with the guidance given in BS 5588: Part 5: 1991.

(b) The proposed layout would mean that the 11 above ground floors, the basement car park and the reception area would all be contained within the firefighting shaft.

(c) The firefighting shaft contains the rising main and fire hose reel points, also the City Council consider that the firefighting lobby should provide a command point for firefighting and rescue operations.

(d) Approved Document B permits a reception area of 10 square metres in a protected stairway providing it is not the only stair serving the building. In this case the proposal is to incorporate a reception/waiting area of 50 square metres within the enclosure of a firefighting stair. The City Council take the view that the size of the reception/waiting area suggests that it could be open to abuse.

The Department's view

What is at issue in this case is whether the removal of the fire separation between the reception/waiting area and the firefighting stairs will:

(a) prejudice safe means of escape (Requirement B1) and
(b) prejudice safe access for the fire service (Requirement B5).

In the Department's opinion Requirement B5 will be the over-riding factor; however, means of escape has been raised as an issue. Approved Document B suggests that in a building where there is more than one stair, then a reception area of up to 10 square metres could be considered acceptable. In this case the reception/waiting area is stated to be approximately 50 square metres and to assess whether this is satisfactory, full details of alternative escape routes would need to be known. You have not given any details of alternative escape routes from the upper floors and therefore on the basis of your submission, the Department is not able to say that the proposals are acceptable in terms of Requirement B1.

With regard to fire service access, BS 5588: Part 5 clearly suggests that the lobby should not be so large as to encourage storage or unauthorized use. A large reception/waiting area open to the firefighting stairs could in the Department's view hinder access for the fire service. This could either be because the route and lobby area had become obstructed or because a fire had occurred within the reception/waiting area itself.

The Department notes from the drawing that an alternative ground floor route is available from the bottom of the stairs to the rear service yard. Therefore any advantage that this originally provided for the fire service would be negated by the removal of the fire separation between the front reception area and the stairs. The Department does not therefore consider the proposals to be acceptable.

The determination

The Secretary of State has given careful consideration to the particular circumstances of this case. He considers that your proposal to remove the fire separation between the ground floor reception/waiting area and the firefighting shaft does not, on the basis of the information given, provide an acceptable standard of escape or provide an acceptable standard of access for the fire service. The Secretary of State therefore determines that your proposals do not comply with Requirements B1 and B5 of the Building Regulations 1991 (as amended).

Loft conversion – means of escape (B.1)

Two new rooms in the roof of a dwelling were to have a means of escape onto a roof walkway and proved not to be acceptable. The solution was a mains-operated alarm in every room opening onto the stairway.

Ref.: 45.1.142   Date: 4 July 1997

In making the following determination the Secretary of State has not considered whether the plans conform to any other relevant requirements.

The proposal

The building works to which this determination relates is a conversion of the roof space of an oasthouse to form two habitable rooms. These are designated on the drawing as a playroom and a study and are shown to have an interconnecting lobby separating each room.

The building was originally a five kiln oasthouse and was converted into five separate three storey dwellings in 1989. The proposal is to make use of two roof spaces above house number 2 to form a new third floor (fourth storey) with two rooms. Each roof space or room is proposed to be linked externally by a new weathered enclosure designated as the connecting lobby on the drawing. The new third floor is stated by the Borough Council to be 7.8 m above ground level; you state that it is 7.8 m at the front and 7.6 m at the rear.

At the time of the original conversion the single stair in each dwelling was required to be protected with 30 minutes' fire resisting enclosure. All doors opening onto the stair enclosure from habitable rooms were made 30 minutes' fire resisting and self-closing. The proposal is to extend this protected stair up to serve the two new rooms on the new third floor. You state that it is proposed to extend the existing fire detection system to serve the new floor level and that this will be linked through to British Telecom on a 24 hour monitoring service.

Alternative escape is provided via a third floor roof window onto a 22 m long walkway between the roofs of the adjoining houses from which you maintain that persons could wait to be rescued. However, the Borough Council have rejected this as not being suitable for alternative escape and it is this that is at issue.

The Applicant's case

You state that it is only point 1 of the Borough Council's decision letter, which relates to the alternative escape route, that you are asking the Secretary of State to determine and that all other issues raised by the Council have been or will be corrected to their satisfaction.

You acknowledge that the proposed third floor presents difficulties in terms of means of escape. However, you consider that you are offering measures that both minimize the risk to occupants and significantly offer more protection than would be the case with a standard loft conversion.

You consider that your proposal to provide accommodation at third floor level should be considered satisfactory and you offer the following points in support of your application:

(i) The protected stair will be extended to serve the new floor and this floor will be separated from the remainder of the accommodation by a fire door.

(ii) Alternative escape will be provided from the protected lobby serving the new third floor via a roof window of the correct opening size. This leads to a walkway between the roofs from which you consider that rescue by the Fire Brigade can be secured. In your view 22 m from the exit widow in open air with adjoining roofs acting as fire barriers, constitutes a place free from the dangers of fire.

(iii) If neighbouring properties construct similar enclosures then this could be beneficial in your view because reciprocal means of escape arrangements could be made through each property.

(iv) The existing fire detection system will be extended to the new floor and provided with a 24 hour monitoring service. You also state that your client would be prepared to consider supplementing this with a mains-operated fire detection and alarm facility and that this could take the form of a condition attached to any approval.

(v) The height of the new third floor in this building is only slightly higher than the height of a second floor in a typical Georgian house. It is also only marginally over the 7.5 m threshold height that requires alternative escape to be provided.

The Borough Council's case

The Borough Council has rejected your proposals on the basis that the alternative escape route across the roof valley of the adjoining properties does not discharge to a final exit and is therefore unsatisfactory. In their view reliance would need to be placed on rescue by the fire brigade and they do not consider that the point at which persons would need to wait to be rescued can be considered as a place free from danger. The Borough Council have also consulted with the local Fire Authority who confirm that rescue from the roof valley would be difficult.

The Borough Council makes reference to paragraph 1.21 of Approved Document B which in turn suggests that the guidance given in clause 4.4 of BS 5588: Part 1: 1990 should be followed for houses with more than one floor over 4.5 m above ground level. Clause 4.4(d) of the British Standard suggests that an alternative escape route should be provided from each storey or level situated 7.5 m or more above the ground or access level.

The Borough Council has noted that the protected stair enclosure and the fire detection system have both been extended to cover the new storey. However, they point out that these are now minimum recommendations for showing compliance with the Building Regulations and although the fire alarm is linked to a monitoring system they do not consider this to be an adequate reason to vary the provisions that satisfy the Regulations.

The Department's view

The Department takes the view that the Secretary of State is being asked to determine whether your proposals to provide additional accommodation at third floor level, will show compliance with Requirement B1 of the Building Regulations.

The Department notes that the height of the new third floor is only 7.8 m above ground level and therefore only slightly in excess of the 7.5 m threshold where alternative escape is suggested. However, what needs to be considered is whether the proposals provide a satisfactory level of escape.

In this case the proposal for alternative escape is along the roof valleys of the adjoining properties and the Department supports the Borough Council's view that this is not satisfactory as an escape route. The use of such a route assumes that it will remain free from obstruction and that persons are fit enough to get out and await rescue by the fire brigade who have stated that such rescue will be difficult.

Because of the low height of the building it would not be unreasonable for persons to expect to either make their escape or be rescued via the internal protected stair. However, for this to be acceptable it is necessary for a suitable early warning system to be provided. You have stated that the alarm system will be extended to cover the new third floor. However, you have not indicated the extent of alarm cover in the building. The Department assumes therefore that the provision of a detection system is limited to the escape route only.

The Department takes the view that for a single escape route to be acceptable in this building, mains-operated detectors should be provided at every level and in every habitable room that opens onto the stairway. The alarm system should be installed to the satisfaction of the Borough Council. However, since this appears not to be the case the Department does not consider the proposals to be satisfactory.

The determination

The Secretary of State has given careful consideration to the particular circumstances of this case. He considers that your proposal to provide habitable accommodation on the new third floor will not provide an acceptable standard for escape in fire. Accordingly the Secretary of State determines that your proposals do not comply with Requirement B1 of Schedule 1 to the Building Regulations 1991 (as amended). However, you may wish to take account of the Department's comments given above.

Extension to dwellinghouse – internal fire spread (B2)

A small addition at the rear of a dwelling is to be lined with t and g boarding. The local authority wanted a Class 1 rating, but it was determined that spread of flame over the new room linings would have little effect upon occupants of other rooms.

Ref.: 45.1.143   Date: 29 July 1997

In making the following determination, the Secretary of State has not considered whether the plans conform to any other relevant requirements.

The proposed work

The building to which the proposed building work relates is an existing private dwellinghouse with existing all-glass conservatory attached at the rear of the premises. The building work involves the demolition of the conservatory and the construction in its place of a new single storey extension. The proposed new room, which has a floor area of 9 square metres, is shown on your drawing to be an extension to the kitchen and is also shown to be fitted out with work surfaces and shelving. A door is shown to separate the new room from the kitchen and a door leading from the new room to the rear garden is also shown to be provided.

The intention is that the new room will be lined both on the walls and ceiling with tongued and grooved timber boarding. You state that the intention is to treat the internal surfaces with one coat of universal primer and one coat of stain topcoat.

These proposals formed the basis of a full plans application which was approved subject to the condition that all internal wall and ceiling finishes were to a Class 1 spread of flame. However, for environmental reasons your client does not wish to use the type of materials which will achieve this class level but you nevertheless believe that compliance with Requirement B2 ('Internal fire spread (linings)') can be achieved. It is in respect of this question that you have applied for a determination.

The Applicant's case

You state that your client has specifically expressed a wish to use only environmentally friendly materials on the internal surfaces of the new room. You accept that the timber treatment that you have specified will not achieve a Class 1 surface spread of flame rating because it is linseed oil based. However, your researches reveal that only non-environmentally friendly treatments, which would be unacceptable to your client, will achieve a Class 1 rating.

You therefore acknowledge that your proposal does not fully comply with the recommendations given in Approved Document 'B' but believe that your proposal demonstrates compliance with Requirement B1 ('Means of escape') because:

(i)   the proposed extension does not form part of an escape route since escape is available via existing french doors; and

(ii) you consider that you have observed the Building Regulations as closely as possible within the remit to produce an extension without using noxious materials.

The Borough Council's case

The Borough Council accepts that your proposed extension is relatively small. However, they point out that the guidance given in Table 10 of Approved Document 'B' suggests that the surface linings of the walls and ceilings in a room of the size that you propose should achieve a surface spread of flame rating of Class 1. In the Borough Council's view the surface treatment that you propose to use may not even achieve the Class 3 rating which would be acceptable for smaller rooms.

The Borough Council do not consider means of escape to be a relevant issue. They also state that they are sympathetic to the use of environmentally friendly materials but consider that the requirements of the Building Regulations must take precedence.

The Department's view

To assess whether your proposals are acceptable, the Department has to gauge them against the Requirements of B2 and the particular section of that requirement which states:

To inhibit the spread of fire within the building the internal linings shall resist the spread of flame over their surfaces.

To satisfy Requirement B2 it is normally expected that the internal surfaces should achieve a surface spread of flame rating that is at least as good as that recommended in Approved Document 'B'. For a room of the size you propose the internal surfaces should therefore achieve a Class 1 rating.

You have stated that your proposed treatment of the internal surfaces will not achieve the recommended classification level given in Approved Document 'B' and what has to be considered is the effect of this on the life safety of the occupants of the dwelling.

The Department notes that the new room is outside the footprint of the existing walls of the building and considers that the speed at which flames could spread over the new room lining will have little effect on the safety of occupants of other rooms in the dwelling. The Department also considers that flame spread in a small room such as the one proposed will have little effect on the safety of the occupants of the new room itself.

The Department therefore considers your proposed timber linings of the new extension to be acceptable in terms of compliance with Requirement B2. In reaching this conclusion the Department has taken account of the room size, its location in relation to the rest of the dwelling, and the alternative exits from the new room.

The determination

The Secretary of State has given careful consideration to the particular circumstances of this case. He considers that your proposal to provide wall and ceiling linings to the new room extension that do not achieve a Class 1 surface spread of flame rating, will provide an acceptable standard. Accordingly the Secretary of State determines that your proposals comply with Requirement B2 of Schedule 1 to the Building Regulations 1991 (as amended).

Internal alterations to offices – means of escape (B1)

A Grade II listed building was involved, and removal of lobby protection on first and second floors was proposed. It was accepted that an adequate smoke detection system would compensate for the loss of protection afforded by the lobbies.

Ref.: 45.1.149   Date: 7 August 1997

In making the following determination, the Secretary of State has not considered whether the plans conform to any other relevant requirements.

The proposed work

The building to which the proposed building work relates is an existing four storey single stair building built around the turn of the century of domestic design and in the 'arts and craft' style. It is listed Grade II and is in office use. As existing the single stair has a lobby approach from the office accommodation on all floors except the top floor. It appears that several alterations have taken place over the years but the current layout dates from 1987.

The proposed building work comprises the removal of the lobby protection to the stair on the first and second floors and some dividing partitions on the first, second and third floors so that additional floor space is made available for office use. On the first and second floors, open plan office accommodation will be provided each side of the stair with the partitions adjacent to the stairway being upgraded so that they achieve 30 minutes' fire resistance. You state that the doors will be fitted with self-closing devices and that a comprehensive smoke detection system will be installed in every room to link through to the existing fire alarm system. What is at issue is the removal of the lobby protection on the first and second floors.

These proposals were the subject of a full plans application which was rejected by the City Council on grounds of non-compliance with Requirement B1 ('Means of escape'). The City Council were not prepared to accept that the removal of the lobby protection on the first and second floors would be in compliance with Requirement B1. However, you believe that this proposal would be in compliance and it is in respect of this question that you have applied for this determination.

The Applicant's case

Your intention is to maximize the usable office space and simplify the door/lobby/corridor configuration so that it remains in keeping with the original building arrangement which was for domestic use. The proposed usable office space in the building would be increased from 165.88 square metres to 193.88 square metres, resulting in an estimated maximum occupancy of 5–6 persons on each floor. The comprehensive smoke detection system will encourage fast evacuation of the building.

You consider that your proposal will ensure the integrity of the building as a functional and efficient interior environment in terms of safety, space and energy.

The City Council's case

The City Council have rejected your proposals on the basis that the revised layouts of the first and second floors show the removal of lobbies designed to protect the downward means of escape and this therefore constitutes a reduction in the protection of the escape route. In reaching this decision the City Council have taken account of your proposal to incorporate a smoke detection system.

The City Council give the following points in support of this rejection:

(i)   Notwithstanding the compactness of the accommodation, the existing building has lobbies which protect the single staircase in accordance with current guidance for means of escape. The City Council take the view that removal of these lobbies will adversely affect the protection to the stairway.
(ii) The City Council recognize that the removal of the lobbies will improve internal circulation but they consider that this would be true of any single staircase building similarly protected in accordance with the Building Regulations.
(iii) The City Council are not aware of any planning or listed building constraints that would restrict the internal layout design.

The Department's view

What needs to be considered in this case is the effect that the removal of the lobby protection to the single stair on the first and second floors will have on the safety of occupants of the building should they have to escape in fire. Any improvement of the layout of the offices by the removal of the lobbies must be achieved within full compliance of Requirement B1.

Approved Document 'B' ('Fire safety') recognizes that there are situations where an escape stair needs the added protection of a protected lobby or protected corridor at all levels except the top storey.

Paragraph 4.26 of Approved Document 'B' suggests this may be necessary where the stair is the only one serving a building (or part of a building) which has more than one storey above or below the ground storey. This is the situation in this case.

The reason for providing lobby protection is to keep the stairway free of smoke whilst persons are making their escape from other areas of the building. If, however, persons are given sufficient warning of the outbreak of fire then they should be able to make their escape before a serious build-up of smoke in the stairway occurs.

You have proposed the installation of a comprehensive smoke detection system to give early warning of fire and this goes beyond the recommendations given in Approved Document 'B'. The Department is of the opinion that the installation of such a smoke detection system in this case can be considered to be an adequate compensatory feature to the extent that the lobby protection to the stairway can be dispensed with. In reaching this conclusion the Department has also taken into account the low occupancy of the building.

You have stated that the partitions separating the office accommodation from the stairway will be upgraded to achieve 30 minutes' fire resistance and the doors will be self-closing. However, you have not been specific about the standard of the fire doors and the City Council have also not raised this as an issue. The proposed standard is directly relevant to achieving compliance and the Department has therefore assumed that the fire doors opening onto the stairway will be FD30S as defined in Appendix B of Approved Document 'B' and self-closing. In addition there would need to be a responsibility on the building management to make sure that these fire doors were kept shut at all times when not in use. Alternatively, hold-open devices linked to the smoke detection system may also be considered to be appropriate.

The determination

The Secretary of State has given careful consideration to the particular circumstances of this case. On the basis of your submitted proposals and the necessary alternative assumptions contained in paragraph 14 above, he has concluded, and accordingly determines, that your proposals to remove the lobby protection to the single stairway on the first and second floor will be in compliance with Requirement B1 of the Building Regulations 1991 (as amended).

Subdivision of large building – internal and external fire spread (B3 and B4)

In this involved situation it was concluded that it appears the proposals would not result in a position which was more unsatisfactory than existed at present.

Ref.: 45.1.117   Date: 8 August 1997

In making his determination the Secretary of State has not considered whether the plans conform to any other relevant requirements.

The proposal

This determination relates to an existing development and concerns two attached buildings, indicated on the site plan as buildings No. 2 and No. 3. Building No. 2 contains Units 1 and 2 and Buildings No. 3 is referred to as Unit 3. In addition to Building No. 2 and No. 3 the site plan shows two other Buildings (No. 1 and No. 4). All four buildings were erected in the 1960s and were originally in the single ownership of a large industrial company. Buildings No. 2 and No. 3 (Units 1, 2 and 3) are now in separate ownership to Buildings No. 1 and No. 4, and a new boundary fence now delineates that ownership.

Building No. 2 was originally erected as a single building of steel portal framed construction and was divided at a later stage into Units 1 and 2. The City Council have expressed some concern that the dividing wall is not a compartment wall due to the provision of doorways within it. However, you state that self-closing fire doors will be provided in the wall and the Department therefore assumes that this is no longer an issue.

Building No. 3 (Unit 3) was constructed later as an attached single storey steel portal framed extension to Units 1 and 2. Unit 1 was open ended and you are now proposing to close the ends; drawing 9144/2 indicates that this is being done by the provision of roller shutter doors. Units 1 and 3 are to be used as warehouses for the storage of wet fruit and jam, and Unit 2 is to be used as a jam production unit. Both units are to be fitted out with storage racks of approximately 4 to 5 metres in height.

Units 1 and 3 have volumes of $12\,600\,m^3$ and $20\,300\,m^3$, respectively. The provisions of Section 65 of the Greater Manchester Act 1981 therefore apply. In order to meet some of the fire precaution matters, the City Council are requiring under section 65(3) of the Greater Manchester Act 1981, you are proposing to incorporate low melt rooflights in both units as an aid to ventilation and also to provide an automatic smoke detection system. However, you are not proposing to provide additional fire resistance to the elements of structure or sprinkler systems in the units which the City Council are requiring under section 65(3). These issues are the subject of a concurrent appeal by you (Ref. No CI/45/4/13).

The City Council appear to have taken the view that your proposed alterations and change of use constitute 'building work' under the Building Regulations 1991 and that as a result they consider requirements B3 and B4 to be applicable. They therefore consider that regard has to be given to the newly constituted boundary resulting from the sale of Buildings No. 2 and No. 3. It is their judgement that fire resistance should be provided to the external walls which have a new boundary condition and the elements of structure which support these walls.

Your full plans application was rejected by the City Council on grounds of non-compliance with Requirement B4 and failure to make provision for certain fire precaution matters contained in section 65(3) of the Greater Manchester Act 1981. Compliance with Requirement B3 has been raised subsequently by the City Council. However, you contend that your proposals to infill the ends of Unit 1 comply with the Building Regulations and that the proposed change of use is not subject to the regulations. You have therefore requested the Secretary of State to determine whether the buildings as existing comply with the Building Regulations 1991, notwithstanding the newly constituted boundary.

The Applicant's case

You have taken the view that the crux of the City Council's argument is that the sale of part of the site and the erection of a boundary fence constitutes a material alteration, and you consider this to be flawed. You do not agree that a boundary change, whether marked by a boundary fence or not, will cause the building to be more likely to spread fire. You consider the interpretation to be unusual in that many buildings are sold without reference to building control.

In addition, you do not accept the City Council's requirement that the boundary wall should be upgraded because the ends of Unit 1 are to be closed. It is your view that Requirement B3 or its equivalent at that time would have been complied with when the building was built.

You point out that to upgrade the fire resistance would involve large areas of external wall in addition to the portal legs and rafters of both units, and that the gable stanchions of Unit 3 would also have to be upgraded.

You state that compartmentation between Units 2 and 3 has been provided. You add that although this is breached by door openings, these can be made self-closing and fire resistant.

The City Council's case

The City Council argue that Units 1, 2 and 3 constituted part of a much larger site under the responsibility of a major industrial company and as such there were two further buildings on the site in close proximity. As a result there existed no boundaries between the buildings, except for the one which defined the extreme curtilage of the site. Unit 1 was effectively a covered way because it was open ended and Unit 3 was vacant, except that at one stage it had been used as a vehicle park and at a later stage was believed to have been used for playing sports. The intention is that all these buildings will now be used for storage/production purposes and, because of the partial sale of the site, will be in close proximity to other industrial buildings.

In respect of Unit 1, the City Council point out that your proposals include the enclosure of the open ends of this building and the provision of fixed storage racks of approximately 4 to 5 metres in height. The portal frame steelwork, whose principal stanchions face the neighbouring site, have no fire protection and provide support to the external masonry walls. There is also no evidence that the foundation bases or steel portal design is acceptable for a boundary condition. The City Council consider that since Unit 1 is to be now used for storage, with the probability of a higher fire load than previously existed, there is a greater risk of fire spread to the adjacent boundary.

In respect of Unit 3, your proposals include the provision of fixed storage racking for goods to a height of approximately 4 to 5 metres. The fire load for this unit has not been quantified and the steel portal frame, including the gable stanchions, are unprotected for the purposes of fire resistance. The new relevant boundary is around 8 metres from the gable elevation and the City Council consider that Requirements B3 and B4 are applicable. They consider the new uses, coupled with the change in boundary conditions, to be a worsening of the existing site conditions and, as such, a worsening of the relevant requirements which they consider constitutes a material alteration.

The Department's view

As the Department understands it, the City Council has approached the questions raised in this determination as follows. With respect to Unit 1 it has regarded the proposed infilling of the existing openings as a material alteration and therefore 'building work' which must demonstrate compliance with Requirement B4. In the Department's view the work is a material alteration. As such it is 'building work' but is not making the existing situation worse in terms of Requirement B4.

With regard to the existing and proposed uses of Units 1 and 3, the City appear to have adopted the view that the new uses and erection of storage racking will constitute a material change of use under the Building Regulations. As such they consider that the existing structure should comply with Requirement B4 having regard to the new boundaries. However, in the Department's view these proposed new uses cannot be defined as material changes of use and are therefore not required to comply with Regulation 6(1). Nor does the Department consider the erection of storage racking to be building work insofar as it will affect Requirement B4.

The particular concern of the City Council is the effects that the unprotected steelwork of Units 1 and 3 will have on the boundary elevations if the steel supports collapse in fire. The Department notes these two units are as existing, except that they are no longer in common ownership with the rest of the site and are to be designated for storage purposes. However, although the Department accepts that the 'building work' within the meaning of the Building Regulations is triggered by the proposed infilling on the existing openings at each end of Unit 1, it does not accept that the change of use falls within the control of the Building Regulations thereby giving a locus to the City Council to examine the compliance of the structure with Requirement B4 having regard to the newly constituted boundary.

The Department accepts that had the units been new, with the same new boundary conditions, then Requirement B4 would not have been satisfied because of the extent of unprotected boundary wall area. In addition, Requirement B3 would not have been satisfied because of the unprotected steelwork supporting the external walls. However, it has to be accepted that buildings and land may be sold resulting in disaggregation of ownerships. The implications for boundary conditions can only be considered in the context of 'building work' as defined in the Building Regulations; and that building work must then fall to be considered on its merits having regard to the particular circumstances of each case.

With regard to the proposed infilling of the existing openings in Unit 1, the Department does not consider that this work will materially alter the situation. The existing openings did not have fire resistance and it would appear that neither do the proposed infill doors. It can therefore be argued that the work, in terms of Regulation 4(2), will be no more unsatisfactory in respect of Requirement B4 than before the work was carried out.

The determination

The Secretary of State has given careful consideration to the particular circumstances of this case. He has concluded that the infilling of the ends of Unit 1 properly fall to be considered under the Building Regulations and determines that your proposal will be no more unsatisfactory than exists at present in respect of Requirement B4.

With regard to the change of use of Units 1 and 3, the newly constituted boundary, and the effect on compliance of these with Requirement B4, he has concluded as follows. On the basis of the evidence provided he does not accept that the new uses and erection of racking in Units 1 and 3 will constitute matters to be controlled under the Building Regulations 1991 and considers that it is therefore inappropriate to examine the compliance of the structures with Requirement B4. Insofar as such an examination would fall to be done in the context of a new boundary, he does not accept the implied proposition that the structure of an existing building should necessarily be examined simply because of a change of boundary.

Loft conversion – protected stairway (B1 and B3)

In question is the fire resistance of the existing first floor, the roof window and the provision of a suitably protected stairway to the front door. The determination went in favour of the applicant.

Ref.: 45.1.154   Date: 17 September 1997

In making the following determination, the Secretary of State has not considered whether the plans conform to any other relevant requirements.

The proposed work

The building work comprises alterations to an existing two storey house to form a new single habitable room in the roof space. There is an 850 mm high and 450 mm wide roof window in the new room which was originally intended to provide an escape route. Because its dimensions do not comply with those recommended in the Approved Document 'B' ('Fire safety') the escape window has been discounted and a fully protected stairway proposed to achieve 30 minutes' fire resistance leading from the new storey down to the front door at ground level. All doors from habitable rooms opening onto the stairway are to be fire resisting and self-closing. The non-load bearing ground and first floor walls which enclose the stairway are offset from each other by approximately 100 mm. The first floor has a modified 30 minutes' fire resistance.

These proposals were the subject of a full plans application which was rejected by the Borough Council on grounds of non-compliance with Requirement B3 ('Internal fire spread (structure)'). The Borough Council considered that the fire resistance of the existing first floor should be upgraded so that it would achieve a full 30 minutes' fire resistance in order to comply with Requirement B3. However, you contend that the existing level of fire resistance is in compliance and it is in respect of this question that you have applied for this determination.

The Applicant's case

Although you believe that the dimensions of the roof window comply with Requirement B1 you have decided to discount its potential and instead have sought to demonstrate compliance with this requirement by showing that there is a fully protected stairway. However, you are not proposing to upgrade the existing fire resistance of the first floor from its current modified 30 minutes' level because you believe that this already complies with Requirement B3. You believe that leaving the fire resistance of the first floor as it exists will not affect the means of escape from the new room and that this will be satisfactory in terms of Requirement B1 ('Means of escape'). You also do not consider that fire spread between the ground and first floor is made worse by the loft conversion.

You make the following points in support of your case:

(i) A protected staircase will be provided from the new room down to the front door of the house.
(ii) The fire resistance of the existing first floor satisfies Requirement B3 for a floor in a domestic loft conversion. You note that there is no specific requirement for a roof escape window.
(iii) Although the enclosing walls are offset between the ground and first floors by approximately 100 mm, the latter is offset towards the staircase and therefore failure of the floor due to a ground floor fire would not jeopardize the staircase.

The Borough Council's case

The Borough Council have rejected your proposals on the basis that you were not prepared to upgrade the existing first floor to the extent that it would achieve a full 30 minutes' fire resistance.

The width of the roof window is 50 mm less than the 500 mm-recommended in Approved Document 'B'. The Borough Council therefore considered that the normal loft conversion rules had not been followed due to what they judged to be the inadequate size of the escape window and that your revised proposal was to provide a fully protected route out of the dwelling instead. That being the case, the Council is of the opinion that the elements of structure, including the existing first floor, should be to a standard that will achieve a full 30 minutes' fire resistance.

In rejecting your proposals the Borough Council took account of the fact that the protected enclosure to the stairway at first floor level was not directly over the wall at ground floor level and so the stair enclosure would rely on the stability of the first floor for its support. In their view, if the floor was not upgraded then premature failure of the floor would negate the provision of the 30 minutes' enclosure to the stairway.

The Department's view

You are proposing not to upgrade the existing first floor so that it will achieve a higher standard of fire resistance. What therefore needs to be considered is the effect that this will have on the occupants of the new room in the roof space if they need to escape or be rescued in a fire situation.

The question of upgrading the fire resistance of the floor would not have arisen if the roof window had not been judged unsuitable by the Borough Council for alternative escape. The width of the window is 450 mm instead of 500 mm as recommended in Approved Document 'B'. Although the width of an escape window is one of the issues that the Department is looking at with regard to revisions of the approved document, the Department accepts the argument put forward by the Borough Council that they can only be expected to take account of current guidance.

You have opted in this case to provide a fully protected stairway instead of relying on the concession given for loft conversions which allows for a partially protected stairway and an escape window of adequate size. The standard expected for a new three storey house is that it will be provided with a fully protected stair and also that its elements of structure, which includes floors, should achieve a full 30 minutes' standard of fire resistance. In this case you are not constructing a new three storey house and the main point at issue is whether the existing first floor, with its reduced level of fire resistance, could prejudice escape via the stairway.

The modified 30 minutes standard requires that the floor must provide for a full 30 minutes in respect of its load bearing capacity (resistance to collapse) but allows for reduced performances with respect to integrity (resistance to fire penetration) and insulation (resistance to the transfer of excessive heat). The existing first floor is expected to achieve 30 minutes' fire resistance with respect to its load bearing capacity and therefore the Department takes the view that the stair enclosure is unlikely to suffer structural collapse during the escape period.

With regard to the reduced resistance to fire penetration of the existing floor, the offset of approximately 100 mm between ground and first floor walls is small and the first floor walls are further into the stairway than those on the ground floor. The Department therefore takes the view that a premature integrity failure of the floor will not, in this instance, unduly prejudice the stairway during the escape period.

The determination

The Secretary of State has given careful consideration to the particular circumstances of this case. He considers that your proposals to retain the existing standard of fire resistance rather than upgrade the existing first floor will provide an acceptable standard. Accordingly the Secretary of State determines that your proposals will comply with Requirement B3; and insofar as compliance with Requirement B3 directly affects means of escape, he determines that your proposals will also comply with Requirement B1 of Schedule 1 to the Building Regulations 1991 (as amended).

Refurbishment of offices – means of escape (B1)

This dispute arose because the Local Authority considered that an interconnecting route within the building would in effect comprise a protected stair. Management procedures proposed secured approval.

Ref.: 45.1.45 Date: 17 September 1997

In making the following determination, the Secretary of State has not considered whether the plans conform to any other relevant requirements.

The proposed work

The building to which the proposed building work relates is an existing rectangular three storey office, conference and laboratory block occupied by a number of separate companies within your client's organization. The ground floor comprises a mix of cellular offices and open plan accommodation, storage facilities, a conference room, and staff accommodation. The first floor similarly comprises cellular offices (which are less in number than on the ground floor) and open plan accommodation together with meeting rooms, a computer suite and payroll/cashier's office. The second floor consists of plant rooms only. The floors are linked by three staircases – two located off centre on both of the longer sides and inter-connected by a corridor (A3G/1) to final exits at ground level; and the other (A3G/52) located off centre on the far, shorter side which has a final exit at ground level.

Your client wishes to divide the ground floor office accommodation into two separate units to be occupied by individual companies. In order to facilitate occasional liaison visits made by their directors, and to avoid a considerable travel distance via corridor A3G/1, your client wishes to ensure a direct means of access between the two units at the opposite end of the building from the corridor.

Because of space constraints and the need to maximize the number of cellular offices, your client does not propose to provide a pass door but proposes to rely on access via and across the landing of stair A3G/1. This will be achieved via an existing door DA3G/58 in one unit which opens onto the stair (and which also provides an altern-

ative escape route via the bottom of stairway A3G/52 to a ground floor final exit) and the installation of a new door (DA3G/41) opening onto the opposite side of the landing from the other unit.

These proposals were the subject of a full plans application which was rejected by the District Council on the grounds that they considered the interconnecting route would form a primary circulation one and would therefore compromise the protected stair. However, you contend that it would not be a primary route and that as such your proposals comply with Requirement B1 ('Means of escape'). It is in respect of this question that you have sought a determination.

The Applicant's case

You disagree with the District Council's interpretation of paragraph 3.12 of Approved Document 'B' ('Fire safety') which states that: 'A protected stairway should not form part of the primary circulation route between different parts of the building at the same level.'

You maintain that stairway A3G/52 and doorway DA3G/41 do not form part of the primary circulation route because the primary circulation route for the ground floor will be via corridor A3G/1 at the opposite end of the building. However, you state that access via this corridor will be infrequent because of the independent nature of the two units.

Your client has requested total separation between the two units with the exception of limited access for directors and security staff which has been provided within the limitation of space and other operational constraints. Door DA3G/41 has been installed to provide this limited access.

In order to try and accommodate the District Council's requirements you made two proposals which they rejected. One was to provide lobby protection to stair A3G/52 by the addition of a 30 minutes' fire resisting door in front of door DA3G/41 (shown on sketch 100). The second proposal was to alarm door DA3G/41 so that a warning bell would be activated should the door be held open for more than a predetermined period.

The District Council's case

The District Council take the view that the protected stair enclosure A3G/52 will form the primary circulation route between the adjacent office accommodation because of the long travel distance involved in using the alternative route you have suggested via corridor A3G/1. The District Council have consulted the local Fire Authority who support their rejection of the proposals.

The District Council have looked at the applicant's alternative proposals but consider that neither overcomes the circulation problem. With respect to providing lobby protection to stair A3G/52, their view is that this will, in effect, exaggerate the possibility of the doors being wedged open. With respect to providing an alarm for door DA3G/41 the District Council consider that if this door is used as a circulation route for a selective number of personnel then it is likely that it will become a primary circulation route for all employees in the long-term. Furthermore they point out that the provision of an alarm to this door will not prevent constant use and will therefore render the self-closing device ineffective.

The Department's view

The intention of paragraph 3.12 in Approved Document 'B' is to prevent escape stairs being prejudiced by the products of combustion if the stairway were to be used as a primary circulation route resulting in fire doors not functioning properly. The Department's view is that if the communicating route between the units via door DA3G/41 and Stairway A3G/52 became the primary circulation route, then it is possible that this stairway could be prejudiced.

However, the Department notes that there is an alternative route of communication between the units via corridor A3G/1 and, although using this route would involve a considerable travel distance, you have indicated that the design philosophy is based on the assumption that communication between the units will be infrequent as they are in effect considered to be for the use of separate companies.

The Department considers that in this case it would be unreasonable to discount your client's stated management procedure for infrequent communication between the units and thereby assume that the route via door DA3G/41 will become the primary circulation route. If this assumption is accepted then the provision of either a lobby or a door alarm is not necessary and the Department therefore considers your original proposal shown on drawing 040 to be acceptable.

However, in reaching this conclusion the Department considers that your client will have a management responsibility to maintain the integrity of the fire protection to stair A3G/52 and to make sure that fire doors are not wedged open.

The determination

The Secretary of State has given careful consideration to the particular circumstances of this case. He considers that your proposal to subdivide the ground floor whilst providing limited access between the two units is acceptable. Accordingly the Secretary of State determines that your proposals comply with Requirement B1 of Schedule 1 to the Building Regulations 1991 (as amended).

Your attention is drawn in particular to paragraphs 16, 17 and 18 and your client's consequential responsibility to maintain the integrity of the fire protection to the stair A3G/52.

Warehouse units – internal fire spread (B3)

Roof lining over a distance of 1.5 m on each side of a compartment wall was not of limited combustibility, and the requirement was not achieved.

Ref.: 45.1.150   Date: 17 September 1997

In making the following determination, the Secretary of State has not considered whether the plans conform to any other relevant requirements.

The proposed work

The proposed building work relates to the construction of a terraced block of five single storey warehouse/industrial units, containing some two storey office accommodation. The units are separated by compartment walls of masonry construction.

The proposed roof construction comprises a continuous roof covering of profiled steel sheets with an external AA designation and a proprietary 40 mm insulation board directly underneath. The insulation board forms the internal lining of the roof. The roof sheeting and insulation are to be fixed on proprietary steel purlins which pass through each compartment wall. The junction between the walls and the roof is to be fire stopped in accordance with Requirement B3 ('Internal fire spread (structure)'). Your client does not propose to install a sprinkler system.

These proposals were the subject of a full plans application which was rejected by the Borough Council on grounds of non-compliance with Requirement B3. In the absence of proposals to underdraw the roof covering with a fire resisting material for a distance of 1.5 m each side of the compartment walls, the Borough Council were not prepared to accept that the proposed fire resistance of the roof would be in compliance with Requirement B3. However, you believe that your proposals would be in compliance and it is in respect of this question that you have applied for a determination.

The Applicant's case

You accept the argument put forward by the Borough Council that any fire protection to the underside of the roof should be applied to the whole roof structure for a distance of 1.5 m each side of the compartment walls and not just to the purlins. However, you are of the opinion that compliance with Requirement B3 is achieved without

the provision of any underdrawing of the roof structure. You have provided copies of the manufacturer's catalogue regarding the proposed insulation board which indicates that the material has an AA rating and may be used without interruption over fire resisting compartment walls in compliance with the Building Regulations.

You accept that the profiled steel roof cladding could buckle in fire but you consider that by the time the fire reached that level of intensity the buildings would have been vacated.

You point out that the buildings are 'speculative' and that the fire loadings are not known. However, in your view onerous fire loadings and any changes in purpose grouping could be taken into consideration at the relevant time.

The Borough Council's case

The Borough Council contend that you have failed to show to the Council's satisfaction that the fire stopping material will be resilient enough to maintain the continuity of the fire resistance of the compartment walls at their junction with the roof. They point out that continuity of fire resistance is a recommendation of Approved Document 'B' ('Fire safety').

To ensure that continuity of fire resistance is provided, the Borough Council require that the roof sheets and supporting structure should be underdrawn with a material that achieves 60 minutes' fire resistance for a distance of 1.5 m on each side of the compartment wall. The Council accept that the roof construction will achieve an external AA rating but believe that the insulating material forming the inner surface cannot be regarded as being of limited combustibility and is therefore unacceptable.

The Department's view

In considering this case the Department referred to the guidance given in Approved Document 'B'. Diagram 24(a) on page 57 suggests that (i) the roof covering over a distance of 1.5 m each side of the compartment wall should be designated AA, AB or AC on a deck of limited combustibility; and (ii) for a compartment to be effective there should be continuity at the junctions of the fire resisting elements enclosing a compartment.

In this case all parties accept that the roof construction will achieve an external AA rating when tested to BS 476: Part 3. However, you have not shown, and the extracts from the manufacturer's catalogue also do not show, that the internal deck of the roof construction – i.e. the insulation board – is a material of limited combustibility as defined in Approved Document 'B'. In view of this, the Department considers that your proposals may not restrict fire spread across the compartment walls. Given that your proposals do not incorporate fire resistance for a distance of 1.5 m each side of the compartment walls, the Department therefore takes the view that they do not comply with Requirement B3.

The determination

The Secretary of State has given careful consideration to the particular circumstances of this case. He considers that because your proposals do not incorporate fire resistance for a distance of 1.5 m each side of the compartment walls they will not provide an acceptable standard for restricting fire spread. Accordingly, the Secretary of State determines that your proposals do not comply with Requirement B3 of Schedule 1 to the Building Regulations 1991 (as amended).

Means of escape from two storey retail unit (B1)

The local authority wanted a fire resistant screen at the head of an accommodation stair. The developer offered a sprinkler and early warning system, and this proved to be acceptable.

Ref.: 45.1.151   Date: 13 January 1998

In making the following determination, the Secretary of State has not considered whether the plans conform to any other relevant requirements.

The proposed work

The proposed building work relates to a two storey retail unit situated in a shopping centre. The first floor comprises the prime trading area as well as containing the staff accommodation. The two floors are linked by a single accommodation stair which is open to both floors. The bottom of this stair is approximately 5 m from the ground floor front entrance door. There is an alternative exit on the ground floor away from the front entrance via a delivery ramp at the rear of the premises, and an alternative escape route from the first floor leading to a protected stairway outside the unit. You state that you have calculated the floor area available to the public within the boundaries of the unit at ground floor level to be less than 90 square metres. However, the Department estimates the total first floor area, excluding staff areas, to be approximately 250 square metres and that of the ground floor to be approximately 150 square metres.

The proposed building work comprises the fitting out of the premises to ensure uniformity with your client's standard corporate image. You wish to retain the open accommodation stair from the ground floor to the prime trading area on the first floor and to omit the usual requirement for a second storey exit leading to a protected staircase. You also wish to retain an inward opening rather than outward opening door leading onto the delivery ramp on the ground floor in order to construct a trolley ramp which will be inclined upwards from the door threshold.

These proposals were the subject of a full plans application which was rejected by the Borough Council on grounds of non-compliance with Requirement B1 ('Means of escape'). The Borough Council were not prepared to accept that the retention of an open accommodation stair would be in compliance with Requirement B1. However, you believe that this proposal would be in compliance and it is in respect of this question that you have applied for this determination. With regard to the inward opening door leading onto the delivery ramp, the Borough Council state that as the occupancy of the ground floor is estimated to be less than 50 persons they accept that this proposal demonstrates compliance with Requirement B1 and that this is a question which does not need to be determined.

The Applicant's case

You take the view that the means of escape from the first floor, which includes the use of the accommodation stair, complies with Requirement B1 ('Means of escape') of the Building Regulations. You point out that the design solution forms part of your client's corporate image which has been accepted by many other building control authorities throughout England and Wales. You also point out that in this particular case it is the first floor of the unit which needs to be seen as the prime trading floor because of its size and therefore you consider it essential to maintain an open staircase.

You accept that in a retail unit of this type there should be two storey exits from each floor which would normally take the form of two staircases. However, you make the following points in support of the acceptability of your proposals:

(i)   Because the unit is part of a shopping centre a sprinkler and fire alarm system are provided, but in addition you propose to offer a smoke detection system on the ground floor as a further compensatory feature. You consider that the sprinkler system will act to control the fire and that the smoke detection system will provide early warning of fire.

(ii)   You refer to the example given in BS 5588: Part 2: 1985 ('Code of practice for shops') of a situation where the use of an accommodation stair may be acceptable and argue that although this is not a small shop you consider the situation to be comparable.

(iii) You accept that persons needing to escape in fire from the first floor via the protected stairway outside the unit will have to traverse the head of the accommodation stair, but you maintain that travel distances are limited and that occupants of the first floor will only be subject to a 10 m travel distance towards

any fire rising through the stair. You argue that this has to be compared with guidance which refers to a permitted distance of 18 m towards a fire, from which point occupants would be able to move to a storey exit.

The Borough Council's case

The Borough Council have rejected the proposals on the basis that they do not show compliance with Requirement B1. They point out that BS 5588: Part 2: 1985 suggests that there should be two protected stairways available from each storey.

When considering your full plans application the Borough Council took the view that because the units in the shopping centre are not large then it would be unreasonable to insist on the provision of two protected stairs from the first floor and accepted an accommodation stair in lieu of one of these. However, the accommodation stair was only acceptable to the Borough Council on condition that a fire resistant screen was provided at the head of the stair to protect the first floor thus allowing a safe passage on the first floor to the protected staircase outside the unit. The Borough Council argue that because the proposals do not show any fire resistant screen, persons using the protected staircase on the first floor as an alternative means of escape would have to pass the head of the accommodation stair which could be obstructed in the event of fire developing on the ground floor.

The Borough Council take the view that they have already accepted a reduced standard of means of escape and they consider that it would be unreasonable to lower the standard any more by accepting the accommodation stair without the protection of a fire resistant screen.

The Department's view

You have requested a determination as to whether the shop fitting of the unit is in compliance with Requirement B1 ('Means of escape') and, in particular, whether the use of an accommodation stair for means of escape purposes complies with the requirement. This determination can only have regard to this specific issue and cannot consider the scheme as a whole – i.e. in the context of the shopping centre.

In this case the accommodation stair is the only stair connecting the ground and first floor. It is necessary to consider whether this stair, together with the alternative protected stair outside the unit, will form a satisfactory means of escape from the first floor. The Department notes that you are offering the benefits of the sprinkler system and an early warning system as compensatory features in lieu of a second fully protected stair. The Department also notes that the provision of a sprinkler system is required because the unit is part of a shopping complex but considers that it would be unreasonable to ignore the advantages that such a system provides with regard to controlling fire growth on the ground floor and, as a consequence, the assisting of escape from the first floor of the unit.

The Department therefore considers that, taking account of the sprinkler system and the early warning system, adequate provision is made for safe escape. However, in reaching this conclusion the Department assumes that adequate fire exit signs are to be provided on the first floor to lead persons needing to escape from the first floor along the alternative escape route via the protected staircase outside the unit. The Department considers that this is important because persons needing to use this alternative escape route will not be familiar with it since it is external to the unit in question.

The determination

The Secretary of State has given careful consideration to the particular circumstances of this case. In considering the acceptability of the accommodation stair he has assumed that adequate fire exit signs are to be provided to the alternative escape route from the first floor via the protected staircase outside the unit. On this basis the Secretary of State has concluded and hereby determines that your proposals, which exclude a fire resistant screen at the head of the accommodation stair, are in compliance with Requirement B1 ('Means of escape') of Schedule 1 to the Building Regulations 1991.

Loft conversion to first floor flat (B1)

The proposal is to form a bedroom and bathroom in the roof over an existing first floor flat, to form a maisonette. Escape measures proposed did not satisfy the Local Authority, and although measures could be taken to overcome the problem, the provision proposed did not comply with B1.

Ref.: 45.1.158   Date: 22 January 1998

In making the following determination, the Secretary of State has not considered whether the plans conform to any other relevant requirements.

The proposed work

The property to which the proposed building work relates is a two storey mid-terrace house which was converted some years ago into a ground and a first floor self-contained flat linked by a single unprotected stairway. A common entrance lobby and passage leads from the front door of the building to the stairway where two internal front doors separate the two flats at ground floor level. Full fire separation exists between ground and first floors and access is available from the existing first floor living room to a balcony at the front of the house.

The proposed building work comprises alterations to the existing roof space to the first floor flat to form a new bedroom with en suite bathroom, thus creating an independent maisonette at first and second floor levels. It is not proposed to provide a separate external entrance at ground level, nor is it proposed to provide any alternative means of escape by way of an external staircase. You state that the fire protection to the first floor elements is being upgraded to meet current Building Regulations and that you have indicated that you would be prepared to install fully linked and hardwired smoke alarms, in accordance with BS 5446: Part 1.

The new second floor accommodation will incorporate an escape rooflight at the rear of the building, overlooking a vehicular service road. In addition to the existing first floor living room access to the external balcony you state that the existing first floor kitchen and bedrooms either do, or will, contain casements providing minimum clear openings of 500 mm and 850 mm in height and whose bottoms are between 900 mm and 1100 mm from the finished floor.

These proposals were the subject of a full plans application which was rejected by the Borough Council on grounds of non-compliance with Requirement B1 ('Means of escape'). The Borough Council were not prepared to accept that the means of escape proposed in conjunction with the provision of smoke alarms would be in compliance with Requirement B1. However, you believe that these proposals would be in compliance and it is in respect of this question that you have applied for a determination.

The Applicant's case

You consider that adequate means of escape has been provided from the new second floor habitable accommodation and make the following points in support of this:

(i)   Additional fire protection is planned to the first floor element as part of the proposal and in response to the Borough Council's questioning of this statement you refer to the '. . . upgrading of inner doors and partitions beyond the possible level required at the time of the original conversion to flats.'

(ii) Escape is available from the existing first floor living room to an external balcony and escape windows of adequate size will be provided to the existing first floor kitchen and bedrooms.

(iii) A rooflight with a clear opening of 850 mm high and 500 mm wide will be provided for escape purposes from the new second floor accommodation.

(iv) The alternative of providing a means of escape via the rear of the loft and an external staircase would be strongly resisted not only by the ground floor occupant but also by the local planning authority.

(v)   A fully linked, hardwired smoke alarm system in accordance with BS 5446: Part 1 has been suggested as part of the proposal and could be made an actual requirement.

The Borough Council's case

The Borough Council have rejected the proposals on the basis that they do not show adequate means of escape and they give the following reasons in support of this:

(i)   The new maisonette will not have the benefit of an independent external ground level entrance and it was considered that the guidance given in paragraph 2.15, and in particular paragraph 2.15 (b), of Approved Document 'B' ('Fire safety') should be followed. The sub-paragraph suggests that an alternative exit from each floor should be provided.

(ii) The Borough Council were prepared to accept the existing first floor alternative escape arrangement onto the balcony but they had suggested that an alternative escape route from the loft area via external escape stairs should be provided.

(iii) The provision at second floor level of an escape rooflight was not considered acceptable at this higher level, and in addition the rooflight faces the rear of the building which is considered unsuitable for quick or easy access.

(iv) The Borough Council are conscious that fire doors to accommodation rooms can be removed or changed, and that if this should happen then the stairs leading from the new second floor could be exposed to a fire in the first floor kitchen or living room.

The Department's view

In this case the roof space of an existing first floor flat is being converted to form habitable accommodation and what is at issue is the safe escape or rescue of the occupants of the new second floor. Although the proposals are similar to a standard loft conversion in that the occupants need a similar level of protection, there is the additional factor that those occupants will have no control over the ground floor flat and in this case will be using the common escape route at ground floor level.

The Borough Council are insisting that an alternative route from the new second floor be provided, such as external escape stairs. The Department accepts that this would satisfy the recommendations given in Approved Document 'B'. However, the Department also accepts that the new second floor accommodation will only consist of a bedroom with en-suite facilities and that the provision of external stairs, even if permitted by the planning authority, may be onerous in this case. The Department also notes that an escape window of the type normally acceptable for loft conversions is being provided in the new second floor accommodation.

It is therefore the Department's view that external stairs need not necessarily be installed, provided adequate compensatory features form part of your proposals. In the Department's view such compensatory features would include the provision of a fully protected internal route of travel from the second floor to the ground floor and an adequate alarm system. You have only stated that two doors marked '++' on the plan will be 30 minutes' fire resisting and self-closing. These are the entrance doors to the ground floor flat and to the new second floor bedroom. However, you have not been specific, even in response to the Borough Council's request for more information, about the upgrading or level of fire resistance provided by the existing first floor doors marked '+' on the plan. A similar problem pertains regarding information about any proposed alarm system. You have stated that it could be made a requirement to provide an interlinked hardwired system but you have not been specific. In both instances the determination can only be based on the proposals as submitted.

In conclusion, although your proposals are similar in many ways to a standard loft conversion there are important differences. These differences may, in the Department's view, be addressed by the provision of a mains-operated alarm system within the first and second floor accommodation installed to BS 5839: Part 6: 1995, and a fully protected internal route of travel from the new second floor to the ground floor with particular attention in this case being paid to the first floor doors.

The determination

The Secretary of State has given careful consideration to the particular circumstances of this case. On the basis of the proposals as presented to him he does not consider that you have made adequate provision for safe escape and therefore determines that your proposals do not comply with Requirement B1 ('Safe escape') of Schedule 1 to the Building Regulations 1991 (as amended).

Conversion of roofspace at doctor's surgery (B1)

The matter at issue revolved around a single open stairway, and it was determined that a ground floor records office, reception area and waiting area open to the stairway did not show compliance with B1.

Ref.: 45.1.157   Date: 26 January 1998

In making the following determination, the Secretary of State has not considered whether the plans conform to any other relevant requirements.

The proposed work

The building to which the proposed building work relates is a large two storey semi-detached dwelling which was converted in 1991 into a doctor's surgery and accordingly incorporates 30 minutes' fire doors throughout. There is a single stairway leading from the first floor to the ground floor where the stair is open to the records office, reception area and waiting space. The ground floor also contains two consulting rooms which are separated from the stairway by single fire doors. The first floor comprises a treatment room, health visitor's room and staff areas. You state that the typical occupancy of each room is 2–3 people, with the exception of the waiting area which is more heavily populated.

The proposed building work comprises an extension to and conversion of the existing roof space to provide additional accommodation space on a new second floor. This additional accommodation space is to comprise three offices and a toilet and will be used for secretarial and general administrative purposes – it is not intended that there should be access by members of the public. The second floor is to be accessed by a continuation of the present stairwell with no lobby protection to the new second floor landing. You propose to upgrade the existing fire doors throughout the building with the provision of smoke seals and in tumescent strips. You also propose to install a fire alarm to BS 5839: Part 1: 1988 to an L1 standard. However, you do not propose to install lobby protection by way of a partition enclosure at ground floor level and lobby doors on the first floor landing.

These proposals were the subject of a full plans application which was rejected by the Borough Council on the grounds of non-compliance with Requirement B1 ('Means of escape'). The Borough Council were not prepared to accept that the installation of a fire alarm, as specified above, in lieu of ground and first floor stairway separation, would be in compliance with Requirement B1. However, you believe that this would be in compliance and it is in respect of this question that you have applied for this determination.

The Applicant's case

You state that the Borough Council were not prepared to accept your proposals because the single staircase will not have two door protection at ground and first floor level. You consider that their require-ment for a partition enclosure at ground floor level and the provision of lobby doors at first floor level to be excessive in terms of both the risk and the alternative provisions proposed. In this respect you point to your proposal to upgrade the existing fire doors throughout with smoke seals and intumescent strips, and your proposal to install a smoke detection system to BS 5839 (L1 standard).

You advise that the building will only be occupied by members of the public when staff are in the building and will often be accompanied by members of staff for the purposes of treatment. You consider that the additional doors and lobbies that the Borough Council wishes to be installed will prove awkward both in terms of general use and circulation within the building. You also consider that they will have a negative effect on the aesthetics of the building.

You understand that the Borough Council's main concern is smoke-logging of the stairwell. However, you point out that the number of persons occupying each room will be relatively low and therefore in your view the quantity of smoke likely to escape into the stairwell, whilst occupants vacate the room via the single door protection, will be minimal. In response to the argument put forward by the Borough Council that the fire protection could be removed at first floor level you contend that this problem could be accentuated if lobby doors were to be installed.

The Borough Council's case

The Borough Council are not prepared to accept your proposals. They consider that they do not show compliance with Requirement B1 ('Means of escape') of the Building Regulations 1991 for the following reasons:

(i)   Lobby protection has not been shown to the stairway at ground and first floor level.
(ii)   The number of storey exits opening onto the stairway enclosure at ground and first floor level is excessive.
(iii)   There is a danger that at first floor level the fire protection to the stairway could be removed due to communication between the different rooms, particularly during the summer months.
(iv)   The records office, reception and waiting space should not be incorporated within the stairway enclosure.
(v)   With regard to the proposed automatic fire detection system, the Borough Council are of the opinion that this will not be an adequate compensating factor for the omission of physical fire protection to the stairway enclosure at ground floor level.

The Department's view

The Department considers that it is the safety of persons using the new second floor office accommodation which has to be considered if they should need to escape in fire. The Borough Council are requesting that lobby protection should be provided at first and second floor levels and the Department understands the concerns expressed by the Borough Council. However, the Department notes that a full detection system is being proposed and is of the opinion that this will give sufficient early warning, when the upgrading of the doors is taken into account, for persons to escape safely without lobby protection. In reaching this conclusion the Department has taken account also of the low numbers of persons stated to be using the premises and the level of management cover.

The Borough Council has raised two further issues, one being that the first floor protection to the stair could be prejudiced and the other being the amount of accommodation open to the stair at ground floor level. With regard to the first issue the Department considers that maintaining the fire separation by making sure that fire doors are properly shut is a management issue. To ask for double door protection in this case on the basis that management is possibly going to fail in its duty to keep the doors shut would, in the Department's view, be unreasonable.

With regard to the stair being open to the ground floor accommodation the Department accepts the Borough Council's judgement on this issue and considers that this could be a fire hazard. Paragraph

4.31 of Approved Document 'B' ('Fire safety') suggests that the facilities incorporated in a protected stairway should be limited to a reception desk or enquiry office area providing it is not the only stair serving the building. In this case the single stair is open to the record office, reception area and waiting space. The Department considers this to be inappropriate.

The determination

The Secretary of State has given careful consideration to the facts and the particular circumstances of this case. He has concluded that the proposed open stair with no lobby protection to any of the three floors is acceptable. However, your proposals at ground floor level incorporate the present arrangements whereby the ground floor records office, reception area and waiting area are incorporated into the single stairway. The Secretary of State has concluded, and hereby determines, that this arrangement will not comply with Requirement B1 of Schedule 1 to the Building Regulations 1991 (as amended).

Roof level bedroom extension (B1)

There was argument regarding the number of storeys in this house and escape measures hinged on whether the property has three or four storeys. It was determined that as there were four storeys the proposals did not comply with B1.

Ref.: 45.1.152   Date: 2 February 1998

In making the following determination, the Secretary of State has not considered whether the plans conform to any other relevant requirements.

The proposed work

The proposed building work relates to a roof extension to an existing three storey, end of terrace, single family dwelling on a corner site. The front door of the property (access level) is at first floor level via steps that lead to the street level. The pavement level is approximately 2.1 m above the ground level adjacent to the property and the Borough Council state that there is a wall on top which extends 1.35 m above pavement level. The gardens of the property are approximately at the level of the ground (lowest) floor which you describe as the basement and an external stairway connects the lowest level to the pavement.

The proposed building work comprises the addition of an extension above the flat roof at the rear of the property, in order to create a new bedroom with a floor area of approximately 24 square metres. This bedroom will be contained within the structure of a new hipped roof which will be integrated into the existing pitched roof to the front of the building. This would render the property a single family four storey dwelling with all levels connected by a single internal stair which as currently exists is unprotected because it is open to the study on the ground/access level and the lower level dining room.

The house is an end-of-terrace corner one and the ground has been dug out on three sides where there are external retaining walls. As a result the front door is effectively on the first floor. However, the plans you have submitted designate the floors as basement, ground floor, first floor and new second floor.

You submitted two proposals in your full plans application – a 'preferred' and an 'alternative' scheme. Both were based on the proposition that the front door level represented the ground level, that a fully protected stair need not be provided, and that inter-connected smoke alarms would provide sufficient compensatory features to ensure compliance with Requirement B1 in either scheme. Both schemes were rejected by the Borough Council on the grounds that they did not show compliance with Requirement B1. However, you believe that both your schemes would be in compliance and it is in respect of this question – and in particular with regard to your 'preferred' scheme – that you have applied for this determination.

The Applicant's case

You accept that because the existing house is three storeys then the concessions given in Approved Document 'B' ('Fire safety') for loft conversions are not strictly applicable, although you conclude that the level of safety provided by the concessions should be considered as reasonable in your case. In your view it would be highly artificial not to take account of the guidance for loft conversions because the front door is on the first floor; and you contend that had the same plans been submitted in respect of an adjacent property where the ground has not been dug out at the front, the Borough Council would have had to approve your alternative scheme notwithstanding the fact that you consider that rescue would be more difficult in the particular circumstances of that property.

You consider that the most important issue is the distance from the new top floor to the final exit and because the final exit is a storey above ground level then you consider it to be wrong to consider the existing house as a three storey one. You point out that the distance from the new top storey to the final exit will be no greater than the same distance in a two storey house with a loft conversion. Your interpretation of paragraph 4.4 of BS 5588: Part 1 is that it defines ground level as being the ground level or access level. You therefore submitted the following two schemes which you consider satisfy the Building Regulations.

The 'preferred' scheme (drawing 259/7):

This arrangement has the following features:

(i)   mains-operated interconnected smoke detectors in every room except bathrooms;
(ii) a fire door on the new second floor bedroom;
(iii) a new partition and fire door separating the entrance hall and stairway (indicated on the drawing as being on the ground floor);
(iv) all doors giving access to the stairway will be made self-closing;
(v)   an escape window in the new bedroom (shown on the drawing as being on the new second floor) which will allow access to rescuers equipped with a ladder.

You consider that the main element of safety in this arrangement is the provision of the fire alarm system and the major difference from the Council's requirements is the absence of fire separation between the ground floor and first floor.

The 'alternative' scheme (drawing 258/8):

This arrangement has the following features:

(i)   a partition and fire door between ground and first floors;
(ii) mains-operated interconnected smoke alarms in circulation areas only;
(iii) a fire door on the new second floor bedroom;
(iv) a new partition and fire door separating the entrance hall and stairway (indicated on the drawing as being on the ground floor);
(v)   all doors giving access to the stairway to be made self-closing;
(vi) an escape window in the new bedroom (shown on the drawing as being on the new second floor) which will allow access to rescuers equipped with a ladder.

You consider that this approach is based on the loft conversion concessions given in paragraphs 1.23 to 1.31 of Approved Document 'B' except that additionally you are providing smoke detection in circulation spaces.

In conclusion you consider that the Secretary of State must decide whether the proposals comply with the Building Regulations and in particular the limitations imposed by Regulation 8 in respect of health and safety, and not simply whether they comply with Approved Document 'B'. You do not consider that the situation appertaining to your case is adequately dealt with in Approved

Document 'B' and that non-compliance with it should not be regarded as non-compliance with the Building Regulations.

The Borough Council's case

The Borough Council are of the opinion that the existing property is not a two storey dwelling and as a consequence of this they consider that the concessions given in paragraphs 1.23 to 1.31 of Approved Document 'B' are not applicable.

The Borough Council also make the following points in support of their rejection of the proposals:

(i) Ladder rescue of persons on the new third floor will be inhibited by the horizontal distance of approximately 3 m from the building to the pavement, the boundary wall/railing and the height of the building in relation to the adjacent ground level.

(ii) The building as proposed will have four storeys and will have more than one floor in excess of 4.5 m above ground level. In these circumstances Approved Document 'B' refers to the guidance given in BS 5588: Part 1: 1990.

(iii) The guidance given in both the British Standard and the Approved Document suggests that no inner rooms are acceptable where the floor level is in excess of 4.5 m above ground level. Therefore the kitchen and dining room at the lowest level and the study at entrance level should be fire separated from the staircase. Without this separation the effects of fire could inhibit or prevent escape from the new third floor. The staircase enclosure should be constructed as a protected stairway with all doors leading onto the stairway being FD30.

(iv) The guidance in BS 5588 which is referred to in Approved Document 'B' also recommends that the new third storey be provided with an alternative escape route. However, the Council recognizes that the provision of a second escape stair will be impractical and are prepared to accept a full alarm and detection system as being suitable compensation for the second stair.

(v) They do not accept that paragraph 4.4 of BS 5588: Part 1 defines the ground level as ground level or access level.

The Department's view

In assessing both your 'preferred' and 'alternative' schemes the Department has to consider the safety of occupants on the new floor should they need to escape or be rescued in a fire situation. The fundamental issue, therefore, is whether the house as converted is a three or four storey property for the purposes of assessing the level of fire precautions that are necessary to satisfy the Building Regulations. This in particular affects the degree of fire protection to the stairway and the required extent of an early warning system.

The Department accepts your points regarding the necessity of your proposals to comply with the Building Regulations, rather than necessarily complying with Approved Document 'B'. There may well be alternative ways of achieving compliance. Thus there is no obligation to adopt any particular solution contained in an Approved Document if you prefer to meet the relevant requirement in some other way. The Department does, however, take the view that Requirement B1 is a life safety matter directly related to health and safety as prescribed by Regulation 8.

The Department accepts that normal access to the dwelling from the street is via steps to the main entrance door at first floor level, but also notes that in accordance with the normal measurement rules the building, as proposed, will consist of four storeys above ground level.

You claim that the concessions applicable to a two storey house with a loft conversion should be acceptable in this case because normal access is at first floor level. However, the Department does not support this view because, as has been pointed out by the Borough Council, it too considers that there would be difficulty in providing safe ladder access from the street to the new third floor room to effect self- or assisted-rescue.

The Department also takes the view that the Borough Council were correct in their interpretation that the dwelling as proposed should be considered as a four storey house. It follows that the loft conversion concessions given in Approved Document 'B' are not appropriate in this case and that the guidance given in both the BS 5588 and the Approved Document 'B' should be followed. This means that a fully protected stair needs to be provided which should not be open to basement/lower ground floor accommodation.

The Department accepts that a smoke detection system has been offered as part of either scheme (albeit that the 'alternative' one restricts the detection system to the circulation areas only) but takes the view that by the time this has been activated escape via an unprotected stairway could have been prejudiced by smoke. Since therefore neither the 'preferred' scheme nor the 'alternative' scheme provide a fully protected stair in accordance with the guidance, the Department considers that neither scheme will provide safe escape in fire.

The determination

The Secretary of State has given careful consideration to the facts and the particular circumstances of this case. He has concluded, and hereby determined, that neither the proposals in your 'preferred' nor those in your 'alternative' scheme comply with Requirement B1 of Schedule 1 to the Building Regulations 1991 (as amended).

Single stair in new office block (B1)

There was to be a single stair at the junction of two wings in this new office block. No protecting screens were proposed to the stairway, and and it was determined that the proposal did not meet B1.

Ref.: 45.1.161   Date: 10 February 1998

In making the following determination, the Secretary of State has not considered whether the plans conform to any other relevant requirements.

The proposed work

The building work comprises the erection of your client's new HQ office block which is to be associated with a new heat treatment factory. The office comprises a ground and first floor with two perpendicular wings joined at right angles. At the juncture of the wings there is a single central staircase, at the base of which is a reception area. Each of the wings is approximately 20 metres in length by 10 metres in width.

The ground floor plan shows seven numbered offices, two other large offices, an archives room, a computer room, a conference room, a plant room, a tea-making room, and toilet facilities. Access to each office is via a central circulation corridor in each wing leading to the juncture of the wings and the central single staircase. A ½ hour fire resisting glazed screen with doors is shown on the plan separating these corridors and circulation areas from the stairwell containing the single stairs, a lift, and a reception area.

The first floor plan shows 11 numbered offices in addition to a boardroom, a tea-making room and toilet facilities. As with the ground floor, access to each office is via a central circulation corridor in each wing leading to the single stair. Access to the stairwell is, again, shown as being through doors in a fire resisting glazed screen.

The conference room/boardroom, tea-making room, and toilet facilities on both floors open into the circulation area between the wings, opposite the fire resisting screens.

The building is equipped throughout with a fire detection and alarm system, including smoke and heat detectors in all areas. All accommodation areas are separated from escape routes by a minimum 30 minutes' fire resisting construction. All doors leading onto escape routes will be provided with smoke seals and self-closing devices. Cloakrooms and storage areas will be provided with the same level of protection. The reception area is to be designated as a 'sterile' space containing no significant quantities of combustible materials.

These proposals were the subject of a full plans application which was rejected by the Borough Council on grounds of non-compliance with Requirement B1 ('Means of escape'). The Borough Council considered that even with fire resistant screens travel distances would be in excess of that recommended in Approved Document 'B' ('Fire safety') and that the omission of fire resisting screens would further increase the travel distances. However, you believe that your proposed fire safety provisions would be in compliance without the screens being in place and it is in respect of this question that you have applied for this determination.

The Applicant's case

You state that your client has specifically requested that the screens forming the circulation route, and separating the stairway, on both floors should not be provided. You refer to the report submitted by your fire consultant and make the following points in support of your submission:

(i)   You accept that the recommended travel distances given in Approved Document 'B' will be exceeded but additional safety features are provided such as the early warning system which you consider will increase evacuation times. The alarm system is understood to comply with BS 5839: Part 1: 1988 (L1 standard).

(ii)   Cross circulation between the wings will not occur in fire because this is an office building and evacuation would be the major priority.

(iii) Various discussions have taken place regarding refuge areas for disabled persons and if the screens were not installed then a suitable disabled refuge will be provided.

(iv)   The reception area is to be kept as a sterile space with reception desks and seating being constructed of materials of limited combustibility, hardwood or stone, with small amounts of padding. Management control procedures will prevent significant quantities of combustibles being placed in this area.

(v)   You consider that the provision of a screen across the stair/reception area would not be of any benefit and may even complicate the exit route and hinder escape.

The Borough Council's case

The Borough Council, after consultation with the Fire Authority, have rejected the proposals on the basis that they are not acceptable and do not show compliance with Requirement B1. They make the following points in support of their decision:

(i)   Provided the fire resisting screens are in place to separate the cross circulation route from the stairway at both ground and first floor levels, then the travel distance to a storey exit is estimated at 25.5 m. Without the screens in place, the maximum travel distance increases to 32 m.

(ii)   Omission of the fire resisting screens which separate the ground and first floor central corridors from the staircase would mean that circulation between the wings of the building would be via the protected stair enclosure. This is contrary to the guidance given in paragraph 3.12 of Approved Document 'B'.

(iii) The guidance given in paragraph 4.31 (c) of Approved Document 'B' limits the use of reception areas within protected stair enclosures but with the proviso that it is not the only stair serving the building. In this case there is only a single stair available for escape.

(iv)   Omission of the screens would increase the risk to disabled persons using the staircase enclosure as a refuge area whilst waiting for assistance with evacuation.

The Department's view

In the Department's view there are two main issues to consider:

(i)   whether it would be permissible in terms of safe means of escape to remove the screens which provide fire separation to the cross circulation route between the wings of the building at ground and first floor levels; and

(ii)   whether it is acceptable to provide a reception area in the single stair enclosure at ground floor level.

(i) *Removal of screens on both floors*   The Department accepts that a full alarm and detection system is to be incorporated but notes that this case relates to quite a large two storey office building comprising two wings on each floor, and with each wing containing a number of offices or other rooms. All these facilities are served by a single centrally-located stair. The Department also notes that even with the protecting screens in place then travel distances exceed the recommended maximum given in Approved Document 'B' and that the travel distance situation is made worse if the screens are removed.

The Department agrees with the Borough Council's view that if the protecting screens are removed then the circulation route from one wing to the other would be via the stair enclosure which is contrary to the recommendations given in Approved Document 'B'. The Department also considers this to be unacceptable in terms of the size of the building and its occupancy. In addition the Department shares the Borough Council's concern regarding the lack of a safe refuge facility for disabled persons on the first floor if the screens are removed; and is also concerned that with the screens removed the tea-making rooms, toilets, and the conference and boardrooms between the wings of the building would open directly into the stair enclosure. The Department therefore considers that the removal of the screens separating the circulation routes on the ground and first floors from the stair enclosure could prejudice safe escape.

(ii) *Reception area in stair enclosure*   The Department accepts that the guidance given in Approved Document 'B' does not not recommend acceptance of a reception area where there is a single stair situation. However, in this case the Department considers that it is reasonable to take account of the proposed early warning system and the management undertaking that this will be a sterile area. Nevertheless, acceptance of the reception area in this location must rely on there being fire separation between the reception area and the circulation route, and is therefore dependent on the protecting screens being in place.

The determination

Your revised proposals to comply with Requirement B1 of Schedule 1 to the Building Regulations 1991 involve the omission of any form of fire separation between the circulation routes and the single stair on both floors, and the retention of a reception area in the single stair enclosure at ground level.

The Secretary of State has given careful consideration to the facts and the particular circumstances of this case. He has concluded, and hereby determines, that your proposals do not comply with Requirement B1.

Complex atrium – internal fire spread (B3)

Alterations are proposed to a new four storey office building. A separate tenant is to occupy the second floor and the Local Authority require fire resistance of service shafts to be 60 minutes. However, the determination allowed a 30 minutes' level of fire resistance.

Ref.: 45.1.60   Date: 8 May 1998

In making the following determination, the Secretary of State has not considered whether the plans conform to any other relevant requirements.

The proposed work

The proposed building work relates to alterations to a new four storey office building. The office accommodation is at ground floor, first floor and second floor levels; with car parking and plant areas at basement level. The building is essentially triangular in footprint with four wings and a concave façade to the hypotenuse of approximately 200 metres in length. In the centre is an atrium at each storey level. Its configuration is complex and involves five linked open spaces extending vertically through the building beneath a glazed roof covering.

The complex atrium design incorporates a natural ventilation system which operates in tandem with a P1 (property protection type) smoke detection system installed in accordance with BS 5839: Part 1: 1988 ('Fire detection and alarm systems for buildings – code of practice for system design, installation and servicing'). The means of escape provisions are based on the assumption of a total evacuation via escape stairs. The perimeter stairways incorporate dry risers, have two hours' fire resistance, and are designated as firefighting stairs.

Your client proposes to accommodate a separate tenant on the second floor. Requirement B3 ('Internal fire spread (structure)') specifies that: 'To inhibit the spread of fire within the building it shall be sub-divided with fire-resisting construction to an extent appropriate to the size and intended use of the building.' The implication of this is that the service shafts passing through and between the first and second floors will need to become protected shafts or alternatively a complete barrier would need to be provided within the service shafts at second floor level. Your client is proposing to adopt the former solution and does not intend to up-grade the current 30 minutes' fire resistance of the protected shafts.

These proposals were the subject of a full plans application which was rejected by the District Council on grounds of non-compliance with Requirement B3. The District Council were not prepared to accept that your proposed level of protection of 30 minutes for the service shafts passing from the first floor to the second would be in compliance with Requirement B3. They cited the guidance, given in Approved Document 'B' ('Fire safety') and contended that there should be a 60 minutes' level of protection. However, you believe that the proposed separate occupancies do not increase the fire risk and that the level of fire separation between the first and second floors of 30 minutes, currently proposed for the building as a single occupancy building, will remain adequate when separate tenancies are introduced to the second floor. Accordingly, you believe that this proposal would be in compliance with Requirement B3 and it is in respect of this question that you have applied for this determination.

The Applicant's case

In support of your case you contend that your proposals to fire-stop small openings in the general areas of the second floor slab and to maintain the existing 30 minutes' fire enclosure of the service shafts at all floor levels, including the second floor, will form an adequate fire separation between the first and second floor. You consider this to be especially so since you have not been asked to provide fire resisting glazing to the circulation areas around the atria or to the external walls.

You make the following additional points in support of your submission:

(i) All parts of the building will be used for office or ancillary purposes and the subdivision of the building into separate tenancies will not affect this. Although with the introduction of more than one tenancy security controls on access within the building will be introduced, the means of escape provisions will not be affected by the sub-division.

(ii) Separate occupation of the second floor will not increase the fire loading of the building; nor can it be considered to have an impact on the overall fire risk. You point out that the smoke control analysis carried out by your fire consultants already assumes that escape from the second floor will be the critical factor.

(iii) The tenancy on the second floor will need to be divided off from the escape stairs in order to maintain the provision for safe escape and this would allow efficient firefighting from any stair enclosure.

(iv) You have agreed with the District Council that the existing curtain walling to atria and courtyards, together with the external glazing, will not require fire resistance. Therefore in your view increasing the fire resistance of the service risers would not appear to have any detrimental effect on fire spread from the lower office areas to the second floor offices.

(v) The alarm and detection systems are automatic throughout and will be monitored by full-time security personnel.

The District Council's case

The District Council consider that it is necessary for adequate compartmentation to be provided between the first and second floors in order to prevent the rapid spread of fire between separate occupancies. In their view reliance upon the existing 30 minutes' fire resistance will not achieve compliance with Requirement B3. On the assumption that adequate compartmentation is in place, and on the basis of the smoke control analysis, they consider the non-fire resisting glazing forming the enclosure to the atrium to be acceptable.

The District Council state that the structural fire precautions for the building have been designed and checked in accordance with Approved Document 'B' ('Fire safety') and in particular Section 8 ('Compartmentation') of that document. They point out that paragraph 8.10 of the Approved Document suggests that a wall or floor provided to divide a building into separate occupancies should be constructed as a compartment wall or floor. The Approved Document further suggests (paragraph 8.30) that if the floor is a compartment floor then the service shafts passing through the floor are deemed to be protected shafts and as such the standard of fire resistance of the service shafts should be uprated to achieve 60 minutes' fire resistance in accordance with Tables A1 and A2 on pages 96–98 of the Approved Document.

The Department's view

The Department agrees that the guidance given in Approved Document 'B' suggests that a floor dividing a building into separate occupancies should be constructed as a compartment floor with the consequence that service shafts passing through that floor should be constructed as protected shafts with the appropriate level of fire resistance, in this case 60 minutes, to maintain the compartmentation. However, the question being put in this particular case is whether the service shafts, which are only provided with 30 minutes' fire resistance, are acceptable.

The Department notes that a full alarm and detection system will have been provided throughout the building which, you state, will be monitored by full-time security control personnel. The Department accepts that the alarm system will form part of the fire engineering package solution for the atria design, although in the Department's view such an alarm system would be of benefit elsewhere in the building in that it will give early warning of fire in other areas such as in the service duct locations. Approved Document 'B' does not give recommendations for alarm systems in office-type buildings. Therefore the installation of such a system could be considered to be a compensatory feature in lieu of the full 60 minutes' fire resistance of the service risers.

In this case you have stated that the second floor of the building is to be tenanted. In the Department's view it appears that the amount of space available for a tenancy could vary. Notwithstanding this the Department considers that the building as a whole will remain in single overall ownership.

After having considered all the relevant aspects of this case, the Department takes the view that the risk of fire spread from the first to the second floor is no greater because of the proposed tenancy than if the building remained in single occupation. The Department therefore considers that it is unreasonable in this case to ask for a higher standard of fire resistance for the protected shafts than the standard of 30 minutes already proposed.

The determination

The Secretary of State has given careful consideration to the facts of this case and the arguments put forward by both parties regarding the circumstances whereby the tenancy will be created. He has concluded, and hereby determines, that your proposals to maintain the 30 minutes' level of fire resistance to the service shafts between the first and second floors are in compliance with Requirement B3 of Schedule 1 to the Building Regulations 1991 (as amended).

PART E: RESISTANCE TO PASSAGE OF SOUND

Floors of converted flats – sound insulation (E2 and E3)

A large three storey Victorian house is to be converted into four flats where two dwellings – one flat and one maisonette – already exist. An amendment of the regulations made in 1995 would apply to this proposal had it not been submitted prior to the amendment coming into effect.

Ref.: 45.1.116   Date: 31 January 1996

The Secretary of State takes the view that he is being asked to determine under section 16(10)(a) whether plans of your proposed work are in conformity with the Requirements E2 and E3 ('Airborne sound (floors and stairs)' and 'Impact sound (floors and stairs)') to the Building Regulations 1991 (as amended). In making his determination the Secretary of State has not considered whether the plans conform with any other relevant requirements.

Your request for a determination concerns the alteration of a large three storey Victorian house currently containing one flat on the ground floor and a maisonette occupying the second and third floors so as to convert the maisonette into three self-contained flats – i.e. two one-bedroom flats on the first floor and one two-bedroom flat on the second floor. The original conversion took place some 30 years ago. The existing floor separating the ground floor flat and the maisonette (i.e. the first floor) is not a sound resisting type.

The Borough Council has rejected your proposals on the grounds that in their view the application constitutes a material change of use as defined by the Building Regulations 1991 (as amended) and that the proposals fail to demonstrate a satisfactory level of sound reduction performance between the two proposed flats at first floor level and the existing ground floor flat in respect of Requirements E2 and E3. You contest this interpretation of a material change of use and have sought a determination based on the assumption that your proposed work will not be required to comply with Requirements E2 and E3.

The Borough Council accepts that the first floor currently separates two residential units but in their opinion the subdivision of the maisonette at first floor level into two flats will give rise to a contravention under Regulation 4.

The Borough Council argue that the formation of the flats at first floor level constitutes a material change of use under Regulation 5(a). They base this on their interpretation of Regulation 5(a) which refers to a change of use as applying when: 'the building is used as a dwelling, where previously it was not' and the fact that Regulation 2 defines a building as including 'part of a building'.

The Borough Council therefore contend that if a material change of use is to occur, Regulation 6(1)(e) will be applicable and that the work would be required to comply with Requirements E1 to E3, and that as a result adequate resistance to the passage of sound should be provided by a floor between dwellings regardless of whether or not the floor existed between a dwelling and part of a dwelling prior to the material change of use.

You contend that your proposals do comply with the relevant Requirement on the basis that the existing floor between the ground floor flat and the first floor after the work is completed will still be performing the same function as before and therefore sound require-

ments do not apply. You argue that the existing first floor both before and after conversion will remain unaltered in terms of its use because it will still be separating one flat from another.

The Department's analysis of this case is as follows. In its view at the date of your application it was not appropriate to define your proposals as a material change of use because no one part of the building had been used as other than a dwelling. the Building Regulations 1991 (as amended) were further amended by the Building Regulations (Amendment) Regulations 1995 which came into effect on 1 July 1995. This amendment introduced Regulation 5(g) ['or (g) the building, which contains at least one dwelling, contains a greater or lesser number of dwellings than it did previously']. In the Department's view this new Regulation would have meant that your proposal would have been defined as a material change of use, had your application been made on or after 1 July 1995.

Apart from any relevant change of use defined in the Building Regulations 1991 (as amended), Part E would also apply if a material alteration had been made. Neither party to this determination has raised this point, but the Department has given consideration to it. In carrying out the conversion work the escape routes from the first and second floors will be altered. On the assumption that at some stage the work would not comply with a relevant Requirement (B1 'means of escape' in this case), then the work would fall within the definition of 'building work' contained in Regulation 3.

Regulation 4(1) requires that 'building work' is carried out so that it complies with the relevant requirements in Schedule 1 and that in complying with them there is no failure to comply with any other such requirements. However, it appears that the building work involved would not affect the structure of the first floor and that therefore this Regulation would not apply to this floor.

Regulation 4(2) requires that after 'building work' has been completed the building either complies with the relevant requirements of Schedule 1 or is no worse than it was before. However, in this case because it appears that the building work would not affect the first floor, the sound insulation provided by the first floor cannot therefore be any worse than it was before. It follows that on this basis the existing construction of the first floor complies with the Building Regulations.

The Secretary of State has given careful consideration to all the facts of this case. In particular he has noted that your application for Building Regulation Approval was made prior to 1 July 1995. The Secretary of State has concluded that Requirements E2 and E3 do not apply to the first floor of your proposals and therefore that your proposals in respect of the continued structural use of the first floor in its unmodified state comply with the Building Regulations 1991 (as amended prior to 1 July 1995).

Sound insulation of a party floor (E2 and E3)

This determination refers to airborne and impact sound insulation in respect of the alteration of a lower ground floor flat to create an additional flat.

The work proposed involves creating two flats where previously one existed and this constituted a material change of use requiring compliance with E1 to E3 for the whole of the floor.

Ref.: 45.1.124   Date: 25 June 1996

The Secretary of State takes the view that he is being asked to determine under section 16(10)(a) whether plans of your proposed work comply with the Requirements E2 ('Airborne sound (floors and stairs)') and E3 ('Impact sound (floors and stairs)') to the Building Regulations 1991 (as amended). In making his determination the Secretary of State has not considered whether the plans conform with any other relevant requirements.

The proposed building work

Your request for a determination relates to the alteration of a single lower ground floor flat into two separate flats; the new one to be the smaller of the two. There are two existing flats on the ground floor above. You state that the elements forming the vertical separation between the lower ground flats will accord with Approved Document 'E' ('Resistance to the passage of sound'). However, you did not propose any upgrading of the sound proofing of the floor construction between the ground and lower ground floor.

The Council rejected your full plans application containing these proposals on the grounds of non-compliance with Requirements E2 and E3 of the Building Regulations 1991 (as amended). Your subsequent request for a relaxation of the Regulations in respect of the requirements to sound proof the existing floor only in the area covering the new, smaller, flat, was viewed by the Council as a request for a dispensation. They considered that the Requirement applied to the whole of the lower ground area, including the existing flat, and therefore refused your request.

At the same time, the Council advised that they were prepared to consider a relaxation based on insulation against airborne sound only being provided in certain rooms and insulation against airborne and impact sound in others. This was based on the vertical alignment of rooms in different types of use. They state that no response was received from you following this offer. You subsequently applied for this determination.

The Applicant's representations

You contend that the balance of the floor area covering the existing flat after the creation of the new, smaller, flat is not required to comply with the Building Regulations because the proposed work would not cause a greater contravention than exists at the present time. However, because you believe that the area of the new flat is materially different given that there will be a flat where previously there was not, you accept that up-grading of the floor over this area is required.

The Borough Council's representations

The Council relies for locus in requiring compliance of all the floor area with Requirements E2 and E3 on the fact that a change of use of part of the building – i.e. the lower ground floor – is proposed. For this they rely upon new regulation 5(g) of the Building Regulations 1991 which was inserted by the Building Regulations (Amendment) Regulations 1995. This added a further definition of a material change of use, namely: '. . . so that after that change the building, which contains at least one dwelling, contains a greater or lesser number of dwellings than it did previously.'

The Council then argue that an additional flat is being created at the lower ground floor level, that a material change of use of part of the building will take place, and that that part of the building must therefore comply with the relevant requirements of Schedule 1 to the Building Regulations. Regulation 6(e) specifies the additional requirements that in the case of a material change of use described in Regulation 5(g) must be complied with – these are requirements E1 to E3.

The Department's view

The Department accepts the Council's view that the proposed work constitutes a material change of use under the Building Regulations (as amended in 1995). However, because the material change of use is in respect of part only of a building it is therefore necessary for the Council to rely upon Regulation 6(2)(b) to require compliance. Because the material change of use transcends the area of both the new and reduced remaining area of the existing flat, in the Department's view it is correct for the Council to interpret the 'part only of the building' as being the whole area of the existing lower ground floor flat.

The Department takes your reference to the 'proposed works not causing a greater contravention than exists at present' in respect of the balance of the floor area over the existing flat, to be a reference to Regulation 4(2). However, to rely upon Regulation 4 it would be

necessary to demonstrate for the purposes of Regulation 4(2)(a) that the building work constitutes a material alteration. A material alteration is defined in Regulation 3(2) and it is the view of the Department that your work as described does not fall within this definition.

The determination

Having given careful consideration to this case, the Secretary of State is content that it is appropriate for the Borough Council to require compliance of the whole floor construction between the ground floor and lower ground floor with Requirements E2 and E3. Accordingly, he determines that your proposals to upgrade the sound insulation of only that area of floor above the new, smaller, flat does not comply with Requirements E2 and E3 of Schedule 1 to the Building Regulations 1991 (as amended).

Additional flat in existing building comprising a dwelling and a shop (E1 and E2)

This is a very useful determination, setting out as it does an explanation of 'change of use' in respect of premises where part of a dwelling was to be converted into a self-contained flat.

Ref.: 45.1.164   Date: 6 February 1998

In making his determination the Secretary of State has not considered whether the plans conform to any other relevant requirements.

The proposed building work

The properties to which the proposed building work relates comprise 2 two storey dwellings occupied and used as one dwelling and a shop. Approximately one half of the ground floor contains a shoe shop and workshop behind, and the other half comprises a two storey dwelling, the first floor of which extends over the top of the shop and workshop on the ground floor. The proposed building work is to convert that part of the first floor of the house which is over the shop and workshop, into a self-contained flat, leaving the original house comprising the remaining part of the first floor and the ground floor area as before. To achieve this conversion additional extension work, including provision of a new external stair, is proposed but is not of relevance to this determination.

The existing wall at ground and first floor level between the shop and the dwelling, and the proposed flat and dwelling, is of $4\frac{1}{2}$ inch common brick with plaster finish to both sides. You amended your plans to include a 32 mm layer of gyprock tri-line to the wall between the flat and the dwelling at first floor level. Your proposals as amended did not contain any provision for additional sound insulation for the wall between the dwelling and the shop; nor do they make provision for additional sound insulation of the floor between the flat and the shop.

These proposals were rejected by the District Council on grounds of non-compliance with Requirement E1 ('Airborne sound (walls)') and Requirement E2 ('Airborne sound (floors and stairs)'). The District Council drew your attention to the existence of paragraph (g) to Regulation 5 ('Meaning of material change of use') of the Building Regulations 1991 (as amended) but you maintained that it was not necessary for sound insulation to be provided to the wall between the shop and the dwelling and to the floor between the proposed flat and shop below in order to achieve compliance with Part E. It is in respect of this question that you applied for a determination.

The Applicant's case

You accept that Regulation 5(g) means that the creation of two dwellings from the one means that a material change of use will occur and that the requirements of Part E should be applied.

However, you do not accept that there is a material change of use between the shop and the flat over, and the shop and the adjacent dwelling. You argue that the use of the building in these particular areas will not change and that therefore no material change of use will occur. It follows that you accept that the requirements should be met in respect of the first floor wall between the existing dwelling and proposed flat, but you do not believe that sound insulation needs to be applied to the wall at ground floor level nor to the existing floor separating the proposed flat from the shop.

The Local Authority's case

The District Council takes the view that the proposals will constitute a material change of use under both Regulation 5(b) and 5(g) – i.e. respectively 'the building contains a flat or flats, where previously it did not'; and 'the building which contains at least one dwelling, contains a greater or lesser number of dwellings than it did previously.' The District Council say that it follows that Regulation 6 requires that where such changes of use have taken place the requirements, among others, of Part E shall apply. On that basis the District Council has requested sound insulation to the existing floor, and the existing walls at ground and first floor level to the standards recommended in Approved Document 'E' ('Airborne and impact sound').

The Department's view

Regulation 2(1) provides that the interpretation of a 'building' includes a reference to part of a building. This provision is subject to the opening line to the Regulation – i.e. 'unless the context otherwise requires'.

Although the Department understands your argument that certain parts of the building are not subject to a change of use in a general sense, the intention of Regulation 5 as set out in its opening phrase is clear. It refers to 'where there is a change in the purposes for which or the circumstances in which a building is used . . .'.

In the Department's view, the reference to 'circumstances' in the context of Regulation 5(g) means that although the same floor area of a building may remain in residential use, the intensity of its use or the juxtapositioning of uses within the dwelling may change (i.e. a bedroom may become a kitchen, or a bathroom may become a bedroom). It is therefore the Department's view that Regulation 5(g) is intended to take account of the effect of a change of use on the building as a whole. In the context of this case the proposed additional dwelling will lead to a change in 'the circumstances in which a building is used'.

You argue, in effect, that only parts of the building are subject to a change of use and that Part E therefore need only be selectively applied to elements of part of the building. However, the Department takes the view that the building as a whole is affected by the proposals and that the reference in Regulation 6(1) to 'such work, if any, should be carried out as is necessary to ensure that the building complies with the relevant requirements . . .' means that whatever work is required, over and above the specific work needed to effect the actual conversion, must be done to achieve compliance with Requirements E1 and E2 within the whole building. The Department would further argue that even if the context for the change of use were to be accepted as being limited to part of the building, Regulation 6(2)(b) would still have the same effect in terms of requiring such work, if any, to be carried out as is necessary.

The determination

The Secretary of State has given careful consideration to the facts of this case and the interpretation of the Building Regulations as he understands them. He has concluded that the proposed work constitutes a change of use within the meaning of Regulation 5(b) and 5(g) of the Building Regulations (as amended) to the whole building and that by virtue of Regulation 6(1)(e) the building as a whole will require such work as may be necessary to ensure compliance with Part E – in this case Requirement E1 and Requirement E2. The Secretary of State therefore determines that your proposed work does not comply with Requirement E1 and Requirement E2 of Schedule 1 to the Building Regulations 1991 (as amended).

PART G: HYGIENE

Water supply, drainage and waste disposal for an experimental dwelling (G and H)

Methods of providing water supply, drainage and waste disposal in this case are unconventional and not covered by approved documents. In these circumstances it is vital to provide adequate supporting evidence that the spirit of building regulations will be achieved.

Inadequate supporting evidence led to the Secretary of State determining that the proposals did not satisfy the regulations.

Ref.: 45.1.126   Date: 13 September 1996

The Secretary of State takes the view that he is being asked to determine under section 16(10)(a) whether plans of the proposed work are in conformity with the requirements of Parts G ('Hygiene') and H ('Drainage and waste disposal') of Schedule 1 to the Building Regulations 1991 (as amended). In making his determination the Secretary of State has not considered whether the plans conform to any other relevant requirements.

The proposed building work

Your request for a determination concerns proposals to construct a single storey dwelling with basement. The dwelling is of timber structure and is to consist of two bedrooms, kitchen, bathroom, living room and conservatory. The size of the dwelling is approximately $100\,m^2$.

Your proposals are for the dwelling to be autonomous and hence the original plans showed no connections to the undertaker's water and sewerage services. Your proposals were as follows:

(i) Rainwater from the roof would be collected and treated for drinking and other water usage.
(ii) Discharge from sanitary conveniences would be collected in a composting chamber which would also receive water discharged from other sanitary appliances. The unit would also deal with macerated kitchen waste and cardboard.
(iii) Effluent from the composting chamber would be drained off by pumping to the garden for disposal either by subsoil drainage or to water garden plants as required.
(iv) Compost from the chamber would be used on the soil either directly or, depending on the moisture content, after a period of drying.

The system is experimental and it is intended to monitor it for information awareness and educational purposes. The house is to be managed by a charity which aims to promote environmental awareness. Your full plans application was originally rejected on the grounds of non-compliance with sections 21, 25 and 26 of the Building Act 1984 and on insufficient details supplied to ascertain compliance with various parts on the Building Regulations 1991, including Part H ('Drainage and waste disposal'). The District Council were made aware that section 26 of the Building Act 1984 had been repealed and superseded by Part G1 (previously G4) of the Building Regulations 1991.

7. Section 21 of the Building Act 1984 requires rejection of the building plans if satisfactory provision for the drainage of the building is not shown. This has been interpreted by the District Council to mean that a drain pipe is required and should be shown on the plans in this case because the building is within 100 feet of a sewer. However, section 21(1)(b) does provide local authorities with the power to dispense with any provision for drainage for a particular building. Section 25 of the Building Act 1984 relates to provision of a water supply which is not covered by Building Regulations.

8. Following a meeting between yourselves and the District Council, it was ascertained that unless supported by the Secretary of State, approval would not be given until it was shown on the drawings that there would be both connections to the water supply and sewer and the provision of a water closet. On this basis you applied for a determination which was subsequently held in abeyance.

9. Since the original design, the National Rivers Authority have advised that the ground is unsuitable for subsoil drainage due to the presence of clay and so a drain connection (to receive the intermittent discharge) was agreed. A connection to the water main to fill up the water storage tank (cistern) should it become empty, was also agreed. The revised plans were submitted to the District Council and you resubmitted a request for determination by the Secretary of State. This was prior to receiving a reply from the District Council following deposit of revised plans.

10. You have requested a determination on two questions. The first is whether a low flush unit (sanitary convenience) sited over a composting tank (from which excess liquid is discharged to the sewer) is acceptable as opposed to the District Council's requirement for a water closet (assumed to be a conventional type) connected directly to the drain. The second question is whether treated rainwater for use as the main domestic supply, with a connection from the water undertaker's main to fill up the water tank if it should run dry, would be acceptable.

11. The District Council, in response to receiving a copy of your request for a determination, have stated in their letter to the Department that they now consider that, in the interests of health and safety, the connections to the sewer and water mains should be fully used.

12. As the Building Regulations 1991 do not cover water supplies, the Secretary of State is unable to determine on this issue. However, advice on water issues is given in paragraph 26 below. Although the Department's earlier letter to the Council refers to a determination in respect of Part G of the Regulations, it is the Department's opinion that Part H of the Building Regulations 1991 is also of relevance in this case.

The Applicant's representations

13. You have adapted your original design to accommodate the Council regarding provision of a water supply and water closet, and also to provide a drain for the discharge of excess liquor from the composting chamber because of problems subsequently found regarding the permeability of the subsoil. Your proposal is to adopt a composting system developed in Australia to receive toilet and other domestic waste water including macerated kitchen waste and cardboard.

In order to conserve as much water as possible, the water closet proposed is of American manufacture and is apparently for use in mobile homes, boats, etc. According to the information supplied, the 'Sealand Traveller' uses 0.5 litres of water per flush. The closet does not incorporate a conventional water seal but a ball which closes onto a Teflon seal is used to prevent odours and insects entering the room. No details of the operation or the size of the discharge connection is given but from information obtained from BRE, it would appear that the flush is operated by raising a foot operated lever which is then depressed to displace the ball from the seal. The outlet diameter is approximately 75 mm. Reference is made in the supporting documentation to an installation of these closets at a National Trust property. The discharge from the closet will go directly to the composting chamber.

In addition, you confirmed the following points in your letter:

(i)   no water from the compost chamber will be discharged onto (or presumably into) surrounding ground;

(ii) the water main will be taken into the basement of the house and if there is a subsequent change of occupancy, the system could be extended to serve the remainder of the house; and

(iii) the monitoring is for information awareness and educational purposes. The house will be managed by a charity that aims to promote environmental awareness through presentation of information.

The District Council's representations

The District Council are adamant that the drainage and water supply should meet the requirements, as they interpret them, of sections 21

and 25 of the Building Act 1984. The original plans submitted for Building Regulation approval were rejected on the grounds that no drain or water supply connection was shown. The District Council also objected because no water closet was shown.

Although your proposals have been amended to include a drain connection (to an existing drain which connects to the undertaker's sewer) and a supply of water from the water undertaker's main, the District Council have stated that, in the interest of health and safety, the connections to the sewer and water mains should be fully used. The Department has taken this to mean that the District Council wish to see all sanitary appliances discharging to drain via discharge pipes and stacks. With regard to the water supply, the Department considers that the District Council would want all water using appliances to be connected to a supply provided by the water undertaker. This could be a direct supply (all at mains pressure) or indirect (served via a cold water storage cistern, except for the sink).

The District Council's letter to the Department also indicates an objection to the calculations given to support the proposed quantity of water storage; the amount of monitoring and treatment required for both the water and sewage systems; and the fact that there is no guarantee that the property will remain in the same ownership. They do not consider that the scheme would provide a safe and healthy environment for people living within and close by.

The Department's views

Requirement G1 of Schedule 1 to the Building Regulations 1991, states that: 'Adequate sanitary conveniences shall be provided in rooms provided for that purpose, or in bathrooms.' The section 'Meaning of terms' in Approved Document G ('Hygiene') sets out the definition of sanitary conveniences to mean 'closets and urinals'. Nowhere does it state that a sanitary convenience in a dwelling shall be a water closet. The guidance at paragraph 1.8 of Approved Document 'G' does, however, state that where a closet is fitted with a flushing apparatus the discharge should be through a trap and discharge pipe into a discharge stack or drain. However, it must be remembered that the guidance contained within Approved Document 'G' was written with only conventional methods of disposal in mind.

The Department is of the opinion that, as Requirement G1 requires adequate (sufficient in numbers) sanitary conveniences to be provided, the type of closet proposed will meet this requirement. However, the requirement of Regulation 7 cannot be ignored and it could well be argued that there is insufficient evidence to prove that the closet will be satisfactory in use from a health point of view. In this context it is noted that from the information provided, there is the option of a hand spray for effective cleaning of the closet.

Requirement H1 of the Building Regulations 1991 states that: 'Any system which carries foul water form appliances within the building to a sewer, a cesspool or septic or settlement tank shall be adequate.' The design as now proposed, incorporates a composting chamber between the appliances and the sewer. Although composting chambers are not mentioned in either the requirement or Approved Document 'H' ('Drainage and waste disposal'), the principle of composting the discharge from closets and grey water recycling systems is not, in the Department's opinion, contrary to the spirit of the Building Regulations 1991. The Department's view is that the discharge side of septic and settlement tanks is not covered by the Building Regulations and that it could therefore be argued that the proposed pumped and gravity drain from the composting chamber is not a matter for the Regulations either. Conversely, as the chamber will be sited within the building and is between the sanitary appliances and the sewer, it could be concluded that it is covered.

The Department has therefore needed to consider whether or not such an arrangement is adequate. To be able to make a judgement as to whether the proposed system will be adequate, the application should provide sufficient evidence that this will be the case. From the documentation provided, there is no third party certification to give such assurance. Indeed, the proposed system is intended to be experimental and as such, it would seem inappropriate for experimental work to be carried out in such a location.

With regard to future change of ownership, the Department does not accept the District Council's view that the proposed system is unacceptable because it cannot be guaranteed that the ownership will not change. The Department does not consider this to be a matter for building control. If a change of ownership is proposed, then it is incumbent on the prospective owner, either directly or through a building surveyor and/or legal adviser, to ascertain the type of house it is and the requirements needed for its everyday running and maintenance.

However, a change of ownership could present the following potential problem. You state in your letter that should the occupation of the dwelling change, then the mains water supply could be connected to the sanitary appliances. The flushing arrangement and pipework, etc., for the sanitary convenience proposed would then have to meet the requirements of Water Byelaws (or future Water Regulations) and there is no information as to whether this would be the case.

The Department's view, therefore, is that the Building Regulations do not prohibit the principle of low water use toilets (or composting toilets) and grey water waste treatment. It is important that the technology and components used for such systems should be shown by third party certification to be adequate for the purpose and thereby meet the requirement of Regulation 7. However, the information provided does not meet with this principle.

As stated at paragraph 12, water supply is not covered by the Building Regulations 1991 and therefore a determination cannot be provided in this respect. However, in the Department's view your proposal to provide a water supply from the water undertaker's main to within the property and for the supply to not be used unless the storage tank (cistern) for the stored rainwater runs dry is unlikely to meet with the approval of the water undertaker. Water Byelaw 59 which, although not specific to this case, is intended to prevent water in any unused pipework becoming stagnant. The maximum period normally permitted for unused water pipework to remain connected to a water undertaker's supply is 60 days.

The determination

The Secretary of State has given careful consideration to all the circumstances of this case, including the location of the proposed dwelling. He has noted your proposals with particular interest in the context of his Department's commitment to a 'green' policy, sustainable development, and Agenda 21 (the action plan for the twenty-first century agreed at the 1992 United Nations Conference on Environment and Development, held in Rio), and would hope to be in a position of being able to encourage rather than discourage such developments and application of new technology. However, given the insufficiency of information on the effectiveness of the proposed systems, the Secretary of State has regrettably concluded that he is unable to determine that your proposals comply with Requirement G1 of the Building Regulations 1991.

PART K: STAIRS, RAMPS AND GUARDS

Loft conversion – new staircase (K1)

This determination relates to the provision of a new staircase to serve a loft conversion containing two bedrooms and a bathroom.

The new staircase is to be similar in pitch, rise and going to the existing stairs from the ground floor. Although not meeting the requirements of K1 it is 'reasonable provision'.

Ref.: 45.1.128   Date: 28 June 1996

The Secretary of State takes the view that he is being asked to determine under section 16(10)(a) of the Building Act 1984 whether the plans of your proposed work are in conformity with Requirement K1 of Schedule 1 to the Building Regulations 1991 (as amended). In making his determination the Secretary of State has not considered whether the plans conform to any other relevant requirements.

The proposed building work

Your request for a determination concerns building work to convert the loft of a 1930s two storey terraced house to provide two small bedrooms and a bathroom. The total area of the new accommodation will be well below 50 square metres resulting in the work qualifying for the relaxation given in Approved Document 'B' ('Fire safety') (see paragraphs 1.23–1.28).

The proposed new staircase will be very similar to the existing stair which gives access from the ground to the first floor, in that it includes a winder and has similar 'rise' and 'going' (204.6 mm and 192 mm respectively). Both the new and existing stairs have a pitch of 46.8°. This contrasts with the guidance in Approved Document 'K' ('Stairs, ramps and guards') for a private stair where the maximum pitch recommended is 42°, and the maximum 'rise' is recommended to be 220 mm and the minimum 'going' to be 220 mm. The new stair is to be accommodated at first floor level by removal of the existing bathroom and the provision of a shower, WC and handbasin in the area remaining. The inward-opening door will be positioned at the base of the new stair.

Your full plans application containing these proposals was rejected on grounds of non-compliance with Requirement K1.

The Applicant's representations

You contend that the existing stair is deemed to be satisfactory and has been for the last 60 years. You argue that as the existing and new stairs will be juxtaposed it would be dangerous and onerous to make the new staircase physically different. You also contend that if the 'rise', and therefore the pitch, is reduced there will not be sufficient headroom over the existing stair; and that if the 'going' is increased, thereby reducing the pitch, the physical size of the door to the proposed shower room would need to be reduced to an almost inadequate size.

The Borough Council's representations

The Borough Council point out that paragraphs 1.21–1.25 of Approved Document 'K' offer guidance on alternatives to normal (42°) stairs for use with loft conversions where a normal stair cannot be accommodated. However, they point out that these alternatives are only for use where there is to be one habitable room and a bathroom but that in this case there will be two habitable rooms plus a bathroom.

The Council considers that the new stair should meet current safe stair standards in order to satisfy requirements B1 and K1. This is because, in this case, there are no unusual space constraints which would prevent construction of a stair meeting the guidance in Approved Document 'K'. To support this they state that examination of similar applications submitted within the locality indicates that compliance can be achieved.

The Department's views

Requirement K1 states that: 'Stairs, ladders and ramps shall offer safety to users moving between levels of the building.' The matter for decision by the Department is whether or not the proposed stair satisfies this requirement in this particular case.

Although the Borough Council do not indicate how, the Department accepts their assertion that a Part 'K' stair (i.e. one which is fully in accord with Approved Document 'K') could be accommodated as part of the conversion works. Taking your proposed layout, the Department equally accepts the limitations identified by you and summarized above. The shower room on the first floor contains the main WC and in the opinion of the Department a further reduction in the size of this door (from 650 mm to 600 mm), and the increased projection of the stair which would be necessary to accommodate a lower pitch, would be a safety hazard – especially for any disabled users.

In the Department's opinion, users of the proposed new stair will already be familiar with the existing stair and this is an important aspect of safety. Consistency between flights is also important. In the Department's opinion these two factors offset the benefit of having a new Part K stair.

Paragraph 1.23 of Approved Document 'K' suggests that the normal design requirements of a stair should only be reduced where the loft conversion comprises one habitable room and a bathroom. In this case, the conversion will comprise two bedrooms and a bathroom. However, both bedrooms are small and only likely to accommodate one single bed.

In summary, the Department accepts that the proposed stair is steeper and has a smaller 'going' than is desirable. However, in the Department's opinion it is reasonable to accept this design in this situation because the disadvantages are offset by familiarity of the users with the similar existing stair, consistency between new and existing juxtaposed flights, and probably only light use of the stair.

The decision

The Secretary of State has given careful consideration to the facts of this case. He has noted that although it is the view of the Council that a Part K stair could be installed, the installation of your proposed stair brings with it consistency of design in use between new and existing stairs, and in addition provides design benefits for your proposals. However, the material consideration must be whether the proposed staircase in the particular circumstances of this case will make reasonable provision for the purposes of Requirement K1. He has concluded, on balance, that it will and accordingly determines that your proposed staircase complies with Requirement K1 of Schedule 1 to the Building Regulations 1991 (as amended).

PART M: ACCESS AND FACILITIES FOR DISABLED PEOPLE

Visual arts building – provision of toilets (M3)

The work involves altering a gymnasium at a school and adding a visual arts centre. The dispute centres around whether or not there should be new toilets for disabled people in the visual arts centre.

It was argued that sufficient facilities were already available with the premises.

It may be noted that although disabled access may be provided it does not necessarily lead to a requirement for the provision of suitable toilets.

Ref.: 45.1.112   Date: 16 February 1996

The Secretary of State takes the view that he is being asked to determine under section 16(10)(a) whether plans of your proposed work are in conformity with the requirements of Requirement M3 ('Sanitary conveniences') of Schedule 1 to the Building Regulations 1991. In making his determination the Secretary of State has not considered whether the plans conform to any other relevant requirements.

Your request for a determination concerns building work at the above school to alter an existing gymnasium and to construct adjacent to this a new building extension for a visual arts centre and to make provision for a WC for disabled persons. The new building extension can be independently approached at ground floor level from within the site and would only be linked to the existing buildings.

The school is an independent, secondary one with approximately 440 pupils, aged between 13 and 18. Construction of the visual arts centre is part of an overall plan to concentrate new developments within the existing complex of older collegiate-style buildings at the school's core, and to allow the gradual renewal and replacement of a number of far-flung prefabricated and temporary buildings. Similarly, sanitary provision has been rationalized over the years; your client's aim being to provide WCs in central locations which are accessible both during and outside the normal academic teaching hours, because many of the pupils are borders.

Your first full plans application was rejected by the Borough Council on the grounds, inter alia, that they did not comply with the requirements of Part G and Part M. In a subsequent letter the Borough Council stated that sanitary accommodation did not need to be installed in the visual arts centre but that there would need to be a WC for disabled persons on the ground and first floor because disabled access was to be provided to and round the building. The letter also stated that disabled access to the first floor could be via a stairlift on the accommodation stair. From the correspondence provided, it appears that it was subsequently agreed by the Borough Council that a WC for disabled persons was not to be required on the first floor.

You then submitted a further full plans application in which it is taken that it was understood that Room G.10 on your plan 1625 (0) 02B of the ground floor was to be designated as a WC for disabled persons. These plans were approved on the 9 January 1995.

Following subsequent discussions and correspondence you submitted a third full plans application showing an alternative proposal on the ground floor for a WC for disabled persons. These plans omitted the provision in the previous full plans application for a WC for disabled persons in the location of Room G.10 but incorporated a cubicle complying with the standards of Approved Document 'M' ('Access and facilities for disabled people') sited at the south end of the Arts School Complex and accessible from the gymnasium corridor – i.e. outwith the new building but within the existing building complex. These plans were rejected on grounds of non-compliance with Part 'M'.

You contend that notwithstanding your preparation and submission of a full plans application, which received approval on the 1 January 1995, Requirement M3 did not oblige your client to make provision for a WC for disabled persons within the visual arts centre itself. It is in respect of this contention that you have sought this determination.

You contend that you are under no obligation to comply with Requirement M3 because paragraph 0.8 of Approved Document: 'M' ('Access and facilities for disabled people') states that: 'In schools or other educational establishments, Requirements M2, M3 and M4 will be satisfied if the provisions comply with . . . Design Note 18 (1984) "Access for Disabled Persons to Educational Buildings", published by the [then] Secretary of State for Education and Science.' You argue that paragraph 5.2. 'Location' of Design Note 18 states that in '. . . junior, middle, secondary schools and colleges, a central location for (WC for disabled persons) . . . is generally preferred and that there is, therefore, no requirement for individual buildings in a school to contain a WC for disabled persons, but rather that the provision should be viewed as an entity and facilities for the disabled provided accordingly. You further point: out that the statutory requirement of Requirement M3 states that: 'If sanitary conveniences are provided in the building reasonable provision should be made for disabled people'; and that by implication since no WCs were to be provided in the visual arts centre, there could be no requirement for a WC for disabled persons.

In a wider context you argue that your clients wish to provide a further WC for disabled persons on the campus to supplement the existing one, but that such a facility would be of much greater benefit to them if it were to be situated outside the confines of the academic buildings (including the visual arts centre) which are locked outside teaching hours. Your clients suggest that reasonable provision is best made by looking at the site as a whole rather than at one new building in isolation, and that this view appears to be supported by Design Note 18.

In your letter you state that the route to the proposed location of the WC for disabled persons proposed in your plans, which were rejected, is through a new covered cloister which forms part of a network of such cloisters through the school campus and that this would be signposted in accordance with good practice. You state that the journey time by wheelchair would be less than one minute from the previously 'mooted' location.

In support of their judgement in rejecting your last full plans application (i.e. that rejected and to which this determination is addressed), the Borough Council refer back to the fact that in their letter they put to you a proposal which allowed for the general sanitary provisions in the new building to be relaxed in this instance, but for the retention of a WC for disabled persons within the new building extension. They point to the fact that your subsequent resubmission accepted these points and on this basis the application was conditionally approved. They refer also to the subsequent rejection of your application which incorporated a WC for disabled persons outwith the new building but within the existing building complex.

The main reasons that the Borough Council put forward for requiring the WC for disabled persons to be within the new extension are:

(a) the fact that they did not insist on provisions within Part G was not to be taken as dismissing the requirement of that Part, but only as accepting that able-bodied students could reasonably travel to existing sanitary units, or could use the disabled sanitary facility which was to be installed within the new building;

(b) the same reasoning for using the existing sanitary units within the school could *not* be applied to disabled students for whom it would clearly be unacceptable and impracticable if they were wheelchair users to vacate the first floor level, leave the new extension, and travel by a covered way to the proposed location of the WC facility;

(c) the time travel to the proposed WC location would be excessive and would be inconvenient and unpleasant in winter months;

(d) The ethos behind Part M is to create an environment which allows disabled persons to access and use a building and to remove a facility designed for a disabled person from that proposed building appears to be illogical;

The Department's analysis of this case is as follows. It seems clear that from the outset your clients did not intend to include, within the visual arts centre, any sanitary accommodation additional to that provided elsewhere in the school. Their grounds were that there was to be no increase in the school population; that the centre was in effect a replacement for an existing art department which was to be demolished; and that sufficient toilets existed elsewhere in the school complex.

Requirement M3 states: 'If sanitary conveniences are provided in the building, reasonable provision shall be made for disabled people.' Because of the 'established WC provisions within the site as a whole', the Borough Council had agreed that there need be no general provision within the visual arts centre. It follows that there is no consequent requirement under Part M to provide sanitary conveniences for disabled people. Moreover, the suggestion by the Borough Council that the provision of appropriate access to a building is a basis on which to require suitable sanitary provision is flawed.

With regard to Design Note 18, the designated provisions therein have the same status as the provisions detailed in Approved Document 'M'. In neither case do they have the statutory status of Requirements of the Building Regulations 1991 (as amended). It is acknowledged that the references in paragraph 0.9 of the Approved Document do not include the general design considerations which accompany those designated requirements from Design Note 18. Nevertheless, in the context of paragraph 5.1 of the Design Note, it would be perverse to ignore the references contained in paragraph 5.2 with respect to the location of WCs suitable for schools and colleges – clearly that paragraph should be read together with paragraph

5.1(d). On this basis it is the Department's view that provision for WCs for disabled persons should be made at a central location.

The Secretary of State has given careful consideration to this case. In particular, he has noted the wording of Requirement M3 and the provisions set out in Design Note 18 1984, 'Access for Disabled Persons to Educational Buildings'. In the particular circumstances of this case he accepts the principle that the WC for disabled persons can be provided outwith the visual arts centre building and has therefore concluded that your plans, which were rejected, comply with Requirement M3.

Extension to club house – access for disabled (M2)

This application for a determination relates to a Local Authority's insistance that wheelchair access should be provided to the changing facilities intended for rugby and soccer teams. This argument was not accepted by the Secretary of State.

Ref.: 45.1.113   Date: 9 May 1996

Your letter requested the Secretary of State to determine under section 16(10)(a) whether plans of the proposed work were in conformity with Requirement M2 ('Access and use') of Schedule 1 to the Building Regulations 1991. In making his determination he has not considered whether the plans conform to any other relevant requirements.

The proposed building work

The proposed building work involves the extension of a single storey sports club. The existing building contains a lounge and associated bar. Directly accessible from the lounge are the club secretary's office, a kitchen and a 'cellar'. At one end of the lounge is an emergency exit. At the other end, there is access to (i) ladies' toilets and changing accommodation; and (ii) a separate corridor which, in turn, provides access to a gentlemen's toilet, a 'unisex' wheelchair WC compartment, and another room (described on registered plan no. 1 simply as a toilet). At the far end of the corridor there is a door which leads out onto the pitches. From the ladies' changing room, a door leads outside to the playing pitches. At the rear of the building, and independently and separately accessible from the lounge, there are two changing rooms and a washroom.

The proposal is to build a single storey extension to the club house, containing four additional changing rooms and associated shower facilities for rugby and football teams. The new accommodation will be accessed at the same level from the existing building via the existing corridor. There will also be a direct entrance for players.

Your full plans application was rejected by the Borough Council on the grounds that they did not comply with Requirement M2 ('Access and use'). In an earlier letter the Borough Council argued that the requirements for Part M of the Building Regulations applied to the extension since the proposal was not governed by any of the limits on application specified in Schedule 1 to the Regulations. They explained that since the proposal had its own extrance, and was not solely approached through the existing building, it should be treated in the same manner as a new building. They referred to the guidance in paragraphs 0.3 to 0.5 of the Approved Document 'M' ('Access and facilities for disabled people') relating to the extension of the existing building. They also argued that the building could be classified as recreational and that paragraphs 3.10 and 3.11 of the Approved Document relating to the provision of changing facilities for the disabled should therefore apply. It is in respect of some of these interpretations and the consequential rejection of your full plans application that you have requested a determination.

Your representations

In support of your contention that your plans do comply with Requirement M2 you draw a clear distinction between the general facilities in the existing club house and those proposed in the extension. You have explained that the facilities provided within the club house are the bar and catering, pool/snooker, card games and such like, and a dance area. You claim that the disabled are catered for in respect of access and toilet facilities.

However, with respect to the proposed changing rooms in the new extension you argue that these are intended for two football and two rugby teams and that this will be a part of the building to which the public will not be permitted. You contend that there is no requirement under the Building Regulations to provide changing rooms for disabled people involved in activities outside the building and state that the Association does not do so for its bowls and tennis sections. Finally, you point out that the Association does not provide, within the building, any sporting or recreational facilities requiring changing facilities, irrespective of who uses or is admitted to the building, and there is therefore no obligation whatsoever to provide changing rooms.

In conclusion, the principle of your argument is that both the existing building and the proposed extension are not themselves recreational buildings providing sporting activities and therefore Requirement M2 should not apply; and that the guidance in paragraphs 3.10 and 3.11 of Approved Document 'M' is not applicable. You further contend that it is not a feasible consideration that a disabled person requiring wheelchair access could be expected to be involved with football or rugby teams.

The Borough Council's representations

In support of their judgement in rejecting your full plans application the Borough Council make the following points:

(a) Part M of the Building Regulations should apply to the building as it is not one of the categories specified in the limits on application in Schedule 1, and that it is an extension which has independent access and is not approached solely through the existing building.

(b) Having in their view established that Part M applies, the question is then whether reasonable provision has been made for disabled people to use the building. The Borough acknowledge that people with wheelchairs do not play rugby or football but argue that the Higham Lane Leisure Association is a multi-purpose organization and, as far as they are aware, do not exclude disabled people from membership and may, either now or in the future, promote activities in which they could participate.

(c) In the Borough's view it would not be impossible to provide changing facilities for the disabled by some re-design of the building, although this would require adequate corridor and doorways and ramped access as well as the changing facility itself.

(d) The Borough do not agree with the applicant's letter explaining that because the club is a members' one this should exempt the building from the requirements of Part M.

(e) Although the Borough applauds the existing provision of a disabled person's toilet in the present building, they argue that notwithstanding what activities disabled persons may or may not be able to do it is certain that they will be excluded if the facilities are not provided to enable them to participate in whatever activities they choose.

The Department's analysis

The Department's analysis of this case is as follows. In the Department's view the Borough Council's judgement that the requirements of Part M apply to the extension because it has a separate entrance, is correct.

The guidance on the performance required of extensions to satisfy Part M is contained in paragraphs 0.3 to 0.5 in Approved Document 'M' and is as follows:

If an existing building is extended, the requirements of Part M apply to the extension provided that it contains a ground storey.

When a building is extended, there is no obligation to carry out improvements within the existing building to make it more accessible to and usable by disabled people than it was before. However, the extension should not adversely affect the existing building with respect to the provisions of the Building Regulations for access to, and use of, the building by disabled people.

An extension should be at least as accessible to and usable by disabled people as the building being extended. Where access to the extension is achieved only through the existing building, it will be subject to the limitations of the existing building, and it would be unreasonable to require higher standards within the extension. On the other hand, it is reasonable that an extension which is independently approached and entered from the boundary of the site should be treated in the same manner as a new building.

The question in this case is what is the reasonable provision that should be made for disabled people to gain access to and use the building. The particular point at issue is whether provision of changing facilities suitable for people who are wheelchair users should be made in the new changing rooms. The Department recognizes the provision made for disabled people to gain access to and use the existing club house. However, as the new extension is to be used solely to provide changing facilities for football and rugby on grass pitches it is considered reasonable not to make provision for changing facilities for people who are wheelchair users as detailed in diagrams 14 and 15 which are referred to in paragraph 3.11 of Approved Document 'M'.

The determination

Having given careful consideration to the above points, the Secretary of State has concluded that reasonable provision for access to, and use of, the planned extension to the club house by disabled people has been made. He therefore determines that your proposals show compliance with Requirement M2 of Schedule 1 to the Building Regulations 1991.

Single storey retail store – wheelchair access to storage platform (M2)

In a number of previous determinations the Secretary of State has emphasized that it is not his remit to decide where disabled people may work. However, in this case he has decided that there is no reason why wheelchair users need access to the storage platform.

Ref.: 45.1.120   Date: 26 June 1996

The Secretary of State takes the view that he is being asked to determine under section 16(10)(a) of the Building Act 1984 whether plans of your proposed work are in conformity with Requirement M2 of Schedule 1 to the Building Regulations 1991 (as amended). In making his determination the Secretary of State has not considered whether the plans conform to any other relevant requirements.

The proposed building work

Your request for a determination concerns building work to provide a mezzanine storage platform to a new single storey retail store as part of fitting out works. The development forms part of a large complex of retail units which have been constructed in two phases; the first phase being the construction of the single storey shell and the second the fitting out for/by the individual occupants. The ground floor area of the unit is approximately 1050 square metres and was approved as part of the first phase application with full access to all parts of the unit fully in accordance with the recommendations contained in Approved Document 'M' ('Access and facilities for disabled people'). It contains all the retail accommodation together with an office, staffroom and toilet facilities including a disabled toilet suitable for wheelchair users.

In this particular case the second phase is to comprise provision of a mezzanine storage platform. It has a floor area of approximately 960 square metres and is accessed by three wide stairs at the rear leading up from the staff accommodation on the ground floor. In addition there are two fire exit stairs at diagonally opposite corners front and rear.

The construction of the mezzanine storage platform was treated by the Borough Council as 'fitting out works' and your full plans application was given conditional approval by them. The condition attaching is that: 'A passenger lift suitable for use by a wheelchair user shall be provided to give access to the first floor of the unit.'

You have resisted implementation of this condition on the grounds that the proposed structure is a storage platform and not, as described in the condition, a 'first floor' and that members of the public will not be allowed onto the platform. It is in respect of the appropriateness of this condition in order to achieve compliance with Requirement M2 that you have sought this determination.

The Applicant's representations

You contend that you have complied with Requirement M2, viz: 'Reasonable provision shall be made for disabled people to gain access to and to use the building.' You confirmed that it was agreed that the disabled facilities available to the ground floor met this requirement but argued that it was inappropriate that disabled access to the storage platform should be provided. The provision on the ground floor includes easy access entrance doors, wide circulation areas, defined clear passageways between displays, disabled toilet, sufficient fire exits and wide areas around the pay and help counter. You confirmed that disabled staff are able to work within the ground floor area but not within the storage areas due to the restricted working requirements. You state that these restrictions are.

(a) that the staff working in the storage areas are required to lift heavy objects or boxed items, either manually or with a manual trolley lift or both. As such your client has to work within the Manual Handling Regulations with regards to methods of lifting heavy objects/bulky items/etc. Furthermore your clients' insurers are quite specific regarding working conditions and any breach of these conditions could result in your clients' insurance being invalidated;

(b) the normal practice is for only two warehouse staff to be in the store at any one time, one of which would be working on the storage platform and the other working in the ground floor storage area. Occasionally there could be two members of staff working on the storage platform area, usually when bulk deliveries are being received;

(c) the Fire Brigade confirms your concerns over wheelchair users being on the storage platform, as there would only normally be one member of staff there at any one time (at certain times of the day the storage room does not have any member of staff in it). Their opinion is that any wheelchair bound person on the storage platform would be at risk in the event of a fire, even with a lift within a protected enclosure.

You conclude that although your clients would like to extend the option of employing disabled staff to work anywhere within their units they have to be practical and realistic about the type of work which certain members of their staff have to carry out. In their view, disabled people cannot be employed to work on the storage platform because of the unnecessary risks that they would be exposed to and for all the other reasons stated above.

The Borough Council's representations

The Borough Council have considered their locus for controlling the construction of the mezzanine storage platform in respect of Part M. They consider that Phase 1 and Phase 2 combined constitute the erection of a building because the construction relating to the two Phases is indivisible and no intervening occupancy has occurred. They are therefore of the view that the Phase 2 work must comply with Requirement M2 and state that in applying that requirement they do not consider they can operate discretion as to whether or not provision should be incorporated for non-ambulant occupants.

The Council argue that the platform is 960 square metres and is therefore in excess of the 280 square metres net floor area criterion given in Approved Document 'M' (paragraph 2.13(a)) for a two storey building. On the basis of this guidance they argue that reasonable access can only be considered to have been provided if a suitable lift is incorporated in the scheme.

The Department's view

You did not seek to challenge the Borough Council's view that Phases 1 and 2 of this development should be considered as the erection of a building for the purposes of the Building Regulations. However, the Department agrees with that view on the basis that there has been no intervening occupancy between the construction of the shell under Phase 1, and the fitting out works being progressed as part of Phase 2. As such the requirements of Part M of the Building Regulations 1991 apply to both the original shell and the fitting out works as if they were being carried out as a single contract. In the Department's view, it follows that the issue which needs to be decided is whether reasonable provision has been made for disabled people to gain access to and to use the facilities within the building, given that no passenger lifts have been provided to enable wheelchair users to reach the mezzanine storage platform.

The Department notes your clients' statement that the mezzanine storage platform is to be used solely for storage purposes and the statement that there are statutory, insurance, and other requirements which restrict the working practices and are such as to prohibit wheelchair users being able to work there. As a consequence in this particular case, the Department concludes that reasonable provision has been made under Requirement M2 for disabled people to gain access to and use the building.

The decision

The Secretary of State has given careful consideration to the facts of this particular case and, in particular, has noted the statutory, insurance and other requirements which are understood to apply and which have the effect of restricting working methods. The Secretary of State would not necessarily seek to take a view on what sort of work can be undertaken by a disabled person. Much will depend on any statutory and other restrictions such as have been identified in this case, the type of disability, the abilities of the individual concerned, and the potential for enterprises to adapt working practices. However, in the circumstances of this particular case he has concluded that it is not reasonable to anticipate disabled wheelchair users requiring access to this storage platform. He therefore determines that your proposals, which exclude any form of lift access to the platform, comply with Requirement M2 of Schedule 1 to the Building Regulations 1991 (as amended).

In coming to this conclusion the Secretary of State has borne in mind the point that if at any time in the future the platform is used to store a different type of stock, and/or different methods of storage are applied, such that it might be feasible to employ a disabled person, then an employer could be obliged to provide suitable access for that disabled person under the Disability Discrimination Act 1995.

New office block – sanitary conveniences for disabled (M1 and M3)

This proposal is to place unisex disabled toilet facilities on the first floor of a three storey building accessible by lift, which upon close examination proved to be in accordance with approved document 'M', and therefore acceptable.

Ref.: 45.1.118   Date: 26 June 1996

In making the following determination the Secretary of State has not considered whether the plans conform to any other relevant requirements.

The proposal

The building work involved concerns the construction of a new three storey office block which it is understood has been built on a speculative basis. The block is L-shaped in plan with the longer external dimensions of the L being approximately 36 and 38 metres in length. In the centre of the L there are service cores containing lifts and sanitary accommodation. Toilets are provided on each floor; and it is proposed that there should be a single disabled unisex WC on the first floor to serve both staff and visitors.

You submitted a full plans application which was rejected in respect of several requirements including M1, M2 and M3. In their rejection letter the Borough Council stated in respect of Requirements M1 and M3 that there was a lack of reasonable provision for disabled toilets. You considered that your proposal to provide a wheelchair WC compartment located on the first floor in the service cores satisfied the interpretation of Requirement M1 and the requirements for provision under M3. You therefore re-submitted a full plans application on stating in your covering letter that a determination would be sought in respect of Requirements M1 and M3.

The Applicant's representations

You have explained that the Borough Council's verbal requirement was for a disabled toilet to be provided on each floor of the three storey office block. You contend that this is beyond what is required under the Building Regulations and believe that your proposal for one unisex WC compartment in accordance with the guidance given in Approved Document 'M' ('Access and facilities for disabled people') (diagram 16 on page 21), located on the first floor adjacent to the non-disabled toilet accommodation and accessed via a common lift lobby, was reasonable provision. You cite a similar proposal where a single unisex WC wheelchair compartment located on the first floor of a three storey office block had been accepted by another Building Control Authority.

In your letter you confirm your view that you have made reasonable provision for disabled staff and any disabled visitors. In particular, you have responded to the Borough Council's view that quick and easy access to a disabled person's WC will not be possible particularly when lifts are being maintained, used for deliveries, or being used by other people. You state that it is unlikely that the occupier will allow both lifts to be out of commission for maintenance purposes at the same time, and that experience shows that in a three storey building of this scale, most non-disabled persons will use any one of the three staircases for internal circulation, thereby leaving the lifts available for disabled persons.

The Borough Council's representations

The Borough Council argue that the performance recommendations contained in the Approved Document 'M' referring to reasonable and convenient provision for staff and visitors are not satisfied by the proposal. They argue that the proposal is neither convenient for visitors nor a reasonable provision for staff. They explain that visitors would have to use a first floor toilet in what would be an often unfamiliar building, which is not only inconvenient in itself but could compromise security systems used by many occupiers. They also argue that staff would face similar difficulties and excessive travel distances from some parts of the building.

The Council considers that your proposal to provide a single disabled person WC on the first floor does not satisfy the requirement with regard to the provision of toilet accommodation under Requirement M3. They argue that the guidance in Approved Document 'M' explains that there are several principles to be considered: the first of which is equal provision; the second is greater need; and finally the number and location of facilities.

With regard to the actual travel distances, where the final arrangement of partitions, furniture and route of travel is not known, the Council argue that the worst case arrangement should be applied. On this basis, and assuming a route of travel with changes of direc-

tion at right angles, they calculate that this would result in 46 metres travel distance on every floor for area 6A, shown on the plans. Moreover, they suggest that if direct distances were to be assumed, then the two thirds rule for 'means of escape' contained in Approved Document 'B' ('Fire safety') would be more appropriate – i.e. 27 metres in this case, rather than the 40 metres suggested in paragraph 4.14 of Approved Document 'M'. However, you have stated that you consider it is incorrect to apply Approved Document 'B' to calculate horizontal travel distances for Part M; and you point out that no distinction is made between direct and indirect travel distances for the purposes of Part M.

The Council state that they have expressed a willingness to consider alternative design solutions to reach a reasonable provision but that these have been declined by you.

The Department's view

The point at issue is whether the single unisex WC for disabled people on the first floor, to serve both visitors and staff, is a reasonable provision under Requirement M3. Requirement M3 states that: 'If sanitary conveniences are provided in the building, reasonable provision shall be made for disabled people.' The objectives of Requirement M3 are stated in paragraph 4.1 of Approved Document 'M': 'In principle, sanitary conveniences should be no less available for disabled people than for able-bodied people. The aim is to provide solutions which will most reasonably satisfy that principle, whilst bearing in mind the nature and scale of the building in which the provision is to be made.' The design considerations given in the Approved Document include the following issues cited on pages 20 and 21:

> 4.3 *The number and location of WCs for disabled people may depend on the size of the building and on the ease of access to the facility. A wheelchair user should not have to travel more than one storey to reach a suitable WC.*

> 4.14 *Requirement M3 will be satisfied by provision for wheelchair users of both sexes on alternate floors: provided that the cumulative horizontal travel distances from a work station to the WC is not more than 40 m and, in a building provided with lift access, the general provision for sanitary conveniences is in areas to which anyone using the building has unrestricted access.*

In this case the Department notes that the sanitary conveniences are provided in the lift lobby areas to which there is unrestricted access. The Department also notes and accepts the points which you have made about the appropriate calculation of travel distances for the purposes of complying with Requirement M2. The Department has therefore concluded that the provision of the single unisex WC for disabled people on the first floor will mean that no one has to travel more than one storey to reach it and that the cumulative horizontal travel distance for staff from a work station to the WC should not be more than the 40 m referred to in Approved Document 'M'.

The determination

The Secretary of State has given careful consideration to the facts of this case. He is satisfied that your clients' proposals have taken properly into account the principles of interpretation contained in Requirement M1, and that your clients' proposals for sanitary conveniences make reasonable provision for disabled people. He accordingly determines that your clients' proposals comply with Requirements M1 and M3.

Laboratory – access to first floor (M2)

> *No lift was to be provided to give access for disabled persons to the first floor in a new building. There was a lift in an adjoining building with a first floor link to the new structure which was considered reasonable provision and this view won the day.*

Ref.: 45.1.135   Date: 7 March 1997

The Secretary of State takes the view that he is being asked to determine under section 16(10)(a) whether plans of your proposed work are in conformity with Requirement M2 ('Access and use') of Schedule 1 to the Building Regulations 1991 (as amended). In making his determination the Secretary of State has not considered whether the plans conform to any other relevant requirements.

The proposed building work

Your request for a determination concerns the construction of a two storey building directly adjacent to the south west side of the headquarters building. The new building is an L-shaped block, with the legs of the L approximately 31 metres and 24.5 metres long and about 11.5 metres wide. The ground floor provides accommodation for two self-contained departments. Each has a separate entrance – one entrance having a designated waiting and reception area. At first floor level there are new laboratory facilities which are intended to be used with, and as an extension to, the existing laboratories on the first floor of the adjacent headquarters building. There is therefore an enclosed link between the first floor levels of both buildings to achieve this operational extension. The size of the first floor of the new building has been calculated by the Department to be approximately 535 m².

The new building contains one staircase which links the laboratories on the first floor with the ground floor entrance lobby of one department but which does not give direct access to any other space on the ground floor. There is also an external fire escape stair to the first floor laboratories at the opposite end of the building from the internal stair. The existing headquarters building contains a lift serving the existing laboratories on the first floor, which is in direct line to the new link corridor to the new laboratories on the first floor of the new building.

Your full plans application for the new two storey block was rejected by the Borough Council for failure to comply with Requirement M2 ('Access and use'). Following an exchange of correspondence, the Borough Council explained that your proposals did not offer adequate access for disabled persons due to the omission of a lift in the new building. However, you believe that although your proposals for the new block do not incorporate a passenger lift, they nonetheless comply with Requirement M2. It is in respect of this question that you have sought a determination.

The Applicant's case

You state that the new block is located adjacent to an existing two storey building which contains two departments: the offices on the ground floor, and laboratories on the first floor. The new building provides for three groups of users; two groups on the ground floor; and an extension to the laboratories on the first floor.

You also state that for the operation of the laboratories a link between the two buildings is being constructed at first floor level only, to allow free movement to personnel and trolleys between the two halves of this Department. This link was planned for in the existing building as a corridor between two rooms and noted as such on the original drawing (M 1612 14) as '1.8 store (Phase II corridor to adjacent building)'.

There is a lift in the existing building which allows access to the first floor and it is a relatively short distance to the link to the new building. You argue that as the link is an essential part of the operation of this one department, which has one section head in control, it seems reasonable that the lift could provide the means of access for a disabled person without constructing one surplus and expensive lift in the new building. You explain that all deliveries of samples will be made via the existing building; the consultant's offices are in the existing building; and the title and site of the existing building on the drawing is proposed office and laboratory accommodation for one department, Phase I. A toilet suitable for the disabled is provided on the first floor in Phase II which will also serve Phase I.

You acknowledge the Borough Council's concern over the difficulty of gaining access from the ground floor departments to the first floor in the new building, but explain that there is no requirement for access between these two floors. You explain that if a disabled person on the first floor wished to visit a department on the ground floor they would have to exit the building, as it is a requirement of your client that one department on the ground floor is shut off from the other department for the sake of privacy. Similarly a visit to the other department on the ground floor would be the same as for any other disabled employee in the building; namely via outside paths.

You maintain that the laboratory is not a public access department; it is fundamentally separate from the ground floor departments because it is the Phase II of the Laboratory Services, whose first phase is located in the existing, adjacent building. The new first floor link is provided primarily to allow the two phases to function together, whilst also providing access to a disabled lift.

You conclude that because there is no obligation to adopt any particular solution contained in an Approved Document these are sufficient reasons to demonstrate that you have satisfied Requirement M2.

The Borough Council's case

In the Borough Council's final rejection of your arguments and proposals they state that the plans have been rejected as they do not offer adequate access for disabled persons because of the omission of a lift. The Borough Council contends that it is wholly unacceptable, particularly when Approved Document 'M' ('Access and facilities for disabled people') recommends a lift for a new building such as this, for disabled persons on the ground floor to exit the building and travel via the existing building to gain access to the first floor. They state that this contravention was pointed out when the proposal was at planning stage.

The Borough Council state that they rejected the plans in respect of Requirement M2 for the following reasons:

(a) The proposal is a new purpose-made building, and therefore should comply fully with the requirements of Part M of the Building Regulations.
(b) The matter of access within the building was discussed prior to an application being made, and further advice was given once the application was lodged and again when construction was in progress (this advice being offered some months before work commenced).
(c) It was suggested that a link at ground floor level would improve the situation, but this was rejected by you.
(d) There is no continuing control by the Borough Council after the building is completed and occupied. This may well result in the building no longer being used for the purpose intended and if both buildings were managed as one department the access for disabled persons would be considerably hindered by the omission of a lift in the new block.

The Department's view

In the Department's view, the requirements of Part M of the Building Regulations apply to the new two storey block. The point at issue in this case is whether reasonable provision has been made for disabled people to gain access to, and to use the facilities within the building, given that no passenger lift is to be provided within the new building to enable wheelchair users to reach the first floor.

The Department notes that provision has been made in terms of toilets for disabled persons on both the ground and first floors of the new building. The Department also notes that the two departments which occupy the ground floor of the new building have been designed as separate self-contained operational units, each with its separate entrance. There is no direct link between these departments and the laboratories at first floor level. You argue that the operation of the laboratories requires that they should be accessed from the existing building and should have no direct links to the new departments on the ground floor. The first floor is dedicated to laboratory use and linked to the existing labs at first floor level, and it is not a department through which public access is permitted. The Department takes the view that your building proposals clearly reflect these operational intentions within the scheme and that it is therefore reasonable to regard the first floor as linked operationally with that of the existing building but separate from the operational functions of the two ground floor departments.

In the Department's view a new building of this scale would normally require provision of a passenger lift in order to demonstrate compliance with Requirement M2. However, the Department notes that there is a lift to the first floor suitable for wheelchair users in the existing building. This would enable wheelchair users who are working in the laboratory to gain access to the first floor extension to those laboratories in the new building. The Department accepts your argument that visits to the ground floor departments would not be a regular part of the operation of the Laboratory department and it is reasonable for access to be provided to these departments for laboratory staff in the same way as for the rest of the staff occupying buildings dispersed around the site – namely, along the normal access routes that link the buildings.

The determination

The Secretary of State has given careful consideration to the particular circumstances of this case. In particular, he has noted the operational requirements of the different departments in the new building and the need for their physical separation. He has concluded that the design of the building reflects these operational requirements and that in that context, and given the availability of a lift in the existing building, you have made reasonable provision for disabled people without the provision of an additional lift in the new building. The Secretary of State accordingly determines that your proposals comply with Requirement M2 ('Access and use') of the Building Regulations 1991 (as amended).

Retail unit – access for disabled (M2)

The local authority considered that a proposed access was not the principal entrance to the shop and therefore discriminated against disabled persons. Building work carried out was determined not to satisfy M2, but an earlier suggestion involving relative levels of floor and pavement would have met the requirements.

Ref.: 48.1.148   Date: 4 August 1997

In making the following determination, the Secretary of State has not considered whether the plans conform to any other relevant requirements.

The proposed building work

Your request for a determination concerns retail unit two at your premises. There are two retail units in the building, each one being independently approached at ground floor level from the existing pavement. There are maisonettes over the retail units but neither these nor retail unit one are the subject of this determination.

The Applicant's case

In requesting a determination with respect to retail unit two, you explain that the proposed access for disabled persons into this unit would, because of the difficulties posed by the fall in the level of the rear of the public footpath across the front of the building, be by the alternative arrangements that follow. From the rear of the shop access would be through doors having a minimum opening of 800 mm and you explain that this direction of approach has the benefit of enabling easy access from the dedicated disabled parking bays in the rear car park. From the front of the shop it would be through a doorway located at the corner of the building, the threshold of which, together with the internal finished floor level, would be set to correspond with the existing level of the rear of the public footpath. You believe that the above facilities more than satisfy the

requirements of M2 as interpreted by the guidance given in paragraph 1.31 of the Approved Document, particularly in respect of subparagraphs (b) and (c).

In submitting evidence from your client, you point out that in the original planning application the entrance to retail unit two was located on the corner. You explain that the problem of access for the disabled has arisen because your client decided to relocate the main entrance from the corner, where the floor level was coincidental with the rear of the public footpath, to the front elevation, without appreciating the consequence for disabled access. Having realized this problem, your client is proposing to reinstate the corner entrance as an alternative entrance to provide access for disabled people. You also explain that the builder had written to the District Council asking for a decision regarding the corner entrance so that they could procure and install the appropriate type of door and make any final modifications to the paving levels, but the builder received no reply. You explain that until the details of the corner entrance have been agreed, a window has been installed to ensure security for the adjoining pharmacy (where Class A drugs are stored).

With your letter you enclosed a copy of a plan which showed the corner door reinstated. Although the earlier plan showed that the floor slab of retail unit two was level with the back of pavement at the site of the proposed corner door, this information could not be determined from the later plan.

The District Council's case

In setting out the background, in their letter the District Council report that you have made three full plans applications for this work, which have all been rejected for not providing details of access for disabled persons as requested. They explain that you made a further submission in relation to retail unit two and which included a letter specifying provision of disabled access to unit two, stating that access had now been provided to both the front and rear of the shop. This application was refused as not complying with Part M.

In commenting on the alternative proposals outlined for retail unit two, the District Council explain that no formal details had ever been submitted with regard to the rear access proposal and they have had no opportunity to comment on this proposal. However, having looked at the situation, including visiting the site, they believe it to be completely unacceptable.

With regard to the rear access, they note that no formal proposals exist on the site or have been submitted with respect to dedicated parking bays and also observe that the car park areas are laid to fairly steep falls. They note also that the general surface finish is unsuitable for wheelchair users as it consists of loose stone to aid surface water drainage and the only tarmac area is the roadway through the site where parking would cause obstruction to delivery trucks, etc. They further argue that once inside the shop the route to the main retail area is via an area which would obviously be used for stock, etc., which could cause the retailer potential security problems with the temptation to lock the door, which would negate disabled access via this route.

The District Council explain that the shopping parade is served by two existing car parks and note that at the front there is space for about 12 cars with one designated disabled bay approximately 50 m from retail unit two.

With regard to the alternative front access by an additional doorway to the corner of unit 2, they conclude that this will not provide adequate access because it is not the principal entrance to the shop and therefore discriminates against disabled persons. They believe that it is impossible to achieve a level landing outside the proposed door as the approach is part of the public footpath which has a fall away from the shop of 1:30. They also note that there is a cross fall across the threshold of 1:20, which makes it extremely difficult for someone using a wheelchair to approach the door safely, comfortably and under control. They believe the access within the shop is likely to be obstructed by retailer displays due to its prominent position. They argue that a photocell automatic door opening device would not be practical in this location as it would be activated by passers-by and the footpath is on a very busy corner.

In addition the District Council provided photographic evidence that the proposed front entrance has in fact been constructed as a window and not a door.

The Department's view

The Approved Document, Part M, gives guidance on how the requirements of Part M of the Building Regulations can be met. In paragraph 1.31 it suggests:

Requirement M2 will be satisfied if:

(a)   *the principal entrance for visitors or customers is accessible and suitable; or*

(b)   *in the event of the space outside the principal entrance being severely restricted, or the site being on sloping ground, an alternative entrance intended for general use is accessible and suitable and there is suitable internal access, available to all who may use the building, from the alternative entrance to the principal entrance; and*

(c)   *where car spaces are provided adjacent to and serving the building, but there are no suitable means of access from them to the principal entrance, an additional entrance, intended for general use, is provided giving suitable internal access to the principal entrance;*

In considering the alternative rear entrance approach the Department considers that this in itself would not provide reasonable access principally because of the inappropriate surface treatment to the car park and the lack of designated parking bays for disabled people. The Department notes that there is a designated car parking bay for disabled people in the front car parking area to the shopping parade.

In considering the proposals for an additional entrance to the front of the shop at the corner of the unit, the Department notes the District Council's concerns over the appropriateness of this entrance with regard to the gradients immediately outside the entrance which the Council considers would not enable a wheelchair user to safely manoeuvre and to manually open the corner entrance. However, the Approved Document suggests, in paragraph 1.9, that a level

approach will satisfy Requirement M2 if it has a gradient not steeper than 1 in 20. The Department thus considers that the gradients at the corner would not preclude an entrance there which would provide reasonable access for disabled people.

The District Council suggested that a photocell door opening device would not be practical given the busy location, but the Department considers that the provision of a suitable, manually operated (not by a PIR or photocell) automatic door that would not be triggered by passers-by on the footpath would provide a reasonable solution.

The Department takes the view that reasonable access for disabled people into the shop at the front would be provided by the location of an entrance door at the corner, as originally designed, and as indicated in the earlier plan with particular reference to the levels indicated for the floor slab and the back of the pavement by this proposed door. However, the work has not been constructed in accordance with these plans in that there is a window at the corner, and not a door.

The Secretary of State's decision

The Secretary of State has given careful consideration to the facts of this particular case and to the points made by both parties. He concludes that reasonable provision has not been made at present for disabled people to gain access into this new shop unit as the rear entrance alone does not provide reasonable access, and the corner entrance has not been provided. He therefore determines that the entry into the new retail unit for customers, as presently proposed and constructed, does not satisfy Requirement M2 of Schedule 1 of the Building Regulations 1991 (as amended). However, had the proposal for a corner entrance set out in the earlier plan been adopted, and the relative levels of the floor slab and the back of the pavement were as in that plan, this proposal, in conjunction with the additional proposals for rear access, would have met requirement M2.

Copies of later decision letters are available on the Internet at the Building Regulations Department website address:
www.construction.detr.gov.uk/br/bro8.htm

Checklists

Errors and omissions may occur in the most careful office, as it is not easy to prepare, or check, a drawing of a proposal and keep in one's memory all the requirements of the building regulations and the approved documents.

Therefore some form of reminder, or checklist, is very useful to have when preparing plans or when checking drawings prior to contract or submission to the local authority or an approved inspector. Equally a checklist in the hands of a building control officer will be an aid to the speedy examination of many submissions.

Examples of checklists are given, and may be adapted to meet the needs of the reader who may vary the series of reminders to current requirements and consultations. It is not wise to leave any blanks on a list. Each aspect of the regulations should be ticked in a column marked 'clear' as the plan is examined. There is provision for raising a query prior to clearing an item.

CHECKLIST NO. 1: GENERAL

Item	Query Clear Comment
Plans	
Building notice	
Full plans	
New building	
Alteration	
Extension	
Works and fittings	
Drawings and information adequate	
Materials and Workmanship	
Reg. 7 – AD	
Fitness	
Past experience	
Agrément certificate	
British Standard	
Independent certificate	
Quality assurance	
Tests and calculations	
Sampling	
Short-lived materials	
Unsuitable materials	
Resistance to subsoil	
Softwood treatment	
CE materials	
Structure – Part A	
Basic stability	
Lateral stability	
Sizes – floor	
ceiling	
roof members	
Walls	
Thickness	
Building proportions	
Loading	
End restraint	
Opening recesses	
Overhang, chases	
Lateral support	

Item	Query Clear Comment
Single leaf walls	
Chimneys	
Strip foundations	
Other foundations	
Codes and standards	
Loading	
Foundations	
Concrete	
Steel	
Aluminium	
Masonry	
Timber	
Disproportionate collapse	
Fire – Part B	
Dwellinghouse	
Flat	
Institutional	
Other residential	
Assembly	
Office	
Shop	
Other non-residential	
Loadbearing elements of structure	
External walls	
Separating walls	
Compartment walls	
Compartment floors	
Protected shafts	
Concealed spaces	
Internal surfaces	
External surfaces	
Stairways	
Loft conversions	
Conversion to flats	
Roof	
Shopping centre	
Steel portal frames	
Raised storage area	
Multi-storey car park	
Petrol pump canopy	
Sprinklers	
Thermoplastic materials	
Junctions	
Wall and roof	
Separating and external walls	
Openings	
Compartment walls	
Compartment floors	
Protected shaft	
Doors	
Pipe protection	
Vent ducts	
Air circulation	
Chimneys and flues	
Cavity barriers	
Fire stopping	
Space separation	
Relevant boundary	

Item	Query Clear Comment	Item	Query Clear Comment
Notional boundary	———————	Hot water storage	———————
Small dwelling	———————	Unvented system	———————
Enclosing rectangle	———————	Controls	
Aggregate notional area	———————	Sanitary conveniences	
Means of escape		Siting	———————
Dwellinghouse of		Design	———————
three or more storeys	———————	Chemical closets	———————
Three or more storeys		Flushing	———————
containing flat	———————	Small bore system	———————
Office		Provisions adequate	———————
Shop		Drainage and Waste Disposal – Part H	
Extended or materially altered		Sanitary pipework	
dwellinghouse of three or	———————	Traps	———————
more storeys		Branch discharge pipes	———————
Three or more storeys and changed		Discharge stacks	———————
to use as a dwellinghouse	———————	Stub stacks	———————
Loft conversion		Ventilation	———————
Self-closing door	———————	Materials	———————
Fire-resistant doors	———————	Watertightness	———————
Wired glazing	———————	Foul drainage	
Stairway	———————	Layout	———————
Enclosure	———————	Pipe cover	———————
Escape window	———————	Gradients and sizes	———————
Number of rooms	———————	Materials	———————
Access for fire service		Beneath building	———————
Access for firefighters		Bedding and backfilling	———————
Venting		Junctions	———————
Firemains	———————	Access	———————
Site Preparation and resistance		Ventilation	———————
to moisture – Part C		Watertightness	———————
Drainage	———————	Cesspools	
Vegetable matter	———————	Siting	———————
Dangerous and offensive		Access	———————
substances	———————	Size	———————
Radon protection	———————	Ventilation	———————
Hardcore	———————	Construction	———————
Damp-proofing	———————	Septic tanks and	
Suspended timber floors	———————	settlement tanks	———————
Walls	———————	Siting	———————
Solid	———————	Access	———————
Cavity	———————	Size	———————
Cladding	———————	Ventilation	———————
Toxic substances – Part D		Construction	———————
Masonry wall	———————	Rainwater drainage	
Construction		Gutters and rainwater pipes	———————
Services		Capacity	
Ventilation	———————	Materials	
Resistance to passage of sound – Part E		Combined system	———————
Walls		Layout	———————
Refuse chutes	———————	Depth	———————
Solid masonry	———————	Gradient and size	———————
Cavity masonry	———————	Bedding and backfilling	———————
Masonry, with lightweight panels	———————	Materials	———————
Timber frame	———————	Watertightness	———————
Junction with external wall	———————	Solid waste	
Roof space	———————	Capacity	———————
Ceiling	———————	Siting	———————
Floors	———————	Design	———————
Junction with partition	———————	Ventilation	———————
Floors		Removal	———————
Floor base with soft covering	———————	Heat-Producing Appliances – Part J	
Concrete base with		Solid fuel up to 45 kW	———————
floating layer	———————	Air supply	———————
Ventilation – Part F		Flue pipes and chimneys	———————
Habitable rooms	———————	Shielding	———————
Kitchens	———————	Factory made chimneys	———————
Common spaces	———————	Hearths	———————
Sanitary accommodation	———————	Placing appliances	———————
Adjoining spaces	———————	Oil-burning output 45 kW	———————
Mechanical	———————	Air supply	———————
Roof space	———————	Factory made chimney	———————
Hygiene – Part G	———————	Flue pipes and chimneys	———————
Bathroom		Balanced flue	———————
Ventilation	———————	Shielding	———————

Item	Query Clear Comment
Placing appliances	_____
Gas burning up to 60 kW	
Air supply	_____
Balanced flue	_____
Chimneys	_____
Flue pipes	_____
Hearths	_____
Shielding	_____
Placing appliances	_____
Protection from falling, collision and impact – Part K	
Stairs	_____
Rise	_____
Going	_____
Headroom	_____
Tapered treads	_____
Width	_____
Length of flights	_____
Handrails	_____
Landings	_____
Guarding	_____
Ramps	_____
Slope	_____
Headroom	_____
Width	_____
Landings	_____
Stepped ramp	_____
Guarding	_____
Protection from falling	_____
Stairs	_____
Ramps	_____
Floors	_____
Balconies	_____
Roofs	_____
Vehicle barriers	_____
Height	_____
Horizontal impact	_____
Construction	_____
Protection from collision and trapping	_____
Conservation of Fuel and Power – Part L	
Dwellings	_____
Roof	_____
Walls	_____
Exposed floor	_____
Ground floor	_____
Windows and rooflights	_____
Elemental method	_____
Target U method	_____
Energy rating method	_____
Thermal bridging	_____
Limiting infiltration	_____
Heating systems controls	_____
Zone controls	_____
Timing controls	_____
Hot water storage systems	_____
Insulation of vessels, pipes and ducts	_____
Conservatories	_____
Other buildings	_____
Residential	_____
Shop	_____
Office	_____
Assembly	_____
Industrial	_____
Roof	_____
Walls	_____
Exposed floor	_____
Ground floor	_____
Windows and rooflights	_____
Elemental method	_____
Calculation method	_____
Energy use method	_____
Heating system controls	_____
Temperature controls	_____
Time controls	_____

Item	Query Clear Comment
Boiler sequence controls	_____
Insulation of vessels, pipes and ducts	_____
Lighting	
Lamp deficiency	_____
Lighting controls	_____
Solar gain	_____
SAP rating	_____
Facilities for Disabled – Part M	
Accommodation	_____
Wheelchair spaces	_____
Ramps and steps	_____
Doors	_____
Internal access	_____
External access	_____
Lifts	_____
Water closets	_____
Fire escape	_____
Glazing Materials and Protection – Part N	
Legislation	_____
Local Act?	_____
1984 Act	
Building over sewer (18)	_____
Provision of drainage (21)	_____
Exits (24)	_____
Water supply (25)	_____
1956 Act	_____
Chimney height (10)	_____
1980 Act – highways	_____
1971 Act	
Conservation area	_____
Listed building	_____
Permitted development	_____
Consultations	_____
Adjoining owner	_____
Building Control Officer	_____
Environmental Health	_____
Electricity Company	_____
Factory Inspector	_____
Fire Officer	_____
British Gas	_____
British Telecom	_____
Structural Engineer	_____
Services Engineer	_____
Water Authority	_____
Police	_____
Party wall	_____
Relaxation Necessary?	_____

CHECKLIST NO. 2: FOUNDATIONS

A1 of Schedule 1 requires that a building must sustain the dead, imposed and wind loads and transmit them to the ground:

- safely
- without causing deflection or deformation of the building, or the ground as will impair the stability of any building.

A2 goes on to require that a building must be constructed so that its stability is not impaired by movements caused by:

- swelling
- shrinkage
- freezing
- reasonably foreseen subsidence or landslip

Item	Query Clear Comment
Site	
Affected by heavy traffic?	_____
Any nearby trees?	_____
Is root penetration possible?	_____
Signs of movement in	_____

Item	Query Clear Comment
adjoining property	
Soil characteristics	
Borehole results	
Is soil corrosive?	
Any soft spots?	
Made up ground	
Clay on waterlogged sand?	
Has previous use	
contaminated soil?	
Is there clay soil on	
sloping site?	
Is ground movement possible?	
Protection against radon	
Choice	
• strip	
• pad	
• raft	
• pile	
• other	
Loading	
Choice appropriate for	
soil conditions	
Ground water pressure	
Existing land drainage affected	
Strip foundations (ground slab	
regarded as 'foundation')	
Should internal walls have	
foundation lower than slab?	
Is reinforcement necessary?	
Are there eccentric loads?	
Thickness	
Depth	
Stepped	
Buttresses	
Chimneys	
Piers	
Volume of concrete	
Mix of concrete	
Depth of fill below slab	
Protection against frost heave	
Where dpcs and dpms	
are incorporated:	
• are materials correct?	
• is bonding adequate?	
• is lapping and sealing adequate?	
• can construction dry out?	
Calculations	
Pad	
Ground not overloaded	
Spacing	
Depth of pads	
Are pads axially loaded	
Reinforcement	
Concrete mix	
Calculation	
Raft	
Is downstand necessary?	
Junction with wall will not	
permit water penetration	
Reinforcement	
Concrete mix	
Calculations	
Pile	
End bearing?	
Friction?	
Spacing	
Specialist advice	
Ground beams	
Calculations	

CHECKLIST NO. 3: EXTERNAL WALLS

A1 of Schedule 1 requires that buildings must sustain the dead, imposed and wind loads and transmit them to the ground safely and without causing deformation of the building.

A3 says that in the event of an accident to a five-storey building the structure must not fall or collapse disproportionately to the cause of the damage.

B2 requires that the materials forming surfaces of walls must resist the spread of flame over the surfaces, and if ignited have a reasonable rate of heat release.

B3 says that buildings must be constructed so that in the event of fire stability will be maintained for a reasonable period.

B4 requires the external walls of buildings to adequately resist the spread of fire over the walls and from one building to another. To have regard to height, use and siting of building.

C4 says that walls must adequately resist passage of moisture to the inside of a building.

D1 requires that steps must be taken to prevent fumes from cavity insulation from getting into a building.

L1 limits the heat loss through the external walls of buildings.

Item	Query Clear Comment
• Cavity	
• Solid	
• Timber frame	
• Other	
Materials	
Loading	
Moisture resistance	
Rain	
Snow	
Wind	
Construction	
Chases	
Combustibility	
Cladding	
Lapping	
Damp-proof courses	
Distance from boundary	
End restraint	
Fire resistance	
Fire spread through wall	
Surface spread of flame	
Cavities – fire stopping	
Area of glazing	
Area of combustible material	
Gables	
Heat recovery systems	
Heat gain	
Insulation	
U value	
Type of insulant	
Toxic emission precautions	
Cold bridges	
Imposed loads	
Loadbearing	
Lateral support by floors	
Lateral support by roof	
Lateral stability	
Moisture resistance	
Damp-roof courses	
Penetrating	
Materials	
Bricks	
Blocks	
Mortar	
Timber	
Concrete	
Other	
Openings	
Overhang	
Parapet	

Item	Query Clear Comment
Loading	
Damp-proof courses	
Height	
Party wall?	
Penetrating pipes	
Pier size and spacing	
Radon protection	
Rendering	
Recesses	
Refuse chutes	
Risk of condensation	
Size	
Length	
Height	
Thickness	
Strapping	
Structural stability	
Structural movement	
Differential loading	
Soil conditions	
Space separation	
Timber frame	
Unprotected areas	
Vapour barriers	
Vertical loads	
Vertical fire spread	
Access for firefighting	
Bay windows	
Calculations	

CHECKLIST NO. 4: ROOFS

A1 requires buildings to sustain the dead, imposed and wind loads and transmit them to the ground safely and without causing deformation of the building.

A3 says that an accident to a five-storey building must not result in failure disproportionate to the cause of the damage.

B3 requires buildings to be constructed so that in the event of fire:

- stability will be maintained for a reasonable period;
- internal fire spread will be inhibited;
- cavities are subdivided and fire stopped;

B4 says that roofs of buildings must adequately resist the spread of fire over the roof or from one building to another.

C4 requires roofs to adequately resist the passage of moisture to the inside of a building.

F2 stipulates that roof voids above insulated ceilings must have reasonable provision to prevent excessive condensation.

H3 requires rainwater to be carried away from roofs to an adequate outfall.

J3 says that flues and chimneys must be installed and constructed so that the risk of a building catching fire from these sources is minimal.

K2 requires roofs in normal use to be provided with barriers.

K3 requires adequate guarding when vehicles have access to a roof.

L1 set standards relating to maximum heat losses through roofs.

Item	Query Clear Comment
Bracing	
Cavities	
Fire stopped	
Compartmentation	
Condensation	
Vapour checks	
Ventilation openings	
Voids	
Covering, type?	
Covering and fixing compatible	
Combustibility	
Distance from boundary	

Item	Query Clear Comment
Edge protection for sheet covering	
Falls	
Fire protection	
Fire stopping	
Flashings	
Flues and chimneys	
Guarding	
Heat gain	
Holding down	
Strapping	
Loading	
Dead	
Wind	
Live	
Lateral support	
Means of escape	
(emergency use of roof)	
Moisture resistance	
Openings	
Rainfall	
Rainwater drainage	
Roof lights, area	
Materials	
Roof and wall junctions	
Structural members	
Size	
Spacing	
Span	
Structural movement	
Size of gutters and downpipes	
Stability	
Separating walls	
Surface spread of flame	
Safe access	
Thermal movement	
Thermal insulation	
Trussed rafters	
Use of plastics	
Underlay	
Upstands protecting openings	
Ventilation	
Vapour barriers	

CHECKLIST NO. 5: INTERNAL WALLS

A1 of Schedule 1 requires a building to sustain the dead, imposed and wind loads and to transmit them to the ground safely without causing deformation of the building.

A3 says that in the event of an accident to a five-storey building must not fail or collapse disproportionately to the cause of the damage.

B2 limits the spread of fire on wall surfaces and if ignited they must have reasonable rate of heat release.

B3 requires that in the event of fire, stability will be maintained for a reasonable period; internal fire spread is to be inhibited, and cavities are to be fire stopped. Separating walls must resist the spread of fire and smoke.

E1 says that walls separating dwellings are to have a reasonable resistance to airborne sound.

J3 requires that heat producing appliances must be installed so that the risk of the building catching fire is minimal.

Item	Query Clear Comment
Additional loads	
Airborne sound insulation	
Bracing other walls	
Construction	
Combustibility	
Chases	
Chimneys	
Cavities	
Compartment	

Item	Query	Clear	Comment
Doors			
Refuse chambers			
Junctions			
Openings			
Pipe junctions			
Refuse chutes			
Lateral stability			
Sound resistance			
Disproportionate collapse			
Damp-proof courses			
End restraint			
Fire-resistance periods			
Fire stops			
Flame spread			
Glazed lobbies			
Height			
Length			
Linings			
Loads			
Dead			
Imposed			
Lateral			
Live			
Temporary			
Vertical			
Loadbearing			
Lateral support by floors			
Lateral support by roof			
Materials			
Mortar			
Non-loadbearing			
Overhangs			
Penetration by			
Pipes			
Ducts			
Services			
Pier sizes			
Pier spacings			
Resistance			
Horizontal			
Oblique			
Vertical			
Recesses			
Structure			
Surfaces			
Separating			
Doors			
Junctions			
Loading			
Pipe junctions			
Refuse chutes			
Complete vertical separation			
Sound resistance			
Structural movement			
Supporting structure			
Thermoplastics			
Thermal movement			

CHECKLIST NO. 6: FLOORS

A1 of Schedule 1 says that buildings must sustain the dead, imposed and wind loads and transmit them safely to the ground without causing deflection or deformation.

A3 relates to disproportionate collapse and says that in the event of an accident to a five-storey building any failure must not be disproportionate to the cause of the damage.

B2 requires that materials forming the surface of a ceiling must adequately resist the spread of flame and if ignited give a reasonable rate of heat release.

B3 says that buildings are to be constructed so that in the event of fire stability will be maintained, fire spread inhibited, and cavities fire-stopped.

C3 requires that floors resist the passage of moisture into a building, or the building's fabric.

E2 says that a floor separating a dwelling from another dwelling or another part of the building not exclusive to the dwelling must have reasonable resistance to airborne sound.

E3 goes on to say that a floor above a dwelling separating one dwelling from another, or another part of the same building, must have reasonable resistance to impact sound.

J3 requires that heat producing appliances, flue pipes and chimneys must be installed so that the risk of a building catching fire from these sources is reduced to a reasonable level.

K2 says that floors to which in normal use people have access must be guarded to prevent users from falling.

K3 has a similar requirement to protect people from falling off floors to which vehicles have access.

Item	Query	Clear	Comment
Chimneys and flues			
Hearths			
Pedestrian guards			
Thermal insulation			
Combustibility			
Openings			
Stairways			
Garages			
Junctions			
Pipe junctions			
Refuse chutes			
Fire resistance			
Floating			
Lateral support by walls			
Moisture resistance			
Damp-proof courses			
Air space			
Ventilation			
Precast			
Timber			
Reinforced			
Hardcore			
Polythene			
Ground covering			
Platform floors			
Sound requirements			
Airborne			
Impact			
Direct transmission			
Pugging			
Weight			
Thickness			
Compartment			
Structural strength			
Loads			
Dead			
Live			
Imposed			
Moving			
Support of walls			
Concentrated			
Impact			
Vibration			
Dynamic			
Wind			
Collapse			
Temporary			
Structural movement			
Supporting structure			
Fire resistance			
Stability			
Integrity			
Insulation			
Services in or through floor			
Suspended ground floor			
Intermediate support			

CHECKLIST NO. 7: PROTECTION FROM FALLING, COLLISION AND IMPACT

B1 of Schedule 1 says that certain buildings are to be provided with means of escape in case of fire.

B3 limits internal fire spread and requires stability to be maintained for a reasonable period.

F1 requires common spaces such as landings and stairways to be ventilated.

K1 says stairs and ramps must afford safe passage for the users of a building.

K2 says that stairs, ramps and floors and balconies together with roofs must be guarded with barriers to stop users from falling.

K3 requires vehicle ramp floors used by vehicles to be guarded.

Vehicle loading bays are to have protection against people colliding with vehicles.

K4 says people moving about must be protected against collision with windows.

K5. There is to be protection against impact from and trapping by doors.

N1. Protect glazing from impact.

N2. Manifestation of glazing.

N3. Safe opening and closing of windows.

N4. Safe access for cleaning windows.

Item	Query Clear Comment
Stairs	
Cavity barriers	
Change of direction	
Connection to wall and floor	
Doors	
Dimensions	
Fire escape	
Fire protection	
Flight lengths	
Flight widths	
Glazed screens	
Going	
Headroom	
Handrails	
Helical	
Incorporation of services	
Landings	
Loading	
Dead	
Live	
Imposed	
Dynamic	
Temporary	
Loft conversions	
Limited combustibility	
Materials	
Openings	
Obstruction	
Protected shaft	
Pitch	
Rise	
Risers in flight	
Sizes	
Spiral	
Steepness	
Structurally sound	
Structural movement	
Thermal movement	
Vibration	
Tapered treads	
Treads level	
Use by disabled people	
Ventilation	
Natural	
Mechanical	
Ramps	
Cavity barriers	

Item	Query Clear Comment
Change of direction	
Connection to wall and floor	
Doors	
Dimensions	
Fire escape	
Fire protection	
Glazed screens	
Headroom	
Handrails	
Incorporation of services	
Kerbs	
Landings	
Loading	
Dead	
Live	
Imposed	
Dynamic	
Temporary	
Limited combustibility	
Materials	
Openings	
Obstruction	
Protected shaft	
Ramp width	
Sizes	
Slope	
Stepped	
Structurally sound	
Structural movement	
Thermal movement	
Vibration	
Use by disabled people	
Ventilation	
Natural	
Mechanical	
Barriers	
Stairs, ramps and	
vehicle parks	
Dimensions	
Guarding	
Balustrades	
Barriers	
Parapet walls	
Construction	
Fixing	
Number	
Siting	
Smooth finish	
Support brackets	
Continuous	
Unclimbable by young children	
Glass blocks	
Laminated glass	
Loading, horizontal	
Protecting disabled people	
Loading bays	
Escape steps	
Refuge	
Windows, skylights and ventilators	
Protection from collision	
Impact or trapping by doors	

CHECKLIST NO. 8: FOUL DRAINAGE

The 1984 Act, section 21, requires proper provision for drainage.

The 1984 Act, section 22, allows buildings to be drained in combination.

H1 of Schedule 1 requires that foul water from a building must be connected to an adequate drainage system.

H2 covers the provision and construction of cesspools, septic tanks and settlement tanks.

Item	Query Clear Comment
Drain	
Private sewer	
Public sewer	
Combined system	
Separate system	
Partially separate system	
Layout	
Access	
Ventilation	
Bends in manholes	
Differential settlement	
Protection	
Through buildings	
Under buildings	
Near buildings	
Depth of cover	
In roads	
In gardens, etc.	
Protected by concrete	
Size	
Gradient	
Capacity	
Materials	
Rigid pipes	
Flexible pipes	
Bedding	
Granular fill	
Selected fill	
Backfilling	
Granular fill	
Selected fill	
Access	
Rodding eyes	
Access fitting	
Inspection chambers	
Manholes	
Siting	
Distances apart	
Manholes; inspection chambers	
Non-ventilating covers	
Step irons	
Half round channels	
Branch connections	
Benching	
Depth	
Width and length	
Construction	
Outfall to	
Private sewer	
Public sewer	
Cesspool	
Septic tank	
Settlement tank	
Other	
Levels	
Cesspool; septic or settlement tank	
Capacity	
Access	
Ventilation	
Construction	

CHECKLIST NO. 8: ACCESS AND FACILITIES FOR DISABLED PEOPLE

Part M now applies to new dwellings as well as other buildings.

M2 reasonable provision is to be made for disabled people to have access to a building.

M3 requires suitable arrangements to be made for the provision of sanitary conveniences for disabled people.

M4 spectator seating is to be provided for disabled people.

B1 requires that satisfactory means of escape be provided.

Item	Query Clear Comment
Approach to building/dwelling	
Access to building/dwelling	
Dwelling threshold	
Entrance doors	
Entrance lobbies	
Access to other storeys	
Access within other storeys	
Corridors and passages	
Internal doors	
Switches and sockets	
Passenger lifts	
Common stairs in flats	
Sanitary conveniences	
Hotels and motels	
Changing facilities	
Audience and spectator seating (space)	
Fire escape facilities	

Index